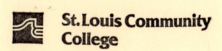

Genetics
and
Molecular
Biology

Genetics and Molecular Biology

ROBERT SCHLEIF
Brandeis University

ADDISON-WESLEY PUBLISHING COMPANY

Reading, Massachusetts • Menlo Park, California • Don Mills, Ontario • Wokingham, England
Amsterdam • Sydney • Singapore • Tokyo • Mexico City • Bogotá • Santiago • San Juan

Library of Congress Cataloging in Publication Data

Schleif, Robert F.
    Genetics and molecular biology.

    Bibliography: p.
    Includes index.
    1. Molecular genetics. I. Title.
QH430.S35   1985   574.87'3282   85-3903
ISBN 0-201-07418-4

ABCDEFGHIJ-HA-898765

p. 433 The fourth equation should be $\dfrac{O}{O_r} = \dfrac{K_D}{K_D + R^2}$

p. 476 Problem 17.7. Line 3 should read ". . . <u>attL</u> ⇄ <u>attR</u>."

p. 491 Illustration at bottom of page should be labeled "tnp A transposase."

p. 498 Line 16 should read ". . . intramolecular resolution takes place."

p. 505

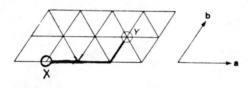

p. 510 Figure 18.20. In Part 2, the horizontal arrow should move from left to right. In Part 3, delete horizontal arrowhead.

p. 532 Line 3 should read "That is, the splicing is not perfect."

p. 535 Problem 19.3. Line 4 should read "and $10J_H$ regions.

p. 545 Last line should read ". . . complement of ribosomal proteins."

p. 555 Figure 20.16.

$Y = (2a, 1b)$ $T = 7$ $S = 420$

p. 557 Figure 20.18.

Preprohead I		Prohead I		Prohead II	
	pC		pB		groE
$(pE + pN_43)$ ⟶	$(pE + pN_43) + pC$ ⟶	$(pE + pN_43) + pC + pB$ ⟶	pE + p		
			pB ⟶ h3		
			pC + pE ⟶ X1 ⟶ X2		

p. 578 At the bottom of the page, ∝-ketoglutarate should appear in text and replace 2 oxogluterate in equation.

p. 581 Last line should read ". . . and a substantial membrane potential exists across . . ."

p. 598 Line 3 should read ". . . changes a glycine to a serine."

p. 598 Last line should read " . . . to the virus RNA and packaged . . ."

p. 606–607 The problems should be numbered for chapter 22.

p. 42   Problem 2.12.   The second line should read ". . . ranging from −3 to +3."

p. 108   Figure 4.23.   7-methylguanine should have a double bond between N+ and carbon.   The ribose of base one should have a $-O(CH_3)$ group, not CH_3.

p. 121   Figure 5.1.   The amino group nitrogen bears the positive charge not the hydrogen.

p. 124   The right-hand peptide bond structure should bear a negative charge on the carbonyl oxygen.

p. 146   Figure 6.2.   The legend should read ". . . at the 3' end of a charged tRNA molecule."

p. 176   Figure 6.26.   Double bonds should be shown on guanine.

p. 192   Figure 7.2.   Bonds should be shown between the hydrogens and nitrogens of the amino groups of guanine and cytosine.

p. 198   Figure 7.10.   The first crossover product shown should be:
        abCDE
        a'b'CDE

p. 268   Line 17 should read ". . . and C's."

p. 271   Figure 10.19.   The four gel lanes should be labeled A G C T respectively.

p. 275   Figure 10.23.   Figure for Problem 10.16.

p. 340   Problem 12.9.   The second line should read ". . . $50mC_i$/m mole."

p. 364   Figure 13.16.   Part (a) should be part (b) and part (b) should be part (a).

p. 407   Line 12 should read " . . . into the fly using the P elements."

p. 408   Problem 15.7, line 2: delete the word diploid.

p. 415   Figure 16.3.   In the region of DNA replication the gene should be designated CII not CIII.

p. 419   Line 14 should read ". . . is located to the left of P_R."

This book evolved from a course in molecular biology that I have taught for the past twelve years to our graduate students and to some undergraduate students. My main theme is that a relatively small number of principles involving cell biology, physical chemistry, genetics, and biochemistry permit the understanding of a large fraction of molecular biology.

This material is intended to encourage thinking and an appreciation for beautiful experiments. Thus the book is selective in the material it presents. Rather than list each fact that is known, I have tried to find a useful subset of the facts. This is necessary, of course, for by now molecular biology is such an ocean of information that no course should touch an appreciable fraction of what is known.

Although the text together with many of the problems forms a self-contained unit, instructors in advanced courses may want to supplement the material contained here with lectures on recent developments and assigned readings of research papers as appropriate to their own course.

Much of the material in this book discusses results found on prokaryotes. This is possible since many of the important principles of molecular biology are most clearly revealed and explored in prokaryotic systems. However, the extension of such results to eukaryotes is discussed, and eukaryotic systems of special interest are presented.

Few undergraduates will have had the ideal preparation for using this book, courses in cell biology, physical chemistry, and biochemistry. To maximize its utility, however, the book is written as though the reader has had partial acquaintance with all three subjects. Consequently, many readers will find familiar material, new material, and a few topics for which reference to other texts will be helpful.

Many of the discussions include numbers and dimensions so that students may begin to develop a reliable intuition as well as a quantitative understanding of biological systems at the molecular level. Also, many experiments are outlined to permit the students to acquire a good idea of the ways in which the facts were learned. This approach appears to facilitate understanding. Although each of the many findings in molecular biology is rather simple, these are remote from direct experience, and hence a large collection of such facts becomes hard to assimilate. Consequently, describing how the knowledge was discovered assists in its learning, because results become less remote, and slight doubts about their reliability are erased.

Many challenging problems are provided at the end of each chapter. Solving two or three of the typical ones per lecture is a reasonable load. A few deal directly with the textual material, but most amplify the textual material or introduce new ideas for which insufficient room existed in the text. Some require access to the literature. The student is urged to work a few of the problems from each chapter and to read all the rest. The particularly difficult ones are marked with a star.

Extensive references to material covered in the text, as well as related material on both prokaryotes and eukaryotes, are provided at the end of each chapter. The Recommended Readings are papers a more advanced student could profitably read in conjunction with the text. The Related Reviews, Books, and Articles list a few of the papers a student can turn to for background information. The papers listed under Deeper Reading permit the advanced student or research worker to find important papers on many topics related to the chapter. The references cover a substantial fraction of the important literature. Naturally, important papers published before those listed will be referred to in these papers. However, important papers published after those listed can also easily be found with the Science Citation Index. This annual reference lists papers and books published in a given year according to the references made in the paper or book. Thus by using this source to find all references to an important earlier work, one can work forward in time from a key paper.

Many people have contributed to this book. I particularly thank Pieter Wensink for penetrating discussions over the years; Donald Green, Thomas Gray, Edward Simon, Robert Vinopal, Philip Perlman, and Mary Ann Jarema for comments and suggestions on the text; James Funston for guidance through the project; and the staff of Addison-Wesley for their excellent production of this book. I would also like to thank Clifford Brunk and Richard L. Bernstein for their opinions on the final manuscript and to my many students for helping me to refine the presentation.

Waltham, Massachusetts   *R.S.*

A NOTE TO THE READER

Since molecular genetics is such a broad subject, many important points have been mentioned only in the problems. Therefore reading all the problems, even if you do not intend to solve them, will help you learn more about what is known. Similarly, some of the references discuss points not dealt with at length in the text, and reading their titles in the lists at the ends of the chapters will add to your knowledge of what is known as well as start you in the direction of more information on a topic.

Many of the problems are of the *"Ah-ha!"* type, so that considerable time may be spent in fully understanding a question or an approach to the answer, and when the insight comes, it often comes in a flash. The problems that are known to be harder are marked with a star, as are sections containing harder material that is not essential to understanding the rest of the book.

C O N T E N T S

■ P A R T   O N E ■

Cell Structure and Cell Constituents

CHAPTER   1   ■   An Overview of Cell Structure and Function   3

Cells Operate with an Immense Amount of Information   3
Rudiments of Prokaryotic Cell Structure   4
Rudiments of Eukaryotic Cell Structure   7
DNA Is Tightly Packed Within Cells   9
Putting Molecules In or Out of Cells   10
Diffusion Is Very Rapid within the Small Volume of the Cell   14
Freely Growing Populations Increase Exponentially   15
Composition Changes Slowly in Exponentially Growing Cells   16
Age Distribution in Populations of Growing Cells   17
Problems   18
Recommended Readings   20
Related Reviews, Books, and Articles   20
Deeper Reading   20

CHAPTER   2   ■   Nucleic Acid Structure   21

The Regular Backbone of DNA   22
Helical Forms of DNA   24
Base-Paired Strands Can Dissociate   26
Polynucleotide Strands Can Specifically Reassociate   27
Specific Reassociation Is a Valuable Tool   28
Dependence of Reassociation Rates on Salt and Temperature   29
Counting DNA Copies by Reassociation Kinetics   29
Topology Introduces Additional Considerations   31

Generating DNA With Superhelical Turns   32
Measuring Superhelical Turns   34
Binding of Proteins and DNA Unwinding   36
Determining Lk, Tw and Wr in Hypothetical Structures   37
Enzymes Can Add or Subtract Superhelical Turns   38
Do Superhelical Turns Have Biological Significance?   39
A Unified Mechanism for DNA Topoisomerases   40
Problems   41
Recommended Readings   43
Related Reviews, Books, and Articles   43
Deeper Reading   43

CHAPTER   3 ■ DNA Synthesis   47

A. ENZYMOLOGY   48
Proofreading, Okazaki Fragments, and DNA Ligase   48
Multiple Types of DNA Polymerases Are Found in Cells   51
Activities Associated With a DNA Replication Fork   53
ϕX174 Replication: A Specific Case   54
Components of Cellular DNA Replication   58
Error and Damage Correction   59

B. PHYSIOLOGICAL ASPECTS   61
DNA Replication Areas in Bacterial Chromosomes   61
DNA Is Synthesized Outward in Both Directions From Origins   63
Escherichia Coli DNA Elongates at About 500 Bases/Sec   65
Constancy of the E. Coli DNA Elongation Rate   67
Keeping DNA Synthesis in Step with Cell Division   69
Making Synchronized Cells   72
How Fast Can DNA Be Replicated?   74
Problems   75
Recommended Readings   77
Related Reviews, Books, and Articles   78
Deeper Reading   78

CHAPTER   4 ■ RNA Polymerase and RNA Synthesis   83

Measuring the Activity of RNA Polymerase   84
Escherichia Coli Contains a Single Type of RNA Polymerase   86
Eukaryotic Cells Contain Three Types of RNA Polymerase   87
RNA Polymerases Contain Multiple Subunits   88
Little of the Cellular Polymerase is Free in the Cytoplasm   91
Specificity by the Sigma Subunit: Heat Shock and Sporulation   92
E. Coli RNA Polymerase Elongates at 60 Nucleotides/Sec   93
The Initiation Process Probably Consists of Many Steps   95
Measurement of Binding and Initiation Rates   96
Relating Abortive Initiations to Polymerase Rate Constants   98
Salt Strongly Affects RNA Polymerase Binding   100

Promoters Are Similar but Not Identical   102
RNA Polymerase Melts about 14 Base Pairs After Binding   103
Transcription Termination Occurs at Specific Sites   105
Some Terminators Require Assistance from Rho Protein   106
Processing Prokaryotic RNAs after Synthesis   107
Processing of Eukaryotic RNAs   107
RNA Splicing   109
Problems   111
Recommended Readings   114
Related Reviews, Books, and Articles   114
Deeper Reading   115

CHAPTER   5 ■ Protein Structure   119

The Amino Acids   120
The Peptide Bond   124
Electrostatic Forces that Determine Protein Structure   125
Hydrogen Bonds and the Chelate Effect   129
Hydrophobic Forces   130
Structures Within Proteins   131
Structure of the Alpha Helix, Beta Sheet, and Beta Turn   132
Thermodynamic Considerations to Protein Structure   134
Prediction of Protein Structure   136
Structure of a DNA-Binding Protein   137
Problems   140
Recommended Readings   141
Related Reviews, Books, and Articles   141
Deeper Reading   141

CHAPTER   6 ■ Protein Synthesis   143

A. CHEMICAL ASPECTS   144
Activation of Amino Acids during Protein Synthesis   144
Decoding the Message   147
Ribosomal RNA Base Pairs with Bacterial Message   150
Experimental Support for the Shine-Dalgarno Hypothesis   151
Eukaryotic Translation Begins with the First AUG   154
Tricking the Translation Machinery into Initiating   154
Protein Elongation   156
Peptide Bond Formation   157
Translocation   158
A Paradox   159
Termination, Nonsense, and Suppression   160

B. PHYSIOLOGICAL ASPECTS   162
Messenger RNA Is Unstable   162
Protein Elongation Rates   162
The Need For Directing Proteins to Cellular Sites   166

The Signal Peptide Model for Protein Excretion   166
Testing the Signal Peptide Model   167
Alternatives to the Signal Hypothesis   170
Expectations for Ribosome Regulation   171
Ribosome Levels Are Proportional to Cell Growth Rates   172
Regulation of Ribosome Synthesis   174
The Stringent-Relaxed Problem of rRNA Regulation   175
Maintaining Balanced Synthesis of Ribosomal Components   177
Problems   179
Recommended Readings   182
Related Reviews, Books, and Articles   182
Deeper Reading   183

■ P A R T   T W O ■

Genetics and Genetic Engineering

C H A P T E R   7 ■ Formal Genetics   189

Mutations   190
Point Mutations, Deletions, Insertions, and Damage   191
Classical Genetics of Chromosomes   193
Complementation, *cis, trans,* Dominant, and Recessive   196
Genetic Recombination   198
Mapping by Recombination Frequencies   199
Mapping by Deletions   202
Heteroduplexes Likely Form during Genetic Recombination   203
Branch Migration and Isomerization   204
Elements of Recombination in *E. coli, RecA, RecBC,* and *Chi*   207
Problems   209
Recommended Readings   211
Related Reviews, Books, and Articles   211
Deeper Reading   212

C H A P T E R   8 ■ Genetic Systems   215

Growing Cells for Genetic Experiments   216
Testing Purified Cultures, Scoring   217
Isolating Auxotrophs, Use of Mutagens, and Replica Plating   217
Genetic Selections   218
Mapping with Generalized Transducing Phage   221
Principles of Bacterial Sex   222
Elements of Yeast Genetics   224
Elements of *Drosophila* Genetics   226

Isolating Mutations in Muscle or Nerve in *Drosophila*   226
Fate Mapping and Study of Tissue-Specific Gene Expression   228
Problems   230
Recommended Readings   232
Related Reviews, Books, and Articles   232
Deeper Reading   232

CHAPTER   9 ∎ **Probabilities and the Luria-Delbrück Fluctuation Test   235**

The Binomial Probability Distribution   236
Derivation of the Poisson Distribution   237
The Gaussian Distribution   238
Measures of Probability Distributions   239
Elements of the Luria-Delbrück Fluctuation Test   240
Mutation Frequencies from Fluctuation Analysis   241
Problems   244
Recommended Readings   246
Related Reviews, Books, and Articles   246
Deeper Reading   246

CHAPTER   1 0 ∎ **Genetic Engineering and Recombinant DNA   247**

The Isolation of DNA   248
The Biology of Restriction Enzymes   250
Cutting DNA with Restriction Enzymes   254
Isolation of DNA Fragments   255
Joining DNA Fragments   256
Vectors Provide for Selection and a Cellular Free Ride   259
Plasmid Vectors   259
A Phage Vector for Bacteria   261
Vectors for Higher Cells   263
Putting DNA Back into Cells   265
Chemical DNA Sequencing   266
Enzymatic DNA Sequencing   269
Problems   273
Recommended Readings   276
Related Reviews, Books, and Articles   276
Deeper Reading   277

CHAPTER   1 1 ∎ **Advanced Genetic Engineering   281**

Cloning from RNA or DNA   283
Plaque and Colony Hybridization for Clone Identification   285
R-Loop Enrichment of DNA Complementary to Purified RNA   285
Walking Along a Chromosome to Clone a Gene   286

Arrest of Translation Assay for DNA of a Gene   287
Tools of Analysis I: Southern Transfers, DNA   289
Tools of Analysis II: Northern and Western Transfers, RNA and Protein   290
Tools of Analysis III: Footprinting and S1 Mapping, DNA   293
Altering Cloned DNA by *in vitro* Mutagenesis   296
Mutagenesis with Chemically Synthesized DNA   300
Detecting Genetic Diseases   302
Doctoring the *E. coli araC* Gene for Hyperproduction   304
Problems   307
Recommended Readings   309
Related Reviews, Books, and Articles   309
Deeper Reading   310

■ PART THREE ■

How Genes Are Regulated

CHAPTER 12 ■ Repression and the *lac* Operon   315

Background of the *lac* Operon   316
The Role of Inducer Analogs in the Study of the *lac* Operon   318
lac Repressor Is a Protein   320
An Assay for *lac* Repressor   321
Wild-Type *lac* Repressor Could Not Be Detected   322
Detection and Purification of *lac* Repressor   324
Repressor Binds to DNA: The Operator Is DNA   326
The Isolation and Structure of Operator   328
Repressor Slides Along DNA to Find the Operator   330
Repressor Binds DNA *in vitro* as Tightly as *in vivo*   331
Repressor's N-Terminus Binds DNA: Genetic Demonstration   332
Isolation and Characterization of 1^{-d} Repressor Mutants   333
Repressor's N-Terminus Binds DNA: Physical Demonstration   334
A Mechanism for Induction   336
RNA Polymerase Binding to the *lac* Promoter   337
Problems   338
Recommended Readings   341
Related Reviews, Books, and Articles   342
Deeper Reading   342

CHAPTER 13 ■ Induction and the *ara* Operon   347

The Sugar Arabinose and Arabinose Metabolism   348
Genetics of the Arabinose System   350

araC Protein Regulates Positively: Detection and Isolation   352
araC Protein Represses as Well as Induces   354
The Promoter for Synthesis of *araC* Protein Is Also Regulated   357
Binding Sites of the *ara* Regulatory Proteins   357
DNA Loops Are Involved in Repression of *araBAD*   360
Electrophoresis to Assay Low Affinity DNA-Binding Proteins   361
Why Positive Regulators Are a Good Idea   364
Problems   366
Recommended Readings   368
Related Reviews, Books, and Articles   368
Deeper Reading   369

CHAPTER   1 4 ▪ **Attenuation and the *trp* Operon**   373

The Aromatic Amino Acid Aynthetic Pathway and Its Regulation   374
Rapid Induction Capabilities of the *trp* Operon: Repression   376
The Serendipitous Discovery of *trp* Enzyme Hypersynthesis   378
Early Explorations of the Hypersynthesis   380
trp Leader RNA Has the Potential to Form Multiple Hybrids   383
Coupling Translation to Attenuation   384
RNA Secondary Structure and the Attenuation Mechanism   386
Other Attenuated Operons   388
Problems   388
Recommended Readings   390
Related Reviews, Books, and Articles   390
Deeper Reading   390

CHAPTER   1 5 ▪ **Developmentally Regulated Genes in Yeast and *Drosophila***   395

Mating Type Conversion in Yeast   396
Cloning the Mating Type Loci in Yeast   397
Transfer of Mating Type Gene Copies to an Expression Site   398
Structure of the Mating Type Loci   399
The Expression and Recombination of Paradoxes   400
DNA Cleavage at the *MAT* Locus   401
Regulated Expression of DOPA Decarboxylase in *Drosophila*   403
Cloning the DOPA Decarboxylase Gene   404
Tissue-Specific Responses: Putting the Cloned Gene Back   405
Tissue-Specific Enhancers for *Drosophila* Yolk Proteins   406
Problems   408
Recommended Readings   408
Related Reviews, Books, and Articles   409
Deeper Reading   409

CHAPTER   16 ▪ Lambda Phage Genes and Regulatory Circuitry   411

A. THE STRUCTURE AND BIOLOGY OF LAMBDA   412
The Physical Structure of Lambda   412
The Genetic Structure of Lambda   413
Lysogeny, Immunity, Integration, and Excision   416
Lambda's Relatives and Lambda Hybrids   417

B. CHRONOLOGY OF A LYTIC INFECTIVE CYCLE   418
Lambda Adsorbs to Cells Via a Maltose Transport Protein   418
Termination Restricts Early Transcription to Genes *N* and *cro*   419
N Protein Prevents Termination of Early Gene Transcription   420
cro Protein Blocks *CI* Repressor Synthesis and Wins the Race   421
Proteins *O* and *P* Initiate DNA Synthesis   422
Proteins *kil, λ, β,* and *exo*   423
Q Protein Blocks Termination to Synthesize Late Proteins   424
Packaging and Lysis   425

C. THE LYSOGENIC INFECTIVE CYCLE AND INDUCTION OF LYSOGENS   426
Chronology of Becoming a Lysogen   426
cro and Repressor Repress and Induce the Same Sites   427
Cooperativity in Repressor Binding and Its Measurement   430
The Need for and Realization of Hair-Trigger Induction   431
Induction from the Lysogenic State   434
Entropy, A Basis for Lambda Repressor Inactivation   436
Problems   437
Recommended Readings   439
Related Reviews, Books, and Articles   439
Deeper Reading   440

▪ P A R T   F O U R ▪

Mobile DNA

CHAPTER   17 ▪ Lambda Phage Integration and Excision   449

Integrated Lambda Maps Like a Chromosomal Gene   450
Chromosomal Deletions Also Delete Lambda Genes   451
DNA Heteroduplexes Prove that Lambda Integrates   452
Gene Order Premutation and the Campbell Model   453
Isolation and Integration-Defective Mutants   455
Isolation of Excision-Deficient Mutants   456
Int and *xis* Are Single Genes and Are Phage-Specific   458
Incorrect Excision Produces *gal* and *bio* Transducing Phage   459
Transducting Phage Carrying Genes Other than *gal* and *bio*   461
Use of Transducing Phase to Study Integration and Excision   462

The Double *att* Phage, *att²*   464
Regulating Integration and Excision: *xis* Is Unstable   466
Retro-Inhibition of *int* Translation   467
In vitro Assay of Integration and Excision   469
Host Proteins Are Involved in Integration and Excision   471
Isolation and Properties of *att* Mutants   472
Structure of the *att* Region   473
Int and *xis* Binding in the *att* Region   475
Problems   476
Recommended Readings   478
Related Reviews, Books, and Articles   478
Deeper Reading   478

CHAPTER   1 8   ■   Transposable Genetic Elements   483

The Discovery of IS Elements in Bacteria   484
The Structure and Properties of IS Elements   486
Discovery of Tn Elements   488
The Structure and Properties of Tn Elements   491
Role of the Repeated Sequence Generated by Transposition   492
Genetic and Physical Mapping of Tn3   494
In vitro Transposition   497
Regulating Flagellin Synthesis by Inverting a DNA Segment   498
Mu Phage Is a Giant Transposable Element   500
Mu Phage Possesses a Segment that Inverts   502
Transposable Elements in Higher Cells   504
Transposons as Genetic Engineering Vectors in Higher Cells   506
Transposition via DNA Intermediates   509
Problems   511
Recommended Readings   513
Related Reviews, Books, and Articles   513
Deeper Reading   513

CHAPTER   1 9   ■   Generating Genetic Diversity: Antibodies   517

The Basic Immune Response   517
Telling the Difference Between Foreign and Self   519
The Number of Different Antibodies Produced   521
Myelomas and Monoclonal Antibodies   521
The Structure of Antibodies   524
Cells Have Many Copies of V Genes and Only a Few C Genes   526
The J Regions   528
The D Regions in H Chains   531
Induced Mutations Add to Antibody Diversity   533
Class Switching of Heavy Chains   533
Enhancers and Expression of Immunoglobulin Genes   534

Problems   535
Recommended Readings   536
Related Reviews, Books, and Articles   536
Deeper Reading   537

■ PART   FIVE ■

Special Topics

CHAPTER   20 ■ Biological Assembly, Ribosomes, and Lambda Phage   541

A. RIBOSOME ASSEMBLY   542
RNase and Ribosomes   542
The Global Structure of Ribosomes   543
Assembly of Ribosomes   545
Experiments With *in vitro* Ribosome Assembly   548
Determining Details of Local Ribosomal Structure   550

B. LAMBDA PHAGE ASSEMBLY   552
General Aspects   552
The Geometry of Capsids   553
The Structure of the Lambda Particle   556
The Head Assembly Sequence   556
Packaging the DNA and Formation of the *cos* Ends   558
Formation of the Tail   560
In vitro Packaging   560
Problems   561
Recommended Readings   563
Related Reviews, Books, and Articles   564
Deeper Reading   564

CHAPTER   21 ■ Chemotaxis   569

Assaying Chemotaxis   570
Fundamental Properties of Chemotaxis   572
Genetics of Motility and Chemotaxis   575
How Cells Swim   576
The Mechanism of Chemotaxis   578
The Energy for Chemotaxis   581
Adaptation and Protein Methylation   582
Problems   585
Recommended Readings   587
Related Reviews, Books, and Articles   588
Deeper Reading   588

CHAPTER   2 2   ■   Oncogenesis, Molecular Aspects   591

Bacterially Induced Tumors in Plants   592
Transformation by Damaging the Chromosome   593
Identifying a Nucleotide Change Causing Cancer   595
Retroviruses and Cancer   598
Cellular Genes Have Retroviral Counterparts   601
Identification of the *src* and *sis* Gene Proteins   603
DNA Viruses and Cancer   604
Directions for Future Research in Molecular Biology   605
Problems   606
Recommended Readings   607
Related Reviews, Books, and Articles   607
Deeper Reading   607

INDEX   ■   611

Genetics
and
Molecular
Biology

Cell Structure and Cell Constituents

An Overview of Cell Structure and Function

I n this book we will be concerned with the basics of cellular processes. For the most part we will investigate processes that occur within cells such as DNA synthesis, protein synthesis, and regulation of gene activity. However, before focusing our study at this level, we should take time for an overview of cell structure and at the same time develop our intuitions about the time and distance scales relevant to the molecules and cells we will be studying.

One of the most widely studied cells is the bacterium *Escherichia coli.* Consequently, much of this book considers information learned about this bacterium. Parallel studies on other bacteria and higher cells have revealed that the basic information that has been found for *E. coli* is generally true for other bacteria and often is true with minor modifications of eukaryotic cells as well.

CELLS OPERATE WITH IMMENSE AMOUNTS OF INFORMATION

Cells face enormous problems in growing. We can develop some idea of the situation by considering a totally self-sufficient toolmaking shop. If we provide the shop with coal for energy and crude ores, analogous to a cell's nutrient medium, then a very large collection of machines and tools is necessary merely to produce each of the parts present in the tools and machines of the shop. Still greater complexity would be added if we required that the shops be totally self-regulating and that each machine be self-assembling. Cells face, and solve, these types of problems. In addition, each of the chemical reactions necessary for growth of cells is carried out in an aqueous environment at near neutral pH. These are conditions that would cripple ordinary chemists.

By the tool shop analogy, we expect cells to utilize large numbers of "parts," and, also by analogy to factories, we expect each of these parts to be generated or interconverted by a specialized machine devoted to production of just one type of part. Indeed, biochemists' studies of metabolic pathways have revealed that an *E. coli* cell contains about 1,000 types of "parts," different small molecules, and that each is generated by a specialized machine, an enzyme. The information required to specify the structure of even one machine is great, a fact made apparent by trying to describe an object without pictures and drawings. Thus, it is reasonable, and indeed it has been found, that cells function with truly immense amounts of information.

DNA is the cell's library in which information is stored in its sequence of nucleotides. Evolution has built into this library the information necessary for cells' growth and division. Because of the extreme value of the DNA library, it is natural that it be carefully protected and preserved. Except for some of the simplest viruses, cells keep duplicates of the information by using a pair of self-complementary DNA strands. Each strand contains a complete copy of the information, and chemical or physical damage to one strand is recognized by special enzymes and is repaired by using information contained on the opposite strand. Higher and more complex cells further preserve their information by possessing duplicate DNA duplexes.

Much of the recent activity in molecular biology can be understood in terms of the cell's information library. Clearly the library contains far too much information for the cell to use at any one time. Therefore mechanisms must have developed to recognize the need for particular portions, "books," of the information and to generate usable copies. In cellular terms, this is the regulation of gene activity.

RUDIMENTS OF PROKARYOTIC CELL STRUCTURE

A typical prokaryote, *E. coli,* is a rod capped with hemispheres (Fig. 1.1). It is $1-3\,\mu$ (10^{-4} cm $= 1\,\mu = 10^4$ Å) long and $0.75\,\mu$ in diameter. Such a cell contains about 2×10^{-13} g of protein, 2×10^{-14} g of RNA that is mostly ribosomal RNA, and 6×10^{-15} g of DNA.

The cell envelope consists of three parts (Fig. 1.2). The outer surface of the outer membrane is largely lipopolysaccharides. These are polysaccharides attached to lipid molecules that constitute the major portion of the outer half of the outer membrane lipid bilayer. The outer membrane consists of these lipopolysaccharides, matrix proteins that form pores for passage of small molecules, and phospholipids. The major shape-determining factor of cells is the peptidoglycan layer or cell wall (Fig. 1.3). It lies beneath the outer membrane and is a single molecule containing many polysaccharide chains crosslinked by short peptides (Fig. 1.4). The outer membrane is attached to the peptidoglycan layer by about 10^6 lipoprotein molecules. One end of a lipoprotein molecule is covalently attached to the diaminopimelic acid

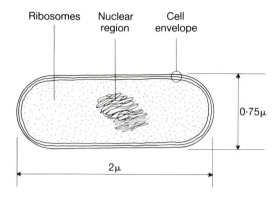

Figure 1.1   The dimensions of a typical *E. coli* cell.

in the peptidoglycan, and the other end, a lipid, is buried in the outer membrane. Finally, the third layer of the cell wall is the inner or cytoplasmic membrane, which consists of phospholipids and proteins. The space between the inner membrane and the outer membrane is known as the periplasmic space. The cell wall and membranes contain about 20% of the cellular protein; after cell disruption by sonicating or grinding, most of this protein is still contained in fragments of wall and membrane and can be easily pelleted by low-speed centrifugation.

The cytoplasm within the inner membrane is a protein solution at about 100 mg/ml, about 10 times more concentrated than the usual cell-free extracts used in

Figure 1.2   Schematic drawing of the structure of the envelope of an *E. coli* cell.

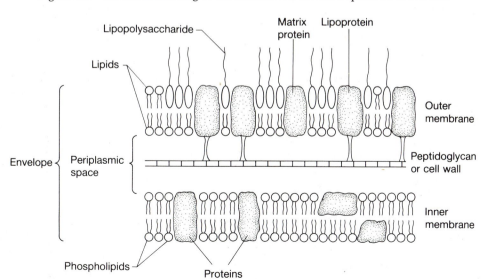

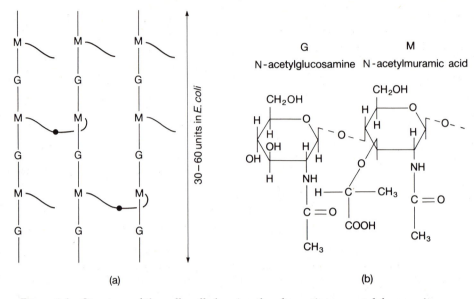

Figure 1.3   Structure of the cell wall showing the alternating N-acetylglucosamine N-acetylmuramic acid units. Each N-acetylmuramic acid possesses a peptide, but only a few are crosslinked in *E. coli*.

Figure 1.4   Structure of the peptide crosslinking N-acetylmuramic acid units. DAP is diaminopimelic acid.

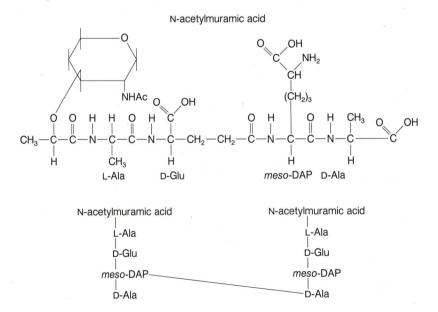

the laboratory. Some proteins in the cytoplasm may constitute as much as 5% by weight of the total cellular protein whereas others may be found at levels as low as 0.01%. The concentrations of many of the proteins vary with growth conditions, and a current research area is the study of the cellular mechanisms responsible for the variations.

Most of the more than 1,000 different types of proteins found within a bacterial cell are located in the cytoplasm. One question to be answered about these proteins is how they manage to exist in the cell without adhering to each other and forming aggregates. Proteins can easily bind to each other. For example, if we denature proteins and expose previously buried portions of their polypeptide chains to the solvent and to other proteins, we produce amorphous precipitates. One might even expect an occasional mutation to inactivate simultaneously two apparently unrelated proteins by the coprecipitation of the mutated protein and some other protein into an inactive aggregate.

The cell's DNA and about 10,000 ribosomes also reside in the cytoplasm. The ribosomes consist of about one-third protein and two-thirds RNA and are roughly spherical with a diameter of about 200 Å. The DNA in the cytoplasm is not surrounded by a nuclear membrane as it is in the cells of higher organisms, but nonetheless it is confined to a portion of the cellular interior. In electron micrographs of cells, the DNA can be seen as a stringy mass in one portion of the cell, and the ribosomes appear as granules uniformly scattered through the cytoplasm.

RUDIMENTS OF EUKARYOTIC CELL STRUCTURE

Like bacteria, eukaryotic cells contain cell membranes, cytoplasmic proteins, DNA, and ribosomes, albeit of somewhat different structure from the corresponding prokaryotic elements (Fig. 1.5). However, eukaryotic cells possess many structural features that even more clearly distinguish them from prokaryotic cells. Within the eukaryotic cytoplasm are a number of structural proteins that form networks of a number of fiber types. Actin, tubulin, and the various intermediate filament proteins form the three main categories of fibers found within eukaryotic cells (Fig. 1.6). Fibers within the cell participate in cell and chromosome movement and also bind the majority of the ribosomes.

The DNA of eukaryotic cells does not freely mix with the cytoplasm but is confined within a nuclear membrane. The DNA itself is tightly complexed with a class of proteins called histones, whose main function appears to be to help it retain a condensed state. Of course, when the cell divides, a specialized apparatus consisting in part of microtubules is necessary to pull the chromosomes into the daughter cells.

Eukaryotic cells also contain specialized organelles such as mitochondria, which perform oxidative phosphorylation to generate the cells' needed chemical energy. In many respects mitochondria resemble bacteria and are thought to have evolved from bacteria. They contain DNA, usually in the form of a circular chromosome like that of *E. coli* and ribosomes that often more closely resemble those found in bacteria than the ribosomes located in the cytoplasm of the eukaryotic cell. Chloroplasts,

Plasma membrane
Mitochondrion
Fibers
Nuclear membrane
Nucleus
Endoplasmic reticulum
Golgi apparatus

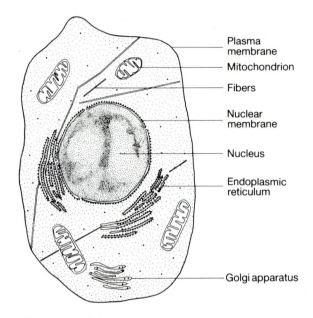

Figure 1.5 Schematic drawing of a eukaryotic cell.

Figure 1.6 Basic structure of the three major cytoskeletal fibers found in most types of eukaryotic cells.

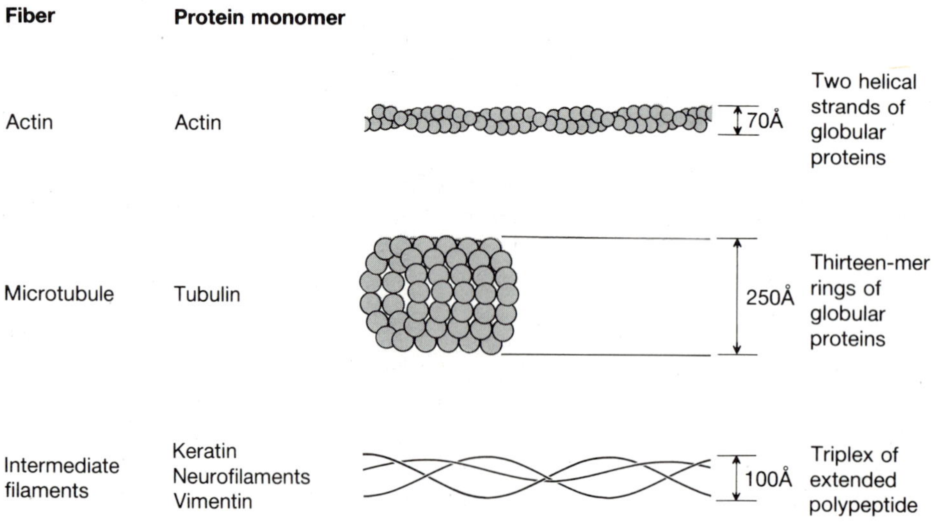

Fiber
Protein monomer
Actin
Actin
70Å
Two helical strands of globular proteins
Microtubule
Tubulin
250Å
Thirteen-mer rings of globular proteins
Intermediate filaments
Keratin
Neurofilaments
Vimentin
100Å
Triplex of extended polypeptide

which carry out photosynthesis in plant cells, are another type of specialized organelle found within eukaryotic cells. Like mitochondria, chloroplasts also contain DNA and ribosomes different from the analogous structures located elsewhere in the cell. Most eukaryotic cells also contain internal membranes. The nucleus is surrounded by two membranes. The endoplasmic reticulum is another membrane found in eukaryotic cells. It is contiguous with the outer nuclear membrane but extends throughout the cytoplasm in many types of cells and is involved with the synthesis and transport of membrane proteins. Similarly, the Golgi apparatus, another structure containing membranes, is involved with modifying proteins for their transport to other cellular organelles.

DNA IS TIGHTLY PACKED WITHIN CELLS

The DNA of the *E. coli* chromosome has a molecular weight of about 2×10^9 and thus is about 3×10^6 base pairs long. Since the distance between base pairs in DNA is about 3.4 Å, the length of the chromosome is 10^7 Å or 0.1 cm. This is very long compared to the 10^4-Å length of a bacterial cell, and the DNA must therefore wind back and forth many times within the cell. Observation by light microscopy of living bacterial cells and by electron microscopy of fixed and sectioned cells shows that, in fact, the DNA is confined to a portion of the interior of the cell with dimensions less than 0.25 μ.

To gain some idea of the relevant dimensions, let us estimate the number of times that the DNA winds back and forth within its confined volume. This will provide an idea of the average distance separating DNA strands and will also give some idea of the proportion of the DNA that lies on the surface of the chromosomal

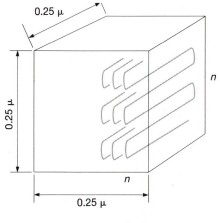

Figure 1.7   Calculation of the number of times the *E. coli* chromosome winds back and forth if it is confined within a cube of edge 0.25 μ. Each of the N layers of DNA possesses N segments of length 0.25 μ.

$$n^2 \times 0.25 \ \mu = 10^3 \ \mu$$
$$n \approx 60$$

mass. The number of times, N, that the DNA must wind back and forth will then be related to the length of the DNA and the volume in which it is contained. If we approximate the path of the DNA as consisting of N layers, each layer consisting of N segments of length 0.25 μ (Fig. 1.7), the total number of segments is N^2, so $2{,}500N^2 = 10^7$ and $N = 60$. We calculate the spacing between adjacent segments of the DNA as $2{,}500/60 = 40$ Å.

The close spacing between DNA strands raises the interesting problem of accessibility of the DNA. RNA polymerase has a diameter of about 100 Å and it may not fit between the strands. Therefore, quite possibly only DNA on the surface of the nuclear mass is accessible for transcription. As we will see later, however, transcription begins on the lactose and arabinose operons as promptly as several seconds after the addition of inducer. Consequently either the nuclear mass is in such rapid motion that any portion of the DNA finds its way to the surface at least once every several seconds, or the RNA polymerase molecules do penetrate to the interior of the nuclear mass and are able to begin transcription of any gene at any time, or possibly the arabinose and lactose operons always reside on the surface of the DNA.

Compaction of the DNA generates even greater problems in eukaryotic cells. Not only is the amount of the DNA very great in some eukaryotic cell types, but the presence of the histones appears to hinder access of RNA polymerase and other enzymes to the DNA. The DNA of many eukaryotic cells is especially contracted before cell division, and at this time it does become inaccessible to RNA polymerase.

PUTTING MOLECULES IN OR OUT OF CELLS

Cells must not leak appreciable quantities of their small-molecule metabolic intermediates to the medium. As a result, the cell membranes are likely to be impermeable to such molecules. Cells must therefore transport certain small molecules like sugars and ions across their membranes. Since the process of concentrating molecules in a volume requires work, metabolic energy of the cells must be coupled to transport and must be consumed in the process.

The amount of work consumed in transporting a molecule into a volume against a concentration gradient may be obtained by considering the simple reaction

$$A_o \rightleftharpoons A_i,$$

where A_o is the concentration of the molecule outside the cell and A_i is the concentration inside the cell:

This reaction can be described by an equilibrium constant

$$K_{eq} = \frac{A_i}{A_o}.$$

The equilibrium constant K_{eq} is related to the free energy of the reaction by the relation

$$\Delta G = -RT \ln K_{eq},$$

where R is about 2 cal/deg $\cdot$ mole and T is 300K (about 25°C), the temperature of many biological reactions. Suppose the energy of hydrolysis of ATP to ADP is coupled to this reaction with a 50% efficiency. Then about 3,500 of the total of 7,000 calories available per mole of ATP hydrolyzed under physiological conditions will be available to the transport system. Consequently, the equilibrium constant will be

$$K_{eq} = e^{-\Delta G/RT} = e^{3,500/600}$$
$$= 340.$$

One interesting result of this consideration is that the work required to transport a molecule is independent of the absolute concentrations; it depends only on the ratio of the inside and outside concentrations.

The transport systems of cells must recognize the type of molecule to be transported, since not all types are transported, and convey the molecule either to the inside or to the outside of the cell. Further, if the molecule is being concentrated within the cell, the system must tap an energy source for the process. Owing to the complexities of this process, it is not surprising that the details of active transport systems are far from being fully understood.

Four basic types of small-molecule transport systems have been discovered. The first of these is facilitated diffusion. Here the molecule must get into or out of the cell on its own, but special doors are opened for it. That is, specific carriers exist that bind to the molecule and shuttle it through the membrane:

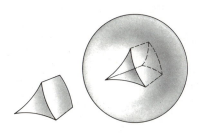

Glycerol enters most types of bacteria by this mechanism. Once within the cell the glycerol is phosphorylated and cannot diffuse back out through the membrane, nor can it exit by using the glycerol carrier protein that carried the glycerol into the cell.

A second method of concentrating molecules within cells is similar to the facilitated diffusion and phosphorylation of glycerol. The phosphotransferase system actively rather than passively carries a number of types of sugars across the cell membrane and, in the process, phosphorylates them (Fig. 1.8). The actual energy for the transport comes from phosphoenolpyruvate. The phosphate group and part of the chemical energy contained in the phosphoenolpyruvate is transferred down a series of proteins, two of which are used by all the sugars transported by this system and two of which are specific for the particular sugar being transported. The final protein is located in the membrane and is directly responsible for the transport and phosphorylation of the transported sugar.

Protons are expelled from the cell during the flow of reducing power from NADH to oxygen:

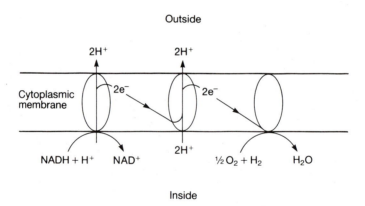

The resulting concentration difference in H^+ ions between the interior and exterior of the cell generates a protonmotive force or membrane potential that can then be coupled to ATP synthesis or to the transport of molecules across the membrane. Active transport systems using this energy source are called chemiosmotic systems. In the process of permitting a proton to flow back into the cell, another small molecule can be carried into the cell or carried out of the cell (Fig. 1.9). Study of all transport systems has been difficult because of the necessity of working with membranes, but the chemiosmotic system has been particularly hard due to the difficulty of manipulating membrane potentials. Fortunately the existence of mutants blocked at various steps of the transport process has permitted partial dissection of the

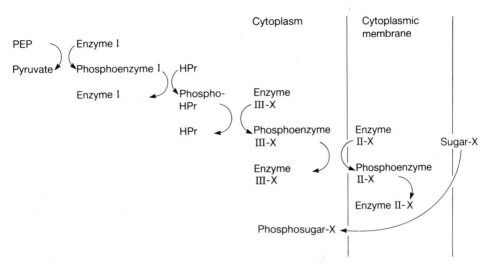

Figure 1.8   The cascade of reactions associated with the phosphotransferase sugar uptake system of *E. coli.*

Figure 1.9   Coupling the excess of H⁺ ions outside a cell to the transport of a specific molecule into the cell; symport, or out of the cell, antiport, by specific proteins that couple the transport of a proton into the cell with the transport of another molecule. The ATPase generates ATP from ADP with the energy deriving from permitting protons to flow back into the cell.

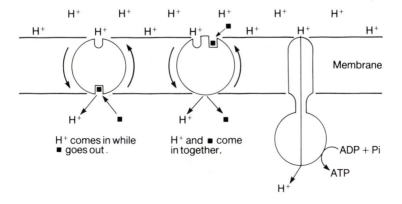

system. However, we are very far from completely understanding the actual mechanisms involved in chemiosmotic systems.

The binding protein systems represent another type of transport through membranes. These systems utilize proteins located in the periplasmic space that specifically bind sugars, amino acids, and ions. Apparently these periplasmic binding proteins transfer their substrates to specific carrier molecules located in the cell membrane. The energy source for these systems is ATP or a closely related metabolite.

Altogether, cells contain a variety of transport mechanisms. Different strains of bacteria and different cell types often transport the same molecule by different systems, one substrate may be transported by several systems, and several systems may even transport the same molecule. The reasons for such nonuniformity and diversity are not at all apparent.

DIFFUSION IS VERY RAPID WITHIN THE SMALL VOLUME OF A CELL

Within several minutes of adding some inducers to bacteria or eukaryotic cells, newly synthesized, active enzymes can be detected. These are the result of the synthesis of the appropriate messenger RNA, its translation into protein, and the folding of the protein to an active conformation. Quite obviously, processes are happening very rapidly within a cell for this entire sequence to be completed in several minutes. We will see that our image of synthetic processes in the cellular interior should be that of an assembly line running hundreds of times faster than usual, and our image for the random motion of molecules from one point to another can be that of a washing machine similarly running very rapidly.

The random motion of molecules within cells can be estimated from basic physical chemical principles. We will develop such an analysis since similar reasoning often arises in the design or analysis of experiments in molecular biology. The

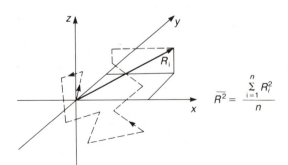

$$\overline{R^2} = \frac{\sum\limits_{i=1}^{n} R_i^2}{n}$$

Figure 1.10   Random motion of a particle in three dimensions beginning from the origin and the definition of the mean squared distance $\overline{R^2}$.

mean squared distance $\overline{R^2}$ that a molecule with diffusion constant D will diffuse in time t is $\overline{R^2} = 6Dt$ (Fig. 1.10). The diffusion constants of many molecules have been measured and are available in tables. For our purposes, we can estimate a value for a diffusion constant. The diffusion constant is $D = KT/f$, where K is the Boltzmann constant, 1.38×10^{-16} ergs/degree, T is temperature in degrees Kelvin, and f is the frictional force. For spherical bodies, $f = 6\pi\eta r$, where r is the radius in centimeters and η is the viscosity of the medium in poise.

The viscosity of water is 10^{-2} poise. Although the macroviscosity of the cell's interior could be much greater, as suggested by the extremely high viscosity of gently lysed cells, the viscosity of the cell's interior with respect to motion of molecules the size of proteins or smaller is more likely to be similar to that of water. This is reasonable, for small molecules can go around obstacles such as long strands of DNA, but large molecules would have to displace the DNA. A demonstration of this effect is the finding that small molecules such as amino acids can readily diffuse through the agar used for growing bacterial colonies, but objects as large as viruses diffuse only slightly.

Since $D = KT/6\pi\eta r$, then $D = 4.4 \times 10^{-7}$ for a large spherical protein of radius 50 Å diffusing in water, and the diffusion constant for such a protein within a cell is not greatly different. Therefore $\overline{R^2} = 6 \times 4.4 \times 10^{-7}t$, and the average time required for such protein molecules to diffuse the length of a cell is $\frac{1}{250}$ second. Analogous reasoning with respect to rotation shows that a protein rotates about $\frac{1}{8}$ radian (about $7°$) in the time it diffuses a distance equal to its radius.

FREELY GROWING POPULATIONS INCREASE EXPONENTIALLY

Reproducibility from one day to the next and between different laboratories is necessary before meaningful measurements can be made on growing cells. Populations of cells that are not overcrowded or limited by oxygen, nutrients, or ions grow freely and can be easily reproduced. Such freely growing populations are almost universally used in molecular biology, and several of their properties are important. The rate of increase in the number of cells in a freely growing population is proportional to the number of cells present, that is,

$$\frac{dN}{dt} = \mu N \quad \text{or} \quad N(t) = N(0)e^{\mu t}.$$

In these expressions μ is termed the exponential growth rate of the cells.

The following properties of the exponential function are frequently useful when manipulating data or expressions involving growth of cells.

$$e^{a \ln x} = x^a$$

$$\frac{d}{dx} e^{ax} = ae^{ax} \qquad \int e^{ax}dx = \frac{1}{a} e^{ax}$$

Table 1.1   The Relation Between
Cell Doublings, Enzyme
Doublings, and Induction Kinetics.

Time	$t = 0$	$t = T_d$	$t = 2T_d$	$t = 3T_d$
Cell mass	1	2	4	8
Total enzyme present if synthesis began long ago	A	2A	4A	8A
Enzyme synthesized during one doubling time		A	2A	4A
Enzyme present if synthesis begins at $t = 0$	0	A	3A	7A

$$e^x = \sum_{n=0}^{\infty} \frac{x^n}{n!}$$

Quantities growing with the population increase as $e^{\mu t}$; throughout this book we will use μ as the exponential growth rate. However, the time required for cells to double in number, T_d, is easier to measure experimentally as well as to think about than the exponential growth rate. Therefore we often need to interconvert the two rates T_d and μ. Note that the number of cells or some quantity related to the number of cells in freely growing populations can be written as $Q(t) = 2^{t/T_d}$, and since $2 = e^{\ln 2}$, $Q(t)$ can also be written as $Q(t) = e^{(\ln 2/T_d)t}$, thereby showing that the relation between T_d and μ is $\mu = \ln 2/T_d$.

COMPOSITION CHANGES SLOWLY IN EXPONENTIALLY GROWING CELLS

In many experiments it is necessary to consider the time course of the induction of an enzyme or other cellular component in a population of growing cells. Suppose that synthesis of an enzyme is initiated at some given time in all the cells in the population and that thereafter the synthesis rate per cell remains constant. What will the enzyme level per cell be at later times?

One way to study induction kinetics is to examine the enzyme synthesis under slightly different conditions. Suppose, instead, that synthesis of the enzyme had begun many generations earlier and thereafter the synthesis rate per cell had remained constant. Since the synthesis of the enzyme had been initiated many cell doublings earlier, by the time of our consideration, the cells are in a steady state and the relative enzyme level per cell remains constant. As the cell mass doubles from 1 to 2 to 4, and so on, the amount of the enzyme, A, also doubles, from A to $2A$ to $4A$, and so on (Table 1.1). The differences in the amount of the enzyme at the different times give the amounts that were synthesized in each doubling time. Now consider the situation if the same number of cells begins with no enzyme but instead begins synthesis at the same rate per cell as the population that had been induced at a much earlier time (see the last row in the table). At the beginning, no enzyme is present, but during the first doubling time an amount A of the enzyme can be synthesized by the cells. In the next doubling time the table shows that the cells can synthesize an amount $2A$ of the enzyme, so after two doubling times the total amount of enzyme present is $3A$. After another doubling time the amount of enzyme present is $7A$.

*AGE DISTRIBUTION IN POPULATIONS OF GROWING CELLS

The cells in a population of freely growing cells are not all alike, for a newly divided cell grows, doubles in volume, and divides into *two* daughter cells. Consequently, freely growing populations contain twice as many cells having just come from division as cells about to divide. The distribution of cell ages present in growing populations is important in a number of molecular biology experiments, one of which is mentioned in Chapter 3. Therefore we will derive the distribution of ages present in such populations.

Consider an idealized case where cells grow until they reach the age of 1, at which time they divide. In reality most cells do not divide at exactly this age, but the ages at which cell division occurs cluster around a peak. To derive the age distribu-

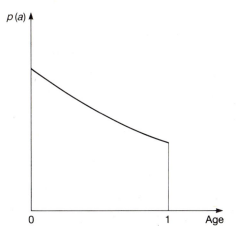

Figure 1.11   Age distribution in an exponentially growing population in which all cells divide when they reach age 1. Note that the population contains twice as many zero-age cells as unit-age cells.

tion, let $N(a, t) \, da$ be the number of cells with age between a and $a + da$ at time t. For convenience, we omit writing the da. Since the number of cells of age a at time t must be the same as the number of zero-age cells at time $t - a$, $N(a, t) = N(0, t - a)$. Since the numbers of cells at any age are growing exponentially, $N(0, t) = N(0, 0)e^{\mu t}$, and $N(a, t) = N(0, t - a) = N(0, t)e^{-\mu a}$. Therefore the probability that a cell is of age a, $p(a)$, is $p(0)e^{-\mu a} = p(0)2^{-a/T_d}$ (Fig. 1.11).

PROBLEMS

1.1. Propose an explanation for the following facts known about E. coli: Appreciable volume exists between the inner membrane and the peptidoglycan layer; the inner membrane is too weak to withstand the osmotic pressure of the cytoplasm and must be supported by a strong, rigid structure; and no spacers have been discovered that could hold the inner membrane away from the peptidoglycan layer.

1.2. If the E. coli interior were water at pH 7, how many H^+ ions would exist at any instant?

1.3. If a population of cells growing exponentially with a doubling time T_d were contaminated at one part in 10^7 with cells whose doubling time is $0.99 \, T_d$, how many doublings will be required until 50% of the cells are contaminants?

1.4. If an enzyme is induced and its synthesis per cell is at a constant rate, show that there is a final upper bound less than 100% of cellular protein that this enzyme can constitute. When the enzyme has reached this level, what is the relation between the rate of synthesis of the enzyme and the rate of dilution of the enzyme caused by increase of cellular volume due to growth?

1.5. In a culture of cells in balanced exponential growth, an enzyme was induced at time $t = 0$. Before induction the enzyme was not present, and at times very long after induction it constituted 1% of cell protein. What is the fraction of cellular protein constituted by this protein at any time $t > 0$? Ignore the 1 min or so lag following induction until the enzyme begins to appear.

1.6. In a culture of cells in balanced exponential growth, an enzyme was fully induced at some very early time, and the level of enzyme ultimately reached 1% of total protein. At time $t = 0$ the synthesis of enzyme was repressed. What fraction of cellular protein is constituted by the enzyme for $t > 0$ (a) if the repressed rate of synthesis is 0 and (b) if the repressed rate of synthesis is 0.01 of the fully induced rate?

*1.7. Suppose the synthesis of some cellular component requires synthesis of a series of precursors P_1, P_2, P_n proceeding through a series of pools S_i.

$$P_1 \longrightarrow P_2 \longrightarrow P_3 \longrightarrow \longrightarrow P_n$$
$$S_1 \longrightarrow S_2 \longrightarrow S_3 \longrightarrow \longrightarrow S_n$$

Suppose the withdrawal of a precursor molecule P_i from pool S_i and its maturation to

S_{i+1} is random. Suppose that at $t = 0$ all subsequently synthesized precursors P_1 are radioactively labeled at constant specific activity. Show that at the beginning, the radioactive label increases proportional to t^n in pool S_n.

*1.8. Consider cells growing in minimal medium. Suppose a radioactive amino acid is added and the kinetics of radioactivity incorporation into protein are measured for the first minute. Assume that upon addition of the amino acid the cell completely stops its own synthesis of the amino acid and that there is no leakage of the amino acid out of the cell. For about the first 15 sec the incorporation of radioactive amino acid into protein increases as t^2 and thereafter as t. Show how this delayed entry of radioactive amino acids into protein results from the pool of free nonradioactive amino acid in the cells at the time the radioactive amino acid was added. Continue with the analysis and show how to calculate the concentration of this internal pool. Use data of Fig. 2 in JMB 27, 41 (1967) to calculate the molarity of free proline in E. coli B/r.

1.9. If the concentration of a typical amino acid in a cell is 10^{-5} M, estimate how long this quantity, without replenishment, could support protein synthesis at the rate that yields 1×10^{-13} g of newly synthesized protein in a cell doubling time of 30 min.

1.10. If a typical protein can diffuse from one end to the other of a cell in $\frac{1}{250}$ sec when it encounters viscosity the same as that of water, how long is required if the viscosity is 100 times greater?

*1.11. Consider a more realistic case for cell division than was considered in the text. Suppose that cells do not divide precisely when they reach age 1 but that they have a probability given by the function $f(a)$ of dividing when they are of age a. What is the probability that a cell is of age a in this case?

*1.12. Show that if $f(a)$ is the probability that a cell of age zero will divide when it reaches age a, show that

$$\frac{f(a)}{\int_a^\infty f(a')\, da'}$$

is the probability that a cell of age a will divide in the next instant.

*1.13. Show that if $f(a)$ is known, μ can be found from the relation $\frac{1}{2} = \int_0^\infty e^{-\mu a} f(a)\, da$. [Hint: Consider that the number of cells of age zero is related to the number of cells of any age.

*1.14. Show that the number of zero-age cells present in a culture at any time is related to the input of zero-age cells to the culture $I(t)$ by the relation $N(0, t) = 2 \int_0^\infty N(0, t - a) f(a)\, da + I(t)$. The Laplace transformation of a function is defined as $F(s) = \int_{-\infty}^\infty e^{-st} f(f)\, dt$, and the function can be recovered by the relation $f(t) = (1/2\pi) \int_{c-i\infty}^{c+i\infty} e^{-st} F(s)\, ds$. Show that $N(s) = I(s)/[1 - 2F(s)]$. By consideration of t approaching infinity or s approaching zero, show that the relation $\frac{1}{2} = \int_0^\infty e^{-\mu a} f(a)\, da$ can be recovered.

RECOMMENDED READINGS

Role of an Electrical Potential in the Coupling of Metabolic Energy to Active Transport by Membrane Vesicles of *Escherichia coli*, H. Hirata, K. Altendorf, F. Harold, Proc. Nat. Acad. Sci. USA *70*, 1804–1808 (1973).

Localization of Transcribing Genes in the Bacterial Cell by Means of High Resolution Autoradiography, A. Ryter, A. Chang, J. Mol. Biol. *98*, 797–810 (1975).

RELATED REVIEWS, BOOKS, AND ARTICLES

Chapter 1, Introduction to Statistical Methods in *Fundamentals of Statistical and Thermal Physics.* F. Reif, McGraw-Hill, New York (1965).

Genetics of the Bacterial Phosphoenolpyruvate Glycose Phosphotransferase System, C. Cordaro, Ann. Rev. of Genet. *10*, 341–359 (1976).

Cellular Transport Mechanisms, D. Wilson, Ann. Rev. of Biochem. *47*, 933–965 (1978).

Bacterial Outer Membranes, ed. M. Inouye, John Wiley & Sons, New York (1979).

Carbohydrate Transport in Bacteria, S. Dills, A. Apperson, M. Schmidt, M. Saier Jr., Microbiol. Rev. *44*, 385–418 (1980).

Biochemistry, L. Stryer, W. H. Freeman and Co., San Francisco (1981).

Growth of the Bacterial Cell, J. Ingraham, O. Maaløe, F. Neidhardt, Sinauer Associates, Inc., Sunderland, Mass. (1983).

Molecular Biology of the Cell, B. Alberts, D. Bray, J. Lewis, M. Raff, K. Roberts, J. Watson, Garland Publishing, Inc., New York (1983).

Biochemistry, G. Zubay, Addison-Wesley Publishing Co., Reading, Mass. (1983).

DEEPER READING

Sugar Transport, I. Isolation of a Phosphotransferase System from *E. coli*. W. Kundig, S. Roseman, J. Biol. Chem. *246*, 1393–1406 (1971).

The Relationship Between the Electrochemical Proton Gradient and Active Transport in *E. coli* Membrane Vesicles, S. Ramos, H. Kaback, Biochem. *16*, 854–859 (1977).

Escherichia coli Intracellular pH, Membrane Potential, and Cell Growth, D. Zilberstein, V. Agmon, S. Schuldiner, E. Padan, J. Bact. *158*, 246–252 (1984).

Nucleic Acid Structure

T hus far we have considered the structure of cells and a few facts about their functioning. In the next few chapters we will be concerned with the structure, properties, and biological synthesis of the molecules that have been particularly important in molecular biology — DNA, RNA, and protein. In this chapter we consider DNA and RNA. The structures of these two molecules make them well suited for their biological roles of storing and transmitting information. This information is fundamental to the growth and survival of cells and organisms because it specifies the structure of the molecules that make up a cell.

Information can be stored by any object that can possess more than one distinguishable state. For example, we could let a stick six inches long represent one message and a stick seven inches long represent another message. Then we could send a message specifying one of the two alternatives merely by sending a stick of the appropriate length. If we could measure the length of the stick to one part in ten thousand, we could send a message specifying one of ten thousand different alternatives with just one stick. Information merely is limiting the alternatives.

We will see that the structure of DNA is particularly well suited for the storage of information. Information is stored in the linear DNA molecule by the particular sequence of four different elements along its length. Furthermore, the structure of the molecule or molecules — two are usually used — is sufficiently regular that generalized tools, called enzymes, can copy, repair, and read out the stored information independent of its content. This duplicated information storage scheme also permits repair of damaged information and a unified mechanism of replication.

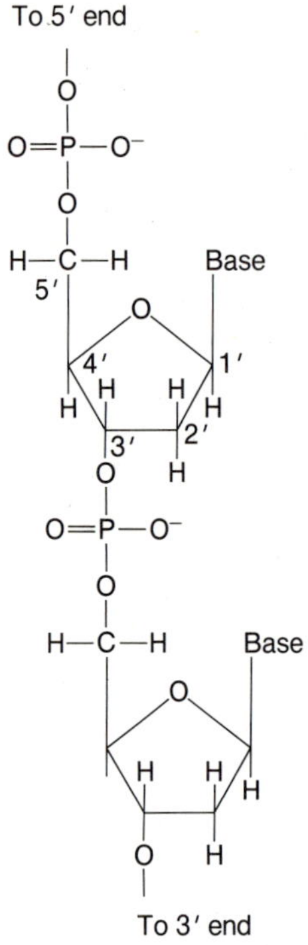

To 5' end

Figure 2.1   The phosphodiester backbone of DNA.

To 3' end

One of the cellular uses of RNA, discussed in later chapters, is as a temporary information carrier. Consequently RNA must also carry information, but ordinarily it need not participate in replication or repair activities.

THE REGULAR BACKBONE OF DNA

The chemical structure of DNA is a regular backbone of 2'-deoxyriboses, ribose lacking the hydroxyl on the 2' carbon, joined by 3'-5' phosphodiester bonds (Fig. 2.1). The information carried by the molecule is specified by bases attached to the 1' position of the deoxyriboses. Four bases are used: the purines adenine and guanine and the pyrimidines cytosine and thymine (Fig. 2.2). The units of base plus ribose or deoxyribose are called nucleosides, and if phosphates are attached to the sugars, the molecules are called nucleotides.

The chemical structure of RNA is similar to that of DNA. The backbone of RNA uses riboses rather than 2'-deoxyriboses, and the methyl group on the 5 position of thymine is absent, leaving the pyrimidine uracil.

Clearly the backbones of DNA and RNA are regular. Can anything be done to

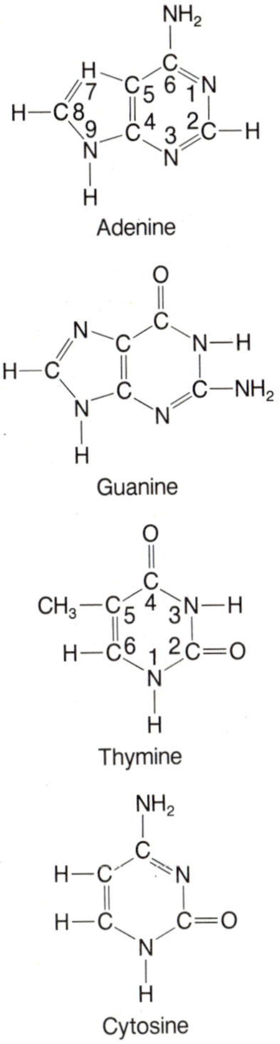

Adenine

Guanine

Thymine

Cytosine

Figure 2.2   The purine bases adenine and guanine and pyrimidine bases thymine and cytosine found in DNA.

make the information storage portion of the molecule regular as well? At first glance this seems impossible because the purines and the pyrimidines are different sizes and shapes. However, as Watson and Crick noticed, pairs of these molecules, adenine-thymine and guanine-cytosine, do possess regular shapes (Fig. 2.3). The deoxyribose residues on both A-T and G-C pairs are separated by the same distance and are at the same relative orientations. Not only are these pairs regular, but they are stabilized by strong hydrogen bonds. The A-T pair can form two hydrogen bonds and the G-C base pair can form three hydrogen bonds. Hydrogen bonds can form when a hydrogen atom can be shared by a donor such as an amino group and an acceptor such as a carbonyl group. The hydrogen bonds between the bases of DNA are strong because in all cases the three atoms participating in hydrogen bond formation lie in nearly straight lines.

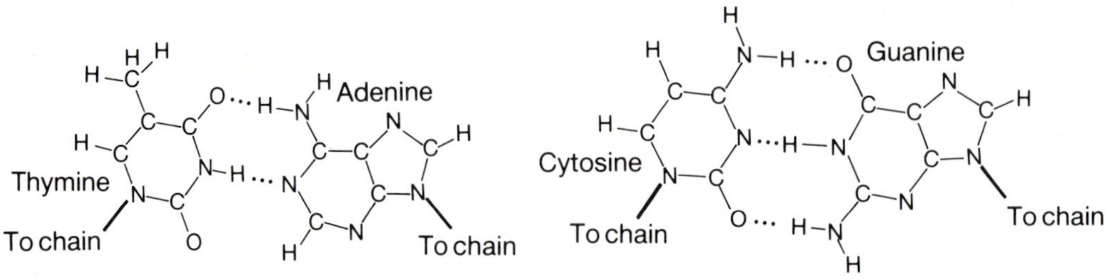

Figure 2.3   Hydrogen bonding between adenine-thymine and guanine-cytosine pairs.

HELICAL FORMS OF DNA

Watson and Crick deduced the basic structure of DNA by using three pieces of information: X-ray diffraction data, the structures of the bases, and Chargaff's findings that the guanine and cytosine and the adenine and thymine compositions of DNA are equal. The Watson-Crick structure is a pair of oppositely oriented DNA strands that wind around one another in a right-handed helix (Fig. 2.4). That is, strands wrap clockwise moving down the axis away from an observer. Base pairs A-T and G-C were proposed to lie on the interior of the helix and the phosphate groups on the outside.

In semicrystalline fibers of native DNA at one humidity as well as in crystals of chemically synthesized DNA, the helix repeat is 10 base pairs per turn. In solution, however, the structure is slightly different, and the helix repeat is 10.4 to 10.6 base pairs per turn. At any of these helical repeats the two phosphate backbones are slightly offset from one another so that the resulting two grooves in the DNA are not equal. The larger groove is known as the major groove, and the smaller is called the minor groove. These basic structures are minor variations in a conformation known as the B form of DNA. A more dramatic structural difference is observed in different sequences of chemically synthesized DNA or in fibers of native DNA at different humidity. The diameters, helical repeat distances, and orientations of the bases in these forms are substantially different from those of the B form. Table 2.1 compares four helical forms.

Studies with models have shown that even left-handed DNA helices are possible. Rich and coworkers have obtained definitive X-ray diffraction evidence for the existence of an unusual left-helical structure that can form from DNA containing an alternating purine-pyrimidine (dG-dC) sequence. Individual base pairs do not occupy equivalent positions; pairs of base pairs do. As a result, the phosphodiester backbone zigs and zags; hence the name Z DNA, shown in Fig. 2.5.

In all of the forms of DNA the two antiparallel DNA strands wrap around one another to form a double helix. The deoxyribose-phosphodiester backbones lie on the outside, and the bases pair between the strands according to the Watson-Crick

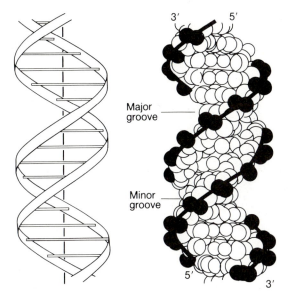

Major
groove

Minor
groove

Figure 2.4   Ribbon-and-stick model and space-filling model of the B form of DNA showing the phosphodiester backbones, antiparallel orientations of the backbones, major and minor grooves, and bases paired across the interior.

rules—A-T, G-C—across the inside of the helix. Which of the various conformations of DNA actually is adopted under physiological conditions? As discussed below, a study of the supercoiling of DNA performed recently provides the best evidence that the conformation of most DNA in solution is close to the B form. Similar experiments indicate that a small fraction of the total DNA in cells might be in the Z conformation.

Table 2.1   Parameters of Some DNA Helices

Helix	Base pairs per turn	Rotation of base pair (degrees)	Rise (Å per base pair)	Diameter (Å)
A	11	32.7	2.56	23
B	10	36	3.38	19
C	$9\frac{1}{3}$	38.6	3.32	19
Z	12	−30	3.63	18

Figure 2.5   Space-filling model of the Z form of DNA.

BASE-PAIRED STRANDS CAN DISSOCIATE

Heating DNA in solution breaks the hydrogen bonds between the A-T and G-C base pairs, unstacks the bases, and destroys the double-helical structure of the DNA. Such a process is called melting. Generally, not all the bonds break at one temperature, and DNA exhibits a transition zone between fully double-stranded DNA and fully melted DNA that often is 15° wide. The midpoint of this melting zone is defined as the melting temperature, which occurs at about 95° in 0.1 M NaCl. However, the actual value of the melting temperature depends on the base composition of the DNA, for the three hydrogen bonds in G-C base pairs provide more stability than the two that can form in A-T base pairs. The ionic composition of the solution also affects melting temperature. The higher the concentration of an ion such as sodium, the greater the shielding between the negatively charged phosphates and the higher the melting temperature. A divalent ion such as magnesium is still more effective in raising the melting temperature of DNA.

Melting can be observed in a variety of ways. One of the simplest is based on the fact that unstacked bases have a higher absorbance in the ultraviolet region of the spectrum than stacked, paired bases (Fig. 2.6). Therefore the absorbance of DNA in the ultraviolet region increases as the DNA melts, and by following the optical

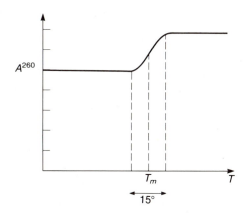

Figure 2.6   Increase in absorbance at 260 nm, A_{260}, of a DNA sample as the temperature is increased through the melting temperature.

density of a DNA solution as a function of temperature, a melting curve can be obtained.

POLYNUCLEOTIDE STRANDS CAN SPECIFICALLY REASSOCIATE

A most remarkable property of denatured DNA is its ability to renature *in vitro* to re-form double-stranded DNA! This re-formation of double-stranded DNA usually is very precise and exactly in register. Two strands may renature to the native double-helical form if their sequences are complementary, that is, if their sequences permit extensive formation of hydrogen-bonded base pairs. The ability of self-com-

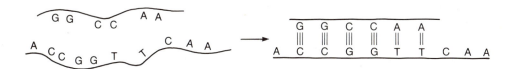

plementary sequences to anneal together and form a double helix is not unique to DNA; RNA can also do this, and it is possible *in vitro* to form RNA-DNA hybrids or RNA-RNA duplexes. The ability of nucleic acids to renature has been extremely important in the development of molecular biology as it has provided ways of detecting the presence of small quantities of specific sequences of RNA or DNA and in some cases to determine their intracellular locations.

SPECIFIC REASSOCIATION IS A VALUABLE TOOL

Let us consider an example illustrating the value of specific renaturation of complementary sequences of RNA and DNA. Suppose we have copies, or a clone, of a gene from *Drosophila melanogaster* fruit fly cells, obtained as described in later chapters, and we have managed to introduce DNA copies of the gene into yeast cells. How can we tell whether the DNA is being expressed by being transcribed to make complementary RNA copies in the cells?

It is not possible to assay for RNA synthesis from the cloned gene simply by measuring the total RNA synthesis in the yeast cells because the overwhelming majority of the RNA they synthesize will not be from the cloned DNA. We need a method for fishing out the RNA complementary to the cloned DNA. Specific reassociation, often called hybridization, is a method for making the measurement.

Figure 2.7 illustrates one method for performing the hybridization. More sophisticated methods are discussed in Chapters 10 and 11 on genetic engineering. The DNA of the cloned gene is denatured and immobilized on special paper. Radioactive RNA extracted from the yeast cells that had been grown on medium containing radioactive RNA precursors is annealed to this DNA, and the amount of radioactivity that becomes bound to the paper by means of its extensive base pairing can be measured and is proportional to the amount of RNA in the sample that was complementary to the immobilized DNA.

Figure 2.7   Schematic of hybridization to quantitate a specific RNA.

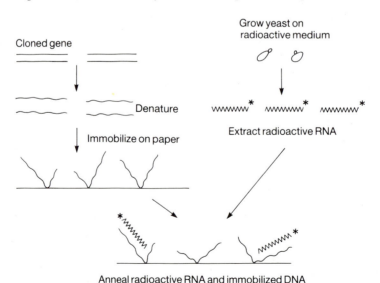

DEPENDENCE OF REASSOCIATION RATES ON SALT AND TEMPERATURE

A little is known about the physical chemistry of the renaturing process itself. At a fixed salt concentration, the rate of renaturation depends on temperature, with the maximum renaturation rate occurring at about 25° below the melting temperature. Varying the salt concentration at a fixed temperature also affects the rate of renaturation, as shown in the graph in Fig. 2.8. As the salt concentration is increased from very low values, the rate of renaturing increases because the negatively charged phosphate groups become better shielded from one another. However, with increasing salt, the renaturation rate reaches a maximum and then declines because high salt concentrations so effectively shield the negatively charged phosphates that imperfectly base-paired regions can form within a single strand of DNA. A molecule folded upon itself in this way is temporarily unavailable for specific reassociation with its complementary strand, and the formation rate of specific complementary strand hybrids decreases.

COUNTING DNA COPIES BY REASSOCIATION KINETICS

The kinetics of the reassociation process can be described by an equation relating the concentrations of the reactants and a constant. If we let $C(t)$, which will be written as just C, stand for the concentration of single-stranded DNA at any time t, C_0 be the initial concentration of DNA, and k_2 be an association rate constant, then the rate of reassociation is given by the following equation:

$$\frac{dC}{dt} = -k_2 C^2.$$

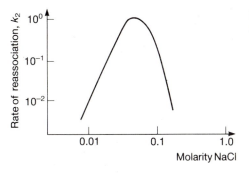

Figure 2.8   Reassociation kinetics of DNA as a function of salt concentration.

This can be solved by rearranging the terms and integrating to yield

$$\frac{C}{C_0} = \frac{1}{1 + k_2 C_0 t}. \tag{2.1}$$

The final result is expressed as $C(t)/C_0$, the relative amount of the total DNA present as single-stranded DNA at time t.

Convenient methods exist for measuring the kinetics of reassociation. One utilizes a column of calcium phosphate crystals, hydroxyapatite, which is capable of separating single and double-stranded DNA. When a mixture of single and double-stranded DNA is applied to the column under the appropriate conditions, only the double-stranded DNA binds. After the unbound single-stranded DNA is flushed out, the double-stranded DNA can be eluted by raising the phosphate concentration in the elution buffer.

The terms C and C_0 as written in the preceding equations refer to concentrations in moles per liter—moles of polymeric single-stranded DNA per liter and not moles of nucleosides per liter. Experimentally it is not difficult to determine C_0 if a well-characterized DNA such as that extracted from a bacteriophage is used. However, it is difficult to know C_0 in experiments that use more complex DNA, such as that extracted from *Drosophila*, because a number of sequences in the genome may be repeated an unknown number of times.

Relation 2.1 can be turned around, however. Either the quantities C_0, C, and t can be used to calculate k_2, or k_2 and $C(t)$ can be measured and used to calculate C_0. This second option permits us to measure the molar concentration of DNA fragments, that is, the molar concentration of polymers of the same sequence. For example, we may know that 10^7 mouse cells contain 1 mg of DNA and we may want to determine how many times per cell a particular sequence is represented. If we measure the hybridization kinetics of our sequence and find that the rate indicates the presence of about 2×10^7 molecules per mg of the DNA, then our sequence is probably present in one copy per chromosome of the cells.

Not only do salt and temperature affect the rate of reassociation, but the lengths of the DNA molecules themselves are also important. The basis for the length effect is as follows. On one hand, as the length of a DNA molecule increases, the random coil of the molecule in solution buries much of the sequence in the inaccessible interior. On the other hand, once two polymers initiate the reassociation, they rapidly "zipper" to completion. The overall effect is to make the rate of reassociation roughly proportional to the square root of the length of the molecule. If DNA concentration is expressed as moles of nucleosides per liter, the salt concentration is 1 M NaCl, and the temperature is about 66°, then

$$k_2 = 3 \times 10^5 (L^{1/2}/N) \text{ liter/mole} \cdot \text{sec,} \tag{2.2}$$

where L is the average length in bases of the DNA and N is the complexity, number of bases before the sequence repeats. Note that when the equation is written in this form, k_2 is not constant: it varies with temperature and salt concentration as well as with the length and complexity of the fragments.

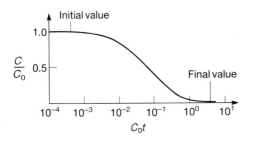

Figure 2.9   Reassociation of lambda phage DNA as a function of $C_0 t$.

It has been found convenient to plot C/C_0 as a function of $\mathrm{Log}(C_0 t)$ because in one experiment both C_0 (the initial concentration of DNA) and the time permitted for reassociating usually are varied over wide ranges. The term $C_0 t$ can therefore vary over a very large range. When a $C_0 t$ measurement is performed with simple DNA such as the 30,000-base-pair DNA extracted from bacteriophage lambda, a simple C/C_0 curve is obtained (Fig. 2.9). The curve is halfway down where $C_0 t = 1/k_2$. This makes it convenient to read off the value of k_2 from a $C_0 t$ curve. Therefore, because the length of the DNA is known, its complexity can be determined from Eq. 2.2. Reassociation curves similar in shape to that for phage lambda are also found when phage T4 DNA or *E. coli* DNAs are used. The points where $C/C_0 = \frac{1}{2}$ or $C_0 t = 1/K_2$ are found to agree with the known sizes of these DNAs, which indicates that these DNAs do not contain an appreciable number of repeated sequences.

Something quite remarkable happens when we measure the reassociation kinetics of DNA that has been extracted from higher organisms and sheared so as to produce a uniform length distribution. The reassociation kinetics of this DNA are complex (Fig. 2.10). First, a fraction of the DNA renatures almost instantaneously. Therefore the genome contains many — often about a million — copies of one or more short sequences. This highly repetitive DNA is usually called satellite DNA because it is found after CsCl equilibrium centrifugation as a satellite to the main band of chromosomal DNA. The next sequences to reassociate are a set of DNA copies repeated on the order of several hundred times per cell. This DNA is called

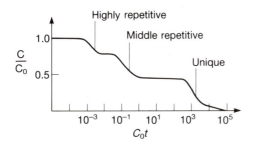

Figure 2.10   Reassociation kinetics of mouse DNA.

middle repetitive DNA. The genes or DNA sequences that appear to be repeated only once in the genome reassociate last.

TOPOLOGY INTRODUCES
ADDITIONAL CONSIDERATIONS

Topology introduces a structural feature in addition to the base-paired and helical aspects of the structure of DNA. The origin of this structure can best be understood by considering a mathematical property of two closed rings. The number of times that one ring encircles the other must be an integral number, and it cannot be changed without physically opening one of the rings.

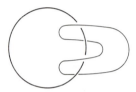

That is, their linking number is a topological invariant. Many types of DNA molecules found in cells are covalently closed circles because each strand is circular. Hence the concept of a linking number applies to DNA molecules obtained from many sources.

The forces tending to hold double-stranded DNA in a right-handed helix with about 10 base pairs per turn add a dimension to the analysis of the structures of covalently closed circles. These forces are sufficiently great that the linking number generally resolves itself into two easily distinguished components: the twist, Tw, which in DNA's usual right-helical form has a value of 1 per each 10.5 base pairs, and the writhe, Wr. If the linking number, Lk, does not equal Tw, then the remainder of Lk is resolved in writhing, a process often called supercoiling or superhelical turns. Therefore, for any covalently closed double-stranded DNA molecule, no matter how it is distorted, unless its phosphodiester backbone is broken, Lk = Tw + Wr. It is curious that the topological invariant Lk equals the sum of two terms, each of which is not invariant.

GENERATING DNA WITH SUPERHELICAL TURNS

To understand how we may experimentally vary Lk, and consequently the degree of supercoiling, let us use the lambda phage DNA. The molecules of this DNA are about 50,000 base pairs long and possess what are called sticky ends; that is, the ends of the DNA duplex are not flush. As shown in Fig. 2.11, the 5′ ends protrude in a

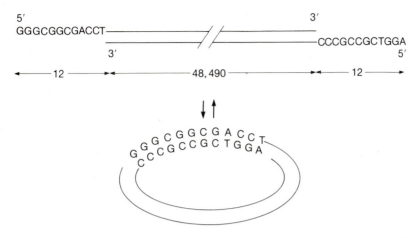

Figure 2.11   Association of the self-complementary single-stranded ends of lambda phage DNA to form a nicked circle.

single-stranded region of 12 bases. The sequence of the left end is complementary to the sequence of the right end. These ends, the sticky ends, can be reassociated together to form a circle, which sometimes is called a Hershey circle after its discoverer. The phosphodiester bonds are not contiguous around the Hershey circle; hence its other name, a nicked circle. Circles having a break in only one of their backbones also are called nicked.

Nicked circles can be covalently closed or sealed with DNA ligase. This enzyme seals the phosphodiester backbone of DNA between nicks that have a 5'-phosphate and a 3'-hydroxyl. Following the ligation, Lk cannot be altered without breaking the backbone of one of the two strands. Hence, the sum of Tw, the right-helical turns, and Wr, the number of superhelical turns, is fixed. If under fixed buffer and temperature conditions we were to anneal the ends of the lambda DNA together and then seal with ligase, the number of superhelical turns would be zero and Lk would be about 5,000. For convenience let us say that the number is exactly 5,000. Furthermore, if we were to introduce distortion or even to wrap this DNA around a protein, the sum of Tw and Wr must still remain 5,000.

Suppose the annealing and sealing had been done in the presence of ethidium bromide. Molecules of this compound are able to slip between two bases, intercalate, and push the bases apart. Consequently the bases do not rotate as much as normal with respect to one another because the phosphate-deoxyribose backbone connecting them cannot lengthen (Fig. 2.12). Hence the amount by which one strand wraps around another is decreased by the intercalation of the ethidium bromide. In the common B form of DNA, the bases are twisted 36° per base, but the intercalation of an ethidium bromide molecule removes 24° of this twist. The number of helical

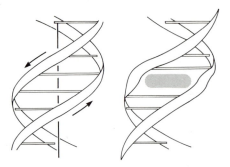

Figure 2.12   Intercalation of a molecule of ethidium bromide between two bases in the DNA duplex reduces their relative twist.

turns in a lambda DNA molecule sealed in the presence of a particular concentration of ethidium bromide might be about 100 less than the number contained in a lambda DNA molecule sealed in the absence of ethidium bromide. Treating with DNA ligase under these conditions would produce a molecule with Lk = Tw = 4,900. If the ethidium bromide were then removed by extraction with an organic solvent, Tw would return to near its standard value of 5,000; but because of the requirement that Tw + Wr = 4,900 be a constant, Wr would become −100 and the circular DNA would writhe. It would have 100 negative superhelical turns.

MEASURING SUPERHELICAL TURNS

Superhelical turns in DNA may introduce important distortions or torsion in the molecules or may result from an important biological process. One way to measure superhelical turns might be to observe the DNA in an electron microscope and see it

twisted upon itself. However, quantitation of the superhelical turns in DNA with more than a few turns is difficult. More convenient measurement methods exist. Consider the DNA molecule just described with 100 negative superhelical turns.

Because it is so twisted upon itself, the molecule is rather compact and sediments in the ultracentrifuge at a high rate. If the sedimentation is done in the presence of a low concentration of ethidium bromide, a few molecules will intercalate into the DNA. This will reduce the number of negative superhelical turns, thereby opening up the DNA, which will sediment more slowly than it would in the absence of ethidium bromide.

Consider a series of sedimentation measurements made in the presence of increasing concentrations of ethidium bromide. At higher and higher concentrations of ethidium bromide, more and more will intercalate into the DNA and unwind the DNA more and more. Consequently the DNA will become less and less compact and sediment more and more slowly (Fig. 2.13). Finally a concentration of ethidium bromide will be reached where the molecule is completely free of superhelical turns. At this concentration the DNA will sediment most slowly. If the centrifugation is done in the presence of still higher concentrations of ethidium bromide, the molecule will be found to sediment more rapidly as the DNA acquires positive superhelical turns and becomes more compact again. The concentration of ethidium bromide required to generate the slowest sedimentation rate can then be related to the number of superhelical turns originally in the DNA via the affinity of ethidium bromide for DNA and the untwisting produced per intercalated ethidium bromide molecule.

Even more convenient than centrifugation for quantitation of superhelical turns has been electrophoresis of DNA through agarose. Under some conditions DNA molecules of the same length but with different linking numbers can be made to separate from one another upon electrophoresis (Fig. 2.14). The separation results from the fact that two molecules with different linking numbers will, on the average during the electrophoresis, possess different degrees of supercoiling and consequently different compactness. Those molecules that are more greatly supercoiled during the electrophoresis will migrate more rapidly. Not only can agarose gels be used for quantitating species with different numbers of superhelical turns, but any

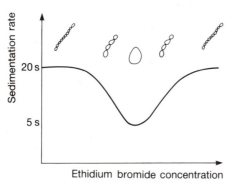

Figure 2.13   Sedimentation rate of a covalently closed circular DNA molecule as a function of the ethidium bromide concentration in the centrifugation solution.

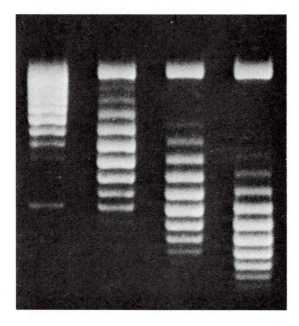

Figure 2.14   Separation of superhelical forms of covalently closed circular DNA species by electrophoresis through agarose. Each of the four samples possessed a different average number of superhelical turns. From Conformational Fluctuations of DNA Helix, R. Depew, J. Wang, Proc. Nat. Acad. Sci. USA 72, 4275–4279 (1979).

particular species can be extracted out of the gel and used in subsequent experiments.

The agarose gels show an interesting result. If DNA is ligated to form covalently closed circles and then subjected to electrophoresis under conditions that separate superhelical forms, it is found that not all of the DNA molecules possess the same linking number. There is a distribution centered about the linking number corresponding to zero superhelical turns, Lk_0. This is to be expected because the DNA molecules in solution are constantly in motion, and a molecule can be ligated into a covalently closed circle at an instant when it possesses a linking number unequal to Lk_0. However, these molecules are frozen in a slightly higher average energy state than those with no superhelical turns. Their exact energy depends on the twisting spring constant of DNA. The stiffer the DNA, the fewer the molecules that will possess any superhelical turns at the time of sealing. Quantitation of the DNA molecules in the bands possessing different numbers of superhelical turns permits evaluation via statistical mechanics of the twisting spring constant of DNA.

BINDING OF PROTEINS AND DNA UNWINDING

The ability to measure accurately the number of superhelical turns in DNA allows a determination of the amount of winding or unwinding produced by the binding of molecules. For example, unwinding measurements first indicated that RNA polymerase melts about 8 bases of DNA when it binds tightly to lambda DNA. Later, more precise measurements have shown that the unwinding is closer to 15 base pairs. This unwinding was shown directly by binding RNA polymerase to nicked circular DNA and then sealing with ligase to form covalently closed circles, removing the RNA polymerase, and measuring the number of superhelical turns in the DNA. The first measurements were done by accurately comparing the sedimentation velocity of the DNA sealed in the presence and in the absence of RNA polymerase. Later experiments have used better DNA substrate and gel electrophoretic measurements of the number of superhelical turns to arrive at the higher values.

Another way to measure the winding produced by binding of a molecule to DNA is to measure the affinity of a molecule for DNA samples containing different numbers of superhelical turns. This method is based on the fact that a protein that introduces negative superhelical turns as it binds to DNA will bind much more tightly to a DNA molecule already containing negative superhelical turns. From the thermodynamics of the situation, this type of approach is very sensitive and has been used to show that the binding of *lac* repressor unwinds DNA about 100°.

*DETERMINING LK, TW, AND WR IN HYPOTHETICAL STRUCTURES

The actual determination of linking number, twist, and superhelical turns in a structure can sometimes be tricky. The following discussion follows the method given by Crick and by Bauer, Crick, and White.

Linking number is the number of times that one DNA strand encircles the other, but it is useful for this number to have a sign that is dependent on the orientation of the strands. Therefore, to determine the linking number of a structure, draw arrows on the two strands pointing in opposite directions. At each point where the two strands cross, assign a + or − value dependent on the orientation. If the upper strand at a crossover can be brought into correspondence with the lower strand with a clockwise rotation, assign a +; if a counterclockwise rotation is required, assign a −.

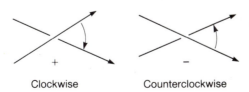

Clockwise   Counterclockwise

The linking number equals the sum of these values divided by 2, as shown in Fig. 2.15.

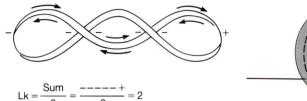

$$Lk = \frac{Sum}{2} = \frac{----+}{2} = 2$$

Figure 2.15   Example of a structure with zero twist but a linking number of 2 and hence containing two superhelical turns.

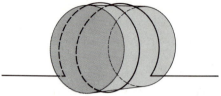

Figure 2.16   Schematic of a nucleosome with about $1\frac{3}{4}$ turns of DNA around it.

Viewing the DNA duplex as a ribbon provides one way to picture twist. Consider a straight line drawn perpendicular to the axis of the ribbon and through both edges. The rotation of this line as it is moved down the axis gives the twist.

For a given structure, the writhing or superhelical turns is most easily given by whatever value Wr must have to make Lk = Tw + Wr.

The ability to compute the superhelical turns in structures has partially answered a paradox raised by the structure of nucleosomes. As mentioned in Chapter 1, the DNA in eukaryotic cells is wrapped around histones. DNA in the B conformation wraps about $1\frac{3}{4}$ times around a core consisting of pairs of the four histones H2A, H2B, H3, and H4 (Fig. 2.16). In the presence of the histone H1, the wrapping is just about two complete turns. Superficially, this structure appears to possess two superhelical turns, and yet when the protein is removed, the DNA is found to possess only one superhelical turn for each nucleosome it had contained. Deeper analysis (see Problem 2.13) reveals, however, that DNA wrapped approximately two turns about each of a series of nucleosomes can possess linking numbers per nucleosome in the range $\pm 0.5 - 2.5$. The value depends on whether the number of turns is a little more or a little less than two and on the path of the DNA to the next nucleosome.

ENZYMES CAN ADD OR SUBTRACT SUPERHELICAL TURNS

Cells have enzymes that alter the linking numbers of covalently closed DNA molecules. Wang, who found the first such protein, called it omega, but now it is often called DNA topoisomerase I. Astoundingly, this enzyme removes negative superhelical turns, one at a time, without hydrolyzing ATP or any other energy-rich small molecule. No nicks are left in the DNA after omega has acted. Purified, covalently closed lambda phage DNA extracted from lambda-infected cells is found to possess about 120 negative superhelical turns. Incubation of omega with this DNA increases

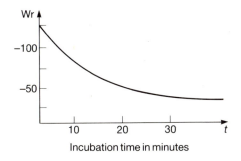

Figure 2.17   Reduction in the number of negative superhelical turns contained in lambda phage DNA as a function of the time of incubation with DNA topoisomerase I.

the linking number until only about 20 negative superhelical turns remain (Fig. 2.17). The enzyme appears to remove twists, and hence superhelical turns, in a controlled way. It binds to the DNA and then interrupts the phosphodiester backbone of one strand by making a high-energy phosphate bond to the enzyme. In this state the enzyme removes one twist and then re-forms the phosphodiester bond in the DNA.

Additional enzymes capable of altering the linking number of DNA have also been discovered. A DNA topoisomerase has been found in eukaryotic cells that can remove positive superhelical turns. Another remarkable enzyme has been found in *E. coli.* This enzyme, DNA gyrase or DNA topoisomerase II, adds negative superhelical turns; as expected, energy in the form of ATP molecules is required for this reaction.

DO SUPERHELICAL TURNS HAVE BIOLOGICAL SIGNIFICANCE?

When covalently closed double-stranded circles are extracted from bacteria or eukaryotic cells, purified and freed of all DNA-binding proteins, and placed in standard buffers, they are found to possess superhelical turns. Is this an indication that DNA within cells actually possesses a torsion acting in the direction that reduces the number of superhelical turns? It is possible that the intracellular DNA is held in a superhelical state by virtue of the binding of molecules such as ethidium bromide or histones. Even though this DNA possesses "formal" superhelical turns, its *in vivo* state would not be altered, as would naked DNA possessing superhelical turns, if it were nicked and then resealed. No "torsion" would be available in this DNA to assist or oppose disruption of base pairs by another protein.

Several experiments suggest that DNA in bacteria not only possesses superhelical turns but also is under a superhelical torsion. The *in vitro* integration reaction of lambda phage, in which a special set of enzymes catalyzes the insertion of covalently closed lambda DNA circles into the chromosome, proceeds only when the lambda DNA possesses negative superhelical turns. In fact, tracking down what permitted the *in vitro* reaction to work led to the discovery of DNA gyrase. Presumably the enzymology of the *in vitro* and *in vivo* integration reactions is the same, and the supercoiling requirement means that *in vivo* the chromosome possesses superhelical turns.

A second experiment also suggests that DNA in normally growing *E. coli* contains superhelical torsion. Adding inhibitors of the DNA gyrase, nalidixic acid or oxolinic acid, which block activity of the A subunit of the enzyme, or novobiocin or coumermycin, which inhibit the B subunit, alters the rates of expression of different genes. The activities of some genes increase while the activities of others decrease. This shows that the drug effects are not a general physiological response and that the DNA must be supercoiled *in vivo*. Yet another indication of the importance of supercoiling to cells is shown by the behavior of DNA topoisomerase I mutants. Such mutants grow slowly, and faster-growing mutants frequently arise. These are found to possess mutations that compensate for the absence of the topoisomerase I by a second mutation that reduces the activity of gyrase, topoisomerase II. Altogether, it seems reasonable to conclude that the DNA in cells not only is supercoiled but also is under a supercoiling torsion.

A UNIFIED MECHANISM FOR DNA TOPOISOMERASES

Studies of the activity of topoisomerases I and II, omega and gyrase, have shown that these enzymes act by a reaction pathway somewhat different from what might have been expected. On first consideration of these enzymes, it would appear most logical for them to proceed by nicking one strand and permitting the other strand to act as a swivel. The broken strand could rotate about the swivel and then be resealed. This is not the way either works, however. Instead, where two DNA duplexes cross, gyrase, with the consumption of an ATP molecule, can permit the duplex lying below to be cut, to pass past the first, and to be rejoined. Topoisomerase I can proceed by the

same sort of pathway, but it passes one strand of a duplex through the other strand, having first denatured a region where strand crossing is to occur (Fig. 2.18). These

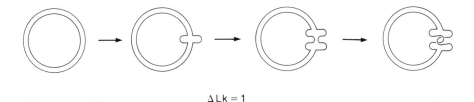

$$\Delta Lk = 1$$

Figure 2.18   The strand inversion pathway as applied by DNA topoisomerase I in which a region of DNA is melted and one strand is broken, passed by the other, and rejoined.

are known as sign-inversion pathways because they change the sign of the linking number that is contributed by a point of crossing of the two DNA strands.

The sign-inversion pathway has a particularly useful property: It permits DNA duplexes to interpenetrate one another. Gyrase is an enzyme that can untangle knots in DNA! Undoubtedly this property is of great value to the cell as the DNA is compressed into such a small volume that tangles seem inevitable.

PROBLEMS

2.1. DNA is stable at pH 11, whereas RNA is degraded by alkali to nucleosides. Look up the reason why.

2.2. In textbooks it is possible to find the phosphodiester backbone of DNA or RNA drawn in either of two ways (Fig. 2.19). Which is correct?

2.3. Look up the reason why RNA cannot easily exist in the B conformation whereas this is the normal conformation of DNA.

2.4. Indicate which portions of the A-T and G-C base pairs lie in the major and minor grooves of the B conformation double helix. Show that the base sequence of a DNA duplex can be read by a protein without denaturing the duplex by contacts made with the DNA in the major groove.

2.5. a) For a solution of denatured DNA at a concentration of 1 μg/ml in 1 M NaCl at 67°, how long would it take to renature 50% of the DNA if it is 1,000 base pairs long and contains a complexity of 1,000 base pairs?

b) How long is required if the DNA is 50,000 base pairs long, like phage lambda, and it contains a complexity of 50,000 base pairs, like phage lambda?

c) How long is required if the lambda DNA of part b is sheared to an average length of 1,000 base pairs?

2.6. Suppose we have three almost identical samples of DNA. The first is linear duplex DNA. The second contains DNA molecules of the same length as the first sample, but the molecules are covalently closed into circles. The molecules in the

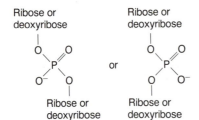

Figure 2.19   Two textbook drawings of the structure of the phosphate in phosphodiester backbones of DNA and RNA.

third sample are circular, but only one strand is covalently closed; the other is nicked. How would these three samples differ in their melting and reassociation properties?

2.7. In a lambda phage DNA molecule, a sequence of nine base pairs was followed by three base pairs, and then the complement of the sequence was repeated in reverse order. Show the results of intrastrand base pairing of these sequences, known as a cruciform. In a covalently closed duplex, how much would the formation of the cruciform, as defined here, change the number of superhelical turns?

2.8. Is the activity of DNA topoisomerase I on negative superhelical DNA and not on positive superhelical DNA proper for this enzyme to act as the protein to unwind twists introduced by DNA replication?

2.9. Show that negative superhelicity but not positive superhelicity assists the melting out of a few base pairs.

2.10. Consider DNA 50,000 base pairs long containing sticky ends, such as phage lambda. If the sticky ends are annealed together and the circle is covalently closed by ligase, several species of superhelically twisted DNA are formed. These differ from one another by single superhelical twists, and they can be separated from one another by electrophoresis through agarose in the presence of moderate concentrations of ethidium bromide. How does the resultant ladder pattern change if lambda DNA containing an internal deletion of five base pairs is used in the experiment? How can this method be extended to determine the helical repeat distance of DNA in solution?

2.11. Show that a sign-inversion pathway such as that proposed for gyrase changes the number of superhelical twists in a covalently closed circular DNA molecule by two.

2.12. Show that approximately two twistless solenoidal turns of DNA can possess a linking number, neglecting the sign, ranging from 0.5 to 2.5. [*Hint:* compute the limit of a series of similar structures and determine the average Lk per structure. Also note that just a little more than two turns may be different than just a little less than two turns and that the DNA coming off the right end may go either to the next solenoid on the right or to the next solenoid on the left. If it goes to the left, strands may cross in two ways.]

2.13. One method for determining the approximate number of superhelical turns in a proposed structure is to build it without twists from a ribbon, to pull the ends taut, and to count the twists. Why does this work?

2.14. Ethidium bromide fluoresces 50 to 100 times more strongly when it has intercalated into DNA. How could this property be used to determine whether DNA from a phage was a covalently closed circle?

2.15. Suppose a biochemist who was studying supercoiling wished to have a large supply of DNA topoisomerase I and gyrase proteins. What would your response be to the biochemist's proposal to generate a bacterial strain that greatly oversynthesized both of these proteins simultaneously?

2.16. Assume that the B form of DNA possesses 10.4 base pairs per turn and the Z form possesses 12 base pairs. Suppose that a large plasmid contains a 156-base-pair insert of alternating dG, dC. Show that the conversion of the insert DNA from the B to the Z form will remove exactly 28 superhelical turns from the plasmid.

2.17. Originally it was thought that the intercalation of one molecule of ethidium bromide reduced the twist of the DNA helix by 12°. Describe experiments that show that the correct value actually is about 24°. How was the first value determined, or why was it incorrect?

2.18. Suppose electrophoresis on agarose is performed such that the twist of DNA is the same as it is in solution. A sample of circular DNA molecules of the same length but possessing -1, 0, and $+1$ superhelical turns is to be electrophoresed. How many bands will appear? How can the $+1$ and -1 species be separated?

*2.19. Davidson (J. Mol. Biol. *66*, 307–309 [1972]) calculated that the affinity of ethidium bromide for DNA containing negative superhelical turns should be about 2.6 times greater than its affinity for linear DNA. This calculation assumed that intercalation of one molecule of ethidium bromide untwists the DNA by 12°. Recalculate using the corrected value of 24°.

RECOMMENDED READINGS

Conformational Fluctuations of DNA Helix, R. Depew, J. Wang, Proc. Nat. Acad. Sci. USA 72, 4275–4279 (1975).
Linking Numbers and Nucleosomes, F. Crick, Proc. Nat. Acad. Sci. USA 73, 2639–2643 (1976).

RELATED REVIEWS, BOOKS, AND ARTICLES

Structure of Chromatin, R. Kornberg, Ann. Rev. of Biochem. *46*, 931–954 (1977).
Biophysical Chemistry, Parts I and III, C. Cantor, P. Schimmel, Freeman, San Francisco (1980).
Supercoiled DNA, W. Bauer, F. Crick, J. White, Sci. American *243*, 118–133 (1980).
DNA Topoisomerases, N. Cozzarelli, Cell 22, 327–328 (1980).

Double Helical DNA: Conformations, Physical Properties, and Interactions with Ligands, M. Record, S. Mazur, P. Melancon, R. Roe, S. Shaner, L. Under, Ann. Rev. of Biochem. *50,* 879–1024 (1981).

Structures of DNA, Cold Spring Harbor Symposia on Quantitative Biology *47,* Cold Spring Harbor Laboratory (1983).

DEEPER READING

Molecular Structure of Deoxypentose Nucleic Acids, J. Watson, F. Crick, Nature *171,* 737–738 (1953).

Kinetics of Renaturation of DNA, J. Wetmur, N. Davidson, J. Mol. Biol. *31,* 349–370 (1968).

Repeated Sequences in DNA, R. Britten, D. Kohne, Science *161,* 529–540 (1968).

Effects of the Conformation of Single-Stranded DNA on Renaturation and Aggregation, F. Studier, J. Mol. Biol. *41,* 199–209 (1969).

Interaction of Closed Circular DNA with Intercalative Dyes, II: The Free Energy of Superhelix Formation in SV40 DNA, W. Aauer, J. Vinograd, J. Mol. Biol. *47,* 419–435 (1970).

Effect of DNA Length on the Free Energy of Binding of an Unwinding Ligand to a Supercoiled DNA, N. Davidson, J. Mol. Biol. *66,* 307–309 (1972).

On the Structure of the Folded Chromosome of *E. coli,* A. Worcel, E. Burgi, J. Mol. Biol. *71,* 127–147 (1972).

Improved Estimation of Secondary Structure in Ribonucleic Acids, I. Tinoco, Jr., P. Borer, B. Dengler, M. Levine, O. Uhlenbeck, Nature New Biol. *246,* 40–41 (1973).

Effects of Microscopic and Macroscopic Viscosity on the Rate of Renaturation of DNA, C. Chang, T. Hain, J. Hutton, J. Wetmur, Biopolymers *13,* 1847–1858 (1974).

Interactions Between Twisted DNAs and Enzymes: The Effect of Superhelical Turns, J. Wang, J. Mol. Biol. *87,* 797–816 (1974).

Electron Microscopic Visualization of the Folded Chromosome of *E. coli,* H. Delius, A. Worcel, J. Mol. Biol. *82,* 107–109 (1974).

Folding of the DNA Double Helix in Chromatin-like Structures from Simian Virus 40, J. Germond, B. Hirt, P. Oudet, M. Gross-Bellard, P. Chambon, Proc. Nat. Acad. Sci. USA *72,* 1843–1847 (1975).

Action of Nicking-Closing Enzyme on Supercoiled and Non-Supercoiled Closed Circular DNA: Formation of a Boltzmann Distribution of Topological Isomers, D. Pulleyblank, M. Shure, D. Tang, J. Vinograd, H. Vosberg, Proc. Nat. Acad. Sci. USA *72,* 4280–4284 (1975).

Determination of the Number of Superhelical Turns in Simian Virus 40 DNA by Gel Electrophoresis, W. Keller, Proc. Nat. Acad. Sci. USA *72,* 4876–4880 (1975).

DNA Gyrase: An Enzyme That Introduces Superhelical Turns into DNA, M. Gellert, K. Mizuuchi, M. O'Dea, H. Nash, Proc. Nat. Acad. Sci. USA *73,* 3872–3876 (1976).

Novobiocin and Coumermycin Inhibit DNA Supercoiling Catalyzed by DNA Gyrase, M. Gellert, M. O'Dea, T. Itoh, J. Tomizawa, Proc. Nat. Acad. Sci. USA *73,* 4474–4478 (1976).

Micrococcus Luteus DNA Gyrase: Active Components and a Model for Its Supercoiling of DNA, L. Liu, J. Wang, Proc. Nat. Acad. Sci. USA *75,* 2098–2102 (1978).

Helical Repeat of DNA in Solution, J. Wang, Proc. Nat. Acad. Sci. USA *76,* 200–203 (1979).

Involvement of Histone H1 in the Organization of the Nucleosome and of the Salt-Dependent Superstructures of Chromatin, F. Thoma, Th. Koller, A. Klug, J. Cell Biol. *83,* 403–427 (1979).

Type II DNA Topoisomerases: Enzymes That Can Unknot a Topologically Knotted DNA Molecule via a Reversible Double-Strand Break, L. Liu, C. Liu, B. Alberts, Cell *19*, 697–707 (1980).

DNA Gyrase Action Involves the Introduction of Transient Double-Strand Breaks into DNA, K. Mizuuchi, L. Fisher, M. O'Dea, M. Gellert, Proc. Nat. Acad. Sci. USA *77*, 1847–1851 (1980).

Torsional Tension in the DNA Double Helix Measured with Trimethylpsoralen in Living *E. coli* Cells: Analogous Measurements in Insect and Human Cells, R. Sinden, J. Carlson, D. Pettijohn, Cell *21*, 773–783 (1980).

Conformational Flexibility of DNA: Polymorphism and Handedness, G. Gupta, M. Bansal, V. Sasisekharan, Proc. Nat. Acad. Sci. USA *77*, 6486–6490 (1980).

Molecular Structure of a Left-Handed Double Helical DNA Fragment at Atomic Resolution, A. Want, G. Quigley, F. Kolpak, J. Crawford, J. van Boom, G. van der Marel, A. Rich, Nature *282*, 680–686 (1980).

Crystal Structure Analysis of a Complete Turn of B-DNA, R. Wing, H. Drew, T. Takano, C. Broka, S. Tanaka, K. Itakura, R. Dickerson, Nature *287*, 755–758 (1980).

Polymorphism of DNA Double Helices, A. Leslie, S. Arnott, R. Chandrasekaran, R. Ratliff, J. Mol. Biol. *143*, 49–72 (1980).

Identification of Two *Escherichia coli* Factor Y Effector Sites Near the Origins of Replication of the Plasmids ColE1 and pBR322, S. Zipursky, K. Marians, Proc. Nat. Acad. Sci. USA *77*, 6521–6525 (1980).

Chromosomes in Living *E. coli* Cells Are Segregated into Domains of Supercoiling, R. Sinden, D. Pettijohn, Proc. Nat. Acad. Sci. *78*, 224–228 (1981).

Catenation and Knotting of Duplex DNA by Type 1 Topoisomerases: A Mechanistic Parallel with Type 2 Topoisomerases, P. Brown, N. Cozzarelli, Proc. Nat. Acad. Sci. USA *78*, 843–847 (1981).

Structure of a B-DNA Dodecamer: Conformation and Dynamics, H. Drew, R. Wing, T. Takana, C. Broka, S. Tanaka, K. Itakura, R. Dickerson, Proc. Nat. Acad. Sci. USA *78*, 2179–2183 (1981).

DNA Flexibility Studied by Covalent Closure of Short Fragments into Circles, D. Shore, J. Langoushi, R. Baldwin, Proc. Nat. Acad. Sci. USA *78*, 4833–4837 (1981).

X-ray Diffraction Study of a New Crystal Form of the Nucleosome Core Showing Higher Resolution, J. Finch, R. Brown, D. Rhodes, T. Richmond, B. Rushton, L. Lutter, A. Klung, J. Mol. Biol. *145*, 757–769 (1981).

Left-Handed DNA in Restriction Fragments and a Recombinant Plasmid, J. Klysik, S. Stirdivant, J. Larson, P. Hart, R. Wells, Nature *290*, 672–677 (1981).

Sequence Dependence of the Helical Repeat of DNA in Solution, L. Peck, J. Wang, Nature *292*, 375–378 (1981).

Sequence-Dependent Helical Periodicity of DNA, D. Rhodes, A. Klug, Nature *292*, 378–380 (1981).

Bent Helical Structure in Kinetoplast DNA, J. Marini, S. Levene, D. Crothers, P. Englund, Proc. Nat. Acad. Sci. USA *79*, 7664–7668 (1982).

Torsional Tension in Intracellular Bacteriophage T4 DNA, Evidence That a Linear DNA Duplex Can Be Supercoiled *in vivo*, R. Sinden, D. Pettijohn, J. Mol. Biol. *162*, 659–677 (1982).

Escherichia coli DNA Topoisomerase I Mutants: Increased Supercoiling Is Corrected by Mutations Near Gyrase Genes, G. Pruss, S. Manes, K. Drlica, Cell *31*, 35–42 (1982).

Escherichia coli DNA Topoisomerase I Mutants Have Compensatory Mutations in DNA

Gyrase Genes, S. DiNardo, K. Voelkel, R. Sternglanz, A. Reynolds, A. Wright, Cell *31*, 43–51 (1982).

Energetics of B-to-Z Transition in DNA, L. Peck, J. Wang, Proc. Nat. Acad. Sci. USA *80*, 6206–6210 (1983).

Energetics of DNA Twisting, I: Relation Between Twist and Cyclization Probability, D. Shore, R. Baldwin, J. Mol. Biol. *170*, 957–981 (1983).

Energetics of DNA Twisting, II: Topoisomer Analysis, D. Shore, R. Baldwin, J. Mol. Biol. *170*, 983–1007 (1983).

Cruciform Formation in a Negatively Supercoiled DNA May Be Kinetically Forbidden Under Physiological Conditions, A. Courey, J. Wang, Cell *33*, 817–829 (1983).

Negatively Supercoiled Simian Virus 40 DNA Contains Z-DNA Segments Within Transcriptional Enhancer Sequences, A. Nordheim, A. Rich, Nature *303*, 674–679 (1983).

Dependence of Z-DNA Antibody Binding to Polytene Chromosomes on Acid Fixation and DNA Torsional Strain, R. Hill, B. Stollar, Nature *305*, 338–340 (1983).

Torsional Rigidity of DNA and Length Dependence of the Free Energy of DNA Supercoiling, D. Horowitz, J. Wang, J. Mol. Biol. *173*, 75–91 (1984).

Characteristic Folding Pattern of Polytene Chromosomes in *Drosophila* Salivary Gland Nuclei, D. Mathog, M. Hochstrasser, Y. Gruenbaum, H. Saumweber, J. Sedat, Nature *308*, 414–421 (1984).

DNA Synthesis

O ne of the fundamental approaches of biochemists is to purify the components of a system to permit analysis without interference from extraneous reactions. The relatively loose association of the proteins required for DNA synthesis has created problems for their study. How can one of the components be purified if all the components must be present for DNA synthesis to occur? We see in this chapter that the problem has largely been solved, but the purification of the many proteins required for DNA synthesis was a monumental task that occupied biochemists and geneticists for many years. By contrast, the machinery of protein synthesis was much easier to study because most of it is bound together in a large structure called a ribosome. Ribosomes can easily be purified and used in studies of protein synthesis.

A basic problem facing an organism is maintaining the integrity of its DNA. Unlike protein synthesis, in which one mistake results in one altered protein molecule, or RNA synthesis, in which one mistake ultimately shows up in the 30 or so translation products of a single messenger RNA, an uncorrected mistake in the replication of DNA can last forever. It affects every descendant every time the altered gene is expressed. Thus it makes sense for the mechanism of DNA synthesis to have evolved to be highly precise. There are two ways to be precise: either be slow or recheck everything for errors a number of times. In view of the fact that bacteria, at least, have evolved to grow rapidly, it is not surprising that they use the rechecking mode. Bacteria use at least two cycles of rechecking to correct errors introduced by replication.

DNA must also maintain its structure against environmental assaults. Damage to the bases of either DNA strand that could lead to incorrect base pairing upon the next round of DNA replication must be repaired. A number of enzymes exist for recognizing, removing, and replacing damaged bases.

DNA synthesis in *E. coli* is regulated by governing the frequency of initiating replication of the chromosome. Although such a regulation system seems difficult to coordinate with cell division, the alternative mechanisms for regulating the rate of DNA synthesis appear more complicated. Regulation of the DNA elongation rate either would require changing the concentrations of many different substrates within the cell, which could be difficult because of their pathways of interconversion, or would require that the activity of the DNA-synthesizing enzyme be variable. Since many cell types can grow at a variety of rates, however, sophisticated mechanisms have developed governing the initiation of DNA replication and the segregation of completed chromosomes into daughter cells.

In this chapter we begin with the process of DNA synthesis itself. After examining the basic problems generated by the structure of DNA, we discuss the enzymology of DNA synthesis. Then we mention the methods cells use to maximize the stability of information stored in DNA. The second half of the chapter concerns physiological aspects of DNA synthesis. Measurement of the number of functioning replication areas per chromosome, the speed of DNA replication, and the coupling of cell division to DNA replication are covered.

A.   Enzymology

PROOFREADING, OKAZAKI FRAGMENTS, AND DNA LIGASE

As we have already seen in Chapter 2, a single strand of DNA possesses a polarity resulting from the asymmetric ribose-3'-phosphate-5'-ribose bonds along the backbone. Most DNA found in cells is double-stranded; a second strand is aligned antiparallel to the first strand and possesses a sequence complementary to the first. This self-complementary structure solves problems in replication because the product of replication is two daughter molecules, each identical to one of its parents. As a result of the structural similarity between the parents and daughters, the mechanisms necessary for readout of the genetic information or replication of the DNA need not accommodate multiple structures. Further, the redundancy of the stored

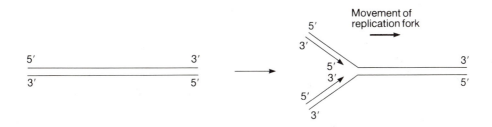

```
                              5'- AAGCTA -3'
5'- AAGCTA -3'                3'-|TTCGAT|-5'
                    ────▶                        ───── Newly synthesized strands
3'- TTCGAT -5'                5'-|AAGCTA|-3'
                              3'- TTCGAT -5'
```

information permits DNA with damage in one strand to be repaired by reference to the sequence preserved in the undamaged, complementary strand.

As is often the case in biology, numerous illustrative exceptions to the generalizations exist. Single-stranded DNA phage exist. These use a double-stranded form for intracellular replication but encapsidate only one of the strands. Apparently what they lose in repair abilities they gain in nucleotides saved.

Generally, however, DNA is double-stranded, and both strands are replicated simultaneously by movement of a replication fork down the DNA. The opposite polarities of the strands requires that, overall, one strand be elongated in a 3'-to-5' direction and the other be elongated in a 5'-to-3' direction. Although cells have been

found to possess 5'-nucleoside triphosphates and not 3'-nucleoside triphosphates, the DNA strands could be elongated nucleotide by nucleotide in both the 5'-to-3' and the 3'-to-5' direction using just 5'-nucleoside molecules (Fig. 3.1). This possibility is excluded by the need for proofreading, however.

The cell's first line of defense against mistakes in DNA synthesis is at the nucleotide incorporation step itself. If the nucleotide most recently incorporated into the elongating strand does not correctly base pair with its complementary strand, the polymerase excises the nucleotide, thereby permitting a second attempt at correct elongation at that position. This proofreading process is useful, but it eliminates the possibility of easily elongating a DNA strand in the 3'-to-5' direction nucleotide by nucleotide.

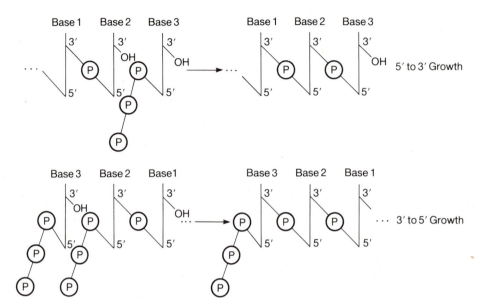

Figure 3.1   DNA may be elongated in both the 5′-to-3′ (top) and the 3′-to-5′ (bottom) directions with the use of only nucleoside 5′ triphosphates.

Proofreading by DNA polymerase to remove a misincorporated base must result in a DNA end precisely like the end that existed before addition of the incorrectly paired nucleotide. Removal of the final nucleotide from the strand growing in the 5′-to-3′ direction can regenerate the 3′-OH that is normally found at the end (Fig. 3.2). However, simple excision of the final nucleotide from a strand growing in a 3′-to-5′ direction cannot regenerate the triphosphate end that existed prior to addition or excision of this final nucleotide.

Both strands being elongated at a replication fork likely grow in a 5′-to-3′ direction. One strand can be synthesized continuously in the 5′-to-3′ direction, but

Figure 3.2   Exonucleolytic removal of the final nucleotide from a chain elongating in the 5′-to-3′ direction, a 3′–5′ exonucleolytic activity, regenerates the 3′-OH.

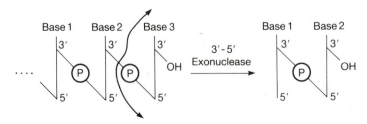

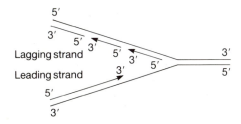

Figure 3.3   Nomenclature of the leading
and lagging strands.

the other strand must be synthesized in the 5′-to-3′ direction in short fragments that
are then linked to yield an intact strand (Fig. 3.3). These fragments are called Okazaki fragments for their discoverer. The strand synthesized continuously is called the
leading strand, and the discontinuously synthesized strand is the lagging strand.

As a result of these considerations we expect cells to contain DNA polymerases
with the following properties. They should use 5′-nucleoside triphosphates to elongate DNA strands in a 5′-to-3′ direction, and they should possess a 3′-to-5′ exonuclease activity to permit proofreading. Additionally, cells should possess an enzyme
to join the fragments of DNA that are synthesized on the lagging strand. The cell
must also solve the problems of initiating DNA replication at the correct time and of
initiating the short DNA fragments from the lagging strand.

MULTIPLE TYPES OF DNA POLYMERASES ARE
FOUND IN CELLS

A purified enzyme is one requirement for careful study of DNA synthesis. In the
early days of molecular biology, important experiments by Kornberg demonstrated
the existence in cell extracts of an enzyme that could incorporate nucleoside triphosphates into a DNA chain. This enzymatic activity could be purified, and the resultant
enzyme was available for biochemical studies. Naturally, the first question to be
asked with such an enzyme was whether it utilized a complementary DNA strand to
direct the incorporation of the nucleotides into the elongating strand via Watson-
Crick base-pairing rules. Fortunately the answer was yes. As this enzyme, DNA pol
I, was more carefully studied, some of its properties appeared to make it unlikely that
the enzyme was directly involved in most cellular DNA replication.

Cairns tried to demonstrate that pol I was not the key replication enzyme by
isolating a bacterial mutant lacking the enzyme. Of course, if the mutant could not
survive, his efforts would have yielded nothing. Remarkably, however, he found a
mutant with much less than normal activity. Such a result appeared to prove that
cells must possess other DNA-synthesizing enzymes, but until a mutant could be
found completely lacking DNA pol I, the demonstration was not complete.

Table 3.1   Comparison of DNA Polymerases from *E. coli*

	Pol I	Pol II	Pol III
Polymerization 5′ → 3′	+	+	+
Exonuclease 3′ → 5′ (proofreading)	+	+	+
Exonuclease 5′ → 3′	+	−	−
Molecules per cell	400	100	10

Another way to show the existence of the DNA polymerases other than DNA pol I is to detect and purify them biochemically. Previously such attempts had proved futile because DNA pol I masked the presence of other polymerases. Once Cairns's mutant was available, however, it was a matter of straightforward bio-chemistry to examine cell extracts for the presence of additional DNA-polymerizing enzymes. Two more such enzymes were found: DNA pol II and DNA pol III. DNA pol III appears to be the main DNA replication enzyme, whereas DNA pol I fills in the gaps in lagging strand synthesis and assists during repair of damaged DNA. No function is known for DNA pol II. Both DNA pol I and DNA pol III possess the 3′-to-5′ exonuclease activity that is necessary for proofreading removal of misincor-porated nucleotides. Table 3.1 compares the three DNA enzymes.

Most eukaryotic cells also possess more than one DNA polymerase. The three main enzymes that have been discovered are α, β, and γ, whose functions are thought to be, respectively, synthesis of the chromosomal DNA in the nucleus, repair of nuclear DNA, and replication of mitochondrial DNA.

The DNA-synthesizing enzymes are not sufficient for the biological replication of DNA. They are able to elongate growing polynucleotide chains, but they cannot initiate synthesis of a chain. This inability is not surprising, however, because initia-tion must be carefully regulated and could be expected to involve a number of other proteins that would not be necessary for elongation.

ACTIVITIES ASSOCIATED WITH A DNA REPLICATION FORK

Since the known DNA polymerases cannot initiate DNA synthesis, primers are necessary to initiate synthesis of the leading strand and to initiate synthesis of each of the short Okazaki fragments on the lagging strand. A hydroxyl group in the correct position is sufficient to prime DNA synthesis. The hydroxyl group can derive from a short stretch of DNA or RNA annealed to one strand, from cleavage of a DNA duplex, or even from a protein, such as the hydroxyl on serine (Fig. 3.4). As we will see in the next section, priming Okazaki fragments is performed by short RNA

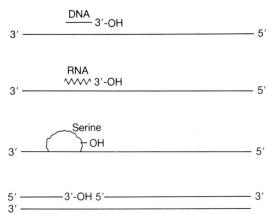

Figure 3.4 Four ways of generating 3'-OH goups necessary for DNA elongation, a deoxyribonucleotide oligomer, a ribonucleotide oligomer, a hydroxyl provided by a protein, and a nick.

molecules. Later one of the enzymes of DNA replication must remove the RNA primers.

After DNA polymerase III has elongated the lagging strand up to the next RNA primer, the enzyme releases from the DNA. Then DNA polymerase I binds and begins excising the ribonucleotides and at the same time adding deoxynucleotides at the 3' end (Fig. 3.5). As it moves, it translates the nick in the 5'-to-3' direction. The ability to substitute radioactive deoxyribonucleotides *in vitro* for nonradioactive ones by nick translation is a convenient method for making DNA radioactive for genetic engineering experiments. When the RNA primer has been completely removed and DNA pol I has released from the DNA, DNA ligase can bind and seal the

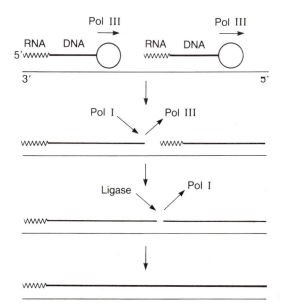

Figure 3.5 Conversion of elongating Okazaki fragments to a completed chain by the actions of DNA pol I and DNA ligase.

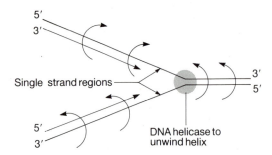

5′
3′

Single strand regions

3′
5′

5′
3′

DNA helicase to
unwind helix

Figure 3.6   The rotations necessary for movement of a DNA replication fork.

nick. As a result of its requirement to release and allow entry of DNA ligase, DNA pol I must not bind to DNA too tightly as it elongates. It must come off from time to time to permit DNA ligase to have a chance to enter, seal the nick, and stop the nick translation process.

Another protein likely to be involved in DNA replication would assist the unwinding and separation of the parental DNA strands. Since such a process would require the breaking of many hydrogen bonds and would therefore consume energy, such a helicase likely would require molecules like ATP (Fig. 3.6). Finally, it is quite reasonable to expect a protein to be found that stabilizes the single-stranded DNA that has been unwound from the double-stranded DNA.

φX174 REPLICATION: A SPECIFIC CASE

Studies of DNA synthesis must provide for initiation because the DNA polymerases lack the ability to initiate. The initiation of DNA replication is confined to specific sites on eukaryotic chromosomes and to a single site on the bacterial chromosome. Template DNA used in the assay of DNA replication must include one of these sites. Trying to assay specific initiation from the origin of DNA replication using whole chromosomal DNA would be a difficult problem. Not only would the total amount of specific synthesis be small, but the background of synthesis deriving from other locations on the template DNA, such as from nicks, is likely to completely mask the specific synthesis. The solution to this problem is to use a DNA template containing a high concentration of origins.

Two sources are apparent for obtaining high concentrations of DNA replication origins. One is to use origin-containing fragments of the chromosome cloned and amplified on plasmids, as described in Chapter 10. Work on DNA synthesis began long before the requisite recombinant DNA technology was available, however. Small DNA phage are another rich source of origins. A phage such as φX174 whose DNA is 5,000 bases long contains an origin of replication on each of its DNA copies. Per milligram of DNA, φX174 DNA contains 500 times as many DNA origins as whole E. coli DNA.

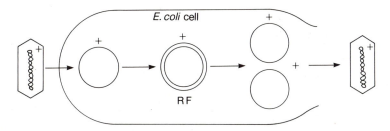

Figure 3.7 The life cycle of a single-strand DNA phage φX174. Plus strand DNA enters the cell and is converted to a double-stranded RF form, which then gives rise to more plus strands that are released as newly encapsidated phage particles.

The DNA contained inside the protein coat of the φX174 phage particle is single-stranded. Each phage particle contains a copy of the same plus strand. Thus the life cycle of the phage requires that the DNA enter the cell, synthesize the complementary minus strand to make the replicative form, RF, and then synthesize and package an excess of the plus strands (Fig. 3.7).

Careful study of the proteins required for *in vitro* synthesis of φX174 DNA has revealed that a large number of proteins is required for conversion of the plus strand to double-stranded DNA. The process can be broken down into initiation and elongation steps. As expected, initiation uses an RNA primer. The protein that actually lays down the RNA primer is the product of the *dnaG* gene. Additionally, priming also uses a set of proteins labeled i, n, n′, n″, *dnaB* protein, *dnaC* protein, and single-stranded binding protein, SSB (Fig. 3.8). The precise roles of these various proteins are not fully known. SSB first binds to most of the DNA holding it in an extended single-stranded form. A portion of the DNA forms a double-stranded hairpin. Protein n′ binds to this region. Then n and n″ bind, followed by *dnaB*, *dnaC*, and i proteins, after which n and i dissociate. Finally the *dnaG* protein binds. Since the complex of proteins forms at a double-stranded hairpin and then moves along

Figure 3.8 Priming of plus strand covered with SSB for conversion to RF. A region of self-complementary sequence forms a double-stranded region to which protein n′ binds. Then n and n″ bind, followed by *dnaB*, *dnaC*, and i proteins. Finally *dnaG* protein binds.

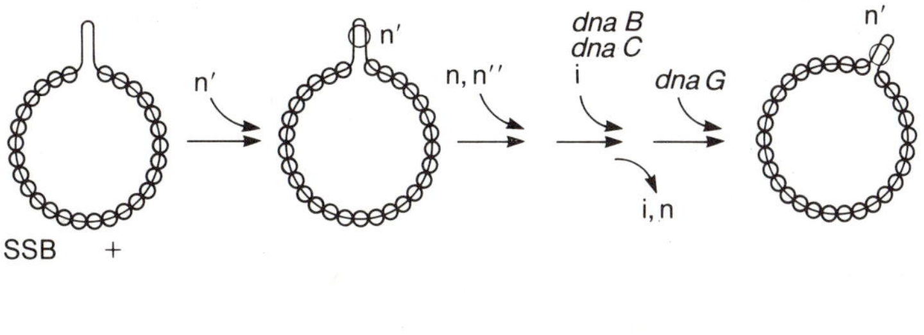

the DNA laying down primers, this complex of proteins has been named a primo-some.

As an aside, how can it be known that all these proteins must be added to assist primer formation? The answer is that the priming reaction can be performed quantitatively. Cell extracts or partially purified proteins from the extracts can be added to the priming reactions to determine if the speed or extent of priming is increased. After an extract is fractionated, for example by passing it over ion exchange columns to separate the various proteins, the priming assay can be used to determine which fractions contain a protein that accelerates the priming reaction. Hence it is possible to detect and purify proteins that speed the priming.

The priming reaction is assayed by incubating the plus strand DNA together with the priming proteins and the four ribonucleoside triphosphates. The priming proteins and the ribonucleoside triphosphates lay down a primer or a number of primers on each of the DNA molecules; when the deoxyribonucleoside triphosphates and DNA polymerase III are added, DNA is rapidly synthesized to yield a covalently closed template and a linear DNA molecule covalently linked to RNA. If less time is provided for primer synthesis, then less DNA is synthesized. This shows that the amount of DNA synthesized is limited by the amount of primer added and that increasing the amount of primer increases the amount of DNA synthesized.

To complete the synthesis of the double-stranded replicative form, RF, from the template and the single-stranded DNA linked to the RNA primer, DNA polymerase I and DNA ligase must be added. DNA pol I with its 5′-to-3′ exonuclease activity excises the RNA primer. Once this has all been removed, DNA ligase can slip in and seal the nick to produce a double-stranded covalently closed circle. This is converted to the supercoiled RF form that is found intracellularly by the inclusion of DNA gyrase in the reaction.

Figure 3.9   Rolling circle conversion of RF to excess plus strands.

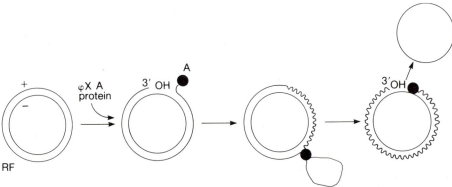

Table 3.2   Requirements for *in vitro* DNA
Replication

Proteins	Phage	
	φX174	G
dnaB	+	−
dnaC	+	−
dnaG	+	+
Single-stranded binding protein, SSB	+	−
i	+	−
n	+	−
n′	+	−
n″	+	−
DNA polymerase III	+	+
DNA polymerase I	+	+
DNA ligase	+	+

The next step in φX174 replication is synthesis of an excess of linear plus strands from the RF double-stranded circles. This process begins when the φX174 gene A protein nicks the plus strand at a specific site (Fig. 3.9). The nick creates a 3′ hydroxyl that serves as a primer. The ensuing replication by DNA pol I or pol III proceeds around and around the template circle. This mode of replication is called the rolling circle. The final problem faced by this mode of DNA replication is the conversion of the long plus strand DNA to unit-length covalently closed plus strand circles. The φX174 gene A protein also is required for this task. As it nicks the plus strand, it covalently bonds to the 5′ end. When the same stretch of DNA is replicated on the next revolution of the rolling circle, the A protein joins the ends of the first circle and bonds to the new 5′ end.

In addition to phage φX174, *in vitro* DNA synthesis directed by a variety of other phage DNAs has also been studied. Curiously, the protein requirements of the phage DNAs for replication are not all the same (Table 3.2). On first consideration it seems likely that the mechanism of initiating DNA replication would be so complicated and specialized that once nature had evolved one solution to the problem, this solution would be used over and over. However, this does not seem to be the case. Some phage require at least ten of the host proteins plus one of their own for replication, whereas other phage need a smaller number of the host proteins.

Table 3.3   Genes and Proteins of DNA Synthesis

Gene or Protein	Identity
Elongation	
dnaB	Component of primosome
dnaE	Alpha subunit of pol III
dnaG	Primase; primes lagging strand
dnaL	
dnaM	
dnaN	Beta subunit of pol III
dnaX	Delta subunit of pol III
dnaZ	Gamma subunit of pol III
gyrA	Subunit of DNA gyrase
gyrB	Subunit of DNA gyrase
SSB	Single-stranded binding protein
Initiation	
dnaA	
dnaC	
dnaI	The genes and proteins are
dnaP	not yet associated. Thus any
	of i, n, n', and n" could be A, C,
i	I, or P.
n	
n'	
n".	
Other or Uncharacterized	
dnaJ	
dnaK	
dnaY	
polA	DNA pol I for repair and gaps
lig	DNA ligase for filling nicks

COMPONENTS OF CELLULAR DNA REPLICATION

The fact that various phage DNA replication systems possess different protein requirements raises the possibility that replication of cellular DNA may use proteins that are different from or in addition to those used by the phage systems. Certainly one additional protein or set of proteins not yet mentioned is that involved in ensuring that the two replicated chromosomes are properly segregated into daughter cells at the time of cell division.

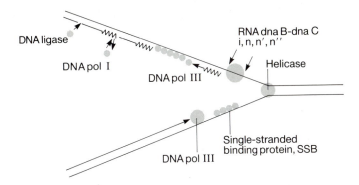

Figure 3.10   Enzymatic activities associated with a replication fork.

Both biochemical and genetic approaches have helped to identify the gene products necessary for cellular DNA replication; these genes and proteins are given in Table 3.3. The genetic approach involves the isolation of bacterial mutants that cannot replicate their DNA. On first consideration this seems impossible because the inability to synthesize DNA should be lethal. The use of temperature-sensitive mutations, however, permits the altered proteins to be active if the cells are grown at low temperatures and then for the protein to be inactivated by raising the temperature of the cells. A sizable number of genes directly involved with DNA replication have been found by this approach. Then the biochemical products of these genes have been identified by reconstitution of *in vitro* DNA replication systems in which extracts from mutant and wild-type cells are compared (Fig. 3.10).

ERROR AND DAMAGE CORRECTION

The bacterial DNA polymerases I and III, but not the eukaryotic DNA polymerases, possess the ability to correct mistakes immediately upon misincorporation of a nucleoside triphosphate. Cells also possess the ability to correct replication mistakes

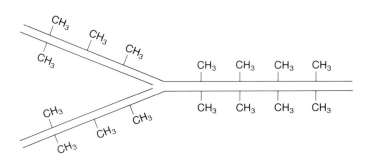

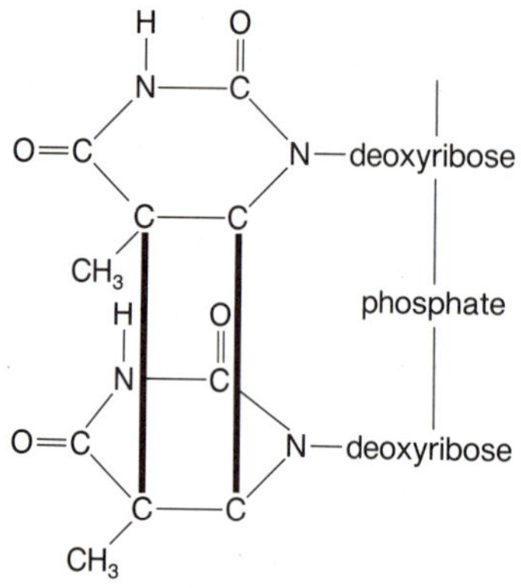

Figure 3.11   Chemical structure of a thymine dimer.

that have eluded the editing activity of the polymerases. Cells contain enzymes that recognize mispaired bases and excise them. The gap thus produced is filled in by DNA polymerase I, or in eukaryotic cells by DNA polymerase β, and sealed with DNA ligase. At first glance, such a repair mechanism would not seem to have much value. Half the time it would correct the nucleotide from the incorrect strand.

How can the cell restrict its repair to the newly synthesized strand? The answer is that the cell marks the DNA strands with methyl groups: Only after a newly synthesized strand has been around for some time does it become methylated and thereby identified as "old." By this means the repair enzymes can tell which of the strands should be repaired. Since the methyl groups involved are not densely spaced along the DNA, the repair enzymes bind to the DNA at appropriate sites and move along the DNA.

The information stored in the DNA can also be degraded after replication. Ultraviolet light damages DNA. The major chemical products of UV irradiation are cyclobutane pyrimidine dimers formed between adjacent pyrimidines, for example between thymine and thymine as shown in Fig. 3.11. These structures inhibit transcription and replication and must be removed. A set of enzymes called the *uvrA*, *uvrB*, and *uvrC* gene products performs this task in *E. coli*. In humans, defects in analogs of these enzymes lead to a condition called Xeroderma pigmentosum, which produces unusual sensitivity to sunlight.

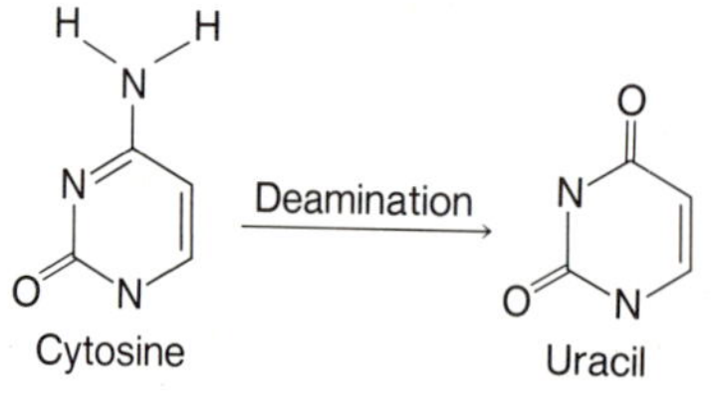

Figure 3.12   Deamination of cytosine produces uracil.

Simple chemical degradation can also compromise the information stored in DNA. The amino group on cytosine is not absolutely stable. Consequently it can deaminate to leave uracil (Fig. 3.12). DNA replication past such a point would then convert the former G-C base pair to an A-T base pair in one of the daughters because uracil has the base-pairing properties of thymine. However, enzymes exist to repair deaminated cytosine. The first enzyme of the pathway is a uracil-DNA glycosidase that removes the uracil. Then the deoxyribose can be removed and the gap that is generated can be filled. These could be the reasons why the cell does not use uracil in

Deamination   Replication

its DNA. If uracil were a normal component of DNA, deaminated cytosine could not be recognized and removed.

An interesting example demonstrates the efficiency of the deaminated cytosine repair system. As mentioned above, and as we examine more carefully in Chapter 10, cells methylate bases found in particular sequences. In *E. coli* one such sequence is CCAGG, which is methylated on the second C. Should this cytosine deaminate, its extra methyl group blocks the action of the uracil-DNA glycosidase, and spontaneous deamination at this position cannot be repaired. As a result, this nucleotide is at least ten times as susceptible to spontaneous mutation as adjacent cytosine residues.

B.   Physiological Aspects

DNA REPLICATION AREAS IN BACTERIAL CHROMOSOMES

After considering the enzymology of the DNA replication and repair processes, we turn to more biological questions. As a first step it is useful to learn the number of DNA synthesis regions per chromosome. To see why this is important, consider the two extremes. On one hand, the chromosome could be duplicated by a single replication fork traversing the entire stretch of a DNA molecule. On the other hand, many replication points per chromosome could function simultaneously. The requisite elongation rates and regulation mechanisms would be vastly different in the two extremes. Furthermore, if many replication points functioned simultaneously, they could be either scattered over the chromosome or concentrated into localized replication regions.

The most straightforward method for determining the number of replication regions on a chromosome is direct examination in an electron microscope. This is possible for smaller phage or viruses, but the total amount of DNA contained in a bacterial chromosome is far too great to allow detection of the few replication regions that would be seen. The situation is even worse for chromosomes from eukaryotes because they contain as much as one hundred times the amount of DNA per chromosome as a bacterium. The solution to this problem is not to look at all the DNA but to look at just the DNA that has been replicated in the previous minute. This can easily be done by autoradiography. Cells are administered highly radioactive thymidine, and a minute later the DNA is extracted and gently spread on photographic film to expose a trail which, upon development, displays the stretches of DNA that were synthesized in the presence of the radioactivity.

The results of such autoradiographic experiments show that cultured mammalian cells contain DNA synthesis origins about every 40,000 to 200,000 base pairs along the DNA. In bacteria the result was different; administration of a short pulse of radioactive thymidine was unnecessary. The entire chromosome of the bacterium could be visualized via the exposed photographic grains, and two startling facts were seen: The chromosome was circular, and it possessed only one or two replication regions (Fig. 3.13).

The existence of a circular DNA molecule containing an additional segment of the circle connecting two points, the theta form, was interpreted as showing that the

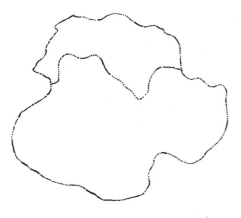

Figure 3.13   Schematic of an electron micrograph of an autoradiograph of an entire partially replicated bacterial chromosome that has been labeled with radioactive precursor for 40 minutes before extraction of DNA and preparation.

chromosome was replicated from an origin by one replication region that proceeded around the circular chromosome. It could also have been interpreted as demonstrating the existence of two replication regions that proceeded outward in both directions from a replication origin. Some of the original autoradiographs published by

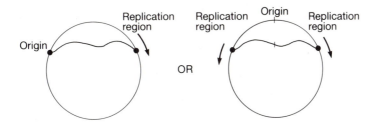

Cairns contain suggestions that the DNA is replicated in both directions from an origin. This clue that replication is bidirectional was overlooked until the genetic data of Masters and Broda provided solid and convincing data for two replication regions in the *E. coli* chromosome.

DNA IS SYNTHESIZED OUTWARD IN BOTH DIRECTIONS FROM ORIGINS

The Masters and Broda experiment is primarily a genetic experiment, but it utilizes the fact discussed in the first chapter that an exponentially growing population contains more young individuals than old individuals. Chromosomes can be considered in the same way. Correspondingly, a population of growing and dividing chromosomes contains more members just beginning replication than members just finishing replication. Since the Cairns autoradiograph experiments show that the chromosome is replicated sequentially, an exponentially growing population will contain more copies of genes located near the origin of replication than genes located near the terminus of replication.

Determining whether the bacterial chromosome is replicated in one direction from a unique origin or in both directions from a unique origin becomes a question of counting gene copies. If we could count the number of copies of genes A, B, C, D, and E, we could determine whether the cell uses monodirectional or bidirectional DNA replication initiating from point X (Fig. 3.14).

Fortunately, nature has provided a rather simple biological way to count the relative number of copies of genes through the use of phage P1. When phage P1 infects cells, it produces about 100 new copies of itself by a process in which about

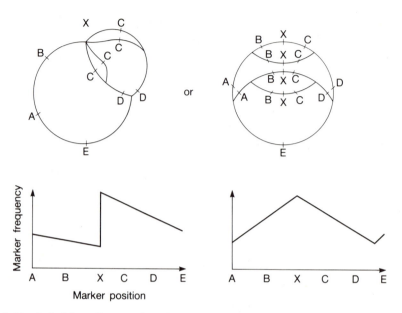

Figure 3.14   Left: Monodirectional replication from point X generates a step plus gradient of marker frequencies. Right: Bidirectional replication originating from X generates two gradients of marker frequencies. In these examples, successive initiations have taken place on the single chromosome.

100 copies of the DNA are synthesized and about an equivalent amount of the phage protein that encapsidates the DNA is also synthesized. The phage DNA packaging process is not altogether accurate, and about 1% of the heads encapsidate a piece of bacterial DNA about the size of the phage P1 DNA. These are defective phage since they carry only bacterial DNA. The defective phage are very useful, however, for they provide a way of injecting their enclosed bacterial DNA into other cells. If a P1 lysate that was prepared on one type of cells is then used to infect a second culture of cells, most of the infected cells will proceed to make new phage P1. Those few cells that are infected with a P1 coat containing *E. coli* DNA from the first cells may be able to recombine that particular stretch of *E. coli* DNA into their chromosomes. They could replace stretches of chromosomal DNA with chromosomal DNA carried by the phage particles. This process is termed transduction.

Transduced cells can be made to reveal themselves as colonies, thereby permitting their accurate quantitation. Consequently the use of phage P1 enabled counting of the relative numbers of copies of various genes in growing cells. The results together with the known genetic map indicated that *E. coli* replicates its chromosome bidirectionally and located on the genetic map the region containing the replication origin.

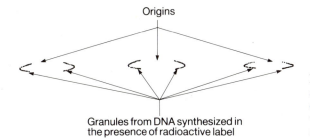

Origins

Granules from DNA synthesized in
the presence of radioactive label

Figure 3.15   Schematic of an electron micrograph of
an autoradiograph prepared from mouse DNA
extracted after 30 seconds of labeling with radioactive thymidine. DNA is replicated bidirectionally
from the multiple origins.

Autoradiographic experiments demonstrated that mammalian DNA also replicates bidirectionally from origins (Fig. 3.15). Multiple replication forks or "eyes"
could be shown to derive from a single DNA duplex. The replication trails indicated
that two forks originated from a single origin. Furthermore, the rate of elongation of
labeled strands showed that the synthesis rate of mammalian DNA is about 200
nucleotides per second.

ESCHERICHIA COLI DNA ELONGATES AT ABOUT 500 BASES/SEC

It seems reasonable that the single bacterial chromosome should be replicated by a
single replication region traversing the chromosome in about one doubling time. If
such a replication region does not contain multiple points of DNA elongation, then
the speed of DNA chain elongation must be on the order of 1,000 nucleotides per
second in order to replicate the chromosome's 3×10^6 bases in 30 minutes, a typical
doubling time for rapidly growing bacteria. Measuring such a rate is very difficult,
but fortunately it is possible to reduce its value by about a factor of 5 by growing the
cells at 20° instead of 37°, the temperature of most rapid growth.

Manor and Deutscher have directly measured the elongation rate of DNA in *E.
coli*. Conceptually their measurement is straightforward. If radioactive DNA precursors are suddenly added to the cell growth medium and growth is stopped soon
thereafter, the total amount of radioactivity incorporated into DNA equals the
product of four factors: a constant related to the specific activity of the label, the
number of growing chains, the elongation rate of a chain, and the time of radioactive
labeling. Similarly, the total amount of radioactivity incorporated into the ends of
elongating chains equals the product of two factors: the same constant related to the
specific activity and the number of elongating chains. If we let T be the total incorporated radioactivity, E be the radioactivity in end nucleotides, c be the constant related

to specific activity, N be the number of elongating chains, and R be their rate, then

$$T = c \times N \times R \times t,$$

$$E = c \times N,$$

and their ratio yields the elongation rate

$$R = \frac{T}{t \times E}.$$

Determination of the T radioactivity is straightforward, but determination of the end radioactivity is not so obvious. Furthermore, careful experimentation would be required so that losses of either T or E samples do not mar the accuracy. These problems were solved by digesting the DNA extracted from the labeled cells with a nuclease that cleaves so as to leave a phosphate on the 3′ position of the deoxyribose. After the digestion, the terminal deoxyribose from the elongating chain lacks a phosphate, whereas the internal deoxyriboses all possess phosphate groups (Fig. 3.16). Hence separation and quantitation of the radioactive nucleosides and nucleotides in a single sample prepared from cells following a short administration of the four radioactive DNA precursors can yield the desired T and E values.

This experiment was difficult. Label could not be provided for too long an interval before synthesis was stopped. One second of synthesis could label several hundred bases, and a separation of nucleosides and nucleotides would then need to be better than one part in several hundred. Also, it is difficult to add label suddenly and then to stop the cells' DNA synthesis quickly. Finally, the specific activity of intracellular nucleoside triphosphate pools does not immediately jump to the same specific activity as the label added to the medium. Fortunately this effect could be

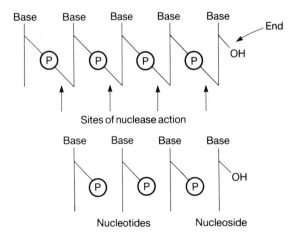

Figure 3.16   Cleavage of DNA between the phosphate and the 5′ position leaves a nucleoside (no phosphate) from the terminal nucleotide of the elongating chain and nucleotides from all other positions.

analyzed mathematically; the result of including a changing specific activity is

$$R = \frac{\ln \dfrac{T(t_2)}{T(t_1)}}{\displaystyle\int_{t_1}^{t_2} \dfrac{E(t)}{T(t)}\,dt}.$$

The sample from the first point that could be taken from the cells after the addition of radioactive label possessed little radioactivity. It had counts of 17–20 cpm in ends and $2-20 \times 10^5$ cpm in total DNA. Samples taken later possessed greater amounts of radioactivity. This experiment yielded elongation rates of 140–250 bases/sec in cells with a doubling time of 150 minutes. This corresponds to a rate of 400 to 800 bases/sec in cells growing at 37° with a doubling time of 45 minutes, and therefore only about two elongation points exist per replication region. The effects on this type of measurement of discontinuous DNA replication on the lagging strand are left as a problem for the reader.

The primary conclusion that should be drawn from the direct measurement of the DNA elongation rate is that a small number of enzyme molecules are involved. The cell does not utilize a factory with many active DNA polymerase molecules in a growing region. What happens when cells grow at 37° but at different growth rates due to the presence of different growth media? Does the DNA elongation rate vary with the growth medium? The Manor-Deutscher experiment could answer this question, but fortunately there are simpler ways to measure the variation of DNA elongation rate with cell growth rate.

CONSTANCY OF THE *E. COLI* DNA ELONGATION RATE

Determining the strategy that cells use to adjust DNA replication to different growth rates once again utilizes measurement of the numbers of gene copies. Consider two cell types, one with a doubling time of 1 hour and one with a doubling time of 2 hours. If each cell type requires the full doubling time to replicate its chromosome, then the distributions of structures of replicating chromosomes extracted from random populations of cells growing at the two different rates would be identical (Fig. 3.17). However, if both cell types replicate their chromosomes in 1 hour, then the cells with the 1-hour doubling time will possess a different distribution of chromosome structures than the cells with the 2-hour doubling time (Fig. 3.18). The problem, once again, is that of counting copies of genes. Measuring the number of copies of genes located near the origin and of genes located near the terminus of replication permits these two possible DNA replication schemes to be distinguished.

Instead of measuring numbers of copies of genes by transduction frequencies, a more precise method — DNA-DNA hybridization — could be used because the locations of the origin and terminus were known. A biological trick was used to obtain

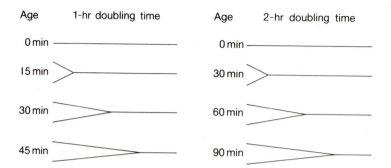

Figure 3.17   States of chromosomes of cells with 1-hour and 2-hour doubling times assuming 1 hour or 2 hours are required to replicate the DNA. The ratio of copies of genes near the replication origin and replication terminus are the same for cells growing with the two rates.

the required samples of DNA that were represented on the chromosome near the origin and terminus. Sequences of DNA were forced to integrate into the chromosomal DNA in these two locations.

Phage lambda can infect cells and proceed through a normal growth cycle followed by lysis of the cells and release of phage. Or it can enter cells, shut down expression of its genes, and, catalyzed by the activities of several enzymes, integrate into a specific site in the chromosome. Integrated in this way, the phage DNA is replicated as though it is part of the bacterial chromosome. Phage Mu is similar to phage lambda, but it integrates nearly anywhere. Therefore the replication experiments utilized a bacterial strain that contained lambda integrated near the terminus and Mu integrated near the DNA replication origin (Fig. 3.19).

DNA was extracted from the cells, denatured, and immobilized on filter paper. Then a mixture of ^{3}H-labeled Mu DNA and ^{14}C-labeled lambda DNA was added

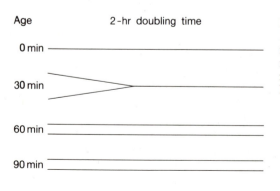

Figure 3.18   States of chromosomes in cells with a 2-hour doubling time, assuming only 1 hour is required for replication of the chromosome. The ratio of the number of copies of genes near the replication origin and terminus is different than if 2 hours were required for replication of the DNA.

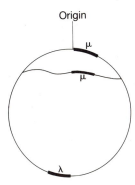

Origin

μ

μ

λ

Figure 3.19   A circular chromosome with phage mu integrated near the origin and phage lambda integrated near the terminus.

and allowed to anneal to the single-stranded DNA on the filters. When all annealing to the immobilized DNA was completed, the filters were washed free of unannealed DNA and the ratio of bound ^{3}H to ^{14}C radioactivity was determined by liquid scintillation counting. This ratio reflects the ratio of the number of origins and termini in the culture of cells from which the DNA was extracted. The ratio could be measured using DNA extracted from cells growing at various growth rates, and it showed that for cell doubling times in the range of 20 minutes to 3 hours the chromosome doubling time remained constant at about 40 minutes.

The constancy of the chromosome doubling time raises new problems, however. How do the cells manage to keep DNA replication and cell division precisely coordinated, and how can DNA replication, which requires 40 minutes, manage to keep up with cell division if cells are dividing in less than 40 minutes?

KEEPING DNA SYNTHESIS IN STEP WITH CELL DIVISION

Helmstetter and Cooper provided an explanation for the problem of maintaining a strain of *Escherichia coli* cells in balanced growth despite a difference between the cellular division time and the chromosome replication time. Even though other strains and organisms may differ in the details of their regulation mechanisms, the model is of great value because it summarizes a large body of data and provides a clear understanding of ways in which cell division and DNA replication can be kept in step.

The model applies most closely to cells growing with a doubling time less than 1 hour. One statement of the model is that a cell will divide $I + C + D$ minutes after the start of synthesis of an initiator substance, which could be a protein. I can be thought of as the time required for the I protein to accumulate to a level such that replication can initiate on all origins present in the cell. In our discussions, we will call this critical level 1. That is, when a full unit of I has accumulated, all chromosomes present in the cell initiate replication, a round of replications begins, and all *I* present at this time is consumed so that its accumulation begins again from a value of zero.

C is the time required for complete synthesis of the chromosome. It is independent of the growth rate of the cells as long as they grow at 37°, and it has a value of 40 minutes. *D* is a constant equal to 20 minutes. This is the time following completion of a round of DNA synthesis until the cell divides. It should be emphasized that in this model a cell must divide *D* minutes after completion of a round of DNA replication. *D* can be considered the time required for segregation of the daughter chromosomes into opposite ends of the cell and for growth of the septum that separates the cells.

The only parameter in the model that is responsive to the growth rate of the cells is the rate at which *I* accumulates. If the growth rate doubles, because of addition of more varieties of nutrients to the growth medium, for example, the rate of accumulation of *I* doubles. The time required for accumulation of a full unit of *I* is the doubling time of the cells.

The Helmstetter-Cooper model also fulfills the essential requirement that such models be stable. If a cell finds itself in an unbalanced state, for example because of accidentally delayed cell division, the model implicitly directs the daughters of such a cell division back into a balanced state.

To illustrate the model, we will consider cells growing with doubling times of 60 minutes and 20 minutes, presented in Figs. 3.20 and 3.21, respectively. For convenience, consider the bidirectional replication of the cell's circular chromosome to be abstracted to a forked line. We use a dot to represent the replication fork, and we enclose the chromosome to represent cell division. Since the model is stable, analysis of a cell division cycle can begin from a point not in a normal division cycle, and continued application of the model's rules should ultimately yield the states of cells growing with the appropriate doubling times.

We begin in Fig. 3.20 from a point at which replication of a chromosome has just begun and the value of *I* is zero. After 40 minutes the chromosome is completely

Figure 3.20   Application of the Helmstetter-Cooper model to cells with a doubling time of 60 minutes. The state of the chromosome, the time, and the level of *I* are shown. Twenty minutes after completion of replication the cell divides.

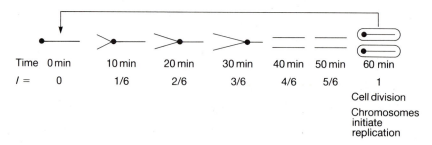

Time	0 min	10 min	20 min	30 min	40 min	50 min	60 min
I =	0	1/6	2/6	3/6	4/6	5/6	1

Cell division

Chromosomes initiate replication

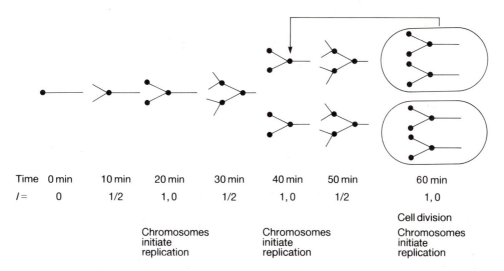

Time	0 min	10 min	20 min	30 min	40 min	50 min	60 min
I =	0	1/2	1, 0	1/2	1, 0	1/2	1, 0

Cell division

Chromosomes initiate replication (under 20 min)

Chromosomes initiate replication (under 40 min)

Chromosomes initiate replication (under 60 min)

Figure 3.21   Application of the Helmstetter-Cooper model to cells with a doubling time of 20 minutes. At 60 minutes, cell division is possible and the resulting state of the cells is the same as cells shown at 40 minutes in the diagram.

replicated and I has reached the value of $\frac{2}{3}$. At 60 minutes, two things happen. The value of I reaches 1, so replication initiates, and I is reset to zero. At this same time, the D interval of 20 minutes has elapsed and the cell can divide. This point of division, with chromosomes initiating replication, is precisely the starting point of the analysis. Therefore, cells proceed through a 60-minute cell cycle as indicated. Newly divided daughters contain one complete chromosome that is just initiating replication.

   The analysis of cells with a doubling time of 20 minutes is also straightforward (Fig. 3.21). At 20, 40, and 60 minutes after the start of replication, full units of I have accumulated, new rounds of replication begin from all the origins present, and the chromosomes become multiforked. The first cell division can occur at 60 minutes after the start of replication, and divisions occur at 20-minute intervals thereafter. The DNA configuration and quantity of I present in cells just after division at 60 minutes is the same as in cells at 40 minutes. Therefore the cell cycle in this medium begins at 40 minutes and ends at 60 minutes. This means that zero-age cells possess two half-replicated chromosomes in this medium.

   The model also easily explains a striking phenomenon that is seen when cells are shifted from a poor medium to a rich medium. Despite the fact that the total cell mass, turbidity, immediately begins to increase at an exponential rate characteristic of the new medium, the division of cells continues at the old exponential rate for 60 minutes (Fig. 3.22). Then it abruptly changes to the new rate. This is simple to understand as a result of the lags built into the model. Immediately following the

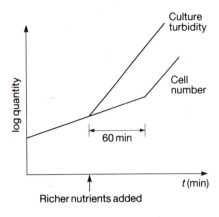

Figure 3.22   A semilogarithmic plot of the culture turbidity, which is roughly proportional to the total cell mass and the cell number. Before the nutritional upshift, cells were in balanced exponential growth, as indicated by the straight-line increase in cell number and turbidity.

upshift, I begins to accumulate at the new, higher rate characteristic of the new medium. Consequently cells begin reaching the full unit of I sooner than they would have before. However, the earliest time that the increased rate at which I reaches unity can be reflected in faster cell divisions is 60 minutes. Forty minutes are required before the upshift is seen in an increase in the rate of completion of chromosome replications, and an additional 20 minutes are required before these can be seen in an increased rate of cell division. During the 60 minutes' lag before adoption of the new division rate, cells have been growing at the new rate. As a result of the lag in division, the average mass of individual cells must increase. This in turn ought to be reflected in a larger cell volume, and indeed it is. The faster *E. coli* grows, the larger its volume.

*MAKING SYNCHRONIZED CELLS

Construction of the Helmstetter-Cooper model required knowledge of the rate of DNA synthesis as a function of cell age from the time of cell division until the next cell division for cells growing with a variety of doubling times. Since the amount of DNA in a single cell is far too small to measure accurately, populations containing many cells all of the same age had to be used. Such populations are called synchronized.

A variety of methods have been used to produce synchronized cells, but most of them traumatize cells so that subsequent growth is abnormal and the resulting data are uninterpretable. The best method of synchronizing bacteria is based on the fact that cells of strain B/r bind to cellulose-acetate cellulose-nitrate filter paper. To use this property, the culture is filtered and the paper inverted and bathed from above with medium that oozes through the paper and washes off all loose cells (Fig. 3.23). Soon the only cells released into the medium dripping off the underside of the paper are those deriving from cell division in which the loose daughter is not bound to the filter.

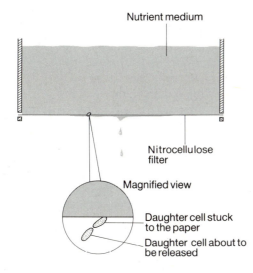

Nutrient medium

Nitrocellulose filter

Magnified view

Daughter cell stuck to the paper

Daughter cell about to be released

Figure 3.23   Schematic and an expanded view of a method to produce synchronized bacteria.

To ascertain that cells have been synchronized properly, an immediate assay is needed. This can be accomplished by recording the electrical resistance through a tiny 30-micron orifice immersed in the medium (Fig. 3.24). Medium with cells is drawn through the orifice at a rate of about 0.01 ml per minute by a slight vacuum. At the instant a cell passes through, it changes the electrical resistance by an amount proportional to its volume because a cell is nearly an electrical insulator. The sizes

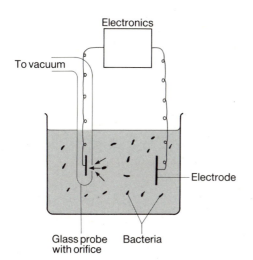

Electronics

To vacuum

Electrode

Glass probe with orifice   Bacteria

Figure 3.24   An electronic method for rapidly measuring the distribution of cell sizes and concentrations of cells.

and numbers of these brief pulses can be electronically quantitated, yielding a rapid measure of the cell densities and individual volumes of 10^5 cells per minute.

Consider the cells that elute from the inverted filter paper in a 2-minute interval. These are "baby" cells. If the population is incubated, the cells will increase in volume and, one doubling time later, divide into twice as many new baby cells. Useful synchrony is lost after about three doublings owing to the variability in the doubling times of individual cells.

*HOW FAST CAN DNA BE REPLICATED?

In the preceding sections we have seen that the bacterial chromosome is replicated by two synthesis forks moving away from a replication origin at a chain elongation speed of about 500 nucleotides per second for cells growing at $37°$. How does this rate compare to the maximum rate at which nucleotides could diffuse to the DNA polymerase? This question is one specific example of a general concern about intracellular conditions. Often it is important to know an approximate amount of time required for a particular molecule to diffuse to a particular site.

If we consider a polymerase molecule to be sitting in a sea of infinite dimensions containing the substrate and if the rate of the enzyme is limited by the diffusion of nucleotides to its active site, then the concentration of substrate is zero on the surface of a sphere of radius r_0 constituting the active site of the enzyme (Fig. 3.25). Any substrate molecules crossing the surface into this region disappear. At great distances from the enzyme, the concentration of substrate remains unaltered. These represent the boundary conditions of the situation, which requires a mathematical formulation to determine the concentrations at intermediate positions.

The basic diffusion equation relates time and position changes in the concentration C of a diffusable quantity. As diffusion to an enzyme can be considered to be

Figure 3.25   A DNA polymerase possessing an active site of radius r_0 into which nucleotides disappear as fast as they can reach the enzyme and the expected nucleotide concentration as a function of distance from the enzyme.

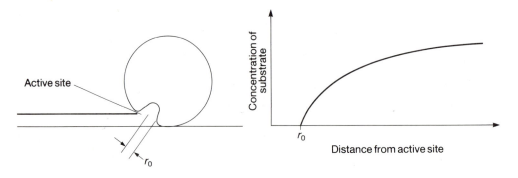

spherically symmetric, the diffusion equation can be written and solved in spherical coordinates involving only the radius r, the concentration C, the diffusion coefficient D, and time t:

$$\frac{dC}{dt} = D \frac{d^2C}{dr^2} + \frac{2D}{r} \frac{dC}{dr}.$$

The solution to this equation, which satisfies the conditions of being constant at large r and zero at $r = r_0$, is

$$C(r) = C_0(1 - r_0/r).$$

The flow rate J of substrate to the enzyme can now be calculated from the equation, giving the flux J:

$$J = -D \frac{dC}{dr}.$$

The total flow through a sphere centered on the enzyme yields

$$\text{Flow} = 4\pi r^2 J$$
$$= 4\pi D r_0 C_0.$$

The final result shows that the flow is independent of the size of the sphere chosen for the calculation. This is as it should be. The only place that material is being destroyed is at the active site of the enzyme. Everywhere else, matter must be conserved. In steady state there is no change of the concentration of substrate at any position, and hence the net amount flowing through the surface of all spheres must be equal.

To calculate the flow rate of nucleoside triphosphates to the DNA polymerase, we must insert numerical values in the final result. The concentration of deoxynucleoside triphosphates in cells is between 1 mM and 0.1 mM. We will use 0.1 mM, which is 10^{-4} moles per liter or 10^{-7} moles per cm^3. Taking the diffusion constant to be 10^{-7} cm^2/sec and r_0 to be 10 Å, we find that the flow is 10^{-20} moles/sec or about 6,000 molecules/sec.

Thus, while the rate of DNA synthesis per enzyme molecule, about 500 nucleoside triphosphates per second, is not at the upper limit of the diffusion rate, it is rather close. Considering that the actual active site for certain capture of the triphosphate is most likely much smaller than 10 Å, the elongation rate of DNA seems remarkably high.

PROBLEMS

3.1. By looking up the appropriate papers, write a brief summary of how the DNA polymerase I mutant in E. coli was found.

3.2. Lark's paper, J. Mol. Biol. 73, 371 (1973), says that Okazaki fragments arise from many regions of the chromosome of a single cell. This would require that there be many nicks scattered about the genome. On the other hand, Worcel and Burgi, J. Mol. Biol. *71*, 127 (1972), claim that there are essentially no nicks in the entire genome of *E. coli.* Can this discrepancy be reconciled?

3.3. From the data of Okazaki et al., Fig. 2, Proc. Nat. Acad. Sci. USA *59*, 598 (1968), calculate the number of Okazaki fragments per replication fork that are being synthesized at any instant. What does your answer mean? In light of the Manor-Deutscher results, was it sensible to conclude that Okazaki fragments are direct replication products? Now read B. Tye et al., Proc. Nat. Acad. Sci. USA *74*, 154 (1977). Comment on the value of a few simple computations (at least in hindsight).

3.4. Show by differentiation that the solution $C = C_0(1 - r_0/r)$ satisfies the diffusion equation in spherical coordinates.

3.5. The Heisenberg uncertainty principle, which can be stated $\Delta E \Delta t$ is greater than or equal to Planck's constant $\hbar$, can also be used to obtain an upper limit for the possible rate of DNA synthesis. Let us suppose that ΔE represents the difference of energy on base-pair binding. From the uncertainty principle, calculate the minimum time that would be necessary to distinguish a correctly paired base from an incorrectly paired base on energetic grounds.

3.6. The method used by Kornberg for detection of host protein components used in formation of the primase involved a system in which most of the host proteins merely accelerated the rate of a reaction that would proceed in their absence. How could one proceed to purify components if each component were absolutely essential for the reaction to proceed at all?

3.7. The synthesis of plus strand ϕX174 DNA requires single-stranded binding protein in quantities stoichometric to the amount of plus strand synthesized. How would this be shown experimentally?

3.8. What would you say about a claim that the only effect of a mutation was an increase in the 3'-5' exonuclease activity of DNA pol I and that this *reduced* the spontaneous mutation frequency?

3.9. How could an autoradiography experiment with *E. coli* be performed to best demonstrate bidirectional replication of the chromosome?

3.10. Masters and Broda, Nature New Biol. *232*, 137 (1970), report that the normalization of gene transduction frequency by P1 is unnecessary; in fact, the raw data support their conclusions better than the normalized data. What is a possible explanation for this result and what experiment could test your proposal?

3.11. In the chromosome replication time measurements of Chandler, Bird, and Caro, a constant replication time of 40 minutes was found. Does their result depend on when in the cell cycle this 40 minutes of replication occurs?

3.12. If the *dam* methylase methylates DNA for use in error correction of newly synthesized DNA strands, why would mutations increasing or decreasing *dam* methylase synthesis be mutagenic?

3.13. A Manor-Deutscher-Littauer experiment yielded the data in the following table. What is the DNA chain elongation rate in these cells?

t (sec)	cpm (nucleosides)	cpm (nucleotides)
5	43	24,000
10	110	91,000
15	140	200,000
20	205	410,000
25	260	650,000

3.14. Why should or shouldn't the existence of Okazaki fragments interfere with a Manor-Deutscher measurement as DNA labeling times become short?

3.15. What biochemical experiments would you do to prove or refute the Helmstetter-Cooper model?

3.16. Cells that have been growing for many generations with a doubling time of 20 minutes are shifted to medium supporting a doubling time of 40 minutes. Assume that there are no nutritional starvations as a result of this shift. Sketch the DNA configurations and indicate cell divisions at 10-minute intervals of cells whose age at the time of the shift was 0 minutes and 10 minutes. Continue the analysis until cells have fully adapted to growth at the slower growth rate.

3.17. The Helmstetter-Cooper model for DNA replication and cell division correctly explains much data. Nonetheless, the model may be incorrect or incomplete. What is the simplest type of new data you can imagine that would definitively show that the model is incorrect?

3.18. Using the Helmstetter-Cooper model for DNA replication and the exponential age distribution, calculate the ratio of the number of copies per cell of the *ara* and *lac* operons for cells with a doubling time of 20 minutes.

3.19. For cells with a doubling time of 60 minutes, what is the ratio of the number of ends of chromosomes to beginnings?

3.20. After loosely bound cells have been washed off the filter and only newly divided cells are being eluted in a synchrony experiment, the concentration of cells in the eluate is found to vary. What is the explanation for this variation? You should assume that the probability that a cell initially binds to the filter is independent of its age.

RECOMMENDED READINGS

The Replication of DNA in *Escherichia coli*, M. Meselson, F. Stahl, Proc. Nat. Acad. Sci. USA *44*, 671–682 (1958).

Chromosome Replication and the Division Cycle of *Escherichia coli* B/r, S. Cooper, C. Helmstetter, J. Mol. Biol. *31*, 519–540 (1968).

Rates of DNA Chain Growth in *E. coli,* H. Manor, M. Deutscher, U. Littauer, J. Mol. Biol. *61,* 503–524 (1971).

The Replication Time of the *E. coli* K-12 Chromosome as a Function of Cell Doubling Time, M. Chandler, R. Bird, L. Caro, J. Mol. Biol. *94,* 127–132 (1975).

E. coli Mutator Mutants Deficient in Methylation-Instructed DNA Mismatch Correction, B. Glickman, M. Radman, Proc. Nat. Acad. Sci. USA *77,* 1063–1067 (1980).

Replication of φX174 DNA with Purified Enzymes, I. Conversion of Viral DNA to a Super-coiled, Biologically Active Duplex, J. Shlomai, L. Polder, K. Arai, A. Kornberg, J. Biol. Chem. *256,* 5233–5238 (1981).

RELATED REVIEWS, BOOKS, AND ARTICLES

DNA Ligase: Structure, Mechanism, and Function, I. R. Lehman, Science *186,* 790–797 (1974).

DNA Repair in Bacteria and Mammalian Cells, P. Hanawalt, P. Cooper, A. Ganesan, C. Smith, Ann. Rev. Biochem. *48,* 783–836 (1979).

Initiation of DNA Synthesis in *Escherichia coli,* J. Tomizawa, G. Selzer, Ann. Rev. Biochem. *48,* 999–1034 (1979).

Discontinuous DNA Replication, T. Ogawa, T. Okazaki, Ann. Rev. Biochem. *49,* 421–458 (1980).

Ultraviolet Light Repair and Mutagenesis Revisited, W. Haseltine, Cell *33,* 13–17 (1983).

DEEPER READING

The Bacterial Chromosome and Its Manner of Replication as Seen by Autoradiography, J. Cairns, J. Mol. Biol. *6,* 208–213 (1963).

On the Regulation of DNA Replication in Bacteria, F. Jacob, S. Brenner, F. Cuzin, Cold Spring Harbor Symp. Quant. Biol. *28,* 329–348 (1963).

Isolation of an *E. coli* Strain with a Mutation Affecting DNA Polymerase, P. De Lucia, J. Cairns, Nature *224,* 1164–1166 (1964).

DNA Synthesis During the Division Cycle of Rapidly Growing *Escherichia coli* B/r, C. Helmstetter, S. Cooper, J. Mol. Biol. *31,* 507–518 (1968).

On the Mechanism of DNA Replication in Mammalian Chromosomes, J. Huberman, A. Riggs, J. Mol. Biol. *32,* 327–341 (1968).

Analysis of DNA Polymerases II and III in Mutants of *Escherichia coli* Thermosensitive for DNA Synthesis, M. Gefter, Y. Hirota, T. Kornberg, J. Wechsler, C. Barnoux, Proc. Nat. Acad. Sci. USA *68,* 3150–3153 (1971).

Evidence for the Bidirectional Replication of the *E. coli* Chromosome, M. Masters, P. Broda, Nature New Biol. *232,* 137–140 (1971).

Mechanism of DNA Chain Growth, XI. Structure of RNA-Linked DNA Fragments of *E. coli,* S. Hirose, R. Okazaki, F. Tamanoi, J. Mol. Biol. *77,* 501–517 (1973).

Genetic and Enzymatic Characterization of a Conditional Lethal Mutant of *E. coli* K-12 with a Temperature-Sensitive DNA Ligase, E. Konrad, P. Modrich, I. Lehman, J. Mol. Biol. *77,* 519–529 (1973).

Genetics and Function of DNA Ligase in *E. coli*, M. Gottesman, M. Hicks, M. Gellert, J. Mol. Biol. *77*, 531–547 (1973).

Bidirectional Growth of the *E. coli* Chromosome, R. Hohlfeld, W. Vielmetter, Nat. New Biol. *242*, 130–135 (1973).

Properties of a Membrane-Attached Form of the Folded Chromosome of *E. coli*, A. Worcel, E. Burgi, J. Mol. Biol. *82*, 91–105 (1974).

Electron Microscopic Visualization of the Folded Chromosome of *E. coli*, H. Delius, A. Worcel, J. Mol. Biol. *82*, 107–109 (1974).

Synthesis of Bacteriophage Lambda DNA *in vitro:* Requirement for O and P Gene Products, H. Shizuya, C. Richardson, Proc. Nat. Acad. Sci. USA *71*, 1758–1762 (1974).

A Conditional Lethal Mutant of *E. coli* K12 Defective in the 5'-3' Exonuclease Associated with DNA Polymerase I, E. Konrad, I. Lehman, Proc. Nat. Acad. Sci. USA *71*, 2048–2051 (1974).

Replication of the *E. coli* Chromosome with a Soluble Enzyme System, T. Kornberg, A. Lockwood, A. Worcel, Proc. Nat. Acad. Sci. USA *71*, 3189–3193 (1974).

Mechanism of DNA Chain Growth, XV. RNA-Linked Nascent DNA Pieces in *E. coli* Strains Assayed with Spleen Exonuclease, Y. Kurosawa, T. Ogawa, S. Hirose, T. Okazaki, R. Okazaki, J. Mol. Biol. *96*, 653–664 (1975).

Transient Accumulation of Okazaki Fragments as a Result of Uracil Incorporation into Nascent DNA, B. Tye, P. Nyman, I. Lehman, S. Hochhauser, B. Weiss, Proc. Nat. Acad. Sci. USA *74*, 154–157 (1977).

The Structure of Histone-Depleted Metaphase Chromosomes, J. Paulson, U. Laemmli, Cell *12*, 817–828 (1977).

An RNA Transcribed from DNA at the Origin of Phage fd Single Strand to Replicative Form Conversion, K. Geider, E. Beck, H. Schaller, Proc. Nat. Acad. Sci. USA *75*, 645–649 (1978).

Modification of DNA by Aflatoxin B_1 Creates Alkali-Labile Lesions in DNA at Positions of Guanine and Adenine, A. D'Andrea, W. Haseltine, Proc. Nat. Acad. Sci. USA *75*, 4120–4124 (1978).

Molecular Basis of Base Substitution Hotspots in *Escherichia coli*, C. Coulondre, J. Miller, P. Farabaugh, W. Gilbert, Nature *274*, 775–780 (1978).

Aphidicolin Prevents Mitotic Cell Division by Interfering with the Activity of DNA Polymerase-α, S. Ikegami, T. Taguchi, M. Ohashi, M. Oguro, H. Nagano, Y. Mano, Nature *275*, 458–460 (1978).

Eucaryotic DNA: Organization of the Genome for Replication, R. Hand, Cell *15*, 317–325 (1978).

dnaA Acts Before *dnaC* in the Initiation of DNA Replication, F. Kung, D. Glaser, J. Bact. *133*, 755–762 (1978).

ATP Utilization by *rep* Protein in the Catalytic Separation of DNA Strands at a Replicating Fork, A. Kornberg, J. Scott, L. Bertsch, J. Biol. Chem. *253*, 3298–3304 (1978).

Map Position of the Replication Origin on the *E. coli* Chromosome, O. Fayet, J. Louarn, Mol. Gen. Genet. *162*, 109–111 (1978).

Mini-Chromosomes: Plasmids Which Carry the *E. coli* Replication Origin, W. Messer, H. Bergmans, M. Meijer, J. Womack, F. Hansen, K. von Meyenburg, Mol. Gen. Genet. *162*, 269–275 (1978).

Nucleotide Sequence of *E. coli* K-12 Replication Origin, K. Sugimoto, A. Oka, H. Sugisaki, M. Takanami, A. Nishimura, Y. Yasuda, Y. Hirota, Proc. Nat. Acad. Sci. USA *76*, 575–579 (1979).

Nucleotide Sequence of the Origin of Republication of the *E. coli* K-12 Chromosome, M. Meijer, E. Beck, F. Hansen, H. Bergmans, W. Messer, K. von Meyenburg, H. Schaller, Proc. Nat. Acad. Sci. USA *76*, 580–584 (1979).

Adenovirus DNA Replication *in vitro*, M. Challberg, T. Kelly, Proc. Nat. Acad. Sci. USA *76*, 655–659 (1979).

An *Escherichia coli* Mutant Defective in Single-Strand Binding Protein Is Defective in DNA Replication, R. Meyer, J. Glassberg, A. Kornberg, Proc. Nat. Acad. Sci. USA *76*, 1702–1705 (1979).

UV-Endonuclease from Calf Thymus with Specificity Toward Pyrimidine Dimers in DNA, E. Waldstein, S. Peller, R. Setlow, Proc. Nat. Acad. Sci. USA *76*, 3746–3750 (1979).

Fidelity of Replication of Phage ϕX174 DNA by DNA Polymerase III Holoenzyme: Spontaneous Mutation by Misincorporation, A. Fersht, Proc. Nat. Acad. Sci. USA *76*, 4946–4950 (1979).

The Delta Subunit of *Escherichia coli* DNA Polymerase III Holoenzyme Is the dnaX Gene Product, U. Hubscher, A. Kornberg, Proc. Nat. Acad. Sci. USA *76*, 6284–6288 (1979).

DNA Sequences and Structural Homologies of the Replication Origins of Lambdoid Bacteriophages, R. Grosschedl, G. Hobom, Nature *277*, 621–627 (1979).

Map Positions of the Replication Terminus on the *E. coli* Chromosome, J. Louarn, J. Patte, Mol. Gen. Genetics *172*, 7–11 (1979).

Mutant Single-Strand Binding Protein of *E. coli* Genetic and Physiological Characterization, J. Glassberg, R. Meyer, A. Kornberg, J. Bact. *140*, 14–19 (1979).

E. coli Mutants Thermosensitive for DNA Gyrase Subunit A: Effects on DNA Replication, Transcription and Bacteriophage Growth, K. Kreuzer, N. Cozzarelli, J. Bact. *140*, 424–435 (1979).

Isolation of a Replication Origin Complex from *Escherichia coli*, K. Nagai, W. Hendrickson, R. Balakrishnan, Y. Yamaki, D. Boyd, M. Schaechter, Proc. Nat. Acad. Sci. USA *77*, 262–266 (1980).

Escherichia coli Mutator Mutants Deficient in Methylation-Instructed DNA Mismatch Correction, B. Glickman, M. Radman, Proc. Natl. Acad. Sci. USA *77*, 1063–1067 (1980).

Nucleotide Sequence of the *Salmonella typhimurium* Origin of DNA Replication, J. Zyskind, D. Smith, Proc. Nat. Acad. Sci. USA *77*, 2460–2464 (1980).

Multienzyme Complex for Metabolic Channeling in Mammalian DNA Replication, G. Reddy, A. Pardee, Proc. Nat. Acad. Sci. USA *77*, 3312–3316 (1980).

DNA Polymerase α Mutants from a *Drosophila melanogaster* Cell Line, S. Ikegami, T. Taguchi, M. Ohashi, M. Oguro, H. Nagano, Y. Mano, Proc. Nat. Acad. Sci. USA *77*, 7049–7053 (1980).

A Fixed Site of DNA Replication in Eucaryotic Cells, D. Pardoll, B. Vogelstein, D. Coffey, Cell *19*, 527–536 (1980).

Assembly of Nucleosomes: The Reaction Involving *X. laevis* Nucleoplasmin, W. Earnshaw, B. Honda, R. Laskey, J. Thomas, Cell *21*, 373–383 (1980).

Supercoiled Loops and Eucaryotic DNA Replication, B. Vogelstein, D. Pardoll, D. Coffey, Cell *22*, 79–85 (1980).

Unique Primed Start of Phage ϕX174 DNA Replication and Mobility of the Primosome in a Direction Opposite Chain Synthesis, K. Arai, A. Kornberg, Proc. Nat. Acad. Sci. USA *78*, 69–73 (1981).

Movement and Site Selection for Priming by the Primosome in Phage ϕX174 DNA Replication, K. Arai, R. Low, A. Kornberg, Proc. Nat. Acad. Sci. USA *78*, 707–711 (1981).

Initiation Sites and Discontinuous DNA Synthesis of Bacteriophage T7, A. Fujiyama, Y. Kohara, T. Okazaki, Proc. Nat. Acad. Sci. USA *78*, 903–907 (1981).

Inhibition of ColE1 RNA Primer Formation by a Plasmid-Specified Small RNA, J. Tomizawa, T. Itoh, G. Selzer, T. Som, Proc. Nat. Acad. Sci. USA *78*, 1421–1425 (1981).

Conservation of the Primosome in Successive Stages of φX174 DNA Replication, R. Low, K. Arai, A. Kornberg, Proc. Nat. Acad. Sci. USA *78*, 1436–1440 (1981).

Enzymatic Replication of the Origin of the *Escherichia coli* Chromosome, R. Fuller, J. Kaguni, A. Kornberg, Proc. Nat. Acad. Sci. USA *78*, 7370–7374 (1981).

Isolation and Characterization of Amber Mutations in Gene *dnaA* of *E. coli* K-12, N. Schaus, K. O'Day, W. Peters, A. Wright, J. Bact. *145*, 904–913 (1981).

Replication of φX174 DNA with Purified Enzymes, II. Multiplication of the Duplex Form by Coupling of Continuous and Discontinuous Synthetic Pathways, N. Arai, L. Polder, K. Arai, A. Kornberg, J. Biol. Chem. *256*, 5239–5246 (1981).

Intermediates of Chromosomal DNA Replication in *Escherichia coli*, M. Raggenbass, L. Caro, J. Mol. Biol. *159*, 273–301 (1982).

Specific Cleavage of the p15A Primer Precursor by Ribonuclease H at the Origin of DNA Replication, G. Selzer, J. Tomizawa, Proc. Natl. Acad. Sci. USA *79*, 7082–7986 (1982).

Binding of the Origin of Replication of *Escherichia coli* to the Outer Membrane, W. Hendrickson, T. Kusano, H. Yamaki, R. Balakrishnan, M. King, J. Murchie, M. Schaecter, Cell *30*, 915–923 (1982).

Primary Structure of the Chromosomal Origins *(oriC)* of *Enterobacter aerogenes* and *Klebsiella pneumoniae*, J. Cleary, D. Smith, N. Harding, J. Zyskind, J. Bact. *150*, 1467–1471 (1982).

Effects of the ssb-1 and ssb-113 Mutations on Survival and DNA Repair in UV-Irradiated ΔuvrB Strains of *Escherichia coli* K-12, T. Wang, K. Smith, J. Bact. *151*, 186–192 (1982).

Structure of Replication Origin of the *Escherichia coli* K-12 Chromosome: The Presence of Spacer Sequences in the *ori* Region Carrying Information for Autonomous Replication, K. Asada, K. Sugimoto, A. Oka, M. Takanami, Y. Hirota, Nucleic Acids Res. *10*, 3745–3754 (1982).

Fidelity of Replication of Bacteriophages φX174 DNA *in vitro* and *in vivo*, A. Fersht, J. Knill-Jones, J. Mol. Biol. *165*, 665–667 (1983).

Contribution of 3'-5' Exonuclease Activity of DNA Polymerase III Holoenzyme from *Escherichia coli* to Specificity, A. Fersht, J. Knill-Jones, J. Mol. Biol. *165*, 669–682 (1983).

Chromosomal Replication Origin from the Marine Bacterium *Vibrio harveyi* Functions in *Escherichia coli*: *oriC* Consensus Sequence, J. Zyskind, J. Cleary, W. Brusilow, N. Harding, D. Smith, Proc. Natl. Acad. Sci. USA *80*, 1164–1168 (1983).

In vitro Synthesis of Bacteriophage φX174 by Purified Components, A. Aoyama, R. Hamatake, M. Hayashi, Proc. Natl. Acad. Sci. USA *80*, 4195–4199 (1983).

Methyl-Directed Repair of DNA Base-Pair Mismatches *in vitro*, A. Lu, S. Clark, P. Modrich, Proc. Natl. Acad. Sci. USA *80*, 4639–4643 (1983).

Partition Mechanism of F Plasmid: Two Plasmid Gene-Encoded Products and a Cis-acting Region Are Involved in Partition, T. Ogura, S. Hiraga, Cell *32*, 351–360 (1983).

ARS Replication During the Yeast S Phase, W. Fangman, R. Hice, E. Chlebowicz-Sledziewska, Cell *32*, 831–838 (1983).

Initiation Site of Deoxyribonucleotide Polymerization at the Replication Origin of the *Escherichia coli* Chromosome, S. Hirose, S. Hiraga, T. Okazaki, Mol. Gen. Genet. *189*, 422–431 (1983).

Changes in DNA Base Sequence Induced by Targeted Mutagenesis of Lambda Phage by Ultraviolet Light, R. Wood, T. Skopek, F. Hutchinson, J. Mol. Biol. *173*, 273–291 (1984).

Effects of Point Mutations on Formation and Structure of the RNA Primer for ColE1 Replication, H. Masukata, J. Tomizawa, Cell *36*, 513–522 (1984).

Bacterial Chromosome Segregation: Evidence for DNA Gyrase Involvement in Decatenation, T. Steck, K. Drlica, Cell *36*, 1081–1088 (1984).

RNA Polymerase and RNA Synthesis

The previous two chapters discussed DNA and RNA structures and the synthesis of DNA. In this chapter we consider the synthesis of RNA. After discussing the structure and activities of RNA polymerases, we will concentrate on the initiation process and mention termination and posttranscriptional modifications of RNA.

Cells must synthesize several types of RNA. The protein synthesis machinery requires tRNA, the two large ribosomal RNAs, and the small ribosomal RNA. Additionally, messenger RNA carries information specifying protein sequence to the ribosomes for translation. Thus cells contain many thousands of copies of the ribosomal and tRNAs and only a few copies each of the many different types of messenger RNAs. Eukaryotic cells use at least three types of RNA polymerase to synthesize the different classes of RNA, whereas *E. coli* uses only one.

Early work with bacteria showed that RNA synthesis in *E. coli* initiates from sites located a little in front of genes. Although the definition of promoter has varied somewhat over the years, we will use the term to mean the sequence required for the transcription of a gene. This will include the nucleotides to which the RNA polymerase binds as well as any others that are necessary for the initiation of transcription. Although the requirements for promoter activity in eukaryotes have not been fully worked out, transcription of eukaryotic messenger RNA also depends on sequences lying ahead of the genes.

Bacterial promoters differ in nucleotide sequence and in the details of their functioning. On some promoters, RNA polymerase by itself is able to bind to and initiate transcription. The promoters in this class vary widely in their RNA initiation rates. Many other promoters require the assistance of one or more auxiliary proteins for RNA polymerase to initiate. Part of this chapter discusses measurements of the differences between promoters because this information ultimately should assist learning the functions of these auxiliary proteins.

MEASURING THE ACTIVITY OF RNA POLYMERASE

Studies on protein synthesis in *E. coli* in the 1960s revealed that a transient RNA copy of DNA is sent to ribosomes to direct protein synthesis. Therefore cells had to contain an RNA polymerase that was capable of synthesizing RNA from a DNA template. This property was sufficient to permit enzymologists to devise assays for detection of the enzyme in crude extracts of cells.

The original assay of RNA polymerase was merely a measurement of the amount of RNA synthesized *in vitro*. The RNA synthesized was easily determined by measuring the incorporation of a radioactive RNA precursor, usually ATP, into a polymer (Fig. 4.1). After synthesis, the radioactive polymer was separated from the radioactive precursor nucleotides by precipitation of the polymer with acid, and the radioactivity in the polymer was determined with a Geiger counter.

While the precipitation procedure that measures total incorporated radioactivity is adequate for the assay of RNA polymerase activity and can be used to guide steps in the purification of the enzyme, this assay is indiscriminate. Only the total amount of RNA synthesized in the reaction tube is quantitated. Since many of the convenient DNA templates contain more than one promoter and since *in vitro* transcription frequently initiates from random locations on the DNA in addition to initiating from the promoters, a higher-resolution assay of transcription is required to study specific promoters and the proteins that control their activities.

Three basic methods are used to quantitate the activities of specific promoters. One is to use DNA that contains several promoters and specifically fish out and quantitate just the RNA of interest by RNA-DNA hybridization (Fig. 4.2). For example, in several cases it has been possible to use a system in which all the biologically

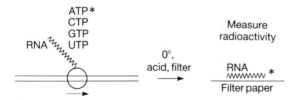

Figure 4.1   Assay of RNA polymerase by incorporation of radioactive nucleotides into RNA in a reaction containing buffer, NaCl, $MgSO_4$, and triphosphates. After transcription, radioactive RNA and radioactive nucleotides remain. These are separated by addition of acid, often trichloroacetic acid, and filtration. The filter paper then contains the RNA whose radioactivity is quantitated in a Geiger counter or liquid scintillation counter.

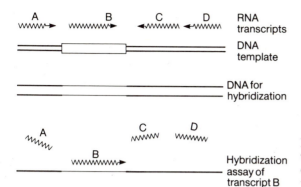

Figure 4.2   Assaying for specific transcription by using a template and DNA for hybridization whose only sequences in common are the region whose transcript is to be assayed.

relevant transcription, that is, the correct transcription, is coded by the same strand of DNA. From these templates any transcription from the other strand must be nonspecific. By assaying the amount of RNA complementary to the two strands, both specific and nonspecific transcription can be quantitated.

The second basic approach for examining the activity of a particular promoter became possible through improvements in recombinant DNA technology. Now small DNA fragments several hundred nucleotides long can be isolated and used as templates in transcription assays. Quantitation of the total RNA synthesized in these

Figure 4.3   Assaying specific transcription by electrophoresis on polyacrylamide gels. The radioactive RNAs of different sizes are synthesized using radioactive nucleoside triphosphates. Then the RNAs are separated according to size by electrophoresis and their positions in the gel found by autoradiography.

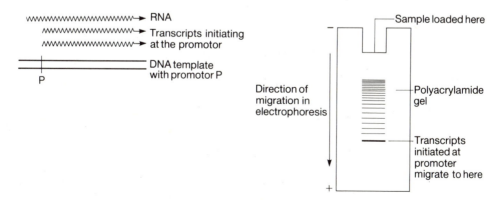

reactions permits assay of a single promoter since the small DNA templates usually contain only one promoter.

Currently, a high-resolution method is often used for examining the activity of promoters. This method permits the simultaneous assay of several promoters as well as nonspecific transcription. Small pieces of DNA 200 to 2,000 base pairs long and containing the promoter in question can be isolated. Transcription initiating from the promoter on these templates begins at the promoter, extends to the end of the DNA, and produces a small transcript of a unique size. Transcription initiating from other sites on the DNA generates other sizes of RNA transcripts. The resultant RNA molecules whose size varies from 10 to 1,000 nucleotides may easily be separated from one another by electrophoresis in polyacrylamide gels in the presence of high concentrations of urea or methyl mercury (Fig. 4.3). These denaturing agents reduce the formation of transient hairpins in the RNA resulting from partial complementarity between portions of the molecules. Therefore the RNA polymers migrate at velocities dependent on their length and independent of their sequences. If α-$^{32}PO_4$ triphosphates have been used during the transcription, the locations of RNA molecules in the gel can subsequently be determined by autoradiography.

ESCHERICHIA COLI CONTAINS A SINGLE TYPE OF RNA POLYMERASE

Once enzymologists could assay and purify an RNA polymerase from E. coli it was important to know the biological role of the enzyme. For example, the bacterial cell might possess three different kinds of RNA polymerase: one for the synthesis of messenger RNA, one for the synthesis of tRNA, and one for the synthesis of ribosomal RNA. If that were the case, much effort could have been wasted in studying in vitro transcription from a gene if the wrong RNA polymerase had been used. Enzymologists' failures to find more than one type of RNA polymerase in E. coli were no proof that more did not exist, for, as we have seen in Chapter 3, detection of an enzyme in cells can be difficult. In other words, what could be done to determine the biological role of the one known RNA polymerase?

Fortunately a way of determining the role of the E. coli RNA polymerase finally appeared. It was in the form of a very useful antibiotic, rifamycin, which blocks bacterial cell growth by inhibiting RNA polymerase.

If many cells are spread on agar medium containing rifamycin, most do not grow. However, a few rifamycin-resistant mutants grow into colonies. Such mutants exist in populations of sensitive cells at a frequency of about 10^{-7}. Examination of the resistant mutants shows them to be of two classes. The first is resistant because the cell membrane is less permeable to rifamycin than the wild-type cells. These are of no interest to us here. The second type is resistant because of an alteration in the RNA polymerase. This can be demonstrated by the fact that the RNA polymerase purified from such rifamycin-resistant cells is indeed resistant to rifamycin in vitro.

As a first step in relating the enzyme studied by enzymologists to cellular

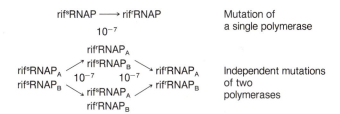

Figure 4.4   Mutation of a single RNA polymerase to rifamycin resistance occurs at a frequency of 10^{-7}, whereas if two independent mutational events are required to mutate two different polymerases to rifamycin resistance, the frequency is $(10^{-7})^2 = 10^{-14}$.

physiology, the rifamycin mutants showed that the RNA polymerase that could be detected and purified is not only a target of rifamycin but is *the* target of rifamycin. Why can't cells contain two RNA polymerases that are inactivated by rifamycin, a missing RNA polymerase, as well as the one purified by enzymologists? Genetics provides the answer. The spontaneous rifamycin-resistant mutants appeared at a frequency of about 10^{-7}, the same frequency as that of spontaneous mutants that lead to the change of a single amino acid in a protein. If both the missing RNA polymerase and the one that could be isolated and purified were mutated to become rifamycin-resistant, two independent mutations would have been required and the frequency of appearance of the mutants would have been approximately $(10^{-7})^2 = 10^{-14}$ (Fig. 4.4).

Thus far, then, we know these facts: The target of rifamycin is a single type of RNA polymerase in bacteria. This RNA polymerase synthesizes at least one essential class of RNA, and this polymerase is the one that biochemists purify. How do we know that this RNA polymerase synthesizes all the RNAs? Careful physiological experiments show that rifamycin addition stops synthesis of all classes of RNA, mRNA, tRNA, and rRNA. Therefore the same RNA polymerase molecule must be used for the synthesis of these three kinds of RNA, and this RNA polymerase must be the one that the biochemists purify.

Unfortunately there is an imperfection in the reasoning leading to the conclusion that *E. coli* cells contain only a single type of RNA polymerase molecule. That imperfection came to light with the discovery, discussed below, that RNA polymerase is not a single polypeptide but in fact contains four different polypeptide chains.

EUKARYOTIC CELLS CONTAIN THREE TYPES OF RNA POLYMERASE

Investigation of eukaryotic cells shows that they do contain more than one type of RNA polymerase. The typical protein fractionation schemes used by biochemists to purify proteins yield three distinct types of RNA polymerase from a variety of higher

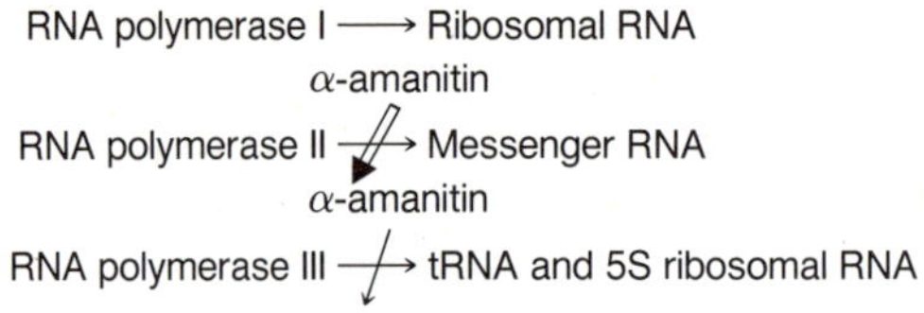

RNA polymerase I ⟶ Ribosomal RNA
α-amanitin
RNA polymerase II ⟶̸ Messenger RNA
α-amanitin
RNA polymerase III ⟶̸ tRNA and 5S ribosomal RNA

Figure 4.5   Three eukaryotic RNA polymerases, their sensitivities to α-amanitin, and the products they synthesize.

cells. These three species are called RNA polymerases I, II, and III for the order in which they elute from an ion exchange column during their purification.

RNA polymerase I appears to synthesize ribosomal RNA, for it is found in the nucleolus, an organelle in which ribosomal RNA is synthesized. Further confirmation of this conclusion is the finding that only purified RNA polymerase I appears capable of correctly initiating transcription of ribosomal RNA *in vitro*. A simple experiment demonstrating this is to use DNA containing ribosomal RNA genes as a template and then to assay the strand specificity of the resulting product. RNA polymerase I synthesizes RNA predominantly from the correct strand of DNA, whereas RNA polymerases II and III do not.

RNA polymerase II has been identified as the polymerase responsible for most synthesis of messenger RNA. *In vitro* experiments show that this polymerase is the most sensitive of the three to a toxin from mushrooms, α-amanitin (Fig. 4.5). The addition of low concentrations of α-amanitin to cells or to isolated nuclei blocks additional synthesis of just messenger RNA.

RNA polymerase III is less sensitive to α-amanitin than is RNA polymerase II, but it is sufficiently sensitive that its *in vivo* function can be probed. The sensitivity profile for synthesis of tRNA and 5S RNA parallels the sensitivity of RNA polymerase III, thus demonstrating the *in vivo* role of the enzyme.

RNA POLYMERASES CONTAIN MULTIPLE SUBUNITS

How can one tell that an enzyme contains multiple subunits? One of the best methods for detection of multiple species of polypeptides in a sample is electrophoresis through a polyacrylamide gel. If the protein has been denatured by boiling in the presence of the detergent sodium dodecyl sulfate, SDS, and the electrophoresis is performed in the presence of SDS, polypeptides separate according to size. This results from the fact that the charged SDS anions that bind to the polypeptides completely dominate the charge as well as force all polypeptides to adopt a rodlike shape whose length is proportional to the molecular weight of the protein. Any two polypeptides of the same size will therefore migrate at the same rate. Following electrophoresis, the positions of proteins in the gel can be visualized by staining.

Figure 4.6   The polypeptide band pattern
found by SDS polyacrylamide gel electrophoresis of purified *E. coli* RNA polymerase.

Each band on a gel derives from a different-sized polypeptide species. When purified *E. coli* RNA polymerase is subjected to SDS polyacrylamide gel electrophoresis,
four distinct bands are seen (Fig. 4.6).

Do all the bands on the gel represent subunits of RNA polymerase or are some of
the bands extraneous proteins that happen to copurify with RNA polymerase? A
reconstitution experiment provides the most straightforward demonstration that the
four different proteins found in RNA polymerase are all essential subunits of the
enzyme. The four bands from an SDS polyacrylamide gel are cut out, the proteins
eluted, and SDS removed. RNA polymerase activity can be regained only if all four
of the proteins are included in the reconstitution mixture.

RNA polymerase from *E. coli* consists of subunits β' and β of molecular weights
155,000 and 151,000, two subunits of α whose molecular weight is 36,000, and one
somewhat less tightly bound subunit, σ, whose molecular weight is 70,000. By
measuring the amounts of each of the four proteins on the polyacrylamide gels it is
clear that the enzyme contains two copies of the α subunit for every single copy of
the others, that is, the subunit structure of RNA polymerase is $\sigma\alpha_2\beta\beta'$.

The subunit structure of the RNA polymerase necessitates a reevaluation of the
rifamycin experiment in which we concluded that *E. coli* contains only one species of
RNA polymerase. The existence of subunits weakens the conclusion. We can say
only that if cells contain more than one type of RNA polymerase, then at least all the
types must have one subunit in common and that subunit is the one inactivated by
rifamycin.

The reconstitution experiments permit pinpointing the actual target of rifamycin. RNA polymerase from rifamycin-sensitive and rifamycin-resistant cells is subjected to SDS polyacrylamide gel electrophoresis. Then reconstitution experiments

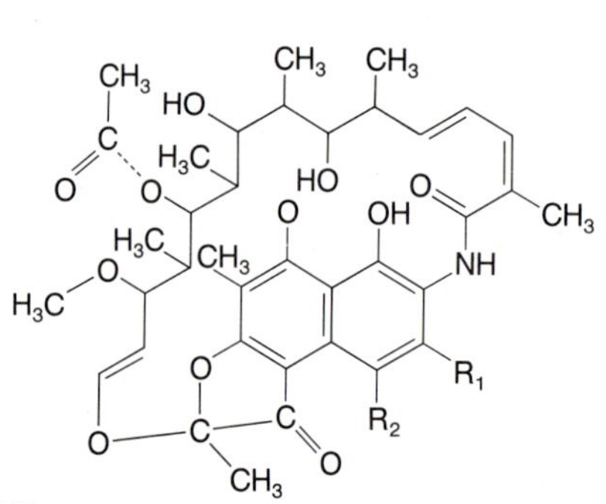

Rifamycin B(R$_1$ = H; R$_2$ = O—CH$_2$—COOH)

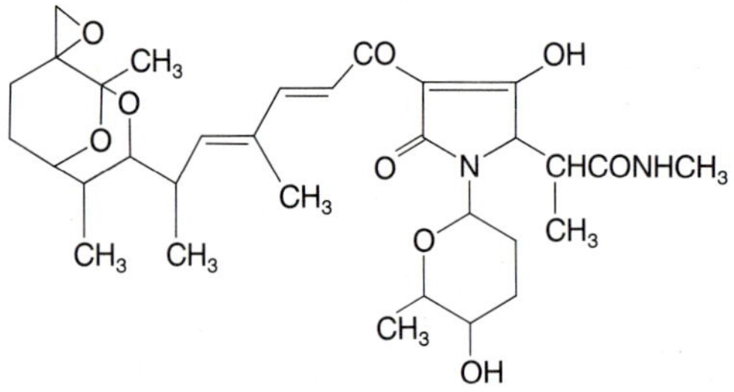

Streptolydigin

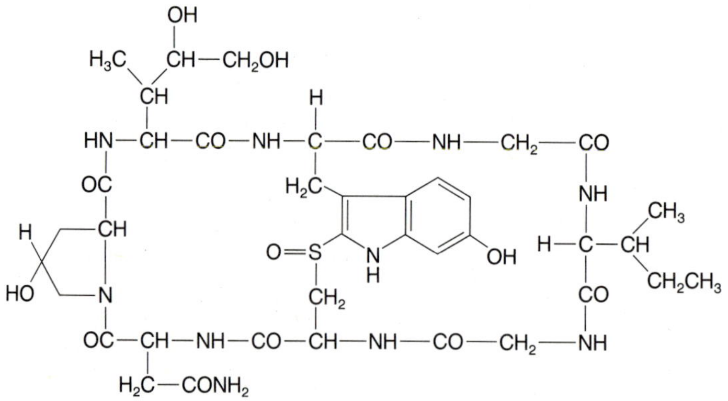

α-amanitin

Figure 4.7   The structures of three RNA polymerase inhibitors, rifamycin, streptolydigin, and α-amanitin.

with the two sets of proteins can be performed in all possible combinations to determine which of the four subunits from the rifamycin-resistant polymerase confers resistance to the reconstituted enzyme. The β subunit was found to be the target of rifamycin.

If we view the RNA polymerase as a biochemical engine, then it is reasonable to expect each different subunit to have a different function. As discussed below, rifamycin inhibits initiation by RNA polymerase, but it has no effect on the steps of elongation of the polynucleotide chain. A different antibiotic, streptolydigin, has also been found to inhibit RNA polymerase (Fig. 4.7). This blocks elongation steps, and therefore we might have expected to find a subunit other than β to be the target of this drug. However, the β subunit is also the target of streptolydigin, and the polymerase begins to look like a complicated machine.

The combined molecular weights of the subunits of RNA polymerase total nearly one half million, but it is not at all clear why the polymerase should be so large. Phage T7, which grows in E. coli, encodes its own RNA polymerase, and this enzyme has a molecular weight of only about 100,000. Apparently the actual RNA initiation and elongation steps do not require an enzyme as large as the E. coli polymerase. Perhaps the additional size of this enzyme permits it to initiate at a wider variety of promoters or to interact with a variety of auxiliary proteins.

The eukaryotic RNA polymerases are even more perplexing. They appear to contain not the four types of subunits seen in the E. coli RNA polymerase but about ten different types of subunits. Possibly some of the differences between the prokaryotic and eukaryotic polymerase are smaller than they first appear. The differences may merely be in the tightness with which subunits cling together. For example, a protein from E. coli that is involved with RNA chain termination, the nusA gene product, could be considered a part of RNA polymerase because it binds to the polymerase after initiation has occurred and after the σ subunit has been released from the core complex of β, β', and α. However, since it does not copurify with the RNA chain elongating activity that is contained in the core, it usually is not so classified. The eukaryotic peptides might just stick together more tightly.

LITTLE OF THE CELLULAR POLYMERASE IS FREE IN THE CYTOPLASM

It is necessary to know the concentration of free intracellular RNA polymerase to design meaningful *in vitro* transcription experiments. One method of determining the concentration utilizes the fact that the β and β' subunits of RNA polymerase are larger than most other polypeptides in the cell. This permits them to be easily separated from other cellular proteins merely by SDS polyacrylamide gel electrophoresis. Consequently a measurement of the amount of protein constituted by these two polypeptides relative to total cellular protein can easily be performed just by electrophoresis and measurement of the fraction of protein contained in the β and β' bands. The results are that a cell contains about 3,000 molecules of RNA polymerase.

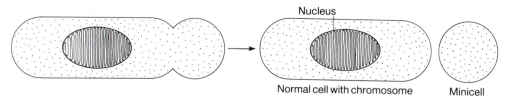

Figure 4.8   The generation of a minicell lacking a chromosome.

A calculation using the cell doubling time and amounts of messenger RNA, tRNA, and ribosomal RNA in a cell leads to the conclusion that about 1,500 RNA molecules are being synthesized at any instant. Hence half the cell's RNA polymerase molecules are synthesizing RNA. Of the other 1,500 RNA polymerase molecules, less than 500 are free of DNA and are able to diffuse through the cytoplasm. The remainder are temporarily bound to DNA at nonpromoter sites.

How do we know these numbers? On first consideration, a direct physical measurement showing that 500 RNA polymerase molecules are free in the cytoplasm seems impossible. However, the existence of a special cell division mutant of *E. coli* makes this measurement straightforward (Fig. 4.8). About once per normal division these mutant cells divide near the end of the cell, producing a minicell that lacks DNA but that contains a sample of the cytoplasm. Hence to determine the concentration of RNA polymerase free of DNA in cells, it is necessary only to determine the concentration of $\beta\beta'$ in the DNA-less minicells. Such measurements show that the ratio of $\beta\beta'$ to total protein has a value in minicells of one-sixth the value found in whole cells. Thus, less than 20% of the RNA polymerase in a cell floats freely through the cytoplasm.

SPECIFICITY BY THE SIGMA SUBUNIT: HEAT SHOCK AND SPORULATION

In vitro transcription experiments with the *E. coli* RNA polymerase have shown that the σ subunit is required for initiation at promoters, but σ is not required for elongation activity. In fact, the σ subunit comes off the polymerase when the transcript is between 2 and 10 nucleotides long. Polymerase lacking the σ subunit, core polymerase, binds randomly to DNA and initiates nonspecifically or from nicks, but it rarely initiates from promoters. These results raise an interesting question: If the σ subunit is required for promoter recognition, then could different σ subunits be used to specify transcription from different classes of genes?

Changes in the σ subunit in *E. coli* and *Bacillus subtilis* have been found. In *E. coli* heat shock stimulates synthesis of a new σ. *Bacillus subtilis* forms spores when growth conditions become poor. Forming a spore requires turning off transcription from many of the genes required for vegetative growth and turning on the sporulation-specific genes. This is an ideal situation for shifting transcription specificity of RNA polymerase by alteration of the σ subunit; indeed, this is just what happens. During sporulation the σ subunit with apparent molecular weight of 55,000, which predominated during normal vegetative growth, is replaced by two other subunits of different molecular weights and the RNA polymerase acquires specificity for transcribing classes of genes that previously were silent.

E. COLI RNA POLYMERASE ELONGATES AT 60 NUCLEOTIDES/SEC

As in the case of DNA synthesis, it is sensible for cells to regulate RNA synthesis at the initiation steps so that the complicated machinery involved in gene regulation need not be built into the basic RNA synthesis module. Once RNA synthesis has been initiated, it likely proceeds at the same rate regardless of growth conditions. Can this be demonstrated? Another need for knowing the RNA elongation rate is in the interpretation of physiological experiments. How soon after the addition of an inducer can a newly synthesized mRNA molecule appear?

RNA elongation rate measurements are not too hard to perform *in vitro*, but they are appreciably more difficult to perform on growing cells. Here we shall explain one method that has been used to determine the *in vivo* RNA elongation rate in *Escherichia coli*.

The measurement used rifamycin to specifically block the initiation of RNA chains. Rifamycin and uridine, a radioactive precursor of RNA, were simultaneously added to cells; thus only those RNA chains that were in the process of elongation at the time of the additions were radioactively labeled (Fig. 4.9). At various times after the rifamycin and uridine addition, samples were taken from the culture and their RNA was separated according to size by electrophoresis on polyacrylamide gels. The interval from the time of uridine and rifamycin additions until no more radioactive uridine was incorporated into a complete class of RNA was the time required for complete synthesis of this RNA.

The ribosomal RNA gene complex was a convenient system for these measurements. This set of genes consist of a promoter where RNA polymerase binds and initiates transcription, the gene for the 16S ribosomal RNA, a spacer region, a tRNA gene, the gene for the 23S ribosomal RNA, and the gene for the 5S ribosomal RNA (Fig. 4.10). The total length of this transcriptional unit is about 5,000 nucleotides. The 16S RNA, spacer tRNA, 23S RNA, and 5S RNA are all generated by cleavage of the growing polynucleotide chain.

The interval between the rifamycin addition and the occurrence of the last RNA polymerase molecule transcribing across the end of the 5S gene is the time required for RNA polymerase to transcribe the distance from the promoter to the end of the ribosomal gene complex. This time is given by the radioactive uridine incorporation

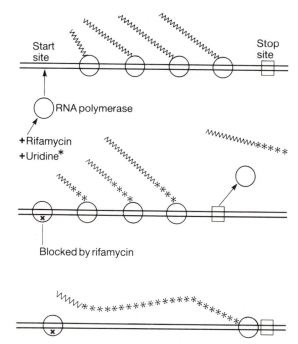

Figure 4.9   Effects of rifamycin addition on transcription of a large operon. Upon the addition of rifamycin, no more RNA polymerase molecules may initiate transcription. Those polymerase molecules that were transcribing continue to the end of the operon. Finally, that polymerase molecule that had initiated transcription just before the addition of rifamycin completes its transcription of the operon.

measurements. Transcription across the 5S gene has ended when the radioactivity in 5S RNA has stopped increasing.

Since the total distance the RNA polymerase must transcribe from the promoter to the end of the 5S gene is about 5,000 bases, the 90 seconds required for transcription of this distance yields an elongation rate of about 60 nucleotides per second (Fig. 4.11). This type of elongation rate measurement has been performed on cells grow-

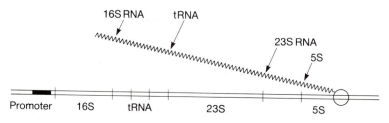

Figure 4.10   Structure of the ribosomal RNA operon used to determine the RNA elongation rate in E. coli.

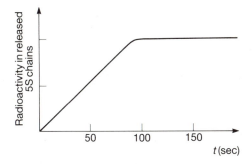

Figure 4.11   Radioactivity incorporation kinetics into 5S RNA following the simultaneous addition of radioactive uridine and rifamycin.

ing at many different growth rates, and the results show that, as expected, the RNA chain growth rate is independent of the growth rate of cells at a given temperature.

THE INITIATION PROCESS PROBABLY CONSISTS OF MANY STEPS

The beginning state for initiation by RNA polymerase is a polymerase molecule and a promoter free in solution, and the end state is a polymerase molecule bound to DNA elongating an RNA chain. Most likely the initiation process that separates these two states is not continuous, but it consists of a number of discrete steps. Can any of these be detected and quantitated and, if so, can measurement of the rates of proceeding from one state to the next during the initiation process provide useful information? Ultimately we would hope that studies like these could explain the differences in activities between promoters of different sequence as well as provide the information necessary to design promoters with specific desired activities.

The first biochemical characterizations of the binding and initiation rates of RNA polymerase on promoters were performed on "strong" promoters. Such a choice for a promoter is natural because it maximizes the signal-to-noise ratio in the data. The data obtained by Chamberlin with these promoters indicated that the initiation process could be dissected into two steps: a rapid step during which polymerase binds to DNA and a slower "isomerization" step in which RNA polymerase shifts to an active form capable of immediately initiating transcription (Fig. 4.12). It was natural to propose that the second step would be the point for regulation of transcription. That is, promoters of different activities would vary in their isomerization rates and if auxiliary proteins were required for initiation at a promoter, they would alter the isomerization rate.

If the regulation of initiation by auxiliary proteins did occur only in the isomerization step, then many polymerase molecules would be nonproductively bound at promoters. There they would be awaiting their activation by auxiliary proteins. This

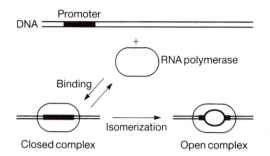

Figure 4.12   Binding of an RNA polymerase to form a closed complex and the near irreversible formation of an open complex.

would be a waste of substantial numbers of polymerase molecules. Hence it seems possible that, instead, the auxiliary proteins assist the first steps of the initiation process. In fact, nature has it both ways. On some promoters, regulation is at the binding step, and on others the regulation occurs at the isomerization step.

MEASUREMENT OF BINDING AND INITIATION RATES

What experiments can be done to resolve unambiguously questions about binding and initiation rates? One of the easiest measurements to perform with RNA polymerase is quantitation of the total RNA synthesized. It is tempting to try to adapt such measurements to the determination of binding and activation rates of RNA polymerase. The execution of the experiments and the interpretation of the data are difficult, however, and many misleading experiments have been done.

Somewhat more direct measurements on the initiation of transcription are possible. Such measurements were first performed by McClure, who discovered that as RNA polymerase initiates *in vitro* it often goes through several cycles of abortive initiation in which a short two, three, or sometimes longer polynucleotide is synthesized. The sequence of these abortive initiation products is the same as the 5′ end of the normal RNA transcript. RNA, like DNA, is elongated in the 5′-to-3′ direction. After one short polynucleotide is made, RNA polymerase does not come off the DNA but remains bound at the promoter, and another attempt is made at initiation. Additionally, two conditions lead to the exclusive synthesis of the short polynucleotides: the presence of rifamycin or the absence of one or more of the ribonucleoside triphosphates (Fig. 4.13). If nucleotides are omitted, those present must permit synthesis of a 5′ portion of the normal transcript. Under these conditions a polymerase molecule bound to a promoter and in the initiation state will continuously synthesize the short, abortive initiation polynucleotides.

Assay of the short polynucleotides synthesized under conditions where longer polynucleotides cannot be synthesized provides a good method of assaying initia-

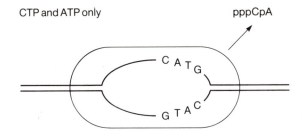

CTP and ATP only   pppCpA

Figure 4.13   Generation of abortive initiation products by provision of only the first two nucleoside triphosphates required by a promoter.

tion by RNA polymerase. The measurement is free of complexities generated by multiple initiations at a single promoter or of the difficulties in interpretation generated by the addition of inhibitors. All that is necessary is the omission of the appropriate nucleoside triphosphates. Once RNA polymerase initiates on a promoter, it begins production of short polynucleotides, but nothing else happens at this copy of the promoter.

Let us first consider the possibility that the time required for RNA polymerase molecules to locate and bind to the promoters is a significant fraction of the total time required for binding and initiation. An experiment can be performed by mixing RNA polymerase, DNA containing a promoter, and two or three of the ribonucleoside triphosphates and measuring the kinetics of appearance of the short polynucleotides that signal initiation. Suppose that the same experiment is repeated but with the RNA polymerase at twice the original concentration (Fig. 4.14). In this case the time required for RNA polymerase to find and bind to the promoter should be half as long as in the first case. Hence the synthesis of the short polynucleotides should begin sooner.

By varying the concentration of polymerase in a set of experiments and appropriately plotting the results, we can extrapolate to determine the results had poly-

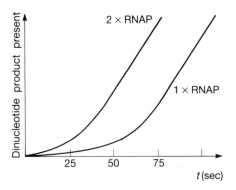

Figure 4.14   The kinetics of synthesis of dinucleotides following the mixing RNA polymerase and DNA at two polymerase concentrations differing by a factor of two.

merase been added at infinite concentration. In such a situation the delay of polymerase binding to promoter should be zero. The residual delay in the appearance of
polynucleotides can only be the time required for conversion of polymerase to the
active state. Knowing this, we can then work back and determine the rate of binding
of RNA polymerase to the promoter.

   We considered above the possibility that the initial binding of RNA polymerase
to promoters was slow. Increasing the concentration of RNA polymerase would then
increase the rate of this binding. A second possibility is that the binding is fast but
that only a fraction of the promoters are occupied by RNA polymerase. The conversion of the bound polymerase to polymerase in an active state would then be the
slow step. Increasing the concentration of RNA polymerase in this situation would
also increase the rate of the initial appearance of the polynucleotides. The reason is
that the initial concentration of polymerase in the bound state is increased owing to
the higher concentration of polymerase in the reaction. The polynucleotide assay
permits quantitation of the apparent binding constant and of the isomerization rate
for this situation as well. The assay may also be used to determine how an auxiliary
protein assists initiation on some promoters.

*RELATING ABORTIVE INITIATIONS TO POLYMERASE RATE CONSTANTS

The binding and initiation reactions explained in the previous section are described
by the following equation:

$$R + P \xrightleftharpoons[k_{-1}]{k_1} RP_c \xrightleftharpoons[k_{-2}]{k_2} RP_0,$$

where R is RNA polymerase free in solution; P is uncomplexed promoter; RP_c is
promoter with RNA polymerase bound in an inactive state, "closed"; and RP_0 is
promoter with RNA polymerase bound in an active state, "open."

   If RNA polymerase and DNA containing a promoter are mixed together, then
the concentration of RP_0 at all times thereafter can be calculated in terms of the initial
concentrations R and P and the four rate constants; however, the resulting solution is
too complex to be of much use. A reasonably close mathematical description of the
actual situation can be found by making three approximations. The first — that R be
much greater than P — is under the experimentalist's control. The second is known
to enzymologists as the steady-state assumption. Frequently the rate constants describing reactions of the type written above are such that during times of interest the
rate of change in the amount of RP_c is small, and the amount of RP_c can be considered
to be in equilibrium with R, P, and RP_0. That is,

$$\frac{dRP_c}{dt} = k_1 R \times P - k_{-1} RP_c - k_2 RP_c + k_{-2} RP_0 = 0.$$

The third assumption is that k_{-2} is much smaller than k_2. Experiments show this to be
a very good approximation. RNA polymerase frequently takes hours or days to

dissociate from a promoter. Straightforward solution of the equations now yields RP_0 as a function of time in a useful form:

$$RP_0 = (P_{initial})(1 - e^{-k_{obs}t}), \tag{4.1}$$

$$k_{obs} = \frac{k_1 k_2 R}{k_1 R + k_{-1} + k_2}, \quad \text{or}$$

$$\frac{1}{k_{obs}} = \frac{1}{k_2} + \frac{k_{-1} + k_2}{k_1 k_2 R}.$$

After starting the binding initiation assay, the total amount of oligonucleotides synthesized by any time can be measured by removing a sample from the synthesis mixture and chromatographically separating nucleoside triphosphates from the short oligonucleotides. Since the rate of oligonucleotide production is proportional to RP_0, the total amount of oligonucleotides synthesized as a function of time is given by

$$\int_0^t RP_0(t')dt' = (P_{initial})(t - 1/k_{obs} + 1/k_{obs}e^{-k_{obs}t}). \tag{4.2}$$

Hence extrapolating the linear portion of the curve to the point of zero oligonucleotides gives $1/k_{obs}$ (Fig. 4.15). As seen above, $1/k_{obs}$ in the limit of high R yields $1/k_2$. At other concentrations, $(k_{-1} + k_2)/k_1 k_2$ is a linear function of $1/R$. Performing the abortive initiation reaction at a variety of concentrations of R and measuring the kinetics of synthesis of oligonucleotides permits straightforward evaluation of k_2 (Fig. 4.16). Most often k_{-1} is much greater than k_2, in which case the ratio k_1/k_{-1}, called K_B, is obtained as well.

A variety of promoters have been examined by these techniques. Highly active promoters must bind polymerase well and must perform the initiation "isomerization" quickly. Less-active promoters are poor in binding RNA polymerase or slow in isomerization.

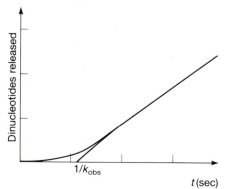

Figure 4.15   Determination of the parameter k_{obs} from the kinetics of incorporation of radioactivity into dinucleotides.

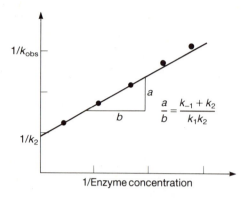

$$\frac{a}{b} = \frac{k_{-1} + k_2}{k_1 k_2}$$

Figure 4.16   Determination of the kinetic parameters describing initiation by RNA polymerase from a series of k_{obs} values obtained at different RNA polymerase concentrations.

SALT STRONGLY AFFECTS RNA POLYMERASE BINDING

The binding of RNA polymerase to DNA is found to be strongly dependent on the concentration of monovalent and divalent cations. The higher the salt concentration, the weaker the binding. Some other proteins that bind to DNA also display similar effects of salt on their binding. Since much of molecular biology revolves around proteins binding to DNA or RNA, we will examine the basis for this finding. Initially it would appear that RNA polymerase could contain a number of positive charges and that attractive forces between these and the negatively charged phosphates on the DNA could contribute to the binding by increasing the rate of association. The presence of high salt concentrations would then shield the attractive forces and weaken the binding. This is a consequence of the fact that the binding constant K_B equals the ratio of the association and dissociation rates:

$$K_B = \frac{k_1}{k_{-1}}.$$

This simple explanation does not seem to hold, however. The salt primarily affects the dissociation rate!

Consider the binding of RNA polymerase to a promoter. It can be written as

$$R + P \rightleftharpoons RP.$$

But if we include an unknown number of sodium ions and, for convenience, let them be represented as being associated with the promoter. The reaction should be written

$$R + Na_n^+P \rightleftharpoons RP + nNa^+,$$

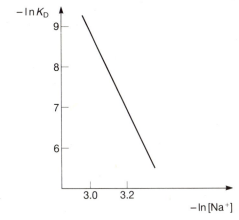

Figure 4.17   A plot of apparent K_D of RNA polymerase from a promoter as a function of the concentration of NaCl in the binding buffer.

and the dissociation constant can be written

$$K_D = \frac{R \times Na_n^+ P}{RP \times (Na^+)^n}, \quad \text{or}$$

$$\log(K_D) = \log(K_0) - n \log(Na^+).$$

The observed variation in the binding of RNA polymerase as a function of the sodium concentration indicates the effective number of ions involved. This number is most easily extracted by plotting the logarithm of the dissociation constant of RNA polymerase from a promoter as a function of the log of the Na^+ concentration. Then the slope equals n. Experimentally this value is found to be about 12 (Fig. 4.17).

The dissociation of RNA polymerase from a promoter would require nearly simultaneous and precise binding of the 12 sodium ions to their former positions on the DNA. The higher the concentration of sodium ions in the solution, the more easily this may be accomplished. Alternatively, the higher the sodium concentration, the weaker the polymerase binding.

How do we know that anions do not also enter the reaction? Their presence is inferred by the finding that polymerase binding is not strongly dependent on the identity of the anion. The dissociation constant measured in the presence of sodium chloride is about the same as the constant measured in the presence of the same concentration of sodium acetate or even sodium sulfate.

It seems most reasonable to assume that the majority of the sodium ions involved with the alteration of the affinity of RNA polymerase with changing concentrations of salt are closely associated with the DNA. Most of the sodium ions associated with the RNA polymerase would be rather loosely bound and poorly localized.

The above explanation can also be couched in thermodynamic terms. The entropic contribution of the sodium ions to the binding-dissociation reaction is large because before the binding of polymerase the ions are localized near the backbone of the DNA. Upon the binding of polymerase, these ions are freed and their entropy change on the binding of polymerase is large. The greater the sodium concentration in the buffer, the smaller the entropy change and therefore the weaker the polymerase binding. Loosely bound, and therefore poorly localized, ions do not make a substantial contribution to the changes in binding as the buffer composition is changed because their entropies do not change substantially during the course of the reaction.

PROMOTERS ARE SIMILAR BUT NOT IDENTICAL

It seems likely that RNA polymerase would recognize a particular nucleotide sequence as a promoter. As the first few bacterial promoters were sequenced, it was a surprise to find so much variation between their sequences that their similarities could not be distinguished. However, when the number of sequenced promoters reached about six, Pribnow noticed a sequence TATAATG, located about 6 bases before the start of messengers, that is, 5'XXXTATAATGXXXXXXAXXXX-3', with the messenger usually beginning with A. This sequence is often called the Pribnow box. Most bacterial promoters also possess elements of a second region of conserved sequence, TTGACA, which lies 35 base pairs before the start of transcription. Protection of promoters from DNAse digestion, chemical modification, and electron microscopy have shown that polymerase indeed is bound to the (-35)-(-10) region of DNA before it begins transcription (Fig. 4.18). Experiments of the type described in Chapters 11 and 12 have permitted direct determination of the bases and phos-

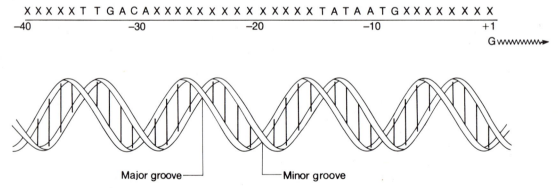

Figure 4.18   The consensus sequences found in *E. coli* promoters and the helical structure of the corresponding DNA, showing the regions of close contact between polymerase and the DNA as determined by mutations and chemical modification experiments. The majority of the contacts are on one side of the DNA.

phates in the promoter region that *E. coli* RNA polymerase contacts. These contact points are clustered in the regions of the conserved bases.

Study of eukaryotic promoters reveals relatively few conserved sequences. Most eukaryotic promoters possess a TATA sequence located between 20 and 60 base pairs before the apparent start of transcription. Removing or altering this sequence markedly reduces promoter activity, but altering other nearby sequences does not have much effect. Eukaryotic promoters often also possess one or more 20-to-30-

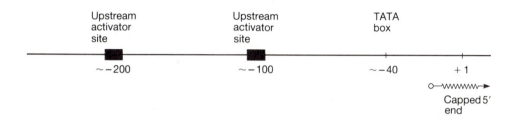

base-pair elements located 100 or 200 base pairs still farther upstream. These elements appear to interact with proteins other than RNA polymerase and to regulate activity of the promoter. In some cases these elements have been moved several hundred nucleotides or even turned around without altering their activities. Elements possessing these latter properties are called enhancers.

RNA POLYMERASE MELTS ABOUT 14 BASE PAIRS AFTER BINDING

One of the first steps of transcription is the binding of RNA polymerase to the proper sequence on the DNA. The best evidence at present supports the notion that RNA polymerase reads the sequence of the DNA and identifies the promoter in a double-stranded unmelted structure. It is theoretically possible that the bases of the growing RNA could be specified by double-stranded unmelted DNA, but it seems vastly easier for these bases to be specified by Watson-Crick base pairing to a partially melted DNA duplex.

Direct experimental evidence shows that during the initiation process, RNA polymerase melts at least 11 base pairs of DNA. Positions on adenine rings normally occupied in base pairs become available for chemical reaction if the pairs are disrupted, and their exact positions along a DNA molecule can then be determined by methods analogous to those used in DNA sequencing. Results obtained from this type of measurement reveal that 11 base pairs of DNA from about the middle of the

Apparent
base pairs
melted

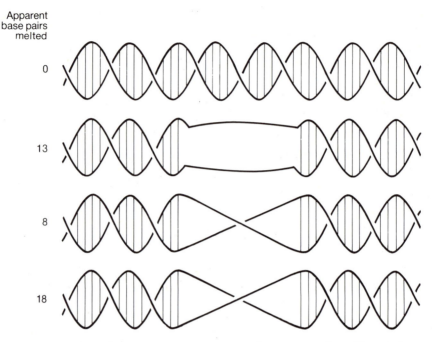

Figure 4.19 Topological measurements cannot give the exact number of base pairs opened by binding of a protein. In each of the three cases, 13 base pairs are broken, but only if the melted region is not twisted does the DNA contain one less twist.

Pribnow box to the start site of transcription are melted when the RNA polymerase binds to a promoter.

A different method has also been used to measure the amount of DNA that is melted by the binding of RNA polymerase. This method consists of binding RNA polymerase to a nicked circular DNA molecule, sealing the nick with polymerase still bound, and determining the change in the supercoiling generated by the presence of the polymerase. If we assume that the melted DNA strands are held parallel to the helix axis, this method yields 17 base pairs melted. However, no method has been developed to determine whether the melted region contains any twist. If it does, then the size of the region that is melted cannot be precisely determined (Fig. 4.19).

The melting of 10 to 15 base pairs of DNA under physiological conditions involves appreciable energy because this length of oligomers anneals together with a very high binding constant. The energy derives from the same source that pushes DNA to base pair — base stacking interactions and hydrogen bonds. Since a large amount of energy is required and a limited amount is available from the binding energy of RNA polymerase, thermal motion in the solution provides the rest. At lower temperatures, and therefore lower thermal motion, a RNA polymerase–DNA duplex is much less likely to possess the requisite energy, and the melting rate at lower temperatures is much reduced. Similarly, the melting rate is affected by the salt concentration. At 0° virtually no RNA polymerase bound to phage T7 DNA is able to initiate in reasonable periods of time, while at 30° almost 100% is able to initiate within a few minutes.

TRANSCRIPTION TERMINATION OCCURS AT SPECIFIC SITES

If transcription of different genes is to be regulated differently, then the genes must be transcriptionally separated. RNA polymerase could just fall off the DNA as it transcribes, but this separation mechanism would require large spaces between genes to eliminate transcriptional readthrough from one gene to the next. A more efficient method to separate transcriptional units is merely to have transcription termination signals at their ends. This is what *E. coli* does; presumably all other organisms also use transcription termination signals.

What is the proof of the existence of transcription termination signals? Two types of demonstration exist. The first is genetic. As will be discussed in more detail later, fusions between transcriptional units, called operons, can be generated. Even though genes may be located close to one another on the chromosome, only by deleting the transcription termination signal at the end of one operon and the promoter at the beginning of an adjacent operon can genes of the second operon be expressed under control of the first promoter (Fig. 4.20).

The second type of demonstration uses *in vitro* transcription. Radioactive RNA is synthesized *in vitro* from a well-characterized DNA template and separated according to size on polyacrylamide gels. Some templates yield a discrete class of RNA transcripts produced from initiation at a promoter and termination at a site before the end of the DNA molecule. Thus the template must contain a transcription termination site.

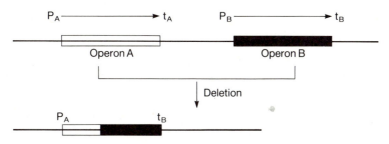

Figure 4.20   Fusion between two operons to place some of the genes of operon B under control of the promoter of operon A. The fusion removes the transcription termination signal t_A and the promoter p_B.

SOME TERMINATORS REQUIRE ASSISTANCE
FROM RHO PROTEIN

The study of a sequence of many transcripts has revealed that the primary nucleotides responsible for determining whether RNA polymerase will terminate transcription lie just at the end of the transcript itself. Several generalizations can be
made about these sequences. One class consists of a region rich in GC bases that can
form a hairpin loop followed by a string of U's.

The class of terminators functions without the need for auxiliary protein factors
from the cell. Most likely as the RNA is elongated past the region rich in GC, it base
pairs with itself to form a hairpin (Fig. 4.21). Then, as the RNA polymerase pauses in
this region, release occurs from the run of U's. A major factor in the release of the
nascent RNA chain from this class of terminators is the exceptional weakness of
oligo (dA:rU) RNA-DNA hybrids. Annealing studies with these oligonucleotides
show that they melt at anomalously low temperatures compared to other oligonucleotide hybrids. Corresponding to this notion is the fact that all terminators that
function *in vitro* without the assistance of auxiliary proteins end in a string of U's.

The second class of terminators is even less understood. This class requires the
presence of rho factor for termination to occur, and its termination activity is stimulated by the presence of the *nusA* protein. During transcription, RNA polymerase
pauses near these termination sequences, most likely aided in pausing by the *nusA*
protein, and then rho terminates the transcription process and releases the RNA and
RNA polymerase by processes yet to be determined.

The discovery of rho factor was accidental. Transcription of lambda phage DNA
in an *in vitro* system produced a large amount of incorrect transcript. This inaccuracy
was revealed by hybridizing the RNA to the two separated strands of lambda phage.
Correct transcripts would have hybridized to only one strand. Apparently the conditions being used for transcription did not faithfully reproduce those existing within
the cell. In a sense, this is a biochemist's dream for it means that something must exist
and is waiting to be found. Therefore Roberts looked for a protein from extracts of
cells that would enhance the fidelity of *in vitro* transcription. He used the enhance-

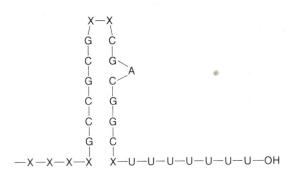

Figure 4.21   Typical sequence and the likely hairpin
structure at the 3'-OH end of the RNA whose
transcription terminates without the assistance of
rho protein.

ment of fidelity of transcription to guide purification of a protein of unknown function. Upon completing the purification and studying the properties of his "fidelity" factor, he discovered that it terminated transcription. We still do not understand how the protein factor, rho, that he found depresses the synthesis of wrong transcripts.

PROCESSING PROKARYOTIC RNAS AFTER SYNTHESIS

It is not altogether surprising that intact transcripts produced by transcription of some operons are not always suited for all the biological roles of RNA. In *E. coli* the ribosomal and tRNAs are processed following transcription. For example, RNAse III first cleaves in a portion of the rRNA that is folded back on itself to form a hairpin, and then other nucleases make additional cuts before the RNA is matured. RNAse III also cleaves the early transcripts of phage T7.

How do we know RNAse III cleaves the ribosomal and phage T7 RNAs? Once again, mutations permit a direct demonstration of the particular biological role played by an enzyme. Investigators isolated a mutant defective in RNAse III by screening extracts prepared from isolated colonies of mutagenized cells for the enzyme RNAse III (Fig. 4.22). About one out of 1,000 colonies yielded cells with greatly depressed levels of the enzyme. Subsequently these cells were found to possess appreciable quantities of uncleaved ribosomal RNA. Furthermore, this RNA could be cleaved *in vitro* with purified RNAse III to yield the smaller ribosomal RNAs.

PROCESSING OF EUKARYOTIC RNAs

Eukaryotic RNAs are processed at their beginnings, their middles, and their ends. Although the structures resulting from these modifications are known, why the modifications occur is not yet clear.

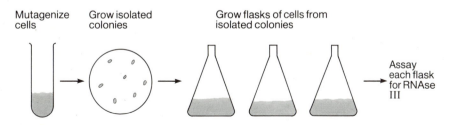

Figure 4.22   An isolation scheme for RNAse III mutants. Mutagenized cells are plated out to yield colonies deriving from single cells. Cultures are grown from these colonies and then tested for the presence of RNAse III.

The 5′ ends of nearly all cellular messenger RNAs but not all viral RNAs contain a "cap," a guanine in a reversed orientation plus several other modifications (Fig. 4.23). More precisely, the cap is a guanine methylated on its 7 position joined through 5′-5′ pyrophosphate linkage to a second methylated guanine followed by a cytosine. These modifications were discovered by tracking down why and where viral RNA synthesized *in vitro* could be methylated. The trail led to the 5′ end whose structure was then determined by Shatkin. The cap appears to assist translation as ribosomes translate cap-containing messenger more efficiently than they translate messenger lacking a cap.

A sizable fraction of the RNA that is synthesized in the nucleus of cells is not transported to the cytoplasm. This RNA is of a variety of sizes and sequences and is called heterogenous nuclear RNA. Within this class of RNA are messenger RNA sequences that ultimately are translated into protein, but, as synthesized, the nuclear RNAs are not directly translatable (Fig. 4.24). They contain extraneous sequences that must be removed by cutting and splicing to form the continuous coding sequence found in the mature messenger located in the cytoplasm. The intervening

Figure 4.23   The structure of the cap found on eukaryotic messenger RNAs. The first base is 7-methylguanylate connected by a 5′-5′ triphosphate linkage to the next base. The 2′ positions on bases 2 and 3 may or may not be methylated.

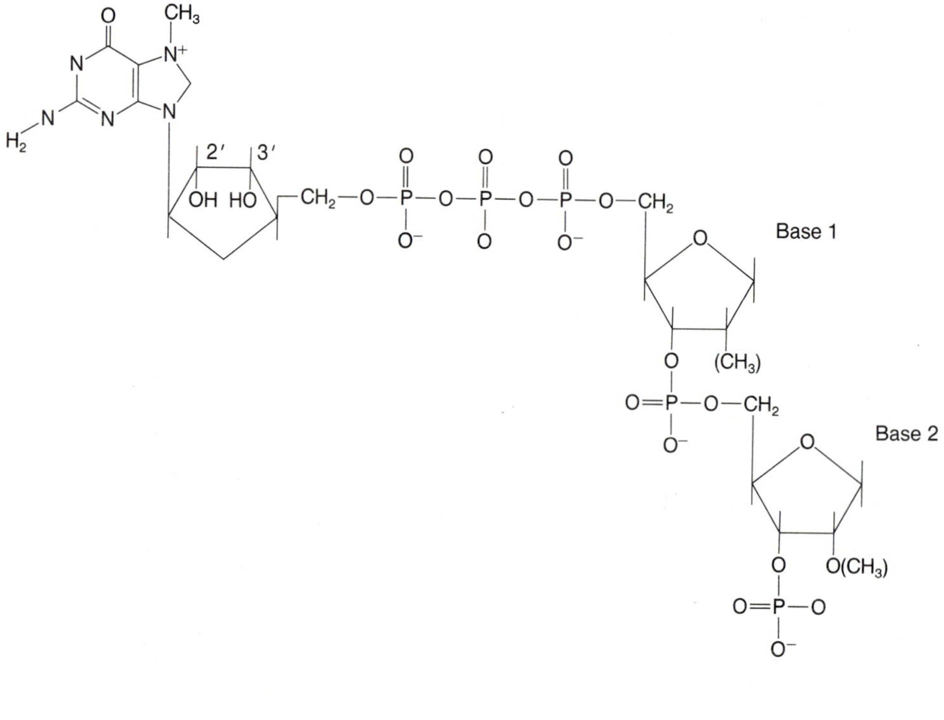

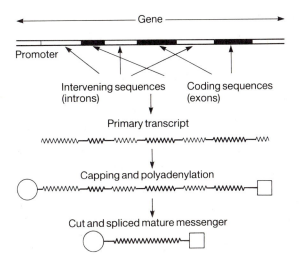

Figure 4.24   Maturation of a primary transcript to yield a translatable messenger.

sequences, which are removed and absent from mature RNA, are called introns; and the other parts of the messenger sequence are often referred to as exons.

The final type of RNA modification found in eukaryotic cells is the addition to the 3' end of 30 to 200 nucleotides of polyadenylic acid. Since the genes that encode the nuclear RNAs containing these poly-A tails do not contain regions complementary to the tails, the poly-A regions must be added posttranscriptionally. Such sequences might have been thought to be timers indicating the time since the RNA was synthesized. That is, from time to time an enzyme could remove one of the adenylic acid residues from an RNA, and, when all had been removed, the RNA could be degraded. However, examination of this question has shown only a poor correlation between the age of an RNA molecule in a cell and the length of its poly-A sequence.

RNA SPLICING

The discovery by Sharp and by Roberts and their coworkers that newly synthesized messenger in eukaryotes could be cut and then spliced back together was a great surprise. Not only is the enzymology of such a reaction unfamiliar to molecular biologists, but the biological need for such a reaction is not apparent.

Intervening sequences do have the property of separating portions of the gene by larger genetic distance; that is, recombination is more likely to occur between coding regions than within them. This permits a coding region to be an independent module. Not surprisingly, then, the portions of proteins that such modules encode often are localized structural domains.

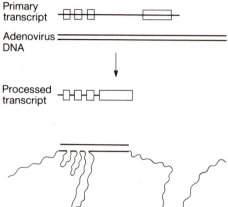

Primary transcript

Adenovirus DNA

Processed transcript

RNA-DNA hybrid between hexon mRNA
and adenovirus DNA

Figure 4.25   The primary and processed transcripts from adenovirus 2 and the structure of an RNA-DNA hybrid between the processed transcript and the viral DNA.

A clear demonstration of messenger splicing used adenovirus RNA. This provided a convenient source of DNA for hybridization reactions as well as viral-specific RNA because adenovirus-infected cells synthesize large quantities of the viral RNA. Electron microscopy of hybrids formed between adenovirus mRNA and fragments of the adenovirus genome demonstrated the existence of the splicing. Messenger that coded for a capsid protein contained sequences at its 5' end originating from three discontinuous regions far upstream of the capsid sequence (Fig. 4.25). The RNA had to have been synthesized, cut several times, and then spliced back together. Splicing of RNA is also seen in the maturation of tRNA in yeast and in many eukaryotic ribosomal RNAs.

Progress in the study of the biochemical mechanisms of splicing initially was slow. Nonetheless, early progress in understanding the mechanism of splicing was greatly assisted by the discovery by Steitz and Flint that splicing *in vitro* could be blocked by antibodies against the U1 class of the abundant small nuclear ribonuclear protein particles, snRNPs. These are particles containing short RNA molecules about 90 nucleotides long and about 10 different proteins. In addition to the inhibition experiments, other clues to the involvement of the U1 snRNPs in splicing are that the sequence at their 5' end is complementary to the consensus sequence at the 5' end of splice junctions, that the snRNP can bind to mRNA containing an intervening sequence and protect it from nuclease digestion *in vitro,* and that the snRNA is found by crosslinking experiments to be associated with nuclear RNA in which intervening sequences have not yet been excised (Fig. 4.26).

Although the proteins of snRNP particles or other proteins could perform the covalent bond breaking and forming required for excision of the intervening se-

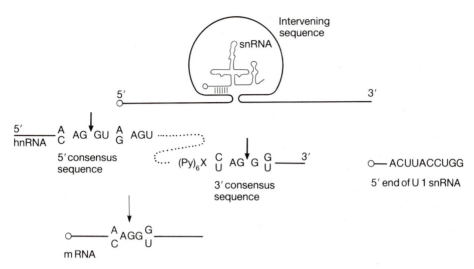

Figure 4.26 Base pairing between the 5′ end of an intervening sequence and the 5′ end of U1 snRNA in which splicing using the consensus sequences generates the resultant mRNA. The small circles on the 5′ ends represent the cap.

quences, the snRNA itself may also play a role. In some intervening sequences of yeast, internal regions are involved in excision and bear some resemblance to portions of the U1 sequences. No protein at all is required for *in vitro* removal of the intervening sequence found in *Tetrahymena* ribosomal RNA. Similarly, RNAse P, which trims tRNA precursors in *E. coli*, contains RNA and protein. The RNA component alone possesses enzymatic RNA cleaving activity.

Considering the importance of RNA to cells, it is not surprising that a multitude of mechanisms have evolved for synthesizing, using, and degrading it. This chapter has considered some of the factors involved in these processes, and later chapters cover in greater depth a number of specific mechanisms found for regulating RNA synthesis.

PROBLEMS

4.1. If RNA polymerase subunits β and β' constitute 0.005 by weight of the total protein in an *E. coli* cell, how many RNA polymerase molecules are there per cell, assuming each β and β' within the cell is found in a complete RNA polymerase molecule?

4.2. A typical *E. coli* cell with a doubling time of 50 minutes contains 10,000 ribosomes. How many RNA polymerase molecules must be synthesizing ribosomal RNA at any instant?

4.3. What is a logical conclusion of the fact that some bases in the -35 region of a promoter are not protected against chemical modification by a bound RNA polymerase molecule and yet their modification prior to addition of RNA polymerase to the DNA prevents open complex formation?

4.4. In Summers, Nature 223, 1111 (1969), what are the experimental data suggesting that the model proposed in the paper is incorrect and instead that phage T7 synthesizes its own RNA polymerase?

4.5. In most *in vitro* transcription experiments using templates such as the *lac* operon on phage lambda, the ratio of the rate of transcription of a gene between the induced and repressed conditions is about 10-fold. *In vivo* the ratio is about 1,000-fold. What are possible reasons for this and what experiments could be done to test your ideas?

4.6. Dennis and Bremer, J. Mol. Biol. 75, 145–159 (1973), present data indicating that the elongation rate of ribosomal RNA is faster than mentioned in this chapter and that it varies with the growth rate of the cells. In light of the much simpler elongation rate measurements utilizing 5S RNA, their results must be incorrect. What is their most likely source of error?

4.7. Suppose radioactive uridine and rifamycin are simultaneously added to growing cells. Assume that both the uridine and the rifamycin instantaneously enter the cells and that the radioactive uridine immediately begins entering RNA chains that are in the process of being elongated. Messenger RNA will be completed and will later decay, whereas ribosomal RNA will be synthesized and will be stable. Sketch the kinetics of radioactivity incorporation into RNA in such an experiment and indicate how it may be used to determine the fraction of RNA synthesis devoted to mRNA and to rRNA.

4.8. Vegetatively growing *B. subtilis* cells appear to contain at least two types of RNA polymerase that differ only in their σ subunits:

	-35 Region Recognized	-10 Region Recognized
σ^{55}	TTGACA	TATAAT
σ^{37}	GGNTNAAA	TATTGTTT

N is any nucleotide and the superscripts are molecular weights in thousands. What astonishing paradox is raised by these data?

*4.9. From the definition of optical density, OD, and the fact that passage of a charged particle through a crystal of silver halide renders it capable of being developed into a silver particle, estimate the number of p^{32} decays that are necessary in an autoradiograph experiment to blacken an area equal to the size of a penny to an OD of 1.0. To solve this it will be necessary to look up some fundamental information on photographic emulsions.

4.10. Why does urea or methyl mercury denature RNA?

4.11. Does RNA polymerase bound to DNA at a nonpromoter site melt out as many bases as RNA polymerase bound at a promoter? [See Nuc. Acids Res. 4, 1225–1241 (1975).]

4.12. It is tempting to conclude that the isomerization step in initiation by RNA polymerase corresponds to the disruption of 10 to 15 base pairs of DNA. How could this correspondence be demonstrated or refuted?

4.13. Suppose *in vitro* transcription from a promoter located on a 400-base-pair piece of DNA requires an auxiliary protein A in addition to RNA polymerase. In experiments to determine A's mechanism of action, order of addition experiments were performed (Fig. 4.27). Propose an explanation consistent with these data and an experiment to confirm your hypothesis.

4.14. Gralla, J. Biol. Chem. *255*, 10423–10430 (1981) claims that most RNA polymerase in a typical *in vitro* transcription experiment is either bound to promoter or nonspecifically bound to other sites, so the concentration of free RNA polymerase is small. Why doesn't this finding negate the work of McClure, Proc. Natl. Acad. Sci. USA *77*, 5634–5638 (1980)?

4.15. Siebenlist's studies of the melting of DNA by binding of RNA polymerase, Nature *279*, 651–652 (1979), directly showed the melting of 11 base pairs, as did the increase in ultraviolet absorbance upon the binding of RNA polymerase [Hsieh and Wang, Nuc. Acids Res. *5*, 3337–3345 (1978)]. Yet Wang and Hsieh's measurements of supercoils with and without bound polymerase, Nuc. Acids Res. *4*, 1225–1241

Figure 4.27 Data for Problem 4.13. Additions are made to the reaction mixtures as indicated on the horizontal time lines, and samples are taken at various times according to the two protocols. The resultant kinetics of RNA chains initiated per minute are indicated.

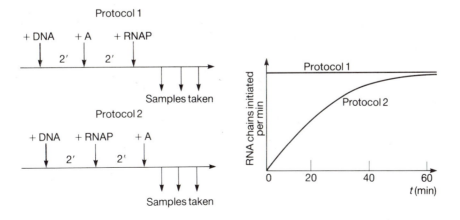

(1977), showed the melting of only about 5 base pairs. Why are these two sets of results not in contradiction?

4.16. How would you go about determining whether rho protein acts by reading DNA sequences, by reading RNA sequences coming out of the RNA polymerase, or by interacting directly with RNA polymerase?

*4.17. Derive equations 4.1 and 4.2. [*Hint:* Compute the amount of RNA polymerase bound to DNA in the closed complex and, from this, the rate of increase in the amount of the open complex as a function of the amount of polymerase not in the closed complex. Consider the amount of RNA polymerase in the closed complex always to be in equilibrium with the free RNA polymerase.]

4.18. An important question about capped eukaryotic messenger RNA is whether the cap location is a processing site produced by cleavage from a longer RNA or whether the cap site is the actual start site of transcription. How could β-labeled ATP be used *in vitro* to answer this question for an RNA that begins GpppApXp?

RECOMMENDED READINGS

Nucleotide Sequence of an RNA Polymerase Binding Site at an Early T7 Promoter, D. Pribnow, Proc. Nat. Acad. Sci. USA 72, 784–788 (1975).

The Molecular Topography of RNA Polymerase Promoter Interaction, R. Simpson, Cell 18, 277–285 (1979).

A Small Nuclear Ribonucleoprotein Is Required for Splicing of Adenoviral Early RNA Sequences, V. Yang, M. Lerner, J. Steitz, J. Flint, Proc. Nat. Acad. Sci. USA 78, 1371–1375 (1981).

Sequence Determinants of Promoter Activity, P. Youderian, S. Bouvier, M. Susskind, Cell 30, 843–853 (1982).

RELATED REVIEWS, BOOKS, AND ARTICLES

Processing of RNA, R. Perry, Ann. Rev. Biochem. 45, 605–629 (1976).

Capping of Eucaryotic mRNAs, A. Shatkin, Cell 9, 645–653 (1976).

"RNA Polymerases," ed. R. Losick, M. Chamberlin, Cold Spring Harbor (1976).

3'-Terminal Addition to HeLa Cell Nuclear and Cytoplasmic Poly(A), S. Sawicki, W. Jelinek, J. Darnell, J. Mol. Biol. 113, 219–235 (1977).

Regulatory Sequences Involved in the Promotion and Termination of RNA Transcription, M. Rosenberg, D. Court, Ann. Rev. Genet. 13, 319–353 (1979).

RNA Processing and the Intervening Sequence Problem, J. Abelson, Ann. Rev. Biochem. 48, 1035–1069 (1979).

Termination of Transcription and its Regulation in the Tryptophan Operon of E. coli, T. Platt, Cell 24, 10–23 (1981).

Snurps and Scyrps, M. Lerner, J. Steitz, Cell 25, 298–300 (1981).

Termination of Transcription in E. coli, W. Holmes, T. Platt, M. Rosenberg, Cell 32, 1029–1032 (1983).

DEEPER READING

A Mammalian System for the Incorporation of Cytidine Triphosphate into Ribonucleic Acid, S. Weiss, L. Gladstone, J. Am. Chem. Soc. *81*, 4118–4119 (1959).

RNA Chain Growth-Rate in *E. coli*, H. Bremer, D. Yuan, J. Mol. Biol. *38*, 163–180 (1968).

Purification and Properties of Ribonuclease III from *Escherichia coli*, H. Robertson, R. Webster, N. Zinder, J. Biol. Chem. *243*, 82–91 (1968).

Termination Factor for RNA Synthesis, J. Roberts, Nature *224*, 1168–1174 (1969).

The Reliability of Molecular Weight Determinations by Dodecyl Sulfate-Polyacrylamide Gel Electrophoresis, K. Weber, M. Osborn, J. Biol. Chem. *244*, 4406–4412 (1969).

Reconstitution of Bacterial DNA Dependent RNA Polymerase from Isolated Subunit as a Tool for Elucidation of the Role of the Subunits in Transcription, A. Heil, W. Zillig, FEBS Letters *11*, 165–168 (1970).

A New Method for Selecting RNA Polymerase Mutants, S. Austin, J. Scaife, J. Mol. Biol. *49*, 263–67 (1970).

Structural Alterations of RNA Polymerase During Sporulation, R. Losick, R. Sorenstein, A. Sonnenshein, Nature New Biol. *227*, 910–913 (1970).

Binding of Dodecyl Sulfate to Proteins at High Binding Ratios: Possible Implications for the State of Proteins in Biological Membranes, J. Reynolds, C. Tanford, Proc. Nat. Acad. Sci. USA *66*, 1002–1007 (1970).

Polyadenylic Acid Sequences in the Heterogenous Nuclear RNA and Rapidly-Labeled Polyribosomal RNA of HeLa Cells: Possible Evidence for a Precursor Relationship, M. Edmonds, M. Vaughn, H. Nakazato, Proc. Natl. Acad. Sci. USA *68*, 1336–1340 (1971).

The Presence and Possible Role of Zinc in RNA Polymerase Obtained from *E. coli*, M. Scrutton, C. Wu, D. Goldthwait, Proc. Nat. Acad. Sci. USA *68*, 2497–2501 (1971).

Amber Mutations of *E. coli* RNA Polymerase, S. Austin, I. Tittawella, R. S. Hayward, J. Scaife, Nature New Biol. *232*, 133–136 (1971).

Studies of the Binding of *E. coli* RNA Polymerase to DNA, V. T7 RNA Chain Initiation by Enzyme-DNA Complexes, M. Chamberlin, J. Ring, J. Mol. Biol. *70*, 221–237 (1972).

Identification of Template Strand in Heteroduplex DNA During Cell-Free Enzyme Synthesis, W. Wetekam, Mol. Gen. Genet. *118*, 57–60 (1972).

ALT: A New Factor Involved in the Synthesis of RNA by *E. coli*, A. Silverstone, M. Goman, J. Scaife, Mol. Gen. Genet. *118*, 223–224 (1972).

Isolation and Characterization of a Ribonuclease III Deficient Mutant of *Escherichia coli*, P. Kindler, T. Keil, P. Hofschneider, Mol. Gen. Genet. *126*, 53–69 (1973).

T7 Early RNAs and *Escherichia coli* Ribosomal RNAs Are Cut from Large Precursor RNAs *in vivo* by Ribonuclease III, J. Dunn, W. Studier, Proc. Nat. Acad. Sci. USA *70*, 3296–3300 (1973).

Studies of Ribonucleic Acid Chain Initiation by *E. coli* Ribonucleic Acid Polymerase Bound to T7 Deoxyribonucleic Acid, I. An Assay for the Rate and Extent of RNA Chain Initiation, W. F. Mangel, M. J. Chamberlin, J. Biol. Chem. *249*, 2995–3001 (1974).

Studies of Ribonucleic Acid Chain Initiation by *E. coli* Ribonucleic Acid Polymerase Bound to T7 Deoxyribonucleic Acid, II. The Effect of Alterations in Ionic Strength on Chain Initiation and on the Conformation of Binary Complexes, W. Mangel, M. Chamberlin, J. Biol. Chem. *249*, 3002–3006 (1974).

The Interaction of Bacterial and Phage Proteins with Immobilized *E. coli* RNA Polymerase, D. Ratner, J. Mol. Biol. *88*, 373–383 (1974).

Reovirus Messenger RNA Contains a Methylated, Blocked 5'-Terminal Structure: m7G(5')ppp(5')GmpCp-, Y. Fukruichi, M. Morgan, S. Muthukrishnan, A. Shatkin, Proc. Natl. Acad. Sci. USA 72, 362–366 (1975).

Sigma Cycle During in vitro Transcription: Demonstration by Nanosecond Fluorescence Depolarization Spectroscopy, C. Wu, L. Yarbrough, Z. Hillel, F. Wu, Proc. Nat. Acad. Sci. USA 72, 3019–3023 (1975).

Identification of a Gene for the α-Subunit of RNA Polymerase at the str-spc Region of the E. coli chromosome, S. Jaskunas, R. Burgess, M. Nomura, Proc. Nat. Acad. Sci. USA 72, 5036–5040 (1975).

A Procedure for the Rapid, Large-Scale Purification of E. coli DNA-Dependent RNA Polymerase Involving Polymin P Precipitation and DNA-Cellulose Chromatography, R. Burgess, J. Jendrisak, Biochem. 14, 4634–4638 (1975).

Bacteriophage T7 Early Promoters: Nucleotide Sequences of Two RNA Polymerase Binding Sites, D. Pribnow, J. Mol. Biol. 99, 419–443 (1975).

The Rho Gene of E. coli Maps at suA, D. Ratner, in "RNA Polymerases," ed. R. Losick, M. Chamberlin, Cold Spring Harbor, 645–655 (1976).

Eukaryotic Nuclear RNA Polymerases, R. Roeder, in "RNA Polymerases," ed. R. Losick, M. Chamberlin, Cold Spring Harbor, 285–329 (1976).

Evidence That Mutations in the suA Polarity Suppressing Gene Directly Affect Termination Factor rho, D. Ratner, Nature 259, 151–153 (1976).

In vivo Distribution of RNA Polymerase Between Cytoplasm and Nucleoid in E. coli, W. Runzi, H. Matzura, J. Bact. 125, 1237–1239 (1976).

Systematic Nomenclature for the RNA Polymerase Genes of Prokaryotes, R. Hayward, J. Scaife, Nature 260, 646–647 (1976).

Identification of a Mutation Within the Structural Gene for the α-Subunit of DNA-Dependent RNAP of E. coli, H. Fujiki, P. Palm, W. Zillig, R. Calendar, M. Sunshine, Mol. Gen. Genet. 145, 19–22 (1976).

Ribosomal RNA Chain Elongation Rates in Escherichia coli, S. Molin, in "Control of Ribosome Synthesis," ed. N. Kjeldgaard, N. Maaloe, Alfred Benzon Symposium IX, Munksgaard, 331–339 (1976).

Transcription Termination at the End of the Tryptophan Operon of E. coli, L. Guarente, D. Mitchell, J. Beckwith, J. Mol. Biol. 112, 423–436 (1977).

A Gene from E. coli Affecting the Sigma Subunit of RNA Polymerase, J. Harris, I. Martinez, R. Calendar, Proc. Nat. Acad. Sci. USA 74, 1836–1840 (1977).

Spliced Segments at the 5' Terminus of Adenovirus 2 Late mRNA, S. Berget, C. Moore, P. Sharp, Proc. Natl. Acad. Sci. USA 74, 3171–3175 (1977).

Promoter Recognition by Phage SP01-Modified RNA Polymerase, C. Talkington, J. Pero, Proc. Nat. Acad. Sci. USA 75, 1185–1189 (1978).

Complementary Sequences 1700 Nucleotides Apart Form a Ribonuclease III Cleavage Site in Escherichia coli Ribosomal Precursor RNA, R. Young, J. Argetsinger-Steitz, Proc. Nat. Acad. Sci. USA 75, 3593–3597 (1978).

Processing of the 5' End of Escherichia coli 16S Ribosomal RNA, A. Dahlberg, J. Dahlberg, E. Lund, H. Tokimatsu, A. Ribson, P. Calvert, F. Reynolds, M. Zahalak, Proc. Nat. Acad. Sci. USA 75, 3598–3602 (1978).

An Amazing Sequence Arrangement at the 5' Ends of Adenovirus 2 Messenger RNA, L. Chow, R. Gelinas, T. Broker, R. Roberts, Cell 12, 1–8 (1977).

Physiochemical Studies on Interactions Between DNA and RNA Polymerase: Unwinding of the DNA Helix by E. coli RNA Polymerase, J. Wang, J. Jacobsen, J. Saucier, Nuc. Acids Res. 4, 1225–1241 (1978).

Physiochemical Studies on Interactions Between DNA and RNA Polymerase: Ultraviolet Absorption Measurements, T. Hsieh, J. Wang, Nuc. Acids Res. 5, 3337–3345 (1978).

A Steady State Assay for the RNA Polymerase Initiation Reaction, W. McClure, C. Cech, D. Johnston, J. Biol. Chem. 253, 8941–8948 (1978).

Contacts Between E. coli RNA Polymerase and Thymines in the lac UV5 Promoter, R. Simpson, Proc. Nat. Acad. Sci. USA 76, 3233–3237 (1979).

Promoter Mutations in the Transfer RNA Gene to tyrT of E. coli, M. Berman, A. Landy, Proc. Nat. Acad. Sci. USA 76, 4303–4307 (1979).

Distinctive Nucleotide Sequences of Promoters Recognized by RNA Polymerase Containing a Phage-Coded "Sigma-like" Protein, C. Talkington, J. Pero, Proc. Nat. Acad. Sci. USA 76, 5465–5469 (1979).

E. coli RNA Polymerase Interacts Homologously with Two Different Promoters, U. Siebenlist, R. Simpson, W. Gilbert, Cell 20, 269–281 (1980).

Binding of E. coli RNA Polymerase Holoenzyme to a Bacteriophage T7 Promoter-Containing Fragment: Selectivity Exists over a Wide Range of Solution Conditions, H. Strauss, R. Burgess, M. Record, Biochem. 19, 3496–3504 (1980).

Binding of E. coli RNA Polymerase Holoenzyme to a Bacteriophage T7 Promoter-Containing Fragment: Evaluation of Promoter Binding Constants as a Function of Solution Conditions, H. Strauss, R. Burgess, M. Record, Biochem. 19, 3504–3515 (1980).

Pentalysine-DNA Interactions: A Model for the General Effects of Ion Concentrations on the Interactions of Proteins with Nucleic Acids, T. Lohman, P. deHaseth, M. Record, Biochem. 19, 3522–3530, (1980).

A Mechanism for RNA Splicing, J. Rogers, R. Wall, Proc. Natl. Acad. Sci. USA 77, 1877–1879 (1980).

Rate-Limiting Steps in RNA Chain Initiation, W. McClure, Proc. Nat. Acad. Sci. USA 77, 5634–5638 (1980).

In vitro Comparison of Initiation Properties of Bacteriophage Lambda Wild-type P_R and x3 Mutant Promoters, D. Hawley, W. McClure, Proc. Nat. Acad. Sci. USA 77, 6381–6385 (1980).

Nucleotide Sequence of a Promoter Recognized by Bacillus subtilis RNA Polymerase, G. Lee, C. Talkington, J. Pero, Mol. Gene. Genet. 180, 57–65 (1980).

DNA-RNA Hybrid Duplexes Containing Oligo (dA : rU) Sequences Are Exceptionally Unstable and May Facilitate Termination of Transcription, F. Martin, I. Tinoco, Jr., Nuc. Acids Res. 8, 2295–2300 (1980).

Evidence That Ribosomal Protein S10 Participates in Control of Transcription Termination, D. Friedman, A. Schauer, M. Baumann, L. Baron, S. Adhya, Proc. Nat. Acad. Sci. USA 78, 1115–1118 (1981).

Deletion Mapping of Sequences Essential for in vivo Transcription of the Iso-1-Cytochrome C Gene, G. Faye, D. Leung, K. Tatchell, B. Hall, M. Smith, Proc. Nat. Acad. Sci. USA 78, 2258–2262 (1981).

Mutations in the Gene Coding for E. coli DNA Topoisomerase I Affect Transcription and Transposition, R. Sternglanz, S. DiNardo, K. Voelkel, Y. Nishimura, Y. Hirota, K. Becherer, L. Zumstein, J. Wang, Proc. Nat. Acad. Sci. USA 78, 2747–2751 (1981).

The nusA Gene Protein of Escherichia coli, G. Greenblatt, J. Li, J. Mol. Biol. 147, 11–23 (1981).

Termination of Transcription by nusA Gene Protein of E. coli, J. Greenblatt, M. McLimont, S. Hanly, Nature 292, 215–220 (1981).

Termination Cycle of Transcription, J. Greenblatt, J. Li, Cell 24, 421–428 (1981).

DNA Strand Specificity in Promoter Recognition by RNA Polymerase, C. Park, Z. Hillel. C. Wu, Nuc. Acids Res. 8, 5895–5911 (1981).

An Estimate of the Frequency of *in vivo* Transcriptional Errors at a Nonsense Codon in *Escherichia coli*, R. Rosenberger, G. Foskett, Mol. Gen. Genet. *183*, 561–563 (1981).

Binding of *Escherichia coli* RNA Polymerase Holoenzyme to Bacteriophage T7 DNA, T. Kadesch, S. Rosenberg, M. Chamberlin, J. Mol. Biol. *155*, 1–29 (1982).

Mechanism of Activation of Transcription Initiation from the Lambda P_{RM} Promoter, D. Hawley, W. McClure, J. Mol. Biol. *157*, 493–525 (1982).

Transcription-Terminations at Lambda t_{R1} in Three Clusters, L. Lau, J. Roberts, R. Wu, Proc. Nat. Acad. Sci. USA *79*, 6171–6175 (1982).

The Yeast *his3* Promoter Contains at Least Two Distinct Elements, K. Struhl, Proc. Nat. Acad. Sci. USA *79*, 7385–7389 (1982).

Transcriptional Control Signals of a Eukaryotic Protein-Coding Gene, S. McKnight, R. Kingsbury, Science *217*, 316–324 (1982).

A Topological Model for Transcription Based on Unwinding Angle Analysis of *E. coli* RNA Polymerase Binary Initiation and Ternary Complexes, H. Gamper, J. Hearst, Cell *29*, 81–90 (1982).

Self-Splicing RNA: Autoexcision and Autocyclization of the Ribosomal RNA Intervening Sequence of *Tetrahymena*, K. Kruger, P. Grabowski, A. Zaug, J. Sands, D. Gottschling, T. Cech, Cell *31*, 147–157 (1982).

A *lac* Promoter with a Changed Distance between −10 and −35 Regions, W. Mandecki, W. Reznikoff, Nuc. Acids Res. *10*, 903–911 (1982).

Laser Crosslinking of *E. coli* RNA Polymerase and T7 DNA, C. Harrison, D. Turner, D. Hinkle, Nuc. Acids Res. *10*, 2399–2414 (1982).

Topography of Transcription: Path of the Leading End of Nascent RNA Through the *Escherichia coli* Transcription Complex, M. Hanna, C. Meares, Proc. Nat. Acad. Sci. USA *80*, 4238–4242 (1983).

Splicing of Adenovirus RNA in a Cell-Free Transcription System, R. Padgett, S. Hardy, P. Sharp, Proc. Nat. Acad. of Sci. USA *80*, 5230–5234 (1983).

Rho-Dependent Termination of Transcription, W. Morgan, D. Bear, P. von Hippel, J. Biol. Chem., 9553–9564 (1983).

The U1 Small Nuclear RNA-Protein Complex Selectively Binds a 5′ Splice Site *in vitro*, S. Mount, I. Pattersson, M. Hinterberger, A. Karmas, J. Steitz, Cell *33*, 509–518 (1983).

Evidence for the Biochemical Role of an Internal Sequence in Yeast Nuclear mRNA Introns: Implications for U1 RNA and Metazoan mRNA Splicing, C. Pikielny, J. Teem, M. Rosbash, Cell *34*, 395–403 (1983).

Yeast Contains Small Nuclear RNAs Encoded by Single Copy Genes, J. Wise, D. Tollervey, D. Maloney, H. Swerdlow, E. Dunn, C. Guthrie, Cell *35*, 743–751 (1983).

The RNA Moiety of Ribonuclease P is the Catalytic Subunit of the Enzyme, C. Guerrier-Tokada, K. Gardiner, T. Marsh, N. Pace, S. Altman, Cell *35*, 849–857 (1983).

Distinctly Regulated Tandem Upstream Activation Sites Mediate Catabolite Repression in the *CYC1* Gene of *S. cerevisiae*, L. Guarente, B. Salonde, P. Gifford, E. Alani, Cell *36*, 503–511 (1984).

Point Mutations Identify the Conserved, Intron-Contained TACTAAC Box as an Essential Splicing Signal Sequence in Yeast, C. Langford, F. Klinz, C. Donath, D. Gallwitz, Cell *36*, 645–653 (1984).

The *htpR* Gene Product of *E. coli* Is a Sigma Factor for Heat-Shock Promoters, A. Grossman, J. Erickson, C. Gross, Cell *38*, 383–390 (1984).

Protein Structure

Proteins carry out most of the interesting cellular processes. Enzymes, structural components of cells, and even excreted cellular adhesives are all proteins. One major property shared by most proteins is their ability to bind only certain other molecules. What gives proteins such a high degree of selectivity? Part of the answer is known and is described in this chapter.

Although it would be satisfying to understand the properties of proteins in general terms, ultimately we want to understand proteins so well that we can design them. That is, our final, distant goal is to be able to specify an amino acid sequence such that, when synthesized, it will assume a desired three-dimensional structure, bind any desired substrate, and then carry out any reasonable enzymatic reaction. Furthermore, if our designed protein is to be synthesized in cells, we must know what necessary auxiliary DNA sequences to provide so that the protein will be synthesized in the proper quantities and at appropriate times. Fortunately, much is already known about the required regulatory sequences, as the information in this book attests.

The most notable advances of molecular biology in the past decade have involved nucleic acids, not proteins. Nonetheless, since DNA specifies the amino acid sequence of proteins, molecular biologists' newfound abilities to alter any desired nucleotides in the DNA sequences of genes mean that the amino acid sequences of proteins can also be specifically altered. Consequently, the pace of research investigating protein structure will dramatically increase. Systematic studies of the structure and activity of proteins resulting from specific amino acid substitutions should soon begin to increase greatly our understanding of protein structure.

In this chapter we examine the fundamentals of protein structure. Much of this information is discussed more completely in biochemistry or physical biochemistry texts. We review the material to develop our intuitions on the structures and properties of proteins so as to have a clearer idea of how cells function. First we discuss the components of proteins, amino acids. Then we consider the linking of amino acids via peptide bonds. A variety of forces are possible between amino acids, and these lead the amino acids along many portions of the polypeptide backbone to adopt relatively simple, specific orientations with respect to one another. These are known as secondary structures, and the most common are alpha helices, beta sheets, and beta bends. In many proteins these secondary structure elements and some random conformations of amino acids are arranged into recognizable domains of amino acids possessing relatively few interactions with amino acids outside the domain. Typically a protein may contain one or more domains.

The secondary structure of many proteins can be rather accurately predicted by empirical rules, which we will discuss. These results are a beginning step toward the ultimate goals of understanding the physical basis of folding of the polypeptide chains and the effects of amino acid substitutions found in mutant proteins. However, little progress has been made in predicting the arrangement of the secondary structure elements into domains or in predicting the complete tertiary structure of a protein.

The final topic of this chapter is an examination of the tertiary structure of repressor protein of phage lambda. This protein binds to specific sequences, termed operators, on the phage DNA and reduces transcription originating from the adjacent promoters. Each subunit of the repressor contains a protruding alpha helix that partly fits into the major groove of the double-stranded DNA. This complementarity in the structure of a protein and the molecule to which it binds is the basis for selectivity in protein binding. The actual DNA-binding structure used by lambda repressor is similar to the DNA-binding portions of other proteins that bind to DNA, and apparently this motif is repeated in a variety of other DNA-binding proteins.

THE AMINO ACIDS

Proteins consist of α-L-amino acids linked by peptide bonds to form polypeptide chains (Fig. 5.1). Near neutrality, the carboxyl group of an amino acid is negatively charged and the amino group is positively charged. In a protein, however, these charges are absent from the interior amino acids owing to the formation of the peptide bonds between the amino groups and carboxyl groups. Of course, the N-terminal amino group of a protein is positively charged and the C-terminal carboxyl group is negatively charged.

Twenty different types of α-L-amino acids are commonly found in proteins (Table 5.1). Except for proline, which technically is an imino acid, these differ from

$$\underset{\underset{\text{H}}{|}}{\overset{\overset{^{+}\text{H} \quad \text{R}}{|\quad\;|}}{\text{H}-\text{N}-\underset{\underset{\text{H}}{|}}{\text{C}_\alpha}-\text{C}}}\overset{\text{O}}{\underset{\text{O}^-}{\diagup\diagdown}}$$

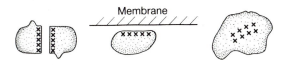

Figure 5.1 An α-L-amino acid with negative charge on the carboxyl and positive charge on the amino group and three amino acids linked by peptide bonds.

one another only in the structure of the side group attached to the alpha carbon. A few other types of amino acids are occasionally found in proteins, resulting from modification of one of the twenty after the protein has been synthesized. Frequently these modified amino acids are directly involved with chemical reactions catalyzed by the protein. Each of the basic twenty must possess unique and invaluable properties since most proteins contain all twenty amino acids.

Even though we must know and understand the individual properties of each of the amino acids, it is convenient to classify the twenty and to understand common properties of the groups. The side groups of the aliphatic amino acids are hydrophobic and prefer to exist in a nonaqueous, nonpolar environment like that found in the contact region between two subunits, in the portion of a protein bound to a membrane, or in the interior of a globular protein. A large fraction of such amino acids on

Membrane

a portion of the surface can make a protein bind to a similar hydrophobic patch on the surface of another protein, as in the case of oligomerization of protein subunits, or it can make the protein prefer to bind to or even dissolve in a membrane. Often hydrophobic amino acids are found on the interior of a protein where they prefer the company of one another to that of water, and they thereby help keep the protein from turning itself inside out.

The basic amino acid side groups possess positive charges at neutral pH. If located on the surface of the protein, such positive charges can assist the binding of a negatively charged ligand, for example DNA. The acidic amino acid side groups possess negative charges at neutral pH. Neutral amino acid side groups possess no

---

Table 5.1   The Amino Acids

Amino Acid	Three-Letter Abbreviation	One-Letter Abbreviation	Character	Side Group
Alanine	Ala	A	Aliphatic	$-CH_3$
Arginine	Arg	R	Basic	$-CH_2-CH_2-CH_2-\overset{H}{N}-C\begin{smallmatrix} NH_2^+ \\ NH_2 \end{smallmatrix}$
Asparagine	Asn	D	Neutral, polar	$-CH_2-C\begin{smallmatrix} O \\ NH_2 \end{smallmatrix}$
Aspartic acid	Asp	D	Acid	$-CH_2-C\begin{smallmatrix} O \\ O^- \end{smallmatrix}$
Cysteine	Cys	C	Neutral	$-CH_2-SH$
Glutamine	Gln	Q	Neutral, polar	$-CH_2-CH_2-C\begin{smallmatrix} O \\ NH_2 \end{smallmatrix}$
Glutamic acid	Glu	E	Acidic	$-CH_2-CH_2-C\begin{smallmatrix} O \\ O^- \end{smallmatrix}$
Glycine	Gly	G	Neutral	$-H$
Histidine	His	H	Aromatic	$-CH_2-C=CH$, H^+N , NH , CH
Isoleucine	Ile	I	Aliphatic	$-\overset{H}{\underset{CH_3}{C}}-CH_2-CH_3$
Leucine	Leu	L	Aliphatic	$-CH_2-\overset{H}{C}\begin{smallmatrix} CH_3 \\ CH_3 \end{smallmatrix}$

---

net charge, and polar amino acid side groups possess separated charges. Separated charges lead to dipole interactions with other amino acids or with ligands binding to the protein.

Cysteine is a notable unique amino acid since in some environments two cysteine residues in a protein can spontaneously oxidize to form a rather stable disulfide bond (Fig. 5.2). In the isolated protein, this bond can be reduced by the presence of an excess of a sulfhydryl reagent to regenerate cysteines.

Table 5.1   The Amino Acids

Amino Acid	Three-Letter Abbreviation	One-Letter Abbreviation	Character	Side Group
Lysine	Lys	K	Basic	$-CH_2-CH_2-CH_2-CH_2-NH_3{}^+$
Methionine	Met	M	Neutral	$-CH_2-CH_2-S-CH_3$
Phenylalanine	Phe	F	Aromatic	$-CH_2-\bigcirc$
Proline	Pro	P	Neutral	
Serine	Ser	S	Neutral, polar	$-CH_2-OH$
Threonine	Thr	T	Neutral, polar	
Tryptophan	Trp	W	Aromatic	
Tyrosine	Tyr	Y	Aromatic	$-CH_2-\bigcirc-OH$
Valine	Val	V	Aliphatic	

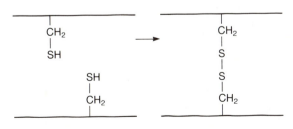

Figure 5.2   Two reduced cysteine residues and their oxidized state, which forms a disulfide bond.

THE PEPTIDE BOND

A peptide bond links successive amino acids in a polypeptide chain. However, the mere linking of amino acids to form a polypeptide chain is insufficient to ensure that the joined amino acids will adopt a particular three-dimensional structure. The peptide bond possesses two extraordinarily important properties that facilitate folding of a polypeptide into a particular structure.

First, as a consequence of the partial double-bond character of the peptide bond between oxygen, carbon, and nitrogen, the unit bounded by the alpha carbon atoms of two successive amino acids is constrained to lie in a plane. Therefore, energy need

$$
\begin{array}{c}
\text{O} \\
\parallel \\
-\text{C}-\text{N}- \\
| \\
\text{H}
\end{array}
\rightleftharpoons
\begin{array}{c}
\text{O} \\
| \\
-\text{C}=\text{N}- \\
| \\
\text{H}
\end{array}
$$

not be consumed from other interactions for "proper" orientation about the C-N bond in each amino acid. Rotation is possible about each of the two bonds from the C_α atom of each amino acid (Fig. 5.3). Angles of rotation about these two bonds are called phi and psi, and their specification for each of the amino acids in a polypeptide completely describes the path of the polypeptide backbone. Of course, the side

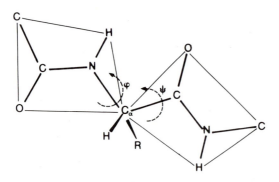

Figure 5.3   Two amino acid units in a polypeptide chain illustrating the planar structure of the peptide bond, the two degrees of rotational freedom for each amino acid unit in a polypeptide, and the angles ϕ and ψ.

chains of the amino acids are free to rotate and may adopt a number of conformations so that the phi and psi angles do not specify the entire structure of a protein.

The second consequence of the peptide bond is that the amide hydrogen from one amino acid may be shared with the oxygen from another amino acid in a hydrogen bond. Since each amino acid in a polypeptide chain possesses both a

$$\overset{\delta^-}{\underset{/}{\overset{}{N}}}\!\!-\!\!H \quad \overset{\delta^+}{\overset{}{O}}\!\!=\!\!\overset{\delta^-}{\underset{\backslash}{C}} \;\rightleftharpoons\; \overset{\backslash}{\underset{/}{N}}\!\!-\!\!H\,\|\|\|\|O\!\!=\!\!\overset{/}{\underset{\backslash}{C}}$$

hydrogen bond donor and an acceptor, many hydrogen bonds may be formed in a polypeptide. Although individual hydrogen bonds are weak, the large number that can form in a protein contributes substantially to maintaining the three-dimensional structure of a protein.

ELECTROSTATIC FORCES THAT DETERMINE PROTEIN STRUCTURE

Proteins barely can survive. If we heat them a little above the temperatures normally found in the cells from which they are isolated, most are denatured. Why should this be? On first consideration it would make sense for proteins to be particularly stable and to be able to withstand certain environmental insults like mild heating. One explanation for the instability is that proteins just cannot be made more stable. A second possibility is that the instability is an inherent part of proteins' activities. It could be that to act as catalysts in chemical reactions or to participate in other cellular activities, proteins must be flexible, and flexibility means that proteins must exist on the verge of denaturing. Future research should illuminate this question. For the present, we will examine the origins of the weak forces that barely manage to give proteins specific shapes.

Whenever possible, it is helpful to think of the interactions between amino acids in terms of forces. Nonetheless, physicists and physical chemists have found great convenience in considering the interactions between objects in terms of potentials (Fig. 5.4). Some of our discussion will be more streamlined if we, too, use potentials. Forces and potentials are easily interconvertible since forces are simply related to potentials. The steepness of an object's potential at a point is proportional to the force on the object while at that point. Alternatively, the potential difference between two points is proportional to the work required to move an object between the two points.

Electrostatics are the basis of several of the forces that determine protein struc-

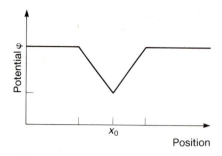

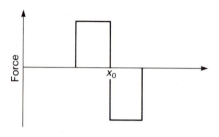

$$\text{Force} = -\frac{d\varphi}{dx}$$

Figure 5.4   A potential ϕ as a function of distance x and the force it generates, which is proportional to the derivative of the potential function.

Figure 5.5   The electrostatic force between two charges of value Q_1 and Q_2 separated by a distance r and the potential ϕ produced by a single charge.

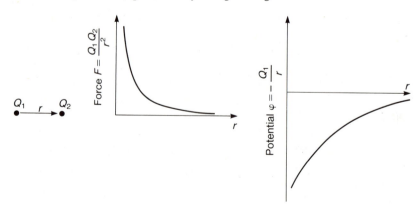

ture. Charges of unlike sign attract each other with a force proportional to each of the charges and inversely proportional to the square of the charge separation (Fig. 5.5). Equivalently, the potential generated at a point by a charge is proportional to the magnitude of the charge and inversely proportional to the distance of the point from the charge.

Since both positively and negatively charged amino acids exist, direct electrostatic attractions are possible in proteins. These generate what are called salt bridges; a number of proteins contain such salt bridges.

When two equal and opposite charges are located near one another, it is convenient to view their effect on other atoms and molecules as a whole rather than summing the results generated by each of the two charges separately. Such a set is known as a dipole. Two dipoles, four charges total, interact with potential possessing a magnitude inversely proportional to the cube of the dipole separation (Fig. 5.6). Some of the amino acids are polar, which means that although they are electrically neutral, their positive and negative charges are not coincident and, consequently, they are dipoles. Many dipole interactions exist in a typical protein.

Dispersion forces, also known as London dispersion forces, are weak, short-ranged attractive forces that exist between all molecules. These are the basis of most of the selectivity in the binding of other molecules to proteins. If the shape of a protein and another molecule are complementary, then many of these attractive forces hold the two molecules tightly together. However, if the molecules are not complementary in shape, only a few of these dispersion forces act and the molecules easily dissociate from one another.

Figure 5.6   An electrical dipole and the potential generated at a distance r from the dipole.

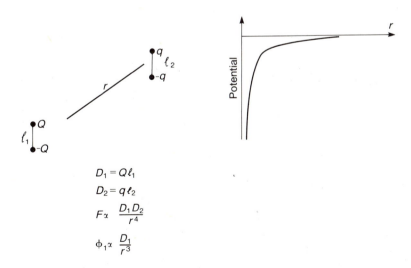

$$D_1 = Q\ell_1$$
$$D_2 = q\ell_2$$
$$F \propto \frac{D_1 D_2}{r^4}$$
$$\phi_1 \propto \frac{D_1}{r^3}$$

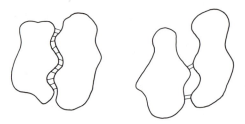

Dispersion forces are a result of electrostatic forces originating from dipoles. They are particularly short-ranged since their attractive force falls off with the sixth power of the distance separating the molecules. Although these forces are best understood in the framework of quantum mechanics, we can understand the origin of the sixth-power dependence. Consider an electrically neutral nonpolar molecule. Thermal fluctuations can generate a momentary separation of its plus and minus charges; that is, a dipole of strength D_1 is briefly generated. This dipole creates an

Dipole, D_1

electric field with a strength inversely proportional to the cube of the distance from the dipole. Such a field can induce a dipole in an adjacent susceptible molecule. The strength of the induced dipole is directly proportional to the strength of the local electric field. That is, the induced dipole has a strength, D_2, proportional to

$$\frac{D_1}{r^3}.$$

Since the interaction energy between two dipoles is proportional to the product of their strength and inversely proportional to the cube of their separation,

$$\text{Energy} = \frac{D_1 D_2}{r^3},$$

the attractive potential between the two dipoles is inversely proportional to the sixth power of their separation,

$$\phi = C\frac{D_1}{r^6}.$$

Due to the inverse sixth-power dependence, the dispersion forces become much stronger as the distance separating two molecules becomes smaller. However, the force cannot become too great, for once the electronic cloud of one molecule begins to interpenetrate the cloud of the other molecule, a very strong repulsive interaction sets in. For many purposes this repulsive potential is approximated with an inverse twelfth power of the atomic separation, and the combination of the two potentials is known as a Van der Waals potential. The radius at which the strong repulsion begins is the Van der Waals radius.

HYDROGEN BONDS AND THE CHELATE EFFECT

A hydrogen atom shared by two other atoms generates a hydrogen bond. This sharing is energetically most important when three atoms are in a straight line and the atom to which the hydrogen is covalently bonded, the hydrogen bond donor,

possesses a partial negative charge and the partner atom, a hydrogen bond acceptor, also possesses a partial negative charge. Then the atoms may approach each other quite closely and the electrostatic attractive forces and the dispersion forces will be appreciable. Since the amide of the peptide bond can be a hydrogen bond donor, and the carboxyl can be a hydrogen bond acceptor, proteins have a potential for forming a great many hydrogen bonds. In addition, more than half the side groups of the amino acids usually participate in hydrogen bonding.

A paradox is generated by the existence of hydrogen bonds in proteins. Studies with model compounds showed that the hydrogen bonds to water are expected to be stronger than the hydrogen bonds beween the amino acids. Why then don't proteins denature and make all their hydrogen bonds to water? A part of the answer is the chelate effect. That is, a great increase in the apparent binding energy between objects occurs if they are held in the correct binding positions even without the binding interaction. In an object such as a protein with a structure that holds amino acids in position, any single bond between amino acids within the protein is stronger than corresponding bonds made to water. A problem not yet understood is how the protein can fold up in the first place so that the amino acids are correctly positioned with respect to one another and so that the hydrogen bonds will be strong.

The chelate effect is important in understanding many phenomena of molecular biology. A different example, explained more fully later, concerns proteins. Much of the work required for two macromolecules to bind to one another is correctly positioning and orienting them. For example, in the interaction between a dimeric protein and another substrate like a protein or DNA, binding of one subunit to the substrate usually ensures that the other is also correctly positioned and oriented. Consequently, the second subunit appears to bind much more tightly than the first. Equivalently, the dimer appears to bind more tightly than might be expected from the binding of the monomer.

HYDROPHOBIC FORCES

The structures of many proteins that have been determined by X-ray diffraction studies reveal that, in general, the polar and charged amino acids are on the surface and the aliphatic amino acids are on the interior. Hydrophobic forces make aliphatic amino acids try to escape from a water environment and to cluster together in the center of a protein away from water.

Figure 5.7   The creation of a water cage around a hydrocarbon in water, when it moves from membrane into water.

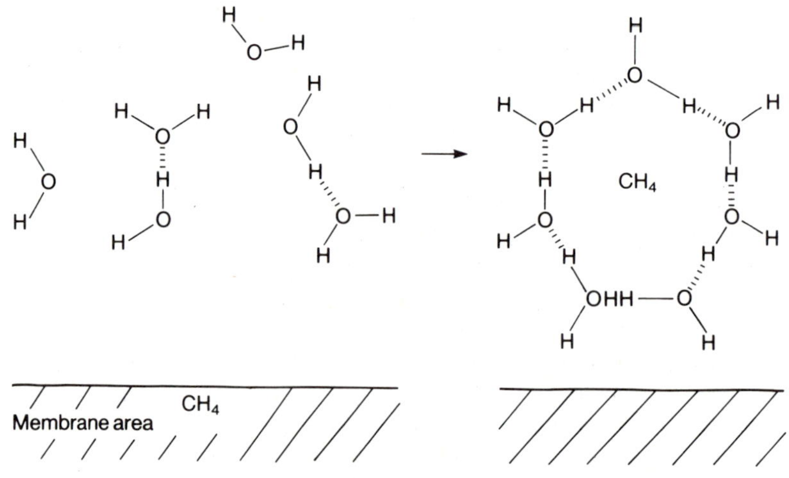

The origin of hydrophobic forces can be understood by considering the energy and entropy change in moving a neutral, nonpolar amino acid out of the interior of a protein and into the surrounding water (Fig. 5.7). The entry of a hydrocarbon into water permits the formation of cages of water molecules around the hydrocarbon molecule. These surround the hydrocarbon but do not significantly interact with it. The energy of formation of these structures actually favors their formation, and if entropy were not a consideration, neutral, nonpolar amino acids would prefer a water environment to the interior of a protein.

The effect of entropy change in the transfer of aliphatic amino acids to water is in the opposite direction and is significantly larger than the effect of the energy change. From considerations at the level presented here, we cannot deduce the magnitude of the effect. That must be determined by measuring the relative solubility of various hydrocarbons in water and organic solvents at various temperatures. However, the fact that an entropy effect exists can be understood. The formation of the cages of water molecules around hydrocarbons greatly reduces the translational and rotational motions of the water molecules involved. As a result, the state of the system in which these cages are absent, the nonpolar amino acid in the interior of the protein, is more probable than the state in which they are present. In thermodynamic terms, there is a substantial entropy decrease on the entry of a hydrocarbon into water from an organic solvent.

STRUCTURES WITHIN PROTEINS

It is useful to focus attention on particular aspects of protein structures. The primary structure of a protein is its linear sequence of amino acids. The local spatial structure of small numbers of amino acids, independent of the orientations of their side groups, generates a secondary structure. The alpha helix, beta sheet, and beta turn are all secondary structures that have been found in proteins. Both the arrangement of the secondary structure elements and the spatial arrangement of all the atoms of the molecule are referred to as the tertiary structure. Quaternary structure refers to the arrangement of subunits in proteins consisting of more than one polypeptide chain.

A domain of a protein is a structure unit intermediate in size between secondary and tertiary structures. It is a local group of amino acids that have many fewer interactions with other portions of the protein than they have among themselves (Fig. 5.8). Interestingly, not only are the amino acids of a domain near one another in the tertiary structure of a protein, but they usually comprise amino acids that lie near one another in the primary structure as well. Often, therefore, study of a protein's structure can be done on a domain-by-domain basis. The existence of semi-independent domains should greatly facilitate the study of the folding of polypeptide chains and the prediction of folding pathways and structures.

Even more useful to the ultimate goal of prediction of protein structure has been the finding that alterations in the structure of proteins tend to be local. This has been

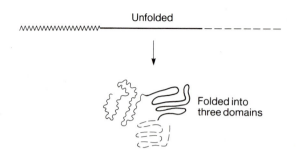

Figure 5.8   Folding of adjacent amino acids to form domains in a protein.

found in studies of the *lac* and lambda phage repressors. In both of these proteins, the majority of the amino acid changes that alter the ability to bind to DNA lie in the portion of the protein that makes contact with the DNA. Similar results can be inferred from alterations in the amino acid sequence of the tryptophan synthetase protein generated by fusing two related but nonidentical genes. Despite appreciable amino acid sequence differences in the two parental types, the fusions that contain various amounts of the N-terminal sequence from one of the proteins and the remainder of the sequence from the other protein retain enzymatic activity. This means that the amino acid alterations generated by formation of these chimeric proteins do not need to be compensated by special amino acid changes at distant points in the protein.

The results obtained with repressors and tryptophan synthetase mean that a change of an amino acid often produces a change in the tertiary structure that is primarily confined to the immediate vicinity of the alteration. This, plus the finding that protein structures can be broken down into domains, means that the majority of the potential interactions between amino acids can be neglected and interactions over relatively short distances of up to 10 Å must play the major role in determining protein structure.

As we will discover later in the book, DNA regions encoding different domains of a protein can be appreciably separated on the chromosome. This permits different domains of proteins to be shuffled so as to accelerate the rate of evolution by building new proteins from new assortments of preexisting protein domains. Domains rather than amino acids become building blocks in protein evolution.

STRUCTURE OF THE ALPHA HELIX, BETA SHEET, AND BETA TURN

The existence of the alpha helix was predicted by Pauling and Cory from careful structural studies of amino acids and peptide bonds. This prediction came before identification of the alpha helix in X-ray diffraction patterns of proteins, much to the chagrin of Bragg, who was director of a laboratory studying protein structure. The helix is found in most proteins and is a fundamental structural element. In the alpha helix, hydrogen bonds are formed between the carbonyl oxygen of one peptide bond and the amide hydrogen of the amino acids located four amino acids away. This

generates a regular helix containing 3.6 amino acids per turn. The side chains of the amino acids extend outward from the helix, and the hydrogen bonds are parallel to the helix axis (Fig. 5.9).

In the beta sheet, a second important structural element of proteins, the polypeptide chains are quite extended (Fig. 5.10). From a top view the peptide backbone is relatively straight, but in a side view the peptide backbone is pleated. The side chains of the amino acids are relatively unconstrained since alternate groups are directed straight up and straight down. The amide hydrogens and the carboxyl groups are directed to either side and are available for hydrogen bonding to another beta sheet lying alongside. This second sheet can be oriented either parallel or antiparallel to the first.

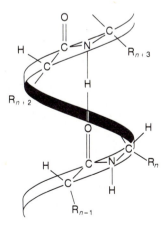

Figure 5.9   A polypeptide chain showing the backbone hydrogen bonds that are formed when the chain is in an alpha helix.

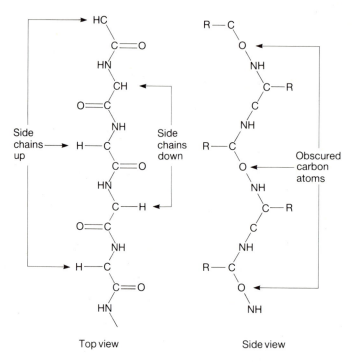

Figure 5.10   A portion of a beta sheet.

The third readily identified secondary structural element is the reverse or beta bend (Fig. 5.11). A polypeptide chain must reverse direction many times in a typical globular protein. The beta bend is an energy-effective method of accomplishing this goal. Four amino acids are involved in a reverse bend.

THERMODYNAMIC CONSIDERATIONS TO PROTEIN STRUCTURE

Thermodynamics provides a useful framework for calculation of equilibrium constants of reactions. This also applies to the "reaction" of protein denaturation. Consider a protein denaturing from a specific conformation, C, to any of a great many nonspecific, random conformations, D. The reaction can be described by an equilibrium constant that relates the amount of the protein found in each of the two states if the system has reached equilibrium.

$$C \rightleftharpoons D,$$

$$K_{eq} = \frac{D}{C}.$$

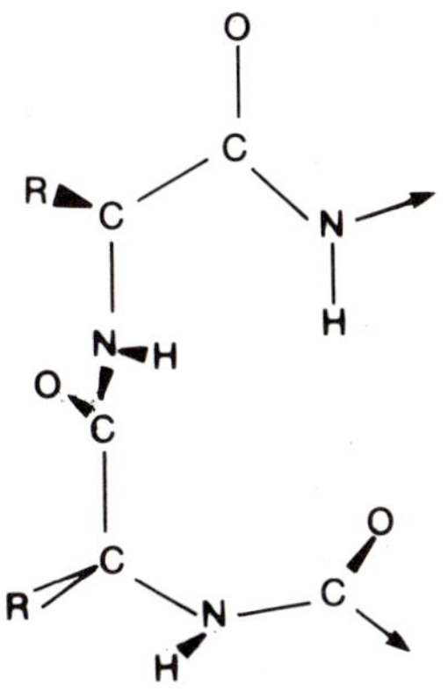

Figure 5.11   Four amino acids forming a beta bend.

Thermodynamics provides a way of calculating K_{eq} as

$$K_{eq} = e^{-\Delta G/RT}$$
$$= e^{-(\Delta H - T\Delta S)/RT},$$

where ΔG is the change in Gibbs free energy; R is the universal gas constant; T is the absolute temperature in degrees Kelvin; ΔH is the enthalpy change of the reaction, which in biological systems is equivalent to binding energy when volume changes can be neglected; and ΔS is the entropy change of the reaction. Entropy measures the number of equivalent states of a system. The state of a protein molecule confined to one conformation without any degrees of freedom possesses much lower entropy than the state of a denatured protein that can adopt any of a great number of conformations all at the same energy. For clarity, we will neglect the contributions of the surrounding water in further considerations, but in physically meaningful calculations these too must be included.

Let us examine why proteins denature when the temperature is raised. If the protein is in the folded state at the lower temperature, K_{eq} is less than 1, that is, $-\Delta G/RT$ is negative. Equivalently,

$$\frac{-\Delta H}{RT} + \frac{T\Delta S}{RT} < 0, \quad \text{or} \quad \frac{\Delta H}{RT} > \frac{\Delta S}{R}.$$

As the temperature increases, neglecting the small temperature-dependent changes that occur in the interaction energies and entropy change, the term $\Delta H/RT$ decreases, eventually becoming less than $\Delta S/R$, and the equilibrium shifts to favor the denatured state.

The temperature dependence of the denaturing of proteins provides the information necessary for determination of ΔH of denaturing. It is very large! This means that ΔS for denaturing is also very large, just as we inferred above, and at temperatures near the denaturing point, the difference of these two large numbers barely favors retention of the structure of the protein. Hence the binding energies of the many interactions that determine protein structure, hydrogen bonds, salt bridges, dispersion forces, and hydrophobic forces just barely overcome the disruptive forces in large part because of the very large number of possible conformations of a denatured protein. Thus we see the value of the peptide bond. If rotations about the C-N bond were not restricted, the increased degrees of freedom available to the protein would be enormous. Then the energy and entropy balance would be tipped in the direction of denatured proteins.

PREDICTION OF PROTEIN STRUCTURE

The tertiary structure a protein will assume in any environment is determined by its amino acid sequence. That is, proteins are capable of folding to their correct conformations without the assistance of any folding enzymes. This is known from the fact that many proteins can be denatured and, under the appropriate conditions, will spontaneously renature or refold to their native conformation. Nature determines the tertiary structure of proteins from their amino acid sequences, but can we? So far the answer is a resounding no.

We can imagine two basic approaches to the prediction of protein structure. The first is simply to consider all the possible atomic interactions in a protein and to determine the state of minimum energy. A closely related approach is to compute a pathway of folding from a denatured state to a state in which the energy is lower than any other easily accessible state. This might more closely resemble the natural problem, for proteins need not be found in their absolutely lowest possible energy state; they merely must be stable for minutes, hours, or days, and the transition from a local energy minimum to the lowest energy state could require a much longer time (Fig. 5.12). Also, considering a pathway of folding from a denatured state may be more realistic than considering just the initial and final folding states because many proteins probably fold up while they are being synthesized. In this case, not all the possible atomic interactions need to be considered since part of the protein folds even before the rest exists. Even with this simplification, the astronomically large

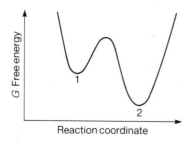

Figure 5.12   A system in state 1 is meta-stable and, when equilibrium is reached, should be found in state 2. Some proteins as isolated in their active form may be in meta-stable states.

numbers of possible conformations that a protein can adopt generally make these types of approaches to structure prediction infeasible.

The second approach is to consider the tertiary structure a code to be broken. General rules in the formulation of the code can then be found by comparing the tertiary structures of many proteins to the amino acids involved. Various degrees of sophistication have been applied to such analyses. Interestingly, most have ended with about the same success in predicting secondary structure and virtually no success in predicting tertiary structure. For simplicity, therefore, we will consider the easiest prediction method based on such analyses.

Chou and Fasman computed structure parameters for each amino acid using 15 proteins whose structures had been determined by X-ray diffraction. These parameters give a probability that an amino acid will be found in alpha helix, beta sheet, or beta turn. For example,

$$P_\alpha(\text{leu}) = \frac{\text{Frequency that leucine is in alpha helix}}{\text{Frequency of all amino acids being in alpha helix}}.$$

The probabilities were calculated for each amino acid in each of the structures alpha helix and beta sheet and in each of the four positions in a reverse or beta turn. Using a table of these probabilities and a few simple rules for combining the probabilities over neighboring amino acid residues, the secondary structures of regions of a protein may be calculated with about 75% reliability. In view of the fact that interactions between nonneighboring amino acids must play a significant role in determining structure, it seems unlikely that further refinement of empirical rules of this nature can significantly improve the predictions. Most likely, a more detailed consideration of the molecular forces will be necessary to predict structures more accurately.

Virtually no success has yet been reported in the prediction of tertiary structures. As mentioned above, the enormous number of possible structures makes evaluation of each one of them an impossible task. Perhaps it will be possible to simplify the problem by solving it as nature does. Most likely the folding of a protein takes place as it is being synthesized, and the folding begins at one or several nucleation points in each domain that are particularly stable regions. Once secondary structure has been determined in these, structure can grow outward much like a crystal grows. This would constitute a folding pathway and would vastly reduce the numbers of conformations that would have to be considered.

STRUCTURE OF A DNA-BINDING PROTEIN

Let us examine a protein whose structure has been determined by X-ray diffraction studies. In it we should see examples of many of the topics covered in this chapter in addition to learning one way proteins can bind to DNA. Interestingly, the binding mode used by the protein we will consider, the N-terminal domain of lambda phage repressor, is used by a sizable fraction of proteins that bind to DNA. In general, each

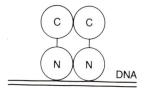

Figure 5.13   Subunit and domain structure of lambda repressor. N and C refer to domains from the N-terminal and C-terminal portions of the polypeptide chain.

of the subunits of these dimeric DNA-binding proteins possesses a pair of alpha helices that make specific contacts with the major groove of the DNA. The two DNA-binding portions of the dimeric protein are separated by about 34 Å, just the distance between adjacent major grooves on one face of the DNA duplex. This means that the most important interactions between these proteins and DNA occur via contacts made on one side of the DNA molecule in contrast to contacts made by wrapping the protein around the cylindrical DNA molecule.

Lambda repressor could not easily be crystallized, but this turned out not to be a

Figure 5.14   A stereo diagram showing the polypeptide backbone structure of the N-terminal dimer of lambda repressor bound to DNA. Helices 2 and 3, which contact the DNA, are darkened. Alongside is shown a sketch of the same complex. The three-dimensional illusion can be generated by fusing the central pair of images that are formed by staring at the drawing as though looking at a distant object. Alternatively, with tubes, restrict the left eye to the left image and the right eye to the right image. Fuse and focus the resulting images. (From The Operator-Binding Domain of Lambda Repressor: Structure and DNA Recognition, C. Pabo, M. Lewis, Reprinted by permission from Nature 298, No. 5873, 443–447. Copyright © 1982 Macmillan Journals Limited.)

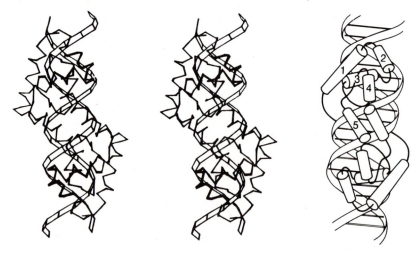

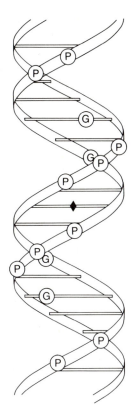

Figure 5.15   The phosphate and guanine contacts made by lambda repressor as it binds to DNA.

problem in the analysis of the structure of lambda repressor. Pabo had shown that each subunit in the dimeric repressor contains two almost independent domains (Fig. 5.13). The N-terminal domain binds to DNA and has only weak interactions with the other N-terminal domain. The C-terminal domain of one monomer tightly binds to the C-terminal domain of another monomer and is mainly responsible for maintaining the dimeric structure of the intact lambda repressor. Upon protease digestion of repressor, the first cleavages occur between the domains. Digestion can be stopped at this point and a 92–amino acid N-terminal domain can be purified and subsequently crystallized.

The N-terminal domain of repressor crystallizes in a dimeric form in which each monomer contains five regions of alpha helix and no regions of beta sheet (Fig. 5.14). The helices enclose a hydrophobic pocket as well as participate in the weak inter-chain interaction. Most interesting, however, was the observation that each subunit contains a region of two protruding alpha helices. These regions are separated by 34 Å, are parallel to one another, and are inclined at an angle just right for fitting into two adjacent major grooves of B form DNA. Such a binding is exactly consistent with chemical modification experiments that had shown that, except for two interactions on the back of the DNA generated by two arms that reach around, lambda repressor makes the majority of its contacts on one side of the DNA (Fig. 5.15). Through this structure, repressor has a shape complementary to DNA and it places a number of amino acids in positions suitable for making contacts with specific hydrogen bond donors or acceptors on the edges of the bases protruding into the major groove of the DNA. In this way the repressor can locate its unique binding site along the DNA.

PROBLEMS

5.1. Which amino acid side chains can serve as hydrogen bond donors and/or acceptors?

5.2. The discussion of the thermodynamics of proteins assumed that protein structures are determined by equilibrium constants and not rate constants. Some proteins spontaneously denature or become inactive. What does this mean about the assumption?

5.3. Hydrophobic forces become stronger with increasing temperature over limited temperature ranges such as 0° to 50°. How can this be? [*Hint:* Consider the thermodynamic relations

$$K = e^{-\Delta G/RT}$$
$$= e^{-(\Delta H - T\Delta S)/RT}.$$

What do you conclude about oligomeric proteins that dissociate into subunits on cooling?

5.4. Why might secondary structure predictions be expected to be more successful with small proteins than with large proteins?

5.5. Why can't proline occur within an alpha helix?

5.6. Suppose a protein contains about 300 amino acids that are mostly in alpha helix form. Estimate a likely value for the number of times the polypeptide chain would make sharp bends so that the protein would be roughly spherical.

5.7. What amino acid side chains might make specific hydrogen bonds within the major groove of DNA to permit detection of specific DNA sequences by proteins?

5.8. In the next chapter we will see that proteins are synthesized from their N-terminal amino acid to their C-terminus. What is the likely significance of the finding that secondary structure predictions are better for the N-terminal half of many proteins?

5.9. Why can't subunits of an oligomeric protein be related to one another by reflections across a plane or inversion through a point as can atoms in crystals?

5.10. Repressor proteins must respond to their environment and must dissociate from their operators under some conditions. However, the DNA-binding domains of repressors constructed like lambda repressor protein could be completely isolated from the domains involved in the binding of small-molecule inducers. Indeed, suppose that the binding of a small-molecule inducer was shown not to alter the structure of the DNA-binding domains at all. What then could cause such a repressor to dissociate from the DNA?

5.11. Roughly, proteins possess a density of 1.3 g/cm^3. Suppose a protein is a dimer with each monomer containing about 300 amino acids. Approximately what overall shape would you expect the protein to possess if when it binds to DNA it makes contact in three adjacent major grooves of the DNA all located on one side of the helix? That is, the protein does not reach around the cylinder to make contacts on the back. Can the protein be spherical or must it be highly elongated?

RECOMMENDED READINGS

The Operator-Binding Domain of Lambda Repressor: Structure and DNA Recognition, C. Pabo, M. Lewis, Nature *298*, 443–447 (1982).
Design, Synthesis and Characterization of a 34-Residue Polypeptide That Interacts with Nucleic Acids, B. Gutte, M. Daümigen, E. Wittschieber, Nature *281*, 650–655 (1979).

RELATED REVIEWS, BOOKS, AND ARTICLES

Protein Crystallography, T. Blundell, L. Johnson, Academic Press, New York (1976).
Empirical Predictions of Protein Conformation, P. Chou, G. Fasman, Ann. Rev. Biochem. *47*, 251–276 (1978).
Principles of Protein Structure, G. Schulz, R. Schirmer, Springer-Verlag, New York (1978).
Biophysical Chemistry, Part I, C. Cantor, P. Schimmel, Freeman, San Francisco (1980).
The Anatomy and Taxonomy of Protein Structure, J. Richardson, Advances in Protein Chemistry *34*, 167–339 (1981).
Protein Purification: Principles and Practice, R. Scopes, Springer-Verlag, New York (1983).
Preparation and Analysis of Protein Crystals, A. McPherson, Wiley, New York (1982).
Biochemistry, G. Zubay, Addison-Wesley, Reading, Mass. (1983).

DEEPER READING

The Theory of Interallelic Complementation, F. Crick, L. Orgel, J. Mol. Biol. *8*, 161–165 (1964).
Handedness of Crossover Connections in β Sheets, J. Richardson, Proc. Natl. Acad. Sci. USA *73*, 2619–2633 (1976).
On the Conformation of Proteins: The Handedness of the β-Strand-α-Helix-β-Strand Unit, M. Sternberg, J. Thornton, J. Mol. Biol. *105*, 367–382 (1976).
The Taxonomy of Protein Structure, M. Rossmann, P. Argos, J. Mol. Biol. *109*, 99–129 (1977).
Logical Analysis of the Mechanism of Protein Folding, IV. Super-Secondary Structures, K. Nagano, J. Mol. Biol. *109*, 235–250 (1977).

β-Turns in Proteins, P. Chou, G. Fasman, J. Mol. Biol. *115*, 135–175 (1977).

Dynamics of Folded Proteins, J. McCammon, B. Gelin, M. Karplus, Nature *267*, 585–590 (1977).

The *lacI* Gene: Its Role in *lac* Operon Control and Its Use as a Genetic System, J. Miller, in *The Operon*, ed. by J. Miller, W. Reznikoff, Cold Spring Harbor Laboratory (1978).

Prediction of β-Turns, P. Chou, G. Fasman, Biophys. J. *26*, 367–384 (1979).

Side-Chain Torsional Potentials: Effect of Dipeptide, Protein, and Solvent Environment, B. Gelin, M. Karplus, Biochemistry *18*, 1256–1268 (1979).

Structure of the *cro* Repressor from Bacteriophage Lambda and Its Interaction with DNA, W. Anderson, D. Ohlendorf, Y. Takeda, B. Matthews, Nature *290*, 754–758 (1981).

Procedure for Production of Hybrid Genes and Proteins and Its Use in Assessing Significance of Amino Acid Differences in Homologous Tryptophan Synthetase α Polypeptides, W. Schneider, B. Nichols, C. Yanofsky, Proc. Natl. Acad. Sci. USA *78*, 2169–2173 (1981).

Thermodynamic and Kinetic Examination of Protein Stabilization by Glycerol, K. Gekko, S. Timasheff, Biochemistry *20*, 4677–4686 (1981).

Structure of the DNA-Binding Region of *lac* Repressor Inferred from Its Homology with the *cro* Repressor, B. Matthews, D. Ohlendorf, W. Anderson, Y. Takeda, Proc. Natl. Acad. Sci. USA *79*, 1428–1432 (1982).

Structural Similarity in the DNA Binding Domains of Catabolite Gene Activator and Cro Repressor Proteins, T. Steitz, D. Ohlendorf, D. McKay, W. Anderson, B. Matthews, Proc. Natl. Acad. Sci. USA *79*, 3097–3100 (1982).

A Simple Method for Displaying the Hydropathic Character of a Protein, J. Kyte, R. Doolittle, J. Mol. Biol. *157*, 105–132 (1982).

The N-Terminal Arms of Lambda Repressor Wrap Around the Operator DNA, C. Pabo, W. Krovatin, A. Jeffrey, R. Sauer, Nature *298*, 441–443 (1982).

Homology Among DNA-Binding Proteins Suggests Use of a Conserved Super-Secondary Structure, R. Sauer, R. Yocum, R. Doolittle, M. Lewis, C. Pabo, Nature *298*, 447–451 (1982).

The Molecular Basis of DNA-Protein Recognition Inferred from the Structure of *cro* Repressor, D. Ohlendorf, W. Anderson, R. Fisher, Y. Takeda, B. Matthews, Nature *298*, 718–723 (1982).

Structure of Catabolite Gene Activator Protein at 2.9 Å Resolution: Incorporation of Amino Acid Sequence and Interactions with Cyclic-AMP, D. McKay, I. Weber, T. Steitz, J. Biol. Chem. *257*, 9518–9524 (1982).

Protein Dynamics in Solution and in a Crystalline Environment: A Molecular Dynamics Study, W. Gunsteren, M. Karplus, Biochem. *21*, 2259–2273 (1982).

Ion-pairs in Proteins, D. Barlow, J. Thornton, J. Mol. Biol. *168*, 867–885 (1983).

The 3 Å Resolution Structure of a D-galactose-binding Protein for Transport and Chemotaxis in *Escherichia coli*, N. Vyas, F. Quicho, Proc. Natl. Acad. Sci. USA *80*, 1792–1796 (1983).

Modular Structural Units, Exons, and Function in Chicken Lysozyme, M. Go, Proc. Natl. Acad. Sci. USA *80*, 1964–1968 (1983).

Cocrystals of the DNA-Binding Domain of Phage 434 Repressor and a Synthetic Phage 434 Operator, J. Anderson, M. Ptashne, S. Harrison, Proc. Nat. Acad. Sci. USA *81*, 1307–1311 (1984).

Specific Amino Acid Substitutions in Bacterioopsin: Replacement of a Restriction Fragment in the Structural Gene by Synthetic DNA Fragments Containing Altered Codons, K. Lo, S. Jones, N. Hackett, H. Khorana, Proc. Nat. Acad. Sci. USA *81*, 2285–2289 (1984).

Protein Synthesis

Now that we have studied the synthesis of DNA and RNA and the structure of proteins, we are prepared to examine the process of protein synthesis. We will first be concerned with the actual steps of protein synthesis. Then, to further develop our understanding of cellular processes, we will discuss the rate of peptide elongation, how cells direct specific proteins to be located in membranes, and how the level of ribosomes in cells is regulated in order to use most efficiently the limited cellular resources. For purposes of this chapter we will take transfer RNA, tRNA, and ribosomes as preexisting cellular components. These molecules are themselves sufficiently complicated and interesting that their own structure and biosynthesis is discussed in later chapters.

In outline, the process of protein synthesis is as follows. Amino acids are activated for protein synthesis by amino acid synthetases which attach the amino acids to their cognate tRNA molecules. The smaller ribosomal subunit and then the larger ribosomal subunit attach to messenger RNA near the initiating codon. Translation then begins at an initiation codon with the assistance of initiation factors. During the process of protein synthesis, the activated amino acids to be incorporated into the peptide chain are specified by three-base codon-anticodon pairings between the messenger and aminoacyl-tRNA. Elongation of the peptide chain terminates on recognition of one of the three termination codons, the ribosomes and messenger dissociate, and the newly synthesized peptide is released.

The actual rate of peptide elongation in bacteria is just sufficient to keep up with transcription; a ribosome can initiate translation immediately behind an RNA polymerase molecule and keep up with the transcription. In eukaryotic cells, however, the messenger is modified and transported from the nucleus to the cytoplasm before it can be translated.

Although the cytoplasm of bacterial cells contains most of the cell's protein, the inner membrane, the periplasmic space, and even the outer membrane contain appreciable protein. Eukaryotic cells also must direct some proteins to organelles and membranes. How do cells do this? One mechanism uses a "signal peptide," the N-terminal 20 amino acids or so of a protein that appear largely responsible for directing the protein away from the cytoplasm into or through the membrane.

Finally, it is necessary to discuss the regulation of the level of ribosomes. Since ribosomes constitute a large fraction of a cell's total protein and RNA, only as many ribosomes as are needed should be synthesized. Consequently, complicated mechanisms have developed coupling protein synthesis to ribosome synthesis.

A.   Chemical Aspects

ACTIVATION OF AMINO ACIDS DURING PROTEIN SYNTHESIS

In Chapter 5 we learned that proteins possess a definite sequence of amino acids that are linked by peptide bonds. Aminoacyl-tRNA participates in the process of protein synthesis in two ways. It brings activated amino acids to the reaction as well as serving as an adapter between the three-base codons in the messenger RNA and the actual amino acids to be incorporated into the growing polypeptide chain (Fig. 6.1). Formation of a peptide bond is energetically unfavorable and is assisted at the step of bond formation by the energy in the amino acid–tRNA bond. This is an ester to the 3' hydroxyl at the invariant —C—C—A end of the tRNA. Aminoacyl-tRNA synthetases, one for each amino acid, form these bonds. The formation of the ester bonds, activation, occurs in two steps. First the enzyme, E, links the amino acid, AA,

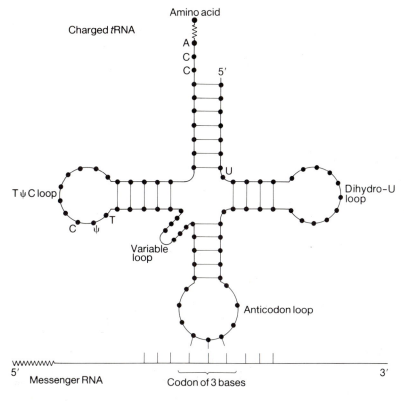

Figure 6.1   The cloverleaf structure of a charged tRNA paired via its anticodon with a codon of the messenger RNA.

to AMP, and then transfers it to the 3′ terminal adenosine of the tRNA (Fig. 6.2).

$$AA + ATP + E \rightleftharpoons E(AMP\text{-}AA) + tRNA + PP \rightleftharpoons E + AMP + AA\text{-}tRNA$$

Some synthetases activate the 2′-hydroxyl of the terminal base of tRNA, some activate the 3′, and some activate both the 2′ and 3′ hydroxyls, but the differences probably do not matter because after the aminoacyl-tRNA is released from the enzyme, the aminoacyl group on the tRNA migrates back and forth.

The aminoacyl-tRNA synthetases are remarkable enzymes since they recognize amino acids and their cognate tRNA molecules and join them together. Inaccuracies in either recognition process could be highly deleterious because choosing the wrong amino acid or the wrong tRNA would ultimately yield a protein with an incorrect sequence. We know, however, that protein synthesis is highly accurate. From measurements on peptides highly purified from proteins of known sequence, the overall amino acid misincorporation frequency is less than about 1/2000.

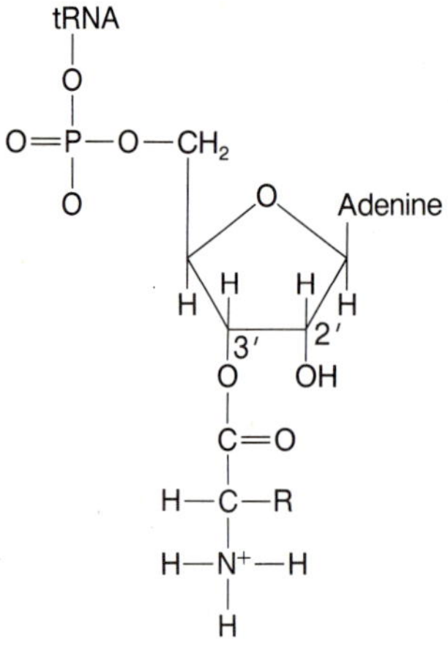

Figure 6.2   The structure of the aminoacyl-ated adenine at the 5' end of a charged tRNA molecule.

Let us first consider the process of choosing the correct amino acid. The greatest difficulty in accurate translation appears to be in discriminating between two highly similar amino acids. Valine and isoleucine are an example since replacing a hydrogen on valine with a methyl group yields isoleucine (Fig. 6.3). The valyl-tRNA synthe-tase probably does not have trouble in discriminating against isoleucine because isoleucine is larger than valine and probably does not fit into a pocket on the enzyme. The reverse situation is more of a problem. Valine will form all of the contacts to the enzyme that isoleucine can form except for those to the missing methyl group. How much specificity could the absence of these contacts provide? Estimates of the differences in binding energy predict about a 200-fold discrimination, but since the actual error rate is found to be much lower, something in addition to binding energy must contribute to specificity. Editing by the synthetase provides the additional accuracy.

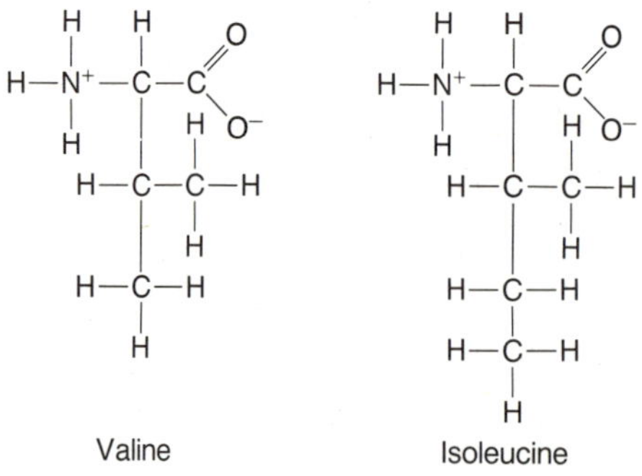

Figure 6.3   The structures of valine and isoleucine.

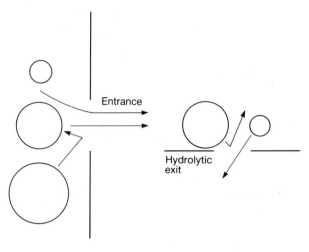

Figure 6.4   A schematic of the double-sieving action of some aminoacyl-tRNA synthetases.

Although isoleucyl-tRNA synthetase can form a valyl adenylate complex, the addition of tRNA^{ile} permits activation of the tRNA^{ile} followed by rapid hydrolysis.

$$E^{ile} + Val + ATP \rightleftharpoons E^{ile}(Val\text{-}AMP) + PP$$

$$E^{ile}(Val\text{-}AMP) + tRNA^{ile} \rightleftharpoons (Val\text{-}tRNA^{ile}) \cdot E^{ile} + AMP$$

$$(Val\text{-}tRNA^{ile}) \cdot E^{ile} \rightleftharpoons Val + tRNA^{ile} + E^{ile}$$

One way to think of this process is that activation is a two-step sieving process (Fig. 6.4). It permits the correct amino acid and similar but smaller amino acids to be activated. Then all amino acids smaller than the correct amino acid have a hydrolytic pathway available for removal of the misacylated amino acid.

A second problem in specificity would appear to be in the selection of the tRNA molecule by the synthetase. In principle this selection could be done by reading the anticodon of the tRNA. A wide variety of experiments have revealed that, although the anticodon is often involved, it is not the sole determinant in the selection of the tRNA, and occasionally it is not involved at all.

DECODING THE MESSAGE

What decodes the messages? On one hand it is apparent that base pairing between a codon of message and an anticodon of the aminoacyl-tRNA decodes the message. However, the aminoacyl-tRNA synthetases are just as important in decoding, for it is also essential that the tRNA molecules be charged with the correct amino acid in the

Table 6.1   The Genetic Code

First Position, 5' End	Second Position				Third Position
	U	C	A	G	
U	Phe	Ser	Tyr	Cys	U
	Phe	Ser	Tyr	Cys	C
	Leu	Ser	End (ochre)	End	A
	Leu	Ser	End (amber)	Trp	G
C	Leu	Pro	His	Arg	U
	Leu	Pro	His	Arg	C
	Leu	Pro	Gln	Arg	A
	Leu	Pro	Gln	Arg	G
A	Ile	Thr	Asn	Ser	U
	Ile	Thr	Asn	Ser	C
	Ile	Thr	Lys	Arg	A
	Met	Thr	Lys	Arg	G
G	Val	Ala	Asp	Gly	U
	Val	Ala	Asp	Gly	C
	Val	Ala	Glu	Gly	A
	Val	Ala	Glu	Gly	G

first place because the ribosome pays no attention to the correctness of the tRNA charging.

Once the amino acids have been linked to their cognate tRNAs, the process of protein synthesis shifts to the ribosome. The "code" is the correspondence between the triplets of bases in the codons and the amino acids they specify (Table 6.1). In a few exciting years molecular biologists progressed from knowledge that there must be a code, to learning that each amino acid is encoded by three bases on the messenger, to actually determining the code. The history and experiments of the time are fascinating, but we cannot discuss them here. Interesting material can be found in *The Eighth Day of Creation* by Horace Freeland Judson and in *Biochemistry* by Lubert Stryer.

In the later stages of solving the code it became apparent that the code possessed certain degeneracies. Of the 64 possible three-base codons, 61 are used to specify the

20 amino acids. From one to six codons may specify a particular amino acid. As shown in Table 6.1, synonyms generally differ in the third base of the codons. In the third position, U is equivalent to C and, except for methionine and tryptophan, G is equivalent to A.

With the study of purified tRNAs, several facts became apparent. First, tRNA contains a number of unusual bases, one of which is inosine, occasionally found in

Inosine   Guanosine

the first position of the anticodon. Second, more than one species of tRNA exists for most of the amino acids. Remarkably, however, the different species of tRNA for any amino acid all appear to be charged by the same synthetase. Third, strict Watson-Crick base pairing is not always followed in the third position of the codon (Table 6.2). Apparently the third base pair of the codon-anticodon complex permits a variety of base pairings (Fig. 6.5). This phenomenon is called wobble. Except for mitochondria, the genetic code and the wobble phenomenon are universal over living things. Mitochondria use a slightly different genetic code, which, by the

Table 6.2   Wobble Base Pairings

First Anticodon Base	Third Codon Base
C	G
A	U
U	A or G
G	U or C
I	U, C, or A

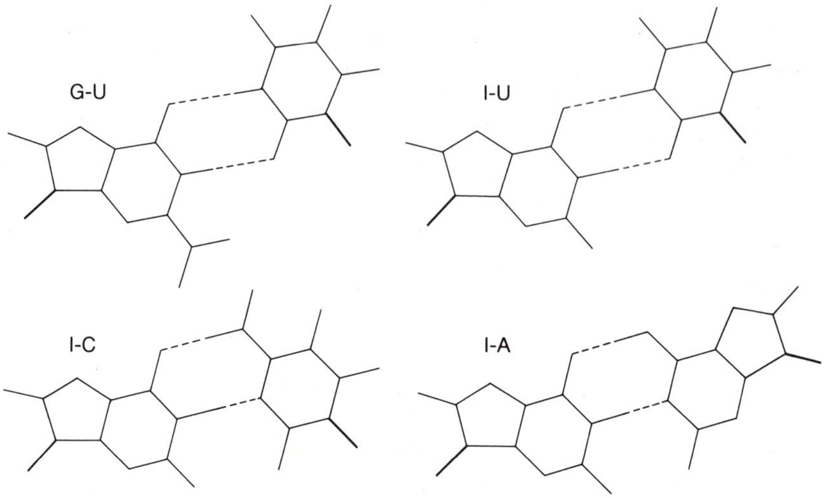

Figure 6.5   Structure of the wobble base pairs.

elimination of a few codons and slightly different-reading rules, permits them to translate all the codons used with only 22 different species of tRNA.

RIBOSOMAL RNA BASE PAIRS WITH BACTERIAL MESSAGE

Ribosomes must recognize a start codon, AUG or GUG, on the messenger RNA to initiate protein synthesis. Characterization of the proteins that are synthesized when the *lac* operon or other operons are induced show that normally only one AUG or GUG of a gene is utilized to initiate protein synthesis. Many of the internal AUG or GUG codons are not used to initiate protein synthesis. Something in addition to the initiating codon itself must signal the point at which translation begins.

Studies of bacterial translation have shown that the first step in initiation is the binding of messenger to the smaller of the two ribosomal subunits, the 30S subunit. Initially it seemed most plausible that a specific sequence on the messenger RNA lying in front of the initiation AUG would be involved. This sequence would be recognized by one of the proteins required for initiation of translation or by one of the ribosomal proteins of the 30S ribosomal subunit. Such a specific protein-RNA interaction was expected to involve relatively inflexible contacts and therefore was expected to utilize a highly conserved sequence in the messenger RNAs. However, the determination of the nucleotide sequence lying in front of a number of genes failed to reveal such a conserved sequence. Instead, a rather weakly conserved sequence was found. At the time this seemed to argue against a specific protein-mRNA interaction, for flexibility in such interactions as are shown by RNA polymer-ase-promoter binding were not then known.

The absence of a strictly conserved sequence preceding start codons suggested that whatever first bound the messenger to the 30S subunit was likely to be an RNA-RNA interaction between mRNA and ribosomal RNA. The originators of this idea, Shine and Dalgarno, were so confident of their proposal that they were not dissuaded by the fact that the then known sequence at the 3′ end of the 16S rRNA which is found in the 30S subunits provided little support for their idea. They repeated the sequencing of the 3′ end of the 16S RNA and discovered that, indeed, the published sequence was incorrect. The correct sequence provided strong support for their idea.

Approximately ten bases before an AUG that serves as a start codon, bacterial messengers contain a sequence that base pairs well with a region near the 3′ end of the 16S ribosomal RNA. This region has been examined in about 70 messenger start

```
                        3′-OH                      5′
        16S rRNA    A U U C C U C C A C U A G · · · ·
                        ‖ ‖‖ ‖‖ ‖
        lacZ mRNA · · · A G G A A A C A G C U A U G · ·
                    5′                              3′
```

sequences; it averages about five base pairs long and is centered about ten nucleotides ahead of the start codon.

EXPERIMENTAL SUPPORT FOR THE SHINE-DALGARNO HYPOTHESIS

Since the time of the original proposal by Shine and Dalgarno, three lines of evidence have provided firm support for the idea that the 3′ end of the 16S rRNA base pairs with a four- to seven-base stretch of the mRNA lying ahead of the translation initiation site. The first line of evidence is the fact that flooding an *in vitro* protein synthesis system with a polynucleotide of sequence very similar to that which is found preceding the AUG of many messengers inhibits mRNA from binding to the ribosome. Presumably this inhibition results from the binding of the polynucleotides near the end of the 16S rRNA, thereby blocking the ability of the ribosomes to bind properly to messenger.

The second line of evidence for the Shine-Dalgarno proposal is direct physical evidence for base pairing between the messenger and the 3′ end of 16S ribosomal

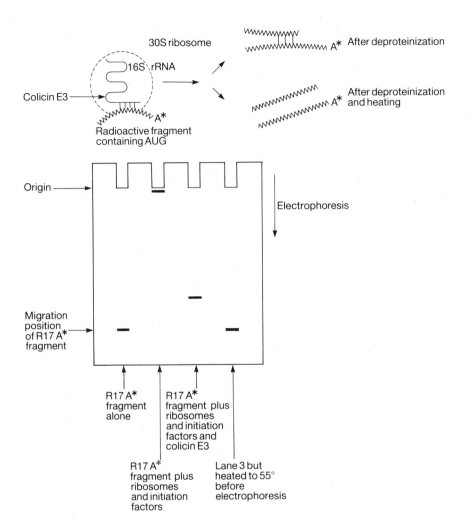

Figure 6.6   (a) The experiment to demonstrate base pairing between the 3′ end of 16S rRNA and messenger RNA. (b) Lanes of a polyacrylamide gel used to separate the messenger, a fragment of R17, and the colicin-cleaved 3′ end of 16S rRNA. Before electrophoresis of all samples, protein was denatured by addition of sodium dodecyl sulfate. When an initiation complex formed, it barely migrated, but if the initiation complex was then cut with colicin E3, the fragment migrated more slowly than fragment alone due to its association with the 40-base piece of rRNA. However, when the hybrid was dissociated by heating, the fragment migrated at its normal rate.

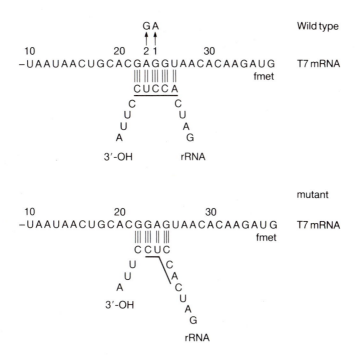

Figure 6.7   The base pairings possible between normal phage T7 messenger and 16S rRNA. Mutation 1 reduces the binding, but mutation 2, a revertant, restores the ability to base pair with the rRNA, but it pairs in a different position.

RNA (Fig. 6.6). This experiment used the ability of a bacteriocidal agent released from some strains of bacteria. The agent, called colicin E3, kills sensitive cells by inactivating their ribosomes by cleaving their 16S rRNA molecules at a position 40 bases from the 3' end.

Jakes and Steitz first bound a fragment of phage R17 messenger *in vitro* to ribosomes in an initiation complex and then cut the ribosomal RNA by the addition of colicin E3. They demonstrated base pairing between the messenger and the end 40 bases of ribosomal RNA by coelectrophoresis of the fragment of R17 RNA with the fragment of 16S rRNA. Repeating the experiment but preventing formation of the mRNA-ribosome initiation complex prevented formation of the hybrid between the two RNAs.

The third line of evidence indicating the correctness of the Shine-Dalgarno hypothesis is from genetics. A base change in phage T7 messenger in the rRNA base-pairing region reduced the translation efficiency of the messenger (Fig. 6.7). The real proof, however, came with the isolation of a revertant of this mutation in which a second base change occurred in the same region. The revertant created a

new, slightly different Shine-Dalgarno sequence, but one that restored high translation efficiency to the messenger.

EUKARYOTIC TRANSLATION BEGINS WITH THE
FIRST AUG

Examination of the sequences preceding the initiation codons in eukaryotic messengers has not revealed extensive homologies to the 18S RNA from the smaller ribosomal subunit. Apparently base pairing between messenger and the ribosomal RNA
is not as important for formation of the initiation complex as it is for bacterial RNAs.
Instead, in the majority of cases it has been found that the *first* AUG of the messenger
is used for initiating translation. The translation efficiency of most, but not all,
messengers in eukaryotic systems is much higher when it contains the cap, as
discussed in Chapter 4. Hence it appears as if translation of most eukaryotic messengers begins with the 40S ribosomal subunit binding to the cap or, less efficiently, to
the 5′ end of the mRNA. The subunit then slides to the first AUG where the large,
60S ribosomal subunit is added and translation begins.

TRICKING THE TRANSLATION MACHINERY
INTO INITIATING

Once messenger has bound to the smaller ribosomal subunit and then a larger
subunit has been added, translation can begin. *Escherichia coli* proteins initiate with
a peptide chain analog in which part of the initiating amino acid looks like a peptide
bond (Fig. 6.8). Perhaps this permits the system to form the first real peptide bond
with essentially the same machinery as it uses to form subsequent peptide bonds.

The initiation analog is N-formyl methionine. The formyl group is added only
after methionine has been transferred to one specialized species of $tRNA^{met}$ called
$tRNA_f^{met}$. Methionine bound to the second $tRNA^{met}$ that is found in cells cannot be
formylated. This scheme appears to prevent the difficulties that would arise from
attempting to initiate a protein with an unformylated methionine or attempting to
put a formylated methionine on the interior of a protein. Now the question is, How
are the two methionine tRNAs distinguished during translation? The AUG codon is
used both at the beginning of proteins and in the interior of proteins.

The met-tRNAs are distinguished by a set of proteins used in the initiation and
elongation processes. These proteins carry the charged tRNA molecules into the
ribosome (Fig. 6.9). The protein IF_2, initiation factor 2, carries the f-met-$tRNA_f^{met}$ into
the site normally occupied by the growing peptide chain, the P site, whereas all other
charged tRNAs, including met-tRNA, are carried into the other site, the A site, by the
elongation factors. Thus N-formyl methionine can be incorporated only at the beginning of a polypeptide.

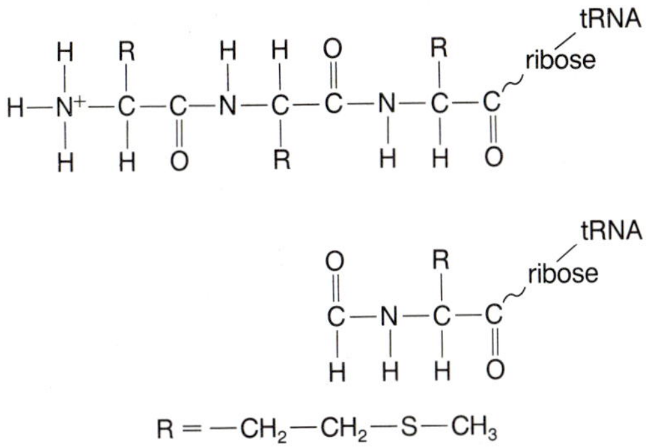

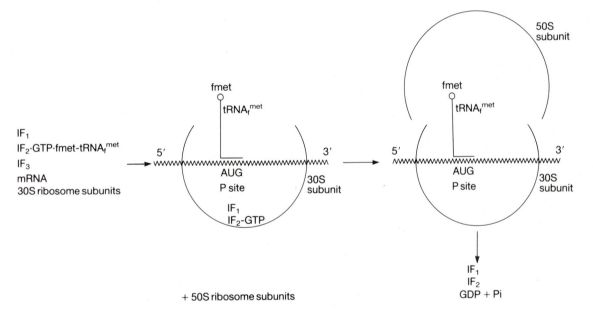

Figure 6.8   The structure of a polypeptide and the similar structure generated by formylation of an amino acid. The initiation amino acid is methionine, whose side group R is shown.

$$R = -CH_2-CH_2-S-CH_3$$

Figure 6.9   Formation of a translation initiation complex.

In addition to the initiation factor IF_2, two other proteins, IF_1 and IF_3, are also used during the initiation steps. Factor IF_1 accelerates the initiation steps but is not absolutely required, and it can be assayed *in vitro* by its acceleration of ^3H-met-tRNA^{met} binding to ribosomes. IF_3 binds to the 30S subunit to assist the initiation process.

The initiation of translation in eukaryotes shows some similarity to the process used in bacteria. Two methionine tRNAs are used, one for initiation and one for elongation, but the methionine on the initiating tRNA is not formylated. However, methionine on this tRNA can be formylated by the bacterial-formylating enzyme. It appears as if the eukaryotic initiation system has evolved from the bacterial system to the point that the formyl group is no longer necessary for protein synthesis, but the tRNA$^{met}_{7f}$ is still sufficiently similar to its progenitor bacterial tRNA$^{met}_f$ that the bacterial-formylating enzymes can recognize and formylate it.

PROTEIN ELONGATION

Although it would appear that the charged tRNAs could diffuse into the ribosome and bind to the codons of mRNA, they are in fact carried into the binding sites on a protein. The protein that serves this function during elongation was originally called Tu (unstable) but is now called EF_1 (elongation factor 1). A rather complex cycle is used for carrying the charged tRNAs to the ribosome A site (Fig. 6.10). First, GTP binds to EF_1, then an aminoacyl tRNA binds, and this complex enters the ribosome A site containing a complementary codon. There GTP is hydrolyzed to GDP, and

Figure 6.10   The cycle by which EF_1 carries GTP and aa-tRNA to the ribosome during protein synthesis.

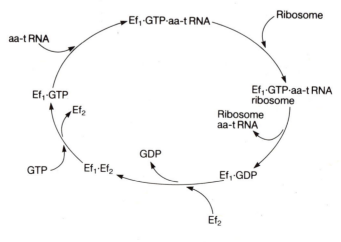

EF$_1$-GDP is ejected from the ribosome. The completion of the cycle takes place in solution. GDP is displaced from EF$_1$-GDP by EF$_2$, which in turn is displaced by GTP. The binding of GDP to EF$_1$ is tight, K_d equals approximately 3×10^{-9} M, and EF$_1$ binds to filters. This permits a simple filter-binding assay to be used to quantitate the protein. Originally the very tight binding for GDP generated confusion because the commercial preparation of GTP contained some GDP that instead was binding.

PEPTIDE BOND FORMATION

Elongation of the polypeptide chain occurs on the ribosome. The growing chain attached to a tRNA occupies the peptidyl, P, site of the ribosome (Fig. 6.11). Alongside is another charged tRNA, also with its anticodon base paired to a codon on the mRNA. This second site is termed the acceptor, A, site since the amino acid here will

Figure 6.11   The process of protein synthesis. (a) The polypeptide occupies the P site, and the incoming aa-tRNA occupies the A site. (b) After formation of the next peptide bond, the polypeptide occupies the A site and the tRNA in the P site is not acylated. (c) The ribosome following translocation, driven by hydrolysis of GTP. The elongated polypeptide now occupies the P site, and the A site is empty, awaiting arrival of EF$_1$ with another aa-tRNA.

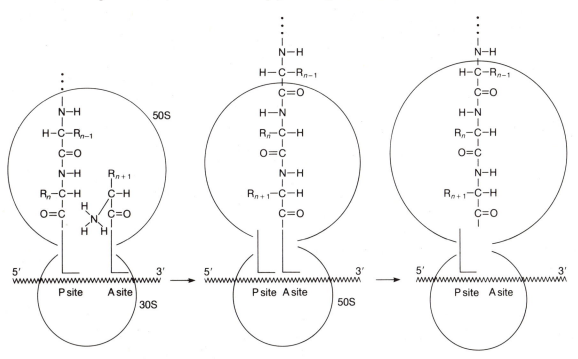

act as an acceptor as the peptide chain is transferred to it. This transfer is catalyzed by the peptidyl transferase, a protein that is an integral part of the 50S ribosomal subunit, and it does not require an external source of energy like GTP.

TRANSLOCATION

Following formation of a peptide bond, the P site of the ribosome contains an uncharged tRNA, and the A site contains a tRNA linked to the growing peptide chain. Translocation is the process of recocking the elongation mechanism. The uncharged tRNA in the P site is ejected, messenger translocates three bases toward the P site, and the tRNA with the peptide chain moves into the P site. The translocation process itself requires hydrolysis of a GTP that has been carried to the ribosome by the EF-G or G factor. Since the E factors are used once for each amino acid added, a large number of molecules of each must be present in the cell to support protein synthesis. It is also logical that their level should parallel the level of ribosomes, and, indeed, as growth rate varies, their levels do keep pace with the levels of ribosomes.

At some time during the growth of the peptide chains the N-terminal group is modified. Approximately 40% of the proteins isolated from *E. coli* are found to begin with methionine, but since all initiate with N-formyl methionine, the remaining 60% must lose at least the N-terminal methionine. Similarly, the 40% of the proteins that do begin with methionine all lack the formyl group. Thus the formyl group must be removed after protein synthesis has initiated. Examination of nascent polypeptide chains on ribosomes shows that the formyl group is missing if they are larger than about 30 amino acids, and therefore the deformylase could well be a part of the

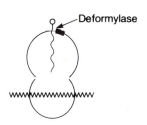

ribosome. It could act when the growing peptide chain is long enough to reach to the enzyme.

The deformylase is a very labile enzyme that is exceedingly sensitive to sulfhydryl reagents. Since many other enzymes isolated from the same cells require the same sulfhydryl reagents for stability, it may be that the deformylase is normally bound to some structure that contributes to its stability, and when it is isolated from extracts and partially purified it is particularly labile in its unnatural environment.

A PARADOX

Hybridization is possible between nucleic acids of complementary sequence. Of course, there is a lower limit to the length of the participating polymers. A single adenine in solution is not normally seen to base pair with a single thymine. The lower length limit for specific hybridization between two nucleic acid molecules in typical buffers containing 0.001 M to 0.5 M salt at temperatures between 10° and 70° is about ten bases. How then can protein synthesis have any degree of accuracy since only three base pairs form between the codon and anticodon?

One part of the answer to how triplet base pairing can provide accuracy in protein synthesis is simply that additional binding energy is provided by contacts other than base pairing. A second part of the answer is that by holding the codon and anticodon rigid and in complementary shapes, binding energy is not consumed in bringing all three bases of the codon and anticodon into correct positions (Fig. 6.12). Once pairing occurs at the first base, the second and third bases are already in position to pair. When two complementary trinucleotides bind, this is not the case. Even after one base pair has formed, the second and third bases of each polymer must be brought into the correct positions before they can base pair. Additional binding energy is consumed in properly orienting them. This is a restatement of the fact that it is difficult to orient molecules properly.

Figure 6.12   Schematic of two complementary trinucleotides in solution and illustrating that base pairing between them produces an ordered structure from the random structures. The anticodons of tRNA molecules are already ordered due to the structure of the entire molecule.

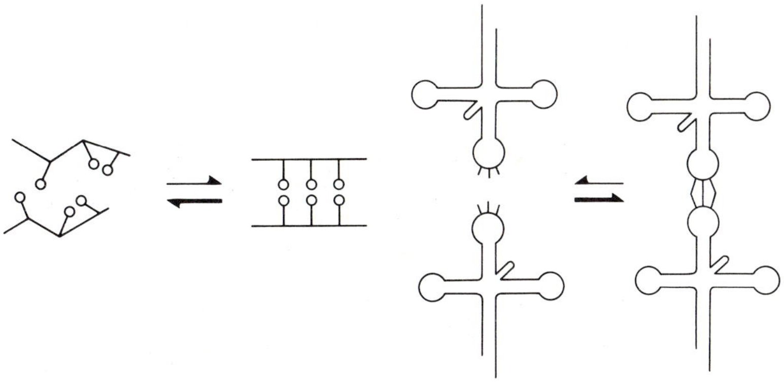

A simple hybridization experiment with tRNA provides a direct demonstration of the consequences of correctly positioning and orienting bases. Two tRNA molecules with self-complementary anticodons will hybridize! Since the structure of the entire tRNA molecule serves to hold the anticodon rigid, the binding energy derived from forming the second and third base pairs need not go into holding these in the proper position. They already are in the correct position. As a result, the two tRNA molecules will hybridize via their anticodons despite the fact that self-complementary trinucleotides will not.

TERMINATION, NONSENSE, AND SUPPRESSION

How is the elongation process ended and the completed polypeptide released from the final tRNA and from the ribosome? The signal for ending the elongation process is any one of the three codons UGA, UAA, and UAG. Of the 64 possible three-base codons, 61 code for amino acids, are "sense"; and 3 code for termination, are "nonsense." In termination, as in the other steps of protein synthesis, specialized proteins come to the ribosome to assist the process. Apparently upon chain termination, the ribosome is unlocked but not immediately released from the messenger. For a short while before it can fully dissociate from messenger it can drift phaselessly forward and backward and can reinitiate translation if it encounters an initiation codon before it dissociates (Fig. 6.13).

Three proteins appear to be involved in termination: a protein factor R_1, which is necessary for termination at UAA and UAG codons; a protein factor R_2, which is

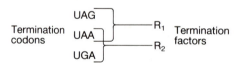

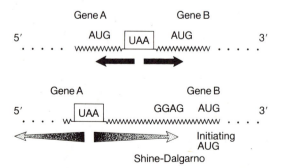

Figure 6.13   Translation termination and reinitiation. (a) A ribosome terminating at the UAA codon could directly reinitiate at either of the nearby AUG codons. (b) A ribosome terminating at the UAA codon would not be able to drift to the AUG codon on the right. A new initiation complex would have to form using the GGAG sequence to bind 16S rRNA.

necessary for termination at UAA and UGA codons; and a protein S, which accelerates the termination process. Cells contain approximately one molecule of R per five ribosomes, a number consistent with the infrequent usage of these molecules.

The existence of chain termination codons is responsible for an interesting phase in the growth of molecular biology. A mutation within a gene can change one of the 61 sense codons into one of the polypeptide chain terminating, nonsense codons. This result shows that only the three bases are necessary to code for chain termination, no others are required, and no special secondary structure of the mRNA is required. As a result of a nonsense codon within a gene, the protein encoded by the mutated gene will be prematurely terminated during translation. Usually the shortened polypeptide possesses no enzymatic activity, and it is frequently degraded by proteases within the cell. Quite surprisingly, however, some bacterial strains were found that could suppress the effects of a nonsense mutation. Although the suppressor rarely restored the levels of the "suppressed" protein to prior levels, the cell often possessed sufficient amounts of the protein to survive. In the case of a nonsense mutation in a phage gene, the suppressor strains permitted the phage to grow and form plaques.

Capecchi and Gussin showed that a suppressing strain inserted a particular amino acid at the site of the nonsense mutation. It did so by "mistranslating" the nonsense codon as a codon for an amino acid. They also showed that the mistranslation resulted from a change in one of the transfer tRNAs for the inserted amino acid. Subsequently, sequencing of suppressor tRNA has shown that, except in a special case, their anticodons have been altered so as to become complementary to one of the termination codons. Two different events can occur when a ribosome reaches a nonsense codon in a suppressing strain. Termination can occur via the normal mechanism, or an amino acid can be inserted into the growing polypeptide chain and translation can proceed.

Suppressors have been found that insert tyr, trp, leu, gln, and ser. The termination codon UAG has come to be called amber and UAA called ochre. No generally used name for the UGA codon exists. Amber-suppressing tRNAs read only the UAG codon, and ochre-suppressing tRNAs read both UAA and UAG codons as a result of the "wobble" in translation. Since the R factors are protein and cannot be constructed like tRNA, it is no surprise that R_2 does not "wobble" and does not recognize the UGG (*trp*) codon.

How do normal proteins terminate in suppressing cells? If a suppressor were always to insert an amino acid in response to a termination codon instead of terminating, then many of the cellular proteins would be fused to other proteins or at least be appreciably longer than usual. The problem of terminating normal proteins could be solved in part by several different termination signals at the end of every gene. Then only the introduction of several different suppressors to a cell could create problems. Not all genes have been found to be ended by tandem translation terminators, however.

The more likely explanation for the viability of nonsense-suppressing strains is that the efficiency of suppression never approaches 100%. Typically it is 10% to

40%. Hence suppression of normal translation termination codons can fuse or lengthen some proteins in a cell, but most terminate as usual. On the other hand, the gene possessing the nonsense mutation would occasionally yield a suppressed instead of terminated protein. A suppression efficiency of 20% could reduce the amount of some cellular proteins from 100% to 80%, relatively speaking, not a substantial reduction. On the other hand, the existence of this suppressor would raise the amount of the suppressed protein from 0% to 20% of normal. Relative to the nonsuppressed level, this is an enormous increase.

B.   Physiological Aspects

MESSENGER RNA IS UNSTABLE

The information for the sequence of amino acids in a protein is carried from the DNA to ribosomes in the messenger RNA. Once the necessary proteins have been synthesized from a messenger, it is necessary that the translation of the messenger cease. In principle, exponentially growing bacterial cells can eliminate unneeded mRNA merely by dilution due to growth of the cells. However, cells that have ceased overall growth, such as eukaryotic cells in a fully grown organism, cannot use this approach.

Complicated mechanisms can be imagined for destruction of messenger once it has been used a fixed number of times. Cells appear not to use strict bookkeeping on translation, however. Once a messenger RNA has been synthesized, it has a fixed probability per unit time of being degraded by nucleases (Fig. 6.14). This random decay process, characterized by a half-life, gives some messenger molecules short lifetimes and others long lifetimes.

Cells that must adapt to a changing environment must vary their enzyme synthesis and consequently possess many messengers with relatively short lifetimes. Most bacterial messengers have half-lives of about two minutes, although some messengers have half-lives of over ten minutes. Some messengers in eukaryotic cells have half-lives of several hours and other eukaryotic messengers have half-lives of several weeks or even longer in stored forms.

PROTEIN ELONGATION RATES

In bacteria the protein elongation rate is about 16 amino acids per second. This value is very close to the corresponding transcription rate of 55 bases per second, so once a ribosome begins translation it can keep up with the RNA polymerase as it tran-

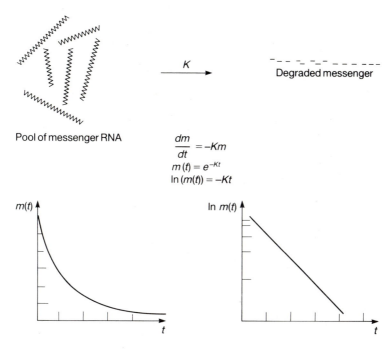

Figure 6.14   Random decay of messenger from a pool characterized by a rate constant K gives rise to an exponential decrease in the amount of messenger in the pool if it is not replenished by new synthesis. Left: Kinetics if plotted on rectilinear coordinates. Right: Kinetics if plotted on semilogarithmic coordinates.

scribes. In several of the better-studied operons, the rate of ribosome attachment is sufficiently fast that the ribosomes are rather closely spaced on the messenger, with about 100 Å of free space between ribosomes. In other operons, however, the spacing is greater.

How is the protein elongation rate measured? One approach would be to use methods analogous to those used to measure RNA elongation rates as described in Chapter 4. However, no good protein initiation inhibitors lacking undesirable side effects exist, and slightly more complicated experiments must be done to measure protein elongation rates. The most general method uses an idea originally developed for measurement of RNA elongation rates before rifamycin was available.

Consider an experiment in which a radioactive amino acid is added to a culture of growing cells. Let us focus our attention on a class of protein of one particular size and consider labeling for intervals that are short compared with the cell doubling time. The number of polypeptide chains in the size class completed after the addition of radioactive label is proportional to the time of labeling. Also, for labeling intervals shorter than the time required to synthesize completely this size of polypeptide chain, the average amount of label incorporated into each chain is proportional to the time that label has been present. Hence during this early period, the total amount

Length of radioactive segment $\alpha\, t$
Number of chains released since
    adding radioactivity $\alpha\, t$
Radioactivity in released chains α Radioactivity per chain
                                                $\times$ Number of chains
                                  $\overline{\phantom{\times \text{Number of chains}}}$
                                                $\alpha\, t^2$

of radioactivity in the particular size class increases in proportion to the number of chains released multiplied by their average radioactivity, both of which are proportional to the time of labeling. Thus the radioactivity in the particular size class of protein increases in proportion to t^2.

After label has been present for the length of time, T, necessary to synthesize completely the polypeptide, the radioactivity per completed peptide chain can no longer increase (Fig. 6.15). After this time, the radioactivity in the particular polypeptide size class can only increase in proportion to the number of chains completed. That is, the radioactivity increase is proportional to t.

For experimental quantitation, it is convenient to compare the radioactivity in a particular size class of polypeptide to the total radioactivity in all sizes of polypeptide. The total amount of radioactivity in all size classes of polypeptide must increase proportional to the time of labeling, t. Therefore the fraction of radioactivity in a particular size class of protein increases in proportion to t until the radioactive label has been present long enough for the entire length of the protein to become radioactively labeled. Thereafter, the fraction of label in the size class remains constant (Fig. 6.16). Hence, determining the transition time from linear increase to being constant, T, provides the synthesis time for the particular size class of polypeptide.

The experimental protocol for performing the elongation rate measurement is

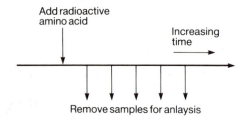

Add radioactive
amino acid

Increasing
time

Remove samples for anlaysis

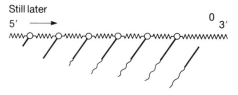

Figure 6.15 Synthesis of a protein by ribosomes traversing a messenger. Ribosomes initiate at the 5′ end of the messenger, and the protein is completed and released at the 3′ end. Addition of radioactive amino acids produces partially labeled polypeptides, indicated by the heavy line, and a completed protein with radioactive amino acids near its carboxy terminus. Longer labeling intervals lead to longer radioactive regions until radioactive amino acids have been present for the length of time to synthesize the complete protein, at which time the released polypeptides are labeled over their entire length.

simply to add a radioactive amino acid to a growing culture of cells. At intervals thereafter, samples are withdrawn and the protein from the entire sample denatured with sodium dodecyl sulfate and electrophoresed on a polyacrylamide gel that provides the requisite size separation of the polypeptides as described in Chapter 4. The results found in *E. coli* and other bacteria are that the protein elongation rate is about 16 amino acids per second when cells are growing at 37°. This value is independent of the length of the polypeptide chain. A similar value is found by measuring the time until appearance of N-terminal label in free completed β-galactosidase or the induction kinetics of β-galactosidase.

Less-satisfactory measurements of protein elongation rates have been made in eukaryotic cells. In reticulocyte cells synthesizing hemoglobin, measurements have been made of times required to label different portions of nascent or completed chains. As these measurements were made at 20° and 25°, their extrapolation to 37°,

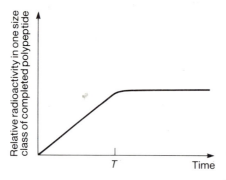

Figure 6.16   The fraction of radioactivity in one size class of polypeptide. The time required to synthesize this peptide is T.

which yields a rate of about 10 amino acids per second, should be considered only an approximation.

THE NEED FOR DIRECTING PROTEINS TO CELLULAR SITES

Cells must direct proteins to several different locations. In addition to the cytoplasm in which most proteins in bacteria are found, proteins are also found in the cell wall. Proteins in eukaryotic cells are found in the cytoplasm and in various cell organelles, as the mitochondria, chloroplasts, lysosomes, and proteins may be excreted as well. Proteins are synthesized in the cytoplasm, but directing them to enter or cross the cell membranes occurs during or shortly after synthesis of the protein. How is this done?

Crucial first observations on protein localization were made on immunoglobulin excretion. There it was observed that ribosomes synthesizing immunoglobulin were bound to the endoplasmic reticulum membranes. Furthermore, when messenger from the membrane-bound ribosomes was extracted and translated *in vitro*, immunoglobulin was synthesized; however, it was slightly larger than the normal protein. This protein possessed about 20 extra amino acids at its N-terminus. When translation was begun *in vivo* and was completed *in vitro*, immunoglobulin of the normal length was synthesized. These observations led Blobel to propose the signal peptide model for excretion of proteins.

THE SIGNAL PEPTIDE MODEL FOR PROTEIN EXCRETION

The signal peptide model utilizes observations on immunoglobulin excretion and has the following parts (Fig. 6.17): first, the N-terminal sequence of a protein to be excreted contains a signal that specifies its transport; second, during translation of messenger coding for an excreted protein, the N-terminal amino acids bind the ribosomes to the membrane; third, during the synthesis of the remainder of the protein, the growing polypeptide chain is directly excreted through the membrane; fourth, often during synthesis the signal peptide is cleaved from the remainder of the protein. This leads to the fifth element of the model: there should exist in or on the membrane a protease to cleave the signal peptide from excreted proteins.

TESTING THE SIGNAL PEPTIDE MODEL

Often it is easier to make observations on eukaryotic cells and then to prove the resulting ideas with experiments done on bacteria. One direct demonstration of concomitant translation and excretion through a membrane was made in bacteria. Chains of a periplasmic protein in the process of being synthesized were labeled by a chemical that was excluded from the cytoplasm by the inner membrane. The experiment was done by adding the labeling chemical to spheroplasts, cells lacking their outer membrane and peptidyl glycan layer, isolating membrane-bound ribosomes, and showing that their nascent polypeptides were labeled.

Does the N-terminal sequence on a protein signal that the remainder of the protein is to be transported into or through a membrane? In principle, this could be

Figure 6.17   The signal peptide model for protein export. As ribosomes begin translation, the hydrophobic signal peptide signals attachment to a site on the membrane. The protein is exported through the membrane as it is being synthesized, and at some point the signal peptide is cleaved.

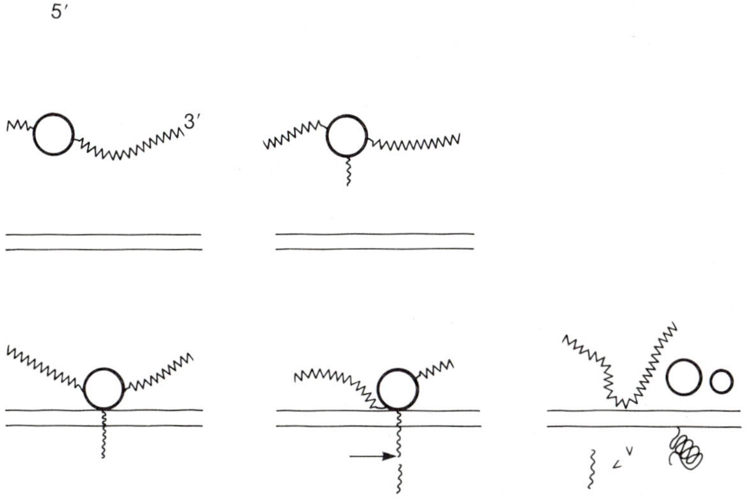

tested by tricking a cell into synthesizing a new protein in which the N-terminal sequence from an excreted protein has been fused to a protein normally found in the cytoplasm. If the hybrid protein is excreted, then the new N-terminal sequence must be signaling export.

Fortunately, *E. coli* has been sufficiently well studied that a number of candidates exist whose N-terminal sequences might be used in such a project. The *malF* gene product, a protein involved in the uptake of maltose into cells and transported through the inner membrane to the periplasmic space, should be an excellent source of an "excretion-coding" N-terminal sequence. The ideal situation to test the excretion hypothesis would be to fuse the N-terminal sequence of *malF* to an easily assayed cytoplasmic protein. One very good candidate for this fusion is β-galactosidase, a protein for which the genetics have also been fully developed.

Remarkably, the fusion of the N-terminal portion of the *malF* gene to β-galactosidase was performed *in vivo* without using recombinant DNA techniques (Fig. 6.18). Through clever genetic manipulations, Silhavy and Beckwith brought the β-galactosidase gene, *lacZ*, near *malF* and then generated a deletion that fused the N-terminal portion of the *malF* gene to β-galactosidase. A postulate of the signal peptide model could therefore be tested. Indeed, a sizable fraction of the β-galactosidase from some of the fusion strains was not located in the cytoplasm but instead was bound to or located within the inner membranes of the cells.

How do you show that a protein is located in the inner membrane of a cell? One

Figure 6.18   Construction of a *malF-lacZ* fusion. The β-galactosidase gene is first brought near the *malF* gene, and then a deletion between the two fuses the N-terminus of *malF* to *lacZ*.

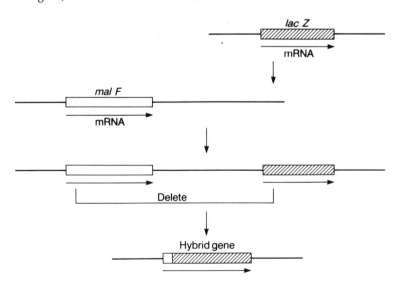

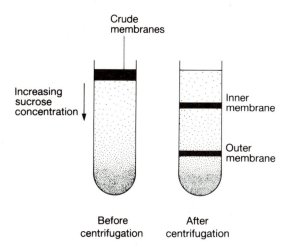

Increasing
sucrose
concentration

Crude
membranes

Inner
membrane

Outer
membrane

Before
centrifugation

After
centrifugation

Figure 6.19   Isopycnic separation of membrane fractions by centrifugation on a preformed sucrose gradient.

method is to disrupt cells, to collect membrane fragments, and then to separate the inner and outer membrane fragments and assay each for the protein. Inner membrane is less dense than outer membrane, and therefore the two can be separated according to density by isopycnic centrifugation in a tube containing a gradient of sucrose concentration (Fig. 6.19). The two membrane fractions sediment down the tube into higher and higher sucrose concentrations until they reach positions where their densities equal the density of the sucrose. There they come to rest. By assay of the fractions collected from such a sucrose gradient, the β-galactosidase from a malF-β-galactosidase fusion could be demonstrated to be located largely in the inner membrane of the cells (Fig. 6.20).

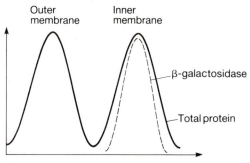

Outer
membrane

Inner
membrane

β-galactosidase

Total protein

Distance from bottom of tube (drop number)

Figure 6.20   Results of isopycnic centrifugation in which outer membrane came to rest near the bottom of the centrifuge tube and inner membrane came to rest closer to the top. The β-galactosidase was associated with the inner membrane fraction.

Is the N-terminal sequence from *malF* actually necessary for export of the β-galactosidase into or through the membrane? Some of the fusions of *malF* to β-galactosidase that were generated in the study yielded proteins that remained in the cytoplasm. Genetic studies showed that the process of producing the fusions in these strains had removed most of the signal peptide of the *malF* gene. Other studies have shown that mutations that abolish export of fusion proteins alter the signal peptide. These experiments show that the signal peptide plays an important role in export. It is clear, however, that the signal peptide is not the whole story. The *malF-lacZ* fusion protein was stuck to the inner membrane. It was not located in the periplasmic space, the final destination of the native *malF* protein. Therefore something in addition to just the leader sequence of *malF* is required for a protein to be vectored to the periplasmic space. In addition, both prokaryotes and eukaryotes appear to stop elongation of the protein after synthesizing the leader, only to resume elongation after binding the ribosome complex to the membrane. Undoubtedly, additional components are involved in these steps.

ALTERNATIVES TO THE SIGNAL HYPOTHESIS

Despite the fact that the signal peptide hypothesis appears to have substantial support, the data discussed above do not constitute proof of the entire hypothesis. For example, nothing excludes the possibility that it is the mere solubility of the

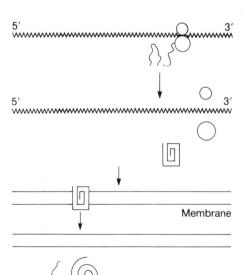

Figure 6.21   An alternative to the signal peptide model for protein export. Following synthesis, the protein adopts a conformation in which it is soluble in the membrane. On cleavage of the signal peptide, the protein adopts a new conformation.

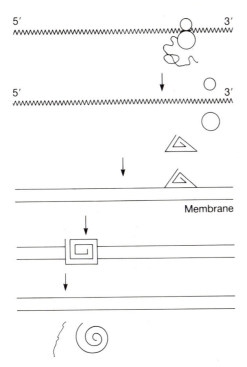

Figure 6.22   The membrane trigger mechanism of protein export. On completion of synthesis, the protein adopts a conformation that is altered to a membrane-soluble form on contact with membrane. Cleavage of the signal peptide leaves the protein in another conformation.

protein in the membrane that specifies its ultimate cellular location (Fig. 6.21). It is possible that the N-terminus is merely the most convenient location to place hydrophobic amino acids that encode membrane solubility. Perhaps a hydrophobic, "membrane solubility" sequence could also be located within the protein or at the C-terminus. An extreme alternative view, but one for which some supporting data exist, is that membrane contact of some exportable proteins triggers them to flip into a membrane-soluble conformation (Fig. 6.22).

EXPECTATIONS FOR RIBOSOME REGULATION

The final topic we consider in this chapter is the regulation of the level of the protein synthesis machinery itself. Not surprisingly, we will find that the synthesis of this machinery is regulated by the intensity of its use.

A ribosome is a large piece of cellular machinery. It consists of about 55 proteins, two large pieces of RNA, and one or two smaller RNAs. Hence it is natural to expect a cell to regulate levels of ribosomes so that they are always used at the highest possible efficiency. Since bacterial cells and some eukaryotic cells may grow with a wide variety of rates, a sophisticated regulation mechanism is necessary to ensure that ribosomes are fully utilized and synthesizing polypeptides at their maximal rate under most growth conditions.

In an earlier section we discussed the finding that proteins in bacteria are elongated at about 16 amino acids per second. In the next sections we will find that the rate of cellular protein synthesis averaged over all the ribosomes is also about 16 amino acids per second. This means that virtually no ribosomes sit idle. All are engaged in protein synthesis.

RIBOSOME LEVELS ARE PROPORTIONAL TO CELL GROWTH RATES

The dramatic ability of bacteria to grow with a wide variety of rates prompts the question of how they manage to maintain balanced synthesis of their macromolecules. In a study of this question, Schaechter, Maaløe, and Kjeldgaard made the discovery that ribosomes are used at constant efficiency, independent of the cell growth rate. To appreciate their contribution fully, it will be helpful first to examine a related question: What is the average rate of protein synthesis per ribosome? As a first step we will estimate this value using typical cellular parameters, then we will calculate this value in a form in which ribosome level and growth rate explicitly enter.

An average bacterial cell with a doubling time of 50 minutes contains about 1×10^{-13} g protein and about 10,000 ribosomes. Approximating the molecular weight of amino acids to be 100,

1×10^{-13} g protein is $1 \times 10^{-13}/10^2 = 10^{-15}$ moles amino acid;

10^{-15} moles amino acid is $6 \times 10^{23} \times 10^{-15} = 6 \times 10^8$ molecules;

10^4 ribosomes polymerize these 6×10^8 amino acids in 50 min or 3×10^3 sec.

Thus the average rate of protein synthesis per ribosome is $6 \times 10^8/3 \times 10^3 = 20$ amino acids per second per ribosome. Compared to the typical turnover number of enzymes, greater than 1,000 per second, this is a low number. However, as we have already seen, the process of addition of a single amino acid to the growing polypeptide chain is complex and involves many steps.

To accurately calculate the rate of protein synthesis per ribosome during steady-state growth, we must include the growth of the cells in the calculation. This can be done in the following way. Define α_r as the relative rate of synthesis of ribosomal protein, that is,

$$\alpha_r = \frac{dP_r/dt}{dP_t/dt}$$

where P_r is ribosomal protein and P_t is total protein. Let $R(t)$ be the number of ribosomes in a culture at time t. Since ribosome number will increase like cell number, $R(t) = R(0)e^{\mu t}$, and, hence, $dR/dt = \mu R(t)$. Also, the rate of ribosome synthe-

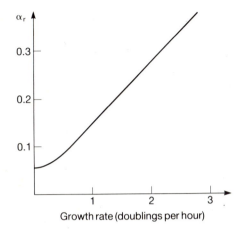

Figure 6.23   The value of α_r as a function of growth rate. Except at slow growth rates, α_r is proportional to growth rate.

sis, dR/dt, equals the rate of ribosomal protein synthesis in amino acids per unit time divided by the number of amino acids in the protein of one ribosome, C:

$$\frac{dR}{dt} = \alpha_r \times \frac{dP}{dt} \times 1/C.$$

Then dP/dt equals the number of ribosomes times the average elongation rate per ribosome, K. That is,

$$\frac{dP}{dt} = KR.$$

Equating these two expressions for dR/dt, we have

$$\frac{dR}{dt} = \mu R(t), \quad \text{and} \quad \frac{dR}{dt} = \frac{\alpha_r K R(t)}{C},$$

which yields $K = \mu C/\alpha_r$, our desired relation. Note that if α_r is roughly proportional to the growth rate, as has been found for bacteria except at the slowest growth rates (Fig. 6.23), then the term μ/α_r is a constant and hence K, average activity of a ribosome, is independent of the growth rate.

For E. coli B/r growing at 37°, with a doubling time of 48 minutes,

$$\mu = \ln 2/Td = 0.693/2.9 \times 10^3 = 2.4 \times 10^{-4} \text{ sec}^{-1},$$

$$C = \frac{\text{Molecular weight of ribosomal protein}}{\text{Average molecular weight of amino acids}}$$

$$= \frac{9.5 \times 10^5}{1.1 \times 10^2} = 8.3 \times 10^3, \quad \text{and}$$

$$\alpha_r = 0.12.$$

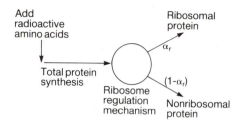

Figure 6.24   Schematic of the ribosomal protein regulation system. The mechanism determines the fraction, α, of total protein synthesis to be devoted to ribosomal protein synthesis.

Therefore

$$K = \frac{2.4 \times 10^{-4} \text{ sec}^{-1} \times 8.3 \times 10^3}{0.12} \text{ amino acids/ribosome}$$

$$= 17 \text{ amino acids per ribosome per second.}$$

This value is close to the elongation rate of polypeptides, showing that most ribosomes in the bacterial cell are engaged in protein synthesis and are not sitting idle.

A measurement of α_r at any time during cell growth can be accomplished by adding a radioactive amino acid to growing cells for a short interval (Fig. 6.24). Then an excess of the nonradioactive form of the amino acid is added. After allowing cells to continue growing until all the radioactive ribosomal proteins have been incorporated into mature ribosomes, the value of α_r is the fraction of radioactivity in ribosomal protein compared to total radioactivity in all the cellular protein. This fraction can be determined by separating ribosomal protein from all other cellular protein and measuring the radioactivity in each sample. Separation of ribosomes from other proteins can most conveniently be performed by agarose gel electrophoresis.

REGULATION OF RIBOSOME SYNTHESIS

We already know that α_r during balanced exponential growth is proportional to the growth rate. Let us consider the response of α_r during a transition between growth rates. Originally this type of data was sought in an effort to place constraints on the possible feedback loops in the ribosome regulation system. By analogy to electronic circuits, the response of the ribosome regulation system could be highly informative. The most straightforward growth rate shift is generated by addition of nutrients supporting faster growth. The resulting response in α_r is a quick increase to the α_r characteristic of the new growth conditions (Fig. 6.25). Superimposed on this shift are small oscillations. The existence of these oscillations shows that a number of cellular components are involved in the regulatory system. The ten-minute period of the oscillations suggests that at least one of the regulating components is not a molecule like ATP that is completely turned over on a time scale appreciably less than ten minutes.

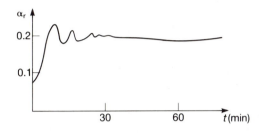

Figure 6.25   The response of α_r to a growth rate increase.

THE STRINGENT-RELAXED PROBLEM OF rRNA REGULATION

In addition to regulating ribosome synthesis in response to differing growth rates, bacterial cells display a second type of ribosome regulation. This is the stringent response, in which the synthesis of ribosomal RNA and tRNA stops if protein synthesis stops. Over the years, study of the stringent-relaxed response of *E. coli* has absorbed much effort, partly because the effect is dramatic. It is always easier to study a large effect because extreme experimental accuracy is unnecessary. A second reason for the interest was the existence of mutants that do not shut off ribosomal and tRNA synthesis on amino acid starvation. These are called relaxed, RC^{rel}, or Rel^- mutants. Their existence facilitated necessary control experiments on the Rel regulation system.

One obvious question about ribosome regulation is whether the stringent response itself is joined to the growth rate response system. Studies have shown that stringent and relaxed cells regulate their ribosome levels identically as a function of growth rate. Consequently, it appears as if these two systems are separate.

The halt in the accumulation of rRNA in amino acid–starved, stringent cells appears to result from decreased initiations by RNA polymerase rather than from degradation of rRNA or blockage of elongation by RNA polymerase. This is not surprising in light of what we have already discussed about regulation of DNA synthesis. One demonstration of this fact is that no radioactive rRNA is synthesized if amino acids, rifamycin, and labeled uridine are simultaneously added to amino acid–starved, stringent cells. The same conclusion is also reached by observing the kinetics of synthesis of the 16S and 23S rRNAs immediately after restoring the missing amino acid to an amino acid requiring strain.

An important clue to the mechanism of the stringent response and perhaps also ribosome growth rate regulation was found by Cashel and Gallant, who discovered that guanosine tetra- and pentaphosphate, ppGpp (Fig. 6.26) and pppGpp, accumulate in amino acid–starved, stringent cells but not in amino acid–starved, relaxed cells. Although accumulation of these compounds, sometimes called magic spots because of their magical appearance on chromatographs of extracts prepared from

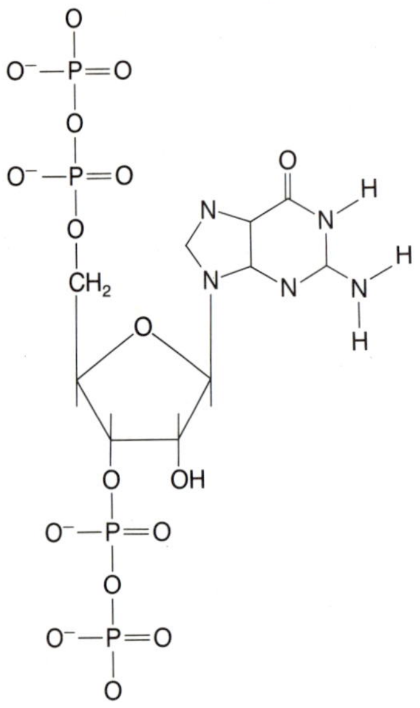

Figure 6.26   The structure of ppGpp.

amino acid–starved cells, could be another of the many secondary responses to amino acid starvation, they were suspected of being the primary effectors of the stringent response because they accumulate in cells before shutoff of rRNA synthesis. Furthermore, the levels of magic spots in cells vary with growth rate. The faster cells grow, and hence the greater the ribosome level, the lower the level of ppGpp. It is as though ppGpp is a corepressor of rRNA synthesis.

Table 6.3   Conditions for Synthesis of ppGpp and pppGpp

Reaction Mixture	Synthesis of ppGpp and pppGpp
—Complete	+
—30S	–
—50S	–
—mRNA	–
—tRNA	–
—Stringent factor	–

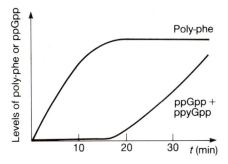

Figure 6.27   *In vitro* synthesis of ppGpp following consumption of phe-tRNA.

Haseltine and Block have shown directly with *in vitro* biochemical experiments the conditions necessary for synthesis of ppGpp. Its synthesis requires a particular protein, ribosomes, messenger, and uncharged tRNAs corresponding to the codon of messenger in the A site of a ribosome (Table 6.3). The required protein is normally rather tightly associated with the ribosomes and is the product of the *relA* gene itself.

The experiment utilized R17 phage RNA, an mRNA of known sequence, purified ribosomal subunits, and additions of various protein synthesis factors. Crucial to these experiments was the complete removal of endogenous messenger and tRNA. This was accomplished by incubating cells with rifamycin for 1.5 hours before extracting the ribosomes. The tRNA was removed from the ribosomes by dialyzing the ribosomes into a buffer containing low magnesium to dissociate the subunits and free the tRNA before centrifuging the ribosomes down a sucrose gradient.

The three proteins encoded by R17 RNA begin with fMetAlaSer, fMetSerLys, and fMetArgAla. Incubating ribosomes with initiation factors and fMet-tRNA$_f^{met}$ did not yield ppGpp and pppGpp synthesis until uncharged tRNA^{ala} or tRNA^{ser} was provided. As expected, if translocation by one codon was allowed by providing the translocation factors and charged ala-tRNA^{ala}, then uncharged tRNA^{ala} was no longer active in stimulating ppGpp or pppGpp synthesis. Also, if limited phe-tRNA was provided to a translation system synthesizing polyphenylalanine coded by polyuridine, ppGpp and pppGpp are synthesized only on exhaustion of the phe-tRNA (Fig. 6.27). In summary, the synthesis of the magic spots is produced by ribosomes in the presence of stringent factor, messenger RNA, and an uncharged tRNA free to react with the codon in the acceptor site of the ribosome.

MAINTAINING BALANCED SYNTHESIS OF RIBOSOMAL COMPONENTS

Although the synthesis of the individual components of ribosomes may be rather well regulated, a slight imbalance in the synthesis of one component could eventually lead to high and potentially toxic levels of that component. Synthesis of the ribosomal RNAs in bacteria are kept in balance by a simple mechanism. As we have already seen in Chapter 4, these RNAs are synthesized in one piece by an RNA polymerase that initiates at a promoter and transcribes across the genes for the three RNAs. A different mechanism is used to maintain balanced synthesis of some of the ribosomal proteins. In a number of the cases, one of the proteins encoded in a ribosomal protein cistron reduces translation of all the proteins in that cistron. This effect is called translational repression.

The finding that a ribosomal protein represses translation of proteins only from the same cistron provides an efficient means for the cell to maintain balanced synthesis of all the ribosomal proteins. Suppose that some ribosomal proteins began to accumulate because their synthesis is a little faster than the other proteins and the rRNA. Then, as the level of these proteins begins to rise in the cytoplasm, they begin to repress their own synthesis, and the system rapidly comes back to a balanced state.

How do we know about translational repression? The main clue came from careful measurements on cells with an increased number of genes coding for some of the ribosomal proteins. The increased copy number might have been expected to increase the synthesis of the corresponding proteins, but it did not. The synthesis of the mRNA for these proteins did increase as expected. It appeared that extra ribosomal proteins in the cell inhibited translation of their own mRNA. Proof of the idea of translational repression came from *in vitro* studies in which levels of individual free ribosomal proteins could be adjusted at will. The addition of DNA containing

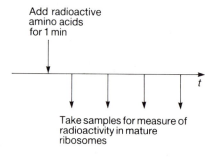

Add radioactive
amino acids
for 1 min

t

Take samples for measure of
radioactivity in mature
ribosomes

Figure 6.28   Determination of the pool size of ribosomal protein by the kinetics of a pulse of radioactive amino acids into ribosomal protein.

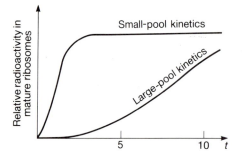

Relative radioactivity in
mature ribosomes

Small-pool kinetics

Large-pool kinetics

5   10   *t*

genes for some of the ribosomal proteins and properly prepared cell extract permits transcription and translation to yield ribosomal proteins synthesized *in vitro*. Nomura found that addition of the appropriate free ribosomal proteins to such a system repressed synthesis of the proteins encoded by the same cistron as the added protein.

Global limits can be placed on the accuracy with which the synthesis of ribosomal components is balanced. A short pulse of radioactive amino acid is provided to the cells, and the total pool of all ribosomal proteins can be determined by measuring the kinetics of incorporation of label into mature ribosomes (Fig. 6.28). The results show that the pool contains less than a five-minute supply of ribosomal proteins. Similarly, the pool size of each individual ribosomal protein can be measured. The results of these experiments show that most of the ribosomal proteins also have a very small intracellular pool of protein awaiting incorporation into ribosomes.

We have seen that the mechanisms regulating ribosome synthesis have good reason to be sophisticated, and, indeed, those aspects that have been investigated have turned out to be complicated. Much of the biochemistry and perhaps even much of the physiology of ribosome regulation remain to be worked out. In bacteria and other single-celled organisms such as yeast, it is likely that most of the regulation mechanism can be dissected by a combination of physiology, genetics, and biochemistry. It will be interesting to see if this potent combination of disciplines proves adequate to solve analogous problems in higher organisms.

PROBLEMS

6.1. Look up the necessary data and calculate the turnover rate of charged tRNA in growing *E. coli*. That is, how long would it take the charged tRNA to become completely uncharged at the normal rates of protein synthesis if charging were suddenly to stop?

6.2. On the basis of finding ribosomes with the enzymatic activity of a nascent but incomplete peptide chain of an induced enzyme, it is sometimes claimed that proteins fold up as they are being synthesized. Why is this reasoning inadequate and what experiments could definitively determine the time between completion of synthesis of a polypeptide and its attainment of enzymatic activity? [*Hint:* Lactose is cleaved to glucose and galactose by the inducible enzyme β-galactosidase. Galactose itself is metabolized by enzymes of another inducible operon.]

6.3. How did N-formyl methionine come to be discovered as the initiating amino acid? Look up the references and describe the experiments.

6.4. In the experiment in which translation of Qβ RNA was blocked by the prior binding of a oligonucleotide homologous to the Shine-Dalgarno sequence, it was found that f-met tRNA could not bind to the ribosomes. However, the trinucleotide, AUG, was able to bind. What is the meaning of this finding and does it contradict the Shine-Dalgarno hypothesis?

6.5. Predict several tRNA synthetases that will activate incorrect amino acids and hydrolytically edit to reduce misacylation. Identify the amino acids likely to be misactivated.

6.6. How do we know of the existence and properties of the A and P sites on ribosomes? Look up the necessary papers and describe the experiments.

6.7. No amber suppressors inserting tryptophan have been found. Amber and UGA mutants at the same site can be found with G to A mutagens. The UGA mutations are suppressible by UGA suppressors but not by suppressors that insert leu or ser. Predict the mutational event producing UGA suppression.

6.8. A frameshift mutation is created by the insertion or deletion of one or more bases that alter the triplet reading frame. The only mutagens that generate mutations suppressing the effects of some frameshift mutations are frameshift mutagens. Some of the resulting suppressors do not lie in the original gene. Propose an explanation and an experiment to test your idea.

6.9. Suppressors for UAA, UAG, and UGA termination codons can be created by single-base changes in the anticodon regions of tRNA genes. What amino acids could be inserted by suppressors to each of these termination signals? Despite energetic searches, not all of these possible suppressors have been found. What are the suppressors that have been found, and what is an explanation for the failure to find the rest?

6.10. With wobble permitted, what is the minimum number of tRNA types that would be required to read all 61 sense codons?

*6.11. At $t = 0$, inducer of an enzyme is added to cells. If T is the time required for transcription-translation-folding, show that the amount of induced enzyme as a function of time is

$$E(t) = \frac{C}{\mu d(\mu + d)} [d(-1 + e^{\mu(t-T)}) + \mu(-1 + e^{-d(t-T)})],$$

where the messenger decay rate is d, that is, messenger decays at a rate proportional to its amount; C is a constant; and μ is the exponential growth rate of the cells. Show that at very small $t - T$, $E(t)$ approximately equals $C(t - T)^2/2$. What is the meaning of the quadratic dependence on time and the lack of dependence on d and μ? Suppose d is much greater than μ. Find T and d from the following data:

t	$E(t)$
0.00	0.00
1.65	0.00
1.70	0.00
1.75	0.00
1.80	0.00
1.85	0.005

1.90	0.011
1.95	0.019
2.00	0.030
2.05	0.043
2.10	0.058
2.15	0.075
2.22	0.095
2.25	0.117
2.50	0.254
2.75	0.436
3.25	0.921
3.75	1.54
4.01	1.89
5.0	3.52
6.0	5.40
7.5	8.53
10.0	14.2
20.0	37.8

6.12. A clever experiment was used to determine the direction and rate of elongation of growing polypeptide chains in reticulocyte cells by Dintzis, Proc. Nat. Acad. Sci. USA 47, 247–261 (1961). The same type of experiment was done by Iwata and Kaji, Proc. Nat. Acad. Sci. USA 68, 690–694 (1971), and yielded the astonishing result that ribosomal protein appears to grow from the C-terminal end. What is the principle of such experiments? Aside from its highly implausible result, what flaws does the second paper contain?

6.13. In view of the fact that ribosomes can translate messenger as fast as it appears, perhaps they could translate still faster if they didn't run up against the RNA polymerase. How do we know this isn't the case?

6.14. If transport of lactose into cells is unnecessary in a strain with β-galactosidase relocated on the outer membrane by fusion to a maltose transport protein, it appears strange that nature even bothers to have a transport system for the *lac* system. Why wouldn't it have been simpler for nature to use a signal peptide to place β-galactosidase on the outer membrane?

6.15. What types of experiments might Silhavy and Beckwith have attempted if they had suspected that the carboxy terminus rather than the amino terminus of a protein specified export through the inner membrane?

6.16. If the level of ribosomes is proportional to the growth rate, show that the rate of synthesis of ribosomes is proportional to the square of the growth rate.

6.17. At the ribosome translation efficiency normally seen in E. coli, what is the absolute maximum growth rate possible?

6.18. From the data of Edlin and Stent, Proc. Nat. Acad. Sci. USA *62*, 475–482, 1969, calculate the intracellular concentrations of nucleoside triphosphate in growing cells and from the data of Lazzarini, Cashel, and Gallant, J. Biol. Chem. *246*, 4381–4385, 1971, the concentration of ppGpp and pppGpp.

6.19. Maaløe has proposed a model for ribosome control as a function of growth rate that should be called a nonmodel. The crux of his idea is that the sum of all α's in the cell must, by definition, equal 1. Thus if everything else is controlled, ribosomes do not need an explicit control. What makes this a plausible model and what critical experiments could be done bearing on its veracity?

6.20. What is the minimum time following a growth rate upshift that would be required for α_r to attain roughly its final value if the 55 ribosomal proteins were distributed into n equal-sized cistrons?

6.21. Sketch a graph of the synthesis of rRNA in amino acid–starved stringent cells on the restoration of amino acids (a) assuming that the block of synthesis was at the RNA polymerase initiation step, and (b) assuming that the block of synthesis was at an RNA chain elongation step and that RNA polymerase molecules are randomly distributed through the rRNA cistrons.

6.22. Overproduction of ribosomal proteins for structural studies was found to make cells so sick that they could hardly grow. What is the most likely explanation?

RECOMMENDED READINGS

Codon-Anticodon Pairing: The Wobble Hypothesis, F. H. C. Crick, J. Mol. Biol. *19*, 548–555 (1966).

Synthesis of Guanosine Tetra- and Pentaphosphate Requires the Presence of a Codon-Specific, Uncharged tRNA in the Acceptor Site of Ribosomes, W. Haseltine, R. Block, Proc. Nat. Acad. Sci. USA *70*, 1564–1568 (1973).

How Ribosomes Select Initiator Regions in mRNA: Base Pair Formation Between the 3′-Terminus of 16S rRNA and the mRNA During Initiation of Protein Synthesis in *E. Coli.*, J. Steitz, K. Jakes, Proc. Nat. Acad. Sci. USA *72*, 4734–4738 (1975).

How Do Eucaryotic Ribosomes Select Initiation Regions in Messenger RNA?, M. Kozak, Cell *15*, 1109–1123 (1978).

RELATED REVIEWS, BOOKS, AND ARTICLES

"Control of Macromolecular Synthesis," O. Maaløe, N. O. Kjeldgaard, Benjamin, New York (1966).

Sense and Nonsense in the Genetic Code, A. Garen, Science *160*, 149–159 (1969).

The Mechanism of Protein Synthesis, Cold Spring Harbor Symposium on Quantitative Biology *34* (1969).

Amino Acyl-tRNA Synthetases: General Features and Recognition of Transfer RNAs, P. R. Schimmel, D. Söll, Ann. Rev. Biochem. *48*, 601–648 (1979).

Transfer RNA: Structure, Properties, and Recognition, ed. P. Schimmel, D. Söll, J. Abelson, Cold Spring Harbor Laboratory (1979).

Five Specific Protein-Transfer RNA Interactions, P. Schimmel, CRC Critical Reviews in Biochemistry, 207–251 (Dec. 1980).

DEEPER READING

Dependency on Medium and Temperature of Cell Size and Chemical Composition During Balanced Growth of *Salmonella typhimurium*, M. Schaechter, O. Maaløe, N. Kjeldgaard, J. Gen. Microbiol. *19*, 592–606 (1958).

Assembly of the Peptide Chains of Hemoglobin, H. M. Dintzis, Proc. Nat. Acad. Sci. USA *47*, 247–261 (1961).

Protein and RNA Synthesis in a Mutant of *E. coli* with an Altered Aminoacyl RNA Synthetase, W. Fangman, F. Neidhardt, J. Biol. Chem. *239*, 1844–1847 (1964).

N-formylmethionyl-sRNA as the Initiator of Protein Synthesis, J. Adams, M. Capecchi, Proc. Nat. Acad. Sci. USA *55*, 147–155 (1966).

In vitro Protein Synthesis: Chain Initiation, R. Webster, D. Engelhardt, N. Zinder, Proc. Nat. Acad. Sci. USA *55*, 155–161 (1966).

Control of Production of Ribosomal Protein, R. Schleif, J. Mol. Biol. *27*, 41–55 (1967).

A Mutant Which Reinitiates the Polypeptide Chain After Chain Termination, A. Sarabhai, S. Brenner, J. Mol. Biol. *27*, 145–162 (1967).

RNA Chain Growth Rate in *Escherichia coli*, H. Bremer, D. Yuan, J. Mol. Biol. *38*, 163–180 (1968).

Control of Haemoglobin Synthesis: Rate of Translation of the Messenger RNA for the α and β Chains, T. Hunt, T. Hunger, A. Munro, J. Mol. Biol. *43*, 123–133 (1969).

Nucleotide Sequence from the Polypeptide Chain Termination Region of the Coat Protein Cistron in RNA Bacteriophage R17, J. L. Nichols, Nature *225*, 147–151 (1970).

Release Factors Mediating Termination of Complete Proteins, M. Capecchi, H. Klein, Nature, *226*, 1029–1033 (1970).

Synthesis of 5S Ribosomal RNA in *E. coli* After Rifampicin Treatment, W. Doolittle, N. Pace, Nature *228*, 125–129

The Nature of the Polypeptide Chain Termination Signal, P. Lu, A. Rich, J. Mol. Biol. *58*, 513–531 (1971).

Removal of Formyl-methionine Residue from Nascent Bacteriophage f2 Protein, D. Housman, D. Gillespie, H. Lodish, J. Mol. Biol. *65*, 163–166 (1972).

Efficiency of Protein and Messenger RNA Synthesis in Bacteriophage T4-infected Cells of *Escherichia coli*, K. Gausing, J. Mol. Biol. *71*, 529–545 (1972).

Accumulation and Turnover of Guanosine Tetraphosphate in *Escherichia coli*, N. Fiil, K. von Meyenburg, J. Friesen, J. Mol. Biol. *71*, 769–783 (1972).

The Frequency of Errors in Protein Biosynthesis, R. Loftfield, D. Vanderjagt, Biochem. J. *128*, 1353–1356 (1972).

Frameshift Suppression: A Nucleotide Addition in the Anticodon of a Glycine Transfer RNA, D. Riddle, J. Carbon, Nature New Biology *242*, 230–234 (1973).

Codon Specific, tRNA Dependent *in vitro* Synthesis of ppGpp and pppGpp, F. Pedersen, E. Lund, N. Kjeldgaard, Nature New Biol. *243*, 13–15 (1973).

Chain Growth Rate of β-galactosidase During Exponential Growth and Amino Acid Starvation, F. Engbaek, N. Kjeldgaard, O. Maaløe, J. Mol. Biol. *75*, 109–118 (1973).

The Anticodon-Anticodon Complex, J. Eisinger, N. Gross, J. Mol. Biol. *88*, 165–174 (1974).

The 3'-Terminal Sequence of *Escherichia coli* 16S Ribosomal RNA: Complementary to Non-sense Triplets and Ribosome Binding Sites, J. Shine, L. Dalgarno, Proc. Nat. Acad. Sci. USA *71*, 1342–1346 (1974).

Ribosomal Protein in *E. coli:* Rate of Synthesis and Pool Size at Different Growth Rates, K. Gausing, Mol. Gen. Genet. *129*, 61–75 (1974).

Transfer of Proteins Across Membranes, G. Blobel, B. Dobberstein, J. Cell Biol. *67*, 835–851 (1975).

Conversion of β-Galactosidase to a Membrane-Bound State By Gene Fusion, T. Silhavy, M. Casadaban, H. Shuman, J. Beckwith, Proc. Nat. Acad. Sci. USA *73*, 3423–3427 (1976).

Extracellular Labeling of Nascent Polypeptides Traversing the Membrane of *Escherichia coli,* W. Smith, P-C. Tai, R. Thompson, B. Davis, Proc. Nat. Acad. Sci. USA *74*, 2830–2834 (1977).

Coding Properties of an Ochre-Suppressing Derivative of *E. coli,* tRNA^{tyr}, S. Feinstein, S. Altman, J. Mol. Biol. *112*, 453–470 (1977).

Characterization of Two mRNA-rRNA Complexes Implicated in Initiation of Protein Biosynthesis, J. Steitz, D. Steege, J. Mol. Biol. *114*, 545–558 (1977).

Mistranslation in *E. coli,* P. Edelmann, J. Gallant, Cell *10*, 131–137 (1977).

Detection of Prokaryotic Signal Peptidase in an *Escherichia coli* Membrane Fraction: Endoproteolytic Cleavage of Nascent f1 Pre-coat Protein, C. Chang, G. Blobel, P. Model, Proc. Nat. Acad. Sci. USA *75*, 361–365 (1978).

Mutations of Bacteriophage T7 That Affect Initiation of Synthesis of the Gene 0.3 Protein, J. Dunn, E. Buzash-Pollert, F. Studier, Proc. Nat. Acad. Sci. USA *75*, 2741–2745 (1978).

Inhibition of Qβ RNA 70S Ribosome Initiation Complex Formation by an Oligonucleotide Complementary to the 3' Terminal Region of *E. coli* 16S Ribosomal RNA, T. Taniguchi, C. Weissmann, Nature *275*, 770–772 (1978).

Patterns of Protein Synthesis in *Escherichia coli:* A Catalog of the Amount of 140 Individual Proteins at Different Growth Rates, S. Pedersen, P. Bloch, S. Reeh, F. Neidhardt, Cell *14*, 179–190 (1978).

The Suppression of Defective Translation by ppGpp and its Role in the Stringent Response, P. O'Farrell, Cell *14*, 545–557 (1978).

Migration of 40S Ribosomal Subunits on Messenger RNA in the Presence of Edeine, M. Kozak, A. Shatkin J. Biol. Chem. *253*, 6568–6577 (1978).

The Accumulation as Peptidyl-Transfer RNA of Isoaccepting Transfer RNA Families in *Escherichia coli* with Temperature-Sensitive Peptidyl-Transfer RNA Hydrolase, J. Menninger, J. Biol. Chem. *253*, 6808–6813 (1978).

Context Effects on Nonsense Codon Suppression in *Escherichia coli,* S. Feinstein, S. Altman, Genetics *88*, 201–219 (1978).

Tandem Promoters Direct *E. coli,* Ribosomal RNA Synthesis, R. Young, J. Steitz, Cell *17*, 225–235 (1979).

Escherichia coli Mutants Accumulating the Precursor of a Secreted Protein in the Cytoplasm, P. Bassford, J. Beckwith, Nature 277, 538--541 (1979).

Efficient Cap-Dependent Translation of Polycistronic Prokaryotic mRNAs Is Restricted to the First Gene in the Operon, M. Rosenberg, B. Paterson, Nature *279*, 696–701 (1979).

Chicken Ovalbumin Contains an Internal Signal Sequence, V. Lingappa, J. Lingappa, G. Blobel, Nature *281*, 117–121 (1979).

Identification of Initiation Sites for the *in vitro* Transcription of rRNA Operons *rrnE* and *rrnA* in *E. coli,* S. Gilbert, H. de Boer, M. Nomura, Cell *17*, 211–224 (1979).

Evidence for the Double-Sieve Editing Mechanism for Selection of Amino Acids in Protein Synthesis: Steric Exclusion of Isoleucine by valyl-tRNA Synthetases, A. Fersht, C. Dingwall, Biochemistry *18*, 2627–2631 (1979).

Secretion of Beta-lactamase Requires the Carboxy End of the Protein, D. Koshland, D. Botstein, Cell *20*, 749–760 (1980).

In vitro Expression of *E. coli* Ribosomal Protein Genes: Autogenous Inhibition of Translation, J. Yates, A Arfstein, M. Nomura, Proc. Nat. Acad. Sci. USA *77*, 1837–1841 (1980).

The Assembly of Proteins into Biological Membranes: The Membrane Trigger Hypothesis, W. Wickner, Ann. Rev. of Biochem. *48*, 23–45 (1979).

Novel Features in the Genetic Code and Codon Reading Patterns in *Neurospora crassa* Mitochondria Based on Sequences of Six Mitochondrial tRNAs, J. Heckman, J. Sarnoff, B. Alzner-DeWeerd, S. Yin, U. RajBhandary, Proc. Nat. Acad. Sci. USA *77*, 3159–3163 (1980).

Different Pattern of Codon Recognition by Mammalian Mitochondrial tRNAs, B. Barrell, S. Anderson, A. Bankier, M. DeBruijn, E. Chen, A. Coulson, J. Drouin, I. Eperon, D. Nierlich, B. Roe, F. Sanger, P. Schreier, A. Smith, R. Staden, I. Young, Proc. Nat. Acad. Sci. USA *77*, 3164–3166 (1980).

Codon Recognition Rules in Yeast Mitochondria, S. Bonitz, R. Berlani, G. Coruzzi, M. Li, G. Macino, F. Nobrega, M. Nobrega, B. Thalenfeld, A. Tzagoloff, Proc. Nat. Acad. Sci. USA *77*, 3167–3170 (1980).

Synthesis, Assembly into the Cytoplasmic Membrane, and Proteolytic Processing of the Precursor of Coliphage M13 Coat Protein, K. Ito, T. Date, W. Wickner, J. Biol. Chem. *255*, 2123–2130 (1980).

The Distal End of the Ribosomal RNA Operon *rrnD* of *E. coli* Contains a tRNA$_1^{thr}$ Gene, Two 5S rRNA Genes and a Transcription Terminator, G. Duester, W. Holmes, Nucleic Acid Res. *8*, 3793–3807 (1980).

The Spontaneous Insertion of Proteins Into and Across Membranes: The Helical Hairpin Hypothesis, D. Engelman, T. Steitz, Cell *23*, 411–422 (1981).

expA: A Conditional Mutation Affecting the Expression of a Group of Exported Proteins in *E. coli* K-12, E. Dassa, P. Boquet, Mol. Gen. Genet. *181*, 192–200 (1981).

Regulation of the Synthesis of *E. coli* Elongation Factor Tu, F. Young, A Furano, Cell *24*, 695–706 (1981).

Identification of Ribosomal Protein S7 as a Repressor of Translation within the *str* Operon of *E. coli*, D. Dean, J. Yates, M. Nomura, Cell *24*, 413–419 (1981).

Lysis Gene Expression of RNA Phage MS2 Depends on a Frameshift During Translation of the Overlapping Coat Protein Gene, R. Kastelein, E. Remaut, W. Fiers, J. van Duin, Nature *295*, 35–41 (1982).

Mechanism of Polypeptide Chain Initiation in Eukaryotes and Its Control by Phosphorylation of the α Subunit of Initiation Factor 2, J. Siekierka, L. Mauser, S. Ochoa, Proc. Nat. Acad. Sci. USA *79*, 2537–2540 (1982).

The Accuracy of Protein Synthesis Is Limited by Its Speed: High Fidelity Selection by Ribosomes of Aminoacyl-tRNA Ternary Complexes Containing GTP (γS), R. Thompson, A. Karin, Proc. Nat. Acad. Sci. USA *79*, 4922–4926 (1982).

A Covalent Adduct Between the Uracil Ring and the Active Site on an Aminoacyl tRNA Synthetase, R. Starzyk, S. Koontz, P. Schimmel, Nature *298*, 136–140 (1982).

Eukaryotic Ribosomes Can Recognize Preproinsulin Initiation Codons Irrespective of their Position Relative to the 5' End of mRNA, P. Lomedico, S. McAndrew, Nature *299*, 221–226 (1982).

Signal Recognition Particle Contains a 7S RNA Essential for Protein Translocation Across the Endoplasmic Reticulum, P. Walter, G. Blobel, Nature *299*, 691–698 (1982).

Diverse Effects of Mutations in the Signal Sequence on the Secretion of β-Lactamase in *Salmonella typhimurium*, D. Koshland, R. Sauer, D. Botstein, Cell *30*, 903–914 (1982).

An Estimate of the Global Error Frequency in Translation, N. Ellis, J. Gallant, Mol. Gen. Genet. *188*, 169–172 (1982).

Effects of Surrounding Sequence on the Suppression of Nonsense Codons, J. Miller, A. Albertinia, J. Mol. Biol. *164*, 59–71 (1983).

Context Effects: Translation of UAG Condon by Suppressor tRNA Is Affected by the Sequence Following UAG in the Message, L. Bossi, J. Mol. Biol. *164*, 73–87 (1983).

Demonstration by a Novel Genetic Technique That Leader Peptidase Is an Essential Enzyme of *Escherichia coli*, T. Data, J. Bact. *154*, 76–83 (1983).

Measurement of Suppressor Transfer RNA Activity, J. Young, M. Capecchi, L. Laski, U. RajBhandary, P. Sharp, P. Palese, Science *221*, 873–875 (1983).

Disassembly and Reconstitution of Signal Recognition Particles, P. Walter, G. Blobel, Cell *34*, 525–533 (1983).

Compilation and Analysis of Sequences Upstream from the Translational Start Site in Eukaryotic mRNAs, M. Kozak, Nucleic Acids Res. *12*, 857–872 (1984).

PART TWO

Genetics and Genetic Engineering

Formal Genetics

T hus far we have covered the structure of cells and the structure, properties, and synthesis of the components of major interest to molecular biologists — DNA, RNA, and proteins. We will now concern ourselves with genetics. Historically, the study and formulation of many genetic principles preceded an understanding of their chemical basis. However, by inverting the order of presentation in this book, major portions of genetics become easier to understand and can be covered in a single chapter.

The chapters in this part of the book cover the mechanics of genetic experiments, some mathematical aspects that frequently enter genetic considerations, and, finally, genetic engineering.

Genetics is central to this book for three reasons. First, the exchange of genetic information, DNA, between cells or organisms and the ability to recombine this DNA by cutting and splicing are widespread in nature. This means that these phenomena must confer high survival value and therefore are of great biological importance. Second, for many years genetics has been at the center of research in molecular biology, first serving as an object of study and later as a helpmate to biochemistry. Third, genetics will become an increasingly valuable tool in enzymology as well as becoming indispensable in the industrial utilization of molecular biology. Genetics will play a particularly important role in the future study of protein structure, in which much attention will be directed toward designing and building proteins to perform specific tasks.

MUTATIONS

Historically, one reason for the study of genetics was to discover the chemical basis of heredity. Naturally, the existence of mutations was necessary to the execution of the classical experiments in genetics, and an understanding of mutations will facilitate our study of these experiments. We have already covered the chemical basis of heredity and the basics of gene expression. Perhaps here we should explicitly state that *gene* refers to a set of nucleotides that specifies the sequence of an RNA or protein. We will now define mutation and in the next section mention the three basic types of mutations. In the following section we will review the classical genetic experiments before turning to recombination.

A mutation is merely a heritable alteration from the normal. It is an alteration in the nucleotide sequence of the DNA or, in the case of RNA viruses, an alteration in the nucleotide sequence of its genomic RNA. We already know that changes in coding portions of DNA may alter the amino acid sequences of proteins and that changes in noncoding regions of DNA have the potential for changing the expression of genes, for example by altering the strength of a promoter. Of course, any cellular process that makes use of a sequence of DNA can be affected by a mutation. The existence of mutations implies that the sequence of DNA in living things, including viruses, is sufficiently stable that most individuals possess the same sequence but sufficiently unstable that alterations do exist and can be found.

The terms wild type, mutant, mutation, and allele are closely related but must be distinguished. Wild type refers to a reference, and it can mean an organism, set of genes, gene, or nucleotide sequence. A mutation then is a change in the nucleotide sequence of the reference that produces a heritable change from the wild type to a mutant organism. A mutant is the organism that carries the mutation. Two mutations are said to be allelic if they lie in the same gene. However, now that many genetic analyses can be performed on a nucleotide by nucleotide basis, in some situations alleles refer to nucleotides rather than to genes.

Until recently, when it became possible to sequence DNA easily, mutations could readily be identified only by their gross effects on the appearance of the cell or organism or its descendants. Some of the most easily studied biological effects of mutations in bacteria and viruses are changes in the colony or plaque morphology. Other easily studied effects of mutations are the inability of cells to grow at low or high temperatures or the inability to grow without the addition of specific chemicals to the growth medium. Such readily observed properties of cells constitute their phenotype. The status of the genome giving rise to the phenotype is called the genotype. For example, an inability to grow on lactose, a Lac⁻ phenotype, can result from mutations in lactose transport, β-galactosidase enzyme, *lac* gene regulation, or the cells' overall regulation of classes of genes that are not well induced if cells are grown in the presence of glucose. Such cells would have negative genotypes in the following genes: *lacY, lacZ, lacI,* or *crp.*

POINT MUTATIONS, DELETIONS, INSERTIONS, AND DAMAGE

The structure of DNA permits only three basic types of alteration or mutation at a site: the substitution of one nucleotide for another, the deletion of one or more nucleotides, and the insertion of one or more nucleotides. A nucleotide substitution at a point is called a transition if one purine is substituted for the other or one pyrimidine is substituted for the other and is called a transversion if a purine is substituted for a pyrimidine or vice versa (Fig. 7.1).

In addition to substitutions of one nucleotide for another in single-stranded DNA or one base pair for another in double-stranded DNA, nucleotides are susceptible to many types of chemical modification. These can include tautomerizations and deamination or more extensive damage such as the complete loss of a base from the ribose phosphate backbone (Fig. 7.2). However, the cellular repair mechanisms remove many such modified bases, and those changes escaping repair cannot themselves be passed on to the next generation because on DNA replication only one of the usual four nucleotides is incorporated into the daughter strand opposite the altered base. A modified base may generate a mutation if the nucleotide incorporated opposite it on the daughter strand is incorrect (Fig. 7.3).

Mutations arise from a variety of sources. As discussed in Chapter 3, point mutations can occur spontaneously during replication of the DNA through the misincorporation of a nucleotide and the failure of the editing mechanisms to correct the mistake or through the chemical instability of the nucleotides. For example, cytosine can deaminate to form uracil, which is then recognized as thymine during DNA replication.

The frequency of spontaneous appearance of point mutations often is too low for convenient experimentation, and mutagens are therefore used to increase the frequency of mutants in cultures 10 to 1,000 times above the spontaneous frequency. A variety of mutagens have been discovered, some by rational considerations and some by chance. Most are either nucleotide analogs that are incorporated

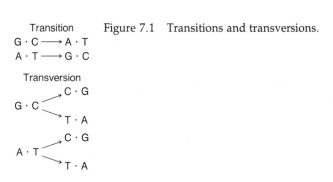

Transition
G · C ⟶ A · T
A · T ⟶ G · C

Transversion

G · C ⟶ C · G
        ⟶ T · A

A · T ⟶ C · G
        ⟶ T · A

Figure 7.1   Transitions and transversions.

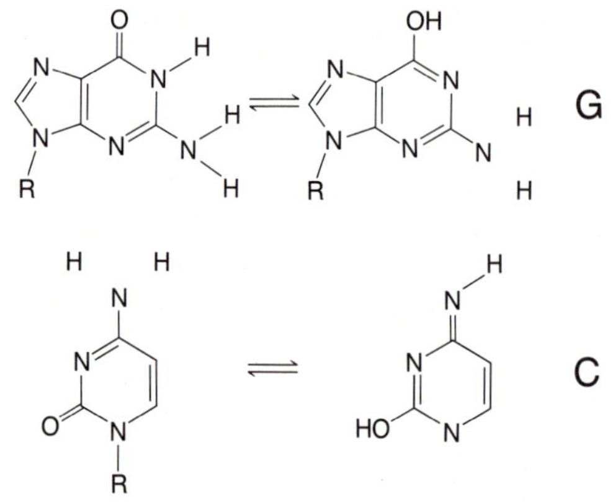

Figure 7.2   Two tautomeric forms of guanine and cytosine.

into the DNA instead of the normal nucleotides, increasing the frequency of mis-pairing in subsequent rounds of DNA replication, or chemically reactive molecules that modify bases in DNA (Fig. 7.4). Ultraviolet light is also a mutagen because it damages DNA, and the cell makes mistakes in correcting the damage. In one way or another, mutagens increase the frequency of mispaired bases escaping repair and thus lead to the change in the original sequence of the DNA.

The mechanisms generating deletions and insertions are not as well understood. Errors in DNA replication provide plausible mechanisms for the generation of one or two base insertions or deletions. Most likely slippage, perhaps stimulated by an appropriate sequence, will permit a daughter strand to possess a different number of bases than the parent strand.

Insertions and deletions larger than a few bases arise by a different mechanism.

Figure 7.3   How replication past a damaged base can introduce a base pair change into the double-stranded duplex.

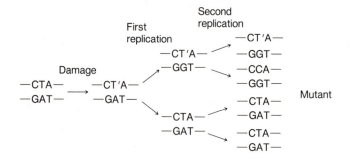

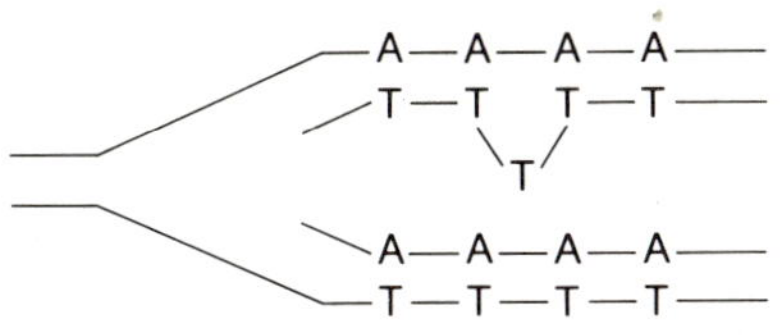

Miller has found that the end points of a number of deletions in bacteria are located at short, repeated or almost repeated sequences. The deletion removes one of the repeats and the intervening sequence. We can picture two plausible events that could create such deletions (Fig. 7.5). The first is a looping of a single chromosome followed by elimination of the material between the two repeats. The second is similar to the first, but it occurs between two chromosomes and transfers the material from one chromosome to the other. One chromosome suffers a deletion and the other an insertion.

The creation of deletions is also stimulated by the presence of some genetic elements called insertion sequences or transposons. These elements transpose themselves or copies of themselves into other sites on the chromsome, thereby creating insertions by mechanisms still not fully understood; we will discuss them more fully in Chapter 18. They often generate deletions in their vicinity as a result of abortive transposition reactions.

CLASSICAL GENETICS OF CHROMOSOMES

We should not proceed to a detailed discussion of molecular genetics without a brief review of Mendelian genetics. Chromosomes in eukaryotes consist mainly of DNA and histones. During some stages of the cell's division cycle in plants and animals, chromosomes can be observed with light microscopes, and they display many beautiful and fascinating patterns. Careful microscopic study of such chromosomes set the stage for subsequent molecular experiments that revealed the exact chemical nature of heredity. We are now approaching an explanation of genetic recombination at a similar depth of understanding.

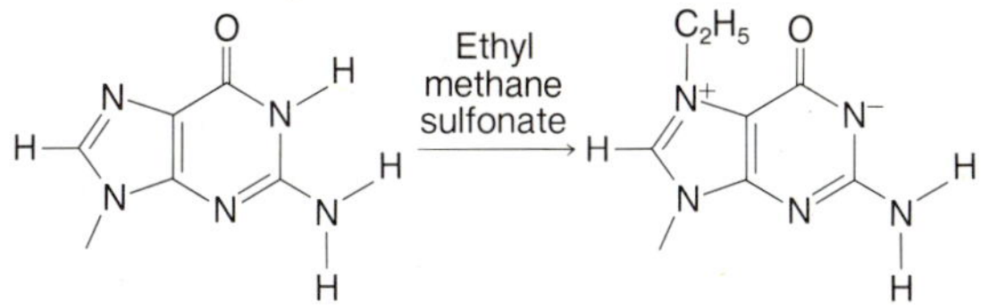

Figure 7.4   A product of ethylation of guanine by ethyl methanesulfonate.

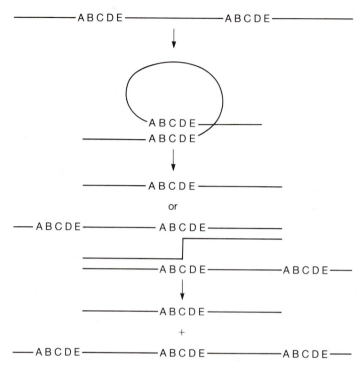

Figure 7.5   Two mechanisms for deletion creation between repeated sequences. The first is looping with recombination between points within a single chromosome, and the second is unequal crossing over between two chromosomes.

The basis of many of the classical studies is that most types of eukaryotic cells are diploid. This means that each cell contains pairs of identical or almost identical homologous chromosomes, one chromosome of each pair deriving from each of the parents. There are some exceptions: Some types of plants are tetraploid or even octaploid, and some variants of other species possess alternate numbers of one or more of the chromosomes.

During normal cell growth and division, the pairs of chromosomes in each dividing cell are duplicated and distributed to the two daughter cells in a process called mitosis. As a result, each daughter cell receives the same genetic information as the parent cell contained. However, the situation must be altered for sexual reproduction. During this process special cells derived from each of the parents fuse and give rise to the new progeny. To maintain a constant amount of DNA per cell from one generation to the next, the special cells, often called gametes, must be haploid instead of diploid. The cell divisions giving rise to the haploid gametes in animals and haploid spores in plants is called meiosis.

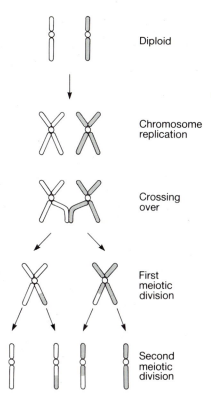

Diploid

Chromosome
replication

Crossing
over

First
meiotic
division

Second
meiotic
division

Figure 7.6   The classical view of meiosis.

During the process of meiosis a pair of chromosomes doubles, genetic recombination may occur between homologous chromosomes, and the cell then divides (Fig. 7.6). Each of the daughter cells then divides again without duplication of the chromosomes. The net result is four cells, each containing only one copy of each chromosome. Subsequent fusion of a sperm and egg cell from different individuals yields a diploid zygote that grows and divides to yield an organism containing one member of each chromosome pair from each parent.

The chromosomes from each of the parents may contain mutations that produce recognizable traits or phenotypes in the offspring. Let us consider just one chromosome pair of a hypothetical organism. Let gene A produce trait A, and, if it is mutant, let it be denoted as gene a and its trait be a. In genetic terminology A and a are alleles. We can describe the genetic state of an individual by giving its genetic composition or genotype. For example, both copies of the chromosome in question could contain the A allele. For convenience denote this as (A/A). Such a cell is called homozygous for gene A. A mating between organisms containing diploid cells of type (A/A) and (a/a) must produce offspring of the type (a/A), which is, of course, identical to (A/a). That

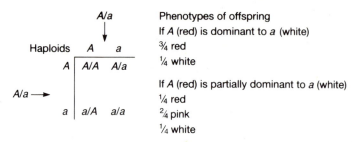

Figure 7.7   The haploids produced from diploid parent cells, the combinations of haploids possible upon their fusion, and the apparent phenotypes if *A* is totally or partially dominant to *a*.

is, the chromosomes in the offspring are copies of each of the parental chromosomes. These offspring are said to be heterozygous for gene *A*.

The interesting results come when two heterozygous individuals mate and produce offspring. A gamete can inherit one or the other of each of the homologous chromosomes from each chromosome pair. This generates a variety of gamete types. However, when large numbers of offspring are considered, many representatives are found of every possible combination of assortment of the chromosomes, and the results become predictable. It is easiest to systematize the possibilities in a square matrix (Fig. 7.7). However, for evaluation of experimental results, the appearance of heterozygotes must be known or deduced. The appearance of a heterozygote (*a/A*) is that of trait A if *A* is dominant, which means automatically that *a* is recessive. Strict dominance need not be seen, and a heterozygote may combine the traits displayed by the two alleles. For example, if the trait of gene *A* were the production of red pigment in flowers and the trait of gene *a* were the absence of production of the pigment, the heterozygote (*a/A*) might produce half the normal amount of red pigment and yield pink flowers.

COMPLEMENTATION, *cis, trans,* DOMINANT, AND RECESSIVE

Complementation, *cis, trans,* dominant, and recessive are commonly used genetic terms. Their meaning can be clarified by considering a simple set of two genes that are transcribed from a single promoter, an operon. Denote the promoter by *P* and the two genes that code for proteins that diffuse through the cell's cytoplasm by *A* and *B*.

Diploids heterozygous for genes of the operon may easily be constructed by genetic crosses in diploid eukaryotes and even in prokaryotes by special tricks. If the genes are on phage genomes, cells may be simultaneously infected with both phage

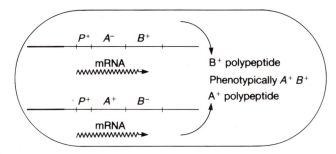

Figure 7.8   Complementation between two genes that produce diffusible products.

types. Consider the possibilities for mutations in genes A and B. If we introduce the operon $P^+ A^- B^+$ into cells that are $P^+ A^+ B^-$, the diploid $P^+ A^- B^+/P^+ A^+ B^-$ will possess good A and B enzymes and phenotypically will appear $A^+ B^+$ (Fig. 7.8). The A^+ gene complements the A^- gene and the B^+ gene complements the B^- gene. Both A^+ and B^+ act in *trans* and are *trans* dominant, or simply dominant. Analogously, A^- and B^- are recessive.

A mutation in the promoter generates more complexity. A P^- mutation also appears to be A^- and B^- even though the A and B genes retain their normal sequences. This results from the fact that no A or B enzyme can be synthesized if its promoter is defective. In genetics terminology, such a strain is *phenotypically* A^- and B^- since, for growth purposes, it behaves as though it lacks the A and B activities. However, it is *genotypically* P^-, A^+, and B^+ since the A and B genes actually remain intact, as may be revealed in other types of experiments. In our example, the P^- gene is *cis* dominant to the A^+ and B^+ genes, that is, its effect is only over genes on the same piece of DNA. Additionally, P^+ is not *trans* dominant to P^- since a partial diploid of the genetic structure $P^- A^+ B^+/P^+ A^- B^+$ appears to be A^- (Fig. 7.9). The

Figure 7.9   Absence of complementation when an intact gene does not produce a diffusible product because its promoter is inactive.

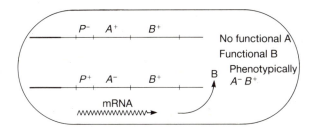

presence of the A^+ gene in a $P^- \, A^+$ chromosome could be revealed by genetic recombination, the subject of the next section.

GENETIC RECOMBINATION

Genetic crosses between two strains, each containing a mutation in the same gene, are occasionally observed to yield nonmutant, that is, wild-type, progeny. This is the result of genetic recombination in which the two parental DNA molecules are precisely broken, exchanged, and rejoined (Fig. 7.10). In this section we describe the phenomenon of recombination and in the next some experiments that make use of recombination to order genetic markers on chromosomes. Finally we discuss what is known about the chemical processes involved in genetic recombination.

Alignment of the pairs of homologous chromosomes during meiosis has long been known through observations with light microscopes. The subsequent movement of identifiable sites on the chromosomes suggested that, indeed, homologous chromosomes were broken and portions exchanged in a process called crossing over or recombination. Although recombination was first observed in eukaryotic cells, it appears to be almost ubiquitous. Even simple bacterial viruses can engage in genetic recombination. The bacterial phage experiments were particularly important because phage provide a simple and small system with few variables, a high sensitivity for recombinants, and a very short generation time so that new important experiments can be done nearly every day.

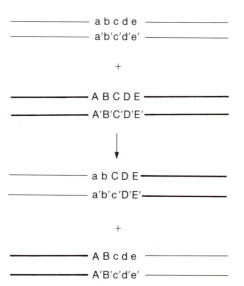

Figure 7.10   Representation of genetic crossover between two DNA duplexes in which a precise cut and splice is made.

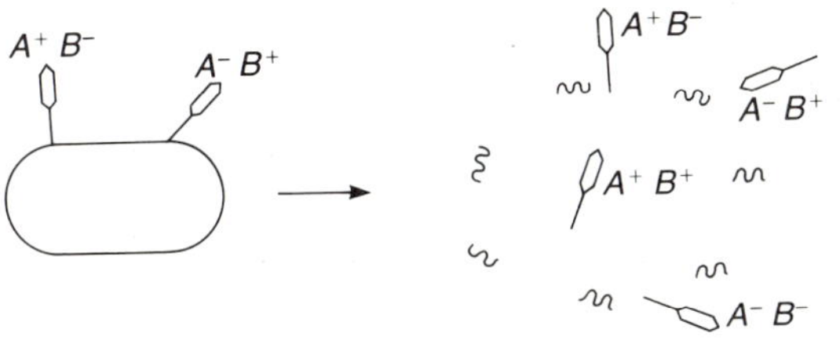

Figure 7.11   Coinfection of a single cell by phage of two different genotypes permits recombination to produce output phage combining genotypes of the parental phage in addition to output phage of the parental types.

Phage mutants can easily be isolated that yield plaque morphologies different from the normal or wild type simply by plating mutagenized phage on cells and locating the occasional different plaque. Nonsense mutations in essential phage genes may be identified as phage that grow only on nonsense-suppressing host cells. Genetic recombination between phage can be revealed by coinfection of cells with two mutants at a sufficiently high multiplicity of infection that each cell is infected with both types of phage (Fig. 7.11). Some of the progeny phage are found to carry alleles from both of the input phage! Since the phage carry only one DNA copy, such progeny have to be recombinants carrying some of the genetic information from each of the parental phage types. The discovery of genetic recombination in phage, by Delbrück and Hershey, opened the way for intensive study of the phenomenon of genetic recombination at the molecular level.

MAPPING BY RECOMBINATION FREQUENCIES

Two or more mutations on any DNA molecule that engages in recombination can be approximately ordered along DNA molecules by measurement of the frequencies of recombination. Let us examine why this is so. Assume that the probability of a recombination or genetic crossover between two points on two almost homologous DNA molecules is a function only of the distance between the points. Then the probability of a crossover between the two points should be linearly proportional to their separation if they are not too far apart. This seems reasonable because increasing the separation increases the number of potential crossover sites. Recombination frequencies should be additive if the frequency is linearly proportional to marker separation. Thus, if $Rf(X, Y)$ is the recombination frequency between markers X and Y, and if strict additivity holds, $Rf(A, C)$ must equal $Rf(A, B) + Rf(B, C)$ (Fig. 7.12).

Unfortunately, the assumption of a linear relation between recombination frequency and distance is not good over extremely short distances. In these cases the specific nucleotides involved can generate profound effects on the recombination frequencies, and additivity fails as a result. These short-distance anomalies are often called marker effects.

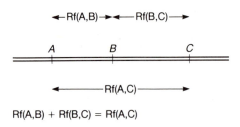

$$Rf(A,B) + Rf(B,C) = Rf(A,C)$$

Figure 7.12   Additivity relations necessary when recombination frequency is linearly proportional to genetic distance.

Another problem arises if the distance between two markers becomes too large. Then more than one crossover is likely in the distance separating them. A molecule having experienced two or any even number of crossovers between the markers appears not to have engaged in crossovers at all (Fig. 7.13). Only an odd number of crossovers between two markers generates recombinants that can be distinguished from the parents. Therefore, as marker separation becomes great and the frequency of crossovers increases, an even number of crossovers becomes as likely as an odd number of crossovers and the recombination frequency approaches 50% as an upper limit. However, if the recombination frequency is not too close to 50%, corrections based on the decreasing likelihood of 3, 5, 7, 9, . . . crossovers can be made. Still larger genetic distances can be measured by subdividing a long interval into smaller intervals and summing their genetic lengths.

Given the vagaries of measuring distances by recombination frequencies, what can be done? Often the major question is merely one of marker order. The actual distances separating the markers are not of great importance. Three-factor genetic crosses are a partial solution. They permit ordering of two genetic markers, B and C, with respect to a third, A, that is known to lie on the outside. That is, the experiment is to determine whether the order is A-B-C, or A-C-B.

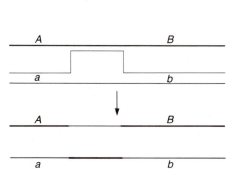

Figure 7.13   If two crossovers occur between markers A and B, then they remain on the same chromosome and the recombination events are not revealed by examining just these markers.

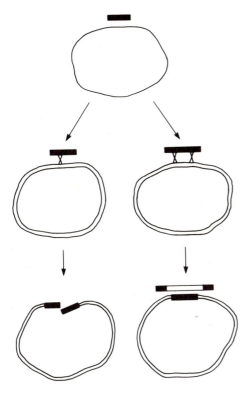

Figure 7.14   In order that a circular chromosome not be opened by genetic recombination, two crossovers between it and a fragment of linear DNA are necessary.

Here the crosses will be described as they are performed in prokaryotes in which two complete chromosomes are not present, but the basic principles apply for the situation of recombination between complete chromosomes. Mating experiments, as described in the next chapter, permit the introduction of a portion of the chromosomal DNA from a donor bacterial cell into a recipient cell. As a result of the inviability of a DNA fragment, if a crossover occurs between the chromosome of the recipient and the incoming chromosome fragment, a second crossover between the two DNAs must also occur for the recipient chromosome not to be broken by the single crossover event (Fig. 7.14). In general, any even number of total crossovers between the incoming DNA and the recipient chromosome will yield viable recombinants.

The genetic markers involved in the genetic cross, A, B, and C, can be different genes or different alleles within the same gene. In principle it matters not, but in practice it could be very difficult to ascertain the different varieties of recombinants if the markers all lie within the same gene.

To determine the gene order, compare the fraction of B^+-C^+ recombinants among all the A^+ recombinants in the two crosses (Fig. 7.15): first between an

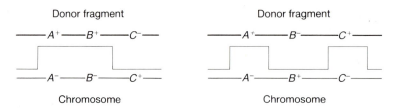

Figure 7.15   Three-factor crosses. The first requires only two crossovers to produce A^+-B^+-C^+ progeny whereas the second requires four.

A^--B^--C^+ chromosome and an A^+-B^+-C^- DNA fragment introduced into the recipient cell, and second between an A^--B^+-C^- chromosome and an A^+-B^--C^+ DNA fragment. If the gene order is A-B-C, then generation of B^+-C^+ recombinants from the first cross requires two crossover events, whereas the second cross requires crossovers in the same two intervals as the first plus an additional two crossovers. Therefore the fraction of the B^+-C^+ recombinants will be much higher from the first cross. Alternatively, if the gene order is A-C-B, the greatest fraction of B^+-C^+ recombinants will be found in the second cross.

MAPPING BY DELETIONS

Deletions may also be used in genetic mapping. The recombination frequency for generating functional genes is not measured in this type of mapping. Instead, all that is asked is whether or not a deletion and a point mutation can recombine to yield a functional gene. If they can, then the deletion must not have removed the nucleotide allelic to the point mutation.

Consider a series of strains each containing a deletion or a point mutation. Suppose the point mutations lie within a gene X and that the deletions all begin beyond the left end of X and extend various distances rightward into X (Fig. 7.16). If a diploid between deletion 1 and point mutation A can yield an X^+ recombinant, then A must lie to the right of the endpoint of 1. If A also fails to yield X^+ recombinants with deletion 2, then deletion 2 ends to the right of A and hence to the right of deletion 1. By this type of reasoning, a completely unordered set of deletions and point mutations may be ordered (see Problem 13).

How close to the end of a deletion can a crossover occur? Measurements in the *lacI* gene by Miller have shown that such crossovers may be detected within three nucleotides of the end of a deletion. Thus if a large number of deletions ending in a gene and point mutations in the gene are available, a precise map can be generated. Also, because the end points of deletions often appear to be randomly distributed, reasonably good comparisons of the sizes of intervals along the DNA in a gene may be made by comparing the numbers of deletions ending in the intervals.

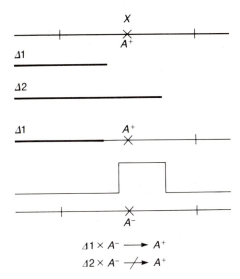

$\Delta 1 \times A^- \longrightarrow A^+$
$\Delta 2 \times A^- \not\longrightarrow A^+$

Figure 7.16   Recombination between a
chromosome containing a point mutation
and a DNA fragment containing a deletion.

HETERODUPLEXES LIKELY FORM DURING
GENETIC RECOMBINATION

Having considered the existence and use of genetic recombination, we are ready to
consider how it comes about. Genetic recombination yields a precise cut and splice
between two DNA molecules (see Problem 10). Even if one DNA molecule were to
have been cut, it is hard to imagine how an enzyme could know where to cut the
other DNA duplex so as to produce the perfect splices that genetics experiments
show occur. The difficulty can be largely overcome by a mechanism utilizing the
self-complementary double-stranded structure of DNA. A denatured portion of one
duplex could anneal to a denatured portion of complementary sequence from the
other duplex (Fig. 7.17). This would hold the two DNA molecules in register while
the remainder of the recombination reaction proceeded.

The life cycle of yeast permits a direct test of the model outlined above. A diploid
yeast cell undergoes recombination during meiosis, and the two meiotic cell divisions yield four haploid spores. These four spores can be isolated from one another
and each can be grown into a colony or culture. In essence, the cells of each colony
are identical copies of one of the original recombinants, and the cells can be tested to
determine the genetic structure of the original recombinants. If one of a pair of
homologous chromosomes contains a mutation and the other does not, generally
two of the four resulting spores will contain the mutation and two will not.

Consider the situation resulting from melting portions of the duplexes and base
pairing between complementary strands of two homologous yeast chromosomes in

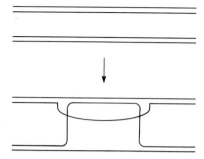

Figure 7.17   One possible mechanism for obtaining cross hybridization between homologous chromosomes during genetic recombination.

the process of genetic recombination. The region of pairing may include the mutation, so a heteroduplex forms that contains the mutant sequence on one strand and the wild-type sequence on the other (Fig. 7.18). As discussed in Chapter 3, mispaired bases are subject to mismatch repair and, if it occurs, the yeast repair system has no apparent reason to choose one strand to repair in preference to the other. Therefore strands may be correctly or incorrectly repaired, so the final outcome can be three copies of the wild-type or mutant sequence and one copy of the other in the meiosis from a single yeast cell. In total, a single yeast cell can produce one or three progeny spores containing the marker from one of the original chromosomes. Without heteroduplex formation and mismatch repair, there is no easy way to generate any ratio other than 2 : 2. This phenomenon is called gene conversion. It is experimentally observed and consequently it is reasonable to expect that pairing between complementary strands of recombinant partners occurs during recombination.

BRANCH MIGRATION AND ISOMERIZATION

Evidence mentioned in the previous section suggests that DNA duplexes engaged in recombination are likely to involve heteroduplexes consisting of one DNA strand from each parent. The problem we will address here is one way these heteroduplexes might be formed and what steps might be necessary to convert them to recombinants. As a glance at any genetics book will show, there are many schemes consisting of more or less reasonable steps that conceivably could be catalyzed by enzymes and that would ultimately lead to the generation of genetic recombination. We will outline one of these.

A nick in one DNA strand could lead to formation of heteroduplexes through DNA polymerization that displaces one strand (Fig. 7.19). This displaced DNA would then be free to anneal with the complementary strand from the other duplex. Indeed, if the other duplex is supercoiled, displacement by the single strand is

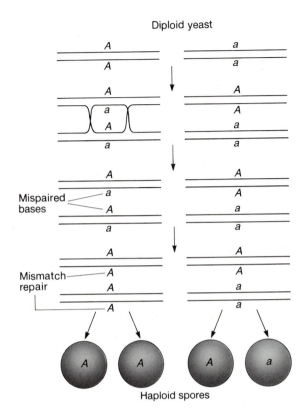

Diploid yeast

Mispaired bases

Mismatch repair

Haploid spores

Figure 7.18   Gene conversion in yeast. A diploid (A/a) undergoes meiosis, which produces heterodu-plexes A-a that are both repaired to A-A.

energetically favored. The displaced loop could be removed and the crossover point could drift in either direction along the DNA. Such drifting is rapid and is isoener-getic since one base pair is formed as another base pair is broken. This is called branch migration.

Before proceeding with the recombination mechanism, we must consider a diversion. For simplicity we will examine a double crossover and then apply the principles to the situation described in the previous paragraph. The strands that appear to cross from one duplex over to the other are not fixed! What is the basis of this remarkable assertion? By a simple reshuffling of the DNA in the crossover region, the other strands can be made to appear to be the ones crossing over (Fig. 7.20). This reshuffling is called isomerization. To understand, consider the more dramatic transformations as shown in the figure. These result in a change in the pair of strands that cross from one duplex to the other. In essence, however, they amount to little more than looking at the DNA from a different angle. Isomerization requires

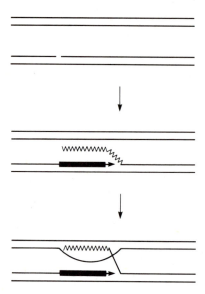

Figure 7.19   One possible mechanism for initiating genetic recombination. A nick is converted to a crossover region by DNA polymerization.

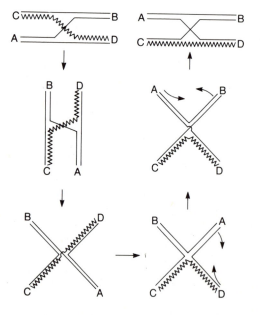

Figure 7.20   DNA crossover region during isomerization. Beginning at the upper right, a series of rotations as indicated ultimately yields the molecule at the upper left, which appears to have altered the DNA strands that cross over from one molecule to the other.

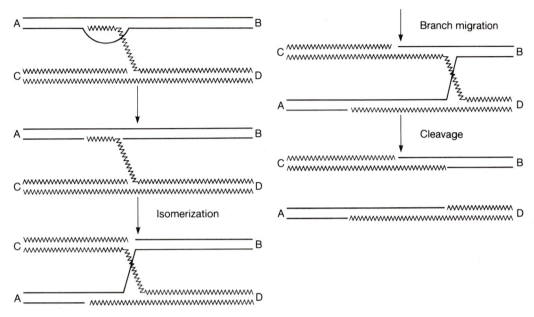

Figure 7.21   One possible pathway for genetic recombination. A nick is converted to a crossover region, which isomerizes and branch migrates, and finally the strands crossing over are cleaved.

only minor structural shifts in the crossover region and therefore is free to occur during genetic recombination.

Now let us return to the crossover mechanism. At the one crossover stage, an isomerization yields a two-strand crossover (Fig. 7.21). Branch migration followed by cleavage of these strands then produces a crossover between the two parental duplexes with a heteroduplex region near the crossover point.

ELEMENTS OF RECOMBINATION IN *E. coli, RecA, RecBC,* AND *Chi*

What types of biochemical evidence can be found in support of the mechanisms of recombination discussed in the previous section? Historically, the elucidation of metabolic pathways was often assisted by the isolation of mutations in enzymes catalyzing individual steps. With similar objectives in mind, mutations have been isolated that decrease or increase the ability of phage, *E. coli*, yeast, and some other organisms to undergo recombination. Subsequently the enzyme products of the recombination genes from *E. coli* have been identified and purified. Of greatest importance to recombination are the *recA* and *recBC* proteins. Additional enzyme activities that might be expected to play roles, such as DNA ligase, DNA polymerases, single-stranded binding protein, and proteins that wind or unwind DNA, have already been discussed and will not be further mentioned here.

RecA mutants are unable to engage in genetic recombination, and the *recA* protein possesses a variety of activities that appear related to recombination pathways. The protein catalyzes the invasion by a single strand of double-stranded DNA. Additionally, the protein is able to bind to single-stranded DNA and catalyzes the renaturing of denatured DNA. For this latter activity, energy from ATP hydrolysis is required. The protein is also able to promote branch migration.

The activities of the *recBC* protein relevant to recombination are also tantalizing, but the actual role in recombination is not apparent. Under some conditions the protein acts as a single-strand and double-strand exonuclease or even as single-strand endonuclease. Under conditions that appear to be closer to those found within cells, the protein acts as a DNA-dependent ATPase and denatures DNA. In some *in vitro* reactions, the protein begins at the end of linear DNA molecules and moves down the DNA at a rate of about 300 nucleotides per second, separating the DNA duplex into single-stranded regions (Fig. 7.22). The enzyme brings the strands back together at a rate of about 200 nucleotides per second. Thus, as the protein moves down the DNA, the enzyme forms single-stranded loops that grow at a rate of

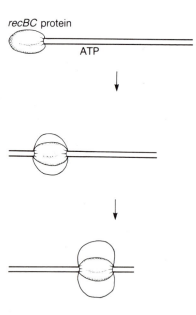

Figure 7.22   Single-stranded bubbles produced by *recBC* protein upon hydrolysis of ATP.

about 100 nucleotides per second. These could participate in the initial event of strand invasion of one duplex by the other.

A DNA element is also necessary for the *recA, recBC* recombination pathway. This is a sequence of eight bases with the name *chi* that is necessary on one of the DNA partners of recombination. Remarkably, *chi* can stimulate recombination at distances up to 10,000 bases away from itself and appears to act only in one direction, that is, it possesses a polarity. Most likely, *chi* acts as an entry site for *recBC* when the DNA is in the circular form.

In summary, this chapter has dealt with classical aspects of genetics, the assortment of chromosomes into progeny, the general principles of genetic mapping, and the biochemical basis of recombination. Although we have good pictures of the processes involved, their molecular and enzymatic details have yet to be worked out.

PROBLEMS

7.1. In many eukaryotes, genetic recombination between chromosomes does not occur during mitosis and is largely restricted to the first stage of meiosis. What is the possible biological value of such a restriction?

7.2. Sperm do not transfer mitochondria to the egg. How could you recognize mitochondrial markers by their genetic behavior?

7.3. Show how backcrossing may be used to determine the genotypes of heterozygotes. [*Hint:* Backcrossing means just about what it sounds like.]

7.4. Invent a scenario by which a single nucleotide change on a chromosome could lead to inactivation of all the genes of an entire chromosome. Note that one of the two X chromosomes in the females of many organisms is inactivated.

7.5. Invent a scenario by which a mutation could be methylation of a particular base on DNA. Note that to qualify as a mutation, the effect must be heritable. That is, if the base is not methylated, it remains nonmethylated, but if it somehow should become methylated, then all descendants will be methylated on the same base.

7.6. Methylation of a base of DNA in a promoter region could inactivate transcription from a gene. Invent genetic data that would lead to this conclusion. That is, the mutation might be *cis* or *trans* dominant, etc.

7.7. How could it be shown that two genetic markers were on the same chromosome in a diploid eukaryotic organism?

7.8. Suppose gene *A* was absolutely known to code for an enzyme and that a mutation inactivating the protein was absolutely known to lie within gene *A*. Why would you be surprised to learn that A^- was dominant to A^+, and what could be an explanation?

7.9. What is the interpretation of the following genetic mapping data? The recombination frequency between markers A and B was 0.002, between markers B and C was 0.002, and the frequency between markers A and C was 0.0035.

7.10. No experiment was described in the text that demonstrates that virtually all recombination events are precise, that is, in register, and do not delete base pairs from one DNA duplex partner and insert them into the other. Design one. [*Hint:* Three nonessential genes A, B, and C are in a row on one chromosome and we have $A^- B^+ C^+$ and $A^+ B^+ C^-$ strains available.]

7.11. Invent genetic recombination data that would lead to the prediction of a two-dimensional array of genetic markers rather than the one-dimensional array of markers that actually is found.

7.12. In mapping a small region by measuring the recombination frequencies between point mutations, the following data were obtained. The numbers show the recombination percentages between the indicated markers. What is the map of the region? In addition to the most likely gene order and the relative distances between the mutations, what else can you say?

		Markers			
Markers	1	2	3	4	5
1	—	—	9	—	4
2	7	0	—	—	5
3	—	—	—	—	—
4	—	—	11	—	—
5	—	—	—	1	0

7.13. The following data set was obtained in an experiment in which deletions entering gene X from the left end and ending somewhere within X were crossed with point mutations in gene X. The ability of the indicated deletion-point mutation pair to yield X^+ recombinants is indicated. What is the best map of the region? Discuss.

	Point Mutation					
Deletion	1	2	3	4	5	6
1	−	+	−	−	−	+
2	−	+	+	+	−	+
3	−	+	−	−	−	−
4	−	+	+	+	−	+
5	+	+	+	+	−	−
6	−	+	−	−	−	+

7.14. Consider a genetic crossing experiment with yeast involving allelic markers *a* and *A*. Why, on first consideration, would you expect that the number of cells yielding one *a* type and three *A* type spores would equal the number yielding three *a* type and one *A* type? What mechanism might be responsible for a serious biasing of the numbers in one direction?

7.15. Why can't gene conversion easily be observed in genetic crosses between bacterial strains?

7.16. What sequences ought to be particularly prone to one- or two-base insertions or deletions?

7.17. What nucleotide would guanine base pair with in DNA replication if it were ethylated on the carboxyl of C-6?

RECOMMENDED READINGS

A General Model for Genetic Recombination, M. Meselson, C. Radding, Proc. Natl. Acad. Sci. USA 72, 358–361 (1975).

RELATED REVIEWS, BOOKS, AND ARTICLES

Special Sites in Generalized Recombination, F. Stahl, Ann. Rev. of Genetics 13, 7–24 (1979).
Genetic Recombination, F. Stahl, W. H. Freeman and Co., San Francisco (1979).
Genetics, U. Goodenough, R. Levine, Holt, Rinehart and Winston, New York (1980).
Chi Hotspots of Generalized Recombination, G. Smith, Cell 34, 709–710 (1983).

DEEPER READING

Induced Mutations in Bacterial Viruses, M. Delbrück, W. Bailey Jr., Cold Spring Harbor Symposium of Quantitative Biology *11*, 33–37 (1946).

Spontaneous Mutations in Bacterial Viruses, A. Hershey, Cold Spring Harbor Symposium of Quantitative Biology *11*, 67–77 (1946).

A Mechanism for Gene Conversion in Fungi, R. Holliday, Genetics Res. *5*, 282–304 (1964).

Isolation and Characterization of Recombination-Deficient Mutants of *E. coli* K-12, A. Clark, A. Margulies, Proc. Natl. Acad. Sci. USA *53*, 451–459 (1965).

A Proposal for a Uniform Nomenclature in Bacterial Genetics, M. Demerec, E. Adelberg, A. Clark, P. Hartman, Genetics *54*, 61–76 (1966).

Formation of Hybrid DNA by Rotary Diffusion During Genetic Recombination, M. Meselson, J. Mol. Biol. *71*, 795–798 (1972).

Molecular Aspects of Genetic Exchange and Gene Conversion, R. Holliday, Genetics *78*, 273–287 (1974).

Mismatch Repair in Heteroduplex DNA, J. Wildenberg, M. Meselson, Proc. Natl. Acad. Sci. USA *72*, 2202–2206 (1975).

Repair Tracts in Mismatched DNA Heteroduplexes, R. Wagner, M. Meselson, Proc. Natl. Acad. Sci. USA *73*, 4135–4139 (1976).

On the Mechanism of Genetic Recombination: Electron Microscopic Observation of Recombination Intermediates, H. Potter, D. Dressler, Proc. Natl. Acad. Sci. USA *73*, 3000–3004 (1976).

Kinetics of Branch Migration in Double-Stranded DNA, B. Thompson, M. Camien, R. Warner, Proc. Natl. Acad. Sci. USA *73*, 2299–2303 (1976).

Hotspots for Generalized Recombination in the *E. coli* Chromosome, R. Malone, D. Chattoraj, D. Faulds, M. Stahl, F. Stahl, J. Mol. Biol. *121*, 473–491 (1978).

ATP-Dependent Renaturation of DNA Catalyzed by the *recA* Protein of *Escherichia coli*, G. Weinstock, K. McEntee, I. R. Lehman, Proc. Natl. Acad. Sci. USA *76*, 126–130 (1979).

Purified *Escherichia coli recA* Protein Catalyzes Homologous Pairing of Superhelical DNA and Single-Stranded Fragments, T. Shibata, C. DasGupta, R. Cunningham, C. Radding, Proc. Natl. Acad. Sci. USA *76*, 1638–1642 (1979).

Initiation of General Recombination Catalyzed in vitro by the *recA* Protein of *Escherichia coli*, K. McEntee, G. Weinstock, I. R. Lehman, Proc. Natl. Acad. Sci. USA *76*, 2615–2619 (1979).

Unwinding and Rewinding of DNA by the *RecBC* Enzyme, A. Taylor, G. Smith, Cell *22*, 447–457 (1980).

Mapability of Very Close Markers of Bacteriophage Lambda, G. Gussin, E. Rosen, D. Wulff, Genetics *96*, 1–24 (1980).

Chi Mutation in a Transposon and the Orientation Dependence of Chi Phenotype, E. Yagil, D. Chattoraj, M. Stahl, C. Pierson, N. Dower, F. Stahl, Genetics *96*, 43–57 (1980).

RecA Protein of *Escherichia coli* Promotes Branch Migration, a Kinetically Distinct Phase of DNA Strand Exchange, M. Cox, I. R. Lehman, Proc. Natl. Acad. Sci. USA *78*, 3433–3437 (1981).

Directionality and Polarity in *recA* Protein-Promoted Branch Migration, M. Cox, I. R. Lehman, Proc. Natl. Acad. Sci. USA *78*, 6018–6022 (1981).

Clustering of Mutations Inactivating a Chi Recombinational Hotspot, D. Schultz, J. Swindle, G. Smith, J. Mol. Biol. *146*, 275–286 (1981).

Structure of *Chi* Hotspots of Generalized Recombination, G. Smith, S. Kunes, D. Schultz, A. Taylor, K. Triman, Cell *24*, 429–436 (1981).

Identity of a Chi Site of *Escherichia coli* and Chi Recombinational Hotspots of Bacteriophage Lambda, K. Triman, D. Chattoraj, G. Smith, J. Mol. Biol. *154*, 393–398 (1982).

Orientation of Cohesive End Site *cos* Determines the Active Orientation of Chi Sequence in Stimulating *recA-recBC*-Mediated Recombination in Phage Lambda Lytic Infections, I. Kobayashi, H. Murialdo, J. Craseman, M. Stahl, F. Stahl, Proc. Natl. Acad. Sci. USA *79*, 5981–5985 (1982).

On the Formation of Spontaneous Deletions: The Importance of Short Sequence Homologies in the Generation of Large Deletions, A. Albertini, M. Hofer, M. Calos, and J. Miller, Cell *29*, 319–328 (1982).

Activation of Chi, A Recombinator, by the Action of an Endonuclease at a Distant Site, M. Stahl, I. Kobayashi, F. Stahl, S. Huntington, Proc. Natl. Acad. Sci. USA *80*, 2310–2313 (1983).

Genetic Systems

A part of an overall investigation of many biological problems consists of isolating mutations, mapping them, and determining their effects in the organism. The previous chapter considered these questions in the abstract. Here we will examine how these questions may be handled in bacteria, yeast, and the fruit fly *Drosophila melanogaster*. The same basic operations used with bacteria or yeast are also used with most other unicellular organisms or with cell cultures from multicellular organisms. Similarly, the principles and genetic operations used with *Drosophila* are similar to many used with other higher organisms, although the *Drosophila* genetic system is much more highly developed than other systems.

Phage and bacteria have been important in molecular biology for many reasons. Among them is the fact that experiments can be performed rapidly, the systems under study appear to be sufficiently simple as to be tractable with reasonable investments of time and effort, genetic experiments can easily be performed, and, most important, biochemical experiments can be performed, mutants can be grown, and their altered genes or gene products can be isolated and definitively tested.

Yeast possess many of the virtues of bacteria. As a simple eukaryote, however, many of the important questions being studied with yeast involve components or processes that are not found in bacteria or phage, such as properties of mitochondria or messenger RNA splicing. One of the most useful properties of yeast is the ease of generating haploids and diploids. The facile generation of mutants requires the haploid form, but complementation studies and genetic mapping require forming of diploids.

The fruit fly *Drosophila melanogaster* has been energetically studied by geneticists since about 1910. It is a eukaryote with differentiated tissues and is relatively easy and inexpensive to study. A large number of mutations have been cataloged and mapped as well as a wide variety of chromosome aberrations such as inversions, substitutions, and deletions. Fortunately *Drosophila* permits the study of many of these rearrangements with the light microscope as the chromosomes in the salivary glands are highly polytene. They contain about 1,000 parallel identical copies. This large amount of DNA and associated macromolecules generates banding patterns characteristic of each region of the chromosome. Perhaps the question of greatest importance presently being investigated with *Drosophila* is tissue-specific gene expression. This includes the differentiation of tissues during development as well as the expression of genes in specific tissues after development. Remarkably sophisticated genetics tools have been developed for investigating these questions. On the other hand, biochemical approaches to these questions are just beginning to be developed, as will be described in following chapters.

GROWING CELLS FOR GENETICS EXPERIMENTS

A culture should be genetically pure before one attempts to isolate a new mutant or to study the properties of an existing mutant. Being genetically pure means that all the cells of a culture are genetically identical. The easiest way to ensure the requisite purity is to grow cultures from a single cell. Then all the cells will be descendants of the original cell, and the culture will be pure unless the spontaneous mutation rate is excessive.

Escherichia coli, the strain of bacteria most widely used in molecular biology, can be simply purified by streaking a culture on a petri plate containing "rich" nutrient medium consisting of many nutrients such as glucose, amino acids, purines, pyrimidines, and vitamins. These plates permit growth of virtually all *E. coli* nutritional mutants and wild type as well as many other types of cells. Streaking is performed by sterilizing a platinum needle, poking it into a colony or culture of cells, and lightly dragging it across an agar surface so that at least a few of the deposited cells are sufficiently isolated that they will grow into isolated and therefore pure colonies.

Essentially, the only modification in the above procedure necessary for other cell types is to use appropriate media. In some cases, however, the cells will grow only in liquid medium. Then purification can be accomplished by diluting a number of cultures to the point that most contain no cells, and therefore the few cultures containing at least a single cell are not likely to contain two or more cells. Often cells from higher organisms display a density-dependent growth and will not divide unless the cell density is above a critical value. Then the isolation of a culture from a single cell requires the use of tiny volumes. One method is to use microdrops suspended from glass cover slips. Evaporation from the drops is prevented by placing the cover slips upside down over small chambers filled with the growth medium.

TESTING PURIFIED CULTURES, SCORING

Once a strain has been purified, its phenotype can be tested. Suppose, for example, that a desired bacterial strain is supposed to be Lac⁻ and Arg⁻, that is, it is unable to use lactose as a sole source of carbon and energy and it is unable to synthesize arginine. What is the simplest way to score the colonies for being Lac⁻ and Arg⁻? One satisfactory growth medium for the mutant would be a medium containing K^+, Na^+, PO_4^{--}, SO_4^{--} Mg^{++}, Ca^{++}, Fe^{++}, NH_4^+, and other trace elements that are present as chemical contaminants of these ingredients, plus glucose at a concentration of 2 g/liter and L-arginine at 0.1 g/liter. This is called a minimal salts medium with glucose and arginine. Each candidate colony could be spotted with sterile wooden sticks or toothpicks onto minimal media containing the following additions: glucose; glucose plus arginine; and lactose plus arginine. The desired strain should grow only on the glucose plus arginine medium (Table 8.1).

ISOLATING AUXOTROPHS, USE OF MUTAGENS, AND REPLICA PLATING

Consider the isolation of a bacterial strain requiring leucine as a growth supplement added to the medium. One method of isolating a leucine-requiring strain would be to dilute cells to a concentration of about 1,000 cells/ml and spread 0.1-ml quantities on the surface of 10,000 glucose plus leucine plates. After these had grown, each of the 1,000,000 colonies could be spot tested for a leucine requirement. By this method, one spontaneously occurring Leu⁻ mutant out of 10^6 cells could be found. This is not a workable method, however, and when faced with problems like this, geneticists devised many shortcuts.

One way to decrease the work of finding a mutant is to increase its frequency of occurrence. Standard techniques for increasing the frequency of mutants in a popu-

Table 8.1 Spot Testing for Arg⁻ and Lac⁻

	Minimal Glucose	Minimal Glucose + Arginine	Minimal Lactose + Arginine
Wild Type (lac⁺, arg⁺)	+	+	+
Arg⁻	−	+	+
Lac⁻	+	+	−
Arg⁻, lac⁻	−	+	−

Table 8.2   Properties of Mutagens

Mutagen	Dominant Effect
N'-methyl-N'-nitro-N-nitrosoguanidine	GC → AT
Ethylmethanesulfonate	GC → AT
Hydroxylamine	GC → AT
Nitrous acid	GC → AT
2-aminopurine	GC → AT
5-bromouracil	GC → AT
UV light	Transversions
Spontaneous	Transversions
MutT mutator gene	AT → GC

---

lation are treatment with chemical mutagens, exposure to UV light, or the use of mutator strains (Table 8.2). The spontaneous mutation frequency in such strains is greatly elevated, for example due to mutations in DNA polymerase. Another short-cut is to reduce the time required to spot many colonies in the scoring steps. Replica plating allows spotting all the colonies from a plate in one operation onto a testing plate. This can be done with a circular pad of sterile velvet or paper that is first pushed against the master plate of colonies. The paper or velvet picks up many cells, which can then be deposited onto a number of replica plates.

In some situations the use of a mutagen might be unwise because of the possibility of introducing more than one mutation into the strain. Therefore a spontaneously occurring Leu⁻ mutant might have to be found. Even with replica plating this could entail much work, and a method of selectively killing all the Leu⁺ cells in the culture would be most valuable.

Penicillin provides a useful reverse selection for bacterial mutants. Ordinarily mutants capable of growing in a particular medium may easily be selected. However, the reverse—the selection for the mutants unable to grow in a particular medium—requires a trick. One is penicilin. This antibiotic blocks formation of the peptide crosslinks in the peptidoglycan layer. As only growing cells synthesize or try to synthesize peptidoglycan, only growing cells in the presence of penicillin make defective peptidoglycan. These burst from their internal osmotic pressure if they are grown in ordinary medium that has a lower osmotic pressure than the cell cytoplasm. Nongrowing cells are penicillin-resistant because they do not attempt to synthesize cell wall. Accordingly, leucine-requiring cells can be selected with penicillin.

A culture of the desired cells is grown on minimal medium containing leucine, then the cells are removed either by filtration or centrifugation. The cells are resuspended in medium lacking leucine, and penicillin is added. The Leu$^+$ cells continue growing and are killed by the penicillin, whereas the Leu$^-$ cells stop growing and remain resistant to the penicillin. A 90-minute treatment kills about 999/1000 of the Leu$^+$ cells. The penicillin is removed and the surviving cells are plated out for spot testing to identify the Leu$^-$ colonies.

GENETIC SELECTIONS

Selective growth of mutants requires the use of conditions in which the desired mutant will grow but the remainder of the cells, including the wild-type parents, will not. This is to be contrasted with scoring, in which all the cells grow and the desired mutant is identified by other means. Often in scoring, all the cells are plated out to form colonies. These are then spotted onto various media on which the mutant may be identified or are grown for assay of various gene products. A simple example of selecting a desired mutant would be isolating a Lac$^+$ revertant from a Lac$^-$ mutant by plating the cells on minimal plates containing lactose as the sole source of carbon. Up to 10^{11} cells could be spread on a single plate in search of a single Lac$^+$ revertant (Fig. 8.1).

A slightly more complicated mutant selection is the use of an agent whose metabolism will create a toxic compound. Cleavage of orthonitrophenyl-β-D-thiogalactoside by the enzyme β-galactosidase yields a toxic compound and cells die. Thus Lac$^+$ cells expressing β-galactosidase in a medium containing glycerol and orthonitrophenyl-β-D-thiogalactoside are killed and only Lac$^-$ cells survive.

Let us examine a more complicated selection, the isolation of a nonsense mutation in the β subunit of RNA polymerase. Such a mutation would be lethal under normal circumstances since RNA polymerase is an essential enzyme and a nonsense mutation terminates translation of the elongating polypeptide chain. However, the desired mutant can be identified if the cells contain a temperature-sensitive nonsense suppressor, Sup$^+$(ts). Such a suppressor results from a mutant tRNA that suppresses nonsense codons but loses its suppressing activity at 42°, a temperature at which *Escherichia coli* normally survives.

Lac$^-$ $\xrightarrow{\text{Select}}$ Lac$^+$ Growth on minimal salts lactose

Lac$^+$ $\xrightarrow{\text{Select}}$ Lac$^-$ Growth on minimal salts glycerol plus orthonitropheny-β-D-thiogalactoside

Figure 8.1   Positive selections for Lac$^+$ and Lac$^-$ cells.

The two key facts in the selection are that rifamycin sensitivity is dominant to resistance and that rifamycin-resistant, Rif^r, mutants possess altered β subunits of RNA polymerase. The first step in selecting the desired mutant is to mutagenize Arg$^-$ Rif^s Sup$^+$(ts) Sm^r (streptomycin-resistant) cells and to grow them at the low temperature of 30°. Under these conditions, a nonsense mutation in the *rif* allele would not be lethal since the suppressor would be active and the complete β chain would be synthesized.

After the first step, a region of the chromosome containing the Arg$^+$ Rif^r genes could be introduced by mating with an appropriate Sm^s strain containing an Arg-Rif episome (see later section for episomes). By selecting for growth in the absence of arginine and the presence of streptomycin, cells diploid for this region could be selected. The majority of the cells that grow will then have the genotype Arg$^+$ Rif^r / Arg$^-$ Rif^s Sup$^+$(ts) Sm^r, in which the genes before the "/" represent episomal genes. Among the cells will be a few of the desired genotype, Arg$^+$ Rif^r / Arg$^-$ Rif^s(amber) Sup$^+$. Since rifamycin sensitivity is dominant to rifamycin resistance, the cells with the amber mutation in the β subunit will be rifamycin-resistant at 42° whereas the others will remain rifamycin-sensitive (Fig. 8.2). The desired cells are able to grow at 42° on minimal medium lacking arginine and containing rifamycin. How could the desired amber mutant be distinguished from a strain that merely

Figure 8.2   Selection for a strain containing a nonsense mutation in the β subunit of RNA polymerase.

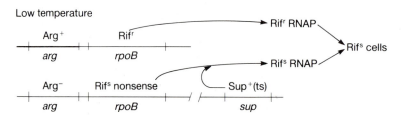

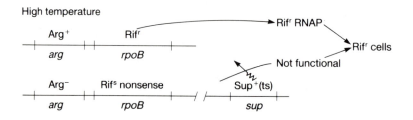

contained a missense mutation in the β subunit? These strains would remain rifamycin-resistant even at 30° whereas the amber mutants would become rifamycin-sensitive!

MAPPING WITH GENERALIZED TRANSDUCING PHAGE

As we have already discussed in Chapter 3, coliphage P1 is imperfect in its encapsulation process, and instead of packaging P1 DNA, a small fraction of the phage package a segment of their host's DNA. Upon infection of sensitive cells with such a lysate, the majority of infected cells are injected with a phage DNA molecule. However, a small fraction of the cells are injected with a segment of the chromosome of the cells on which the phage lysate was prepared. This DNA segment cannot permanently survive in the cells because it lacks a DNA replication origin and, furthermore, exonucleases would degrade it. However, it is free to engage in genetic recombination before it is degraded. This type of transfer of genetic markers via a phage is called generalized transduction.

The ability to transduce cells with a phage like P1 facilitates fine-structure genetic mapping. Three-factor genetic crosses may be performed with phage P1 to determine the order of genetic markers regardless of the sex of the cells. Even more useful, however, are the consequences of the fact that the length of DNA that a P1 phage particle carries is only 1% of the size of the bacterial chromosome. Therefore the frequency with which two genetic markers are simultaneously carried in a single transducing phage particle is high if they are close to one another. The frequency falls if the markers are more widely separated, and it becomes zero if they are too widely separated, about 1% of the size of the chromosome (Fig. 8.3). As a result, the frequency of cotransduction of two genetic markers provides a good measurement of their separation.

Cotransduction frequency often is easy to measure. For example, bacterial conjugation experiments could have indicated that the genes for synthesis of leucine and the genes permitting utilization of the sugar arabinose were closely spaced on the chromosome. P1 mapping can measure their separation more accurately. P1 could be grown on an Ara$^+$Leu$^+$ strain and used to transduce an Ara$^-$Leu$^-$ strain to

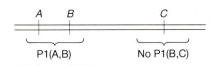

Figure 8.3   A portion of a bacterial chromosome showing genetic markers A, B, and C and the length of DNA that can be accommodated within one full phage head of P1. A and B can be copackaged, whereas B and C cannot.

Ara$^+$ by spreading the P1 infected cells on agar containing minimal salts, arabinose as a carbon source, and leucine. Then the Ara$^+$ transductants could be scored by spot testing for the state of their leucine genes.

PRINCIPLES OF BACTERIAL SEX

Lederberg and Tatum discovered that some bacterial strains can transfer DNA to recipient cells, that is, they can mate. This discovery opened the doors for two types of research. First, genetic manipulations could be used to assist other types of studies involving bacteria; throughout this book we will see many examples of the assistance genetics provides to biochemical, physiological, and physical studies. Second, the mechanism of bacterial mating itself was interesting and could be investigated. In this section we will review the actual mechanism of bacterial mating.

Male cells possess a mating module called the F-factor, for fertility, consisting of DNA containing about 25 genes. One of the F genes codes for the F-pilus, an appendage essential for DNA transfer. Others code for additional parts of the F-pilus, its membrane attachment, and DNA replication of F-factor DNA. The system also contains at least one regulatory gene.

F-pilus contact with a suitable female activates the mating module. As a result, a break is made within the F sequences of DNA, and a single strand of DNA synthesized via the rolling circle replication mode is transferred into the female (Fig. 8.4). Immediately on entry of the single strand into the female, the complementary strand is synthesized. If no breakage occurs during transfer, all the F-factor DNA, including the portion initially left behind at the break, can be transferred into the female cell. The F-factor thus codes for transfer into female cells of itself and also any DNA it is connected to. If the F-factor is located in the chromosome, then the entire chromosome can be mobilized for transfer into a recipient cell.

A little more than 100 minutes are required for transfer of the entire *E. coli* chromosome to a female. Genes located anywhere on the chromosome can be mapped by determining the timing of their transfer to female cells. Mating is initiated by mixing male and female cells, and genetic markers on one side of the F-factor will be transferred in a few minutes, while markers on the opposite side of the

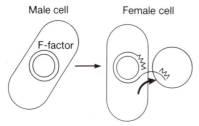

Male cell   Female cell

F-factor

Figure 8.4   Transfer of one strand of the double-stranded F-factor from male cells into female cells.

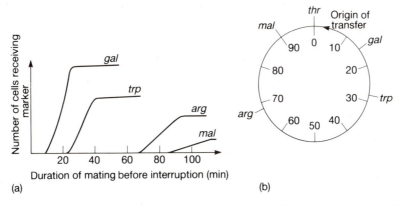

Figure 8.5   (a) The number of recombinants of the types indicated as a function of the duration of mating. (b) The approximate chromosomal locations of the markers used in the mating.

chromosome will not be transferred for more than half an hour (Fig. 8.5). At any time, the further transfer of markers can be stopped merely by vigorously shaking the culture to separate the mating couples.

Cells that transfer their chromosomal genes to recipient cells are called Hfr for high frequency of recombination; the cells that contain F-factors separate from the chromosome are called F or F′, and the females are called F⁻. Unfortunately, random breakage of the DNA during mating from an Hfr to an F⁻ often interrupts the transfer so that only infrequently can the complete chromosome be transferred. This is a blessing in disguise, however, for in addition to the time of transfer, the frequency of a marker's transfer indicates its chromosomal position with respect to the integrated F-factor. Fortunately, more than one integration site of F has been found, so a variety of origins of transfer are available and the entire chromosome can be easily mapped.

A mating experiment might be performed in the following way. Male Arg⁺ Gal⁺ Leu⁺ Smˢ (streptomycin-sensitive) and female Arg⁻ Gal⁻ Leu⁻ Smʳ cells would be grown to densities of about 1×10^8 per ml and mixed together in equal portions. The cells would then be shaken very gently. At intervals, a sample of cells would be taken, vigorously shaken so as to separate mating couples, and dilutions spread on streptomycin-containing medium so as to select for Arg⁺ or Gal⁺ or Leu⁺ Smʳ recombinants.

F-factors facilitate genetic study with E. coli in a second way. They need not be associated with the chromosome. Instead they can be autonomous DNA elements existing alongside the chromosome. The transfer properties remain the same for an F-factor in this position, but the genetics are slightly altered in two ways. First, the extrachromosomal F-factor can have some chromosomal genes associated with it, in

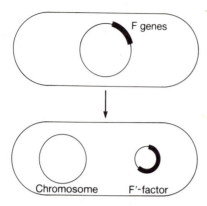

Figure 8.6   Formation of an F'-factor carrying part of the chromosome.

which case it is called an F'-factor. Presumably these genes were picked up at an earlier time by an excision of the integrated F-factor from the chromosome of an Hfr cell (Fig. 8.6). The second alteration of an extrachromosomal F-factor is that often the entire mating module and attached genes are transferred to a female. As a result, the female acquires a functional F-factor with its associated genes and becomes diploid for a portion of the chromosome as well as capable of transferring copies of the same F' onto other female cells.

Little is known about the F-factor DNA replication system. Chromosomal DNA replication in an Hfr remains under control of its own genes and begins at its own origin, *oriC*, despite the presence of the integrated F-factor. However, the DNA replication system of an F-factor is activated when it excises from the chromosome, for as an extrachromosomal element it must initiate DNA replication from its own origin.

ELEMENTS OF YEAST GENETICS

Yeast contain about 1.5×10^7 base pairs of DNA per cell, about five times the amount per bacterium. In baker's yeast, *Saccharomyces cerevisiae*, the number of chromosomes is about 17, so most of the chromosomes are substantially smaller than the *E. coli* chromosome.

As already mentioned, yeast can be found as haploids or diploids. This property greatly facilitates mutation isolation and genetic analysis and, combined with the small chromosome size, makes yeast a good choice for experiments that require eukaryotic cells. A diploid yeast cell can grow in culture much like a bacterium, although yeast divide not in half like bacteria, but they generate daughter buds that enlarge and finally separate from the mother cell. In contrast to *E. coli,* but like some other bacteria, yeast can sporulate. This occurs if they are starved of nitrogen in the presence of a nonfermentable carbon source like acetate. In this 24-hour process, a single diploid yeast cell undergoes meiosis and forms an ascus containing four spores. On incubation in rich medium, the spores will germinate and grow into haploid yeast cells (Fig. 8.7).

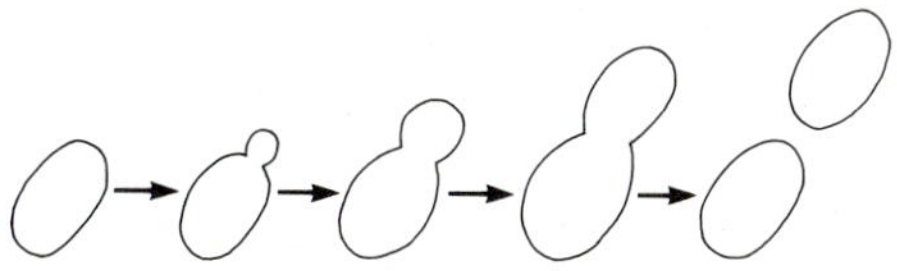

If one of the spores is isolated from an ascus and is grown up separately from other spores, the resulting culture is different from the parental culture. The cells are haploid and they cannot sporulate. In fact, an ascus contains two types of spores. These are designated as mating types a and α. Both of these generate haploid cultures and cannot sporulate. However, if a and α cultures are mixed, cells of opposite mating type adhere and form mating pair aggregates. First the cytoplasms of these pairs fuse, and then the nuclei fuse to generate diploid cells that grow and remain diploids. The diploids then can sporulate and regenerate the haploids of mating types a and α.

Recombination can be used for genetic mapping in yeast much as it is used in bacteria. However, since no yeast phage are known, genetic transfer must be accomplished either by the fusion of haploids as described above or by direct DNA transfer as described in a later chapter on genetic engineering. Most genetic recombination occurs during meiosis, although mitotic recombination can be detected.

Figure 8.7   The yeast cell cycle.

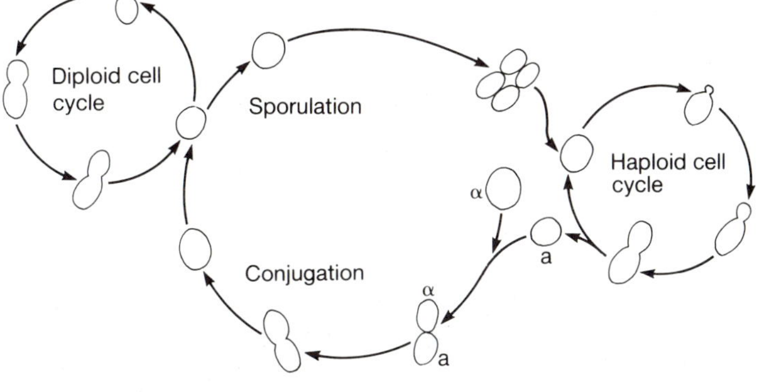

ELEMENTS OF *DROSOPHILA* GENETICS

Genetics can profitably be applied to the study of the development of tissues and tissue-specific gene expression as well as behavior, vision, muscle, and nerve function in *Drosophila.* The *Drosophila* genome of about 1.65×10^8 base pairs of DNA is contained in four chromosomes. Chromosomes II, III, and IV are normally found as pairs, while in males the constitution of the fourth chromosome is XY and in females it is XX. Unlike yeast, there is no simple way to make haploid fruit flies, and therefore the genes that are located on chromosomes II, III, and IV cannot easily be studied. However, genes located on chromosome X may easily be studied, for males are haploid for the X chromosome and females are diploid. Therefore recessive mutations in genes located on the X chromosome will be expressed in males, and the ability of X-located genes to complement can be tested in females.

Once a mutation is generated in the DNA of a bacterium, one of the daughter cells usually is capable of expressing the mutation. The analog is also true in *Drosophila.* However, the equivalent of a cell division in bacteria is the next generation in *Drosophila.* An adult fly can be mutagenized, but many of its genes are expressed only in development. Therefore mutagenized adults must be mated and their progeny must be examined for the desired mutation. One straightforward way to mutagenize flies is to feed them a 1% sucrose solution containing ethylmethanesulfonate (EMS). If male flies are mutagenized and mated with females, four types of progeny are obtained. In the first generation only females could contain a mutagenized X chromosome. If these females are collected and mated again with unmutagenized

$$
\begin{array}{c}
\text{Mutagenesis} \\[4pt]
\underset{\substack{\text{Male} \qquad \text{Mutagenized} \\ \text{male}}}{\text{XY} \longrightarrow \text{X'Y}} \quad \times \quad \underset{\text{Female}}{\text{XX}} \longrightarrow
\left\{
\begin{array}{l}
\text{X'X} \\
\text{X'X} \\
\text{YX} \\
\text{YX}
\end{array}
\right.
\end{array}
$$

males, in the second generation half of the females have a chance of receiving a mutagenized X chromosome. Clearly, considerable tedious sorting of males and females could be required in the detection of rare mutants.

ISOLATING MUTATIONS IN MUSCLE OR NERVE IN *DROSOPHILA*

How could mutations in muscle or nerve be isolated? Since these mutations could be lethal, they ought to be conditional. That is, the mutation ought to be expressed only under special conditions, for example at elevated temperature. Suzuki developed

$$X'Y \times \hat{X}XY \longrightarrow \begin{cases} X'\hat{X}X & \text{lethal} \\ X'Y & \text{male} \\ Y\hat{X}X & \text{female} \\ YY & \text{lethal} \end{cases}$$

Figure 8.8   Use of the attached X chromosome to generate special populations of flies. Although an XY fly is male, an $\hat{X}XY$ fly is female.

ingenious methods for the isolation of temperature-sensitive paralytic mutations. The flies ought to be perfectly normal at low temperature, be paralyzed at high temperature, and recover rapidly when returned to low temperature. Undoubtedly such mutations would be exceedingly rare, and great numbers of flies would have to be screened to find a few candidate mutants. Such large numbers necessitated the use of tricks to eliminate the need for sorting males and females.

The first trick used an attached X chromosome. This is an inseparable pair of X chromosomes. Males mated with attached X chromosome females yields the expected four types of offspring (Fig. 8.8). Both the $X'\hat{X}X$ and YY are lethal, and therefore only mutagenized males or nonmutagenized females result from this mating. If the attached X chromosome contains a temperature-sensitive lethal mutation, the females can be killed by a brief temperature pulse, leaving only the desired mutagenized males as a pure stock. The male and female stocks required for the first mating can be generated by this same technique with the result that sorting according to sex of the fly is eliminated.

The second problem was the actual selection for the temperature-sensitive paralytics. This was done by introducing up to 10^4 flies into a cubical box about two feet on a side. The temperature in the box was raised, and the box was given a bang on a tabletop to make the flies fly upward. Any temperature-sensitive paralytics remained at the bottom of the box. These were collected by rotating the box so that they were collected on a ledge. Then the flies were anesthetized by adding carbon

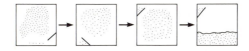

dioxide or ether, and those flies that had been able to fly fell to the bottom of the box where they were killed by the addition of detergent and acetic acid.

Like most mutant selection schemes, several additional phenotypes were found in addition to those sought. One of these was *rex*, for rapid exhaustion. After running around for few minutes a *rex* mutant shudders a bit and falls over in a paralysis that lasts a few minutes. It can then get up and is perfectly normal for about an hour. Another mutant was the *bas*, for bang-sensitive. The desired mutants fell into three types: par^{ts}, for temperature-sensitive paralytic; sts^{ts}, for stoned; and shi^{ts} from the Japanese word for paralyzed. Here the genetics have greatly outrun the biochemistry, and so far little is known about the biochemical bases for these mutations.

FATE MAPPING AND STUDY OF TISSUE-SPECIFIC GENE EXPRESSION

One obvious way to examine the tissue specificity of gene expression is to isolate the tissues and assay each for the protein or gene product in question. A slight modification of this approach is quite reasonable. The synthesis of messenger from the various tissues of a fly can be measured approximately by DNA-RNA hybridization. As we will see in Chapter 11, DNA from desired genes can be obtained and then used in such *in situ* hybridization experiments. Remarkably, genetics experiments called fate mapping can also locate the tissues in which an altered gene is expressed. This approach does not require knowledge of the gene involved and therefore it is useful in initial studies. For example, a fly may be unable to flap its wings because of a defective wing, a defective wing muscle, a defective nerve to the muscle, or defective neurons in the brain. Fate mapping can determine which tissue is responsible for such altered behavior.

Fate mapping relies on the developmental pathway of the fly. A fertilized *Drosophila* egg contains a single cell whose nucleus undergoes about nine divisions. These nuclei then migrate to the surface of the egg to form the blastula stage, and

Figure 8.9   Typical body color patterns encountered in fate mapping in which haploid and diploid tissues can be distinguished by body color.

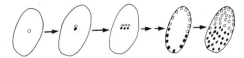

three more divisions occur before cell walls are laid down. At this stage, different cells on the surface ultimately become different parts of the adult fly, but cells lying near each other frequently develop into adjacent parts on the fly. Therefore a map can be drawn on the egg of the parts of the adult fly that each of the cells will become. If it were possible to associate a particular phenotype in an adult with a particular location on the egg, then the tissue responsible for the adult phenotype would be determined. This is possible!

The association of tissues in the mature fly with positions on the blastula utilizes selective chromosome loss during development of the blastula. Fly development is not greatly altered in a female egg cell if one of the X chromosomes contains a defect so that it begins the first nuclear replication a little late. Consequently this chromosome often is not segregated into one of the two resultant daughter nuclei resulting from the first nuclear division in the egg. The final result of this chromosome loss is that about half of the cells of the blastula will be diploid XX and the others will be haploid X. Since the spatial orientation of the first cleavage with respect to the egg shell is not uniform from egg to egg and because there is little mixing of the nuclei or cells during subsequent divisions, different sets of cells will be XX and X in different blastulas. Suppose that these two types of cells can be distinguished. This can be done by placing a recessive body color marker gene, for example yellow, on the X chromosome that is not lost. Then cells of the fly possessing the XX genotype will be black and the X genotype will be yellow. The adult fly will have a mottled appearance (Fig. 8.9).

The probability that two different body parts possess different colors will be proportional to the distance separating their corresponding ancestor cells in the blastula state. The greater their distance apart, the greater the probability that the dividing line between the two cell types will fall between them. If they are close together, there is little chance they will be of different cell type and therefore there is little chance they will possess different body color.

A collage appears when the body parts of the adult fly are mapped to the blastula (Fig. 8.10). The map can then be used as follows to locate the tissue in which a recessive mutation is expressed. If the mutation is located on the X chromosome that is not lost during the development, then those tissues that can express the mutant phenotype will be haploid. For example, if the mutant phenotype always and only appears in flies with haploid second left legs, then it can be concluded that the tissue in which the mutation is expressed is the second left leg. More generally, however, the frequency of association of the mutant phenotype with a number of landmarks gives the distance on the blastula between the landmarks and the tissue

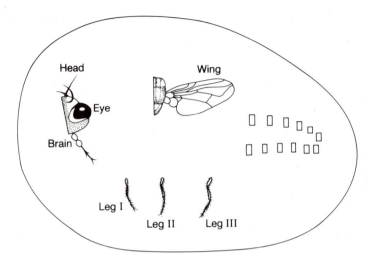

Figure 8.10   Locations of cells on one side of the blastoderm that give rise to the adult fly body parts indicated.

in question. Transferring these distances to the blastula fate map then reveals the tissue in question.

PROBLEMS

8.1. Why is streptomycin sensitivity dominant to streptomycin resistance? [*Hint:* Streptomycin-sensitive ribosomes synthesize toxic peptides in the presence of the drug.]

8.2. Why is rifamycin sensitivity dominant to rifamycin resistance?

8.3. In the usual protocol for phage P1 transduction of most genetic markers, the P1-infected cells are immediately spread on plates selective for the marker being transduced. Why can't this protocol be used for the transduction of the streptomycin and rifamycin resistance alleles? Note that these alleles are recessive to the sensitive alleles.

8.4. In the isolation of leucine-requiring strains with the use of a penicillin selection, why are the cells grown in leucine-containing medium before the addition of penicillin?

8.5. Nitrosoguanidine mutagenizes the growing points of the bacterial chromosome. Explain how this property could be used to isolate mutations concentrated in the vicinity of a selectable genetic marker. For example, if you possessed a Leu⁻ mutation, how could you isolate many mutations near the leucine genes?

*8.6. Nitrosoguanidine mutagenesis of *E. coli* growing rapidly on a rich yeast extract medium yielded double mutants that were simultaneously rifamycin-resistant and arabinose-minus. These markers lie one-third of the chromosome apart. These double mutants were not seen if the mutagenesis was performed on cells growing on minimal medium. Why? [*Hint:* See Problem 8.5.]

8.7. Expand on the following cryptic statement: Tritium suicide was used to isolate temperature-sensitive mutations in the protein synthesis machinery.

8.8. Physiological and *in vitro* measurements showed that the RNA polymerase from streptolydigin-resistant cells was resistant to the drug. It was then desired to map the mutation. Most likely this mutation would map near the rifamycin resistance mutation. The most straightforward way to show that the two markers were close to each other would have been to start with Stl^r Rif^s strain and to transduce it to Rif^r from Rif^{r}Stl^s and determine if it became Stl^s. Presumably this would show that the Rif and Stl alleles cotransduce. Why is this reasoning incorrect?

8.9. What are the expected effects on spontaneous mutation frequencies of hyper and hypo synthesis of the *dam* methylase? This protein is thought to mark "old" DNA strands by methylation of adenine residues in the sequence d(GATC).

8.10. Why are mutant reversion studies more sensitive than forward mutation studies in the detection of mutagenic properties of compounds? What are the limitations of determining mutagenic specificity by reversion studies?

8.11. What is a reasonable interpretation of the following? Ultraviolet light irradiation increases the frequency of mutations in phage only if *both* phage and cells are irradiated.

8.12. Look up and sketch the structural basis for the mutagenic actions of nitrous acid, hydroxylamine, ethylmethane sulfonate, 5-bromouracil, and 2-aminopurine.

8.13. What can you conclude about synthesis of the outer surface components of *E. coli* from the fact that male-specific and female-specific phage exist?

8.14. Yeast ordinarily contain 10 to 50 mitochondria. Since mitochondria contain DNA and code for some of their own proteins, mutations may be isolated in these organelles. How can mitochondrial genetic markers be recognized?

8.15. In yeast, how can mitotic and meiotic genetic recombination be distinguished?

8.16. Why is it not an omission to neglect consideration of the behavior of mutations located on the *Drosophila* Y chromosome?

8.17. How could a stock of female *Drosophila* be generated using the properties of the attached X chromosome?

8.18. How could flies unable to detect bright light be isolated?

8.19. Could fate mapping work if the orientation of the plane of the first nuclear division with respect to the egg were not random? Could it work if there were no mixing of cells during subsequent divisions? Could it work if there were complete mixing?

8.20. In fate mapping of a behavioral mutation in *Drosophila*, the following data were obtained by screening 200 flies. The mutation lies on the X chromosome, which also carries a recessive mutation that leads to a yellow body color. Indicate where on the blastoderm the behavioral mutation could map to.

	Behavior			Behavior	
Eye	Normal	Mutant	Wing	Normal	Mutant
Black	41	9	Black	39	11
Yellow	11	39	Yellow	12	38

RECOMMENDED READINGS

Genetic Studies of the *lac* Repressor, IV. Mutagenic Specificity in the *lacI* Gene of *Escherichia coli*, C. Coulondre, J. Miller, J. Mol. Biol. *117*, 577–606 (1977).

Dominant Mutators in *Escherichia coli*, E. Cox, D. Horner, Genetics *100*, 7–18 (1982).

Mapping of Behavior in *Drosophila* Mosaics, Y. Hotta, S. Benzer, Nature *240*, 527–534 (1972).

RELATED REVIEWS, BOOKS, AND ARTICLES

Sexuality and the Genetics of Bacteria, F. Jacob, E. Wollman, Academic Press, New York (1961).

Experiments in Molecular Genetics, J. Miller, Cold Spring Harbor Laboratory (1972).

The Microbial World, R. Stanier, E. Adelberg, J. Ingraham, Prentice-Hall, Englewood Cliffs, N.J. (1976).

The Conjugation System of F-like Plasmids, N. Willetts, R. Skurray, Ann. Rev. of Genetics *14*, 41–79 (1980).

The Biology of the Yeast Saccharomyces, Life Cycle and Inheritance, Ed. J. Strathern, E. Jones, J. Broach, Cold Spring Harbor Laboratory (1981).

The Molecular Biology of the Yeast Saccharomyces, Metabolism and Gene Expression, Ed. J. Strathern, E. Jones, J. Broach, Cold Spring Harbor Laboratory (1982).

Yeast Genetics, Fundamental and Applied Aspects, Ed. J. Spencer, D. Spencer, A. Smith, Springer-Verlag, New York (1983).

DEEPER READING

Novel Genotypes in Mixed Cultures of Biochemical Mutants of Bacteria, J. Lederberg, E. Tatum, Cold Spring Harbor Symposium of Quantitative Biology *11*, 113–114 (1946).

Transduction of Linked Genetic Characters of the Host by Bacteriophage P1, E. Lennox, Virology *1*, 190–206 (1955).

Selecting Bacterial Mutants by the Penicillin Method, L. Gorini, H. Kaufman, Science *131*, 604–605 (1960).

Optimal Conditions for Mutagenesis by N-Methyl-N'-Nitro-N-Nitrosoguanidine in *Escherichia coli*, E. Adelberg, M. Mandel, G. Chen, Biochem. Biophys. Res. Communications *18*, 788–795 (1965).

Behavioral Mutants of *Drosophila* Isolated by Countercurrent Distribution, S. Benzer, Proc. Natl. Acad. Sci. USA *58*, 1112–1119 (1967).

Temperature-Sensitive Mutations in *Drosophila melanogaster*, D. Suzuki, Science *170*, 695–706 (1970).

Localized Mutagenesis of Any Specific Small Region of the Bacterial Chromosome, J. Hong, B. Ames, Proc. Natl. Acad. Sci. USA *68*, 3158–3162 (1971).

Amber Mutations of *Escherichia coli* RNA Polymerase, S. Austin, I. Tittawella, R. Hayward, J. Scaife, Nature New Biology *232*, 133–136 (1971).

Pedigrees of Some Mutant Strains of *Escherichia coli* K-12, B. Bachmann, Bacteriol. Rev. *36*, 525–557 (1972).

Escherichia coli K-12 F-Prime Factors, Old and New. K. Low, Bacteriol. Rev. *36*, 587–607 (1972).

Rapid Mapping of Conditional and Auxotrophic Mutations in *Escherichia coli* K-12, B. Low, J. Bact. *113*, 798–812 (1973).

Culture Medium for Enterobacteria, F. Neidhardt, P. Bloch, D. Smith, J. Bacteriol. *119*, 736–747 (1974).

Fate Mapping of Nervous System and Other Internal Tissues in Genetic Mosaics of *Drosophila melanogaster*, D. Kankel, J. Hall, Dev. Biol. *48*, 1–24 (1976).

Genetic Engineering in vivo Using Translocatable Drug-Resistance Elements, N. Kleckner, J. Roth, D. Botstein, J. Mol. Biol. *116*, 125–159 (1977).

Genetic Studies of the *lac* Repressor VII: On the Molecular Nature of Spontaneous Hotspots in the *lacI* Gene of *Escherichia coli*, P. Farabaugh, J. Miller, J. Mol. Biol. *126*, 847–863 (1978).

Export Without Proteolytic Processing of Inner and Outer Membrane Proteins Encoded by F Sex Factor *tra* Cistrons in *Escherichia coli* Minicells, M. Achtman, P. Manning, C. Edelbluth, P. Herrlich, Proc. Natl. Acad. Sci. USA *76*, 4837–4841 (1979).

A Genetic Analysis of F Sex Factor Cistrons Needed for Surface Exclusion in *Escherichia coli*, M. Achtman, P. Manning, B. Kusecek, S. Schwuchow, N. Willetts, J. Mol. Biol. *138*, 779–795 (1980).

Mutagenic Deamination of Cytosine Residues in DNA, B. Duncan, J. Miller, Nature *287*, 560–561 (1980).

Linkage Map of *Escherichia coli* K-12, 6th ed., B. Bachmann, K. Low, Microbiol. Rev. *44*, 1–56 (1980).

Conditioning of Leg Position in Normal and Mutant *Drosophila*, R. Broker, W. Quinn, Proc. Natl. Acad. Sci. USA *78*, 3940–3944 (1981).

Genetic and Sequence Analysis of Frameshift Mutations Induced by ICR-191, M. Calos, J. Miller, J. Mol. Biol. *153*, 39–66 (1981).

Mosaic Analysis of a *Drosophila* Clock Mutant, R. Konopka, S. Wells, T. Lee, Mol. Gen. Genet. *190*, 284–288 (1983).

Probabilities and the Luria-Delbrück Fluctuation Test

A nalogous to the finding that most of the interesting astronomical objects are faint, most of the interesting genetics experiments seem to involve infrequent events. Fortunately, the power of genetic selections permits them to be studied relatively painlessly. Nonetheless, the numbers of events examined in typical experiments, that is, the number of cells having generated a particular genetic crossover, often range from a few to several thousand. The numbers could be made larger, but the time and expense would not justify the effort. Consequently, more than casual analysis of the results often is required to extract reliable conclusions from data of this nature. Consider a simple but common example where analysis of small numbers is required. Suppose a cell culture was diluted and plated, and yielding 100 colonies per plate — corresponding to an original cell density of 1×10^8 cells/ml. What are the chances that the culture was actually at 1.1×10^8, or 1.5×10^8, or even 2×10^8 cells/ml? Conversely, how many duplicate plates would have to be averaged to determine with a reliability of 90% that the average was within 5% of the true value?

In this chapter we will consider two types of statistics questions. The first concerns three widely encountered probability distributions: the binomial distribution, the Poisson distribution, and the Gaussian distribution. For simplicity, we will derive the binomial distribution and then consider the remaining two as special cases of the first. The second question considered in this chapter is closely related but much deeper mathematically. It concerns analysis of heritable random events that occur in growing populations and therefore is a frequently occurring topic in biology.

THE BINOMIAL PROBABILITY DISTRIBUTION

The calculation of probabilities in many situations merely is a question of counting the numbers of equally likely events. For example, a flipped coin can come down heads or tails with equal likelihood, and therefore the probability of throwing a heads in a single toss is $1/2$.

Consider the probabilities involved in flipping a weighted coin such that each time it is flipped the probability of coming up heads is p. What is the probability of tossing n heads out of a total of N tosses? First note that the probability of coming up tails must be $(1 - p)$ because the probability that the coin comes up one way or the other must be 1. Also, the chances on any toss are independent of the results of all previous tosses. Under these conditions the probability of any particular sequence of heads and tails is just the product of the probabilities for the individual tosses. For example, the probability of throwing n consecutive heads followed by $(N - n)$ tails is

$$\underbrace{p \times p \times p \times \cdots \times p}_{n \text{ factors}} \times \underbrace{(1 - p) \times (1 - p) \times \cdots \times (1 - p)}_{(N - n) \text{ factors}},$$

which equals $p^n(1 - p)^{N-n}$. This is one particular way of throwing the n heads — to have them all at the beginning.

Other patterns containing a total of n heads and $(N - n)$ tails can also occur. Any other particular pattern also has a probability of $p^n(1 - p)^{N-n}$. Since we are concerned merely with the number of heads and tails, not with their order of occurring, we must determine the total number of different patterns of n heads and $N - n$ tails. If this number of patterns is X, the probability of n heads is then $X \times p^n(1 - p)^{N-n}$.

Calculating the number of different orders in which n heads and $(N - n)$ tails appear in a total of N tries is equivalent to counting the number of ways of arranging n objects of one type and $(N - n)$ objects of another type into N positions in a line. To calculate this, assume for the moment that each of the objects is distinguishable. Then the total number of orders is $N!$, where $N! = N \times (N - 1) \times (N - 2) \times \cdots \times 1$, and is called N factorial. This results from the fact that any of the N objects can occupy the first position. For each of these N possibilities, any of the $(N - 1)$ remaining objects can occupy the second position. For each of these $N \times (N - 1)$ possibilities, any of the $(N - 2)$ remaining objects can occupy the third position, and so on. Now we will remove the constraint that each of the objects is distinguishable. As far as this problem is concerned, the N heads are completely indistinguishable among themselves. Therefore for the $N!$ total orders, the indistinguishability of n heads permits them to have $n!$ indistinguishable reorderings among themselves. These reorderings were counted in the total $N!$ and must be removed. Similarly, the indistinguishable reorderings of the $N - n$ objects must be removed. The final result is that there are a total of

$$\frac{N!}{n! \times (N - n)!}$$

orders of the two types of objects.

In the original problem the probability, $P(n)$, of throwing n heads and $N - n$ tails is therefore

$$P(n) = \frac{N!}{n! \times (N - n)!} \times p^n \times (1 - p)^{N-n}.$$

This is known as the binomial probability distribution.

DERIVATION OF THE POISSON DISTRIBUTION

The Poisson distribution arises frequently — from the scores in soccer games to the number of Prussian soldiers fatally kicked by horses in a year (Table 9.1) to the numbers of radioactive decays in an interval of time. It also describes the probabilities of spreading different numbers of cells on petri plates when identical plates are being prepared, the various numbers of phage that attach to different bacteria when a culture is infected, or the number of plasmids taken up by transformed cells.

The Poisson distribution is nothing but a special case of the binomial distribution. This special case occurs so often, however, that it is convenient to use a formation of the binomial distribution that avoids factorials and at the same time

Table 9.1   Number of Prussian Cavalry Killed Each Year by the Kick of a Horse, 1875–1894

Casualties in a Year in One Troop	Number of Occurrences of n Deaths in a Troop in a Year	
	Actual	Theoretical
0	144	139
1	91	97.3
2	32	34
3	11	7.9
4	2	1.39
≥5	–	0.196

Source: L. Bortkiewiecz, *Das Gasetz der kleinen Zahlen,* Teubner, Leipzig (1898), quoted by H. Jeffreys, *Theory of Probability,* 2nd ed. Oxford, (1948).

involves only the variables apparent in the experimental situation. The special case occurs when $n \ll N$, and $p \ll 1$. Approximating the binomial, we have

$$\frac{N!}{(N-n)!} = \underbrace{(N-n+1) \times (N-n+2) \times \cdots (N)}_{n \text{ factors}}$$

$$\approx N^n,$$

and, since

$$e^x = 1 + x + \frac{x^2}{2!} + \frac{x^3}{3!} + \cdots$$

for very small p,

$$e^{-p} \approx 1 - p, \quad \text{and}$$
$$(1-p)^{N-n} \approx (1-p)^N$$
$$\approx (e^{-p})^N$$
$$\approx e^{-Np}, \quad \text{and}$$

$$P(n) = \frac{N^n}{n!} p^n e^{-Np}.$$

Since the average or mean number events that will occur is pN, define pN to be m. Then, $P(n) = (m^n/n!)e^{-m}$. It is left as a problem to show that this is correctly normalized and hence that the sum of the probabilities of all possible results, from 0 to N, is 1, as indeed it must.

Note that the Poisson approximation has buried N, but it is usually unnecessary to recover it. Suppose, for example, that on the average a manuscript contains five mistakes per page. What is the probability that a page contains no mistakes? Since $m = 5$, $P(0) = e^{-5} = 0.0067$.

*THE GAUSSIAN DISTRIBUTION

The Gaussian distribution is the familiar bell-shaped curve. It is an approximation to a binomial distribution that applies when N is very large. Since the binomial distribution could be sharply peaked in the vicinity of its maximum, approximating the logarithm of $P(n)$ rather than $P(n)$ itself yields better results. This is so because a large change in the value of $P(n)$ will produce only a small change in the value of $\ln P(n)$. Approximating $\ln P(n)$ in a Taylor series around the point $n = N_0$ where $\ln P(n)$ assumes its maximum value, we have

$$\ln P(n) = q(N_0) + (n - N_0)q'(N_0) + \frac{(n - N_0)^2}{2} q''(N_0) + \cdots, \quad \text{and}$$

$$q(N_0) = \ln P(n), \qquad \text{when } n = pN,$$

$$q'(N_0) = \frac{d \ln P(n)}{dn} = 0, \qquad \text{when } n = pN,$$

$$q''(N_0) = \frac{d^2 \ln P(n)}{dn^2} = \frac{-1}{Np(1-p)}, \qquad \text{when } n = pN.$$

Thus,

$$\ln P(n) \approx q(N_0) - \frac{(n - N_0)^2}{2} \times \frac{1}{Np(1-p)}, \qquad \text{or}$$

$$P(n) = e^{q(N_0) - [-(n - pN)^2/2Np(1-p)]}.$$

Since $e^{q(N_0)}$ is a constant, and since the approximated $P(n)$ must be normalized so that the sum of the probabilities of all possible values of n is 1, the approximation for $P(n)$ finally becomes

$$P(n) = \frac{1}{\sqrt{2\pi\sigma^2}} e^{-[(n - m)^2/(2\sigma^2)]},$$

where $m = pN$ and $\sigma^2 = Np(1 - p)$.

MEASURES OF PROBABILITY DISTRIBUTIONS

The three types of commonly occurring probability distributions derived in the preceding sections possess maximum values and characteristic widths (Fig. 9.1). Although most experimental situations are closely described by these distributions, some are not. In these cases some understanding of the situation can be gained by

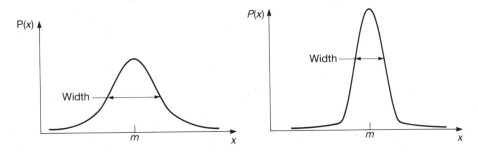

Figure 9.1   Means and widths of probability distributions. A demonstration that both parameters are important for characterization of a distribution.

Table 9.2   Characteristics of the Binomial, Poisson, and Gaussian Distributions

Distribution		Mean m	Variance σ^2
Binomial	$\dfrac{N!}{n!(N-n)!}\, p^n(1-p)^{N-n}$	Np	$Np(1-p)$
Poisson	$\dfrac{m^n}{n!}\, e^{-m}$	m	m
Gaussian	$\dfrac{1}{\sqrt{2\pi\sigma^2}}\, e^{-[(n-m)^2/(2\sigma^2)]}$	m	σ^2

learning the point of maximum value and the width or dispersion of the distribution. A convenient mathematical measure of width is the variance σ^2, which is defined as

$$\sigma^2 = \int (x-m)^2 P(x)dx \qquad \text{for continuous functions}$$

or

$$\sigma^2 = \sum_i (x_i - m)^2 P(x_i) \qquad \text{for discrete functions,}$$

where m is the mean value. The more sharply peaked a probability distribution is, the smaller the value of σ^2 (Table 9.2).

ELEMENTS OF THE LURIA-DELBRÜCK FLUCTUATION TEST

The Luria-Delbrück fluctuation test was developed to answer a question on the origin of mutations. Two possibilities were considered: that mutants were adaptations of a "plastic" organism generated in response to the environment or that mutants were generated regardless of the environment and the environment merely selected for their survival. Now, of course, we understand the source of the mutants as being random changes of the DNA sequence that are produced independently of the environment.

The solution of this problem was provided by Luria and Delbrück in one of the first papers of modern genetics. They were concerned with the origin of bacterial mutants resistant to a phage. A secondhand story of the origin of this discovery is that Luria was measuring the mutation frequency to phage resistance and was about to publish, but his data were not reproducible. Naturally he repeated the experiment to make the data better, but despite his best efforts to narrow in on the mutation frequency, his data always showed large fluctuations and did not settle on one value. Delbrück helped analyze this situation, and the result was both a method for determining the mutation frequency from the data and a proof that exposure to the phage was not generating the mutations.

Remarkably, it is not possible to write down a function giving the probability of finding a culture containing n mutants. The best mathematics can do is to provide various measures of the distribution, but it never provides the distribution itself. Fortunately, what can be provided, the mutation rate, is the reason geneticists studied the question in the first place.

The demonstration that phage-resistant cells existed in the culture before the addition of phage was straightforward. Experiments demonstrated that the probability of a mutation to phage resistance was small and that large populations had to be used to find such mutants. Therefore if the mutations were generated at the time of exposure of the populations to the phage, the numbers of phage-resistant cells in a number of parallel cultures should be Poisson distributed, that is,

$$P(n) = \frac{m^n}{n!} e^{-m},$$

for the probability that a particular population contains n mutants if the average number of mutants per culture is m. Thus, $\sigma^2 = m$.

Suppose, on the other hand, that the mutations were generated well before addition of the phage to the cells and that the mutations were Poisson distributed at the time of their generation. Thus $\sigma^2 = m$ at the time of creation of the mutants. For simplicity, consider the case that all the mutations were created t' minutes before exposure to the phage. Then at the time of addition of the phage, the new mean $m' = e^{\mu t'}m$, and a little calculus shows that the new variance $\sigma'^2 = (e^{\mu t'})^2 m$. Hence $\sigma'^2 \neq m'$, and at the time of addition of phage the numbers of phage-resistant cells in the cultures are no longer Poisson distributed. In conclusion, if numbers of phage-resistant mutants originally were Poisson distributed, but the culture grew after their generation, their numbers will no longer be Poisson distributed.

When the actual experiment was done and the numbers of phage-resistant cells were measured in each of 50 to 100 independent cultures, the variance was found not to equal the mean. The resistant cells had to exist in the culture before the addition of the phage, and cells were spontaneously mutating to phage resistance during growth.

*MUTATION FREQUENCIES FROM FLUCTUATION ANALYSIS

In this section we will derive the Luria-Delbrück equation for calculation of the actual mutation frequency. This analysis follows Luria and Delbrück's paper but fills in the mathematical details. To parallel their paper, let m stand for the number of mutants and a be the mutation frequency. Also use time units such that cell growth can be described as e^t, that is, $\mu = 1$. Then the equation describing the number of mutants in the population is

$$\frac{dm}{dt} = m + aN(t).$$

The term m says that mutants, once they exist, grow and divide normally, that is, if the entire population were mutant, $dm/dt = m$ and $m(t) = e^t$. This is just the exponential growth term. The second term, $aN(t)$, comes from the generation of new mutants. At any time, the contribution from this source is proportional to the number of cells present at this time and the mutation frequency a. It is a that is to be determined from experimental data on the numbers of mutants in cultures.

Before solving this equation it must be rewritten in standard form:

$$\frac{dm}{dt} - m = aN(t).$$

As usual in solving this type of equation, the solution consists of the general and particular solutions. The general solution is defined as that function $m(t)$ that satisfies the equation with the right-hand term set equal to zero:

$$\frac{dm}{dt} - m = 0.$$

Such a function $m(t)$ can always be added to the solution $m(t)$ that satisfies the equation when the right-hand term is not set equal to zero. By substitution it is clear that the general solution is $m = Ae^t$, where A has any value.

The particular solution must satisfy the equation

$$\frac{dm}{dt} - m = aN(t).$$

The cells grow exponentially, so $N(t) = e^t$. Thus

$$\frac{dm}{dt} - m = ae^t.$$

Substitution shows that a solution to this equation is

$$m = ate^t$$
$$= atN(t).$$

Hence the complete solution is

$$m(t) = Ae^t + atN(t),$$

where A can be set to any desired value. This freedom is used to set the value of m to the initial conditions.

The original equation, $dm/dt = m + aN(t)$, is valid only when $m \geq 1$. Therefore the solution derived above applies only after the first mutant is likely to have appeared in one of the cultures. Let the time at which the first such mutant is likely to have appeared be t_0. Then the value of A must be determined to fit this condition. At time t_0 the number of mutants m in the population is 1 and hence

$$1 = Ae^{t_0} + at_0N(t_0).$$

Utilizing the fact that $N(t_0) = e^{t_0}$, we have

$$1 = Ae^{t_0} + at_0e^{t_0},$$

and

$$A = e^{-t_0} - at_0.$$

Now substituting this into the solution given above for $m(t)$, we have

$$m(t) = (e^{-t_0} - at_0)e^t + atN(t),$$

and, since $N(t) = e^t$,

$$m(t) = a(t - t_0)N(t) + e^{t-t_0}.$$

The term e^{t-t_0} turns out to be negligible compared to the first term, as can be verified later, and will be dropped. Thus

$$m(t) = a(t - t_0)N(t).$$

The remaining problem is to determine the time, t_0, at which the first mutation is likely to have occurred in the cultures. To find this time, consider the original differential equation. Before any mutants exist, $m = 0$ and the equation becomes

$$\frac{dm}{dt} = aN(t).$$

Following the procedure already used in the solution of the first equation, we have

$$m(t) = B + ae^t$$
$$= B + aN(t),$$

where B is the general solution and can be adjusted so that the entire solution fits the initial conditions.

At $t = -\infty$ there are no mutants, or $m(-\infty) = 0$, and $0 = B + aN(-\infty)$. Also $N(-\infty) = 0$, and $B = 0$. Therefore $m(t) = aN(t)$.

If the experiment is being performed with C identical cultures, the solution for $m(t)$ is applicable from the time that there is likely to be a single mutant somewhere in the C cultures. N still stands for the number of cells in one culture. Thus,

$$1 = aCN(t_0),$$

$$1 = aCe^{t_0}.$$

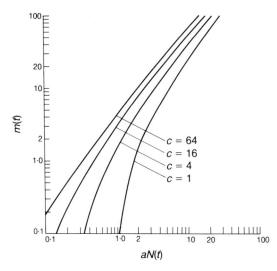

Figure 9.2   Four plots of $m(t)$ as a function of $aN(t)$ for different values of C. Given a mean value $m(t)$ of the number of mutants in C cultures, the value of $aN(t)$ may be read from the graph, and this can be divided by the number of cells in a culture, $N(t)$, to determine a, the mutation frequency.

Looking back, it is apparent that it will be more convenient to have an expression for $t - t_0$. Therefore, if

$$1 = aCe^{t_0},$$

$$e^{-t_0} = aC,$$

$$e^t \times e^{-t_0} = aCe^t,$$

$$e^{t-t_0} = aCN(t),$$

$$t - t_0 = \ln aCN(t).$$

And finally, substituting yields

$$m(t) = aN(t) \ln aCN(t).$$

Experimentally, the average number of mutants per culture at time t and the number of cells per culture at time t can be measured. C is just the number of cultures, and therefore the mutation rate a can be determined. Figure 9.2 shows the relation between m, a, N, and C.

PROBLEMS

9.1. Using the binomial distribution, calculate the probability of throwing a total of 15 heads in a total of 20 tries.

9.2. Show algebraically that the binomial distribution is correctly normalized, that is, that the sum of all probabilities of $p(n)$ from $n = 0$ to $n = N = 1$. Note that this is just the probability that something happens and it had better equal 1. [*Hint:* Consider $(1 - p + p)^N$.]

9.3. Show that $\sigma^2 = m$ for a Poisson distribution.

9.4. Assume that 0.01 ml aliquots are plated with perfect precision from a 10-ml culture at 10,000 cells/ml. Show that this situation satisfies the approximation used in deriving the Poisson distribution and hence that the numbers of cells per plate will be Poisson distributed. What will the variance equal?

9.5. Design an experiment that uses mutagenic techniques to determine the approximate number of essential genes in *E. coli.*

9.6. Individual cells were grown to 3×10^8 and spread on plates selective for a character x^+. The number of x^+ colonies per plate was

Number of x^+ per Plate	Number of Plates
0	27
1	16
2	5
3	6
4	2
5	3
7	1
9	1
11	1
20	1
28	1

Do the plates induce the mutation? What is the mutation rate as read off from Fig. 9.2?

9.7. Mutagenesis can be directed at small regions of the *E. coli* chromosome by the use of transducing phage P1. The phage stock is grown on a wild-type strain and then mutagenized *in vitro* by treatment with hydroxylamine. The recipient bacterial strain that contains an auxotrophic marker in the region of interest is transduced to repair the auxotrophy by the P1 stock. Consequently, a high fraction of the transduced recipients also acquire closely linked mutagenized DNA. Suppose that the hydroxylamine treatment produces 0.001 survival of the plaque-forming P1. What is the average number of lethal hits per phage? If lethal hits were the only effect of the treatment, by how much would the transducing titer for an average-size gene have fallen? Consider that P1 contains about 50 average-size genes.

9.8. Derive a relation between UV dose and the fraction of surviving cells if the cells possess n independent nuclei. Assume that UV damages only DNA and that if one nucleus is not damaged, the cell will survive. What is the most appropriate way to plot experimental data to determine n?

9.9. Assuming equal utilization of all sense codons, what is the average spacing between possible nonsense mutations created by a single base change from the wild-type sequence? What is the probability that x amino acids lie between two possible nonsense mutation locations?

9.10. Show how the theoretical numbers contained in Table 9.1 were calculated. See J. M. Keynes, *A Treatise on Probability*, (MacMillan: London, 1921), for a detailed breakdown of these numbers, but not the correct calculations.

RECOMMENDED READINGS

Mutations of Bacteria from Virus Sensitivity to Virus Resistance, S. Luria and M. Delbrück, Genetics *28*, 491–511 (1943).

RELATED REVIEWS, BOOKS, AND ARTICLES

Calculus and Analytic Geometry, 6th ed., G. Thomas, R. Finney, Addison-Wesley, Reading, Mass. (1984).
An Introduction to Probability Theory and Its Applications, W. Feller, Wiley, New York (1968).

DEEPER READING

We Wuz Robbed, Editorial, Nature *211*, 670 (1966).
The Frequency Distribution of Spontaneous Bacteriophage Mutants as Evidence for the Exponential Rate of Phage Reproduction, S. Luria, Cold Spring Harbor Symp. Quant. Biol. *16*, 463–470 (1951).

Genetic Engineering and Recombinant DNA

The terms "genetic engineering" and "recombinant DNA" refer to techniques in which DNA may be cut, rejoined, its sequence determined, or a segment altered in sequence to suit an intended use. For example, a DNA fragment may be isolated from one organism, spliced to other DNA fragments, and put into a bacterium or another organism. This process is called cloning because many identical copies can be made of the original DNA fragment. In another example of genetic engineering, a stretch of DNA, often an entire gene, may be isolated and its nucleotide sequence determined, or its nucleotide sequence may be altered by *in vitro* mutagenic methods. These and related activities in genetic engineering have two basic objectives: to learn more about the ways nature works and to make use of this knowledge for practical purposes.

Before 1975 the most detailed studies of biological regulatory processes were restricted to genes of small phage or bacterial genes that could be placed on the phages. Only by beginning with such phage could DNA or regulatory proteins be obtained in quantities adequate for biochemical study. Also, only with phage or phage-carrying bacterial genes could variant DNA sequences be easily generated for the study of the effects of altering the DNA-binding sites of proteins or of altering the proteins themselves. The isolation of specialized transducing phage that carried the genes of the *lac* operon were particularly important advances in this era. These phage provided an instantaneous 100-fold enrichment of the *lac* genes compared to the chromosomal DNA and stimulated a wide variety of important studies that greatly furthered our understanding of gene regulation. Now genetic engineering permits the same sorts of studies to be carried out on virtually any gene from virtually any organism.

The second major reason for interest in genetic engineering is the "engineering" it makes possible, such as the synthesis of a number of rare and expensive hormones or antibiotics in bacteria or yeast at low cost. Such hormones can be modified by genetic engineering techniques to alter and perhaps improve their biological activities. Many chemical processes may be substantially improved by the use of enzymes obtained at low cost through genetic engineering. Similarly, genetic engineering is facilitating the production of antigens for immunization of humans and animals against a variety of diseases and parasites. Much research in genetic engineering is also directed at plants with the hope of improving on traditional genetic methods of crop modification or of adding nitrogen fixation abilities to plant cells to reduce crop fertilization costs.

Genetic engineering of DNA requires that the following steps be readily performed. The DNA for study must be in pure form. It must be possible to cut this DNA reproducibly at specific sites so as to produce fragments containing genes or parts of genes. Next, it must be possible to rejoin DNA fragments to form hybrid DNA molecules. Vectors must exist to which fragments can be joined and then introduced into cells by the process called transformation. The vectors must have two properties: first, they must provide for autonomous DNA replication of the hybrid in the cells, and second, they must permit selective growth of only those cells that have received the vectors. This chapter discusses these fundamental steps of genetic engineering in addition to one nonessential but highly useful operation, that of determining the sequence of DNA fragments. The next chapter discusses many of the more specialized and exotic operations that constitute genetic engineering.

THE ISOLATION OF DNA

Cellular DNA, chromosomal or nonchromosomal, is the starting point of many genetic engineering experiments. Such DNA can be extracted and purified by the traditional techniques of heating the cell extracts with detergents and removing proteins by phenol extraction. If polysaccharides contaminate the sample, they can be removed by equilibrium density gradient centrifugation in cesium chloride.

As we will see, many experiments require the purification of vector DNA. Two types of vectors are commonly used: plasmids and phage. A plasmid is a DNA element similar to an episome that replicates independently of the chromosome (Fig. 10.1). Usually plasmids are small, 3,000 to 25,000 base pairs, and circular. Generally lambda phage or closely related derivatives are used for phage vectors in *Escherichia coli*, but for cloning in other bacteria, like *Bacillus subtilis*, other phage are used. In some cases a plasmid can be developed that will replicate autonomously in more than one host organism. These "shuttle" vectors are of special value in studying genes of eukaryotes; we consider them later.

The complete purification of plasmid DNA generally requires several steps.

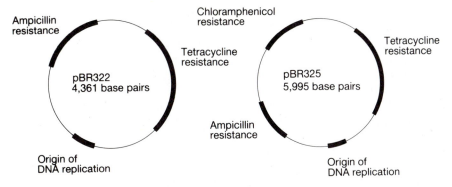

Figure 10.1   Structure of two plasmids commonly used in genetic engineering.

After the cells are opened using lysozyme and detergents, most of the chromosomal DNA is removed by centrifugation. Then, to separate the plasmid DNA from all the remaining chromosomal DNA, proteins, and most of the large RNA molecules, the DNA is centrifuged to its equilibrium position in equilibrium density gradient centrifugation performed in the presence of ethidium bromide. Any chromosomal DNA remaining with the plasmid will have been fragmented and will be linear, whereas most of the plasmid DNA will be covalently closed circles. As we saw in Chapter 2, a covalently closed DNA molecule can intercalate less ethidium bromide than a linear molecule can. Since ethidium bromide possesses a lower density than DNA, the linear and circular DNA molecules acquire different densities, and they are separated by equilibrium density gradient centrifugation. Following the centrifugation, the two bands of DNA are easily observed by shining UV light on the tube (Fig. 10.2).

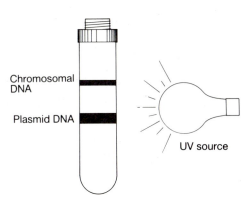

Figure 10.2   The separation of covalently closed plasmid DNA from nicked plasmid DNA and fragments of chromosomal DNA on CsCl equilibrium density gradient centrifugation in the presence of ethidium bromide. The bands of DNA fluoresce upon illumination by UV light.

The natural fluorescence of ethidium bromide is enhanced 50 times by intercalation into DNA, and the bands glow a bright cherry red under UV light.

Lambda phage are easily purified by utilizing their unique density of 1.5 g/cm³, which is halfway between the density of protein, 1.3, and the density of DNA, 1.7. The phage may be isolated by equilibrium density gradient centrifugation in which the density halfway between the top and bottom of the centrifuge tube is 1.5. They too may be easily observed in the centrifuge tube as a bluish band resulting from their preferential scattering of shorter wavelengths of light — the Tyndall effect, the reason the sky is blue and sunsets are red.

THE BIOLOGY OF RESTRICTION ENZYMES

In this section we digress into the biology of restriction enzymes and then return to their use in cutting DNA. In the past fifteen years a large number of enzymes have been found, primarily in bacteria, that cut DNA at specific sites. These enzymes are called restriction enzymes because in the few cases that have been carefully studied, the cleaving enzyme is part of the cell's restriction-modification system.

The phenomenon of restriction-modification in bacteria is a small-scale immune system for protection from infection by foreign DNA. In contrast to higher organisms in which identification and inactivation of invading parasites, bacteria, or viruses can be performed extracellularly, bacteria can protect themselves only after foreign DNA has entered their cytoplasm. For this protection, many bacteria specifically mark their own DNA by methylating bases on particular sequences with modifying enzymes. DNA that is recognized as foreign by its lack of methyl groups on the appropriate sequences is cleaved by the restriction enzymes and then exonucleases degrade the DNA to nucleotides. Less than one phage out of 10^4 wrongly

Table 10.1   Plating Efficiencies of Phage Grown on *E. coli* Strains C, K, and B When Plated on These Bacteria

Phage	Strain		
	C	K	B
lambda-C	1	$<10^{-4}$	$<10^{-4}$
lambda-K	1	1	$<10^{-4}$
lambda-B	1	$<10^{-4}$	1

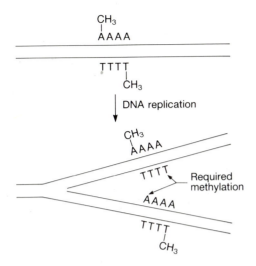

Figure 10.3   Methylation of an asymmetrical sequence necessitates recognition and methylation of two different sequences on the daughter strands on DNA replication.

methylated infecting phage is able to grow and lyse an *E. coli* protected by some restriction-modification systems.

Arber studied restriction of lambda phage in *E. coli* and found that *E. coli* strain C does not contain a restriction-modification system. Strain B has one restriction-modification system, and yet a different one recognizes and methylates a different nucleotide sequence in strain K-12. Phage P1 also specifies a restriction-modification system of its own, and this can be superimposed on the restriction-modification system of a host in which it is a lysogen.

Let the notation lambda-C represent lambda phage that has been grown on *E. coli* strain C. Infection of strains B, K-12, and C with lambda grown on the various strains yields different efficiencies of plaque formation (Table 10.1). Passage of the phage through a host of one type modifies the DNA so that it is recognized as "self" and plates at high efficiency if the phage reinfects that same strain and is recognized as "foreign" and plates at low efficiency if the phage infects a strain with a different restriction-modification system.

Possession of a restriction-modification system introduces complexities to the process of DNA replication. Imagine that the double-stranded DNA contains methyl groups on both strands of the DNA at a recognition sequence. DNA replication creates a new duplex in which the new daughter at first lacks the modification. This half-methylated DNA must not be recognized as foreign DNA and cleaved, but instead must be recognized as "self" and methylated (Fig. 10.3). Therefore, the restriction-modification system functions like a microcomputer, recognizing three different states of methylation of its recognition sequence and taking one of three different actions. If the sequence is unmethylated, the enzymes cleave. If the DNA is

Some palindromic DNA sequences

```
    5'          3'
      —AGCT—
      —TCGA—
    3'          5'

    5'          3'
      —AGNCT—        N is any base
      —TCN'GA—       N' is complementary to N
    3'          5'

    5'          3'
      —ACGCGT—
      —TGCGCA—
    3'          5'
```

Figure 10.4   Palindromic DNA sequences. Because DNA strands possess an orientation, the reverse of the sequence is contained on the opposite strand. In palindromes with an odd number of bases, the central nucleotide is irrelevant.

methylated on one of the two strands, the modification system methylates the other strand; if the DNA is methylated on both strands, the enzymes do nothing.

A palindromic recognition sequence streamlines operation of the restriction-modification system. (A palindrome is a sequence that reads the same forward and backward, such as *repaper* and *radar*.) Palindromes, of course, can be of any size, but most utilized as restriction-modification recognition sequences are four, five, or six bases (Fig. 10.4). By virtue of the properties of palindromes, both daughter duplexes of replicated palindromic sequences are identical, and thus the modification enzyme must recognize and methylate only one type of substrate (Fig. 10.5). The use of

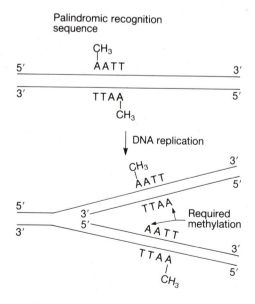

Palindromic recognition sequence

Figure 10.5   Methylation of a palindromic sequence permits recognition and methylation of only one sequence during DNA replication.

nonpalindromic recognition sequences would require that the modification enzyme recognize two different sequences. Presumably dimeric proteins are used to recognize palindromic sequences.

Restriction enzymes are divided into two easily distinguishable classes, I and II. Those in class II cleave in or near their recognition sequence. The enzymes in class I cleave at sites far distant from the recognition sequence and will not be further discussed here even though they were the first to be discovered.

A restriction enzyme within a cell is a time bomb because physical-chemical principles limit the enzyme's specificity for binding only to the recognition sequence. If a restriction enzyme were to bind to a wrong sequence, and a typical bacterium contains about 4×10^6 such sequences, most likely the sequence would not be methylated and the enzyme could cleave. This would break the chromosome, and the cell would die. However, the experimental observation is that cells containing restriction enzymes do not noticeably die any faster than cells without restriction enzymes. How, then, is the extraordinarily high specificity of the restriction enzymes generated?

The requisite high specificity can be obtained if cutting the DNA duplex is a two-step process (Fig. 10.6). An enzyme could bind to the recognition sequence, cleave one strand, wait a while, then cleave the other strand. This has the effect of utilizing the recognition sequence twice for each cleavage, for if the enzyme binds at a site other than the recognition sequence it rapidly dissociates before cleaving the second strand. Therefore, restriction enzymes are likely to produce nicks in the DNA

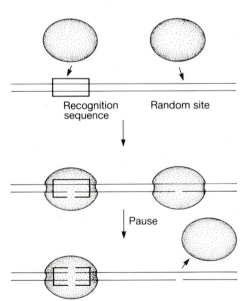

Figure 10.6   Cleavage by a restriction enzyme at a recognition sequence and at a random sequence. The enzyme binds and cleaves one strand but can remain bound long enough to cleave the second strand only if it has bound to a recognition sequence.

at sites other than the recognition sequence, but these nicks can be repaired with
DNA ligase and the cell will not be harmed in the process. No restriction enzymes are
likely to be found that cleave both strands of the DNA in a concerted process.

CUTTING DNA WITH RESTRICTION ENZYMES

The restriction enzymes provide a necessary tool for cutting fragments of DNA out
of larger molecules. Their exquisite specificity permits very great selectivity, and
because more than one hundred different restriction enzymes are known (Fig. 10.7),

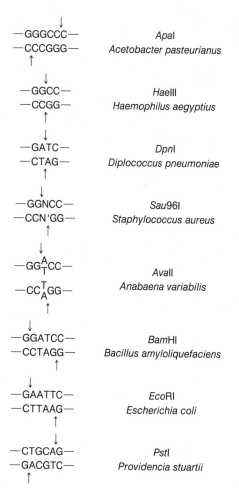

Figure 10.7   The recognition sites and
cleavage sites of several restriction enzymes.

their wide variety permits a large degree of choice in the cleavage sites utilized. Often, fragments may be produced with cuts located within 20 base pairs of any desired location.

One of the more useful restriction enzymes for genetic engineering is part of the restriction-modification system produced by the *E. coli* plasmid R. The corresponding restriction enzyme is called *Eco*RI. Instead of cleaving in the center of its palindromic recognition sequence, this enzyme cleaves off-center and produces four-base self-complementary ends. These "sticky" ends are enormously useful in

recombinant DNA work as they can be reannealed like the "sticky" ends of phage lambda, thereby permitting efficient joining of DNA fragments during DNA ligation steps. In some situations, DNA fragments can even be arranged to have different types of sticky ends so that their insertion into other DNA can be forced to proceed in one particular orientation.

ISOLATION OF DNA FRAGMENTS

Following cleavage of DNA by restriction enzymes or other exotic manipulations to be discussed later, DNA fragments frequently must be isolated. Fortunately, fractionation according to size is particularly easy because DNA possesses a constant charge-to-mass ratio and double-stranded DNA fragments of the same length have the same shape and therefore migrate during electrophoresis at a rate nearly independent of their sequence. Generally, the larger the DNA, the slower it migrates.

Remarkable resolution is obtainable in electrophoresis. With care, two fragments whose sizes differ by 0.5% can be separated if they lie within a range of 2 to 50,000 base pairs. No single electrophoresis run could possess such high resolution over this entire range. A typical range for adequate size separation might be 5 to 200 base pairs or 50 to 1,000 base pairs, and so on. The material through which the DNA is electrophoresed must possess special properties. It should be inexpensive, easily used, and uncharged, and it should form a porous network. Two materials meet the requirements: agarose and polyacrylamide.

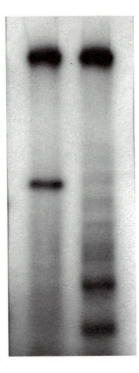

Figure 10.8   Separation and identification of DNA fragments by electrophoresis. (a) The bands generated by cleavage of plasmid pBR322 with different concentrations of a restriction enzyme. (b) The identification of radioactively labeled DNA fragments by autoradiography following electrophoresis. (Courtesy of S. Hahn.)

Following electrophoresis, bands formed by the different-sized fragments may be located by several means (Fig. 10.8). Low quantities of radioactive DNA may be easily detected by autoradiography if the DNA was radiolabeled before the separation. Usually $^{32}PO_4$ is a convenient label because phosphorus emits particularly energetic electrons and phosphate-containing materials can be made highly radioactive owing to the short half-life of the phosphorus. Often, sufficient DNA is present that it may be detected directly by staining with ethidium bromide. The enhanced fluorescence from ethidium bromide intercalated in the DNA compared to its fluorescence in solution permits detection of as little as 5 ng of DNA in a band. Following electrophoresis and detection, the DNA can be isolated from the gel and impurities removed to yield pure DNA fragments.

JOINING DNA FRAGMENTS

Having discussed how DNA molecules can be cut and purified, it is now necessary to discuss the joining of DNA molecules. *In vivo*, the enzyme DNA ligase repairs nicks in the DNA backbone. This activity may be utilized *in vitro* for the joining of two

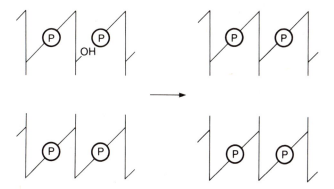

DNA molecules. Two requirements must be met: First, the molecules must be the correct substrates, that is, they must possess 3'-hydroxyl and 5'-phosphate groups; and second, these groups on the molecules to be joined must be properly positioned with respect to one another. The method for generating the proper positioning has two variations: either to hybridize the fragments together via their sticky ends or, if flush-ended fragments are to be joined, to use such high concentrations of fragments that from time to time they are spontaneously in the correct positions.

Hybridizing DNA fragments possessing self-complementary sticky ends generates the required alignment of the DNA molecules. The DNA ligase seals the nicks and joins the molecules. Restriction enzymes such as *Eco*RI produce four-base sticky

Figure 10.9   Joining two DNA fragments by poly-dA and poly-dT tails.

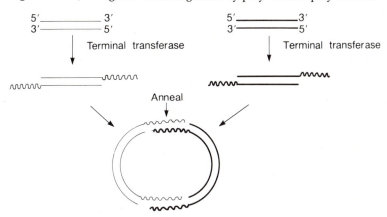

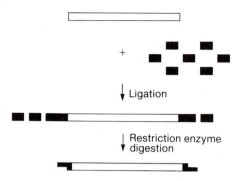

↓ Ligation

| Restriction enzyme
▼ digestion

Figure 10.10   Addition of linkers by ligation and their conversion to sticky ends by restriction enzyme digestion.

ends. These ends will hybridize and can be ligated together at temperatures on the order of 12–14°C. Other restriction enzymes generate flush ends. One method of converting such flush-ended molecules to sticky-ended molecules is to use the enzyme terminal transferase to synthesize a poly-dA tail on the 3′ ends of one of the fragments and to synthesize a poly-dT tail on the 3′ ends of the other fragment (Fig. 10.9). The two fragments can then be hybridized together by virtue of their self-complementary ends and ligated together. Or, if the tails are long enough, the complex can be directly introduced into cells, where the gaps and nicks will be filled and sealed by the cellular enzymes.

Joining of flush-ended DNA molecules is much less efficient than joining of sticky-ended molecules. Not only are the numbers of molecules joined low, but to obtain joining at all, high DNA and ligase concentrations must be used. The high concentrations increase the chances that two DNA molecules occupy the active site of the enzyme at the same time.

Linkers can also be used to generate self-complementary single-stranded molecules (Fig. 10.10). Linkers are short, flush-ended DNA molecules containing the recognition sequence of a restriction enzyme that produces sticky ends. The ligation of linkers to DNA fragments proceeds with reasonably high efficiency because high molar concentrations of the linkers in the presence of ligase may easily be obtained. After the linkers have been joined to the DNA segment, the mixture is digested with the restriction enzyme, which cuts the linkers and generates the sticky ends. In this way a flush-ended DNA molecule is converted to a sticky-ended molecule that may easily be joined to other DNA molecules.

The ability of ligase to join flush-ended DNA molecules can also be used for direct cloning of a fragment into a vector. While this method is straightforward, it suffers from two drawbacks: It requires high concentrations of DNA and ligase for the reaction to proceed, and even then the ligation efficiency is low; and it is difficult later to excise the fragment from the vector.

VECTORS PROVIDE FOR SELECTION AND A CELLULAR FREE RIDE

Cloning a piece of DNA requires that it be replicated when it is put back into cells. Hence the DNA to be cloned must itself be a replicon or must be joined to a replicon. Additionally, since the efficiency of introduction of DNA into cells is much below 1, transformed cells must be readily identified. In fact, since only about one cell out of 10^5 is transformed, selections must usually be included to permit only the transformed cells to grow.

Vectors must fulfill the two requirements described above, replication in the host cell and growth selection of the cells receiving the transforming DNA. As mentioned earlier, two basic types of vectors are used: plasmids and phage. Plasmids contain bacterial replicons that can coexist with the normal cellular DNA and at least one selectable gene. Usually it is a gene conferring resistance to an antibiotic. Phage, of course, carry genes for replication of their DNA; since they enter cells effectively, selectable genes usually are unnecessary.

PLASMID VECTORS

Most plasmids are small circles that contain the elements necessary for DNA replication, one or two drug-resistance genes, and a region of DNA into which foreign DNA may be inserted without damage to essential plasmid functions. One of the most widely used plasmids, pBR322, carries genes coding for resistance to tetracycline and β-lactamase (Fig. 10.11). The latter confers resistance to penicillin and related analogs by cleaving in the lactam ring, which renders them biologically inactive. Chloramphenicol, tetracycline, and kanamycin resistance are other selectable drug-resistance genes commonly carried on plasmids.

In a typical cloning experiment a plasmid is cut in a nonessential region with a restriction enzyme, say *Eco*RI, foreign *Eco*RI-cut DNA is added, and the single-stranded ends are hybridized together and ligated. Only a fraction of the resulting plasmids will contain inserted DNA, however; the rest will have recircularized without insertion of foreign DNA. How can transformants whose plasmids contain inserted DNA be distinguished from plasmids without inserted DNA? Of course, in some conditions a genetic selection can be used to enable only transformants with the desired fragment of inserted DNA to grow. More often this is not possible, and so an efficient method of rapidly identifying candidates containing inserted DNA is used.

One method for identifying candidates relies on insertional inactivation of a drug-resistance gene. For example, within the ampicillin-resistance gene in pBR322 exists the only plasmid cleavage site of the restriction enzyme *Pst*I. Fortunately *Pst*I cleavage generates sticky ends and DNA may readily be ligated into this site, whereupon it inactivates the ampicillin-resistance gene. The tetracycline-resistance gene

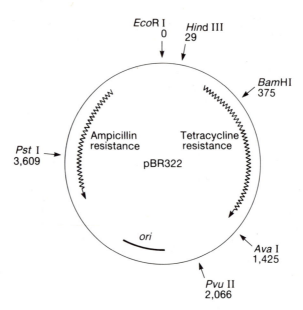

Figure 10.11   A map of the plasmid pBR322 showing the ampicillin- and tetracycline-resistance genes, the origin, and a few restriction enzyme cleavage sites.

remains intact and can be used for selection of the cells transformed with the
recombinant plasmid. The resulting colonies can be spot tested for their sensitivity to
ampicillin. Only the ampicillin-sensitive, tetracycline-resistant transformants contain foreign DNA in the plasmid. The ampicillin-resistant transformants derive from
plasmid that recircularized without insertion of foreign DNA.

Efficient genetic engineering necessitates that plasmid DNA be obtainable in
high quantities. Some plasmids maintain only three or four copies per cell, whereas
other plasmids have cellular copy numbers of 25 to 50. Many of the high-copy-number plasmids can be further amplified because their DNA replication is under relaxed
control. The plasmid continues to replicate after protein synthesis and cellular DNA
synthesis have been blocked by the addition of chloramphenicol. After 20 hours of
drug treatment, a cell might contain as many as 3,000 copies of a plasmid with
relaxed DNA replication.

Nature did not deliver plasmid vectors ready-made for genetic engineering. The
most useful plasmid vectors were themselves constructed by genetic engineering.
The starting materials were R plasmids, which are plasmids or autonomously replicating DNA elements that carry one or more drug-resistance genes. The R plasmids
are the cause of serious medical problems, for various bacteria can acquire R plasmids and thereby become resistant to the normal drugs used for treatment of infections (Table 10.2). The conversion of an R plasmid into a useful vector requires

Table 10.2   Various R-Factors Found in *E. coli*

R Plasmid	Antibiotic Resistances*	Molecular Weight ($\times 10^b$)
R1	Am, Cm, SM, Su, Km	65
R6	Su, Sm, Cm, Tc, Km	65
R6K	Am, Sm	26
R28K	Am	44
R15	Su, Sm	46

* Am = ampicillin; Cm = chloramphenicol; Sm = streptomycin; Su = sulfonamides; Km = kanamycin.

elimination of large amounts of extraneous DNA because the smaller the plasmid the greater its copy number following amplification by chloramphenicol treatment. Additionally, for convenient cloning, at least one of the restriction enzymes should possess only a single cleavage site in the plasmid, and such a site must be located in a nonessential region.

In the construction of cloning vectors, R plasmids were digested with various restriction enzymes, and the resulting mixture of DNA fragments was hybridized together via their self-complementary ends and were ligated to produce many combinations of scrambled fragments. These were transformed into cells. Only new plasmids containing at least the DNA segments necessary for replication and drug resistance survived and yielded colonies. The desirable plasmids containing only single cleavage sites for some restriction enzymes could be identified by amplification and purification of the DNA followed by test digestions with restriction enzymes and electrophoresis to characterize the digestion products. The plasmid pBR322 possesses single *Hind*III, *Eco*RI, *Pst*I, and *Pvu*II sites.

A PHAGE VECTOR FOR BACTERIA

Phage vectors and phage-derived vectors are useful for two reasons. Phage can carry larger inserted DNA fragments than plasmids, so substantially fewer transformed candidates must be examined to find a desired clone, and the efficiency of infecting repackaged phage DNA into cells is much greater than the efficiency of transforming plasmid DNA into cells. Once a desired segment of DNA has been cloned on a phage, however, the convenience of manipulating plasmids, due in part to their smaller size, dictates that the segment be subcloned to a plasmid. Lambda was an ideal choice for a phage vector because it is well understood and easy to work with. Most important, the phage contains a sizable nonessential internal region flanked by *Eco*RI cleavage sites.

Before *Eco*RI-cleaved lambda DNA could be used for cloning, it was necessary to eliminate cleavage sites that are located in essential regions of the lambda genome.

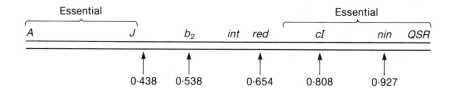

First, a lambda hybrid phage was constructed by *in vivo* genetic recombination. This lacked the three *Eco*RI cleavage sites in the nonessential region — positions 0.438, 0.538, and 0.654 — and retained the two sites in the essential regions. Then the two remaining *Eco*RI cleavage sites were eliminated by mutation and selection. In the selection the phage were cycled between hosts lacking and containing the *Eco*RI restriction-modification systems. Any phage with cleavage sites mutated so as to be nonrecognizable by the *Eco*RI system would have a greater probability of escaping the restriction enzymes upon growth in the second host. After 10 to 20 cycles of this selection scheme, Davis found a phage that had lost one of the two RI sites, and after an additional 9 to 10 cycles, he found a mutant that had lost the other site. This mutant phage was then made into a useful cloning vector by recombination with wild-type lambda to restore the three *Eco*RI cleavage sites located in the central portion of the genome.

The DNA isolated from lambda phage particles is linear, and it can be cleaved by *Eco*RI to yield the smaller central fragments and the larger left and right arms. *Eco*RI-cleaved DNA fragments to be cloned can then be hybridized together and ligated to purified right and left arms. This DNA either can be used as it is to transfect cells made competent for its uptake or can be packaged *in vitro* into phage heads and used as a normal phage lysate to infect cells. The efficiency of the latter path is much greater than the former and is used when the fragment to be cloned is present in only a few copies.

A trick can be used to eliminate the background from lambda arms that have rejoined without the insertion of foreign DNA. Lambda phage are inviable if the encapsulated DNA is less than 73% of the length of the wild-type genome. Therefore a special lambda derivative was constructed so that, following removal of the central section, its length was a little less than 73% of the wild-type length. Then a reconstructed lambda would be viable and form plaques only if foreign DNA was inserted, bringing its length to greater than 73% of that of wild-type lambda. The arms that rejoin without insertion of foreign DNA are inviable.

VECTORS FOR HIGHER CELLS

Cloning DNA in higher cells poses the same problems as cloning in bacteria. The vectors must permit simple purification of sizable quantities of DNA, must permit selection of transformed cells, and must have space for inserted DNA. Shuttle vectors, which have been extensively used for cloning in yeast, are a neat solution to these requirements. In addition to containing the normal bacterial cloning-plasmid elements, they contain a yeast replicon and a genetic marker selectable in yeast (Fig. 10.12). As a result, large quantities of the vectors can be obtained by growth in *E. coli* and then transformed into yeast for the actual research. The ability to shuttle between bacteria and yeast saves much time and expense in genetic engineering experiments.

Yeast chromosomal DNA replication origins and yeast 2-μ circles, a yeast plasmid-like element with unknown function, have been used as the yeast replicons, and nutritional markers such as uracil, histidine, leucine, and tryptophan biosynthesis have been used as selectable genes in auxotrophic yeast. As yet, no dominant selectable marker like a drug-resistance gene has found wide use in shuttle vectors.

Viruses form a basis for many vectors useful in higher plant and animal cells. For example, one of the simplest vectors for mammalian cells is the simian virus SV40. It permits many of the same cloning operations as phage lambda.

The terminology used with mammalian cells can be confusing. "Transformation" can mean either that cells have been "transformed" by a tumor-causing virus

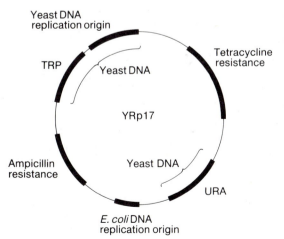

Figure 10.12   The structure of a vector for shuttling between *E. coli* and yeast. It contains genes permitting DNA replication and selection in both organisms.

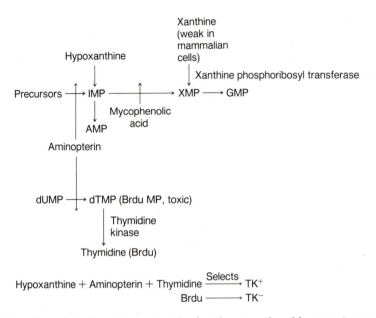

Figure 10.13   The metabolic pathways involved with some selectable genes in mammalian cells. IMP-inosine monophosphate, XMP-xanthine monophosphate, GMP-guanosine monophosphate, dUMP-deoxyuridine monophosphate, dTMP-deoxythymidine monophosphate, Brdu-bromodeoxyuridine, Brdu-MP, bromodeoxyuridine monophosphate. Aminopterin blocks tetrahydrofolate reductase, which is required for synthesis of IMP and dTMP, and mycophenolic acid blocks synthesis of XMP.

like SV40 to lose their contact inhibition and continue growing past the stage of a confluent cell monolayer, the point at which normal mammalian cells cease growth, or that foreign DNA has been introduced into the cells. Be alert! Although loss of contact inhibition could be useful in identifying cells that have incorporated the SV40 DNA or SV40 hybrids, this property is of limited use because constructions involving SV40 could eliminate the "transforming" genes. Therefore, selectable genetic markers suitable for mammalian cells are required.

One of the most useful genes in genetic selections in mammalian cells has been the thymidine kinase gene because TK$^+$ cells can be selected by growing them in medium containing hypoxanthine, aminopterin, and thymidine, or TK$^-$ cells can be selected by growing them in medium containing bromodeoxyuridine (Fig. 10.13). Furthermore, virologists had previously discovered that the herpes simplex virus codes for its own thymidine kinase. Therefore the viral genome can be used as a concentrated source of the gene in an expressible form for initial cloning experiments.

Although the thymidine kinase gene has been useful in selecting cells that have taken up foreign DNA, a selectable gene that does not require the prior isolation of mutants in the cell lines would also be valuable. The *E. coli* enzyme xanthine-guanine phosphoribosyl transferase gene appears to meet these requirements. The protein product of the gene functions in mammalian cells and permits selective growth of nonmutant cells that contain the enzyme. The required growth medium contains xanthine, hypoxanthine, aminopterin, and mycophenolic acid. Other dominant genes useful for the selection of transformed cells are mutant dihydrofolate reductase that is resistant to methotrexate, a potent inhibitor of the wild-type enzyme, and kanamycin-neomycin phosphotransferase. The latter is an enzyme derived from a bacterial transposon and confers its drug resistance on many cell types: bacterial, yeast, plant, and mammalian. Of course, for proper expression in the higher cells the gene must be connected to an appropriate transcription unit and must contain the required translation initiation and polyadenylation signals.

PUTTING DNA BACK INTO CELLS

Once the DNA has been joined to the appropriate vector, the hybrid must be transformed into cells for biological amplification. The phenomenon of transformation of pneumococci by DNA has been known since 1944. Once the desirable genetic properties of *E. coli* were realized, it too was tried in transformation experiments, unsuccessfully for many years. Suddenly, and without the motivation of doing genetic engineering, a method for transforming *E. coli* was discovered. This occurred at a most opportune time, for the developments in the enzymology of DNA cutting and joining were almost ready to be used in a system of putting the DNA that had been altered *in vitro* back into cells. Reintroducing a DNA molecule containing a replicon into a cell permits a biological amplification of greater than 10^{12}, so copies of a single molecule may be prepared in sufficient quantities that physical experiments using milligrams can be performed.

The key to transformation of *E. coli* was treatment of the cells with calcium ions. Such treatment permitted incorporation of bare plasmid or phage DNA. The term for transformation with phage DNA that then yields an infected cell is transfection. Yeast may also be transformed after a treatment that includes incubation in the presence of calcium ions. Some types of mammalian cells, mouse L cells for example, can be transformed merely by sprinkling them with a mixture of the DNA and calcium phosphate crystals. Here the mechanism of transformation appears to be uptake of the DNA-calcium-phosphate complex.

Direct manual injection of small volumes of DNA into cells has been highly useful for the study of cloned DNA fragments because it eliminates the need for a eukaryotic replicon or a selectable gene. Microinjection into the oocytes of the frog *Xenopus laevis* has yielded much information, and microinjection into culture mammalian cells is also possible. DNA injected into *Xenopus* cells is transcribed for many hours and translated into easily detectable amounts of protein. As a result of these properties, a DNA segment can be cloned onto a plasmid such as pBR322, manipu-

lated *in vitro,* and injected into the cells for examination of its new biological properties. Recently it has become possible to microinject into a fertilized mouse embryo.
The embryo can then be reimplanted to develop into a mouse. Since the fragments of
injected DNA often will recombine into the chromosome, the mice are transformed
by the DNA.

CHEMICAL DNA SEQUENCING

Two efficient techniques have been developed for DNA sequencing: chemical and
enzymatic. Both utilize electrophoresis at high temperatures in the presence of urea
so as to denature the DNA. Under these conditions the single-stranded molecules
migrate at rates independent of their sequence and dependent only on their length.
In such gels two single-stranded DNA fragments differing in length by a single base
can be resolved from one another if their length is less than 200 to 400 bases.

    The basic principle of DNA sequencing by the chemical or enzymatic method is
to generate a set of radioactive DNA fragments covering a region. The sizes of these
fragments indicate the nucleotide sequence of the region. How is this possible?
Consider a large number of identical single-stranded molecules (Fig. 10.14). Suppose that in this population of molecules, some were cleaved at the first A residue
from a particular end, some were cleaved at the second A residue, some were cleaved
at the third A residue, and so on. That is, the population of cleaved DNA molecules
would consist of some cleaved at each of the A's. If the first few A residues were
located four, five, and nine nucleotides from the end, then fragments containing the
original end produced by cleavage at A's would be four, five, and nine nucleotides
long. The sizes of these fragments could be determined by electrophoresis.

    In the same way that A-specific cleavages could be used to determine the
distances of A residues from the end, other base-specific cleavages could be used to
determine the distances from the end of the DNA molecules to occurrences of the

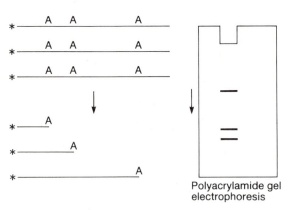

Polyacrylamide gel
electrophoresis

Figure 10.14   DNA sequencing. Modification at an
average of one A per molecule in the population
followed by cleavage at the modification sites
generates a population of molecules ending at the
former positions of A's. Their electrophoresis
generates the ladder pattern indicated.

other bases. Note that a whole set of distances is determined by these means, not just the distance to the first occurrence of a base. The distance to the second occurrence would also be determined since some of the molecules are cleaved not at the first occurrence but at the second occurrence.

Practical considerations slightly complicate the above procedure. First, the amounts of DNA that must be identified are too small to be detected by staining with ethidium bromide and observing the fluorescent DNA bands. Second, the chemical method generates extraneous fragments in addition to those originating from the one end under consideration. These other fragments could interfere with the identification of the desired fragments. Both problems are solved by radioactively labeling the DNA at the end from which the sequencing is being done. Autoradiography of the gels then provides adequate sensitivity, and the only fragments that are radioactive and generate bands on the autoradiographs are precisely those that include the end from which the sequencing begins.

Obtaining a DNA fragment labeled only at one end is not difficult. Each strand in a double-stranded segment can be labeled on its 5′ end with the enzyme T4 phage polynucleotide kinase. Then the two strands can be separated by denaturing the DNA and electrophoresis under partially denaturing conditions. Frequently the two strands migrate at different rates and can be separated. If so, either can be used in the chemical sequencing. A second way to obtain the DNA fragment with a radioactive label on only one end also begins with end-labeled double-stranded duplex (Fig. 10.15). This duplex is next cleaved with a restriction enzyme and if the two fragments are unequal in size, they can be separated by electrophoresis. Either of the fragments is suitable for DNA sequencing because only one of the strands in each fragment is then radioactive, and it is labeled on only one end.

To sequence a stretch of DNA, four populations of radiolabeled DNA are made, one partially cleaved at each of the four bases. These four populations are then subjected to electrophoresis in four adjacent lanes on the denaturing gels. The resulting four ladder patterns permit directly reading of the sequence of the bases on the DNA.

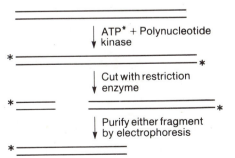

Figure 10.15   A protocol for labeling DNA on one end.

The cleavages at the four bases are made chemically rather than enzymatically. An average of about one cleavage per several hundred bases is optimal for most sequencing. Maxam and Gilbert discovered that a way of generating the requisite base specificity was to break the procedure into two parts. The first part introduces a highly base-specific chemical modification under controlled and mild conditions. Then harsh conditions are used to generate the actual cleavages at all modified positions.

Dimethyl sulfate provides a highly specific methylation of guanines (Fig. 10.16). The introduction of the methyl group permits subsequent depurination followed by cleavage of the sugar-phosphate backbone at the position of the missing base by a reaction with piperidine. Slightly altered conditions yield methylation by dimethyl sulfate of guanines, and to a lesser extent, methylation of adenines. Hydrazine is used for pyrimidine-specific reactions. Both thymine and cytosine react with hydrazine unless high concentrations of salt are present. Then only cytosine reacts; so in reality, four different base-specific reactions are not used in sequencing. For example, one commonly used set of reactions yields bands for G's, G's + A's, C's + T's, and T's. These are sufficient for sequence determination.

Figure 10.16   Basis of the G-specific reaction in Maxam-Gilbert sequencing. The final product is strand scission at the former position of a guanosine.

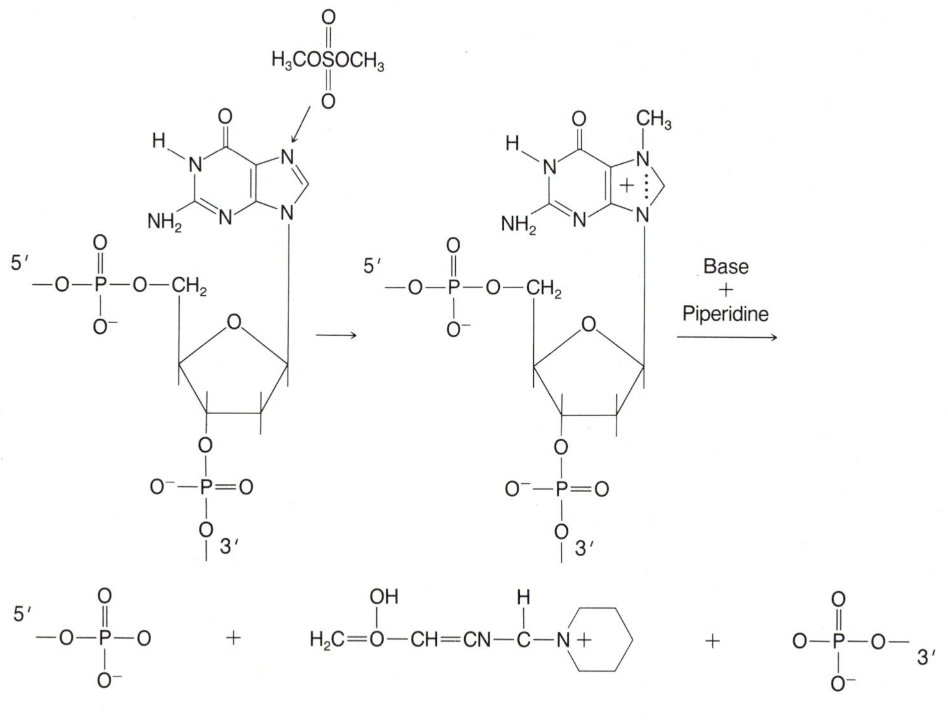

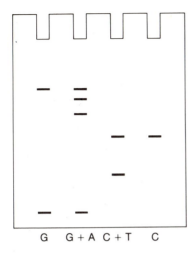

Figure 10.17   A sequencing gel in which G, G + A, C + T, and C-specific reactions were used. The sequence reads G-T-C-A-A-G.

Consider, for example, that the DNA sequence near the 5′ end of the fragment was 5′-GTCAAG-3′ and the fragment was labeled on its 5′ end. Then electrophoresis of the four reaction products yields a ladder pattern from which the sequence may be read by proceeding upward on the gel from lane to lane (Fig. 10.17). From 200 to 400 bases may be read from a single gel. Since several fragments may be processed at once, under good conditions, and after the DNA has been purified, it is possible to sequence 1,000 nucleotides of DNA in a week.

ENZYMATIC DNA SEQUENCING

Sanger and his colleagues developed an enzymatic method for generating the set of DNA fragments necessary for sequence determination. It relies on the fact that elongation of a DNA strand by DNA polymerase cannot proceed beyond an incor-

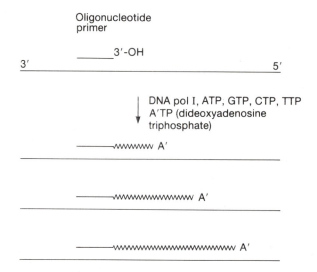

Figure 10.18   The Sanger sequencing method. Primer extension by DNA pol I in the presence of dideoxyadenosine triphosphate, A'TP, creates a nested set of oligomers ending in A'.

porated 2',3'-dideoxyribonucleotide, for the absence of the 3'-OH on such an incorporated nucleotide prevents chain extension. The chain terminates once a single dideoxyribonucleotide has been incorporated. Thus, DNA pol I elongation reactions are performed in the presence of the four deoxyribonucleotide triphosphates plus sufficient dideoxyribonucleotide triphosphate that about one of these nucleotides will be incorporated per hundred normal nucleotides. The DNA synthesis must be performed so that each strand initiates from the same nucleotide and from the same strand. This is accomplished first by using only a single-stranded template and second by hybridizing an oligonucleotide to the DNA to provide the priming 3'-OH necessary for DNA elongation by DNA polymerases (Fig. 10.18).

In sequencing, four different elongation reactions are performed, each containing one of the four dideoxyribonucleotides. Radioactive label is introduced to the fragments by the use of either labeled nucleotides or labeled primer. The fragments generated in the elongation reactions are analyzed after electrophoresis in the same manner as the fragments generated by chemical cleavage (Fig. 10.19).

The main difficulties in the Sanger method of sequencing are generating the single-stranded DNA template and obtaining the necessary primer. Elegant methods have been developed for solving both problems. One trick is to open the double-stranded circular form of a plasmid with a restriction enzyme that makes a single cut and digest with exonuclease III, which removes only from the 3' ends and leaves only a little of the DNA in double-stranded form (Fig. 10.20). An oligonu-

ACGT

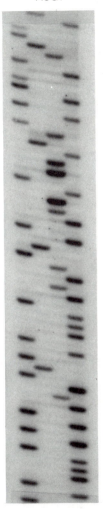

Sequence is

ATATTTATATATGT
ACATATTTATGTGA
CTATGGTAGGTCGA
TATAATAGCAATC

Figure 10.19   A gel from Sanger sequencing; the region beyond the α-tubulin gene
of *Drosophila melanogaster*, (Courtesy
W. Theurkoff.)

cleotide primer for the sequencing can be hybridized to the resulting single-stranded
DNA.

The virus M13 is another source of the single-stranded DNA. Only the plus
strand of the virus is packaged and secreted from virus-infected cells. Purification of
this material from as little as 2 milliliters of growth medium yields sufficient DNA for
sequencing, and sizable fragments of DNA can be inserted into nonessential regions
of the virus genome without loss of viability. Furthermore, M13 has no rigid structural limitations on the length of DNA that can be inserted because its DNA is
packaged as a rod during its extrusion through the cell membrane. Although the

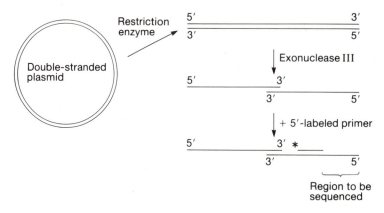

Figure 10.20   Generation of single-stranded DNA for Sanger sequencing by cleavage with a restriction enzyme, digestion with exonuclease III, and hybridization of a primer.

viral form of the virus is single-stranded, the necessary molecular hybrids between M13 and foreign DNA may easily be constructed using the double-stranded intracellular replicative form of the virus DNA. This DNA can then be transformed into cells just as plasmid DNA is.

To facilitate cloning into the M13 virus, the N-terminal portion of the *E. coli* β-galactosidase gene, *lacZ*, has been inserted into a nonessential region of the virus genome (Fig. 10.21). The virus can be infected into special cells that synthesize the remainder of the β-galactosidase protein. The two polypeptides bind to each other and, as a result, generate functional β-galactosidase. By plating infected cells on medium containing substrate of β-galactosidase, which when cleaved generates an intense blue dye, the infected cells yield blue "plaques." In reality a plaque is not formed because the virus does not kill infected cells, it merely makes them grow more slowly than noninfected cells. The particular utility of the special M13 virus is that a region containing a number of unique restriction enzyme cleavage sites not present elsewhere in the virus has been placed within the β-galactosidase gene. When foreign DNA is inserted into this region, the N-terminal portion of β-galactosidase is no longer synthesized in an active form, and such a virus yields white "plaques." An oligonucleotide that hybridizes near the insertion site is used as a primer in the sequencing. Therefore only one primer can be used to sequence any DNA that is inserted into the virus. Of course, if the inserted DNA exceeds 300 to 500 nucleotides, subfragments must be generated before the entire sequence can be determined.

Needless to say, the vast explosion in the amount of sequenced DNA has stimulated development of highly efficient methods for analysis. Computer programs have been written for comparing homology between DNA sequences, searching for symmetrical or repeated sequences, cataloging restriction sites, writing

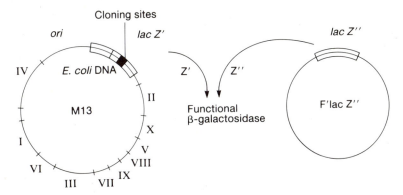

Figure 10.21   The structure of M13 with a portion of the *lacZ* gene inserted suitable for cloning and Sanger sequencing.

down the amino acid sequences resulting when the DNA is transcribed and translated into protein, and performing other time-consuming operations on the sequence. The amount of information known about DNA and the ease in manipulating it has greatly extended our understanding of many biological processes and is now becoming highly important in industrial and medical areas as well.

PROBLEMS

10.1. What is a reasonable interpretation of the fact that while many different strains of bacteria possess virtually identical enzyme activities for the metabolic pathways, the nucleotide sequence specificities of their restriction-modification systems are drastically different?

10.2. What is the transformation efficiency per DNA molecule if 5×10^5 transformants are produced per microgram of 5,000-base-pair plasmid?

10.3. How could it be determined whether bacteria that were transformed under a particular set of conditions took up more than one molecule of plasmid DNA?

10.4. Suppose you are constructing a plasmid vector from an R plasmid. How would you select for the elimination of a fragment containing the only cleavage site for a particular enzyme? How would you select for a plasmid that contains only a single copy of a particular restriction site?

10.5. The *Eco*RI and *Bam*HI cleavage sites on lambda are as shown at the top of page 274. If another enzyme cleaves lambda DNA once, show how its position of cleavage may be unambiguously determined merely by using various combined digestions of the enzymes and accurate measurement of the fragment sizes.

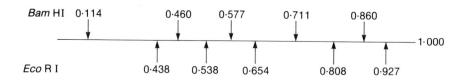

10.6. How many times is the *Hae*III recognition sequence GGCC likely to occur in 50,000 base pairs of DNA with GC content 30%, 50%, and 70%?

10.7. How frequently should the recognition sequence of *Hind*II, GTPyPuAC occur in 50,000 base pairs of DNA if AT = GC = 50%.

*10.8. Suppose phage T4 DNA ligase is used to perform end-to-end intermolecular joining of DNA molecules of different length. At a fixed molar concentration of DNA molecules, how should the rate of circularization vary with the length of the DNA? [*Hint:* What is the concentration of one end in the presence of the other as a function of the length of the DNA?]

10.9. How would you proceed to put a small fragment of DNA that has been generated by *Hae*III cleavage into the *Eco*RI cleavage site on a plasmid such that the fragment could be cut out of the plasmid (after cloning and growth) by *Eco*RI?

*10.10. Suppose the cleavage sites for a restriction enzyme were randomly located in the *E. coli* genome. (a) Neglecting for a moment the fact that the genome is of finite size, calculate the length distribution for restriction enzyme cleavage products if their average size is 10,000 base pairs. (b) Use the length distribution you just calculated to estimate the number of fragments from the *E. coli* chromosome whose size lies between 6,000 and 9,000 base pairs. (c) Using the fact that two DNAs of 6,000 and 6,050 base pairs can just be resolved from one another in agarose gel electrophoresis, calculate or closely estimate how the region 6,000 to 8,000 base pairs of digested *E. coli* DNA should look after electrophoresis. Should it form discrete and well-separated bands, a smear, or something in between?

10.11. Suppose a gene has been cloned using *Eco*RI sticky ends. Suppose also that it is necessary to remove one of the sticky ends so that the other end may be modified. How could you rather easily alter the sequence of one of the *Eco*RI ends for this purpose? [*Hint:* Consider partial digestions and the use of another enzyme.]

10.12. When sequencing by the Maxam-Gilbert method, how would you recognize a position that is methylated *in vivo* by a modification enzyme?

10.13. A hybrid plasmid was formed by inserting foreign DNA via the polydA-polydT method into the single *Eco*RI site contained on the parent plasmid. Sketch the DNA heteroduplex formed between one strand of DNA from the hybrid plasmid, opened at a random location by shear forces, and one strand from the parent plasmid, opened by *Eco*RI cleavage.

10.14. *Eco*RI cleaves as shown and a hypothetical restriction enzyme *Hyp*I

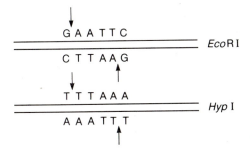

EcoRI

HypI

Figure 10.22   Figure for Problem 10.14.

cleaves as shown in Fig. 10.22. Why would or wouldn't you expect to be able to form molecular hybrids using sticky ends generated by this pair of restriction enzymes?

10.15. Suppose you wished to clone the enzyme from *Haemophilus aegyptius* coding for the restriction enzyme *Hae*. How would you select or screen for cells having received plasmids containing the genes coding for the desired restriction enzyme?

*10.16. Lambda phage DNA was digested with a certain restriction enzyme that yields more than 30 fragments of the phage. These fragments were labeled on their 5′ ends with polynucleotide kinase and sequenced by the Gilbert-Maxam procedure. Let us suppose that four specific base cleavage reactions were performed and it is claimed that the autoradiogram in Fig. 10.23 was obtained. What do you conclude?

10.17. Suppose the gene coding for the restriction enzyme *Hae*III can be cloned in *E. coli*, but the gene for *Hae*II cannot be cloned despite the fact that both proteins contain only a single polypeptide chain. What is likely to be going on?

10.18. Will the following method work, in principle, for DNA sequencing?

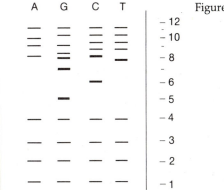

Figure 10.23   Figure for Problem 10.16.

a) Label the DNA by reaction with T4 DNA kinase and γ-^{32}PO$_4$-ATP, cut into two unequal pieces, and isolate one of the pieces.

b) Lightly nick with DNase.

c) Nick translate in four reactions with DNA pol I using all XTPs with one of each of the four dideoxytriphosphates (none radioactive).

d) Run on sequencing gels and read the sequence.

10.19. Restriction enzymes *Bam*HI and *Bcl*I cleave GGATCC and TGATCA, respectively. Why can DNA fragments produced by *Bcl*I digestion be inserted into *Bam*HI sites? If a plasmid contains a single *Bam*HI cleavage site, how could you select for plasmids containing inserts of *Bcl*I fragments?

10.20. How would you select for a clone carrying the methylase partner of a restriction enzyme?

10.21. Plasmids carrying the *Eco*RI restriction-modification system may be mated into recipient cells with 100% viability. What does this tell us?

10.22. How could the DNA replication origin of the *E. coli* chromosome be cloned?

RECOMMENDED READINGS

Cleavage of DNA by RI Restriction Endonuclease Generates Cohensive Ends, J. Mertz, R. Davis, Proc. Natl. Acad. Sci. USA *69*, 3370–3374 (1972).

Viable Molecular Hybrids of Bacteriophage Lambda and Eukaryotic DNA, M. Thomas, J. Cameron, R. Davis, Proc. Natl. Acad. Sci. USA *71*, 4579–4583 (1974).

A System for Mapping DNA Sequences in the Chromosomes of *Drosophila melanogaster*, P. Wensink, D. Finnegan, J. Donelson, D. Hogness, Cell *3*, 315–325 (1974).

A New Method for Sequencing DNA, A. Maxam, W. Gilbert, Proc. Natl. Acad. Sci. USA *74*, 560–564 (1977).

DNA Sequencing with Chain-Terminating Inhibitors, F. Sanger, S. Nicklen, A. Coulson, Proc. Natl. Acad. Sci. USA *74*, 5463–5467 (1977).

RELATED REVIEWS, BOOKS, AND ARTICLES

Host-Controlled Modification of Bacteriophage, W. Arber, Ann. Rev. Microbiol. *19*, 365–378 (1965).

Recombinant DNA (an issue devoted mainly to the topic), Science *196*, no. 4286, 159–221 (1977).

Recombinant DNA, Methods in Enzymology *68*, ed. R. Wu, Academic Press, New York (1979).

Recombinant DNA (an issue devoted to the topic), Science *209*, no. 4463 (1980).

Nucleic Acids, Methods in Enzymology *65*, ed. L. Grossman and K. Moldave, Academic Press, New York (1980).

Industrial Microbiology (an issue devoted to the topic), Scientific American (September 1981).

Applications of Computers to Research on Nucleic Acids (an issue devoted to the topic), Nucleic Acids Res. *10*, no. 1 (1982).

DEEPER READING

The Inosinic Acid Pyrophosphorylase Activity of Mouse Fibroblasts Partially Resistant to 8-Azaguanine, J. Littlefield, Proc. Natl. Acad. Sci. USA *50*, 568–576 (1963).

A Dye-Buoyant-Density Method for the Detection and Isolation of Closed Circular Duplex DNA: The Closed Circular DNA in Hela Cells, R. Radloff, W. Bauer, J. Vinograd, Proc. Natl. Acad. Sci. USA *57*, 1514–1521 (1967).

Molecular Weight Estimation and Separation of Ribonucleic Acid by Electrophoresis in Agarose-Acrylamide Composite Gels, A. Peacock, C. Dingman, Biochem. *7*, 668–674 (1968).

Enzymatic Breakage and Joining of Deoxyribonucleic Acid, B. Weiss, A. Jacquemin-Soblon, T. Live, G. Fareed, C. Richardson, J. Biol. Chem. *243*, 4543–4555 (1968).

Supercoiled Circular DNA-Protein Complex in *Escherichia coli*: Purification and Induced Conversion to an Open Circular DNA Form, D. Clewell, D. Helinski, Proc. Natl. Acad. Sci. USA *62*, 1159–1166 (1969).

Calcium-Dependent Bacteriophage DNA Infection, M. Mandel, A. Higa, J. Mol. Biol. *53*, 159–162 (1970).

Enzymatic Joining of Polynucleotides, P. Modrich, I. Lehman, J. Biol. Chem. *245*, 3626–3631 (1970).

Polynucleotide Kinase from *Escherichia coli* Infected with Bacteriophage T4, C. Richardson, Proc. Nucleic Acid Res. *2*, 815–828 (1971).

Enzymatic Oligomerization of Bacteriophage P22 DNA and of Linear Simian Virus 40 DNA, V. Sgaramella, Proc. Natl. Acad. Sci. USA *69*, 3389–3393 (1972).

DNA Nucleotide Sequence Restricted by the RI Endonuclease, J. Hedgpeth, H. Goodman, H. Boyer, Proc. Natl. Acad. Sci. USA *69*, 3448–3452 (1972).

Nature of ColE1 Plasmid Replication in *Escherichia coli* in the Presence of Chloramphenicol, D. Clewell, J. Bact. *110*, 667–676 (1972).

Enzymatic End-to-End Joining of DNA Molecules, P. Lobban, A. Kaiser, J. Mol. Biol. *78*, 453–471 (1973).

A Suggested Nomenclature for Bacterial Host Modification and Restriction Systems and Their Enzymes, H. Smith, D. Nathans, J. Mol. Biol. *81*, 419–423 (1973).

Construction of Biologically Functional Bacterial Plasmids *in vitro*, S. Cohen, A. Chang, H. Boyer, R. Helling, Proc. Natl. Acad. Sci. USA *70*, 3240–3244 (1973).

A New Technique for the Assay of Infectivity of Human Adenovirus 5 DNA, F. Graham, A. Van der Eb, Virology *52*, 456–467 (1973).

Determination of a Nucleotide Sequence in Bacteriophage f1 DNA by Primed Synthesis with DNA Polymerase, F. Sanger, J. Donelson, A. Coulson, M. Kossel, D. Fischer, J. Mol. Biol. *90*, 315–333 (1974).

Manipulation of Restriction Targets in Phage Lambda to Form Receptor Chromosomes for DNA Fragments, N. Murray, K. Murray, Nature *251*, 476–481 (1974).

Fusion of the *Escherichia coli lac* Genes to the *ara* Promoter: A General Technique Using Bacteriophage Mu-1 Insertions, M. Casadaban, Proc. Natl. Acad. Sci. USA *72*, 809–813 (1975).

Colony Hybridization: A Method for the Isolation of Cloned DNAs That Contain a Specific Gene, M. Grunstein, D. Hogness, Proc. Natl. Acad. Sci. USA *72*, 3961–3965 (1975).

A Rapid Method for Determining Sequences in DNA by Primed Synthesis with DNA Polymerase, F. Sanger, A. Coulson, J. Mol. Biol. *94*, 441–448 (1975).

Ligation of *Eco*RI Endonuclease-Generated DNA Fragments into Linear and Circular Structures, A. Dugaiczyk, H. Boyer, H. Goodman, J. Mol. Biol. *96*, 171–181 (1975).

Chain Length Determination of Small Double-and-Single-Stranded DNA Molecules by Polyacrylamide Gel Electrophoresis, T. Maniatis, A. Jeffrey, H. Van de Sande, Biochem. *14*, 3787–3794 (1975).

Injected Nuclei in Frog Oocytes Provide a Living Cell System for the Study of Transcriptional Control, J. Gurdon, E. Robertis, G. Partington, Nature *260*, 116–120 (1976).

*Eco*RI Endonuclease, Physical and Catalytic Properties of the Homogeneous Enzyme, P. Modrich, D. Zabel, J. Biol. Chem. *251*, 5866–5874 (1976).

Terminal Labeling and Addition of Homopolymer Tracts to Duplex DNA Fragments by Terminal Deoxynucleotidyl Transferase, R. Roychoudhury, E. Jay, R. Wu, Nucleic Acids Res. *3*, 101–116 (1976).

Packaging Recombinant DNA Molecules into Bacteriophage Particles *in vitro*, B. Hohn, K. Murray, Proc. Natl. Acad. Sci. USA *74*, 3259–3263 (1977).

Chemical Synthesis of Restriction Enzyme Recognition Sites Useful for Cloning, R. Scheller, R. Dickerson, H. Boyer, A. Riggs, K. Itakura, Science *196*, 177–180 (1977).

Screening lambda gt Recombinant Clones by Hybridization to Single Plaques in situ, W. Benton, R. Davis, Science *196*, 180–182 (1977).

Transfer of Purified Herpes Virus Thymidine Kinase Gene to Cultured Mouse Cells, M. Wigler, S. Silverstein, L. Lee, A. Pellicer, Y. Cheng, R. Axel, Cell *11*, 223–232 (1977).

Mapping Adenines, Guanines, and Pyrimidines in RNA, H. Donis-Keller, A. Maxam, W. Gilbert, Nucleic Acids Res. *4*, 2527–2538 (1977).

Construction and Characterization of New Cloning Vehicles, I. Ampicillin-Resistant Derivatives of the Plasmid pMB9, F. Bolivar, R. Rodriguez, M. Betlach, H. Boyer, Gene *2*, 75–93 (1977).

Transformation of Yeast, A. Hinnen, J. Hicks, G. Fink, Proc. Natl. Acad. Sci. USA *75*, 1929–1933 (1978).

Cosmids: A Type of Plasmid Gene-Cloning Vector That Is Packageable *in vitro* in Bacteriophage Lambda Heads, J. Collins, B. Hohn, Proc. Natl. Acad. Sci. USA *75*, 4242–4246 (1978).

Transformation of Yeast by a Replicating Hybrid Plasmid, J. Beggs, Nature *275*, 104–108 (1978).

pBR322 Restriction Map Derived from the DNA Sequence: Accurate DNA Size Markers up to 4361 Nucleotide Pairs Long, J. Sutcliffe, Nucleic Acids Res. *5*, 2721–2728 (1978).

Transformation of Mammalian Cells with Genes from Procaryotes and Eucaryotes, M. Wigler, R. Sweet, G. Sim, B. Wold, A. Pellicer, E. Lacy, T. Maniatis, S. Silverstein, R. Axel, Cell *16*, 777–785 (1979).

Cloning in Single-Stranded Bacteriophage as an Aid to Rapid DNA Sequencing, F. Sanger, A. Coulson, B. Barrell, A. Smith, B. Roe, J. Mol. Biol. *143*, 161–178 (1980).

High Efficiency Transformation by Direct Microinjection of DNA into Cultured Mammalian Cells, M. Capecchi, Cell *22*, 479–488 (1980).

Sequencing End-Labeled DNA with Base-Specific Chemical Cleavages, A. Maxam and W. Gilbert, in Methods in Enzymology *65*, ed. L. Grossman and K. Moldave, 499–559, Academic Press, New York (1980).

Selection for Animal Cells that Express the *Escherichia coli* Gene Coding for Xanthine-Guanine Phosphoribosyl-Transferase, R. Mulligan, P. Berg, Proc. Natl. Acad. Sci. USA *78*, 2072–2076 (1981).

Involvement of Outside DNA Sequences in the Major Kinetic Pathway by Which *EcoRI* Endonuclease Locates and Leaves Its Recognition Sequence, W. Jack, B. Terry, P. Modrich, Proc. Natl. Acad. Sci. USA *79*, 4010–4014 (1981).

Propagation of Foreign DNA in Plants Using Cauliflower Mosaic Virus as Vector, G. Gronenborn, R. Gardner, S. Schaefer, R. Shepherd, Nature *294*, 773–776 (1981).

Construction and Characterization of New Cloning Vehicles, V. Mobilization and Coding Properties of pBR322 and Several Deletion Derivatives Including pBR327 and pBR328, L. Covarrubias, L. Cervantes, A. Covarrubias, X. Soberón, I. Vichido, A. Blanco, Y. Kuparsztoch-Portnoy, F. Bolivar, Gene *13*, 25–35 (1981).

Somatic Expression of Herpes Virus Thymidine Kinase in Mice Following Injection of a Fusion Gene into Eggs, R. Brinster, H. Chen, M. Trumbauer, A. Seneac, R. Warren, R. Palmiter, Cell *27*, 223–231 (1981).

Introduction of Rat Growth Hormone Gene into Mouse Fibroblasts via a Retroviral DNA Vector: Expression and Regulation, J. Doehmov, M. Barinaga, W. Vale, M. Rosenfeld, I. Verma, R. Evans, Proc. Natl. Acad. Sci. USA *79*, 2268–2272 (1982).

A Systematic DNA Sequencing Strategy, G. Hong, J. Mol. Biol. *158*, 539–549 (1982).

Transformation of Mammalian Cells to Antibiotic Resistance with a Bacterial Gene Under Control of the SV40 Early Region Promoter, P. Southern, P. Berg, J. Molec. Appl. Genet. *1*, 327–341 (1982).

Restriction and Modification Enzymes and Their Recognition Sequences, R. Roberts, Nucleic Acids Res. *10*, r117–r144 (1982).

Isolation of Z-DNA-Containing Plasmids, R. Thomae, S. Beck, F. Pohl, Proc. Natl. Acad. Sci. USA *80*, 5550–5553 (1983).

The Alternate Expression of Two Restriction and Modification Systems, S. Glover, K. Firman, G. Watson, C. Price, S. Donaldson, Mol. Gen. Genet. *190*, 65–69 (1983).

Site-Specific Cleavage of DNA at 8- and 10-Base-Pair Sequences, M. McClelland, L. Kesler, M. Bittner, Proc. Nat. Acad. Sci. USA *81*, 1480–1483 (1984).

Genomic Sequencing, G. Church, W. Gilbert, Proc. Nat. Acad. Sci. USA *81*, 1991–1995 (1984).

DNA Polymorphic Loci Mapped to Human Chromosomes 3, 5, 9, 11, 17, 18, and 22, S. Tayler, A. Sakaguchi, D. Barker, R. White, T. Shows, Proc. Nat. Acad. Sci. USA *81*, 2447–2451 (1984).

Immunity and Protection Against Influenza Virus by Synthetic Peptide Corresponding to Antigenic Sites on Hemagglutinin, M. Shapira, M. Jibson, G. Muller, R. Arnon, Proc. Nat. Acad. Sci. USA *81*, 2461–2465 (1984).

Advanced Genetic Engineering

The previous chapter described the fundamentals of genetic engineering: cutting, splicing, vectors, transformation, and DNA sequencing. This chapter describes more advanced manipulations, which, for the most part, are technological aspects of genetic engineering. While of some interest in themselves, their main value is in the sophisticated studies they facilitate. Research described in the remainder of the book, particularly that involving eukaryotic cells, made extensive use of the techniques described in this and the previous chapter.

Since higher cells possess large genomes, cloning a gene that is present as a single copy poses a serious logistic problem. Many hundreds of thousands of candidates may have to be scored to find the desired clone. Fortunately, a number of strategies have been devised to ease this burden; several are discussed in this chapter.

The study of cloned DNA segments has revealed many unexpected sequence properties of the DNA from higher cells. Several techniques based on electrophoresis that greatly facilitate studies of these properties are described here. Among them are Southern transfers and the use of nuclease digestions and electrophoresis. These have shown that the moderately repeated DNA sequences mentioned in Chapter 2 are scattered about the genome and that the protein-coding sequences in many of the genes in higher organisms are not contiguous but are interrupted by inserted sequences of unknown function that are spliced out of the corresponding RNAs after transcription.

Table 11.1   Enzyme Properties for Use in Genetic Engineering

Enzyme	Activity
Bal 31 exonuclease	Digests both strands of double-stranded DNA
S1 nuclease	Single-strand specific, digests SS RNA and DNA
Exonuclease III	Digests only from 3'-OH end of DNA in duplex form
DNA pol I, large fragment	Has polymerizing activity of DNA pol I but lacks its $5' \rightarrow 3'$ exonuclease activity
T4 DNA ligase	Forms phosphodiester bond between 5' phosphate and 3' hydroxyl ends in duplex DNA
Restriction enzymes	Cleave DNA at specific sequences

Once investigators have determined the sequence, locations of coding regions, and the transcription start and stop sites of a cloned segment of DNA, they frequently turn to a study of its function with *in vitro* or even *in vivo* experiments. Many of these manipulations utilize properties of enzymes that modify DNA (Table 11.1). Sequence alterations of the cloned DNA permit "molecular dissections" of regions of DNA and thus are a method for systematic study. One highly useful tool for generating nucleotide modifications is *in vitro* mutagenesis. Nucleotide insertions and deletions can easily be made at the ends of DNA fragments, and random base changes can be directed to any desired small portion of a gene. Additionally, specific base changes can be made at a specific point in a gene by utilizing chemically synthesized DNA. In fact, advances in the art of DNA synthesis have reached the point that entire genes are now synthesized for some applications.

Before the final topic of this chapter, two direct medical applications of genetic engineering are mentioned: the prenatal detection of the sickle cell trait by hybridization, and the mapping of the gene responsible for the neurodegenerative disease, Huntington's chorea, to one of the human chromosomes and the possibility of a prenatal test for this genetic disease. The final topic of this chapter integrates some of the techniques previously described into a sequence of steps that actually were used to construct a plasmid for the hypersynthesis of a bacterial regulatory protein by the introduction of an active promoter and efficient ribosome binding site in front of the gene.

CLONING FROM RNA OR DNA

The isolation of clones carrying some genes of *Escherichia coli* is straightforward because mutant strains can be used as recipients of the cloning vectors. Only those cells that receive a vector carrying the gene to be cloned overcome the mutant's growth deficiency and become able to grow (Fig. 11.1). Many times however, the gene to be cloned is not from *E. coli* and therefore may not be properly expressed in this organism or simply cannot be used as the basis of a growth selection.

How can clones carrying nonselectable genes be detected? One of the simplest ways to improve the odds is to begin with DNA that is enriched for the desired genes. Occasionally this is straightforward because multiple copies of the DNA may exist in certain tissues of higher cells at certain developmental stages. More often, however, one is forced to begin with messenger RNA that is enriched for transcripts of the gene.

Figure 11.1   Cloning DNA from *E. coli* in which a selection for X⁺ transformants permits direct selection of the desired clones.

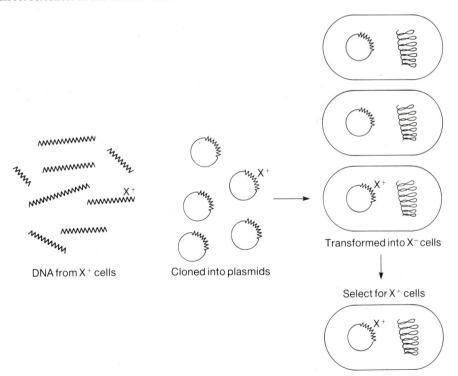

DNA from X⁺ cells   Cloned into plasmids

Transformed into X⁻ cells

Select for X⁺ cells

Extraction of RNA from cells yields a preponderance of ribosomal RNA. However, messenger RNA can easily be separated from the ribosomal RNA since most messenger RNA from most higher organisms contains a poly-A tail at the 3' end. This tail can be used for isolation by passing a crude fraction of cellular RNA through a cellulose column to which poly-T has been linked. Under hybridizing conditions, the poly-T linked to the column and the poly-A tails of messenger molecules hybridize and bind the messenger RNAs to the column, whereas the ribosomal RNA molecules flow through the column. The messenger RNAs are eluted by lowering the salt concentration or by raising the temperature. Such a purification step frequently provides a several-hundred-fold enrichment for messenger RNA and eliminates most ribosomal RNA. This procedure, in combination with the choice of a particular tissue at a specific developmental stage as a source of the RNA, often yields sufficiently pure mRNA.

The single-stranded RNA obtained by the steps described above cannot be cloned directly. Either the RNA can be transcribed to DNA via a complementary strand, cDNA, or the RNA can be used to aid identification of a clone containing the complementary DNA sequence. To generate a cDNA copy of the poly-A-containing messenger, several steps are performed (Fig. 11.2). First, a poly-dT primer is hybridized to the messenger. Next, reverse transcriptase is used to elongate the primer to yield a DNA copy. This enzyme, which is found within the free virus particle of some

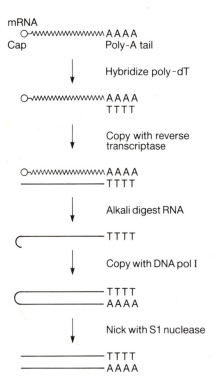

Figure 11.2   Steps in the replication of an mRNA molecule into a double-stranded DNA copy suitable for cloning.

animal viruses, synthesizes DNA using an RNA template. At this point the sequence exists as an RNA-DNA hybrid, but the RNA can be specifically removed by digestion in alkali to leave a cDNA strand. The end of the DNA strand left by reverse transcriptase often folds over on itself and therefore can provide the necessary 3'-OH to prime synthesis by DNA polymerase I of the strand complementary to the first DNA strand. Finally, the looped single strand of DNA can be converted to two strands of DNA by brief digestion with an enzyme that cleaves single-stranded nucleic acid, S1 nuclease. This resulting double-stranded DNA can be cloned using linkers or the poly-A, poly-T tailing methods.

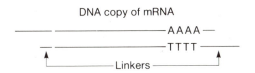

PLAQUE AND COLONY HYBRIDIZATION FOR CLONE IDENTIFICATION

A variety of direct screening techniques exist for detecting a cloned gene. In these approaches, cloning the foreign DNA into a lambda vector is convenient because the phage can accommodate inserted fragments of large size and many lambda phage may be screened on a single petri plate. The candidate phage collection, called a lambda bank or library, containing inserted DNA segments is plated out at thousands of plaques per plate. Then a replica copy of the phage plaques is made by pressing a filter paper onto the plate. After the paper is removed, it is immersed in alkali. By these steps the DNA is denatured and fixed onto the paper. Then radioactively labeled RNA or DNA, called a probe, is hybridized to any complementary DNA sequences in plaque images on the paper. The location of the areas with bound RNA are then determined by autoradiography, and viable phage containing the desired insert can be isolated from the corresponding position on the original plate. Analogous techniques exist for screening colonies containing plasmids.

R-LOOP ENRICHMENT OF DNA COMPLEMENTARY TO PURIFIED RNA

Instead of screening many plaques for the few containing a rare sequence, in some situations the DNA to be cloned can first be enriched for the gene. One such technique of DNA purification is based on RNA-looping, usually shortened to R-looping. Ordinarily, double-stranded DNA possesses greater stability than an

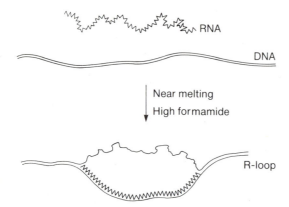

Figure 11.3   An R-loop. Single-stranded DNA is more squiggly than double-stranded nucleic acid.

RNA-DNA duplex. However, near the melting temperature of RNA-DNA duplexes and in the presence of high concentrations of formamide, RNA-DNA duplexes are more stable than DNA-DNA duplexes. In these conditions an RNA molecule will displace one of the DNA strands of a duplex to form an R-loop (Fig. 11.3). The physical properties of the R-looped DNA are sufficiently different from the input RNA and the input DNA that it may be purified on the basis of either its density or its size. The RNA can then be removed from the R-loop and the DNA cloned.

WALKING ALONG A CHROMOSOME TO CLONE A GENE

Another method of cloning a gene is walking. Suppose a randomly isolated clone has been shown to lie within several hundred thousand bases of the DNA segment to be cloned. Of course, demonstrating such a fact is not trivial in itself. However, in the case of the fruit fly *Drosophila melanogaster* the demonstration is straightforward.

As explained in Chapter 8, the polytene chromosomes in cells of several *Drosophila* tissues possess characteristic bands that serve as chromosomal landmarks. Therefore the approximate amount of DNA involved in rearrangements or deletions can readily be determined. Also, highly radioactive DNA originating from a cloned segment of *Drosophila* chromosome can be hybridized *in situ* to *Drosophila* polytene chromosomes. The position to which the fragment hybridizes can then be determined by autoradiography (Fig. 11.4). Hence the chromosomal position of a cloned fragment can be determined. If it lies near a gene of interest, walking may be in order.

A chromosomal walk begins with a restriction map of the cloned piece of DNA. The right- and left-end terminal restriction fragments are used as probes of a lambda bank. Several clones are picked that hybridize to the right-hand fragment and not to the left-hand fragment. Then restriction maps of these clones are made and again the

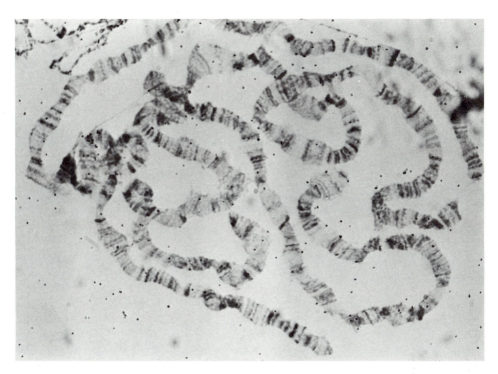

Figure 11.4   The polytene chromosomes from the salivary gland of *Drosophila melanogaster* stained and used for *in situ* hybridization with radioactive DNA. The radioactive DNA hybridized to a portion of one chromosome near the bottom center of the photograph. (Courtesy J. Hirsh.)

right- and left-hand fragments are used to find new clones that hybridize only to the right-hand fragments (Fig. 11.5). The successive lambda transducing phage identified permit walking to the right, and when a sufficient distance has been covered, *in situ* hybridization permits determination of which direction on the cloned DNA corresponds to moving toward the desired gene.

ARREST OF TRANSLATION TO ASSAY FOR DNA OF A GENE

In vitro translation of mRNA forms the basis of one technique for identification of clones containing DNA of a particular gene. The technique requires that the enriched messenger RNA of the gene, such as the mRNA obtained from an oligo-dT column, be translatable *in vitro* to yield a detectable protein product of the desired

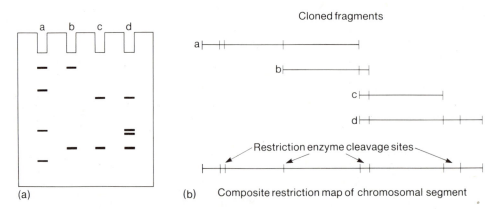

(a)   (b)   Composite restriction map of chromosomal segment

Figure 11.5   (a) A gel showing DNA fragments resulting from a restriction enzyme digestion of a set of four partially overlapping DNA sequences. (b) The four DNAs and their composite.

gene. To perform the identification, DNA for a candidate clone is denatured and hybridized to the RNA used in the translation mixture. If the DNA contains sequences complementary to the mRNA, the two will hybridize together, the messenger will become unavailable for *in vitro* translation, and the gene product will not be synthesized (Fig. 11.6). DNA from a clone not possessing the DNA sequence of the gene will not interfere with translation of the messenger. Hence DNA carrying

Figure 11.6   Hybridization arrest of translation as an assay for DNA of a gene whose mRNA can be translated into protein.

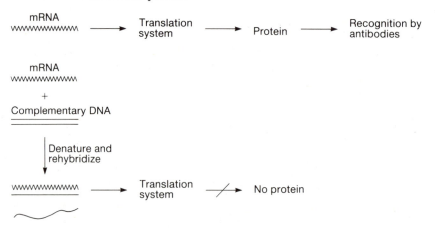

sequences complementary to the messenger can be detected by hybridization arrest of translation.

TOOLS OF ANALYSIS I: SOUTHERN TRANSFERS, DNA

Many examples have already been mentioned of the utility of detecting sequence complementarity between two nucleic acids. In this section more examples will be given. The main point of the present discussion is that powerful and simple techniques have been developed to permit sophisticated questions raised by cloning experiments to be answered with ease.

Consider the problem of learning whether a segment of DNA suffered a deletion or rearrangement on cloning. One method of answering this question would be to determine the size of several large restriction fragments within the cloned DNA and the size of the corresponding fragments obtained by digestion of DNA extracted from the original organism. If all the fragments obtained from both sources were of the same size, then a deletion is unlikely to have occurred. The experimental difficulty of this approach is measuring the size of the chromosomal restriction fragments.

One method of measuring the size of the restriction fragments would be to electrophorese the DNA on agarose. Then the gel could be sliced up, and DNA extracted from each of the slices could be hybridized to radioactively labeled fragment obtained from the clone. This would identify the slice containing the complementary DNA and permit comparison of the electrophoretic mobilities of the chromosomal DNA and the cloned fragments.

Southern devised a hybridization method for DNA separated by electrophoresis that eliminates the need for any slicing and eluting of DNA. Following electrophoresis, the DNA is denatured while in the gel and transferred *in toto* to a sheet of nitrocellulose paper (Fig. 11.7). Recall that generally only single-stranded DNA binds to such paper. One way to transfer the DNA is to place the gel with the denatured DNA on top of a stack of paper placed in a reservoir of buffer. The nitrocellulose sheet is placed on top of the gel, and a stack of dry absorbent paper is placed on top of the nitrocellulose sheet. Capillary action draws the buffer through the gel, carrying the fragments of DNA from the agarose into the nitrocellulose sheet, where they are bound. Another transfer method is electrophoretic. The gel and nitrocellulose paper or other membrane that binds single-stranded DNA are placed in contact, and a transverse electric field moves the DNA to the paper or membrane. The key to both of these transfer processes is that they preserve the original separation pattern of the DNA fragments within the gel.

The nitrocellulose paper with the pattern of immobilized DNA can then be used for hybridization with the radioactive probe. If the probe is RNA, the paper can be used as it is, but if the probe is DNA, the paper must be treated to prevent the additional nonspecific binding of single-stranded DNA. Soaking in dextran, bovine

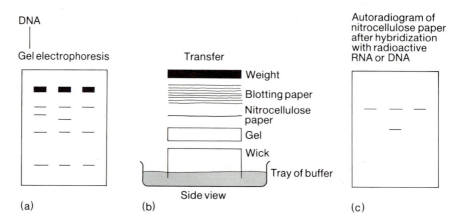

Figure 11.7   Southern transfer methodology. (a) Gel electrophoresis separates DNA fragments. (b) The denatured fragments are blotted to nitrocellulose paper. (c) Hybridizing the paper with radioactive probe and autoradiography identifies fragments homologous to the probe.

serum albumin, polyvinylpyrollidone, sodium dodecyl sulfate, and a foreign DNA such as salmon sperm DNA before the hybridization step usually prevents nonspecific binding of DNA to the paper. Then only DNA complementary to DNA immobilized on the paper will bind. To hybridize, the paper is simply incubated under hybridizing conditions in a solution containing the probe. After the incubation, free probe is rinsed off the nitrocellulose paper, and the positions of bound probe are visualized by autoradiography (Fig. 11.8). Many different DNA samples can be hybridized to a single probe in a single experiment by this method.

TOOLS OF ANALYSIS II: NORTHERN AND WESTERN TRANSFERS, RNA AND PROTEIN

Northern transfers are the converse of Southern transfers in that it is RNA rather than DNA that is separated by electrophoresis, transferred, and immobilized on paper to preserve the original pattern. Although under normal conditions only single-stranded DNA binds to nitrocellulose paper, RNA can be made to bind by transferring in the presence of high (3 M) concentrations of salt and baking. Paper with immobilized RNA can then be used in hybridization just like paper with immobilized DNA.

What kinds of questions can be answered with immobilized RNA? One question concerns the *in vivo* state of various RNAs. Transient precursors of a mature RNA molecule can easily be detected because they will be larger than the mature RNA and

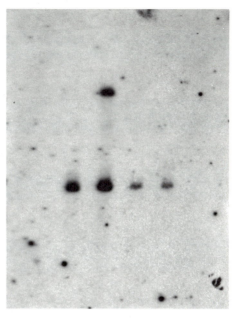

Figure 11.8   A Southern transfer experiment with *Drosophila melanogaster.* Flies were ground up, the DNA extracted and digested with *Hind* III, and the DNA subjected to electrophoresis on an agarose gel. (a) A lane of size standards and four lanes of experimental material. The DNA was stained with ethidium bromide and photographed with UV illumination. (b) The DNA was denatured and transferred to nitrocellulose paper. The bound DNA was then hybridized with a radioactive DNA probe that hybridized to one size of fragment in the first, third, and fourth lanes and to two sizes of fragment in the second lane. (Courtesy M. C. Hung and M. Garabedian.)

will be separated during electrophoresis. This permits following the maturation process of an RNA molecule. Not only can the changing sizes of the maturing species be monitored, but the specific regions that are lost during the maturation can be determined by using radioactive probes from the regions in question.

Specific RNAs or DNAs can be purified by use of Northern or Southern transfers. Either single-stranded RNA or single-stranded DNA can be bound to the paper. Then either RNA or DNA fragments complementary to the immobilized RNA or DNA can easily be isolated from a mixture by hybridization followed by elution. As an application, messenger RNA eluted from such immobilized DNA can be translated *in vitro* to provide a definitive identification of a candidate clone for a specific gene.

Western transfers involve proteins, not nucleic acids. The principle is the same as for Northern and Southern transfers. A pattern of proteins separated by electro-

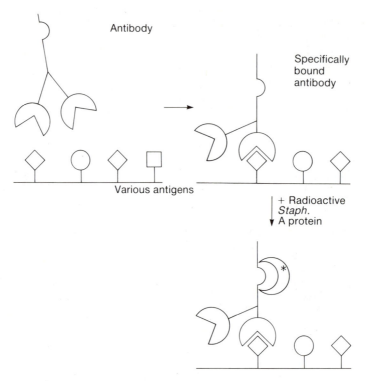

Figure 11.9   The use of radioactive *Staphylococcus* A protein to identify antibody-antigen complexes in Western transfers.

phoresis is transferred to nitrocellulose paper and then specific proteins are visualized. A type of protein that can easily be located is one capable of binding to a specific nucleic acid. The paper with the immobilized proteins is incubated with the radioactive nucleic acid and washed, and an autoradiograph is prepared.

Antibodies also provide a method of detecting specific immobilized proteins in the Western transfer method. Molecules of one particular antibody type bind highly selectively to just one particular shape found in some other macromolecule, their antigen. Almost any protein can be used as an antigen to elicit the synthesis of antibodies. Thus antibodies provide highly selective agents for the detection of specific proteins, their antigens. The antibody selectivity corresponds to the hybridization selectivity of nucleic acids.

Although radioactive antibody could be used to detect antigen on the Western transfer paper, it is not efficient, for different antibodies would then have to be made radioactive for the detection of different proteins. The A protein from *Staphylococcus aureus* provides a more general detection method. This protein binds to a portion of the antibody molecule so that one sample of radioactive *Staphylococcus aureus* A protein suffices for the detection of any and all antibody-protein complexes (Fig. 11.9). As in the case of Northern transfers, one of the applications of the method is the detection of precursor molecules.

TOOLS OF ANALYSIS III: FOOTPRINTING AND S1 MAPPING, DNA

The preceding section described methods that can be used to determine whether a protein binds to a particular DNA fragment. Often the next level of investigation requires precise determination of the binding location of a protein on DNA or the transcription start or splice site of an RNA. One method of making this determination for a protein is to bind the protein to the DNA fragment, extensively digest with DNase, then determine the sequence of whatever was protected from digestion. Of course, this method has drawbacks. It is difficult to digest completely the DNA not covered with protein without digesting much of the covered DNA. Also, any inhomogeneity in digestion of the ends complicates the analysis. Finally, the protected piece of DNA may be too small for convenient analysis.

Galas and Schmitz developed an elegant solution to the problem of determining the location at which a protein binds to DNA. Their method is called footprinting and is based on the principles behind Maxam-Gilbert DNA sequencing. It can detect the binding of a protein up to about 300 nucleotides from the end of a DNA fragment.

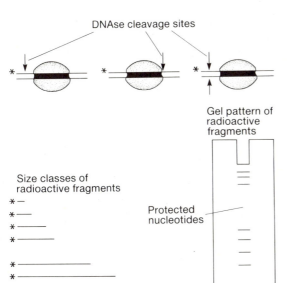

Figure 11.10   The principles of DNase footprinting. Light digestion by DNaseI nicks each of the labeled DNA molecules about once. No nicking occurs under the bound protein and therefore no fragments appear of the size corresponding to the distance from the protein to the labeled end of the DNA.

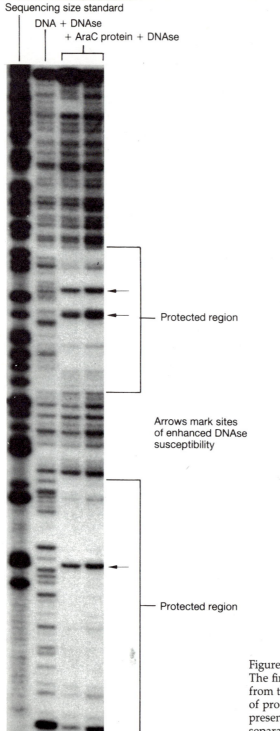

Sequencing size standard

DNA + DNAse

+ AraC protein + DNAse

Protected region

Arrows mark sites
of enhanced DNAse
susceptibility

Protected region

Figure 11.11   Autoradiograph of a footprinting gel.
The first lane is a size standard; the second, DNA
from the ʟ-arabinose operon digested in the absence
of protein; the third and fourth, digested in the
presence of *araC* protein, which binds to two sites
separated by about 15 base pairs. (Courtesy W.
Hendrickson.)

The DNA fragment to be investigated must first be labeled. This can be done with polynucleotide kinase, with terminal transferase, which adds nucleotides to the 3'-OH end of DNA, or even with DNA pol I, which fills out sticky ends left by many restriction enzymes. As in DNA sequencing, the label must be present at only one point in the sequence. The usual method of achieving this is to cleave the end-labeled DNA into fragments of unequal length with a restriction enzyme and to separate them by electrophoresis.

The actual DNA footprinting begins with double-stranded DNA labeled on one end. After the binding of a protein to a particular site on the DNA, the complex is briefly treated with DNase. The duration of this enzyme digestion is adjusted so that about one cleavage occurs per DNA molecule. Consequently, the population of molecules will contain examples of phosphodiester bond breakage at all positions except those covered by the protein (Fig. 11.10). Then electrophoresis of the denatured DNA fragments on a sequencing gel produces a radioactive ladder pattern in

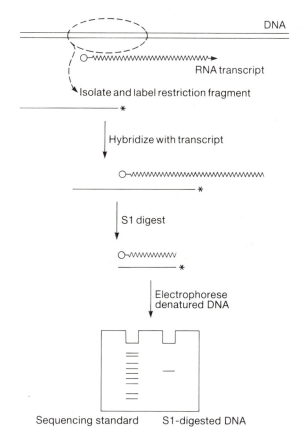

Figure 11.12   The principles of S1 nuclease mapping. Hybridization of RNA to an appropriately labeled DNA fragment protects against S1 digestion. The size of the radioactive segment is then determined on a DNA sequencing gel.

which rungs are present at each position except those corresponding to nucleotides protected by the protein.

The experiment can also be performed with the protein protecting from chemical attack rather than enzymatic attack. Dimethylsulfate can methylate guanine residues except some of those protected by the protein. After the methylation, the DNA can be cleaved at each of the methylated guanines, and the denatured, labeled fragments can be subjected to electrophoresis on a sequencing gel. Both DNaseI and dimethylsulfate are imperfect in that their reaction with unprotected DNA is somewhat base-specific. Consequently, DNase footprinting or dimethylsulfate protection reactions are done in parallel, one with the binding protein and one without the binding protein. The differences in the rung patterns are attributable to the protein (Fig. 11.11).

S1 mapping, developed by Berk and Sharp, can easily determine the end points of RNA molecules. This is done by techniques similar to protein footprinting. Consider the problem of determining the *in vivo* transcription start point of a particular gene. RNA is isolated from the organism, freed of excess protein and DNA, then hybridized to end-labeled, single-stranded DNA that includes the transcription start point. After hybridization, the single-stranded RNA and DNA tails are digested with S1 nuclease. The exact size of the DNA that was protected from nuclease digestion can then be determined by electrophoresis on a DNA sequencing gel (Fig. 11.12). This size gives the distance from the transcription start point to the end of the DNA fragment used in the hybridization.

ALTERING CLONED DNA BY *in vitro* MUTAGENESIS

Detailed study of DNA-related biological mechanisms requires more than the DNA and associated proteins. It often requires alteration of the components. Not only does variation of the relevant parameters reveal more about the working mechanism, but the ability to test variants permits definitive proof of theories. Mutants have been used in molecular biology almost from its origins, first in the elucidation of biochemical pathways and now prominently in structural studies of the mechanisms by which proteins function as enzymes or recognize and bind to specific nucleotide sequences on DNA.

The efficient isolation of mutations has always posed a problem in molecular biology. Suppose mutations are desired in a particular gene or DNA sequence. If the entire organism must be mutagenized, then to obtain a reasonable number of alterations in the desired target, many more alterations will inevitably occur elsewhere on the chromosome. Often these other mutations will be lethal, so the necessary alterations in the target cannot easily be found. A method is needed for directing mutations just to the target gene. *In vitro* mutagenesis of cloned DNA fragments is a solution. DNA of only the target sequence can be heavily mutagenized and then put back into cells.

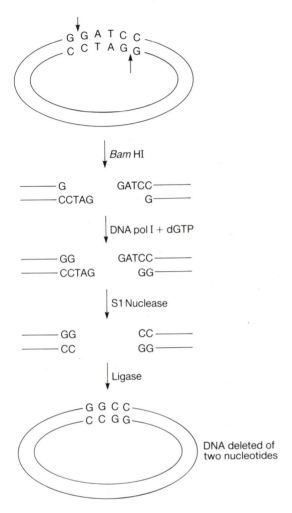

Figure 11.13   Creating a two-base-pair deletion at a *Bam*HI site by filling out two of the four-base single-stranded ends and removing the other two with S1 nuclease.

Insertions and deletions can easily be generated at the cleavage site of a restriction enzyme. For example, a four-base insertion can be generated at the cleavage site of *Bam*HI by filling in the four-base single-stranded ends with DNA pol I and ligating the flush ends together. Similarly, a four-base deletion can be generated by digesting the single-stranded ends with the single-stranded specific nuclease S1 before ligation. Variations on these themes are to use DNA pol I in the presence of only one, two, or three of the nucleotides to fill out part of the single-stranded ends before nuclease treatment and ligation (Fig. 11.13). Mixing and matching entire restriction fragments from a region under study is another closely related method of

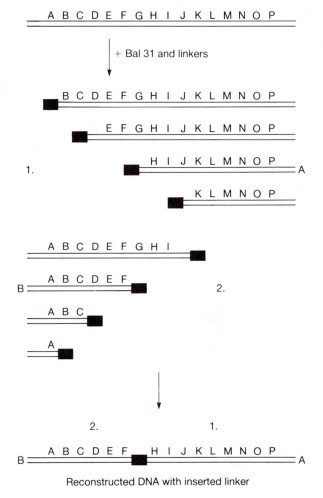

Figure 11.14   Digestion with Bal 31 from either direction and addition of linkers generates
a set of molecules that can be rejoined via the linkers to yield a molecule essentially like
the original wild type but with a drastic insertion comprising the linker molecule itself.

changing portions of DNA binding sites or substituting one portion of a protein for
another.

More extensive deletions from the ends of DNA molecules can be generated by
double-stranded exonuclease digestion. The nuclease Bal 31, isolated from the culture medium of the bacterium *Alteromonas espejiana,* is particularly useful for this
purpose. With it, a set of clones with progressively larger deletions into a region can
easily be isolated. The addition of linkers after Bal 31 digestion permits targeted

substitution of a set of nucleotides or a change in the number of nucleotides between two sites. Deletions are isolated entering the region from both directions. Before recloning, a restriction enzyme linker is added. After these steps, a pair of deletions can be easily joined via their linkers to generate a DNA molecule identical to the wild type except for the alteration of a stretch comprising the linker (Fig. 11.14). If different pairs of deletions are chosen, then distances also can be altered.

Bases within DNA fragments can be changed with chemical *in vitro* mutagenesis. Hydroxylamine will effectively mutagenize the cytosines in denatured DNA fragments, which can then be renatured and recloned. Alternatively, mutagenesis can be directed to particular regions. One method is to generate a single-stranded region by nicking one strand as a result of digestion with a restriction enzyme in the presence of ethidium bromide and then briefly digesting with exonuclease III. The mutagenesis is then done with a single-stranded specific reagent such as sodium bisulfite, which mutagenizes cytosines and ultimately converts them to thymines, or by compelling misincorporation of bases during repair of a gap.

The final method of mutagenesis to be mentioned in this section is insertional inactivation of a gene. Suppose a yeast gene is to be inactivated by mutation and that the cloned copy of the gene is available. The gene could be mutagenized *in vitro* and reincorporated, but that would not inactivate the resident chromosomal gene. The following is a way to inactivate the chromosomal gene. A vector can be constructed from a bacterial replicon, bacterial selectable gene, the yeast gene fragment, and a yeast selectable gene such as uracil synthesis. Uracil-requiring yeast cells are trans-

Figure 11.15   Insertional inactivation of a yeast gene.

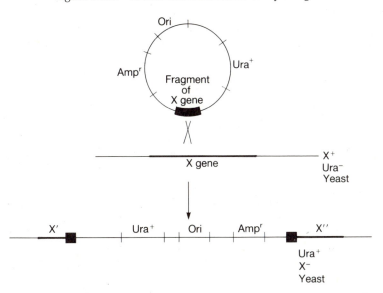

formed, and selection is performed for cells able to grow without exogenously added uracil. Since the vector lacks a yeast replicon, it must integrate into the yeast chromosome to survive. Furthermore, it must integrate by means of a recombination in a region of DNA homology between the plasmid and the chromosome. Two areas exist for this: the region of the gene coding for uracil synthesis and the region of the gene fragment. Use of an appropriate recipient with a deletion of the gene coding for uracil synthesis can eliminate the possibility of recombination in this region. Recombination within the fragment of the gene splits the chromosomal gene into two incomplete parts (Fig. 11.15). The nonhomologous chromosomes can then be segregated by growth so that homozygous yeast result or haploids can be generated by sporulating the diploids. If the gene with the insertion is completely inviable, only two spores from each tetrad would be viable. The structure of the parental diploid yeast could be demonstrated by an appropriate Southern transfer.

MUTAGENESIS WITH CHEMICALLY SYNTHESIZED DNA

Before 1965 no researchers would have had a good idea of what to do with the sequence of an entire chromosome if it were presented to them. We are hardly in that situation now. Similarly, before 1975 there seemed to be little reason to try to synthesize DNA chemically. Not only were relatively few interesting sequences known, but the fraction of the synthesized material that would possess the desired sequence was likely to be too small to be of use. However, with the development of cloning since 1975 and the overall increase in our knowledge of biological mechanisms, the picture has changed. Now it is conceivable, when necessary, to synthesize a gene chemically and to clone it. Its sequence can be checked and, thereafter, large quantities can be prepared. Less ambitious projects can use chemically synthesized oligomers for modification of cloned genes, as hybridization probes for the detection of clones, or for sequencing.

Khorana laid the groundwork for chemical synthesis of DNA. He developed techniques to form the phosphodiester bond between nucleotides while at the same time preventing the reactive amino, hydroxyl, and other phosphorus groups from reacting. With these techniques he and his co-workers then synthesized a complete tRNA gene. Originally many person-years were required for the synthesis of 80 nucleotide oligomers. However, continued development by many research groups has improved oligonucleotide synthesis to the point that as many as 15 nucleotides may be joined in specific sequence in a day.

Blocking groups are placed on the reactive groups that are not to participate in the condensation to form a phosphodiester, and the oligonucleotide is gradually built (Fig. 11.16). After synthesis of the complete oligonucleotide, all the blocking groups are removed. If the desired oligonucleotide is particularly long, blocks of short, overlapping oligonucleotides can be synthesized, hybridized, ligated, and finally cloned.

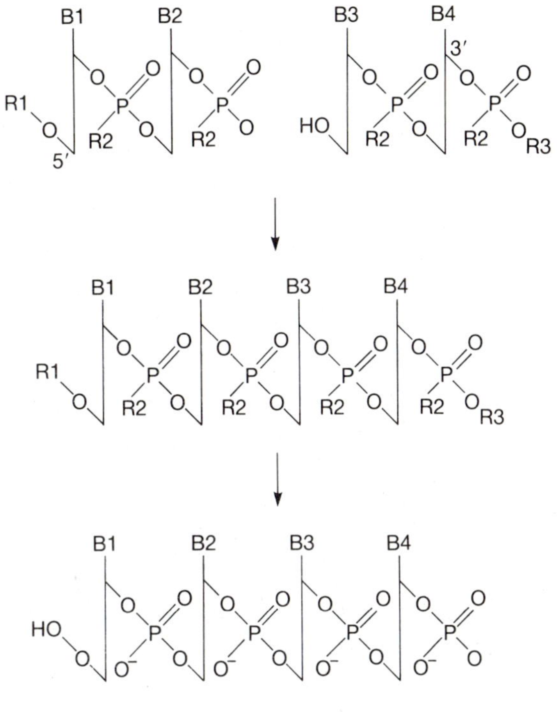

Figure 11.16 General reagents for chemical synthesis of DNA. The blocking groups R1 and R2 can be removed under acid or base conditions to yield units, as shown, that can be condensed. Finally, all the blocking groups can be removed to yield the desired oligomer.

With the ability to synthesize oligomers 15 to 20 nucleotides long, how can a gene be mutated? Of course, a complete gene containing the desired mutation could be synthesized from overlapping oligonucleotides, but easier methods exist for using an oligonucleotide to generate a specific mutation. An oligonucleotide containing the desired mutation, insertion, or deletion will hybridize to complementary wild-type single-stranded DNA and can serve as a primer for DNA pol I (Fig. 11.17). The resulting double-stranded DNA contains one wild-type strand and one mutant strand. Upon replication in cells, one of the daughter duplexes is wild type and the other is mutant. Therefore, following transformation and segregation, a mutant gene can be obtained.

Oligonucleotide-directed mutagenesis requires not only the primer but also a single-stranded template. One source of the template is the single-stranded phage M13. If this source of DNA is to be used, the gene must be cloned into the phage. Alternatively, the necessary single-stranded DNA template can be generated from double-stranded plasmid by cleavage with a restriction enzyme and digestion with exonuclease III. The plaques or colonies containing the mutant sequence can be distinguished from the wild-type sequence by using the oligonucleotide primer itself as a probe since hybridization conditions can be found such that a probe will discriminate between two DNAs containing single-base differences.

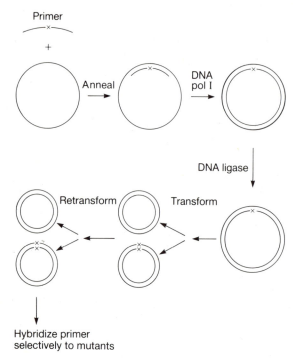

Figure 11.17   Mutagenesis with chemically synthesized DNA. The oligomer hybridizes, except for the mispaired base. Extension of the primer with DNA pol I and ligation yields heteroduplex molecules that can be transformed. Following DNA replication in the transformants, the two types of DNA molecules segregate to yield wild-type or mutant homoduplexes. Retransformation yields plaques containing entirely mutant or wild-type DNA, which can be identified by hybridization with radioactive mutant oligomer.

DETECTING GENETIC DISEASES

Huntington's chorea is a neurodegenerative disease that becomes apparent later in life, after the carrier is likely to have had children. The mutation is dominant, and prenatal detection of the condition or knowledge that one is a carrier could be of great value in a decision to have children. Since no biochemical alteration was known as a result of the mutation and no genes linked to the condition were known, progress appeared unlikely. But in 1983 striking progress was made.

The human chromosome contains a sizable number of polymorphisms. These reveal themselves as size differences between individuals of specific fragments resulting from restriction enzyme digestion. They are detected merely by digesting chromosomal DNA with a restriction enzyme, subjecting the fragments to electrophoresis, and hybridizing a radioactive probe derived from a clone of a human segment. Among a small set of clones that yielded polymorphisms — that is, size differences between individuals in a population — one restriction enzyme fragment polymorphism has been found that is tightly linked to Huntington's disease. This now permits the genetic mapping, cloning, and detection of the gene responsible for the condition.

Sickle cell anemia is another genetic disease whose easy detection would be

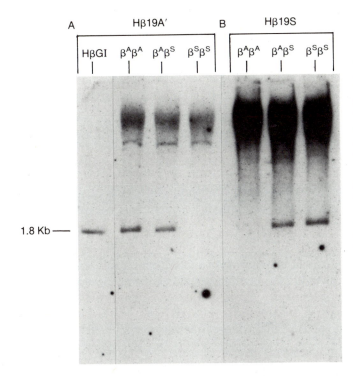

A   Hβ19A'   B   Hβ19S

HβGI   βᴬβᴬ βᴬβˢ βˢβˢ   βᴬβᴬ βᴬβˢ βˢβˢ

1.8 Kb —

Figure 11.18   Autoradiographs of Southern transfer experiments using whole human chromosomal DNA digested with *Bam*HI from normal cells, cells haploid for the sickle cell allele, and cells diploid for the sickle cell allele. The leftmost lane contains a size standard. The lanes contain *Bam*HI-cut DNA of the types indicated, β^A for normal and β^S for sickle. After electrophoresis and transfer to nitrocellulose, the left four lanes were hybridized to a 19-base-pair probe exactly homologous to the β^A sequence but containing a single-base mismatch to the β^S sequence. The right three lanes were hybridized with radioactive 19-base-pair probe with a single-base mismatch to the β^A sequence and exactly homologous to the β^S sequence. (From Detection of Sickle Cell β^S-Globin Allele by Hybridization with Synthetic Oligonucleotides, B. Conner, A. Reyes, C. Morin, I. Itakura, R. Teplitz, R. Wallace, Proc. Nat. Acad. Sci. USA *80*, 278–282 [1983]).

valuable. In this case the basis of the disease is known. As a result of an A to T substitution in the gene coding for β-globin, valine is substituted for glutamic acid in the protein. This permits the deoxy form of the globin to form long polymers that elongate the red blood cells so that capillaries are blocked.

Remarkably, the single nucleotide changes responsible for the sickle cell condition can be detected by hybridization. As little as 10 μg of cellular DNA can be digested with a restriction enzyme and subjected to Southern transfer. An oligonucleotide homologous to the nonmutant sequence does not hybridize to the mutant sequence and vice versa (Fig. 11.18). This approach shows great promise for detection of genetic defects.

DOCTORING THE *E. coli araC* GENE FOR HYPERPRODUCTION

Once a gene has been cloned, usually it must be modified to facilitate study of the biochemical system. An outline is given here of the steps that could be used to increase the *in vivo* synthesis of the product of a cloned *araC* gene. This protein is required for induction of the arabinose operon.

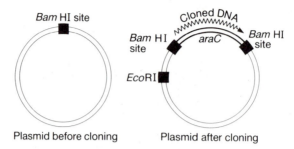

Figure 11.19   Generation of two *Bam*HI cleavage sites by cloning a *Bam*HI-ended molecule in a *Bam*HI-cut plasmid.

Suppose the *araC* gene has been cloned into the *Bam*HI site of a plasmid by digesting chromosomal DNA with *Bam*HI, transforming, and selecting AraC$^+$ cells (Fig. 11.19). Further, assume that by sequencing or by restriction enzyme digestion of different clones, one clone, in which the proximal end of the *araC* gene is toward the *Eco*RI site of the plasmid, has been chosen for further study. Finally, assume that the sequencing showed that this end of the *araC* gene was a little over 200 nucleotides from the *Bam*HI site.

The *Bam*HI site then is a good place to begin modification of the *araC* gene. However, if the *araC*-plasmid DNA is digested with *Bam*HI, the sites at both ends of the gene will be cleaved. Therefore the cleavage site at the distal end of the gene must be removed. This can be done by partial *Bam*HI digestion of the plasmid so that on average only one of the two *Bam*HI sites per plasmid is cut, filling out of the sticky

ends with DNA pol I to make them double-stranded, ligation of the DNA back to circular form, and transformation (Fig. 11.20).

Some of the plasmids resulting from the above treatment will have the *Bam*HI site at the distal end of the *araC* gene inactivated by the partial cutting and filling process. These plasmids are then identified by double restriction enzyme digestion. *Bam*HI plus any other enzyme that makes multiple cuts in the plasmid can be used, and the pattern of cuts can be analyzed to detect the plasmid in which the functional *Bam*HI site is at the proximal end of the *araC* gene.

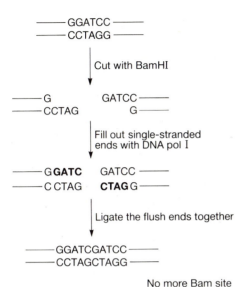

```
——— GGATCC ———
——— CCTAGG ———
```

              Cut with BamHI

```
——— G            GATCC ———
——— CCTAG             G ———
```

              Fill out single-stranded
              ends with DNA pol I

```
——— G GATC        GATCC ———
——— C CTAG        CTAG G ———
```

              Ligate the flush ends together

```
——— GGATCGATCC ———
——— CCTAGCTAGG ———
```

              No more Bam site

Figure 11.20 Elimination of a *Bam*HI
cleavage site by treatment with DNA pol I
and ligase.

Then the extra nucleotides at the proximal end of the gene must be removed. This is done by cutting the plasmid with *Bam*HI, which now will cut only at the proximal end of the *araC* gene. Then the cut DNA can be lightly digested with nuclease Bal 31 or digested with exonuclease III followed by digestion with nuclease S1. Either of these treatments can be adjusted to remove about 175 nucleotides of the DNA. This treatment will remove the 175 nucleotides in front of the *araC* gene and about the same number of nucleotides from the plasmid. After this treatment, *Bam*HI linkers are ligated onto the DNA.

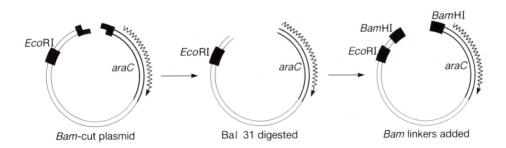

Bam-cut plasmid   Bal 31 digested   *Bam* linkers added

The next step in the process is cutting the DNA with *Eco*RI. This frees a small and useless fragment of plasmid that is lost in subsequent steps (Fig. 11.21). Then a

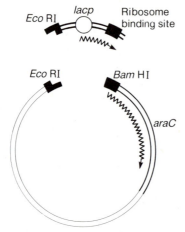

Figure 11.21   Structure of the plasmid after *Eco*RI cleavage and the structure of the *lac* fragment with an *Eco*RI sticky end, *lac* promoter, *lac* ribosome binding site, and a flush end that is ligated to the *Bam*HI linker.

special fragment of DNA about 90 base pairs long is ligated into the plasmid. This fragment will have an *Eco*RI sticky end and a flush end and will contain the *lac* promoter and Shine-Dalgarno ribosome binding site. The fragment does not include the ATG sequence corresponding to the AUG initiation codon of the *lacZ* gene. After ligation of this fragment into the plasmid, cells are transformed.

At this point a good promoter and a ribosome binding site have been placed ahead of the *araC* gene. However, it is unlikely that the distance between the Shine-Dalgarno sequence and the beginning of the *araC* gene is optimal. Most likely this distance must be shortened somewhat. By sequencing across the inserted *lac* promoter ribosome binding site sequence into the *araC* gene, the exact distance from the ribosome binding site to the ATG sequence can be found. This sequencing can be

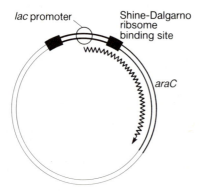

Figure 11.22   Structure of the plasmid after cleaving with *Bam*HI, light digestion with Bal 31, and ligation to re-form the circle.

simply done by cutting the plasmid with *EcoRI*, treating the DNA with polynucleo-tide kinase and radioactive ATP so as to place a radioactive phosphate at the 5' ends of the DNA, digesting the DNA with an enzyme that cuts the plasmid a number of times, isolating the *lac* promoter–*araC* radioactive fragment, and then performing the Maxam-Gilbert chemical degradation sequencing.

The sequencing reveals the amount of DNA that must be removed to generate the 6-to-10-base optimal distance between the Shine-Dalgarno ribosome binding sequence and the beginning of the *araC* gene. Now the DNA can be cut with *BamHI*, digested again with Bal 31 or exonuclease III and S1, cut with *EcoRI*, and ligated with an excess of fresh 90-base *lac* promoter–ribosome binding site DNA (Fig. 11.22). Transformants from this DNA can be screened for their overproduction of *araC* protein. One simple screening is merely to subject crude lysates of transformants to SDS polyacrylamide gel electrophoresis and look for those that synthesize large quantities of a protein of the correct molecular weight.

In this final example we have seen that a sizable number of rather simple steps usually are involved in genetic engineering constructions. One of the virtues of biological engineering is that many of the intermediates in the long series of steps can be amplified biologically, that is, the plasmids can be transformed into cells and large quantities of the modified plasmid grown and purified before attempting subse-quent steps.

PROBLEMS

11.1. Describe how a gene could be cloned without the use of any restriction enzymes.

11.2. In cloning DNA, often the vector is treated with alkaline phosphatase before the ligation step. What does such a treatment achieve?

11.3. What problems are generated if, in walking along a chromosome, a stretch of middle repetitive DNA is encountered, and what could be done to jump over it?

11.4. Suppose an insert in M13 is 700 bases, all of which must be sequenced, but resolution limits accurate reading to only 400. Without recloning, how could you arrange to read the sequence of the inserts in the "other" direction and still use the Sanger method?

11.5. How could you determine the approximate physical location of the origin of DNA replication of a phage by making use of restriction enzymes?

11.6. Suppose you isolated what could be a previously unknown restriction enzyme. Without sequencing, how might you determine the recognition sequence of your enzyme?

11.7. What type of selection could be used in isolating plasmids containing the *E. coli* origin of DNA replication?

11.8. Consider a gene that contains two intervening sequences. Suppose that these regions have been removed in the processed messenger. What will the R-loop between the genomic DNA and the processed RNA look like?

11.9. A concern in the early days of recombinant DNA work was that gross rearrangements would occur in cloned DNA at high rates. Suppose you have a cloned stretch of DNA. How could you determine, using Southern transfers, whether it has retained the same relative positions of restriction sites as DNA extracted directly from the organism?

11.10. Putting a particular DNA sequence from a higher organism into *E. coli* as is done in cloning might make the recipient cell become sick or die. How might you test for the possibility that a sizable fraction of the genome from the higher organism cannot be easily cloned because of these problems?

11.11. Comment on the following statement: "Intervening sequences exist to frustrate molecular biologists by preventing the use of the hybridization arrest translation assay system in screening clones."

11.12. To reduce the number of colonies that must be screened in cloning a gene, it is useful to clone DNA fragments of as large a size as possible. Both *Bam*HI (from *Bacillus ameloliquaefaecens*) and *Bst*I (from *Bacillus stearothermophilus*) recognize the same cleavage sequence. Why would fragmentation of DNA with *Bst*I at 65°, a temperature at which *Bam*HI is not active, be expected to yield larger DNA fragments than cleavage of exactly the same DNA with *Bam*HI at 37°?

11.13. Researchers involved with genetic engineering have been concerned that the spontaneous mutation rate in *E. coli* would lead to the rapid degeneration of any special gene for which there was no selection for the wild-type gene sequence. What properties of *E. coli* can you cite to put their minds at ease?

11.14. DNA can be injected into a mouse egg and implanted in a pseudopregnant female from which baby mice are ultimately obtained that contain copies of the injected DNA. How would you explain the finding that sometimes the resultant mice are mosaics, that is, some of the cells contain the DNA and others do not? How could you make use of these findings to determine useful information about development of the mouse?

11.15. The homology between two sequences S1 and S2 can be compared by the dot matrix method (see, for example, Cell *18*, 865–873 [1979]) as follows: A dot is placed in the ith row and jth column of the comparison matrix (generated by computer) if the sequences S1 and S2 possess n out of m identical bases beginning with the ith base of S1 and the jth base of S2. Typical operating parameters would be to require identity in 8 out of 16 bases, that is, $n = 8$ and $m = 16$. Regions of homology appear as lines of dots running downward to the right. How would you use the dot matrix method to look for inverted repeated sequences — palindromes — in a sequence?

11.16. Suppose you wished to use directed mutagenesis against a portion of a gene coding for an essential enzyme in haploid yeast. How could you simultaneously inactivate the resident chromosomal gene and substitute a portion of the gene that you had modified *in vitro?*

*11.17. The restriction enzyme *Xba*I cleaves the sequence T↓CTAGA and *Hind*III cleaves the sequence A↓AGCTT. Suppose *Hind*III cleaves a plasmid once in a non-essential region. By what simple procedure not involving linkers or similar oligonucleotides could you clone *Xba*I-cut restriction fragments into *Hind*III-cut plasmid and at the same time eliminate any problems from reclosure of the plasmid without insertion of the *Xba*I-cut DNA?

11.18. Suppose you have a vector that contains a promoter, Shine-Dalgarno sequence, and the very beginning of any gene. Suppose also that translation in *E. coli* of a portion of a gene from another organism often generates sufficient protein that it may be detected by antibodies. How could these suppositions be put together to yield an efficient scheme for the cloning of any gene for which you have the protein and can make antibodies?

11.19. Why does a DNA molecule n nucleotides long resulting from an S1 digestion experiment migrate at a different rate than a DNA molecule n nucleotides long resulting from Maxam-Gilbert sequencing? That is, to determine the exact size of a DNA molecule resulting from an S1 mapping experiment, what correction must be done when using size standards resulting from Maxam-Gilbert sequencing?

RECOMMENDED READINGS

Detection of Specific Sequences Among DNA Fragments Separated by Gel Electrophoresis, E. Southern, J. Mol. Biol. *98*, 503–517 (1975).

Screening lambda gt Recombinant Clones by Hybridization to Single Plaques in situ, W. Benton, R. Davis, Science *196*, 180–182 (1977).

Transcriptional Control Signals of a Eukaryotic Protein-Coding Gene, S. McKnight, R. Kingsbury, Science *217*, 316–324 (1982).

RELATED REVIEWS, BOOKS, AND ARTICLES

Recombinant DNA (an issue devoted mainly to the topic), Science *196*, no. 4286, 159–221 (1977).

Recombinant DNA, Methods in Enzymology *68*, ed. R. Wu, Academic Press, New York (1979).

Recombinant DNA (an issue devoted to the topic), Science *209*, no. 4463 (1980).

Nucleic Acids, Methods in Enzymology *65*, ed. L. Grossman, K. Moldave, Academic Press, New York (1980).

Methods of DNA and RNA Sequencing, ed. S. Weissman, Praeger Publishers, New York (1983).

Biological Frontiers (an issue mainly on applications of genetic engineering), Science *222*, no. 4625 (1983).

DEEPER READING

Transposition of R Factor Genes to Bacteriophage Lambda, D. E. Berg, J. Davies, B. Allet, J-D. Rochaix, Proc. Natl. Acad. Sci. USA *72*, 3628–3632 (1975).

Colony Hybridization: A Method for the Isolation of Cloned DNA's That Contain a Specific Gene, M. Grunstein, D. Hogness, Proc. Natl. Acad. Sci. USA *72*, 3961–3965 (1975).

Nucleic Acid Hybridization Using DNA Covalently Coupled to Cellulose, B. Noyes, G. Stark, Cell *5*, 301–310 (1975).

Hybridization of RNA to Double-Stranded DNA: Formation of R-Loops, M. Thomas, R. White, R. Davis, Proc. Natl. Acad. Sci. USA *73*, 2294–2298 (1976).

A Simple Method for DNA Restriction Site Mapping, H. Smith, M. Birnstiel, Nucleic Acids Res. *3*, 2387–2398 (1976).

Labeling DNA to High Specific Activity *in vitro* by Nick Translation with DNA Polymerase I, P. Rigby, M. Dieckmann, C. Rhodes, P. Berg, J. Mol. Biol. *113*, 237–251 (1977).

Packaging Recombinant DNA Molecules into Bacteriophage Particles *in vitro*, B. Hohn, K. Murray, Proc. Natl. Acad. Sci. USA *74*, 3259–3263 (1977).

Method for Detection of Specific RNA's in Agarose Gels by Transfer to Diazobenzyloxy-methyl-Paper and Hybridization with DNA Probes, J. Alwine, D. Kemp, G. Stark, Proc. Natl. Acad. Sci. USA *74*, 5350–5354 (1977).

Sizing and Mapping of Early Adenovirus mRNAs by Gel Electrophoresis of S1 Endonuclease-Digested Hybrids, A. Berk, P. Sharp, Cell *12*, 721–732 (1977).

Expression in *Escherichia coli* of a Chemically Synthesized Gene for the Hormone Somatostatin, K. Itakura, T. Hirose, R. Crea, A. Riggs, H. Heyneker, F. Bolivar, H. Boyer, Science *198*, 1056–1063 (1977).

Mapping Adenines, Guanines, and Pyrimidines in RNA, H. Donis-Keller, A. Maxam, W. Gilbert, Nucleic Acids Res. *4*, 2527–2538 (1977).

Bacteriophage T4 RNA Ligase: Preparation of a Physically Homogeneous Nuclease-Free Enzyme from Hyperproducing Infected Cells, N. Higgins, A. Geball, T. Snopek, A. Sugino, N. Cozzarelli, Nucleic Acids Res. *4*, 3175–3186 (1977).

Cosmids: A Type of Plasmid Gene-Cloning Vector That Is Packageable *in vitro* in Bacteriophage Lambda Heads, J. Collins, B. Hohn, Proc. Natl. Acad. Sci. USA *75*, 4242–4246 (1978).

Phenotypic Expression in *E. coli* of a DNA Sequence Coding for Mouse Dihydrofolate Reductase, A. Chang, J. Nunberg, R. Kaufman, H. Erlich, R. Schimke, S. Cohen, Nature *275*, 617–624 (1978).

Local Mutagenesis: A Method for Generating Viral Mutants with Base Substitutions in Preselected Regions of the Viral Genome, D. Shortle, D. Nathans, Proc. Natl. Acad. Sci. USA *75*, 2170–2174 (1978).

DNase Footprinting: A Simple Method for the Detection of Protein-DNA Binding Specificity, D. Galas, A. Schmitz, Nucleic Acids Res. *5*, 3157–3170 (1978).

A General Method for Maximizing the Expression of a Cloned Gene, T. Roberts, R. Kacich, M. Ptashne, Proc. Natl. Acad. Sci. USA *76*, 760–764 (1979).

The Evolution and Sequence Comparison of Two Recently Diverged Mouse Chromosomal β-Globin Genes, D. Konkel, J. Maizel, P. Leder, Cell *18*, 865–873 (1979).

Plasmid Vehicles for the Direct Cloning of *E. coli* Promoters, G. An, J. Friesen, J. Bact. *140*, 400–407 (1979).

The Use of R-Looping for Structural Gene Identification and mRNA Purification, J. Woolford, M. Rosbash, Nucleic Acids Res. *6*, 2483–2497 (1979).

Improved Methods for the Formation and Stabilization of R-Loops, D. Kaback, L. Angerer, N. Davidson, Nucleic Acids Res. *6*, 2499–2517 (1979).

Filter Replicas and Permanent Collections of Recombinant DNA Plasmids, J. P. Gergen, R. H. Stern, P. C. Wensink, Nucleic Acids Res. *7*, 2115–2136 (1979).

Total Synthesis of a Gene, H. Khorana, Science *203*, 614–625 (1979).

Targeted Deletions of Sequences from Cloned Circular DNA, C. Green, C. Tibbetts, Proc. Natl. Acad. Sci. USA *77*, 2455–2459 (1980).

Unusual α-Globin-Like Gene That Has Cleanly Lost Both Globin Intervening Sequences, Y. Nishioka, A. Leder, P. Leder, Proc. Natl. Acad. Sci. USA *77*, 2806–2809 (1980).

A Mouse α-Globin-Related Pseudogene Lacking Intervening Sequences, E. Vanin, G. Goldberg, P. Tucker, O. Smithies, Nature *286*, 222–226 (1980).

Improved Methods for Maximizing Expression of a Cloned Gene: A Bacterium That Synthesizes Rabbit β-Globin, L. Guarente, G. Lauer, T. M. Roberts, M. Ptashne, Cell *20*, 543–553 (1980).

Directed Deletion of a Yeast Transfer RNA Intervening Sequence, R. Wallace, P. Johnson, S. Tanaka, M. Schold, K. Itakura, J. Abelson, Science *209*, 1396–1400 (1980).

Chemical DNA Synthesis and Recombinant DNA Studies, K. Itakura, A. Riggs, Science *209*, 1401–1405 (1980).

The Detection of DNA-Binding Proteins by Protein Blotting, B. Bowen, J. Steinberg, U. Laemmli, H. Weintraub, Nucleic Acids Res. *8*, 1–20 (1980).

Construction of a General Vector for Efficient Expression of Mammalian Proteins in Bacteria: Use of a Synthetic Ribosome Binding Site, G. Jay, G. Khoury, A. Seith, E. Jay, Proc. Natl. Acad. Sci. USA *78*, 5543–5548 (1981).

A DNA Fragment with an α-Phosphorothioate Nucleotide at One End Is Asymmetrically Blocked from Digestion by Exonuclease III and Can Be Replicated *in vivo*, S. Putney, S. Benkovic, P. Schimmel, Proc. Natl. Acad. Sci. USA *78*, 7350–7354 (1981).

Total Synthesis of a Human Leukocyte Interferon Gene, M. Edge, A. Greene, G. Heathcliffe, P. Meacock, W. Schuch, D. Sunlon, T. Atkinson, C. Newton, A. Markham, Nature *292*, 756–762 (1981).

Propagation of Foreign DNA in Plants Using Cauliflower Mosaic Virus as Vector, B. Gronenborn, R. Gardner, S. Schaefer, R. Shepherd, Nature *294*, 773–776 (1981).

The Use of Synthetic Oligonucleotides as Hybridization Probes, II. Hybridization of Oligonucleotides of Mixed Sequence to Rabbit β-Globin DNA, R. Wallace, M. Johnson, T. Hirose, T. Miyake, E. Kawashima, K. Itakura, Nucleic Acids Res. *9*, 879–894 (1981).

A Systematic DNA Sequencing Strategy, G. Hong, J. Mol. Biol. *158*, 539–549 (1982).

Efficient Expression of *Escherichia coli* Galactokinase Gene in Mammalian Cells. D. Schumperli, B. Howard, M. Rosenberg, Proc. Natl. Acad. Sci. USA *79*, 257–261 (1982).

Segment-Specific Mutagenesis: Extensive Mutagenesis of a *lac* Promoter/Operator Element, H. Weiher, H. Schaller, Proc. Natl. Acad. Sci. USA *79*, 1408–1412 (1982).

Gap Misrepair Mutagenesis: Efficient Site-Directed Induction of Transition, Transversion, and Frameshift Mutations *in vitro*, D. Shortle, P. Grisafi, S. Benkovic, D. Botstein, Proc. Natl. Acad. Sci. USA *79*, 1588–1592 (1982).

Introduction of Rat Growth Hormone Gene into Mouse Fibroblasts via a Retroviral DNA Vector: Expression and Regulation, J. Doehmer, M. Barinaga, W. Vale, M. Rosenfeld, I. Verma, R. Evans, Proc. Natl. Acad. Sci. USA *79*, 2268–2272 (1982).

Oligonucleotide-Directed Mutagenesis as a General and Powerful Method for Studies of Protein Function, G. Dalbadie-McFarland, L. Cohen, A. Riggs, C. Morin, K. Itakura, J. Richards, Proc. Natl. Acad. Sci. USA *79*, 6409–6413 (1982).

Cloning the Double-Stranded RNA Genes of Reovirus: Sequencing of the Cloned S2 Gene, L. Cashdollar, J. Esparza, G. Hudson, R. Chmelo, P. Lee, W. Joklik, Proc. Natl. Acad. Sci. USA *79*, 7644–7648 (1982).

Chromosomal Localization of Human Leukocyte, Fibroblast, and Immune Interferon Genes by Means of *in situ* Hybridization, J. Trent, S. Olson, R. Lawn, Proc. Natl. Acad. Sci. USA *79*, 7809–7813 (1982).

Transcriptional Control Signals of a Eukaryotic Protein-Coding Gene, S. McKnight, R. Kingsbury, Science *217*, 316–324 (1982).

Diverse Mechanisms in the Generation of Human β-Tubulin Pseudogenes, C. Wilde, C. Crowther, N. Cowan, Science *217*, 549–552 (1982).

An Alternative Method for Synthesis of Double-Stranded DNA Segments, J. Rossi, R. Kierzek. T. Huang, P. Walker, K. Itakura, J. Biol. Chem. *257*, 9226–9229 (1982).

Genetic Transformation of *Drosophila* with Transposable Element Vectors, G. Rubin, A. Spradling, Science *218*, 348–353 (1982).

Site-Specific Mutagenesis of *Agrobacterium* Ti Plasmids and Transfer of Genes to Plant Cells, J. Leemans, C. Shaw, R. Deblaere, H. DeGreve, J. Hernalsteens, M. Maes, M. Van Montegu, J. Schell, J. Mol. Appl. Genet. *1*, 149–164 (1982).

Transformation of Mammalian Cells to Antibiotic Resistance with a Bacterial Gene Under Control of the SV40 Early Region Promoter, P. Southern, P. Berg, J. Mol. Appl. Genet. *1*, 327–341 (1982).

Detection of Sickle Cell β^S-Globin Allele by Hybridization with Synthetic Oligonucleotides, B. Conner, A. Reyes, C. Morin, I. Itakura, R. Teplitz, R. Wallace, Proc. Natl. Acad. Sci. USA *80*, 278–282 (1983).

Efficient Isolation of Genes Using Antibody Probes, R. Young, R. Davis, Proc. Natl. Acad. Sci. USA *80*, 1194–1198 (1983).

Expression and Regulation of a Human Metallothionein Gene Carried on an Autonomously Replicating Shuttle Vector, M. Karin, G. Cathala, M. Nguyen-Huu, Proc. Natl. Acad. Sci. USA *80*, 4040–4044 (1983).

Isolation of Z-DNA-Containing Plasmids, R. Thomae, S. Beck, F. Pohl, Proc. Natl. Acad. Sci. *80*, 5550–5553 (1983).

Chromosomal Walking and Jumping to Isolate DNA from *Ace* and *rosy* Loci and the Bithorax Complex in *Drosophila melanogaster*, W. Bender, P. Spierer, D. Hogness, J. Mol. Biol. *168*, 17–33 (1983).

A Polymorphic DNA Marker Genetically Linked to Huntington's Disease, J. Gusella, N. Wexler, P. Conneally, S. Naylor, M. Anderson, R. Tanzi, P. Walkins, K. Ottina, M. Wallace, A. Sakaguchi, A. Young, I. Shoulson, E. Bonilla, J. Martin, Nature *306*, 234–238 (1983).

Enzymatic Techniques for the Isolation of Random Single-Base Substitutions *in vitro* at High Frequency, P. Abarzúa, K. Marians, Proc. Nat. Acad. Sci. USA *81*, 2030–2034 (1984).

Two-Dimensional S1 Nuclease Heteroduplex Mapping: Detection of Rearrangements in Bacterial Genomes, T. Yee, M. Inouye, Proc. Nat. Acad. Sci. USA *81*, 2723–2727 (1984).

A Colorimetric Method for DNA Hybridization, M. Renz, C. Kurz, Nucleic Acids Res. *12*, 2435–3444 (1984).

How Genes Are Regulated

Repression and the *lac* Operon

H aving discussed genetics and the structure and synthesis of proteins and nucleic acids in the previous chapters, we are now prepared for consideration of biological regulatory mechanisms. What exactly is meant by "mechanism of regulation"? The term refers to the means by which the expression of a specific gene or set of genes is selectively increased or decreased, that is, induced or repressed. Thus a protein that binds only under some conditions to the promoter of one particular gene could regulate expression of this gene, whereas ATP would not be considered a regulator despite the fact that its presence is necessary for expression of the gene. Changes in ATP levels would be expected to affect expression of all genes similarly.

This section of the book describes the *lac, ara, trp,* and lambda phage sets of genes found in *Escherichia coli* as well as systems from eukaryotic cells; the mating type system in yeast; and regulation of dopa decarboxylase and yolk protein synthesis in *Drosophila melanogaster.* The prokaryotic genes were chosen because much is known about their regulation and because each is regulated by a dramatically different mechanism. The eukaryotic systems represent clear examples of regulation and are good examples of the use of recombinant DNA technology being applied to this type of problem.

Much is known about the four prokaryotic systems, in part because intensive study was begun on them well before similar studies became possible on the eukaryotic systems. Even before the development of genetic engineering, investigators of the bacterial systems devised means of combining genetics, physiology, and physical-chemical studies. The depth to which the studies has penetrated has revealed general principles that are likely to be operant in cells of any type. The diversity of regulation mechanisms used in the systems considered here ranges from a simple competition between lactose repressor and RNA polymerase for binding to DNA of the promoter region for the *lac* operon, to a series of proteins binding to the DNA and assisting RNA polymerase to initiate transcription of the *ara* operon, to involvement of the translation apparatus in determining whether transcription of the *trp* operon will continue once begun, and finally to highly intricate cascade systems regulating transcription of the genes of phage lambda.

The general principles used in regulating the gene systems considered here cover a broad range, but nature's diversity in gene regulation mechanisms is certainly not exhausted by these. Other regulation systems display both minor and major differences. Some, whose study is too preliminary to include here, seem to use entirely different mechanisms of gene regulation.

BACKGROUND OF THE *lac* OPERON

The initial studies on the *lac* system, like those on most other bacterial systems, were genetic. At the Pasteur Institute in Paris during World War II, Monod began a study of the process of adaptation of *E. coli* to growth on medium containing lactose. This first led to studies on the origin of the enzymes that were induced in response to the addition of lactose to the medium and then to studies on how the induction process was regulated. Research on the *lac* operon at the Pasteur Institute flourished and spread around the world, and for many years the most active research area in molecular biology was the lactose system.

By now, the essential regulatory properties of the lactose system have been fully characterized as a result of extensive physiological, genetic, and biochemical analysis. Research on the lactose system recently has turned to the more fundamental questions of how RNA polymerase recognizes promoters and initiates transcription and how proteins fold during their synthesis, recognize their substrates, and bind to other proteins or to specific sequences on DNA.

The lactose system consists of four genes that enable *E. coli* to grow on lactose as a source of carbon and energy (Fig. 12.1): *lacI*, which codes for the *lac* repressor, a regulatory protein; *lacZ*, which codes for β-galactosidase, an enzyme that cleaves

Figure 12.1   The *lac* operon of *E. coli* showing the regulatory gene *lacI*, the promoters p_{lac} and p_I, operator O, and the genes *lacZ*, *lacY*, and *lacA*.

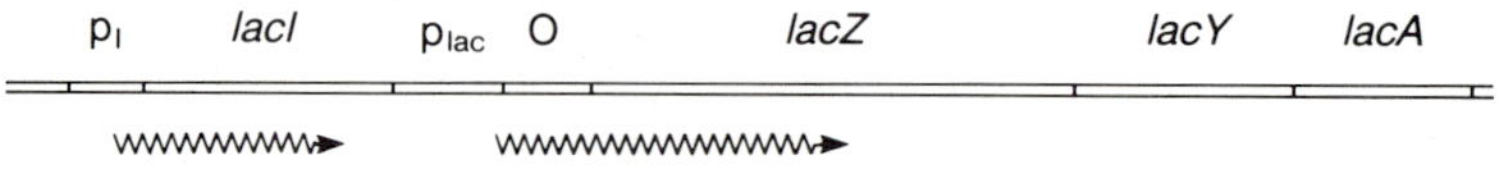

lactose to glucose and galactose; *lacY*, which codes for a protein required for active transport of lactose into cells; and *lacA,* which codes for an enzyme that transfers acetyl groups to some galactosides, thereby reducing their toxicity. The three genes *lacZ, lacY,* and *lacA* constitute an operon by the strict definition of the word since a single promoter serves more than one gene. We will also use the word "operon" to refer to a transcriptional unit plus its related regulatory sequences even if it contains a single gene. The use of a separate word such as "regulon" or "transcripton" for a situation in which a single gene is served by a single promoter seems unnecessarily neologistic.

Figure 12.2 shows the picture that has been derived for the mechanism of regulation of the *lac* operon in *E. coli.* The *lacI* gene codes for a repressor molecule that binds to a specific site in the promoter region, termed the operator or O site. While bound to the operator, it prevents the binding of RNA polymerase to the promoter and blocks transcription of the Z, Y, and A genes.

If lactose is added to a growing culture, the synthesis rate of the *lac* enzymes is increased about a thousandfold. Curiously, induction requires activity of the very enzyme that is being induced, β-galactosidase. The small uninduced level of β-galactosidase cleaves most of the lactose that leaks into the cells or that enters via the basal levels of *lacY* protein. A side reaction of β-galactosidase generates a molecule related to lactose—allolactose—from lactose, and this binds to repressor. Such binding causes the repressor to reduce its affinity for operator by about a thousand-

Figure 12.2   Schematic drawing of the *lac* operon in a repressed state and in an induced state in the presence of cAMP.

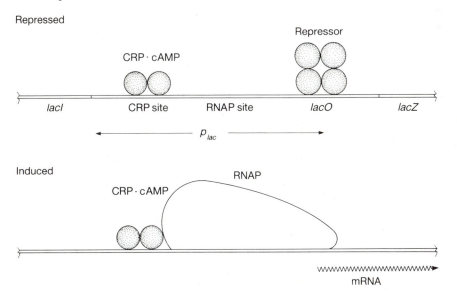

fold and to be released from the DNA. Transcription of the operon can begin, and induction occurs. Much more lactose is then transported into the cells and cleaved to glucose and galactose.

The efficient initiation of transcription by RNA polymerase at the promoter of the *lac* operon requires not only the absence of bound *lac* repressor but also the presence of cAMP and a protein that binds cAMP, cyclic AMP receptor protein, often called CRP. This auxiliary induction requirement is thought to result from the fact that the carbon and energy requirements of *E. coli* are most efficiently met by catabolizing glucose rather than other sugars. Consequently, cells have evolved a way to shut down the possibly inefficient use of other carbon utilization pathways if glucose is present, a phenomenon known as the glucose effect or catabolite repression. The glucose effect is generated in two ways: by excluding inducers of some operons from the cell and by reducing the inducibility of some operons. When glucose is present and is being metabolized, the concentrations of cyclic AMP are low, and few CRP proteins contain bound cyclic AMP. Only when CRP has bound cAMP that occurs when glucose is absent can the protein specifically bind to DNA and assist RNA polymerase to initiate transcription of the CRP-dependent operons such as *lac*.

THE ROLE OF INDUCER ANALOGS IN THE STUDY OF THE *lac* OPERON

Compared to progress on many other problems in molecular biology, research on the *lac* operon proceeded rapidly. In part, this was due to the fact that measuring the response of the *lac* operon to changes in the cell's environment was easy. Addition of lactose induces the operon a thousandfold, and the assay of β-galactosidase is particularly simple. Another important reason for the rapid progress is that many useful analogs of lactose can readily be synthesized. These analogs facilitate assay of the *lac* operon proteins as well as simplify the isolation of useful mutants.

The properties of several analogs are shown in Table 12.1. ONPG is particularly useful as its cleavage yields galactose and orthonitrophenol, which when ionized at basic pH is bright yellow. This provides a simple and sensitive assay for β-galactosidase requiring only an optical density measurement of the product produced in a given time. IPTG is useful in inducing the operon for two reasons. First, IPTG induces by binding directly to the repressor, so functional β-galactosidase is unnecessary; and second, it is not metabolized, and therefore its use for induction does not disturb the metabolic pathways of the cell as might be the case for induction by lactose. Phenylgalactose is another useful analog. It is not an inducer, but it can be metabolized if β-galactosidase is present to hydrolyze it. Therefore only permanently derepressed cells grow on medium containing phenylgalactose as the carbon and energy source. Such permanently induced mutants are called constitutive and result either from defective repressor that is unable to bind to the operator or from mutant operator that is no longer recognized and bound by repressor.

Table 12.1 Properties of Lactose Analogs

Analog	Inducer	Transported	β-galactosidase Substrate	Comments
Lactose	−	+	+	Not an inducer
Allolactose	+	?	+	True inducer
Phenyl-β-D-galactose	−	+	+	Selects nonrepressing mutants
Orthonitro-phenyl-β-D-galactose (ONPG)	?	+	+	When cleaved, produces yellow nitrophenol
Isopropyl β-D-thio-galactose (IPTG)	+	+	−	Unmetabolized inducer
Orthonitro-phenyl-β-D-fucoside	−	−	−	Inhibits induction
X-gal, 5-bromo-4-chloro-3-indolyl-β-D-galactoside	−	−	+	Produces blue dye on hydrolysis

lac REPRESSOR IS A PROTEIN

The phenomenon of repression in the *lac* operon was characterized by genetic experiments that showed that a product of the *lacI* gene diffused through the cell and shut off expression of the *lac* genes. Basically these were complementation experiments that showed that *lacI* acted in *trans* to repress. Although a LacI⁻ LacZ⁺ strain constitutively expresses β-galactosidase in the absence or presence of inducers, in an F′LacI⁺ ΔLacZ/LacI⁻ LacZ⁺, the chromosomal LacZ gene is repressed except in the presence of *lac* inducers.

Research on the *lac* operon then shifted to learning what repressor was — whether RNA or protein — and how it acted to block expression of the *lac* enzymes. Since the target of the repressor was a sequence that could be genetically mapped, two likely possibilities existed. The operator could be a region on the DNA to which repressor bound to reduce the intracellular levels of functional *lac* messenger. Or the functional operator that was recognized by repressor was a region on the *lac* messenger to which repressor bound to reduce the translation efficiency. Either of these possibilities required recognition of a specific nucleotide sequence by the *lacI* gene product. To many investigators, such an ability seemed most reasonable for an RNA molecule. Indeed, early results obtained with RNA and protein synthesis inhibitors led to the conclusion that it was not protein. Gilbert and Müller-Hill, however, reasoned that an RNA molecule was unlikely to possess all the regulatory properties required of the *lacI* product. Therefore Müller-Hill designed a simple genetic experiment that indicated that repressor contained at least some protein.

He began with a strain containing amber mutations in both the *gal* and *his* operons. This strain was then made LacI⁻ by selection of constitutives. Most of these mutants were missense, but a few were nonsense I⁻ mutants. Each of the constitutives was then reverted simultaneously to Gal⁺ and His⁺. The only way a cell could perform this double reversion was to become Su⁺ and simultaneously suppress both nonsense mutations. These nonsense-suppressing revertants were then tested for

$$\text{Gal}^-_{\text{nonsense}}\text{His}^-_{\text{nonsense}} \xrightarrow{\text{Select LacI}^-} \text{Gal}^-\text{His}^-\text{LacI}^- \xrightarrow[\text{His}^+]{\text{Select Gal}^+} \text{Gal}^+\text{His}^+(\text{Su}^+) \begin{cases} \text{LacI}^+, \text{ nonsense I}^- \\ \text{LacI}^-, \text{ missence I}^- \end{cases}$$

lac constitutivity, and a few were found to have simultaneously become I⁺. Consequently, these were nonsense I mutations that were now being suppressed. This proves that the I gene product contains protein, but it does not prove that the repressor is entirely protein. The repressor had to be purified before it could be shown to be entirely protein.

AN ASSAY FOR *lac* REPRESSOR

Genetic and physiological experiments investigating properties of the *lac* operon provided information from which a number of regulatory mechanisms were proposed. These ranged from the logical mechanism of *lac* repressor binding to DNA and inhibiting transcription to complicated translational control mechanisms utilizing tRNA molecules. Clear demonstration of the regulation mechanism required purification of its components and *in vitro* reconstruction of the *lac* system.

The most important step in the reconstruction of the *lac* regulatory system was the ability to detect repressor. *Lac* repressor, of course, had to be highly purified from lysed cells. If regulation of the *lac* operon were efficient—and that is the main reason for the existence of regulation—then the cell should contain far fewer molecules of repressor than of the induced gene products. Furthermore, since *lac* repressor possessed no known enzymatic activity, no easy and sensitive assay was available. Without the ability to detect repressor, its purification was impossible, for the fractions containing repressor that would be generated during purification could not be identified.

Repressor's only known property was that it bound inducer, IPTG being one. Therefore Gilbert and Müller-Hill developed an assay of *lac* repressor based on the protein's ability to bind to inducer molecules. Equilibrium dialysis can detect a protein that binds a particular small molecule. The protein solution to be assayed is placed in a dialysis sack and dialyzed against a buffer that contains salts to maintain the pH and ionic strength and the small molecule that binds to the protein (Fig. 12.3). In the case of repressor, radioactive IPTG was used. After equilibrium has been attained, the concentration of free IPTG inside the sack and outside the sack is equal. However, the total concentration of IPTG inside the sack equals the concentration of free IPTG plus the concentration of IPTG bound to repressor. Both the inside and outside concentrations of IPTG can be determined by measuring the amount of

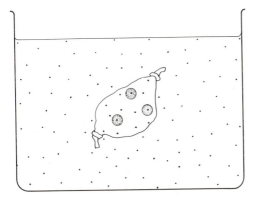

Figure 12.3   Representation of equilibrium dialysis in which the concentration of free IPTG inside and outside the sack is equal, but repressor in the sack binds additional IPTG inside the sack.

radioactivity contained in samples of known volumes taken from outside and inside the dialysis sack.

Does an equilibrium dialysis assay possess sufficient sensitivity to detect the small amounts of *lac* repressor that are likely to exist in crude extracts of cells? The binding reaction between repressor and IPTG is closely approximated by the reaction

$$R_f + IPTG \rightleftharpoons R \cdot IPTG$$

where R_f is the concentration of free repressor, IPTG is the concentration of free IPTG, and R-IPTG is the concentration of the complex between repressor and IPTG. A dissociation constant K_D describes the relations between the concentrations:

$$K_D = \frac{R_f \times IPTG}{R \cdot IPTG}.$$

Substituting the conservation equation, $R_f + R\text{-}IPTG = R_t$, where R_t is the total amount of repressor, and rearranging yields the relation we need. Biochemists have many different names for the equivalent algebraic rearrangements of this equation but usually call the phenomenon Michaelis-Menten binding:

$$R \cdot IPTG = \frac{R_t \times IPTG}{K_D + IPTG}.$$

The ratio of radioactivity in the samples obtained from inside and outside the sack is

$$\frac{IPTG + R \cdot IPTG}{IPTG} = 1 + \frac{R_t}{K_D + IPTG}.$$

Normally in liquid scintillation, counting a 5% difference between samples with more than 100 cpm can be readily determined. Thus the quantity

$$\frac{R_t}{K_D + IPTG}$$

must be greater than 0.05 for detection of *lac* repressor by this assay.

WILD-TYPE *lac* REPRESSOR COULD NOT BE DETECTED

Before trying to detect *lac* repressor, Gilbert and Müller-Hill estimated the signal that could be expected in the equilibrium assay and decided that they were unlikely to detect wild-type repressor. We examine such a calculation. Two quantities are needed: the dissociation constant of repressor for IPTG and the concentration of repressor in cell extracts.

To make a crude guess of the dissociation constant of IPTG from *lac* repressor, assume that *in vivo* the repression of the *lac* operon is directly proportional to the

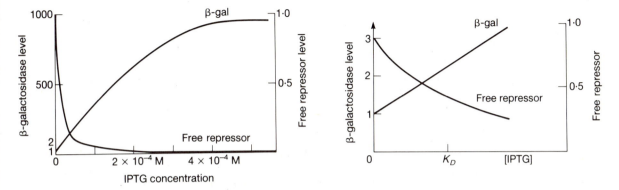

Figure 12.4   The relation between free repressor concentration and the induction level of the *lac* operon. This indicates the relation between the dissociation constant of repressor K_D and the IPTG concentration giving half-maximal induction of the operon.

concentration of repressor uncomplexed with IPTG. Consequently, if half the repressor is bound to IPTG and half is free of IPTG, the basal level of the *lac* operon will be doubled. The concentration of IPTG at which the 50-50 binding occurs and the enzyme level is twice the basal level equals the dissociation constant. Although this concentration of IPTG could be measured in reasonably straightforward experiments, it can be calculated from data already known. The *lac* operon can be induced a thousandfold and is half-maximally induced by 2×10^{-4} M IPTG (Fig. 12.4). Roughly, then, at an IPTG concentration of $(2 \times 10^{-4})/500 = 4 \times 10^{-7}$ M IPTG, the level of expression of the *lac* operon will be twice the basal level. Consequently an estimate of the dissociation constant of *lac* repressor and IPTG is 4×10^{-7} M.

The volume of a cell is 10^{-12} cm^3 = 10^{-15} liter. Using this value, a packed cell pellet is $10^{15}/(6 \times 10^{23})$, or about 10^{-9} M in cells. If a cell contains 10 repressor molecules, the concentration of repressor in a packed cell pellet is 10^{-8} M. A cell lysate cannot easily be made at a higher concentration than that obtained by opening cells in a packed cell pellet. Hence a reasonable estimate for the concentration for repressor in the equilibrium dialysis assay is 10^{-8} M.

If high specific activity radioactive IPTG is available, it can be used in the assay at a concentration well below the K_D estimated above (see Problems 12.9 and 12.10). Therefore the ratio of the radioactivities contained in samples of equal volumes taken from inside and outside the sack is

$$\frac{\text{Counts inside sack}}{\text{Counts outside sack}} = 1 + \frac{R_t}{K_D}$$

$$= 1 + \frac{10^{-8}}{4 \times 10^{-7}}$$

$$= 1 + 0.025,$$

which is less than can be reliably detected.

DETECTION AND PURIFICATION OF *lac* REPRESSOR

The previous section showed that wild-type *lac* repressor in crude extracts of cells was not likely to produce a detectable signal in the equilibrium dialysis assay. Therefore Gilbert and Müller-Hill isolated a mutant repressor that bound IPTG more tightly than the wild-type repressor. Crude extracts made from this strain showed an excess of counts in the dialysis sack! The excess was barely detectable; nonetheless it was statistically significant, and fractionation of the extract immediately yielded a protein sample with an easily detectable excess of counts.

Once the assay of *lac* repressor detected something, it was of great importance to prove that the origin of the signal was repressor and not something else. The proof used the tight binding mutant. First, the tight binding mutant was used to develop a partial purification of repressor so that a fraction could be obtained in which the signal was large. Then this same purification procedure was used to obtain a similar fraction from wild-type cells. This too generated a significant signal. The proof came with the demonstration that the apparent dissociation constant for IPTG in the fraction from the mutant and the wild type were different. This was simply done by performing the dialysis on a series of samples at different concentrations of IPTG. The sample obtained from the mutant bound IPTG more tightly, that is, had a smaller K_D, than the wild type (Fig. 12.5). As the only difference between the mutant and the wild type was a mutation in the *lacI* gene, the signal in the assay had to result from *lac* repressor.

The definitive detection of repressor opened the door to biochemical studies. First, with an assay, the repressor could in principle be purified and used in biochemical studies probing its mechanism of action. Second, it was possible to attempt to isolate mutants that synthesized elevated quantities of repressor so as to ease the burden of purification. With an assay, such candidates could be identified.

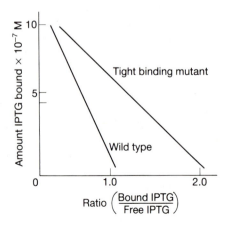

Figure 12.5   Results of equilibrium dialysis at different IPTG concentrations of wild-type repressor and a tight-binding mutant repressor. Rearranging the binding equation derived in the text yields a form convenient for plotting data,
$R \cdot I = R - [K \times (R \cdot I)/I]$. In this form the value of R · I when $(R \cdot I)/I = 0$ yields the concentration of R molecules capable of binding I and the slope of the binding curve gives K.

A mutation rendering *lac* repressor temperature-sensitive was used in the isolation of mutants possessing higher levels of repressor. Cells were grown at a temperature just high enough to inactivate most of the temperature-sensitive repressor. Consequently the *lac* operon was no longer repressed, and the cells expressing the *lac* operon were then killed. The survivors, which were able to repress the operon, could be of two types. Either the repressor could be altered so that it could repress at the elevated temperature or more repressor could be synthesized. With the equilibrium dialysis assay, the two types of mutants could easily be distinguished and an overproducing mutant was identified.

The selection for the loss of constitutivity in the scheme described above used yet another lactose analog, TONPG (o-nitrophenyl-1-thio-β-D-galactoside). This inhibits growth when it is cleaved by β-galactosidase, but it is not an inducer. Mutant cells unable to cleave this compound grow in its presence. Three types of mutants have this property: the necessary mutants as well as *lacZ* and *lacY* mutants. Both of the undesired mutant types were easily eliminated subsequently by requiring the mutants to grow on lactose. The selection scheme was successful, and mutants were found that contained elevated amounts of *lac* repressor.

$$\text{I}^{-\text{ts}} \xrightarrow[\substack{\text{t-ONPG} \\ \text{glycerol} \\ \text{high temperature} \\ (43°)}]{} \substack{\text{I}^+ \text{ phenotype} \\ \text{Z}^- \\ \text{Y}^-} \xrightarrow{\text{lactose}} \text{I}^+ \text{ Z}^+ \text{ Y}^+$$

The isolation of the *lacI* overproducer was the first clear example of the successful isolation of a promoter mutation and was itself a breakthrough. The resulting I^Q (Q for quantity) mutation mapped at the beginning of the I gene, as expected for a promoter mutation, and generated a 10-fold increase in the level of repressor. Although such an increase was helpful to the biochemical studies, the levels of *lac* repressor were still low. An additional increase in repressor levels was obtained by putting the *lac* gene with the I^Q mutation onto a lambda-type phage. Growth of this phage in cells yielded 50 copies of the *lacI* gene instead of the usual two to four, and the overall effect was to increase the level of *lac* repressor up to 50-fold.

Normally, lambdoid phage stop macromolecular synthesis in cells about an hour after induction and lyse the cells soon thereafter. However, a defect in the phage S gene prolongs the period of macromolecular synthesis in two ways. First, the cells do not lyse because the phage-encoded lytic enzymes do not reach the peptidolglycan layer in cells infected with S$^-$ lambda. Second, the defect leads to retention within the cell of the small molecules necessary for macromolecular synthesis. The use of S-defective phage multiples the repressor yield another factor of two to five so that altogether the gene dosage effect, I^Q mutation, and S-defect provide cells in which 0.5% of the protein is *lac* repressor. Later a better promoter mutation, I^{SQ}, was isolated and, in combination with the phage, yielded several percent of the cellular protein as *lac* repressor. The combination of these tricks to overproduce repressor and facilitate the biochemical study are a real achievement for the precloning era. Now the generation of such an overproduction is reasonably straightforward, as discussed in the preceding two chapters.

REPRESSOR BINDS TO DNA: THE OPERATOR IS DNA

As mentioned in the introduction to this chapter, we now know that repressor acts by binding to the operator and preventing RNA polymerase from binding to the *lac* promoter. The first experiments to show that repressor binds to *lac* DNA used ultracentrifugation of radioactive repressor and the DNA from lambda-*lac* phage (Fig. 12.6). The DNA sediments at 40 S, whereas repressor sediments at 7 S. If repressor bound so tightly to DNA that it did not come off during the centrifugation, it would sediment at about 40 S. Indeed, *lac* operator–containing DNA carried *lac* repressor along with it down the centrifuge tube, but only if inducers of the *lac* operon were not present (Fig. 12.7). These are the properties expected of repressor if it were to regulate by binding to DNA to prevent transcription.

A considerably easier assay of DNA binding by *lac* repressor was developed later by Riggs and Bourgeois. Pure lambda DNA passes through cellulose nitrate filters, but repressor, like many other proteins, binds to the filters. It is bound by molecular interactions and not by filtration. Most surprisingly, if repressor is bound to DNA, the DNA molecule will not pass through the filter either. Thus if the DNA is made radioactive, the retention of *lac* DNA on the filter can easily be detected, providing a simple assay for repressor binding to operator (Fig. 12.8).

Figure 12.6   Initial configuration of the centrifuge tube in the experiment to show that repressor binds to operator-containing DNA.

lac DNA
+ repressor

lac DNA + repressor
+ IPTG

Non-*lac* DNA
+ repressor

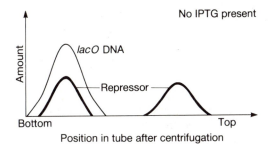

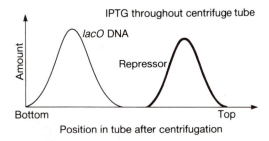

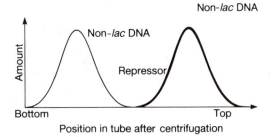

Figure 12.7   Repressor binds to operator-containing DNA only in the absence of IPTG.

Figure 12.8   Schematic of the filter binding assay for *lac* repressor.

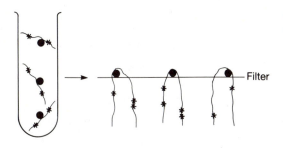

The filter binding assay is very sensitive because the DNA can be made highly radioactive with $^{32}PO_4$: The long DNA molecule contains very many phosphate groups, and the half-life of ^{32}P is short. This assay allows convenient measurement of the rates of binding and dissociation as well as a determination of the equilibrium constant for binding.

The rate of *lac* repressor dissociation from operator is easily measured by mixing repressor with radioactive DNA so that the majority of repressor is bound to operator. Then the solution is diluted and a large excess of nonradioactive *lac* DNA is added. Any repressor that subsequently dissociates from operator either remains free in solution or has a much greater chance of binding to nonradioactive DNA. At intervals after the dilution, the mixture is filtered to determine the fraction of repressor still bound to the radioactive *lac* operator–containing DNA. The kinetics in the reduction in the amount of radioactive DNA bound to the filters with increasing time give the dissociation rate of repressor from operator.

THE ISOLATION AND STRUCTURE OF OPERATOR

After the isolation of *lac* repressor and the demonstration that it bound to DNA, interest turned to the structure of operator. The first questions concerned its size and sequence. Answering such questions is now straightforward with the application of genetic engineering techniques such as those discussed in the previous two chapters. However, the questions about the *lac* operator came to the fore before the era of genetic engineering. Because ingenious techniques were developed and the work directly led to the development of some of the genetic engineering techniques now used, we will review the techniques, now antique, used to isolate and sequence the *lac* operator.

Before developing their chemical sequencing method, Gilbert and Maxam sequenced the operator, primarily by direct DNA sequencing methods, whereas somewhat later Reznikoff, Barnes, Abelson, and co-workers sequenced the entire *lac* regulatory region by purifying and sequencing RNA copies of it (Fig. 12.9). Gilbert's method required isolation of pure operator itself. DNA was isolated from phage carrying the *lac* genes. This DNA was sonicated to fragments about 1,000 base pairs long, repressor was added, and the mixture was passed through a cellulose nitrate filter. The DNA fragments containing operator were bound by repressor and held on the filter. Then the operator-containing fragments were specifically released by adding IPTG to the rinse buffer. Finally, repressor was again bound to these fragments, and the portions of the fragments not protected by repressor were digested by DNase. At the time these steps seemed like a rapid and convenient procedure. The operator was about 20 base pairs long, and it had been purified 2000-fold from the phage or 20,000-fold from the *E. coli* DNA!

The method for sequencing the entire *lac* regulatory region via RNA did not depend on the tight binding of any proteins to the DNA or RNA. Although this method was more general than the method for isolating *lac* operator and using

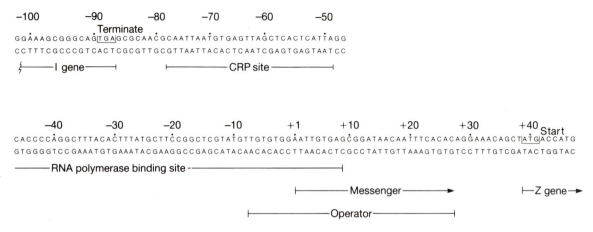

Figure 12.9   The DNA sequence of the *lac* regulatory region and the binding sites of the proteins that bind there.

repressor to protect against DNase, the method required an enormous backlog of genetic work to provide the necessary deletions. First, radioactive RNA was synthesized from one lambda-*lac* phage, then the RNA was hybridized to another lambda phage to remove all phage-specific sequences, and then free RNA in solution was hybridized to another lambda-*lac* phage (Fig. 12.10). The two lambda-*lac* phage

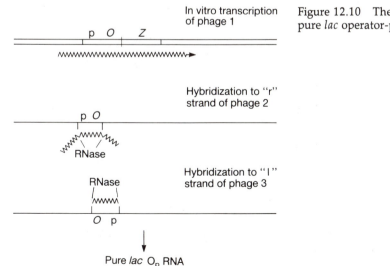

Figure 12.10   The sequence of steps used to obtain pure *lac* operator-promoter DNA for sequencing.

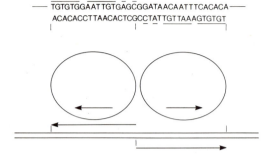

TGTGTGGAATTGTGAGCGGATAACAATTTCACACA
ACACACCTTAACACTCGCCTATTGTTAAAGTGTGT

Figure 12.11   The symmetry in *lac* operator and how a protein consisting of two identical subunits could bind to it.

were chosen such that the only *lac* sequences they possessed in common were the desired regulatory sequences of the operon. The resulting RNA was then sequenced by the methods that had been developed by Sanger for sequencing RNA, which relied on the enzymes that cleave RNA adjacent to certain bases.

The *lac* operator was found to possess a symmetrical sequence. This suggested that two of the repressor subunits symmetrically made contact with it (Fig. 12.11). To probe the structure of *lac*-operator-repressor complexes, Gilbert and Maxam turned to DNA protection studies. They bound repressor to operator and investigated the reaction of dimethylsulfate with bases of the operator in the complex. Indeed, a symmetrical protection pattern was obtained, but, more important, they noticed that the method could easily and rapidly be adapted to yield the sequence of a DNA fragment.

REPRESSOR SLIDES ALONG DNA TO FIND THE OPERATOR

Measurements of the repressor-operator binding rates have shown that the association reaction is very fast, indeed faster than predicted for a simple binding of repressor to a target the size of operator. The speed of the binding reaction when the operator is contained in a longer piece of DNA is dependent on the salt concentration in which the reaction is performed. Within limits (see Problem 25), the lower the salt concentration, the faster the binding rate (Fig. 12.12).

$$\text{Repressor} + \text{Operator} \cdot 10 \, \text{Na}^+ \rightleftharpoons$$

$$\text{Repressor} \cdot \text{Operator} + 10 \, \text{Na}^+$$

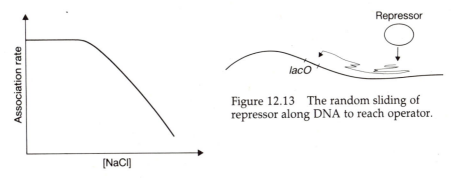

Figure 12.13   The random sliding of repressor along DNA to reach operator.

Figure 12.12   The association rate of *lac* repressor for operator as a function of salt concentration.

The effect of salt concentration on the repressor binding rate is understandable in terms of the sodium ions that must be displaced from operator as repressor binds. The reverse reaction, the dissociation of repressor from DNA, requires that the sodium ions be returned to their prior positions near the phosphate groups on the DNA. At low sodium concentrations in the solution, this replacement is difficult and infrequent, but at high sodium concentrations, it readily proceeds. Thus at low salt concentrations repressor can bind to nonoperator DNA, cannot easily come off, and can slide an appreciable distance (Fig. 12.13). The lower the salt concentration, the further repressor can slide along the DNA. Hence the DNA on either side of the operator acts as an antenna to feed repressor to operator. Such a salt effect is seen only when the DNA fragment containing operator is appreciably larger than the operator so that sliding to operator is possible. The salt concentrations *in vivo* permit the antenna effect to significantly accelerate repressor binding to the operator.

REPRESSOR BINDS DNA *in vitro* AS TIGHTLY AS *in vivo*

Filter binding experiments also allow a direct measurement to be made of the dissociation constant of repressor for operator. The results agree with the dissociation constant that can be calculated from the association and dissociation rates. The dissociation constant obtained by either of these methods can also be compared to *in vivo* data. A correspondence in the numbers suggests, but does not prove, that the *in vitro* reactions and conditions closely mimic the *in vivo* environment.

An estimation of the *in vivo* dissociation constant of repressor for the operator uses the inducibility of the *lac* operon. Assume that the operator is completely free of

repressor when the operon is fully induced and that the basal level of $\frac{1}{1000}$ the fully induced level results from the operator being free of repressor $\frac{1}{1000}$ of the time.
    Since

$$K_D = \frac{R \times O}{RO}$$

$$= 10^{-8}\,\frac{O}{RO}$$

$$= 10^{-8}\,\frac{1}{1000}$$

$$= 10^{-11}\ \text{M}.$$

This result closely agrees with the values found *in vitro* in buffers with salt concentrations equal to the *in vivo* value of about 0.15 M.

REPRESSOR'S N-TERMINUS BINDS DNA: GENETIC DEMONSTRATION

The study of *lac* repressor and lambda phage repressor binding to their respective operators and RNA polymerase binding to promoters made it clear that proteins are capable of binding to specific DNA sequences. The research questions then turned from finding what produced repression to determining how repression was accomplished, for example, how proteins recognize sequences. Like so many scientific questions, the fundamental problem was too difficult to attack directly. Instead, much simpler questions were investigated that led toward answering the major one. Genetic experiments were ideally suited to such studies.
    Work by Miller and collaborators showed that nonsense mutations lying early in the *lacI* gene did not abolish expression of the entire gene. Although the ribosomes terminate translation on reaching the nonsense codon, often they do not dissociate from the mRNA before reinitiating translation at a site within the *lacI* gene. As a result, they synthesize a repressor molecule that lacks up to 61 of the amino acids normally found at the NH_2-terminus. Most surprisingly, these truncated repressor polypeptides fold, associate as usual to form tetramers, and bind IPTG. However, they are incapable of binding to DNA as shown *in vivo* by their inability to repress and *in vitro* by the DNA filter binding assay. The simplest explanation of their inability to bind to DNA is that the amino acids missing from their NH_2-terminus are the part of the wild-type repressor that recognizes and binds to operator. In addition, more detailed genetic analysis of the repressor have shown that the overwhelming majority of missense mutations affecting DNA binding by the repressor map in the region of the I gene coding for the first 60 amino acids.
    The amino-terminal truncated repressors possess a curious genetic property. A typical LacI⁻ mutant is recessive to the wild-type I gene. That is, a *lacI⁺/lacI⁻* diploid

is repressible and can be induced. The same is not true of the repressor mutants missing their N-terminal amino acids. They are dominant I^-, denoted I^{-d}, and the diploid $lacI^+/lacI^{-d}$ behaves like an I^-. The effect is more pronounced when an I^Q is mutated to nonsense in the early amino acids to produce an I^{Q-d}. (See Problem 12.1.)

The dominance of I^{-d} mutations to the I^+ allele results from the tetrameric structure of repressor. During the synthesis of the two types of repressor in a LacI$^+$/LacI^{-d} cell, the probability that four newly synthesized wild-type repressor subunits will associate to form a wild-type tetramer is low. Instead, most of the wild-type subunits are sequestered in oligomers containing at least one mutant repressor subunit. That is, the good subunits are titrated out by the bad subunits, and the probability of forming a repressor with four wild-type subunits or even two wild-type subunits is small. Note that the dominance of the I^{-d} allele necessitates that either the bad subunits coerce the good subunits into inactive conformations or that more than one subunit is involved with the binding of the tetrameric repressor molecule to operator.

ISOLATION AND CHARACTERIZATION OF I⁻ᵈ REPRESSOR MUTANTS

The steps to combine to isolate a particular mutant sometimes are not apparent, so we will list and explain a series of genetics steps directed to the isolation of an I^{-d} mutant.

1. Isolate I^- mutations using phenyl-β-D-galactoside plates. Since phenyl-gal is not an inducer, but is a substrate of β-galactosidase, any cells growing on these plates must be constitutive.

2. Score to determine which are nonsense I^- mutations. The mutations can be isolated on an episome, F'Lac$^+$ Pro$^+$, and to test, the episome can be mated into an F$^-$ su$^+$ strain. This strain should be deleted of lac and pro genes. To determine whether the episome in the suppressing strain is I^+ or I^-, the plates can include 5-bromo-4-chloro-3-indolyl-β-D-galactoside (X-gal). If X-gal is included in minimal glycerol plates, the constitutives will form deep blue colonies because the hydrolysis of this substrate by β-galactosidase produces an insoluble blue dye.

3. Map the locations of the I^- mutations using the ability of strains containing the mutation on an episome and a deletion into the I gene on the chromosome to reconstruct an I^+ gene (Fig. 12.14). A point mutation is able to recombine with deletion 1 but not deletion 2 to form I^+ recombinants. Although this type of mapping only positions the mutation to the left or to the right of the end point of the deletion, an ordered set of deletions can be used to locate the mutation to within about 10 base pairs.

How can the selective plates be arranged to permit growth of only the cells that can reconstruct a functional I gene? The plates must prevent growth of each of the

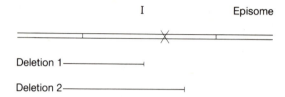

Figure 12.14   Deletion mapping of *lacI* mutations. Only the strain containing deletion 1 can recombine with the mutation indicated to reconstruct a functional *lacI* gene.

parent types of cells as well as mated cells that are unable to form a wild-type I gene. This can be done by using phenyl-gal and glycerol in plates. The deletion mapping can be done in a GalE⁻ strain. The resulting defective galactose epimerase gene renders the cells sensitive to galactose. The I⁻ cells are constitutive and cleave the phenyl-gal to produce galactose, which prevents their growth. However, an I⁺ recombinant avoids this problem and grows on the glycerol without cleaving the phenyl-gal.

4. Test candidates for being *trans*-dominant repressor negative by mating the episome into an I⁺ strain by selecting for transfer of the *pro* marker. This strain is tested by spotting onto X-gal plates as before.

5. Physical studies of the repressor from the I⁻ᵈ mutants can test whether the repressor is a tetramer, and SDS gels can show that the mutated repressor is shorter than wild type.

Protein sequencing can provide the final proof that the repressor synthesized in the nonsense I⁻ᵈ mutants results from translational restarts. The mutant repressor can be purified by precipitation with antibody, separated from the antibody by SDS gel electrophoresis, eluted from the gel, and the N-terminal amino acids sequenced.

REPRESSOR'S N-TERMINUS BINDS DNA: PHYSICAL DEMONSTRATION

The physical experiments showing that the N-terminal amino acids of repressor are directly involved in binding to operator were simple. On limited digestion in the presence of high concentrations of glycerol and Tris or at low temperatures, proteases like trypsin or chymotrypsin cleave repressor once about 50 amino acid residues from the N-terminus. The "headpiece" and the much larger core can each be isolated by passing the digest through a gel filtration column that separates proteins according to size. Purified core binds IPTG normally but does not bind to DNA, whereas purified headpiece binds to *lac* operator specifically.

The combination of nuclear magnetic resonance (NMR) and genetics provides the second demonstration that the N-terminal region of repressor specifically binds to operator. Conventional NMR monitors the electronic environment of nuclei

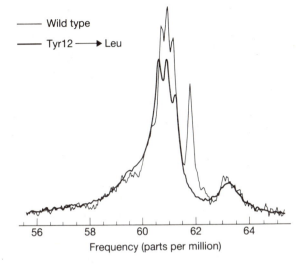

—— Wild type

—— Tyr12 ——▶ Leu

Frequency (parts per million)

56   58   60   62   64

Figure 12.15   A nuclear magnetic resonance (NMR) spectrum of fluorotyrosine substituted wild-type *lac* repressor and the spectrum from a mutant lacking tyr12. (Courtesy P. Lu.)

(atoms) containing nonzero spins. The isotopes ^{1}H, ^{13}C, and ^{19}F, all produce usable NMR signals. The NMR spectrum of a protein generally possesses many overlapping resonance peaks. If each resonance peak of repressor could be resolved from the others and be identified with a particular atom in the sequence of the protein, then, since the environment of only a few atoms ought to be altered by the binding to operator, only a few resonance peaks ought to be shifted. If this were found, NMR could identify the atoms of repressor that come close to operator.

The inclusion of fluorotyrosine in cells' growth medium substitutes this molecule for the eight tyrosines in each repressor subunit or the four tyrosines in the headpiece. The resulting repressors bind to operator normally. The reason for including fluorotyrosine is that the NMR spectrum of fluorine possesses a resonance peak well separated from the carbon and hydrogen peaks of the rest of the repressor molecule. Therefore the environments of just these four or eight amino acids in the protein can be monitored.

Genetics assisted the unambiguous identification of the tyrosine-fluorotyrosine residues in the *lac* repressor sequence responsible for the peaks (Fig. 12.15). This was done with nonsense mutations located in the tyrosine codons. Suppression of each of these mutations generates a repressor missing a tyrosine from a known position, unless Su$_{III}$ is used. Comparison of the NMR spectra from wild type and the suppressed nonsense mutant then identifies one resonance peak. These approaches have been used with both fluorine-substituted and normal repressor and headpiece proteins and have shown that isolated headpiece retains its native structure and that it binds to operator. In addition, the chemical environments of only a few of the tyrosines, most notably the tyrosine at position 7, are altered on operator binding.

A MECHANISM FOR INDUCTION

How do inducers reduce repressor's affinity for operator? It is possible that they bind near the operator binding site and merely interfere with repressor's correct binding to operator. This possibility seems unlikely in view of the genetic data and the

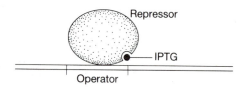

separability of the IPTG binding and operator binding substructures of repressor discussed above. It seems more likely that IPTG merely causes the subunits of repressor to alter positions slightly with respect to one another. Why should this drastically weaken the binding? The headpiece experiments contain the answer.

The headpiece binds to operator with a much lower affinity than wild-type repressor. Such reduced affinity is expected. The wild-type tetrameric repressor molecule possesses a relatively rigid structure in which pairs of the N-terminal regions are held in positions appropriate for binding to a single operator. That is, the binding of one of the N-terminal regions to half of the symmetrical operator perforce brings a second N-terminal region into position for its binding to the other half of the operator, and most of the interaction energy goes into tightly binding this second subunit in place (Fig. 12.16). This is another example of the chelate effect. Overall, the result is that the oligomeric repressor tightly binds to operator. The same is not true of the isolated headpieces. The binding of one headpiece does not automatically bring another into position for binding to the other half of operator. As a result of their independent binding, the apparent affinity of headpiece for operator is low.

The chelate effect also streamlines the explanation of induction: Binding of IPTG shifts pairs of subunits away from optimal relative positions for headpiece binding to operator. Consequently, the affinity of repressor for operator is greatly

Figure 12.16   The binding of one subunit of repressor to half of the operator correctly positions the other subunit for binding to the other half of operator.

reduced, and repressor dissociates. Eventually, direct experiments may be able to test such ideas. In the meantime, two types of repressor mutations are consistent with this point of view. Repressor mutants can be found that are not located in the N-terminal region and that result in much tighter or much weaker repressor-operator binding. These mutants possess no discernible structural alterations. Most likely these types of mutation merely shift the positions of the subunits slightly with respect to one another. The tighter-binding mutant must bring the subunits into closer complementarity with operator, and the weaker-binding mutants must shift the subunits away from complementarity.

RNA POLYMERASE BINDING TO THE *lac* PROMOTER

One straightforward way to study the repressor-operator interaction would be through the three-dimensional structure of the repressor bound to operator. Similarly, knowing the structure of the CRP-RNA polymerase-promoter interaction could tell how polymerase initiates. Presently these questions exceed the capabilities of X-ray crystallography, and no other method can provide the needed structural information. Partial understanding of these questions has come with the determination of the sequence of the *lac* operon regulatory region, the determination of the regions protected by binding of repressor, CRP, or RNA polymerase from chemical or enzymatic attack, footprinting, and the sequences of many other promoters. Before this, electron microscopy was able to detect repressor binding to operator and RNA polymerase binding to the promoter (Fig. 12.17). The microscopy showed that

Figure 12.17   An electron micrograph of *lac* repressor bound to operator on a 203-base-pair piece of DNA and a micrograph of RNA polymerase bound to the promoter on this same-sized fragment.

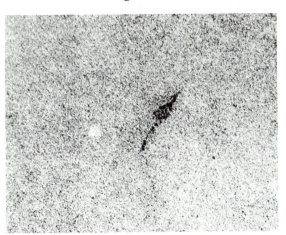
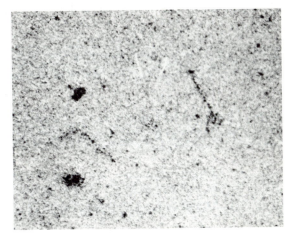

repressor and polymerase both bound to the same area and therefore would likely complete for binding, a result confirmed biochemically. Furthermore, these experiments could show that effective RNA polymerase binding to wild-type DNA required the presence of CRP. The abortive initiation assay as described in Chapter 4 has been used to show that CRP accelerates the binding rate of RNAP to DNA. CRP has also been found to occlude the binding of RNA polymerase to a second promoter in *lac*. RNA polymerase bound to this second site both initiates transcription very slowly and also blocks RNA polymerase from binding to the normally functional promoter. CRP therefore stimulates *lac* transcription in two ways: by preventing clogging and by assisting initiation from the correct promoter.

An enormous amount has been learned about the *lac* operon, and only a fraction could be mentioned in this chapter. Active research on the mechanism of regulation of the *lac* operon seems nearly over as most, but not all, of the observed physiological phenomena have been explained biochemically. Attention is now turning to more fundamental questions of protein function and protein–nucleic acid interaction. An amino acid sequence has been recognized both in the lambda cro protein and CRP protein that binds to DNA. A similar amino acid sequence is found in the headpiece of *lac* repressor, and much interest is directed to how the basic sequence can encode DNA binding and yet permit sufficient variability to provide each of these proteins with specificity for its own DNA sequence.

PROBLEMS

12.1. Consider the repressor that is synthesized in cells containing both the I^{-d} and I^+ alleles. If subunits mix randomly during assembly of the tetrameric repressor, what is the fraction of tetrameric molecules having the composition I_4^+, $I_3^+I_1^{-d}$, $I_2^+I_2^{-d}$, $I_1^+I_3^{-d}$, I_4^{-d}? Consider that the synthesis of the two types of subunits is equal. Repeat, assuming that the synthesis of I^{-d} subunits is in 10-fold excess over I^+ subunits.

*12.2. Assuming both that *lac* repressor messenger is translated by about 30 ribosomes and that repressor subunits do not exchange between tetramers, and in light of the low numbers of repressor molecules per cell, is a paradox generated by the existence of dominant I^- mutations?

12.3. Define temperature-sensitive synthesis (TSS) mutations as those that lead to the protein's not being able to fold at high temperatures but that allow a protein that folded at low temperature to be stable at high temperature. These are to be distinguished from TS mutations, which lead the protein to be heat-unstable regardless of the temperature at which it folded. Invent a selection for TSS Z^- mutations and for TSS I^- mutations.

12.4. Design the simplest experiment you can that could answer the question, now hypothetical, of whether repressor blocks the binding or the transcriptional progress of RNA polymerase.

12.5. Transcription of a region of DNA could alter the probability of recombina-

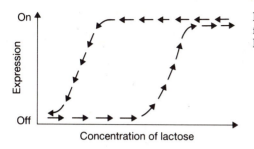

Figure 12.18 Behavior of the *lac* system as a function of the lactose concentration in the medium. Induction follows the arrows.

tion in that interval. Design an experiment to measure the effect of transcription on recombination. What precautions would be needed?

12.6. The induction curve of the *lac* operon versus IPTG concentration is approximately quadratic, that is, cooperative. However, the Gilbert–Müller-Hill paper shows that the IPTG binding to repressor obeys the equation derived in the text, that is, it is Michaelis-Menten. Is there a contradiction here? Why or why not?

12.7. The events at the preceding initiation by RNA polymerase are important and need appropriate definitions. One term used in this area is "promoter." Observe the varying usages of the term in the papers J. Mol. Biol. *69*, 155–160 (1972); J. Mol. Biol. *38*, 421 (1968); Nature *217*, 825 (1968), and others these papers lead you to. Comment on the uses and make a suitable definition of your own.

*12.8. The presence of the active transport system for the *lac* operon allows the system to behave as a bistable flip-flop inducing at a high lactose concentration and then remaining induced at a lactose concentration well below that required for induction (Fig. 12.18). Let C_i be the concentration of IPTG inside cells and C_o its external concentration. Assume that the degree of derepression (induction) of the

Figure 12.19 Hypothetical data giving the ability of the *lac* transport system to concentrate inducer intracellularly to a concentration C_i from an outside concentration C_o and the level of derepression of the *lac* operon as a function of the intracellular inducer concentration.

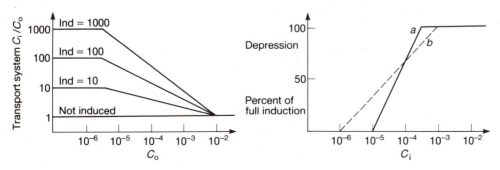

transport system as a function of C_i and the ability of the transport system to concentrate IPTG as a function of C_o are as shown in Fig. 12.19. Show that the response curve *a* produces a system in which fully induced cells that are put into 10^{-6} M IPTG remain induced indefinitely or, if they were not induced, when put into 10^{-6} M IPTG they do not induce despite minor fluctuations in the levels of enzymes and the active transport capabilities. Show that the curve *b* does not produce this bistable response.

12.9. a) If cells contain 100 repressor monomers and a cell extract at 50 mg/ml is dialyzed against 50 mC_i/mM IPTG at 1×10^{-6} M, what are the cpm per 0.1 ml sample inside and outside the sacks. Assume $K_D = 4 \times 10^{-6}$ M.

b) What is the ratio if the IPTG concentration is lowered to 1×10^{-7} M?

c) Is it of any use to use very high specific activity IPTG?

12.10. If the dissociation constant of IPTG from *lac* repressor, instead of being 2×10^{-7} M,

a) were 1×10^{-8} M, could repressor have been detected in the original experiments?

b) were 2×10^{-5} M, but the specific activity of IPTG had been 10 times higher, could *lac* repressor have been detected in the original experiments?

*12.11. Would it be reasonable that the NMR peaks originating from fluorotyrosines located in the portion of the repressor responsible for subunit-subunit association should be appreciably altered on repressor-operator or repressor-inducer binding? Why?

12.12. Like many proteins, *lac* repressor binds to nitrocellulose filters. Quite unexpectedly, however, if IPTG is bound to repressor, it is not easily released when repressor has stuck to the filters. Thus if radioactive IPTG is used, only *lac* repressor can bind radioactivity to the filters, and this constitutes a convenient assay for *lac* repressor. The assay is limited to the use of small quantities of protein, however, since about a monolayer of protein on the filters prevents further protein binding. What purity of repressor would be needed to bind 200 cpm of 25 mC_i/mm IPTG to filters 2 cm in diameter?

12.13. What are the biological advantages of symmetrical, palindromic DNA binding sites for proteins?

12.14. Why might missense mutations in the N-terminal region of *lac* repressor not be *trans* dominant?

12.15. Are the CRP and repressor binding sites the only palindromes in the *lac* regulatory region? What do you conclude about the biological significance of palindromes?

*12.16. How could Müller-Hill's genetic experiment be correct but repressor be purely RNA?

12.17. What dye is produced by β-galactosidase cleavage of X-gal, and why is its insolubility important for its use as an indicator in plates?

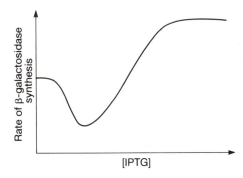

Figure 12.20   Data for Problem 12.24.

12.18. What is the relation between the association and dissociation rates and the equilibrium binding constant for a reaction like *lac* repressor binding to operator?

12.19. After examination of the appropriate literature, describe how allolactose was found to be the true inducer of the *lac* operon.

12.20. What will the kinetics of β-galactosidase synthesis look like in the mated culture if $F'I^+Z^+Y^+$ cells are mixed with $F^-\Delta lac$ cells in the absence of *lac* operon inducers? How will the kinetics be altered if the male strain is $F'I^QZ^+Y^+$?

12.21. How could you prove that only two of the four subunits of *lac* repressor are required for binding of the *lac* repressor to operator?

12.22. Devise a method for the isolation of a *lacI* mutation in which repressor binds IPTG less tightly than the wild-type repressor.

12.23. Suppose it was observed during footprinting of *lac* repressor that the footprint of repressor was not removed by the inclusion of IPTG in the buffer during the DNase digestion. The same experiment performed with the same components and with the same amounts of DNA and repressor, but in a volume 100-fold larger, yielded the expected result. What is the likely explanation?

12.24. A mutant *lac* repressor shows the bizarre induction behavior graphed in Fig. 12.20. The purified repressor was found to bind to DNA, both operator and nonoperator, with about 100 times the affinity of normal repressor. Explain these data.

RECOMMENDED READINGS

Isolation of the Lac Repressor, W. Gilbert, B. Müller-Hill, Proc. Natl. Acad. Sci. USA *56*, 1891–1898 (1966).

lac Repressor-Operator Interaction, I. Equilibrium Studies, A. Riggs, H. Suzuki, S. Bourgeois, J. Mol. Biol. *48*, 67–83 (1970).

The Nucleotide Sequence of the *lac* Operator, W. Gilbert, A. Maxam, Proc. Natl. Acad. Sci. USA *70*, 3581–3584 (1973).

RELATED REVIEWS, BOOKS, AND ARTICLES

The Lactose Operon, ed. J. Beckwith, D. Zipser, Cold Spring Harbor Laboratory (1970).
The Operon, ed. J. Miller, W. Reznikoff, Cold Spring Harbor Laboratory (1978).

DEEPER READING

Suppressible Regulator Constitutive Mutants of the Lactose System in *Escherichia coli*, B. Müller-Hill, J. Mol. Biol. *15*, 374–379 (1966).

Transposition of the *lac* Region of *E. coli*, II. On the Role of Thiogalactoside Transacetylase in Lactose Metabolism, C. F. Fox, J. R. Beckwith, W. Epstein, E. Singer, J. Mol. Biol. *19*, 576–579 (1966).

The Lac Operator Is DNA, W. Gilbert, B. Müller-Hill, Proc. Natl. Acad. Sci. USA *58*, 2415–2421 (1967).

DNA Binding of the *lac* Repressor, A. Riggs, S. Bourgeois, R. Newby, M. Cohn, J. Mol. Biol. *34*, 365–368 (1968).

Mutants That Make More Lac Repressor, B. Müller-Hill, L. Crapo, W. Gilbert, Proc. Natl. Acad. Sci. USA *59*, 1259–1264 (1968).

A Mechanism for Repressor Action, W. Reznikoff, J. Miller, J. Scaife, J. Beckwith, J. Mol. Biol. *43*, 201–213 (1969).

The *lac* Repressor-Operator Interaction, III. Kinetic Studies, A. Riggs, S. Bourgeois, M. Cohn, J. Mol. Biol. *53*, 401–417 (1970).

Inactivation and Degradation of Messenger RNA from the Lactose Operon of *E. coli*, T. Schwartz, E. Craig, D. Kennell, J. Mol. Biol. *54*, 299–311 (1970).

Lac Repressor Binding to DNA Not Containing the *Lac* Operator and to Synthetic Poly dAT, S. Lin, A. Riggs, Nature *228*, 1184–1186 (1970).

Mutational Inversion of Control of the Lactose Operon of *E. coli*, G. Meyers, J. Sadler, J. Mol. Biol. *58*, 1–28 (1971).

The Nature of Lactose Operator Constitutive Mutations, T. Smith, J. Sadler, J. Mol. Biol. *59*, 273–305 (1971).

Mapping of the Lactose Operator, J. Sadler, T. Smith, J. Mol. Biol. *62*, 139–169 (1971).

Translational Restarts: AUG Reinitiation of a *lac* Repressor Fragment, T. Platt, K. Weber, D. Ganem, J. Miller, Proc. Natl. Acad. Sci. USA *69*, 897–901 (1972).

Transposition of the *lac* Region to the *gal* Region of the *E. coli* Chromosome: Isolation of Lambda *lac* Transducing Bacteriophages, K. Ippen, J. A. Shapiro, J. R. Beckwith, J. Bact. *108*, 5–9 (1971).

Regulation of *lac* mRNA Synthesis in a Soluble Cell-Free System, B. de Crombrugghe, B. Chen, M. Gottesman, I. Pastan, H. Varmus, M. Emmer, R. Perlman, Nature New Biol. *230*, 37–40 (1971).

Role of Cyclic AMP 3′,5′ and Cyclic AMP 3′,5′ Receptor Protein in the Initiation of *lac* Transcription, B. de Crombrugghe, B. Chen, W. Anderson, M. Gottesman, R. Perlman, I. Pastan, J. Biol. Chem. *246*, 7343–7348 (1971).

Lac Repressor Binding to Operator Analogues: Comparison of Poly(d(A-T)), Poly (d(A-BrU)), and Poly (d(A-U)), S. Lin, A. Riggs, Biochem. Biophys. Res. Comm. *45*, 1542–1547 (1971).

Lac Repressor-Operator Interaction, VI. The Natural Inducer of the *lac* Operon, A. Jobe, S. Bourgeois, J. Mol. Biol. *69*, 397–408 (1972).

Lac Repressor Binding to Non-Operator DNA: Detailed Studies and a Comparison of Equilibrium and Rate Competition Methods, S. Lin, A. Riggs, J. Mol. Biol. *72*, 671–690 (1972).

Distribution of Suboptimally Induced β-ᴅ-Galactosidase in *E. coli:* The Enzyme Content of Individual Cells, P. Maloney, B. Rotman, J. Mol. Biol. *73*, 77–91 (1973).

The Nucleotide Sequence of the Lactose Messenger Ribonucleic Acid Transcribed from UV5 Promoter Mutant of *E. coli*, N. Maizels, Proc. Natl. Acad. Sci. USA *70*, 3585–3589 (1973).

Lac Repressor, Specific Proteolytic Destruction of the NH₂-Terminal Region, and Loss of the DNA Binding Activity, T. Platt, J. Files, K. Weber, J. Biol. Chem. *248*, 110–121 (1973).

A New Class of Promoter Mutations in the Lactose Operon of *E. coli*, J. Hopkins, J. Mol. Biol. *87*, 715–724 (1974).

Translational Reinitiation: Reinitiation of *lac* Repressor Fragments at Three Internal Sites Early in the *lacI* Gene of *Escherichia coli*, J. Files, K. Weber, J. Miller, Proc. Natl. Acad. Sci. USA *71*, 667–670 (1974).

Photochemical Attachment of *lac* Repressor to Bromodeoxyuridine-Substituted *lac* Operator by Ultra Violet Radiation, S. Lin, A. Riggs, Proc. Natl. Acad. Sci. USA 71, 947–951 (1974).

Non-Specific DNA Binding of Genome Regulating Proteins as a Biological Control Mechanism, I. The *lac* Operon: Equilibrium Aspects, P. von Hippel, A. Revzin, C. Gross, A. Wang, Proc. Natl. Acad. Sci. USA *71*, 4808–4812 (1974).

Measurements of Unwinding of *lac* Operator by Repressor, J. Wang, M. Barkley, S. Bourgeois, Nature *251*, 247–249 (1974).

Regulation of the Synthesis of the Lactose Repressor, P. Edelman, G. Edlin, J. Bact. *120*, 657–665 (1974).

Genetic Fusions Defining *trp* and *lac* Operon Regulatory Elements, D. Mitchell, W. Reznikoff, J. Beckwith, J. Mol. Biol. *93*, 331–350 (1975).

Initiation of *in vitro* mRNA Synthesis from the Wild-Type *lac* Promoter, J. Majors, Proc. Natl. Acad. Sci. USA *72*, 4394–4398 (1975).

Interaction of Effecting Ligands with *lac* Repressor and Repressor-Operator Complex, M. Barkley, A. Riggs, A. Jobe, S. Bourgeois, Biochem. *14*, 1700–1712 (1975).

Genetic Regulation: The *lac* Control Region, R. Dickson, J. Abelson, W. Barnes, W. Reznikoff, Science *187*, 27–35 (1975).

Conformational Transitions of the *lac* Repressor from *E. coli*, F. Wu, P. Bandyopadhyay, C. Wu, J. Mol. Biol. *100*, 459–472 (1976).

Mutations Affecting the Quaternary Structure of the *lac* Repressor, A Schmitz, U. Schmeissner, J. Miller, P. Lu, J. Biol. Chem. *251*, 3359–3366 (1976).

Thiogalactoside Transacetylase of the Lactose Operon as an Enzyme for Detoxification, K. Andrews, E. Lin, J. Bact. *138*, 510–513 (1976).

Genetic Studies of the *lac* Repressor, I. Correlation of Mutational Sites with Specific Amino Acid Residues: Construction of a Colinear Gene-Protein Map, J. Miller, D. Ganem, P. Lu, A. Schmitz, J. Mol. Biol. *109*, 275–301 (1977).

Genetic Studies of the *lac* Repressor, II. Fine Structure Deletion Map of the *lacI* Gene, and Its Correlation with the Physical Map, U. Schmeissner, D. Ganem, J. Miller, J. Mol. Biol. *109*, 303–326 (1977).

Interaction of *lac* Repressor with Inducer: Kinetic and Equilibrium Measurements, B. Friedman, J. Olson, K. Matthews, J. Mol. Biol. *111*, 27–39 (1977).

Nucleotide Sequence Changes Produced by Mutations in the *lac* Promoter of *E. coli*, R. Dickson, J. Abelson, P. Johnson, W. Reznikoff, W. Barnes, J. Mol. Biol. *111*, 65–75 (1977).

Transcription and Translation Initiation Frequencies of the *E. coli lac* Operon, D. Kennell, H. Riezman, J. Mol. Biol. *114*, 1–21 (1977).

Genetic Studies of the *lac* Repressor, III. Additional Correlation of Mutational Sites with Specific Amino Acid Residues, C. Coulondre, J. Miller, J. Mol. Biol. *117*, 525–575 (1977).

Genetic Studies of the *lac* Repressor, IV. Mutagenic Specificity in the *LacI* Gene of *E. Coli*, C. Coulondre, J. Miller, J. Mol. Biol. *117*, 577–606 (1977).

Contacts between the *lac* Repressor and Thymines in the *lac* Operator, R. Ogata, W. Gilbert, Proc. Natl. Acad. Sci. USA *74*, 4973–4976 (1977).

Minimal Length of the Lactose Operator Sequence for the Specific Recognition by the Lactose Repressor, C. Bahl, R. Wu, J. Stawinsky, S. Narang, Proc. Natl. Acad. Sci. USA *74*, 966–970 (1977).

Binding of Synthetic Lactose Operator DNAs to Lactose Repressors, D. Goeddel, D. Yansura, M. Caruthers, Proc. Natl. Acad. Sci. USA *74*, 3292–3296 (1977).

Nonspecific DNA Binding of Genome-Regulating Proteins as a Biological Control Mechanism: Measurement of DNA-Bound *E. coli lac* Repressor *in vivo*, Y. Kao-Huang, A. Revzin, A. Butler, P. O'Conner, D. Noble, P. von Hippel, Proc. Natl. Acad. Sci. USA *74*, 4228–4232 (1977).

Isolation of the Amino-Terminal Fragment of Lactose Repressor Necessary for DNA Binding, N. Geisler, K. Weber, Biochemistry *16*, 938–943 (1977).

Molecular Parameters Characterizing the Interaction of *E. coli lac* Repressor with Non-Operator DNA and Inducer, A. Butler, A. Revzin, P. von Hippel, Biochem. *16*, 4757–4768 (1977).

Direct Measurement of Association Constants for the Binding of *E. coli lac* Repressor to Non-Operator DNA, A. Revzin, P. von Hippel, Biochem. *16*, 4769–4776 (1977).

Nonspecific Interaction of Lac Repressor with DNA: An Association Reaction Driven by Counterion Release, P. deHaseth, T. Lohman, M. Record, Biochem. *16*, 4783–4791 (1977).

Genetic Studies of the *lac* Repressor, V. Repressors Which Bind Operator More Tightly Generated by Suppression and Reversion of Nonsense Mutations, A. Schmitz, C. Coulondre, J. Miller, J. Mol. Biol. *123*, 431–456 (1978).

Genetic Studies of the *lac* Repressor, VII. On the Molecular Nature of Spontaneous Hotspots in the *lacI* Gene of *E. coli*, P. Farabaugh, U. Schmeissner, M. Foffer, J. Miller, J. Mol. Biol. *126*, 847–863 (1978).

Sequence of the *lacI* Gene, P. Farabaugh, Nature *274*, 765–769 (1978).

Correlation of Nonsense Sites in the *LacI* Gene with Specific Codons in the Nucleotide Sequence, J. Miller, C. Coulondre, P. Farabaugh, Nature *274*, 770–775 (1978).

Molecular Basis of Base Substitution Hotspots in *E. coli*, C. Coulondre, J. Miller, P. Farabaugh, W. Gilbert, Nature *274*, 775–780 (1978).

Tight-Binding Repressors of the *lac* Operon: Selection System and *in vitro* Analysis, M. Pfahl, J. Bact. *137*, 137–145 (1979).

Characteristics of Tight Binding Repressors of the *lac* Operon, M. Pfahl, J. Mol. Biol. *147*, 1–10 (1981).

Mapping of I Gene Mutations Which Lead to Repressors with Increased Affinity for *lac* Operator, M. Pfahl, J. Mol. Biol. *147*, 175–178 (1981).

Interaction of the cAMP Receptor Protein with the *lac* Promoter, R. Simpson, Nucleic Acids Res. *8*, 759–779 (1981).

Genetic Assignment of Resonances in the NMR Spectrum of a Protein: Lac Repressor, M. Jarema, P. Lu, J. Miller, Proc. Natl. Acad. Sci. USA *78*, 2707–2711 (1981).

Salt Dependence of the Kinetics of the *lac* Repressor-Operator Interaction: Role of Nonoperator DNA in the Association Reaction, M. Barkley, Biochem. *20,* 3833–3842 (1981).

lac Repressor: A Proton Magnetic Resonance Look at the Deoxyribonucleic Acid Binding Fragment, K. Arndt, F. Boschelli, P. Lu, J. Miller, Biochem. *20,* 6109–6118 (1981).

lac Repressor–*lac* Operator Interaction: NMR Observations, H. Nick, K. Arndt, F. Boschelli, M. Jarema, M. Lillis, J. Sadler, M. Caruthers, P. Lu, Proc. Natl. Acad. Sci. USA *79,* 218–222 (1982).

Repressor-Operator Interaction in the *lac* Operon, II. Observations at the Tyrosines and Tryptophans, H. Nick. K. Arndt, F. Boschelli, M. Jarema, M. Lillis, H. Sommer, P. Lu, J. Sadler, J. Mol. Biol. *161,* 417–438 (1982).

Secondary Structure of the *lac* Repressor DNA-Binding Domain by Two-Dimensional ¹H Nuclear Magnetic Resonance in Solution, E. Zuiderweg, R. Koptein, K. Wuthrich, Proc. Natl. Acad. Sci. USA *80,* 5837–5841 (1983).

A Perfectly Symmetric *lac* Operator Binds *lac* Repressor Very Tightly, J. Sadler, H. Sasmor, J. Betz, Proc. Natl. Acad. Sci. USA *80,* 6785–6789 (1983).

Dual Promoter Control of the *Escherichia coli* Lactose operon T. Malon, W. McClure, Cell *39,* 173–180 (1984).

Induction and the *ara* Operon

M olecular biologists were greatly surprised to discover that nature regulates gene activity in more than one way. Perhaps this conservatism was a carryover from physics since many of the early molecular biologists had been trained as physicists. In physics the near universality of laws had made research similar in spirit to searching for the Holy Grail. Consequently, once the lactose operon was characterized as negatively regulated, evidence that a different gene system might not be regulated similarly was received by some with hostility. Now, of course, we understand that nature uses a number of gene regulation mechanisms, among which are negative regulation, discussed in the previous chapter, and positive regulation, discussed in this chapter.

Genetic study of the arabinose operon of *Escherichia coli* was begun by Gross and Englesberg and pursued for many years by Englesberg. The work began as a straightforward genetic mapping exercise using phage P1. However, as data began to accumulate that indicated that the regulation mechanism might not be a simple variation of that used in the *lac* operon, Englesberg began a more intensive characterization of the system. To this day the mechanism of regulation of the arabinose operon is yet to be fully understood. Consequently, this chapter will explain how the genetic data indicated that the operon is positively regulated and then consider the *in vitro* experiments that proved that the system is positively regulated. Next, the unexpected findings that indicated that the *ara* system is also negatively regulated will be discussed. These led to the discovery of an "action at a distance" phenomenon, which is explained as the result of looping the DNA so as to bring together two sites that are separated by 200 base pairs.

THE SUGAR ARABINOSE AND ARABINOSE METABOLISM

The pentose L-arabinose occurs naturally in the walls of plant cells. The bacterium *E. coli*, but not humans, can use this sugar as a source of carbon and energy. Before arabinose can be metabolized by the intracellular arabinose enzymes, it must be transported from the growth medium through the inner membrane to the cytoplasm. This is done by two independent arabinose transport systems, products of the *araE* and *araFG* genes. Arabinose, like many sugars, exists in solution predominantly in a ring form in either of two conformations, which interconvert with a half-time of about 10 minutes (Fig. 13.1). Therefore it is possible that one of the transport systems is specific for the α anomeric form of arabinose and the other transports the β form. The β form is a substrate for the first enzyme of the catabolic pathway, but it is not known whether the other ring form or the liner form, which exists in only trace concentrations, might be the true inducer and play a role analogous to that of allolactose in the *lac* operon.

The duplication of arabinose uptake systems has hindered their genetic and physiological study because mutants are difficult to identify. A defect in either of the systems is masked by the activity of the other system. The genes coding for the enzymes required for arabinose catabolism, however, have been easier to map and study. They are located near the top of the genetic map as it is usually drawn (Fig. 13.2). Arabinose is first converted to L-ribulose by arabinose isomerase, the *araA*

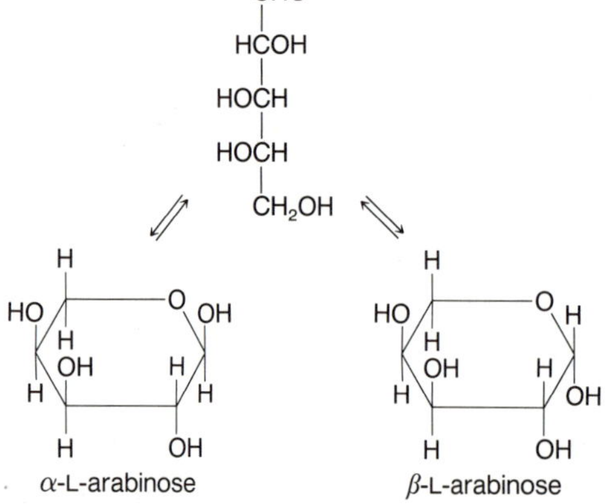

Figure 13.1   The open chain and pyranose ring forms of L-arabinose.

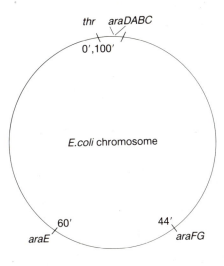

Figure 13.2   The locations on the *E. coli* chromosome of the genes specifically required for uptake and metabolism of L-arabinose.

gene product, then ribulose is phosphorylated by ribulokinase, the product of the *araB* gene, to yield L-ribulose-5-phosphate (Fig. 13.3). The ribulose phosphate is then converted to D-xylulose-5-phosphate by the *araD* gene product, ribulose phosphate epimerase. Xylulose phosphate enters the pentose phosphate shunt, and the enzymes subsequently involved are not induced specifically by the presence of arabinose.

The two arabinose active transport systems map at two different regions of the chromosome separate from the genes required for the metabolism. Therefore three sets of genes have been discovered whose activities are regulated by arabinose. Why

Figure 13.3   The path of catabolism of L-arabinose and the gene products catalyzing the conversions.

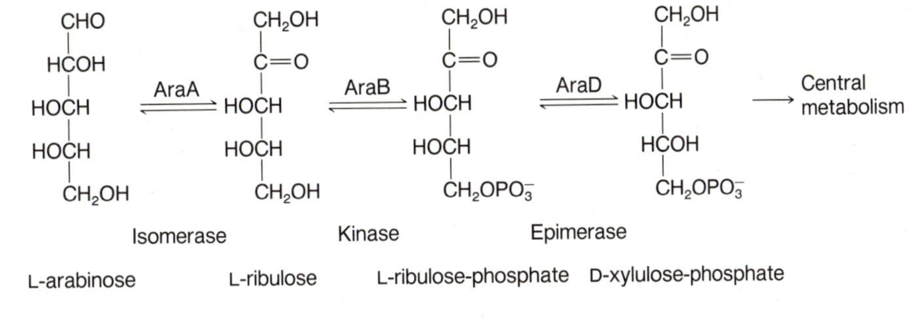

should this be? Differential regulation of the genes is one reason, and speed of induction is another reason not to have all the genes in a single long operon.

At low arabinose concentrations a high-affinity, but perhaps energy-inefficient or low-capacity, uptake system might be necessary. On the other hand, in the presence of high concentrations of arabinose, a different uptake system might be more useful. These different needs would necessitate splitting the corresponding genes into separate operons so that they could be differentially regulated. Alternatively, suppose arabinose were suddenly presented to bacteria, as it might be in the gut. It is to the cell's great advantage to begin metabolizing this new nutrient as soon as possible. However, if all the genes for the uptake and metabolism of arabinose were in a single long operon, then from the time induction of the arabinose operon began until RNA polymerase could reach the end of the operon would be about three minutes. Any strain that divided its arabinose operon into two or three separately transcribed units could induce all its *ara* enzymes more quickly and begin appreciable arabinose metabolism a minute or two sooner than cells with an undivided arabinose operon. The time saved in induction could be of enormous selective value over evolutionary time scales because cells might have to induce as often as once per day. Presumably, then, a balance must be reached between the speed of enzyme induction and the burden of carrying extra nontranslated sequences such as promoters.

Induction of the arabinose operon requires the presence of cyclic AMP and the cyclic AMP receptor protein CRP. As in the *lac* operon, the main role of this protein could be to enable induction of the arabinose operon only in the absence of glucose. This would prevent the cell's attempting to utilize arabinose when the better carbon source glucose is present.

GENETICS OF THE ARABINOSE SYSTEM

The phenotypes resulting from most mutations in the arabinose operon are not exceptional. The *araB, araA,* and *araD* genes lie in one transcriptional unit served by promoter p_{BAD} (Fig. 13.4). Mutations in the genes act as expected: They abolish the activity of the enzyme in question and leave the cells arabinose-negative, that is, unable to metabolize arabinose. However, *araD* mutations have the useful property that they make the cells sensitive to the presence of arabinose. This results from such cells' accumulation of ribulose phosphate. This type of sensitivity is not an isolated case, for high levels of many sugar phosphates in many types of cells are toxic or growth inhibitory.

The arabinose sensitivity of AraD$^-$ cells provides a simple way to isolate mutations in the arabinose genes. AraD$^-$ mutants that have further mutated to become resistant to arabinose contain mutations preventing the accumulation of ribulose phosphate. These are isolated by spreading large numbers of AraD$^-$ cells onto plates containing arabinose and some other source of carbon and energy, such as glycerol or yeast extract plus tryptone. Each cell capable of growing into a colony must

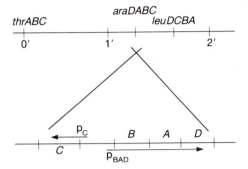

Figure 13.4   Fine structure of the *E. coli* genome in the region of the *araBAD* genes, the *ara* promoters, and the directions of transcription of the genes. In the top the genes are drawn in the conventional orientation of the *E. coli* genetic map, and in the bottom the arabinose genes have been inverted to conform with their conventional usage, in which the major transcription is rightward.

contain a secondary mutation in the arabinose operon in addition to the AraD⁻ mutation. This scheme does not yield transport-negative mutants since both transport systems would have to be inactivated before the cells would be arabinose-resistant.

The behavior of mutations in one of the arabinose genes was paradoxical. These mutations made the cells arabinose-negative, but no enzymatic activity could be associated with the gene. In fact, none seemed necessary because the other arabinose gene products performed all the required metabolic conversions. Additionally, cells with mutations in this gene, *araC*, had the strange property of not possessing any of the arabinose-induced enzymes or active transport systems (Table 13.1). The *araC* product clearly was a protein, however, because nonsense mutations existed in the *araC* gene.

Formally, the behavior of the *araC* mutants was consistent with several regulatory mechanisms. First, contrary to expectations prevalent at the time of this work, *araC* could code for a positive regulator, one whose presence is necessary for synthe-

Table 13.1   Induction of the Arabinose Operons

Genotype	Isomerase Levels		Transport Genes	
	Basal	Induced	Basal	Induced
C^+	1	1000	1	150
C^-	1	1	1	1
C^+/C^-	1	1000	1	150
$F'C^+A^-/C^-A^+$	1	1000	1	150

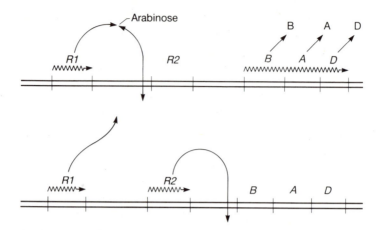

Figure 13.5   A double-negative regulation system in which R1 looks like an activator in genetic and physiological tests.

sis of the other *ara* gene products. Second, *araC* could code for a component directly involved with the uptake of arabinose into cells. Its absence would mean that the intracellular level of arabinose would never become high enough to permit induction of the arabinose genes. Third, *araC* could be part of a double-negative regulation system (Fig. 13.5). That is, *araC* protein in the presence of arabinose could repress synthesis of the true repressor of the arabinose operon. In this case, if *araC* were inactive or arabinose were absent, then the repressor would be synthesized and the arabinose enzymes would not be synthesized.

araC PROTEIN REGULATES POSITIVELY: DETECTION AND ISOLATION

No matter how clever, genetic and physiological experiments rarely seem able to provide rigorous proofs of proposals for mechanisms of gene regulation. Proof of a model often requires purification of the system's individual components and *in vitro* reconstruction of the system. How then could *araC* protein be purified for the biochemical demonstration that it is a positive regulator as well as for biochemical studies of its mechanism of action? Detection of the *lac* repressor was difficult enough, and repressor was known to bind inducer tightly. Not even this handle was available for detection of *araC* protein. *In vivo* experiments measuring the induction level of arabinose enzymes in cells deficient in arabinose transport as a function of arabinose concentration had shown that the affinity of *araC* protein for arabinose was too low to permit its detection by binding to arabinose.

Many different detection schemes for *araC* protein were tried, but the one that worked was based on the protein's only known activity, that of inducing synthesis of the arabinose enzymes. Working with the *lac* operon, Zubay had developed a means of preparing a partially fractionated cell lysate in which added DNA could be transcribed and translated. When a concentrated source of *lacZ* DNA is added to this system, *lac* mRNA is synthesized and translated into active β-galactosidase. This system was adapted to the arabinose genes to search for the *araC* protein.

The necessary coupled transcription-translation system consists of a cell extract from which much cellular DNA and all mRNA has been removed (Fig. 13.6). Then DNA template, salts, amino acids, and enzyme cofactors are added. The cell extracts must be made from *araB* mutants so that the small quantities of the *araB* enzyme, ribulokinase, which will be synthesized *in vitro*, can be detected. Also, the extracts must be free of *araC* protein so that it can be detected on its addition to the extracts. Of course, it is sensible that the source of *araC* protein be as concentrated as possible as well as lack ribulokinase. Finally, as concentrated a source of *araB* DNA as possible must be added to the synthesis extract.

It was possible to meet the various requirements of the *in vitro* system. The extract for *in vitro* synthesis was prepared from cells deleted for the *araCBAD* genes. The source of *araC* protein was the ΔaraCBAD cells infected with a phage carrying the *araC* gene but not an active *araB* gene. The source of *araB* DNA was a phage carrying the arabinose genes. Of course, similar experiments done at the present time would utilize plasmids carrying the desired genes. When the ingredients were mixed, *araB* gene product was synthesized only if both arabinose and *araC*-containing extract were added to the reaction! This result rigorously excludes the possibility that *araC* protein acts positively in induction via active transport of arabinose into cells. It also makes the possibility of a double repressor regulation system exceedingly remote.

Figure 13.6   Schematic of the coupled transcription-translation system.

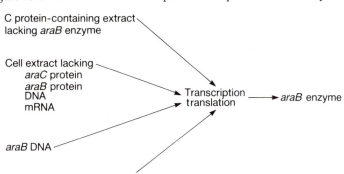

Of course, the last word on the positive nature of the regulation system came with a completely pure *in vitro* transcription system consisting of just cyclic AMP receptor protein, RNA polymerase, *ara* DNA, and *araC* protein. *Ara*-specific messenger is synthesized if and only if these components and arabinose are present in the reaction. The fact that this system works also shows that the regulation is exerted at the level of initiation of transcription rather than via degradation of mRNA or even at a translational level.

Although the assay was cumbersome, the coupled transcription-translation system permitted quantitation of *araC* protein in fractions obtained from purification steps. Micrograms of the protein could be purified. Only when the *araC* gene had been fused to the *lacZ* promoter and placed on a high-copy-number plasmid, as described in Chapter 11, did the level of *araC* protein reach values permitting straightforward purification. Now it is possible in just a few days to purify 50 milligrams of the protein so that physical experiments examining the basis of the protein's activities can be done.

araC PROTEIN REPRESSES AS WELL AS INDUCES

In the presence of arabinose, *araC* protein induces expression of the metabolic and active transport genes. That is, it is a positive regulator. Most unexpectedly, *araC* protein also appears to repress expression of these genes. Three types of experiments demonstrate the repression exerted by *araC* protein. The simplest utilizes Ara$^+$ revertants that are isolated in strains deleted of the *araC* gene. These mutations are called I^c, they lie in the RNA polymerase binding site, and they permit a low rate of polymerase binding and initiation in the absence of *araC* protein. Repression is revealed by the fact that the constitutive promoter activity of I^c mutants is reduced by the presence of *araC* protein. The protein can act in *trans* to repress just like the *lac* repressor.

Experiments with the I^c mutations also show that at least part of the site required for repression of p$_{BAD}$ lies upstream from all of the sites required for induction. This is shown by the properties of strains containing the two deletions, Δ1 or Δ2. Δ1 extends just to the end of the *araC* gene and Δ2 ends beyond the *araC* gene, so that half of the region between the *araC* and *araBAD* genes has been deleted. The promoter p$_{BAD}$ in both of the strains is undamaged by the deletions because it remains fully inducible

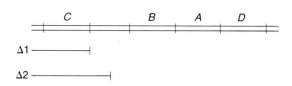

Table 13.2   Repression of I^c
Depends on an Upstream Site

Episome	Chromosome	
	$\Delta 1\ I^c$	$\Delta 2\ I^c$
No C gene	30	30
F'C$^+\Delta$BAD	1	30

when *araC* protein is provided in *trans*. Repression of the I^c mutation by *araC* protein occurs only in $\Delta 1$ (Table 13.2). Therefore at least part of the site required for repression has been deleted by $\Delta 2$.

The repression exerted by *araC* might conceivably occur only in the I^c mutants. Therefore experiments will be described that do not use the I^c mutations. The experiments use the same two deletions as described above, $\Delta 1$ and $\Delta 2$. AraC$^+\Delta$ (BAD) episomes are introduced into the deletion strains. When these cells are grown in the absence of arabinose, the basal level in the strain containing $\Delta 1$ is normal but the basal level in the strain containing $\Delta 2$ is up to 30 times normal. This result confirms the repression effect and also shows that in the absence of arabinose, a little of the *araC* protein is in the inducing state and can weakly induce p_{BAD} if the region defined by $\Delta 2$ has been deleted.

A third demonstration of repression in the *ara* system utilizes *araC* constitutive mutations, *araC*^{c}. This type of mutation causes the arabinose enzymes to be induced even in the absence of the normal inducer, L-arabinose. Diploids containing both *araC*^{c} and *araC*$^+$ mutations are surprising, for the C$^+$ allele is dominant to the C^c allele (Table 13.3). The diploids possess nearly the normal *uninduced* level of arabinose enzymes in the absence of arabinose and nearly the fully induced level of enzymes in the presence of arabinose. In light of the other experiments showing repression, these results are most simply explained as resulting from repression by the C$^+$ protein despite the presence of the C^c protein. However, these results are also consistent with *araC* protein being an oligomer in which the *in vivo* dominance of C$^+$ results from subunit mixing.

The *araC*^{c} mutants mentioned above are easily isolated with the aid of the arabinose analog 5-methyl-L-arabinose, which is also known as D-fucose. D-fucose cannot be metabolized by *E. coli*, but it does interact with *araC* protein to inhibit the normal induction by arabinose. Mutants able to grow on arabinose in the presence of fucose are called *araC*^{c}.

Table 13.3   Dominance of C$^+$ to C^c

Genotype	Expression of BAD Genes
C$^+$	1
C^c	1000
C^c/C$^+$	50

The results described above indicate that *araC* protein can induce or repress the initiation of transcription of the arabinose operons in *E. coli*. *AraC* protein must

H
HO ———O
H
OH   H H,OH
H
H   OH
L-arabinose

CH$_3$
HO ———O
H
OH   H H,OH
H
H   OH
D-fucose

therefore exist in at least two states, a repressing and an inducing state, and arabinose must drive the population of *araC* protein molecules in a cell toward the inducing state (Fig. 13.7).

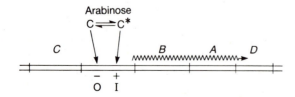

Figure 13.7   The presence of arabinose drives *araC* protein into the inducing state, which acts at a site downstream from the repression site at which uncomplexed *araC* protein acts.

THE PROMOTER FOR SYNTHESIS OF *araC* PROTEIN IS ALSO REGULATED

The *araC* gene is oriented oppositely to the *araBAD* genes, and its promoter, p_C, is adjacent to the regulatory sites for the BAD operon. This location raises the possibility that the activity of p_C might be regulated by the same proteins that regulate the activity of p_{BAD}.

Since the measurement of *araC* protein is difficult, lengthy, and imprecise, Casadaban chose not to study regulation of the p_C promoter by measuring *araC* protein itself but instead to fuse the promoter to a much more easily measured enzyme, β-galactosidase, and to measure it (Fig. 13.8). Once again, modern techniques would facilitate a more direct approach to the problem. Now the Berk-Sharp S1 nuclease mapping method permits a straightforward characterization of the promoter activity under a variety of conditions. Through a series of intricate genetic operations the β-galactosidase gene was brought near the *araC* gene and then intervening DNA was deleted so that the structural gene of β-galactosidase was fused to p_C. Hence the measurement of β-galactosidase became a measurement of the activity of p_C.

The first finding was that p_C is stimulated about threefold by the presence of cyclic AMP-CRP. A more surprising finding was that p_C is about six times more active in the absence of *araC* protein than in its presence. That is, *araC* protein represses its own synthesis. A third finding was that on the addition of arabinose to cells, the level of *araC* messenger increases about fourfold in several minutes and then slowly falls back to its prior level (Fig. 13.9). The mechanism for this transient derepression is not known.

BINDING SITES OF THE *ara* REGULATORY PROTEINS

Now that the regulatory phenomena of the *ara* operon have been laid out, what can be said about its mechanism of regulation? The nucleotide sequences of the promoters p_{BAD}, p_E, p_{FG}, and another *ara*-responsive promoter called p_H show three

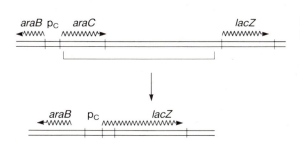

Figure 13.8   Deletion of intervening DNA fuses p_C to a transposed *lacZ* gene.

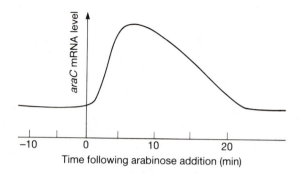

Figure 13.9   Transient derepression of *araC* messenger level following the addition of arabinose to growing cells.

distinct regions of homology among themselves. One of these is the RNA polymerase binding site. Another is the cyclic AMP receptor protein binding site, and the third is the *araC* protein binding site. Inexplicably, the *araBAD* RNA polymerase binding site is quite similar to the RNA polymerase sites of promoters that do not require auxiliary proteins for activity. Since p_{BAD} requires the auxiliary proteins *araC* and CRP for its activity, the polymerase binding sequence might have been expected to be markedly different from the consensus RNA polymerase binding sequence.

The identity of the protein binding sites was established by DNase footprinting. In *araBAD* the *araC* protein binding site is just upstream from RNA polymerase, and the CRP binding site is just upstream from that of *araC* (Fig. 13.10). The *araC* binding site is called *araI* for induction. These three sites are required for induction of p_{BAD}.

A second *araC* protein binding site called $araO_1$ lies another 60 nucleotides upstream of *araI*. This site overlaps the RNA polymerase binding site of p_C, and since occupancy of $araO_1$ by *araC* protein sterically interferes with the binding of RNA polymerase at p_C, $araO_1$ is an operator of p_C. $AraO_1$ is not directly involved with

Figure 13.10   Schematic of the binding sites of *araC* protein, CRP, and RNA polymerase on the *araBAD*. $AraO_1$ is the operator and *araI* site required for induction.

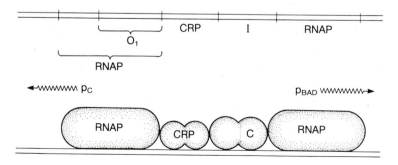

Table 13.4 p_C Activity as a Function
of C Level

Level of C Protein	Promoter Activity	
	p_C	p_{BAD}
No C	50	1
Wild type, C^+	10	1000
$2 \times$ Wild type	1	1000
$60 \times$ Wild type	0.1	1000

repression of p_{BAD}. A simple demonstration of this is the behavior of p_C and p_{BAD} as the level of *araC* protein is increased in a series of strains containing plasmids with *araC* fused to promoters of different strength. As the level of *araC* protein is increased, p_C shuts down but p_{BAD} remains fully inducible (Table 13.4).

Deletion analysis showed that one of the sites required for repression of p_{BAD} lies beyond $araO_1$. It is called $araO_2$, and it lies 270 base pairs upstream from the p_{BAD}

Position with respect to p_{BAD} transcription start

transcription start point. Footprinting has confirmed that *araC* protein binds to this site, and mutation analysis has shown that a single nucleotide change in this region can eliminate repression of p_{BAD}. Deletion of this site is the reason that the original deletion that was isolated by Englesberg could no longer repress p_{BAD} but could induce. His deletion extended through $araO_2$ into $araO_1$, and during the several-year interval between the discovery of O_1 and the discovery of O_2, $araO_1$ was thought to be required for repression of p_{BAD}.

The location of the CRP site required for stimulation of the p_{BAD} promoter is most surprising. In the *lac* operon, the CRP site lies at positions -48 to -78 with respect to the start of transcription, just next to RNA polymerase. In *ara*, the *araC* protein occupies this site and CRP is beside *araC*, in positions -80 to -110. If CRP makes specific contacts with RNA polymerase in functioning on the *lac* promoter, does it make a different set of specific contacts with *araC* protein, and does *araC* protein make yet another set of contacts with RNA polymerase? Most likely not, and therefore specific contacts between CRP and RNA polymerase are not the basis of the activity of CRP in all operons. *AraC* protein is interposed between CRP and RNA polymerase in the p_{BAD} operon. However, CRP is adjacent to RNA polymerase on p_C, and CRP functions to stimulate p_C as well as p_{BAD} from this site.

DNA LOOPS ARE INVOLVED IN REPRESSION OF *araBAD*

How can repression of p_{BAD} be generated from the $araO_2$ site that is located more than 200 base pairs upstream? There are three possibilities (Fig. 13.11). A signal could be transmitted through the DNA, for example by changing the angle of the base tilt; something could polymerize along the DNA starting at $araO_2$ and finally cover I; or the DNA could bend back and bring the $araO_2$ region near p_{BAD}. An indirect demonstration of the looping is the effect of altering the number of base pairs between $araO_2$ and p_{BAD}. If five base pairs, half a helical turn, are added between these two sites, the ability to repress is greatly diminished, but if 11 base pairs are added, the ability to repress is restored. An addition of 15 base pairs eliminates repression, and a longer addition of 31 base pairs restores repression.

These results show that the absolute distance between the two sites is not important but that the angular orientation between the two is. This implies that the DNA loops back to form contacts between *araC* protein bound at $araO_2$ and some-

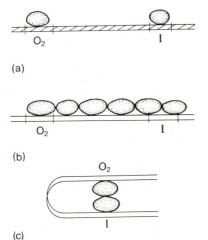

(a)

(b)

(c)

Figure 13.11   Three possible mechanisms for repression from a site several hundred nucleotides from a promoter. (a) Properties of the DNA can be altered, (b) protein can polymerize along the DNA, or (c) loops can form.

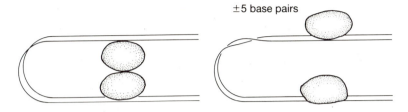

Figure 13.12   How a five-base-pair insertion can introduce half a helical turn and prevent formation of a loop.

thing in the promoter region. If the protein at $araO_2$ is on the opposite side of the DNA (Fig. 13.12), the additional energy required to twist the DNA half a turn is too great and the loop required for repression does not form. The loop appears to form between $araC$ bound at $araO_2$ and $araC$ bound at $araI$ because some point mutations in $araI$ eliminate repression.

Two steps are required for induction of the arabinose operon. These are reversing the repression generated by the DNA loop and then initiating RNA polymerase transcription. *In vivo* and *in vitro* experiments have shown that overcoming repression requires both CRP and *araC* protein plus arabinose. Inducing requires only *araC* protein plus arabinose because the operon can be induced in the absence of CRP if the $araO_2$ site has been deleted.

ELECTROPHORESIS TO ASSAY LOW AFFINITY DNA-BINDING PROTEINS

Biochemical experiments on the mechanism of *ara* regulation require assaying the binding of CRP, *araC* protein, and RNA polymerase to DNA. Although many proteins, such as *lac* repressor, bind to nitrocellulose filter paper, some, like *araC* protein, release DNA as they bind. Quantitation of such protein's DNA binding can be performed by DNase footprinting and similar protection experiments, but these methods are awkward and difficult, so a different and convenient assay of DNA binding is required. Crothers's and Revzin's laboratories developed an electrophoretic assay of DNA binding in their work with *lac* repressor and CRP that is applicable to many DNA-binding proteins. (This assay has the additional virtue that a protein need not be mostly pure before it can be assayed. The filter-binding assay has such a requirement; see Problem 12.12.)

The gel electrophoresis assay of DNA-binding proteins involves binding the protein to a DNA molecule of length 100 to 1000 base pairs and electrophoresis through an acrylamide or agarose gel. A protein molecule remains bound to a DNA

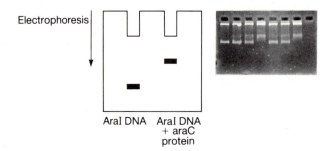

Figure 13.13   An acrylamide gel after electrophoresis of a 440-base-pair DNA fragment of the *ara* regulatory region in the first lane and the fragment plus increasing quantities of *araC* protein in the second through fourth lanes and fifth through seventh lanes.

molecule during the electrophoresis run and retards the migration rate of the DNA. Therefore DNA without bound protein migrates at one rate and forms a band, and DNA with the protein bound migrates at another rate and forms a different band on the gel (Fig. 13.13). The binding of two or even three proteins to a DNA molecule can be quantitated because additional binding retards the migration rate still more. By quantitating the amounts of the two bands, the fraction of the DNA with a bound protein can be determined. Either nonradioactive DNA can be used with ethidium bromide fluorescence to locate and quantitate the DNA bands or radioactive DNA can be used with autoradiography.

The gel electrophoresis assay requires that the protein remain bound to the DNA during the entire electrophoresis run. The fact that many DNA-binding proteins remain on the DNA less than one minute in solution does not generate a problem, for two reasons. First, as mentioned earlier, many proteins displace cations from the DNA as they bind. The dissociation of protein from the DNA then requires that cations bind to the DNA in place of the protein. However, if the electrophoresis is performed in the presence of very low concentrations of ions, this replacement of the cations on the DNA is infrequent and the protein remains bound to the DNA much longer than it would in normal buffers.

The second reason for a prolonged lifetime of protein-DNA complexes in the gel results from the fact that the dissociation constant equals the ratio of the rate of binding and the rate of dissociation (Fig. 13.14). Assume that the equilibrium constant for the binding of protein to DNA is not greatly altered by the presence of the gel. However, the gel greatly reduces the forward-rate constant because the gel interferes with the diffusion of the protein to its binding site. If the forward-rate constant is much reduced, the reverse-rate constant must also be greatly reduced so that the equilibrium constant remains unaltered. Consequently, proteins that would dissociate in less than 30 seconds in solution remain bound for more than half an hour in the gel.

$$P + DNA \underset{k_{-1}}{\overset{k_1}{\rightleftharpoons}} P \cdot DNA$$

At equilibrium, $\dfrac{d}{dt}([P \cdot DNA]) = 0$

$\dfrac{d}{dt}(P \cdot DNA) = k_1 \times [P] \times [DNA] - k_{-1}[P \cdot DNA] = 0$

or $\dfrac{[P] \times [DNA]}{[P \cdot DNA]} = \dfrac{k_{-1}}{k_1} = K_D$

Figure 13.14   The relation between the forward-rate and reverse-rate constants of a reaction and the equilibrium constant.

The gel electrophoresis assay works well for *araC* protein and has been used to learn the buffer conditions that maximize the stability of this labile protein and then to determine the association and dissociation rates and examine the effects of CRP and *araC* protein on the binding of RNA polymerase. The assay permitted a useful variation of the footprinting class of experiments.

Figure 13.15   Schematic of a chemical modification experiment. DNA labeled on one end is lightly modified on bases or phosphates, protein is able to bind to DNA molecules not containing modifications in critical sites, DNA with and without bound protein is separated, DNA is cleaved at the sites of modification and subjected to electrophoresis on a DNA sequencing gel.

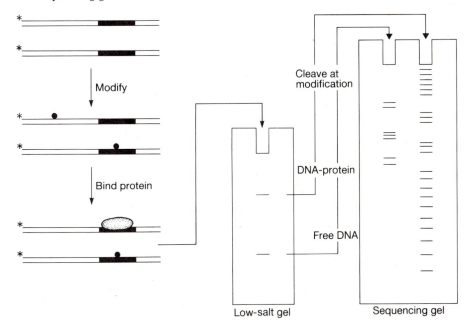

Modify

Bind protein

Cleave at modification

DNA-protein

Free DNA

Low-salt gel   Sequencing gel

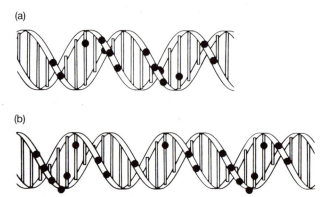

Figure 13.16   Guanines and phosphates where chemical modification interferes with protein binding *araC* protein or lambda repressor binding. (a) The *araI* site; (b) an operator of phage lambda.

In footprinting the protein is bound to DNA and the sites protected from DNase or dimethylsulfate are determined. The experiment can also be performed in reverse. The DNA can be modified and then the protein added and the DNA separated into two classes (Fig. 13.15). These are the molecules for which the modifications do not lie in essential regions and which therefore bind *araC* protein and the molecules that have modifications in essential regions and that prevent binding of *araC* protein. The two classes of modified DNA molecules are easily separated with the gel electrophoresis assay, and then the sites that were modified can be determined by cleaving at those sites, as is done in DNA sequencing and then electrophoresing on a sequencing gel. This modify-and-bind approach is particularly useful for determining the points of close protein-phosphate approach because the conditions required for modification of the phosphate inactivate most proteins, and modification therefore must be done in the absence of the protein.

The chemical modification experiments with *araC* protein showed an unexpected finding. The protein makes contacts with guanines in three adjacent major grooves instead of the two grooves as seen with *lac* repressor, phage repressors, and CRP protein (Fig. 13.16). *AraC* protein must therefore use a different DNA binding motif than these other proteins.

WHY POSITIVE REGULATORS ARE A GOOD IDEA

A positive regulation mechanism makes good biological sense because it can easily be adapted to permit rapid accumulation of enzymes following induction. This ability ought to be particularly valuable to an organism that must live in a changing environment. If cells of one strain can adapt to the presence of a new nutrient a little

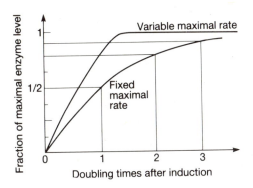

Figure 13.17   Induction kinetics of an enzyme synthesized under control of a promoter that turns on at a fixed rate and under control of a promoter that turns on at a high initial rate and then reduces to a lower rate.

more quickly than their competitors, then the strain may survive better on an evolutionary time scale.

Suppose that, on induction, a promoter turns fully on and stays on at that rate thereafter. Then, as mentioned in Chapter 1, one doubling time later the cells will possess half as much of the induced enzymes per cell mass as they would have possessed had the promoter been induced for many generations (Fig. 13.17). This situation is less than optimal, for one of two reasons. Either the submaximal specific activities that exist for the first doubling time or two after induction partially limit the growth rate of the cells or, if such submaximal specific activities are sufficient for maximum growth, then after four or five doubling times of induction, an excess of enzyme is present and the cells are wasting their resources synthesizing this excess.

For purposes of rapid adaptation to a changing environment, it is more sensible to turn the promoter on to a high rate at the beginning and then, as the enzyme level per cell reaches an optimum value, to turn the promoter rate down to a value just sufficient to maintain this value. The arabinose operon is capable of doing this. The apparent induction of p_C resulting from the addition of arabinose approximately doubles the intracellular level of $araC$ protein in 20 minutes, but after two hours the level of $araC$ protein per cell has been diluted back to its preinduction value. If the activity of p_{BAD} were not fully saturated by the normal level of $araC$ protein, the temporary increase in $araC$ protein immediately following arabinose induction would be reflected in a temporary increase in the activity of the promoter—just what is required to bring the catabolic enzymes rapidly to optimum intracellular concentrations. CRP in the lac operon could serve the same function, for the metabolism of lactose ought to create partial catabolite repression and reduce the activity of p_{lac} after the metabolism of lactose has begun.

Another reason for the existence of positive regulatory systems applies with greater strength to higher cells that have several hundred thousand genes to regulate. Let us use the lac regulatory system as the basis for comparison. At any instant, more than half of the lac repressor in a bacterial cell is nonspecifically bound at nonoperator sites. If the cell contained a thousand times as much DNA, then at least a thousand times as much *lac* repressor would be required to maintain full repression of the *lac* operon. If the cell contained 100,000 operons regulated by repressors with similar characteristics, then the nucleus could not accommodate the total amount of regulatory proteins required to regulate these operons. Of course, it may be possible for a repressor to possess greater selectivity than *lac* repressor for its operator site, but there has to be an upper limit.

Positive regulatory systems do not require high levels of repressor to maintain low basal expression rates since a positively regulated promoter is naturally off. For many genes, the occupancy of only a fraction of the promoter regions by positive regulators may be sufficient to provide sufficient activity of the promoter when it must be turned on.

PROBLEMS

13.1. How could subunit mixing explain the dominance of C^+ to C^c without requiring that C^+ protein be a true repressor?

13.2. Frequently it is claimed that phosphorylation of sugars prevents their leakage out of cells. If this were the case, it would seem reasonable to phosphorylate arabinose before the other enzymatic conversions. Is there a biochemical reason for not phosphorylating arabinose before the isomerization step?

13.3. Suppose that two different missense mutations in the arabinose isomerase were found to complement one another. How could this be possible? Suppose that a point mutation in the *araC* gene could nonetheless yield Ara$^+$ cells when a particular CRP$^-$ mutation was introduced into the cell. What could you infer about the functioning of the system, and what could you conclude definitely?

13.4. What experiment would be the best to determine whether α- or β-L-arabinose is actively transported by cells? What value would it be to further research to know which it is?

13.5. Why is it of interest to know whether AraI^C promoters can be repressed in strains lacking active CRP?

13.6. Suppose that dominant Ara$^-$ mutations are very rare but do exist. Devise an efficient selection method for their isolation.

13.7. The PaJaMa experiment, named for Pardee, Jacob, and Monod, consists of mating F'I$^+$Z$^+$ episomes into F$^-$ cells deleted of the *lac* region. Immediately after entry of the episome into the female, synthesis of *lacZ* protein begins, but its synthesis is turned off after about 30 minutes because of the accumulation of *lac* repressor. What could a PaJaMa experiment tell about the arabinose operon and what strains would be needed?

13.8. Devise an efficient method for the accurate mapping of C^c mutations.

13.9. "Localized folding" is the tendency for a pair of amino acids that lie near each other in the primary sequence to be near each other in the three-dimensional structure of the properly folded protein. A proper definition and some further examples are given in "Nucleation, Rapid Folding, and Globular Interchain Regions in Proteins," D. Wetlaufer, Proc. Natl. Acad. Sci. USA *70*, 697–701 (1973). The arabinose binding site of the C protein and the DNA-binding site(s) of the *araC* protein might be constructed from amino acids from only restricted portions of the C protein polypeptide. Devise genetic experiments to test this prediction.

13.10. The claim has been made that the existence of temperature-sensitive mutations in the *araC* gene proves that the gene product is a protein. Why is such reasoning faulty?

13.11. In the paper Purification of the *araC* Protein, G. Wilcox, K. Clemetson, D. Santi, E. Englesberg, Proc. Natl. Acad. Sci. USA *68*, 2145–2148 (1971), the claim is made that *araC* protein is detected and purified by affinity chromatography.

a) How could identification of the DNA-binding activity as *araC* protein be in error?

b) What experiments could provide definitive proof as to the identity of the DNA-binding protein and *araC* protein?

c) Data in the paper show that 1 μg of "purified" C protein containing approximately 0.25 μg of "C band" bound approximately 0.01 μg of DNA to filters. The molecular weight of lambda DNA is approximately 30×10^6 and of *araC* protein is approximately 60,000. What fraction of the 1 μg in fact was active in binding DNA?

d) What do you say about identifying any band on gel electrophoresis as *araC* protein under these conditions?

e) If there were 20 molecules of C protein per typical cell containing 2×10^{-13} g of protein, what fraction of total protein was C protein? Was any purification accomplished at all?

*13.12. Knowledge of the affinity of arabinose for the *araC* protein is important for the proper execution of experiments, and a number of determinations of this constant have been made. Most of these measurements had to be made indirectly because insufficient *araC* protein was available for a direct measurement. Therefore the measurements have observed the ability of the *araC* protein to induce synthesis of the arabinose enzymes as a function of the arabinose concentration.

a) Show why a measurement of an apparent K_m in these systems could yield a "binding constant" indicating a much higher affinity of arabinose or *araC* protein than actually is the case. The published data indicate a broad spread of measured values. The first published data, J. Mol. Biol. *46*, 197–199 (1969), yielded a high nonlinear induction curve and a K_m of about 10^{-4}. Although this provided the needed answer—that searching for *araC* protein by equilibrium dialysis was pointless—the data were flawed by the undetected presence of an active transport system. A later paper in which similar measurements are reported, J. Bact. *110*, 56–65 (1972), presents a linear induction dependence on arabinose and a K_m of about 6 mM.

b) What must be the matter with the first paper? When it became possible to perform similar experiments *in vitro*, the first K_m measured was about 3 mM, Nature

New Biol. *233*, 166–170 (1971). However, later *in vitro* measurements, J. Biol. Chem. *249*, 2946–2952 (1974), yielded a K_m of about 20 mM.

c) Which of the two *in vitro* measurements is more likely to be incorrect and why?

d) Comment on the *in vivo* measurement of $K_m = 60$ mM, J. Mol. Biol. *80*, 433–444 (1973). You may be interested to know that direct physical measurements using the electron microscope also support the value of 60 mM.

13.13. How could you test the possibility that the role of CRP in induction of p_{BAD} is to help break repression? Does this make biological sense?

13.14. Design an experiment to show whether or not the turndown in the activity of p_C 10 minutes after the addition of arabinose is due to catabolite repression generated by the metabolism of arabinose.

*13.15. Cells growing on either a rich broth medium or on glucose minimal medium were treated with two mutagens. Each of the resulting rifamycin-resistant mutants was then tested for its Ara phenotype. What is an explanation for the appearance of Rif^r-Ara^- double mutants only when cells were treated with one type of mutagen and then only when they had grown on rich medium?

13.16. The maltose operon is regulated in a purely positive fashion by the *malT* product. Predict the consequences on regulation of the *mal* operon of a vast excess of *malT* protein in the cell.

13.17. *Lac* UV5 is a mutation that makes activity of the *lac* promoter independent of cyclic AMP receptor protein. Suppose promoter minus mutants of *lac* UV5 occur at a frequency of 10^{-8}. Suppose also that promoter minus mutations in the arabinose operon occurred only at a frequency of 10^{-10}. Not only that, but on analysis, all of these turn out to be little insertions or deletions, whereas the *lac* mutations were primarily base changes. What would be the best interpretation of these data?

RECOMMENDED READINGS

The L-arabinose Operon in *Escherichia coli* B/r: A Genetic Demonstration of Two Functional States of the Product of a Regulator Gene, E. Englesberg, C. Squires, F. Meronk, Proc. Natl. Acad. Sci. USA *62*, 1100–1107 (1969).

Arabinose C Protein: Regulation of the Arabinose Operon *in vitro*, J. Greenblatt, R. Schleif, Nature New Biol. *233*, 166–170 (1971).

RELATED REVIEWS, BOOKS, AND ARTICLES

The Operon, ed. J. Miller, W. Reznikoff, Cold Spring Harbor Laboratory (1978).

DEEPER READING

Determination of the Order of Mutational Sites Governing L-arabinose Utilization in *E. coli* B/r by Transduction with Phage P1bt, J. Gross, E. Englesberg, Virology *9*, 314–331 (1959).

Direct Selection of L-arabinose Negative Mutants of *E. coli* Strain B/r, H. Boyer, E. Englesberg, R. Weinberg, Genetics *47*, 417–425 (1962).

Positive Control of Enzyme Synthesis by Gene C in the L-arabinose System, E. Englesberg, J. Irr, J. Power, N. Lee, J. Bact. *90*, 946–957 (1965).

The L-arabinose Permease System in *E. coli* B/r, C. Novotny, E. Englesberg, Biochim. et Biophys. Acta *117*, 217–230 (1966).

Further Evidence for Positive Control of the L-arabinose System by Gene *araC*, D. Sheppard, E. Englesberg, J. Mol. Biol. *25*, 443–454 (1967).

An Analysis of "Revertants" of a Deletion Mutant in the C Gene of the L-arabinose Complex in *E. coli*, B/r: Isolation of Initiator Constitutive Mutants (I^c), E. Englesberg, D. Sheppard, C. Squires, F. Meronk, J. Mol. Biol. *43*, 281–298 (1969).

Directed Transposition of the Arabinose Operon: A Technique for the Isolation of Specialized Transducing Bacteriophages for Any *E. coli* Gene, S. Gottesman, J. Beckwith, J. Mol. Biol. *44*, 117–127 (1969).

Induction of the L-arabinose Operon, R. Schleif, J. Mol. Biol. *46*, 197–199 (1969).

Arabinose-Leucine Deletion Mutants of *E. coli* B/r, D. Kessler, E. Englesberg, J. Bact. *98*, 1159–1169 (1969).

Selection of *araB* and *araC* Mutants of *E. coli* B/r by Resistance to Ribitol, L. Katz, J. Bact. *102*, 593–595 (1970).

Nonsense Mutants in the Regulator Gene *araC* of the L-arabinose System of *E. coli* B/r, J. Irr, E. Englesberg, Genetics *65*, 27–39 (1970).

Dual Control of Arabinose Genes on Transducing Phage Lambda d*ara*, R. Schleif, J. Greenblatt, R. Davis, J. Mol. Biol. *59*, 127–150 (1971).

L-arabinose Operon Messenger of *E. coli*, R. Schleif, J. Mol. Biol. *61*, 275–279 (1971).

Hyperinducibility as a Result of Mutation in Structural Genes and Self-Catabolite Repression in the *ara* Operon, L. Katz, E. Englesberg, J. Bact. *107*, 34–52 (1971).

Anomeric Specificity and Mechanism of Two Pentose Isomerases, K. Schray, I. Rose, Biochem. *10*, 1058–1062 (1971).

Initiator Constitutive Mutants of the L-arabinose Operon (OIBAD) of *E. coli*. B/r, L. Gielow, M. Largen, E. Englesberg, Genetics *69*, 289–302 (1971).

Fine Structure Deletion Map of the *E. coli* L-arabinose Operon, R. Schleif, Proc. Natl. Acad. Sci. USA *69*, 3479–3484 (1972).

Induction of the *ara* Operon of *E. coli* B/r, M. Doyle, C. Brown, R. Hogg, R. Helling, J. Bact. *110*, 56–65 (1972).

Different Cyclic AMP Requirements for Induction of the Arabinose and Lactose Operons of *E. coli*, J. Lis, R. Schleif, J. Mol. Biol. *79*, 149–162 (1973).

In vivo Experiments on the Mechanism of Action of L-arabinose C Gene Activator and Lactose Repressor, J. Hirsh, R. Schleif, J. Mol. Biol. *80*, 433–444 (1973).

Induction Kinetics of the L-arabinose Operon of *E. coli*, R. Schleif, W. Hess, S. Finkelstein, D. Ellis, J. Bact. *115*, 9–14 (1973).

Novel Mutation to Dominant Fucose Resistance in the L-arabinose Operon of *E. coli*, N. Nathanson, R. Schleif, J. Bact. *115*, 711–713 (1973).

Direction of Transcription of the Regulatory Gene *araC* in *E. coli* B/r, G. Wilcox, J. Boulter, N. Lee, Proc. Natl. Acad. Sci. USA *71*, 3635–3639 (1974).

The Arabinose C Gene Product of *E. coli* B/r Is Hyperlabile in a Cell-Free Protein Synthesis System, D. Steffen, R. Schleif, Mol. Gen. Genet. *128*, 93–94 (1974).

Fusion of the *E. coli lac* Genes to the *ara* Promoter: A General Technique Using Bacteriophage Mu-1 Insertions, M. Casadaban, Proc. Natl. Acad. Sci. USA *72*, 809–813 (1975).

The Isolation and Characterization of Plaque-Forming Arabinose Transducing Bacteriophage Lambda, J. Lis, R. Schleif, J. Mol. Biol. *95*, 395–407 (1975).

The Regulatory Region of the L-arabinose Operon: Its Isolation on a 1000 Base-Pair Fragment from DNA Heteroduplexes, J. Lis, R. Schleif, J. Mol. Biol. *95*, 409–416 (1975).

The Regulatory Region of the L-arabinose Operon: A Physical, Genetic and Physiological Study, R. Schleif, J. Lis, J. Mol. Biol. *95*, 417–431 (1975).

Paucity of Sites Mutable to Constitutivity in the *araC* Activator Gene of the L-arabinose Operon of *E. coli*, N. Nathanson, R. Schleif, J. Mol. Biol. *96*, 185–199 (1975).

The General Affinity of *lac* Repressor for *E. coli* DNA: Implications for Gene Regulation in Procaryotes and Eucaryotes, S. Lin, A. Riggs, Cell *4*, 107–111 (1975).

Regulation of the Regulatory Gene for the Arabinose Pathway *araC*, M. Casadaban, J. Mol. Biol. *104*, 557–566 (1976).

Electron Microscopy of Gene Regulation: The L-arabinose Operon, Jay Hirsh, Robert Schleif, Proc. Natl. Acad. Sci. USA *73*, 1518–1522 (1976).

Hypersensitivity to Catabolite Repression in the L-arabinose Operon of *E. coli* B/r Is trans Acting, D. Sheppard, M. Eleuterio, J. Bact. *126*, 1014–1016 (1976).

The *araC* Promoters: Transcription, Mapping, and Interaction with the *araBAD* Promoter, J. Hirsh, R. Schleif, Cell *11*, 545–550 (1977).

Overproducing *araC* Protein with Lambda-arabinose Transducing Phage, D. Steffen, R. Schleif, Mol. Gen. Genet. *157*, 333–339 (1977).

In vitro Construction of Plasmids Which Result in Overproduction of the Protein Product of the *araC* Gene of *E. coli*, D. Steffen, R. Schleif, Mol. Gen. Genet. *157*, 341–344 (1977).

Nucleotide Sequence of the L-arabinose Regulatory Region of *E. coli* K-12, B. Smith, R. Schleif, J. Biol. Chem. *253*, 6931–6933 (1978).

The *E. coli* L-arabinose Operon: Binding Sites of the Regulatory Proteins and a Mechanism of Positive and Negative Regulation, S. Ogden, D. Haggerty, C. M. Stoner, D. Kolodrubetz, R. Schleif, Proc. Natl. Acad. Sci. USA *77*, 3346–3350 (1980).

DNA Sequence of *araBAD* Promoter Mutants of *E. coli*, A. Horowitz, C. Morandi, G. Wilcox, J. Bact. *142*, 659–667 (1980).

Identification of *araC* Protein on Two-Dimensional Gels: Its *in vivo* Instability and Normal Level, D. Kolodrubetz, R. Schleif, J. Mol. Biol. *149*, 133–139 (1981).

Regulation of the L-arabinose Transport Operons in *Escherichia coli*, D. Kolodrubetz, R. Schleif, J. Mol. Biol. *151*, 215–227 (1981).

L-arabinose Transport Systems in *Escherichia coli* K-12, D. Kolodrubetz, R. Schleif, J. Bact. *148*, 472–479 (1981).

A Gel Electrophoresis Method for Quantifying the Binding of Proteins to Specific DNA Regions: Application to Components of the *Escherichia coli* Lactose Operon Regulatory System, M. Garner, A. Revzin, Nucleic Acids Res. *9*, 3047–3060 (1981).

Equilibria and Kinetics of *lac* Repressor Operator Interactions by Polyacrylamide Gel Electrophoresis, M. Fried, D. Crothers, Nucleic Acids Res. *9*, 6505–6525 (1981).

Is the Amino Acid Sequence but Not the Nucleotide Sequence of the *Escherichia coli araC* Gene Conserved? C. Stoner, R. Schleif, J. Mol. Biol. *154*, 649–652 (1982).

Arabinose-Inducible Promoter from *Escherichia coli*, Its Cloning from Chromosomal DNA, Identification as the *araFG* Promoter and Sequence, B. Kosiba, R. Schleif, J. Mol. Biol. *156,* 53–66 (1982).

Hyperproduction of *araC* Protein from *Escherichia coli*, R. Schleif, M. Favreau, Biochem. *21,* 778–782 (1982).

Transcription Start Site and Induction Kinetics of the *araC* Regulatory Gene in *Escherichia coli* K-12, C. Stoner, R. Schleif, J. Mol. Biol. *170,* 1049–1053 (1983).

The *araE* Low Affinity L-arabinose Transport Promoter, Cloning, Sequence, Transcription Start Site, and DNA Binding Sites of Regulatory Proteins, C. Stoner, R. Schleif, J. Mol. Biol. *171,* 369–381 (1983).

In vivo Regulation of the *Escherichia coli araC* Promoter, S. Hahn, R. Schleif, J. Bact. *155,* 593–600 (1983).

An *araBAD* Operator at −280 Base Pairs That Is Required for p_{BAD} Repression: Addition of DNA Helical Turns Between the Operator and Promoter Cyclically Hinders Repression, T. Dunn, S. Hahn, S. Ogden, R. Schleif, Proc. Nat. Acad. Sci. USA *81,* 5017–5020 (1984).

Regulation of the *Escherichia coli* L-arabinose Operon Studied by Gel Electrophoresis DNA Binding Assay, W. Hendrickson, R. Schleif, J. Mol. Biol. *174,* 611–628 (1984).

Upstream Repression and Catabolite Sensitivity of the *Escherichia coli* L-arabinose Operon, S. Hahn, T. Dunn, R. Schleif, J. Mol. Biol. *180,* 61–72 (1984).

Attenuation and the
trp Operon

In the previous two chapters we have seen examples of regulation of gene expression by control of RNA polymerase initiation frequency. The main subject of this chapter is yet another means that cells use to regulate transcription of some operons: controlling transcription termination at a special point after the mRNA has been initiated but before the RNA polymerase has entered the operon's structural genes. Either the RNA synthesis terminates and no messenger for the structural genes is synthesized or termination does not occur and functional messenger is synthesized. The major regulator of this RNA transcription attenuation mechanism is the secondary structure of the RNA itself. Not only does a similar mechanism operate in a number of other amino acid biosynthetic operons, but related mechanisms also appear to regulate ribosomal protein operons and other operons whose enzyme levels vary with growth rate.

No cells are continuously exposed to ideal growth conditions. Indeed, the majority of the time *Escherichia coli* cells are exposed to adverse conditions, and growth is slowed or stopped. However, from time to time, spurts of growth are possible when nutrients suddenly appear or when dense cells are diluted. Therefore cells must possess regulatory mechanisms that turn on and turn off at appropriate times the synthesis of enzymes like those of the tryptophan biosynthetic pathway. Three important cellular states should be considered:

1. Tryptophan is absent, but otherwise cells may be capable of synthesizing protein.

2. Tryptophan is absent, and no protein synthesis may be possible.

3. Tryptophan is present.

Only in the first state is it sensible for cells to synthesize *trp* messenger RNA. In both the second and third states it is energy-efficient for cells not to synthesize *trp* messenger RNA. The main lesson we will learn from examination of the *trp* operon is that an exceptionally clever regulatory mechanism has evolved that accomplishes these goals.

Although a number of other similarly regulated systems are now known, this chapter uses the *trp* operon of *E. coli* as an example. For completeness, we first briefly consider regulation of the aromatic amino acid pathway of which the tryptophan pathway is one part.

THE AROMATIC AMINO ACID SYNTHETIC PATHWAY AND ITS REGULATION

The first chemical reaction common to the synthesis of tryptophan, tyrosine, and phenylalanine is the condensation of erythrose-4-P and phosphoenolpyruvate to form 3-deoxy-D-arabino-heptulosonate-7-P, DAHP, a reaction catalyzed by DAHP synthetase (Fig. 14.1). Due to its position at the head of the aromatic acid biosynthetic pathway and to the fact that this reaction is irreversible, it is logical that this reaction should be a key point of regulation. Indeed, this expectation is met. *Escherichia coli* regulates both the amount and activity per molecule of DAHP synthetase as a function of the intracellular levels of the aromatic amino acids.

Figure 14.1   Outline of the aromatic amino acid biosynthetic pathway. Each arrow represents an enzymatic step. Tryptophan feedback inhibits one of the DAHP synthetases as well as the first enzyme of the pathway committed to tryptophan synthesis.

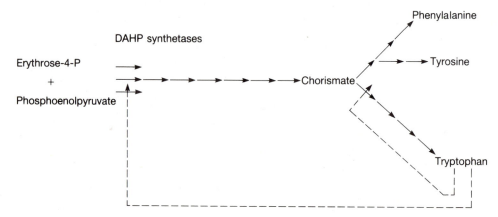

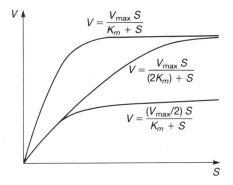

$$V = \frac{V_{max}\, S}{K_m + S}$$

$$V = \frac{V_{max}\, S}{(2K_m) + S}$$

$$V = \frac{(V_{max}/2)\, S}{K_m + S}$$

Figure 14.2   Two possibilities for feedback inhibition of an enzyme. The top curve is enzyme velocity as a function of substrate concentration, and the equation is the activity of an uninhibited enzyme as a function of its K_m, V_{max}, and substrate concentration S. The middle curve and equation are for an inhibited enzyme with a doubled K_m, and the bottom curve and equation are for an inhibited enzyme with a halved V_{max}.

A double regulation of total DAHP synthetase activity is logical. While regulation of the enzyme level minimizes unnecessary consumption of amino acids and energy in the synthesis of the enzyme, this type of regulation is incapable of producing appreciable changes in the enzyme levels or total enzyme activity on time scales shorter than minutes. A much more rapidly responding regulation mechanism is also necessary to adjust the synthesis rates of tryptophan, tyrosine, and phenylalanine on time scales of seconds.

In addition to stabilizing the activity of the synthetic pathway against random fluctuations, a rapidly responding regulation would fine-tune the biosynthetic flow rates of the aromatic amino acids and would be able to respond rapidly to growth rate changes generated by changes in the nutrient medium. Feedback inhibition of an enzyme's activity meets the requirements, as this mechanism can alter an enzyme's activity in milliseconds.

Feedback inhibition is an example of an allosteric interaction in which accumulation of the product of the pathway leads to inhibition of the activity of the enzyme: The enzyme is feedback-inhibited. This is a specific example of an allosteric interaction in which a molecule dissimilar in shape to the substrates of an enzyme can bind to the enzyme, usually at a site on the enzyme far from the active site, and can generate conformational changes that alter the catalytic activity of the enzyme. Feedback inhibition can reduce an enzyme's activity in either of two fundamental ways (Fig. 14.2). The tryptophan-sensitive DAHP synthetase is feedback-inhibited largely as a result of a change in its V_{max}, whereas the first enzyme of the pathway used solely for tryptophan synthesis, anthranilate synthetase, is an example of the other possibility. It is feedback-inhibited by tryptophan via a change in its K_m.

Bacillus subtilus possesses a single DAHP synthetase whose synthesis and activity is regulated. *Escherichia coli,* however, possesses three different DAHP synthetases. The activity of one, the *aroH* protein, is feedback-inhibited by tryptophan, another is feedback-inhibited by tyrosine, and the third is feedback-inhibited by phenylalanine. Only if the cell's growth medium possesses all three amino acids is all DAHP synthetase activity within a cell inhibited.

This is an example of the fact that different microorganisms possess different overall schemes for the regulation of tryptophan synthesis. On one hand, it is possible that the different evolutionary niches occupied by different microorganisms require these different schemes. On the other hand, perhaps one scheme is no better than another, and the different ones just happened to have evolved that way. Either way, this diversity means that the overall scheme for regulation of tryptophan synthesis in *E. coli* is not the only one that works.

RAPID INDUCTION CAPABILITIES OF THE *trp* OPERON: REPRESSION

The tryptophan operon consists of five genes — *trpE, D, C, B,* and *A* — that code for the enzymes unique to the synthesis of tryptophan (Fig. 14.3). In addition, the *trpR* gene codes for a repressor that helps regulate expression of these genes. *Trp* repressor bound to the *trp* operator blocks access of RNA polymerase to the promoter. Since this repressor binds to *trp* operator far better in the presence of its corepressor tryptophan than in its absence, the transcription of the *trp* genes is repressed in the presence of excess tryptophan and derepressed during times of tryptophan deficiency.

The classical repressor regulatory mechanism of the *trp* operon is adequate for meeting part of the regulatory requirements, as discussed in the first section of this chapter. However, repression may not adequately block *trp* messenger synthesis when protein synthesis is not possible. The next section begins discussion of the regulatory mechanisms that shut down *trp* messenger synthesis under these conditions.

Figure 14.3   The *trp* operon that codes for the enzymes necessary for the conversion of chorismate to tryptophan. Shown is the o, operator; p, promoter; *trpL,* leader region; and the positions within the operon of the five *trp* structural genes.

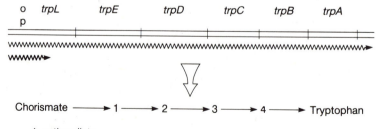

1. anthranilate
2. phosphoribosyl anthranilate
3. carboxyphenylamino-deoxyribulose-phosphate
4. indoleglycerol phosphate

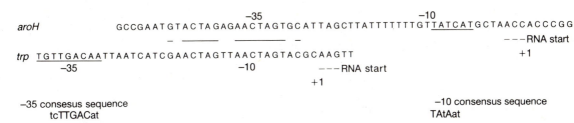

```
                                    -35                              -10
aroH        GCCGAATGTACTAGAGAACTAGTGCATTAGCTTATTTTTTTGTTATCATGCTAACCACCCGG
            _ _____  _____ _                                    ---RNA start
                                                                     +1
trp TGTTGACAATTAATCATCGAACTAGTTAACTAGTACGCAAGTT
     -35                          -10            ---RNA start
                                   +1
```

-35 consesus sequence   -10 consensus sequence
    tcTTGACat   TAtAat

Figure 14.4   DNA sequences of the *aroH* and *trp* operon RNA polymerase binding sites aligned according to the homologies between the *trpR* binding sites. Identical bases are shown with a bar. Shown below are the consensus −35 and −10 RNA polymerase sequences.

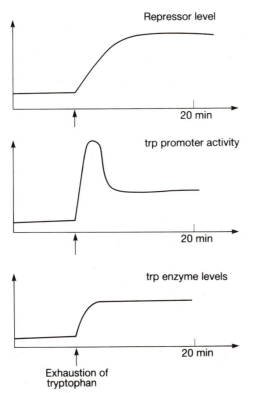

Repressor level

20 min

trp promoter activity

20 min

trp enzyme levels

20 min

Exhaustion of tryptophan

Figure 14.5   Repressor level, *trp* promoter activity, and *trp* enzyme levels following exhaustion of tryptophan from the medium.

The *trpR* repressor represses synthesis of the *aroH*, DAHP synthetase, messenger as well as the *trp* operon messenger by blocking binding of RNA polymerase to the two promoters in the presence of tryptophan. However, an interesting variation is shown between the two promoters. In the *trp* operon, the operator is centered around the -10 region of the RNA polymerase binding site, whereas in the *aroH* operon, the repressor binding site is located around the -35 region. Since both of these operators are quite similar, the -10 region of the *trp* promoter and the -35 region of the *aroH* promoter only weakly resemble the promoter consensus sequences for these two regions (Fig. 14.4). Apparently to compensate for this drastic alteration in part of the RNA polymerase binding site, the other portions of the RNA polymerase binding sites of these two promoters are highly homologous to RNA polymerase consensus promoter sequences for the two regions.

In addition to regulating the *trp* operon and *aroH*, trpR repressor in the presence of tryptophan also represses its own synthesis. As we will see, the consequences of this self-repression are that the cellular levels of the *trp* enzymes can rapidly increase to optimal levels following tryptophan starvation. The *lac* and *ara* operons can use positive-acting regulation mechanisms to generate a rapid induction response followed by a lower steady-state response, whereas the *trp* operon can do the same with only negative-acting elements.

How is the rapid enzyme induction accomplished? Consider cells growing in the presence of excess tryptophan. In such a case the *trp* operon, *trpR*, and *aroH* genes are all repressed. On tryptophan starvation, these genes are all derepressed, and the gene products are synthesized at a high rate (Fig. 14.5). However, as the intracellular concentration of *trpR* repressor and tryptophan both increase, appreciable repression can set in, and transcription of the three sets of genes can decrease. Finally, when steady state has been reached in minimal medium lacking tryptophan, the *trp* operon is 90% repressed, in part because the level of *trp* repressor is much higher than it is in the presence of tryptophan.

THE SERENDIPITOUS DISCOVERY OF *trp* ENZYME HYPERSYNTHESIS

Polarity, the decrease in the expression of a gene downstream in an operon from a nonsense mutation, commanded the attention of many molecular biologists from 1970 to about 1976. Experiments by Yanofsky and Jackson at this time were designed to study the phenomenon of polarity by generating a set of deletions internal to the *trp* operon that eliminated the polar effects of a mutation near the beginning of the operon on the expression of a gene near the end of the operon. As deletions are rare to begin with, special selections and scorings must be used to identify cells containing the desired one. The basic selection method was to use conditions in which polarity reduced expression of a promoter-distal *trp* gene to the extent that cells could not grow and then to select for secondary mutants that could grow. Among them would be deletions removing the cause of the polarity. Scoring by replica plating onto petri plates spread with lawns of phage carrying parts of the *trp* operon then identified the candidates containing internal deletions.

The last two genes of the tryptophan operon, *trpA* and *trpB*, code for the α and β subunits of tryptophan synthetase, which normally has an $\alpha_2\beta_2$ structure. The synthetase from a *trpA* mutant has a β_2 structure. This dimer possesses part of the enzyme's normal activity and catalyzes the conversion of indole and serine to tryptophan and water at about 3% the rate that the wild-type enzyme catalyzes the conversion of serine and indole-3-glycerol phosphate to tryptophan and water. The activity of the β_2 complex is such that if indole is present in the growth medium, tryptophan can be synthesized at a rate sufficient to supply the cells' tryptophan requirements. However, the reduction in expression of the *trpB* gene in a *trpA* mutant strain containing a polar *trpE* mutation leaves the cells unable to convert indole to tryptophan at a rate adequate for growth (Fig. 14.6). Under these conditions, a deletion of the *trpE* mutation retaining the *trp* promoter and the *trpB* gene relieves the polarity on *trpB* expression and permits cells to satisfy their tryptophan requirements with indole.

Candidate colonies that could grow on indole were selected then scored for loss of the *trpD* and *trpC* genes with special phage carrying point mutation copies of these genes. The phage can infect cells, but many cells are not lysed. Instead, the phage can either recombine into the chromosome or simply remain in the cell for several generations until diluted away by cell growth. While present in the cells, the phage can recombine with homologous DNA on the host chromosome and reconstruct a functional *trpD* or *trpC* gene if the chromosome is not deleted of these genes. Thus, simply by replicating the indole⁺ colonies onto lawns of the *trp* phage and picking colonies that did not yield TrpD⁺ or TrpC⁺ recombinants, Jackson and Yanofsky isolated many internal deletions in the *trp* operon that relieved polarity. We might have expected that fusions of the *trpB* gene to other genes of the chromosome might also have occurred, but, curiously, none were found.

Out of 34 internal deletions that had removed polarity on the expression of the *trpB* gene, all were still regulated by the *trp* repressor. Most unexpectedly, two deletion strains hypersynthesized the *trpB* protein by a factor of three to ten; each of

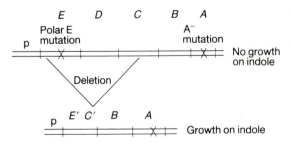

Figure 14.6   How an internal *trp* deletion can relieve polarity generated by a polar *trpE* mutation and increase expression of the *trpB* gene.

	trp repression	β_2 specific activity
Wild type	−	10
	+	0.5
Δ1	−	10
	+	0.5
Δ2	−	40
	+	2
Δ3	−	40
	+	0.5

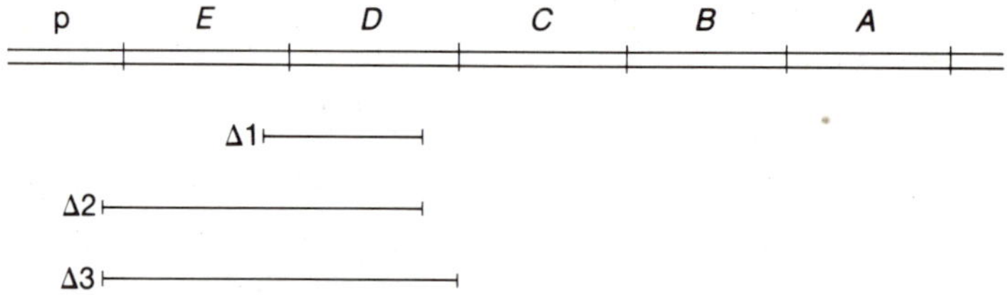

Figure 14.7   The specific activity of the β_2 protein of the *trp* operon in the presence and absence of *trp* repressor in wild-type cells and in cells containing the three different deletions indicated.

these was deleted of all of the *trpE* gene (Fig. 14.7). The source of this hypersynthesis was then examined more carefully, as described in the next section.

EARLY EXPLORATIONS OF THE HYPERSYNTHESIS

The experiments described in this section were performed before the development of modern genetic engineering. It is interesting to see how Yanofsky and his collaborators cleverly utilized the technology available at the time to learn the cause of the hypersynthesis and thereby discover the phenomenon of attenuation. The first question in their careful investigation was whether the hypersynthesis in the internal deletion strains mentioned above resulted from an alteration in the levels of messenger. Either more messenger was present in the deletions or messenger was being translated more efficiently. The choice of subsequent experiments depended on which it was.

Messenger RNA in cells was briefly labeled with tritiated uridine, extracted, and hybridized to complementary *trp* DNA immobilized on paper. *Trp*-containing DNA was available in the form of $\phi80$ and lambda phage in which part of the phage genome was substituted with *trp* sequences. Use of such phage containing all or part of the *trp* operon substituted for the use of fragments that today would be isolated by restriction enzyme cleavage of an appropriate plasmid and subsequent purification by gel electrophoresis. The hybridization revealed that in all strains, the levels of *trpB* messenger paralleled the levels of *trpB* enzyme. Therefore the cause of *trp* enzyme hypersynthesis in some of the internal deletion strains was elevated messenger levels.

Some of the internal deletion strains hypersynthesized the *trpB* protein and messenger and some did not. Hence it was important to learn whether the effect could be correlated with the extent of the deletion. This was particularly important because sequencing of *trp* messenger had revealed that the *trp* operon contained an unexpectedly long leader of 162 bases between the start of transcription and the translation start of the first *trp* enzyme.

The deletion end points were determined by isolating *trp* messenger synthesized *in vivo* by the deletion strains. Total RNA from the deletion strains was isolated and hybridized to wild-type *trp* DNA. The region of complementarity between the RNA and DNA was resistant to RNase. After the digestion, this region was then melted off the DNA and the exact extent of *trp*-specific sequence obtained from each deletion was apparent on RNA sequencing. RNA sequencing was used since DNA sequencing techniques had not yet been invented. It was found that only those deletions that removed a site located 20 nucleotides before the *trpE* gene hypersynthesized *trpB*.

Conceivably, the deletions caused hypersynthesis by altering the activity of the *trp* promoter, although this possibility seemed unlikely because some deletions ended as far as 100 nucleotides downstream from the promoter. More likely, the deletions removed a site located 20 bases ahead of the start of the *trpE* gene, which

Figure 14.8   Two-step hybridization for determination of the relative amounts of *trpL* and *trpB* mRNA in cells. Messenger from the *trp* operon as indicated on the top line is hybridized to a phage DNA carrying part of the *trp* operon. RNA eluted from the first phage was then hybridized to phage containing sequences homologous to either the leader region or the *trpB* region of the operon.

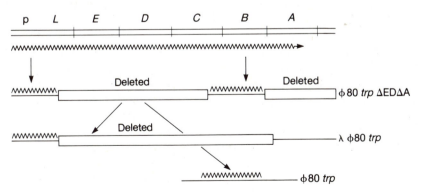

terminated most transcription before it entered the *trp* structural genes. This idea was tested by comparing the amount of *trp* messenger specified by a sequence lying upstream of the *trpE* gene and upstream of the potential termination site to the amount of messenger specified by sequences downstream from the site (Fig. 14.8). The downstream messenger contained the *trpB* region of the operon. A two-step hybridization permitted quantitation of messenger RNA in cells from each of these regions. This experiment provided the telling clue. Wild-type repressed cells as well as repression-negative cells contained an eight-to-ten-fold molar excess of messenger from the leader region preceding the *trpE* gene over messenger from the *trpB* region. However, in cells deleted of only the critical region just ahead of the *trpE* gene, the amount of *trpB* messenger was elevated and was nearly comparable to the former. These results prove that, indeed, termination occurs at the site 20 bases ahead of the *trpE* gene and that under normal conditions a majority of the transcription does not enter the *trp* structural genes. The transcriptional termination that occurs ahead of the *trpE* gene is called attenuation and the site responsible for it is called the attenuator, *a*.

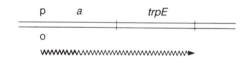

The obvious and most important question to ask at this point was whether transcriptional attenuation, which ordinarily permitted only one in eight to ten polymerases to transcribe into the *trp* structural genes, was a regulatory element. Starvation for tryptophan, generated by addition of indolylacrylic acid that inhibits tryptophan synthesis as well as tryptophanyl-tRNA synthetase, did decrease attenuation. Starvation for other amino acids did not decrease attenuation in the *trp* operon, proving that attenuation does play a regulatory role in the *trp* operon.

In vitro transcription of the *trp* operon DNA was the logical next step in these studies. *trp* RNA from such experiments could be quantitated by its synthesis from $\phi80$ and lambda *trp* phage and appropriate RNA-DNA hybridizations. These showed that virtually all transcripts ended at the point identified in the *in vivo* experiments as responsible for attenuation. The sequence of this region showed that it resembles other sequences at which RNA transcription terminates (Fig. 14.9). It contains a G-C-rich region that can form a hairpin followed by a string of eight U's.

The leader region also contains an AUG codon and can bind ribosomes just like the beginning of authentic genes. This was shown by the straightforward but technically difficult method of isolating radioactive leader region RNA from cells, binding ribosomes to it in the presence of initiation factors and f-met-tRNA^fmet, lightly

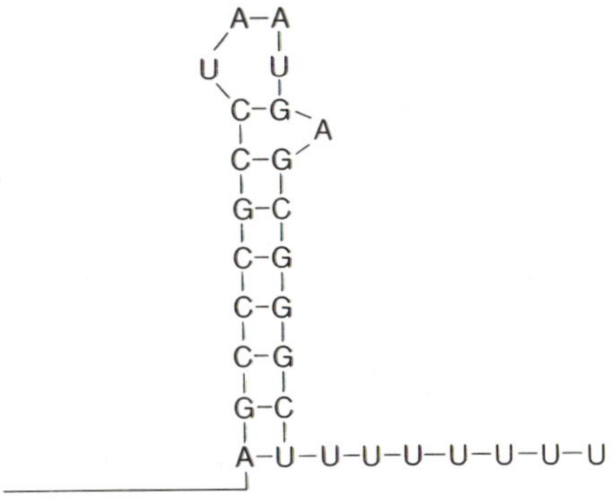

```
      A—A
     /    |
    U      U
     \    |
      C—G
      |    \
      C—G    A
      |    /
      G—C
      |  |
      C—G
      |  |
      C—G
      |  |
      C—G
      |  |
      G—C
      |  |
      A—U—U—U—U—U—U—U
```

Figure 14.9   The 3-4 hybrid of the *trp* leader RNA.

digesting with RNase, and sequencing the RNA that was protected from digestion. Following the AUG codon are 13 more codons before a translation termination codon. Clearly it was of great interest to determine whether this leader peptide is synthesized *in vivo*. Vigorous attempts to isolate it failed, however. Therefore an indirect experiment was used to prove that the leader is translated. The leader peptide was fused by deletion to *lac* repressor, and the fusion protein was found to be synthesized at a high rate. Since the leader peptide contains two tryptophans, attenuation appeared to be coupled to translation. This suspicion was deepened by the findings that *trp* regulation is altered in some *trp* synthetase mutants.

trp LEADER RNA HAS THE POTENTIAL TO FORM MULTIPLE HYBRIDS

Once the hairpin secondary structure in the RNA immediately preceding the *trp* attenuation site was suspected of playing a role in transcription termination, several critical experiments were apparent. One was to change the sequence of the region. Mutations changing the stability of the hairpin might be expected to alter attenuation. A second and easier experiment was to examine naturally occurring sequence variants. Because related bacterial strains must have evolved from a common predecessor, they probably share the same basic mechanism of *trp* regulation. If the leader sequences of their *trp* operons possess structural features in common, these features are likely to be important in the regulatory mechanisms. Indeed, this was found. The *trp* operon in *Salmonella typhimurium* and *Serratia marcescens* also contain leader regions. Both code for a leader peptide containing several tryptophans, and both mRNAs are capable of forming a hairpin just before a string of U's located about 20 bases ahead of the *trpE* gene.

Close examination of the leader region of the *trp* operon of *E. coli*, the *trp* operons from other strains of bacteria, and other amino acid biosynthetic operons has re-

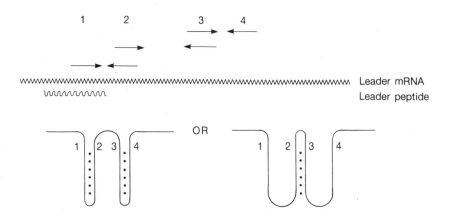

Figure 14.10   The two possible structures that the second half of the *trp* leader region is capable of assuming. Above is indicated the leader region with the complementary regions. Opposed arrows at the same height above the leader are homologous and can base pair. The wavy line below the leader mRNA is the region encoding the leader peptide.

vealed two additional facts. First, the peptides encoded by the leader regions always contain one to seven of the amino acid residues synthesized by that operon. Second, the leader mRNAs possess at least *four* regions that can form intermolecular base-paired hairpins (Fig. 14.10). They are numbered 1, 2, 3, and 4. The pairing of 3 and 4 forms the hairpin discussed above that is required for termination of transcription. Additionally, region 2 can base pair either with region 3 or with region 1. Likewise, region 3 can base pair with region 2 or 4. The particular hybrids that form on a leader molecule determine whether transcription of that molecule will terminate as RNA polymerase passes the attenuation region. As we will see, the pairing choice is regulated by the position on the leader of a ribosome translating the leader peptide.

COUPLING TRANSLATION TO ATTENUATION

The location of the leader peptide with respect to the four different regions capable of hybrid formation in the *trp* leader region provides a simple mechanism for regulating attenuation. If the 3-4 hybrid forms during transcription of this region, attenuation is possible because this hybrid is the "termination" loop. Conversely, if the 2-3 hybrid forms during transcription of leader, then formation of the "termination" loop is prevented. Finally, if the 1-2 hybrid forms, then region 2 is not available for formation of the 2-3 hybrid, but as RNA polymerase transcribes regions 3 and 4, they are free to base pair, and transcription terminates.

How can the presence or absence of tryptophan affect formation of the 1-2 or 2-3 hybrid? Ribosomes translating the leader region in the absence of charged

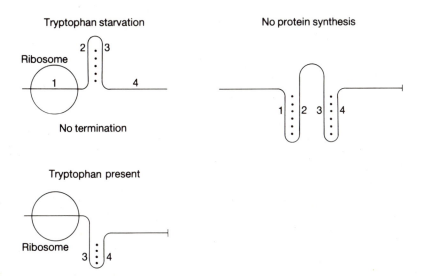

Figure 14.11   Possible structures of the leader mRNA in the presence of ribosomes under the three conditions: tryptophan starvation, no protein synthesis, and tryptophan excess.

tRNA^{trp} will stall at the *trp* codons and, owing to their location, they will block the formation of the 1-2 hybrid (Fig. 14.11). Formation of the 2-3 hybrid is not blocked, and as soon as these regions of the mRNA are synthesized, this hybrid forms. Consequently, the 3-4 hybrid does not form in time to terminate transcription, and attenuation does not occur.

In the absence of any protein synthesis, ribosomes cannot bind to the leader region, and the 1-2 hybrid forms. In turn, this permits formation of the 3-4 hybrid. Attenuation follows. This happens in an *in vitro* transcription system, which, as we saw above, terminates transcription with a high probability at the attenuation site.

Finally, what happens in the presence of excess tryptophan? In cells containing adequate levels of tryptophan, much initiation, of course, will be blocked by the *trp* repressor. However, the transcription that is initiated will be largely terminated, for ribosomes either complete translation of the leader peptide and permit hybrids 1-2 and 3-4 to form, or ribosomes remain awhile at the termination codon. From this position they block formation of the 1-2 hybrid. In either case, the 3-4 hybrids form, and termination at the attenuator occurs.

It is necessary to note that the relative thermodynamic stability of the various leader hybrids is not important to attenuation. The factor determining whether attenuation will occur is which structures are not blocked from forming. The kinetics of formation of the base-paired structures should be on time scales less than milliseconds so that if they are not blocked from forming, they should form while the RNA is being synthesized. Then several seconds after initiating transcription, RNA poly-

Free Energy of Leader Hairpins (From Fig. 4.10)

Hairpin	ΔG
1–2	−11.2 Kcal/mol
2–3	−11.7 Kcal/mol
3–4	−20 Kcal/mol

merase reaches the attenuation site and terminates or not depending on whether the 3-4 hybrid has formed. The interval between initiation and termination is much shorter than the interconversion time of many hybrid structures, and they therefore will not have had time to interconvert and adopt their lowest-energy conformation. The factor determining termination is which of the hybrid structures is capable of forming as the RNA is synthesized, not which is the most stable.

The system is not yet fully understood. During tryptophan starvation, how does the system manage to ensure that ribosomes initiate translation as soon as the messenger has been synthesized? If ribosomes do not promptly initiate translation of the leader, premature termination at the attenuator will result. One way to synchronize transcription and translation would be for RNA polymerase to stall a short distance after the AUG codon of the leader and not to proceed until a ribosome has begun translation of the leader and nudged the RNA polymerase back into action.

RNA SECONDARY STRUCTURE AND THE ATTENUATION MECHANISM

Point mutations provide excellent support for the attenuation mechanism as outlined above. Changing the AUG codon of the leader to an AUA has the expected effect (see Problem 14.2). A base change abolishing a base pair in the 2-3 hybrid, but one having no effect on the base pairs of hybrids 1-2 or 3-4 prevents the leader from forming the 2-3 hybrid. Consequently, the 3-4 hybrid always forms, and attenuation inevitably results. Mutations that reduce the stability of the 3-4 hybrid reduce transcription termination, both *in vitro* and *in vivo*.

Direct experiments also provide evidence for the existence of secondary structure in the leader RNA. Light digestion of the RNA with RNase T1 cleaves only at those guanosines postulated not to be in the hybrids (Fig. 14.12).

A second experiment also suggests that base pairing in the leader region is important. As mentioned above, in an *in vitro* transcription system with purified DNA and RNA polymerase, virtually all transcripts are terminated at the attenuation site. However, the substitution of inosine triphosphate (ITP) for GTP eliminates the premature termination. While it is possible that specific contacts between the inosine and RNA polymerase or between the inosine and the DNA are responsible for the lack of premature termination, it seems more likely that the failure of the substituted RNA to form its 3-4 hybrid is the reason. Whereas G-C base pairs possess three hydrogen bonds, I-C base pairs have only two because inosine lacks an amino group

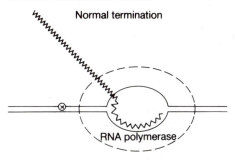

Cytosine-guanine   Cytosine-inosine

that is present on guanosine. The two hydrogen bonds do not provide sufficient energy to form the usual hybrids, and thus termination does not occur. Analogs incorporated into the DNA generate much smaller perturbations on attenuation, also suggesting that it is the base-paired structure of the transcribed RNA that is necessary for termination at the attenuation site. The use of DNA molecules hetero-duplex in the leader region also yields data consistent with the attenuation mecha-

Figure 14.13 The use of heteroduplex template DNAs for demonstrating that con-formation of the product RNA determines whether termination occurs.

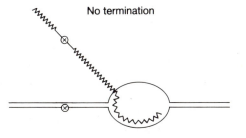

Figure 14.12 The points of RNase T₁ cleavage of the leader RNA with respect to the 1-2 and 3-4 hybrid structure that likely forms on naked *trp* leader RNA *in vitro*.

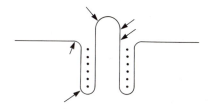

met-thr-arg-val-gln-phe-lys-HIS-HIS-HIS-HIS-HIS-HIS-HIS-pro-asp

met-thr-ala-LEU-LEU-arg-VAL-ILE-ser-LEU-VAL-ILE-ser-VAL-VAL-VAL-ILE-ILE-ILE-pro-pro-cys-gly-ala-ala-leu-gly-arg-gly-lys-ala

met-ser-his-ile-val-arg-phe-thr-gly-LEU-LEU-LEU-LEU-asn-ala-phe-ile-val-arg-gly-arg-pro-val-gly-gly-ile-gln-his

met-lys-his-ile-pro-PHE-PHE-PHE-ala-PHE-PHE-PHE-thr-PHE-pro

met-lys-arg-ILE-ser-THR-THR-ILE-THR-THR-THR-ILE-THR-ILE-THR-THR-gly-asn-gly-ala-gly

met-lys-ala-ile-phe-val-leu-lys-gly-TRP-TRP-arg-thr-ser

Figure 14.14   The sequences of leader peptides from some amino acid operons.

nism. DNA molecules can be constructed that are wild type on one strand and mutant in a single base in the leader region on the other (Fig. 14.13). Only when the mutation is on the strand copied by RNA polymerase and therefore producing an altered RNA product molecule does the mutant nucleotide alter attenuation. If the RNA product is wild type, so is transcription and termination.

Finally, direct physical evidence for the attenuation model has also been obtained. In an *in vitro* transcription system, the usual termination at the attenuator can be blocked by including a large excess of an oligonucleotide complementary to region 1. Apparently this oligonucleotide anneals to region 1 as it is synthesized and prevents its pairing with region 2. Consequently, the 2-3 hybrid can form, and the 3-4 hybrid cannot form. Termination is therefore blocked.

OTHER ATTENUATED OPERONS

Study of the sequences of a number of amino acid biosynthetic operons has revealed that they are likely to be regulated by attenuation. Their leader peptides contain dramatic runs of the amino acid whose synthesis the operon codes for (Fig. 14.14). In summary, the attenuation mechanism seems to be an exceptionally efficient method for regulating the amino acid biosynthetic operons because the necessary regulation is obtained by the properties of just 160 nucleotides of RNA. In the case of the *trp* operon, the double-barreled regulation provided by *trp* repressor and attenuation provides up to a 700-fold regulation range, 70-fold coming from repression and another 10-fold from attenuation. The autoregulation of *trpR* permits rapid accumulation of optimal enzyme level in cells on tryptophan starvation followed by a lower rate of enzyme synthesis when steady-state conditions have been reached.

PROBLEMS

14.1. Fusidic acid interferes with elongation of peptide chains and slows the progress of ribosomes in translation. How would "low" and "high" fusidic acid concentrations affect regulation of the tryptophan operon?

14.2. What would the effect be of changing the AUG codon of the *trp* leader peptide to AUA or of damaging the Shine-Dalgarno sequence?

14.3. If the leader region of the *trp* operon of the messenger RNA has signaled the RNA polymerase passing the attenuation region whether or not it is useful to terminate, it would be sensible for the attentuation region to no longer be involved in binding ribosomes and synthesizing a short polypeptide. How could this be accomplished?

14.4. Suppose the enzymes of the *trp* operon were highly unstable, that is, had a lifetime of about five minutes. What types of regulation would then be necessary or unnecessary?

14.5. What is likely to happen to regulation of the *trp* operon in a mutant *E. coli* in which RNA polymerase elongates at twice the normal rate?

14.6. Suppose *E. coli* were being modified to permit maximum synthesis of tryptophan. Suppose also that the signal that has been chosen to induce this maximal synthesis is starvation or semistarvation for leucine. What cellular modifications would be necessary?

14.7. In view of the rates discussed in earlier chapters, estimate the average time between initiations by ribosomes on the *trp* leader sequence. What would be the effect on *trp* attenuation if the ribosome initiation frequency were limited only by the peptide elongation rate?

14.8. How could an attenuation mechanism be used for regulation of synthesis of purines or pyrimidines?

14.9. The data presented in this chapter on the *trpB* enzyme levels in the different deletions are anomalous. What is unusual and what is a likely explanation?

14.10. Considering that a DNA sequence that encodes transcription termination contains a region rich in G's and C's that can base pair followed by a string of about six U's, devise several mechanisms that could use this information to signal the RNA polymerase to terminate transcription. For example, one might be that the separated DNA strands at the transcription point themselves form hairpins in addition to the RNA, and the three hairpins bind to sites within the polymerase to cause it to terminate.

14.11. Suppose, in the days before definitive experiments could be performed with purified components, you wished to determine whether tryptophan, trp-tRNA^{trp}, or tryptophanyl-tRNA-synthetase was involved in repression of the *trp* operon. Invent data you could obtain with crude extracts or partially purified extracts, perhaps using appropriate mutants or tryptophan analogs, that would prove that tryptophan was the corepressor of the *trp* operon.

14.12. In describing the experiments of Yanofsky, no mention was made in this chapter of testing whether the hypersynthesizing deletions might be the effect of a diffusible product in the cells. Design an experiment and invent results that would prove that the hypersynthesis was a *cis* effect.

14.13. How do you reconcile the existence of suppressors of polarity that function on polar mutants in *lac* and *trp* with the fact that *in vitro* transcription of *trp* terminates nearly 100% at the attenuator even in the absence of rho?

14.14. Consider the use of hybridization to quantitate *trp* leader and distal mRNA resulting from *in vitro* transcription. What would the results look like and what mistaken conclusion would be drawn if the DNA used for hybridization were not present in molar excess over the RNA?

14.15. Recombination with *trp* phage permits approximate location of the end points of *trp* internal deletions. How could the end points be mapped more accurately?

14.16. The paper Transcription Termination at the *trp* Operon Attenuators of *E. coli* and *S. typhimurium:* RNA Secondary Structure and Regulation of Termination, F. Lee, C. Yanofsky, Proc. Natl. Acad. Sci. USA *74,* 4365–4369 (1977), did not present the complete attenuation model. What critical information was missing?

RECOMMENDED READINGS

The Region Between the Operator and First Structural Gene of the Tryptophan Operon of *E. coli* May Have a Regulatory Function, E. Jackson, C. Yanofsky, J. Mol. Biol. *76,* 89–101 (1973).
Transcription Termination *in vivo* in the Leader Region of the Tryptophan Operon of *E. coli,* K. Bertrand, C. Squires, C. Yanofsky, J. Mol. Biol. *103,* 319–337 (1976).

RELATED REVIEWS, BOOKS, AND ARTICLES

Amino Acid Biosynthesis and Its Regulation, H. Umbarger, Ann. Rev. Biochem. *47,* 533–606 (1978).
Regulation of Gene Expression in the Tryptophan Operon of *Escherichia coli,* T. Platt, in The Operon, ed. J. Miller, W. Reznikoff, Cold Spring Harbor (1978).
Attenuation in the Control of Expression of Bacterial Operons, C. Yanofsky, Nature *289,* 751–758 (1981).

DEEPER READING

Nonsense Codons and Polarity in the Tryptophan Operon, C. Yanofsky, J. Ito, J. Mol. Biol. *21,* 313–334 (1966).
The Internal Low Efficiency Promoter of the Tryptophan Operon of *E. coli,* D. Morse, C. Yanofsky, J. Mol. Biol. *38,* 447–451 (1968).
Transcription of the Operator Proximal and Distal Ends of the Tryptophan Operon: Evidence That *trpE* and *trpA* Are the Delimiting Structural Genes, J. Rose, C. Yanofsky, J. Bact. *108,* 615–618 (1971).

Expression of the Tryptophan Operon in Merodiploids of *E. coli*, H. Stetson, R. Somerville, Mol. Gen. Genet. *111*, 342–351 (1971).

Polarity and Enzyme Functions in Mutants of the First Three Genes of the Tryptophan Operon of *E. coli*, C. Yanofsky, V. Horn, M. Bonner, S. Stasiowski, Genetics *69*, 409–433 (1971).

Internal Promoter of the Tryptophan Operon of *E. coli* Is Located in a Structure Gene, E. Jackson, C. Yanofsky, J. Mol. Biol. *69*, 307–313 (1972).

Internal Deletions in the Tryptophan Operon of *E. coli*, E. Jackson, C. Yanofsky, J. Mol. Biol. *71*, 149–161 (1972).

Hyper-Labile Messenger RNA in Polar Mutants of the Tryptophan Operon of *E. coli*, S. Hiraga, C. Yanofsky, J. Mol. Biol. *72*, 103–110 (1972).

Detection and Isolation of the Repressor Protein for the Tryptophan Operon of *E. coli*, G. Zubay, D. Morse, W. Schrenk, J. Miller, Proc. Natl. Acad. Sci. USA *69*, 1100–1103 (1972).

In vitro Synthesis of Enzymes of the Tryptophan Operon of *E. coli*, P. Powels, J. Van Rotterdam, Proc. Natl. Acad. Sci. USA *69*, 1786–1790 (1972).

Structure and Evolution of a Bifunctional Enzyme of the Tryptophan Operon, M. Grieshaber, R. Bauerle, Nature New Biol. *236*, 232–235 (1972).

Normal Repression in a Deletion Mutant Lacking Almost the Entire Operator-Proximal Gene of the Tryptophan Operon of *E. coli*, S. Hiraga, C. Yanofsky, Nature New Biol. *237*, 47–49 (1972).

Isolating Tryptophan Regulatory Mutants in *E. coli* by Using a *trp-lac* Fusion Strain, W. Reznikoff, K. Thornton, J. Bact. *109*, 526–532 (1972).

Nucleotide Sequences from Messenger RNA Transcribed from the Operator-Proximal Portion of the Tryptophan Operon of *E. coli*, P. Cohen, M. Yaniv, C. Yanofsky, J. Mol. Biol. *74*, 163–177 (1973).

Nucleotide Sequences from Tryptophan Messenger RNA of *E. coli:* The Sequence Corresponding to the Amino-Terminal Region of the First Polypeptide Specified by the Operon, M. Bronson, C. Squires, C. Yanofsky, Proc. Natl. Acad. Sci. USA *70*, 2335–2339 (1973).

Tryptophanyl-tRNA and Tryptophanyl-tRNA Synthetase Are Not Required for *in vitro* Repression of the Tryptophan Operon, C. Squires, J. Rose, C. Yanofsky, H. Yang, G. Zubay, Nature New Biol. *245*, 131–133 (1973).

Regulation of *in vitro* Transcription of the Tryptophan Operon by Purified RNA Polymerase in the Presence of Partially Purified Repressor and Tryptophan, J. Rose, C. Squires, C. Yanofsky, H. Yang, G. Zubay, Nature New Biol. *245*, 133–137 (1973).

Synthesis of Tryptophan Operon RNA in a Cell-Free System, D. McGeoch, J. McGeoch, D. Morse, Nature New Biol. *245*, 137–140 (1973).

Interaction of the Operator of the Tryptophan Operon with Repressor, J. Rose, C. Yanofsky, Proc. Natl. Acad. Sci. USA *71*, 3134–3138 (1974).

Interaction of the *Trp* Repressor and RNA Polymerase with the *trp* Operon, C. Squires, F. Lee, C. Yanofsky, J. Mol. Biol. *92*, 93–111 (1975).

An Intercistronic Region and Ribosome-Binding Site in Bacterial Messenger RNA, T. Platt, C. Yanofsky, Proc. Natl. Acad. Sci. USA *72*, 2399–2403 (1975).

Punctuation of Transcription *in vitro* of the Tryptophan Operon of *E. coli*, H. Pannekoek, W. Brammer, P. Pouwels, Mol. Gen. Genet. *136*, 199–214 (1975).

Dual Control of the Tryptophan Operon Is Mediated by Both Tryptophanyl-tRNA Synthetase and the Repressor, D. Morse, A. Morse, J. Mol. Biol. *103*, 209–226 (1976).

Regulation of Transcription Termination in the Leader Region of the Tryptophan Operon of *Escherichia coli* Involves Tryptophan or Its Metabolic Product, K. Bertrand, C. Yanofsky, J. Mol. Biol. *103*, 339–349 (1976).

Nucleotide Sequence of the 5′ End of Tryptophan Messenger RNA of *Escherichia coli*, C. Squires, F. Lee, K. Bertrand, C. Squires, M. Bronson, C. Yanofsky, J. Mol. Biol. *103*, 351–381 (1976).

Termination of Transcription *in vitro* in the *Escherichia coli* Tryptophan Operon Leader Region, F. Lee, C. Squires, C. Squires, C. Yanofsky, J. Mol. Biol. *103*, 383–393 (1976).

Ribosome-Protected Regions in the Leader-*trpE* Sequence of *Escherichia coli* Tryptophan Operon Messenger RNA, T. Platt, C. Squires, C. Yanofsky, J. Mol. Biol. *103*, 411–420 (1976).

Polarity Suppressors Defective in Transcription Termination at the Attenuator of the Tryptophan Operon of *E. coli* Have Altered Rho Factor, L. Korn, C. Yanofsky, J. Mol. Biol. *106*, 231–241 (1976).

Nucleotide Sequence of Region Preceding *trp* mRNA Initiation Site and Its Role in Promoter and Operator Function, G. Bennett, M. Schweingruber, K. Brown, C. Squires, C. Yanofsky, Proc. Natl. Acad. Sci. USA *73*, 2351–2355 (1976).

Mutations Affecting tRNA[Trp] and Its Charging and Their Effect on Regulation of Transcription Termination at the Attenuator of the Tryptophan Operon, C. Yanofsky, L. Söll, J. Mol. Biol. *113*, 663–677 (1977).

Transcription Termination at the *trp* Operon Attenuators of *E. coli* and *S. typhimurium*: RNA Secondary Structure and Regulation of Termination, F. Lee, C. Yanofsky, Proc. Natl. Acad. Sci. USA *74*, 4365–4369 (1977).

Nucleotide Sequence of the Promoter-Operator Region of the Tryptophan Operon of *E. coli*, G. Bennett, M. Schweingruber, K. Brown, C. Squires, C. Yanofsky, J. Mol. Biol. *121*, 113–137 (1978).

Nucleotide Sequence of the Promoter-Operator Region of the Tryptophan Operon of *Salmonella typhimurium*, G. Bennett, K. Brown, C. Yanofsky, J. Mol. Biol. *121*, 139–152 (1978).

RNA Polymerase Interaction at the Promoter-Operator Region of the Tryptophan Operon of *E. coli* and *Salmonella typhimurium*, K. Brown, G. Bennett, F. Lee, M. Schweingruber, C. Yanofsky, J. Mol. Biol. *121*, 153–177 (1978).

Sequence Analysis of Operator Constitutive Mutants of the Tryptophan Operon of *E. coli*, G. Bennett, C. Yanofsky, J. Mol. Biol. *121*, 179–192 (1978).

Nucleotide Sequence of the Attenuator Region of the Histidine Operon of *Escherichia coli* K-12, P. DiNocera, F. Blasi, R. Frunzio, C. Bruni, Proc. Natl. Acad. Sci. USA *75*, 4276–4282 (1978).

DNA Sequence from the Histidine Operon Control Region: Seven Histidine Codons in a Row, W. Barnes, Proc. Natl. Acad. Sci. USA *75*, 4281–4285 (1978).

The Regulatory Region of the *trp* Operon of *Serratia marcescens*, G. Miozzari, C. Yanofsky, Nature *276*, 684–689 (1978).

Translation of the Leader Region of the *E. coli* Tryptophan Operon, G. Miozzari, C. Yanofsky, J. Bact. *133*, 1457–1466 (1978).

Regulation of the Threonine Operon: Tandem Threonine and Isoleucine Codons in the Control Region and Translational Control of Transcription Termination, J. Gardner, Proc. Natl. Acad. Sci. USA *76*, 1706–1710 (1979).

leu Operon in *Salmonella typhimurium* Is Controlled by an Attenuation Mechanism, R. Gemmil, S. Wessler, E. Keller, J. Calvo, Proc. Natl. Acad. Sci. USA *76*, 4941–4945 (1979).

Attenuation in the *E. coli* Tryptophan Operon: Role of RNA Secondary Structure Involving the Tryptophan Codon Region, D. Oxender, G. Zurawski, C. Yanofsky, Proc. Natl. Acad. Sci. USA *76*, 5524–5528 (1979).

Alternative Secondary Structures of Leader RNAs and the Regulation of the *trp, phe, his, thr,* and *leu* Operons, E. Keller, J. Calvo, Proc. Natl. Acad. Sci. USA *76*, 6186–6190 (1979).

The Effect of an *E. coli* Regulatory Mutation on tRNA Structure, S. Eisenberg, M. Yarus, L. Soll, J. Mol. Biol. *135*, 111–126 (1979).

E. coli Tryptophan Operon Leader Mutations Which Relieve Transcription Terminators Are cis-Dominant to *trp* Leader Mutations Which Increase Transcription Termination, G. Zurawski, C. Yanofsky, J. Mol. Biol. *142*, 123–129 (1980).

E. coli RNA Polymerase and *trp* Repressor Interaction with the Promoter-Operator Region of the Tryptophan Operon of *Salmonella typhimurium*, D. Oppenheim, G. Bennet, C. Yanofsky, J. Mol. Biol. *144*, 133–142 (1980).

Functional Analysis of Wild-Type and Altered Tryptophan Operon Promoters of *Salmonella Typhimurium* in *E. coli*, D. Oppenheim, C. Yanofsky, J. Mol. Biol. *144*, 143–161 (1980).

Model for Regulation of the Histidine Operon of Salmonella, H. Johnson, W. Barnes, F. Chumley, L. Bossi, J. Roth, Proc. Natl. Acad. Sci. USA *77*, 508–512 (1980).

Nucleotide Sequence of *ilvGEDA* Operon Attenuator Region of *Escherichia coli*, F. Nargang, C. Subrahmanyam, H. Umbarger, Proc. Natl. Acad. Sci. USA *77*, 1823–1827 (1980).

Multivalent Translational Control of Transcription Termination at Attenuator of *ilvGEDA* Operon of *Escherichia coli* K-12, R. Lawther, G. Hatfield, Proc. Natl. Acad. Sci. USA *77*, 1862–1866 (1980).

Structure and Regulation of *aroH*, the Structural Gene for the Tryptophan-Repressible 3-deoxy-D-arabinose-heptulosonic acid-7-phosphate Synthetase of *E. coli*, G. Zurawski, R. Gunsalus, K. Brown, C. Yanofsky, J. Mol. Biol. *145*, 47–73 (1981).

DNA Sequence Changes of Mutations Altering Attenuation Control of the Histidine Operon of *Salmonella typhimurium*, H. Johnston, J. Roth, J. Mol. Biol. *145*, 735–756 (1981).

Tandem Termination Sites in the Tryptophan Operon of *E. coli*, A. Wu, G. Christie, T. Platt, Proc. Natl. Acad. Sci. USA *78*, 2913–2917 (1981).

Antitermination Regulation of *ampC*, Ribosome Binding at High Growth Rate, B. Jaurin, T. Grundstrom, T. Edlund, S. Normark, Nature *290*, 221–225 (1981).

Pausing of RNA Polymerase During *in vitro* Transcription of the Tryptophan Operon Leader Region, M. Winkler, C. Yanofsky, Biochem. *20*, 3738–3744 (1981).

Effects of DNA Base Analogs on Transcription Termination at the Tryptophan Operon Attenuator of *Escherichia coli*, P. Farnham, T. Platt, Proc. Natl. Acad. Sci. USA *79*, 998–1002 (1982).

Transcription Termination at the Tryptophan Operon Attenuator Is Decreased *in vitro* by an Oligomer Complementary to a Segment of the Leader Transcript, M. Winkler, K. Mullis, J. Barnett, I. Stroynowski, C. Yanofsky, Proc. Natl. Acad. Sci. USA *79*, 2181–2185 (1982).

trp Aporepressor Production Is Controlled by Autogenous Regulation and Inefficient Translation, R. Kelley, C. Yanofsky, Proc. Natl. Acad. Sci. USA *79*, 3120–3124 (1982).

Transcript Secondary Structures Regulate Transcription Termination at the Attenuator of *S. marcescens* Tryptophan Operon, I. Stroynowski, C. Yanofsky, Nature *298*, 34–38 (1982).

Superattenuation in the Tryptophan Operon of *Serratia marcescens*, I. Stroynowski, M. van Cleemput, C. Yanofsky, Nature *298*, 38–41 (1982).

Nucleotide Sequence of Yeast *LEU2* Shows 5′- Noncoding Region Has Sequences Cognate to Leucine, A. Andreadis, Y. Hsu, G. Kohlhaw, P. Schimmel, Cell *31*, 319–325 (1982).

Attenuation Control of *purB1* Operon Expression in *Escherichia coli* K-12, C. Turnbough, K. Hicks, J. Donahue, Proc. Natl. Acad. Sci. USA *80*, 368–372 (1983).

Purification and Characterization of *trp* Aporepressor, A. Joachimiak, R. Kelley, R. Gunsalus, C. Yanofsky, Proc. Natl. Acad. Sci. USA *80*, 668–672 (1983).

Transcription Analyses with Heteroduplex *trp* Attenuator Templates Indicate That the Transcript Stem and Loop Structure Serves as the Termination Signal, T. Ryan, M. Chamberlin, J. Biol. Chem. *258*, 4690–4693 (1983).

Attenuation of the *ilvB* Operon by Amino Acids Reflecting Substrates or Products of the *ilvB* Gene Product, C. Hauser, G. Hatfield, Proc. Natl. Acad. Sci. USA *81*, 76–79 (1984).

Use of Complementary DNA Oligomers to Probe *trp* Leader Transcript Secondary Structures Involved in Transcription Pausing and Termination, R. Fisher, C. Yanofsky, Nucleic Acids Res. *12*, 3295–3302 (1984).

Developmentally Regulated Genes in Yeast and *Drosophila*

R apid progress is being made in learning the basic structure and properties of a great many regulated genes and gene families of eukaryotic cells. Much of the progress is attributable to the power of genetic engineering, which has made the task of cloning and sequencing many eukaryotic genes relatively straightforward. Once a gene has been cloned, the sequences involved with its regulation can be defined by the standard operations of deleting and altering sequences, and then, in the most amenable cases, the DNA can be reintroduced into the original organism to examine the consequences of the alterations.

Once the effects of DNA modifications on the expression of a gene have been explored, the effect on the organism of the altered regulation can be investigated, and the biochemical mechanisms underlying the regulation can be more easily studied. The newness of these research areas means that the understanding of regulation in higher organisms is not yet at the same depth as it is in the bacterial systems we have considered in the earlier chapters.

This chapter examines regulated systems in two organisms: the mating type system in yeast and two gene systems in the fruit fly *Drosophila melanogaster*. Chapter 8 briefly mentioned the yeast cell cycle in which haploid *a*- or α-type cells are generated by sporulation of diploid a/α cells. A haploid yeast cell contains both *a*- and α-type genetic information, but it expresses only one or the other. This, then, is a simple example of cell differentiation. *Drosophila* too is a good subject for examination of regulated genes because it is a higher eukaryote, much genetic information is known about it, and the fly follows a developmental pathway to generate different tissues. The proteins considered here are the enzyme dopa decarboxylase and the egg yolk proteins. These proteins are regulated with respect to both the tissues and the times in the life cycle at which they are synthesized.

MATING TYPE CONVERSION IN YEAST

The *a* and *α* cell types of yeast are not absolute and permanent. Yeast possessing the *ho* allele switch from mating type *a*, which is written as *MATa*, to *MATα* or vice versa at a frequency of about 10^{-6} per doubling time, whereas yeast possessing the *HO* allele, the form most commonly found in nature, switch mating types as frequently as once each generation.

The status of the mating type of a clone or colony can easily be determined by mating. Colonies to be checked should possess at least one nutritional requirement for growth. These colonies can be replica plated onto rich plates. Then tester cells of mating type *MATa* and of mating type *MATα* that possess different nutritional requirements can be streaked across them. After several hours of mating, the plates are replicated onto minimal medium plates lacking the nutritional requirements of both the colonies to be checked and the tester strains. For example, cells of mating type *a* mate only with the *α* mating type tester strain and yield a patch of cells able to grow on the minimal plates. The cell type of yeast that switch mating type nearly every generation can be measured by micromanipulation to permit observation of individual cell-to-spore or cell-to-cell interactions.

As in the study of many biological phenomena, the study of mutants revealed hidden details of the mating type system. *Saccharomyces* isolated directly from nature that possessed defects in mating type or mutants in mating type that were isolated in the laboratory showed an astonishing fact. If a mata⁻ mating type mutant is switched to *MATα* mating type and then switched back to *MATa* mating type, the original defect, the mata⁻ defect, disappears. The mata⁻ allele cannot be found no matter how many times the *a*-*α* conversion is performed. Analogous experiments with matα⁻ mutations also reveal the same properties. The only reasonable conclusion is the conclusion first postulated by Oshima and Takano — that the cells contain nonexpressed reference copies of the *MATa* and *MATα* mating type information on

Figure 15.1   Reference and expression copies of yeast mating-type genes. Information is copied from *HML* or *HMR* into *MAT*.

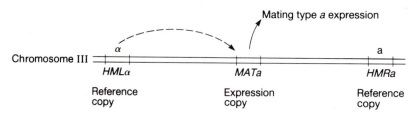

the chromosome in addition to the expressible forms of the *MATa* or *MATα* mating type information. As explained below, direct physical data have confirmed this conclusion. The reference sites are named *HML* and *HMR,* and the expression site is named *MAT* (Fig. 15.1). Usually, but not always, *HML* contains α mating type information, and *HMR* contains *a* type information.

CLONING THE MATING TYPE LOCI IN YEAST

How could the existence of reference copies of mating type information at *HML* and *HMR* and an expression copy at *MAT* be proven? The most straightforward proof is simply to clone DNA copies from the three loci and use DNA hybridization to demonstrate the transfer of the *HM* DNA sequences to the *MAT* locus.

To clone the mating type loci, a yeast shuttle vector was used that permitted selection of both transformed yeast and transformed bacteria. The mating type genes could not conveniently be cloned directly into yeast because the transformation efficiency of yeast is too low. Vast quantities of yeast DNA digested with a restriction enzyme, plasmid, and ligase would have been required. A solution to this problem was to generate smaller quantities of the hybrid plasmids, transform into *E. coli,* grow up large quantities of cells with plasmid DNA, extract the plasmid DNA, and transform this into yeast (Fig. 15.2).

To clone the mating type locus, the yeast DNA that was ligated into the shuttle vector was first extracted from a strain able to mate. The plasmid was then transformed into a yeast strain possessing a defect in the *MAT* locus. Therefore, after the transformation, the yeast able to mate would be those that had been transformed with a vector containing a copy of the functional *MAT* locus.

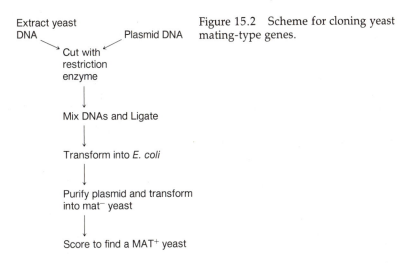

Extract yeast DNA

Plasmid DNA

Cut with restriction enzyme

Mix DNAs and Ligate

Transform into *E. coli*

Purify plasmid and transform into mat⁻ yeast

Score to find a MAT⁺ yeast

Figure 15.2   Scheme for cloning yeast mating-type genes.

TRANSFER OF MATING TYPE GENE COPIES TO AN EXPRESSION SITE

Once the *MATα* mating type clone had been obtained by Nasmyth and Tatchell and by Hicks, Strathern, and Klar, experiments testing DNA transposition to the *MAT* locus were possible. Southern transfers were performed on yeast DNA cut with various restriction enzymes and separated according to size by electrophoresis. The location of the α-specific or *a*-specific sequences were determined by hybridization using the cloned α or *a* sequences as probe.

Figure 15.3   Southern transfer of mating-type genes probed with *MATα* sequences. The restriction enzyme used did not cleave in the mating-type genes. *HML* was on the largest fragment and *MAT* was on the smallest. (Courtesy S. Stewart and J. Haber.)

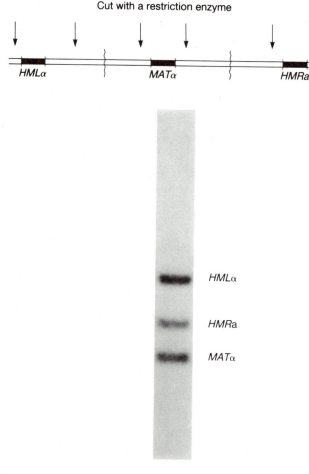

DNA from

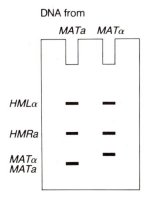

Figure 15.4   Southern transfer showing that *MATa* sequences are about 100 base pairs shorter than *MATα* sequences.

We might have expected the *a* mating type locus to possess no homology to the *α* mating type locus, in which case an *α*-specific probe would have been homologous to two different sizes of restriction fragment in DNA extracted from *α* mating type cells, *HMLα* and *MATα*. This same probe would have been homologous to only one restriction fragment in DNA extracted from *a* mating type cells, *HMLα*. Surprisingly, a different answer was seen. The *α* sequence also possesses homology to *a* mating type sequences. Therefore three different sizes of restriction fragments with homology to either probe were observed in *MATa* and in *MATα* cells (Fig. 15.3). These were sequences at *HML, HMR,* and *MAT.*

The fact that the *a*-specific and *α*-specific sequences were partially homologous permitted a direct and simple demonstration of the basis of mating type conversion. The genetic data indicated that sequences from *HML* or *HMR* were copied into *MAT,* where they were expressed. Because the *a* mating type sequence is about 100 nucleotides shorter than the *α* mating type, the identity of the sequence occupying the *MAT* locus can be directly determined on Southern transfers (Fig. 15.4). The experiments showed, as expected, that mating type *a* cells possessed the *a* sequences at *MAT,* and that mating type *α* cells possessed the *α* sequences at *MAT.* For example, the Southern transfer of a diploid yeast that possesses a *MATa/MATα* genotype yields two bands originating from the *MAT* locus that differ in size by 100 base pairs. Clones of the four spores of a diploid can be grown, their mating type determined, and their DNA extracted for Southern transfer analysis. The *a*-type clones possess only the shorter *MAT* sequence and the *α*-type clones possess only the longer *MAT* sequence.

STRUCTURE OF THE MATING TYPE LOCI

Cloning the mating type sequences permitted direct determination of their structures. DNA-DNA heteroduplexes revealed the overall structural similarities and differences, while DNA sequencing was used to examine the detailed aspects of the

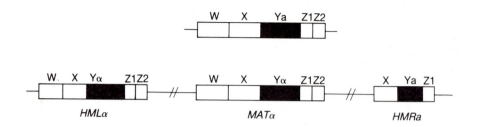

structures. Remarkably, the a^- and α-specific sequences are relatively small, 642 and 747 base pairs, respectively. In addition to these regions, which are called Ya and Yα, the *HML, MAT,* and *HMR* loci are flanked by common sequences called W, X, Z1, and Z2.

The RNA transcripts of the *MAT* region originate from near its center and extend outward in both directions. These transcripts were identified by S1 mapping by

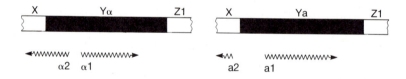

extracting RNA from cells, hybridizing it to end-labeled DNA fragments, and digesting that RNA and DNA not in a duplex. Measurement of the size of the DNA that was protected from digestion by the RNA and knowledge of the locations of the ends of the DNA fragment give the transcription start points. On the basis of genetic complementation tests, the $\alpha 1$, $\alpha 2$, and a1 transcripts appear to be translated into protein, but the a2 transcript may not be translated since no mutations have been found in the a1 region and the a1 transcript lacks a good open reading frame preceded by an AUG codon.

THE EXPRESSION AND
RECOMBINATION PARADOXES

Two interesting questions are raised by the structure of the mating type loci and the locations of transcripts in *MATa* and *MATα*. First, how is transposition always forced to proceed from *HML* or *HMR* to *MAT* and never the reverse, and, second, how is expression of a or α from *HML* or *HMR* prevented?

Relatively little is known about the biochemistry of the transposition reaction. However, the available genetic data suggest that a direct transfer of DNA does not

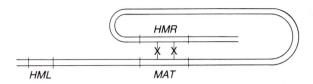

Figure 15.5 Folding the chromosome to bring *HMR* near *MAT* so sequence information could be passed directly from *HMR* to *MAT*.

occur but that mating-type conversion instead is a result of gene conversion (Fig. 15.5). In part this conclusion is based on the failure to find free DNA copies of sequences from *HML* or *HMR* in yeast.

One simple method for expressing sequences at *MAT* and not expressing the sequences at *HML* or *HMR* would have been for the region around *MAT* to provide a promoter that specifies transcription across whatever sequence has been inserted at *MAT.* Clearly nature has chosen a different solution, for both the *a* and *α* mating type genes possess internal promoters. This in turn makes the second question even more acute. What prevents expression of the *a* or *α* mating-type genes from *HML* or *HMR*?

Genetics provides part of the answer to how the mating type genes are regulated. Mutations in any of four genes that are called *MAR* and *SIR* permit expression of the mating type genes from *HML* and *HMR*. Since these mutations are recessive, the *MAR* and *SIR* gene products likely code for repressors. Nonetheless, the mecha-

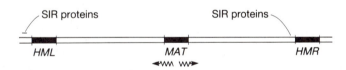

nism by which these proteins prevent expression of *HML* and *HMR* loci but permit expression of the genes in the *MAT* locus must be complex. One possible mechanism, for which some evidence exists, is that supercoiling density of the DNA is different in the different regions.

DNA CLEAVAGE AT THE *MAT* LOCUS

One way to initiate the transfer of a sequence from *HML* or *HMR* by a gene conversion mechanism would be to cleave a target sequence in *MAT*. A specific endonuclease much like a restriction enzyme could generate a double-stranded break in DNA

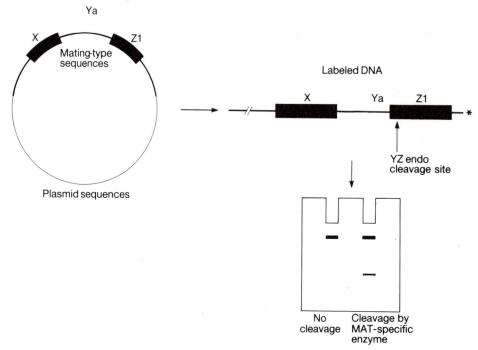

Figure 15.6   Scheme for sensitively detecting an endonuclease that cleaves specifically in *MAT* sequences.

at the *MAT* locus. How could such an enzyme be sought? Restriction enzymes can be detected by their cleavage of DNA, such as lambda phage or plasmid DNA. However, if the mating type endonuclease is specific, then only DNA carrying the *MAT* locus and possibly the *HML* and *HMR* loci would be substrates. Also, the enzyme generating the cleavages should be found in haploid *HO* yeast but probably would not be found in *ho* or in diploid yeast since these latter types do not switch mating type.

A particularly sensitive assay was used to search for the predicted cleavage (Fig. 15.6). A plasmid containing the mating type region was cut and end-labeled with $^{32}PO_4$. Incubation of this DNA with extracts prepared from various yeast strains should cut the DNA somewhere within the *MAT* locus. Such cleavage would generate a smaller radioactive fragment, which could be detected on electrophoresis of the digestion mixture and autoradiography. As expected, *HO* haploids contain a cleaving activity, and, furthermore, this activity is present at only one part of the cell cycle, just before the period of DNA synthesis. Using this cleavage assay, the enzyme has been partially purified, and biochemical studies on it are now possible.

The site of DNA cleavage lies just within the Z region of the *MAT* locus. Some of the mutations that act in *cis* to prevent the normal high frequency of mating-type conversion do not possess the double-stranded break *in vivo,* and, as expected, they prevent·cleavage *in vitro* and lie within 10 base pairs of the cleavage site.

REGULATED EXPRESSION OF DOPA DECARBOXYLASE IN *DROSOPHILA*

As a fruit fly develops from an egg to an adult, it molts two times as a larva, metamorphoses to the pupal state, and finally emerges as an adult fly. At each of these stages, the cuticular exoskeleton must be grown and then hardened and pigmented. Although the biochemistry of the tanning and hardening of the cuticle that occurs at these developmental stages is not fully understood, one of the essential components in the process is known to be N-acetyldopamine. This compound is synthesized from tryosine. Tanning and hardening the cuticle is not the full extent of

$$\text{Tyrosine} \longrightarrow \text{Dopa} \xrightarrow[]{\substack{\text{Dopa decarboxylase} \\ \text{(DDC)}}} \text{Dopamine} \longrightarrow \text{N-acetyldopamine}$$

this compound's use by the fly, for it is also found in the brain and nervous system and may be closely related to a neurotransmitter.

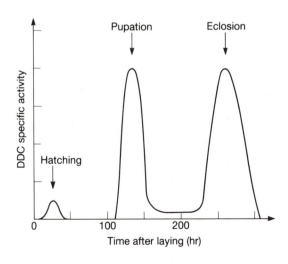

Figure 15.7   Dopa decarboxylase activity per milligram of protein in the egg, larva, and pupa as a function of time after the egg was laid.

As expected from the requirements for N-acetyldopamine, one of the enzymes required for its synthesis, dopa decarboxylase, appears in the hypoderm, the cells located just below the surface, just at the times of molting, pupation, and eclosion (Fig. 15.7). The enzyme, abbreviated DDC, is also synthesized in the brain at other times. Therefore, the mechanisms governing the dramatic time- and tissue-specific expression of DDC make the regulation of this enzyme an ideal subject for analysis.

CLONING THE DOPA DECARBOXYLASE GENE

As we have seen in the previous chapters, one of the key ingredients to a penetrating study of a gene regulation mechanism is the ability to manipulate the DNA of the gene. This requires cloning the gene first. Chapters 10 and 11 considered general methods for cloning of genes. Here another specific example is described.

Three basic facts were used to clone the dopa decarboxylase gene. First, the enzyme is synthesized at high rates only at certain times, primarily in the hypodermis; second, mRNA from the genes must, of course, code for the protein; and third, the genetic location of the gene on the *Drosophila* polytene chromosome was known.

DNA extracted from embryos was easily cloned into lambda vectors. The problem was determining which phage carried the dopa decarboxylase gene. This was done by searching for a lambda phage to which poly(A) mRNA extracted from larval integument hybridized in preference to poly(A) mRNA extracted from embryos. Although the poly(A) RNA itself could have been used as a probe, to increase the amount of radioactivity in the hybridization probe radioactive cDNA was made from the poly(A) mRNA, and this was hybridized to the lambda phage.

To increase further the hybridization signal, Hirsh had first partially purified the mRNA for dopa decarboxylase by separating the RNA according to size by sedimentation. Assay of the various RNA fractions for the dopa decarboxylase–specific mRNA was done by *in vitro* translation followed by precipitation with antibody against dopa decarboxylase and two-dimensional gel electrophoresis in which isoelectric focusing was done in one dimension and size fractionation was done in the other dimension (Fig. 15.8).

Clones that selectively hybridized cDNA sequences enriched for dopa decarboxylase sequences were then screened to determine from which chromosome and which part of the chromosome they originated. The necessary *in situ* hybridizations were done with radioactively labeled RNA made by *in vitro* transcription of the clones. One was found that hybridized to the region known to contain the dopa decarboxylase gene.

The approximate cytologic location of the dopa decarboxylase gene had been determined by classical *Drosophila* genetic methods by Hodgetts and by Wright. They had measured the levels of dopa decarboxylase in flies haploid, diploid, and triploid for many regions of the chromosome. Only for the region known as band 37C-37D did ploidy parallel the enzyme level.

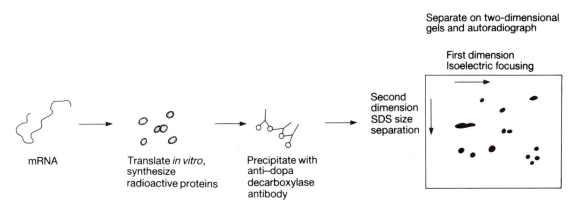

Figure 15.8   Detection of mRNA sequences that encode dopa decarboxylase.

The final proof that the candidate clone contained the dopa decarboxylase gene was provided by purifying larval mRNA that hybridized to the clone and then eluting the RNA and translating it with an *in vitro* translation system. This RNA yielded authentic dopa decarboxylase. The DNA from the phage could then be used for the typical experiments, such as determining the location of the gene by R-looping, confirming that the *in vivo* synthesis rate of dopa decarboxylase paralleled the level of the mRNA, and for the detection of intervening sequences within the gene.

TISSUE-SPECIFIC RESPONSES: PUTTING THE CLONED GENE BACK

To study the DNA elements involved in regulation *in vivo*, investigators must be able to put modified copies of the cloned dopa decarboxylase gene back into flies. Remarkably, a modification of the direct method actually works. Fertilized eggs can be injected with a DNA solution of plasmid containing the gene at a stage before the nuclei are surrounded by cell walls. A copy of the dopa decarboxylase DNA and surrounding DNA can then be transposed to the fly chromosome in some of the nuclei. If the transfer occurs to nuclei that become germ cells, then a copy of the transposed sequence will appear in the eggs or sperm deriving from this nucleus. Consequently, some of the next generation of flies can contain this sequence. If these flies are chosen for propagation, all of the next generation will contain copies of the dopa decarboxylase gene.

The transposition of the dopa decarboxylase gene from the plasmid into the chromosome was catalyzed by DNA element called transposon. The particular transposon used in this case is called a P element. These will be covered more fully in Chapter 18, but it suffices to say here that the P element codes for enzymes that

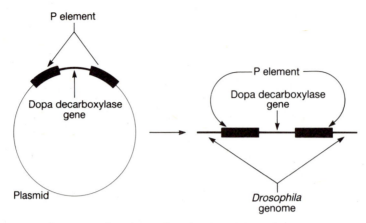

Figure 15.9   P element mediated transfer of a cloned dopa decarboxylase gene to the genome of *Drosophila*.

catalyze the transposition of the P element into new DNA sites. If the dopa decarboxylase gene has been inserted into the P element, then this DNA can be transferred into the fly chromosomes (Fig. 15.9).

How can the presence of the reintegrated dopa decarboxylase gene be detected? Fortunately, Wright had isolated temperature-sensitive mutations in the gene. Mutant flies grown at low temperature contain reduced levels of the enzyme, but if grown at high temperature, so little enzyme is present that they die. Consequently, transformants grown at 25° containing the injected gene possess nearly normal levels of the enzyme and can be identified by their slightly different cuticle color and also by the fact that they are viable and develop into adults at restrictive temperatures. With the ability to reintroduce the cloned gene and measure its regulation properties, study of the effects of DNA alterations became possible. Remarkably, removing part of the DDC upstream sequence altered the pattern of expression of the enzyme without eliminating expression.

TISSUE-SPECIFIC ENHANCERS FOR *DROSOPHILA* YOLK PROTEINS

The *Drosophila* genome contains three genes that code for yolk proteins. Two of these genes, *yp1* and *yp2*, are divergently oriented and separated by 1,200 base pairs. These genes are transcribed only in females and only at specific times and in specific tissues, mainly in the ovaries and in the fat body cells located primarily in the abdomen. After synthesis in these tissues, the proteins are transported into the eggs, where they serve as food for late developmental stages of the embryo.

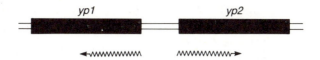

Like the dopa decarboxylase gene, the yolk protein gene system provides an ideal opportunity for study of tissue and time specificity in gene expression in developing organisms. Such genes could possess elements that regulate activity of the gene's promoter, one element for the tissue in which the gene is to be expressed and perhaps another that responds to developmental stage. Then the gene could be expressed only if both its tissue and time specificity elements were on. Such a scheme of gene regulation would raise at least two questions. First, how do the elements regulate activity of the promoter, and, second, what causes particular cells to differentiate to the point that they activate only certain of the tissue-specific elements?

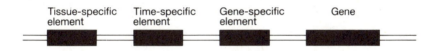

Remarkably, tissue-specific elements have been detected and localized for *yp1* and *yp2*. A clone of *yp1* and *yp2* was cut between the genes, and each of the separated genes was reintroduced into the fly using the p elements. When *yp1* and *yp2* are not separated, both are expressed in both the fat body and the ovary. However, Wensink and Garabedian have shown that after splitting, *yp1* is expressed only in the fat body and *yp2* is expressed only in the ovaries (Fig. 15.10). This means that there are at least

Figure 15.10   Expression pattern of the native and split yolk protein genes of *Drosophila*.

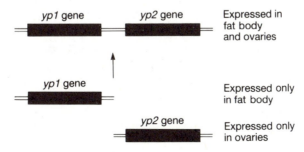

two independent tissue-specific expression elements acting on each gene. One specifies expression in the ovaries, the other in the fat body. The linear topology of the normal two-gene situation means that the stimulation by at least one of the elements leapfrogs the other. That is, the elements possess some of the characteristics of tissue-specific enhancers.

PROBLEMS

15.1. What can we infer about the biological molecules involved in mating type determination and function in light of the fact that mating type can be detected to change nearly every generation in *HO* cells?

15.2. Describe how the yeast transformant carrying a mating type gene could be detected.

15.3. How can dominance and recessivity of *MAR* and *SIR* genes be tested since mating type can ordinarily be determined for haploid yeast only?

15.4. A mutation known as *rad52* is lethal in haploid *HO* strains, apparently because of the double-stranded break that is generated at the *MAT* YZ site. How could the properties of this mutation be used to infer that the mating system generates no similar breaks elsewhere in the yeast chromosomes?

15.5. How would you test whether the absence of the mating-specific endonuclease in diploid and *ho* yeast results from its failure to be synthesized or from the synthesis of an inactivator of the enzyme?

15.6. Draw a scheme by which a double-stranded cleavage in the Z1 locus could initiate a gene conversion in which the Y region at *MAT* is altered to that of *HML* or HMR.

15.7. In seeking the enzyme that makes the double-stranded cut in the *MAT* locus, it would be wise to use diploid *HO* cells. What problem would this raise and how would you overcome it?

15.8. Why is it logical that the mating-type conversion be initiated by a cut in the DNA at *MAT* rather than a cut at *HML* or *HMR?*

15.9. What can be inferred if, of the enzymes catalyzing the synthesis of N-acetyldopamine from tyrosine, only dopa decarboxylase is developmentally regulated?

15.10. Suppose that a temperature-sensitive mutation were not available in the dopa decarboxylase gene. How could the cloned gene have been modified so that its transcription could have been assayed when it was reintroduced into flies?

15.11. How would you seek a protein that specifically binds to an enhancer?

RECOMMENDED READINGS

Transposable Mating Type Genes in *Saccharomyces cerevisiae*, J. Hicks, J. Strathern, A. Klar, Nature *282*, 478–483 (1979).

The Structure of Transposable Yeast Mating Type Loci, K. Nasmyth, K. Tatchell, Cell *19*, 753–764 (1980).

The Cloned Dopa Decarboxylase Gene Is Developmentally Regulated When Reintegrated into the *Drosophila* Genome, S. Scholnick, B. Morgan, J. Hirsh, Cell *34*, 37–45 (1983).

RELATED REVIEWS, BOOKS, AND ARTICLES

The Molecular Biology of the Yeast Saccharomyces, ed. J. Strathern, E. Jones, J. Broch, Cold Spring Harbor Laboratory (1981).

Molecular Genetics of Yeast Mating Type, K. Nasmyth, Ann. Rev. Genet. *16*, 439–500 (1982).

DEEPER READING

A New Method for Hybridizing Yeast, C. Lindegren, G. Lindegren, Proc. Nat. Acad. Sci. USA *29*, 306–308 (1943).

Mating Types in *Saccharomyces:* Their Convertibility and Homothallism, Y. Oshima, I. Takano, Genetics *67*, 327–335 (1971).

The Response of Dopa Decarboxylase Activity to Gene Dosage in *Drosophila:* A Possible Location of the Structural Gene, R. Hodgetts, Genetics *79*, 45–54 (1975).

The Genetics of Dopa Decarboxylase in *Drosophila melanogaster*, I. Isolation and Characterization of the Structural Locus and the α-Methyl Dopa Hypersensitive Locus, T. Wright, R. Hodgetts, A. Sherald, Genetics *84*, 267–285 (1976).

The Genetics of Dopa Decarboxylase in *Drosophila melanogaster*, II. Isolation and Characterization of Dopa Decarboxylase Mutants and Their Relationship to the α-Methyl Dopa Hypersensitive Mutants, T. Wright, G. Bewley, A. Sherald, Genetics *84*, 287–310 (1976).

Mapping of the Homothallism Genes *HMα* and *HMa* in *Saccharomyces* Yeasts, S. Harashima, Y. Oshima, Genetics *84*, 437–451 (1976).

Asymmetry and Directionality in Production of New Cell Types During Clonal Growth: The Switching Pattern of Homothallic Yeast, J. Strathern, I. Herskowitz, Cell *17*, 371–381 (1979).

MAR-1-A Regulator of *HMa* and *HMα* Loci in *Saccharomyces cerevisiae*, A. Klar, S. Fogel, K. MacLeod, Genetics *93*, 37–50 (1979).

A Suppressor of Mating Type Locus Mutations in *Saccharomyces cerevisiae:* Evidence for and Identification of Cryptic Mating Type Loci, J. Rine, J. Strathern, J. Hicks, I. Herskowitz, Genetics *98*, 837–901 (1979).

The Structure and Organization of Transposable Mating Type Cassettes in *Saccharomyces* Yeast, J. Strathern, E. Spatola, C. McGill, J. Hicks, Proc. Nat. Acad. Sci. USA *77*, 2839–2843 (1980).

Induction of Translatable mRNA for Dopa Decarboxylase in *Drosophila:* An Early Response to Ecdysterone, G. Kraminsky, W. Clark, M. Estelle, R. Gietz, B. Sage, J. O'Connor, R. Hodgetts, Proc. Nat. Acad. Sci. USA *77*, 4175–4179 (1980).

The Isolation and Characterization of *Drosophila* Yolk Protein Genes, T. Barnett, C. Pachl, J. Gergen, P. Wensink, Cell *21*, 729–738 (1980).

Developmental Relationship Between Dopa Decarboxylase, Dopamine Acetyl Transferase, and Ecdysone in *Drosophila*, J. Marsh, T. Wright, Dev. Biol. *80*, 379–387 (1980).

Regulation of Transcription in Expressed and Unexpressed Mating Type Cassettes of Yeast, A. Klar, J. Strathern, J. Broach, J. Hicks, Nature *289,* 239–244 (1981).

The Sequence of the DNAs Coding for the Mating-Type Loci in *Saccharomyces cerevisiae,* C. Astell, L. Ahlstrom-Jonasson, M. Smith, K. Tatchell, K. Nasmyth, B. Hall, Cell *27,* 16–23 (1981).

Isolation and Characterization of the Dopa Decarboxylase Gene of *Drosophila melanogaster,* J. Hirsh, N. Davidson, Mol. Cell. Biol. *1,* 475–485 (1981).

Transcript Maps of *Drosophila* Yolk Protein Genes, M. Hung, T. Barnett, C. Woolford, P. Wensink, J. Mol. Biol. *154,* 581–602 (1982).

The Regulation of Yeast Mating-Type Chromatin Structure by SIR: An Action at a Distance Affecting Both Transcription and Transposition, K. Nasmyth, Cell *30,* 567–578 (1982).

Homothallic Switching of Yeast Mating Type Cassettes Is Initiated by a Double-Stranded Cut in the *MAT* Locus, J. Strathern, A. Klar, J. Hicks, J. Abraham, J. Ivy, K. Nasmyth, C. McGill, Cell *31,* 183–192 (1982).

Genetic Transformation of *Drosophila* with Transposable Element Vectors, G. Rubin, A. Spradling, Science *218,* 348–353 (1982).

Deletions and Single Base Pair Changes in the Yeast Mating-Type Locus That Prevent Homothallic Mating-Type Conversions, B. Weiffenbach, D. Rogers, J. Haber, M. Zoller, D. Russell, M. Smith, Proc. Nat. Acad. Sci. USA *80,* 3401–3405 (1983).

Molecular Analysis of a Cell Lineage, K. Nasmyth, Nature *302,* 670–676 (1983).

Sequence and Structure Conservation in Yolk Proteins and Their Genes, M. Hung, P. Wensink, J. Mol. Biol. *164,* 481–492 (1983).

Genetic Dissection of Monamine Neurotransmitter Synthesis in *Drosophila,* M. Livingstone, B. Tempel, Nature *303,* 67–70 (1983).

Correct Developmental Expression of a Cloned Alcohol Dehydrogenase Gene Transduced in the *Drosophila* Germ Line, D. Goldberg, J. Posakony, T. Maniatis, Cell *34,* 59–73 (1983).

A Site-Specific Endonuclease Essential for Mating-Type Switching in *Saccharomyces cerevisiae,* R. Kostriken, J. Strathern, A. Klar, J. Hicks, F. Heffron, Cell *35,* 167–174 (1983).

Molecular Cloning of Hormone-Responsive Genes from Yeast *Saccharomyces cerevisiae,* G. Stetler, J. Thorner, Proc. Nat. Acad. Sci. USA *81,* 1144–1148 (1984).

A Conserved DNA Sequence in Homoeotic Genes of the *Drosophila* Antennapedia and Bithorax Complexes, W. McGinnis, M. Levine, E. Hafen, A. Kuroiwa, W. Gehring, Nature *308,* 428–433 (1984).

Independent Control Elements that determine Yolk Protein Gene Expression in Alternative *Drosophila* Tissues, M. Garabedian, M. Hung, P. Wensink, Proc. Nat. Acad. Sci. USA *82,* 1396–1400 (1985).

Lambda Phage Genes and Regulatory Circuitry

A s a result of its size and biological properties, lambda phage has been intensively studied for many years. Lambda is neither so small that each of its genes must play multiple roles in phage development nor so large that there are too many genes to study or understand. Furthermore, lambda has an interesting dual mode of existence. On one hand, a lambda phage particle can infect a cell, grow vegetatively to produce a hundred copies of itself, and lyse the cell. On the other hand, a lambda phage particle can infect a cell and enter a quiescent phase. In this lysogenic state only three phage genes are expressed, and both daughters of cell division are similarly lysogenic. Although it is highly stable and can be passed for many generations to descendant cells, lysogeny need not be permanent. The lambda within a lysogenic cell can be induced to enter its vegetative mode and will then multiply and lyse the cell.

Two fundamental questions are raised by lambda. First, how does lambda regulate its growth? Lytic growth alone poses regulation problems because genes must be expressed at the right time and to the correct extent to maximize viable phage yield. A second aspect to lambda's regulation problems is lysogeny. The phage existing as a lysogen must be able to switch most of its genes back on upon induction. The regulation system involved with the two states must be reasonably stable because the lambda in a lysogenic cell only rarely induces spontaneously to begin vegetative growth.

The second fundamental question is raised by the existence of the lysogenic state. In such a state, lambda DNA has been integrated into the host chromosome, and, upon induction, lambda excises from the chromosome. How are these steps accomplished? What enzymes and DNA substrates are required for these reactions? How is the reaction forced to proceed in the direction of integration following infection, and how is it forced to go in the direction of excision upon phage induction?

The research that has been concentrated on lambda has told us much about this phage, particularly the research that combined genetic analysis and physiological studies. One unexpected practical result of these studies was the ability to use lambda as a vector in genetic engineering. Now most research on lambda is concentrated on the more difficult questions of how things happen rather than what happens, where the genes are, and what their products accomplish for the phage.

This chapter explains the structure and scheme of gene regulation in phage lambda. Part 4 describes mobile DNA, beginning with a chapter on the integration and excision of lambda DNA from the host chromosome.

A.   The Structure and Biology
of Lambda

THE PHYSICAL STRUCTURE OF LAMBDA

Lambda phage particles possess an isometric head of diameter 650 Å and a tail 1,500 Å long and 170 Å wide (Fig. 16.1). A tail fiber extends an additional 200 Å from the tail. This tail fiber makes a specific contact with a protein of the host cell, and the phage DNA is injected through the tail into the cell.

The DNA within the phage particle is double-stranded and linear, with a length such that if it were in its circular form it would be exactly 48,502 base pairs long. This is roughly 1% the size of the *Escherichia coli* chromosome. Intracellularly, lambda DNA exists as monomeric circles or polymeric rolling circle forms, but during encapsidation, unit-length lambda linear genomes are generated by two cuts of the DNA strands offset from one another by 12 bases (Fig. 16.2). Thus the ends of the encapsidated phage DNA are single-stranded and self-complementary. The arrangement of DNA within the head suggests that the DNA is pushed into the head through a hole and then the tail is attached, and proposals have been made for such insertion mechanisms.

Roughly half the mass of a lambda particle is DNA and half is protein. Conse-

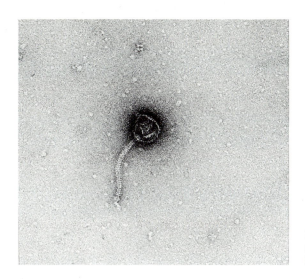

Figure 16.1   Electron micrograph of a negatively stained lambda phage particle. (Courtesy T. Kirchhausen.)

quently, the particle has a density in CsCl halfway between the density of DNA, 1.7 gm/cm³, and the density of protein, 1.3 gm/cm³. This density facilitates purification of the phage since isopycnic banding in CsC1 density gradients easily separates the phage from most other cellular components.

THE GENETIC STRUCTURE OF LAMBDA

The first mutations isolated and mapped in phage lambda were those that changed its plaque morphology. Ordinarily, lambda plaques are turbid or even contain a minicolony of cells in the plaque center. These result from growth of cells that have become immune to lambda infection. Consequently, lambda mutants that produce clear plaques may easily be identified amid many turbid plaques. Kaiser isolated and

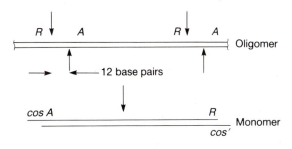

Figure 16.2   Cleavage at the two *cos* sites generates sticky-ended lambda monomers from oligomers during encapsidation of lambda DNA.

mapped such clear plaque mutants of lambda. These fell into three complementation groups, which Kaiser called *CI, CII,* and *CIII,* with the C standing for clear.

A   J b_2 int   β   γ   *CIII*   N   *CI*   cro   *CII*   O   P   Q   S   R R_z

One specific mutation in the *CI* gene is particularly useful. It is known as CI_{857}, and the mutant *CI* product is temperature-sensitive. At temperatures below 37° the phage forms normal turbid plaques, but at temperatures above 37° the mutant forms clear plaques. Furthermore, lysogens of lambda CI_{857} can be induced by shifting their temperature above 37°.

Many additional lambda mutations were isolated and mapped by Allan Campbell. These were not restricted to nonessential genes, as he used conditional mutations. On one host bacterial strain, the phage grow because the mutation is not expressed, but on another strain they do not grow because the mutation is expressed and the mutant gene product is inactive. Nonsense suppressing–nonsuppressing pairs of hosts provide a good method for isolation and study of mutations in phage. The mutants are isolated by plating mutagenized phage on a nonsense-suppressing strain. Plaques deriving from phage containing nonsense mutations may be identified by their inability to grow after being spotted onto the nonsuppressing strain.

Phage within nonsense mutations in various genes may then be studied by first preparing phage stocks of the nonsense mutants on suppressing hosts. The phage can then be used in a variety of studies. For example, pairs of mutants can be crossed against one another, and the frequency of generation of wild-type phage by recombination between the two mutations can be quantitated by plating on nonsuppressing strains. The nonsense mutants also facilitate study of phage gene function. Nonsuppressing cells infected with a nonsense mutant phage stock progress only partway through an infective cycle. The step of phage development and maturation that is blocked by the mutation can be determined by use of radioactive isotopes to quantitate protein, RNA, and DNA synthesis or electron microscopy to determine which phage macromolecules or structures are and which are not synthesized.

Campbell named the genes he found and mapped *A* through *R,* left to right. The genes identified and mapped after his work are identified by the remaining letters of the alphabet, by three-letter names, and by other symbols.

The sequence of the entire lambda DNA molecule has been determined. A few of the known genes could be identified in the sequence by the amino acid sequences of their products or by mutations that changed the DNA sequence. Many others could be identified with a high degree of confidence by their position and the existence of a good Shine-Dalgarno ribosome-binding sequence preceding a sequence potentially coding for a protein of appropriate size (Fig. 16.3). The identifica-

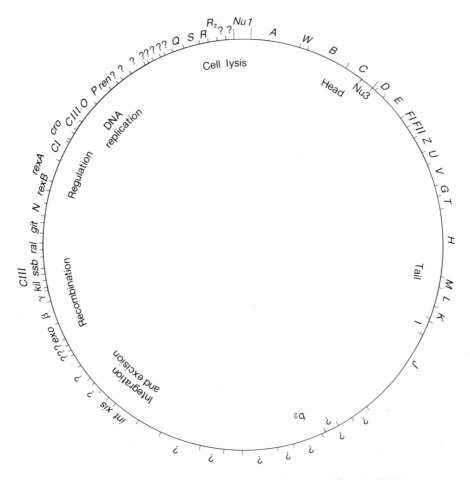

Figure 16.3 The complete genetic map of lambda determined from its DNA sequence. The sizes of genes are indicated along with the functions of various classes of proteins.

tion of such open reading frames was greatly assisted by examination of codon usage. An open reading frame that is not translated into protein usually contains all sense codons at about the same frequency, whereas the reading frames that are translated into protein tend to use a subset of the codons. That is, many proteins use certain codons with a substantially higher frequency than other codons. Unexpectedly, applying these criteria to the DNA sequence of phage lambda revealed 14 possible genes not previously identified genetically or biochemically. It remains to be shown how many of these actually play a role in phage growth and development.

The DNA sequencing revealed a second unexpected property of the lambda genes. Many are partially overlapped. The function of this overlap could be to conserve coding material, although it may also play a role in translation. For example, overlapping a translation initiation codon AUG with a termination codon UGA in the sequence AUGA could eliminate the need for a ribosome-binding site for the second gene. Since ribosomes drift phaselessly forward and backward after encoun-

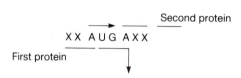

tering a nonsense codon and before dissociating from the mRNA, this overlapping can reduce the requirement for a Shine-Dalgarno ribosome-binding sequence for the second gene.

Genes of related function are clustered in the lambda genome. The genes *A* through *F* are required for head formation, *Z* through *J* for tail formation, the genes in the b_2 region are not essential for phage growth and may be deleted without material effect. Since such deletion phage still possess the normal protein coat but lack about 10% of their DNA, they are less dense than the wild-type phage; that is, they are *buoyant density mutants*. The genes *int* and *xis* code for proteins that are involved with integration and excision and will be discussed in the next chapter. The genes *exo*, *β*, and *γ* are involved with recombination. The proteins from genes *CIII N, CI, cro,* and *CII* regulate the expression of phage genes. The genes *O* and *P* code for proteins required for initiation of lambda DNA replication. The *Q* gene protein turns on synthesis of the entire set of genes required for formation of the head and tail as well as the genes *S*, *R*, and R_z, which are required for cell lysis.

LYSOGENY, IMMUNITY, INTEGRATION, AND EXCISION

When lambda lysogenizes a cell, it inserts its DNA into the DNA of the host. There the lambda DNA is passively replicated by the host just as though it were host DNA. Such a lambda lysogen is highly stable, and lysogeny can be passed on for hundreds of generations. Upon induction of the phage, which may be either spontaneous or caused by exposure to inducing agents such as UV light, lambda excises itself from the chromosome and enters a lytic cycle. About one hundred phage are produced, and the cell lyses.

The advantages to the phage and cell of lysogeny would largely be lost if a lysogenic cell could be lytically infected by another lambda phage. Indeed, lambda lysogenic cells cannot be lytically infected with more phage. The cells are said to possess lambda immunity. The superinfecting phage can absorb to immune cells and inject their DNA into the cell, but the superinfecting phage DNA does not function. This DNA is not replicated or integrated into the host chromosome and therefore is diluted away by cell growth since it is passed to only one of the two daughter cells on cell division.

Lambda repressor protein, the product of the CI gene, confers immunity. Repressor encoded by the lysogenic phage diffuses throughout the cytoplasm of the cell. When another lambda phage injects its DNA into this cell, the repressor binds to specific sites on the superinfecting DNA and inactivates the promoters necessary for the first steps of vegetative phage growth. This same repressor activity also prevents the lysogenic phage from initiating its growth. However, upon induction of the lysogen, the repressor is destroyed and the phage can begin a lytic growth cycle.

LAMBDA'S RELATIVES AND LAMBDA HYBRIDS

Lambda does not uniquely occupy its ecological niche. A variety of both close and distant relatives of lambda are known (Table 16.1). The near relatives all possess the same sticky ends of the chromosome and are of almost the same size and genetic structure as lambda. Some possess the same immunity as lambda, which means that they cannot grow in lambda lysogens, but others possess different immunities. That is, they are heteroimmune. Remarkably, the lambda relatives can form recombinants between one another to form hybrids. DNA heteroduplexes between them show that their genes tend to be either closely homologous or quite dissimilar. It is as though nature possesses a few fundamental lambda-type phage and can inter-

Table 16.1   Properties of a Number of Lambdoid Phage

Phage	Immunity	Position When Integrated in Host	Sticky Ends
λ	λ	gal	λ
21	21	trp	λ
ϕ80	ϕ80	trp	λ
ϕ81	ϕ81	gal	λ
82	82	gal	−
434	434	gal	λ

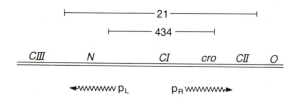

Figure 16.4   The regions of hybrid lambda imm[21] and lambda imm[434], which derive from lambda and phage 21 or phage 434, respectively.

change their parts to produce the large number of different lambdoid phage that are observed.

A study of the similarities and differences between lambda's relatives has been fruitful in understanding lambda. Phage 21, ϕ80, and 434 have been particularly useful in the research on phage lambda. They have highlighted the crucial invariants of lambda growth and regulation. Recombinant hybrids have also been constructed between lambda and both 21 and 434 (Fig. 16.4). These were constructed so as to retain the immunity of the heteroimmune phage but to contain as much of the rest of lambda as possible. These lambda hybrids are called lambda imm[434], or lambda i[434], and lambda i[21]. They permit recombination and complementation studies that otherwise would be impossible because of the existence of immunity.

B.   Chronology of a Lytic Infective Cycle

LAMBDA ADSORBS TO CELLS VIA A MALTOSE TRANSPORT PROTEIN

Phage particles ought not to attempt wantonly injecting their DNA into anything. For the highest survival value, they should release their encapsidated DNA only after they have made stable contact with a suitable host cell. Not unexpectedly, then, the lambda tail fiber makes a specific contact with an *E. coli* outer membrane protein. Only when this protein is present can lambda adsorb and inject.

Many different membrane structures are used by different phage in the adsorption and injection process. Lambda uses a protein specified by the maltose operon. It is named the *lamB* protein because its first known function was lambda adsorption and only later was it discovered to be part of the maltose operon. The normal cellular function of the *lamB* protein is to create a pore through the outer membrane somewhat larger than the pores usually found there. This maltose-inducible pore is necessary for the diffusion of maltodextrins, and maltose if it is present only at low concentrations, to the periplasmic space. Induction of the maltose operon increases the levels of this protein and speeds the adsorption process; eliminating synthesis of this protein or mutationally altering it makes cells resistant to lambda and can leave them Mal⁻.

TERMINATION RESTRICTS EARLY TRANSCRIPTION TO GENES *N* AND *cro*

Within a minute after injection into the host, the sticky ends of the lambda DNA anneal to each other and are covalently joined by the DNA ligase of the host. However, transcription of the lambda DNA begins promptly upon infection and probably does not require this circularization. The major transcription begins from the two promoters p_L and p_R. These are located on either side of *CI*. A third promoter, p_{RM}, possesses less activity and is located farther to the right of *CI*.

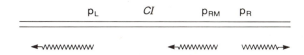

As on all promoters, initiation of transcription from the promoters p_L and p_R is followed by the dissociation of the σ subunit of RNA polymerase after the polynucleotide chains have reached a length of 6 to 15 nucleotides. Then a host protein, the *nusA* product, binds to RNA polymerase. *Nus* stands for N utilization substance. Most likely it binds in the site previously occupied by the σ subunit because only one or the other of these proteins can bind to RNA polymerase at once.

The RNA polymerase molecules continue past the point where the σ and *nusA* subunits exchange and they transcribe through the *N* gene on the left and the *cro* gene on the right. However, this is all that happens at this stage of the lambda infection, for just beyond these genes the RNA polymerase pauses, assisted by the *nusA* protein, and the host transcription terminator, *rho*, releases the RNA chain and RNA polymerase from the DNA (Fig. 16.5). This pausing and termination occurs at the first transcription termination sites in the p_L and p_R operons, t_{L1} and t_{R1}. Additional terminators with slightly different properties, t_{L2} and t_{R2}, exist farther down-

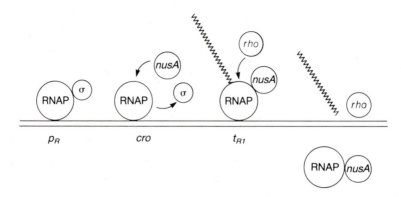

Figure 16.5   The σ subunit dissociates from RNA polymerase and is replaced by the *nusA* protein. At the terminator, the *rho* protein interacts with the complex and releases RNA and RNA polymerase.

stream in both operons. As a result of t_{L1} and t_{R1}, the first stage of lambda vegetative growth is confined to the accumulation of N and *cro* proteins.

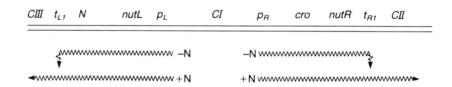

N PROTEIN PREVENTS TERMINATION OF EARLY GENE TRANSCRIPTION

Several minutes after infection, N protein reaches sufficiently high concentrations in the cytoplasm to be able to bind to RNA polymerase. This binding occurs as the RNA polymerase crosses a special phage DNA sequence called *nut*, from *N-utilization*. Without a functional *nut* sequence, or without the *nusA* protein already bound to RNA polymerase, the N protein does not bind to the polymerase. RNA polymerase with N protein bound to it no longer terminates transcription at the sites t_{L1} and t_{R1} and instead continues across these sites to synthesize messenger for genes *CII* and *CIII* as well as the more distal genes in these operons (Fig. 16.6).

It should be mentioned at this time that a number of *E. coli* mutants exist in which wild-type lambda does not grow. Some of these behave as though the lambda

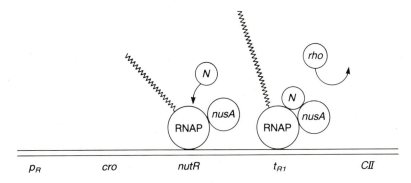

Figure 16.6   A representation of how lambda *N* protein could antagonize the action of *rho* protein at the terminators.

possessed a mutation in the *N* protein. These are called *nus* mutants, and the *nusA* protein is the product of one such gene. Presumably, *nusA* protein can be altered so that it still fulfills its normal cellular function, but lambda *N* protein does not properly interact with it. Consequently, termination always occurs at t_{L1} and t_{R1} in *nus* mutants. Other *nus* mutations lie in RNA polymerase and in ribosome genes.

Cro PROTEIN BLOCKS *CI* REPRESSOR SYNTHESIS AND WINS THE RACE

Lambda development may proceed either down the vegetative pathway or down the lysogenic pathway. The main protagonists in this decision are *cro* protein and the

Figure 16.7   *Cro* protein first inactivates p_{RM} and at higher concentrations reduces activity of p_L and p_R.

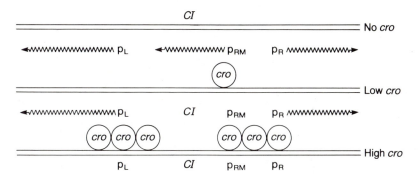

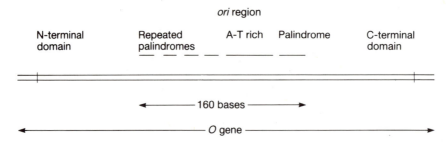

Figure 16.8   The structure and location of the *ori* region within the lambda *O* gene.

CI protein, both of which are synthesized initially, *cro* from p_R and *CI* from p_{RM}. After several minutes the balance tips ever so slightly one way or the other, and *cro* or *CI* commits the phage to one of the possibilities. In a lytic response, the *cro* protein gains the upper hand and blocks the synthesis of *CI*. This is accomplished in two ways (Fig. 16.7). First, *cro* binds to an operator adjacent to the promoter for the synthesis of *CI*, the p_{RM} operator. Thus, no *CI* is made via this route. Later, *cro* represses transcription from p_L and p_R by binding to operators adjacent to the promoters. This secondary effect reduces the synthesis of the phage *CII* and *CIII* proteins, which, when present at sufficiently high concentrations, stimulate the synthesis of *CI* from a promoter called p_{RE}. The p_{RE} promoter is explained more fully in a later section.

PROTEINS *O* AND *P* INITIATE DNA SYNTHESIS

Transcription originating from p_R leads to the accumulation of the *O* and *P* proteins. These, plus transcription into or near the *ori* site, direct phage DNA replication to begin. When the *O* protein has reached appropriate levels, it binds to a series of four repeats of a palindromic DNA sequence located at *ori* (Fig. 16.8). The *P* protein binds to *O* protein, and the host protein *dnaB* binds to *P*. The additional host proteins, *dnaE*, *dnaG*, *dnaJ*, *dnaK*, *dnaZ*, RNA polymerase, and DNA gyrase, are also required for the DNA replication, which then proceeds outward in both directions from *ori*.

Curiously, *ori* is located within the lambda *O* gene itself! The coding regions of the *O* gene on either side of *ori* specify amino acids whose secondary structure, as predicted by the Chou-Fasman folding rules mentioned in Chapter 5, yields well-defined domains. However, the amino acids encoded by the portion of *O* containing *ori* — which consists of the four repeats, an A-T-rich region, and a palindrome — are not predicted to possess much secondary structure. Such a picture is consistent with the experimental findings that the N-terminal portion of the *O* protein contains the phage-specific DNA determinants and that the carboxy-terminus of *O* binds the lambda *P* protein. The *O* protein itself is highly unstable *in vivo*, suggesting that the unstructured central portion of the protein is susceptible to proteolysis.

Mutations in the *ori* site itself support the structure described above. *Ori* mutants are recognized because they are *cis*-dominant mutations affecting DNA replication and they generate very tiny plaques. The sequence of several such dominant mutations has shown that they lie in the *ori* section of *O*. Some are small deletions within the *ori* region, but the sizes of all preserve the reading frame; that is, they are multiples of three bases. Therefore, the amino acids encoded by the *ori* portion of the *O* gene are unimportant to the functioning of *O* protein.

Ordinarily the requirement for transcriptional activation of *ori* is met by transcription originating from p_R. However, lambda mutants may be isolated that create other promoters that activate *ori*. Most curiously, their transcription does not have to cross *ori*. One promoter that activates lambda phage DNA replication lies 95 base pairs away, and its transcription is directed away from *ori*. One plausible role for the transcription requirement is to bring or hold the *ori* region out of the highly concentrated mass of chromosomal DNA. As discussed in Chapter 1, chromosomal DNA is so highly concentrated that many proteins do not have free access to it. Since transcription likely occurs by RNA polymerase located on the surface or outside this mass, transcription near *ori* could expose *ori* to proteins necessary for DNA replication.

PROTEINS *kil*, γ, β, AND *exo*

Leftward transcription beyond *CIII* leads to the accumulation of the *kil*, γ, β, and *exo* proteins in addition to others whose functions are not known but that are nonessential. *Kil* stops cell division. The γ protein shuts off the host recombination pathway, and the *exo* and β proteins open a new pathway for genetic recombination.

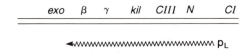

Turning off the host recombination pathway is essential to the phage. As mentioned above, lambda DNA is first replicated bidirectionally from the *ori* region. Later in its growth cycle, however, lambda's replication shifts to a rolling circle mode. This generates the oligomers of lambda DNA that are obligatory for encapsidation. However, few if any rolling circles can be generated in the presence of active *recBC* enzyme. Apparently this enzyme recognizes an intermediate in the generation of a rolling circle as a recombination intermediate. Therefore, to prevent the cell from blocking the process of making rolling circles, lambda inactivates *recBC* with a protein of its own, γ (Fig. 16.9). However, lambda still has need for recombination,

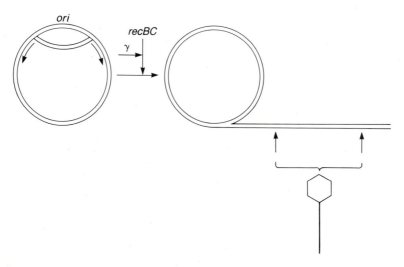

Figure 16.9   Conversion of theta replicative forms to rolling circles is blocked by *recBC*, whose activity is itself blocked by *γ*.

and it therefore potentiates another pathway for genetic recombination by synthesizing the *exo* and *β* proteins.

Q PROTEIN BLOCKS TERMINATION TO SYNTHESIZE LATE PROTEINS

The final genes in the early operons are the *int* and *xis* genes on the left and the *Q* gene on the right. The *int* and *xis* genes are important in the lysogenization process and are discussed in the next chapter. The *Q* protein functions as an antiterminator of a promoter located immediately to its right. Without *Q* protein, polymerase

Figure 16.10   Action of *Q* protein in preventing premature termination at $t_{R'}$.

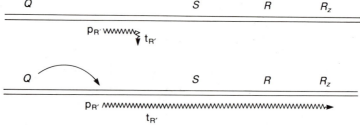

initiates at this promoter and terminates 190 bases later (Fig. 16.10). However, when *Q* protein is present, this termination is abolished, and transcription proceeds across the *S* and *R* genes and into the late genes. The late genes code for the head and tail structures of the phage. Since transcription of the late genes occurs at a time when about 40 copies of the DNA exist in the cell, large quantities of the phage late proteins are synthesized.

PACKAGING AND LYSIS

One problem the phage must solve is lysing the cells at the right moment. Most likely this time is a compromise that has been worked out over the eons between maximizing the number of completed phage particles released and minimizing the interval after infection or induction until some completed phage are released. Although infection can begin by slipping lambda's DNA molecule into the cell through a small hole, release of newly assembled phage particles requires something more drastic. At the least, holes large enough for the phage head must be punched in the rigid peptidoglycan layer, and the inner and outer membranes must be ruptured as well.

Three lambda proteins are known to participate in the lysis process. They are all late proteins synthesized under control of the *Q* gene product. The first to be identified is the product of the lambda *R* gene. Originally this was called the lysozyme for its ability to lyse cells, then for a while it was mistakenly called an endopeptidase or endolysin for the specific bonds in the peptidoglycan the enzyme was thought to cleave, and now it is correctly known to be a transglycosylase (Fig. 16.11). Another lambda-encoded protein also degrades the peptidoglycan layer of the cell. It is the product of the R_z gene, and it is an endopeptidase. The third protein required for lysis is the product of the *S* gene. Experiments indicate that this protein forms a pore

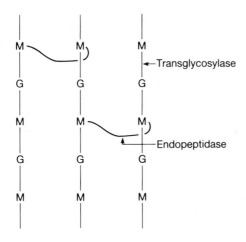

Figure 16.11   The structure of the peptidoglycan layer and the bonds cleaved by the *R* and R_z proteins.

through the inner membrane so that the R and R_z products, which are cytoplasmic proteins, are provided access by S protein to their substrate the peptidoglycan.

The behavior of cells infected with S^- phage is consistent with the idea that the S gene codes for a pore. As expected, the R and R_z products accumulate in the cytoplasm of such cells, and lysis does not occur unless the inner membrane is damaged. Chloroform treatment or freezing and thawing are two methods for disrupting the integrity of the inner membrane. Protein synthesis, DNA synthesis, and respiration do not shut off in cells infected with S^- phage as they normally do 40 minutes after infection with S^+ phage. Therefore, the shutoff of macromolecular synthesis that normally occurs 40 minutes after infection results from leakage of crucial intracellular components, as at this time the cell loses its ability to concentrate small molecules intracellularly.

As mentioned in Chapter 12, S^- mutants facilitate work with lambda. First, since macromolecular synthesis does not shut off at 40 minutes, phage continue to be made and the phage yield per infected or induced cell is raised from about 100 to about 500. Second, phage may easily be harvested from the cell growth medium by centrifuging cells full of phage into pellets, resuspending in small volumes, and lysing by the addition of chloroform. As a result, large quantities of highly concentrated phage are easily obtained.

C.   The Lysogenic Infective Cycle and Induction of Lysogens

CHRONOLOGY OF BECOMING A LYSOGEN

We now examine how an infecting lambda phage can shut off transcription of all but three of its own genes. The beginning steps are the same as in a lytic infection: DNA is injected into the cell, and the N and cro proteins, which are sometimes called the immediate early genes, begin to accumulate. Then the remainder of the proteins synthesized under control of the p_L and p_R promoters, including CII and $CIII$, begin to be synthesized. These latter proteins are crucial to the lysogenic response in two ways. First, they are necessary if CI is to win the race with cro, and, second, they are required for adequate synthesis of the one phage protein required for integration of phage DNA into the host DNA, the int protein.

CII and $CIII$ proteins activate two phage promoters. One, p_{RE} is located to the right of cro and is oriented opposite p_R. Thus the initial part of the RNA synthesized under control of this promoter is anti-cro messenger. The complementarity of this

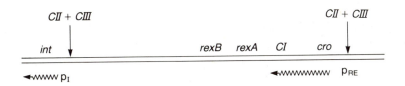

RNA to *cro* messenger is not known to have any biological importance, however. The important fact is that the transcription initiated at p_{RE} continues into the *CI* gene, and therefore lambda repressor, the *CI* gene product, as well as two other proteins in this operon, *rexA* and *rexB*, accumulate at a high rate. The *CII* and the *CIII* proteins also assist RNA polymerase to initiate transcription from the p_I promoter. This promoter lies partly in the *xis* gene and just in front of the *int* gene.

As the lambda repressor accumulates, resulting from the highly active p_{RE}, it soon reaches a level sufficient to bind to the operators that overlap p_L and p_R. This binding shuts off synthesis of *N*, *cro*, and the other early gene products. Hence, although a little *exo*, β, γ, *O*, *P*, and *Q* have accumulated, their levels are insufficient to sustain growth. Lambda destined to become a lysogen may even undergo several rounds of DNA replication, but then the early genes are shut off and the instability of *O* prevents further development. The *int* protein, which also has been synthesized at high rates under control of p_I, participates with a number of host proteins in the integration of one copy of the lambda DNA into the host.

After high levels of lambda repressor have effectively shut off the phage early promoters, three phage promoters can still produce functional messenger, p_{RM}, p_{RE}, and p_I. However as *CII* and *CIII* lose activity and are diluted away with cell growth, p_{RE} and p_I shut down and leave only p_{RM} active. This occurs whether or not lambda has been successful in integrating itself into the host chromosome. If integration has occurred, then each daughter cell inherits a copy of the lambda genome and is lysogenic. If integration has not occurred, then lambda is passed to only one daughter and soon is lost by dilution.

Cro AND REPRESSOR REPRESS AND INDUCE FROM THE SAME SITES

Regulation of the early promoters by *cro* and repressor is more complicated than has been indicated. At both p_L and p_R are three binding sites for each of these proteins, O_{L1}, O_{L2}, and O_{L3} on the left and O_{R1}, O_{R2}, and O_{R3} on the right. Repressor bound at O_{L1} is sufficient to inactivate p_L. Repressor and *cro* binding on the right is more interesting and complicated because these proteins regulate the activities of both p_R and P_{RM}. Repressor binding to the operators on the right acts as a repressor for p_R and a stimulator or, at higher concentrations, as a repressor for p_{RM} (Table 16.2). These different activities are accomplished by the following means. Repressor bound at O_{R1}

$$p_L \qquad\qquad CI \qquad p_{RM} \qquad p_R$$

$$O_{L1} \ O_{L2} \ O_{L3} \qquad\qquad O_{R3} \ O_{R2} \ O_{R1}$$

inactivates p_R, but repressor bound at O_{R1} and O_{R2} simultaneously represses p_R and activates p_{RM}. Repressor bound at O_{R3} inactivates p_{RM}. *Cro* protein bound at O_{R3} represses p_{RM}, and *cro* bound to O_{R1} or O_{R2} represses p_R (Table 16.3).

Despite the fact that *cro* and repressor bind to virtually the same sequences, at least as assayed by DNase protection and the behavior of mutations lying in the three operator sites on the right, their binding is not the same. As the level of *cro* begins to rise in cells, for example during a lytic infective cycle, *cro* binds first to O_{R3} and shuts off the synthesis of *CI* repressor. Only later, after the level of *cro* has risen still higher, does it bind to O_{R2} and O_{R1} and shut down the activity of p_R. On the other hand, during a phage developmental cycle that will result in the production of a lysogen, as repressor begins to accumulate it first binds to O_{R1} and O_{R2} and shuts off p_R and turns on p_{RM}. At still higher concentrations, repressor binds to O_{R3} and shuts off p_{RM}.

We owe the preceding picture to a series of clever *in vivo* and *in vitro* experiments by Ptashne and his collaborators. Two operations had to be performed to examine the *in vivo* effects of *cro* and repressor on the activities of p_R and p_{RM}. The synthesis of the two proteins had to be decoupled and a means had to be found of varying the

---

Table 16.2   Effects of increasing concentrations of lambda repressor, its binding to operators, and its effects on the promoters p_{RM} and p_R

Activity of p_{RM}	Operator Occupancy by Repressor			Activity of p_R	
	O_{R3}	O_{R2}	O_{R1}		
Weak	−	−	−	On	Increasing repressor concentrations
On	−	+	+	Off	
Off	+	+	+	Off	

$$O_{R3} \qquad\qquad O_{R2} \qquad\qquad O_{R1}$$

⟵wwwwwww p_{RM} wwwwwww⟶ p_R

Table 16.3   Effects of increasing concentrations of *cro* protein, its binding to operators, and its effects on the promoters p_{RM} and p_R

Activity of p_{RM}	Operator Occupancy by Cro Protein Cro on Operator			Activity of p_R	
	O_{R3}	O_{R2}	O_{R1}		
Weak	−	−	−	On	
Off	+	−	−	On	Increasing *cro* concentrations
Off	+	+	−	On	
Off	+	+	+	Off	

level of one protein while examining the activity of each of the promoters. Genetic engineering came to the rescue. In one case *cro* protein synthesis was placed under control of the *lac* promoter via a p_{lac}-*cro* fusion. Since the *lac* promoter was still regulated by *lac* repressor, the level of *cro* in cells could be varied by varying the concentration of *lac* inducer, IPTG, added to the culture medium. Quantitating the activities of p_R and p_{RM} was facilitated by fusing either promoter to the β-galactosi-

Figure 16.12   Decoupling of *cro* and repressor from their promoters and reconstruction to facilitate examination of their interactions via the inducibility of the *lac* promoter and the assay of β-galactosidase.

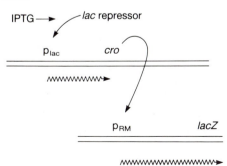

Figure 16.13   Response of the p_R and p_{RM} promoters to increasing concentrations of *cro* protein.

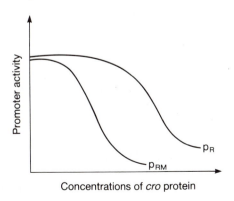

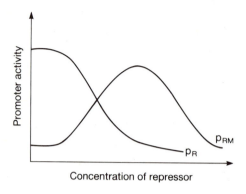

Figure 16.14   Response of the p_R and p_{RM} promoters to increasing concentrations of repressor protein.

dase gene (Fig. 16.12). Hence, although IPTG was added to cells and β-galactosidase was measured, the results were about behavior of the lambda early gene regulation mechanism.

As the intracellular concentration of *cro* was increased, first p_{RM} and then p_R was repressed (Fig. 16.13) and, as indicated above, these promoters showed quite different responses to repressor (Fig. 16.14). At the same concentrations, p_R was shut off and p_{RM} was turned on. Then, at higher concentrations, p_{RM} was shut off. *In vitro* transcription experiments using DNA fragments several hundred base pairs long yielded the same results.

Quantitation of the binding of *cro* and repressor to the three operators was done with DNase footprinting. At the lowest *cro* concentrations, only O_{R3} is occupied, but as *cro* levels are increased, O_{R2} and O_{R1} are occupied. The results with repressor are the reverse. O_{R1} has the highest apparent affinity for repressor, being 50% occupied at a concentration of 3 nM *in vitro*. At twice this concentration, O_{R2} is 50% occupied, but 25 times this concentration is required for O_{R3} to be 50% occupied.

COOPERATIVITY IN REPRESSOR BINDING AND ITS MEASUREMENT

As seen in the preceding section, repressor does not bind with equal affinity to the three operators on the right. More detailed experiments also reveal that repressor molecules do not bind independently to the three operators (Table 16.4). Not surprisingly, repressors bound at the operators interact with one another and change the overall binding energy of repressor for an operator. Most unexpectedly, however, the results show that a repressor molecule bound at the middle operator can interact either with a repressor bound at O_{R1} or with a repressor bound at O_{R3}, but it does not simultaneously interact with both.

Operator mutants were used to measure the intrinsic affinity of repressor for the

Table 16.4   The binding energies of repressor dimers to the various operators and the interaction energies that generate cooperative binding of repressor dimers

Components	Binding Energy, ΔG (Kcal)
Protein-DNA	
$R\text{-}O_{R1}$	-12
$R\text{-}O_{R2}$	-10
$R\text{-}O_{R3}$	-10
Protein-Protein	
$R_{OR1}\text{-}R_{OR2}$	-2
$R_{OR2}\text{-}R_{OR3}$	-2

three operator sites as well as to reveal interactions between adjacently bound repressor molecules. By eliminating repressor binding to O_{R2}, the binding energy of repressor to O_{R1} and to O_{R3} could be measured without complications caused by interactions with adjacent repressors. Similarly, eliminating binding at O_{R1} or O_{R3} permitted determination of the other binding energies. Ultimately, this series of measurements provided enough data to permit calculation of the interaction energies of adjacent repressor molecules. The measurements showed that repressor binding at O_{R2} has a dramatic effect on the binding of adjacent repressors. Once all the binding energies could be calculated, statistical mechanics could be used to calculate the exact behavior of the system. The results closely agreed with the *in vivo* results.

THE NEED FOR AND THE REALIZATION OF HAIR-TRIGGER INDUCTION

Why does lambda possess regulation systems as complicated as those described above? Part of the answer is that lambda must switch between two states, lytic and lysogenic. Even so, this does not explain why a simple repressor like that found in the *lac* operon could not suffice for lambda's needs. In part the answer lies in the fact that lambda must regulate its gene activities over a much greater range than *lac*. The basal level of the *lac* operon is 0.1% of the fully induced level, and there is no apparent reason why a lower basal level would serve any useful purpose. However, a similar basal level of repression of lambda early genes would be disastrous, for lambda would then have a high rate of spontaneous induction from the lysogenic state. Typically fewer than one cell out of 10^6 spontaneously cures of lambda or induces. This low spontaneous induction rate means that the basal rate of expression of the lambda early genes is very low indeed.

Suppose the basal level of expression of a repressible operon similar to the *lac* operon is to be generated by increasing the concentration of repressor. To reduce the basal level to $\frac{1}{1000}$ of normal, the repressor level would have to be raised 1000-fold. Not only would this be wasteful of repressor, but even worse are the implications for induction. With this much repressor in the cell, inducing the early genes to greater than 50% of maximal would require inactivating more than 99.999% of the repressor. Although lambda phage may like to hitch a free ride in healthy cells, once viability of the cells is in question because of damage to the host DNA, lambda bails out and induces. The lambda in very few cells indeed would be capable of inducing if 99.999% of the repressor within a cell had to be inactivated before early genes could be turned on efficiently.

One of lambda's solutions to the problem of being either repressed or induced has been to evolve a nonlinear induction response. At normal levels of repressor, about 100 dimers per cell, lambda is fully repressed. However, if 90% of repressor has been inactivated, the p_R promoter of lambda is 50% of fully induced. For comparison, inactivation of 90% of *lac* repressor induces the *lac* operon only 3% (Table 16.5). The highly cooperative binding of repressor to O_{R1} and O_{R2} is largely responsible for lambda's nonlinear response. This can be understood quantitatively as follows. Since the high cooperativity in repressor binding means that most often O_{R1} and O_{R2} are either unoccupied or simultaneously occupied, it is a good approximation to assume that two repressors bind to operator at the same time. Combining

Table 16.5   Derepression of the lambda promoter p_R and the *lac* promoter as a function of the amount of repressor remaining active

Concentration of Repressor	Activity	
	p_R	p_{lac}
100%	<0.01	<0.01
10%	0.5	0.03
1%	–	0.5

the binding equation that defines the dissociation constant with the conservation equation and solving yields the ratio of free operator to total operator, that is, the relative amount of derepression as a function of repressor concentration:

$$O + 2R \rightleftharpoons O \cdot 2R,$$

$$K_D = \frac{O \times R^2}{O \cdot R_2},$$

$$O + O \cdot 2R = O_T,$$

$$\frac{O}{O_T} = \frac{K}{K_D + R^2}.$$

As the concentration of free repressor falls from a value a bit above $K_D^{1/2}$ to a value below $K_D^{1/2}$, promoter activity, which is proportional to the fraction of operator unoccupied by repressor, O/O_T, increases rapidly due to the R^2 term. Hence very sharp changes in operator occupancy can be produced with relatively small changes in the concentration of repressor.

Lambda repressor uses another mechanism in addition to cooperative binding to the operators to increase the nonlinearity in its response to repressor concentration. Only dimers of the repressor polypeptide are able to bind to lambda operators under physiological conditions. The dissociation constant governing the repressor monomer-dimer reaction is such that a small reduction from the normal intracellular concentration of repressor sharply reduces the concentration of active repressor dimers. The combination of repressor monomer association-dissociation, cooperativity in repressor binding, and the need for sustained synthesis of N protein for lambda induction and development produces a very nonlinear induction response to changes in repressor level. At one concentration of repressor, lambda is highly stable as a lysogen, but at a lower concentration, lambda almost fully derepresses p_L and p_R (Fig. 16.15).

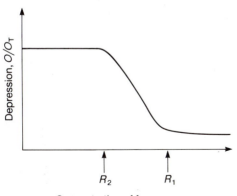

Figure 16.15   Slightly exaggerated example of the nonlinearity in the repression of p_R as a function of repressor concentration. Operator is fully occupied and off at concentration R_1; at a repressor concentration about half of R_1, the operator is unoccupied and the promoter is fully derepressed.

INDUCTION FROM THE LYSOGENIC STATE

In this section we discuss how lambda escapes from the lysogenic state and enters a lytic cycle. This occurs when DNA of the host cell is damaged and undergoes repair. The signal a phage uses for detection of this damage is a product of DNA repair, oligonucleotides excised from the damaged sites. These oligonucleotides activate *recA* protein to stimulate the cleavage both of lambda repressor and of another cellular repressor, the *lexA* protein (Fig. 16.16). These cleaved repressors are no longer able to repress. It seems most mysterious that a protein that normally is involved with binding to single-stranded DNA and also is an ATPase should be a catalyst for a proteolytic activity and that this catalysis should occur as a result of the presence of an oligonucleotide.

In a nonlysogenic cell the genes that are normally repressed by *lexA* protein are the *lexA* gene itself, the *recA* gene, and a set of about 15 others that are called the SOS system. Known functions of this system are to repair damaged DNA and to postpone cell division until repair is completed. Once repair is completed, the oligonucleotides are no longer available to activate *recA*. Consequently, newly synthesized *lexA* protein is no longer cleaved and it therefore represses synthesis of *recA* and the other proteins of the SOS system. In a normal nonlysogenic cell, the SOS system switches off when it is no longer needed.

The proteolytic activity catalyzed by *recA* protein disconnects the N-terminal domain of the lambda repressor from the C-terminal domain (Fig. 16.17). A number

Figure 16.16   *LexA* protein represses its own synthesis as well as that of *recA* protein. Cleavage of *lexA* and lambda repressor activated by *recA* derepresses the *lexA, recA,* and lambda operons.

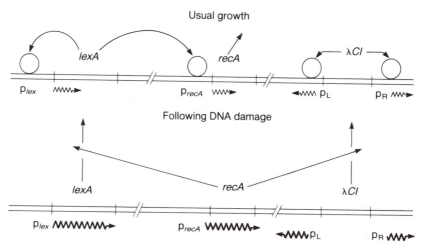

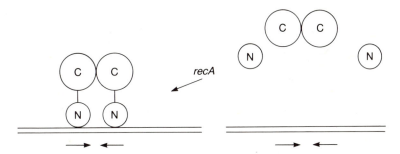

Figure 16.17   Representation of the structure of lambda repressor and the effects of cleavage by *recA* protease.

of physical experiments with protease-digested repressor or with N-terminal fragments of repressor produced from nonsense mutations have revealed that the N-terminal half of the protein folds up to form a compact domain that can bind to operator. The C-terminus also folds to form a compact domain, but this domain is responsible for the dimerization of repressor. The C-terminal domains lacking N-terminal domain still dimerize, whereas the N-terminal fragments do not.

As required for induction, the affinity of the N-terminal domains for operator is much lower than that of the intact dimeric repressor. Furthermore, these N-terminal domains do not show cooperativity in their binding to adjacent operators. In the next section we will see why the proteolytic cleavage and elimination of dimeric DNA-binding domains greatly reduce the affinity of repressor for operator. As a result of this reduced affinity, repressor comes off the DNA and the phage induces.

The N-terminal domain of repressor makes protein-protein contacts with RNA polymerase when it stimulates transcription from the p_{RM} promoter (Fig. 16.18). The stimulation provided by this positive-acting factor should be compared to the stimulation provided by CRP protein on the *lac* operon. Lambda repressor on O_{R1} and O_{R2} accelerates the isomerization rate by RNA polymerase after it has bound to p_{RM},

Figure 16.18   The activation of RNA polymerase at p_{RM} by two repressor dimers bound at O_{R1} and O_{R2}.

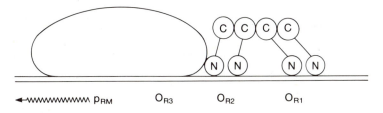

whereas CRP stimulates transcription by increasing the binding of RNA polymerase to the *lac* promoter. Positive-acting transcription factors can function by either of these two possible mechanisms.

Two types of evidence indicate that it is the N-terminal domain of repressor that contacts RNA polymerase. The first evidence is mutations in lambda repressor that eliminate the positive stimulation of p_{RM} by repressor when it occupies O_{R1} and O_{R2}. These lie in the portion of the protein immediately adjacent to the helices of the N-terminal domain that contact operator. The second evidence is *in vivo* experiments using nonsense mutations in repressor protein. The N-terminal domain of repressor that is synthesized by a hyperproducing fusion from a nonsense mutation in *CI* is capable of stimulating p_{RM}.

*ENTROPY, A BASIS FOR LAMBDA REPRESSOR INACTIVATION

The binding of dimeric repressor to operator can be described by a dissociation constant that is related to the change in the free energy via the standard thermodynamic relationship:

$$K_D = e^{-\Delta G/RT}$$
$$= e^{-(\Delta H - T\Delta S)/RT},$$

where K_D is the dissociation constant of the dimer from operator, ΔG is the change in free energy, R is the gas constant, T is the temperature, ΔS is the change in entropy, and ΔH is the change in enthalpy.

It is natural to assume that if monomeric repressor were binding to operator, then roughly half as many contacts would be formed between repressor and DNA and roughly half as many water molecules and ions would be displaced from DNA and the protein as the repressor bound. Therefore we might write for the dissociation constant of the monomer

$$K_M = e^{-[\Delta H/2 - (T\Delta S)/2]/RT},$$

or $K_M = K_D^{1/2}$. However, this is not correct. To make this clear, let ΔS be written as the sum of the entropy changes involved with the contacts and displacement of water, $\Delta S_{\text{interaction dimer}}$, plus the change in entropy involved with immobilization and orientation of the dimeric repressor, $\Delta S_{\text{intrinsic dimer}}$:

$$\Delta S_{\text{dimer}} = \Delta S_{\text{interaction dimer}} + \Delta S_{\text{intrinsic dimer}}.$$

The same type of equation can be written for the monomer.

$$\Delta S_{\text{monomer}} = \Delta S_{\text{interaction monomer}} + \Delta S_{\text{intrinsic monomer}}.$$

Roughly, the monomer makes half as many interactions as the dimer and displaces half as many water molecules and ions as it binds. Therefore

$$\Delta S_{\text{interaction monomer}} = \Delta S_{\text{interaction dimer}}/2.$$

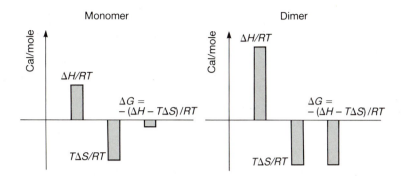

Figure 16.19   Why a dimer can bind more tightly than a monomer. Almost all the additional ΔH provided by the second monomer's binding can go into increasing ΔG, whereas for binding of a monomer, almost no ΔG is left over to contribute to binding.

However, the same is not true of the intrinsic entropies. The entropy change associated with bringing repressor monomer to rest on operator by correctly positioning and orienting it is nearly the same as the change associated with the dimeric repressor. Thus

$$\Delta S_{\text{monomer}} \neq \Delta S_{\text{dimer}}/2,$$

and detailed calculations based on statistical mechanics or experimental estimation of the inequality show that the inequality can be severe. In other words, a sizable fraction of the total entropy change involved with repressor binding to DNA is associated with its correct positioning. Roughly the same entropy is required to orient a monomer or dimer, but in the case of lambda repressor, the presence of the dimer adds twice as much to the binding energy. Most of this additional energy can go into holding the dimer on the DNA, and hence K_D is much less than K_M^2 (Fig. 16.19).

PROBLEMS

16.1. How would you perform the crosses between lambda and phage 434 to produce a lambda imm^{434} phage?

16.2. The *rex* genes of lambda are in the *CI* operon, and *rex* protein gives lambda lysogens immunity to a class of T4 mutants possessing defects in the r_I or r_{II} genes. Wild-type T4 grows well on lambda lysogens. How would you isolate *rex* mutants of lambda?

16.3. An interesting cloverleaf structure was proposed for the *ori* region with bound *O* protein in "DNA Sequences and Structural Homologies of the Replication Origins of Lambdoid Bacteriophages," R. Grosschedl, H. Hobom, *Nature* 277, 621–

627 (1979). What precisely does the DNase footprinting of *O* bound to *ori,* say about this model? See "Purified Bacteriophage Lambda O Protein Binds to Four Repeating Sequences at the Lambda Replication Origin," T. Tsurimoto, K. Matsubara, Nucleic Acids Res. *9,* 1789–1799 (1981).

16.4. How would you isolate nonsense mutations in the *S* gene of lambda?

16.5. Why does lambda possess a linear genetic map when in fact the DNA molecules undergoing recombination are circular?

16.6. Does the paper "Isolation of the Operators of Phage Lambda," Pirrotta, Nature New Biol. *244,* 13–16 (1973), contain data that could have led to the discovery described in "Multiple Repressor Binding at the Operators in Bacteriophage Lambda," T. Maniatis, M. Ptashne, Proc. Nat. Acad. Sci. USA *70,* 1531–1535 (1973)? Explain.

16.7. How could one show genetically that v_1v_3, a double-point mutation in the rightward operators, makes p_R insensitive to repression? Similarly, how could one show genetically that v_2, a mutation in the leftward operators, makes p_L repressor insensitive?

16.8. C17 is a repressor-insensitive new promoter for genes *CII, O,* and *P.* Why might lambda $C17CI^-$ be virulent (capable of forming plaques on a lambda lysogen) but lambda $C17CI^+$ be not virulent?

16.9. What is a reasonable interpretation of the fact that cells infected with S^+ phage and treated with metabolic inhibitors, such as cyanide, or protein synthesis inhibitors, such as chloramphenicol, later than 15 minutes after infection promptly lyse?

16.10. The lambda *R* gene product was once identified as an endopeptidase in "Endopeptidase Activity of Phage Lambda-Endolysin," A. Taylor, Nature New Biol. *234,* 144–145 (1971). What is a likely reason for the misidentification and what control experiments would you have performed?

16.11. A lysogen of lambda $CI_{857}N^-O^-P^-$ is grown at 42° for many generations and then grown at 35°. Infection of the cells grown in this manner with lambda yielded clear plaques and a lysogenization frequency much lower than seen on infection of a similarly treated nonlysogenic strain. Why?

16.12. Simultaneous infection of a lambda lysogen with lambda and lambda i^{434} produced a burst containing only lambda i^{434}. Why? How would you isolate lambda mutants that would grow in a lambda lysogen when coinfected with lambda i^{434}, and where would you expect these to map?

16.13. Write a few intelligent comments about the locations of genes and their sites of action in phage lambda.

16.14. The two papers "Protein X Is the Product of the *recA* Gene of *E. coli,*" K. McEntee, Proc. Nat. Acad. Sci. USA *74,* 5275–5279 (1977), and "Identification of the *recA (tif)* Gene Product of the *E. coli,*" L. Gudas, D. Mount, Proc. Nat. Acad. Sci. USA *74,* 5280–5284 (1977), both propose explicit models for regulation of the *recA* protein levels. How do these models differ? From the data presented, are either of the models preferred? From data alluded to, is either of the two models preferred?

16.15. Conceivably a *nus* mutation might act by causing a host protein to inactivate the *N* protein of phage lambda. Although a complicated biochemical experiment could be used to test this idea, what simple genetic experiment could disprove the proposal?

16.16. Devise a selection or scoring method for the isolation of *CI* mutations that repress normally but do not stimulate p_{RM}.

16.17. If the *lamB* porin possesses larger holes than the two porins normally present in the outer membrane, why not have just the *lamB* porin and forget the others? Note that none of the porins is ion-selective.

16.18. Given the genes in which *nus* mutations have been found, invent a mechanism of antitermination.

16.19. How would you isolate lambda mutants that have bypassed their requirement for transcriptional activation from p_R for activation of DNA replication?

*16.20. (a) How do the lambda promoters respond if the intracellular repressor levels become too high? (b) Contrast the wisdom of maintaining a low intracellular level of repressor via making repressor messenger infrequently but translating it many times versus making messenger often but translating each molecule only a few times. (c) With respect to the answer in part b, comment on the fact that the 5′ ribonucleotide of *CI* messenger is the A of the initiation AUG codon.

RECOMMENDED READINGS

Isolation of the Lambda Phage Repressor, M. Ptashne, Proc. Nat. Acad. Sci. USA *57*, 306–313 (1967).

Control of Lambda Repressor Synthesis, L. Reichardt, A. Kaiser, Proc. Nat. Acad. Sci. USA *68*, 2185–2189 (1971).

Positive Control of Endolysin Synthesis *in vitro* by the Gene N Protein of Phage Lambda, J. Greenblatt, Proc. Nat. Acad. Sci. USA *69*, 3606–3610 (1972).

Altered Reading of Genetic Signals Fused to the N Operon of Bacteriophage Lambda: Genetic Evidence for Modification of Polymerase by the Protein Product of the N Gene, N. Franklin, J. Mol. Biol. *89*, 33–48 (1974).

Interactions Between DNA-Bound Repressors Govern Regulation by the Lambda Phage Repressor, A. Johnson, B. Meyer, M. Ptashne, Proc. Nat. Acad. Sci. USA *76*, 5061–5065 (1979).

RELATED REVIEWS, BOOKS, AND ARTICLES

The Bacteriophage Lambda, ed. A. Hershey, Cold Spring Harbor Laboratory (1971).

Autoregulation and Function of a Repressor in Bacteriophage Lambda, M. Ptashne, K. Bachmann, M. Humayun, A. Jeffrey, R. Maurer, B. Meyer, R. Sauer, Science *194*, 156–161 (1976).

Control of Transcription Termination, S. Adhya, M. Gottesman, Ann. Rev. Biochem. *47*, 967–996 (1978).

How the Lambda Repressor and *Cro* Work, M. Ptashne, A. Jeffrey, A. Johnson, R. Maurer, B. Meyer, C. Pabo, T. Roberts, R. Sauer, Cell *19*, 1–11 (1980).

Lambda II, ed. R. Hendrix, J. Roberts, F. Stahl, R. Weisberg, Cold Spring Harbor Laboratory (1983).

DEEPER READING

Cohesive Sites on the DNA from Several Temperate Coliphages, R. Baldwin, P. Barrand, A. Fritsch, D. Goldthwait, F. Jacob, J. Mol. Biol. *17*, 343–357 (1966).

Specific Binding of the Lambda Phage Repressor to Lambda DNA, M. Ptashne, Nature *214*, 232–234 (1967).

Regulation of Bacteriophage Lambda DNA Replication, M. Green, B. Gotchel, J. Hendershott, S. Kennel, Proc. Nat. Acad. Sci. USA *58*, 2343–2350 (1967).

Mutations in Bacteriophage Lambda Affecting Host Cell Lysis, A. Harris, D. Mount, C. Fuerst, L. Siminovitch, Virology *32*, 553–569 (1967).

Electron-Microscopic Visualization of Deletion Mutations, R. Davis, N. Davidson, Proc. Nat. Acad. Sci. USA *60*, 243–250 (1968).

Control of Development in Temperate Bacteriophages, II. Control of Lysozyme Synthesis, C. Dambly, M. Couturier, R. Thomas, J. Mol. Biol. *32*, 67–81 (1968).

New Mutations in the S. Cistron of Bacteriophage Lambda Affecting Host Cell Lysis, A. Goldberg, M. Howe, Virology *38*, 200–202 (1969).

Strains of Phage Lambda in Current Use, W. Dove, Virology *38*, 349–351 (1969).

A Site Essential for Expression of All Late Genes in Bacteriophage Lambda, I. Herskowitz, E. Signer, J. Mol. Biol. *47*, 545–556 (1970).

Genetic Expression in Bacteriophage Lambda, III. Inhibition of *E. coli* Nucleic Acid and Protein Synthesis During Lambda Development, S. Cohen, A. Chang, J. Mol. Biol. *49*, 557–575 (1970).

Intracellular Pools of Bacteriophage Lambda Deoxyribonucleic Acid, B. Carter, M. Smith, J. Mol. Biol. *50*, 713–718 (1970).

Bypassing a Positive Regulator: Isolation of a Lambda Mutant That Does Not Require N Product to Grow, N. Hopkins, Virology *40*, 223–229 (1970).

Entropic Contributions to Rate Accelerations in Enzymic and Intramolecular Reactions and the Chelate Effect, M. Page, W. Jencks, Proc. Nat. Acad. Sci. USA *68*, 1678–1683 (1971).

Bacterial Mutants in Which the Gene N Function of Bacteriophage Lambda Is Blocked Have an Altered RNA Polymerase, C. Georgopoulas, Proc. Nat. Acad. Sci. USA *68*, 2977–2981 (1971).

Lysis Defective Mutants of Bacteriophage Lambda: Genetics and Physiology of S Cistron Mutants, R. Reader, L. Siminovitch, Virology *43*, 607–622 (1971).

Relationship Between the N Function of Bacteriophage Lambda and Host RNA Polymerase, A. Ghysen, M. Pironio, J. Mol. Biol. *65*, 259–272 (1972).

Escape Synthesis of the Biotin Operon in Induced Lambda b2 Lysogens, K. Krell, M. Gottesman, J. Parks, J. Mol. Biol. *68*, 69–82 (1972).

Note on the Structure of Prophage Lambda, P. Sharp, M. Hsu, N. Davidson, J. Mol. Biol. *71*, 499–501 (1972).

The Topography of Lambda DNA: Insertion of Ordered Fragments and the Physical Mapping of Point Mutations, J. Egan, D. Hogness, J. Mol. Biol. *71*, 363–381 (1972).

Isolation and Properties of rex⁻ Mutants of Bacteriophage Lambda, G. Gussin, V. Peterson, J. Virology *10*, 760–765 (1972).

The Specificity of Lamboid Phage Late Gene Induction, R. Schleif, Virology *50*, 610–612 (1972).

Regulation of the Expression of the N Gene of Bacteriophage Lambda, J. Greenblatt, Proc. Nat. Acad. Sci. USA *70*, 421–424 (1973).

Multiple Repressor Binding at the Operators in Bacteriophage Lambda, T. Maniatis, M. Ptashne, Proc. Nat. Acad. Sci. USA *70*, 1531–1535 (1973).

Purification and Properties of the γ-Protein Specified by Bacteriophage Lambda: An Inhibitor of the Host Rec BC Recombination Enzyme, Y. Sakaki, A. Karu, S. Linn, H. Echols, Proc. Nat. Acad. Sci. USA *70*, 2215–2219 (1973).

Replication of Bacteriophage Lambda DNA Dependent on the Function of Host and Viral Genes, I. Interaction of *red, gam,* and *rec,* L. Enquist, A. Skalka, J. Mol. Biol. *75*, 185–212 (1973).

Role of the *cro* Gene in Bacteriophage Lambda Development, H. Echols, L. Green, B. Oppenheim, A. Oppenheim, A. Honigman, J. Mol. Biol. *80*, 203–216 (1973).

Gene N Regulator Function of Phage Lambda imm21: Evidence That a Site of N Action Differs from a Site of N Recognition, D. Friedman, G. Wilgus, R. Mural, J. Mol. Biol. *81*, 505–516 (1973).

Structure of the Lambda Operators, T. Maniatis, M. Ptashne, Nature *246*, 133–136 (1973).

Isolation of the Bacteriophage Lambda Receptor from *E. coli,* L. Randall-Hazelbauer, M. Schwartz, J. Bact. *116*, 1436–1446 (1973).

Genetic Characterization of a prm⁻ Mutant of Bacteriophage Lambda, K. Yen, G. Gussin, Virology *56*, 300–312 (1973).

An Endonuclease Induced by Bacteriophage Lambda, M. Rhoades, M. Meselson, J. Biol. Chem. *248*, 521–527 (1973).

Promoters Are in the Operators in Phage Lambda, R. Maurer, T. Maniatis, M. Ptashne, Nature *249*, 221–223 (1974).

Release of Polarity in *Escherichia coli* by Gene N of Phage Lambda: Termination and Antitermination of Transcription, S. Adhya, M. Gottesman, B. deCrombrugghe, Proc. Nat. Acad. Sci. USA *71*, 2534–2538 (1974).

Transcription Termination and Late Control in Phage Lambda, J. Roberts, Proc. Nat. Acad. Sci. USA *72*, 3300–3304 (1975).

Lambda Repressor Turns Off Transcription of Its Own Gene, B. Meyer, D. Kleid, M. Ptashne, Proc. Nat. Acad. Sci. USA *72*, 4785–4789 (1975).

Control of Bacteriophage Lambda Repressor Synthesis After Phage Infection: The Role of N, cII, cIII and *cro* Products, L. Reichardt, J. Mol. Biol. *93*, 267–288 (1975).

Control of Bacteriophage Lambda Repressor Synthesis: Regulation of the Maintenance Pathway by the *cro* and *cI* Products, L. Reichardt, J. Mol. Biol. *93*, 289–309 (1975).

Rec-Mediated Recombinational Hot Spot Activity in Bacteriophage Lambda, III. Chi Mutations Are Site-Mutations Stimulating Rec-Mediated Recombination, F. Stahl, J. Craseman, M. Stahl, J. Mol. Biol. *94*, 203–212 (1975).

Studies on the Late Replication of Phage Lambda: Rolling-Circle Replication of the Wild-Type and a Partially Suppressed Strain *Oam29 Pam80,* D. Bastia, N. Sueoka, E. Cox, J. Mol. Biol. *98*, 305–320 (1975).

Reversible Interaction Between Coliphage Lambda and Its Receptor Protein, M. Schwartz, J. Mol. Biol. *99*, 185–201 (1975).

Recognition Sequences of Repressor and Polymerase in the Operators of Bacteriophage Lambda, T. Maniatis, M. Ptashne, H. Backman, D. Kleid, S. Flashman, A. Jeffrey, R. Mauer, Cell *5*, 109–113 (1975).

The Kil Gene of Bacteriophage Lambda, H. Greer, Virology *66*, 589–604 (1975).

The Nature and Origin of a Class of Essential Gene Substitutions in Bacteriophage Lambda, D. Henderson, J. Weil, Virology *67*, 124–135 (1975).

Purification and Properties of a DNA-Binding Protein with Characteristics Expected for the *Cro* Protein of Bacteriophage Lambda: A Repressor Essential for Lytic Growth, A. Folkmanis, Y. Takeda, J. Simuth, G. Gussin, H. Echols, Proc. Nat. Acad. Sci. USA *73*, 2249–2253 (1976).

Sequence of *cro* Gene of Bacteriophage Lambda, T. Roberts, H. Shimatake, C. Brady, M. Rosenberg, Nature *270*, 274–275 (1977).

Amino Acid Sequence of *Cro* Regulatory Protein of Bacteriophage Lambda, M. Hsiang, R. Cole, Y. Tekada, H. Echols, Nature *270*, 275–277 (1977).

Inactivation and Proteolytic Cleavage of Phage Lambda Repressor *in vitro* in an ATP-Dependent Reaction, J. Roberts, C. Roberts, D. Mount, Proc. Nat. Acad. Sci. USA *74*, 2283–2287 (1977).

Completed DNA Sequences and Organization of Repressor-Binding Sites in the Operators of Phage Lambda, Z. Humayun, A. Jeffrey, M. Ptashne, J. Mol. Biol. *112*, 265–277 (1977).

The Essential Role of the *Cro* Gene in Lytic Development by Bacteriophage Lambda, A. Folkmanis, W. Maltzman, P. Mellon, A. Skalka, H. Echols, Virology *81*, 352–362 (1977).

Cro Regulatory Protein Specified by Bacteriophage Lambda, Structure, DNA Binding and Repression of RNA Synthesis, Y. Takeda, A. Folkmanis, H. Echols, J. Biol. Chem. *252*, 6177–6183 (1977).

Sites of Contact Between Lambda Operators and Lambda Repressor, Z. Humayun, D. Kleid, M. Ptashne, Nucleic Acids Res. *4*, 1595–1607 (1977).

Nucleotide Sequence of *cro*, cII, and Part of the *O* Gene in Phage Lambda DNA, E. Schwarz, G. Scherer, G. Hobom, H. Kossel, Nature *272*, 410–414 (1978).

The Relationship Between Function and DNA Sequence in an Intercistronic Regulatory Region in Phage Lambda, M. Rosenberg, D. Court, H. Shimatake, C. Brady, Nature *272*, 414–423 (1978).

Mechanism of Action of *Cro* Protein of Bacteriophage Lambda, A. Johnson, B. Meyer, M. Ptashne, Proc. Nat. Acad. Sci. USA *75*, 1783–1787 (1978).

Interaction of Bacteriophage Lambda Repressor with Nonoperator DNA Containing Single-Strand Gaps, R. Sussman, J. Resnick, K. Calame, J. Baluch, Proc. Nat. Acad. Sci. USA *75*, 5817–5821 (1978).

Coliphage Lambda nutL⁻: A Unique Class of Mutants Defective in the Site of Gene N Product Utilization for Antitermination of Leftward Transcription, J. Salstrom, W. Szybalski, J. Mol. Biol. *124*, 195–221 (1978).

Specificity Determinants for Bacteriophage Lambda DNA Replication, II. Structure of *O* Proteins of Lambda-ϕ80 and Lambda-82 Hybrid Phages and of a Lambda Mutant Defective in the Origin of Replication, M. Furth, J. Yates, J. Mol. Biol. *126*, 227–240 (1978).

Nucleotide Sequence of the O Gene and of the Origin of Replication in Bacteriophage Lambda DNA, G. Scherer, Nucleic Acids Res. *5*, 3141–3156 (1978).

Regulatory Functions of the Lambda Repressor Reside in the Amino-Terminal Domain, R. Sauer, C. Pabo, B. Meyer, M. Ptashne, K. Backman, Nature *279*, 396–400 (1979).

Location of the Regulatory Site for Establishment of Repression by Bacteriophage Lambda, M. Jones, R. Fischer, I. Herskowitz, H. Echols, Proc. Nat. Acad. Sci. USA *76*, 150–154 (1979).

The Lambda Repressor Contains Two Domains, C. Pabo, R. Sauer, J. Sturtevant, M. Ptashne, Proc. Nat. Acad. Sci. USA *76*, 1608–1612 (1979).

Transcription of the *int* Gene of Bacteriophage Lambda New RNAP Binding Site and RNA Start Generated by *int*-Constitutive Mutations, R. Fischer, Y. Takeda, H. Echols, J. Mol. Biol. *129*, 509–514 (1979).

Receptor for Bacteriophage Lambda of *E. coli* Forms Larger Pores in Black Lipid Membranes Than the Matrix Protein, B. Boehler-Kohler, W. Boos, R. Dieterle, R. Benz, J. Bact. *138*, 33–39 (1979).

N-Independent Leftward Transcription in Coliphage Lambda: Deletions, Insertions, and New Promoters Bypassing Termination Functions, J. Salstrom, M. Fiandt, W. Szybalski, Mol. Gen. Genet. *168*, 211–230 (1979).

Lambda Encodes an Outer Membrane Protein: The *lom* Gene, J. Reeve, J. Shaw, Mol. Gen. Genet. *172*, 243–248 (1979).

L Factor That Is Required for β-galactosidase Synthesis Is the *nusA* Gene Product Involved in Transcription Termination, J. Greenblatt, J. Li, S. Adhya, D. Friedman, L. Baron, B. Redfield, H. Kung, H. Weissbach, Proc. Nat. Acad. Sci. USA *77*, 1991–1994 (1980).

Site Specific Recombination Functions of Bacteriophage Lambda: DNA Sequence of Regulatory Regions and Overlapping Structural Genes for *Int* and *Xis*, R. Hoess, C. Foeller, K. Bidwell, A. Landy, Proc. Nat. Acad. Sci. USA *77*, 2482–2486 (1980).

Promoter for the Establishment of Repressor Synthesis in Bacteriophage Lambda, U. Schmeissner, D. Court, H. Shimatake, M. Rosenberg, Proc. Nat. Acad. Sci. USA *77*, 3191–3195 (1980).

Control of Transcription Termination: A Rho-Dependent Termination Site in Bacteriophage Lambda, D. Court, C. Brady, M. Rosenberg, D. Wulff, M. Behr, M. Mahoney, S. Izumi, J. Mol. Biol. *138*, 231–254 (1980).

Gene Regulation at the Right Operator (O_R) of Bacteriophage Lambda, I. O_R3 and Autogenous Negative Control by Repressor, R. Maurer, B. Meyer, M. Ptashne, J. Mol. Biol. *139*, 147–161 (1980).

Gene Regulation at the Right Operator (O_R) of Bacteriophage Lambda, II. O_R1, O_R2 and O_R3: Their Roles in Mediating the Effects of Repressor and *cro*, B. Meyer, R. Maurer, M. Ptashne, J. Mol. Biol. *139*, 163–194 (1980).

Gene Regulation at the Right Operator (O_R) of Bacteriophage Lambda, III. Lambda Repressor Directly Activates Gene Transcription, B. Mayer, M. Ptashne, J. Mol. Biol. *139*, 195–205 (1980).

Structure of the *Cro* Repressor from Bacteriophage Lambda and Its Interaction with DNA, W. Anderson, D. Ohlendorf, Y. Takeda, B. Mathews, Nature *290*, 754–758 (1981).

Purified Lambda Regulatory Protein *c*II Positively Activates Promoters for Lysogenic Development, H. Shimatake, M. Rosenberg, Nature *292*, 128–132 (1981).

Purified *lexA* Protein Is a Repressor of the *recA* and *lexA* Genes, J. Little, P. Mount, C. Yanisch-Perron, Proc. Nat. Acad. Sci. USA *78*, 4199–4203 (1981).

Mechanism of Action of the *lexA* Gene Product, R. Brent, M. Ptashne, Proc. Nat. Acad. Sci. USA *78*, 4204–4208 (1981).

SOS Induction and Autoregulation of the *himA* Gene for the Site-Specific Recombination in

Escherichia coli, H. Miller, M. Kirk, H. Echols, Proc. Nat. Acad. Sci. USA *78,* 6754–6758 (1981).

The nusA Gene Protein of *Escherichia coli,* Its Identification, and a Demonstration That It Interacts with the Gene N Transcription Anti-Terminating Protein of Bacteriophage Lambda, J. Greenblatt, J. Li, J. Mol. Biol. *147,* 11–23 (1981).

Interaction of the Sigma Factor and the *nusA* Gene Protein of *E. coli* with RNA Polymerase in the Initiation-Termination Cycle of Transcription, J. Greenblatt, J. Li, Cell *24,* 421–428 (1981).

Multilevel Regulation of Bacteriophage Lambda Lysogeny by *E. coli himA* Gene, H. Miller, Cell *25,* 269–276 (1981).

Cell Lysis by Induction of Cloned Lambda Lysis Genes, J. Garrett, R. Fusselman, J. Hise, L. Chiou, D. Smith-Grillo, J. Schulz, R. Young, Mol. Gen. Genet. *182,* 326–331 (1981).

The R Gene Product of Bacteriophage Lambda Is the Murein Transglycosylase, K. Bienkowska-Szewczyk, B. Lipinska, A. Taylor, Mol. Gen. Genet. *184,* 111–114 (1981).

A Fine Structure Map of Spontaneous and Induced Mutations in the Lambda Repressor Gene, Including Insertions of IS Elements, M. Leib, Mol. Gen. Genet. *184,* 364–371 (1981).

Purified Bacteriophage Lambda O Protein Binds to Four Repeating Sequences at the Lambda Replication Origin, T. Tsurimoto, K. Matsubara, Nucleic Acids Res. *9,* 1789–1799 (1981).

Downstream Regulation of *int* Gene Expression by the b2 Region in Phage Lambda, C. Epp. M. Pearson, Gene *13,* 327–337 (1981).

DNA Sequence of the *att* Region of Coliphage 434, D. Mascarenhas, R. Kelley, A. Cambell, Gene *15,* 151–156 (1981).

The N-Terminal Arms of Lambda Repressor Wrap Around the Operator DNA, C. Pabo, W. Krovatin, A. Jeffrey, R. Sauer, Nature *298,* 441–443 (1982).

The Operator-Binding Domain of Lambda Repressor: Structure and DNA Recognition, C. Pabo, M. Lewis, Nature *298,* 443–447 (1982).

Posttranscriptional Control of Bacteriophage Lambda *int* Gene Expression from a Site Distal to the Gene, G. Guarneros, C. Montanez, T. Hernandez, D. Court, Proc. Nat. Acad. Sci. USA *79,* 238–242 (1982).

Quantitative Model for Gene Regulation by Lambda Phage Repressor, G. Ackers, A. Johnson, M. Shea, Proc. Nat. Acad. Sci. USA *79,* 1129–1133 (1982).

Mutant Lambda Phage Repressor with a Specific Defect in Its Positive Control Function, L. Guarente, J. Nye, A. Hochschild, M. Ptashne, Proc. Nat. Acad. Sci. USA *79,* 2236–2239 (1982).

Replication of Lambda dv Plasmid *in vitro* Promoted by Purified Lambda *O* and *P* Proteins, T. Tsurimoto, K. Matsubara, Proc. Nat. Acad. Sci. USA *79,* 7639–7643 (1982).

Specificity Determinants for Bacteriophage Lambda DNA Replication, III. Activation of Replication in Lambda ri^c Mutants by Transcription Outside of *ori,* M. Furth, W. Done, B. Meyer, J. Mol. Biol. *154,* 65–83 (1982).

Control of Transcription by Bacteriophage P22 Repressor, A. Poteete, M. Ptashne, J. Mol. Biol. *157,* 21–48 (1982).

Regulation of Bacteriophage Lambda *int* Gene Expression, A. Oppenheim, S. Gottesman, M. Gottesman, J. Mol. Biol. *158,* 327–346 (1982).

Lambda Phage *cro* Repressor, DNA Sequence-Dependent Interactions Seen by Tyrosine Fluorescence, F. Boschelli, K. Arndt, H. Nick, Q. Zhang, P. Lu, J. Mol. Biol. *162,* 251–266 (1982).

The Phage Lambda Q Gene Product: Activity of a Transcription Antiterminator *in vitro,* E. Grayhack, J. Roberts, Cell *30,* 637–648 (1982).

Analysis of *nutR*: A Region of Phage Lambda Required for Antitermination of Transcription, E. Olson, E. Flamm, D. Friedman, Cell *31*, 61–70 (1982).

Control of Phage Lambda Development by Stability and Synthesis of *cII* Protein: Role of the Viral *cIII* and *hflA, himA* and *himD* Genes, A. Hoyt, D. Knight, A. Das, H. Miller, H. Echols, Cell *31*, 565–573 (1982).

Repressor Structure and the Mechanism of Positive Control, A. Hochschild, N. Irwin, M. Ptashne, Cell *32,* 319–325 (1982).

Effect of Bacteriophage Lambda Infection on Synthesis of *groE* Protein and Other *Escherichia coli* Proteins, D. Drahos, R. Hendrix, J. Bact. *149,* 1050–1063 (1982).

A Second Function of the S Gene of Bacteriophage Lambda, D. Wilson, A. Okabe, J. Bact. *152,* 1091–1095 (1982).

Nucleotide Sequence of Bacteriophage Lambda DNA, F. Sanger, A. Coulson, G. Hong, D. Hill, G. Petersen, J. Mol. Biol. *162,* 729–773 (1982).

The *Rex* Gene of Bacteriophage Lambda Is Really Two Genes, K. Matz, M. Schandt, G. Gussin, Genetics *102,* 319–327 (1982).

Bacteriophage Lambda Protein *cII* Binds Promoters on the Opposite Face of the DNA Helix from RNA Polymerase, Y. Ho, D. Wulff, M. Rosenberg, Nature *304,* 702–708 (1983).

S Gene Product: Identification and Membrane Localization of a Lysis Control Protein, E. Altman, R. Altman, J. Garrett, R. Grimaila, R. Young, J. Bact. *155,* 1130–1137 (1983).

Evidence That a Nucleotide Sequence, "*boxA*," Is Involved in the Action of the *NusA* Protein, D. Friedman, E. Olson, Cell *34,* 143–149 (1983).

Autodigestion of *lexA* and Phage Lambda Repressors, J. Little, Proc. Nat. Acad. Sci. USA *81,* 1375–1379 (1984).

PART FOUR

Mobile
DNA

Lambda Phage Integration and Excision

A s our understanding of molecular biology has grown, our reverence for DNA has steadily declined. Once DNA was immutable and not subject to change by humans. For a while, the use of mutagens and the isolation of mutations were in vogue. Then we found that chromosomes are subject to gross rearrangements, something classical *Drosophila* geneticists tried to tell us years earlier. Finally, as described in earlier chapters, manipulating DNA sequences *in vitro* by cutting, joining, and synthesizing has become commonplace and a possible pathway to biological and financial miracles.

The next three chapters examine *in vivo* recombination events between DNA molecules that lack the sequence homologies necessary for generalized recombination. Such recombination processes require special proteins or enzymes and usually involve a special DNA site on at least one of the participating DNA molecules. Chapter 17 treats lambda phage, whose DNA molecule readily enters and exits from the *Escherichia coli* chromosome by a path of specialized recombination. A wealth of information is known about this process as it has been extensively studied from the early years in molecular biology up to the present. Chapter 18 treats other DNA segments that integrate into or excise from chromosomes. One of the most striking findings is that integrating sequences, now called transposons, are ubiquitous. They have been found in all organisms that have been carefully examined. Chapter 19 treats the generation of diversity of binding specificities in antibody molecules. In part this is accomplished by shuffling DNA sequences within the chromosomes of cells in the immune system. Hence, in contrast to the situation with transposons, which can be viewed as parasites, the ability to move DNA sequences about may also be of great positive value to an organism.

INTEGRATED LAMBDA MAPS LIKE A CHROMOSOMAL GENE

How do we know lambda phage associates with the chromosome when it lysogenizes a cell? Lambda lysogens could possess copies of the phage DNA freely floating in the cytoplasm. These molecules could be randomly inherited by daughter cells on division. However, to explain the lambda's low rate of spontaneously becoming nonlysogenic, cells would have to possess large numbers of the phage genomes. A second method to ensure proper inheritance of lysogeny would be for lambda to possess a special segregation mechanism. Then only low numbers of the lambda genome would have to exist in cells. The simplest segregation mechanism would be for lambda to attach to or integrate into the host chromosome. Then it would be replicated and segregated into daughter cells with the host chromosome.

Elegant genetic experiments showed not only that lambda associates with the chromosome but that it associates with a specific site on the chromosome. Now, of course, a simple Southern transfer experiment could settle the issue. The original data suggesting that the lambda DNA was associated with a particular region of the chromosome came from mating experiments between lysogenic male bacteria and nonlysogenic female bacteria.

Transfer of genetic markers lying between the origin of transfer and the *gal* region showed no differences between lysogenic and nonlysogenic males. However, the frequency of female recombinants incorporating genetic markers transferred after the *gal* genes revealed a discrepancy (Fig. 17.1). Their frequency was much

Figure 17.1   The transfer frequency of markers from lambda lysogens or nonlysogens. Marker X^+ is located near the origin of transfer, and markers Y^+ and Z^+ are transferred after the point of lambda insertion.

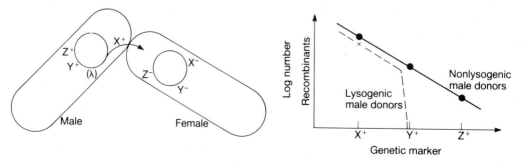

lower if the DNA was transferred from lysogenic males than if the DNA came from nonlysogenic males. This can be understood as follows: If lambda is attached to the host genome in the vicinity of *gal*, then when it enters a nonlysogenic female it finds itself in an environment lacking lambda repressor. This is analogous to infection, and the cell normally proceeds through a lytic cycle. This process of a phage inducing upon its transfer into a nonlysogenic female is called zygotic induction.

A minor paradox is raised by the results discussed above. The DNA is transferred to females in a linear fashion beginning at a particular point. Therefore it would seem that if a genetic marker lying ahead of the integrated lambda were transferred to females, the lambda itself eventually would also be transferred. Why are such cells not killed by zygotic induction?

The resolution to the paradox is that the majority of recipient cells that receive a *gal*-proximal marker do not also receive *gal*. The DNA strand being transferred frequently breaks and conjugation stops. Even though genetic markers lying near the origin are transferred at high efficiency, markers lying farther away and therefore transferred later in the mating process are transferred with substantially lower efficiencies. Consequently, most cells receiving a marker proximal to lambda or *gal* never receive lambda or *gal*, so zygotic induction occurs in only a tiny fraction of the cells that receive markers transferred before *gal*.

CHROMOSOMAL DELETIONS ALSO DELETE LAMBDA GENES

The simplest method of attaching lambda to the host chromosome would be to insert it directly into the DNA. The simplest ideas are not always correct, however, and in the mid-60s when the question was being considered, DNA seemed sacrosanct. Experiments were designed to determine whether lambda did integrate into the chromosome or whether it merely was stuck onto a special place on the chromosome.

The genetic demonstration of lambda's insertion into the host chromosome utilized deletions. The conclusion was that lambda is integrated into the chromosome if a deletion of host genes extends into the lambda and removes some, but not all, lambda genes. Often the identification of deletions is difficult. However, in this case their identification and isolation were easy.

The nitrate reductase complex permits *E. coli* to use nitrate as an electron sink in the absence of oxygen. This enzyme complex is not entirely specific, and it will reduce chlorate as well. However, the product, chlorite, is toxic. Such a situation is a geneticist's dream, for if wild-type cells can be killed, then mutants that are not killed by the chlorate can easily be isolated. Although it would be appropriate to name the genes involved after nitrate, the genetic locus is frequently called *chl* for chlorate resistance.

The deletions in the various *chl* loci frequently extend into adjacent genes. If chlorate-resistant mutants are isolated in a lambda lysogen, some of the deletions are

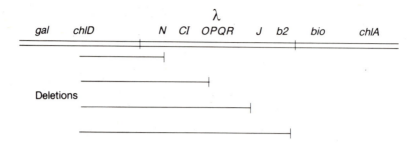

Figure 17.2 Deletions from *chlD* into lambda genes that are consistent with lambda being directly integrated into the host chromosome.

found to be missing some but not all lambda genes. The pattern of lambda genes remaining or deleted always is consistent with the lambda genome being linearly incorporated into the chromosome (Fig. 17.2).

If some of the lambda genes have been deleted, how does one show that some lambda genes remain? The cell certainly cannot produce infective lambda! A method called marker rescue answers the question. Suppose we wish to test whether the deleted lysogen still has an intact *J* gene. A lawn of these cells is poured in soft agar on a plate and a small volume of a lambda *imm⁴³⁴ susJ* (nonsense mutation in *J*) stock is spotted onto the lawn. Being heteroimmune, the superinfecting phage is not repressed even if the deleted lysogen still possesses immunity. Some of these phage can exchange their defective *J* gene by recombination for the good one existing on what remains of the partially deleted lysogenic lambda. Sufficient phage can make this replacement so that the resulting *sus⁺* phage will grow and lyse all cells in the spot, producing a clearing after 12 hours of growth. If no *J* gene is present on the partially deleted chromosome, the lambda *imm⁴³⁴ susJ* cannot grow, and the spot is turbid. In some cases, a gene on the partially deleted lysogen will be able to complement the superinfecting phage, but the mapping results are unaltered. Only if the gene is present can the superinfecting phage grow.

Marker rescue experiments on *chl* deletions revealed many deletions that extended partway into the lambda genes. Certainly the simplest explanation for this result is that the DNA of lambda is contiguous with the host DNA.

DNA HETERODUPLEXES PROVE THAT LAMBDA INTEGRATES

Davidson and Sharp directly demonstrated the insertion of lambda DNA into the genome using electron microscopy. They used a procedure developed to examine sequence homologies between strands from different DNA molecules. After denaturation to separate strands of the two input DNA populations, the strands are allowed to anneal. Then the DNA is mixed with a cytochrome C solution to coat the

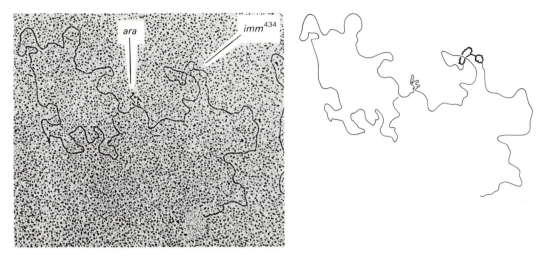

Figure 17.3   A DNA heteroduplex formed with one strand of lambda *imm*[434] and one strand of an *ara* transducing lambda. The bush in the middle is the single-stranded *ara* sequence, and the two bubbles in the immunity region are from the nonhomologous sequences that have partial homology over a stretch. This portion hybridized together.

molecules and increase their diameter as well as improve their staining properties. To spread out the long, snarled DNA molecules, a small volume of the DNA plus cytochrome C is layered on top of a buffer. As the protein spreads and forms a monolayer, the DNA molecules are stretched out and can be picked up on an electron-transparent support and stained. Single- and double-stranded DNA visualized by this procedure is easily distinguished because single-stranded DNA is curlier and more flexible than double-stranded DNA.

When two homologous DNA strands reanneal, a simple double-stranded molecule results. However, a heteroduplex between a strand deleted of a stretch of sequence generates a single-stranded "bush" on the other strand, and if the two strands possess a stretch of sequence without homology, a "bubble" is formed (Fig. 17.3). The heteroduplexes formed between an F′-factor containing an integrated lambda genome and lambda DNA showed the structure expected if lambda were to insert itself into the chromosome (see Problem 17.4).

GENE ORDER PERMUTATION AND THE CAMPBELL MODEL

Determination of the lambda phage gene order in lysogens by deletion mapping exposed an anomaly: The gene order is a particular circular permutation of the order found by recombination frequencies between genes for vegetatively growing

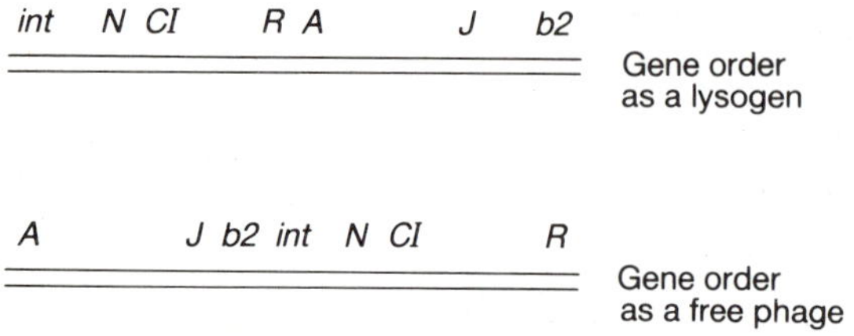

int N CI R A J b2
Gene order as a lysogen
A J b2 int N CI R
Gene order as a free phage

Figure 17.4 The gene orders of lambda phage determined for lysogens by deletion mapping and for lytically infected cells in which the DNA has not integrated into the chromosome by recombination frequencies.

lambda (Fig. 17.4). An explanation for this and other problems concerned with integration was found by Campbell and is known as the Campbell model. The idea is that the lambda DNA, which is linear within the virion, circularizes upon entering a cell and then opens its genome at another point as it reciprocally recombines into the chromosome of the host. The results of the DNA heteroduplexes between DNA

Figure 17.5 The Campbell model for lambda integration. Linear phage DNA circularizes soon after infection by association of the *A* and *R* ends. Integration occurs by means of a site-specific recombination event in the *O* regions of the phage and bacterial DNA to generate the integrated phage. *P* and *P′* and *B* and *B′* are the phage and bacterial DNA sequences required for this process.

A J b2 int N CI QSR
A R S Q
λ
CI
J
POP′
N
b2 int
X
BOB′
Chromosome
BOP′ int N CI QSRA J b2 POB′
Integrated λ

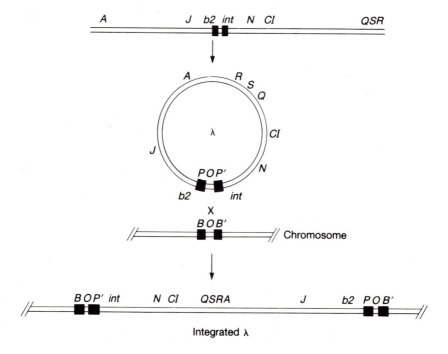

extracted from lambda phage particles and an episome containing an integrated lambda fit this model.

A simple notation has developed for the DNA elements involved in integration and excision. The region of the host DNA where lambda inserts is labeled *BOB'*, where *B* denotes the left portion of the bacterial sequence into which lambda inserts, *O* is the crossover point or region of the recombination between phage and chromosomal DNA, and *B'* is the right portion of the bacterial sequence. The entire region *BOB'* is called *attB*. Similarly, the phage regions are called *POP'* and *attP*. The products of integration are *BOP'*, which is also called *attL*, and *POB'*, which is called *attR* (Fig. 17.5). Note that this notation does not prejudice the discussion by assuming that *B* and *P* are homologous or that *B* is the same as *B'*. In fact, we can predict that the system would not operate with significant homologies between *BOB'* and *POP'* because the *recABC*-mediated general recombination system could then recombine to insert or excise a phage genome at any time.

ISOLATION OF INTEGRATION-DEFECTIVE MUTANTS

A regulated *BOB'-POP'* specific recombination system would solve lambda's problem of restricting integration and excision reactions to the right time and right sites on the phage and host genomes. That lambda possesses such a recombination system could be proved by the isolation of lambda mutants unable to lysogenize because of the absence of an integration enzyme. The main difficulty in such a proof is a common problem in genetics — that of identifying the desired mutant.

At the time (about twenty years ago) that integration mutants were being sought, a lambda mutant was already known that permitted good guesses to be made about the properties of the desired integration minus mutants. This lambda mutant is a deletion of the lambda *b2* region, which extends into the *POP'* region and leaves the phage deficient in its integration abilities. Lambda *b2* phage can establish repression without integrating and therefore can form turbid plaques similar to those of wild-type lambda. If cells from the turbid center of a lambda *b2* plaque are streaked on solidified medium and allowed to proceed through additional generations of growth in the absence of superinfecting phage, the lambda *b2* are diluted away since they do not replicate by their own *ori*, and they are not integrated into the chromosome to be replicated by the host. Therefore the resultant cells are nonlysogenic. Thus, seeking mutants defective in the ability to integrate merely required seeking phage that could form turbid plaques but that could not stably integrate.

The two steps of growing cells to dilute away the nonintegrated lambda and testing for lambda immunity can be combined. The candidate cells are streaked on plates containing the pH indicators eosine yellow and methylene blue and are spread with 10^7 lambda CI^- mutants. The cells spread on the plate grow, divide, and occasionally encounter a lambda CI^-. If a cell infected by a lambda CI^- particle does not possess lambda immunity, the phage is not repressed, and it grows and lyses the

Strain 152
infected with λ

Colonies on EMBO plate spread with
λb2c after 20 hr at 31°C

(1) b2+ int+

(2) b2

(3) int6

(4) int 41

(5) int 41 & b2

(6) int 41 & int 6

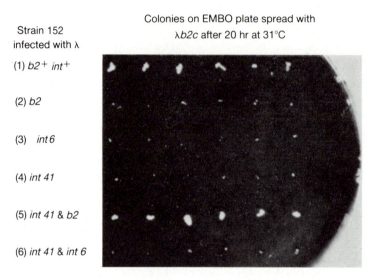

Figure 17.6   Spot test for lysogenicity. Small numbers of cells infected as indicated were spotted on the plate spread with 10^9 λb2c and incubated. Rows 5 and 6 show complementation between λint41 and λb2 but no complementation between λint41 and λint6. (From Integration-Negative Mutants of Bacteriophage Lambda, M. Gottesman, M. Yarmolinsky, J. Mol. Biol. 31, 487–505 [1968].)

cell. This releases some acid. Adjacent cells may also be infected and lysed, but because the colony stops growing before all cells can be lysed, some cells remain in the colony (Fig. 17.6). A colony of nonlysogenic cells spotted on these plates therefore grows into a ragged and purple colony, whereas colonies of immune cells yield smooth and pink colonies.

Using a brute-force pick-and-spot technique based on the knowledge of the behavior of lambda b2 mutants and the indicating plates, Echols isolated lambda mutants unable to lysogenize. These are called int mutants. Some of them were nonsense mutants, proving that the phage encodes a protein that is required for integration.

ISOLATION OF EXCISION-DEFICIENT MUTANTS

If an int mutant has been helped to lysogenize by complementation, it is found to excise poorly. Hence int protein is required for excision. Although it would seem that the process of integration should be readily reversible with the same components that are used for integration, another phage-encoded protein has been found to be required for the excision process to proceed at an appreciable rate.

A phenomenon known as heteroimmune curing, loss of the lysogenic phage, forms the basis of a simple demonstration that excision requires a protein in addition to *int* protein. After lambda lysogens have been infected with the heteroimmune phage lambda *imm*[434], many of the survivors are found to have been cured of the

$$\lambda imm^{434} + \text{Lysogen} \longrightarrow \text{Some nonlysogens}$$

lambda. Apparently the superinfecting heteroimmune phage provide diffusible products that increase excision of the lambda phage. Some Int⁻ phage can promote curing when they superinfect lysogens of different immunity. Such mutants must be providing something other than *int* protein to generate the curing.

The isolation of a nonsense mutation in the gene required for excision proved that it coded for a protein. This isolation used heat-pulse curing. If a lysogen of lambda CI_{857} growing at 32° is heated to 42° for five minutes and grown at 32°, the heat-sensitive CI_{857} repressor is first denatured, and phage growth begins. Then, after the cells are cooled to 32°, the repressor renatures and further phage development ceases before sufficient phage products have accumulated to kill the cells. However, in the five minutes of derepression, sufficient phage proteins are synthesized so that lambda can excise from the chromosome. After further cell growth, the excised lambda genome is diluted away, and daughter cells appear at high frequency that are free of lambda.

Heat-pulse curing was used to isolate excision-defective mutants in the following way (Fig. 17.7). A mutagenized stock of lambda CI_{857} was used to lysogenize cells. This step selected for phage retaining the ability to lysogenize. The lysogenic cells resulting from this step were grown at 32° and then replicated onto three petri plates. One was incubated at 32°, one was incubated at 42°, and the third was incubated at 42° for five minutes, at 32° for two hours, and then at 42° overnight.

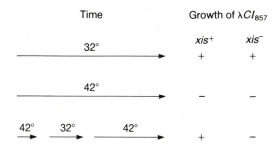

Figure 17.7   Temperature protocols for the identification of *xis*⁺ and *xis*⁻ by the inability of the latter to excise following a short heat induction.

The first plate kept viable copies of all the colonies. The second plate showed which colonies were infected with lambda, as growth at 42° would induce the phage and kill lysogens. Growth on the third plate indicated which colonies could heat-pulse cure. On this plate, the lambda lysogens would heat-pulse cure, and the cured cells in such a colony would be capable of growing into colonies during the subsequent growth at 42°. Any colony whose cells possessed excision-defective phage, *xis* mutants, could not heat-pulse cure and would therefore be killed by the subsequent growth at 42°.

Later, a simpler method for detecting excision-defective mutants was devised. This scheme uses cells in which lambda has mistakenly integrated into a site within the *gal* genes that resembles the authentic *attB* site. The cells are Gal⁻ as a result. Infection of these cells by phage able to provide the *xis* function in *trans* catalyzes excision of the phage from the *gal* genes, and the cells become Gal⁺. By plating infected cells on MacConkey galactose-indicating plates, those deriving from Xis⁺ phage are red because the cured cells are Gal⁺ and turn the dyes red, while the plaques deriving from Xis⁻ phage are clear or white. With this convenient assay, the excision abilities of many different phage can be assayed on a single galactose indicator plate.

Using this plate assay, Enquist and Weisberg performed a thorough genetic analysis of the *att-int-xis* region of the lambda chromosome. They isolated and characterized hundreds of *int* and *xis* point mutations and mapped many of them against a set of deletions ending in the region. The large number of mutants studied permitted a reasonable estimation of the sizes of the genes. The *int* gene appeared to be about 1,240 base pairs long, and the *xis* gene was very small, only 110 base pairs long. No additional phage genes acting in *trans* and directly involved in the integration or excision process were discovered.

Int AND *xis* ARE SINGLE GENES AND ARE PHAGE-SPECIFIC

How many phage-encoded proteins are required for integration and excision? If *int* and *xis* proteins will function in *trans*, then complementation tests can be performed to determine the number of *int* genes and the number of *xis* genes on the phage. Experiments showed that the *int* protein of one phage will indeed help another phage integrate, for cells coinfected with a lambda *int* mutant and a heteroimmune lambda *imm*⁴³⁴ phage yield some lambda lysogens in addition to lambda *imm*⁴³⁴ lysogens and double lysogens. However, coinfection of cells with two different

$$\lambda imm^{434} + \lambda int_x^- \xrightarrow{\text{infect}} \lambda \text{ and } \lambda imm^{434} \text{ lysogens}$$

lambda *int* mutants does not yield lysogens at the frequencies normally observed for complementation. All the *int* mutations tested therefore must lie in the same gene.

$$\lambda int_x^- + \lambda int_y^- \xrightarrow{\text{infect}} \text{No lysogens}$$

Analogous experiments with *xis* mutants show that they too are confined to a single gene.

Int and *xis* mutations are not complemented by most phage other than lambda.

$$\lambda int_x^- + \phi 21 \xrightarrow{\text{infect}} \phi 21 \text{ lysogens only}$$
$$\lambda int_x^- + \phi 80 \xrightarrow{\hspace{1cm}} \phi 80 \text{ lysogens only}$$

Although phage 21 possesses its own *int* and *xis* proteins, these do not complement lambda *int* or *xis* mutants. Thus, *int* and *xis* can be phage-specific.

INCORRECT EXCISION PRODUCES *gal* AND *bio* TRANSDUCING PHAGE

Lambda phage can act as a specialized transducer of certain host genes. This means that each phage particle carries a copy of the same bacterial DNA and can transduce infected cells. Specialized transduction by lambda should be contrasted to generalized transduction in which only a small fraction of the phage particles contain nonphage DNA, and only a small fraction of these contain DNA of any particular bacterial gene.

Transducing phage once were necessary for the enrichment and ultimate isolation of regions of DNA involved in regulation of bacterial genes. Also, cells infected with transducing phage often hypersynthesize the proteins they encode owing to the presence of multiple phage genomes during much of the phage growth cycle. Without this gene dosage effect, the proteins often would be present in quantities too low for biochemical study. The enormous value of transducing phage in research on bacterial gene regulation undoubtedly stimulated efforts to develop genetic engineering techniques so that DNA, proteins, and gene regulation systems from other organisms could also be studied biochemically.

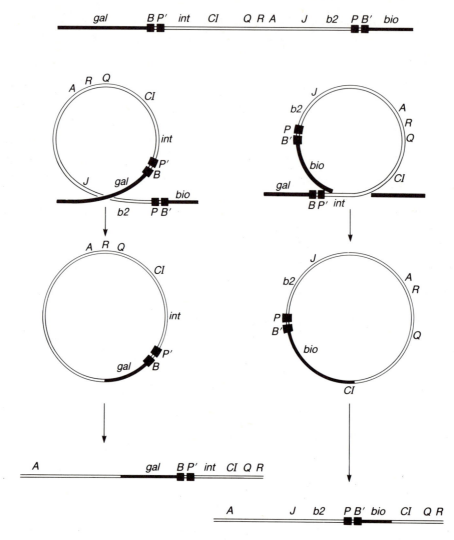

Figure 17.8   Production of *gal* and *bio* transducing phage by recombination events located to the left or to the right of the end of an integrated lambda.

Upon induction of a lysogen, most phage excise uneventfully. However, at a frequency of 10^{-5} to 10^{-7}, phage excise incorrectly from the host chromosome and pick up bacterial DNA adjacent to the phage integration site. These incorrect excision events produce transducing phage. One of the joys and powers of genetics is that exceptionally rare events such as these may be captured and perpetuated for further study.

When lambda excises incorrectly, presumably utilizing *int* and *xis* but performing the site-specific recombination event at a point other than *BOP'* and *POB'*, then bacterial DNA on one side or the other of the integrated lambda DNA is picked up by the phage. Looping out to the left produces a phage with *gal* genes substituting for the *b2* region (Fig. 17.8). That is, *gal* DNA in the phage is flanked on both sides by lambda genes. Also, in place of the phage attachment region *POP'* is *attL, BOP'*.

The *gal* genes are sufficiently far from *att* lambda that phage having picked up *gal* genes must, of necessity, have left at least the *J* gene behind so that the transducing phage genome is small enough to be packaged in a lambda coat. These phage are defective because they do not yield viable phage upon infection of cells. They are propagated by coinfecting them with phage that provide their missing *J* gene product. Such defective *gal* transducing phage are called *λdgal*.

Phage that excise the other way pick up the biotin genes. These substitute for the nonessential *int, xis,* and other early genes under control of p_L. Usually these phage are not defective and hence are called *λpbio* for their ability to form plaques and transduce biotin genes. They possess *attR, POB'*.

TRANSDUCING PHAGE CARRYING GENES OTHER THAN *gal* AND *bio*

Isolation of *gal* or *bio* transducing phage is straightforward because the lambda attachment site is near these genes. However, for genes located farther away, a simple improper excision event cannot produce transducing phage. Genes located perhaps as far as a minute away on the genetic map, about 40,000 base pairs, from an integrated phage can be picked up by a combination of a deletion and improper excision event. However, the combined probabilities for these are so low that large cell cultures must be used to find the transducing phage. What can be done to isolate phage that carry genes located more than a minute from *attB* of lambda?

Two approaches are used to bring an integrated phage and a particular gene close enough together to permit generation of transducing phage. Either the gene is moved close to the integrated phage genome or the phage is moved close to the gene.

One method of moving a gene X^+ close to a phage requires isolating an episome containing the gene, then forcing the episome to recombine right into the middle of a gene near a phage attachment site. $\phi80$, a relative of lambda, is suitable for these purposes. The gene coding for the $\phi80$ receptor protein lies adjacent to $\phi80$ *att*, so the episome can be forced to integrate into the receptor gene and then $\phi80$ transducing phage can be isolated that have picked up the X gene, which is now located close to the phage attachment site (Fig. 17.9).

Finding the cell in which the episome has inserted into the gene encoding the receptor protein for $\phi80$ requires multiple selections because the event is rare. The episome is made temperature-sensitive in its replication, and X^- cells are used. Therefore, selecting for $\phi80$-resistant, temperature-resistant X^+ cells yields the cell containing the desired insertion. Candidates for the insertion can be confirmed by the ability of the integrated F' to mobilize the host chromosome, thereby converting the strain to an Hfr male in which the origin of transfer is near the $\phi80$ *att* region.

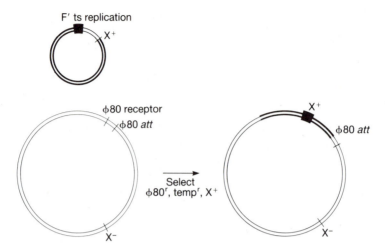

Figure 17.9   Selection of cells in which an episome unable to replicate itself at high temperature and carrying the gene X^+ recombines into the gene encoding the $\phi 80$ receptor protein.

However, the resulting $\phi 80^r$ cells must be infected and lysogenized with $\phi 80$ before the final transducing phage can be made. This is done by infecting with a host range–variant phage called $\phi 80h$, which binds to a structure different from the $\phi 80$ receptor protein and can infect $\phi 80^r$ cells.

The second way of moving lambda and a particular gene close to one another is to bring the phage to the gene. This is easily done by selecting for lysogens following infection of a host deleted of the phage attachment region *attB*. At very low but usable frequencies, lambda will then integrate into sites that possess some similarity to the normal chromosomal integration site. These secondary integration sites are so widely scattered over the chromosome that lambda can be forced, albeit at very low frequencies, to integrate into or adjacent to just about any gene. Either the desired integrant can be selected by genetic means or the entire population of cells with phage integrated at many different sites can be used as a source of phage. Some of the phage will excise incorrectly from their abnormal positions and transduce adjacent bacterial genes.

USE OF TRANSDUCING PHAGE TO STUDY INTEGRATION AND EXCISION

One use of transducing phage is to demonstrate that lambda normally integrates and excises at precisely the same point. Of course, lambda can be integrated and excised many times from the bacterial *att* region, and the region apparently suffers

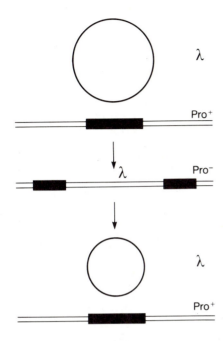

Figure 17.10 Integration of lambda phage into a pseudo *att* site in a gene inactivates that gene, but excision restores the original nucleotide sequence and the gene is reactivated.

no harm. However, how do we know, without sequencing, that bases are not inserted or deleted in the process? The integration of lambda into secondary *att* sites provided the proof that, as far as the sequence of the host chromosome is concerned, excision is the exact opposite of integration.

When lambda integrates into a secondary *att* site, the gene into which lambda has inserted is disrupted and therefore inactivated. However, when the lambda is induced and excises from these sites, the majority of the surviving cells possess a perfectly normal gene. A very few of the lambda improperly excise and produce transducing phage as described in the previous section. That is, except for the products of the rare improper excision events, no nucleotides are inserted or deleted at the *att* site. Therefore it is reasonable to infer that the integration and excision cycle at the normal *att* site similarly does not alter its sequence. For example, one site for secondary lambda integration is a gene coding for a protein required for proline synthesis. The insertion of lambda makes the cells Pro⁻, but heat-pulse curing leaves the cells Pro⁺ (Fig. 17.10).

A second use of transducing phage is in the study of the biochemistry of the integration and excision reactions themselves. These site-specific recombination events take place at the *att* regions, but the partners need not be confined to a phage and host chromosome. For example, the excision reaction can be performed between two phage, a

$$BOP' + POB' \longrightarrow BOB' + POP'$$

λdgal and a λpbio to form a lambda and a λdgal-bio. To detect these recombination products, the input phage must be genetically marked. This can be done using nonsense mutations located in the A and R genes, which are at opposite ends of the lambda DNA. The cross might therefore be between $\lambda dgalA^-R^+ \times \lambda pbioA^+R^-$, and the frequency of production of wild-type lambda, those able to form plaques on su⁻ cells, can be measured. Crosses between all combinations of att regions can be performed by similar approaches. For example, lambda with BOB' can be constructed by isolating a gal-bio phage from the above cross. Of course, to prevent generalized recombination from producing similar products, both the cells and phage must be made recombination-defective.

The main results of studies examining the site-specific recombination events catalyzed by int and xis proteins are that all combinations of att regions will recombine, but at different rates, and that the xis protein is required only for the excision type of reaction.

THE DOUBLE att PHAGE, att²

The preceding sections have described a little of what has been learned from genetic experiments about lambda's ability to integrate and excise. One way to study the biochemistry of integration and excision is to construct an in vitro system that mimics the in vivo reaction. The first requirement for integration or excision is to bring the two participating molecules close together. In vivo, this requirement is partially met as a simple consequence of the fact that both DNA molecules are confined to the volume of the cell and therefore are held in close proximity. The in vitro integration or excision reaction should be greatly speeded if the two att regions similarly can be forced close to one another. One way to accomplish this is to place both att sites on the same DNA molecule. Then the concentration of one att site in the vicinity of the other is, of necessity, high.

This section describes the isolation and properties of such a double att phage, and the following sections describe its use and the use of a similar phage to study the integration and excision reactions more deeply both in vivo and in vitro.

Within the size limits of DNA that can be packaged in a lambda phage head, the locations of the crossover points that produce gal or bio transducing phage appears random. From this it should not be inferred that all DNA except the attachment region are equivalent, because farther from the lambda integration site could lie sequences that recombine with one another at higher frequencies catalyzed either by the int-xis site-specific system or by the general recombination system. Indeed, on either side of lambda are two sites that the host recombination system will recombine with each other to excise a phage that has picked up host DNA from both sides of the phage integration site (Fig. 17.11). Ordinarily this event cannot be detected because the resulting DNA is too long to be packaged in a lambda coat. However, if several deletions have first been put into lambda, this particular excision product will fit into the head, and viable phage are formed.

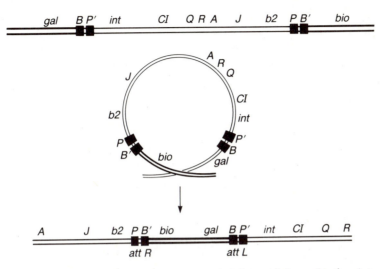

Figure 17.11   Recombination between two points beyond the ends of an integrated lambda phage generates the lambda *att²* phage containing *attR* and *attL*.

The structure of the phage produced by recombination between two sites flanking lambda is most interesting. The phage contains two *att* regions, *attL* and *attR*, and is called *att²*. Such an *att²* phage can lose the extra bacterial DNA by a reaction analogous to normal excision:

$$POB' + BOP' \longrightarrow POP' + BOB'.$$

This process requires the *int* and *xis* proteins and produces a viable lambda phage genome of normal size, except for its original deletions, and a minicircle (Fig. 17.12).

Figure 17.12   An excision reaction between the two *att* regions on *att²* generates a viable phage with a single *att* region and a minicircle.

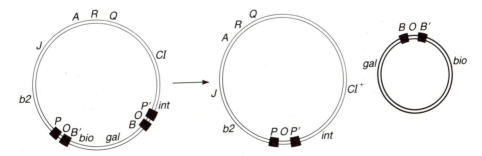

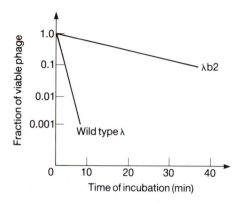

Figure 17.13   Killing of lambda $b2$ and wild-type lambda by heating in the presence of a Mg^{++} chelator.

Study of the excision reaction requires quantitating the input phage lambda att^2 as well as the product phage. Fortunately, the two many be readily distinguished. The concentrations of λatt^2 and the lambda generated by excision can be assayed in a mixture of the two by first separating the phage on the basis of density in equilibrium centrifugation. The λatt^2 phage is more dense than lambda as a result of its additional DNA.

A simpler assay of the two types of phage makes use of the sensitivity of lambda to heat and chelators of Mg^{++}. Removal of Mg^{++} ion from the phage reduces the charge neutralization between phosphates of the DNA backbone, and as a result the DNA expands and can burst the phage coat. Phage particles possessing less than the usual amount of DNA are more resistant to heating and removal of Mg^{++}. Combinations of Mg^{++} chelators and elevated temperatures can be found in which wild-type lambda is killed by factors of 10^3 to 10^4, but the lambda $b2$ deletion mutant is unharmed (Fig. 17.13). Slightly different conditions can be used to kill the λatt^2 but permit its excision product to form plaques. A phage stock containing λ and λatt^2 can be titered on plates with and without pyrophosphate, which chelates magnesium. Both the phage and the excision product grow on the normal plates, but only the excision product phage can grow on the plates containing pyrophosphate.

REGULATING INTEGRATION AND EXCISION:
Xis IS UNSTABLE

When lambda infects a cell, both *int* and *xis* proteins are synthesized, so the phage can repeatedly integrate and excise until the proteins are diluted away by growth. Why is the last event an integration rather than an excision? Part of the explanation of how integration wins out is that *int* protein is synthesized in preference to *xis* under control of the *CII*-activated promoter, p_I. Despite this, an infected cell will contain some *xis* protein. One simple mechanism for ensuring integration is for *xis*

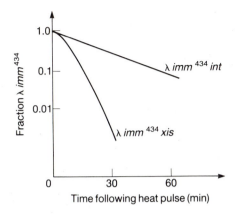

Figure 17.14   The fraction of lambda *imm*⁴³⁴ *att*² Int⁻ or Xis⁻ as a function of time that is converted to lambda *imm*⁴³⁴ by the *int* and *xis* proteins induced at zero time.

protein to be less stable than *int*. Gottesman used the *att*² phage to simplify testing this hypothesis with an *in vivo* assay of the stability of *int* and *xis* proteins.

The test used a lysogen of $\lambda Cl_{857}S^-$ briefly heat-pulsed to denature the thermolabile repressor and induce a burst of *int* and *xis* protein synthesis. The repressor then renatured, and the phage-encoded protein synthesis stopped. At times thereafter, portions of the culture were additionally infected with λimm^{21} *att*² Xis⁻ or λimm^{21} *att*² Int⁻. At early times after the heat pulsing the *int* and *xis* protein already present from the first phage complemented the *int* or *xis* defect of the *att*² and catalyzed the excision reaction to produce "wild-type" λimm^{21} from the λimm^{21} *att*². At later times, however, the *int* or *xis* synthesized by the first phage had decayed, and therefore fewer wild-type λimm^{21} were produced (Fig. 17.14). Simply infecting at various times after the heat pulse and assaying for the number of non-*att*² lambda that were produced allowed the kinetics of decay of the *int* and *xis* protein to be measured. The half-life of *int* was greater than an hour, whereas the half-life of *xis* was less than half an hour.

RETRO-INHIBITION OF *int* TRANSLATION

A lambda phage not integrated in the host chromosome and proceeding down the lytic pathway would do well to minimize its synthesis of *int* protein. On the other hand, an induced lambda located in the chromosome has great need for *int* protein to assist in its excision. Thus the phage can find itself in two nearly identical situations, but in only one should it restrict its production of *int* protein. How can the phage know whether it is integrated and adjust its *int* protein synthesis accordingly? In both situations, the *int* specific promoter p_I is off and the synthesis and regulation of *int* and *xis* proteins must derive from the p_L promoter.

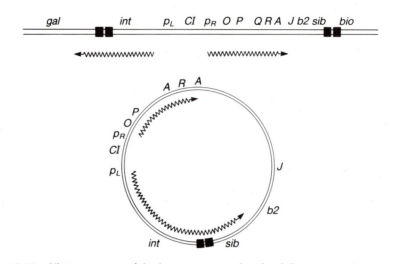

Figure 17.15   *Sib* is not a part of the *int* messenger when lambda is transcribed as a lysogen.

The topology of the phage signals whether it is integrated in the chromosome or free in the cytoplasm! When the phage has excised, it is circular, and the *b2* region is adjacent to the *int* gene (Fig. 17.15). However, when the phage is integrated, host DNA is adjacent to the *int* gene. These two situations have different effects on the translation efficiency of *int* messenger deriving from the p_L promoter. A sequence called *sib* lying in the *b2* region is transcribed and becomes part of the *int* messenger only when the phage is not integrated. This sequence inhibits translation of the *int* message.

The synthesis of *int* and *xis* proteins was characterized in lytically infected cells that had been exposed prior to infection with a low dose of UV light to block host-encoded protein synthesis. These cells were then infected with phage, and radioactive amino acids were added. At various times, samples were taken and subjected to SDS polyacrylamide gel electrophoresis. The *int* and *xis* proteins were easily detected and quantitated by autoradiography. Alternatively, no UV light was used, and instead the *int* or *xis* proteins were precipitated with antibody, further separated from the other proteins by the electrophoresis, and finally quantitated from autoradiographs.

Following infection by wild-type lambda, the synthesis of *int* protein is low, but infection by lambda *b2* yields high synthesis of *int* protein. Point mutations and deletions have also been isolated that yield high levels of *int* protein following infection. These narrowly define a region in the *b2* region called *sib*, which is responsible for the altered synthesis rate of *int*. The deletions and point mutations have a strictly *cis* effect on *int* protein synthesis. Therefore the effect is called retro-inhibition because it acts backward on expression of the preceding gene. Most likely retro-inhibition acts through the secondary structure of the messenger RNA.

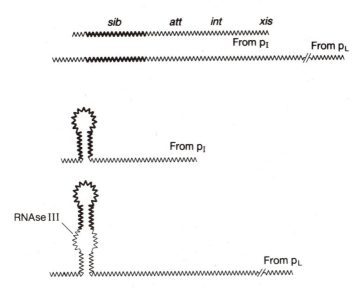

Figure 17.16   Secondary structures of the *sib* region in transcripts originating from p_I and p_L. The promoter p_I lies just ahead of *xis*, so the transcription start point does not include all of *xis*.

The *sib* region contains a typical transcription termination sequence, and the *sib* mutations reduce base pairing within the transcript in the termination hairpin (Fig. 17.16). Transcripts initiated at p_I normally terminate near the end of this region, but the transcripts initiated at p_L ignore termination signals because of the phage N protein bound to RNA polymerase, and the transcription continues past this region. The transcripts are capable of forming an additional base-paired region that is sensitive to RNase III, an enzyme that cleaves special double-stranded RNA structures, whereas the normally terminated transcript that initiated at p_I is not. Most likely, then, cleavage by RNase III leads to degradation of the adjacent *int* gene messenger. As expected, the *sib* effect is absent in RNase III–defective mutants.

In vitro ASSAY OF INTEGRATION AND EXCISION

Although ingenious genetic experiments can reveal much about biochemical systems, ultimately the details must be studied biochemically. Therefore it has been of great interest to isolate the *int* and *xis* proteins. Originally, attempts were made to isolate them without regard to their biochemical activities. The proteins encoded by lambda were labeled with radioactive amino acids. This was done in a double label experiment in which ^{14}C-amino acids were used to label cells infected with wild-type lambda and ^{3}H-amino acids were used to label cells infected with *int* or *xis* nonsense mutants. The labeled extracts were mixed, and proteins were fractionated by column chromatography, electrophoresis, or centrifugation. The *int*-containing fractions were identified by their elevated ratios of ^{14}C to ^{3}H. This assay guided a partial purification of the *int* product, but it provided no indication of whether the protein was biologically active or even of how to measure its activity.

A second type of assay that has been used to purify *int* protein of lambda was based on the expectation that the protein should bind specifically to the *att* region of lambda. Like so many assays in molecular biology, this one utilized the fact that double-stranded DNA, but not most proteins, freely passes through cellulose acetate – cellulose nitrate filters. However, a protein bound to a DNA molecule can prevent the passage of the DNA through the filter. Therefore retention of radioactive DNA on filters can easily be quantitated. Lambda *att*$^+$ and *att*$^-$ DNAs used in the same binding reaction were labeled with different radioactive isotopes to provide a built-in specificity measure of any binding that was observed. This assay was an improvement over the first, but it too could yield a protein without the full activity of *int* protein.

The most reliable assay for *int* protein activity is to seek a protein that carries out the entire integration reaction *in vitro*. As mentioned above, one likely requirement for such a reaction to proceed *in vitro* is a high concentration of *attB* and *attP* regions. The easiest way to obtain these is to place both on the same DNA molecule.

Can *attB* and *attP* be put on the same molecule, as are *attL* and *attR* of λatt^2? Nash constructed such a phage by forcing a recombination to occur in a region of nonhomology between a λgal-*bio* containing *attB* and a lambda (Fig. 17.17). Similar to λatt^2, his construct, $\lambda attB$-*attP,* can undergo an integration type of reaction that removes the *bio* region and leaves the phage considerably more resistant to Mg^{++} chelation or heat. Hence, the parental phage and derivatives that have undergone an integration reaction and have become smaller can easily be distinguished.

Initial tests with the $\lambda attB$-*attP* phage showed that an *in vitro* integration reaction would work. The experiment was performed by incubating the $\lambda attB$-*attP* DNA with a cell extract prepared from lambda-infected cells. Then the DNA was extracted from the mix and used to infect cells that had been made capable of taking up naked

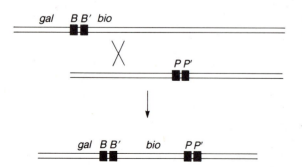

Figure 17.17   The illegitimate recombination by which the lambda *attB-attP* phage was generated.

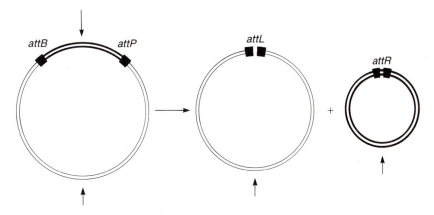

Figure 17.18   Digestion with a restriction enzyme generates different fragments from the lambda *attB-attP* phage and its integration "product," lambda and the minicircle.

lambda DNA, supporting growth of the phage, and lysing. This provided an extremely sensitive assay for the integration reaction. The infected spheroplasts could be spread on a lawn of lambda-sensitive cells on a plate containing pyrophosphate to yield plaques only from spheroplasts infected with products of the integration reaction. Later, as the assay conditions were improved, the integration reaction could be assayed merely by physical quantitation of the DNA products. The locations of cleavage sites in the DNA rearrange by the integration reaction, so new sizes of restriction fragments are generated (Fig. 17.18). These are detected by separating the resulting fragments by electrophoresis.

The *in vitro* integration reaction required Mg^{++}, ATP, and spermidine. More careful examination showed that, normally, linear DNA was not a substrate for integration and that the use of supercoiled DNA eliminated the requirement for ATP. The assays permitted the purification and characterization of biologically active *int* protein, and analogous experiments have been done with the λatt^2 phage for the purification of *xis* protein.

HOST PROTEINS ARE INVOLVED IN INTEGRATION AND EXCISION

As mentioned above, the integration reaction requires supercoiled DNA or relaxed circular DNA plus ATP. This suggests that the cell extracts contain a protein that can supercoil DNA using ATP for energy. Indeed, this proved to be so. The use of either the ability of an extract to perform the integration reaction or merely the production of supercoiled DNA as assayed by electrophoresis allowed identification and purification of the enzyme DNA gyrase. As explained in Chapter 2, this enzyme introduces negative superhelical twists in covalently closed DNA circles at the expense of ATP.

Further study of the *in vitro* integration reaction revealed the requirement for yet another host protein. Extensive purification and characterization of this protein, which is called IHF, for integration host factor, showed it to be a dimer of rather small subunits, 11,000 and 9,500 molecular weight. The genes coding for these peptides, *himA* and *himD* or *hip*, had been identified genetically from host mutations that block phage integration. It is surprising that cells should possess a nonessential protein that is required for integration of lambda, but even more surprising is the fact that synthesis of *him* protein is under control of the *lex* protein. That is, *him* protein is a member of the cell's SOS system that is induced when DNA is damaged.

ISOLATION AND PROPERTIES OF *att* MUTANTS

The properties of mutations in the *att* region could reveal much about the integration and excision reactions. Lambda *att*² provides an elegant method for the isolation of such mutants. The desired *att* mutants should not be substrates for the conversion of

Figure 17.19   Integration of lambda carrying a mutation consisting of a change in a base pair of the *O* region of an *att* sequence. If the DNA strand cleavages by which integration occurs bracket the base change, then the change is placed at each of the *O* regions surrounding an integrated phage. However, mismatch repair can correct either of these to leave the mutation in one or the other or neither of these sites.

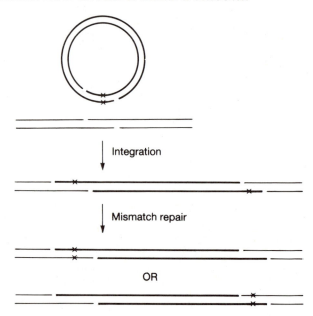

lambda *att*^{2} to lambda in the presence of *int* and *xis*. Therefore the desired mutants can be isolated by infecting cells already containing the *int* and *xis* proteins with mutagenized lambda *att*^{2} and selecting the lambda *att*^{2} that do not excise their inserted sequence to yield the minicircle and the lambda circle.

As expected for mutations in the *att* region, those that were found were *cis*-dominant. That is, they were not mutations in diffusible and hence complementable *trans*-acting proteins like *int* or *xis*. Unexpectedly, however, the mutations did not lie in the *B*, *B'*, *P*, or *P'* regions; instead, they were in the *O* portion of the *BOP'* or *POB'* regions. This was very interesting, for it showed that the *O* region was not merely the space "between" two nucleotides. Instead, it must consist of at least one nucleotide. Finally, the fact that heterozygotes of the mutations could be isolated almost certainly means that the cleavages of the two DNA strands that occur in the *O* region upon integration and excision must be staggered rather than flush (Fig. 17.19).

STRUCTURE OF THE *att* REGION

Historically the ability to sequence DNA came after geneticists had isolated the *att* mutants and had determined much about the integration and excision reactions. Therefore, obtaining the sequence would permit checking the genetic inferences. Sequencing the *att* region was straightforward because phage carrying the four possible *att* regions — *POP'*, *BOB'*, *BOP'*, and *POB'* — were available. Landy identified the appropriate restriction fragments from these phage, isolated them, and sequenced them. As expected from the genetic experiments, all four contained an identical core sequence. The size of the core could not have been determined from the genetic experiments, but it turned out to be 15 base pairs long (Fig. 17.20).

The recognition regions flanking the cores are of different lengths. By deletion analysis, the sequences on either side of *O* of the bacterial *att* region that are neces-

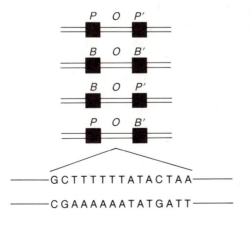

Figure 17.20   Nucleotide sequence of the *O* region of *att*.

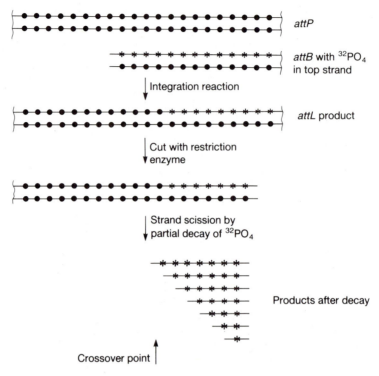

Figure 17.21   Scheme for determining the site of strand transfer in the integration reaction. Small circles represent nonradioactive phosphates and small *x*'s represent radioactive phosphates. More precision is gained by placing radioactive phosphates adjacent to only one of the four bases by using the appropriate α-labeled triphosphates when the strand is labeled. The strand is labeled by using separated strand containing *attB* and a primer sequence to provide the 3'-OH for DNA polymerase.

sary for integration have been shown to be only about five bases. By contrast, *attP* is about 240 base pairs long.

The sites of *int*-promoted cleavage of the *O* sites have been determined exactly using radioactive phosphate both as a label and as an agent to cleave the DNA. An *in vitro* integration reaction was performed between an *attP* and a short DNA fragment containing *attB* that had been extensively labeled with radioactive phosphate on either strand. All phosphates of the product *attL* or *attR* that lie beyond the transfer point derive from *attB* are therefore radioactive. The positions of these radioactive phosphates were then determined exactly by permitting some to decay. The decay event cleaves the phosphate backbones, so subsequent electrophoresis of the denatured fragments on a sequencing gel yields bands corresponding to each position that had a radioactive phosphate (Fig. 17.21). The cleavages are separated from one another by seven base pairs.

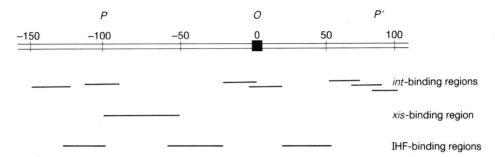

Figure 17.22   The regions in *attP* to which *int*, *xis*, and IHF proteins bind.

Int AND *xis* BINDING IN THE *att* REGION

One approach to the study of the integration and excision reactions is to determine where purified *int, xis,* and IHF proteins bind in the *att* regions. Since the *att* sites have been cloned on plasmids, their DNA is readily available for footprinting and chemical protection experiments.

The proteins have a complex binding pattern (Fig. 17.22). For example, *int* protein binds to four sites in *attP:* one in the common core, two on the *P* arm, and one on the *P'* arm. IHF binds to three sites, and *xis* protein binds to at least one site. Together these binding sites cover nearly the entire region from −150 to +90 with respect to the center of the common core. Indeed, when the complex of the various *att* regions and the *int, xis,* and IHF proteins is examined in the electron microscope, a compact and topologically complicated structure is seen. This structure is more complicated than might be imagined because it appears to contain six to ten molecules of *int* protein. Such a large complex is much different from the one-to-one binding of repressors to operators or RNA polymerase to promoters. This would make us fear that it could be an artifact except that a complex of *int* plus *xis* proteins bound at *attR* specifically pairs with a similar complex formed at *attL* on a different DNA molecule. Since the bacterial and phage *att* regions possess no sequence homology except for their common core sequences, the bound proteins must be holding the pair of DNA molecules together in much the same way as they must be held for the excision reaction.

We end this chapter at the point where molecular biologists, biochemists, and physical chemists have discovered what elements are involved in the lambda phage integration and excision reactions. It remained to be determined what happens at the atomic scale, what makes it happen, and how we might design and build a system with the properties of being able to recognize a specific site on DNA and integrate another DNA into it.

PROBLEMS

17.1. Suppose lambda DNA molecules freely float in the cell cytoplasm and are randomly segregated into daughter cells. How many DNA molecules per cell would be required to generate a spontaneous curing rate of 10^{-6} per generation?

17.2. Breakage of a DNA strand during conjugation prevents transfer of genetic markers lying after the breakage point. If the probability of breakage per unit length of chromosome transferred to female cells is a constant b, show that the probability of transfer of a marker located at position l from the origin of transfer is e^{-bl}.

17.3. Using the Southern transfer technology (J. Mol. Biol. 98, 503–517 [1975]), how would you show that lambda DNA in a lysogenic strain is directly inserted into the E. coli genome?

17.4. Sketch the expected structure of the heteroduplex between F' (lambda) and lambda DNA.

17.5. If lambda and lambda $b2$ coinfect cells, the only $b2$ lysogens found are double lysogens simultaneously possessing lambda and lambda $b2$ integrated into their chromosomes. Why?

17.6. How would the following data be interpreted?

Infection by	Gives bursts consisting of
lambda N$^-$ + lambda i^4	50 lambda N$^-$ + 50 lambda i^4
lambda N$^-$ + lambda i^{21}	50 lambda N$^-$ + 50 lambda i^{21}

Induction of the lysogen	plus infection by	Gives bursts of consisting of
(lambda N$^-$)	lambda i^4	50 lambda N$^-$ + 50 lambda i^4
(lambda N$^-$)	lambda i^{21}	0 lambda N$^-$ + 100 lambda i^{21}

17.7. Design a simple experiment in which the frequency of integration into an episome and chromosome would be capable of proving that $attB \neq attP$ and/or $attB' \neq attP'$.

17.8. Phage mu is much like lambda except that it appears to be completely indiscriminate with respect to the host attachment site. (a) Show how insertion of mu into a gene produces an effect just like a point mutation on the activity of the gene product and that in mapping the mu insertion maps like a point mutation. (b) How would you show that mu phage could insert in either direction within a gene?

17.9. Suppose that Int$^-$ lambda cannot lysogenize at all. It is found, however, that mu phage can help lambda integrate. It is also found that the mu-assisted integration of lambda is not altered if the *POP'* of lambda is deleted. Predict the genetic structure of mu-assisted lambda lysogens.

17.10. The integration-excision reaction is asymmetric in that *int* is required for integration and both *int* and *xis* are required for excision. How can this be true without violating the fundamental physical law of microscopic reversibility?

17.11. Lambda's requirement of supercoiled DNA for integration may merely be a way to detect whether its DNA is circular. Why would integration of linear lambda DNA be lethal? Which substrate would be more likely to possess a requirement for supercoiling, the bacterial or the phage DNA?

17.12. The superhelical twists required for the lambda excision reaction could be required to place the DNA in a conformation that is recognized by the *int* and *xis* proteins, or the strain put into the DNA could assist part of the biochemistry, melting out the crossover region, for example. What experiments could you devise that could distinguish between these two possibilities?

17.13. The ability to cross *att* mutations from the phage onto the chromosome need not prove the existence of staggered cuts. How could this be?

17.14. Why would lambda *imm*^{434} but not lambda *imm*^{21} be expected to heteroimmune-cure a wild-type lambda lysogen?

17.15. A *λpara* phage was produced by insertion of lambda into the *araB* gene followed by excision to produce a phage carrying the *araC* gene and part of the *araB* gene. What would be the interpretation of the fact that induction of a single lysogen of this phage yielded only a few phage per cell unless arabinose was also present in the medium? Then it would produce normal quantities of phage.

17.16. How could you make a lambda lysogen of a Gal$^-$ strain that is also deleted of the maltose gene activator so that it is lambda-resistant? You might try infecting with lambda phage containing ϕ80 tail fibers, but the cells are ϕ80-resistant as well as resistant to all other lambdoid phage and their host range mutants. The conditions of the problem are to make the lysogen in one step so that transducing to make them Mal$^+$ and then lysogenizing them is forbidden.

17.17. When lambda is forced to integrate into the chromosome at sites other than the normal lambda attachment site, it does so at low frequencies that vary widely from site to site. What correlation would you expect between integration frequency at a site and the fraction of excised phage from sites that transduce adjacent bacterial genes?

17.18. Look up and describe the basis of the genetic selection Nash used in the construction of the lambda *attB-attP*.

17.19. In studying integration or excision events from *att*^{2} types of substrates, the two *att* regions can be in parallel or opposite orientations on the DNA molecule. How do the products of the reaction differ in these two cases? What use might be made of either of these?

RECOMMENDED READINGS

Prophage Lambda at Unusual Chromosomal Locations, I. Location of the Secondary Attachment Sites and the Properties of the Lysogens, K. Shimada, R. Weisberg, M. Gottesman, J. Mol. Biol. *63*, 483–503 (1972).

A Genetic Analysis of the *att-int-xis* Region of Coliphage Lambda, L. Enquist, R. Weisberg, J. Mol. Biol. *111*, 97–120 (1977).

Viral Integration and Excision: Structure of the Lambda *att* Sites, A. Landy, W. Ross, Science *197*, 1147–1160 (1977).

RELATED REVIEWS, BOOKS, AND ARTICLES

The Bacteriophage Lambda, ed. A. Hershey, Cold Spring Harbor Laboratory (1971).

Integration and Excision of Bacteriophage Lambda: The Mechanism of Conservative Site-Specific Recombination, H. Nash, Ann. Rev. Genet. *15*, 143–167 (1981).

Retroregulation: Control of Gene Expression from Sites Distal to the Gene, M. Gottesman, A. Oppenheim, D. Court, Cell *29*, 727–728 (1982).

Lambda II, ed. R. Hendrix, J. Roberts, F. Stahl, R. Weisberg, Cold Spring Harbor Laboratory (1983).

DEEPER READING

Mutants of Bacteriophage Lambda Unable to Integrate into the Host Chromosome, R. Gingery, H. Echols, Proc. Nat. Acad. Sci. USA *58*, 1507–1514 (1967).

A Deletion Analysis of Prophage Lambda and Adjacent Genetic Regions, S. Adhya, P. Cleary, A. Campbell, Proc. Nat. Acad. Sci. USA *61*, 956–962 (1968).

Integration-Negative Mutants of Bacteriophage Lambda, M. Gottesman, M. Yarmolinsky, J. Mol. Biol. *31*, 487–505 (1968).

Directed Transposition of the Arabinose Operon: A Technique for the Isolation of Specialized Transducing Bacteriophages for Any *Escherichia coli* Gene, S. Gottesman, J. Beckwith, J. Mol. Biol. *44*, 117–127 (1969).

Ter, a Function Which Generates the Ends of the Mature Lambda Chromosome, S. Mousset, R. Thomas, Nature *221*, 242–244 (1969).

Evidence for a Prophage Excision Gene in Lambda, A. Kaiser, T. Masuda, J. Mol. Biol. *47*, 557–564 (1970).

New Mutants of Bacteriophage Lambda with a Specific Defect in Excision from the Host Chromosome, G. Guarneros, H. Echols, J. Mol. Biol. *47*, 565–574 (1970).

Specialized Transduction of Galactose by Lambda Phage from a Deletion Lysogen, K. Sato, A. Campbell, Virology *41*, 474–487 (1970).

Deletion Mutants of Bacteriophage Lambda, I. Isolation and Initial Characterization, J. Parkinson, R. Huskey, J. Mol. Biol. *56*, 369–384 (1971).

Deletion Mutants of Bacteriophage Lambda, II. Genetic Properties of *att*-Defective Mutants, J. Parkinson, J. Mol. Biol. *56*, 385–401 (1971).

Deletion Mutants of Bacteriophage Lambda, III. Physical Structure of *att*, R. Davis, J. Parkinson, J. Mol. Biol. *56*, 403–423 (1971).

Isolation of Transducing Particles of φ80 Bacteriophage That Carry Different Regions of the *E. coli* Genome, R. Press, N. Gladsdorff, P. Miner, J. DeVries, R. Kadner, W. Maas, Proc. Nat. Acad. Sci. USA *68*, 795–798 (1971).

Specialized Transduction of the *Salmonella hut* Operons by Coliphage Lambda: Deletion Analysis of the *hut* Operons Employing Lambda-p*hut*, G. Smith, Virology *45*, 208–223 (1971).

Concatemers in DNA Replication: Electron Microscopic Studies of Partially Denatured Intracellular Lambda DNA, A. Skalka, M. Poonian, P. Bartl, J. Mol. Biol. *64*, 541–550 (1972).

Regulation of Lambda Prophage Excision by the Transcriptional State of DNA, R. Davis, W. Dove, H. Inokuchi, J. Lehman, R. Roehrdanz, Nature New Biol. *238*, 43–45 (1972).

Isolation of Plaque-Forming, Galactose-Transducing Strains of Phage Lambda, M. Feiss, S. Adhya, D. Court, Genetics *71*, 189–206 (1972).

Prophage Lambda at Unusual Chromosomal Locations, II. Mutations Induced by Bacteriophage Lambda in *E. coli* K12, K. Shimada, R. Weisberg, M. Gottesman, J. Mol. Biol. *80*, 297–314 (1973).

Attachment Site Mutants of Bacteriophage Lambda, M. Shulman, M. Gottesman, J. Mol. Biol. *81*, 461–482 (1973).

Thermal Asymmetry of Site-Specific Recombination by Bacteriophage Lambda, G. Guarneros, H. Echols, Virology *52*, 30–38 (1973).

Int-Constitutive Mutants of Bacteriophage Lambda, K. Shimada, A. Campbell, Proc. Nat. Acad. Sci. USA *71*, 237–241 (1974).

Purification of Bacteriophage Lambda Int Protein, H. Nash, Nature *247*, 543–545 (1974).

Quantitative Aspects of Gene Expression in a Lambda-*trp* Fusion Operon, J. Davison, W. Brammer, F. Brunel, Mol. Gen. Genet. *130*, 9–20 (1974).

Specialized Transducing Phages Carrying Fusions of the *trp* and *lac* Regions of the *E. coli* Chromosome, W. Schrenk, J. Miller, Mol. Gen. Genet. *131*, 9–19 (1974).

Lambda *att*B-*att*P: A Lambda Derivative Containing Both Sites Involved in Integrative Recombination, H. Nash, Virology *57*, 207–216 (1974).

Elements Involved in Site Specific Recombination in Bacteriophage Lambda, S. Gottesman, M. Gottesman, J. Mol. Biol. *91*, 489–499 (1975).

Integrative Recombination in Bacteriophage Lambda: Analysis of Recombinant DNA, H. Nash, J. Mol. Biol. *91*, 501–514 (1975).

Prophage Lambda at Unusual Chromosome Locations, III. The Components of the Secondary Attachments Sites, K. Shimada, R. Weisberg, M. Gottesman, J. Mol. Biol. *93*, 415–429 (1975).

Specialized Transducing Phages for Ribosomal Protein Genes of *E. coli*, S. Jaskunas, L. Lindahl, M. Nomura, Proc. Nat. Acad. Sci. USA *72*, 6–10 (1975).

Integrative Recombination of Bacteriophage Lambda DNA *in vitro*, H. Nash, Proc. Nat. Acad. Sci. USA *72*, 1072–1076 (1975).

Packaging of Prophage and Host DNA by Coliphage Lambda, N. Sternberg, R. Weisberg, Nature *256*, 97–103 (1975).

Constitutive Integrative Recombination by Bacteriophage Lambda, H. Echols, Virology *64*, 557–559 (1975).

Restriction Assay for Integrative Recombination of Bacteriophage Lambda DNA *in vitro*: Requirement for Closed Circular DNA Substrate, K. Mizuuchi, H. Nash, Proc. Nat. Acad. Sci. USA *73*, 3524–3528 (1976).

DNA Gyrase: An Enzyme That Introduces Superhelical Turns into DNA, M. Gellert, K. Mizuuchi, M. O'Dea, H. Nash, Proc. Nat. Acad. Sci. USA *73*, 3872–3876 (1976).

The Use of Specialized Transducing Phages in the Amplification of Enzyme Production, A. Moir, W. Brammer, Mol. Gen. Genet. *149*, 87–99 (1976).

Characterization of the Integration Protein of Bacteriophage Lambda as a Site-Specific DNA-Binding Protein, M. Kotewicz, S. Chung, Y. Takeda, H. Echols, Proc. Nat. Acad. Sci. USA *74*, 1511–1515 (1977).

Involvement of Supertwisted DNA in Integrative Recombination of Bacteriophage Lambda, K. Mizuuchi, M. Gellert, H. Nash, J. Mol. Biol. *121*, 375–392 (1978).

Interaction of Int Protein with Specific Sites on Lambda *att* DNA, W. Ross, A. Landy, Y. Kikuchi, H. Nash, Cell *18*, 297–307 (1979).

An *Escherichia coli* Gene Product Required for Lambda Site-Specific Recombination, H. Miller, D. Friedman, Cell *20*, 711–719 (1980).

The Lambda Phage *att* Site: Functional Limits and Interaction with *int* Protein, P. Hsu, W. Ross, A. Landy, Nature *285*, 85–91 (1980).

Lambda *alt* SF: A Phage Variant That Acquired the Ability to Substitute Specific Sets of Genes at High Frequency, D. Friedman, P. Tomich, C. Parsons, E. Olson, R. Deans, E. Flamm, Proc. Nat. Acad. Sci. USA *78*, 410–414 (1981).

Retroregulation of the *int* Gene of Bacteriophage Lambda: Control of Translation Completion, D. Schindler, H. Echols, Proc. Nat. Acad. Sci. *78*, 4475–4479 (1981).

SOS Induction and Autoregulation of the *himA* Gene for Site-Specific Recombination in *Escherichia coli*, H. Miller, M. Kirk, E. Echols, Proc. Nat. Acad. Sci. USA *78*, 6754–6758 (1981).

Direct Role of the *himA* Gene Product in Phage Lambda Integration, H. Miller, H. Nash, Nature *290*, 523–526 (1981).

Multilevel Regulation of Bacteriophage Lambda Lysogeny by E. *coli himA* Gene, H. Miller, Cell *25*, 269–276 (1981).

Structure and Function of the Phage Lambda *att* Site: Size, Int-binding Sites, and Location of the Crossover Point, K. Mizuuchi, R. Weisberg, L. Enquist, M. Mizuuchi, M. Buraczynska, C. Foeller, P. Hsu, W. Ross, A. Landy, Cold Spring Harbor Laboratory Symposium of Quantitative Biology *45*, 429–437 (1981).

Purification and Properties of the *E. coli* Protein Factor Required for Lambda Integrative Recombination, H. Nash, C. Robertson, J. Biol. Chem. *256*, 9246–9253 (1981).

DNA Sequence of the *att* Region of Coliphage 434, D. Mascarenhas, R. Kelley, A. Campbell, Gene *5*, 151–156 (1981).

Downstream Regulation of *int* Gene Expression by the b2 Region in Phage Lambda, C. Epp, M. Pearson, Gene *13*, 327–337 (1981).

Posttranscriptional Control of Bacteriophage Lambda *int* Gene Expression from a Site Distal to the Gene, G. Guarneros, C. Montanez, T. Hernandez, D. Court, Proc. Nat. Acad. Sci. USA *79*, 238–242 (1982).

Site-Specific DNA Condensation and Pairing Mediated by the *int* Protein of Bacteriophage Lambda, M. Better, C. Lu, R. Williams, H. Echols, Proc. Nat. Acad. Sci. USA *79*, 5837–5841 (1982).

Bacteriophage Lambda *int* Protein Recognizes Two Classes of Sequence in the Phage *att* Site: Characterization of Arm-Type Sites, W. Ross, A. Landy, Proc. Nat. Acad. Sci. USA *79*, 7724–7728 (1982).

Regulation of Bacteriophage Lambda *int* Gene Expression, A. Oppenheim, S. Gottesman, M. Gottesman, J. Mol. Biol. *158*, 327–346 (1982).

Biochemical Analysis of *att*-Defective Mutants of the Phage Lambda Site-Specific Recombination System, W. Ross, M. Shulman, A. Landy, J. Mol. Biol. *156*, 505–529 (1982).

Control of Phage Lambda Development by Stability and Synthesis of cII Protein: Role of the Viral cIII and *hflA* and *himD* Genes, M. Hoyt, D. Knight, A. Das, H. Miller, H. Echols, Cell *31*, 565–573 (1982).

Purification of the Bacteriophage Lambda *xis* Gene Product Required for Lambda Excisive Recombination, J. Biol. Chem. K. Abremski, S. Gottesman, J. Biol. Chem. *257*, 9658–9662 (1982).

Deletion Analysis of the Retroregulatory Site for the Lambda *int* Gene, D. Court, T. Huang, A. Oppenheim, J. Mol. Biol. *166*, 233–240 (1983).

Knotting of DNA Caused by a Genetic Rearrangement, Evidence for a Nucleosome-like Structure in Site Specific Recombination of Bacteriophage Lambda, T. Pollock, H. Nash, J. Mol. Biol. *170*, 1–18 (1983).

Site-Specific Recombination of Bacteriophage Lambda: The Change in Topological Linking Number Associated with Exchange of DNA Strands, H. Nash, T. Pollock, J. Mol. Biol. *170*, 19–38 (1983).

Role of DNA Homology in Site-Specific Recombination, R. Weisberg, L. Enquist, C. Foeller, A. Landy, J. Mol. Biol. *170*, 319–342 (1983).

Role of the *xis* Protein of Bacteriophage Lambda in a Specific Reactive Complex at the *attR* Prophage Attachment Site, M. Better, W. Wickner, J. Auerbach, H. Echols, Cell *32*, 161'–168 (1983).

The Mechanism of Lambda Site-Specific Recombination: Site-Specific Breakage of DNA by *Int* Topoisomerase, N. Craig, H. Nash, Cell *35*, 795–803 (1983).

Patterns of Lambda *Int* Recognition in the Regions of Strand Exchange, W. Ross, A. Landy, Cell *33*, 261–272 (1983).

Resolution of Synthetic att-site Holliday Structures by the Integrase Protein of Bacteriophage Lambda, P. Hsu, A. Landy, Nature *311*, 721–726 (1984).

Transposable Genetic Elements

T ransposable elements are special DNA sequences ranging in length from several hundred to about 5,000 base pairs and, in the case of a phage that is a transposable element, to about 50,000 base pairs. These sequences can spread within a cell by being copied into new DNA locations as well as within a population through transformation, transduction, or conjugation.

The phenomenon of transposing elements was first described in maize in the elegant genetic studies of McClintock. However, general interest in the subject lay dormant until a physical basis for it was stimulated by genetic and physical studies with *Escherichia coli*. Now, with the development of genetic engineering, the detection, characterization, and study of transposing elements is much easier than before, and such elements have been found in a number of other organisms. This chapter considers the discovery, structure, properties, and use of transposable elements that are found in the bacteria *E. coli* and *Salmonella typhimurium*, the yeast *Saccharomyces*, and the fruit fly *Drosophila*.

The cell's chromosome would be an ideal home for a parasitic sequence of nucleotides. There it would enjoy almost the same treatment as an integrated lambda phage genome. Furthermore, the copying of such a sequence into a new chromosomal location likely would require only biochemical activities already present or easily synthesized in the host cell. Given such an environment, it is not surprising that parasitic sequences have evolved and now exist. A deeper question is why such sequences do not constitute most of the cell's DNA. Most likely, the tendency of such sequences to proliferate is countermanded by evolution on a grander scale. A cell line or organism that is weighted down by an excessive number of unused DNA sequences would be at a survival disadvantage compared to others containing fewer such sequences. Consequently, evolution will constantly select for cells without too many transposable sequences, and, overall, an unhappy balance will exist between the sequence and the host cell line. This is not to say that transposable elements are without value to their hosts. As we will see, the presence of repeated DNA sequences can facilitate chromosome rearrangements. This reshuffling of genetic material may greatly speed evolution and aid cells containing repeated sequences.

THE DISCOVERY OF IS ELEMENTS IN BACTERIA

Nonsense mutations in many operons have polar effects, that is, they reduce the expression of downstream genes. In efforts to study these effects more carefully, in the late 1960s Malamy, Shapiro, and Starlinger each isolated and characterized strongly polar mutations in the *lac* and *gal* operons. In addition to the expected classes of nonsense mutations, another type was also found. This new type was highly polar but not suppressible by known nonsense suppressors. Since the mutations reverted, albeit at very low rates, it seemed that they could be insertions of any number of nucleotides or small deletions. Therefore it became of great interest to find out just what these mutations were. Nowadays this question would probably be answered by cloning a portion of the DNA containing the mutation and sequencing it to determine the exact nature of the mutation. However, such a procedure was impossible at the time, and less direct experiments were performed.

One likely cause of the polarity was the insertion of extraneous nucleotides into the *lac* and *gal* genes. Fortunately, this possibility could be tested by the techniques then available. If the insertions were larger than a few hundred nucleotides, they

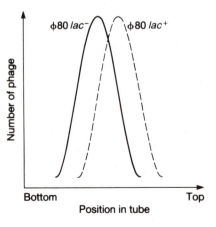

Figure 18.1   The positions of *lac*$^+$ and strongly polar *lac*$^-$ phage particles in the centrifuge tube after CsCl equilibrium gradient centrifugation. The polar mutation makes the phage more dense.

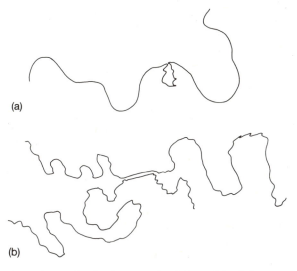

Figure 18.2   (a) DNA heteroduplex between lambda and lambda containing IS2. The single-stranded bubble is the DNA of IS2. (b) DNA heteroduplex formed between two r strands of lambda, each containing IS3 but in opposite orientations.

could be detected directly by generating a change in the density of *lac* or *gal* transducing phage. Therefore the mutations were crossed onto λd*gal* and φ80λ*lac* phage and the densities of phage both with and without the mutations were carefully compared by equilibrium gradient centrifugation (Fig. 18.1). The results showed that transducing phage carrying the polar mutations were denser than the corresponding transducing phage not carrying the mutations. Therefore the mutations were caused by the insertion of DNA elements that are called IS, for insertion sequences.

At first, the insertions were suspected of being merely random sequences that somehow had been inserted into the genes being studied. However, the mechanism of their generation is not so easy to imagine. Point mutations, insertions and deletions of a nucleotide or two, and sizable deletions may all be created by simple reactions likely to occur in cells. Not so for sizable insertions.

Additional study of the IS sequences began to clarify their origin. One particularly valuable technique for their study was the formation of DNA heteroduplexes (Fig. 18.2). DNA isolated from the *gal* and *lac* transducing phage with and without the IS sequences was used in these experiments. When necessary, the strands of these phage could even be separated and purified. The lengths of IS sequences fell into classes of particular sizes, and any particular IS sequence was found to insert into a gene in either of the two possible orientations. The existence of the discrete-length classes suggested that a small number of IS sequences might be responsible

Table 18.1   Sizes of various bacterial IS
sequences and the number of base pairs of host
DNA they duplicate upon insertion

IS Element	Size of Element (base pairs)	Size of Host Duplication (base pairs)
IS1	768	9
IS2	1,327	5
IS3	1,300	3, 4
IS4	1,426	11, 12
IS5	1,195	4
IS10-R	1,329	9
IS50-R	1,531	9

for all the insertions. Indeed, less than ten different IS sequences have been found in *E. coli* (Table 18.1). The demonstration that IS elements were specific sequences did not explain how they were inserted into different sites in the cell's chromosome, but by analogy to lambda phage, the elements were expected to code for one or more proteins involved in the process. By now the IS elements have been obtained on phage and plasmids and they have been sequenced and characterized using *in vitro* mutagenesis as described below. From these studies a comprehensive picture of their structure and function has emerged.

THE STRUCTURE AND PROPERTIES OF IS ELEMENTS

Most of the bacterial IS sequences contain two characteristic elements. First, a sequence of 10 to 40 base pairs at one end is repeated or nearly repeated in an inverse orientation at the other end. This is called IR, for inverted repeat. Second, the IS sequences contain one or two regions that appear to code for protein because they possess promoters, ribosome-binding sequences, ATG (AUG) codons, and sense codons followed by peptide chain termination codons. If proteins were made under control of these sequences, they would likely be involved with the process of copying the IS into a new position, transposition. Indeed, one of the gene products is generally called the transposase. In IS5, the shorter reading frame plus its associated promoter and ribosome-binding sequence is oriented opposite to the first and is contained entirely within the first reading frame as well as being in the same codon register (Fig. 18.3)! Such an astonishing degree of information compaction naturally raises the question of whether both reading frames actually are utilized.

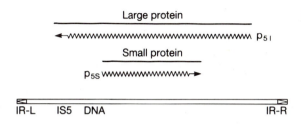

Figure 18.3   The organization of IS5. The large and small proteins are in the same codon register.

The activity of the promoter sequences within IS5 has been examined by fusing them to the gene encoding galactokinase, *galK*. Since the assay for galactokinase is sensitive and easy to perform, the activities of the IS5 promoters can readily be measured. The promoter for the large protein is active *in vivo* and *in vitro*, but for unknown reasons the *in vitro* activity of this promoter is much less than the activity of the promoter for the small protein.

A high expression of the IS genes most likely would stimulate a high rate of transposition into new chromosomal sites. Since IS transposition rates appear low, something must protect IS genes from readthrough transcription and translation if

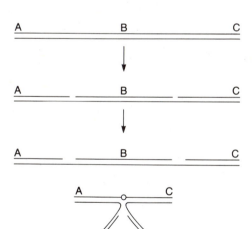

Figure 18.4   A mechanism by which nicking DNA can stimulate generation of deletions.

the sequence has inserted into an active gene of the host cell. IS5 possesses six peptide chain termination codons flanking the protein genes — three at each end of the transposon, one in each reading frame. Additionally, the left end of the IS5 contains a transcription termination sequence oriented so as to prevent transcription from adjacent bacterial genes from entering the IS in the reading direction of the transposase.

Not only do IS sequences copy themselves into new DNA locations, but they also stimulate the creation of deletions located near the IS elements. These deletions have one end precisely located at the end of an IS element and are therefore thought to arise in abortive transposition attempts that begin with transposase nicking the DNA at the boundaries of the IS element (Fig. 18.4). Such a facile creation of deletions in any selected region is highly useful to geneticists, for, as we have seen, a set of such deletions can be used for fine structure mapping.

DISCOVERY OF Tn ELEMENTS

While IS elements were under investigation, another category of genetic elements was also being studied. These were plasmids, some of which can conjugate like F-factors, that carry genes encoding proteins that confer resistance to antibiotics. They are called R-factors.

Not long after antibiotics began to be widely used, many bacterial isolates from infections were found to be resistant to one or more of the drugs. These isolates were not resistant by virtue of a mutation altering the cellular target of the antibiotics; instead, they synthesized specific proteins or enzymes, which conferred resistance on the cells either by detoxifying the antibiotics or by blocking their entry into the cell. For example, penicillin-resistant strains were found to synthesize a β-lactamase that opens the lactam ring of penicillin and renders it harmless to the cells (Fig. 18.5).

Although many R-factors were fertile and could transfer themselves to other cells, the speed with which different drug resistances appeared on R-factors was

Figure 18.5   The structure of the lactam ring of penicillins and related antibiotics and their structure following ring opening by β-lactamase. Different penicillins possess different R groups.

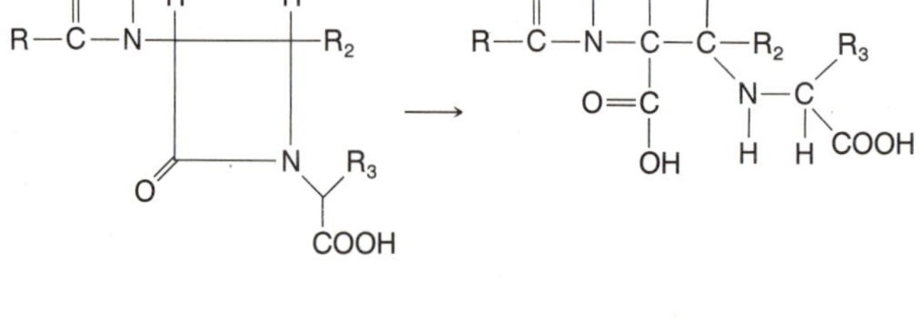

astonishingly rapid. Soon after the introduction of a new antibiotic, R-factors that carried genes conferring resistance to the new antibiotic would be found in many geographic locations. Unfortunately this has created great problems in the treatment of infections. The drug-resistance factors rapidly pick up genes encoding resistance to the most commonly used drugs, and soon treatment of infections becomes diffi-cult. Now physicians are using restraint in prescribing antibiotics to slow the spread of drug-resistance genes. Only antibiotics known to be effective against a bacterium are used, doses are high enough to be effective quickly, and the duration of treat-ment is sufficiently long that the infecting cells most likely are all killed. Further-more, a number of drugs are withheld from general use so that they may be reintro-duced at a later time when the bacteria have shed the relevant drug-resistance gene through loss of either the R-factor or the gene on the R-factor responsible for antibiotic resistance. Such a loss can be a selective advantage to cells not growing in the presence of the antibiotic.

The ability of drug-resistance genes to spread rapidly suggested that they could hop from one R-factor to another or to other DNA sequences such as phage or the bacterial chromosome. One demonstration of this property came from an attempt to construct a lambda phage carrying the kanamycin-resistance gene from an R-factor.

Cells deleted of *attB* and carrying a kanamycin-resistance R-factor were infected with lambda. This lambda phage was deleted of part of the *b2* region so that the DNA of the resulting kanamycin transducing phage would not be too large to be packaged. However, to permit the phage to form lysogens, the deletions chosen did not extend into the *att* region. As seen in the previous chapter, a small fraction of the resulting lysogens should then contain a lambda inserted into DNA adjacent to the kanamycin gene(s) by virtue of integrating into sites that weakly resemble lambda's normal integration site. Upon induction, a small fraction of the phage were expected to excise incorrectly, pick up the adjacent kanamycin-resistance genes, and be capa-ble of transducing kanamycin resistance to other cells.

Indeed, kanamycin transducing phage were isolated. However, the phage possessed several unexpected properties. First, the additional DNA in these phage was not located immediately adjacent to the *attP*. Second, kanamycin-resistant transformants obtained by infecting cells with the *λkan* phage were lysogenic for lambda only if they contained an *attB*. These properties can be understood as follows. The DNA coding for kanamycin resistance originally hopped or was copied onto lambda by a recombination event rather than arriving there after an improper phage excision of the type that produces *gal* and *bio* transducing particles. Therefore the foreign DNA could be located anywhere and need not be immediately adjacent to *attP*. Second, a copy of the sequence encoding the kanamycin resistance could also hop or be copied from the phage onto the chromosome. This is the easiest way that cells that were blocked from becoming lysogens at normal frequencies by their absence of a phage *att* region could become drug-resistant after infection with a *λkan*.

The DNA responsible for the kanamycin resistance was examined by DNA heteroduplexes. It was found to be 5,400 base pairs long and flanked by inverted

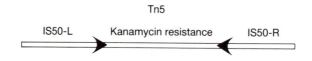

repeated sequences 1,500 base pairs long. Some of the drug-resistance genes studied on the R-factors themselves also possessed similar flanking sequences. Further study showed that the repeated sequences flanking the drug-resistance genes were trans-ferred when the drug-resistance gene hopped. Finally, it was found that the repeated elements at the ends of many of the drug-resistance genes were of essentially identical base sequence to known types of IS elements. Thus it appears that two IS elements can flank a gene and then facilitate copying the composite element into other DNA locations. That is, studies have shown that in bacteria these elements do not excise like lambda phage does, but they transpose by generating a new copy of themselves in the target sequence. Sequences with these general properties and structure are called transposons, or Tn elements (Table 18.2). The element described in this section is called Tn5, and the IS sequences associated with it are called IS50 so as not to be confused with IS5.

Table 18.2   Selected transposons, genes they carry, sizes, associated IS elements, and the orientation of repeated sequences

Transposon	Genes Carried	Size (kilobases)	IS Element	Repeats
Tn1	Ampicillin[R]	5	–	–
Tn3	Ampicillin[R]	4.957	–	–
Tn4	Ampicillin[R], Streptomycin[R], Mercuric[R] ion	20.5	– IS50	–
Tn5	Kanamycin[R]	5.7	IS50	Inverted
Tn6	Kanamycin[R]	4.2	–	–
Tn7	Trimethoprim[R]	13.5	–	–
Tn9	Chloramphenicol[R]	2.5	IS1	Direct
Tn10	Tetracycline[R]	9.3	IS10	Inverted

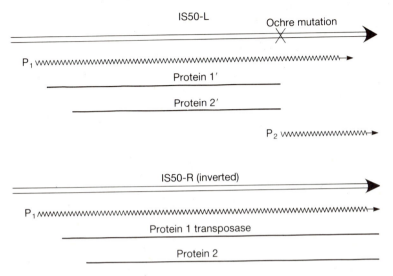

IS50-L

Ochre mutation

P₁

Protein 1'

Protein 2'

P₂

IS50-R (inverted)

P₁

Protein 1 transposase

Protein 2

Figure 18.6   The organizations of the IS50-L and IS50-R of Tn5.

THE STRUCTURE AND PROPERTIES OF Tn ELEMENTS

Transposons can be grouped into three broad structural classes. Members of the class containing Tn5 and Tn10 possess a pair of IS elements on either side of a DNA sequence that is transposed. Although the right- and left-hand IS elements are nearly identical, the right-hand IS elements encode a transposase that is required for transposition, whereas the left IS elements do not. The difference between the left element of Tn5, which is called IS50-L, and the right element, IS50-R, is a single nucleotide change (Fig. 18.6). This creates an ochre mutation in the genes both for the transposase and for a second protein that is translated in the same reading frame. The nucleotide change also greatly strengthens a promoter oriented toward the DNA inserted between the IS elements.

Transposition by members of the Tn5-Tn10 class of transposons appears to require at least one protein encoded by the transposon and only DNA sites located

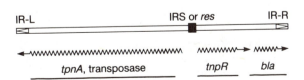

IR-L   IRS or *res*   IR-R

tpnA, transposase   *tnpR*   *bla*

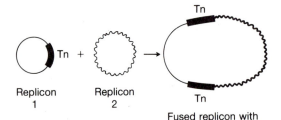

Figure 18.7   Incomplete transposition can fuse two replicons with the duplication of the transposition element.

very near the ends of the element. That is, if only internal sequences are deleted and if the necessary proteins are provided in *trans,* the deleted element will transpose.

Tn3 belongs to a second class of transposons. Members of this group do not contain flanking IS sequences and possess only short inverted repeats at their ends. Their transposition also requires two transposon-encoded proteins — a transposase and another protein. However, transposition by these elements requires more than the two proteins and the ends of the transposon. A DNA site in the middle of the transposon is also required. Without this site, the transposon does not complete transposition events; it copies itself into new locations but connects the old location to the new (Fig. 18.7). Consequently, a transposon lacking the internal site fuses the donor and target DNA replicons when it attempts transposition. These properties are a clear demonstration that transposition occurs by copying rather than by excision and reintegration.

The third class of transposon is represented by phage Mu. Superficially this can be thought of as a transposon that happens to have picked up a phage. Therefore it replicates not by freeing itself from the chromosome but by copying itself into multiple locations. This property is particularly useful because instead of transposing at a frequency of perhaps 10^{-3} per generation like many transposons, when induced it transposes at a frequency of 10^2 per cell doubling time.

ROLE OF THE REPEATED SEQUENCE GENERATED BY TRANSPOSITION

One obvious question about IS and Tn elements is how the elements choose the sequences into which they will insert. For example, they could be like lambda phage and choose a specific target sequence. Then, of course, sequencing the boundaries of the elements would reveal the equivalents of recognition and crossover points. Such a possibility seemed highly likely with the discovery that both the IS and Tn elements are flanked by repeated sequences deriving from the host of a small number of bases, five or nine being common. However, this idea became less appealing with the finding that the sequences flanking many different examples of insertion by one

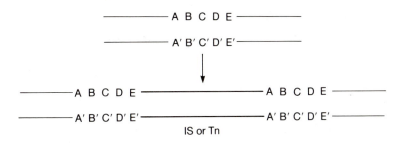

type of element were all unrelated. Nonetheless, the possibility still existed that the repeated sequences reflect a fundamental feature of the transposition process.

Lambda phage was used to demonstrate that the identity of the sequences flanking Tn10 is not required for transposition. Kleckner isolated two different insertions of Tn10 into the *CI* gene of lambda. By sequencing, she showed that these possessed different nine-base-pair flanking repeats. Then, by performing a cross between the two phage, she found a recombinant from a crossover within the Tn10 (Fig. 18.8). Restriction enzyme digests of the DNAs from the two parents and the recombinant proved that the crossover had occurred within the Tn10 and not within the adjacent phage DNA. Of course it was most likely that the crossover should occur within the Tn10 because this element is 9,300 base pairs long and constituted the majority of the DNA between the genetic markers in the *N* and *O* genes used in the cross. The resultant transposon contained different flanking sequences at its ends, and yet it could transpose into other locations as well as a Tn10 containing identical flanking sequences, thus proving the point.

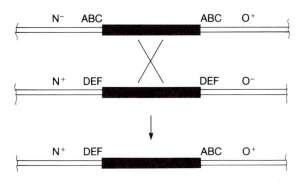

Figure 18.8   Construction of Tn10 with nonidentical repeated flanking host sequences by recombination between two independent insertions.

GENETIC AND PHYSICAL MAPPING OF Tn3

Although the nucleotide sequence of a transposon could readily be determined, our knowledge does not yet permit identification of genes and determination of their activities just from the nucleotide sequence. Therefore a functional analysis of the components of a transposon had to be coupled to a genetic analysis. In principle this could have been done by first mutagenizing cells containing a plasmid carrying a transposon. Plasmids from different clones could then have been sequenced and their transposing properties could have been characterized and correlated with the nucleotide changes. Clearly, this is a most inefficient strategy.

Heffron devised an efficient method of isolating and mapping mutations in the transposon. Purified plasmid was digested with DNase I under conditions in which it makes nearly flush double-stranded scissions at random locations in the DNA (Fig. 18.9). This reaction was stopped before any appreciable number of plasmid molecules was cleaved more than once. Before proceeding, an *EcoRI* site within the

Figure 18.9   Procedure for mutagenesis by insertion of *EcoRI* linkers at random positions.

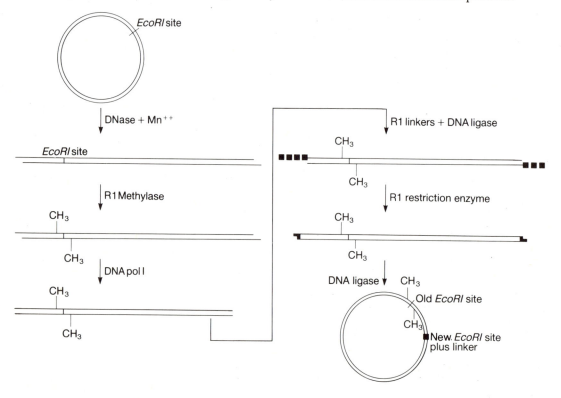

plasmid was made nonreactive by methylating it with the *EcoRI* methylase. After overhanging ends resulting from the DNase I digestion were converted to flush ends by filling out with DNA pol I, *EcoRI* restriction enzyme linkers were ligated on. The flush-ended linkers were cut with *EcoRI* to make them sticky-ended and to remove any multiple linkers that may have been generated by the ligation. Then the linear molecules were converted to covalently closed circles with DNA ligase, which joined the *EcoRI* sticky ends.

The overall effect of the steps described above was to introduce an *EcoRI* linker at a random location within the plasmid. The linker's eight nucleotides constitute a small insertion that should inactivate genes encoding for proteins or recognition sites of proteins on DNA. More important, however, the location of the linker in the transposon may be demonstrated by restriction enzyme digestion. Should knowledge of the exact location of the linker be necessary, only a small fraction of the transposon must be sequenced.

A genetic test was used to quantitate the transposition ability of the mutated Tn3 elements. The basis of the test was that some plasmids are unable to replicate in DNA pol I mutants. Therefore two types of episomes were placed in pol I$^+$ cells. One was capable of conjugation as well as replication in pol I mutants, and the second was unable to conjugate or to replicate in a pol I mutant. The second plasmid contained the mutant Tn3 factors. Cells containing both of the plasmids were mated with female DNA pol I mutant cells, and the growth medium was chosen to select for female cells that had received the fertile episome as well as the drug-resistance gene carried by Tn3 (Fig. 18.10). Cells growing under these conditions could result only from male cells in which the Tn3 had transposed to the fertile episome and had subsequently been transferred to the female by conjugation.

Figure 18.10   Method for testing transposition capabilities of Tn3 by its ability to transpose to a fertile plasmid that is capable of replication in a *recA* female.

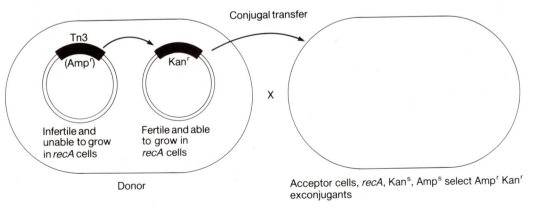

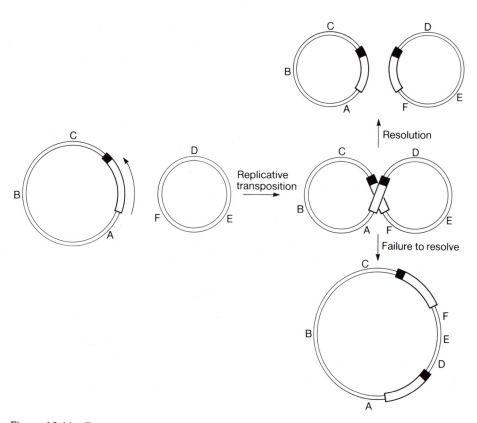

Figure 18.11   Top: transposition into a second plasmid with recombination between the two *res* sites; bottom: transposition into a second plasmid without recombination between the two *res* sites.

The linker mapping technique described above showed that two genes and three DNA sites are required for Tn3 transposition. That is, in addition to the two transposon ends, a third site in the middle of Tn3 is required for normal transposition. When the third site is damaged, Tn3 forms cointegrates between the donor and recipient DNA molecules containing two copies of the Tn3 in the same orientation (Fig. 18.11). Resolution of the structure by a recombination event between the two Tn3 factors restores the original structure and yields the acceptor replicon with a Tn3. In the absence of the site required for the resolution step — the internal resolution site, *res* — the transposition process stops at the cointegrate stage. One of the proteins encoded by Tn3 was shown to be the transposase, and the other was a protein that functions both as a repressor of transposase synthesis and as a protein catalyzing the resolution reaction. It is called resolvase.

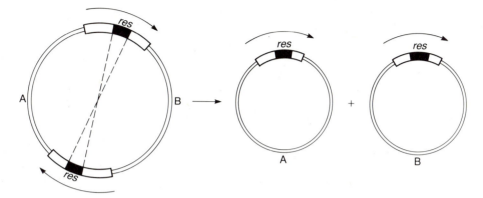

Figure 18.12   Recombination between two *res* sites on a plasmid can generate two product molecules.

In vitro TRANSPOSITION

Learning the biochemical details of transposition requires use of *in vitro* transposition reactions. A plasmid with duplicated internal resolution, *res,* sites facilitated study of a transposition reaction, just as the use of lambda phage with duplicated *att* sites was useful for biochemical study of the integration reaction. Such a plasmid

Figure 18.13   Threading DNA through resolvase to ensure that both *res* sites have the same orientation.

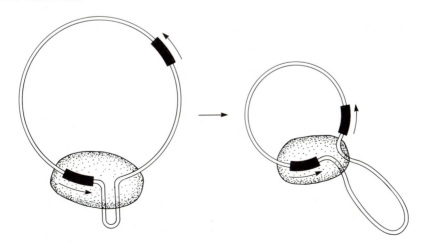

permitted study of the second step of transposition, the recombination between two *res* sites that resolves the donor and acceptor replicons. Resolution eliminates the DNA between the two internal resolution sites of such a plasmid (Fig. 18.12).

Experiments with the double *res* site plasmid showed that resolution would proceed *in vivo* in the presence of a Tn3 with a functional resolvase gene. The reaction would also take place *in vitro* if resolvase was synthesized at much higher than usual levels in cells and the partially purified enzyme added to the reaction. However, the reaction occurred only if the plasmid DNA was supercoiled. Relaxed DNA would not work. No other source of energy was required, and host proteins such as the *him* product that are required for lambda integration are not required for resolution. Resolvase works only when the two *res* sites on the same plasmid are correctly oriented. If one is turned around with respect to the other, no resolution occurs. This means that DNA is threaded out from one *res* site until the second is reached and then resolution occurs (Fig. 18.13). The consequences of such a property are that resolution does not occur between two different DNA molecules. Only intermolecular resolution takes place.

REGULATING FLAGELLIN SYNTHESIS BY INVERTING A DNA SEGMENT

Later, a full chapter will be devoted to chemotaxis, the ability of cells to sense and move toward attractants or away from repellents. Here, however, we will be concerned with a form of regulation of chemotaxis that involves inverting a DNA segment. *Salmonella* can propel itself by rotating a bundle of its flagella. The bacte-

rium possesses two genes encoding the protein component of flagella, flagellin. Depending on the orientation of an invertible controlling DNA element, only one or the other of the flagellin genes is expressed.

Considering one habitat of *Salmonella,* the gut, considerable selective advantage accrues to the strain that can express more than one flagellin type. Antibody, which is secreted by patches of special cells on the walls of the intestine, could bind to a flagellum and block rotation of the flagella and therefore stop chemotaxis. Hence, to survive and swim in the presence of antiflagellin antibody, it is to a cell's advantage to be able to synthesize a second type of flagellin against which the host does not yet possess antibody. However, it is likely that simultaneous synthesis of two types of flagellin within a cell would be harmful. If both types of flagellin were incorporated into flagella, then either of two types of antibody could block chemotaxis rather than just one. Consequently, if two flagellin genes are present, one should be completely inactive while the other is functioning.

How can an appropriate flagellin synthesis flip-flop circuit be constructed using regulatory elements available to a cell? Only one of the flagellin genes should possess any activity at one time, but reversing the activity pattern should be readily possible. These requirements are not easily met using regulatory elements of the type we have thus far considered. For example, although the lambda phage early gene system possesses the necessary selectivity in having two states, fully off and fully on, this system is not reversible. Alternatively, genes regulated like *lac*, *trp*, and *ara* either do not possess large differences between their on and off states or they are hard to flip from one state to another. In general, the regulation systems we have considered in previous chapters do not permit construction of simple circuits possessing two states of dramatically different activities and at the same time permit simple switching between the two states.

An invertible controlling segment of DNA solves *Salmonella*'s problem. In one orientation, only one type of flagellin is synthesized, and in the other orientation, only the other type is synthesized. The apparatus responsible for flipping the DNA segment can be completely independent of flagellin or expression of the flagellin genes. That is, the protein or proteins necessary for inversion of the DNA segment can be synthesized independently at a rate such that the segment will invert at the optimum rate for overall survival of the bacteria.

The structure of the invertible segment in *Salmonella* is similar to IS elements and a portion of Mu phage called the G segment. It is 995 base pairs long, it has inverted repeats of 14 bases at its ends, and its central region encodes a protein that is necessary for inversion. At its right end the element contains a promoter oriented to the right. Just to the right of the invertible segment lies one of the flagellin genes, *H2*, and another gene called *rhl* (Fig. 18.14). The *rhl* protein is a repressor of the second

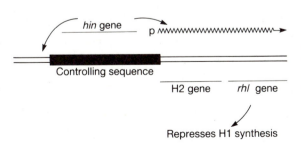

Figure 18.14   Structure of the DNA at the flagellin controlling sequence in *Salmonella*. The *hin* gene codes for a protein that can invert the region. The region also contains a promoter for synthesis of *H2* and *rhl* messenger when the region is in one orientation.

flagellin gene, *H1*, located elsewhere on the chromosome. As a result of this structure, one orientation of the invertible element leads to synthesis of *H2* protein and *rhl*
repressor. In this state the *rhl* repressor blocks synthesis of *H1* flagellin. When the
controlling element is in the other orientation, *H2* and *rhl* protein are not synthesized
because they have been separated from their promoter, and therefore synthesis of
H1 is not repressed.

The strategy of switching the antigenic properties of crucial surface structures is
not unique to bacteria. African trypanosomes, unicellular flagellated cells, can evade
the immune systems of vertebrates by switching their synthesis of surface proteins.
These protozoa have hundreds of different genes for the same basic surface protein.
However, the process of switching on the synthesis of any one of them resembles
more closely the mating-type switch of yeast than the orientation switching involved with the switching of *Salmonella* flagellin synthesis.

MU PHAGE IS A GIANT TRANSPOSABLE ELEMENT

A transposon could carry DNA with phage-like functions rather than a simple
antibiotic resistance gene. Such a transposon would be able to travel freely from cell
to cell as a virus does and would not be limited by the transfer rate of host DNA.
However, bacterial transposons do not excise from the chromosome, and a
transposon-phage would need some means for excising phage units from the chromosome for packaging.

Phage Mu first drew attention by its ability to generate mutations in infected cell
populations. These mutations were generated by its insertion into various bacterial
genes. Careful mapping of many Mu phage insertions into the β-galactosidase gene
showed that Mu inserted without high specificity in the target sequence. Thus, Mu
behaved like a lambda phage with no sequence specificity in its choice of the
bacterial *att* site.

Two facts, however, indicate that Mu is more like a transposon than like phage
lambda. First, Mu duplicates five bases of chromosomal DNA upon its insertion.
Second, Mu does not excise from the chromosome and replicate in the cytoplasm.
Even though an induced cell may yield a hundred Mu phage upon lysis, never
during the lytic cycle are any free Mu DNA molecules observed. The Mu DNA is
replicated only by transposition, and packaging of the Mu takes place directly on the
integrated DNA. A dramatic demonstration of replication by transposition is provided by Southern transfers of DNA taken from a Mu lysogen before and after
induction of the phage. Before induction, only a single restriction fragment of the
host DNA contains sequences homologous to a restriction fragment including the
end of Mu, but half an hour after induction of Mu growth, many restriction fragments contain these sequences (Fig. 18.15).

Mu packages its DNA right out of the chromosomal insertions. Part of the head
structure recognizes a sequence near the left end of the DNA, reaches out beyond the
phage 50 to 150 base pairs, and begins packaging. A headful of DNA is packaged,

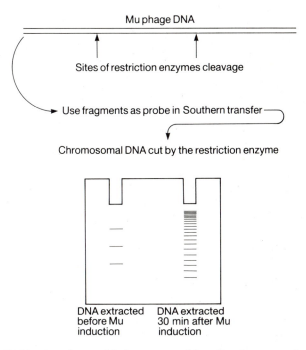

Figure 18.15   DNA taken from a Mu lysogen used in a Southern transfer. When Mu occupies one chromosomal location, only one restriction fragment of host DNA contains the left end of lambda. After induction of the phage, many fragments contain this sequence, and the Southern transfer shows many bands homologous to this probe.

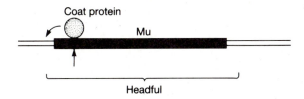

and the remaining DNA is then cleaved and the tail is attached. Since a Mu phage headful of Mu DNA is slightly larger than the genome of the phage, the packaged DNA usually extends beyond the right end of the phage and includes about 3,000 base pairs of bacterial DNA as well. The headful hypothesis explains the fact that a Mu phage carrying an insertion packages a smaller amount of bacterial DNA roughly equal in amount to the size of the insertion.

The sequence heterogeneity that results from the headful mode of packaging can be seen very dramatically when heteroduplexes of phage genomes are analyzed in the electron microscope. Since it is highly unlikely that the two strands from the same phage will rehybridize upon formation of the heteroduplex, two unrelated strands will associate. The phage DNA portions of the strands will, of course, be complementary and will form a duplex. However, the strands of bacterial DNA on the right are unlikely to be complementary, and these will remain single-stranded and are observed as "split ends." The similar split ends from the left end are too small to be clearly observed.

MU PHAGE POSSESSES A SEGMENT THAT INVERTS

The split-ended heteroduplexes described in the preceding section are observed with DNA that is obtained from Mu-infected cells. However, Mu DNA obtained from an induced lysogen yields heteroduplexes containing not only the split end, but half of them also contain an internal single-stranded region of 3,000 base pairs, a bubble (Fig. 18.16). This bubble results from the fact that the region, called G, is inverted in about half of the phage.

The ability of Mu to invert G can be abolished by a small deletion at the end of G or by a point mutation in the adjacent region of the DNA called *gin*. This point mutation can be complemented, thereby proving that *gin* encodes a protein that participates in inversion of G. Curiously, phage P1 also contains the same G loop, and P1 can complement Mu *gin* mutants.

The G loop codes for proteins that determine the host range of Mu. Phage induced from lysogens with G in the plus orientation are able to absorb to *E. coli* K-12, but not to *Citrobacter freundii* or *E. coli* C. When G is in the minus orientation, the abilities of the resultant phage to adsorb to these strains are reversed. This phenomenon provides an explanation of the heteroduplexing experiments. Those Mu able to adsorb to *E. coli* K-12 must be in the plus orientation. Since the rate of inversion is low, the G region of most of the phage in the resulting lysate will still be in the plus orientation, and heteroduplexes prepared from this DNA will not contain

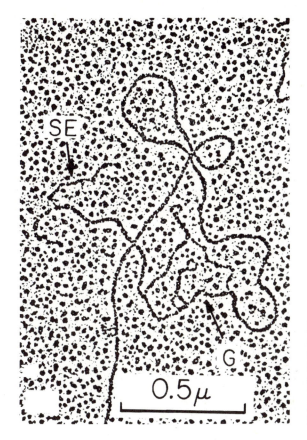

Figure 18.16   Electron micrograph of a portion of a denatured and rehybridized Mu phage DNA duplex showing the split ends, SE, and the G loop. Under these conditions, sequences within the G loop are showing partial self-hybridization. (From Electron Microscope Heteroduplex Study of the Heterogeneity of Mu Phage and Prophage DNA, M. Hsu, N. Davidson, Virology *58*, 229–239 [1974]).

G bubbles. Lysogens are different. Although the phage originally entered cells with the G loop in the plus orientation, during the many generations of growth from a single lysogen, G can invert, and eventually about half the cells in the population will contain Mu with G in the plus orientation and half will contain Mu with G in the minus orientation. Induction of the phage in this cell population and preparation of DNA heteroduplexes will yield half the molecules with G bubbles.

The mutants with G stuck in one orientation because of the deletion at the end of G or the point mutation in *gin* permit a particularly simple demonstration of the infectivity properties. Mu lysogens with G stuck in the minus orientation yield phage particles, but these are not infective on *E. coli* K-12. Mu lysogens containing G stuck in the plus orientation are fully infective when plated on *E. coli* K-12.

Genetic analysis of the G region of the phage shows that two sets of genes are involved: S and U, and S' and U'. S and U are expressed when G is in the plus orientation and S' and U' are expressed when G is in the minus orientation (Fig.

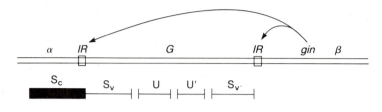

Figure 18.17   Structure of the G region of phage Mu. Transcription of the S and U or S′ and U′ genes initiates at the promoter p in the α region. The *gin* gene product acts at the inverted repeats, *IR*, to invert the G region at a low rate.

18.17). The promoter for these genes and the initial portion of the S gene is contained in the α region adjacent to G. Thus, G contains a variable portion of S and S′ and the intact U and U′ genes. This is an interesting example of a situation in which a given DNA sequence can be used to specify different products.

TRANSPOSABLE ELEMENTS IN HIGHER CELLS

How might the transposable elements of *E. coli* have been discovered if they had not carried drug-resistance genes? Southern transfers would have been one method. Upon cloning various regions of the chromosome, someone would have found a DNA fragment that hybridized to many sizes of restriction fragments in the digest of whole chromosomal DNA. Transposable elements in yeast were discovered in almost this way. In an investigation of a DNA restriction fragment that varied in size from one yeast strain to another, Davis found that the region responsible for the variability contained a DNA segment that hybridized to about 35 different restriction fragments of the yeast chromosome. This segment had many of the properties of transposable elements in bacteria and was named Ty1.

Hybridization and sequencing of Ty1 and the regions into which it inserts have revealed its structure. It duplicates five bases upon insertion and consists of two flanking regions, called delta, of 330 base pairs oriented as direct repeats around a 5,600-base-pair central region. Not all delta elements found in yeast are identical, nor are the central regions identical, for some Ty elements are able to block expression of nearby yeast genes while others stimulate expression of genes near the point of integration.

A recombination event between the two delta elements deletes the central region and one delta element. Hence it is not surprising that the yeast chromosome contains about 100 of these solo delta elements. Recombination between different Ty elements can create various chromosomal rearrangements. Although the consequences of recombination can be determined easily in yeast, similar chromosome rearrangements catalyzed by recombination between transposons must occur in other organisms. Consequently transposons may be of positive value to an organism because they may speed chromosome rearrangements that may directly generate new proteins and new schemes of gene regulation.

Retrovirus particles contain single-stranded RNA that is translated upon infection. This RNA can also be converted to an RNA-DNA hybrid and finally to double-stranded DNA through the action of a virally specified enzyme called reverse transcriptase, which is carried within the virus particle. The resulting DNA molecule can then be integrated into the host cell's chromosome, in which form it is called a provirus.

The sequence arrangements of proviruses are similar to those of transposable elements. Upon their insertion into the chromosome, retroviruses duplicate a small number of bases as a result of generating staggered nicks in the target sequence. At each end of the retrovirus sequence is an inverted repeat of about 10 base pairs that is

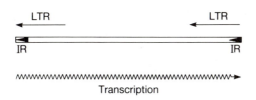
Transcription

part of a few-hundred-base-pair direct repeat. Between the direct repeats, which are named long terminal repeats or LTRs, are sequences of about 5,000 base pairs that code for viral coat protein and other proteins. Another similarity between transposable elements and retroviruses is their transcription pattern. Transcription begins near the end of one LTR, proceeds through the central region, and ends within the other LTR.

Some of the middle repetitive DNA in *Drosophila* has been analyzed, and its properties suggest that it is related to proviruses. One of these families is called *copia*. Small virus-like particles can be isolated from the nucleus of *Drosophila* cells containing RNA homologous to *copia* DNA sequences. This RNA is translatable into one of the proteins that coats the RNA. Since the *copia* sequences are not frozen in place within the chromosome of the fly, either they transpose much like bacterial transposable elements or they can copy themselves into other positions via the RNA route.

Retroviruses have been much harder to demonstrate in humans than in other animals. Nonetheless, they have been found. A more easily observed element in the human genome is the *Alu* family of sequences. Humans contain 100,000 to 500,000 copies of this sequence. The name derives from the fact that the restriction enzyme *Alu* cleaves more than once within the sequence. Consequently, digestion of human DNA with *Alu* yields 100,000 identical fragments, which upon electrophoresis generate a unique band in addition to the faint smear generated by the heterogeneity of the remainder of the DNA. The *Alu* sequences look like direct DNA copies of mRNA molecules because they contain a stretch of poly deoxyadenosine at their 3′ ends. Like transposons, the *Alu* sequences also are flanked by direct repeats of chromosomal sequences of 7 to 20 base pairs.

It is not possible to tell from their structure whether retroviruses and *Alu* sequences evolved from transposable elements or the reverse. It does seem clear, however, that sequences with similar structures and properties as the transposable elements that are found in bacteria are found in many other cell types as well.

TRANSPOSONS AS GENETIC ENGINEERING VECTORS IN HIGHER CELLS

One would think that transposons would form ideal vectors for genetic engineering in higher cells. DNA to be transposed into a chromosome merely would have to be added to a nonessential site within a transposon active in the desired cells, and then the transposon DNA would have to be injected into the cells. Indeed, such constructs have been made in bacterial transposons. Tn3 has been cut in its β-lactamase gene, *bla*, and a DNA fragment encoding tetracycline resistance has been inserted. This new structure transposes as well as the former. Analogous reconstructions have been performed with Tn5 and Tn10.

From what has been mentioned above, one source of transposon in the fruit fly *Drosophila* could be one of the middle repetitive sequences. One way to isolate such a sequence would be to locate a sequence heterogeneity between two strains. For example, the nucleotide sequences of *Drosophila melanogaster* Oregon R and Canton S are virtually identical over long stretches but are punctuated by the insertion into one of the two sequences of a stretch of several thousand nucleotides. Apparently this is a transposon. A short sequence of the host sequence, that is, five bases, is even duplicated at each end of the inserted sequence. The use of these transposons for genetic engineering could generate two problems. They might be inactive because of the accumulation of nucleotide changes after they transposed, or they might transpose at far too slow a rate to be useful.

The P element transposons in *Drosophila* appeared to be a much better starting material for a genetic engineering vector. When a male P^+ line is mated with a female

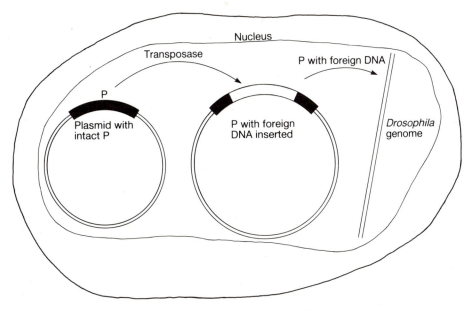

Figure 18.18   Using a plasmid with a functional *Drosophila* P element to provide trans-posase so that the P element containing foreign DNA can transpose into the *Drosophila* genome.

non-P line, a short interval of intense transposition in the developing egg follows, and many genes are inactivated by the insertion of P elements. Afterward, the transposition activity ceases, and the resultant strains are genetically stable. The reverse — mating a P^+ female with a non-P male — does not stimulate transposition. By analogy to the bacterial transposons, the P elements behave as though they encode a repressor and enzymes necessary for transposition. Upon entry of a P element into a new cytoplasm, the temporary absence of repressor permits the transposition enzymes to be synthesized, and transposition likely occurs. One important difference, however, between the *Drosophila* P elements and the *coli* IS and Tn elements is that P elements appear to be stimulated to excise from their former locations as well as to copy themselves into new locations because mutations generated by insertions of P elements can be induced to revert by the same mating process that stimulates P element mutations. The IS and Tn elements are not readily excised, although rare recombination events remove them from an integrated position. In their ability to integrate and excise, the P elements resemble lambda phage.

A P element would form a useful vector for inserting DNA into a chromosome of a fruit fly. By analogy with the bacterial transposons, nearly any sequence of foreign DNA could be inserted into the middle of the P element. Then, for transformation of the foreign DNA into a cell's chromosome, if this P element plus a normal P element

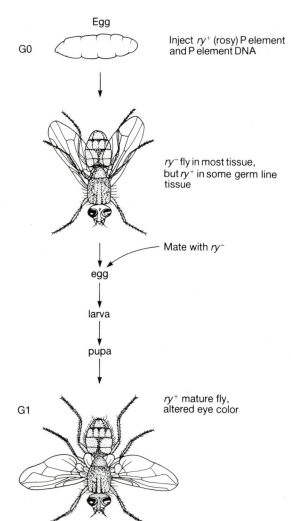

Egg

G0

Inject *ry*⁺ (rosy) P element
and P element DNA

ry⁻ fly in most tissue,
but *ry*⁺ in some germ line
tissue

Mate with *ry*⁻

egg

larva

pupa

G1

ry⁺ mature fly,
altered eye color

Figure 18.19   Schematic of the transformation
process in *Drosophila*. Some cells deriving from one
end of the egg including the germ line cells of the
G0 generation may be transformed. Only in the G1
generation can all the cells in a fly contain copies of
the transforming DNA.

were introduced into a non-P egg cell, transposition enzymes should be synthesized
and move or copy the altered P element into the cellular DNA (Fig. 18.18).

Rubin and Spradling found that all these expectations were correct and that P
elements do form a basis for efficient transformation of DNA into *Drosophila*. They
inserted the *rosy* gene coding for xanthine dehydrogenase into the middle of an
internally deleted P element that had been cloned on a plasmid. Then they injected
both plasmid DNA containing the P element with inserted DNA and plasmid DNA
containing an intact P into developing *rosy* negative *Drosophila* eggs. The injection

was into a portion of the egg in which the germ line cells develop and at a time when the egg was multinucleate but before cell walls had developed. Many of the injected eggs survived and yielded adults whose progeny possessed a functional *rosy* gene in all the fly's tissues (Fig. 18.19). These flies were easily detected because their eye color was different from that of the untransformed *rosy* negative mutants.

The sequence of events in transformation is as follows. The injected P element DNA is transcribed in the nucleus, and the resulting RNA is transported to the cytoplasm where it is translated into transposition enzymes that return to the nucleus. There they catalyze the transposition of the P element containing the *rosy* gene into a *Drosophila* chromosome. In the meantime, cell walls form, and the egg develops to a larva and finally to an adult. Some of the cells of its germ line contain the functional *rosy* gene carried in the P element. When this fly matures and mates with a mutant *rosy* strain, its progeny that receive egg or sperm containing the *rosy* gene on the P element will then express functional xanthine dehydrogenase.

TRANSPOSITION VIA DNA INTERMEDIATES

Two basic methods exist for inserting a transposable element or its copy into a new site. In one method, staggered cuts in both strands of the target site are made and a strand of the element is transferred to the target sequence (Fig. 18.20). Continuous DNA replication in the 5'-to-3' direction first across the element in one direction and then back in the opposite direction is followed by transfer of the original strand of the transposable element back to the donor molecule. A minor variation of this method is for the second strand of the host element to be copied discontinuously, net 3' to 5', while the target region is replicated 5' to 3' and then to be followed by various strand exchanges to create either fused donor and target molecules or a transposition. This method requires that only one end of the transposable element be held in a complex at a time.

The second basic method of transposition is for both ends of the transposable element to be held in a complex at the same time and for single strands to be nicked and then replicated (Fig. 18.21). If the process is stopped here, the donor and target replicons will be fused. This is analogous to the structures generated by transposition of Tn3 when the internal resolution site has been damaged. The final step in a normal transposition is this recombination event between the two copies of an internal resolution site that separates the two replicons. Considering the variability that nature displays in the schemes of gene regulation, it would be surprising if examples were not found for each of these basic transposition modes. However, the experiments to demonstrate the transposition mode are not simple. Not only are the transposition enzymes normally synthesized in low quantity, but they appear to be highly unstable.

This chapter has considered transposable elements that are found in a variety of organisms. The next chapter will continue the theme of *in vivo* cutting and splicing of DNA. In this chapter we saw that cutting and splicing served as a convenient method

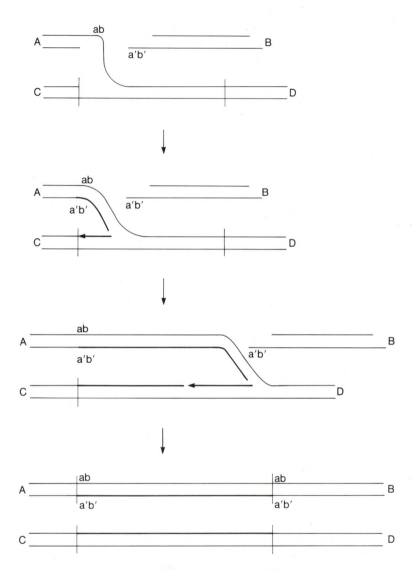

Figure 18.20   One mode of transposition. The transposable element located on the C-D segment is nicked, and one strand is transferred to the A-B duplex, which is cut on both strands. Replication in the 5′-to-3′ direction from the bottom strand of A and replication on the upper strand of C-D followed by strand exchange upon completion of replication yields a copy of the transposable element in C-D. If the strands are exchanged differently, the A-B and C-D replicons are fused.

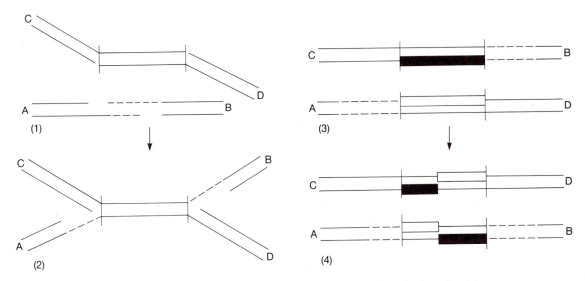

Figure 18.21   A second mode of transposition. Nicks are made at both ends of the transposon and on both strands of the acceptor molecule, A-B. Joining and DNA replication yield a copy of the replicon, but the replicons are joined. Recombination within the replicon generates the separate A-B and C-D replicons.

for moving DNA around and presumably increasing the viability of the transposition sequence. Cutting and splicing also served as a convenient method for switching expression from one to another gene as in the case of *Salmonella* and phage Mu. Cutting and splicing will be seen in the next chapter to perform something much different than the mere regulation of gene activity. It also generates new genes.

PROBLEMS

18.1. Describe how R-factors or R-factors and F'- factors, instead of lambda phage as was described in the text, could be used to demonstrate transposition by drug-resistance transposons.

18.2. Consider the problem of comparing the densities of lambdoid transducing phage carrying polar mutations to the densities of phage without the mutations. What sequence of steps could be used to cross a mutation on the chromosome onto a transducing phage such that the only difference between the progeny and parent phage is the mutation?

18.3. Demonstration of the identity of insertion sequences in different lambda phage required the separation of the DNA strands of the phage. Look up how the

DNA strands, sometimes called r for right and l for left or Watson and Crick, of lambda can be separated.

18.4. Why are oppositely oriented overlapping genes likely to be in the same codon register?

18.5. Either invent a selection or look up the genetic selection used by Kleckner that will yield lambda phage with insertions of Tn10 into the *CI* gene of lambda.

18.6. Show how deletions and inversions can be generated by the insertion of a Tn factor into its own replicon.

18.7. How could you determine the genetic locations of IS sequences that are naturally resident on the chromosome of *E. coli*?

18.8. What is a straightforward genetic method of looking for replicon fusion generated by a transposon?

18.9. The left split ends of DNA heteroduplexes of phage Mu are too short to be easily seen in electron microscopy of DNA heteroduplexes. How could you precisely measure the distribution of sizes of host DNA located at the left end?

18.10. Show how denaturing and reannealing DNA containing two inverted repeated sequences separated by some other DNA will create a stem and loop structure, a "lollipop."

18.11. What problem, unaddressed in the text, is raised by the following fact? Density labeling Mu phage and infecting show that the very DNA molecule that was packaged in the phage coat, not a copy of this DNA molecule, is integrated into the chromosome of a newly formed Mu lysogen.

18.12. Consider a transposon in the Tn5-Tn10 class that is flanked by inverted IS sequences. During transposition, the outside ends of the IS elements function as ends. Using a plasmid, how could you test whether the opposite ends could be used instead? That is, instead of transposing with the outside ends of the IS sequences, perhaps the inside ends could be used.

18.13. To make use of its broad host range, Mu has a great need to circumvent an infected cell's restriction-modification system. How might it do this and how would you test your idea?

18.14. Without the use of mutants unable to invert the G segment, stuck or *gin* mutants, how could you demonstrate the variable host range specificities of Mu phage?

18.15. Look up and describe how the invertible segment of *Salmonella* that controls flagellin synthesis was discovered.

18.16. Based on the reasons for shifting flagellin type in *Salmonella*, estimate sensible limits for the frequency of inverting the flagellin controlling segment.

18.17. Consider sequences such as transposable elements in the chromosome of a cell. Show how recombination events between these portable regions of homology can create deletions, duplications, inversions, and, in cells with more than one chromosome, translocations.

18.18. Give a plausible sequence of events by which a transposable element such as Tn10 could evolve to become a retrovirus.

18.19. How could a strong promoter oriented outward from a site within a transposon reduce translational expression of the transposase from transcriptional readthrough entering the transposase as well as generate repression in *trans* of transposase expression?

RECOMMENDED READINGS

Genetic Engineering *in vivo* Using Translocatable Drug-Resistance Elements, N. Kleckner, J. Roth, D. Botstein, J. Mol. Biol. *116*, 125–159 (1977).

In vitro Mutagenesis of a Circular DNA Molecule by Using Synthetic Restriction Sites, F. Heffron, M. So, B. McCarthy, Proc. Nat. Acad. Sci. USA *75*, 6012–6016 (1978).

Recombinational Switch for Gene Expression, J. Zeig, M. Silverman, M. Hilmen, M. Simon, Science *196*, 170–172 (1978).

Invertible DNA Determines Host Specificity of Bacteriophage Mu, P. van de Putte, S. Cramer, M. Giphart-Gassler, Nature *286*, 218–222 (1980).

G Inversion in Bacteriophage Mu: A Novel Way of Gene Splicing, M. Gilphart-Gassler, R. Plasterk, P. Van de Putte, Nature *297*, 339–342 (1982).

Genetic Transformation of *Drosophila* with Transposable Element Vectors, G. Rubin, A. Spradling, Science *218*, 348–353 (1982).

RELATED REVIEWS, BOOKS, AND ARTICLES

DNA Insertion Elements, Plasmids, and Episomes, ed. A. Bukhari, J. Shapiro, S. Adhya, Cold Spring Harbor Laboratory (1976).

Transposable Elements, M. Calos, J. Miller, Cell *20*, 579–595 (1980).

Transposable Elements, N. Kleckner, Ann. Rev. Genet. *15*, 341–404 (1981).

Mobile Genetic Elements, ed. J. Shapiro, Academic Press, Orlando (1983).

DEEPER READING

The Origin and Behavior of Mutable Loci in Maize, B. McClintock, Proc. Nat. Acad. Sci. *36*, 344–355 (1950).

Bacteriophage-Induced Mutation in *Escherichia coli*, A. Taylor, Proc. Nat. Acad. Sci. USA *50*, 1043–1051 (1963).

Frameshift Mutations in the Lactose Operon of *E. coli*, M. Malamy, Cold Spring Harbor Symp. Quant. Biol. *31*, 189–201 (1966).

Strong Polar Mutations in the Transferase Gene of the Galactose Operon of *E. coli*, E. Jordan, H. Saedler, P. Starlinger, Mol. Gen. Genet. *100*, 296–306 (1967).

O^0 and Strong Polar Mutations in the *gal* Operon Are Insertions, E. Jordan, H. Saedler, P. Starlinger, Mol. Gen. Genet. *102*, 353–363 (1968).

Mutations Caused by the Insertion of Genetic Material into the Galactose Operon of *Escherichia coli,* J. Shapiro, J. Mol. Biol. *40,* 93–105 (1969).

Mu-1 Promoted Integration of a *lambda-gel* Phage in the Chromosome of *E. coli,* M. Faelen, A. Toussaint, M. Couturier, Mol. Gen. Genet. *113,* 367–370 (1971).

Mutations in the Lactose Operon Caused by Bacteriophage Mu, E. Daniell, R. Roberts, J. Abelson, J. Mol. Biol. *69,* 1–8 (1972).

Insertion Mutations in the Control Region of the Galactose Operon of *E. coli,* II. Physical Characterization of the Mutations, H. Hirsch, H. Saedler, P. Starlinger, Mol. Gen. Genet. *115,* 266–276 (1972).

Polar Mutations in *lac, gal,* and Phage lambda Consist of a Few IS-DNA Sequences Inserted with Either Orientation, M. Fiandt, W. Szybalski, M. Malamy, Mol. Gen. Genet. *119,* 223–231 (1972).

Connecting Two Unrelated DNA Sequences with a Mu Dimer, A. Toussaint, M. Faelen, Nature New Biology *242,* 1–4 (1973).

Similarity of Vegetative Map and Prophage Map of Bacteriophage Mu-1, C. Wijffelman, G. Westmaas, P. Van de Putte, Virology *54,* 125–134, (1973).

Electron Microscope Heteroduplex Study of the Heterogeneity of Mu Phage and Prophage DNA, M. Hsu, N. Davidson, Virology *58,* 229–239 (1974).

Mutagenesis by Insertion of a Drug-Resistance Element Carrying an Inverted Repetition, N. Kleckner, R. Chan, B. Tye, D. Botstein, J. Mol. Biol. *97,* 561–575 (1975).

Transposition of R Factor Genes to Bacteriophage lambda, D. Berg, J. Davies, B. Allet, J. Rochaix, Proc. Nat. Acad. Sci. USA *72,* 3628–3632 (1975).

Electron Microscope Heteroduplex Studies of Sequence Relations Among Plasmids of *Escherichia coli:* Structure of F13 and Related F-Primes, S. Hu, E. Ohtsubo, N. Davidson, J. Bact. *122,* 749–763 (1975).

Occurrence of Insertion Sequence (IS) Regions on Plasmid DNA as Direct and Inverted Nucleotide Sequence Duplications, K. Ptashne, S. Cohen, J. Bact. *122,* 776–781 (1975).

IS1 Is Involved in Deletion Formation in the *gal* Region of *E. coli* K-12, H. Reif, H. Saedler, Mol. Gen. Genet. *137,* 17–28 (1975).

Location of the "Variable End" of Mu DNA Within the Bacteriophage Particle, R. Inman, M. Schnos, M. Howe, Virology *72,* 393–401 (1976).

State of Prophage Mu DNA upon Induction, E. Ljunquist, A. Bukhari, Proc. Nat. Acad. Sci. USA *74,* 3143–3147 (1977).

DNA Sequence at the Integration Sites of the Insertion Element IS1, M. Calos, L. Johnsrud, J. Miller, Cell *13,* 411–418 (1978).

IS1 Insertion Generates Duplication of a Nine Base Pair Sequence at Its Target Site, N. Grindley, Cell *13,* 419–426 (1978).

Role of the G. Segment in the Growth of Phage Mu, N. Symonds, A. Coelho, Nature *271,* 573–574 (1978).

The Invertible Segment of Bacteriophage Mu DNA Determines the Adsorption Properties of Mu Particles, A. Bukhari, L. Ambrosio, Nature *271,* 575–577 (1978).

Inversion of the G DNA Segment of Phage Mu Controls Phage Infectivity, D. Kamp, R. Kahmann, D. Zipser, T. Broker, L. Chow, Nature *271,* 577–580 (1978).

Phase Variation in *Salmonella:* Genetic Analysis of a Recombinational Switch, M. Silverman, J. Zieg, M. Hilmen, M. Simon, Proc. Nat. Acad. Sci. USA *76,* 391–395 (1979).

Molecular Model for the Transposition and Replication of Bacteriophage Mu and Other Transposable Elements, J. Shapiro, Proc. Nat. Acad. Sci. USA, *76,* 1933–1937 (1979).

DNA Sequence Analysis of Tn10 Insertions: Origins and Role of 9 bp Flanking Repetitions During Tn10 Translocation, N. Kleckner, Cell *16,* 711–720 (1979).

Physical Structures of Tn10-Promoted Deletions and Inversions: Role of 1400 bp Inverted Repetitions, D. Ross, J. Swan, N. Kleckner, Cell 16, 721–731 (1979).

Evidence for Transposition of Dispersed Repetitive DNA Families in Yeast, J. Cameron, E. Loh, R. Davis, Cell 16, 739–751 (1979).

DNA Sequence Analysis of the Transposon Tn3: Three Genes and Three Sites Involved in Transposition of Tn3, F. Heffron, B. McCarthy, H. Ohtsubo, E. Ohtsubo, Cell 18, 1153–1163 (1979).

Nucleotide Sequences at the Attachment Sites of Bacteriophage Mu DNA, R. Kahmann, D. Kamp, Nature 280, 247–250 (1979).

Identification of the Protein Encoded by the Transposable Element Tn3 Which Is Required for Its Transposition, R. Gill, F. Heffron, S. Falkow, Nature 282, 797–801 (1979).

Transposition Protein of Tn3: Identification and Characterization of an Essential Repressor-Controlled Gene Product, J. Chou, P. Lemaux, M. Casadaban, S. Cohen, Nature 282, 801–806 (1979).

Transposition of DNA Inserted into Deletions of the Tn5 Kanamycin Resistance Element, R. Meyer, G. Boch, J. Shapiro, Mol. Gen. Genet. 171, 7–13 (1979).

Analysis of the Nucleotide Sequence of an Invertible Controlling Element, J. Zieg, M. Simon, Proc. Nat. Acad. Sci. USA 77, 4196–2000 (1980).

Regulation of Tn5 Transposition in Salmonella typhimurium, D. Biek, J. Roth, Proc. Nat. Acad. Sci. USA 77, 6047–6051 (1980).

Molecular Consequences of Deletion Formation Mediated by Transposon Tn9, M. Calos, J. Miller, Nature 285, 38–41 (1980).

Phase Variation: Genetic Analysis of Switching Mutants, M. Silverman, M. Simon, Cell 19, 845–854 (1980).

DNA Rearrangements Associated with a Transposable Element in Yeast, G. Roeder, G. Fink, Cell 21, 239–249 (1980).

Insertion of the Drosophila Transposable Element copia Generates a 5 Base Pair Duplication, P. Dunsmuir, W. Brorein, Jr., A. Simon, G. Rubin, Cell 21, 575–579 (1980).

A Mechanism of DNA Transposition, R. Harshey, A. Bukhari, Proc. Nat. Acad. Sci. USA 78, 1090–1094 (1981).

Resolution of Cointegrates Between Transposons Gamma-Delta and Tn3 Defines the Recombination Site, R. Reed, Proc. Nat. Acad. Sci. USA 78, 3428–3433 (1981).

Transposon Tn3 Encodes a Site-Specific Recombination System: Identification of Essential Sequences, Genes, and Actual Site of Recombination, R. Kostriken, C. Morita, F. Heffron, Proc. Nat. Acad. Sci. USA 78, 4041–4045 (1981).

Nucleotide Sequence of Terminal Repeats of 412 Transposable Elements of Drosophila melanogaster, B. Will, A. Bayev, D. Finnegan, J. Mol. Biol. 153, 897–915 (1981).

Genetic Organization of Transposon Tn10, T. Foster, M. Davis, D. Roberts, K. Takeshita, N. Kleckner, Cell 23, 201–213 (1981).

Heterogeneous Host DNA Attached to the Left End of Mature Bacteriophage Mu DNA, M. George, A. Bukhari, Nature 292, 175–176 (1981).

Homology Between the Invertible Deoxyribonucleic Acid Sequence That Controls Flagellar-Phase Variation in Salmonella sp. and Deoxyribonucleic Acid Sequences in Other Organisms, E. Szekely, M. Simon, J. Bact. 148, 829–836 (1981).

DNA Sequence Organization of IS10-Right of Tn10 and Comparison with IS10-Left, S. Halling, R. Simons, J. Way, R. Walsh, N. Kleckner, Proc. Nat. Acad. Sci. USA 79, 2608–2612 (1982).

Inverted Repeats of Tn5 Are Transposable Elements, D. Berg, L. Johnsrud, L. McDivitt, R. Ramabhadran, B. Hirschel, Proc. Nat. Acad. Sci. USA 79, 2632–2635 (1982).

Conservative Integration of Bacteriophage Mu DNA into pBR322 Plasmid, J. Liebart, P. Ghelardini, L. Paolozzi, Proc. Nat. Acad. Sci. USA *79*, 4362–4366 (1982).

Expression of Two Proteins from Overlapping and Oppositely Oriented Genes on Transposable DNA Insertion Element IS5, B. Rak, M. Lusky, M. Hable, Nature *297*, 124–128 (1982).

Control of Tn5 Transposition in *Escherichia coli* Is Mediated by Protein from the Right Repeat, J. Johnson, J. Yin, W. Reznikoff, Cell *30*, 873–882 (1982).

Transposition of Cloned P Elements into *Drosophila* Germ Line Chromosomes, A. Spradling, G. Rubin, Science *218*, 341–347 (1982).

Identification of a Transposon Tn3 Sequence Required for Transposition Immunity, C. Lee, A. Bhagwat, F. Heffron, Proc. Nat. Acad. Sci. USA *80*, 6765–6769 (1983).

Transcription Initiation of Mu *mom* Depends on Methylation of the Promoter Region and a Phage-Coded Transactivator, R. Plasterk, H. Vrieling, P. Van de Putte, Nature *301*, 344–347 (1983).

Retrovirus-like Particles Containing RNA Homologous to the Transposable Element *copia* in *Drosophila melanogaster*, T. Shiba, K. Saigo, Nature *302*, 119–124 (1983).

DNA Sequence of the Maize Transposable Element *Dissociation*, H. Doring, E. Tillman, P. Starlinger, Nature *307*, 127–130 (1983).

Tn10 Transposase Acts Preferentially on Nearby Transposon Ends *in vivo*, D. Morisato, J. Way, H. Kim, N. Kleckner, Cell *32*, 799–807 (1983).

Site-Specific Relaxation and Recombination by the Tn3 Resolvase: Recognition of the DNA Path Between Oriented *res* Sites, M. Krasnow, N. Cozzarelli, Cell *32*, 1313–1324 (1983).

Three Promoters Near the Termini of IS10: pIN, pOUT, and PIII, R. Simons, D. Hoopes, W. McClure, N. Kleckner, Cell *34*, 673–682 (1983).

Translational Control of IS10 Transposition, R. Simons, N. Kleckner, Cell *34*, 683–891 (1983).

Isolation of the Transposable Maize Controlling Elements Ac and Ds, N. Fedoroff, S. Wessler, M. Shure, Cell *35*, 235–242 (1983).

Mapping of the *pin* Locus Coding for a Site-Specific Recombinase That Causes Flagellar-Phase Variation in *Escherichia coli* K-12, M. Enomoto, K. Oosawa, H. Momota, J. Bact. *156*, 663–668 (1983).

Mechanism of Transposition of Bacteriophage Mu: Polarity of the Strand Transfer Reaction at the Initiation of Transposition, K. Mizuuchi, Cell *39*, 395–404 (1984).

Generating Genetic Diversity: Antibodies

The restriction-modification systems discussed in earlier chapters provide bacteria with the ability to identify and destroy foreign DNA. Multicellular organisms also must be able to identify and destroy foreign invaders. Such invaders include not only viruses but also bacteria and yeasts as well as multicelled parasites. Finally, higher organisms must have a means of protection against themselves. From time to time, a few of their cells lose their control and begin uncontrolled growth. Often these runaway cells are stopped by the immune system, but when they are not, the result is cancer.

Similar to a bacterium's "immunity" to infection by phage lambda, an organism can be resistant to infection by a parasite. The immune response of vertebrates can be considered to consist of two steps: first, recognizing a macromolecule that is foreign to the organism; and second, doing something with the recognized macromolecule. Because of the complexity of the immune system, this chapter will be concerned primarily with only a portion of the first part of the system, that involving recognition. Rearrangements of DNA segments and alterations in DNA sequences generate a multitude of genes, each slightly different from the others, that code for the many different proteins that recognize the foreign molecules.

THE BASIC IMMUNE RESPONSE

Vertebrates can detect and dispose of most foreign macromolecules. Foreign macromolecules that can stimulate the immune system to produce a response are called antigens. These can include protein, nucleic acids, polysaccharides, and lipids that originate in a vertebrate from bacterial, viral, or parasitic infections, from injections of foreign materials, or from the transformation of a normal cell into a cell type that is not growth-inhibited. The transformed cells usually possess altered cell surfaces that often are recognized by the immune system.

Evolution has chosen a compromise between the ability of an animal to respond immediately to any one of millions of foreign antigens and the enormous burden that would result from being able to respond quickly. Young vertebrates possess the ability to recognize and respond to all foreign invaders, but this ability begins at a very low level for most antigens. Only some time after exposure to an antigen does the animal's ability to respond to that antigen begin to be significant. Thereafter, the animal can respond more quickly to the reappearance of a particular antigen. As a result of such exposure to invaders or to antigens related to them, animals acquire immunity.

In most cases, the immune system can mobilize to deal with a new antigen before the presence of its source can harm the animal. Speeding and amplifying the immune response is the objective of medical immunization. In such immunizations, antigens are injected to prime the immune system. However, to avoid transmitting the disease associated with the antigen, antigenic material extracted from the organism is injected, or the organism is killed before injection, or a relatively harmless variant of the organism is injected. Therefore the animal can generate an immune response without suffering the actual dangerous infection. Often, a second or third injection is made after several weeks to boost the animal's response.

Two responses begin to be apparent within a week of the introduction of a new antigen into an animal. New proteins appear in the serum. These are called antibodies and are able to bind specifically to the antigen and not to other molecules. Also, a subpopulation of the cell type called T lymphocytes appears. T lymphocytes, which ordinarily are found in the blood, spleen, and lymph nodes, have the ability to bind the antigen to their surface.

In some cases, the mere binding of antibodies or T cells to an antigen is enough to protect the animal from the source of the antigen. As an example, the binding of antibody to the flagella of *Salmonella,* discussed in the previous chapter, may sufficiently immobilize the cell that it cannot survive in the intestinal tract. Usually, however, an animal's immune response includes more than just antibody molecules or T cells binding to the antigen. Additional proteins and cell types may participate in elimination of antigens. These recognize complexes of antibody or T cells bound to antigen and act to kill or digest or eliminate the foreign material as well as to modulate the additional synthesis of antibodies and specific T cells. One such array of proteins is called the complement system.

The antibodies found in the serum are synthesized by B-type lymphocytes. These cells are found in bone marrow, blood, spleen, and lymph nodes. All the antibody molecules synthesized by a particular B cell are of a single type and are capable of binding to only one antigen. Similarly, each T cell is capable of binding only one type of antigen to its surface. One of the problems of immunology is how the different B or T cells acquire the ability to synthesize proteins with different binding specificities. As we will see in more detail later, immature lymphocytes possess the ability to recognize different antigens. The presence of one of these antigens can trigger one of the pre – B cell lymphocytes to begin growth and division. After about twelve divisions, this expanded clone of identical cells can synthesize and secrete appreciable quantities of antibody.

TELLING THE DIFFERENCE BETWEEN FOREIGN AND SELF

What keeps a vertebrate from synthesizing antibodies against its own macromolecules? In principle, one solution to this problem would be to have the entire repertoire of an animal's antibody specificities explicitly encoded in the genome and expressed during development. While the encoded specificities could avoid synthesis of antibodies against the animal, this would be inefficient for at least two reasons First, as we will see in the next section, the very large number of specificities expressed by the immune system would use an inordinate amount of the coding potential of the genome. Second, such an inflexible system would be highly dangerous, for the moment an invader mutated and developed surface molecules not neutralized by the host cells' immune system, the invader could freely grow in the host animal, and the entire population of host animals could be killed. It would be as though they didn't possess any immune system at all, for the mutated invader would encounter no opposition.

For these reasons we expect a considerable amount of randomness, even in genetically identical animals, in the generation of antibody-binding specificities so that at least a few animals in any population may be capable of recognizing any antigen. However, it is this randomness that generates the problem of identifying only foreign macromolecules. If the generation of the ability to recognize diverse antigens is random and not explicitly encoded in the DNA, the same randomness will generate antibodies against the organism itself, and these self-recognizing antibodies must be eliminated if the animal is to survive.

The problem of preventing self-recognition is solved by killing the lymphocytes that recognize macromolecules of the organism itself. This, of course, must be done after development of the immune system but before the organism is exposed to foreign macromolecules. For purposes of the molecular biology of the immune system, it is useful to consider the following simplified picture of the development of an animal's lymphocytes (Fig. 19.1). Eventually in the cells deriving from a fertilized egg, there arises a single cell that can be considered the precursor of the T and B cells. Subsequent divisions of this cell produce only T and B cells. Each of the T or B cells develops the ability to recognize only one antigen. In mammals, at about the time the animal is born, those lymphocytes are killed that are capable of synthesizing anti-

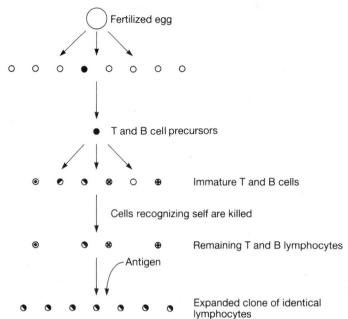

Figure 19.1   Growth and differentiation of a fertilized egg to yield the immature T cells, some of which recognize self and are killed. The remainder recognize other antigens and reside in the animal until the binding of antigen to their surface triggers them to begin growth, after which "clonal expansion" yields a population of size sufficient to synthesize appreciable quantities of immunoglobulin.

bodies to antigens present in the animal. The lymphocytes capable of synthesizing antibodies to antigens not present at this time survive and form the pool of immature T and B lymphocytes that are capable of responding to foreign antigens.

The consequences of incomplete killing of lymphocytes that recognize "self" or the subsequent alteration of specificity so that "self" is recognized are predictable. In either case the immune system turns against the animal, and autoimmune conditions result. Autoimmunity is not at all uncommon. Forms of diabetes mellitus result from antibodies directed against part of the pituitary gland. Myasthenia gravis, a condition in which antibodies are directed against muscle-nerve connections, and systemic lupus erythematosus, a condition in which antibodies are directed against many cellular components, are also autoimmune diseases, and many other examples are known. Autoimmune conditions also appear to result from loss of the ability to repress activation of T or B cells or from mutation in lymphocytes so that their antigen specificity changes and the antibodies interact with macromolecules of the animal.

THE NUMBER OF DIFFERENT
ANTIBODIES PRODUCED

How many different antibodies can be generated by a single mouse or human? If the number is small, then all the information for their synthesis could be explicitly stored in the genome, however unwise this would be, but if it is very large, some other mechanism must be used for generation of the diversity. A straightforward physical measurement made with mice provides an estimation of the lower limit to the number of different antibodies animals are capable of synthesizing.

After immunization of mice with phosphocholine linked to a macromolecule, antibodies are induced that are capable of binding to phosphocholine. The levels of these antibodies are increased at least a factor of 100 compared to their level before immunization. Together these induced phosphocholine-binding antibodies constitute about 0.5% of the total serum protein of the mice. The total number of different antibody varieties in this 0.5% of the serum is more than 100 because isoelectric focusing shows more than 100 different bands of phosphocholine-binding antibody.

These numbers lead to the estimate that mice can synthesize antibodies with more than 2×10^6 different specificities. The reasoning is as follows. Let N be the number of different antibodies present in a mouse. If these are all synthesized in nearly the same levels, each constitutes about $1/N$ of the total serum. If one particular antibody is induced by a factor of 100, that one then constitutes $100/N$ of the total serum. The 100 different phosphocholine-specific antibodies then constitute $100 \times (100/N)$ of the serum. This was $1/200$ of the total serum. That is, $100 \times (100/N) = 1/200$, or $N = 2 \times 10^6$. If the mechanisms of antibody synthesis required one gene for each different antibody, these numbers would necessitate that a sizable portion of the genome be used to code for the immune system.

$$2 \times 10^6 \text{ genes} \times 10^3 \frac{\text{nucleotides}}{\text{gene}} = 2 \times 10^9 \text{ nucleotides}$$

Typical haploid DNA content of vertebrate $= 3 \times 10^9$ base pairs

MYELOMAS AND MONOCLONAL ANTIBODIES

The tremendous diversity in antibody specificities present in the serum of an animal precludes any simple study of homogeneous antibody. Although purifications based on binding to an antigen might be attempted for obtaining antibody with one specificity, the resultant antibodies would still be highly heterogeneous, for many different antibodies normally are induced by a single antigen. Even a molecule as simple as phosphocholine can induce hundreds of different antibodies. A typical protein could induce thousands of different antibodies. Instead of pursuing the purification of a single antibody out of the thousands that are induced, researchers have made use of a cancer of the immune system to obtain homogeneous antibodies.

A lymphocyte with unregulated cell division generates a cancer called myeloma. The serum of patients with myeloma contains high concentrations of one single antibody. Myelomas can also be induced in mice, and the cancerous cells can be transferred from mouse to mouse. The ascites fluid collected from such mice is a solution of almost-pure homogeneous antibody. Thus a particular myeloma can be indefinitely propagated, and its corresponding antibody can be easily purified and studied.

The major problem with the use of myelomas is that their antigens are unknown. Of course, it is possible to screen hundreds of myeloma antibodies against thousands of potential antigens to find a molecule the antibodies bind to, but this has not proved highly successful. It would be ideal to be able to induce a myeloma for any desired antigen. This, in fact, can be done! Milstein and Kohler discovered that the immortality of a myeloma cell line can be combined with the selectable antigen specificity of antibodies that are obtained from immunized animals. The result is a

Figure 19.2   Scheme for the production of cell lines synthesizing monoclonal antibodies. Neither of the parental cell types can grow in the hypoxanthine aminopterin medium.

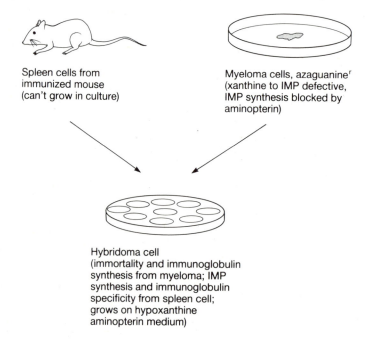

Spleen cells from
immunized mouse
(can't grow in culture)

Myeloma cells, azaguanine[r]
(xanthine to IMP defective,
IMP synthesis blocked by
aminopterin)

Hybridoma cell
(immortality and immunoglobulin
synthesis from myeloma; IMP
synthesis and immunoglobulin
specificity from spleen cell;
grows on hypoxanthine
aminopterin medium)

cell line that synthesizes a single type of antibody, one with a desired binding specificity. The cell line can be grown *in vitro* or injected into mice, giving them myeloma but providing a convenient source of large quantities of their antibody product. The antibodies produced by these means are called monoclonal because they derive from a single cell or clone.

Mouse cell lines synthesizing monoclonal antibodies are isolated by fusing myeloma cells with lymphocytes extracted from the spleen of a mouse about five days after its immunization with the desired antigen (Fig. 19.2). The cell fusion is accomplished merely by mixing the two cell types in the presence of polyethylene glycol. This agent induces fusion of the cell membranes; after an interval of unstable growth in which a number of chromosomes are lost, relatively stable chromosome numbers, called karyotypes, are established in the population. To reduce the background deriving from growth of unfused parental cell types, a genetic selection against both of the parental cell types is used so that only a fusion product can grow. The fused cells are then diluted to a concentration of a few cells per well of a culture tray, and after they have had time to grow and secrete antibody to the growth medium, the individual cultures from the wells are screened for the presence of antibody with the desired specificity. To avoid generating clones that synthesize two types of antibodies, those synthesized by the original myeloma and those specified by the lymphocytes, nonsecreting mutant myelomas are used in the cell fusions.

Monoclonal antibodies find wide use in diagnostic medicine and in many types of basic research. In addition to providing an almost limitless supply of antibody with unchanging characteristics, the monoclonal technology permits antibodies to be made against materials that otherwise cannot be completely purified. For example, monoclonal antibodies can be made against the cell walls of neurons from an animal. The neurons are dissected out, cell walls isolated and injected into a mouse, and the fusion of spleen cells to the myeloma line performed. Even unfractionated brain could be used for the initial injections. The antibodies synthesized by the fused cells can be screened for desired properties. Some of the resulting monoclonal

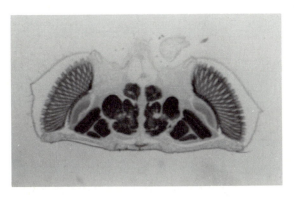

Figure 19.3   Reaction of monoclonal antibody to a central nervous system antigen with a brain section of *Drosophila melanogaster.* After binding of the antibody, a second antibody to which peroxidase was linked against the first was added. Then, a peroxidase substrate was added. The peroxidase generates an insoluble product from the substrate and this indicates the regions to which the monoclonal antibody bound. (Courtesy K. White.)

antibodies have been found to bind to all cells, some bind only to neurons, and others bind to just a few neurons (Fig. 19.3). Presumably neurons of the latter class are closely related to one another in ancestry or function.

THE STRUCTURE OF ANTIBODIES

Determining the amino acid sequences and three-dimensional structures of antibodies has greatly helped our understanding of how they are encoded by a reasonable number of genes and how they function. Although eight major classes of human antibody are known, this chapter deals primarily with IgG, the most prevalent antibody in serum. The other antibodies, IgM, the first detectable antibody synthesized in response to an antigen, as well as IgD, IgG_3, IgG_1, IgG_{2b}, IgG_{2a}, IgE, and IgA, have slightly different structures and apparently are specialized for slightly different biological roles. IgG contains four polypeptide chains, two identical light,

Figure 19.4   Schematic of the structure of IgG showing the inter- and intrachain disulfide bonds, hinge region, and the fragments produced by papain digestion. The antigen-binding site is at the amino terminal end. The molecule contains four types of structurally distinct domains: V, C_H1, C_H2, and C_H3. The region in the vicinity of the papain cleavage site is called the hinge region since the molecule is highly flexible here.

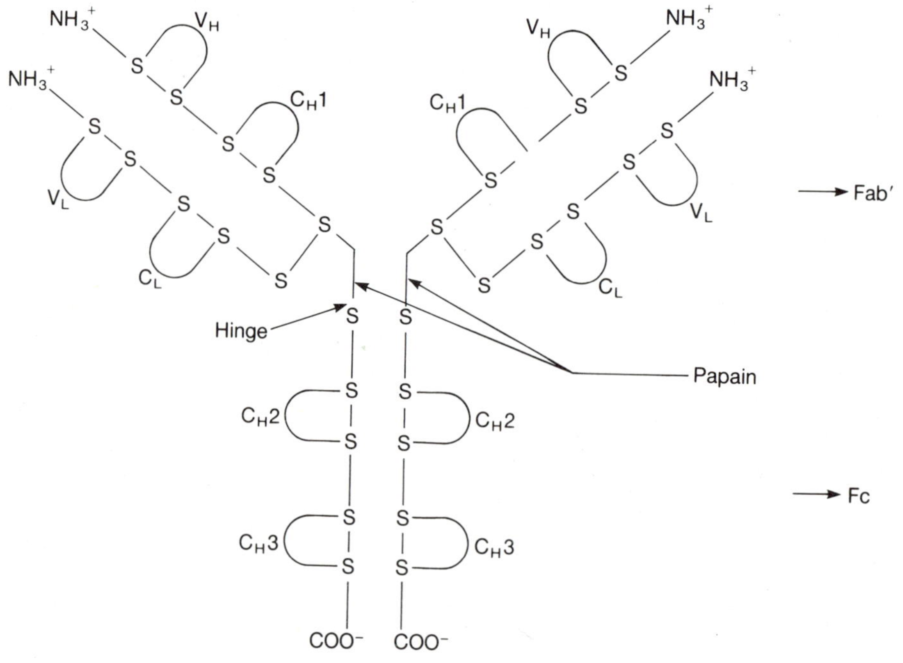

or L, chains of 22,500 molecular weight, and two identical heavy, or H, chains of about 50,000 molecular weight, giving a structure H_2L_2. Two subclasses of light chains are known in mouse and humans, kappa and lambda, whereas the eight major immunoglobulin classes are determined by the eight types of heavy chain, M, D, G3, G1, G2b, G2a, E, and A.

The chains of IgG are linked by inter- and intrachain disulfide bonds into a Y-shaped structure (Fig. 19.4). Papain digestion of one mole of IgG yields two moles of Fab' fragments and one mole of Fc fragment. The ab' indicates that these fragments possess the same binding specificity as the intact IgG antibody, and the c indicates that these fragments are readily crystallizable. Analysis of fragments of IgG has been highly important because fragments are much more easily studied than the intact IgG molecule.

The amino acid sequences of both the L and the H chains of IgG isolated from myelomas revealed an astonishing fact. About the first 100 amino acid residues of each of these chains are highly variable from antibody to antibody, whereas the remainder of the chains is constant (Fig. 19.5). Even more remarkable, however, is the finding that three or four regions within the variable portion of the protein are hypervariable. That is, most of the variability of the proteins lies in these regions. In view of their variability, such regions are likely to constitute the antigen-binding site of the antibody molecule. This suspicion was confirmed by affinity labeling of antibodies with special small-molecule antigens that chemically link to nearby amino acid residues. They were found to link to amino acids of the hypervariable portions of the L and H chains, just as expected.

Proteins from myelomas also assisted the X-ray crystallography of antibodies. Without the homogeneity provided by monoclonal antibodies, the crystal structures of antibodies could never have been determined. The structure of Fab' fragments confirms most of the postulates that the amino acid sequence raised. The hypervariable portions of the L and H chains together form the antigen-binding site at the end of each arm of the antibody molecule in a domain formed by the variable regions

Figure 19.5   The variability of the light and heavy chains as a function of residue position. The hypervariable regions show as regions with significantly greater variability than the average.

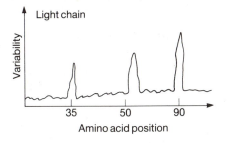

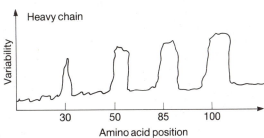

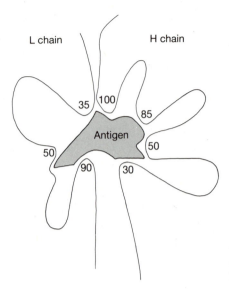

Figure 19.6   Schematic of the structure of the antigen-binding region of the V domain showing the confluence of the hypervariable regions to form the antigen-binding pocket.

(Fig. 19.6). In addition, the constant region of the L chain and the first constant region, C_H1, of the H chains form another domain (shown earlier in Fig. 19.4). The constant regions C_H2 and C_H3 of the two heavy chains form two additional domains of similar structure.

The fact that the variable portions of the L and H chains together form the antigen-binding site greatly reduces the number of genes required to specify antibodies. If any one of 1,000 L chains could combine with any one of 1,000 different H chains, then a total of 2,000 different genes could code for antibodies, with a total of 1×10^6 different binding specificities. The next sections describe how B lymphocytes combine small numbers of segments of the L and H genes to generate large numbers of different L and H genes, which then generate still larger numbers of antibodies because almost any L gene can function with almost any H gene. Techniques similar to those used for study of the antibody genes from B cells have recently been applied to the identification and analysis of clones containing the genes encoding the T cell antigen receptors. These proteins also contain two polypeptides and variable plus constant regions that are rearranged during the ontogeny of the organism.

CELLS HAVE MANY COPIES OF V GENES AND ONLY A FEW C GENES

The variability of the first 100 residues of the L and H chains and the constancy of the remainder of these chains suggested that the genes for L and H proteins were split into two parts. The V regions could then be encoded by any one of a set of different V region segments; upon differentiation into lymphocytes, one particular V region segment would be connected to the appropriate constant region segment to form a gene coding for an entire L or H chain. Such a proposal should be readily verifiable by the techniques of genetic engineering.

Cloning DNA coding for the V and C regions was not difficult. Since antibodies are secreted, they are synthesized by ribosomes attached to the endoplasmic reticulum of lymphocytes. Therefore, isolation of poly-A RNA from membranes of mouse myeloma cells yielded an RNA fraction greatly enriched for antibody messenger. This was then converted to cDNA with reverse transcriptase and cloned. Sequencing of the clones provided positive identification of those that contained the immunoglobulin genes. While this approach provided DNA copies of the RNA that is translated to form immunoglobulin, it did not directly indicate the status of the portion of the genome that encoded the RNA.

Southern transfer experiments done by Tonegawa and by Leder using fragments of cloned cDNA as a probe showed three facts about the genes involved with immunoglobulin synthesis. First, any single V region sequence possesses detectable but variable homology with up to about 30 different V region segments in the genome. The total number of V region coding segments in the genome can then be estimated from the numbers of such segments sharing homology to more than one V region probe. Results of such experiments indicate that the genome of the mouse contains from 100 to 1,000 kappa V region genes, with the most likely number being about 300. The second result from the Southern transfer experiments was that the genome contains much smaller numbers of the C region genes. The mouse genome contains four of the lambda class C region L genes and only one of the kappa class L region gene. The third result obtained from Southern transfers was that the genome in the vicinity of a V and C region gene undergoes a rearrangement during differentiation from a germ line cell to a lymphocyte. That is, the size of DNA fragment generated by digestion with a restriction enzyme containing homology to the V region differed between DNA extracted from embryos and DNA extracted from the myeloma source of the cDNA clone.

Answering deeper questions about the structure and rearrangements of the genes involved with antibody synthesis required additional cloning. Copies of undifferentiated V and C region genes could be obtained by cloning DNA isolated from mouse embryos. The differentiated copies could be obtained by cloning from myelomas. Although millions of clones must be screened to find the few containing the desired genes, such a search is straightforward because DNA fragments from the cDNA clones can be used as hybridization probes for identification of clones carrying the desired regions. Therefore, DNA extracted from mouse embryos and myelomas was cloned in lambda phage, and the plaques were screened for those containing DNA homologous to the V or C region from the cDNA clone. The results of these experiments in conjunction with the Southern transfer experiments showed that in the germ line cells the C and V region genes are situated far from one another. In a myeloma cell, however, a DNA rearrangement has placed the expressed V and C region genes beside one another. After the rearrangement, they are separated by

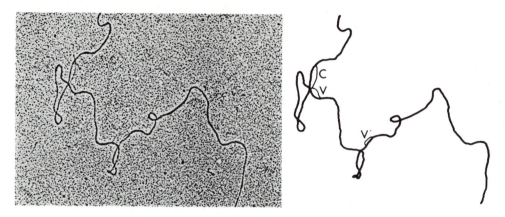

Figure 19.7   An R-loop between light chain messenger and myeloma cell DNA showing the intervening sequences between V' and V and between V and C. (From Two Immunoglobulin Genes Are Expressed in Myeloma S107, S. P. Kwan, E. Max, J. Seidman, P. Leder, M. Scharff, Cell *26*, 57–61 (1981), Copyright MIT.)

only about 1,000 base pairs. An R-loop between light chain mRNA and the rearranged DNA shows the V and C region genes separated by an intervening sequence plus an additional portion of V region gene called V' (Fig. 19.7). This codes for the leader sequence that is required for secretion of the polypeptide.

THE J REGIONS

The segments of embryonic DNA encoding the V regions of both the lambda and kappa light chains of mouse IgG do not code for about 13 amino acids normally considered to be part of the variable region of the protein. The sequence of the embryonic C region DNA does not solve the problem, for it contains just the amino acids that are constant from antibody to antibody. The missing amino acids have been found to be encoded by a segment of DNA lying from 1 to 3 kilobases (Kb) upstream from the 5' end of the C region gene. This region is called the J region because it is joined to the V region in the process of DNA rearrangement that connects a V and C region pair.

Of course, the most direct demonstration of the existence of a J region was provided by sequencing. But Tonegawa had first performed R-looping to locate the J region approximately. Light chain messenger from myelomas was hybridized to C region DNA cloned from embryos (Fig. 19.8). The R-looped structures of the lambda chain located a short region with the necessary coding properties about 1,000 to 3,000 nucleotides upstream from the C region. By contrast, kappa light chain mRNAs extracted from a series of myelomas indicated the existence of four different

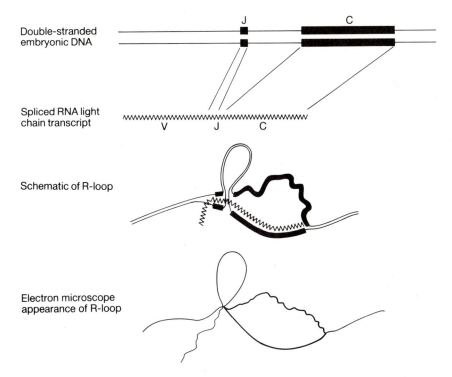

Figure 19.8   Genesis of an R-loop demonstrating the existence and location of a J region.

J regions ahead of the kappa C segment. Thus the presence of these J regions adds another means for generation of antibody diversity. Any of the several hundred V segments can be linked to the C segment by any one of the four J regions. Since the J regions of the light chains lie just beyond the third hypervariable region of the antibody, variability in this position has a direct effect on the structure of the antigen-binding site of the resultant antibody.

While nucleotide sequencing revealed a total of five kappa chain J regions, sequencing of myeloma proteins showed that one of the J regions is not functional.

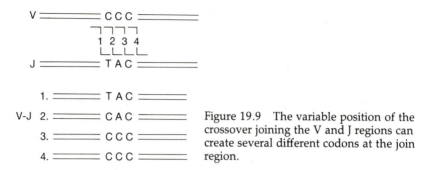

Figure 19.9   The variable position of the crossover joining the V and J regions can create several different codons at the join region.

This finding is in keeping with the absence of a GT sequence alongside the region in the position almost always occupied by this dinucleotide at the start of an intervening sequence.

Nucleotide sequencing by Leder demonstrated that additional diversity was also generated in connection with the J regions. The DNA crossover site that connects the V segment to one of the J regions is not fixed. Slight variability in its position creates codons not present on either the V or the J segment being joined (Fig. 19.9).

The J region and the surrounding sequences are special. Immediately upstream of the J region itself must lie signals that are used in the DNA splicing of a V region to the J region (Fig. 19.10). Then, immediately after the J region are the nucleotides that

Figure 19.10   The process of genome rearrangement and transcript splicing that finally yields a processed mRNA molecule capable of being translated into an intact immunoglobulin light chain.

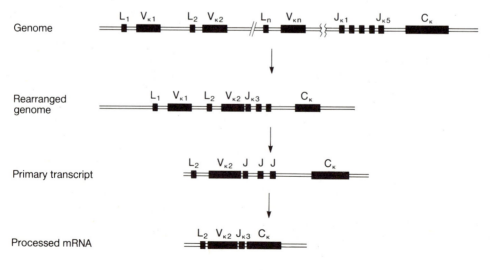

signal the mRNA processing machinery so that the intervening sequence of about 2,500 bases between the J region and the beginning of the C region can be removed and the messenger can be made continuous for translation. Later sections explain two additional functions of this region: activating transcription and serving as one end of deletion that changes the C region joined to the V-J region.

THE D REGIONS IN H CHAINS

The heavy chains of mice and humans have also been examined to determine whether they are spliced together from segments as are the light chains. Indeed, the heavy chains are. About 300 variable H chain segments exist as well as four J regions in mice and ten in humans, and a single C region of the G class. However, these segments do not encode all the amino acids found in the heavy chains. Thus it seemed likely that an additional segment of DNA is spliced in to generate the complete immunoglobulin heavy chain. These segments are called D, for diversity, because they lie in the third hypervariable region of the heavy chain.

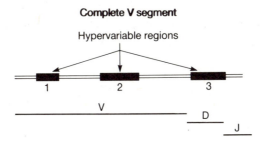

Almost the same strategy as was used to find the J regions worked in finding the missing segments of the heavy chains. Tonegawa had observed that the J regions in some myelomas were rearranged even though they were not connected to V segments. Most likely the rearrangements involved the D segments. This hypothesis proved to be true. By cloning a rearranged J region and using the D region it contained, the D regions were located on the embryonic DNA ahead of the J regions.

One of the interesting features of the D regions was that they possessed flanking sequences of seven and nine bases just like those found alongside the V and J regions of the light chain (Fig. 19.11). Similar sequences also exist behind the heavy chain V regions and in front of the heavy chain J regions. These flanking sequences signal the processing system and properly locate the splicing reactants.

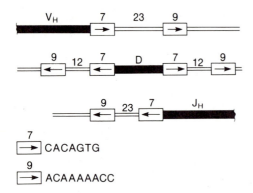

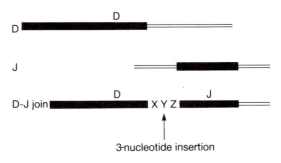

Figure 19.12   An insertion of three bases generated upon splicing a D and J segment.

Figure 19.11   The V, D, and J regions and the positions, sizes, and sequences of the flanking recognition sequences used in the processing to produce an intact immuno-globulin gene.

Not only does introduction of one of the ten or so different D regions into a V-D-J-C gene generate diversity, but imperfections in the splicing process itself involving the D regions generate additional diversity. That is, the slicing is not perfect. Often deletions or insertions of several bases are made as the D segment is spliced in (Fig. 19.12).

The variable region of the H chain that is generated by the fusions and rear-rangements of different DNA regions is joined to the C region. The C region of the heavy chain itself contains three intervening sequences. One lies between each

stretch of 100 amino acids. These groups of 100 amino acids fold together to form the domains of the protein that interact with the portions of the immune system that dispose of foreign molecules bound to antibody molecules. The fact that intervening sequences separate well-defined structural domains of a protein led to the prediction that most intervening sequences in other proteins would also be found to separate the nucleotides coding for functional and structural domains of the protein. Such a spacing increases the chances that sequences coding for intact domains could be shuffled between proteins in the course of evolution.

INDUCED MUTATIONS ADD TO
ANTIBODY DIVERSITY

We have seen that lymphocytes use reasonably small numbers of V, J, and D segments to generate a large number of different genes coding for antibody polypeptide chains. This, plus the fact that antigenic specificity is generated by the combination of two antibody chains, permits an animal to synthesize millions of different antibodies. As mentioned earlier, however, survivability of a species dictates that some antibody specificities be randomly determined. Furthermore, it is likely that within the population of animals of one species, the genes involved with antibody diversity should themselves be considerably more variable than other genes. Such variability at a genetic locus is known as a genetic polymorphism and indeed is observed in the antibody genes.

The evidence for randomness beyond that introduced at the ends of D segments and the variable crossover points in the splicing of J segments is direct. In one study, myelomas whose IgG products bound phosphocholine were examined. In contrast to most variable regions, the light chain variable region RNA from the myelomas was found to hybridize to only one genomic V region. Upon sequencing a number of these V regions cloned from mouse myelomas, a fraction was found to possess exactly the same V region sequence as the homologous V region cloned from embryonic mouse DNA. However, several of the cloned V regions differed from the embryonic sequence in one or more locations. Since all the V region genes originated from the same germ line V region and the corresponding sequence in nonlymphocyte cells is unaltered, the slight sequence differences represent mutations that were selectively introduced into the gene. Some of the differences encoded different amino acids, whereas others were neutral. Furthermore, the changes were localized to within about 2 Kb of the V region gene. These results mean that the mechanism responsible for generating the mutations makes mutations randomly but confines the changes to the immediate vicinity of the V region gene.

CLASS SWITCHING OF HEAVY CHAINS

As mentioned above, the first detectable antibody synthesized in response to an antigen is IgM. Later, the other classes, IgA, IgD, IgG, IgA, and IgE, can be detected. Remarkably, myelomas in culture also can switch the class of antibody they are synthesizing, but without changing the antigen specificity. Since the class of an antibody is determined solely by the constant region of the heavy chain, class switching must be generated by a change in the C segment that is connected to the V, D, and J segments.

Just as expected, sequencing has shown that after the J segment, the chromosome contains a series of C region segments beginning with the one coding for the M class of immunoglobulin. This array of eight C region segments is spread over 200 Kb of DNA and possesses an order consistent with the sequence of possible class

switching that has been observed. Class switching often replaces one of the down-stream C regions for the M constant region by means of a deletion in which inter-vening DNA is lost. Hence, the intervening sequence between the J and C_μ region contains signals that participate in class switching.

ENHANCERS AND EXPRESSION OF IMMUNOGLOBULIN GENES

Up to now in this chapter we have ignored the problem of regulation of expression of the immunoglobulin genes. Consider the 300 or so kappa-type V regions. The cell would waste considerable resources on the synthesis of messenger, and perhaps its translation, if each of these gene segments were transcribed. How then can the fusion of a V segment to a C segment activate transcription of the spliced gene and only the spliced gene? The explanation is provided by enhancer sequences. As mentioned earlier, these sequences, in conjunction with tissue-specific and develop-ment-specific proteins can stimulate the activity of promoters lying within one or two thousand bases in particular tissues at particular times (Fig. 19.13).

Figure 19.13   An enhancer in the region between the J and C regions stimulates a promoter lying in front of a V region segment after the V region has been joined to the C region.

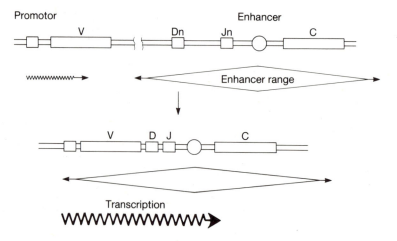

Enhancers have been found on most eukaryotic viruses, where they are essential for growth of the virus. Enhancers have also been found in the intervening sequence that lies between the J segment and the C segment of the heavy and light immunoglobulin genes. Each V region contains a promoter, which is only weakly active, without an enhancer nearby. Therefore, not only does gene splicing generate a substantial portion of the diversity in antibodies, but it also activates transcription of a spliced gene, a neat two-for-one maneuver.

PROBLEMS

19.1. Why are some antigens precipitated by IgG, but never by Fab' fragments?

19.2. With the introduction of as few additional postulates as possible, explain why the new antibodies elicited by repeated boostings of the same animal often possess higher and higher affinities for the antigen.

19.3. Estimate a theoretical maximum number of different kappa antibody specificities an animal could synthesize merely by splicing — that is, without mutation, deletion, or insertion — if the animal possessed 300 V_L, 10 J_L, 300 V_H, 10 D, and 10 C_H regions. Again in the absence of diversity generated by variable position of splicing, deletions or insertions, and introduction of mutations, why could this theoretical maximum not be achieved?

19.4. Invent data using DNA probes homologous to just 30 different V regions that would indicate that the total mouse genome contains 200 different V region segments.

19.5. How would you interpret data obtained from sequencing a large number of V regions of the L and H chains that showed that about 10 amino acid residues of each chain were absolutely invariant?

19.6. Sketch the structure of an R-loop formed by annealing IgG heavy chain mRNA and its genomic DNA.

19.7. Suppose you have a cDNA clone containing a V region and you wish to identify a clone containing its embryonic V region gene. Considering that a number of genomic V genes will possess appreciable homology to the V region on the cDNA clone, how could a clone containing the desired V region be identified without sequencing?

19.8. Devise a method using known enzyme activities by which lymphocytes could hypermutate DNA in the vicinity of the appropriate V region segment. The mutations must be confined to the V region area, and only the one V region that has been connected to the C region should be mutated.

19.9. The mutations in a V region gene seen confined to the one linked to the C region gene. Why is it logical that IgM not possess these mutations and that only the classes that are generated after IgM possess them?

19.10. Hypothesize a mechanism for splicing the V, J, D, and C regions together based on the sequences and spacings between the flanking conserved nucleotides of these regions.

19.11. What is the importance of knowing the orientation of the V segments on the chromosome relative to the D, J, and C segments?

19.12. No mention was made of the fact that mouse and human cells are diploid. What is a likely reason and explanation for the fact that lymphocytes synthesize only one antibody type despite the fact that with two chromosomes they could synthesize two or more types?

19.13. Sketch the heteroduplex that would be formed between the mRNA and the genomic DNA for the heavy chain.

19.14. How would you look for the DNA that encodes the carboxy terminal amino acids that anchor the membrane-bound forms of antibody?

19.15. Without sequencing or cloning, how could you test the idea that class switching results from RNA processing and not from DNA rearrangements?

19.16. About what fraction of the human or mouse genome would be required to code for 2×10^6 average-sized genes?

RECOMMENDED READINGS

Two mRNAs Can Be Produced from a Single Immunoglobulin mu Gene by Alternative RNA Processing Pathways, P. Early, J. Rogers, M. Davis, K. Calame, M. Bond, R. Wall, L. Hood, Cell 20, 313–319 (1980).

Somatic Mutation of Immunoglobulin Light-Chain Variable-Region Genes, E. Selsing, U. Storb, Cell 25, 47–58 (1981).

Identification of D Segments of Immunoglobulin Heavy Chain Genes and Their Rearrangement in T Lymphocytes, Y. Kurosawa, H. vonBoehmer, W. Haas, H. Sakano, A. Trauneker, S. Tonegawa, Nature 290, 565–570 (1981).

A Tissue-Specific Transcription Enhancer Element Is Located in the Major Intron of a Rearranged Immunoglobulin Heavy Chain Gene, S. Gillies, S. Morrison, V. Oi, S. Tonegawa, Cell 33, 717–728 (1983).

RELATED REVIEWS, BOOKS, AND ARTICLES

Three Dimensional Structure of Immunoglobulins, D. Davies, E. Padlan, D. Segal, Ann. Rev. Biochem. 44, 639–667 (1975).

Three-Dimensional Structure, Function and Genetic Control of Immunoglobulins, R. Poljak, Nature 256, 373–376 (1975).

Somatic Generation of Antibody Diversity, S. Tonegawa, Nature 302, 575–581 (1983).

Immunology: The Science of Self-Nonself Discrimination, J. Klein, Wiley & Sons (1984).

DEEPER READING

The Participation of A and B Polypeptide Chains in the Active Sites of Antibody Molecules, H. Metzger, L. Wofsy, S. Singer, Proc. Nat. Acad. Sci. USA *51*, 612–618 (1964).

The Molecular Basis of Antibody Formation: A Paradox, W. Dreyer, J. Bennett, Proc. Nat. Acad. Sci. USA *54*, 864–869 (1965).

An Analysis of the Sequences of the Variable Regions of Bence Jones Proteins and Myeloma Light Chains and Their Implications for Antibody Complementarity, T. Wu, E. Kabat, J. Exp. Med. *132*, 211–250 (1970).

The Three-Dimensional Structure of the Fab′ Fragment of a Human Myeloma Immunoglobulin at 2.0-Å Resolution, R. Poljak, L. Amzel, B. Chen, R. Phizackerley, F. Saul, Proc. Nat. Acad. Sci. USA *71*, 3440–3444 (1974).

Binding of 2,4 Dinitrophenyl Compounds and Other Small Molecules to a Crystalline Lambda-Type Bence-Jones Dimer, A. Edmundson, K. Ely, R. Girling, E. Abola, M. Schiffer, F. Westholm, M. Fausch, H. Deutsch, Biochem. *13*, 3816–3827 (1974).

Continuous Cultures of Fused Cells Secreting Antibody of Predefined Specificity, G. Kohler, C. Milstein, Nature *256*, 495–497 (1975).

A Complete Immunoglobulin Gene Is Created by Somatic Recombination, C. Brack, M. Hirama, R. Lenhard-Schuller, S. Tonegawa, Cell *15*, 1–14 (1978).

The Arrangement and Rearrangement of Antibody Genes, J. Seidman, P. Leder, Nature *276*, 790–795 (1978).

Sequences of Five Potential Recombination Sites Encoded Close to an Immunoglobulin kappa Constant Region Gene, E. Max, J. Seidman, P. Leder, Proc. Nat. Acad. Sci. USA *76*, 3450–3454 (1979).

Domains and the Hinge Region of an Immunoglobulin Heavy Chain Are Encoded in Separate DNA Segments, H. Sakano, J. Rogers, K. Huppi, C. Brack, A. Traunecker, R. Maki, R. Wall, S. Tonegawa, Nature *277*, 627–633 (1979).

Sequences at the Somatic Recombination Sites of Immunoglobulin Light-Chain Genes H. Sakano, K. Huppi, G. Heinrich, S. Tonegawa, Nature *280*, 288–294 (1979).

A Kappa-Immunoglobulin Gene Is Formed by Site-Specific Recombination Without Further Somatic Mutation, J. Seidman, E. Max, P. Leder, Nature *280*, 370–375 (1979).

Isolation of Variants of Mouse Myeloma X63 That Express Changed Immunoglobulin Class, A. Radbruch, B. Liesegang, K. Rajewsky, Proc. Nat. Acad. Sci. USA *77*, 2909–2913 (1980).

An Immunoglobulin Heavy Chain Variable Region Gene Is Generated from Three Segments of DNA: V_H, D, J_H, P. Early, H. Huang, M. Davis, K. Calame, L. Hood, Cell *19*, 981–992 (1980).

Amino Acid Sequence of Homogeneous Antibodies to Dextran and DNA Rearrangements in Heavy Chain V-Region Gene Segments, J. Schilling, B. Clevinger, J. Davie, L. Hood, Nature *283*, 35–40 (1980).

The Joining of V and J Gene Segments Creates Antibody Diversity, M. Weigert, R. Perry, D. Kelly, T. Hunkapiller, J. Schilling, L. Hood, Nature *283*, 497–499 (1980).

Complete Nucleotide Sequence of Immunoglobulin γ2b Chain Gene Cloned from Newborn Mouse DNA, Y. Yamawaki-Kataoka, T. Kataoka, N. Takahashi, M. Obata, T. Honjo, Nature *283*, 786–789 (1980).

A Monoclonal Antibody for Large-Scale Purification of Human Leukocyte Interferon, D. Secher, D. Burke, Nature *285*, 446–450 (1980).

Deletions in the Constant Region Locus Can Account for Switches in Immunoglobulin Heavy Chain Expression, S. Cory, J. Jackson, J. Adams, Nature *285*, 450–456 (1980).

Two Types of Somatic Recombination Are Necessary for the Generation of Complete Immunoglobulin Heavy-Chain Genes, H. Sakano, R. Maki, Y. Kurosawa, W. Roeder, S. Tonegawa, Nature *286*, 676–683 (1980).

DNA Sequences Mediating Class Switching in α-Immunoglobulins, M. Davis, S. Kim, L. Hood, Science *209*, 1360–1365 (1980).

Repetitive Sequences in Class-Switch Recombination Regions of Immunoglobulin Heavy Chain Genes, T. Kataoka, T. Miyata, T. Honjo, Cell *23*, 357–368 (1981).

A Single V_H Gene Segment Encodes the Immune Response to Phosphorylcholine: Somatic Mutation Is Correlated with the Class of the Antibody, S. Crews, J. Griffin, H. Huang, K. Calame, L. Hood, Cell *25*, 59–66 (1981).

Gene Segments Encoding Transmembrane Carboxyl Termini of Immunoglobulin γ Chains, J. Rogers, E. Choi, L. Souza, C. Carter, C. Word, M. Kuehl, D. Eisenberg, R. Wall, Cell *26*, 19–27 (1981).

Antibody Diversity: Somatic Hypermutation of Rearranged V_H Genes, S. Kim, M. Davis, E. Sinn, P. Patten, L. Hood, Cell *27*, 573–581 (1981).

Structure of the Human Immunoglobulin mu Locus: Characterization of Embryonic and Rearranged J and D Genes, J. Ravetch, U. Siebenlist, S. Korsmeyer, T. Waldmann, P. Leder, Cell *27*, 583–591 (1981).

Monoclonal Antibodies Distinguish Identifiable Neurones in the Leech, B. Zipser, R. McKay, Nature *289*, 549–554 (1981).

Identification and Nucleotide Sequence of a Diversity DNA Segment (D) of Immunoglobulin Heavy-Chain Genes, H. Sakano, Y. Kurosawa, M. Weigert, S. Tonegawa, Nature *290*, 562–565 (1981).

Organization, Structure, and Assembly of Immunoglobulin Heavy Chain Diversity DNA Segments, Y. Kurosawa, S. Tonegawa, J. Exp. Med. *155*, 201–218 (1982).

Clusters of Point Mutations Are Found Exclusively Around Rearranged Antibody Variable Genes, P. Gearhart, D. Bogenhagen, Proc. Nat. Acad. Sci. USA *80*, 3439–3443 (1983).

Structure of Genes for Membrane and Secreted Murine IgD Heavy Chains, H. Cheng, F. Blattner, L. Fitzmaurice, J. Mushinski, P. Tucker, Nature *296*, 410–415 (1982).

A Lymphocyte-Specific Cellular Enhancer Is Located Downstream of the Joining Region in Immunoglobulin Heavy Chain Genes, J. Banerji, L. Olson, W. Schaffner, Cell *33*, 729–740 (1983).

Somatic Mutations of Immunoglobulin Genes Are Restricted to the Rearranged V. Gene, J. Gorski, P. Rollini, B. Mach, Science *220*, 1179–1181 (1983).

A Human T Cell–Specific cDNA Clone Encodes a Protein Having Extensive Homology to Immunoglobulin Chains, Y. Yanagi, Y. Yoshikai, K. Leggett, S. Clark, I. Aleksander, T. Mak, Nature *308*, 145–149 (1984).

Isolation of cDNA Clones Encoding T Cell–Specific Membrane-Associated Proteins, S. Hedrick, D. Cohen, E. Nielsen, M. Davis, Nature *308*, 149–153 (1984).

Complete Primary Structure of a Heterodimeric T-Cell Receptor Deduced from cDNA Sequences, H. Saito, D. Kranz, Y. Takogaki, A. Hayday, H. Eisen, S. Togegawa, Nature *309*, 757–762 (1984).

Special Topics

Biological Assembly, Ribosomes, and Lambda Phage

We have already dealt with biological assembly in the folding of polypeptide chains. In that case, although we do not understand the process well enough to predict how a sequence of amino acids will fold, we have the basic principles well in hand. Polypeptide folding in the environment of the cell or in physiological buffers appears to be completely specified by the peptide itself. No external folding engine or scaffolding is required. Is the same true of much larger biological structures, or does the assembly of larger structures involve more than just the substructures?

Two fundamentally different types of larger structures are considered in this chapter: the structure and assembly of ribosomes and the structure and assembly of lambda phage. As we saw in Chapter 6, ribosomes of bacteria consist of two nonidentical subunits, each containing one or two RNA molecules and many different ribosomal proteins. The great majority of these proteins are present in each ribosome as a single copy. Thus, ribosomes are irregular and asymmetric. On the other hand, lambda phage, and most other viruses, possess a highly regular coat that covers DNA or RNA molecules. The coat of lambda consists of many copies of a few proteins that form an icosahedral head in addition to a number of other proteins that form a tail.

Both in the case of ribosomes and in the case of virus assembly, one important fundamental question is the order of assembly.

A.   Ribosome Assembly

RNase AND RIBOSOMES

Originally, much difficulty was encountered in the study of ribosomes and their constituent rRNAs from *E. coli* because of the activity of RNaseI, which contaminated most preparations of ribosomes. One might ask how the ribosomes could function *in vivo* without degradation if they are so quickly degraded *in vitro*. The answer is that RNaseI is located in the cell's periplasmic space. There it does no harm to the ribosomes until the cells are lysed and it is released, whereupon it adventitiously binds tightly to ribosomes and degrades their RNA, particularly at room temperature or above, but even at temperatures near 0°.

Molecular biologists used two sensible approaches to solve the RNase problem. The first was a classic case of applying genetics to solve a biochemical problem: Gesteland isolated an RNaseI⁻ mutant. This was not a trivial task because no genetic selection was apparent for permitting the growth of just RNaseI⁻ mutants, nor was any physiological trait likely to reveal the desired mutants. The only known characteristic of the desired mutants was their lack of RNaseI in the periplasmic space. The obvious solution to the problem of isolating the desired mutant under the circumstances, but apparently not one used before, was merely to use a brute-force approach and score several thousand candidate colonies from a heavily mutagenized culture for absence of the enzyme.

To minimize the work of scoring, the mutagen had to be highly effective. Fortunately, nitrosoguanidine can induce a number of mutations into each cell. As a result, any mutant lacking a nonessential gene activity can be found in a population of a few thousand candidates from nitrosoguanidine-mutagenized cultures. The work of doing conventional RNaseI assays on several thousand different cultures is great. Therefore Gesteland devised two simple scoring methods. In one, the whole cells from individual colonies grown from a mutagenized culture were resuspended at 42° in buffer containing radioactive ribosomal RNA and EDTA. The high temperature and the EDTA released the RNase from the cells without lysing them. Then, after an incubation in the presence of radioactive RNA, the undegraded RNA was precipitated by addition of acid, and its radioactivity was determined. Several hundred colonies per day could be assayed for the ability of their RNaseI to degrade the RNA. The second scoring method used a clever plate technique in which duplicate plates contained the colonies to be tested. One was overlayed with several milliliters of agar containing a high concentration of tRNA and EDTA and was incubated at 42°. During a few hours of incubation, those colonies containing RNaseI digested the tRNA to short oligonucleotides. The mutant colonies lacking

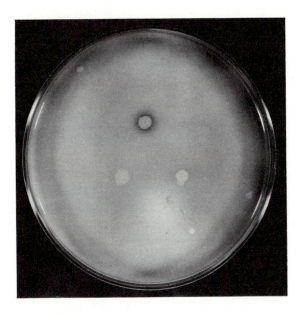

Figure 20.1   Photograph of a petri plate containing an RNaseI plus colony and an RNaseI⁻ colony with a dark halo. (Courtesy R. Gestland.)

RNaseI could not digest the tRNA in their immediate vicinity. Then the plate was flooded with concentrated HCl. The acid precipitated the tRNA, leaving the plate opaque except in the areas surrounding colonies containing RNaseI. The desired RNaseI⁻ colonies lacked cleared halos and could easily be detected (Fig. 20.1).

The second approach for elimination of RNaseI problems was biological. A number of different bacterial strains were examined for this enzyme. A satisfactory strain was found, MRE600, that lacked the enzyme, so it was used as a source of ribosomes in some structural and assembly studies.

THE GLOBAL STRUCTURE OF RIBOSOMES

The topic of *in vivo* assembly or *in vitro* reassembly cannot be studied without some knowledge of the structure being assembled. Learning the structure of ribosomes has been a tantalizing problem. Most of the interesting details of ribosomes are too small to be seen by electron microscopy, but ribosomes are too large for application of the conventional physical techniques such as nuclear magnetic resonance or other spectroscopic methods. Ribosomes are also too large for their structures to be solved in detail by X-ray crystallography.

One of the first questions to answer about ribosome structure is the folding of rRNA. It could be locally folded, which means that all nucleotides near each other in the primary sequence are near each other in the tertiary structure (Fig. 20.2). Recall

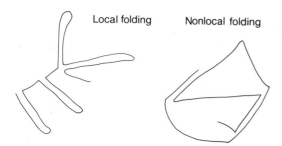

Local folding   Nonlocal folding

Figure 20.2   Examples of local and nonlocal folding.

that many proteins, IgG for instance, are locally folded and contain localized domains, each consisting of a group of amino acids contiguous in the primary sequence. Ribosomal RNA possesses a sequence that permits extensive folding, most
of which is local but a part of which is nonlocal. Many stretches of the RNA can form
double-helical hairpins, but other portions form large loops separating complementary sequences (Fig. 20.3).

Several lines of evidence support the secondary structure, which is predicted
primarily on the basis of sequence. One of the most compelling is interspecies
sequence comparisons. Frequently, a stretch A_1, which is postulated to base pair
with stretch B_1, is altered between species. In such a case, the sequences of A_2 and B_2
in the second species are both found to be altered so as to preserve the complemen-

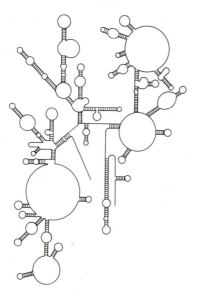

Figure 20.3   Secondary folding of 16S
rRNA based on sequence, RNase susceptibility, and interspecies comparisons.

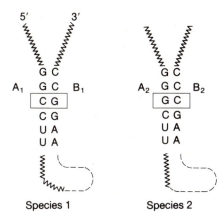

Figure 20.4 How base differences in two regions of RNA strengthen proposals for base pairing between the regions involved.

tarity (Fig. 20.4). Additional evidence supporting the secondary structure predictions is the locations of sites susceptible to cleavage by RNases or to crosslinking. The cleavage sites of RNaseA lie in loops or unpaired regions. Crosslinking portions of the rRNA with psoralen joins regions postulated to be paired and also connects nucleotides that are in close proximity because of the tertiary structure of the ribosome.

The electron microscope has permitted determination of gross details of the ribosome. Not only can a shape be discerned, albeit barely, but within this shape specific ribosomal proteins can be located. Finding the placement of ribosomal proteins on the surface of the ribosome utilizes antibodies against individual proteins (Fig. 20.5). IgG is bivalent, so it will join two ribosomal subunits. Owing to the asymmetry of the subunits, the position at which the two are linked by the antibody can often be ascertained. Not only can many ribosomal proteins be located (Fig. 20.6), but the exit site of nascent polypeptide chains has been found through the use of antibodies to a nonribosomal protein whose synthesis can be highly induced.

ASSEMBLY OF RIBOSOMES

How are ribosomes formed? We would like to have an answer to this question concerning *in vivo* assembly because we want to know how things work. We also would like to be able to carry out *in vitro* assembly so that sophisticated experiments on the structure and function of ribosomes could be done.

Study of the *in vivo* assembly of ribosomes is difficult because the experimenter can do so little to alter the system. One or two ribosomal precursors can be distinguished from mature ribosomes on the basis of their sedimentation velocities. They contain only a subset of the complement or ribosomal proteins. Additionally, since

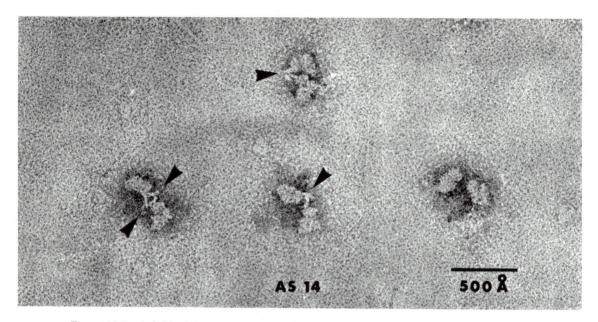

Figure 20.5   A field of 30S ribosomal subunits reacted with antibodies against protein S14. The IgG antibodies are indicated by arrows. Protein S14 is exposed, so it is possible for two IgG molecules to bind to the ribosomal subunit simultaneously. (Reprinted with permission of the authors from Lake, Pendergast, Kahan, and Nomura, Proc. Nat. Acad. Sci. USA 71, 4688–4692 [1974].)

rRNA is methylated in a number of positions, the degree of methylation of RNA extracted from the precursors can be measured. Not surprisingly, such studies have not greatly illuminated the subject of ribosome assembly.

Nomura has made the most significant contributions to our understanding of *in vitro* ribosome assembly. As a first step in the attack on this difficult problem, he removed a few proteins from ribosomes and learned how to replace them and regain the ribosome's ability to synthesize protein. The proteins of the smaller ribosomal subunit were separated into two classes: those few that split off the 30S subunit when it is centrifuged to equilibrium in CsCl and the remaining cores that contain the 16S rRNA and the majority of the smaller subunit's proteins. The assay of reassembly was by sedimentation velocity of the products or their ability to function in an *in vitro* translation assay.

The next step in the study of ribosome assembly was the total reconstitution from isolated rRNA and ribosomal proteins. The RNA was purified by phenol extraction to remove the proteins and then extensively dialyzed to remove the phenol. The ribosomal proteins were obtained by extracting purified 30S subunits with urea and lithium chloride. This harsh treatment denatures the proteins but

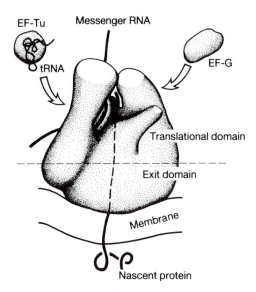

EF-Tu   Messenger RNA

tRNA

EF-G

Translational domain

Exit domain

Membrane

Nascent protein

Figure 20.6   Diagrammatic representation of the exit and translational domains of the ribosome and their orientations with respect to the membrane binding site. The nascent protein is shown as an unfolded, extended chain during its passage through the ribosome. (Adapted with permission of the authors from Bernabeu and Lake, Proc. Nat. Acad. Sci. USA *79*, 3111–3115 [1982].)

solves one of the harder technical problems in the study of ribosomal proteins, that of their insolubility. Much time was invested by many investigators to find ways of solubilizing these proteins. Why the proteins should be poorly soluble under some conditions is not at all clear; perhaps they cannot achieve their correct conformation in the absence of the correct rRNA binding site and, as a result, the proteins easily aggregate via hydrophobic patches on their surfaces.

To reconstitute ribosomes, a solution containing ribosomal proteins is slowly added to a stirred buffer containing the rRNA. In the solution it often is a race between reconstitution and aggregation. However, a substantial number of intact 30S functional ribosomal subunits are formed. Once functional 30S subunits could be formed, optimal reconstitution conditions could be found. Not surprisingly, they were similar to *in vivo* conditions. They are about 0.3 M KCl, pH 6 to 8, and Mg^{++} greater than 10^{-3} M. Soon it became possible to reconstitute ribosomes from purified 16S rRNA and the individually purified ribosomal proteins. These results definitively proved that ribosome assembly requires no additional scaffolding or assembly proteins that are not found in the mature particle. The remarkable self-assembly process can be likened to making watches by shaking their parts together in a paper bag.

One major reason Nomura succeeded where many others failed was that he tried to reconstitute the ribosomes in buffers closely resembling conditions found *in vivo* and at temperatures at which cells grow well, 37°. Most other workers had attempted the reconstitution at temperatures near freezing as a result of experiences with ribosomes contaminated with RNaseI or proteins contaminated with proteases.

EXPERIMENTS WITH *in vitro* RIBOSOME ASSEMBLY

More sophisticated experiments on ribosome structure and assembly became possible with the ability to isolate individual ribosomal proteins and to reconstitute ribosomes from them. For example, the concentration of each protein could be varied in the reconstitution experiments. This allowed determination of the kinetic order of the reaction, that is, the number of components that had to interact simultaneously for the reaction to proceed. It is certainly to be expected that the assembly process would be sequential, with proteins being added one after another; the alternative case — all 21 of the 30S ribosomal subunit proteins and the rRNA coming together simultaneously in a reaction involving 22 components — would have far too low an assembly rate. Indeed, the probability that even a few of the components come together at once is low. Quite surprisingly, however, the most rate-limiting step in ribosome assembly is not the coming together of two components, it is an intramolecular rearrangement of a structure already formed (Fig. 20.7). That is, the rate-limiting reaction is unimolecular, and its rate relative to the amount of components present cannot be speeded by increasing the concentration of any components. Such a unimolecular step is analogous to the isomerization of RNA polymerase to form an active complex at a promoter.

The rate of *in vitro* ribosome assembly is strongly temperature-dependent. From this dependence, the activation energy can be calculated to have the unusually high value of 38 Kcal/mole. This high value suggested that *in vivo* ribosome formation also requires high energy. If so, the activation energy might be increased in some ribosome mutants. This has been found. A sizable fraction of cold- sensitive mutants in *E. coli* are defective in ribosome assembly. At low temperatures, such mutants accumulate precursors of ribosomes that contain some, but not all, of the ribosomal proteins. Raising the temperature allows these precursors to mature.

Reconstitution of ribosomal subunits from cold-sensitive mutants permitted pinpointing of the target of the mutations. They were in the ribosomal proteins. A similar approach can also be used to locate the altered component in other types of ribosome mutations. For example, the protein that is altered in streptomycin-resistant mutants is a protein designated S21. The ribosomal proteins are designated by S for proteins of the smaller, 30S subunit and L for proteins of the larger, 50S subunit. They are numbered in the order of their locations on a two-dimensional gel in which they all are separated from one another (Fig. 20.8).

Figure 20.7   Assembly of 30S subunits. A subset of 30S proteins and 16S rRNA form an RI complex that then must be activated by heat to form RI*, to which the remainder of the ribosomal proteins bind to form the complete 30S subunit.

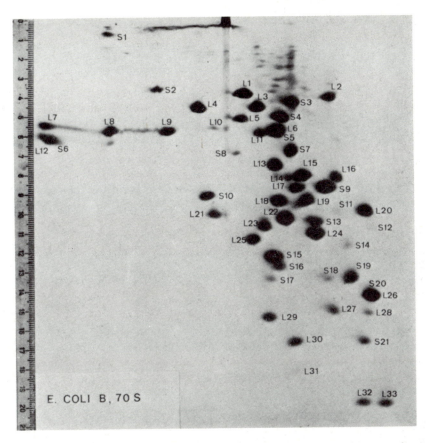

Figure 20.8   A photograph of the proteins from 70S subunits separated on two-dimensional gel electrophoresis. Proteins are numbered left to right in rows as a page is read. (From Ribosomal Proteins XII, E. Kaltschmidt, H. G. Wittmann, Proc. Nat. Acad. Sci. USA *67*, 1276–1282 [1970].)

The sequence in which individual ribosomal proteins bind to the maturing ribosome can also be determined by *in vitro* assembly experiments. The binding of the isolated radioactive ribosomal proteins to rRNA can be measured by sedimentation of the RNA after incubation. Only four of the 30S proteins are capable of binding to the naked 16S rRNA. The complexes thus formed can then be used to determine which proteins can bind next. Building in this fashion, a complete assembly map of the ribosomal subunits has been constructed (Fig. 20.9). In general it shows that any of several proteins can bind to the structure at any stage in its assembly. Also, many of the proteins that interact strongly during assembly are encoded by genes in the same ribosomal protein operon (Fig. 20.10).

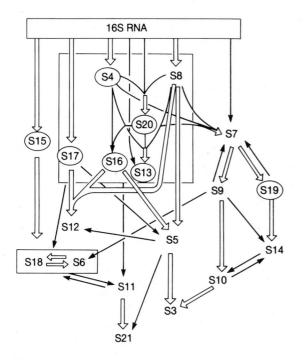

Figure 20.9   Assembly map of the 30S ribosomal subunit. (From Assembly Mapping of 30S Ribosomal Proteins from *E. coli*, W. Held, B. Ballow, S. Nizushima, M. Nomura, J. Biol. Chem. *249*, 3103–3111 [1974].)

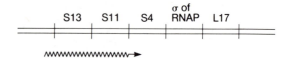

Figure 20.10   Gene structure of the ribosomal protein operon containing S13, S11, and S4. This operon also contains the gene for the sigma subunit of RNA polymerase and the ribosomal protein L17.

DETERMINING DETAILS OF LOCAL RIBOSOMAL STRUCTURE

Consider the fundamental question of determining which proteins are close neighbors in the ribosome. One direct approach to this question is to crosslink two proteins on the intact ribosome with bifunctional crosslinking reagents. If two ribosomal proteins are connected by the reagent when they are in a ribosome, but not when they are free in solution, it can be concluded that the proteins are near one another in the ribosome.

Many of the proteins that are crosslinked to each other in this way are proteins that depend on one another during assembly of the ribosomal subunit. Some of these proteins are encoded in the same operons; proteins that are adjacent to one another in the ribosome tend to derive from adjacent genes in the chromosome. For example, ribosomal proteins S4, S11, and S13 lie in the same operon, S13-S11-S4. S4 and S13 and S13 and S11 crosslink, S4 and S13 interact during assembly, and together they interact with S11 during assembly.

The ability to reassemble ribosomes from their isolated components greatly facilitates structural studies. A ribosome can be partially assembled, for example, and then antibody against a component in the immature ribosome can be added. If the presence of the antibody blocks the subsequent association of a ribosomal protein added later, it is reasonable to expect that the antibody directly blocks access of the protein to its site.

If all ribosomal proteins were spherical, their complete spatial arrangement would be determined by knowing the distances between the centers of proteins. Some of the requisite measurements can be made with fluorescence techniques or slow neutron scattering. Fluorescent molecules possess an absorption spectrum such that illumination by photons within this wavelength band excites the molecule, which then emits a photon of longer wavelength within what is called the emission spectrum of the molecule (Fig. 20.11). This property can be used to measure distances separating two ribosomal proteins if one protein is labeled with a fluorescent molecule possessing an emission spectrum overlapping the excitation spectrum of a second fluorescent molecule located on the second ribosomal protein.

In vitro assembly of ribosomes can be used to construct a ribosome in which two of the proteins contain the fluorescent probes. By illuminating the rebuilt ribosomes with light in the excitation spectrum of the first molecule and measuring the strength of the fluorescence in the wavelength of the emission spectrum of the second molecule, the distance between the two fluorescent molecules can be determined. The amount of light in the second emission spectrum varies as the sixth power of the distance separating the molecules:

$$E = \frac{R_o^6}{R_o^6 + R^6},$$

Figure 20.11   Spectra used in measuring distances separating ribosomal proteins. Dotted line is the excitation and emission spectrum of fluorescent molecule 1 and the solid line is the excitation and emission spectra for molecule 2.

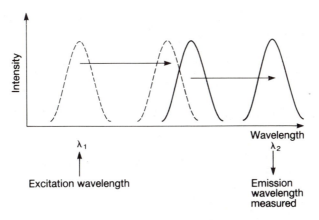

where R is the distance between the fluorescent molecules and R_0 is a constant that depends on the orientations of the molecules, the spectral overlap of the fluorescent emission and excitation spectra, and the index of refraction of the medium separating the molecules. The method yields the most reliable data for proteins separated by 25 to 75 Å; that is, the method is best at determining the distances of nearest neighbors in the ribosome.

Neutron diffraction is another method of measuring distances between ribosomal proteins. It, too, relies on reassembly of ribosomal subunits. Two proteins in the ribosome are replaced by their deuterated equivalents. These proteins are obtained from cells grown on deuterated medium. Since the neutron scattering properties of hydrogen and deuterium are different, an interference pattern is generated by the presence in the ribosome of the two proteins with different scattering properties. The angular separation in the peaks of the interference pattern can be related to the distance separating the two altered proteins in the reconstituted ribosome. Overall, the results of crosslinking, assembly cooperativity, immune microscopy, fluorescence transfer, and neutron scattering give a consistent picture for the locations within the ribosomal subunits of the ribosomal proteins.

B.   Lambda Phage Assembly

GENERAL ASPECTS

The assembly of lambda, and most other virus particles as well, can be broken down into three basic problems: How is the coat assembled? How is the nucleic acid gotten inside the coat? How is the nucleic acid released from the virus particle into the appropriate cell? At present, partial answers are known to these questions, but it is not yet possible to design parts of a virus such that, when present in proper concentrations in an environment similar to that found in a cell, they will properly assemble and then, when appropriately triggered, release their DNA.

Not only is the structure of lambda phage dramatically different from that of ribosomes, but its assembly process also is notably different. First, as in the case of many larger viruses, one virus-encoded protein is used during assembly but is not present in the final particle. Second, a host protein participates enzymatically in the assembly process. Third, a number of proteins are cleaved or covalently joined during assembly.

The most interesting problem involved in phage maturation is how the DNA is

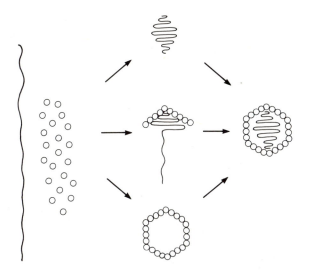

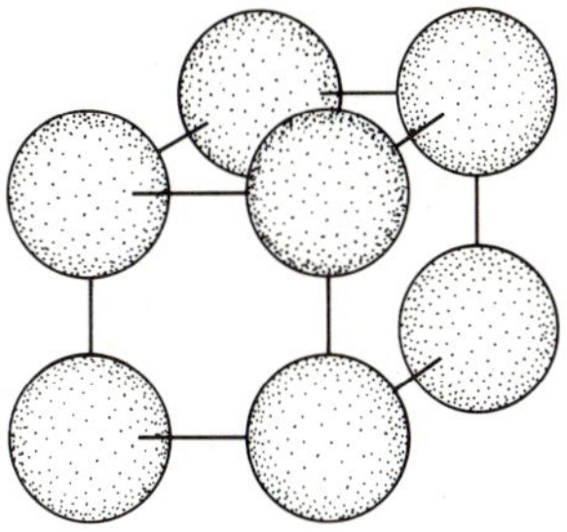

Figure 20.13   A cube with subunits at each vertex.

Figure 20.12   Three schemes for maturation of head subunits and the lambda DNA molecule into an assembled head with the DNA inside.

packaged in the head. There are three possible models for maturation of the DNA into the head (Fig. 20.12). The first is condensation of the DNA followed by maturation of the protein capsid around it. The second is a concerted condensation of DNA and capsid components, and the third is insertion of the DNA into a preformed capsid. Of these three possibilities, phage lambda matures by the third.

THE GEOMETRY OF CAPSIDS

Previously, we saw that lambda's genes are crowded together and even overlapping in places. Other bacterial virus genomes are similarly squeezed, and it is likely that survival of many types of viruses depends on their packing as much information as possible into as short a genome as possible. This being the case, it is likely that as few genes as possible will be used to code for DNA encapsidation. Consequently, it is reasonable to expect that, primarily, the virus coat will be constructed from multiple copies of one or two virus-encoded proteins. What general principles might then apply in virus assembly?

A virus coat could be constructed like a regular polyhedron. For example, eight identical subunits could form a cube with one subunit at each corner (Fig. 20.13). If four subunits occupied each face, then the coat would consist of 24 subunits, and a larger volume would be enclosed. Note that each of the 24 subunits would make

Figure 20.14   A regular icosahedron.

identical contacts with its neighbors. The maximum volume that could be enclosed by a set of subunits, each making identical contact with its neighbors, is based on the 20-sided icosahedron (Fig. 20.14). Each face could consist of three subunits. Thus the maximum number of subunits that can be utilized to enclose a volume in which each subunit is exactly equivalent to any other is 60. Experimentally, however, many viruses, lambda included, are found to be approximately icosahedral but to possess more than 60 subunits in the coat.

Caspar and Klug have investigated the structures that can be constructed when the constraint that each subunit is equivalent to any other is weakened to permit them to be only quasiequivalent. That is, each of the subunits will have nearly the same shape and will make nearly the same contacts with its neighbors. Using the same contact points over and over necessitates that the final structure be symmetrical. Maximizing the similarity of contacts with neighbors necessitates that the structure be a variation on an icosahedron.

The restriction to icosahedral symmetry can be most easily understood by considering why some other symmetry is less favored. For example, why is a large cubic structure unlikely? A face of such a cube could consist of a minimum of nine subunits making identical contacts with one another. Suppose, in fact, that a large planar network of such subunits is to be converted to a cube. This can be done by converting a point of fourfold symmetry to a point of threefold symmetry by removing a semi-infinite quadrant of subunits (Fig. 20.15). This generates the necessary vertex of the cube, but note that the subunits engage in two types of interaction — four together and three together. The differences between the two types of interactions necessitate that the individual subunits be distorted.

A similar analysis can be performed if a plane is covered with equilateral triangles and some points of sixfold symmetry are converted to points of fivefold symmetry. Of course, if the fivefold symmetric vertices are adjacent to one another, the regular icosahedron is generated. However, if the fivefold vertices are not adjacent so as to enclose a larger volume, then a structure is generated in which subunits engage in two types of interactions. Some subunits are adjacent to vertices with

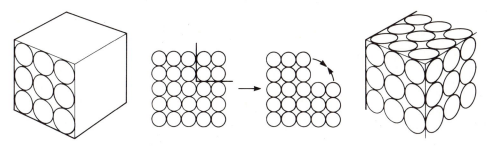

Figure 20.15   A cube with nine subunits on each face. Conversion of a planar net of such subunits to a cube by removing a sector of the plane and bringing the edges together.

sixfold symmetry, and some are adjacent to vertices with fivefold symmetry. The difference between these two types of interactions introduces the smallest strain possible and therefore it is likely that nature often will choose to construct virus coats on this principle. Often the units of five or six subunits can be isolated intact or observed as a unit in the electron microscope. These polymers are called pentons and hexons.

One way to view some of the higher-order structures is that they subdivide the equilateral triangles of the regular icosahedron. The triangles can be subdivided into 4, 9, 16, 25, and so on, subtriangles. However, other icosahedra are also possible. Consider a plane net of equilateral triangles and unit vectors **a** and **b** (Fig. 20.16). Satisfactory icosahedra can be constructed if one of the fivefold vertices is placed at an origin and the second at position $(n\mathbf{a}, m\mathbf{b})$, where n and m are integers. If we define $T = (n^2 + nm + m^2)$, then the regular icosahedron has $T = 1$ since $n = 1$ and $m = 0$. The number of subunits, S, in all the icosahedra is $S = 60T$. Although lambda phage's coat is based on a $T = 7$ icosahedron, it intersperses two types of coat subunits. In addition, smaller numbers of other proteins are found in the head. Some of these form the connection between the head and the tail.

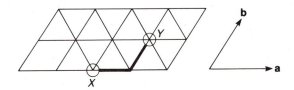

$Y = (2\mathbf{a}, 1\mathbf{b})$ $T = 7$ $S = 420$

Figure 20.16   Coordinates for locating vertices for icosahedra constructed with quasiequivalent subunits. X is one vertex and Y is a second vertex located 2 units in the **a** direction and 1 unit in the **b** direction. The structure that would be formed possesses a T number of 7 and 420 subunits.

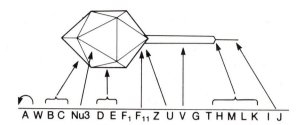

Figure 20.17   The genetic map of some genes involved with forming the head and tail of lambda and the locations of the gene products in the physical structure of the head and tail.

THE STRUCTURE OF THE LAMBDA PARTICLE

The physical structure of lambda phage and the organization of the genetic map of lambda show the same tendency as the physical structure of the ribosome and the organization of the genetic map of ribosomal proteins. The genes for proteins that lie near one another in the particles often lie near one another in the genetic map (Fig. 20.17). Remarkably, however, the structural genes of lambda phage are translated at vastly different efficiencies so that synthesis from the different morphogenic genes closely parallels the numbers of the different types of protein molecules that are found in the mature phage particle.

The A protein, often denoted pA, is responsible for producing the cohesive ends of the DNA. Proteins pD and pE are the major capsid proteins. Protein pE plus Nu3 followed by pB and pC is able to form a precursor of the head to which pD is added later. In this structure pB is cleaved, and pC is fused to pE. In the final structure, protein pD is interspersed with equal numbers of protein pE, and it should be noted that the genes for these proteins lie next to each other. Proteins D and E are the two major capsid proteins, and since E has a molecular weight of 38,000 and D has a molecular weight of 12,000, molecules of D probably help fill the gaps left between molecules of E. Protein F_{II} as well as protein W is involved with steps preparing the head assembly for receipt of the tail. F_{II} provides a specificity for the type of tail that is attached to the head. Protein V is the major structural protein of the tail. Proteins pH, pM, pL, and pK are all involved in the transition from the major tail to the tail fiber at the end, protein J.

THE HEAD ASSEMBLY SEQUENCE

Three discernible head precursors of lambda can be isolated by centrifugation and can thereby be studied. The first can be called preprohead I and consists of pE and pNu3 (Fig. 20.18). Without pE, nothing forms, and without pNu3, only large amorphous shapes appear. Hence pNu3 is a shape-determining protein. Next, proteins pC and then pB are added, and the resulting precursor is called prohead I. Upon the action of the bacterial protein, the *groE* gene product, pNu3 leaves the structure, pB is

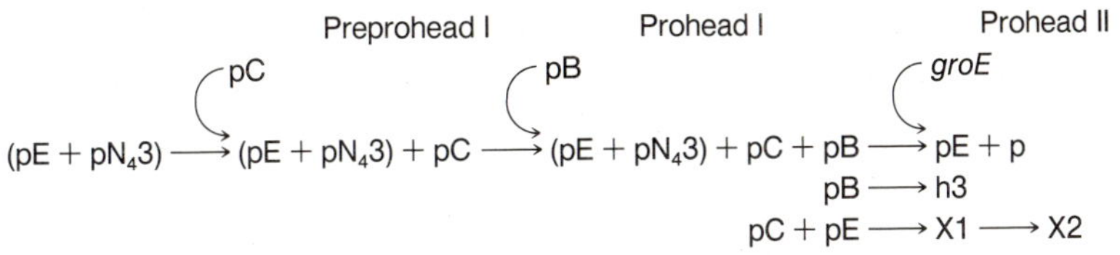

Preprohead I   Prohead I   Prohead II

$$\begin{array}{ccccc}
& \overset{pC}{\curvearrowright} & & \overset{pB}{\curvearrowright} & & \overset{groE}{\curvearrowright} & \\
(pE + pN_43) & \longrightarrow & (pE + pN_43) + pC & \longrightarrow & (pE + pN_43) + pC + pB & \longrightarrow & pE + p \\
& & & & pB & \longrightarrow & h3 \\
& & & & pC + pE & \longrightarrow & X1 \longrightarrow X2
\end{array}$$

Figure 20.18   Maturation scheme of lambda heads.

cleaved, resulting in a new protein called h3, and pC and a few of the copies of pE are fused to form X1, some of which is cleaved to form X2. These reactions leave a structure called prohead II. The final head is formed by the addition of lambda DNA, its cleavage to unit-length molecules containing the 12-base single-stranded *cos* ends, and the addition of pD on an equimolar basis to pE. Proteins pW and pF_{II} prepare the head for attachment of the tail. One of the functions of such a protein is to act as an adapter between the hexagonal tail and the pentagonal vertex to which the tail is attached.

The origins of h3, X1, and X2 were determined by making precursor heads very radioactive and purifying pB, h3, X1, and X2. The purification was simply done by electrophoresis on SDS gels. The proteins were eluted from the gels, digested with trypsin, and subjected to electrophoresis in one dimension and chromotography in the second dimension. This two-dimensional separation of the tryptic digest yields approximately 15 radioactive peptides from each protein. The tryptic fingerprints were identical for pB and h3 proteins except for two peptides missing from h3 but present in pB. Thus the h3 protein is a truncated version of the pB protein. Similarly, pC and pE were found to be covalently joined in the phage coat to yield X1, some of which is cleaved to yield X2.

The protein Nu3 occupies a role predicted for many other proteins but thus far found rather infrequently. It appears to be a structural protein and is used during the assembly of the phage particle but is not present in the assembled phage particle. It is synthesized in high quantities and forms a part of the scaffolding during formation of the particle. It is not cleaved proteolytically nor is it fused to any other protein, and it is continuously reused for the maturation of particles.

The protein product of the host cell *groE* gene has been conjectured to be a membrane protein that is involved in the synthesis of cell wall or cell membrane. Mutants with altered *groE* protein frequently fail to support the growth of not only phage lambda but also related phage and even phage T4. In addition to failing to cleave pB, they fail to package DNA. However, mutants of lambda can be isolated that overcome the *groE* defect. These mutants are usually found to be altered in the lambda E gene. Some of the rarer *groE* mutations can be overcome by compensating mutations in the lambda B gene. These results strongly suggest that *groE* protein is responsible for the cleavage and actually touches the E and B proteins during the maturation of the phage.

The *groE* mutation joins the list of other cellular mutations that may block maturation of the phage. In addition to *groE,* there also are the *groP* mutations in the host cell, which block DNA replication of the phage but which can be overcome by mutation in the lambda P gene. The *groP* mutations lie within a subunit of DNA polymerase. Some *groN* mutations lie in the RNA polymerase and make the phage appear N^- unless they are overcome by a compensating mutation in the phage N gene.

PACKAGING THE DNA AND FORMATION OF THE *cos* ENDS

Up to this point, we have not considered the system that converts circular lambda DNA, which is found in cells, into the sticky-ended linear molecules that are found in phage particles. The *ter*mini-producing protein, the A gene product, generates the required staggered nicks while packaging the DNA. The generation of these nicks can easily be assayed by using a double lysogen in which both prophage are Int⁻ or Xis⁻. Normal phage excision is not possible from such a double lysogen, but the *ter* system can clip an intact lambda genome out of the middle of the two excision-defective lambda genomes (Fig. 20.19).

In experiments designed to study effects of duplication of portions on lambda DNA, a gratuitous duplication of the cohesive end of phage lambda occurred. A study of this mutant has yielded appreciable insight into the packaging of phage DNA. The starting phage for generation of the duplications was lambda deleted of a sizable fraction of the *b2* region of the phage, *b221,* and containing an amber mutation in the *red* gene. These were grown, and selection was made for phage containing duplications by selecting denser phage after separation according to density on equilibrium centrifugation in CsCl. As expected, most of the resulting dense phage contained duplications as shown by the following criteria: Heteroduplexes between strands from these phage and strands from wild-type phage contained a bubble; the phage were denser than the parental type; and, upon growth in Rec⁺ cells or in Su⁺ cells to suppress the *red* mutation, unequal crossing over around the duplication yielded both triplications and phage lacking the duplication altogether.

If the duplication phage were grown on cells unable to recombine and unable to

Figure 20.19   Cutting a mature full-length lambda out of a tandem double lysogen in which each individual phage is excision-defective.

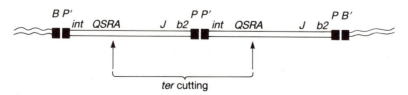

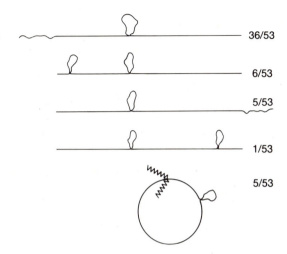

36/53

6/53

5/53

1/53

5/53

Figure 20.20 The structures of heteroduplexes formed between a lambda containing a duplication on the left arm and a lambda containing a duplication of the *cos* site.

suppress the phage *red* mutation, no change occurred in most of the duplication phage. However, one duplication phage was unstable under these conditions. At an appreciable frequency it segregated phage lacking the duplication. Upon heteroduplex formation, an even more startling result was found. The duplication could appear at either end of the phage! To distinguish the ends in these experiments, the heteroduplexes were formed between this strange phage and another phage containing a duplication in the left arm. In all cases, one duplication bubble was formed in the left half of the molecule and an additional bubble, that form the duplication in the strange phage, was found near the right or left end.

A number of types of heteroduplexes were found (Fig. 20.20). Two important conclusions can be drawn from these experiments: the phage contains a duplicated *cos* site, but both *cos* sites need not be cleaved for the DNA to be packaged, and packaging of the DNA proceeds from left to right. The reasoning for this is left to a problem at the end of the chapter.

42/48

6/48

The finding of a polarized left-to-right packaging of the lambda DNA is consistent with the *in vitro* packaging experiments investigating the nature of the DNA required for packaging. These *in vitro* packaging experiments found that a molecule containing only a single genome's equivalent of lambda DNA, a monomer, was not capable of being packaged. Only polymers of the lambda genome were capable of being packaged. *In vivo*, the necessary polymers would derive from the rolling circle mode of DNA replication or from recombination between circular molecules. The obvious experiments were designed to discover the minimum DNA capable of being packaged. The minimum is a lambda monomer containing a sticky left end and a right *cos* end covalently joined to a left *cos* end. These results all strongly suggest that the left end of lambda is packaged first and the right end enters last. Since the tail is put on after packaging of the DNA, it is natural to expect that the right end of lambda would be contained either within the tail or just at the union of the tail to the head; indeed, when tailless lambda phage are isolated that contain lambda DNA and are lightly treated with DNase, it is the right end of lambda that is attacked by the DNase. All of this strongly suggests that upon infection, the right end of lambda should be injected first into the cells.

FORMATION OF THE TAIL

The tail of lambda is formed independently of the head and is then attached to the head. The most striking property of tail maturation is a mechanism that grows the tail to a precise length. First, proteins G, H, I, J, K, L, and M interact to form an initiator of polymerization of the major tail protein, V. The V protein polymerizes on this initiator to form the tail tube. Finally, the terminator of polymerization, U, is added then functions. After this, Z protein functions, and the tail is attached to the head.

If mutations in the tail genes are present, abnormal tails may be formed. For example, if the terminator of polymerization, pU, is absent, then normal tails are formed, but the Z protein cannot function and the phage is largely inviable. Upon prolonged incubation of U⁻ extracts, *in vivo* or in an *in vitro* reconstitution system, the tails will extend beyond their normal length to form what is called a polytail. This structure is sufficiently abnormal that the Z protein can function on it, and the polytail is attached to the head to yield a phage particle, but one of very low infectivity. Of course, if the V protein is absent, no tail polymerization occurs. Similarly, mutations in any of the genes G through M block initiator formation and therefore no tail is formed. The H protein of the tail undergoes proteolytic cleavage during maturation to yield a protein called H_{II}. Apparently the H protein acts much like a tape measure in determining the length of the tail. Internal deletions in H yield phage with shorter tails.

In vitro PACKAGING

In vitro packaging of lambda DNA is possible and, indeed, is widely used to package restructured DNA in genetic engineering operations. Such packaging is done by preparing and mixing two highly concentrated extracts of cells, each infected with lambda mutants incapable of forming complete lambda particles or synthesizing DNA. Together these extracts contain all the required proteins for packaging. *In vitro* packing proceeds well if one of the extracts contains prohead II as a result of mutations in genes D or F. If such an extract is mixed with an extract prepared from bacteria growing lambda defective in the E gene and lambda DNA is added, mature lambda are formed. This *in vitro* packaging requires ATP, as would be expected from the fact that DNA somehow is stuffed into the head.

In this chapter we have seen two extremes for biological assembly. The "unique" structure of the ribosome is generated with ribosomal RNA and, except for one duplicated protein, with single copies of each of the ribosomal proteins. On the other extreme, many phage coats are virtually crystalline in that they contain mainly one protein used over and over in a regular array. Although we understand many facts about assembly, in reality we understand very little about the actual process. We cannot look at the structure of a protein like the lambda coat protein and predict that it is capable of forming a regular icosahedron, nor do we know anything about the changes in the shapes of proteins that occur when they assemble. We are also quite in the dark about understanding how and why nucleic acid is inserted into phage coats. Far more remains to be learned in this area than is already known.

PROBLEMS

20.1. a) How would you prove that an antibiotic that blocked protein synthesis did so by interfering with the functioning of ribosomes?

b) How would you show that its site of action is restricted to the 30S or 50S subunits?

c) If the effect were on the 30S subunit, how would you tell whether a mutation providing resistance to the antibiotic altered the RNA or protein?

d) Suppose parts a, b, and c above showed that the target was 16S RNA. Suppose sequencing showed that all rRNA from strains resistant to the drug were altered identically. How could this be possible in light of the fact that *E. coli* has seven genes for ribosomal RNA?

20.2. Devise a method for the isolation of nonsense mutations in the gene coding protein S12. You might wish to use the facts that some changes in this protein produce streptomycin resistance and that streptomycin sensitivity is dominant to resistance.

20.3. If the 30S ribosomal subunit were to be characterized by a set of distances between the centers of mass of the individual proteins, what is a minimum number of distances generally required to specify the topography of 21 proteins? How many different arrangements are compatible with this set of distances?

20.4. Draw to scale the outline of a ribosomal subunit, its RNA as a straight line, and a ribosomal protein of molecular weight 20,000, assuming it is spherical.

20.5. Demonstrate the improbability that ribosome assembly requires the simultaneous occurrence of all 21 proteins in a small volume. Estimate the probability that in a solution at 10 micrograms per ml in each of the proteins whose weight may be taken as 20,000, all 21 are present simultaneously in a sphere of radius twice that of a ribosome.

20.6. Suppose antibodies were made against a "purified" ribosomal protein and used in electron microscopy to locate the ribosomal protein on the ribosome. The microscopy located two sites 100 Å apart to which the antibodies bound. We are left concluding either that the protein has a highly asymmetric shape in the ribosome, or . . . ? What simple experiments could resolve the issue?

20.7. How would you determine which proteins lie on the interface between the 30S and 50S subunits?

20.8. What ideas would you have and what experiments would you perform if the nucleotide sequence at the end of the gene for ribosomal protein S11 and the beginning of the gene for S4 were XYZUGAXXXXXXAUG?

20.9. Of what evolutionary value is it for proteins that are adjacent in the ribosome to derive from genes that are adjacent on the chromosome?

20.10. How could an experiment be done to test for *ter*-mediated interchange of *cos* sites?

20.11. The geodesic dome (Fig. 20.21) was designed according to the same principles as those that govern icosahedral virus assembly. How many triangular facets would the entire sphere contain?

20.12. Where does trypsin cut peptides? What fraction of peptides cleaved from random protein by trypsin could be expected to contain sulphur? If cells are labeled at 10 mCi/μM in sulphur, how many counts per minute would exist in a typical tryptic spot of lambda phage E protein derived from 10 micrograms of phage?

20.13. It was once proposed that a protein found in lambda tails was a product of the A gene. How, with labeling but with no use of peptide maps, could this be disproved? Remember that an A$^-$ mutant makes no phage.

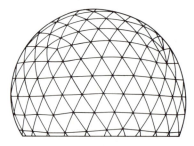

Figure 20.21   A geodesic dome constructed according to the principles of quasiequivalence.

20.14. If the binding of lambda phage to *E. coli* triggers the injection of lambda DNA, there could exist *E. coli* mutants capable of absorbing lambda but incapable of triggering DNA injection. How would you isolate such a mutant?

20.15. What is the origin of the circular structures with single-stranded tails observed in the heteroduplexes containing a duplication of the *cos* site?

20.16. How do the data presented on the heteroduplexes of the *cos* duplication lead to the conclusion that packaging proceeds from left to right?

20.17. There exist bacterial *groE* mutants that are able to grow but that do not support the growth of lambda phage. How could a lambda transducing phage be isolated carrying the *groE* gene?

20.18. The single-stranded regions in heteroduplexes between lambda and lambda containing small duplications are not found at unique sites; rather, they "wander." Why? How can such a wandering bubble be forced to appear at a unique location with another duplication?

20.19. The products of genes F_{II} and W prepare lambda heads for attachment of the tails. How would you determine whether these proteins must act in a particular order, and, if they do, how would you determine which acts first?

*20.20. In a *λpara* transducing phage, the *E. coli* DNA replaced the *int* gene of lambda. A double lysogen of this phage and wild-type lambda was made, but the two phage types could not be separated by equilibrium density gradient centrifugation because they were the same density. Therefore a double lysogen was made with the *ara* transducing phage and lambda *b2*. Induction of this double lysogen yielded transducing phage of *b2* density and plaque forming nontransducing phage of wild-type density. How did the density interchange of the two phage come about?

*20.21. Show that the number of subunits in the icosahedral viruses is proportional to $n^2 + nm + m^2$, where n and m are as described in the text.

*20.22. Consider the construction of a cube from a piece of paper by successively cutting out quadrants as illustrated in Fig. 20.15. Define the angular deficit as the angle between the two lines defining the quadrant. In the case of a cube, the angular deficit required to form each vertex is 90°, so the total angular deficit of the cube is 720°. Generalize to the other regular polyhedra, irregular convex polyhedra, convex and concave polyhedra, and continuously curved, closed surfaces.

RECOMMENDED READINGS

Assembly Mapping of 30S Ribosomal Proteins from *Escherichia coli*, S. Mizushima, M. Nomura, Nature *226*, 1214–1218 (1970).

Bacteriophage Lambda Derivatives Carrying Two Copies of the Cohesive End-Site, S. Emmons, J. Mol. Biol. *83*, 511–525 (1974).

Testing Models of the Arrangement of DNA Inside Bacteriophage Lambda by Crosslinking the Packaged DNA, R. Haas, R. Murphy, C. Cantor, J. Mol. Biol. *159*, 71–92 (1982).

Site-Directed Mutagenesis of Ribosomal RNA, Construction and Characterization of Deletion Mutants, R. Course, M. Stark, A. Dahlberg, J. Mol. Biol. *159*, 397–416 (1982).

Length Determination in Bacteriophage Lambda Tails, I. Katsura, R. Hendrix, Cell *39*, 691–698 (1984).

RELATED REVIEWS, BOOKS, AND ARTICLES

Ribosomes, ed. M. Nomura, A. Tissieres, P. Lengyel, Cold Spring Harbor Laboratory (1974).

Virus Assembly, S. Casjens, J. King, Ann. Rev. Biochem. *44*, 555–611 (1975).

Head Morphogenesis of Complex Double-Stranded DNA Bacteriophages, H. Murialdo, A. Beaker, Microbiol. Rev. *42*, 529–576 (1978).

Secondary Structure of 16S Ribosomal RNA, H. Noller, C. Woese, Science *212*, 403–410 (1981).

Structure, Assembly, and Function of Ribosomes, K. Nierhaus, Current Topics in Microbiology and Immunology *97*, 81–155, Springer-Verlag, Berlin (1982).

Architectural Design of Spherical Viruses, A. Klug, Nature *303*, 378–379 (1983).

Lambda II, ed. R. Hendrix, J. Roberts, F. Stahl, R. Weisberg, Cold Spring Harbor Laboratory (1983).

DEEPER READING

Physical Principles in the Construction of Regular Viruses, D. Caspar, A. Klug, Cold Spring Harbor Symposium on Quantitative Biology *27*, 1–24 (1962).

Isolation and Characterization of Ribonuclease I Mutants of *Escherichia coli*, R. Gesteland, J. Mol. Biol. *16*, 67–84 (1966).

Ter, a Function Which Generates the Ends of the Mature Lambda Chromosome, S. Mousset, R. Thomas, Nature *221*, 242–244 (1969).

Intracellular Pools of Bacteriophage Lambda DNA, B. Carter, M. Smith, J. Mol. Biol. *50*, 713–718 (1970).

Studies of the 16 and 23S Ribosomal RNA of *E. coli* Using Composite Gel Electrophoresis, A. Dahlberg, A. Peacock, J. Mol. Biol. *55*, 61–74 (1971).

Purification and Characterization of 50S Ribosomal Proteins of *E. coli*, G. Mora, D. Donner, P. Thammana, L. Lutter, C. Kurland, G. Craven, Mol. Gen. Genet. *112*, 229–242 (1971).

Spatial Arrangement of Ribosomal Proteins: Reaction of the *E. coli* 30S Subunit with bis-Imidoesters, T. Bickle, J. Hershey, R. Traut, Proc. Nat. Acad. Sci. *69*, 1327–1331 (1972).

Mechanism of Kasugamycin Resistance in *E. coli*, T. Helser, J. Davies, J. Dahlberg, Nature New Biology *235*, 6–9 (1972).

Head Assembly Steps Controlled by Genes F and W in Bacteriophage Lambda, S. Casjens, T. Hohn, A. Kaiser, J. Mol. Biol. *64*, 551–563 (1972).

Identity of the Ribosomal Proteins Involved in the Interaction with Elongation Factor G, J. Highland, J. Bodley, J. Gordon, R. Hasenbank, G. Stoffler, Proc. Nat. Acad. Sci. USA *70*, 147–150 (1973).

Host Participation in Bacteriophage Lambda Head Assembly, C. Georgopoulos, R. Hendrix, S. Casjens, D. Kaiser, J. Mol. Biol. *76*, 45–60 (1973).

In Vitro Assembly of Bacteriophage Lambda Heads, A. Kaiser, T. Masuda, Proc. Nat. Acad. Sci. USA *70*, 260–264 (1973).

Mapping *E. coli* Ribosomal Components Involved in Peptidyl Transferase Activity, N. Sonenberg, M. Wilchek, A. Zamir, Proc. Nat. Acad. Sci. USA *70*, 1423–1426 (1973).

Evidence That the Cohesive Ends of Mature Lambda DNA Are Generated by the Gene A Product, J. Wang, A. Kaiser, Nature New Biol. *241*, 16–17 (1973).

The Capsid Structure of Bacteriophage Lambda, M. Bayer, A. Bocharov, Virology *54*, 465–475 (1973).

Chemical Linkage of the Tail to the Right-Hand End of Bacteriophage Lambda DNA, J. Thomas, J. Mol. Biol. *87*, 1–9 (1974).

Capsid Structure of Bacteriophage Lambda, R. Williams, K. Richards, J. Mol. Biol. *88*, 547–550 (1974).

Bacteriophage Lambda F_{II} Gene Protein: Role in Head Assembly, S. Casjens, J. Mol. Biol. *90*, 1–23 (1974).

Comments on the Arrangement of the Morphogenetic Genes of Bacteriophage Lambda, S. Casjens, R. Hendrix, J. Mol. Biol. *90*, 20–23 (1974).

Protein Fusion: A Novel Reaction in Bacteriophage Lambda Head Assembly, R. Hendrix, S. Casjens, Proc. Nat. Acad. Sci. USA *71*, 1451–1455 (1974).

Determination of the Location of Proteins L14, L17, L18, L19, L22, and L23 on the Surface of the 50S Ribosomal Subunit of *Escherichia coli* by Immune Electron Microscopy, G. Tischendorf, H. Zeichhardt, G. Stoffler, Mol. Gen. Genet. *134*, 187–208 (1974).

Protein Cleavage in Bacteriophage Lambda Tail Assembly, R. Hendrix, S. Casjens, Virology, *61*, 156–159 (1974).

Assembly Mapping of 30S Ribosomal Proteins from *Escherichia coli*, W. Held, B. Ballou, S. Mizushima, M. Nomura, J. Biol. Chem. *249*, 3103–3111 (1974).

Processing of Bacteriophage Lambda DNA During Its Assembly into Heads, M. Syvanen, J. Mol. Biol. *91*, 165–174 (1975).

Assembly of Bacteriophage Lambda Heads: Protein Processing and Its Genetic Control in Petit Lambda Assembly, R. Hendrix, S. Casjens, J. Mol. Biol. *91*, 187–199 (1975).

Singlet Energy Transfer Studies of the Arrangement of Proteins in the 30S *Escherichia coli* Ribosome, K. Huang, R. Fairclough, C. Cantor, J. Mol. Biol. *97*, 443–470 (1975).

Identification by Diagonal Gel Electrophoresis of Nine Neighboring Protein Pairs in the *Escherichia coli* 30S Ribosome Crosslinked with Methyl-4-Mercaptobutyrimidate, A. Sommer, R. Traut, J. Mol. Biol. *97*, 471–481 (1975).

Petite Lambda: A Family of Particles from Coliphage Lambda Infected Cells, T. Hohn, F. Flick, B. Hohn, J. Mol. Biol. *98*, 107–120 (1975).

Ribosomal Proteins S5, S11, S13, and S19 Localized by Electron Microscopy of Antibody-Labeled Subunits, J. Lake, L. Kahan, J. Mol. Biol. *99*, 631–644 (1975).

Neutron Scattering Measurements of Separation and Shape of Proteins in 30S Ribosomal Subunit of *Escherichia coli*: S2-S5, S5-S8, S3-S7, D. Engelman, P. Moore, B. Schoenborn, Proc. Nat. Acad. Sci. USA *72*, 3888–3892 (1975).

Ribosome Structure Determined by Electron Microscopy of *Escherichia coli* Small Subunits, Large Subunits, and Monomeric Ribosomes, J. Lake, J. Mol. Biol. *105*, 131–159 (1976).

Morphogenesis of Bacteriophage Lambda Tail, Polymorphism in the Assembly of the Major Tail Proteins, I. Katsura, J. Mol. Biol. *107*, 307–326 (1976).

Lambda Head Morphogenesis as Seen in the Electron Microscope, A. Zachary, L. Simon, S. Litwin, Virology, *72*, 429–442 (1976).

Triangulation of Proteins in the 30S Ribosomal Subunit of *E. coli*, P. Moore, J. Langer, B. Schoenborn, D. Engelman, J. Mol. Biol. *112*, 199–234 (1977).

Significant Changes in 16S RNA Conformation Accompanying Assembly of the 30S Ribosome *in vitro*, H. Hochkeppel, G. Craven, J. Mol. Biol. *113*, 623–634 (1977).

Early Events in the *in vitro* Packaging of Bacteriophage Lambda DNA, A. Becker, M. Marko, M. Gold, Virology *78*, 291–305 (1977).

Symmetry Mismatch and DNA Packaging in Large Bacteriophages, R. Hendrix, Proc. Nat. Acad. Sci. USA *75*, 4779–4783 (1978).

Selection for *Escherichia coli* Mutants with Proteins Missing from the Ribosome, E. Dabbs, J. Bact. *140*, 734–737 (1979).

Packaging of the Bacteriophage Lambda Chromosome: A Role for Base Sequences Outside *cos*, M. Feiss, R. Fisher, D. Siegele, B. Nicols, J. Donelson, Virology *92*, 56–67 (1979).

DNA Packaging by the Double-Stranded DNA Bacteriophages, W. Earnshaw, S. Casjens, Cell *21*, 319–331 (1980).

Ribosomal Proteins S3, S6, S8, and S10 of *E. coli* Localized on the External Surface of the Small Subunit by Immune Electron Microscopy, L. Kahan, D. Winkelmann, J. Lake, J. Mol. Biol. *145*, 193–214 (1981).

Localization of 3' Ends of 5S and 23S rRNAs in Reconstituted Subunits of *Escherichia coli* Ribosomes, M. Stoffler-Meilicke, G. Stoffler, O. Odom, A. Zinn, G. Kramer, B. Hardesty, Proc. Nat. Acad. Sci. USA *78*, 5538–5542 (1981).

Arrangement of the Subunits in the Ribosome of *Escherichia coli:* Demonstration by Immuno-electron Microscopy, B. Kastner, M. Stoffler-Meilicke, G. Stoffler, Proc. Nat. Acad. Sci. USA *78*, 6652–6656 (1981).

Topography of 16S RNA in 30S Subunits and 70S Ribosomes, Accessibility to Cobra Venom Ribonuclease, S. Vassilenko, P. Carbon, J. Ebel, C. Ehresman, J. Mol. Biol. *152*, 699–721 (1981).

Positions of Proteins S6, S11, S15 in the 30S Ribosomal Subunit of *Escherichia coli,* V. Ramakrishnan, S. Yabuki, I. Sillers, D. Schindler, D. Engelman, P. Moore, J. Mol. Biol. *153*, 739–760 (1981).

Structure of the Scaffold in Bacteriophage Lambda Preheads: Removal of the Scaffold Leads to a Change of the Prehead Shell, P. Kunzler, H. Berger, J. Mol. Biol. *153*, 961–978 (1981).

Topography of the *E. coli* 5S RNA-Protein Complex as Determined by Crosslinking with Dimethylsuberimidate and Dimethyl-3, 3' Dithiobispropionimidate, T. Fanning, R. Traut, Nucleic Acids Res. *9*, 993–1004 (1981).

Assembly Map of the Large Subunit (50S) of *Escherichia coli* Ribosomes, R. Rohl, K. Nierhaus, Proc. Nat. Acad. Sci. USA *79*, 729–733 (1982).

Nascent Polypeptide Chains Emerge from the Exit Domain of the Large Ribosomal Subunit: Immune Mapping of the Nascent Chain, C. Bernabeu, J. Lake, Proc. Nat. Acad. Sci. USA *79*, 3111–3115 (1982).

Bacteriophage Lambda DNA Packaging: Scanning for the Terminal Cohesive End Site During Packaging, M. Feiss, W. Widner, Proc. Nat. Acad. Sci. USA *79*, 3498–3562 (1982).

Ribosomal Protein S4 Is an Internal Protein: Localization by Immunoelectron Microscopy on Protein-Deficient Subribosomal Particles, D. Winkelmann, L. Kahan, J. Lake, Proc. Nat. Acad. Sci. USA *79*, 5184–5188 (1982).

Three-Dimensional Reconstruction and Averaging of 30S Ribosomal Subunits of *Escherichia coli* from Electron Micrographs, V. Knaver, R. Hegerl, W. Hoppe, J. Mol. Biol. *163*, 409–430 (1983).

Three-Dimensional Reconstruction and Averaging of 50S Ribosomal Subunits of *Escherichia coli* from Electron Micrographs, H. Oettl, R. Hegerl, W. Hoppe, J. Mol. Biol. *163*, 431–450 (1983).

Tests of Spool Models for DNA Packaging in Phage Lambda, J. Widom, R. Baldwin, J. Mol. Biol. *171*, 419–437 (1983).

Packaging of DNA into Bacteriophage Heads, A. Model, S. Harrison, J. Mol. Biol. *171*, 577–580 (1983).

Separate Sites for Binding and Nicking of Bacteriophage Lambda DNA by Terminase, M. Feiss, I. Kobayashi, W. Widner, Proc. Nat. Acad. Sci. *80*, 955–959 (1983).

Structure of *E. coli* 16s RNA Elucidated by Psoralen Crosslinking, J. Thompson, J. Hearst, Cell *32*, 1355–1365 (1983).

Positions of Proteins S14, S18, S20 in the 30 S Ribosomal Subunit of *Escherichia coli*, V. Ramakrishnan, M. Capel, M. Kjeldgaard, D. Engelman, P. Moore, J. Mol. Biol. *174*, 265–284 (1984).

Proteinase Sensitivity of Bacteriophage Lambda Tail Proteins gpJ and pH* in Complexes with the Lambda Receptor, C. Roessner, G. Ihler, J. Bact. *157*, 165–170 (1984).

Chemotaxis

In this chapter we will see that genetics, physiology, and biochemistry, the tools of molecular biology that have been so useful in the study of relatively simple questions, can also be applied to the study of biological phenomena as complicated as behavior. Of course, to utilize the full power of genetics, the organism ought to be a phage, bacterium, yeast, or fruit fly, and to facilitate biochemical work, it ought not to be the fruit fly. Phage do not respond to their environment or appear to learn anything, but bacteria do and therefore have been used in such studies.

Bacteria respond to their environment by swimming toward some chemicals like sugars, amino acids, and oxygen and away from others like phenol. This response, called chemotaxis, involves a number of steps. The bacteria must be able to sense the chemical in question, choose the appropriate direction to swim, and be able to swim. That is, they must contain elementary sensors, an analog of a nervous system, and motors. How such elements can be constructed from simple bacterial components is one interest in such studies. A second interest is that understanding the systems in bacteria is likely to help our understanding of the analogs of such systems in higher animals.

In the last century, microbiologists observed chemotactic behavior of bacteria in the microscope. In contrast to nonmotile bacteria strains, which vibrate slightly due to Brownian motion, motile bacteria swim rapidly. Chemotactic bacteria are motile and direct their motility to move themselves toward attractants and away from repellents. The swimming is accomplished by moving long, slender flagella that are attached to the cell wall.

To swim in the correct direction, bacteria must be capable of detecting differences in the concentrations of attractants or repellents. This can be done in two ways. At any instant they could make a measurement of attractant concentration at two points, that is, compare concentrations at their two ends. However, the difficulty of this approach is that it requires a highly precise concentration measurement. For example, if an attractant changed in concentration from zero to one molar over a distance of one centimeter, over the length of a cell the concentration would change by only 10^{-4} M. Although telling the difference between zero and 10^{-4} is not hard, telling the difference between 0.5 and .5001, $0.5 + 10^{-4}$, is. An alternative scheme for detecting a concentration gradient is for the bacteria to measure the attractant concentration at one location, remember it, swim to another location, make a second measurement, and compare the two values. Despite the apparent complexity of the second method, this is the one used.

ASSAYING CHEMOTAXIS

Sensitive and simple characterization of a cell's chemotaxis ability is necessary for efficient study of the phenomenon. One straightforward assay is the motility plates or swarm plates developed by Adler (Fig. 21.1). In one application, the medium in

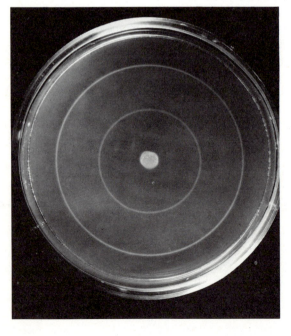

Figure 21.1   The appearance of a minimal-salts plate containing galactose and glucose several hours after placing a drop of cells in the center. (By permission, J. Adler, Science 153, 708–716, 1966, © 1966 AAAS.)

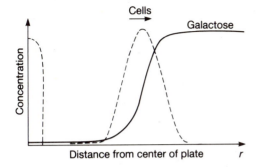

Figure 21.2   The concentrations of galactose (solid line) and cells (dashed line) as a function of distance r from the center of the plate at 5 hours after spotting cells.

these plates contains the usual salts necessary for cell growth, a low concentration of galactose, and a concentration of agar such that the medium has the consistency of vichyssoise. Several hours after a drop of chemotactic cells is placed in the center, the plate will contain the spot of cells in the center and a ring about two centimeters in diameter surrounding the center. This ring expands outward with time and ultimately reaches the edge of the plate. Usually, a second ring also forms and moves behind the first.

The origin of the rings is straightforward. Sensitive chemical tests show that the cells spotted in the center of the plate gradually consume the galactose where they are placed (Fig. 21.2). However, just beyond the edge of the spot, galactose remains in its original concentration. That is, the consumption of galactose creates a concentration gradient in galactose at the edge of the spot. The bacteria on the edge of the spot detect this gradient, swim toward higher galactose concentrations, and, at the same time, consume the galactose in the new position. Consequently, the cells move outward pursuing and consuming the galactose. The second ring that forms on the galactose plates derives from bacteria utilizing their endogenous energy sources to swim toward oxygen. If tryptone medium is substituted for galactose, then multiple rings form, the first ring from bacteria swimming toward the most active attractant, the second toward a second attractant, and so on. Also in this case, one of the rings is formed by bacteria pursuing oxygen.

The capillary tube assay is more sensitive and quantitative than the swarm plate assay. Instead of requiring the bacteria to produce their own gradient of attractant concentration, the gradient is produced by diffusion of the substrate. A short capillary tube is filled with medium containing an attractant and is sealed at one end. The open end is pushed into a drop of medium lacking attractant but containing cells, and the tube is allowed to rest there several hours. As attractant diffuses out of the mouth of the capillary, a gradient of attractant is produced in the region (Fig. 21.3). The chemotactic bacteria in the drop detect and swim up the gradient a short distance into the capillary. At the end of the assay the capillary tube is removed and rinsed from the outside, and the bacteria inside are blown out into a tube. These are

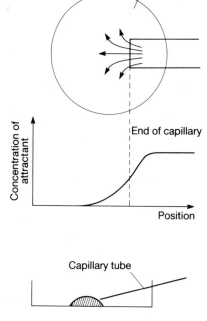

Figure 21.3   Diagram of the capillary tube assay in which the capillary is inserted in a drop of medium in a petri plate, a top view showing diffusion out of a capillary to generate a gradient, and the effective concentration as a function of position of the attractant after several hours of diffusion.

diluted and plated on petri plates. Since the number of bacteria having entered the capillary can be accurately measured by counting the number of colonies on the plates after incubation, the chemotactic ability of the cells can be quantitated. Typically less than 100 nonmotile bacteria enter the capillary. Several thousand motile but nonchemotactic bacteria normally enter a capillary with or without attractant, and as many as 500,000 chemotactic bacteria will enter a capillary tube that contains attractant.

FUNDAMENTAL PROPERTIES OF CHEMOTAXIS

The capillary tube assay permits several simple measurements delimiting methods by which cells accomplish chemotaxis. The first is that metabolism of attractant is not required for chemotaxis. An analog of galactose, fucose, is not metabolized by *Escherichia coli,* and yet this serves as an attractant in the capillary tube assay. Another line of evidence leading to this same conclusion is that some mutants that are unable to metabolize galactose are still able to swim toward galactose.

Since cells can swim toward or away from a wide variety of chemicals, it is likely that they possess a number of receptors, each with different specificity. Therefore the question arises as to how many different types of receptors a cell possesses. This question can be answered by a cross-inhibition test. Consider the case of fucose and galactose. Since their structures are similar, it seems likely that cells detect both

D-galactose   D-fucose

chemicals with the same receptor protein. This conjecture can be tested by using a high concentration of fucose in both the tube and the drop so as to saturate the galactose receptor and blind cells to the galactose that is placed only in the tube. Fucose does blind cells to a galactose gradient, but it does not blind them to a serine gradient. These findings prove that galactose and fucose share the same receptor.

By blinding experiments and the use of mutants, nine receptors have been found for sugars and three for amino acids (Table 21.1). The three amino acid receptors are not highly selective, and they allow chemotaxis of E. coli toward 10 different amino acids. Except for the glucose receptor, synthesis of the sugar receptors is inducible, as is the proline receptor. The glucose, serine, and aspartate receptors are synthesized constitutively.

The capillary assay also allows a convenient determination of the sensitivity of the receptors. By varying the concentration of attractant within the capillary, the ranges over which the receptor will respond can be easily determined. Typically, the lowest concentration is about 10^{-7} M, and the highest concentration is about 10^{-1} or 10^{-2} M (Fig. 21.4). It is not at all surprising that a sugar-detecting system is no more sensitive than 10^{-7} M, for at about 10^{-6} M the rate of diffusion of a sugar to a bacterium is just adequate to support a 30-minute doubling time if every sugar molecule reaching the cell is utilized. The cell could survive with normal growth rates in the presence of still lower concentrations of sugar by rapidly swimming through the solution. However, it would be surprising if it could achieve much more than 10-fold improvement.

The actual receptor proteins for the different sugars have been sought, and several have been identified biochemically. The receptors for galactose, maltose, and ribose are found in the periplasm. That is, they are located outside the inner membrane but inside the peptidoglycan layer. As a group, these and other periplasmic proteins can be removed from cells by osmotic shock.

Table 21.1   Partial List of
Attractants, Repellants, and
Their Sensors

Sensor	Response
N-acetyl glucosamine	Attractant
Fructose	Attractant
Galactose	Attractant
Glucose, fucose	
Glucose	Attractant
Mannose	Attractant
Glucose	
Mannitol	Attractant
Ribose	Attractant
Sorbitol	Attractant
Trehalose	Attractant
Aspartate	Attractant
Glutamate	
Methionine	
Proline	Attractant
Serine	Attractant
Cysteine	
Alanine	
Glycine	
Asparagine	
Threonine	
Alcohols	Repellent
Fatty acids	Repellent
Hydrophobic amino acids	Repellent
Indole	Repellent
Skatole	

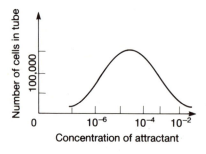

Figure 21.4   A typical response curve
showing the number of cells that have entered the capillary tube as a function of the
concentration of attractant placed in the
tube.

No periplasmic binding proteins have been found for the glucose, mannitol, and trehalose systems. Instead, these systems use receptors that are tightly bound to or located in the inner membrane. These receptors serve a double purpose as they also function in the group translocation of their substrates into the cell via the phospho-transferase transport system.

GENETICS OF MOTILITY AND CHEMOTAXIS

Mutants permit dissection of the chemotaxis system, both by allowing observation of the behavior of cells containing damaged components of the system and by facilitating biochemical isolation and study of the components. One method for isolation of chemotaxis mutants utilizes the swarm plates. About 100 candidate mutant cells are diluted into a plate containing low agar concentration, minimal salts, any necessary growth factors, and a metabolizable attractant such as galactose. As each cell grows into a colony, it generates an outward-moving ring if the cells are able to consume the attractant and swim toward higher concentrations. A motile but nonchemotactic cell makes a moderately large colony with diffuse edges, whereas a nonmotile cell makes a compact colony with sharp edges.

Several types of galactose-specific chemotactic mutants have been found with the swarm plates. One type possessed an altered galactose-binding protein. The protein had a greater dissociation constant for galactose, as did the galactose transport system, proving that the same protein is involved with both galactose chemotaxis and galactose transport. In another class, active transport became inactive, but the periplasmic binding protein was not altered.

A different type of genetic selection proved useful for the isolation of deletions and point mutations in flagellin genes. It is based on the fact that antibodies against flagella stop their motion and block motility. Bacterial mutants with different antigenic determinants on their flagella are resistant to the antibody. Imagine the situation in a merodiploid cell where an episome possesses genes for the mutant flagellin and the chromosome codes for the nonmutant flagellin. Such cells will not be chemotactic in the presence of the antibody because their flagella will contain both types of flagellin and their action will be blocked by the antibody. However, any

Flagellum of a and b type flagellin

mutants synthesizing only the flagellin encoded by the episome will be chemotactic because their flagella will be resistant to the antibody. As a result, the only cells able to swim out of a spot on a swarm plate containing the antiflagellin antibody must be ones whose chromosomal flagellin genes *only* are not functional. Apparently deletions are the most common method for accomplishing this.

Another straightforward method for the isolation of mutants in flagella employs the phage chi, which appears to use moving flagella to reach the cell. Perhaps the tail fibers in the phage wrap around the flagella and the phage is drawn to the cell by their motion. Therefore cells may become resistant to the phage either by not possessing a flagellum or by not moving it.

HOW CELLS SWIM

Escherichia coli propels itself through liquid by rotating its flagella. Normally the flagella are left-handed helices and their rotation generates a thrust that moves the cell. In this section we consider the structure of flagella, how we know they rotate, how the rotation is created, and how the several flagella present on a single cell can function together.

Flagella are too thin to be easily seen by ordinary light microscopes. They can be visualized by light microscopes operating in the dark-field mode or by electron microscopes, where they appear curly or helical. Careful isolation of flagella shows that they are attached to a hook-shaped structure connected to a set of rings that is embedded in the cell's membranes. The rings have the appearance of a motor that rotates the flagella (Fig. 21.5). The hook is a flexible connector between the basal structure and the flagella. Such a universal joint is necessary because in *E. coli* the

Figure 21.5   The appearance of a flagellum at low magnification in the electron microscope and the structure of the basal body, the motor, at high magnification showing the rings and the membranes of the cell wall.

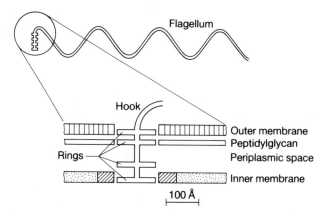

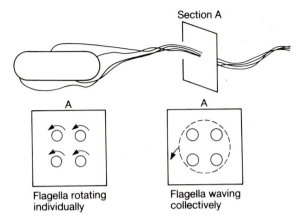

Section A

A

A

Flagella rotating
individually

Flagella waving
collectively

Figure 21.6   A bacterium with flagella sprouting from various locations bending via the
hook portion and coming together in a bundle. The section taken at point A shows flagella
rotating individually and their movement if the bundle as a whole waves.

flagella sprout from random points on the cell's surface, but the flagella join together
in a bundle to propel the cell.

Because of the size of the flagella, indirect means must be used to demonstrate
that they rotate. One simple experiment uses the phenomenon that antibody against
flagellin can block motility. More precisely, bivalent antibody blocks motility, but
monovalent antibody does not (Fig. 21.6). This result can be understood if the
flagella form a bundle and each flagellum rotates within this bundle. A bivalent
antibody molecule can link different flagella and prevent their rotation. However, if
flagella waved or rotated as a group, bivalent antibodies would have an effect no
different from monovalent antibodies.

The most graphic demonstration that flagella rotate is also the basis of many
other important experiments on chemotaxis. Simon used one mutation to block
flagellin synthesis and another to permit greater than usual growth of the hook. As
the synthesis of flagella is sensitive to catabolite repression, growth of cells in glucose
reduced the number of the resulting polyhooks from about five per cell to about one.
These cells could be bound to a microscope slide by means of antihook antibody that
had bound to the hook and nonspecifically bound to the glass as well (Fig. 21.7).
Chemotactic cells immobilized in this way rotated at two to nine revolutions per
second. This leads to the conclusion that the hook normally rotates, but when it was
tied down, the cell rotated instead. Nonmotile cells did not rotate. Of course, motile
but nonchemotactic mutants did rotate because they can swim but are incapable of
swimming up a gradient.

The immobilization experiment shows that a single flagellum rotates and that
the rotation is associated with chemotaxis. Dark-field microscopy has shown that
the bundle of flagella on a cell is stable as long as the flagella rotate counterclockwise.
If the flagella reverse their direction, their left-helical structure compels the bundle

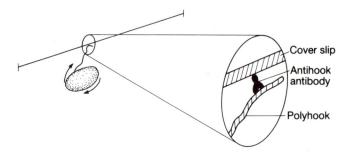

Figure 21.7   Demonstration of the hook's rotation. After attachment of the hook or a short flagellum to a coverslip, the cell rotates, and this can easily be observed.

to fly apart temporarily. Furthermore, if the reversal is sufficiently vigorous, the flagella snap into a right-helical conformation. This further ensures that the bundle of flagella disperses.

THE MECHANISM OF CHEMOTAXIS

As viewed in the microscope, chemotactic bacteria appear to swim smoothly for 50 to 100 bacterial lengths in about 1 second, to tumble for about 0.1 second, and then to swim smoothly in another direction. How this run-tumble-run behavior could be converted to overall swimming toward increasing concentrations of an attractant was one of the major problems in chemotaxis.

Berg constructed an elaborate tracking microscope that quantitated the movement of a chemotactic bacterium in three dimensions. This showed that, indeed, the impression obtained from simple visual observation was correct. Following a tumble, a cell's subsequent run was in a random direction. More important, however, was the behavior of bacteria under conditions simulating the presence of a gradient in attractant. Since it is difficult to control a gradient in the concentration of an attractant around the cell and since it is likely that a cell detects not a spatial gradient but the change in the concentration of attractant from one time to the next, Berg devised a system to vary the concentration of attractant with time.

Alanine aminotransferase catalyzes the interconversion of alanine and α-ketogluterate to pyruvate and L-glutamate. Both alanine and glutamate are attractants, and therefore the reaction cannot be used to create or destroy attractant. However, by using a mutant that is blind to alanine but not blind to glutamate, only

Alanine aminotransferase
$$\text{Alanine} + 2 \text{ oxogluterate} \rightleftharpoons \text{Pyruvate} + \text{L-glutamate}$$

ATTRACTANTS   REPELLENTS

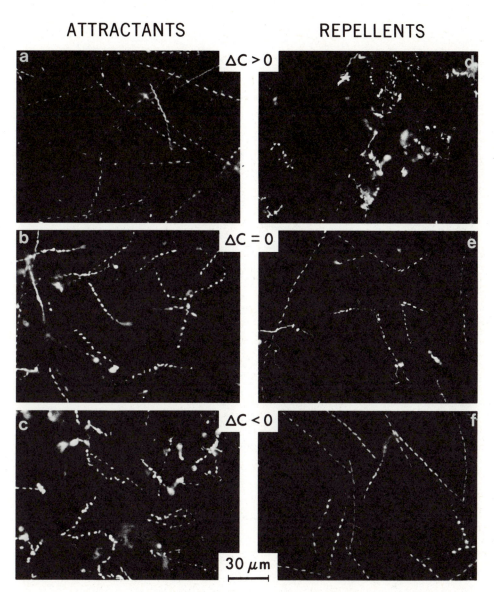

Figure 21.8   Time-lapse photographs of chemotactic bacteria rapidly exposed to increases or decreases in attractant or repellant concentrations. The smooth path of a swimming bacterium looks dashed in the blinking light source used in the experiment. (Courtesy D. Koshland.)

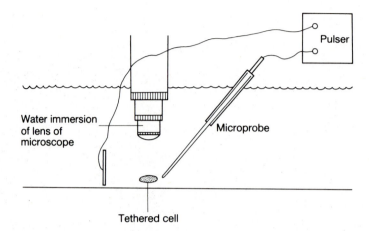

Figure 21.9 The method by which Berg generates small impulses of attractant and observes the cell's responses. The microscope contains additional electronics that record the rotation of the bacterium.

Figure 21.10 An impulse is the limit of a pulse of finite area but width approaching zero. It can produce a response as shown in some circuits. When the circuit elements are linear, the response to two impulses is the sum of the individual responses.

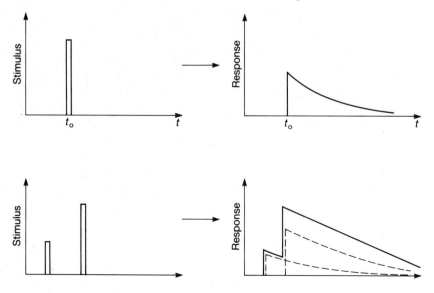

glutamate is an attractant and the enzyme-catalyzed reaction can be used to vary the concentration with time of attractant surrounding a cell. If alanine, α-ketoglutarate, and alanine aminotransferase are put in the medium bathing the cell being observed, then the concentration of attractant, glutamate, increases with time. In this case, the runs of the bacteria are longer than average. Alternatively, when the glutamate concentration is decreasing as a result of adding glutamate, pyruvate, and alanine aminotransferase, the runs are shorter than average.

Once it was clear that cells made decisions about swimming on the basis of comparisons in time, simpler equipment than a tracking microscope could be used for study of the phenomenon. Koshland developed a stopped-flow apparatus in which cells are rapidly diluted into a solution containing a different concentration of attractant or repellent and then, after an adjustable interval, photographed using a time exposure of about 1 second. In this duration of exposure, swimming cells leave streaks and tumbling cells generate irregular blobs (Fig. 21.8). Good statistics of probabilities of runs and tumbles can be obtained using the many bacteria that can be quantitated at one time.

The technique of immobilizing a cell via its hook or flagellum has been highly useful in characterizing chemotaxis responses. After such an immobilization, a microprobe containing attractant or repellent can be positioned near a cell, and application of a brief electrical pulse will drive a known amount of the chemical into the medium in the immediate area (Fig. 21.9). If the concentration changes are within the linear response range of the cell, the response to a brief pulse of attractant contains all the information necessary to predict the response of a cell to any other function of attractant as a function of time. Basically any other function can be considered to be broken up into a series of impulses of appropriate magnitude, and the responses to each of these can be summed (Fig. 21.10). Such a technique of circuit analysis is well known to electrical engineers, but it has not been widely applied to biochemical systems.

All the experiments indicate that cells modulate the duration of their intervals of smooth swimming. Such a modulation is sufficient for the cells to achieve a net drift up a gradient of attractant. As long as a cell is swimming up a gradient in attractant concentration, it decreases the chances of its tumbling. However, if the attractant is decreasing in concentration, the cell is more likely to tumble and try swimming in a new direction. These properties mean that cells drift up gradients in attractant concentration.

THE ENERGY FOR CHEMOTAXIS

Application of genetic and biochemical methods has allowed determination of the energy source required for chemotaxis. On one hand it seems logical that ATP would be the direct source of the mechanical energy required for swimming since most energy transductions in higher cells appear to use ATP. However, the flagella originate in the cell membrane, and a substantial proton gradient exists across the inner

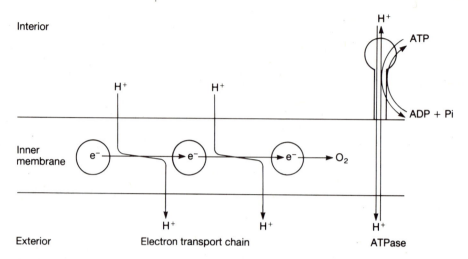

Figure 21.11   A schematic of the inner membrane of a cell showing that electron flow down the electron transport chain leads to export of protons and ultimate transfer of electrons to oxygen. Reentry of protons to the cell through the ATPase generates ATP. Conversely, ATP can be hydrolyzed by ATPase to pump protons out of the cell.

membrane under most growth conditions. Therefore the direct source of energy could also derive from the membrane potential.

ATP and the protonmotive force across the inner membrane are normally inter-convertible by means of the membrane-bound ATPase (Fig. 21.11). The membrane potential generates ATP, and, conversely, ATP can be used to generate a membrane potential. Therefore, attempting to block formation of either ATP or the membrane potential is likely to have an effect on the other. However, cells can be grown in the presence of arsenate to block ATP formation both by glycolysis and by ATPase. This treatment preserves the membrane potential. Such cells are motile, but they do not swim up gradients of attractants. Next, to investigate the role of the membrane potential, cells can be grown anaerobically. This blocks the usual means of genera-tion of the membrane potential because the electron transfer chain becomes inactive owing to the lack of a terminal electron acceptor. To prevent energy from ATP from being used to create a membrane potential, an ATPase-negative mutant was used. Anaerobically grown ATPase mutants are not motile, leading to the conclusion that the motors that drive the flagella are run by the cell's membrane potential.

ADAPTATION AND PROTEIN METHYLATION

Many sensory systems in animals adapt to a stimulus. That is, excitation with a particular stimulus will evoke a response, but the response disappears during con-

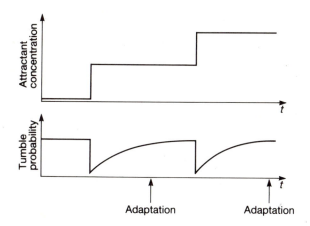

Figure 21.12   Adaptation to increases in attractant concentration. Before the first increase in concentration, cells had a fixed probability of shifting their flagella from counterclockwise to clockwise rotation. Immediately after the first step in attractant, the probability of reversing the direction of rotation fell but then returned to its initial value. Following the second step in attractant, the same adaptation occurred.

tinued application of the stimulus. An increase in the exciting stimulus will then produce a new response, and, eventually, a new adaptation occurs (Fig. 21.12). Such an adaptation phenomenon is also observed in bacterial chemotaxis and is likely inherent in the machinery a cell uses for comparing attractant concentrations at two different times. Tethered bacteria will rotate counterclockwise for up to about 10 seconds after the attractant concentration in their medium is suddenly increased over biologically meaningful concentration ranges. Then they will adapt to the attractant and punctuate their counterclockwise spins with clockwise spins at the same frequency as before the addition of attractant. When whopping concentration changes are applied, the cells will spin counterclockwise for up to several minutes before they adapt.

The phenomena of excitation and adaptation can be simply generated by comparing the sensor's present signal to some average of the signal over the recent past (Fig. 21.13). Such averaging is equivalent to a slow response, and this provides the cell's "memory." Hence the instantaneous conditions surrounding a cell must be compared to an average of the conditions taken over the previous few seconds. If the concentration of attractant has increased during that interval, the instantaneous signal will exceed the average signal, and the comparator should produce a signal that suppresses clockwise rotations. However, if the concentration of attractant has not increased over an interval of a few seconds, the instantaneous and averaged signals will be equal and the comparator signal will fall. Consequently, the motor will be permitted to reverse direction and the cells will tumble and as a result will randomize their swimming direction.

The central component in the cell's adaptation has been found. Adler had observed that methionine is necessary for chemotaxis. That is, a methionine auxotroph could be grown in the presence of methionine in which it would be chemotactic, that is, would swim up gradients of attractant. However, upon removal of the

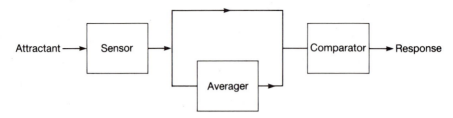

Figure 21.13   One circuit that will generate a response usable for chemotaxis. The comparator response regulates the direction of flagellum rotation.

methionine, the cells remained motile for hours, but they were no longer chemotactic. Later, when the mechanism of chemotaxis was understood to be modulation in the frequency of runs and tumbles, it was sensible to ask what methionine did. Experiments showed that methionine-starved cells are unable to tumble. The next obvious question was whether methionine starvation destroyed the ability to tumble or whether it eliminated the signal to tumble. At first this question seems unapproachable. However, genetics came to the rescue. Among the many types of non-chemotactic mutants known are some that incessantly tumble. The question then was whether these tumbling mutants continued to tumble during methionine starvation. All that was necessary to answer this question was to make these tumbling mutants methionine auxotrophs and to starve them of methionine. After this treatment they continued to tumble. Thus methionine is necessary for signaling wild-type cells that tumbling is necessary, and it is not essential for the act of tumbling.

How might methionine be required? Methionine via s-adenosylmethionine is a known source of methyl groups in metabolism. Most likely, then, the methionine requirement says that methylation is involved in the tumbling signal. Consistent with this notion is the fact discussed above that the addition of arsenate to cells, which blocks ATP formation and hence s-adenosylmethionine synthesis, also blocks chemotaxis but not motility.

With such clues for the involvement of methylation in chemotaxis, it is natural to look for methylated proteins. The level of methylation of several membrane proteins has been found to be correlated with chemotaxis behavior for some attractants. The addition of one of these attractants to cells leads to preferential methylation of one of a set of four membrane proteins, products of genes named *tsr, tar, tap,* and *trg* (Fig. 21.14). Methyl groups are transferred from s-adenosylmethionine to glutamyl residues in these proteins to make carboxy methyl esters. Each of these proteins may receive as many as five methyl groups.

Increase in the concentration of an attractant must lead to a conformation change in one of these membrane-bound proteins. This change then signals the flagellar motor. Subsequent methylation of the protein reverses the protein's ability to send the signal; that is, by the time the methylation level of the protein approaches

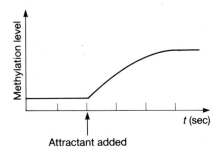

Figure 21.14   Methylation as function of time after attractant addition of one of the proteins in the inner membrane that behaves like the signal averager.

its final value, the protein has returned to its normal conformation and no longer signals the motor. This reversal of the conformation change by methylation is the actual adaptation. These methylatable proteins are transducers in the chemotaxis system. Either they directly bind the attractant or repellent or they couple with proteins that do, and they send a signal on to the motor.

For methylation to be a satisfactory "memory," at least two enzymes are required: an enzyme to methylate and an enzyme to demethylate the transducer proteins. More precisely, a methyltransferase and a methylesterase must exist. Both of these have been found. They act on the proper membrane proteins and are encoded by genes in the set of eight or nine genes that are required for chemotaxis but are not required for motility. That is, these genes are involved with the regulation of tumbles.

Chemotaxis is remarkable. A single cell manages to sense attractants with receptors located outside the cytoplasm, transmit the information to the interior, remember the information it has received so as to adapt if the attractant or repellent concentration is no longer changing, build and energize a motor, and synthesize a bundle of flagella that smoothly rotate together or fly apart when necessary. The biochemical basis for parts of the system are known. A few of the receptors have been purified, and the center of the adaptation phenomenon, the methylation of three membrane proteins, is partially understood. Whether the biochemical and physical basis for operation of the signal transmission system, the motor, and the flagella can be worked out in the years ahead is an interesting question.

PROBLEMS

21.1. Suppose a strain of chemotactic cells with no endogenous energy source is starved of all possible sources of metabolic energy. Why should they swim for a minute or two after dilution from high potassium to low potassium containing valinomycin (a potassium ionophore)?

21.2. What information do we learn about the chemotaxis system by the finding that the steeper the gradient in attractant concentration, the faster the bacteria swim up the gradient?

21.3. How would you isolate chemotactic mutants (a) that blind to the usual attractant, (b) that are temperature-sensitive and blind, (c) with an altered K_m for an attractant?

21.4. What would be the behavior of a singly flagellated cell when the flagellum changes from counterclockwise to clockwise rotation?

21.5. How fast do bacteria move when they swim?

21.6. In the case of repellent versus attractant, how could you determine whether the response to the situation of repellent and attractant is just sometimes being attracted and other times being repelled or an algebraic summation of both desires to produce a uniform response?

21.7. Through an ordinary light microscope, does a right-handed screw thread look right-handed or left-handed, that is, do bacteria rotating clockwise actually appear to be rotating clockwise when observed through the microscope?

21.8. Is the model proposed on the basis of physiological information in Sensory Transduction in *Escherichia coli:* Role of a Protein Methylation Reaction in Sensory Adaptation, M. Goy, M. Springer, J. Adler, Proc. Natl. Acad. Sci. USA 74, 4964–4968 (1977), essentially the model proposed in Transient Response to Chemotactic Stimuli in *E. coli,* H. Berg, P. Tedesco, Proc. Natl. Acad. Sci. USA 72, 3235–3239 (1975)? Explain how they are fundamentally the same or fundamentally irreconcilable.

21.9. How would you determine from which end the flagellum, exclusive of the basal parts and hook, grows? Why would anyone want to know?

21.10. In terms of the amount of carbohydrate consumed by a bacterial cell in one doubling time, how much carbohydrate metabolism is required to provide the energy necessary for a cell to swim 5 cm?

21.11. Polymerized flagellin exists in at least two conformations. Must any bonds be made or broken during a transition between conformations? If so, what experiments could confirm their existence?

21.12. In light of the Macnab paper, Proc. Natl. Acad. Sci. USA 74, 221–225 (1977), is it still reasonable to consider phage chi to be brought to the cell surface rotation of the flagella as a nut is drawn to the head of a bolt by rotation?

21.13. The chapter has discussed run-tumble, counterclockwise-versus-clockwise flagellar rotation for variations in attractant concentrations. What must the responses be for increases and decreases in repellent concentrations?

21.14. Show that the diffusion of sugar to the surface of a bacterium at 10^{-6} M is just adequate to support a 30-minute cell doubling time.

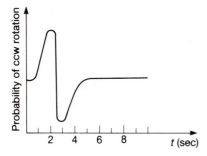

Figure 21.15   One possible probability of counterclockwise rotation of flagella following an impulse of attractant.

21.15. Look up Weber's law concerning the magnitudes of stimulation required to evoke responses. Consider a system in which the ratio of the increase in the magnitude of the stimulus to the magnitude itself must be greater than or equal to some constant value for a response to be evoked. Why does or does not such a system follow Weber's law?

21.16. Why is the separation of both the *tsr* and the *tar* proteins according to their methylation on SDS polyacrylamide gels unexpected, and what is a likely explanation for the separation?

21.17. Both the methyltransferase minus and methylesterase minus mutants are not observed to methylate the *tsr* and *tar* proteins in using the standard protocols of treating cells with chloramphenicol and then adding radioactive methionine. Why is this surprising on first consideration, and what is the likely explanation?

21.18. What is the likely chemotaxis behavior of methyltransferase minus and methylesterase minus mutants?

21.19. Suppose the response to a small, short pulse of attractant at $t = 0$ produced the change in rotation probabilities shown in Fig. 21.15. What modifications to the circuit shown in Fig. 21.13 would be required to generate the delay in the response and the symmetric output? Why is or is not this particular response biologically sensible?

RECOMMENDED READINGS

Flagellar Rotation and the Mechanism of Bacterial Motility, M. Silverman, M. Simon, Nature 249, 73–74 (1974).

Change in Direction of Flagellar Rotation Is the Basis of the Chemotactic Response in *Escherichia coli*, S. Larsen, R. Reader, E. Kort, W. Tso, J. Adler, Nature 249, 74–77 (1974).

Dynamic Properties of Bacterial Flagellar Motors, H. Berg, Nature 249, 77–79 (1974).

Sensory Transduction in *Escherichia coli*: Role of a Protein Methylation Reaction in Sensory Adaptation, M. Goy, M. Springer, J. Adler, Proc. Nat. Acad. Sci. USA 74, 4964–4968 (1977).

RELATED REVIEWS, BOOKS, AND ARTICLES

Chemotaxis in Bacteria, J. Adler, Ann. Rev. Biochem. *44*, 341–356 (1975).

Protein Methylation in Behavioural Control Mechanisms and in Signal Transduction, M. Springer, M. Goy, J. Adler, Nature *280*, 279–284 (1979).

Biochemistry of Sensing and Adaptation in a Simple Bacterial System, D. Koshland, Jr., Ann. Rev. Biochem. *50*, 765–782 (1981).

Bacterial Chemotaxis, A. Boyd, M. Simon, Ann. Rev. Physiol. *44*, 501–517 (1982).

DEEPER READING

Effect of Amino Acids and Oxygen on Chemotaxis in *Escherichia coli*, J. Adler, J. Bact. *92*, 121–129 (1966).

Chemotaxis in Bacteria, J. Adler, Science *153*, 708–716 (1966).

Nonchemotactic Mutants of *Escherichia coli*, J. Armstrong, J. Adler, M. Dahl, J. Bact. *93*, 390–398 (1967).

Chemoreceptors in Bacteria, J. Adler, Science *166*, 1588–1597 (1969).

Role of Galactose Binding Protein in Chemotaxis of *E. coli* Toward Galactose, G. Hazelbauer, J. Adler, Nature New Biol. *230*, 101–104 (1971).

Fine Structure and Isolation of the Hook-Basal Body Complex of Flagella from *Escherichia coli* and *Bacillus subtilis*, M. DePamphilis, J. Adler, J. Bact. *105*, 384–395 (1971).

Attachment of Flagellar Basal Bodies to the Cell Envelope: Specific Attachment to the Outer, Lipopolysaccharide Membrane and the Cytoplasmic Membrane, M. DePamphilis, J. Adler, J. Bact. *105*, 396–407 (1971).

The Gradient-Sensing Mechanism in Bacterial Chemotaxis, R. Macnab, D. Koshland, Proc. Nat. Acad. Sci. USA *69*, 2509–2512 (1972).

Chemotaxis in *E. coli* Analysed by Three-Dimensional Tracking, H. Berg, D. Brown, Nature *239*, 500–504 (1972).

Quantitative Analysis of Bacterial Migration in Chemotaxis, F. Dahlquist, P. Lovely, D. Koshland, Nature New Biol. *236*, 120–123 (1972).

Chemotaxis Toward Amino Acids in *E. coli*, R. Mesibov, J. Adler, J. Bact. *112*, 315–326 (1972).

Bacteria Swim by Rotating Their Flagellar Filaments, H. Berg, R. Anderson, Nature *245*, 380–382 (1973).

Genetic Analysis of Flagellar Mutants in *E. coli*, M. Silverman, M. Simon, J. Bact. *113*, 105–113 (1973).

Chemotaxis Toward Sugars in *E. coli*, J. Adler, G. Hazelbauer, M. Dahl, J. Bact. *115*, 824–847 (1973).

Genetic Analysis of Bacteriophage Mu-Induced Flagellar Mutants in *Escherichia coli*, M. Silverman, M. Simon, J. Bact. *116*, 114–122 (1973).

Chemomechanical Coupling Without ATP: The Source of Energy for Motility and Chemotaxis in Bacteria, S. Larsen, J. Adler, J. Gargus, R. Hogg, Proc. Nat. Acad. Sci. USA *71*, 1239–1243 (1974).

Temporal Stimulation of Chemotaxis in *Escherichia coli*, D. Brown, H. Berg, Proc. Nat. Acad. Sci. USA *71*, 1388–1392 (1974).

Phosphotransferase-System Enzymes as Chemoreceptors for Certain Sugars in *E. coli* Chemotaxis, J. Adler, W. Epstein, Proc. Nat. Acad. Sci. USA *71*, 2895–2899 (1974).

Bacterial Motility and Chemotaxis: Light Induced Tumbling Response and Visualization of Individual Flagella, R. Macnab, D. Koshland, Jr., J. Mol. Biol. *84*, 399–406 (1974).

Bacteria Can Swim Without Rotating Flagellar Filaments, C. Calladine, Nature *249*, 385 (1974).

Data Processing by the Chemotaxis Machinery of *Escherichia coli*, J. Parkinson, Nature *252*, 317–319 (1974).

Positioning Flagellar Genes in *E. coli* by Deletion Analysis, M. Silverman, M. Simon, J. Bact. *117*, 73–79 (1974).

Isolation and Complementation of Mutants in Galactose Taxis and Transport, G. Ordal, J. Adler, J. Bact. *117*, 509–516 (1974).

Properties of Mutants in Galactose Taxis and Transport, G. Ordal, J. Adler, J. Bact. *117*, 517–526 (1974).

"Decision"-Making in Bacteria: Chemotactic Response of *Escherichia coli* to Conflicting Stimuli, J. Adler, W. Tso, Science *184*, 1292–1294 (1974).

Quantitation of the Sensory Response in Bacterial Chemotaxis, J. Spudich, D. Koshland, Proc. Nat. Acad. Sci. USA *72*, 710–713 (1975).

Transient Response to Chemotactic Stimuli in *E. coli*, H. Berg, P. Tedesco, Proc. Nat. Acad. Sci. USA *72*, 3235–3239 (1975).

Role of Methionine in Bacterial Chemotaxis: Requirement for Tumbling and Involvement in Information Processing, M. Springer, E. Kort, S. Larsen, G. Ordal, R. Reader, J. Adler, Proc. Nat. Acad. Sci. USA *72*, 4640–4644 (1975).

Bacterial Behavior, H. Berg, Nature *254*, 389–392 (1975).

Evidence for an s-adenosylmethionine Requirement in the Chemotactic Behavior of *Salmonella typhimurium*, D. Aswad, D. Koshland, Jr., J. Mol. Biol. *97*, 207–233 (1975).

Isolation, Characterization, and Complementation of *Salmonella typhimurium* Chemotaxis Mutants, D. Aswad, D. Koshland, Jr., J. Mol. Biol. *97*, 225–235 (1975).

cheA, cheB, and cheC Genes of *E. coli* and Their Role in Chemotaxis, J. Parkinson, J. Bact. *126*, 758–770 (1976).

Bacterial Flagella Rotating in Bundles: A Study in Helical Geometry, R. Macnab, Proc. Nat. Acad. Sci. USA *74*, 221–225 (1977).

Identification of a Protein Methyltransferase as the *cheR* Gene Product in the Bacterial Sensing System, W. Springer, D. Koshland, Jr., Proc. Nat. Acad. Sci. USA *74*, 533–537 (1977).

A Protonmotive Force Drives Bacterial Flagella, M. Manson, P. Tedesco, H. Berg, F. Harold, C. van der Drift, Proc. Nat. Acad. Sci. USA *74*, 3060–3064 (1977).

Sensory Transduction in *Escherichia coli*: Two Complementary Pathways of Information Processing That Involve Methylated Proteins, M. Springer, M. Goy, J. Adler, Proc. Nat. Acad. Sci. USA *74*, 3312–3316 (1977).

Normal to Curly Flagellar Transitions and Their Role in Bacterial Tumbling: Stabilization of an Alternative Quaternary Structure of Mechanical Force, R. Macnab, M. Ornston, J. Mol. Biol. *112*, 1–30 (1977).

Identification of Polypeptides Necessary for Chemotaxis in *Escherichia coli*, M. Silverman, M. Simon, J. Bact. *130*, 1317–1325 (1977).

A Protein Methylesterase Involved in Bacterial Sensing, J. Stock, D. E. Koshland, Jr., Proc. Nat. Acad. Sci. USA *75*, 3659–3663 (1978).

Inversion of a Behavioral Response in Bacterial Chemotaxis: Explanation at the Molecular Level, S. Khan, R. Macnab, A. DeFranco, D. Koshland, Proc. Nat. Acad. Sci. USA *75*, 4150–4254 (1978).

Complementation Analysis and Deletion Mapping of *Escherichia coli* Mutants Defective in Chemotaxis, J. Parkinson, J. Bact. *135,* 45–53 (1978).

Energetics of Flagellar Rotation in Bacteria, M. Manson, P. Tedesco, H. Berg, J. Mol. Biol. *138,* 541–561 (1980).

The Steady-State Counterclockwise/Clockwise Ratio of Bacterial Flagellar Motors is Regulated by Protonmotive Force, S. Khan, R. Macnab, J. Mol. Biol. *138,* 563–597 (1980).

Multiple Methylation in Processing of Sensory Signals During Bacterial Chemotaxis, A. DeFranco, D. Koshland, Jr., Proc. Nat. Acad. Sci. USA *77,* 2429–2433 (1980).

Genetic and Biochemical Properties of *Escherichia coli* Mutants with Defects in Serine Chemotaxis, M. Hedblom, J. Adler, J. Bact. *144,* 1048–1060 (1980).

Signal Processing Times in Bacterial Chemotaxis, J. Segall, M. Manson, H. Berg, Nature *296,* 855–857 (1982).

Flagellar Hook Structures of *Caulobacter* and *Salmonella* and Their Relationship to Filament Structure, T. Wagenknecht, D. DeRosier, J. Mol. Biol. *162,* 69–87 (1982).

Adaptation in Bacterial Chemotaxis: CheB-Dependent Modification Permits Additional Methylations of Sensory Transducer Proteins, M. Kehry, F. Dahlquist, Cell *29,* 761–772 (1982).

Impulse Responses in Bacterial Chemotaxis, S. Block, J. Segall, H. Berg, Cell *31,* 215–226 (1982).

Asynchronous Switching of Flagellar Motors on a Single Bacterial Cell, R. Macnab, D. Han, Cell *32,* 109–117 (1983).

Isotope and Thermal Effects in Chemiosmotic Coupling to the Flagellar Motor of *Streptococcus,* S. Khan, H. Berg, Cell *32,* 913–919 (1983).

Adenylate Cyclase Is Required for Chemotaxis to Phosphotransferase System Sugars by *Escherichia coli,* R. Black, A. Hobson, J. Adler, J. Bact. *153,* 1187–1195 (1983).

Adaptation Kinetics in Bacterial Chemotaxis, S. Block, J. Segall, H. Berg, J. Bact. *154,* 312–323 (1983).

Correlation Between Bacteriophage Chi Adsorption and Mode of Flagellar Rotation of *Escherichia coli* Chemotaxis Mutants, S. Ravid, M. Eisenbach, J. Bact. *154,* 604–611 (1983).

Successive Incorporation of Force-Generating Units in the Bacterial Rotary Motor, S. Block, H. Berg, Nature *309,* 470–472 (1984).

Direction of Flagellar Rotation in Bacterial Cell Envelopes, S. Ravid, M. Eisenbach, J. Bact. *158,* 222–230 (1984).

Structure of the Trg Protein: Homologies with and Differences from Other Sensory Transducers of *Escherichia coli,* J. Ballinger, C. Park, S. Harayama, G. Hazelbauer, Proc. Nat. Acad. Sci. USA *81,* 3287–3291 (1984).

Oncogenesis, Molecular Aspects

The cells of multicellular organisms face different growth problems from bacteria. They must differentiate, proceed through the proper number of divisions, and then stop growing despite the presence of ample nutrients. How can these requirements be met? On one hand, a cell's lineage could regulate the processes. The first twenty generations of descendants of a particular cell could, for example, become a liver and then the cells could stop dividing. This type of scheme operates in a nematode, *Caenorhabditis elegans*, which is being carefully studied by geneticists and molecular biologists, and portions of developmental pathways in higher organisms make use of this principle as well. Chemically or electrically signaling to one another is a second approach cells can take. By this they can learn their positions within an organism before developing into tissues appropriate to their locations. The ability of many tissues to heal following a wound or to regenerate following partial excision indicates that this mechanism also functions in some organisms.

We have already seen that the elementary regulatory mechanisms found in bacteria can be damaged by mutations. The regulatory mechanisms of higher cells can also be damaged. A breakdown in the cellular growth regulators can lead to uncontrolled growth in an organism—cancer. Such a state is expressed *in vitro* in cell culture as the failure of the cells to stop growing when they have grown to form a confluent monolayer. The cells are said to be transformed to a noninhibited state. That is, these cells are no longer contact-inhibited. It is not hard to imagine that alterations in regulator sites, promoter activity, translation efficiency, structure of any of a variety of proteins in a cell, or introduction of a new gene into a cell could all result in uncontrolled growth. The experimental question is determining which is the cause of a particular type of cancer as well as determining how the alteration changes the cell.

This chapter focuses on the application of principles and techniques discussed in earlier chapters to the determination of molecular alterations that lead to the cancerous state in plants and animals. The information resulting from such studies is providing valuable windows for an examination of the workings of normal cells and indicates that about 30 genes play highly important roles in regulating the growth of animal cells. However, it is not yet clear whether the information gained from these studies will be directly applicable to the treatment of cancer patients.

BACTERIALLY INDUCED TUMORS IN PLANTS

Plants appear to be simpler than mammals and yet they are susceptible to tumors just like mammals. Therefore, some of the lessons we learn about plant tumors may be helpful in the analysis of animal tumors. One feature that makes the study of plant cells and plant tumors valuable in research on oncogenesis is that with some kinds of plants, cell cultures can be maintained indefinitely *in vitro,* and, when desired, these undifferentiated cells can be induced to differentiate back into normal plants that reproduce sexually. Such techniques permit detailed analysis of causes of oncogenesis. In addition, these approaches have the potential for yielding valuable mutants by using techniques similar to those used with bacteria for the isolation of single mutant cells from large cultures and then regenerating complete mutant plants.

The bacterium *Agrobacterium tumefaciens* can induce the growth of masses of undifferentiated cells, called crown galls, in susceptible plants. This transformation to the undifferentiated state requires a 200-Kb plasmid carried by the bacterium. In the transformation process at least 8 to 10 Kb of DNA from the plasmid are transferred into the plant cells. There the DNA is integrated into the chromosome of the plant, where it is replicated along with the cellular DNA. As a result, all cells of the crown gall contain the DNA originating from the plasmid. Part of the integrated

Figure 22.1   Structures of octopine and nopaline.

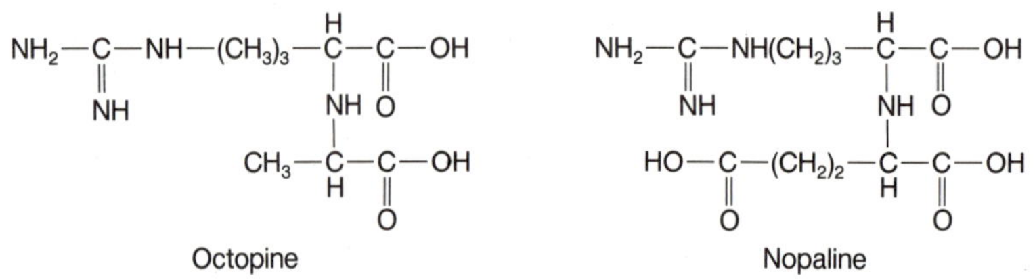

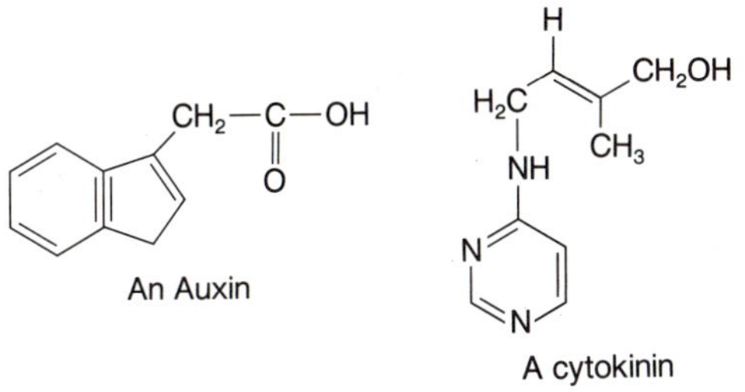

Figure 22.2   Indole 3-acetic acid, an auxin, and 6-(4-hydroxy-3-methyl-trans-2-buten-ylamino)purine, a cytokinin.

plasmid DNA directs the plant to synthesize and excrete the compounds octopine or nopaline (Fig. 22.1). In turn, these compounds can be catabolized by the *Agrobacter* bacteria in the crown gall. Few other bacteria or parasites can utilize these compounds for growth. Thus, one *Agrobacterium* subverts part of the plant to produce nutrients for a large bacterial colony.

In contrast to the nontransformed plant cells, the cells from crown galls do not require the growth factors auxin and cytokinin for their continued growth in culture (Fig. 22.2). Ordinarily, medium for plant cells requires the presence of these two small-molecule growth factors in addition to a variety of other metabolites. With the ratio of auxin and cytokinin at one value, plant cells in culture remain undifferentiated, but if the ratio of auxin to cytokinin is increased, stems and leaves tend to develop. Conversely, if the auxin to cytokinin ratio is decreased, roots tend to develop, and without either auxin or cytokinin present, the cells do not grow. In the whole plant, auxin is synthesized in the stem tips and cytokinin in the root tips. A concentration gradient in these molecules from the top to bottom of the plant helps cells identify their positions and develop appropriately. Plant cells transformed with *Agrobacter* do not require either auxin or cytokinin for growth. It is likely that the DNA that was acquired from the plasmid by the transformed cells directs or induces synthesis of auxin and cytokinin-like substances that substitute for these chemicals both in the crown gall and in cell culture. By comparison to the crown galls in plants, it is likely that many types of animal cancers will involve alterations in the cell's synthesis of or response to growth factors.

TRANSFORMATION BY DAMAGING THE CHROMOSOME

Epidemiologists have long recognized that certain chemicals induce cancer, and Ames has shown that many bacterial mutagens are carcinogens. The mutagenicity of chemicals, as detected by the reversion of a set of histidine mutants, correlates well with their carcinogenicity. Therefore chromosome damage in the form of mutations or small insertions and deletions is one cause of cancer. The next section explains

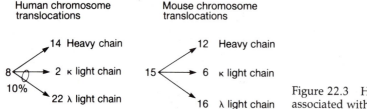

Figure 22.3   Human and mouse translocations associated with cancers of the immune system.

how the exact base change responsible for induction of one particular cancer has been determined.

In addition to small lesions, chromosomal rearrangements can alter either the expression or the structure of RNAs or proteins. Such chromosomal rearrangements are frequently observed in a class of human cancers called Burkitt's lymphomas and in mouse cancers called plasmacytomas. Just as areas of the *Drosophila* polytene chromosomes can be identified by their banding pattern in the light microscope, so also can areas of individual human chromosomes. However, the bands of the mouse or human chromosomes that appear after trypsin treatment and staining with Giemsa stain are smaller than the *Drosophila* polytene bands, and fewer are formed or can be recognized. Nonetheless, the chromosomes involved in some translocations can be identified. For example, the majority of Burkitt's lymphomas contain cells with part of chromosome 8 translocated to chromosome 14 (Fig. 22.3).

Lymphomas are cancers of immunoglobulin-excreting cells, and therefore it is likely that the chromosome rearrangements in Burkitt's lymphomas actually involve the immunoglobulin genes. Indeed, this is the case. The actual site of the chromosome translocation from a Burkitt's lymphoma has been identified by cloning the

Figure 22.4   Schematic of the heavy chain region of human chromosome 14 and the approximate location of the material from chromosome 8 that is translocated to chromosome 14. Transcripts originating in the region are also shown.

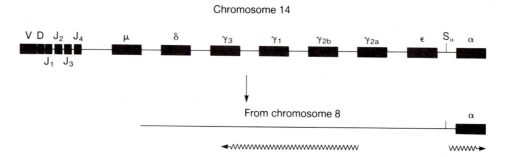

heavy chain constant region from lymphoma and nonlymphoma cells. The lymphoma cells contained an insertion of DNA from chromosome 8 within the switch region of the heavy chain S_α (Fig. 22.4). The junction of the insertion sometimes contains a sequence resembling the DNA it replaced, implying that a recombination event, or an immunoglobulin switch-related event, was responsible for the translocation. The DNA that has been translocated to the immunoglobulin region likely codes for one of the cell's growth factors. As a result of the translocation, its expression is increased and the cells are transformed. One way the expression could be increased would be for the region into which the growth factor gene has been translocated to contain an enhancer. This could be, for example, the enhancer found between the J region and the C region of the immunoglobulin heavy chain.

IDENTIFYING A NUCLEOTIDE CHANGE CAUSING CANCER

As mentioned above, since mutagens are carcinogenic, it is likely that some cancers are caused by mutation. How could we determine the nucleotide change or changes that cause a human cell to be transformed? If such a change could be detected, it would be important to know if the gene that has been altered, called an oncogene, codes for a cellular growth factor. In principle, comparing the nucleotide sequences of all the chromosomes from a normal cell and a transformed cell would reveal the change. However, even if such an ill-advised attempt to sequence about 6×10^9 base pairs of DNA could be accomplished, the important change would likely be masked by the multitude of extraneous nucleotide changes resulting from either random mutation or from the mutagenic treatment that generated the base change that caused the cancer in the first place.

To reduce to manageable proportions the problem of locating the nucleotide change that causes a cancer, a powerful selection technique for obtaining the critical DNA fragment is required. Cells growing in culture can be used in screening DNA for its ability to induce the cancerous state. For example, one line of mouse cells, NIH 3T3 cells, has become adapted for continuous growth in culture, in contrast to many cell lines that grow for only a limited number of generations. The NIH 3T3 cells grow well in culture and can be considered to be partway along the pathway leading to transformation, but they do display contact inhibition. However, if one of these cells is transformed, it loses its contact inhibition, and its descendants continue growing when the other cells in the monolayer have stopped. The resulting pile of cells is easily detected and is called a focus. Weinberg and others have found that DNA extracted from 10% to 20% of different human tumors can transform NIH 3T3 cells to form foci.

Although the focus assay for transforming DNA fragments could be used as the basis for a straightforward biochemical purification of the desired DNA fragments, genetic engineering methods permitted more rapid isolation of the DNA. The basic trick was to tag the desired DNA fragment with a sequence not normally found in the

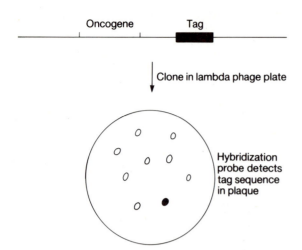

Figure 22.5   The use of a tag to permit detection of an oncogene when cloning into *E. coli*. The desired plaque contains material complementary to the tag known to be associated with the oncogene.

cultured mouse cells or in *Escherichia coli* (Fig. 22.5). Then the human-derived sequence could be distinguished from mouse or even from *E. coli* DNA. Suppose for a moment that such tagging has been accomplished. Then a gene library from the transformed cells can be made in a lambda phage or plasmid vector. One of the lambda phage or plasmids will then contain the fragment that causes the cancer, and alongside the sequence will be the tagging sequence. This particular phage or plasmid can be immediately identified by hybridization using an appropriate probe to recognize the tag.

Two methods can be used for tagging the oncogene isolated from cancerous cells. As mentioned in Chapter 18, one of the highly repetitive sequences in human DNA is called the *Alu* sequence because it is cleaved more than once by the *Alu* restriction enzyme, and therefore the resulting fragment readily shows up upon electrophoresis of *Alu*-restricted human DNA (Fig. 22.6). The mouse genome lacks this sequence. Simply because of its high repetition number, an *Alu* sequence was likely to lie near the mutation responsible for transformation. Indeed, one did, and

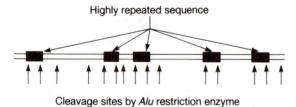

Figure 22.6   A large number of fragments of exactly the same length is generated by *Alu* restriction enzyme cleavage of human DNA.

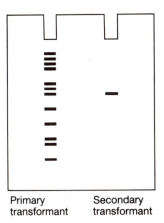

Primary   Secondary
transformant   transformant

Figure 22.7   Appearance of a Southern transfer of DNA extracted from primary and secondary mouse transformants in which the radioactive probe would be the human *Alu* sequence.

Weinberg was able to use this as a tag of the human mutated sequence in the mouse cells or in recombinant lambda phage containing the gene.

A second method for tagging the oncogene is direct and general. DNA isolated from a tumor can be digested with a restriction enzyme, and the desired tagging sequence can be ligated to it. Then the mixture of DNAs can be added to the cultured mouse cells for selection of the fragment containing the oncogene.

A minor technical point must be mentioned. The cultured cells that take up the human DNA usually take up many molecules. Therefore, although a focus of transformed cells will contain the desired sequence that is responsible for the transformation, it is likely to contain a number of other irrelevant human sequences as well. Using DNA extracted from this primary round of transformation for a second round of transformation should eliminate these extraneous human sequences from the transformants. Indeed, after the first cycle, a Southern transfer prepared from DNA of a transformed cell revealed many sizes of restriction fragments containing the human-derived *Alu* tagging sequence (Fig. 22.7). However, after a second cycle of transformation, all the transformants could easily be shown to contain only a single copy of the tagging sequence. Presumably this was directly linked to the stretch of DNA that is responsible for the transformation.

The techniques described above permitted the cloning and isolation of DNA fragments 5 to 10 Kb long that contained the cellular oncogene from bladder carcinoma cells, which could transform NIH 3T3 cells. Once these fragments were obtained, they could be used as probes in screening genomic libraries made from DNA isolated from noncancerous cells. Thus the predecessor of the oncogene could also be obtained. Then the question was merely to determine the difference between the two DNAs. Again, they could have been totally sequenced, but cleverness reduced the work. Restriction fragments between the two were exchanged to find a small DNA segment that was necessary for transformation (Fig. 22.8).

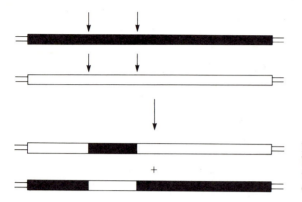

Figure 22.8   Restriction fragment interchange between the cellular homolog of the oncogene and the oncogene itself, which can identify the fragment containing the critical base change.

A single nucleotide difference separates the normal cellular prototype and its oncogenic variant. The change occurs in the thirteenth position of a coding region and changes a glycerine to a serine. Since both Northern transfers of total cellular RNA and immune precipitations of the actual gene product involved show that neither transcription of the gene nor its translation efficiency is appreciably altered by the mutation, the structural change in the protein must be the cause of its oncogenicity.

RETROVIRUSES AND CANCER

Just as bacteria can induce a type of cancer in plants, bacteria or viruses could cause cancer in animals. One method to determine whether these agents can do so in animals is to induce cancer in animals using a cell-free or bacteria-free extract prepared from tumor cells. One of the first such isolations was made by Peyton Rous from a cancer of connective tissue — a sarcoma — isolated from a chicken. Hence the virus is called Rous sarcoma virus, or RSV. The procedure has revealed that RSV is a member of a class of RNA viruses that cause sarcomas in birds, mice, cats, monkeys, and other animals. These viruses are called murine, feline, and simian sarcoma viruses. Similar techniques have been used to detect cancers of various types of blood cells from nonhumans, and the causative viruses are called leukemia viruses. Detection of a human tumor virus has been much more difficult.

The sarcoma and leukemia viruses are both retroviruses. That is, the nucleic acid within the virus particle is single-stranded RNA, and yet this RNA can be duplicated into double-stranded DNA and integrated into the genome of some infected cells (Fig. 22.9). The enzyme reverse transcriptase that is packaged within the virus particle performs the replication. Initiation of the replication, much like replication from DNA templates, requires a free hydroxyl for initiation. Remarkably, a tRNA molecule hybridized to the virus DNA and packaged within the virus provides the

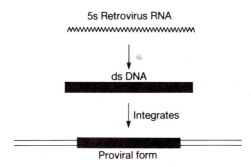

5s Retrovirus RNA

ds DNA

Integrates

Proviral form

Figure 22.9   Retrovirus converts from single-stranded RNA in a virion to double-stranded DNA in proviral form found in the chromosome.

necessary hydroxyl (Fig. 22.10). The conversion to double-stranded DNA slightly rearranges and duplicates the ends of the genome to generate the long terminal repeats, LTRs, from the U3 and U5 regions. This process most likely makes use of the two genetically identical RNA molecules that are encapsidated within the virion. After its duplication, the linear double-stranded DNA migrates to the nucleus, where a few molecules circularize and perhaps one integrates with little sequence specificity into the chromosome. Such an integration into a germ line cell permits retroviruses to be passed vertically from one generation to the next in addition to the spread of the virus horizontally by infection.

Nondefective retroviruses are able to grow within cells without the assistance of a helping virus. Three essential viral genes are found in nondefective retroviruses: *gag, pol,* and *env.* In Rous sarcoma virus these three genes encode, respectively, four

Figure 22.10   Initiating the reverse transcription process from tRNA and copying to generate the LTRs from the R, U3, and U5 sequences.

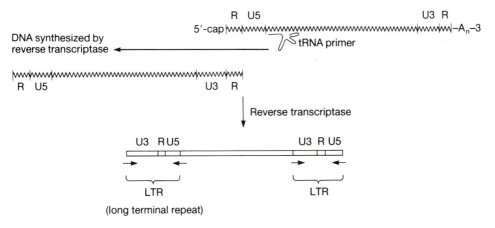

structural proteins found within the virus; the reverse transcriptase, its associated RNase H, an enzyme that degrades RNA in an RNA-DNA duplex, as well as a DNA endonuclease; and two virus envelope glycoproteins. The virus uses two methods to synthesize this large group of different proteins from just three genes and one major promoter.

Eukaryotes do not use polycistronic messengers. As a rule, just the first AUG in a messenger is used as the translation start point; in some cases a slightly more elaborate signal is required, but multiple translation initiations do not occur on a single messenger. One method to generate multiple proteins from one transcription unit is to cleave a long precursor-polypeptide into the required polypeptides. This strategy is used in the *gag* gene to generate the four internal viral proteins.

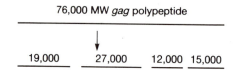

A more elaborate scheme is used for translation of the *pol* and *env* genes. For synthesis of the *pol* gene product, the retrovirus mRNA is spliced so that the N-terminal portion of *gag* is fused to the *pol* gene. After translation of this hybrid messenger, the *gag*-encoded portion is cleaved off. Splicing also places the *env* gene at the 5' end of messenger so that it can be translated. Its product also is cleaved to generate two polypeptides.

The integrated or proviral form of the retrovirus is transcribed to yield capped and polyadenylated mRNA, which is both spliced as described above and then

Figure 22.11   Transcription of the proviral form yields messenger for translation or RNA for encapsidation. Transcription initiates or is processed to the left edge of the R site and terminates or is cleaved at the right edge of the second R site.

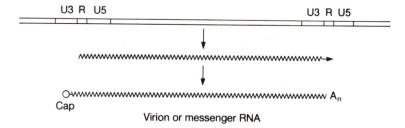

translated or which is encapsidated and incorporated into new virions (Fig. 22.11). Therefore the transcription start point lies at the beginning of the R region of the virus. The promoter that must therefore lie in the U3 region exists in two copies, one at each end of the virus. The right-hand promoter is available to transcribe whatever cellular DNA the virus has integrated near. Therefore a retrovirus can act like a portable promoter, usually causing no damage when it turns on a host gene. However, when it turns on certain genes, the cell can be transformed and lose normal growth regulation.

Acutely transforming retroviruses are capable of inducing tumor formation in several weeks to several months after injection into animals. These carry an additional gene, an oncogene. Such acutely transforming viruses usually are defective, however, and lack functional *pol* or *env* genes. Therefore they must be grown in association with their nondefective relatives, analogous to complementation by lambda of a defective transducing phage. RSV is an exception to the general case, as this virus is nondefective and carries an oncogene, called *src*. Translation of the *src*

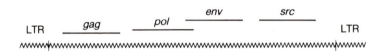

transcript is accomplished by a splicing reaction that places the AUG codon for this protein near the 5′ end of the RNA molecule.

Retroviruses lacking oncogenes are also capable of inducing tumors in susceptible animals, but more slowly than the acutely transforming retroviruses. These slow retroviruses often take a year or more to generate tumors. Apparently, out of a great number of virus-infected cells, a small number, often only one, contain a retrovirus inserted adjacent to a cellular gene involved with growth regulation. The nearby location of the retrovirus increases expression of the gene, either from readthrough of a viral transcript or from increased expression from the cellular promoter as a result of an enhancer sequence within the virus. The virus may also promote rearrangement of the gene so that variant protein is synthesized.

CELLULAR GENES HAVE RETROVIRAL COUNTERPARTS

Are the viral oncogenes of acutely transforming viruses related to any cellular genes? This is a good question, for a number of reasons. Since chromosomal rearrangements point to specific genes as being involved in transformation to the cancerous state, genes related to these may be carried on the retroviruses. Also, since cancer can be induced by mutagens, it would seem likely that the mutations responsible could create genes closely similar to those carried on the acutely transforming retroviruses. Finally, occasionally a slow retrovirus becomes acutely transforming by the substitution of part of its genetic material by host material. Most likely the new gene originates from the chromosome of the cells in which the virus has been growing.

The question as to whether the viral oncogenes are related to cellular genes can easily be answered by Southern transfers. Undue fears of dangers from cloning such DNAs have passed, and viral oncogenes may now be handled without elaborate precautions. The viral oncogenes from a number of retroviruses have been cloned by making cDNA from the virus and ligating it into plasmid or phage vectors. Then the appropriate restriction fragments have been used to probe Southern transfers of DNA extracted from noncancerous animals. Most surprisingly, chickens, mice, and humans all possess a sequence of DNA with high homology to the src gene. Clearly this gene has not evolved rapidly in the time since these animals diverged from one another during the course of evolution. This implies that the cellular function of this protein is frozen and most likely closely tied to other cellular functions.

The viral form of the src gene is called v-src and the cellular form is c-src. Homology measurements similar to those carried out for Rous sarcoma virus have been done for other acutely transforming retroviruses (Table 22.1). Most of the viral oncogenes in these viruses possess cellular counterparts as well. About 30 different viral and cellular forms of the oncogenes are known. This means that a reasonably small number of genes are involved and that there is some hope of being able to deduce how the products of these genes function in normal and transformed cells.

The mutated gene that Weinberg found in the human bladder carcinoma cells is carried in a slightly modified form of a rat-derived Harvey murine sarcoma virus.

Table 22.1   Oncogenes Found on Some Retroviruses

Gene	Virus	Likely Origin
src	Rous sarcoma virus	Chicken
myc	Avian myelocytomatosis virus	Chicken
erbA	Avian erythroblastosis virus	Chicken
erbB	Avian erythroblastosis virus	Chicken
myb	Avian myeloblastosis virus	Chicken
ras	Kirsten murine sarcoma virus	Rat
sis	Simian sarcoma virus	Monkey

The gene, named c-*myc,* which translocates into the heavy chain locus of the immunoglobulin genes in Burkitt's lymphoma, has also been identified on a retrovirus.

IDENTIFICATION OF THE *src* AND *sis* GENE PROTEINS

Eukaryotic cells contain thousands of proteins. Most likely only a few of these are involved in regulation of growth, and these probably are synthesized in small quantities. Therefore it would seem like a difficult task to identify a protein as the product of either c-*src* or v-*src* and virtually impossible to determine the enzymatic activity of such a protein. Nonetheless, Erikson accomplished both for v-*src.*

The starting point of this work was the fact that animals with sarcomas or lymphomas frequently synthesize antibodies against the retroviral proteins. The v-*src* protein might also induce antibody synthesis in animals if it were sufficiently different in structure from all other cellular proteins. Therefore rabbits were infected with an avian sarcoma virus that could induce sarcomas even in some other types of animals. The same virus was also used to transform chick embryo fibroblasts *in vitro.* These chicken cells ought to synthesize the same viral proteins as the rabbit. As a result, these proteins would be recognized by the rabbit antibodies. In addition, the chicken embryo fibroblasts ought to contain very few of the other proteins that would be recognized by the rabbit antibodies. Therefore, the v-*src* protein ought to be one of the very few proteins recognized by the antibodies.

Extracts from radioactively labeled RSV-transformed chick cells were incubated with the serum from the Rous-infected rabbits. Since the quantities of proteins bound by the antibodies were low, an antigen-antibody lattice was unlikely to form. Another method had to be used for selectively isolating the antigen bound to antibody. Erikson used whole cells of *Staphylococcus aureus* because one of its surface proteins, protein A, specifically binds to IgG. After incubation of rabbit serum with the radioactive chick proteins, the *Staphylococcus* cells were added and the mixture

Protein A   IgG   pp60^{src}

centrifuged to separate antibody-bound radioactive proteins from those not bound to antibody.

About 20 radioactive proteins were precipitated by the rabbit serum. However, among them was one that was present only if the labeled proteins originated from transformed chick cells. Furthermore, this protein was not present if the cells had been infected with a nonacutely transforming retrovirus. In addition, the protein was the same molecular weight as a protein synthesized *in vitro* using the oncogene portion of RSV as a messenger. In this way, the v-*src* product was identified as a 60,000-molecular-weight protein, which is called pp60*src*.

An enzymatic activity associated with the v-*src* protein was discovered on the basis of an educated guess. Since the protein probably is involved with regulating growth and since phosphorylation of proteins is involved in regulating many activities in eukaryotic cells, the *src* product could be a protein kinase. Therefore radioactive ATP and potential kinase substrate proteins were included with the *Staphylococcus*-precipitated IgG-precipitated *src* protein. Indeed, a kinase activity was detected. It was proved to be the product of the *src* gene because it was absent from cells grown at high temperature that had been transformed with mutants of RSV that do not transform at high temperature. Together, these pieces of evidence provide solid proof that the kinase was the product of v-*src*. Another related question is whether the antibodies against the v-*src* gene product could detect the c-*src* protein. They could, even though the protein is synthesized at a small fraction of the level at which the viral homolog is synthesized. The cellular protein also is a kinase. Similar studies have permitted partial purification of a number of other viral and cellular oncogene products. The cellular locations of these proteins can also be determined in many cases.

The actual cellular function of the oncogene carried on a simian retrovirus, v-*sis*, has also been identified. This was not done via antibodies and enzyme assays. Instead, the nucleotide sequence of its oncogene identified its product as a protein that has been purified for entirely different reasons from blood plasma, the platelet-derived growth factor. This protein is released by platelets at the sites of wounds and stimulates growth of fibroblasts, smooth muscle cells, and glial cells. Here is one clear case of computers being of great help to molecular biologists. When the amino acid sequence for the growth factor was entered into a data base and homologies were sought, the simian virus oncogene immediately was revealed as encoding an extremely close relative. Almost certainly the retrovirus gene derives from the growth factor gene.

DNA VIRUSES AND CANCER

DNA viruses are less closely associated with cancer than are the retroviruses because their infection does not usually yield tumors and, except for papilloma viruses, which cause warts, DNA viruses cannot readily be found in tumors. However, papovaviruses, including SV40, and the very closely related polyoma virus, adeno-

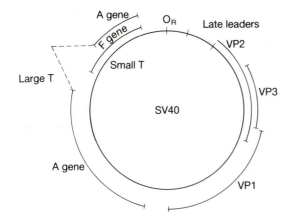

Figure 22.12   Genes of SV40 showing the splicing pattern that generates the oncogenes large T and small t.

viruses, and herpes viruses are capable of transforming cells in culture so that they lose their contact inhibition or are able to grow into colonies from isolated cells. Such transformed cells can then be injected into animals, where they cause tumors.

SV40 and polyoma both encode two or three proteins, which are collectively known as T antigens (Fig. 22.12). These have been identified immunologically in the same way the retroviral oncogene proteins were identified. The antibodies obtained from animals carrying SV40 and polyoma-induced tumors have been used to show that the T antigen genes are expressed early after infection and that the T antigens bind to proteins in the nucleus, to DNA, and to other cellular structures. The presence of one of the T antigens alone is usually insufficient to transform cells, suggesting that in general the process of transformation requires a number of steps. This conclusion is in accord with the fact that the NIH 3T3 cell foci-inducing DNAs isolated from tumors are not capable of transforming most other cell lines.

The activity of the T antigens has been a subject of great interest, but the low levels at which they are synthesized *in vivo* has greatly hampered their study. They are, therefore, perfect subjects for genetic engineering. Indeed, by properly fusing the DNA encoding the different proteins to suitable Shine-Dalgarno ribosome-binding sequences, these proteins have been hypersynthesized in bacteria, and the proteins can be used for biochemical studies.

DIRECTIONS FOR FUTURE RESEARCH IN MOLECULAR BIOLOGY

It is clear that knowledge of the eukaryotic cell is far from our understanding of lambda phage or *E. coli,* but our understanding of even these simpler objects is far from complete. Therefore we can perceive three general areas for future research in genetics and molecular biology. The systems like *lac,* lambda phage, and a few others will be studied in ever greater depth so that we may understand as much as possible about the physics, chemistry, and biochemistry of these systems. An ultimate goal of this direction of work is the ability to predict an amino acid sequence such that a polypeptide chain would fold up and bind to any desired sequence on a nucleic acid or possess any reasonable enzymatic activity.

A second general area of work will be understanding cellular processes. This includes study of the mechanisms used in nature for regulating gene and enzyme activity; the complicated interactions between metabolic pathways; and the mechanisms of many functions such as movement, active transport, RNA splicing, cell division, and receptor function. A third area of future work will be that of understanding organisms as a whole. This will include how cells and tissues develop and differentiate and how they signal to one another and then respond.

PROBLEMS

21.1. Recalling that either of the X chromosomes in each cell of a woman is inactivated, how would you determine whether a carcinoma derived from transformation of a single cell as contrasted to transformation of multiple cells?

21.2. Rous sarcoma virus contains two identical single-stranded RNA molecules, each of which contains all the genetic information of the virus. Devise an experiment to test the conjecture that recombination between the two genomes, not necessarily in their RNA form, frequently generates an intact genome from two damaged ones.

21.3. What control experiments are necessary to ensure reliability in the series of experiments in which serum from an animal with a sarcoma was used to precipitate proteins from cultured transformed cells and thereby identify the *src* gene product?

21.4. In the experiments examining the kinase activity of v-*src*, what experiments would you do next if you observed that protein phosphorylation was nearly instantaneous and that the rate was independent of the volume of the reaction?

21.5. In experiments to determine which portion of the mutated gene conferred tumorgenicity, restriction fragments were exchanged between the oncogenic and nononcogenic forms. This sort of experiment is fraught with danger, however, for false positives could arise as a result of incomplete restriction enzyme digestion, minor DNA contamination, and so on. What controls can be put into this experiment to warn of such pitfalls?

21.6. Why is injection of a retrovirus or DNA virus into a newborn or young animal more likely to generate a tumor than injection into a mature animal?

21.7. Suppose you had a collection of tumors whose DNA was capable of transforming cells in culture. What is the most efficient way to determine, not with absolute certainty but with a high degree of confidence, whether the oncogenes involved are identical?

21.8. In transforming susceptible cells in a search for the DNA sequences involved, one might worry that several disconnected DNA segments may be required. How could this possibility be excluded by varying the transformation conditions?

21.9. Remarkably, monoclonal antibody against the oncogene product responsible for the EJ bladder carcinoma was available for comparing the translation efficiency of the cellular prototype and the oncogenic version. By searching through the literature, find how this antibody was obtained.

21.10. Exposure to UV light can induce cancers that can ultimately be traced to retroviruses. Propose a hypothesis to explain this finding and an experiment to test your idea.

RECOMMENDED READINGS

Carcinogens Are Mutagens: A Simple Test System Combining Liver Homogenates for Activation and Bacteria for Detection, B. Ames, W. Durston, E. Yamasaki, F. Lee, Proc. Nat. Acad. Sci. USA *70*, 2281–2285 (1973).

Activation of a Cellular *onc* Gene by Promoter Insertion in ALV-Induced Lymphoid Leukosis, W. Hayward, B. Neel, S. Astrin, Nature *290*, 475–480 (1981).

Dietary Carcinogens and Anticarcinogens, B. Ames, Science *221*, 1256–1264 (1983).

The Embryonic Cell Lineage of the Nematode *Caenorhabditis elegans*, J. Sulston, E. Schierenberg, J. White, Dev. Biol. *100*, 64–119 (1983).

RELATED REVIEWS, BOOKS, AND ARTICLES

Molecular Biology of Tumor Viruses, ed. J. Tooze, Cold Spring University Laboratory (1973).

Molecular Biology of Tumor Viruses, Second Edition, DNA Tumor Viruses, ed. J. Tooze, Cold Spring Harbor Laboratory (1980).

Molecular Biology of Tumor Viruses, Second Edition, RNA Tumor Viruses, ed. R. Weiss, N. Teich, H. Varmus, J. Coffin, Cold Spring Harbor Laboratory (1982).

Cellular Oncogenes and Retroviruses, J. Bishop, Ann. Rev. Biochem. *52*, 301–354 (1983).

Growth Factors: Mechanism of Action and Relation to Oncogenes, C. Heldin, B. Westermark, Cell *37*, 9–20 (1984).

DEEPER READING

Rous Sarcoma Virus: A Function Required for the Maintenance of the Transformed State, G. Martin, Nature *227*, 1021–1023 (1970).

Banding in Human Chromosomes Treated with Trypsin, H. Wang, S. Fedoroff, Nature New Biology *235*, 52–54 (1971).

Identification of a Polypeptide Encoded by the Avian Sarcoma Virus *src* Gene, A. Purchio, E. Erikson, J. Brugge, R. Erikson, Proc. Nat. Acad. Sci. USA *75*, 1567–1571 (1975).

Post-Embryonic Cell Lineages of the Nematode *Caenorhabditis elegans*, J. Sulston, H. Horvitz, Dev. Biol. *56*, 110–156 (1977).

Protein Kinase Activity Associated with the Avian Sarcoma Virus *src* Gene Product, M. Collett, R. Erikson, Proc. Nat. Acad. Sci. USA *75*, 2021–2024 (1978).

Nucleotide Sequences Related to the Transforming Gene of Avian Sarcoma Virus Are Present in DNA of Uninfected Vertebrates, D. Spector, H. Varmus, J. Bishop, Proc. Nat. Acad. Sci. USA *75*, 4102–4106 (1978).

A 32,000 Dalton Nucleic Acid–Binding Protein from Avian Retrovirus Cores Possesses DNA Endonuclease Activity, D. Grandgenett, A. Vora, R. Schiff, Virology *89*, 119–132 (1978).

Uninfected Vertebrate Cells Contain a Protein That Is Closely Related to the Product of the Avian Sarcoma Virus Transforming Gene *(src)*, H. Opperman, A. Levinson, H. Varmus, L. Levintow, J. Bishop, Proc. Nat. Acad. Sci. USA *76*, 1804–1808 (1979).

Purification of Human Platelet-Derived Growth Factor, H. Antoniades, C. Scher, C. Stiles, Proc. Nat. Acad. Sci. USA *76*, 1809–1813 (1979).

Synthesis of Simian Virus 40 t Antigen in *Escherichia coli*, T. Roberts, I. Bikel, R. Yocum, D. Livingston, M. Ptashne, Proc. Nat. Acad. Sci. *76*, 5596–5600 (1979).

Nonrandom Chromosome Changes Involving the Ig Gene-Carrying Chromosomes 12 and 6 in the Pristine-Induced Mouse Plasmacytomas, S. Ohno, M. Babonits, F. Wiener, J. Spira, G. Klein, M. Potter, Cell *18*, 1001–1007 (1979).

The Protein Encoded by the Transforming Gene of Avian Sarcoma Virus (pp60^{src}) and a Homologous Protein in Normal Cells Are Associated with the Plasma Membrane, S. Courtneidge, A. Levinson, J. Bishop, Proc. Nat. Acad. Sci. USA *77*, 3783–3787 (1980).

Detection and Isolation of Type C Retrovirus Particles from Fresh and Cultured Lymphocytes of a Patient with Cutaneous T-Cell Lymphoma, B. Poiesz, F. Ruscetti, A. Gazdar, P. Bunn, J. Minna, R. Gallo, Proc. Nat. Acad. Sci. USA *77*, 7415–7419 (1980).

Chemical Synthesis of a Polypeptide Predicted from Nucleotide Sequence Allows Detection of a New Retroviral Gene Product, J. Sutcliffe, T. Shinnick, N. Green, F. Liu, H. Niman, R. Lerner, Nature *287*, 801–805 (1980).

Identification of a Functional Promoter in the Long Terminal Repeat of Rous Sarcoma Virus, Y. Yamamoto, B. de Crombrugghe, I. Pastan, Cell *22*, 787–797 (1980).

Tumor DNA Structure in Plant Cells Transformed by *A. tumefaciens*, P. Zambryski, M. Holsters, K. Kruger, A. Depicker, J. Schell, M. Van Montagu, H. Goodman, Science *209*, 1385–1391 (1980).

Transforming Activity of Human Tumor DNAs, T. Krontiris, G. Cooper, Proc. Nat. Acad. Sci. USA *78*, 1181–1184 (1981).

Activation of Related Transforming Genes in Mouse and Human Mammary Carcinomas, M. Lane, A. Sainten, G. Cooper, Proc. Nat. Acad. Sci. USA *78*, 5185–5189 (1981).

Unique Transforming Gene in Carcinogen-Transformed Mouse Cells, B. Shilo, R. Weinberg, Nature *289*, 607–609 (1981).

Human-Tumor-Derived Cell Lines Contain Common and Different Transforming Genes, M. Perucho, M. Goldfarb, K. Shimizu, C. Lama, J. Fogh, M. Wigler, Cell *27*, 467–476 (1981).

Structure and Functions of the Kirsten Murine Sarcoma Virus Genome: Molecular Cloning of Biologically Active Kirsten Murine Sarcoma Virus DNA, N. Tsuchida, S. Uesugi, J. Virol. *38*, 720–727 (1981).

Viral *src* Gene Products Are Related to the Catalytic Chain of Mammalian c-AMP-Dependent Protein Kinase, W. Barker, M. Dayhoff, Proc. Nat. Acad. Sci. USA *79*, 2836–2839 (1982).

Transforming Genes of Human Bladder and Lung Carcinoma Cell Lines Are Homologous to the *ras* Genes of Harvey and Kirsten Sarcoma Viruses, C. Der, T. Krontiris, G. Cooper, Proc. Nat. Acad. Sci. USA *79*, 3637–3640 (1982).

Isolation and Preliminary Characterization of a Human Transforming Gene from T24 Bladder

Carcinoma Cells, M. Goldfarb, K. Shimizu, M. Perucho, M. Wigler, Nature 296, 404–409 (1982).

Monoclonal Antibodies to the p21 Products of the Transforming Gene of Harvey Murine Sarcoma Virus and of the Cellular ras Gene Family, M. Furth, L. Davis, B. Fleurdelys, E. Scolnick, J. Virol. 43, 294–304 (1982).

Cellular myc Oncogene Is Altered by Chromosome Translocation to an Immunoglobulin Locus in Murine Plasmacytomas and Is Rearranged Similarly in Human Birkitt Lymphomas, J. Adams, S. Gerondakis, E. Webb, L. Corcoran, S. Cory, Proc. Nat. Acad. Sci. USA 80, 1982–1986 (1983).

Sarcoma Growth Factor from Conditioned Medium of Virally Transformed Cells Is Composed of Both Type α and Type β Transforming Growth Factors, M. Anzano, A. Roberts, J. Smith, M. Sporn, J. De Larco, Proc. Nat. Acad. Sci. USA 80, 6264–6268 (1983).

A Point Mutation Is Responsible for the Acquisition of Transforming Properties by the T24 Human Bladder Carcinoma Oncogene, E. Reddy, R. Reynolds, E. Santos, M. Barbacid, Nature 300, 149–152 (1983).

Mechanism of Activation of a Human Oncogene, C. Tabin, S. Bradley, C. Bargmann, R. Weinberg, A. Papageorge, E. Scolnick, R. Dhar, D. Lowy, E. Chang, Nature 300, 143–149 (1983).

Activation of the T24 Bladder Carcinoma Transforming Gene Is Linked to a Single Amino Acid Change, E. Taparowsky, Y. Suard, O. Fasano, K. Shimizu, M. Goldfarb, M. Wigler, Nature 300, 762–765 (1983).

Homology Between Human Bladder Carcinoma Oncogene Product and Mitochondrial ATP-Synthase, N. Gay, J. Walker, Nature 301, 262–264 (1983).

Platelet-Derived Growth Factor Is Structurally Related to the Putative Transforming Protein p28^{sis} of Simian Sarcoma Virus, M. Waterfield, G. Scrace, N. Whittle, P. Stroobant, A. Johnson, Å. Wasteson, B. Westermark, C. Heldin, J. Huang, T. Deuel, Nature 304, 35–39 (1983).

Activation of a Translocated Human c-myc Gene by an Enhancer in the Immunoglobulin Heavy-Chain Locus, A. Hayday, S. Gillies, H. Saito, C. Wood, K. Wiman, W. Hayward, S. Tonegawa, Nature 307, 334–340 (1983).

Simian Sarcoma Virus onc Gene, v-sis, Is Derived from the Gene (or Genes) Encoding a Platelet-Derived Growth Factor, R. Doolittle, M. Hunkapiller, L. Hood, S. Devare, K. Robbins, S. Aaronson, H. Antoniades, Science 221, 275–277 (1983).

Nucleotide Sequence of Rous Sarcoma Virus, D. Schwartz, R. Tizard, W. Gilbert, Cell 32, 853–869 (1983).

Structure and Sequence of the Cellular Gene Homologous to the RSV src Gene and the Mechanism for Generating the Transforming Virus, T. Takeya, H. Hanafusa, Cell 32, 881–890 (1983).

Regeneration of Intact Tobacco Plants Containing Full Length Copies of Genetically Engineered T-DNA, and Transmission of T-DNA to R1 Progeny, K. Barton, A. Binns, A. Matzke, M. Chilton, Cell 32, 1033–1043 (1983).

Size, Location, and Polarity of T-DNA-Encoded Transcripts in Nopaline Crown Gall Tumors; Common Transcripts in Octopine and Nopaline Tumors, L. Willmitzer, P. Dhaese, P. Schreier, W. Schmalenbach, M. Van Montagu, J. Schell, Cell 32, 1045–1056 (1983).

Genetic Analysis of T-DNA Transcripts in Nopaline Crown Galls, H. Joos, D. Inze, A. Caplan, M. Sormann, M. Van Montagu, J. Schell, Cell 32, 1057–1067 (1983).

The Product of the Avian Erythroblastosis Virus erbB Locus Is a Glycoprotein, M. Privalsky, L. Sealy, J. Bishop, J. McGrath, A. Levinson, Cell 32, 1257–1267 (1983).

The Adenovirus Type 5E1A Transcription Control Region Contains a Duplicated Enhancer Element, P. Hearing, T. Shenk, Cell *33*, 695–703 (1983).

A Tissue-Specific Transcription Enhancer Element Is Located in the Major Intron of a Rearranged Immunoglobulin Heavy Chain Gene, S. Gillies, S. Morrison, V. Oi, S. Tonegawa, Cell *33*, 717–728 (1983).

Proviruses Are Adjacent to c-*myc* in Some Murine Leukemia Virus–Induced Lymphomas, D. Steffen, Proc. Nat. Acad. Sci. USA *81*, 2097–2101 (1984).

Trans-acting Transcriptional Activation of the Long Terminal Repeat of Human T Lymphotropic Viruses in Infected Cells, J. Sodroski, C. Rosen, W. Haseltine, Science *225*, 381–385 (1984).

Index

A site, ribosome, 156–159, 180
Abortive RNA initiations, 96–99, 338
Absorbance, UV, DNA, 26, 27
Accuracy in protein synthesis, 145–147
Acidic amino acids, 121–123, 130, 131
Actin, 7, 8
Action at a distance, 401, *see also* enhancer
Activation of promoters, 95–98
Active transport, 10–14, 181
   arabinose, 348–351
   binding system, 14
   lactose, 317, 319, 339
Acutely transforming retrovirus, 601
Adaptation, *ara*, 365
Adaptation, chemotaxis, 582–585
Adaptation, *lac*, 316
Adapter, lambda, head–tail, 557
Adenine, structure, 23
Adenovirus, 604, 605
Adenovirus, mRNA splicing, 110
Adsorption, lambda, 418, 419
Affinity labeling, 525
Affinity, *araC* protein-arabinose, 352, 361–364
   araC protein-DNA, 361–364
   lac repressor-operator, 331, 332
   lac repressor-IPTG, 321, 322
   RNA polymerase–DNA, 100, 101
Agarose gels, 255, 256
Age distribution, cells, 17–19
   DNA, 67–72
Aggregation, proteins, 7
   ribosomal proteins, 546, 547
Agrobacter, 592
Aliphatic amino acids, 121–123, 130, 131
Allele, 190, 201
Allelic exclusion, 536
Allolactose, 317, 319, 341
Allosteric interaction, 375, 376
Alpha helix, 131–134, 139, 140
Alu sequences, 506, 596–598
Amanitin, 88, 90, 91
Amber codon, 160–162
Amino acid, 120–123, 140
   activation, 144–147
   diffusion, 15

intracellular concentration, 19
   modification, 121
Aminoacyl-tRNA, 144–147
Aminopterin, 264
Aminopurine, 218
Ampicillin resistance, 259–261
Anaerobic growth, 582
Annealing, *see* hybridization
Annealing kinetics, 29–31
Anomers, arabinose, 348
Antibiotics, 488–491
Antibody, 498, 517–536
   blocking, 551
   detection of protein, 292
   effect on chemotaxis, 575
   structure, 524–526
Anticodon, tRNA, 143, 147–150
Antigen, 517–536
Antigen binding site, 525, 526
Antigenic switching, 500
Antiport, 13, 14
Anuclear cells, 92
ara genes, 348, 349
ara operon, 347–368
ara phage, 453
Arabinose
   catabolism, 348, 349
   operon, 10, 304–307
   structure, 348
araC protein, 351–368
araC^c, 354
Aromatic amino acids, 374, 375
ARS, yeast autonomous replicating sequence,
   263–265
Arsenate, 582
Ascites fluid, 522
Ascus, 225
Assay of protein-DNA binding, 293–296, 326–
   328, 340, 341, 358, 361–364
Assembly map, ribosomes, 549, 550
Assembly order, ribosomes, 549
Assembly, biological, 541–563
Association of nucleic acids, 29–31, 258, 296–273,
   283–307, 383–388
ATP, energy of hydrolysis, 11

ATPase, 13, 582
att mutants, 472, 473
att region, lambda, 473, 474
Attached X chromosome, 227
*att*B, 454, 455
Attenuation, 373–390
*att*L, 454, 455
*att*P, 454, 455
*att*R, 454, 455
Autogenous regulation, 357
   trp, 376–378
Autoimmunity, 519, 520
Autoradiography, 285–307
   chromosome, 62, 63, 76
   DNA, 255, 256, 267
   transcription quantitation, 85, 86, 105
Auxin, 593
Auxotroph, 217–219

B cells, 519–536
Backcrossing, 209
Bacterial mating, 222–224, 320, 333, 334, 341,
   451, 476
Bal 31 deletion, 298–300, 304–307
BAP, treatment in cloning, 307
Basal level expression, *ara*, 351, 352
   lac, 323, 332, 432
   lambda, 432
   transposase, 487, 488
   trp, 376–378
Base pair, structure, 23, 24
   wobble, 150
Basic amino acids, 121–123, 130, 131
Beta sheet, 131–134
Beta turn, 131–134
Beta-galactosidase, 168–170, 315–345
Beta-lactamase, 488
Bidirectional DNA replication, 63–65, 422, 423
Binding constant, *see* dissociation constant
Binding energy, cooperative, lambda, 431
Binomial distribution, 236, 237, 240, 244
Biotin transduction, 460–464
Bisulfite mutagenesis, 299
Bivalent antibody, 524–526, 577
Blastula, 228–230
Blind scoring, genetics, 542, 543
Blinding assay, chemotaxis, 573
Blood, 518
BOB', 454, 455
Boltzman constant, 15
Bone marrow, 518
Branch migration, 204–209, 401

Brdu, 264
Bromodeoxyuridine, 264
Bromouracil, 218
Brownian motion, 14, 15, 569
Brute force, genetics, 542, 543
   RNAseIII mutant, 107
Budding, yeast, 224, 225
Burkitt's lymphoma, 594, 603

Caenorhabditis, 591
Calcium phosphate, 30
   use in transformation, 265
cAMP, 318, see CAP
Campbell, model, 454, 455
Cancer, 517, 521–523, 591–607
CAP, CRP, 318, 329, 338, 350, 357, 361, 364, 577
Capillary tube assay, 571–573
Capsid structure, 552–563
Carcinogen, 593
Carrier protein, 11, 12
Casette mechanism, yeast, 396–403
Catabolite repression, 318, 338, 361, 364
   flagella, 577
cDNA cloning, 284
Cell, age distributions, 17–19
   culture, 592–607
   cycle, yeast, 224, 225
   doubling time, 16
   fusion, 523
   growth, 71, 72
   growth, coordinating DNA synthesis, 48, 69–72
   lineage, 591
   membrane, permeability, 86
   physiology, synchronized cells, 72–74, 77
   structure, eukaryotic, 7–9
   structure, prokaryotic, 7–9
   wall, 5
Cellular oncogenes, 601–603
Chain termination, polypeptide, 160–162
Chelate effect, 129, 130, 140, 141, 159, 160,
   434–437
   lac operon, 336, 337
Chelation, Mg^{++}, 466
Chemical DNA sequencing, 266–269, 274–276
Chemiosmotic transport system, 12–14
Chemotaxis mutants, 584
Chemotaxis, 498–500, 569–587
   assay, 570
   energy source, 581, 582
chi sequence, recombination, 207–209
Chi, phage, 576
Chimeric proteins, 132

Chloramphenicol resistance, 259–261
Chlorate reductase, 451, 452
Chlorate resistance, 451
Chloroplast, 7–9
Chromosome, breakage, lambda, 451
  circular, 7
  damage, 593–607
  eukaryotic, 7
  genetics, 193–196
  loss, fate mapping, 229–230
  rearrangements, 505
CI, lambda, 414, 415
cis, genetics, 196–198
cis-dominant mutation, 196–198, 423
Class switching, 533, 534
Clear plaques, 413, 414
Cloning, 247–304, 404, 405
  antibody genes, 527–536
  DNA enrichment, 283
  M13, 272, 273
  oncogene, 595–598
  RNA enrichment, 283–284
Closed complex, RNA polymerase, 98, 99
Cloverleaf, tRNA, 143
Code, genetic, 148
Codon usage, 415
Cohesive ends, lambda, 32–34, 412, 413, 556,
    558–560
Cold-sensitive mutants, 548
Colicin E3, 152
Colony hybridization, 285
Combinatorics, 236
Complement system, 518
Complementation, 196–198, 215, 458, 459
  lac operon, 320
Complexity, DNA, 30, 31
Concentration measurement, chemotaxis, 570
Concerted cleavage of DNA, 26, 27, 254
Conformational changes, protein, 375, 376
Conjugation, 450, 451, 486, 495
  bacteria, 222–224, 320, 333, 334, 476
  yeast, 224, 225
Constant regions, 525–537
Constitutive mutations, araI^c-cip, 354
  lac operon, 320
Contact inhibition, 264, 591, 592, 595, 605
Convolution, impulse, 580, 581, 587
Cooperativity, lambda repressor binding, 428–430
copia, 505, 506
Core proteins, ribosomes, 546
Core, lac repressor, 334
Cot, 29–31

Cotransduction, P1, 221, 222
Coumermycin, 40
Counting gene copies, 29–31
Coupled transcription-translation, 353–354,
    384–390
cro protein, 420–422, 427–430
Cross-inhibition, chemotaxis, 573
Crosslinking, ribosomes, 550
  rRNA, 545
Crossover, genetic, 203–209
Crown gall, 592, 593
CRP, 318, 329, 350, 357, 361, 364, 577
CsCl density gradient, 249, 250, 413, 466
Cubic symmetry, 554, 555
Curing, lambda, 450, 451, 456–458
Cuticle, fruit fly, 403–406
Cyclic AMP receptor, 318, 329, 338, 350, 357, 361,
    364, 577
Cyclobutane, 60
Cytokinin, 593
Cytoplasm, 5, 6
Cytoplasmic membrane, 5–6
Cytosine, structure, 23

D regions, antibody, 531–536
Damage repair, DNA, 49, 59–61
DDC, 403–406
Deamination of cytosine, 60, 61, 191
Defective transducing phage, 461
Defective retrovirus, 601
Deformylase, protein synthesis, 158, 159
Delayed early genes, lambda, 426, 427
Deletion, 191–194, 282, 297–300, 304, 307, 461
  antibody, 534
  generation, 487
  isolation, 378, 379
  lambda, 451, 452
  mapping, 202, 203
  mapping, lac, 334
Delta elements, 504, 505
Denaturation, protein, 125, 134–137
Denaturing DNA, 26, 27
Density separation, phage, 466
Deoxyribose, structure, 22
Detergent, SDS, 88–91
Detoxification, 488, 489
Development, 395–408
Diabetes mellitus, 520
Diagnosis, monoclonal antibodies, 523
Dideoxy sequencing, 269–273
Differential equation, solving, 242–244

Differentiation, 592, 593
  cellular, 216, 226–230, 403–406
Diffusion constant, 15
Diffusion, 14, 15, 19, 330, 331, 571–573
  equation, 75
  in DNA synthesis, 74–76
Dihydrofolate reductase, 264
Dimethylsulfate, 296, 364
Dimethylsulfate, use in DNA sequencing, 268
Diploid, 194, 193–198, *see lac, ara, trp*
Diploid, yeast, 395–403
Dipole interaction, 127–129
Directed mutagenesis, 296–302, 304–308
Directed transposition, 400, 401
Dispersion force, 127–129
Dissociation constant, *ara*C protein-arabinose, 352
  *ara*C protein, DNA, 361–364
  lac repressor-DNA, 331, 332
  lac repressor-IPTG, 321, 322
  lambda repressor-DNA, 436, 437
  RNA polymerase, 100, 101
Disulfide bonds, antibody, 524, 525
Diversity segments, 531–536
DNA, absorbance, 26, 27
  affinity of proteins for, 37, 100, 101, 331, 332,
      361–364
  binding domain, *lac* repressor, 332–335
  binding, *int*, 470
  binding, motif, 137–140, 332–335, 364
  chemically synthesized, 300–302
  complexity, 30, 31
  damage and error correction, 48–53, 59–61, 76,
      416, 417
  degradation, 250
  diameters, 24–26
  electrophoresis, 255, 256
  electrophoresis of superhelical form, 35, 36
  elongation direction, 49
  elongation rate, 65–75
  error correction, 49, 76
  Escherichia coli, molecular weight, 9
  exonuclease, 50, 52, 53, 76
  footprinting, 293–296
  fragment isolation, 255, 256
  grooves, 24–26
  gyrase, 39, 471, 472
  helical structures, 24–26
  helicase, 54
  Hershey circle, 33
  heteroduplex, 203–209, 452, 455, 476, 485
  host repair, 191
  initiation of synthesis, 48, 69–72

  injection, 563
  ionic effects on, 26, 27
  isolation, 248–250, 255, 256
  labeling fragments, 295
  lagging strand, 51
  leading strand, 51
  ligase, 37, 256–258
  ligase, in recombination, 208
  linking number, 32, 33
  loops, 360, 361
  melting, 26, 27
  methylation, 59–61, 76, 250–255
  nicked circle, 33
  nuclear strand separation, 9–10
  nucleus, 7
  Okazaki fragments, 49–54, 77
  origins, 328, 422, 423, 437, 455
  packaging, 552, 553
  physiology of repliction, 61–75
  pol I, 282, 297–309
  polymerases, 51–53, 59, 76
  polymorphisms, 302
  protein contacts, 41, 100, 101, 137–140,
      293–296, 331, 332, 363, 364, 475
  reassociation, 27–31, 67–69, 159, 160, 285–307
  rearrangements, 517–536
  repeated sequences, 30, 31
  replication origins, 54–56, 62–65, 422, 423, 437,
      455
  replication, 42, 62–65
  replication, bidirectional, 63–65, 422, 423
  replication, lambda, 422–424, 427
  replication, rolling circle, 222
  satellite, 31
  sedimentation of superhelical, 35, 36
  segregation, 58
  sequencing, 190, 266–271
  sequencing, origin, 328
  strand separation, 30
  structure, 21–41
  superhelical turns, 32–41
  synthesis, 47–85
  synthesis, chemical, 282
  synthesis, enzymology of, 47–61
  synthesis, primer, 51, 53–57
  theta form, 62
  topology, 31–41
  torsion, 39, 40
  twist, 32–38, 360, 361
  twisting spring constant, 36
  uracil glycosidase, 61
  writhe, 32–38

DNAse I, 494
Domains, lambda repressor, 435
Domains, proteins, 109–111, 120, 131, 524–526, 532
Dominance, rifamycin sensitivity, 220, 221
Dominant mutations, 196, 230, 366
Dopa decarboxylase, 396, 403–406
Dot matrix homology, 308
Double helix, DNA, 24–26
Doubling time of cells, 16
Drosophila, genetics, 226–230, 404–408
Drug resistance, 259–261, 488–491
Duplication, lambda, 558, 559

Eclosion, 403, 404
EcoRI cleavage sites, lambda phage, 262
EcoRI, restriction enzyme, 255, 274
Editing, DNA, 49–50, 191
    by aminoacyl-tRNA synthetase, 146–147, 180
    misacylation, 146, 147, 180
Egg, 195, 196
Electron acceptor, 382, 451
Electron microscopy, *lac*, 337
Electron transport, 12
Electrophoresis, DNA fragments, 255, 256
    proteins, 164–166
    RNA, 85, 86, 105
    SDS, 88–90
    superhelical DNA, 35, 36
Electrostatic forces, 125–129
Elongation, protein, 156, 157
Elongation, protein synthesis factors, 156, 157
    rate, DNA, 65–75
    rate, RNA, 93–95
    rate, protein, 162–166
EMBO plates, 456
Embryo, 226–230
    antibody genes, 527
    Encapsidation, geometry, 552–563
    lambda DNA, 262, 425
    phage P1, 64
Endolysin, 425, 438
Endonuclease, 401
Endopeptidase, 425, 438, 600
Endoplasmic reticulum, 8, 9, 166, 167, 527
Enhancer, 103, 406, 407, 534, 535, 595, 601
Enthalpy, 436, 437
Entropy, 436, 437
    cation displacement, 101, 102
    hydrophobic forces, 131
    lac repressor, 336, 337

of molecule orientation, 159, 160
    salt binding, 331
env, retrovirus, 599–601
Envelop glycoprotein, 600, 601
Enzymatic DNA sequencing, 269–273
Enzyme, induction kinetics, 16, 17, 350, 364–366, 376
Episome, 220–224, 461
    maintenance, 495
Equilibrium constant, 11, *see* affinity
Equilibrium density gradient centrifugation, 168–170, 248–250, 413, 466
Equilibrium dialysis, 321–324, 340
Error repair, 203, 204, 472, 473
Error frequencies, protein synthesis, 145–147
Erythrose, 374
Escherichia coli, genetics, 216–224
    structure 1–7
Ethidium bromide, 33–36, 249, 256, 267, 362
Ethylmethanesulfonate, 193, 218, 226
Eukaryotic cell, structure, 7–9
Evolution, 532
    ara, 364–366
    trp, 376
Excision, mutants, 456–458
    lambda, 412, 416, 417, 449–478
ExoIII, use in DNA sequencing, 270, 282
Exon, 106–111, 308, 528–536
Exonuclease, 50, 52, 53, 76, 270, 282
Exoskeleton, 403
Exponential, cell growth, 71, 72
    function, 15, 16
    growth, 15–18
Expression site, mating type, 397–403

F-factor, 222–224, 461, 495
F-pilus, 222
Fab' fragments, 524, 525
Facilitated diffusion, 11, 12
Factorial, 236
Fat body, 406, 407
Fate mapping, *Drosophila*, 228–231
Feline sarcoma, 598
Filter binding assay, 326–328, 361, 470
Filter hybridization, 28, 285–307
Flagella, 498–500, 569–587
    helix, 576, 577
    structure, 576–578
Flip-flop regulatory circuits, 339, 499, 500
Fluctuation test, 235–245
Fluorescence transfer, 551
Fluorotyrosine, 335

Focus, 595
Folding pathways, protein, 136
Footprinting, 293–296, 341, 358, 363, 364
Formylation, methionine, 154–156
Frameshift, protein synthesis, 180
Free energy, 11
Fruit fly, *see Drosophila*
Fucose, 355, 572
Fusidic acid, 388
Fusion proteins, *trp*, 383
Fusion, gene, 105, 167–170, 357, 427–430

G loop, Mu phage, 502–504
G segment, Mu phage, 499
gag, retrovirus, 599–601
gal transduction, 460–464
Galactose, 317
Galactose chemotaxis, 571–576
Galactose-binding protein, 575
Gamete, 194–196
Gaussian distribution, 238–240
Gel binding assay, 361–364
Gene, 190, 201
    activation, lambda, 430
    conversion, 204, 205, 211, 401
    expression, assay of, 28
    expression, increasing, 304–307
    fusion, 105, 168, 170, 357, 427–430
    rearrangements, 517–536
Generalized transduction, 461, 462
Genetic, code, 148
    code, mitochondria, 149, 150
    diversity, 517–536
    engineering, 247–276, 281–309, 429, 459,
        506–509, 560, 561
    map, lambda, 415, 438, 454, 556
    mapping, 198–209
    mapping, Hfr, 222–224
    mapping *lac* repressor, 333, 334
    mapping Tn3, 494–496
    phage P1, 221, 222
    polymorphism, 519
    selection, 219–221
Genetics, 189–230
    deletion mapping, 202, 203
    Drosophila, 226–230
    genotype, 190, 195–199
    mutations, 191–194
    phenotype, 190
Genotype, 190, 195–199
Geometry, capsids, 553–555, 563
Germ line, 509

Giemsa stain, 594
gin, Mu phage, 502
Globin polymers, 303
Glucose, 317
Glucose effect, 318, 338
Glycerol transport, 12
Glycosidase, uracil-DNA, 61
Golgi apparatus, 8, 9
*gro*E, 557, 558, 563
Growth cycle, lambda, 417
Growth factors, 592, 593, 598
Growth inhibition, 591, 592
Growth rate, cells, 16, 173, 174
    DNA, 65–75
    protein, 162–166
    RNA, 93–95
Growth requirements, 217
Growth, exponential, 15–18
Guanine, structure, 23
Guanosine tetra-, penta-phosphate, 175–177
Guard sequences, 487, 488
Gyrase, 39, 56, 471, 472

Hairpins, RNA, 106, 107, 383–390
    rRNA, 544
Haploid, yeast, 224, 225, 395–403
HAT medium, 264
Head assembly, lambda, 556, 557
Headfull mechanism, 501
Headpiece, *lac* repressor, 334, 335
Heat induction, 457
Heat pulse curing, 457, 458
Heat shock, *E. coli*, 92, 93
Heavy chain, 524–536
Helical repeat, DNA, 24
Helicase, DNA, 54
Helix-turn-helix, DNA binding domain, 139–140,
    332–335, 364
Helmstetter-Cooper model, 69–72, 77
Heredity, 189–193
Herpes virus, 264, 605
Hershey circle, 33
Heteroduplex, DNA, 203–209, 452, 455, 476, 485
Heterogenous nuclear RNA, 106–111
Heteroimmune curing, 457, 477
Heteroimmune phage, 417, 452
Heterozygote, 196–198
Hfr, bacterial mating, 223, 229
him protein, 472, 498
Hinge region, 524
hip, 472
Histones, 10, 38

HML, yeast mating type, 396–403
HMR, yeast mating type, 396–403
HO, ho, yeast, 396–403
Homology, V region genes, 527
Homozygote, 195, 196
Hook, flagella, 576, 577
Horizontal propagation, 599
Host proteins, lambda integration, 471, 472
Host range, φ80, 462
Host range, Mu, 502–504
Host repair, 191
Huntington's Chorea, 282, 302
Hybrid arrest of translation, 287–289
Hybrid, RNA-DNA, 27–31
    RNA-DNA, strength, 106
Hybridization, assay of RNA polymerase, 84, 85
    assay of RNA, 28–31
    detection of single base differences, 301–303
    kinetics, nucleic acids, 29–31
    nucleic acids, 27–31, 285–307
    RNA-DNA, 67–69
    trinucleotides, 159, 160
Hydrazine, use in DNA sequencing, 268
Hydrogen bond, nucleic acids, 23, 24
    proteins, 125, 129, 130
Hydrophobic amino acids, 121–123, 171
Hydrophobic forces, 130, 131, 140
Hydroxyapatite, 30
Hydroxylamine mutagenesis, 218, 299
Hyperchromicity, 26, 27
Hypermutation, antibody genes, 533, 535
Hypersynthesis, 459
    genetic engineering, 304–307
    lac repressor, 324, 325
Hypervariable regions, antibodies, 525–536
Hypoderm, 404

I⁻ᵈ, lac, 332–334, 338
Icosahedron, 554, 555
IgG, IgM etc, 524–526
IgG and ribosome structure, 545
Immediate early genes, lambda, 426, 427
Immune system, 517–536
    cancers, 594, 595
Immuno-microscopy, 545
Immunity, lysogen, 416, 417
Immunization, 518–521
Immunoglobulin, 517–536
    signal peptide, 166–171
Impulse response, 580, 581, 587
In situ hybridization, 286, 287
In vitro, lambda excision, 464–466, 469–471

lambda integration, 470, 471
mutagenesis, 296–302, 304–308
packaging, 262, 560, 561
protein synthesis, 287–289, 353, 354
ribosome assembly, 545–552
Incorrect excision, lambda, 460–466
Incorrect insertion, lambda, 462
Indicating plates, 319, 333, 334, 455, 456, 458
    tRNA-HCl, 542, 543
Indole, 379
Induced dipole, 128, 129
Induced mutation, antibody, 533
Inducer exclusion, 318
Induction kinetics, of enzymes, 16
    ara, 350, 364–366
    trp, 376
Induction, ara, 351–368
    lac operon, 315, 336, 337, 345,
    lambda, 428
Infective cycle, lambda, 418–426
Information redundancy, 48, 49
Information storage, 21, 22
Information, 1, 2
Initiating translation, 150–156
Initiation factors, protein synthesis 154–159
Initiation of DNA synthesis, 48, 69–72
Injection, Drosophila, 508
Inner membrane, 5, 12, 13, 168, 170
Inosine base pairing, 386–387
Insertion, genetic, 191–194, 282, 297, 405–408,
    484–488, 504–509
Insertional inactivation, 299, 300
Insolubility, ribosomal protein, 546, 547
Instability, int, xis, 466, 467
int binding, 475, 476
int, 456–459
Integration host factor, 472
Integration mutants, 455, 456
Integration, lambda, 411, 412, 416, 417, 449–478
Intercalation, DNA, 33–36
Intermediate filaments, 7, 8
Intermolecular resolution, 497, 498
Interspecies comparisons, 383, 384, 544, 545
Intervening sequence, 106–111, 308, 528–536
Intron, 106–111, 308, 528–536
Inverted repeat, 486–488
Invertible elements, 498–504
Ion localization, 102
Ionic effects, DNA melting, 26, 27
    nucleic acid renaturation, 29–31
    RNA polymerase binding, 100–103
IPTG, 318, 319

Iq, *lac*, 325
IS elements, 484–515
Isomerization, by RNA polymerase, 95–99
    in DNA recombination, 204–209
Isometric, 412
Isopropyl-thio-galactoside, 318, 319
Isopycnic centrifugation, 248–250, 413, 466
    membranes, 169

J regions, antibody, 528–536

Kanamycin resistance, 259–261, 489, 491
Kappa chains, 525–536
Karyotype, 523
Keratin, 8
Kinase, *src*, 604
Kinetic order, ribosome assembly, 548
Kinetics, of RNA initiations, 98, 99
Klenow fragment, 282
Knots, untangling, 41

Labeling DNA fragments, 53, 295
lac, induction, 431–433
    operon, 10, 202, 315–345, 429
    basal level, 323, 332
    mutant repressor, 324, 325
    promoter, use in gene expression, 304–307
    repressor hypersynthesis, 324, 325
Lactose analogs, 318, 319
Lactose operon, 10, 202, 315–345, 429
Lagging strand, DNA, 51
*lam*B, 418, 419
Lambda chains, 525–536
Lambda gene bank, cloning, 281, 285, 286
Lambda phage, adsorption, 418, 419
Lambda phage, att², 464–467
    assembly, 552–563
    chromosome integration, 68, 69
    cloning vector, 261–263
    delayed early genes, 426, 427
    DNA, 31, 38, 39, 412
    genes, 416
    genetic map, 415, 438
    immediate early genes, 426, 427
    immunity, 413–437
    induction, 434–436
    integration, 39, 40
    late genes, 424
    repressor, structure, 137–140
    sticky ends, 32–34, 412, 413, 556, 558–560
    transcription, 106
Lambda-*bio*, 460–464

Lambda-*gal*, 460–464
Lambda-hybrid phage, 418
Lambdoid phage, 417, 418
Larva, 403
Late genes, lambda, 424
Leader peptide, *trp*, 381–383
Leader sequence, protein, 166–170, 528
Leading strand, DNA, 51
Leukemia virus, 598
*lex*A, 434
Libraries, cloning, 285
Life cycle, φX174 phage, 55
    yeast, 203
Ligase, DNA, 37, 51–56, 208, 256–258
Light chain, antibody, 524–536
Linker, cloning, 285
    in DNA joining, 258
    mutagenesis, 494–496
    scanning, 298
Linking number, 31–41
Localized folding, 366, 377
    rRNA, 543, 544
Localized mutations, antibody, 533
Lollipops, 512
London force, 127–129
Long terminal repeats, 505, 599
Loops, DNA, 360, 361
LTR's, 505, 599
Lupus, 520
Luria-Delbruck test, 235–245
Lymph nodes, 518
Lymphocytes, 518–536
Lysis, by lambda, 425, 438
Lysogenic response, 422–437
Lysogeny, 411, 412, 416, 417
Lysozyme, lambda, 425, 438
Lytic infection cycle, lambda, 418–426

M13 virus, use in DNA sequencing, 271–273
MacConkey plates, 458
Magic spot, 175–177
Major groove, DNA, 24–26
Maltose, cell uptake, 418, 419
Mapping, genetic, 198–209
    Hfr, 222–224
    integrated lambda, 450, 451
    Tn3, 494–496
MAR, repressor, mating type, 401
Marker effect, 199
Marker rescue, 452
MAT, yeast mating type, 396–403
Mating type conversion, 396–403

Mating type, yeast, 225, 395–403
Mating, bacteria, 220, 222–224, 450, 451, 476
Matrix proteins, 6
Maturation, RNA and protein, 290–293
Maxam-Gilbert sequencing, 266–269, 274–276,
    293–296
Mean, of probability distribution, 240
Meiosis, 194–196, 204–209, 225
Meiotic recombination, 231
Melting DNA, 26, 27, 103–105
Membrane, disruption, 426
    permeability, 10
    potential, 12–14, 581, 582
Memory, 585
Mendel, 193–196
Messenger RNA, decay, 162, 163
    purification, 284
    self-splicing, 111
    splicing, 106–111, 530–536, 600
    synthesis, RNA, see RNA
    stability, 468
Meta-stable state, proteins, 136
Metamorphosis, 403
Methotrexate, 265
Methyl mercury, denaturing RNA, 86
Methyl transferase, 585
Methylase, 495
Methylation, DNA, 59–61, 76, 209, 231, 250–255,
    274
    protein, 582–585
    RNA, 106
    rRNA, 546
Methylesterase, 585
Michaelis-Menton, 322
Microinjection, 265, 266
Microtubule, 8
Middle repetitive DNA, 31, 505, 506
Minicells, 92
Minicircles, 465, 466
Minor groove, DNA, 24–26
Misacylation, 146, 147, 180
Mismatch repair, 203, 204, 472, 473
Misrepair mutagenesis, 299
Mitochondria, 7, 209, 215, 231
    genetic code, 149, 150
Mitosis, 194–196, 203–209
Mitotic recombination, 225, 231
Molecular weight determination, RNA, 88–90
Monoclonal antibodies, 521–523
Monomer-dimer interactions, lambda, 433
Mosaics, sexual, 228, 231
Motility, 572–587

mRNA, see messenger RNA
Mu phage, 476, 492, 500–504
    chromosome integration, 68, 69
Murine sarcoma, 598
Muscle mutants, *Drosophila*, 226–231
Mushroom toxin, 88
Mutagen, 191, 192, 217–219
Mutagenesis, 192, 218
    in vitro, 296–302, 304–308
    oligomers, 301, 302
Mutant, 190
    chemotaxis, 575
Mutation rate, 241–246, 308
Mutation, 190–193
    antibody, 533
    causing cancer, 595–598
    deamination, 61
    spontaneous, 61
    temperature sensitive, 190, 227, 228
Mutator gene, 218
Myasthenia gravis, 520
myc, 603
Mycophenolic acid, 264
Myelomas, 521–537

N protein, lambda, 420, 421
N-acetyldopamine, 403–406
N-acetylglucosamine, 6
N-acetylmuramic acid, 6
N-formyl methionine, 154, 156, 179
Nalidixic acid, 40
Negative regulation, see *ara, lac,* lambda, *trp*
Neomycin phosphotransferase, 265
Nerve mutants, *Drosophila*, 226–231
Nervous system, 226–231, 302, 404, 523, 569
Neurodegeneration, 302
Neurofilaments, 8
Neurons, 523
Neurotransmitter, 403–406
Neutron diffraction, 552
Nick translation, 53
Nicked circle, 33
NIH 3T3 cells, 595–598
Nitrate reductase, 451, 452
Nitrocellulose, 289, 290
Nitrosoguanidine, 218, 230, 542
Nitrous acid, 218
NMR, *lac* repressor, 334, 335
Nonlinear induction, lambda, 120, 131, 431–433,
    524–526, 532
Nonlocal folding, 366, 367, 544

Nonsense mutation, 199, 414, 436
　　lac repressor, 332–334
　　protein synthesis, 160–162, 180
　　RNA polymerase, 219–221
　　suppression, 160–162, 180, 320
Nonspecific binding, *lac*, 365
Nopaline, 592, 593
Northern transfers, 290–293
Novobiocin, 40
Nuclear magnetic resonance, 334, 335
Nuclear membrane, 8, 9
Nucleation, protein structure, 137
Nucleic acids, association, 29–31, *see also*
　　hybridization
Nucleoside, 22
Nucleosomes, 38
Nucleotide, 22
Nucleotide pools, 66, 67
Nucleus, 8–10
Number of antibodies, 521
nus proteins, 419–421
nut, lambda, 420, 421
Nutritional requirements, testing 217

Ochre codon, 160–162, 491
Octopine, 592, 593
Okazaki fragments, 49–54, 75–77
Oligo-T column, 284
Oligomer, DNA, 300–302
Oligomer, *lac* repressor, 333
Oligomerization, protein subunits, 121
Oligonucleotide, chemically synthesized, 300–302
Oligonucleotide-directed mutagenesis, 301, 302
Omega protein, topoisomerase, 38, 39
Oncogene, 595–607
Oncogenesis, 591–607
ONPG, 318, 319
Open complex, RNA polymerase, 98, 99
Operator, *ara*, 358, 359
　　lac, 316, 317, 326–332
　　lambda, 427–437
Operon, *see ara, lac, trp*
Operon, definition, 317
Optical density, DNA, 26, 27
ori, lambda, 422, 423, 437, 455
Orientation entropy, 436, 437
Orientation of peptide bond, 124, 125
Origin, DNA replication, 54–56, 62–65, 422, 423, 437, 455
　　DNA replication, cloning, 276
Orthonitrophenol, 318, 319
Orthonitrophenyl-beta-D-thiogalactoside, 219
Osmotic pressure, cells, 4, 218, 219

Outer membrane, 5–7, 10
Ovary, 406–407
Overlapped genes, 415, 416
Oxidation, cysteine, 122, 123
Oxidative phosphorylation, 7
Oxolinic acid, 40

P element, 405, 406, 506–509
P site, ribosome, 156–159, 180
Packaging, lambda, 262, 425
　　Mu, 500
PaJaMa experiment, *lac*, 341
　　ara, 366
Palindromes, DNA, 252–255, 340
Papain cleavage, 524
Papilloma virus, 604, 605
Papovavirus, 604, 605
Pausing, RNA polymerase, 386, 419
Penicillin, 230, 488
Penicillin, action, enrichment, 218, 219
Penicillin resistance, 488
Pentose phosphate shunt, 349
Peptide bond, 124, 125, 144
　　formation, 157–159
Peptidoglycan layer, 5, 218, 219, 425, 438, 325, 573
Periplasm, 5, 573
Periplasmic protein, 14, 542, 575
Permutation, circular, 453–455
Petri plate, 216
pH indicator, 455
Phage assembly, 552–563
Phage Chi, 576
Phage DNA replication, host requirements, 57, 422, 423, 437
Phage excision, *ter*, 558
Phage G, 57
Phage maturation, 552–563
Phage P1, 63, 64
　　genetic mapping with, 221, 222
Phage ϕX174, 54–56
Phage R17, 153, 154
Phage receptor gene, 461, 462
Phage relatives of lambda, 417
Phage T7, 153, 154
Phage vectors, 248–250, 261–263
Phage, specialized transducing, 247,
　　ara, 353, 563
　　bio, 459–464
　　gal, 459–464, 484–486
　　kan, 489–490
　　lac, 325–328, 484–486
　　trp, 379
Phenol extraction, 248

Phenotype, 190, 195–199, 217
Phenotype, *ara*, 350–352
Phenylgalactose, 318, 319
Phi-psi angles, 124, 125
Phosphocholine, 521, 522
Phosphodiester, DNA backbone, 22
Phosphoenolpyruvate, 12, 13, 374
Phosphorylated sugars, 349, 366
Phosphorylation, *src*, 604
Phosphotransferase system, 12, 13, 575
Phylogenetic comparison, 383, 384, 544, 545
Physiological state, supercoiled DNA, 39, 40
Physiology, cell growth rate, 181
    free RNA polymerase, 91, 92
    functions of RNA polymerase, 86–88
    protein synthesis, 162–179
Piperidine, use in DNA sequencing, 268
Plaque hybridization, 285
Plaque morphology, 413, 455
Plasmid vectors, 238–250, 259–261
Plasmids, use in studying transposons, 486–515
Plasmocytoma, 594
Platelet derived growth factor, 604
Ploidy, 194
Point mutation, 191
Poise, 15
Poisson distribution, 237, 238, 240, 245, 246
PolI mutants, 495
pol, retrovirus, 599–601
Polacrylamide gels, DNA, 255, 256
Polar amino acids, 122, 123, 127–131
Polar mutations, 484–486
Polarity, 378–380
Poly(A) mRNA, 109, 284, 404, 600, 601
Polyacrylamide gels, SDS, protein, 164–166
Polyadenylation of RNA, 109, 284, 404, 600, 601
Polyhedron, 553–555
Polyhook, 578
Polymerization rate, DNA, 65–75
    protein, 162–166
    RNA, 93–95
Polymorphisms, DNA, 302
Polynucleotide kinase, 267
Polypeptide chain termination, 160–162
Polyprotein cleavage, 600
Polytene chromosome, 216, 286, 287, 404, 594
Pool size, charged tRNA, 179
    magic spot, 182
    nucleoside triphosphates, 182
    ribosomal protein, 179
POP', lambda, 454, 455
Pore, membrane, 425, 426
Porin, maltose, 418, 419

Positive regulation, 347–368
Potential energy, 125–129
ppGpp, 175–177
Precipitation, ribosomal proteins, 546, 547
Precursors, RNA and protein, 290–293
Predicting protein structure, 136, 137
Premodification experiments, 363, 364
Preprohead, 557
Pribnow box, 102
Primer, DNA, 51, 53, 56, 57
    in dideoxy sequencing, 270, 271
    retrovirus, 598, 599
Primosome, 56
Probability, 235–245
Programmed differentiation, 591
Prohead, 557
Prokaryotic cell structure, 4–7
Promoter, 84–86, 95–99, 190, 407, 491,
    consensus sequence, 102, 103
    definition, 83
    genetics, 197–198
    heat shock, 92, 93
    lac, 316, 317, 329, 337, 339
    lambda, 419–436
    mutation, 325
    recognition by sigma subunit, 92, 93
    retrovirus, 600, 601
    trp, 378
Proofreading, 48–51, 76
Protease, 547
Protection, protein-DNA binding, 293–296, 341,
    358, 363, 364
Protein A, *Staphlococcus*, 603, 604
Protein, aggregation, 7
    elongation, 156, 157
    domains, 109–111, 210, 131, 524–526, 532
    excretion, 166–171, 527
    fusions, 132
    precursors, 290–293
    prediction, 136, 137
    purification, 47
    secretion, 166–171, 527
    structure, 119–140
Protein synthesis, 143–183
    accuracy, 145–147
    chain termination, 160–162
    decoding, 147–150
    deformylase, 158, 159
    elongation, 156, 157
    elongation rate, 162–166
    folding rates, 179
    initiation factors, 154–156
    peptide bond formation, 157–159

ribosome-mRNA binding, 150–154
signal peptide, 166–171
stable RNAs, 83
targeting, 166–171
translocation, 158, 159
Protein-DNA binding, 41, 100, 101, 137–140,
    293–296, 331–332, 363, 364, 475
Proton flow, 12–14
Proton gradient, 581, 582
Proton-motive force, 12–14
Provirus, 505, 600
Pseudo *att* sites, 462, 463
Psoralen, 545
Pupa, 403
Purification, *araC* protein, 354
    cells, 216, 217
    int, 469–471
    lac repressor, 321–326
    transposase, resolvase, 497–498
Purine, 23
Pyranose, arabinose, 348
Pyrimidine, 23
Pyrophosphate, chelation, 466

Q protein, lambda, 424, 425
Quantitating gene copies, 29–31
Quasiequivalence, 554, 555

R plasmids, 260, 261, 488–491, 511
R-factors, 260, 261, 488–491, 511
R-loop, 285, 286, 528, 529
Random motion, 14, 15
Randomness, antibody, 533
Reading frames, lambda 415
Readthrough, transcription, 105
Rearrangements, DNA, 517–536
Reassembly, ribosome, 546
Reassociation of RNA, 27–31, 106, 159, 160
    DNA, single base differences, 301–303
    RNA-DNA, 67–69
Reassociation, kinetics, 29–31
RecA protein 207–209, 434
RecBC protein, 207–209
RecBC, lambda inactivation, 423, 424
Receptor gene, 461, 462
Receptor, chemotaxis, 572–575
Recessive, genetic trait, 196, 230
Recombinant DNA, 247–276, 281–309
Recombination, 193–209, 414, 452, 453, 558, 559
    between chromosomes, 195, 196
    in genetic mapping, 221–225
    in transduction, 64

lambda, 423, 424
    meiotic, 225
    mitotic, 225
Reconstitution, ribosome, 546–552
Regular polyhedron, 553
Regulation, *see* specific object
Regulation, ribosome synthesis, 174–179
Relaxed replication, plasmid, 260
Relaxed RNA response, 175–177
Release factors, protein synthesis, 160–162
Renaturation of proteins from SDS, 89–91
Renaturation, nucleic acids, 27–31
Repair, mismatch, 204, 205, 472, 473
Repeated sequences, 30, 31
Repeated sequences, transposition, 490, 492, 493,
    500, 504
Replica plating, 217–219
Replication by transposition, Mu, 500
Replication fork, DNA, 51, 53, 54, 63–65
    rotations, 54
Replication origins, 54–56, 62–65, 422, 423, 437,
    455
Replicon, 259, 265, 300, 496
Replicon fusion, 492, 496, 509
Repression, *ara,* 354–356
    lac, 315–345
Repressor, *lac,* 315–345, 428, 430
    lambda, structure, 137–140, 413–437
    sliding, 330, 331
    trp, 373–390
    yeast mating type, 401
res site, Tn3, 492, 496–498
Resolution, transposition, 496, 509
Resolvase, 491, 492
Restriction enzyme specificity, 253, 254
Restriction enzymes, 250–255
Restriction-modification, 250–255, 262
Retro-inhibition, 467–469
Retrovirus, 505, 513, 598–604
Reverse bend, see beta turn
Reverse transcriptase, 284, 285, 505, 598, 599
Revertant, 219
rex genes, lambda, 437
RF, ϕX174 phage, 54–58
rhl, repressor, 499, 500
Rho protein, 106, 107, 419–421
Ribonuclear protein particles, 110, 111
Ribose, structure, 22
Ribosome, structure and assembly, 541–552
    A and P sites, 156–159, 180
    average activity, 173
    dissociation, 177

gene complex, 93–95
in protein synthesis, 143–179
level variation with growth rate, 172–174
numbers of, 7
pairing with messenger, 152–154
prokaryotic, 7
regulation, 171–179
termination, 416
Ribulose, 349
Rifamycin resistant mutants, 86–91
Rifamycin, 175, 177
    RNA polymerase, 86, 220, 221
    structure, 90
RNA polymerase, 83–105
    abortive initiation, 98, 99
    antibiotic sensitivity, 90, 91
    assay, 84–86
    diameter, 10
    DNA contacts, 102, 103
    electrophoresis, 88–92
    elongation rate, 93–95
    inhibitors, 90, 91
    initiation of RNA, 95, 96, 435, 436
    isomerization, 96–99
    lac operon, 315, 345
    melting DNA, 36, 37, 103–105
    mutation, 86, 87, 219–221
    pausing, 386, 419
    salt effects, 100, 101
    sigma subunit, 92, 93
    stuttering, 386
    subunit structure, 87–90
    types in cells, 86–88
RNA, 5′ cap, 106
    alkali degradation, 41
    digestion by alkali, 285
    elongation rate, 181
    hairpins, 106, 107, 383–390, 544
    intervening sequence, 106–111
    methylation, 106
    precursors, 290–293
    processing, 107–111, 215, 293–296
    secondary structure, 386–390
    self-splicing, 111
    sequencing, 328–330
    splicing, 106–111, 215, 293–296
    synthesis, stringent-relaxed response, 175–177
    structure, 21–41
RNA-DNA duplex, stability, 286
RNAseP, 111
RNAseH, 600
RNAseI, 542, 543, 547

RNAseIII, 107, 469
Rolling circle DNA replication, 57, 58, 222, 412, 560
rosy, Drosophila, 508, 509
Rotation, flagella, 576–578
Rous sarcoma virus, 598–604
rRNA, 544–552
Run-tumble, chemotaxis, 578–581

S gene, lambda, 325
S1 mapping, 293–296, 357
S1 nuclease, 282, 293–297
Salivary chromosome, 216
Salivary gland, Drosophila, 287
Salt bridge, 127–129
Sanger sequencing, 269–273
Sarcoma, 598
Satellite DNA, 31
Scaffolding, 547, 556, 557
SDS protein gels, 164–166
Secondary integration sites, lambda, 462, 463
Secondary structure, proteins, 120, 131–134
    RNA, 386–390
    rRNA, 544, 545
Segregation, genetic, 300, 450
Selection, genetic 219–221
Self assembly, ribosomes, viruses, 541–563
Self splicing RNA, 111
Self-recognition, antibody, 519
Selfish DNA, 483, 484
Sequencing DNA, 266–276
Serum, 522
Sexual mosaics, 228–231
Shielding, ionic, 29
Shine-Dalgarno hypothesis, 151–154
Shine-Dalgarno sequence, 306, 307, 416
Shuttle vector, 248, 263–265, 397
sib, lambda, 469
Sickle cell trait, 282, 303
Sigma subunit, RNA polymerase, 419, 420
Sign-inversion, topoisomerases, 41
Signal averaging, 583, 584
Signal peptide, 166–171
Simian sarcoma, 598
Single strand binding protein, 57–59, 76, 208
Single-stranded phage, 49
SIR, repressor, mating type, 401
sis, oncogene, 604
Sliding repressor, 330, 331
Slow retrovirus, 601
snRNP, 110, 111
Sodium dodecyl sulfate, polyacrylamide gel, 88–91
SOS repair, 434

Southern transfer, 290, 300, 304, 389
Specialized transducing phage, 247,
    ara, 353, 563
    bio, 459–464
    gal, 459–464, 484–486
    kan, 489–490
    lac, 325–328, 484–486
    trp, 379
Specificity, *int* and *xis,* 458, 459
    of restriction enzymes, 253, 254
Sperm, 195, 196
Spermidine, 471
Spheroplast, 471
Spleen, 518
Splicing RNA, 106–111, 530–536, 600
Split ends, Mu, 502, 503
Split genes, antibody, 526–536
Split proteins, ribosomes, 546
Spontaneous induction, lambda, 432
Spontaneous mutation, 218, 240–245
Spore, yeast, 203–206, 396, 399
Sporulation, 194–196, 300
    B. subtilis, 92, 93
    yeast, 224, 225
src, 601–604
    identification, 603, 604
Stability, *int, xis,* 466, 467
Staphylococcus A protein, 292, 293
Statistics, 235–245
Sticky ends, DNA, 255
    lambda DNA, 32–34, 412, 413, 556, 558–560
Strand invasion, 204–209
Strand scission, DNA sequencing, 268
Strand separation, 267, 511, 512
Streptolydigin, 90, 91
Streptomycin resistance, 548
Stringent RNA response, 175–177
Structural domains, proteins, 109–111, 120, 131,
    524–526, 532
Structure prediction, Chou-Fasman, 137
Structure, DNA, 21–41
    protein, 119–140
    RNA, 119–140
Stuttering, transcription, 386, 419
Subunit mixing, 355, 366
Sugar phosphate sensitivity, 350
Supercoil, in recombination, 204, 205
Supercoiled DNA, 56, 401, 477
    in vivo, 25
    assay of, 34–36, 42, 43
Supercoiling, DNA melting by RNA polymerase, 104
Superinfection, phage, 417

Superposition theorem, 580, 581, 587
Suppression, nonsense, 160–162, 180, 199, 320,
    414, 436,
    temperature sensitive, 219
Sus mutations, 414
SV40, 263–265
Swarm plates, chemotaxis, 570, 571
Switch region, antibody genes, 594
Switch, yeast mating type, 396
Symmetry, 553–556
Symport, 13, 14
Synchronized cells, 72–74, 77
Synthesis rate, DNA, 65–75
    protein, 162–166
    RNA, 93–95
Systemic lupus erythematosus, 520

T antigen, 605
T cells, 518–536
T4 phage, DNA, 31
Tail, lambda, 412
Tailing, use in cloning, 256–258, 285
Tanning, 403–406
tap, chemotaxis, 584
tar, chemotaxis, 584,
Targeted mutagenesis, 296–302, 304–308
TATA box, 103
Tautomerization, 191, 192
Temperature dependence, ribosome, 548
Temperature sensitive, mutation, 190, 227, 228
    repressor, lambda, 414
    synthesis, 338
ter, lambda, 558
Terminal transferase, 258
Termination, transcription, 382–390, 419–422, 469
    translation, 148, 160–162
Tertiary structure, 120, 131–134
Tester strain, 396
Tethered bacteria, 577–583
Tetracycline resistance, 259–261
Tetrahydrofolate reductase, 264
Tetrahymena RNA, self splicing, 111
Thermal motion, 104
Thermodynamics, energy relations, 11
    entropy, 101, 102, 131, 159, 160, 331, 336, 337,
        436, 437
    proteins, 134–136, 140
Theta form of DNA, 62
Three-factor cross, 200–202
Thymidine kinase, 264
Thymine, structure, 23
Time-lapse photography, 579

Tissue-specific gene expression, 403–408
Tn elements, 488–515
TONPG, *lac* operon, 325
Topoisomerases, 38–41
Topology, lambda DNA, 468
    supercoiling, 31–41
Torsion of DNA, 39, 40
Tracking microscope, 578–581
trans dominant, 332–334
trans, genetics, 196–198
Transacetylase, 317
Transcription, *ara*, 10, 354
    attenuation, 373–390
    elongation rate, 181
    factors, 435, 436
    hybridization assay, 28
    initiation, 83
    lac, 10
    pausing, 386
    termination, 105, 107, 469
    termination, lambda, 419–422
    termination, *trp*, 382–390
Transducing phage, 379, 459–464, 484–486,
        489–490, 563
Transduction, by P1, 63, 64, 221, 222
Transfection, 265, 266
Transfers, Southern, Northern, Western, 289–293
Transformation, 263–266, 273, 283–289
    cancer, 591–607
    yeast, 397
Transient derepression, 357, 358
Transition, genetic, 191
Translation, assay for cloning, 287–289
    efficiency, 380–382, 468, 556
    eukaryotic initiation, 154
    in vitro, 546
    initiation, 150–156
Translational repression, ribosome, 178, 179
Translocation, chromosome, 594
Translocation, protein synthesis, 158, 159
Transport, active, 10–14, 181
    arabinose, 348, 349, 350, 351
    lactose, 317, 319, 339
Transposable elements, 483–515
Transposase, 486–492, 496–498
Transposition, 405, 406
Transposition mechanism, 509, 510
Transposon, 193, 483–511
Transposon vectors, 506–509
Transversion, genetic, 191
trg, chemotaxis, 584
Triplication, lambda, 558

tRNA, charging, 111, 144–147
tRNA, maturation, 111
    rare bases, 149
trp operon, 373–390
Tryptic digests, 557
tsr, chemotaxis, 584
TSS mutation, 338
Tubulin, 7, 8
Tumble, chemotaxis, 578–581
Tumor virus, 263–265, 598–606
Turbid plaques, 413, 414, 455
Twiddle, chemotaxis, 578–581
Twist, DNA, 32–38, 360, 361
Ty elements, 504–506
Tyndall effect, 250

Ultracentrifugal assay of superhelical DNA, 34–36
Ultraviolet light, mutagenesis, 192
Unilinear inheritance, 427
Universal joint, 576, 577
Upstream activators, 103, *see also* enhancers
Upstream repression, 354, 355
Urea, denaturing RNA, 86
UV damage to DNA, 60, 61, 192, 231
UV repair, 416, 417

V region genes, antibodies, 527–536
Van der Waals potential, 129
Variable regions, 525–536
Variable splicing, 530–532
Variance, statistics, 240
Vectors, genetic engineering, 248–250, 412,
        506–509
Vegetative pathway, lambda, 421
Vertical propagation, 599
Vichyssoise, 571
Vimentin, 8
Viral oncogenes, 601–603
Virulent mutants, 438
Virus, coat, 555
    herpes simplex, 264
    lambda, 411–477
    lambda structure, 552–561
    M13, 271–273
    P1, 64–65, 221, 222
    principles of structure, 552–563
    SV40, 263–265
Viscosity, 15

Walking, cloning, 286, 287, 307
Water cage, 130
Watson-Crick structure, 24–26, 149

Weber's law, 587
Western transfers, 290–293
Wild type, 190
Wobble, 149
Work, of active transport, 10, 11
Writhe, 32–38, 42, 43

X-gal, 318, 319
X-ray diffraction, 132
Xanthine dehydrogenase, 508, 509
Xenopus leavis, 265

Xeroderma pigmentosum, 60, 61
xis binding, 475, 476
xis, 456–459
Xylulose, 349

Yeast, 203–206, 224, 225, 263–265
    genetics, 203–206
Yolk proteins, 406–408

Zygote, 195, 196
Zygotic induction, 451, 507–509

SALES TRAINING HANDBOOK

A Guide to Developing Sales Performance

Sponsored by the American Society for Training and Development

Robert L. Craig, Editor

Leslie Kelly, Assistant Editor

PRENTICE HALL
Englewood Cliffs, New Jersey 07632

Prentice-Hall International (UK) Limited, *London*
Prentice-Hall of Australia Pty. Limited, *Sydney*
Prentice-Hall Canada, Inc., *Toronto*
Prentice-Hall Hispanoamericana, S.A., *Mexico*
Prentice-Hall of India Private Limited, *New Delhi*
Prentice-Hall of Japan, Inc., *Tokyo*
Simon & Schuster Asia Pte. Ltd., *Singapore*
Editora Prentice-Hall do Brasil, Ltda., *Rio de Janeiro*

© 1990 *by*

PRENTICE-HALL, Inc.

Englewood Cliffs, NJ

10 9 8 7 6 5 4 3 2 1

Library of Congress Cataloging-in-Publication Data

Sales training handbook : a guide to developing sales performance /
 Robert L. Craig, editor, Leslie Kelly, assistant editor.
  p. cm.
  "Sponsored by the American Society for Training and Development."

 Includes bibliographical references.
 ISBN 0-13-788175-4
  1. Sales personnel—Training of—Handbooks, manuals, etc.
I. Craig, Robert L. II. Kelly, Leslie. III. American Society for
Training and Development.
HF5439.8.S35 1990   89-16356
658.3 1245—dc20   CIP

ISBN 0-13-788175-4

PRENTICE HALL
BUSINESS & PROFESSIONAL DIVISION
A division of Simon & Schuster
Englewood Cliffs, New Jersey 07632

PRINTED IN THE UNITED STATES OF AMERICA

FOREWORD

Sales trainers are a peculiar breed. Peculiar in that they defy being categorized in nice, neat little boxes, or lumped into predictable groups. The only term that is really applicable is unique. For this reason, the Sales and Marketing Professional Practice Area of the American Society for Training and Development decided that a handbook for new sales trainers was needed.

Most sales trainers come from the field in their profession. They don't begin in personnel or training and are then assigned to training as is often true in other training areas. They are often superstars in their own right—having sold successfully for a number of years. Their organizations recognize their uniqueness, their talent, their energy, and their inner motivation and hope that by making them the sales trainers, somehow they will clone themselves and ensure that an entire sales force will match their success.

Selling successfully and training successfully are two distinct tasks. We hear cries for help frequently from sales personnel, turned sales trainers. Help is needed in all areas from understanding sales as a discipline to evaluating the final product with many steps in between.

Up until now, no book has been on the market that is expressly for the new sales trainer. No book was available to help the sales person turned sales trainer begin at ground zero and build a successful program. This book is such a book. Even the experienced sales trainer will find the book a wonderful reference tool and an excellent tool for training new sales trainers.

The book is a labor of love by ASTD's Sales and Marketing Professional Practice Area's Executive Committee. Begun nearly three years ago, the book has been written by experts in the field who have donated their talents to help make sales trainers' jobs easier.

The book would not have been possible without the support of the American Society for Training and Development—Curt Plott, Executive Vice President; Nancy Olson, Director of Publications; Fred Voss, Director, Technical Training Programs; Greta Kotler, Manager, Professional Product Development; Kermit Boston and Lynn Sellers, the Chairs of the Executive Committee during the project; and Leslie Kelly's office staff. A special thanks goes to Dick Miller of Merck who went above and beyond his role as a member of the Editorial Review Board, Jan Blough, Isabel Kersen, and Dave Schultze.

Finally, a project never succeeds without the devotion of an Acquisitions Editor, in our case, Olivia Lane. She believed in our project; she gave us guidance and important direction; she displayed patience and dedication to its completion. For her help, we are eternally grateful.

We dedicate this to Melvin Kallett, a member of our original Editorial Advisory Board. His untimely death shortly after the start of this project touched us all.

To our readers we say: Enjoy!

Executive Committee–Sales and Marketing
Professional Practice Area
American Society for Training and Development

WHAT THIS HANDBOOK WILL DO FOR YOU

The success of a business enterprise depends directly on its sales performance, which, in turn, depends directly on the competence of the sales force. Today, constantly changing job knowledge and skill, as well as a continuous movement of sales personnel, demand a new order of effective sales training. It is an essential corporate function.

This SALES TRAINING HANDBOOK brings together a wealth of successful sales training experience through the medium of a contributed reference work. It is a collection of the know-how of a specially selected group of outstanding authorities in sales training. It is a convenient, single source of basic information about results-oriented sales training.

It is designed for use by anyone responsible for training sales staff—full-time sales trainers, part-time sales trainers who may have other human resource or sales responsibilities, and for sales managers who have the total responsibility for sales performance.

The book covers the entire range of sales training activities and includes . . .

- how to determine what training is needed and when
- how to design and develop the right training programs and materials for your specific needs
- how to deliver training efficiently to the sales force
- how to manage sales training
- how to select and train the sales training staff
- how and when to use the various training methods—from the lecture to computer-based and interactive video training
- how to measure and evaluate the results of your training
- where and how to buy outside training materials and services

In short, this Handbook is a practical, quick resource of the basics for those charged with improving sales results through training.

As Editor, I particularly want to express my sincere appreciation to all the contributors who have labored through the extensive effort of putting this work together.

My special thanks and respect go to the members of our Editorial Advisory Board whose ongoing advice and reviews were so essential in producing the Handbook—they are:

Donald R. Botto, Goodyear International; Earl D. Honeycutt, Jr., The University of North Carolina; Brenda W. Hahne, ODI; Renee A. Huss, Southwestern Bell Corporation; Richard S. Miller, MSD-AGVET (Merck); Linda J. Segall, Ball Corporation; Lara Steinel, *USA TODAY*; and Jack D. Wilner, Jack Wilner & Associates, Inc.

Robert L. Craig, *Editor*
Fairfield, Connecticut

EDITORIAL ADVISORY BOARD

CONTENTS

SECTION I THE SALES TRAINING FUNCTION

**Chapter 1 Developing Professional Selling Skills:
The Role of the Sales Trainer**
James F. Evered 3

Where Sales Training Fits into the Company Structure 4

The Sales Trainer's Impact on Corporate Image, Growth, and
Earnings 5

Your Responsibility to Your Students 6

Your Responsibility to Your Company 6

How Sales Training Has Evolved in Corporations 8

The Changing Role of the Sales Trainer 9

The Sales Trainer and Corporate Planning: Be Supportive 10

Budget Field Trips to Bolster Communications 11

You have to *Sell* as Well as Teach 12

Try for a Revolving Sales Training Program 13

Importance of Follow-up 14

Training Objectives: Know What You Want to Accomplish 14

Evaluate Trainess after the Program 15

Maintain a State-of-the-Art Program 16

Make Self-training Materials Available 17

Use Internal Resources 17

External Resources: Don't Try to Do It All Yourself 18

The Future of Sales Training 19

Recognize Your Responsibility 21

Now It's Up to You! 21
Suggestions for Further Reading 21

Chapter 2 The Manager's Role in Sales Training
            Charles H. Kinney 27

Sales Manager Role Versus Training Role 29
Establish Clear Agreements Between Sales Manager and
Trainer 30
Learn How to Define Needs 31
How to Prepare for Your Training Program 32
Delivery: On-the-Job Training is the Best Choice 36
Follow-up 40
Special Tips for Sales Trainers 41

**Chapter 3 Sales Training Management: How to Develop
            Performance Leaders   *William E. Law* 42**

Strategic Skills: What Sales Training Managers Need to
Know 44
How to Establish a Successful Training Management
System 49
Strategy Formation 49
The Management Process 51
Organization Structure 62
Staff 64
Summary 67

**Chapter 4 Staffing for Sales Training: How to Improve Your
            Selection Techniques and Predict
            Performance   *Dennis S. McGurer* 68**

How to Assess Prospective Personnel *Before* You Hire
Them 70
Job Positions and Staffing Alternatives 76
Managing the Training Staff 81
Training and Developing the Staff 85
Evaluating Training Staff Performance 86
The Training Professional of the Future 88

Chapter 5 Preparing Sales Trainers *Roger E. Wenschlag* **90**

Four Ways Sales Trainers Positively Influence the
Organization 91
Tips on Selecting the Sales Trainer 92
What Sales Trainers Do 94
How to Prepare the New Sales Trainer 102
Summary 104

**SECTION II PROGRAM DEVELOPMENT AND
ADMINISTRATION**

**Chapter 6 How Adults Learn: Instructional Guidelines for Sales
Training** *W. Franklin Spikes III* **107**

Adult Learning Characteristics 109
Andragogy: The Art of Teaching Adults 120
How to Apply Adult Learning Skills to Your Sales Training
Program 123
Summary 124

**Chapter 7 Performance Systems Design: How to Develop an
Instructional System in Eight Steps**
Kenneth A. McClung, Jr. **125**

Performance Systems Design: How to Develop a Performance
System in Eight Steps 126
Performance Systems Design Overview 126
How Performance Systems Design Can Help Your
Company 127
Performance Systems Design Team: Eight Key Roles 128
How to Develop a Project Plan 130
PSD Step 1: Analyze the Problem/Requirement 133
PSD Step 2: Design a Strategy or Approach to Solve the
Problem 137
PSD Step 3: Develop Support Materials 144
PSD Step 4: Test the Materials 147
PSD Step 5: Produce/Reproduce the Final Materials 149
PSD Step 6: Implement the Program 149

PSD Step 7: Evaluate the Results 150

PSD Step 8: Maintain and Periodically Update 151

Summary 152

Chapter 8 Determining Sales Training Needs
  Robert C. Immel **154**

Nine Benefits of Conducting a Needs Analysis 155

Is It a Training Problem? 156

Suggestions for Collecting and Reporting Data 158

Seven Methods for Assessing Sales Training Needs 163

Sales Manager Needs and Their Impact on Your Analysis 174

District Meetings 175

Summary 176

**Chapter 9 The Big Picture: How to Create a Theme for Your
  Sales Training Program** *Diane Hessan
  and Thomas C. Keiser* **177**

What Is the Big Picture? 180

Ten Reasons Why Creating an Overall Message Can Help
Improve Your Sales Training Delivery 180

How to Craft the Best Message for Your Sales Program in
Four Steps 182

How to Incorporate the Big Picture Theme in Your Sales
Training Program 187

Summary 192

**Chapter 10 Developing and Writing Successful Sales Training
  Programs** *Warren Kurzrock* **193**

Formulating a Content Overview 194

Developing Content: Four Key Methods 195

Selecting a Training Method 200

Learning Objectives for Structuring Your Workshop 206

Before You Write Your Program: Two Critical Outlines to
Develop 207

Specific Formats for Developing and Writing Programs 213

Chapter 11 The Learning Environment and Its Impact on Your Sales Training   *Isabel L. Kersen* 217

How Environment Affects the Learning Process 218
Seven Major Elements of the Learning Environment 220
On-the-Job Training: Choosing the Best Learning Environment Possible 238
The Ideal Classroom Environment: A Ten-Point Checklist 240
Summary 241

Chapter 12 Administering Sales Training   *Gerry V. Marx* 242

Positioning Sales Training Managers As "Counselers" 243
Skills That Can Be Moldeled 245
Where to Locate the Sales Training Department/Training Activities 247
Records, Budgets, and Purchasing Recordkeeping 249
Activities to Consider Prior to Trainee Arrival 260
W.I.I.F.M. 265
Operations During the Training Program 265
Follow-up/Evaluation—Postprogram Activities 267
Summary 269

SECTION III IMPLEMENTING SALES TRAINING

Chapter 13 Coaching and On-the-Job Sales Training   *Nicholas H. Ward and Kathryn S. Wolfson* 273

How Top-Performing Salespeople Learn 274
Performance Assessment and Monitoring 275
How Sales Data Analysis Helps to Pinpoint Training Needs 278
Altering Comfortable Patterns of Sales Activity 279
The Trainer's Role in Building One-on-One Coaching Effectiveness 283
Day-to-Day Coaching of the Call Process 284

Performance Management Coaching 296
Sales Meetings: Group Coaching Opportunities 304
Training Sales Managers to Be Coaches and On-the-Job
Training Reinforcement 305

Chapter 14 Self-study and Self-development, Flexibility, and
            Learner Initiative for Your Sales Training
            System   *Greta Kotler* 306

Self-study 309
Self-development 323
Summary 333

Chapter 15 Lecture Method and Classrooms: The Most Frequent
            Method in Sales Training   *Robert W. Pike* 336

How the Lecture Method is Used in Sales Training 338
Handling Questions and Answers 341
The "Kilis" Method for Assessing the Audience 344
Developing Your Presentation Using the Four-Step AIDA
Formula 346
Techniques for Building Support in Your Presentation 349
Other Techniques for Supporting the Lecture 351
Tips for Communicating Sales Training Ideas Effectively 355
How to Present an Effective Opening 357
Closing the Lecture 359
Room Arrangement for the Lecture Method 360
Summary 363

Chapter 16 Using Role Playing for Sales Training
            Scott B. Parry 364

Role Play—A Misunderstood Opportunity for Training
Programs 365
Two Main Sales Training Purposes of Role Plays 366
How It All Started 367
Demonstration: Key to the Three-Stage Learning Model 368
How to Handle the Demonstration in Sales Training 369

Guidelines for Selecting Role Plays 377

The Three Objectives in Writing Role Plays 382

Preparing the Trainees: Two Training Tools for Practicing
What Is Learned 384

Conducting the Role Play 391

Leading the Postenactment Discussion 393

Group Variations: Six Techniques 394

Summary 402

**Chapter 17 The Case Method: Teaching Sales Managers to Make
Better Decisions   *Thomas J. Newman* 403**

The Case Method: A System of Instruction Based on
"Guided" Discovery 405

Designing Your Own Cases with In-House Resources 406

Using Specialized Professional Writers to Develop Cases 409

Maintaining Quality in Case Selection 410

When to Use the Case Method 410

How to Write a Case in Five Steps 412

How to Use the Case Method in an In-House Training
Seminar 414

Four Methods of Measuring the Impact of the Case
Method 418

Sources of Case Material for Your Training Program 420

Summary 420

**Chapter 18 Gaming and Simulation in Sales Training
J. Clayton Lafferty with Tracy L. Range 422**

How Gaming and Simulation Are Used in Everyday Learning
Experiences 424

Simulation: Three Key Types to Consider in the Training
Environment 425

When to Use a Simulation or Game 427

How to Use Prepared Simulation Materials 429

Creating Your Own Simulation Materials 435

Summary 437

**Chapter 19 Effective Audio Visual Media in Sales
             Training *Rick Wills* 439**

Audio Visual Media: Support Materials That Appeal to Sales
Trainees 440

How Audio Visual Media Help Sales Trainees to Learn 441

Types of Audio Visual Media to Use in Sales Training 442

Six Criteria for Deciding Which Audio Visual Aid to Use in
Your Training Program 464

Guidelines for Preparing and Using Audio Visual Aids 465

Working with Audio Visual Suppliers to Choose the Best
Equipment for Your Needs 466

Summary 467

**Chapter 20 How Video and Film Can Improve Your Sales
             Training Presentations *James A. Baker* 469**

Instructional Guidelines for Using Training Materials 471

The Advantages and Disadvantages of Video and Film 472

Video: Seven Reasons for Its Popularity with Training
Managers 475

Developing a Sales Training Program with Video Tape 478

Purchasing Videotaped Sales Training Materials 480

How to Make Your Own Sales Training Video in Six
Steps 481

Marketing Your Sales Training Video Program 489

Summary 490

**Chapter 21 Computer-Based and Interactive Video Training
             Technology *Patrick A. Cataldo, Jr., and
             Daniel S. Cooper* 492**

Computer-Based and Interactive Video Training: Fast,
Effective, and Flexible 494

How CBT and IVT Differ from Other Forms of Media-Based
Instructions 495

Five Examples of CBT and IVT Programs Used in Sales
Training 496

How CBT and IVT Compare with Other Sales Training
Methods 498

Practical Applications 499

Advances in CBT and IVT Technology 506

The Impact of Computer-Based and Interactive Video
Training on Sales Trainers 509

Summary 510

**Chapter 22 How to Improve Your Sales Training
          Meetings *Homer Smith* 512**

Five Reasons for Conducting a Sales Training Meeting 513

How to Make Your Sales Training Meeting a "Group Sale"
Using the AIDA Formula 514

Setting the Meeting Objective 515

Four Basic Training Methods for Sales Meetings 516

Seven Steps for Planning a Successful Sales Meeting 518

National Versus Regional Meetings 519

Tips for Planning Large Sales Meetings 521

Sources for Meeting Speakers 522

Preparing an Effective Meeting Room Environment 524

**SECTION IV MEASURING AND EVALUATING SALES
           TRAINING**

**Chapter 23 Practical Guidelines for Measuring What Trainees
          Learn *Thomas R. Currie* 433**

Five Reasons for Measuring Learning in Sales Training 534

Learning Objectives: The Basics for Measurement 537

Individual and Group Problem Solving 543

Action Planning: How to Measure Specific Areas of
Learning 543

Learner Evaluations: Encouraging People to Think About
What They've Been Taught 545

Summary 546

Chapter 24 Measuring Training Results: Behavioral Change and Performance Improvement *C. E. (Gene) Hahne, Robert E. Lefton, V. R. Buzzotta* **548**

How Behavioral Change Training Works 551

Three Kinds of Objectives for a Successful Sales-Behavior Training Program 552

Three Questions for Determining If Your Sales-Behavior Objectives Have Been Met 553

Two Basic Measuring Principles 554

Precipitating Events: Why Sales Training Programs Don't "Just Happen" 555

How Sales Trainers Determine the Need for Sales-Behavior Training 555

How to Win Management's Commitment by Linking Various Corporate Interests 557

Designing Measures of Behavioral and Performance Change 558

When to Measure 559

How to Measure 569

Obtaining Qualitative or Quantitative Measurements 570

How Numbers Help in Compiling Results 573

Designing Pre- and Postmeasures 574

The Need for Statistical Checks 575

Presenting the Findings 575

Summary 577

SECTION V SELECTING AND USING SALES TRAINING RESOURCES

Chapter 25 Sources of Sales Training Information *Edward R. Del Gaizo and Derwin A. Fox* **583**

Product Training 585

Selling Skills: Professional, Commercial, and Practitioner Sources 589

Selling Skills: The Academic Approach 604
Summary 606

**Chapter 26 Sales and Marketing Education: Programs Offered
by Colleges and Universities
Steward W. Husted 607**

How Sales and Marketing Education on Campus Began 608
Types of Programs Offered on Campus Today 609
Continuing Education on Campus 612
Continuing Education and Services off Campus 618
Industry Involvement with Marketing ED/HRD Academic
Programs 622
Evaluation of College and University Programs 628
Summary 629

**Chapter 27 Consulting with External Sales Trainers: A Four-
Phase Process   *Chip R. Bell* 630**

Consulting with External Sales Trainers 631
Sales Consultant—Specialists in the Selling Process 631
How an External Sales Trainer Can Help You 633
The Consulting Process and External Sales Trainers 634
Phase 1: Entry—The Initial Contact 635
Phase 2: Diagnosis—Examining the Problem 645
Phase 3: Response—The Action Part of the Consulting
Process 646
Phase 4: Closure–Evaluation and Follow-through 648

**Chapter 28 When and How to Use Packaged Sales Training
Programs   *Richard C. McCullough* 650**

The Trend Toward Packaged Sales Training Programs 651
Guidelines for Making Your Build or Buy Decision 653
Identifying Packaged Programs for Sales Training 658
Choosing the Right Sales Training Packaged Program 662
Working with the Vendor of Your Choice 669
Summary 670

Chapter 29 Public Seminars as a Sales Training Resource
**   *Don M. Schrello* 672**

Public Seminars and Their Role in Sales Training 673
When to Use Public Seminars in Sales Training 679
Estimating the Costs and Payback of a Public Seminar 682
How Public Sales Training Seminars Operate 683
Sources of Information on Public Seminars 690
Tips on Selecting the Right Seminar 695
Getting the Most out of a Public Seminar 700
Summary 705

Chapter 30 Professional Networking for Sales Training
**   *Judith C. Quinn* 707**

How to Network Successfully in Sales Training 708
What Are Networks? 710
How to Start Networking 711
How to Use a Network 713
The Benefits of Networking in Sales Training 714
Summary 715

Index 719

section one

THE SALES TRAINING FUNCTION

Chapter 1 Developing Professional Selling Skills: The Role of the Sales Trainer

Chapter 2 The Manager's Role in Sales Training

Chapter 3 Sales Training Management: How to Develop Performance Leaders

Chapter 4 Staffing for Sales Training: How to Improve Your Selection Techniques and Predict Performance

Chapter 5 Preparing the Sales Trainers

chapter one

DEVELOPING PROFESSIONAL SELLING SKILLS: THE ROLE OF THE SALES TRAINER

JAMES F. EVERED

James F. Evered is president of HRD Services, Inc., a sales and management consulting firm in Denton, Texas. His background includes more than 30 years of experience as a trainer of over 100,000 sales and sales management personnel for more than 500 companies throughout the world, many among the *Fortune* 500. He is past president and honorary life member of the National Society of Sales Training Executives. He has been awarded ten Medals of Excellence by the Society, including four coveted Gold Medals. He is a member of ASTD and has received the Society's Professional Contributions Award for his outstanding work in sales and marketing. Evered also holds the Certified Speaking Professional certification from the National Speakers Association. An accomplished salesman, manager, author, trainer and speaker, he is recognized by his professional colleagues as one of the nation's top trainers.

Sales trainers have one unifying purpose: to develop the professional selling skills of all sales personnel which will, in turn, assist them in reaching corporate marketing goals. The sales trainer's responsibilities cover a wide spectrum, from determining training needs, program development, making proposals to management, designing and delivering instruction, measuring training results, and supporting field sales management in their training efforts to providing resources for self-development of sales personnel.

The professional development of salespeople, those who move the merchandise from the farm to the market, from the mill to the factory, from the factory to the distributor, to the dealer and to the ultimate consumer has never been more essential. In today's fast-changing world with fierce international competition, computerization, highly educated and sophisticated buyers, and rapid changes in the marketplace, the professional training of salespeople has become a vital ingredient in corporate survival and success. Whether you are a professional sales trainer or a field sales manager, your training responsibilities have never been more important. Training salespeople has multiple objectives, that is, to develop salespeople to their highest possible level of professionalism, increase sales, increase productivity, reduce turnover, develop stronger and better customer relations, decrease selling costs, and improve morale.

WHERE SALES TRAINING FITS
INTO THE COMPANY STRUCTURE

The sales trainer's position on the organizational chart covers a full spectrum, depending on the company, its philosophy of training, its size, and the size of the sales department. The trainer usually reports either to the director of sales or the director of personnel. In some cases, the trainer and his or her department report directly to corporate management.

In smaller companies, the sales trainer is usually a one-person department, responsible for all functions, that is, determining training needs, program design, proposing the program to management, writing training materials, developing audio/visual (A/V) aids, delivering the instruction and measuring results. The one-person department may also be responsible for other activities within the company, especially if the sales force is small. In other words, sales training may be a part-time function.

In many larger companies the sales training department may have a full staff of writers, instructors, A/V specialists, and so on, all reporting to a departmental manager. Many of these larger departments provide a spectrum of training for the company, for example, for sales managers, supervisors, executives, and the company's customers. In high-technology companies the sales trainer may even be responsible for teaching product knowledge, assisted by product managers.

Many companies do not even have a sales trainer per se, and the responsibility is left fully in the hands of field sales managers. In such cases, however, the responsibilities remain the same. The sales manager must budget a portion of time for this important function. The development of salespeople is a high priority. Often, however, training is interrupted by other activities of the manager, and the training loses continuity. In a centralized training atmosphere this is not a problem.

THE SALES TRAINER'S IMPACT ON CORPORATE IMAGE, GROWTH, AND EARNINGS

Few employees within a company have the enormous impact on corporate image, growth, and earnings as that of the sales trainer. To paraphrase the great Peter Drucker, "Selling is the only thing that contributes to earnings—everything else contributes to costs." Salespeople *are* the revenue producers of the organization, and the revenue they produce correlates directly with the quality of training they receive. Salespeople carry the company image and acceptance into the marketplace.

Field sales managers and corporate sales trainers who have responsibility for the professional development of salespeople have a key role in the future of the company.

There is no such thing as a *born* salesperson. Professional salespeople are *developed*. While some people have the personality to help them succeed, many have limited sales experience, some may even have extensive sales experience, but it will be your job to bring them all to the highest possible level of performance. Time and time again, I have seen salespeople fail, or suffer mediocre incomes, simply because they were not adequately trained. Most of them had to learn selling techniques by observation and osmosis. What a waste of careers! Many companies hire "experienced" salespeople, thinking they don't need training.

When such people are hired, the company is buying the bad habits with the good, and bad habits are often difficult to change. As a sales trainer, it will also be your responsibility to change these bad habits.

YOUR RESPONSIBILITY TO YOUR STUDENTS

Most sales trainers feel a real responsibility *for* their students while in the learning environment or the training program. You will also be responsible *to* your students. You will have a direct influence on the salesperson's life-style, earnings, growth, future, and standard of living. This responsibility must not be taken lightly. If you do a mediocre job of training, that person will likely have a mediocre future. If you do an excellent job of training, that person will likely have an excellent future. Many top corporate executives have risen from the ranks of sales. In every case, I have found that they had outstanding trainers and managers who had a personal dedication to "giving their best." They recognized the needed skills, taught those skills, coached the skills, and gave of themselves, with a personal dedication to helping their people grow. Positive reinforcement and performance feedback is a way of life with good trainers and managers.

The salesperson should always know where he or she stands, and understand what is needed for improvement. Unfortunately, many managers and trainers are reluctant to provide effective feedback to their salespeople. It seems to be the old "no news is good news" syndrome. As criminal and immoral as it may be, many salespeople are notified for the first time, on the day of their termination, that they aren't performing satisfactorily. Whose fault is it? The fault can be laid directly at the door of the manager, the trainer, or both.

Stop and think about it—the amount of merchandise or services salespeople sell, the earnings they receive, the life-style and the standard of living they enjoy, the career success they achieve, and job satisfaction they get is heavily dependent upon the training they receive. To repeat, your responsibility is no small one. The importance you put on your job, the professionalism of your performance, your constant willingness to improve, and your dedication to seeing others succeed will have great impact.

YOUR RESPONSIBILITY TO YOUR COMPANY

Due to the sales trainer's impact on corporate earnings, growth, and image, he or she bears a great responsibility to the company, and that responsibility manifests itself in RESULTS. The trainer will, or should, be measured by the results produced by the training program. As a sales trainer, you must develop your company's revenue producers to the highest level possible through ongoing sales development for both the novice and veteran. Your responsibilities are the same, whether you are

- A career or long-term trainer.
- In a stepping-stone position grooming for management.
- A one-person or a part-time training department.
- A training manager with a full staff.
- Responsible for sales training only.
- Responsible for both sales and sales management training.
- Responsible for training customers' salespeople.
- A field sales manager with 100 percent training responsibilities.

Training is not a one-shot deal, although many perceive it as such. The development of professional salespeople is ongoing. Skills deteriorate, we forget, we go through periods of professional degeneration, products and markets change, and constant sales training keeps skills honed to a fine edge. It is your responsibility to see that your company has a continuing program.

The routine of selling frequently takes its toll. Salespeople get tired of making the same presentation over and over. Without realizing it, salespeople begin to take shortcuts, and their presentations deteriorate. They get away from the basic fundamentals and techniques. As a sales trainer, you must provide constant refresher training and/or reinforcement on the job for your company's existing salespeople to keep the selling skills sharp.

Although, when assigned to the training job, you may have "inherited" an existing training program, you must constantly analyze the program to be certain it is on target and results-oriented. Further, many sales trainers become victims of the "training fad syndrome"; that is, they look for ways to upgrade their programs and lose sight of the basic needs of salespeople. Repetition takes its toll among trainers as well as salespeople. Many times they look for the fads simply to have something different to do. Any time you make a change in your training program, be certain it's for the benefit of your students, not for you. As a professional rose grower once told me, "Prune for the rose's sake, not the pruner's sake."

Even today, sophistication of the training language continues. This has been an evolution in itself, and space will not be devoted here for a full discussion. However, the modern language of training has taken on some interesting facets. Consider some of the current terminology we see in publications crossing our desks:

Sophisticated terms used to replace "sales," "salesman," and "selling" include

Organizational behavior modification

Behavior modification specialist

Functional analysis

Intervention strategies

Response frequencies

Self-modification for personal adjustment

Behavior engineering systems training

Behavior modification technology

When I study these trends, I continue to ask myself, "Why?" "What are we doing?" "Do these terms more clearly define what we are doing?" "Is it mainly an effort to professionalize?" "Are we groping for something different that might work?" Beware the "fad" trap!

HOW SALES TRAINING HAS EVOLVED
IN CORPORATIONS

The corporate sales trainer is gradually taking his or her rightful place in the organization. Over the past several years, companies have finally begun to realize the true value of their sales trainer and the impact he or she has on the organization. We have seen the trainer move higher and higher in the corporate hierarchy, many becoming vice presidents. The income level of sales trainers, too, has risen, commensurate with their contribution to the organization. Many have become the head of the human resources department, with added responsibilities for all personnel functions. But who are the trainers who have made such strides in these past few years? They are the ones who kept their training programs current, kept them on target, kept them results-oriented, did a professional job of teaching, produced measurable results, and reported those results to their management. They proved their worth to their companies.

On the other hand, it is saddening to see the number of companies which have not yet recognized the valuable contribution their sales trainers are making and are keeping the job at a level far below that which is appropriate. This situation, naturally, keeps the job at a stepping-stone level, and the trainer cannot afford to stay there long. The result is a regular turnover in the job and a heavy loss of accumulated experience with each turn. This weakens the effectiveness of the position and hinders its growth.

Why, then, does the sales training department become a prime target for the corporate axe? Simply because management does not see enough value in the function to retain and intensify it. I'm sure they have a "gut feeling" that it's doing some good—but how much good? Therein lies the problem. When management looks around for excess weight to throw overboard, they look in terms of cold, hard dollar figures. Unfortunately, the only figures they usually have regarding sales training is the "cost" of the program, that is, salaries; facilities; materials; equipment; and the fortune they have spent on meeting rooms, meals, and travel expense. Sadly, these people are looking at only half an operating statement. It's like judging the health of a business by the operating expenses alone.

Why, then, don't they analyze the other side of the statement? They don't have the other half, that's why. And it is the sales training department that failed to set up a system of measuring its own worth. When you talk to a sales trainer about his or her failure to establish a measuring system, you hear a host of old familiar cop-outs like, "there are just too many variables, or we're dealing with personalities," or "No matter how well I teach, if field management doesn't support it, you can't——." Hogwash! If you can't measure your results, you better get out of the business of training sales personnel.

We'll find measurable results for almost any sales training program. But these usually turn out to be measured against objectives set by the sales trainer, in terms of what he or she plans to accomplish, that is, number of meetings held, number of people trained, new programs to be developed, budgetary controls, and so on. While such objectives are most commendable and we all have them (or should), that's not enough.

"Measuring the *results* of what we taught" (trainee performance) is what really counts. To clarify things, I'm not talking about the tests you give or the demonstrations your trainees make at the end of the course to see whether you accomplish your *instructional objectives*. I'm talking about the results the trainees achieve back in the field *because* of your teaching. This is where you prove your worth as a sales trainer, and this is where you strengthen your position by improving sales, which, incidentally, brings trainer job security. The company simply can't afford to lose you.

Why, then, don't many sales training managers have a system of measurement set up? Often, perhaps, they are sincerely afraid to put themselves into a measurable situation because of the inherent threat that goes along with it. They are surviving by "gut feelings," testimonials, and impressive platform performances. I wonder whether they have enough faith in their own competence to stake their reputation on it. Will they lay their job on the line, protected only by professional ability? Perhaps they don't know how to set up a measuring system.

THE CHANGING ROLE OF THE SALES TRAINER _____

Years ago, the sales training slot was often filled with a "burned-out" sales representative. Although the sales rep was burned out on the territory, the company felt obligated to keep him or her on, due to longevity, and the sales training position was a logical "pigeonhole." Can you imagine the negative effects these trainers had on their sales trainees? Fortunately, this practice is seldom seen today.

As a professional sales trainer, you simply must not stagnate. You must be in a constant state of evolution, keeping your training program up to date using new technologies (e.g., computer-assisted instruction) and responding to the changing roles of sales personnel, changing markets, changing customers, new products, changing management philosophies, and so on. Today's customers are more educated, more sophisticated, and more business-minded. You must develop salespeople capable of dealing with today's customers. That means constantly analyzing your program to see whether it is truly meeting the needs. Your company's customers can be a great source of research to assess needed alterations in your training.

As the sales trainer position moves higher and higher in the organization, we're seeing larger and larger training budgets, larger training facilities, and larger training staffs. This is visible evidence that companies are beginning to realize the value of sales training. They are seeing trangible results on the bottom line. They are beginning to see training as an investment, not an expense. As the sales trainer, you are responsible for seeing that the investment pays a healthy return to your company. The trainer can be a company leader in good times and the company savior during hard times.

THE SALES TRAINER AND CORPORATE PLANNING: BE SUPPORTIVE

Few corporate changes are made that do not, in some way, affect sales, and the sales trainer must become a part of corporate planning. As the trainer, you play an important role in helping management meet its goals. If you are not presently a part of planning meetings, *ask* to be invited to sit in. All marketing plans must be implemented by field sales personnel, promotional programs must be implemented by them, and new products must be introduced by them. You play an important role in getting these jobs done.

Sales trainers often develop a special rapport with their students and this rapport with field personnel can be especially useful in implementing of marketing plans. This rapport seems to continue, even after the sales representatives have been out in the field for a long time. Thus, the sales trainer can help bring the perspective of the field sales staff to corporate planning.

The field sales manager will always be 100 percent responsible for developing his or her sales personnel. Your job as sales trainer is to supplement the manager's efforts and to help him or her train the sales representatives.

Many sales managers send their representatives to sales training sessions, thinking they will come back high-producing professionals. They expect the sales trainer to do the entire job for them. This just won't happen! The field sales manager *must* follow up to reinforce the skills taught, to keep the skills alive. In most cases, you will have to train the sales manager *how* to follow up on your training. The manager may be sending his or her salespeople to training experiences the manager has never seen. How can they possibly follow up on something they have never seen? They can't. You will have to tell them, specifically, what needs to be done.

You are also a trainer for your sales managers. Let's face it, not all good managers are good teachers. Doing a job well and teaching someone else to do it well require two different sets of skills. You will have to "train the trainer" in most cases. You may have to assist the sales manager in developing training programs, writing outlines and leader's guides, developing pass-out materials, and producing audio/visual aids. Don't do it *for* the manager; rather do it *with* him or her. The chapter on "How Adults Learn" can be of great help to you and your field managers.

BUDGET FIELD TRIPS TO BOLSTER COMMUNICATIONS

Your time must not be limited to your office and the training center. You need to budget a lot of time for the field, both traveling with field sales representatives and working with field sales managers. You may be asked to travel with a "problem" sales representative, even providing coaching. You may wish to travel with the company's "star" sales representative to find out what he or she is doing right so the information can be fed into your sales training program. You may wish to travel with several sales reps to determine their training needs. You may even wish to travel with them to get tangible evidence of the results of your training.

Above all, you should budget time in the field with your company's field sales managers for maintaining communications and gaining field support for your training. The field sales manager should definitely be a part of your research on training needs. It's a great way to get the manager to "buy into" your program. If the sales managers feel they have a say in what goes into the program, they are a lot more likely to support the program. Let's face it, these sales managers live with their sales representatives on a day-to-day basis. They know their needs. They know where their weaknesses and their strengths lie. They can be of enormous help in helping you develop (or change) a training program that best serves the company's needs. This also provides a great opportunity to discuss what the sales manager must do to follow up on your training to keep the skills alive.

While working with the field sales managers, let them know they should not expect miracles. It's going to be impossible for the trainees to put into practice everything you taught them immediately after your sales training sessions. Psychologically, none of us is able to handle that much change at once. Help your sales manager determine those skills he or she wants to see practiced at once, limiting it to two or three skills. After the sales representatives become comfortable with these changes, bring in two or three more. You don't paint a barn by throwing the paint on by the bucketful. You put the paint on "stroke by stroke." The same is true in implementing new selling skills. Give them what they can handle and then give them some more. Eventually, you and the sales manager will have many skills at work. You will see tangible results from this approach. The sales representatives are perfectly normal people, and they will tend to avoid techniques with which they feel uncomfortable. They will tend to "retreat to the familiar." Don't give them more than they can handle at one time.

Never, never, never make a field sales trip with a sales representative without first notifying the sales representative and the sales manager. Don't pull any surprise visits. That will surely put the sales representative on the defensive, he or she will resent it, and little will be accomplished. It should be a *planned* trip, with all parties knowing why you are there.

YOU HAVE TO *SELL* AS WELL AS TEACH

If we are to be effective sales trainers, we must practice what we teach. We have to *sell* our students on what we teach. If the student doesn't "buy" the material being presented, you can rest assured he or she will never use it. Actually, in the classroom you are in the role of a salesperson, and your student is in the role of customer. Just like a customer, the student is silently asking the most important question in all of selling, "What's in it for me?" "Why should I do it this way?" "Will it make my job of selling easier?" "Will I make more money?" "Will it enhance my future?"

In the classroom, as a sales trainer, you are constantly selling your students on what is being taught. Presenting isn't enough—everything has to be sold. Just like a product, "if they buy it, they will use it." Your reputation as a professional sales trainer will be determined not only by what you *teach*, but by what you *sell*.

That's why it's important that every segment of your program be relevant to your students. But, just as with a customer, the student is not always aware of his or her needs. Don't *assume* relevancy—*establish* it. Throughout your program, constantly ask yourself, "Is this really relevant to my students?" Also, throughout your training sessions, constantly remind your students of the *why* as well as the *what*.

At the end of my training sessions, I use an interesting method of determining whether I have sold the students on the material presented. I allow about 30 minutes at the end of each seminar for this little exercise. I ask each student to take a piece of blank paper and write the answers to this question: "Here's what I plan to do as a result of attending this seminar." Then, we go around the room, reading every one of them aloud. When the reading is finished, I wrap it up with a remark like this: "If each of you do what you just said you will do, think of the bottom-line impact that will have on the company's operating statement—and on your income!" They have, in fact, committed themselves to what they are going to do. That way, I can tell whether I have "closed the sale."

TRY FOR A REVOLVING SALES TRAINING PROGRAM

Wouldn't it be nice if we could do all our sales hiring on January 1 and begin the training program on January 2? How simple that would make it. But it just doesn't work that way. Naturally, it is important to bring a new hire into the program as soon as possible, but that can't always be done. It can't even be done on a quarterly basis. What do we do with the new hire until the next sales training seminar is scheduled?

There are two basic philosophies of training. One is, "Throw them to the wolves on a territory and let them get a little experience before bringing them in for training." The other is, "train them before we throw them to the wolves." I won't argue in favor of either system, but the problem exists. Personally, I want to bring them into a training seminar as quickly as possible so I can teach them the good habits before they have a chance to develop bad ones. Throwing them to the wolves too early takes a heavy toll in loss of potentially good salespeople. I want an early opportunity to get them sold on the idea that selling is a great profession, that high earnings are possible, that they can be proud to be a salesperson, that they can enjoy success if they go at it right, that is, using professional selling skills.

Having them kill time until the next sales training class by asking them to travel with various salespeople has serious drawbacks. They will pick up both bad and good habits, and bad habits are hard to break. This will happen, no matter how carefully you select the salespeople with whom they travel.

Then, too, you should have courses that veteran sales representatives can return to for a refresher in certain skills. I often find veteran salespeople who feel a need for a refresher course in closing, or some other selling skill. A program must allow them to come in for "parts" of it. Veterans can also be supportive for the new trainee in the class. He or she can share successful ideas with your newer people and substantially reinforce your program.

One of your toughest jobs will be to train salespeople who have had considerable experience with another company. They will have a lot of things to "unlearn," and that's difficult. They will have habits with which they are perfectly comfortable, some of which are incongruent with what you are teaching. Here you have a real selling job. Changing behavior is difficult. You can't "demand" change, but you can "sell" change. Your training program may even need customizing to accommodate the person with sales experience.

IMPORTANCE OF FOLLOW-UP

Salespeople are often considered trained simply because records show they have attended a course, yet their attitude toward other people and toward quality of selling skills and sales productivity show that their training has not been effective. Why? Well, there could be several reasons, some of which can be laid right at the door of the sales trainer. Perhaps poor teaching skills can be blamed. Perhaps the trainer failed to establish relevancy. Perhaps the trainer failed to "sell" the participants on the importance of the training. Perhaps the cause is a lack of follow-up by the field sales manager. If this is the case, the sales trainer can do something about it.

Your program must be "on target," that is, designed to meet the specific needs of the student, must be results-oriented, loaded with "how-to," heavy on participation and practice, and in accord with good adult learning practices. Above all, the students must "buy" what is being presented. That's a selling job on your part.

TRAINING OBJECTIVES: KNOW WHAT YOU WANT TO ACCOMPLISH

Later chapters will cover measuring learning, and it is important that trainees take away from the program those things you want them to take away. From the very outset of developing the program, you must decide (in behavioral terms) what you want the trainee to be able to *do* as a result of the training. That is the only way you can build a meaningful program. You must "sell" the relevancy of the material, and you must do it throughout the entire program.

Training objectives (behavioral outcomes) must be clearly defined for the student *prior* to beginning training. In other words, you must provide the trainees a list of "here's what you will be able to do as a result of this program," prior to the course. As you provide the list, you must sell them on the importance of doing it that way. That establishes the relevancy. If they don't understand that relevancy, believe me, they won't buy it.

After you have established relevancy and advised them what they will be able to *do* after the course, you must measure, at the end of the course, whether they can *do* what you taught. That means demonstration of the skills taught.

Chapter 16 covers the basics of role play that can be used to determine whether they can actually perform the skills taught. You can't assume they learned the skills; you must have them demonstrate the skills.

After my training sessions, no sales manager has ever been able to say to me, "My reps don't know how to close a sale." I know better! If I had not observed them closing sales properly, they would never have left the classroom. Every one of them had demonstrated the ability to close sales. But here is where follow-up by the sales manager is essential, that is, keeping those skills alive until they become habits.

EVALUATE TRAINEES AFTER THE PROGRAM

Sales trainers can determine, through role play and demonstration, that trainees are capable of displaying the skills taught. But what happens after they leave the training environment? The greatest measure of your effectiveness is what happens back on the job after they leave your program, the results gained in the field as a result of your training efforts. Here's where you must have full support of the field sales manager, that is, keeping the skills alive on the job where bottom-line results are evident.

Some of the measurable results which will enhance the validity of your training, whether with novice or veteran salespeople, are

- Call/sales ratio on prospects.
- Call/sales ratio on existing accounts.
- Average dollar volume per sale on prospects.
- Average dollar volume per sale on existing accounts.
- Dollar sales per mile traveled.
- Returned merchandise (monthly average).
- Customer complaints (monthly average).
- Expense to sales ratio
- Average calls per day on prospects.
- Average calls per day on existing accounts.
- Report inaccuracies (monthly average).
- Late report frequency (average number of days).
- Forecasting accuracy.
- Collections (average dollars uncollected per month).
- Total sales volume.
- New accounts obtained.

- New markets developed.
- Promotional programs conducted or implemented.

Through the cooperation of your field sales managers, it is imperative that training results be reported to your management. Remember, you *must* gain the support of management for your training program to make your full contribution to the organization. Keep your management thoroughly informed of your training results.

MAINTAIN A STATE-OF-THE-ART PROGRAM

Avoid the temptation of the "training fads" as something exotic to impress management. If it won't add to the essential skills of your students, don't do it! Remember the old KISS rule: "Keep It Simple, Salesman."

To keep your program up to the "state of the art," it will be important to do a lot of reading, interpreting, evaluating the myriad materials that cross your desk. Weed them out carefully, but watch for those things that will truly enhance your efforts.

It is also important that you participate actively in organizations which will add to your professional growth, for example, the American Society for Training and Development's (ASTD) Sales and Marketing Professional Practice Area, the National Society of Sales Training Executives (NSSTE), the National Speakers Association, and other organizations, particularly within your own industry. They can add immeasurably to your professional growth. All industries have their associations, and you should take full advantage of their programs.

You should also take advantage of other sources of professional growth. There is an "orchard" out there if you will go after it, for example, cassette programs, books, community college courses, extension courses, and trade publications. Take advantage of all. The concluding section of this chapter will give you several sources to which you can turn for your own professional development, as well as for the development of your salespeople and sales managers.

Throughout elementary school, high school, and college, we were guided by textbooks to supplement teachers' efforts. Why not in your sales training program? Select an excellent book on selling skills which agrees with your philosophies and provide it to your students. It not only supplements your efforts, but it also provides material for night assignments and follow-up reinforcement after your program. Nearly all books on selling skills are generic, but you can extract many examples from the book and adapt them to your specific product or service.

MAKE SELF-TRAINING MATERIALS AVAILABLE _____

You can't do it alone! Training of sales personnel must go far beyond the classroom. There are vast resources of materials available for salespeople who have a sincere desire to continue their professional growth. As the sales trainer, you should maintain a library of material available to your field salespeople. Among the items to be included are (see the suggestions for further reading at the end of this chapter)

- Books relating to selling skills.
- Books on management for the salespeople who have aspirations of moving into management positions.
- Manuals, both from within the company and from without.
- Trade publications.
- Audio cassette programs.
- Video cassette programs.
- A recommended reading list for sales personnel.
- The curricula from colleges, universities, and community colleges relating to sales and sales management.

To keep the library current will require a lot of research on your part. But just having a library available is not enough. You must constantly encourage your field salespeople to take full advantage of it, even maintaining a log of the people using it and how frequently. That tells you a lot about the potential of your salespeople. Naturally, it is equally important to maintain such a library for your field sales managers. Self-development is just as important for them. Again, constantly encourage the use of these resources. It can substantially reinforce your own efforts.

USE INTERNAL RESOURCES _____

I invariably make the following statement in every seminar I conduct: "Ladies and gentlemen, there is more genius sitting in this room than there is standing." It relates to the synergistic effect, that is, "None of us is as smart as all of us." Never fall into the trap of thinking that the sales trainer knows it all, and his or her job is to deliver all that genius. We go around the room asking each participant how many years of experience they have in selling, and you would be amazed at the total. For example, over the previous 30 years of training salespeople, I have been exposed to approximately 400,000 years of sales experience. If a trainee can't draw something from that much experience, he or she doesn't need to be in the training seminar—he or she should be in therapy.

Within your company you will have many people who can make a significant contribution to your program. You have product managers, engineers, sales managers, executives, and a host of others who can strengthen your program. You should take advantage of this accumulated experience to reinforce the skills you are teaching. You may even want to take advantage of your company's "star" salespeople to share their successful experiences. Using such people in your program gains valuable "buy-in" to your program, and you will find more healthy support for your program.

But, a strong word of caution in using such people. Don't fall into the trap of having a "parade of executives" on the program, presenting a lot of boring and irrelevant material for the sole purpose of salving executive egos.

If you are not careful, you will end up with a lot of what I call "y'oughtas" on the program. For example, while we have all these salespeople gathered, y'oughta have the personnel department talk to them for a while, y'oughta have R&D spend some time with them, y'oughta have the credit manager give them a couple of hours, and y'oughta have the warehouse manager tell them how the products are shipped. If you let this happen, you'll have little time left to teach the basics necessary to succeed on the job. Keep these "y'oughtas" out of your program. To put it bluntly, y'oughta stop letting management do this to you. Don't allow anyone on the program who can't make a significant contribution to the skills and knowledge necessary to succeed in selling.

EXTERNAL RESOURCES: DON'T TRY TO DO IT ALL YOURSELF

Don't try to go it alone! Even though you may have all the experience and genius to do a terrific job of training, you need reinforcement. Outside resources can do a tremendous job of reinforcing everything you are teaching. Although later chapters will cover the use of outside resources, a few words of wisdom are appropriate at this point. The outside consultant comes from a neutral corner, is nonbiased regarding company procedures, and can bring in a host of new approaches and/or techniques. He or she can provide a needed "change of face" and freshen your program. You, too, will learn many things from an outsider that you can use in your own programs. You will pick up new techniques, skills, and perceptions. You will also see new audio/visual aids you can use in your program. The chapter on "Using Consultants" will give you important guidelines on how to select the consultant you want to use. Be careful here! Don't be misled with "oversophistication." Look for relevancy and those things which will reinforce what you are teaching. Be sure you get what you want and what you need. Be sure you get a consultant who can "express," not "impress." And when you come to the cost of an outside consultant, please remember a sign I saw in the conference room of a large company for which I did some training—"If you think training is expensive, consider the cost of ignorance!" Weigh carefully whom you select, and be certain he or she understands thoroughly what you want taught, what you want reinforced, what you want avoided, and so on. Be sure he or she does the job *you* want done. And be careful of a thing called "consultant dependency," that is, a program so sophisticated that only the consultant can deliver it. Be certain the program he or she delivers can be left operational in your hands—unless, that is, you want it otherwise.

THE FUTURE OF SALES TRAINING

The job of corporate sales trainer isn't what it used to be. The trainer will continue to rise in the organizational structure and in stature so long as

- The program is based on real needs (not theoretical).
- The program is behavior-oriented and results-oriented.
- The program produces measurable results.
- Those results are reported to management in bottom-line terms.
- The program changes with the times and needs and doesn't stagnate.
- The trainer continues to develop professionally.

The training job is often kept at an unreasonably low level primarily because management has failed to recognize the contribution of the trainer. This is the direct fault of the trainer for not measuring and reporting training results. By the same token, higher-level trainers often jeopardize their positions during hard economic times for the same reason, that is, failing to prove their track record in terms of bottom-line contributions.

What's ahead for the trainer? I see two very significant trends: (1) trainers will find it necessary (and feasible) to expand their services, and (2) trainers will be required to answer hard-line questions from their management, for example,

"How was sales performance improved?"

"What is training's return on investment (ROI)?"

"What is the turnover record of the people you have trained?"

"What is the track record of people you have trained, vis-à-vis, promotions, earnings, contributions, and so on?"

In answering these hard-line questions, the trainer will no longer be able to cop out with, "Once I have trained them, it's no longer my problem—it's up to the sales manager." That's like a production line employee saying, "I built the darn thing. If it doesn't work, it's not my problem."

Trainers should take note of a couple of frequently seen mistakes:

1. Refusal to measure and report results beyond the "end of the course." I will never underestimate the need for measuring whether instructional objectives were met (and set). Frankly, I can't get too enthused about the "happy report" at the end of the course. I'm not particularly concerned whether they were satisfied with the facilities, the food, the maid service, or the temperature of the coffee. I'm not convinced that these factors have an important effect on learning, but a lot of trainers are. They spend enormous amounts on expensive hotels to satisfy these "happy" needs—but that's their business. I do not believe in cow psychology, that is, "Keep 'em contented and you'll keep 'em productive." Personally, I feel the reciprocal is true: "Keep 'em productive and you'll keep 'em contented."

   True results must be measured back on the job where it counts, but many trainers are afraid of the risk in such measurements. They exist on "happy reports," gut feelings, and testimonials following a "great course." These trainers will be in trouble when management asks, "What did we get for what we spent?"

2. Failure to grow in their jobs. Many trainers are teaching exactly what they were teaching five years ago, no more, no less. Perhaps they inherited the course and are doing no more than perpetuate it until they move out of the job. These people are hardly trainers—they are instructors.

If the trainer is to survive in today's business world, the job must be loaded vertically, that is, adding responsibilities such as sales management training, supervisory training, and so on, in other words, enlarging the trainer's contribution to the organization. As trainers, we must expand our services within our organization to serve a broader spectrum of needs. We must get involved in company planning to make more contributions toward meeting company objectives.

Several years ago, a great many members of the NSSTE and ASTD were training sales representatives only. Today, many of these people have widened their services up to and including complete human resource development. But this broad-spectrum service was not thrust upon them by management—they studied to prepare for it—and *asked* for it. They wanted to serve every facet of their companies. These are the trainers whose positions have been significantly elevated, both organizationally and financially. I believe this is the career growth route for today's professional sales trainer.

If you are not yet so qualified, I strongly suggest study of supervisory skills, management, human resources, and so on as a means of getting ready for advancement. Once you are ready, you can honestly tell your management that you are being "underspent" and that you want to serve the training needs of the entire organization. That is precisely the course I took with my former company— and succeeded. When I joined the company I was responsible for training sales personnel only. Later, the training function was established as a separate department and the training was extended to executive management, supervisors, plant general managers, production managers, materials managers, and quality control managers, as well as sales representatives, sales managers, and customers and their sales personnel.

RECOGNIZE YOUR RESPONSIBILITY

The future of a lot of people rests squarely in your hands. The future of your company, its growth, image, and earnings also rest, to a large degree, in your hands. You are in a position of enormous influence. Be certain you meet that responsibility in a professional manner. The rewards can be handsome.

NOW IT'S UP TO YOU!

This chapter has been an overview of the purpose and role of the sales trainer. The rest of the book is the "how-to." Study each chapter as needed until you understand it thoroughly. Write down your "here's what I'm going to do" notes as you go through each chapter—then make them happen. It is not my intent to encroach on the material presented by other chapter authors; rather it is to establish the importance of what each has presented.

Each chapter has been written by a dedicated and experienced professional who has a sincere desire to strengthen your sales training capabilities. Each chapter is experience based. The best way to succeed in any job is to emulate those who have been highly successful *in* that job. Those are your authors. Now the rest is up to *you*.

SUGGESTIONS FOR FURTHER READING

This section is designed to be a starting point for building a library of resources for your salespeople, your sales managers, and yourself. It is only a starting point. Feel free to add to it or delete from it as you wish. The addresses of the publishers are included following the reading lists.

TITLE	AUTHOR	PUBLISHER

Reading List for Sales Personnel

TITLE	AUTHOR	PUBLISHER
How to Become a Master Sales Builder	Gschwandtner	Prentice Hall
Consultative Selling	Hanan, Cribben & Heiser	AMACOM
Selling Through Negotiation	Smith	Marketing Education Associates
Zig Ziglar's Secrets for Closing Sales	Ziglar	F. H. Revell Co.
Superachievers	Gschwandtner	Prentice Hall
How I Raised Myself from Failure to Success in Selling	Bettger	Prentice Hall
The Disciplines of Selling	Evered	JEFCO Publ. Co.
Keys to Human Relations in Selling	Zion	Prentice Hall
How to Sell Against Tough Competition	Kuesel	Prentice Hall
Successful Salesmanship	Ivy/Horvath	Prentice Hall
Magic of Creative Selling	Shields	Prentice Hall
Psychocybernetics	Maltz	Prentice Hall
Back to Basic Selling	Taylor	Prentice Hall
The Best Seller	Willingham	Prentice Hall
Strategic Selling	Miller/Heimas	William Morrow & Co.
Winning Moves	Delmar	Warner Books
How to Get Control of Your Time and Your Life	Lakein	Signet (New American Library)

Subscription Publications

Personal Selling Power	Gschwandtner	Personal Selling Power
Master Salesmanship	various	Clements Communications

Reading List for Sales Managers

Analyzing Performance Problems	Pipe/Mager	Fearon
From Selling to Managing	Brown	AMACOM
Management by Objectives	Odiorne	Prentice Hall
Basic Principles of Supervisory Management	AMA	AMACOM
The Success System That Never Fails	Stone	Prentice Hall
The Sales Manager as a Trainer	NSSTE	Addison-Wesley
Shirt-Sleeves Management	Evered	AMACOM
The I.B.M. Way	Rodgers	Harper & Row
The Management of Sales Training	NSSTE	NSSTE
Strategic Sales Management	Hughes/Singler	Addison-Wesley
How to Succeed in Middle Management	Lumsden	AMACOM
The Vital Difference	Harmon	AMACOM
In Search of Excellence	Peters	Harper & Row
The Disciplines of Selling	Evered	JEFCO

Publishers' Addresses

Addison-Wesley
Reading, MA 01867

American Society for Training and Development
1630 Duke St.
Box 1443
Alexandria, VA 22313

AMACOM
135 W. 50th St.
New York, NY 10020

Clements Communications
Concord, PA 19331

Fearon Publishers, Inc.
Belmont, CA 94002

Harper & Row, Publishers
10 E. 53rd St.
New York, NY 10022

JEFCO Publishing Co.
P.O. Box 1450
Denton, TX 76202

Marketing Education Associates
4004 Rosemary Street
Chevy Chase, MD 20015

William Morrow & Company
105 Madison Avenue
New York, NY 10016

Personal Selling Power
P.O. Box 5467
Fredericksburg, VA 22403

National Society of Sales Training Executives
203 E. Third St.
Sanford, FL 32771–1803

Prentice Hall
Englewood Cliffs, NJ 07632

F. H. Revell Company
Old Tappan, NJ 07675

Signet (New American Library)
1633 Broadway
New York, NY 10019

Warner Books
666 Fifth Avenue
New York, NY 10103

Sources of Free Catalogs for Audio and Video Cassette Programs on Sales and Sales Management

AMACOM
135 W. 50th St.
New York, NY 10020

American Society for Training and Development
1630 Duke St.
Box 1443
Alexandria, VA 22313

The Dartnell Corporation
4660 Ravenswood Avenue
Chicago, IL 60640

JEFCO Publishing Co.
P.O. Box 1450
Denton, TX 76202

Learn Incorporated
113 Gaither Drive
Mount Laurel, NJ 08054–9987

National Speakers Association
4747 N. 7th St., #310
Phoenix, AZ 85014

Nightingale-Conant Corporation
7300 N. Lehigh Avenue
Chicago, IL 60648

Personal Selling Power
P.O. Box 5467
Fredericksburg, VA 22403

Prentice Hall
Prentice Hall Audio
Business and Professional Publishing
Englewood Cliffs, NJ 07632

Simon & Schuster, Inc.
Simon & Schuster Audio Publishing
1230 Avenue of the Americas
New York, NY 10020

Success Motivation Institute
P.O. Box 7614
Waco, TX 76710

Tape Rental Library
1 Cassette Center
Covesville, VA 22931

Thompson-Mitchell & Associates
3384 Peachtree Road, NE
Atlanta, GA 30326

Weldon, Joel & Associates
7975 N. Hayden Road, #D-147
Scottsdale, AZ 85258

Dave Yoho Associates
10803 W. Main St.
Fairfax, VA 22030

Zig Ziglar Corporation
3330 Earhart Drive
Carrollton, TX 75006

chapter two ————————————

THE MANAGER'S ROLE IN SALES TRAINING

———————————————— *CHARLES H. KINNEY*

Charles Kinney is recently retired from the position of National Training Manager for Pitney-Bowes, Inc., responsible for overseeing the implementation of sales training by field managers to the 7,000-person sales and service organization. Prior to this, Kinney was manager of sales training, where he and his staff designed, developed,and implemented the training for salespeople at the company's National Training Center in Aberdeen Woods, Peachtree City, Georgia. Before entering the field of training, Kinney spent over 15 years as a sales manager and sales representative. He is a member of the American Society for Training and Development and the National Society of Sales Training Executives. Kinney currently is working as a consultant in the field of sales and sales management training.

This chapter will assist the sales trainer in establishing training team work with the sales manager. The information will also be useful to the sales manager who must work alone. It will list ideas that can be used by both managers and trainers and will relate methods for imparting the information from the trainer to the manager in the field. We will show how the differing roles and objectives of the two managers can interact for the best training results. The best training results will lead to the ultimate goal of increased sales at a more profitable level.

It is easier for trainers to understand the manager's role in training if they understand the job of a sales manager. The ultimate goal is the attainment of team quota on a consistent basis. For this to happen, the sales manager has to do many things:

- Hire and fire salespeople.
- Assign sales territories.
- Establish sales standards.
- Establish individual team member quotas.
- Develop strategic sales plans.
- Verify orders, call reports, itineraries, expense accounts, and so on.
- Assure proper product mix.
- Communicate with home office.
- Plan and conduct meetings.
- Introduce new products.
- Maintain competitive information.
- Maintain marketing information.
- Maintain product information.

- Work in the field with salespeople.
- Assure that proper salaries and commissions are paid.
- Handle complaints.
- Perform public relations.
- Lead a personal life.
- Motivate the sales force.
- Train the sales force.
- Evaluate salespeople.

Is it any wonder that the sales manager does not place the highest priority on training? Consider that in most companies the most important factors affecting the manager's compensation are sales dollars and profits.

Most definitions of management include the statement, "Getting the work done through others." It is impossible to get salespeople to perform if they do not know what they are supposed to do or how it is supposed to be accomplished. Training to impart that information becomes a very important part of any manager's job, whether they realize it or not.

SALES MANAGER ROLE VERSUS TRAINING ROLE

It is important that trainers understand the difference between line and staff. The staff role of the trainer is to support the line position by providing the tools and resources necessary to get the job done. This is especially difficult for the person who moves into a trainer position from a field or line position. Many such trainers have problems adapting to what they consider a secondary position. The sales manager gets all the glory and usually is better compensated. Hence, there may be an animosity on the part of the trainer toward the manager that prevents the full cooperation necessary to get the job done. As trainers, we must recognize this and do everything in our power to eliminate this potential danger to full cooperation.

Consider the fact that the manager's job is very demanding and that the glory of a successful sales month or quarter is short-lived. The first day of the next month or quarter, the manager starts at zero.

A recent appointee to sales trainer from the ranks of sales management states, "Training is a lot like baseball, Mom, and apple pie. Nearly everyone believes in it!"

"Unfortunately, training is often the first activity curtailed at 'crunch time,' even though it's acknowledged and most needed. Crunch time is as simple as too busy or as difficult as no budget. I have found it helpful to acknowledge this reality and respond by staying flexible and on call with shorter, modulized programs that can be delivered in the field."

A recent conversation with four sales managers from four different industries revealed many differences in the way that training is approached. Some companies require the sales managers to do it all, while others have a very long, formal training program before the salesperson is introduced to the manager. What is important to remember here is that every one of the sales managers interviewed placed a high importance on training. They gave it a high priority, yet one person stated it soon becomes a decreased priority for implementing.

Reasons that managers give for not implementing training as planned vary. The trainee or salesperson may have his or her own agenda that requires immediate attention or the training requested by the home office is not relevant. They also stated that one reason for not using the training department or requesting help from them is the fact that they need to react more quickly, that training had to be more timely, and that the trainers need to get more field feedback.

Another consideration in the role of the sales manager is the method used by many companies to select sales managers. Too many companies promote the best salespeople to the manager position without regard as to the person's ability to manage. Many of these sales managers received little, if any, sales training as salespeople. They are self-starters and self-learners. They have the desire, drive, and ability to succeed, and succeed they do, often in spite of their managers. Too often this type of sales manager does not realize the real need for training. They think that everyone is like them. They think that all salespeople will learn it on their own.

Our perceptions of what training is, and should be, vary according to our backgrounds. The competent, self-made sales manager may well feel that "telling is training." If I tell them what I want, it will be done. The sales trainer on the other hand is versed in needs analysis, program development, delivery, and all the time-consuming elements that go into training. The sales manager wants and needs it now, while the trainer may be more deliberate and methodical.

Once the sales goals have been established, the next step is to determine how to reach these goals. This is where the trainer enters the picture. In this situation, a good trainer can act as a consultant with the sales manager in deciding on the training needs. Our goal is to establish a common meeting ground where we can feel comfortable. The training must be relevant and effective.

ESTABLISH CLEAR AGREEMENTS BETWEEN SALES MANAGER AND TRAINER

One of the most important needs is the need for clear agreements between the sales manager and the sales trainer. We need to know who is to do what and when it is to be done. Many variables in any training venture need to be resolved before we can have a successful program:

- What is the actual training need?
- Who will prepare the training?
- Who will deliver the training?
- What specific help does the sales manager need?
- What resources are available?
- Who needs to be trained?
- Where will the training occur?
- Who has the budget?
- How will we measure the results?

While trainers may be better at preparing training programs and possibly at delivering programs to large groups, sales managers may be better at delivering to small groups and in one-on-one instruction. The most common situation that sales trainers will run into is a request for a special program that will be delivered by the sales manager. It is important that we use as many of the manager's ideas as possible since that will give him or her pride and ownership in the program and will ensure better implementation. As you continue to read the balance of this chapter, keep in mind the necessity for clear contracts with the manager. It will avoid problems in the end.

LEARN HOW TO DEFINE NEEDS

Too often a sales manager or product manager will call the training department with a request and the training department in turn sends back a 10- or 20-page request form that is too detailed. Usually if the manager could answer all the questions on the form there would be no problem. As a result, the manager ignores the form for the present and the training department in the future.

Sales managers usually are not as specific in defining needs and setting objectives as are trainers. They have goals and/or problems, not objectives. It is the trainer's job to evaluate the problem and suggest the solution. Once a manager has outlined a goal or problem, it has been our experience that the training department should write the specific objectives. This will make it easier to get agreement sooner.

Usually when sales of a particular product fall below quota for a period of time the first reaction is "We need a training program." We need to address two areas to define the true training needs. First, is it a training problem? Is the reason that sales have fallen off because the salespeople are not trained, or is there some other reason? For example, the commission rate may be too low compared to other products. Competition may have a better product at a better price. The product may simply be obsolete, and we have not been willing to acknowledge that fact. This is a good example of the need for the trainer to act as a consultant. Once we have determined that it is a training problem, we need to determine the specific training needs and write the objectives. What exactly does a salesperson need to know and be able to do to sell the product? The objectives must be based on the true needs. If the problem is not a training problem, the trainer should be honest and suggest alternative solutions.

Other chapters of this book deal in detail how to define needs; you should refer to these chapters for more specific information.

HOW TO PREPARE FOR YOUR TRAINING PROGRAM

Once the training needs have been established and the objectives written, it is time to build or write the program. Generally the training department will have the responsibility for preparing or developing the program. Once again the subject of program development is covered elsewhere in this book so we will limit our discussion to a few rules that trainers should follow when preparing programs for managers to deliver. We will also list a few ideas for managers to use in preparing programs.

As a trainer,

- Always include everything needed to deliver the training. Do not ask the person delivering the program to locate and/or refer to other texts, old bulletins, memos, and so on.
- Keep it simple and easy to understand.
- Develop case studies or exercises where possible that permit the trainees to do it themselves.
- Permit the person delivering the program a degree of flexibility—they may know more about the subject than you do.
- Provide a way for the deliverer to determine if the material was absorbed.

Sales managers that prepare their own training programs are usually confronted with needs in one of four areas.

- Sales skills
- Product or competitive knowledge
- Markets and applications
- Policies and procedures

Sales managers must differentiate between skill and knowledge. Skill is the ability to use knowledge. It is a learned power of doing something competently. Knowledge is the fact of having information. We acquire skill differently from the way in which we acquire knowledge. So it follows that a training program that teaches sales skills will be different from one that teaches knowledge.

There are many ways to teach a skill, but probably one of the easiest for a sales manager to prepare involves role modeling by the instructor and practice by the participant(s).

A simple model for skill building is

- Give reason for skill need.
- Receive acknowledgment.
- Outline a model (if there is one).
- Demonstrate the skill.
- Engage participants to perform skill.
- Critique; measure against the model.
- Practice correct performance.
- Commit.
- Follow-up—observe performance in fields.

Figure 2–1 is an outline a sales manager might prepare to teach the skill of "handling objections."

Figure 2–1 may go into more detail than a sales manager would prepare. Use as much or as little outline as needed to conduct the meeting. But be sure that you can perform the role models well in any skill-building session.

Both trainers and sales managers should note that a training session to teach product or market knowledge will use a different format from that of the role model and practice just described. One advantage of knowledge training is that you can give a trainee the resources to acquire the knowledge without interrupting selling time. A simple test will determine if the trainee has acquired the knowledge. Knowledge can be transferred via video or print or in today's marketplace by computer. Unfortunately, most of these methods are too time consuming and expensive for a sales manager to prepare. A study of the section of this book on program development will give the sales manager many good methods of preparing training sessions to transfer knowledge.

It is always a good idea to put the burden of learning on the trainees by giving them the resources and verify the acquisition of the knowledge by testing. The key is in the preparation of the test. The test must answer two questions:

Figure 2–1 Sample outline for handling objections

Establish Need (give reason for skill need)
   Ask: "Has anyone had a prospect raise an objection that they could not
       answer or actually caused you to lose the deal?" Expand with other
       questions as necessary.

Acknowledgment
   Ask: "For agreement, do we all agree that the ability to answer objections
       is critical?" (or similar) Be sure and get agreement from all participants.

Model Outline
   Use easel or overhead:

   • Acknowledge.
   • Refocus.
   • Answer.
   • Trial close.

Discuss and get agreement on value and use of model; that is,

   • Track to run on
   • Letting prospect know we are working together rather than jumping in to
     overcome or overwhelm the objection

Role Model (demonstrate the skill)
   Ask: "For most prevalent objection (NOTE: Be prepared to role model any,
       but 95 percent of the time the objections will be "It costs too much!")."

Acknowledge: "I see."

Refocus: (In this case) "Usually when people feel that a product costs too much
it is because of one of two reasons: they may feel that the seller is making too
much profit, in plain words, the product is overpriced, or they may feel that
the product is priced correctly, but that it simply does not offer enough value
to their particular need or application. Which *is* the case here?" (Be prepared
to answer either.) Practice before meeting.

If overpriced, review:

   • Company history, profit picture

- Various costs going into product
- Service and so on

Example: "Mr. Prospect, I am sure that you feel that it is important to make a profit to stay in business and serve customers in the future. As a businessperson you are aware of the costs involved with administration, research, production, distribution, advertising, marketing, and in my case, direct selling. Our company has been in business over X years, and our profits are a matter of public record. They have not been excessive, but they do assure you that we will be here to serve you in future years."

Trial Close: "Does this seem fair to you?"

Participants(s) Perform Skill

- Have volunteer(s) perform skill.

Critique

- Ask volunteer to critique self.
- Have participants critique.
- Engage in instructor critique (be sure it follows model).

Practice:

- Assign teams of three (one salesperson, one prospect, and one observer).
- Practice. Each team has different objectives.
- As the instructor, rotate, observe, critique.

Commitment:

- Request two or three demonstrations by participants.
- Get agreement that participants will perform in field.

Follow Up

- Ask for example of success at next meeting.
- Observe performance as you work with people in field.

1. Is it relevant?
2. Is the answer in the resource material?

DELIVERY: ON-THE-JOB TRAINING IS THE BEST CHOICE

What is the most effective way to deliver training to sales representatives? Does it require any special techniques? Who is most effective at delivering training to sales representatives?

These questions are difficult to answer, but a recent discussion with a group of new salespeople indicated strongly that they prefer to be taught by people who have had successful sales experience. Further questioning of this group revealed that not all areas of training require sales experience. Product training, corporate policies, and paperwork procedures were mentioned as areas not requiring sales experience to teach. The message is loud and clear to trainers that, if we do not have sales experience, we should team teach with sales managers or highly respected sales representatives whenever possible.

Much has been written about delivering training, but the fact remains that the best training a salesperson can receive is on the job. This type of training can be most beneficial to the sales manager as well, since it offers the opportunity to train while performing other functions. Too often sales managers feel that they do not have time to train. In reality they are training every minute that they are with the salesperson.

Managers are role models for sales representatives, and in this way they are sending messages to the sales representatives at all times. The message can be positive or negative. The boss sets the tone for the office. If the boss arrives early, the people will arrive early, and if the boss leaves early, the people leave early. If the boss wears casual clothes, the people will wear casual clothes. It is extremely important for sales managers to realize the role they are playing. This becomes even more important when the manager and the sales representative are making a joint sales call. The joint sales call offers the opportunity to reinforce as well as implement training. It also offers the opportunity to observe the sales representative and determine other areas of training need.

The method of delivering on-the-job training will vary depending upon whether or not the sales representative is new (a trainee) or experienced.

If the sales representative is new, the manager should make the first call and possibly more if necessary to show the proper procedure. Most managers would prefer to discuss the call first, usually stressing the one facet of the call that is to be trained. The point-reason-example-point technique can be used in any training situation. Simply

State the *point* you want to make.

Give a *reason*.

Give an *example*.

Give *feedback* that indicates knowledge of the *point*.

The dialogue might go like this:

Point: "It is essential that we give the prospect three important facts in our approach. They are 'Who we are,' 'Why we are here,' and 'The possible benefit to the prospect.'"

Reason: "The reasons that this is important are, first, it provides a model, a track to follow. Second, it reduces time spent in unimportant chit-chat and lets us get down to business. Third, the fact that we give the prospect a potential benefit will assist us in continuing the call. Fourth, most businesspeople are busy and do not care for idle conversation. Fifth, it is more professional."

Example: "Mr. or Mrs. Prospect, my name is William Sales, and I represent the Bayluck Manufacturing Company. The reason for my call today is to exchange information on shipping procedures. A next-day delivery procedure that will assure shippers' customer satisfaction is being introduced, and I feel it will benefit your business. May we talk?"

Point: To trainee: "Do you see how these three statements in the approach will be beneficial to you? Do you understand how to use them? Once again, what are the three facts?"

Following this discussion of the approach and a discussion of the call objectives, the manager should make the call. The trainee should observe and think of possible improvements or other verbage that can be discussed after the call.

On-the-job training of an experienced sales representative will require a different approach. Actually, it is in the form of coaching. One proven method of training for improvement is to make a positive critique immediately following the call, and to hold back any negative critique or suggestion until just before making the next call. At that time, a suggestion relating to the inappropriate performance on the last call could be addressed by saying "By the way, how do you think it might work with the new prospect if we tried it this way?" and state the proper performance. Ask the sales representative to practice the proper performance and obtain agreement to try it on the prospect.

A sales manager who uses these two techniques for training both new and senior sales representatives during sales call will increase valuable training time and will also see results.

Guidelines for an Effective Training Session

Much has been written about platform skills and how to sway an audience using spellbinding techniques—how to stand, making eye contact, and the like—but I can guarantee that a skill more important in training sales representatives is the skill of being a catalyst. Making things happen in a training session—getting the trainees to do it themselves, making the trainees participate and practice—are solid rules for an effective session. The following ideas can be used by either a sales manager or a trainer.

Probing. Select a particular product or a particular market and ask the group, "What questions can we ask to uncover needs for this product?" or "What questions can we ask in this market to uncover needs for our products or services?" All questions are listed on the easel or board. After all questions are listed, the second part of the session is to ask the group, "Why would we ask that question?" This can be followed by discussions of how to make questions less directive and how to ask the questions so that it does not sound like an attorney interrogating a witness on the stand. The lists of questions can then be put in some logical order, transcribed, and distributed to the group. Does learning take place? You bet it does. Another advantage of this type of training session is that it requires very little preparation.

New Product Introduction. In this case, all the product information is delivered to the sales representatives before the meeting. This can be done from two days to two weeks ahead of time depending upon the complexity of the information. Their assignment is to study it and retain all the information that they feel is relevant to selling the product. The meeting itself is held by breaking the group into small teams and asking each team to develop a sales or marketing plan to sell the new product. The role of the meeting leader is simply that of a catalyst. There are several ways that the presentations can be made. If the product is complex, teams might have different assignments: one to explore markets, another to develop demonstration and presentation techniques, and so on. This type of training can be expanded or shortened as needed. One absolute requirement at the end of this or any other training session is to get a commitment from each attendee as to what action will be taken. Commitment can vary from the number of sales within a certain time frame to a list of prospects showing time frames for the first call. The key is participation. In all cases, the written commitment should be delivered to the appropriate sales manager.

Demonstration Contests

Another exciting training session is in the form of a demonstration contest. The contest can be by teams or individuals. Notify the participants in advance. Learning takes place in the preparation, and further learning occurs as each attendee picks up an idea or technique from the other attendees. You must have a method of judging the contest that is fair. This requires a set of standards or criteria for measurement. Assume the following definition.

Give a selling demonstration that

- Follows the model.
- Answers at least one objection.
- Relates three benefits directly to prospect needs (hot buttons).
- Provides a smooth operation.
- Involves the prospect.

To measure the demonstration, provide a checkoff sheet. The checkoff sheet can list the steps and give points for each item completed. The checkoff sheet can be broken into detail.

The demonstration contest is fun and very participative.

Competitive Presentations. Another training session that has high participation is one in which a team or teams are preassigned to study a particular competitive product(s) or service. At the training session the assigned team will make a competitive presentation while another team makes a presentation of your products.

A winner can be declared and discussions held that will enhance the attendees' competitive knowledge and eliminate the fear of competition.

Case Studies. The use of case studies in sales training is highly effective. They are even more effective when the salespeople develop their own cases from live situations. The method is especially effective if used in a classroom session over several days where successive role plays involving probing, defining applications, presenting, closing, and so on are held. Over a five-day period the attendees will help each other solve their own case.

The foregoing are but a few of many delivery ideas that involve the trainee.

Role Play. Role play is practice. It should be used to improve performance, not as a judgment as to how well the trainee performed. Role plays can be conducted one-on-one with no observers to practice short segments of a sales call. They can be used in front of other class members with full video encompassing the entire sales call or anywhere in between these two extremes.

The trainer or sales manager should play the part of the customer whenever possible because too often sales representatives will present a unique situation that they experienced that is not likely to occur again. Role-play situations should be as normal as possible.

In a classroom situation you should assign trainees a particular area for their observation and comments. Try dividing the class into pros and cons so that one group watches for good performance while the other group watches for performance that can be improved. You can further divide individuals to watch for number of open probes versus closed probes, handling objections, trial classes, or any number of specific areas.

The normal rules for roleplay apply when critiquing. Ask participant first: "How do you feel you did?" "What would you change?"

Ask observers for

Positive areas.

Improvement areas.

Offer your own critique last, stating positive areas first.

Always try to be creative in delivering training. The more the participants do themselves, the better the training. Be sincere and creative and always build in a commitment section so that the training can be measured.

FOLLOW-UP

The sales manager is the one person with the best opportunity to follow up and measure the results of training programs. Previously in this chapter, we discussed the advantages of on-the-job sales training of sales representatives by sales managers. One additional advantage is the opportunity of the manager to witness firsthand the sales representative in action and observe if past training has been effective. Did the sales representative use the skills learned in the training session, and if not, why not? It also offers the opportunity for remedial training on the spot where the use of the trained skill has slipped.

If training is to be effective it must be followed up and reinforced constantly. This is the reason that the last part of any sales training session include some commitment in writing from the attendee. The sales manager then has a tool to determine if the training is valuable. The follow-up does require effort on the part of the manager, but it has twofold results. It will get incremental business as well as measure the training if used correctly.

SPECIAL TIPS FOR SALES TRAINERS

There are a few bits of information that do not fall under any category that may be helpful to a sales trainer or sales manager. Some of these are obvious and some of them have been discussed previously, but need to be reinforced.

1. *Telling Is Not Training.* Do not assume that they know it just because you told them.

2. *All Deficient Performance Cannot Be Solved by Training.* There are four basic reasons why salespeople do not perform, and only one can be solved by training. If they do not know how, you have a training problem. The other three reasons? They don't want to, they do not know what they are supposed to do, and finally, the manager or the system gets in their way.

3. *Practice, Practice, Practice.*

4. *Get "Ego" Out of the Way.* We are referring here to the ego of the manager or the trainer. Do not be so carried away with your own performance that you feel that standing in front of a group and entertaining them is training. Likewise, do not take over every call or office duty to show them that *you* know how to do it better than they do.

It does not matter if we are trainers or managers. Our ultimate goal has to be to help the sales representative get the order. Training is the critical element in helping us to reach this goal.

chapter three —————————————

SALES TRAINING MANAGEMENT: HOW TO DEVELOP PERFORMANCE LEADERS

WILLIAM E. LAW

William E. Law has experienced more than twenty-five years in the corporate world, mostly with AT&T and mostly in sales, sales management, and the management of sales training. In 1984, he established his own consulting firm, Law & Associates, Inc., and currently works with clients in the areas of strategic management of training leadership development, strategic thinking, and organizational change. Law is active in the National Society of Sales Training Executives and the National Society for Performance and Improvement. He is an adjunct faculty member at Xavier University. His trademark is the practical application of innovative ideas to today's challenges.

This chapter will concentrate on two essential issues: (1) identifying and applying "strategic skills" to leading the training function and (2) designing and implementing a strategic management system for training.

More and more companies are recognizing the critical contribution that training and development can make to their survival and profitability. Increasing world competition, rapid technological innovation, continued domestic economic volatility, changing work force quality, and shifting demographics are best managed by having highly trained and capable people at all levels in the business. As a result, they are placing greater emphasis on the training and development of everyone from top management to clerical employees.

Tom Peters says it well: "What have you done today to enhance (or at least ensure against the decline of) the relative overall useful skill level of your work force vis-à-vis competitors? Work force training must become a corporate (and indeed national) obsession. And it is on this variable that the outcome of the overall competitive struggle may most strongly depend."[1]

Another perspective is offered by Larry Wilson as he talks of the need to change the way we sell to clients. According to Wilson, the world in which our customers live is experiencing radical change. As a result, salespeople will need to change the way they sell by "creating the future of selling by *anticipating* where our clients are going, developing *innovative* responses and then *strategically* changing—changing the game of selling."[2]

As the role of training assumes increased importance and as pressures increase for a new perspective on the way we sell, the challenge for sales training managers becomes quite clear. Training organizations **must** *become part of their*

---

[1] Thomas J. Peters, *Thriving on Chaos, Handbook for a Management Revolution* (New York: Alfred A. Knopf, 1987), pp. 323–324, 328.

[2] Larry Wilson, *Changing the Game: The New Way to Sell* (New York: Simon & Schuster, 1987), p. 28.

corporation's key strategic direction. They need to accomplish this for survival—their own and their company's. More important, they must have the vision and foresight to equip their clients—the sales training population—with the right kinds of skills that will enable them to confront and manage these volatile marketplace challenges.

STRATEGIC SKILLS: WHAT SALES TRAINING MANAGERS NEED TO KNOW

The first major area in this chapter emphasizes the skill needs of sales training managers. We have already said that the rapidity of change in the marketplace is creating a demand for a new paradigm for selling with emphasis on innovation, anticipation, and strategic positioning. Salespeople need to become leaders with their clients by offering innovative solutions to business problems, by being on the forefront of thinking and problem recognition, and by being part of their clients' planning process. These same characteristics apply doubly to sales training managers.

The evolving paradigm for sales training managers requires a new set of leadership skills in addition to the traditional management skills which will be reviewed shortly.

Doing the Right Thing versus Doing Things Right

Traditional skills include planning, organizing, staffing, directing, evaluating, and controlling. These skills tend to focus more on actions such as cutting costs, maintaining stability, making things run smoothly, improving efficiency, and employing other measures to control processes and procedures. In the words of Warren Bennis, they refer to "doing things right."[3]

The new leadership skills, which will be the subject in this discussion, focus on "doing the right things."[4] If we think about the changes that are affecting business today, a set of higher-level skills is obviously needed. These skills, just like those of salespeople, need to be more innovative, more strategic, and more future-oriented.

What Training Managers Must Do to Solve Critical Performance Problems

If a training group is to be considered part of a company's strategic process and part of the solution to critical performance problems, the manager of that

---

[3] Warren, Bennis, *Leaders* (New York: Harper & Row, 1985), pp. 26–27.
[4] Ibid.

training group must, and I emphasize *must*, proactively seek involvement with senior management, must provide the business with innovative solution options to performance issues, must anticipate and head off performance situations before they become problems, must initiate activities which provide business solutions to performance issues rather than waiting to be asked (or worse, being told) to do something, and must provide *value* to the business in return for the investment in training and development.

EXAMPLE

Dunkirk Industries discovered that their salespeople were having a hard time with their new S773 product line. The cause(s) were not easily identified. To create as nonthreatening an environment as possible, the sales training staff decided to set up a series of mock sales calls with the training staff role playing all the major customers. In addition the decision was made to video tape the session and debrief the tapes one-on-one with the sales managers. Consequent activity revealed that the product knowledge training had been incomplete and supporting technical information weak. Early intervention prevented a potential failure of the new product line.

Five Key Questions That Can Improve Your Leadership Skills

Conduct a quick survey of your own situation by answering the following questions as candidly and objectively as you can.

- "Am I frequently asked by senior management to discuss performance issues and options for dealing with them?"
- "Do I devote a higher percentage (30 to 40 percent) of my time to anticipating my client group performance needs?"
- "Is the performance of the training group evaluated by line managers' perceptions of improved job performance?"
- "Is an assignment in the sales training group actively sought by top performers?"
- "Does my training group operate from a plan which stresses the achievement of business results?"

If you answered a resounding "yes" to these questions, then you are well on your way to being a true leader within your business and to positioning sales training as a strategic asset to your company.

If you hedged on answering or answered "no" to any, then you need to think seriously about what you are doing and why—in other words, "what is your *real* role?"

The checklist of questions that follows (see Figure 3–1) is meant to assist

Figure 3–1 A checklist for positioning sales training

• *POSITIONING*: The organization occupies a position of respect and trust within the company structure; possesses position power, but, because of interactions with key stakeholders, has high levels of personal power (expert, information, connection, and referent); is perceived as being influential; and fosters partnering relationships with clients and external networking resources.

Please use the following scale to respond to survey questions.

     5 – Completely accurate description
     4 – Mostly accurate description
     3 – Undecided
     2 – Mostly inaccurate description
     1 – Completely inaccurate description

Training and Development . . .

Is frequently called on by senior executives to offer suggestions for improving performance.	5	4	3	2	1
Is frequently asked to justify staff and budget levels.	5	4	3	2	1
Conducts management training and development.	5	4	3	2	1
Is frequently called upon by line managers to discuss performance problems and options for solution.	5	4	3	2	1
Sees itself as a catalyst for organizational change.	5	4	3	2	1
Is clearly perceived by senior management as a catalyst for change.	5	4	3	2	1
Conducts senior executive development.	5	4	3	2	1
Spends all its time and resources on nonsupervisory training and development.	5	4	3	2	1

Source: Excerpted from Performance Level Survey, Copyright 1988, Law & Associates, Inc.

you in identifying those areas which are "weak" and which are "strong" indicators of your positioning.

Appendix or Artery, Which Describes Your Training Group?

I have often used a metaphor to describe training groups as functioning as an appendix or as an artery within their companies.

A training group that concentrates on the number of training sessions delivered and on the number of students trained, which frequently operates in a crisis mode while reacting to demands from a variety of sources, which is in the posture of continual justification of its existence, which seldom plans beyond the current training schedule is an "appendix" to its company. At the first sign of the onset of an "inflammation" like budget reductions, staff reductions, and restricted internal travel, this organization is likely to be snipped out. Just like an appendix!

Conversely, the training group which initiates activities to identify performance problems and to offer solutions, which concentrates on doing those things which improve business results, which keeps close to the future direction of the business and the industry, which seeks innovative solutions to performance issues is functioning as a "performance artery" for their company. This organization will be actively sought as a participant in seeking ways to improve overall business performance—it will be valued and supported!

The difference between being an appendix and an artery is a function of the leadership capabilities of the training manager.

Five Valuable Training Leadership Skills for Managers

When we think about leadership and about those we consider to be leaders, we are likely to recall situations involving innovation, significant change, and perhaps even chaos.

Leaders seem to be those individuals who introduce the new way of working with a particular situation, who manage to introduce change without the turmoil normally associated with it, or who rescue an organization from the brink of disaster.

Several researchers have recently published results of their studies of leadership, among them Warren Bennis. In his book, *Leaders*,[5] Bennis identifies four

---

[5] Ibid.

characteristics that were found to be common among those individuals who are considered to be leaders. They tended to

- Create a vision of what can be.
- Share and communicate the meaning of that vision.
- Inspire trust and commitment through their behavior.
- Empower people to grow and achieve as they work on accomplishing their part of the vision.

These traits have special significance to sales training managers. They are fundamental to achieving strategic positioning of sales training as a corporate asset. These traits translate into a set of training leadership skills, namely, visioning, focus, positioning, networking, and innovating.

1. *Visioning*. Visioning is the ability to imagine some future situation and to develop a plan for making it become reality. A visionary has a clear idea of what he or she will be doing at some point in the future, what impact he or she will have, and how he or she will be functioning. The preference is for creating the future by managing conditions which are likely to generate the desired situation. As Peter Drucker once said, "The best way to predict the future is to create it."

2. *Focusing*. This leadership skill is described as the ability to concentrate on issues which have the greatest impact. It is the ability to concentrate resources and attention on achieving results rather than on performing activities. It is doing those things that positively impact business results rather than merely conducting training programs.

3. *Positioning*. Call it power or influence, but the positioning enjoyed by sales training is crucial to overall success. Positioning is the capacity to mobilize people and resources to get something done. When this definition is applied to the sales training function, it is the capacity to influence the behavior of line sales and marketing staff to implement the skills and knowledge received through learning experiences.

4. *Networking*. The ability to access either information which is considered to be valuable to others or the connections with people who are considered influential is networking. It is a type of influence based on information and resources that are not formally attached to the position of sales training management. (See Chapter 30, "Professional Networking.")

5. *Innovating*. Innovation is the ability to seek out and implement new approaches to learning and development. It is the ability to incorporate the new with the old and to produce improved results.

EXAMPLE

An innovative approach might be developing a computer based, self-study program to update salespeople on product line refinements. Rather than spend the money to fly everyone in for a day or two of instruction, the same result is accomplished by a computer training program that is mailed to the field or sent via modems.

The role of sales training managers, as has been repeatedly pointed out, is to serve their companies by providing critical skills and knowledge through training and development. The challenge to them is to provide that support in such a way that their participation is valued by key senior management.

Performing traditional management functions is important and vital; however, the ability of the sales training manager to assume a position of leadership and influence is based on a higher-level set of skills—those of leadership.

HOW TO ESTABLISH A SUCCESSFUL TRAINING MANAGEMENT SYSTEM

A fundamental element for any successful organization is a framework that defines the work environment. A training management system provides direction and meaning for those inside the training group. It also affords those outside the training organization with awareness and understanding of the purpose of training and how its involvement contributes to the overall attainment of corporate and departmental results. A training management system is also the structure through which individual actions become integrated with one another and result in senior management support of overall training directions.

An effective system for managing sales training contains four key components: strategy formation, management processes, organization structure, and staff. Each of these elements is essential by itself, but training success is achieved through the complex interaction of strategy with the other four elements.

STRATEGY FORMATION

A frequent topic of discussion among training managers is the linkage of sales training activities to the overall direction of the business in general and the marketing/sales department specifically. This concern, however, addresses only a small part of the issue. Linkage is certainly needed; but a strong, almost symbiotic, relationship with the ethics, values, goals, objectives, and purpose of the business is unconditionally *required*.

Strategy formation provides the training manager with a clear focus on those issues which are likely to have the greatest impact on the business. When these issues are *based on facts*, have *key management involvement*, and take *implementation into account*, meaningful training strategies which address them can be generated.

Look for Changes That May Impact on Your Sales Effort

Strategy formation does not mean a formal, annual planning process. Instead, it refers to a persistent process of scanning the business horizon for changes or indications of changes which are likely to have an impact on the business and ultimately on the sales effort. When this scanning process identifies issues which can be influenced by sales training, the training manager needs to begin gathering factual data on the change and its impact. Senior management should be informed and their support sought for developing and implementing strategies which address the effects of the change on sales force skills and abilities.

Functioning in this manner will help to position the training manager as a knowledgeable businessperson rather than being seen as "just a trainer." It will also begin to pave the way to link the training function directly to the strategic direction of the business.

Determine Where Your Organization Is Now and Where It Needs to Be

It is important to understand the significant difference between strategy formation and implementation.

Strategy formation is determining *where* your organization is now and *where* it *needs to be* in the future. With this definition, it is obvious that strategies are future focused and long range, consider available options, and emphasize results or outcomes.

Implementation, on the other hand, is deciding *how* to move your organization from where it is to where it needs to be. The focus is more tactical and short term and is based on performing activities.

Identify Where Training Can Positively Affect Performance

Strategy formation is the proactive catalyst that links the training group to the needs of the sales organization and the company. Successful strategies, as previously mentioned, are fact based, focus on implementation, contain manageable risk, and can be supported by key management stakeholders. It is the responsibility of the training manager to assume a proactive role in identifying those business areas where training can have a positive effect on business performance.

The training manager who seems to be in a rather constant state of reacting to crises only becomes involved in an endless stream of activity which does little to enhance the value that quality sales training can have on the profits of the business. When budgets get tight, this manager is fighting for survival. The losers when this happens are the training group, the sales organization, and the business.

Visible indicators of well-developed strategies are the mission, goal, and objective statements developed and implemented by the training organization. The next section will discuss management processes as ways to turn strategies into realities.

THE MANAGEMENT PROCESS

How to Offer Direction and Monitor Performance

Management processes are the tools used to implement strategies. These tools, as previously mentioned, are directly related to the other elements of the training system and include business planning, curriculum planning, training scheduling, marketing planning, evaluation, and budgeting. They help training management establish direction and expectations and provide the systems to monitor performance against those expectations. Primary elements in the management process include mission, goal, and objective statements and the process to monitor attainment of performance against these expectations.

Develop a Mission Statement. Mission statements provide the direction, aim, or focus for a corporation, department, or group. The corporate mission statement, aside from providing general guidelines for strategic planning, has special relevance to the formation of individual program strategies and the nature of the business. Mission statements determine the competitive arena in which a business operates. They determine how attention will be channeled to different areas of opportunity. They keep people in the organization from spinning their wheels working on strategies and plans that might be considered irrelevant or inappropriate by top management.

A mission statement for a sales training group should do basically the same thing, except that the focus is narrowed. It provides direction for both training staff and internal clients to help them understand how the role of sales training supports the corporate and the sales and marketing department missions.

An example of a sales training mission statement might read

> The Sales Training Group will positively impact client buying behavior by providing professional sales training and support programs which allow our sales force to be clearly differentiated by the added value it brings to the client relationship.

In addition to providing training direction, this statement also identifies the value placed on the efforts of those in the training group. More will be mentioned about value later.

Set Goals. Next are goals, broad statements of focus, which identify specific areas for the training group to apply its talents. Examples of some goal areas which support the previous value example might be "client focus," "sales force professionalism," "profitability," "optimizing employee resources."

A sample goal statement, with sales force professionalism as the goal area, might read

> Develop and deliver sales training programs which focus on the technical and professional selling and sales management skills required by the sales force for maximum sales attainment. All programs shall emphasize the integrity of the sales process and the value of individual attitudes in a client relationship. This will increase the level of sales force professionalism and permit client differentiation of our sales force in the marketplace.

Goals drive activities and direct resource allocation toward those issues which are most important. Activities and demands for training effort which falls outside the identified goal areas *must* be challenged to determine whether they should be performed. If the decision is made to proceed, then that decision is made with thought and not by default. Unnecessary effort is avoided, and the focus of training activities is kept on track with company needs. A rule for training managers is, "If it doesn't fit the goal areas, we don't do it." This should be translated to say that demands for training should be challenged to make sure that they are the right things to use training resources on.

State Objectives. The final direction-setting elements are objectives. Too often, the terms goals and objectives are used interchangeably. To avoid this confusion, we will define objectives as *specific, measurable, time-bound,* and *simple statements.*

Continuing with the previous example of sales force professionalism, objectives which support both the group value statement and a specific goal area might include some or all of the following:

> Conduct advanced sales training programs for 85 percent of sales representatives with more than two years in the job;
>
> Maintain a training effectiveness rating of 90 percent from line managers;

Review all sales programs with line managers to ensure that appropriate skills and knowledge are being taught.

The issue of establishing a strong link to corporate direction is important for more reasons than just linkage alone. Developing a strong value or mission statement along with supporting goals and objectives provides the training manager with valuable tools for communicating with his or her organization, it affords an opportunity and a means for establishing expectations, and it becomes the means for tracking and evaluating performance.

Gain Commitment of Others. Coupled with developing mission statements, goals, and objectives is the need for some way to gain the commitment of those who will be implementing these direction-setting elements.

Much has been written during the past several years about the value of having members of an organization be involved in setting group goals and objectives. Gaining acceptance of training direction and objectives is certainly much easier when members of the group have the opportunity to actively participate in the goal and objective-setting process. That does not mean that some objectives are not directed downward from the manager to the group. It simply means that, when it is appropriate for group members to participate, they are involved and their input is incorporated.

Dictating versus negotiating. Using the previous example for our value statement (sales force professionalism) and goal areas, there are several possible objectives that might be "dictated" and many that are negotiable. For example, you might indicate that the minimum number of weeks of training delivery be 75, that the minimum overall course evaluation index be maintained at 90 percent, and that 90 percent of existing courses offered be reviewed for content relevancy. While these objectives have been "dictated," there is still room for each member of the training staff to negotiate their individual objectives for the upcoming planning period. This option will make securing commitment to overall direction easier and, at the same time, more palatable for the staff. In addition to gaining commitment to personal goal and objective numbers, this process will enable you to communicate more of the "why's" associated with required or minimum levels of performance. Experience has indicated that people are usually more willing to support a direction if they have some idea why performance levels have been set as they are.

To illustrate how this process might be implemented, consider a situation where a training manager receives a list of goals and objectives from his or her boss. Before committing to the objectives, the manager shares the list with the training group along with a request for each member of the group to identify the portion of the overall objectives they are willing to commit to as their personal objectives. When these individual lists are received, the manager can then compare them with overall group objectives to identify which goal areas need a higher level of support and commitment. This analysis can then be used as the basis for negotiating individual performance commitments. (Note: This process typically produces a higher level of performance commitment than does a top-down direction alone.) When individual commitments are finalized, the manager can then submit a total package of goals and objectives which has been reviewed and committed to all members of the training group.

Other areas not directly related to overall group performance or which are internal to the training group can then be negotiated with each individual. These areas typically focus on personal development and performance improvement issues.

Commitment gaining is a critical element in successful management of a training group. It serves as the communications vehicle for explaining "why" minimum levels of performance are needed and for obtaining group and individual "buy-in" to these levels. It is also the method through which you can encourage and monitor individual development and improvement.

Strategic Tools of the Management Process

Business Planning. The first of the management processes is business planning. This component establishes the link between the company's overall direction and that of the sales training group. Corporate and departmental mission and goal statements should provide the direction and framework for determining what is expected of sales training. Critical business issues are normally contained in the mission and goal statements, and all sales training activity should support attainment of these key business goals. Business planning for sales training is the process of identifying those areas where sales training can have the greatest impact on business success.

EXAMPLE _____

Sample Business Planning Questions for New Products

The sales training manager has the opportunity to talk with the product development manager. Several key areas are explored:

What new products are being planned for introduction in the next twelve to eighteen months?

When these products are introduced, how much training or retraining will need to be done with the salespeople to ensure a good, solid technical knowledge of the product?

Will your engineers be available to make sales calls with my people again this year to ensure continuity with our larger clients?

Will the pricing be similar or different than our present pricing structure?

What kind of a budget should I plan on for training support materials that will fully cover your technical needs?

What type of coordination will need to be done with marketing?

What kind of planning have you done with marketing up to this point?

What type of sales projections are being made for break even? Normal profit level? Smash hit success?

Will we be coordinating our activities with our product distribution centers again? Have they indicated an interest in participating in the training again this year?

At what point will we need to intervene with the sales force if the initial training does not provide a complete enough background for their comfort level? Will your engineers be able to alert us to problems as early as they did last time?

What visual materials and models will be needed during your product introduction portion of the training? Do we need to go out into the plant in the line as we did last time?

Will you be able to provide a quality survey for our salespeople to give to customers, as before, after they are using the new products so we can monitor problems?

Even if your company does not have a formalized mission statement and set of goals identified, you must concentrate on identifying those business issues which dominate senior management's attention. When talking with management, it is important to get beyond the point of asking what training needs they have. It is more meaningful to talk about business conditions. Ask about those things that are going well and those that are hurting. Ask what the plans are for the future. Ask these people how they plan to get where they need to go with their part of the business. Only when answers to these types of questions are obtained can you really know what training issues to concentrate on. Questions like these will lead to business discussions which identify specific business needs. They focus attention on performance issues rather than on delivering more training programs.

Your sales training department must have a working system for assessing company needs and a process in place for responding to those needs which are identified. Sound business planning facilitates this process and helps change the perception that training is little more than another special program on closing techniques, overcoming objections, or negotiating skills.

Always remember that sales training is about change, strategies, and organization systems—in our own and in client organizations. These things take time to influence, and only through persistent emphasis on the business issues surrounding them can sales training become part of their solution.

Components of a strategic training plan should include

- An Executive Summary
- A Situational Analysis
- A Statement of Key Business Issues
- Training Mission, Goals, and Objective Statements
- A Curriculum Plan
- Budget Plans
- A Delivery Schedule.

The Executive Summary is a synopsis of the plan. It provides senior management with the salient points of the plan in a one-page overview. It should include highlights of the key issues, training mission statement, and key goal areas. The Executive Summary should promote the key role that sales training plays in supporting sales performance.

The Situational Analysis is both an examination of the current skill and knowledge levels for the target training population and a description of critical weaknesses.

It is simply determining "where your organization is" compared to "where it needs to be" with respect to the needs of the business. A guide to making this determination is obtaining answers to the following questions from key people in your training population:

Do you feel that training supports your ability to meet the expectations placed on your organization?

Are we addressing performance issues quickly enough?

Do you rely on training to help you meet your goals?

The Key Business Issues section provides the link between sales training and those issues that are of most concern to senior and line managers. This category clarifies the relationship of capability and performance in those areas which are training related.

The other components are reviewed elsewhere and will not be discussed further here.

Curriculum Planning. This component provides the link between training activity and the business issues identified through the business planning process. If we define the role of sales training as helping the sales force to establish a clear difference between themselves and the competition through the training and development we provide, then the "training things" we do must directly support that role. The amount, sequence, and types of programs offered must be planned and implemented for no other reason than establishing a clear differ-ence between our sales force and the competition.

The Curriculum Plan is the muscle, sinew, and nerve system for the "beast" we call sales training. Overlayed on the skeletal structure (the Business Plan), a Curriculum Plan provides form and substance. It gives meaning to the training business plan by linking acquisition of necessary skills and knowledge to those conditions being faced by the business.

SIX ELEMENTS OF A CURRICULUM PLAN

The elements of a curriculum plan include

- Skill and knowledge requirements by job title.
- Skill and knowledge requirements grouped by experience or course.
- Expected levels of performance following training.
- The sequence of courses and on-the-job experiences.
- Requirements for progressing through the curriculum.
- A review process for ensuring that content is relevant to organizational needs.

Once a Curriculum Plan has been established and committed to by key managers, specific courses, programs, and learning experiences can then be de-signed to ensure that each salesperson and sales manager reaches the expected level of performance within the time allocated.

In addition to identifying the sequence and types of learning experiences that are required by each job, there must also be a process in place for periodic review of the curriculum. This review procedure must involve key stakeholders, including line sales managers, marketing staff, and training staff. The outcome of the review is to reexamine the curriculum to ensure that content is appropriate and relevant to existing and expected conditions, that the sequence is suitable to line and staff needs, and that the emphasis within the various experiences is pertinent to sales and marketing needs.

This review process should be done on a regular basis, at least annually and prior to the annual budget planning cycle for your company. It should also include key managers from line sales, marketing staff, and training.

Training Schedule. Training schedules are a channel for communicating be-tween the training population and the training group. The schedule is a visual indicator of the level of training activity and, at the same time, a very subtle signal to the sales force which calls attention to the types of skills and knowledge they need to develop and refine if they are to be successful.

Timing the publication of a delivery schedule is an issue which has subtle ramifications. An annual schedule informs the sales force of the training that will be available during the entire year, but it also says that there is little flexibility to meet evolving needs. Like any annual plan, you need to be able to accurately predict needs that will exist 6 to 12 months in the future. It is doubtful that many of us possess such wisdom.

A procedure should be put in place which provides the sales training manager with input from key people—sales managers, marketing staff, senior management. The input should enable the training manager to identify the types of training which will be needed and the numbers of students who need each type of training. The result will be a firm training schedule.

The suggested time frame for this schedule is a "rolling" three or six months. With this approach, a schedule which has a firm three (or six) months and a tentative three (six) months is published. Each month, the first firm month becomes the current schedule, the first tentative month becomes fixed, and a sixth, tentative month is added. As the schedule is reviewed each month, the month being added to the firm category can be reviewed and modified, if necessary, to reflect current needs. Looking at the last month, either sixth or twelfth, on a regular basis forces the sales training manager to keep a constant eye on the business horizon and to stay in touch with the changes taking place in the environment and within the company.

Maintaining a schedule which has this short-/long-term perspective built into it, will very subtly communicate a sense of anticipation of and attention to evolving needs while providing solutions to current needs.

Keeping in touch with key line sales and staff marketing people and asking them the types of business-driven questions discussed earlier will indicate to them that sales training is more than a list of training programs which people attend in random fashion. The training schedule can help communicate that meaning.

Training Budget. Rensis Likert and Eric Flamholtz have both suggested that business must begin to account for its human resources with the same attention that other assets are tracked. The implications of their theories to sales training managers is not lost.

The training budget represents the funding needed for staff, equipment, and facilities to implement successfully the training business plan during a fiscal year. Budgeting must be developed in support of the overall sales and marketing plans and should be for the same time frame. It must be based on addressing business issues, not on the number of training programs to be delivered. It must be based on helping sales personnel achieve expected levels of performance rather than on the number of weeks of training that will be delivered. In addition, it should emphasize the value of investing in people as business assets. A resource for further examination of human resource accounting is *Human Resource Accounting* by Eric Flamholtz.[6]

A critical point to be made regarding budgets is that they are basically a short-term measurement tool. Because of this focus, it is easy for managers to be trapped into placing too much emphasis on short-term activities at the expense of attention to longer-term business issues.

A budget typically represents *only* the expense side of the value equation, and it is the responsibility of the sales training manager to emphasize continually the entire equation which includes attaching value to people as business assets.

Budget should be developed following establishment of goals and objectives and creation of a delivery schedule. The sequence should not be reversed.

Evaluating Performance

The need to identify and promote the value of sales training has been emphasized throughout this chapter. The capacity to do this, however, is the direct result of having a method to evaluate and track performance. It is through tracking training activities and evaluating their impact on improving job performance that a training manager is able to establish the value of training and development.

Eight Reasons for Tracking Performance. Performance tracking or evaluation is conducted for several reasons:

- Measuring individual learning
- Providing information for individual development planning
- Evaluating group performance
- Measuring skill and knowledge acquisition
- Measuring effectiveness of training design and delivery
- Measuring transfer of learning to the job
- Measuring the relevance of training to job and company needs
- Identifying the contribution of training to individual and group performance

Three Levels of Training Performance to Consider. A viable evaluation system would consist of tracking at least three levels of training-related activity.

First is the *reactive level*. It consists of the traditional "smile test" which

---

[6] Eric G. Flamholtz, *Human Resource Accounting* (San Francisco: Jossey-Bass, 1985).

seeks student reaction to the course and the instructor. Such input is important for establishing trends for instructor performance or acceptance of the program, but has marginal value in truly evaluating learning effectiveness and relevance to the job.

There are several reasons for this perspective. Students, especially those who are newer employees, are not always the best judges of what they really need to be successful on their jobs. Instructors can easily manipulate student input by emphasizing certain styles of delivery and content at the expense of others. Also, the questions used in reaction surveys are typically not designed to measure learning and therefore do not provide input on the actual effectiveness of the program.

The second level of evaluation is *on-the-job follow-up*. Input is gathered from students who have attended a particular program and from their managers. The process is to solicit information through personal interviews (in person or by telephone) or through questionnaires. Students who have attended a program and subsequently had time to apply what they learned in training to their job are asked to identify the relevance and applicability of key elements of the program to actual conditions in the field. Both good and bad input is sought and used to revise program design and content to improve job performance.

In addition, input is sought from the student's manager to identify examples of improved job performance which relate to skills or knowledge acquired during the program. This is an important element because it involves the manager in the training and development process and emphasizes the fact that training is a process, not an event. Without follow-up and continual reinforcement on the job, little is gained from attendance at a training program.

Level 2 evaluation provides data for program modification and enhancement. It also produces the data which can be used to establish the value of the training effort.

This is a critical factor in a training manager's ability to position training as a vital corporate asset. Given the difficulty of clearly identifying the impact of training on bottom-line performance, the *most practical option* for gauging contribution seems to be the opinion of sales managers. Training performance needs to be monitored against a set of standards which indicate value for training clients (sales managers) as well as for upper management. It is therefore imperative that training managers work with their clients and stakeholders (line sales managers, marketing managers, senior corporate executives) to identify those issues which are considered important and to observe them very closely. The most simple approach is to ask, *"How are we doing?"* on a frequent basis. Level 2 evaluation provides opportunities to ask the question.

The third level of evaluation is the *curriculum review* process which was described earlier. A curriculum review provides the opportunity to obtain valid, objective input on the relevance and applicability of training to the needs of the sales effort.

Figure 3–2. Strategic training management model

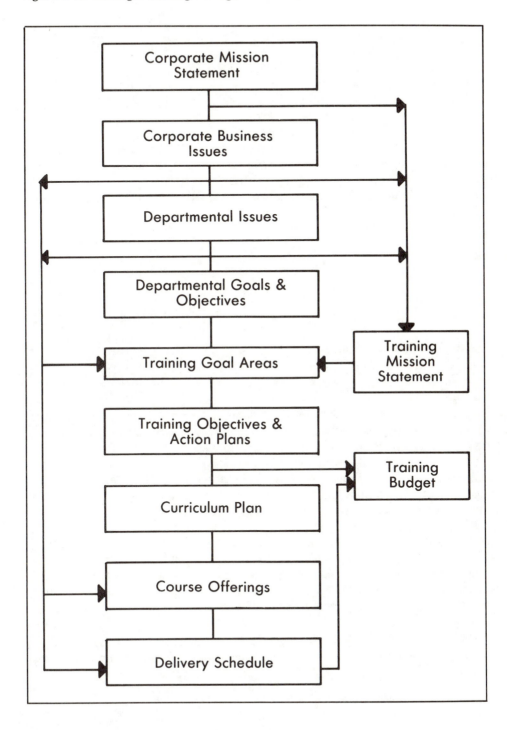

The review process is usually conducted by a Training Review Board which is comprised of key sales managers, marketing managers, and training management. Their purpose is to review current training programs to determine content relevance, to determine whether the amount of delivery is adequate, and to establish future training needs.

Combining the three levels of tracking into an evaluation system will provide a sales training manager with the information to identify those critical elements which demonstrate value and support.

Management processes as shown in Figure 3–2 are critical to the overall management and leadership of sales training. They provide the structure and the procedures which enable the sales training manager to plan, direct, control, and organize the function for optimal performance. They also provide a foundation for direction and leadership.

ORGANIZATION STRUCTURE

Deciding Between a Centralized and Decentralized Structure

The question of the structure of training often revolves around the issue of centralized versus decentralized. While this is an issue best addressed within the specific parameters of a company, there are some key points to be considered.

Customer Convenience Versus Fragmentation. First, as companies attempt to follow the latest prescriptions of contemporary management gurus and move decision making down organizational lines and closer to the customer, there has been a marked emphasis on moving training activities to regional and district locations. While this fits one prescription for closeness to the customer, it runs counter to another—optimizing resources. Decentralized functions almost always mean that more resources—people, funding, facilities, and so on—will be needed to perform those functions.

In addition, moving to a decentralized structure before there is a high level of readiness on the part of all managers across all functional boundaries will almost always lead to fragmentation and wholesale attention to purely local concerns and solutions. This occurs at the expense of overall corporate direction and commitment to a single vision for the business.

Develop a "Delivery Mix" Strategy. Second, the real issue is often whether students need to travel to a central location for training. A "delivery mix" strategy which addresses the appropriateness of various types of training delivery must be developed. This delivery mix should determine those types of training delivery which are best suited to the kinds of training which is being conducted. A delivery mix should consider

- The type of training (skill versus knowledge, basic versus advanced, etc.).
- Geographic locations.
- Numbers of students.
- Delivery schedules.
- Using appropriate training technologies.

When these elements have been evaluated, a delivery strategy should be developed which includes remote, central, regional, and local delivery. Of these options, central, regional, and local delivery all refer to traditional classroom delivery. Remote delivery refers to either teleconferenced/teletrained or satellite delivery.

Remote delivery requires some investment in facilities to permit transmission of two-way audio to simultaneous locations. This investment can range from inexpensive (speakers and microphones for each location plus long-distance usage) to expensive (high-quality microphones and speakers, acoustically conditioned rooms, electronic writing pads, personal computers with image capture boards, satellite networks, etc.). The level of investment in remote delivery technology must be carefully considered in view of the learning requirements of the training population and the types of content to be delivered. Several sources, including AT&T, can offer assistance in the design of cost-effective remote training options.

The Reporting Structure. A third issue concerns reporting structure—to Marketing or Sales or to Personnel or HRD. This is a touchy issue, but the conclusion of virtually every sales training manager who has been asked is that they would rather report to the functional organization, Sales or Marketing, rather than to Personnel or HRD. The reasons are basic.

Reporting to a function which is not part of Sales and Marketing places a potential barrier between the training group and the internal client population. This barrier can be due to departmental emphasis, operating against a different set of goals and objectives, and competition for resources from other functional units.

Organization structure is sometimes a financial issue and sometimes a political issue within a company. From the perspective of optimizing training performance and resource utilization, the best structure for sales training is one which has an unencumbered relationship with the training population. The structure should enable the training manager to work freely with all management levels in the client department. It should not impede the ability to carry out the training mission statement effectively through artificial barriers.

Staff

Other frequent topics for discussion are the size and composition of the training staff.

Four Guidelines to Consider When Evaluating Staff Size. Size is an issue because of the budgetary implications and because of the manner in which training budgets are typically developed. Because the value of training is often suspect, the staff is kept to a very low number, regardless of the needs of the client population. Unfortunately, a guideline for calculating staff size is not available. There are four factors, though, which might be useful to consider:

1. Product or service diversity, complexity, changes to product offerings, frequency of new product introductions.
2. The amount of travel needed to provide training to widely dispersed student locations.
3. The extent to which the staff is required to update or develop new materials.
4. The number and variety of courses offered. A large number or a big variety of training will typically require a larger staff.

The fundamental issue here is the level of credibility and the value that senior management has in training. The higher the level, the more likely it is that *adequate* staff size will be supported.

How to Achieve a Balanced Staff Makeup. The other question is staff composition. Should the staff be comprised of line people or professional training people? There are pros and cons to either position.

A training staff totally composed of line personnel certainly has a strong feel for the pulse of the marketplace. This is a strong argument for recruiting directly from field sales into training positions because of experience with competition, clients, and product/service applications.

On the other hand, professional sales types usually know little about learning processes, about designing and developing training for optimal learning, about developing realistic valuative measures, or about training technology in general.

In addition, they tend to emphasize their personal experiences which may not be broad enough to meet the needs of other salespeople.

Using professional training people can address the issue of training competence, but it begs the issue of practical knowledge and experience with the "real world." This lowers their credibility with line managers and salespeople. It can also place a communications barrier between the sales force and the training staff because the two do not talk the same language. Trainers use technical training terms like "needs assessment," "terminal objectives," "governing objectives," "task relevance," and the list goes on and on.

Salespeople talk in their own language as well. They say things like "carrying a bag," "price objections," "closing," and "prospecting," and this list goes on and on.

The point is that to be successful, the training staff must talk the same language as their clients. If they don't, they are seldom listened to.

The suggestion for staff makeup is a balanced group with people from both disciplines. This can be accomplished by using professional training people for design, development, and evaluation activity and line sales people for delivery. They can learn from each other and, over time, can exchange some responsibilities.

Another suggestion is to plan for the rotation of staff, especially those doing delivery. This will keep a stream of fresh perspectives on what is happening in the marketplace. It will also help to manage the problem of training "burnout" which occurs when instructors deliver a high volume of training over a long period of time. For example, an instructor delivering 35 to 40 weeks of training over a 15- to 18-month period is likely to lose enthusiasm and become very regimented in his or her style. Realistic rotation of about two years would forestall this type of problem.

Regardless of the type of staff and the time frame of their training involvement, a sales training manager should have a workable and realistic plan for developing and maintaining a staff that has the skills and knowledge to implement the training plan effectively. The key to this plan is the level of support available to the training manager and the positioning the group enjoys.

Marketing Plan

The final element of a training management system is a marketing plan which packages and promotes training as a crucial service for the sales organization, and perhaps for other functional areas of the company. A marketing plan has several purposes. It should stimulate awareness of the value of training, create interest in the training process, promote the value of sales training activities to sales and marketing clients, and enable the training manager to secure commitment and support for continued training involvement and participation.

Marketing of sales training is closely tied to the tracking and evaluation process. Minus a workable mechanism to promote training's contribution, the challenge of demonstrating the value of training becomes increasingly more difficult.

When we realize that training is an intangible service, the most important thing to remember is that the client often doesn't know what he's getting until he doesn't get it. Only when the result is not consistent with expectations does the client begin to question the original support and commitment to training.

How a Marketing Plan Can Help Your Training Group. The intent of a marketing plan for training is to position sales training as a viable, necessary element in the company's strategic planning process. The fundamental issue is to remind clients regularly of what they are receiving from this service called training and how overall company direction is supported. Without a conscious effort to employ a marketing strategy, clients will tend to forget the value they are receiving.

Three Steps for Developing a Marketing Plan for Training Managers. Because training management is essentially a marketing issue, it involves developing and maintaining relationships with a variety of stakeholders including line sales, sales management, and marketing staff. A marketing plan designed to nurture and sustain those relationships must include some familiar steps.

First, training managers need to generate *awareness* among their training clients by identifying performance problems, showing that those problems have costs, that there is opportunity for solution, and that the opportunity has benefits to the company.

Second, training management needs to determine, through a *situation analysis*, where the company now stands and what is needed to obtain the desired results. This step enables the sales training group to demonstrate the application of technical training skills to real-time business issues. It also affords the opportunity to provide a valuable service to the training client.

Third is the responsibility of the sales training manager to assume responsibility and *accountability* for maintaining client relationships. Training managers will find that relationships will deteriorate unless appropriate energy is directed toward maintaining and enhancing them.

Because providing training is an intangible service, it is important to create surrogates or metaphors for tangibility—how we dress, speak, write, design and present proposals, work with prospects, respond to inquiries, initiate ideas—all show how well we understand the client's business.

The most valuable asset a training organization can have is its relationships with its clients—line sales. It is not just "who you know" that is important, but how well you are known to them that makes the difference. That difference is a function of the nature and quality of your relationships with them—how well you have managed those relationships.

SUMMARY

As we look at the issue of sales training management, it becomes obvious that a new paradigm of training leadership is emerging. The paradigm is a combination of the traditional management skills of planning, organizing, staffing, directing, evaluating, and controlling plus the leadership skills of visioning, focus, positioning, networking, and innovation.

The challenges of contemporary business demand that the human resource be more adaptable, more creative, and more willing to innovate. Sales training organizations, as role models and as the organizational function charged with implementing change, must also change. Training managers must move their organizations forward—get them going in the right direction. Training managers must search for opportunities, look for ways to radically alter the status quo, to lead their organizations where they have never been before. They must become *performance leaders*. James MacGregor Burns stated in *Leadership*, that

> the essence of leadership . . . is the recognition of real need, the uncovering and exploiting of contradictions among values and between values and practice, the realigning of values, the reorganization of institutions where necessary, and the governance of change. Essentially, the leader's task is consciousness-raising on a wider plane.[7]

What better definition of the role of the manager of sales training can we develop?

The sales training manager who is waiting for management to dictate the company's sales training needs will find that either management will not come to him or her or that they will come with the wrong things. The challenge to today's sales training manager is to be proactive, to anticipate changes that will impact the sales force's skills, and to be ready with innovative performance improvement solutions when the need is greatest. The time for training management to be involved is in the problem recognition phase, not after the fact—not after someone else has defined the problem and determined the training solution without input from the training manager.

The sales training manager of the future *must* become a sales training leader and position sales training as a strategic performance artery for the business.

---

[7] James MacGregor Burns, *Leadership* (New York: Harper Torchbooks, 1978), pp. 43–44.

chapter four

STAFFING FOR SALES TRAINING: HOW TO IMPROVE YOUR SELECTION TECHNIQUES AND PREDICT PERFORMANCE

DENNIS S. McGURER

Dennis S. McGurer is vice president, general manager of marketing, at DISCUS Electronic Training, a Kodak company. He recently accepted this position after a successful term as director of educational development and planning for Eastman Kodak Company's Marketing Education Services Organization. McGurer has given many presentations on the development of sales training personnel to local and national business groups. He holds a B.S. degree in Marketing and Speech from Bowling Green University and is a member of the American Management Association, American Society for Training and Development, the National Society of Sales Training Executives, and the National Society of Pharmaceutical Sales Trainers.

The success of a training organization, like all service groups, is primarily dependent on the quality of its personnel. This chapter examines five essential aspects of personnel in the training organization:

1. Selection of potentially superior performers
2. Staffing for maximum efficiency
3. Management of creative professionals
4. Training and motivation
5. Appraising performance

It also looks at major trends in training personnel:

- Increased use of competency-based standards for selection
- Increased availability of academically qualified professionals
- Development of multiservice training organizations with smaller staffs of generalists who act as consultants

Professionalism has always been central to training. But over the last decade, the selection, education, and development of training personnel have undergone fundamental changes. Today, a new training professional has emerged—one who is highly educated, more specialized, and ultimately more influential in the corporate structure. This new breed fits well in the era of professionalism that characterizes modern training organizations.

The future of training and development may be influenced most by academia. The most important recent development has been the acknowledgment by colleges and universities that training deserves academic recognition. The result has been a growing number of degree programs which produce highly skilled training developers. Degree programs have attracted talented students who forecast a training community of the future that will be better trained and more career-oriented than in the past.

Personnel selection techniques have also improved. Competency research into the skills, knowledge, and motivation required for superior performance has given administrators a more reliable approach to selection. New types of intern programs provide departments with ways of assessing potential personnel *before* hiring them.

HOW TO ASSESS PROSPECTIVE PERSONNEL *BEFORE* YOU HIRE THEM

The Job Competency Approach

Degrees, grades, and even past jobs are unreliable predictors of future performance. For that reason businesses are turning to the job competency approach for selection. Job competency is based on the concept that significantly superior performers have certain distinct skills, knowledge, or motivation that account for their increased effectiveness. A job competency assessment compares superior with average performers and determines the behaviors unique to the superior group. The model can then be used as a basis for selection, not only of those individuals who possess the competencies, but also of individuals who are most likely to use them on the job.

For organizations large enough to justify the cost of their own job competency assessment, the results can be extremely valuable. However, there is a second and less costly alternative.

Fifteen Key Training and Development Roles. In 1983, the American Society for Training and Development (ASTD) published a sophisticated competency study of 15 key training and development roles. Because the study was based on a general population, it looked at outputs and behaviors required for various roles in the training and development field—but not solely those of superior performers. The publication, *Models for Excellence*,[1] provides a workable foundation for selection based on universal competencies found in the 15 key roles. It also lists definitions and critical outputs expected of each role. An additional benefit of the ASTD study is the definition of the expertise expected from basic, intermediate, and advanced training practitioners. The 15 roles identified are the following:

---

[1] Patricia A. McLagan and Richard C. McCullough, *Models for Excellence*: *The Conclusions and Recommendations of the ASTD Training and Development Competency Study* (Washington, DC: American Society for Training and Development, 1983).

1. *Evaluator*: Identifies the extent of a program, service, or product's impact.

2. *Group Facilitator*: Manages group discussions and group process so that individuals learn and group members feel the experience is positive.

3. *Individual Development Counselor*: Helps individuals assess personal competencies, values, and goals; identifies and plans development and career actions.

4. *Instructional Writer*: Prepares written learning and instructional materials.

5. *Instructor*: Presents information and directs structured learning so that individuals learn.

6. *Manager of Training and Development*: Plans, organizes, staffs, controls operations or projects, and links training and development operations with other organization units.

7. *Marketer*: Sells training and development viewpoints, learning packages, programs, and services to target audiences outside own work unit.

8. *Media Specialist*: produces software and uses audio, visual, computer, and other hardware-based technologies for training and development.

9. *Needs Analyst*: Defines gaps between ideal and actual performance and specifies causes of gaps.

10. *Program Administrator*: Ensures that facilities, equipment, materials, participants, and other components of a learning event are present and that program logistics run smoothly.

11. *Program Designer*: Prepares objectives, defines content, and selects and sequences activities for specific programs.

12. *Strategist*: Develops long-range plans for what the training and development structure, organization, direction, policies, programs, services, and practices will be to accomplish the training and development mission.

13. *Task Analyst*: Identifies activities, tasks, subtasks, and human resource and support requirements needed to accomplish specific results in a job or organization.

14. *Theoretician*: Develops and tests theories of learning, training, and development.

15. *Transfer Agent*: Helps individuals apply learning after the learning experience.

How Competency Models Help in Selection and Development. In addition to the critical outputs and competencies for the 15 key training and development roles, the ASTD study identified the knowledge/skills areas important in virtually all major activities in the training and development field. While a wealth of valuable information, in actual practice the knowledge/skills are not unique jobs but aspects of several jobs. Often they are duplicated in more than one real job category. Using the materials for selection interviewing requires in-depth knowledge of the tasks involved in the jobs involved. Once determined, the interviewer may find it helpful to develop specific questions centering on the knowledge/skills and their related competencies.

Preparing interview questions on easily defined skills is relatively simple. Most could be used as written or verbal questions.

EXAMPLE

Evaluator

A training program has just been completed that teaches new salespeople how to use electronic cash registers. How would you evaluate the program in terms of (1) learning immediately after the presentation (2) long-term effectiveness (3) needed revisions, if any?

Media Specialist

Explain the process involved in converting a slide-tape program to videotape.

However, for competencies that involve less well-defined skills, a more sophisticated questioning approach is required. This is covered in the next section.

Developing Competency Models to Fit Specific Jobs. By necessity the ASTD study presents the broadest form of a competency model. A study specific to an organization is highly desirable. It will provide not only the knowledge/skills required for superior performance but also the less easily determined competencies of motivation and attitude.

A note of caution. Sometimes specific job models are published in journal articles or industry reports. "Borrowing" such models and applying them on the basis of a generic job title can be misleading. Among other things, the actual tasks involved for the same job title in different organizations often vary significantly. For example, in one organization the job of instructor might primarily involve competencies related to classroom skills in a single subject area. In another it may include program development in a variety of areas as well as classroom skills in many subjects.

Specific competency assessments are usually essential for each critical job or "family of jobs" with closely related duties. The resulting models would have four parts:

1. *Basic Competencies*: Those that must be exhibited by an individual to do an acceptable but average job.

2. *Essential Competencies*: Those which superior performers possess more often

than average performers. These tend to be competencies that are difficult or impossible to develop in a person within a reasonable period of time. Examples include persistence, assertiveness, and systematic and forward thinking.

3. *Enhanceable Competencies*: Those that could be enhanced through training experience. Normally there should be some evidence of the competencies in the individual because developing them from scratch would be difficult. Enhanceable competencies include political sensitivity, concern for positive impact, and situational control.

4. *Trainable Competencies*: Those in which a person could be trained with little or no prior knowledge or experience. For example, learning about a product line or organization.

Creating a model of hiring criteria is facilitated through use of the following form:

In the absence of a more definitive job competency study, a tasks based model is frequently used. The tasks required to perform the job and the skills required to do them are determined and listed as the basis for candidate selection. The model is deficient in that the competencies required to do the tasks in a superior way are not included. This deficiency can be reduced, but not eliminated, by having superior performers review the tasks and determine what knowledge and motivation, in addition to the skills, they believe each requires to complete the tasks satisfactorily. Skills and knowledge can often be taught but motivation and personal traits are difficult to develop or change. The importance of each to performing the job should be considered in the selection process.

Hiring Model

Position

I. Primary Tasks: Basic Skill or Knowledge Required

_____

_____

_____

_____

_____

_____

II. Motivation or personal trait required to perform each task in addition to the skill requirement:

Motivation	Personal Trait	Should this be:	
		Hired For	Enhanced By training
_____		( )	( )
_____			
_____			

In addition to the above what qualities are required to be a superior performer in this job:

The model could be used for a variety of purposes, including performance expectations, development opportunities, and overall assessment. But its first use is usually in selection interviews for new personnel.

The competency-based interview would use a special questioning technique that probes past experiences for behavioral detail. Basically, the candidate is asked to relate extremely successful and unsuccessful experiences. Used correctly, the technique can provide a balanced picture of an individual's skills, knowledge, motivation, self-image, and other traits that can be compared with the competency model. While seemingly simple, the technique requires considerable practice to do well.

Since all models are an ideal, few candidates will precisely match each of the competencies. However, only those possessing all the *basic* and a high number of *essential* competencies should be considered for a position. They would have the greatest potential for superior performance. Possession of additional *enhanceable* competencies would further indicate potential success.

Procedures for developing competency studies and questioning skills are referenced in books[2,3] or can be purchased from competency consulting organizations.

---

[2] George O. Klemp, *Job Competency Assessment: Defining Success Factors of Job Performance*, ASTD *Research Series*, Vol. 8 (Washington, DC: American Society for Training and Development, 1982).

[3] George O. Klemp and David C. McClelland, "What Characterizes Intelligent Functioning Among Senior Managers," in *Practical Intelligence Concepts of Competence in the Everyday World*, edited by R. Sternbert and R. Wagner (New York: Cambridge Publishers, 1986).

Six Alternate Approaches for Selecting Personnel

Multiple Interviews. Regardless of the basic approach to interviewing, multiple interviews, often using staff members, are a desirable selection technique. If staff members are used, the interviews can help to determine how well the candidate will fit in with the group, as well as give the individual a practitioner's understanding of the scope and expectations of the job. Multiple interviews can help to verify competencies and provide a check on any first impressions that might be overly negative or positive.

Résumés. Résumés are primarily useful as a record of past formal and informal education, and serve as a basic job record. Because their wording tends to be vague, it is often useful to ask candidates for specific duties in past jobs and for an explanation of listed accomplishments.

Transcripts of Grades. Transcripts will verify graduation and are useful as an overview of subjects taken. Unfortunately, course titles and course contents may have little in common. Studies repeatedly show that grades are poor predictors of future performance.

Telephone Screening. Screening conducted by telephone is sometimes used both by the interviewer and by the candidate. Screening should be used to determine genuine interest, to answer questions triggered by the résumé, and to ensure that essential competencies are present. This step can save a lot of time by eliminating interviews of unqualified candidates.

The following form will help capture important information in telephone screening

Telephone Screen

Position _____

Candidate _____ Telephone number _____

Date of call _____ Interviewer _____

1. Does person indicate an understanding of the job ( ) Yes (No)
2. Questions raised by resume: Resolved
_____ ( ) Yes ( ) No

3. Background, experience or interests not indicated on resume that may be significant:

4. Basic skills and competencies required for job:

_____ Quotes indicating candidate

_____ does or does not possess

_____

_____

_____

_____

5. Questions asked by candidate that need to be answered by   ( ) Mail
                                                                                                 ( ) Phone

6. Personal interview ( ) Yes ( ) No

       If YES: Date _____

                    Time _____

                    Location _____

7. Comments:

References. Because of the fear of lawsuits, most companies do not provide information on past employees other than to verify employment. Personal references are seldom considered reliable, but may be useful to verify some information.

Samples of Work. Some organizations request samples of work or provide a standardized work exercise packet which the candidate is encouraged to complete. Evaluation of the work samples may include a personal interview in which the individual is questioned about the development of a particular sample, its evaluation, and other details. The interview is also useful to determine how much of the work was individual or a cooperative effort.

JOB POSITIONS AND STAFFING ALTERNATIVES _____

The positions that make up a training staff depend largely on the size and duties of the organization. In large organizations, the staff may consist of hundreds of people, each with specific duties. In small organizations, training may be an additional duty of a single individual and that person would be expected to do everything from analysis to instruction. The most frequently required duties, and those traditionally assigned to them, are presented on page 71.

Permanent versus Temporary Personnel

Most organizations employ both permanent and temporary personnel. Permanent personnel, whether one person or hundreds, are necessary to add stability to the training group. They best understand the company's and the client's basic needs, plans, and goals. They are normally assigned major projects.

Temporary personnel can be a wide variety of individuals from vendors to contract designers, writers, media producers, and editors. In general, these individuals are usually either employees of outside agencies or freelancers. In either case, they are paid by the hour, by the project, or by some other form of short-term contract. Normally they are awarded no additional benefits other than the salary they receive from their contractor agency or employer.

Departments with tight personnel restraints may be comprised entirely of temporary employees. The flexibility of this arrangement has increased its popularity. Some of the advantages are

- Only the personnel actually needed are working.
- Talent can be chosen that best fits a need.
- If selected individuals do not produce, they are more easily dismissed.
- In slow periods, personnel reduction is quickly and easily accomplished.

The major considerations when using temporary personnel are the cost of acquainting them with the company, making certain that quality is maintained, and potential legal implications. While the initial cost of company orientation may be high, it is a one-time outlay if the individual is used on a frequent basis. The quality problem will be overcome if an adequate quality control system is in place. Legal implications depend on the locality.

Intern Programs: Cost-effective and Helpful to Both Company and Student

Many training departments use intern programs to provide on-the-job experience for prospective new hires. Most of these programs are coordinated with colleges and universities. A typical program is conducted by Eastman Kodak Company's Marketing Education and Support Organization. In this program students in their last year of advanced study are awarded six-month internships based on recommendations from their schools, personal interviews, and competitive work-related exercises. During their tenure they are assigned to actual projects with marketing divisions. Under the guidance of a mentor, they perform all phases of analysis and development of training programs up to the instruction stage. The internship does not guarantee the student a job offer, but it provides both the intern and the company a chance to evaluate potential opportunities.

Internships can be cost-effective while providing help for college students. Potential new hires can be evaluated in real job situations; they, in turn, can determine their best fit in the training profession.

Instructors: Four Key Types

In small organizations, instructors are often the same people who develop training programs; in larger groups, there may be a staff that instructs only. Some companies like to use managers, salespeople, or other experienced personnel in the instructor role. There are advantages and disadvantages to each of these ideas, both in effectiveness and in their reception by those being trained.

Permanent Instructors. A permanent, full-time, instructional staff is almost essential when there is a consistently large volume of training. Highly trained and experienced individuals are valuable assets in the classroom. However, the wise administrator will pay close attention to the training staff to make certain that they are continually challenged by their classroom role. This can be accomplished by rotation of job assignments, training seminars which develop new presentation skills, and assignments to completely different activities at periodic intervals.

Developer-Trainers. The major advantage of combining the program developer/instructor role is that the instructor will have intimate knowledge of the subject matter. But knowledge of training design or content is not necessarily an indication of a good instructor. Many developers prefer not to be in the classroom, and this should be considered before assigning them to it.

Experienced Practitioners. Managers, salespeople, engineers, and others may seem logical instructors. They are assumed to know the materials; plus they bring an aura of reality and credibility to the classroom. But it is unfair both to the individuals and the students to put subject-matter experts in the classroom without proper instructor training in classroom techniques and in the exact materials to be taught. These instructors can be very valuable, especially in sales training, but they must be trained to be effective trainers.

Contract Instructors. Professional instructors, such as moonlighting college professors or public school teachers, can often be a good source of employee instructors. Again, it is essential that their capabilities in the classroom not be taken for granted. It is especially important to consider the change in the type of people they are used to teaching. Good public school teachers may not do as well with adults, and some college professors may not feel the urgency that is required in the industrial classroom.

Internal Consultants

While the term "consulting" has become a popular one in training, it is not consistently defined. Some organizations use "consultant" as a general title for all training developers, regardless of their roles. Ideally, "consulting" is the use of trainers as internal advisors on any aspect of the organization that may be influenced by training or training-related activities. The true potential for this consulting role lies with the abilities, respect, and knowledge of organizational problems that reside in the training department. It is not a matter of suddenly shifting from a "training organization" to a "consulting group" overnight. The role of consultant is one that is earned through demonstrated ability.

Training Coordinators

A coordinating role is a growing practice in companies that prefer to maintain smaller training staffs. It places a few key specialists in the position of coordinators who understand the training development process but contract out the actual development and production of training. Often these trainers are called "consultants" because they begin each client contact by performing a front-end analysis and then make training or other intervention recommendations. To do this successfully, most staff members need to be educated to perform the coordinating role. One of the most important skills they should learn is how to maintain quality control. Since their work load is usually heavy, good control systems, such as computer-assisted scheduling, are often useful. Their education should also include selecting, contracting, and working with vendors as well as budget development and cost control. In some cases, these staff members must also learn to sell training services. This total "business attitude" is essential for the training coordinator concept to succeed. (See Figures 4–1 and 4–2.)

Figure 4–1 Training coordinator's typical quality control checklist

1. Does the analysis provide the information needed to produce a quality product?

   In clear concise language, what are the objectives?
   What is the definition of quality for this project? If necessary list each aspect of the project such as video, writing, artwork, and other standards.
   What, if any, restraints are there to producing the quality level desired?

Figure 4–1 Con't.

Is there sufficient time to produce the quality desired?
Is there sufficient money to produce the quality desired?
Is there sufficient talent available to produce quality desired?
What is the profile of the audience? Is the quality level sufficient to match the audience expectations?
What certainty is there that the approach can meet the objectives both in effectiveness and quality? If low certainty, who assumes the risk of failure?

2. During Production

Are there planned and significant quality checks during the development?
Are *all* aspects of the project checked for quality, for example, writing, audio, video, and packaging.
What provision is there for correction if quality is below desired level?
Who decides on the quality of the various parts of the project? Is there agreement on this in writing?

3. Prototype Stage

Will the project be tested on a group accurately representing that of the real audience?
Is the prototype planned early enough for significant changes to be made if necessary?
Is there adequate budget remaining to make quality corrections?

Figure 4–2 Vendor selection checklist

1. Is there a past record of quality work in the needed area?
   What proof do you have of this?
      Your own
      Highly reliable source
      Vendor's say-so
      Products
         Are products recent?
         Are people who produced them still with the vendor?
         Were the products actually produced by the vendor, co-produced with someone else, or farmed out?
2. Is there a record of reliability?

Figure 4–2 Con't.

3. Will the vendor guarantee quality to meet your standards?
   Will the vendor agree to your standards or only his or hers?
   What happens if standards are not met?
   Is there a financial obligation to the vendor if products are rejected for quality? Other reasons?

4. Does the vendor have the staff, facilities, and other capabilities to produce the needed materials?
   If not what assurance is there that they are available?

5. What provision does the vendor have for backup in critical projects?

6. Is the vendor the only one who possesses a specific talent, facility, technique, or other unique quality needed for the project?

7. Is the vendor easy to work with?

8. Does the vendor tend to take on too many projects for the staff or facilities available?

9. Is the vendor financially stable, especially if project requires substantial up-front investment?

10. Is this the vendor of choice or lower on the desired list?

(These checklists are not meant to be all-inclusive but rather to provide suggestions.)

MANAGING THE TRAINING STAFF

Good management of a training staff is not substantially different from good management in any other staff. The basic duties are

- Setting guidelines for staff selection.
- Setting policy for promotion and career development.
- Establishing and monitoring adherence to budgets.
- Communicating a sense of belonging, purpose and recognition.
- Meeting the goals and objectives of the organization and staff.
- Ensuring that quality control is an integral part of the entire operation.

People who are attracted to the training profession are often those with strong creative perspectives. In addition, training positions are usually less structured than jobs in other areas of the company. These factors can present problems to managers who are used to more formal approaches to work. However, training professionals are no less productive than other employees. "Appropriate management" in training organizations usually means making certain that employees understand the goals and standards of management and then helping them meet those goals and standards through appropriate support.

How Management Philosophy Determines Staffing

While the size of the company may dictate, to some extent, the size of the training organization, it is the philosophy of management that will ultimately determine the number of positions and duties required.

Proactive versus Reactive Departments. One aspect of that management philosophy is the proactive versus reactive nature of the training organization.
The proactive department

- Actively solicits business. It does not wait for business to come in.
- Takes a broad view of "training," using its full resources to even out seasonal training demand.
- Performs a management consulting role.
- Is closely related to the personnel or human resources department, helping in decisions on policy and selection.
- Is actively involved in problem solving for all areas of the company. Becomes a part of the "strategy process."
- Aims ultimately to be a profit center or at least a "zero-budget" operation.

The reactive department

- Primarily responds to requests of others.
- Depends on budgets set by others and works within them.
- Has seasonal work load with slow periods used for staff development or other internal activities.

While these statements make it clear that a proactive department is the better of the two models, being proactive is not a traditional role of training. It may be resisted by both training management and upper management. Over the long run, however, proactive departments tend to be ones that attract employees who see training as an exciting profession in which their efforts will gain the greatest prestige and rewards for the department and the company.

Separating Departmental Functions to Improve Efficiency. The second philosophical concern is the separation of roles and duties. Some managers prefer personnel with "total competencies"—those who can take a training problem from analysis to development to production and into the classroom. Individuals who can do all these jobs with equal competence are rare, but the demand for them is great, especially in smaller organizations. Even in larger groups, the separation of the major functions is often minimal. The reasons for this vary from budget considerations to a desire for greater continuity. This idea is sometimes promoted by developers who feel most comfortable when in total control of every step in the training process.

The opposite concept is the division of the development and production activities into two (or more) separate and distinct positions. Some managers feel that the use of specialists promotes both efficiency and better results. However, when the jobs are separated, perception and prestige may make one job more appealing than the other. Content developers may see media producers in the role of skilled artisans rather than in the role of more academically trained "professional" developers. This perception can be heightened in the not unusual situation of a staff made up of degree holders in content development and nondegree personnel in media. The problem is made even more difficult if media people feel that the only way to bridge the gap is to take on more content development. An administrator should be aware of these possibilities and make an effort to treat each position equally in respect, salary and advancement.

Keeping Staff to a Minimum. The third management philosophy affects the basic number of personnel in the department. There is a growing trend toward smaller "coordinating" departments. Using a minimal permanent staff, these departments emphasize the use of vendor or contract labor as a means of cost control without a loss of quality or output. This concept is covered more fully under the next section, "Administering the Training Department."

Administering the Training Department

The type of organization a department chooses will have a significant impact on the administration of the department and requirements for personnel. Organizational structures of training departments traditionally match those of other support groups.

However, an organizational structure which closely resembles that of a manufacturing group may be more realistic in personnel utilization. Manufacturing groups start with a defined need. Then specialists design a product to meet that need. The product is produced using a planned series of operations, each one of which adds to the product's value and determines its quality. The total group is responsible for customer satisfaction with the product. This is basically the same process used by most training groups to develop and produce training programs. While the "product" may be customized and intellectual, training organizations, especially larger ones, are by their nature "production houses," utilizing basic manufacturing systems whether they care to acknowledge this or not.

The concept of training departments as production houses not only allows an administrator to understand and control the flow of projects better, from concept through to delivery, but promotes the use of modern management tools. Through the use of production and quality control systems, most training organizations can cut costs, increase efficiency, and improve quality. By using standardized procedures, reducing the number of suppliers, bidding on contracts, and using personnel more efficiently and effectively, they improve the products, the service, and the financial bottom line. All this tends to emphasize the training organization's value to the parent company.

There is usually some opposition to the use of production and quality control systems in creative organizations. For example, employee output expectations in creative positions are a universal problem for managers. This is primarily because measurement seems difficult. However, analysis of often-performed operations such as analysis, script writing, and audio visual products can provide reliable time and cost estimates. These enable managers to use such data in scheduling and cost management systems like PERT (Program Evaluation and Review Technique).

Resistance by personnel to the use of such systems is usually toward record keeping and perceived reduction in the creative atmosphere. As long as the record keeping is kept simple and the time estimates are based on reality over a significant type and number of projects, these objections become minimal. Their advantages both to the individual and the organization can be dramatic. A training manager and his or her organization which can predict with reasonable accuracy the cost and time required to produce quality products becomes invaluable to their clients and their company.

Administrators of training departments should center their efforts on personnel selection and development, administrative functions such as providing standards and budget control, and directing efforts toward meeting corporate goals. Generally, managers should be well informed about projects but not actually involved in them. In smaller departments, the administrator may be more heavily involved in projects by necessity. In either position the skill of the administrator in understanding company culture, comprehending the real versus the stated needs of the clients, and appreciating the competition may be the determining factor in the department's success.

In some companies the training administrator is a pivotal position used as a developmental assignment before higher office. Because the training department tends to represent all parts of the company, the idea is frequently a good one. However, if this is the policy, there should be a strong administrative assistant position to maintain the stability the organization needs for quality and productivity.

TRAINING AND DEVELOPING THE STAFF _____

Training should be an ongoing process and integrated to the vision of the department. What the administrator and the staff see as the future should be the goal of staff development. Individual development should be based on the actual, rather than the assumed, level of knowledge. Too often, entire groups are trained in specific areas when only a few need the education. This is wasteful in terms of time and money and is often regarded as a negative by those who do not need the information. Group training should be aimed at group needs, such as new developments in delivery systems, new approaches to standard procedures, and maintaining knowledge of the industry or profession.

Outside Sources of Development

Outside sources include seminars, workshops, and institutes. There is virtually no area of training and development that is not covered by an outside training program. The major question is usually not "what subject" but "when," "where," "what cost," and "how good" is the program. Computer-based "catalogs" are available on-line that provide updated material on a daily basis. Some provide a review by former participants. The most reliable means of determining the value of a program is talking directly to a previous attendee whose opinion is valued.

Development activities should follow the rules of any good training program by stating objectives, having logical development, and providing for evaluation. Information on all these, plus references for evaluation, should be available from those offering the program.

In-House Resources

Inside resources are often overlooked as a valuable source of personnel development. Large organizations may have formal training in areas not covered by your own department. For example, manufacturing may have a basic course that introduces scheduling concepts and structure to the training department. Don't overlook internal experts who could give presentations on their specialties.

Your own staff is another source of experience and expertise which is often ignored. Presentations about projects and their development can be valuable sources of information, especially to new people. Explanations of new training delivery systems, validation and evaluation, and new trends in training and development are often useful. Reviews of conferences can also provide useful information and could be a required follow-up for those who attend the conferences.

Encourage New Career Path Opportunities

Career paths in training and development have traditionally been limited. People often came into training after careers in other areas, or they joined a training staff when employed by the company and stayed there. This has been a major management failing in the past. Fostered by the idea that trainers were one dimensional, the tendency was to overlook the training and development employees when opportunities became available.

The *new professionalism*, and notably the *proactive* nature of many training departments, has helped to make great strides in breaking this pattern. In departments that have assumed a consulting role within the whole company, the proven capabilities of trainers have opened doors that were previously closed.

Because career opportunities within a training organization are often restricted by size or limited turnover, opportunities within the whole organization must provide both incentive and motivation to move into other areas. Managers should work with the staff on a long-range plan that will make the full capabilities of trainers well known throughout the organization. The consulting role is one part of such a plan; another might be assigning trainers to committees, product launch groups, and other high-profile activities. The objective is to prove that the developers and instructors are educated, highly flexible, and well suited to other job opportunities within the company.

While some trainers (perhaps most) prefer to work in jobs much like their present ones, simply being informed of other opportunities outside of the training department can be a positive motivator.

EVALUATING TRAINING STAFF PERFORMANCE

Performance appraisal in modern companies is based on goals and objectives established well in advance of the appraisal period and reviewed periodically before the end of it. This same approach should be applied both to individual trainers and to the department as a whole.

How to Make Your Appraisal System Work

There has never been and probably never will be an entirely satisfactory appraisal system. All appraisals are subjective to some degree, but the goal is to make them as unbiased as possible. Many companies use a standard appraisal system that encourages an employee and the manager to establish work goals and expectations early in the year. Training departments may set standards in terms of number of programs completed, hours devoted to certain functions, budget control, or numbers of people taught. There are often additional objectives, which may be termed "stretch goals." These are achievements above and beyond normal expectations and often are accompanied by financial or other motivational rewards.

If the goals are clear-cut and viewed as achievable by both parties, evaluation is relatively simple, especially if the goals are met. If the goals are not met and there are obvious reasons why, the process still works well.

But too often goals are not clear-cut, or they are made and forgotten by both parties until the end of the year. Then appraisal becomes a virtual negotiation, with both parties feeling that they are on the losing side. For the system to work, three things have to happen:

1. Goals must be clear to both manager and employee.
2. Goals must be mutually agreeable and either literally or figuratively contracted to be met.
3. Goals must be renewed at set intervals. If they are not being met, adjustments either to the goals or to performance must be made.

Other areas of evaluation which may be outside of absolute and easily definable goals include such things as learning new skills, eliminating waste, and ensuring client satisfaction.

Two important areas almost always overlooked are quality of the developers' products and how well instructors actually train. Most approaches to evaluation of trainers tend to assess the total product, not individual development or presentation. Another avoided area is the return on investment of a training program. This is usually left to the client because of the difficulty in measuring the economic impact of training. But both these considerations are rapidly becoming more important to customers of training departments and will likely have increased importance to the appraisal of individuals in the future.

Evaluating the Staff and Department

Did the department meet its objectives? Were projects on time? Were they on or under budget? These three questions are normally asked by upper management in a general appraisal of the training department. Answers are fairly easy to obtain, and if these were the only questions the training administrators asked, life would be considerably easier.

But additional and much harder questions are increasingly being asked, for example,

- Was the best use of resources (people, plant, and materials) made? How do you know?
- Was there an adequate return on investment?
- Were the activities of the training department essential to the goals of the company?

To answer these questions, administrators must have continuing methods of evaluating what the department is doing and how well it is doing it. Managers must also know, with the same certainty, that their people fit exactly into what the company expects from training in terms of goals, costs, and return on investment. These questions—and especially the last one—can be extremely difficult to determine.

Smart administrators use the same basis for answering these questions that they use in answering personnel questions.

They make certain that there is a meeting of the minds with upper management about goals, that they know what will be accepted as proof of achievement, and that their personnel are working under the basic guidelines established by the company.

THE TRAINING PROFESSIONAL OF THE FUTURE

While they have recognized the potential of training and development and have agreed to its value, colleges and universities are still in the developmental stage in producing the training professional of the future. The schools have recognized the demand for training "specialists" and are graduating educational designers, media producers, and other professionals.

In spite of this growth, a major problem exists in the lack of central focus among teaching institutions. Each school has developed its own unique philosophy, which may not mesh with those of other institutions. Debates continue over "industrial technology" versus "performance technology" and whether students should be true specialists or educated in all phases of training. While most degree content is based on well-accepted principles, the new hire may find the philosophies of employee training substantially different from those of his or her school. A department staffed by people from different schools may be a menagerie of different approaches to design and could be difficult to manage as a team.

While the future may see training departments staffed entirely with college-trained specialists, a long transitional period will continue with departments having a mix of people: college-trained specialists, generalists, and experienced people without formal training or education. This growth of the new academic emphasis on preparation for a career in training will maintain its momentum for only as long as the results match the expectations.

Overall, the need for training and development personnel has increased so rapidly that specialists are in short supply. Technology, such as video conferencing and computer-based training, will have a tremendous impact on sales training personnel in the future. As new delivery systems become more sophisticated and demands increase for faster, more effective results, the challenge to the energies of available talent will increase even more.

Much of training management will continue indefinitely to be managers by experience rather than training professionals who have moved up through the ranks. The new era "professionals" are still too young to have reached the upper ranks of management. That will eventually change, except in those cases where training management positions remain pivotal developmental positions, as many are today.

chapter five

PREPARING SALES TRAINERS

ROGER E. WENSCHLAG

Roger E. Wenschlag is the vice president of marketing and sales support at McLagan International, St. Paul, Minnesota, a firm that offers performance management and human resource development products and consulting services. He recently joined McLagan after several years of sales training and product management experience in the training industry. Wenschlag holds a B.S. and a M.A. degree in Marketing Education from the University of Minnesota. He is a member of the American Marketing Association and the American Society for Training and Development.

Of all the sales support functions, there is none more visible or important than that of sales training. In the most positive sense, those who are employed to serve as sales trainers are partners with their sales managers. They have the responsibility to help members of the sales force learn and apply the skills and knowledge needed to help their customers solve problems and achieve their goals.

Since sales trainers are so important, the methods by which we prepare them deserve our close attention. The purpose of this chapter is to explore some of those methods. After first showing how the sales trainer can positively influence the organization, we will explore some tips on selecting people for this important role. Then we will examine the many potential roles the sales trainer can play and the different training competencies that may be required in doing so. Finally, we will look at specific ways to prepare the new trainer for his or her new role.

FOUR WAYS SALES TRAINERS POSITIVELY INFLUENCE THE ORGANIZATION

Depending on the sales organization, the function of sales training may be perceived in different ways. Some see it as an integral part of an organization's ability to compete and grow: others see it as a perfunctory service needed to avoid too much turnover in the ranks of new hires. Obviously, these are extreme points of view on a continuum, and we are likely to find ourselves somewhere between them. Where is your organization on this issue? What do you believe?

Taking the broad view, let us look briefly at some of the ways sales trainers can positively influence the organization. If you are in a situation where sales training is perceived in its narrowest light, this could be of value to you.

Promoting Leadership

When the sales training staff works in partnership with sales managers, they can often exert a force of leadership in the sales organization. Because of their unique position they can bring perspective, new ideas, and vision to sustain sales managers in their daily grind of fighting it out in the "trenches." Some examples of leadership the sales trainer could contribute are (1) helping to shape a sales meeting agenda, (2) creating practical in-field coaching aids, (3) creating ideas for better sales productivity management, and (4) helping to assure that training is actually applied on the job. Obviously, such contributions must stand the test of reality, or the credibility of the sales trainer could suffer.

Encouraging Sales Force Productivity

From the proper initial training of new hires to the ongoing development of veterans, the sales trainer plays a key role in sales force productivity. Salespeople need to understand how to sell, how best to solve the customers' problems, how to manage themselves and their account base, and how to identify and use company resources.

Achieving Strategic Objectives

When the sales trainer is strongly linked to the broad mission and function of the sales organization, we know that training can directly support very critical goals. Here we might include the skills and concepts needed to achieve strategic growth objectives and to establish competitive advantage in the marketplace.

Developing Growth of Salespeople

Aside from just helping to make salespeople more effective in selling and serving customers, the sales trainer also plays a big role in the career development and satisfaction of salespeople. Adjusting to the job, self-management, personal assessment, and development planning are some of the more important things that can be taught to salespeople to help them grow.

TIPS ON SELECTING THE SALES TRAINER

As with any other function in your organization, you would want to find and place the best qualified individuals in the sales training job. As we know, the sales trainer may have a profound impact on your company and its salespeople.

Chapter 4, "Staffing for Sales Training," covers selection in depth. But, to set the context for this chapter, we will briefly discuss here some key elements of selection.

Qualities to Look for in Potential Sales Trainers

Aside from sales experience and any training expertise the individual brings to the job, there are several personal qualities we should expect in the sales trainer. Chief among these is interpersonal sensitivity and versatility because training is, above all, a "people" business. A caring kind of person in the job will contribute greatly to the growth and satisfaction of training participants.

Another important quality of the sales trainer is flexibility and the ability to live with a fair degree of ambiguity. Today's selling environment dictates no less. Constant change caused by competition and your own moves in the marketplace is an ongoing challenge for the sales trainer who must respond to the needs of the sales organization. The kinds of changes one can expect in one's own organization might include such things as new product rollouts, pricing actions, reorganization, reassignment of a well-cultured training allies, new customer service policies, and new ways of selling. One who lives in the past and is unwilling to change may not have a future.

Career versus "Passthrough" Trainers

A subject of ongoing debate in sales training circles is the issue of staffing the sales training function with "career" trainers or "passthrough" trainers from the sales force. There are certainly pros and cons to each staffing method. Ideally, in a department large enough, you would want a mix of both types. Let's examine each.

The chief advantage of the career trainer is that he or she lends training expertise and stability and continuity to the sales training function. It is a good career promotion option for situations where you have "passthrough" trainers who want to stay in training. The chief disadvantage of career trainers is the potential for a program to get stale and out of phase with the needs of the sales organization. There are several ways to prevent this: (1) have an ongoing evaluation program with input from sales managers, (2) use a field advisory committee to help manage the program, and (3) assure that all trainers stay current by spending time in the field with salespeople and customers.

The chief advantages of the "passthrough" trainer are instant credibility, fresh ideas, and thorough knowledge of how the organization's products and services are used in the marketplace. Ultimately, "passthrough" trainers can make other beneficial impacts on the organization because of the skills they take with them and the support for training that spreads throughout the company. Trainer experience is excellent background for future sales managers, product managers, marketing staff, and other support jobs in the company that interface with salespeople. Recent estimates indicate that "passthrough" trainers is a preferred pattern by about 40 percent of sales organizations.

On the downside, "passthrough" trainers can pose some disadvantages as well. The quality of training offered could suffer because of disruption and lack of continuity and the new trainer's lack of training skills.

A training assignment also offers some potential problems for the would-be trainer coming in from the field. There is often a disparity between the pay and perquisites of selling and training jobs. Trainers may also run the risk in some organizations of being the victim of corporate retrenchment programs where training is often targeted for cuts.

WHAT SALES TRAINERS DO

Fifteen Roles Performed by Sales Trainers

One of the most crucial questions we must ask ourselves when selecting and preparing the sales trainer is, "What will the person be expected to do in this sales training job?" In a smaller company, where one trainer might be sufficient, the demands of the job will require a broad range of training competencies. In a larger operation, with more staff involved, we are likely to see a division of labor where individuals would tend to specialize and thus require fewer overall competencies.

To identify clearly the training roles and competencies of the sales trainer we will use *Models for Excellence* (see McLagan and McCullough), a study recently done by the American Society for Training and Development (ASTD). Its purpose was to define the roles, functions, and competencies of people who work in training and development.

The first major part of the ASTD study is a description of the 15 distinct roles that exist in the human resources development (HRD) field. The value of these role descriptions lies in their use in understanding the job of a sales trainer because to varying degrees, the sales trainer may have elements of all these roles as part of his or her job. Following are brief descriptions of each of the roles, and in a sales training context, some typical outputs one might expect to see for each. As you review them think about your own situation and ask, "To what extent will our sales trainer(s) perform these roles?" "Will some be more important than others?"

Marketer. The role of marketing and contracting for HRD viewpoints, programs, and services. Some typical outputs are

- Marketing and/or sales presentations.
- Customer/client needs identification.
- Contracts and agreements.

Needs analyst. The role of identifying ideal and actual performance and performance conditions and determining causes of discrepancies. Some typical outputs are

- Strategies for analyzing needs.
- Needs assessment tools.
- Skill, knowledge, and attitude needs identification.

Theoretician. The role of identifying, developing, or testing new information and using it to improve individual or organizational performance. Some typical outputs are

- Research designs.
- Research reports and articles.
- Action research.

Media specialist. The role of producing material for and using audio, video, computer, and other hardware for HRD-related activities, programs, or services. Some typical outputs are

- Graphics for training materials.
- Video-based material and broadcasts.
- Media production plans.

Group facilitator. The role of managing group or team discussions, processes, and interventions. Some typical outputs are

- Sales team intervention plans.
- Group process feedback.
- Group planning outcomes.

Instructor. The role of presenting information and directing structured learning experiences. Some typical outputs are

- Lectures, presentations, demonstrations.
- Feedback to course developers.
- Evaluation of and feedback to students.

Program designer. The role of preparing objectives, defining content, and selecting and sequencing activities for a specific intervention. Some typical outputs are

- Learning objectives.
- Program plans and designs.
- Instructional plans.

Manager. The role of supporting and leading a group's work and linking that work to the total organization. Some typical outputs are

- Training unit plans and direction.
- Performance management.
- Resource acquisition and allocation.

Administrator. The role of providing coordination and support services for the delivery of HRD programs and services. Some typical outputs are

- Facilities and equipment selection and schedules.
- Program and client records.
- User support and client service.

Career and development advisor. The role of helping individuals to assess personal competencies, values, goals, and to identify and plan development and career goals and actions. Some typical outputs are

- Third-party referrals.
- Development and career planning resources.
- Career analysis/assessment.

Strategist. The role of establishing long-range HRD direction to support the overall business goals. Some typical outputs are

- HRD policies.
- Future scenarios.
- HRD strategies.

Materials developer. The role of preparing material for HRD interventions. Some typical outputs are

- Sales aids.
- Facilitator guides.
- Participant materials.

Evaluator. The role of identifying the impact of an intervention on individual or organizational effectiveness. Some typical outputs are

- Evaluation instruments and tools.
- Evaluation plans and programs.
- Reports of program/product/process impact.

Task analyst. The role of identifying activities, tasks, human resources, and other requirements necessary to accomplish specific results in a job or organization. Some typical outputs are

- Job designs.
- List of key job/unit tasks.
- List of competency requirements of a job/unit.

Transfer agent. The role of helping individuals apply learning after the learning experience. Some typical outputs are

- Individual action plans for on-the-job application.
- Jobs aids to support learning.
- On-the-job environment modified to support the transfer of learning.

Critical Competency Skills

The foregoing role descriptions are helpful in getting the big picture of what the sales trainer may do. We can get an even clearer view if we look at another part of the ASTD competency study which outlines specific critical knowledges and skills for excellent performance in the trainer's job. While all 31 of these competencies are too numerous to list in this chapter, it might be helpful to illustrate how several of them go together to show examples of typical "profiles" of various sales training jobs.

Instructor/Program Designer Skills

Let's suppose that we have a sales trainer in a small department with two staff. The principal roles of our trainer are that of "Instructor" and "Program Designer." Given these roles we are likely to want our trainer to possess, at the very least, these critical competencies from the ASTD competency study:

1. *Adult Learning Understanding.* Knowing how adults acquire knowledge, skills, and attitudes; understanding individual differences in learning.
2. *A/V Skill.* Selecting and using audio/visual hardware and software.
3. *Competency Identification Skill.* Identifying the knowledge and skill requirements of jobs, tasks, and roles.
4. *Feedback Skill.* Communicating opinions, observations, and conclusions such that they are understood.
5. *Group Process Skill.* Influencing groups to both accomplish tasks and fulfill the needs of their members.
6. *Intellectual Versatility.* Recognizing, exploring, and using a broad range of ideas and practices. Thinking logically and creatively without undue influence from personal biases.
7. *Objectives Preparation Skill.* Preparing clear statements which describe desired outputs.
8. *Performance Observation Skills.* Tracking and describing behaviors and their effects.
9. *Presentation Skills.* Verbally presenting information such that the intended purpose is achieved.
10. *Relationship Versatility.* Adjusting behavior to establish relationships across a broad range of people and groups.
11. *Training and Development Techniques Understanding.* Knowing the techniques and methods used in training; understanding their appropriate uses.
12. *Writing Skills.* Preparing written material which follows generally accepted rules of style and form, is appropriate for the audience, is creative, and accomplishes its intended purposes.

Manager/Needs Analyst Skills. Given the same sales training department as just outlined, let us suppose that the other individual performs the principal roles of "Manager" of the department and "Needs Analyst." We would be likely to want this person to have the following critical competencies from the ASTD competency study:

1. *Adult Learning Understanding.* See earlier description.
2. *Cost-Benefit Analysis Skill.* Assessing alternatives in terms of their financial, psychological, and strategic advantages and disadvantages.
3. *Data Reduction Skill.* Scanning, synthesizing, and drawing conclusions from data.
4. *Delegation Skill.* Assigning task responsibility and authority to others.
5. *Facilities Skill.* Planning and coordinating logistics in an efficient and cost-effective manner.
6. *Futuring Skill.* Projecting trends and visualizing possible and probable futures and their implications.
7. *Negotiation Skill.* Securing win-win agreements while successfully representing a special interest in a decision situation.
10. *Organization Behavior Understanding.* Seeing organizations as dynamic, political, economic, and social systems which have multiple goals; using this larger perspective as a framework for understanding and influencing events and change.
11. *Organization Understanding.* Knowing the strategy, structure, power networks, financial position, and systems of a specific organization.
12. *Personnel/HR Field Understanding.* Understanding issues and practices in other HR areas such as selection and staffing and human resource planning.
13. *Relationship Versatility.* See earlier description.
14. *Research Skills.* Selecting, developing, and using methodologies and statistical and data collection techniques for a formal inquiry.
15. *Training and Development Field Understanding.* See earlier description.
16. *Training and Development Techniques' Understanding.* See earlier description.
17. *Writing Skills.* See earlier description.

As you can see, many of the competencies required for the Instructor/Program Designer are also needed by the Manager/Needs Analyst. This is because those who manage the training function need this broader mix of knowledges and skills to be able to judge whether or not their subordinates are doing their work according to acceptable HRD practices. Those who manage a training department and have little knowledge of HRD practices are at a distinct disadvantage.

Another common situation where a broad base of competencies is needed is in the one-person department. At one moment this individual may find himself writing next year's department plans and an hour later could be planning the learning design of a product information workshop scheduled for two weeks away.

Now that you have seen some typical sales training jobs and their competency requirements, you should do further analysis of your own situation. Given the training roles that are needed in your organization, what do you think are the most critical competencies needed? Using the "Pretraining Assessment of Sales Trainer Competency," below, you can conduct an assessment of your training needs. When completed, the profile you create will be useful in assessing the qualifications of job candidates and in planning developmental experiences for the new trainer(s).

Table 5–1 Pretraining assessment of sales trainer competency

The Competency	Job Importance			Current Level of Expertise			
	N/A	Lo	Hi	None	Basic	Intermed.	Advanced
Adult Learning Understanding							
A/V Skill							
Career Development Knowledge							
Competency Identification Skills							
Computer Competence							
Counseling Skill							
Data Reduction Skill							

Table 5–1 Con't.

The Competency	Job Importance			Current Level of Expertise			
	N/A	Lo	Hi	None	Basic	Intermed.	Advanced
Delegation Skill							
Facilities Skill							
Feedback Skill							
Futuring Skill							
Group Process Skills							
Industry Understanding							
Intellectual Versatility							
Library Skill							
Model-Building Skill							
Negotiating Skill							
Objectives Preparation Skill							
Organization Behavior Understanding							
Organization Under-standing							
Performance Observa-tion Skills							
Personnel/HR Field Un-derstanding							
Presentation Skills							
Questioning Skills							
Records Management Skill							
Relationship Versatility							
Research Skill							
T&D Techniques' Un-derstanding							
Writing Skills							

Note: For a complete description of behaviors related to the Basic, Intermediate, and Advanced levels refer to the complete ASDT competency study, *Models of Excellence.*

HOW TO PREPARE THE NEW SALES TRAINER _____

As with first-time entry to any other technical job, the new sales trainer needs to be adequately prepared too. There is too much at stake for the organization and the individual to leave this to chance.

Five Trainer Preparation Options to Consider

Since you now know from your analysis the role(s) your trainer will perform, and the most critical competencies in your environment, you need to consider how you will prepare the new trainer. The development you provide will likely range from quite general training concepts and skills to very specific leader training for programs you have purchased from outside suppliers.

There are many possible sources and suppliers of general training concepts and skills. Be sure to know what you want before committing to a course of action. It's very common to see flyers and brochures come through the mail all offering "Train the Trainer" courses. Yet, when you examine the content and skills they offer, they may vary considerably. Following are samples of the kind of courses you are likely to see offered through the many sources:

- "Beginning Classroom Instruction"
- "Designing and Writing Effective Training Programs"
- "Advanced Classroom Techniques"
- "Group Process Facilitation"
- "Effective On-the-Job Training"
- "Teaching Technical Subjects"
- "Managing a Training Department"

Your best bet in choosing courses would be to use your competency analysis as the criteria. Obviously, the more sales focused they are, the more relevant. Following are several trainer preparation options you may want to consider:

- New trainer workshops conducted by your local chapter of ASTD and national programs offered by the Sales and Marketing Professional Practice Area of ASTD. Information about these programs is available through the ASTD national office at (703) 683-8100, 1630 Duke Street, Box 1443, Alexandria, VA 22313.
- The "Train the Trainer" course offered periodically by the National Society of Sales Training Executives, whose national office may be reached at (407) 322-3364, 203 E. Third Street, Sanford, FL 32771.

- The trainer development programs (often several in a series) offered by many colleges and universities. An example are the programs (short term, not regular courses) of the University of Michigan's Management Education Center at Ann Arbor.

- Nationally known seminar/workshop vendors like Practical Management Associates, Inc., American Management Association, and Friesen Kay, and Associates, who offer basic and advanced trainer preparation courses. You can learn more about these kinds of resources by consulting the advertisements in training periodicals, reviewing direct-mail advertising brochures, and accessing ASTD's "TRAINET" database on training programs.

- Self-study programs and trainer reference books on a variety of subjects. The ASTD HRD book list is an outstanding source for trainers to begin looking for printed resources. From "performance analysis" to "test construction," you will find a wide variety of titles covering virtually everything the new trainer needs to know. One ASTD resource of great value is *Sales and Marketing Training Resources in Print*, prepared by Leslie Kelly. It lists books, tapes, journals, films, video, and periodical abstracts.

Sample Training Plan for a New Trainer

Suppose that you are soon about to begin work as a new sales trainer, or you are about to replace one who is due for rotation in your organization. Let us develop a sample training plan for a trainer who is to perform the roles of "Instructor/Program Designer." In completing this plan, you will help assure that you or another gets off to a good start and learns the most critical competencies required for this important assignment. Realistically, with course development and teaching responsibilities, such a plan would take at least a year to accomplish:

- *Competency Model*. Using Table 5-1, identify the most critical competencies you want the new person to be able to perform. As guidelines for coaching the new person, use the examples in the study of the three levels of expertise for each competency. Prioritize the competencies for development purposes.

- *Presentation Skills*. Attend a three-day "skills" course which focuses on basic classroom presentation techniques. These are generally widely available.

- *Leader Training*. If your organization uses training programs sold by a training supplier that requires trainer certification you would want to get the new person leader trained in as many courses as are appropriate.

- *Instructional Design*. After the initial flurry of presentation courses, it would be appropriate to give the new person solid grounding in the basics of instructional design. These are generally widely available.

Once the new sales trainer is on the job we have the responsibility to monitor and support his or her development. We need to give the appropriate feedback, coaching, and support necessary for continued growth.

SUMMARY

As we have seen, the sales trainer is important to our organizations. They can have a great influence on the productivity and satisfaction of salespeople. Depending on circumstances, the sales trainer may perform a variety of roles such as manager, instructor, instructional designer, and needs analyst. And given these roles, we would expect our sales trainer to have a mix of competencies appropriate for our situation. The ASTD Competency Study is a key resource in understanding more about trainer roles and competencies.

When we select the new sales trainer, we know there are both pros and cons when it comes to choosing either the "career" trainer or the "passthrough" trainer. The important thing is that the sales trainer be up to date and responsive to the needs of the sales organization. There are many options to prepare the new sales trainer, including public and private workshops, seminars, and printed resources. Regardless of developmental need, we should outline the kind of competencies we want and support the new trainer with the appropriate feedback and coaching.

REFERENCES

Geber, Beverly. "Who Should Do the Sales Training," *Training*, May 1987, pp. 69–76.

Kelly, Leslie. *Sales and Marketing Training Resources in Print*. Washington, DC: Sales and Marketing Professional Practice Area, ASTD, Spring 1987 (mimeographed).

McLagan, Patricia A., and Richard C. McCullough. *Models for Excellence: The Conclusions and Recommendations of the ASTD Training and Development Study*. Washington, DC: ASTD, 1983.

Mitchell, Frank G. "Selecting the Training Department Staff." In *The Management of Sales Training*, ed. Jared F. Harrison. Orlando, FL: National Society of Sales Training Executives, 1983.

section two ———————————————

PROGRAM DEVELOPMENT AND ADMINISTRATION

Chapter 6   How Adults Learn: Instructional Guidelines for Sales Training

Chapter 7   Performance Systems Design: How to Develop an Instructional System in Eight Steps

Chapter 8   Determining Sales Training Needs

Chapter 9   The Big Picture: How to Create a Theme for Your Sales Training Program

Chapter 10   Developing and Writing Successful Sales Training Programs

Chapter 11   The Learning Environment and Its Impact on Your Sales Training

Chapter 12   Administering Sales Training

chapter six

HOW ADULTS LEARN: INSTRUCTIONAL GUIDELINES FOR SALES TRAINING

W. FRANKLIN SPIKES III

W. Franklin Spikes III is associate dean and associate professor of business administration in the School of Business Administration at the University of Missouri-St. Louis. He is responsible for managing ongoing management development and supervisory training activities for a variety of corporate clients of the university. Previously, Dr. Spikes was manager of Sales Training Strategy & Marketing for Southwestern Bell publications. He has written extensively in the field of training administration and adult and continuing education. Spikes received his doctorate from Northern Illinois University and has performed postdoctoral study at the University of Texas and at Harvard University.

This chapter examines two of the more important themes of adult learning: the characteristics of the adult learner, including an introduction to the ideas surrounding self-directed learning and learning styles, and the concept of andragogy, that is, the art and science of helping adults learn.

American corporations are spending billions of dollars on training new and existing employees. Retraining of workers to assume and perform new roles in an increasingly complex work environment is becoming an ever-more-important aspect of today's business world. In one nationwide study it was estimated that in 1987 more than 38 million American workers received "formal, employer sponsored training."[1] Other research has indicated that more than 95 percent of the adult population is annually engaged in some type of learning project.[2] In many colleges and universities adult students outnumber traditional-aged students. At the same time adult women are returning to school at an ever-increasing rate. Another study, performed by the College Board, has found that more than half of the adults involved in the research had studied at least two topics within a year.[3]

With this increase in participation and level of investment comes a clear and undeniable need on the part of educators and trainers to understand how adults learn. Over the years a great deal of information has become available about the adult learning process, instructional techniques that can be used with adult learners and other matters related to teaching and training adults.

---

[1] C. Lee, "Where the Training Dollars Go," *Training*, 24 (10) (1987), pp. 51–66.

[2] A. M. Tough, *The Adults' Learning Projects* (Toronto: Ontario Institute for Studies in Education, 1971).

[3] C. Aslanian and H. M. Brickell, *Americans in Transition: Life Changes as Reasons for Adult Learning* (Princeton, NJ: The College Board, 1980).

ADULT LEARNING CHARACTERISTICS _____

Six Ways That Life-Change Events Contribute to Learning

In an article in TRAINING magazine entitled "30 Things We Know for Sure About Adult Learning,"[4] Ron and Susan Zemke wrote about those elements concerning adult learning that are both present in the current common body of knowledge and upon which some agreement exists. They subdivided these "30 Things" into three categories of information about: (1) adult learners and their motivation, (2) designing curriculum for adults, and (3) working with adults in the classroom. Space limitations obviously prohibit reproducing this article in its entirety here. However, it is useful to highlight the more significant portion of the information that is found in it, as this work provides a very useful place to begin a discussion of the characteristics of adult learners.

Zemke and Zemke have suggested that there are six concepts that are commonly held about adult learning and motivation:

1. Adults seek out learning experiences to cope with special life-change events. Marriage, divorce, a new job, a promotion, being fired, retiring, losing a loved one, and moving to a new city are examples.

    For example, new sales people are eager to learn. They are like sponges, but need to learn at a pace that will ensure long term retention.

2. The more life-change events an adult encounters, the more likely he or she is to seek out learning opportunities. Just as stress increases as life-change events accumulate, the motivation to cope with change through engagement in a learning experience increases. Since the people who most frequently seek out learning opportunities are people who have the most overall years of education, it is reasonable to guess that for many of us learning is a coping response to significant change.

    For example, the introduction of new products produces high stress with new sales people so training program content needs to accommodate their high need for control and understanding.

3. The learning experiences adults seek out on their own are directly related—at least in their own perception—to the life-change events that triggered the seeking. Therefore, if 80 percent of the change being encountered is work related, then 80 percent of the learning experiences sought should be work related.

---

[4] TRAINING magazine, Vol. 6 (1981). Reprinted with permission. R. Zemke and S. Zemke, "30 Things We Know for Sure About Adult Learning," pp. 115–117.

For example, the advent of computers and modems for field sales staffs in recent years has triggered a demand for extensive computer training.

4. Adults are generally willing to engage in learning experiences before, after, or even during the life-change event. Once convinced that the change is a certainty, adults will engage in any learning that promises to help them cope with the transition.

   Sales people will do anything necessary to retain their air of authority in product knowledge. However, early notice of impending change, the type of training that will take place, and follow-on support services mean a great deal to their success.

5. Although adults have been found to engage in learning for a variety of reasons—job advancement, pleasure, love of learning, and so on—it is equally true that for most adults, learning is not its own reward. Adults who are motivated to seek out a learning experience do so primarily because they have a use for the knowledge or skill being sought. Learning is a means to an end, not an end in itself.

   Because every minute away from one's sales territory may mean a loss of money, training must be relevant, focused, skill building, and income enhancing for many sales people. How to do his or her job better is a real important feature of training for a sales person.

6. Increasing or maintaining one's sense of self-esteem and pleasure are strong secondary motivators for engaging in learning experiences. Having a new skill and enriching current knowledge can be both, depending on the individual's personal perceptions.[5] Because successful sales people have big eyes, all training needs to enhance a sales person's public image and reinforce highly successful skills.

It becomes obvious from the preceding information that one of the major characteristics of adult learners is the strong and close tie that exists between changes in life events and the desire to take part in learning activities. Aslanian and Brickell[6] describe this situation in the following way in reporting on their research:

> Regardless of their demographic characteristics, almost all the adult learners interviewed pointed to their own changing circumstances as their reason for learning. . . . it is being in transition from one status in life to another that causes most adults to learn.

---

[5] Ibid.

[6] Aslanian and Brickell, *Americans in Transition.* pp. 161–62.

As life circumstances change, individuals experience role transitions which require learning to take place in order for them to adapt effectively to their new role. In such instances there is both a clear cause-and-effect relationship between these three elements of a person's life and a direct correlation between an adult's life circumstances and the degree and extent of learning that takes place in his or her life.

New sales managers are particularly concerned about training for their new responsibilities. Often they have been very successful field sales people, earning substantial incomes, and enjoying numerous sales awards, and the company feels that if they can "clone" themselves or teach others their successful sales techniques, the company will gain a great deal. Unfortunately many training programs don't focus on the major transition issues which include moving from a technical line position to a people manager position. An ill conceived program for training new sales managers can send the new managers scampering for outside help or, even worse, be responsible for their failure. Such failures lead to some very talented people leaving companies to return to field sales positions.

Pragmatism and Learning Orientation

A second important characteristic of adult learners concerns the pragmatism with which they approach a learning experience. Over the years evidence has shown that adults participate in learning activities for various reasons. In *The Inquiring Mind*,[7] Cyril Houle studied a small group of adult students to determine why they participated in learning activities. In this first work of its kind, he identified three types of learners:

1. Goal-Oriented Learners—those who use education as a means of accomplishing fairly clear-cut objectives.

   Good sales people might be highly goal oriented. A 10 percent increase in sales translates into learning any method that will allow them to sharpen their skills to accomplish this.

2. Activity-Oriented Learners—those who take part because they find in the circumstances of learning a meaning which has no necessary connection, and often no connection at all, with the content or announced purposes of the activity.

   Sales people who are activity oriented might find they sharpen their skills best by "doing" and have others observing, evaluating, and feeding back. Activity alone stimulates learning.

---

[7] C. P. Houle, *The Inquiring Mind* (Madison: University of Wisconsin Press, 1961).

3. Learning-Oriented—those who seek knowledge for its own sake.

Many sales people might take training, some unrelated to sales, to stimulate their creative thinking processes, to pick up one new idea, to broaden their perspective, or to better understand a particular aspect of business. These individuals are highly oriented towards personal growth.

A good amount of research has taken place since this trifold typology was proposed in the early 1960s. While Houle's notion has been modified and to some extent expanded over time, its essence still holds true. Moreover, it lends credence to the concept that for the most part learning becomes a means to an end for many adults. Adults learn because they have a goal to achieve or perhaps because they enjoy the activities associated with learning. Learning for learning's sake, now as then, still comes in third place in a three-horse race of reasons why adults participate in educational activities.

How Experience Contributes to Self-worth and Desire to Learn

A third characteristic of adult learners concerns the individual experience that each learner brings to the classroom. Judy Arin-Krupp[8] has suggested that "no adult is the same as any other adult, because each has a unique past and a unique perception of that past."

It is evident from both years of practice and evidence in the literature that the role of experience of adult learners is an important factor in how they approach a learning experience. Many salespeople see their yearly sales awards meetings and training sessions as a catalyst for the coming year. Ideas and techniques are exchanged, new approaches are introduced, the latest sales and marketing are reviewed. Although exhausted from long, jam packed days, they come away renewed, refreshed, and motivated.

More specifically,

> Adults obviously enjoy the potential benefits and drawbacks of more accumulated life experience than children. Much of this experience is qualitatively different from that of children.[9]

Eli Lilly based in Indianapolis approaches the replacement of sales people very carefully. Often the new person will work on territory with the present

[8] J. Arin-Krupp, "A Holistic View of Adult Learners," in *Methods and Materials in Adult and Continuing Education*, edited by C. Klevins (Los Angeles: Klevins, 1987), pp. 230–241.

[9] R. M. Smith, *Learning How to Learn: Applied Theory for Adults* (New York: Cambridge Publishers, 1982), p. 40.

sales person for up to 18 months. The new person thus has the cumulative experience of the seasoned person for a productive period before taking over.

It is clear that as an adult grows, ages, and learns, the experience gained from these activities affects the way in which he or she views new learning experiences. It forms the basis upon which new learning is built and often prohibits new learning from occurring. By the mere passage of time, an adult's life will include a wider variety of life experiences than will a child's. It is safe to say that an adult does not leave these experiences at the classroom door. Like any other individual physiological or psychological characteristic, past experience accompanies the adult student in any learning activity. In the same way, the self-concept of an adult is clearly related to past experience. Thus learning can easily be a vehicle which can build up or threaten an individual's feeling of self-worth.

A final thought about the importance of the relationship of experience to learning is appropriate here. Simply put, the experience of adult learners in the classroom is perhaps one of the best resources for learning that is available. Carefully utilized, this rich resource can dramatically and positively affect most types of learning activities. Conversely, as an adult can be viewed as a sum of his or her experiences, by denying the merit and value of these experiences, the adult himself or herself feels a clear sense of denial and loss of self-worth. Most sales people learn as much from one another in a sales training session as they do from the sales trainers. Such learning is often unplanned and unrecognized. However, programs need to allow time for such informal learning.

Participation in Adult Learning Activities

A number of studies have looked at the patterns of participation among adult learners in educational activities. As early as 1965 Johnstone and Rivera[10] began to examine participation patterns of adults. During the 1970s the Commission on Non-Traditional Study presented the results of their analysis of participation in adult learning programs.[11] They sought to determine why people wanted to further their education, and if they did not, what factors prohibited them from doing so. Even though the data are now more than ten years old, they provide an interesting and useful view of why adults do and do not continue their education.

Tables 6–1 and 6–2 show the findings of this research.

[10] J. W. C. Johnstone and R. J. Rivera, *Volunteers for Learning* (Chicago: Aldine, 1965).

[11] A. Carp, R. Peterson, and P. Roelfs, "Adult Learning Interests and Experiences," in K. P. Cross, J. R. Valley, and Associates, as found in K. P. Cross, *Adults as Learners* (San Francisco: Jossey-Bass, 1981), pp. 89,99.

Table 6–1 Reasons for learning

Reasons	Percentage of Would-be Learners Checking "Very Important"	Percentage of Learners Checking Why They Participated
Knowledge		
Become better informed	56%	55%
Satisfy curiosity	35	32
Personal goals		
Get new job	25	18
Advance in present job	17	25
Attain degree	21	14
Community goals		
Understand community problems	17	9
Become better citizen	26	11
Work for solutions to problems	16	9
Religious goals		
Serve church	12	19
Further spiritual well-being	19	13
Social goals		
Meet new people	19	18
Feel sense of belonging	20	9
Escape goals		
Get away from routine	19	19
Get away from personal problems	11	7
Obligation fulfillment		
Meet educational standards	13	4
Satisfy employer	24	27
Personal fulfillment		
Be better parent, spouse	30	19
Become happier person	37	26
Cultural knowledge		
Study own culture	14	8
Other reasons	4	2
No response or other response	14	3

Source: A. Carp, R. Peterson and P. Roelfs, "Adult Learning Interests and Experiences," in K. P. Cross, J. R. Valley and Associates, as found in K. P. Cross, *Adults As Learners* (San Francisco, Jossey-Bass, 1981), p. 89.

Table 6–2 Perceived barriers to learning

Barriers	Percentage of Potential Learners
Situational barriers	
Cost, including tuition, books, child care, and so on	53%
Not enough time	46
Home responsibilities	32
Job responsibilities	28
No child care	11
No transportation	8
No place to study or practice	7
Friends or family don't like the idea	3
Institutional barriers	
Don't want to go to school full time	35
Amount of time required to complete program	21
Courses aren't scheduled when I can attend	16
No information about offerings	16
Strict attendance requirements	15
Courses I want don't seem to be available	12
Too much red tape in getting enrolled	10
Don't meet requirements	6
No way to get credit or a degree	5
Dispositional barriers	
Afraid that I'm too old to begin	17
Low grades in past, not confident of my ability	12
Not enough energy and stamina	9
Don't enjoy studying	9
Tired of school, tired of classrooms	6
Don't know what to learn or what it would lead to	5
Hesitate to seem too ambitious	3

Source: A. Carp, R. Peterson and P. Roelfs, "Adult Learning Interests and Experiences," in K. P. Cross, J. R. Valley and Associates, as found in K. P. Cross, *Adults as Learners* (San Francisco, Jossey-Bass, 1981), p. 99.

Boiled down to its essence, this study found that a high relationship existed among job, career, and vocationally related educational programs and the respondents' expressed willingness to participate in such activities. More than one-third (35 percent) of the participants indicated that they participated in "vocationally oriented" training programs. They also indicated that the acquisition of practical and useful knowledge was a primary motivator of furthering their education. More than half (56 percent) indicated that furthering their knowledge was the most important reason for pursuing additional educational opportunities.

The implications of these data for the sales trainer are clear and can be simply stated. Training that focuses on knowledge and skill acquisition, training that is practically oriented, training that is able to show a clear relationship to job-related activities will attract (and perhaps retain) adult learners more readily than will training programs that are not developed in such a focused manner.

Other Considerations Concerning Adult Learning

At least three other major issues should be considered when beginning to develop an understanding of the characteristics of adult learners and how they learn. The first two of these issues, self-directed learning and learning styles, will be discussed briefly in the remainder of this portion of this chapter. The other, the concept of andragogy, its distinctive concepts and characteristics, will be presented in the next section of this chapter.

Self-directed Learning. In recent years there has been an emerging interest in the concept of self-directed learning among adults. Traditionally, learning has been thought of as an institutionally based phenomenon. Students go to school where the learning process is directed and controlled. Employees attend training programs where their learning is planned and directed by a trainer or teacher. Learning thus becomes inexorably intertwined with institutions.

In the early 1970s the idea of examining the extent and scope of noninstitutionally based learning was put forth by a Canadian researcher, Alan Tough. In his study he developed the notion of self-directed learning projects, that is, those activities which are planned, controlled, and implemented by an adult with regard to learning a skill, acquiring a concept, or increasing knowledge about a given topic or topics. This research reported that a staggering amount of learning goes on outside the traditionally institutionally oriented environment.[12]

While there is a continuing discussion among researchers and theoreticians about the nature of self-directed learning and how to define and describe this concept, it is clear that adults do not just depend on institutions as sources of learning. Thought of as "individuals who are able to plan, initiate, and evaluate their own learning experiences with or without the assistance of others,"[13] self-directed learners clearly make up a significant portion of today's adults who are continuing their education. The learning that takes place among these adults often is viewed as a specific process having certain steps through which a learner moves or perhaps in which certain responsibilities are adopted by the learner rather than an institution.[14] Still others have discussed self-directed learning

---

[12] A. M. Tough, *The Adults' Learning Projects*, 2nd ed. (Austin, TX: Learning Concepts, 1979).

[13] R. S. Cafarella and E. P. Caferella, "Self-directedness and Learning Contracts in Adult Education," *Adult Education Quarterly*, 36(4) (Summer 1986), pp. 226–234.

[14] M. S. Knowles, *Self-directed Learning: A Guide for Learners and Teachers* (Chicago: Follett, 1975).

in relationship to selected aspects of personality development. Oddi[15] has suggested that by examining personality characteristics of adults and linking them to the process of self-directed learning, a greater understanding of this concept could be achieved.

Sales training programs have shifted to a number of self-directed learning programs. Computer-based programs with topics such as product knowledge, financial analysis, new forms and reporting, generic sales letters and order samples are quickly becoming the norm.

Companies are even producing in house video tapes that are sent to field sales people to supplement training manuals.

With the explosion of new technology on the scene in the past few years companies have been able to produce very innovative training programs that are self-directed.

Regardless of which view is taken or perhaps which conceptual foundation is utilized to think about the process of self-directed learning, it is crucial for educators and trainers to remember that learning does not just take place in the classroom. Learning is not limited to the eight-hour day in which trainees learn about products or increase their sales skills and competencies. Rather, the evidence is clear that more adults will take responsibility for some form of their learning during their lifetime. They will bring this desire for self-direction into the classroom. They will also bring the information gained in this process with them as well. Trainers must recognize the significance of the self-directing learning process and realize that the employees they face on a daily basis have not only the capability but the desire and intent to seek other sources and procedures for their own individual growth and learning. Alan Tough[16] has perhaps encapsualized the notion of the adult learner in relation to self-directed learning activities when he said

> The adult learner of the future will be highly competent in deciding what to learn and planning and arranging his own learning. He will successfully diagnose and solve almost any problem or difficulty that arises. He will obtain appropriate help competently and quickly, but only when necessary.

Many companies are developing sales training libraries to accommodate a tremendous number of varying learning needs.

Three Major Learning Styles. In addition to the interest in self-directed learning among trainers and educators of adults, a great deal of information about

[15] L. F. Oddi, "Perspectives on Self-directed Learning," *Adult Education Quarterly*, 38(1) (Fall 1987), pp. 21–30.

[16] Tough, *The Adults' Learning Projects*, appears in Smith, 1982, p. 94.

adult learning styles is now available in the literature. Bob Smith in his book, *Learning How to Learn—Applied Theory for Adults*, defined learning style as "the individual's characteristic ways of processing information, feeling, and behaving in learning situations."[17] In this same work Smith described the relationship of an individual's learning style to the total learning process. He suggested that there are three factors, an individual's learning style being the first of these, that should be examined when attempting to understand the learning process. The other two elements that are related to learning style are *needs*, that is, "what people need to know and be able to do for success in learning"[18] and *training*, that is, "organized activity directed toward increasing competencies in learning."[19]

Needs assessment and training design are dealt with elsewhere in this book. However, a reference is made to these concepts here so that the reader can begin to see how these three concepts are related to each other as well as to learning styles.

There are three major components of an individual's learning style. They are (1) cognitive factors, (2) affective factors, and (3) environmental factors. Each of these elements affects an adult's learning in some way. Each of these elements also has subelements or categories, some of which are described in the paragraphs that follow.

Cognitive factors: how to "think through" your problem. Cognitive factors are those "patterns of perceiving, remembering, thinking and problem solving"[20] that allow an individual to think through and cope with issues in his or her environment. Several indices can be measured to determine the nature of an adult's cognitive orientation. These include

- *Field independence versus field dependence.* This is the tendency of a learner to look at an issue from the viewpoint which looks at it in pieces (field independent) versus looking at the same issue from the total perspective (field dependent).

    Some sales people can focus in on corporate needs easily. Others are locked into the perspective of their own territory. All will learn easier if that perspective is known.

- *Conceptualizing and categorizing.* An adult who has a categorizing style is one who uses concepts of similarity and puts bits of information together. Someone with a different conceptual style, on the other hand, acts on the

---

[17] Smith, *Learning How to Learn*.

[18] Ibid.

[19] Ibid.

[20] Ibid.

basis of external and physical attributes of information. Kirby[21] has described the differences between these two elements of style in this way:

> Confronted with three items—meat, cheese and bread . . . the [categorizing] person might describe them as a sandwich, while the [conceptualizing] person might call them edibles.[22]

A categorizing sales person would need to know in introducing a product change what's the difference with the old product and what's the change with the new product. A conceptual sales person, on the other hand, would be more interested in the functional aspects of a product and would learn about it accordingly.

- *Reflectivity versus impulsivity.* This dimension translates simply to a matter of accuracy (reflectivity) versus speed (impulsivity). Someone with a reflective style will respond in a carefully thought out and cautious manner while an impulsive person is more of a risk taker with responses being given more quickly.

- *Experiencing and organizing information*—People tend to think about and organize information in three ways: (1) A kinesthetic orientation, that is, one who learns through physical experiences. These sales people need to see, touch and use a product; (2) A visual orientation, that is, someone who learns through figurative or spatial thinking. These sales people need notebook materials and visual reinforcement. Slides and overhead transparencies are valuable; and (3) An auditory orientation, that is, learning through verbally related thinking. These sales people can have someone describe an item or process and understand/learn about it.

Failure to recognize this last time can result in a program that misses the mark. Most sales training programs contain a real balance of these elements.

Affective and environmental considerations. The second major are of consideration that is related to learning style assessment is related to affective considerations. In examining this dimension of style such things as the need for structure and authority versus autonomy and collaboration, and expectations and motivation, of the learner must be examined.

Environmental issues. The third and final component of learning style relates to environmental issues. In this case the environment can be viewed from the perspective of something as simple to control as the light and temperature

---

[21] P. Kirby, *Cognitive Style, Learning Style and Transfer Skill Acquisition* (Columbus, OH: National Center for Research in Vocational Education, 1979).

[22] Smith, *Learning How to Learn.*

of the training environment to factors such as the need for affiliation or individuality desired by the learner.

Ultimately the concept of learning-style assessment becomes a useful one when planning and conducting training programs. It is safe to say that programs that utilize information about individual learning styles can clearly be more effective than are similar programs that do not consider such participant data. Such information has been and can continue to be utilized for such purposes as placement of participants, diagnosis of learning problems, program development and planning, selection of training methodologies, and training environment analysis and development.

ANDRAGOGY: THE ART OF TEACHING ADULTS

Recently I had the opportunity to conduct a multipart training program for employees of a credit union associated with one of the nation's *Fortune* 500 organizations. A part of this program was designed to help first-line supervisors as well as their managers understand how adults learn. Ultimately these employees were to use this information to conduct on the job training sessions with their subordinates.

After the second session of the program, a long-time employee who had been very quiet yet involved in her own way in the first session of the program came to me and said, "I've been studying the materials that you gave us last time. I've read all of them. I've gone to the dictionary, several dictionaries in fact, and I still don't know what this andragogy stuff is. Can you please, please tell me!"

The concluding section of this chapter is designed to answer the question of what is this "andragogy stuff" anyway, both in terms of describing some of the more important aspects of this concept as well as discussing some of the ways that this notion impacts upon both trainees and trainers.

The Concept of Andragogy in Sales Training

Malcolm Knowles has generally been given credit as being the "father" of the concept of andragogy in the United States. Over the last 20 years he has written extensively about this concept.[23-27] Over time he has applied this frame-

---

[23] M. S. Knowles, "Andragogy Not Pedagogy," *Adult Leadership*, 16 (1968), pp. 350–352, 386.

[24] M. S. Knowles, *The Modern Practice of Adult Education* (New York: Association Press, 1970).

[25] M.S. Knowles, *The Adult Learner: A Neglected Species* (Houston, TX: Gulf, 1973).

[26] M.S. Knowles, "Andragogy Revisited, Part II," *Adult Education*, 30 (1979), pp. 52–53.

[27] M. S. Knowles, *Andragogy in Action* (San Francisco: Jossey-Bass, 1984).

word to a variety of settings and has lectured extensively on its implementation in a variety of instructional settings. In his work *The Modern Practice of Adult Education: Andragogy Versus Pedagogy*, Knowles defined andragogy as being the "art and science of helping adults to learn." In this work the concept of andragogy was compared and contrasted with pedagogy, "the art and science of teaching children."[28]

Recently the history of this concept has been examined more thoroughly with its derivation being traced back to Germany in the 1830s.[29] Over the last six decades evidence exists that the idea of andragogy has appeared in a variety of contexts. These include references to it in Europe in 1921,[30] in America in 1926,[31] and again in 1927,[32,33] and in the mid-1960s in the United States.[34] The issue is not who came first, Knowles, his European colleagues, or his American predecessors. Rather the more important concept is that the teaching of adults using some of the elements espoused by Knowles is an old one which has a history of successful practice over time and in various settings.

Despite this history, some scholars have argued that the idea of andragogy as a theory for basing instructional strategies and practices is not a sound one. Since it first appeared in Knowles's works, many others have discussed its merits from both a theroretical and practical point of view.

Some suggest that the idea of andragogy is not really a sound theory,[35–38] that it has not been well researched,[39] or that it is not clearly defined.[40] Despite

---

[28] Knowles, *The Modern Practice of Adult Education*.

[29] J. Davenport, "Is There Anyway Out of the Andragogy Morass?" *Lifelong Learning: An Omnibus of Practice and Research*, 11(3) (1987), pp. 17–21.

[30] Ibid.

[31] E. Lindeman, *Worker's Education*, 1926, cited in D. W. Stewart, "Perspectives," *Lifelong Learning: An Omnibus of Practice and Research*, 9(2) (1986).

[32] M. Anderson and E. Lindeman, *Education Through Experience* (New York: Workers Education Bureau, 1927(a)).

[33] M. Anderson and E. Lindeman, "Education Through Experience," *Workers Education: A Quarterly Journal*, 4 (1927(b)) pp. 33–34.

[34] Knowles, *Andragogy in Action*.

[35] C. O. Houle, *The Design of Education* (San Francisco: Jossey-Bass, 1972).

[36] J. London, "Review of *The Modern Practice of Adult Education: Angragogy Versus Pedagogy*," *Adult Education*, 24 (1973), pp. 60–70.

[37] J. Elias, "Andragogy Revisited," *Adult Education*, 29 (1979), pp. 252–256.

[38] A. Hartree, "Malcolm Knowles' Theory of Andragogy: A Critique," *Internal Journal of Lifelong Education*, 3 (1984), pp. 203–210.

[39] A. Clardy, "Andragogy: Adult Learning and Education at Its Best?" cited in J. Davenport, "Is There Anyway Out of the Andragogy Morass?" *Lifelong Learning: An Omnibus of Practice and Research*, 11(3) (1987), pp. 17–21.

[40] Davenport, "Is There a Way Out of the Andragogy Morass?"

this criticism in the academic community, many of the basic principles described by Knowles with regard to andragogy have been put into practice effectively. This latter issue is addressed in the remainder of this section.

Four Ways That Teaching Children and Adults Differ

Four assumptions differentiate the concept of andragogy from pedagogy.[41] These assumptions are related to changes in self-concept, the role of experience in learning, readiness to learn, and learning orientation. A description of each of these assumptions follows.

Changes in Self-concept. "This assumption is that as a person grows and matures his [her] self-concept moves from total dependence [in childhood] to one of increasing self-directedness."[42] The sales person often is highly self-directed because they work alone, unsupervised. They also often have good self-management skills.

The Role of Experience. "This assumption [suggests] that as an individual matures he accumulates an expanding reservoir of experience that causes him to become an increasingly rich resource for learning, and at the same time provides him with a broadening base to relate new learnings."[43] Sales people are very dependent on their intuition. The accumulation of experiences is key to this.

Readiness to Learn. The third assumption of andragogy suggests that "as an individual matures, this readiness to learn is decreasingly the product of his biological development . . . and is increasingly the product of the developmental tasks required for the performance of his social roles."[44] Sales people like to look good, so well-conceived training is highly valued.

Orientation to Learning. This assumption suggests that adults have a problem-centered orientation to learning with a time perspective that requires the learning of today to be put in place and used tomorrow. Sales people value well directed training that helps them do their job better.

Table 6–3 shows how the characteristics/assumptions of andragogy compare with those of pedagogy.[45]

---

[41] Knowles, *The Modern Practice of Adult Education.*

[42] Knowles, *The Adult Learner*, p. 55.

[43] Ibid., p. 56.

[44] Ibid., p. 57.

[45] Ibid., p. 58.

Table 6–3 A comparison of the assumptions of pedagogy and andragogy

	Pedagogy	Andragogy
Self-concept	Dependency	Increasing self-directiveness
Experience	Of little worth	Learners are a rich resource for learning
Readiness	Biological development social pressure	Developmental tasks of social roles
Time perspective	Postponed application	Immediacy of application
Orientation to learning	Subject centered	Problem centered

From *The Adult Learner*, Third Edition, by Malcolm Knowles. Copyright © 1984 by Gulf Publishing Company, Houston, TX. Used with permission. All rights reserved.

HOW TO APPLY ADULT LEARNING SKILLS TO YOUR SALES TRAINING PROGRAM

This chapter has presented a discussion of a number of factors related to the characteristics of adult learners, to self-directed learning, to learning styles, and to the concept of andragogy. What, then, given this base of information, are the implications for developing effective sales training programs? Several suggestions follow.

1. Sales training programs should clearly take advantage of the collective experience of the trainees. Keep in mind that to deny the worth of the learner's experience is, in effect, to deny him or her. Trainers should seek to build upon the related experience of the trainees rather than assume that they (the trainers) possess the only relevant experience to share.

2. Continuing efforts should be made to assess the learning styles of trainees and to incorporate this information into the training materials and into the classroom methodologies of sales trainers.

3. When possible, training programs should involve as many experientially based learning activities as possible. These activities should be developed on a problem-centered rather than content-centered basis.

4. Sales training programs, in the aggregate and with regard to individual learning components, should be developed so that the information learned can be immediately used.

5. Sales training programs should be offered at those times when critical life events of adults will be focused upon the content of the programs and when there is a high level of readiness to learn.

6. Ongoing efforts should be made to tie formal learning in the training classroom to the related learning that occurs in an informal, self-directed manner among sales trainees.

7. When possible sales training activities and program planning strategies should be mutually developed with front-end input from past trainees, line managers, and training staff members being used as the basis for improving learning activities.

SUMMARY

This chapter has described several facets of the adult learning process. Characteristics of adult students, learning styles, self-directed learning, and other issues have been discussed. It has further provided an examination of one of the major conceptual foundations of the field, andragogy. This information has been accompanied by charts, graphs, and tables that summarize the narrative in visual form. Finally, seven recommendations for using this information in developing effective sales training programs have been provided.

By using the information that has been offered here in concert with the information in other portions of this handbook, sales trainers and developers should be able to develop more effective training programs that make maximum use of what is known about adult learners and their learning patterns.

chapter seven ——————————————————————

PERFORMANCE SYSTEMS DESIGN: HOW TO DEVELOP AN INSTRUCTIONAL SYSTEM IN EIGHT STEPS

—————————————— *KENNETH A. McCLUNG, JR.*

Kenneth A. McClung, Jr., is a senior partner with the Instructional Design Group, Inc. (IDG), a performance technology consulting firm located in Morristown, New Jersey. Over the past 14 years he has assisted corporations in the development of more than 150 successful sales, management, and technical training programs. Prior to joining IDG, McClung had experience as an independent developer and 11 years as an Army officer, including three years on the faculty of the U.S. Military Academy at West Point. He has a B.S. degree from North Georgia College and an M.S. and Ed.D. in Instructional Technology from the University of Southern California. He is the co-author of a five-book series on microcomputers for professionals published by John Wiley and Sons.

PERFORMANCE SYSTEMS DESIGN: HOW TO DEVELOP AN INSTRUCTIONAL SYSTEM IN EIGHT STEPS

There are several goals for this chapter.

- To introduce you to the Performance System Design (PSD) process. This chapter provides an overview of the process; following chapters provide details on specific steps.
- To illustrate the importance of Performance Systems Design to the development of sales training.
- To provide you with practical tools and advice for implementing Performance System Design within your organization.

PERFORMANCE SYSTEMS DESIGN OVERVIEW

Performance Systems Design involves eight steps:

1. Analyze a problem or requirement.
2. Design a strategy/approach to solve the problem.
3. Develop materials to support the strategy.
4. Test the materials.
5. Produce/reproduce the final materials.
6. Implement the program.
7. Evaluate the results.
8. Maintain as required. . . .

. . . eight seemingly simple steps to which numerous volumes have been dedicated. Since its launch during World War II, the foregoing process has been variously referred to as

- Educational Technology
- Instructional Technology
- Performance Technology
- Instructional Design
- Instructional Systems Design
- Performance Systems Design

I have chosen to call this process by the last title, Performance Systems Design. I think it most accurately portrays the tasks which sales trainers are called upon to do in the course of their jobs.

Sure, sales training is, by definition, instructional. But what about new product introductions that are designed to motivate or inform rather than instruct? or job aids (such as documenation or checklists) that guide salespeople through a process but don't require retention of skills or knowledge? Although job aids, new product introductions, and many other facets of your job may not be instructional, they are all designed to increase human performance.

HOW PERFORMANCE SYSTEMS DESIGN CAN HELP YOUR COMPANY

Why should you and your company be concerned with PSD? The answer can be summarized in five words: quality, cost, efficiency, consistency, and satisfaction.

Quality means different things to different people. Here I have chosen to use Phil Crosby's definition: "quality is conformance to requirements." You can have a single one-page handout or a five-day video workshop; if both meet their respective requirements, they are both quality. PSD can help you determine your requirements, as well as whether you've fulfilled them.

Cost can be looked at from several perspectives:

- What are the cost and benefit of developing and implementing a program?
- What is the potential cost impact if that program doesn't meet its objectives?
- Are there potential cost savings in a noninstructional solution?

All companies want to maximize the benefits of their training dollars. PSD can help to do this.

Efficiency is measured by comparing production output with cost. By routinely applying PSD to your sales training development, you can get maximum productivity for your resource expenditures.

Consistency is very important to long-term success in sales training. You can succeed without PSD, but your chances of having consistent successes are much better with PSD.

Satisfaction is also very important. By using PSD, you can ensure that your audiences receive the performance solutions which best meet their needs. Knowing that you've successfully met these needs is a very good feeling, indeed.

THE PERFORMANCE SYSTEMS DESIGN TEAM: EIGHT KEY ROLES

PSD is a process which requires several different roles. Depending upon the size of your organization, the scope of the project, and your individual experience, you might be called upon to perform one or more of these roles:

- Project manager
- Instructional designer
- Writer/developer
- Subject-matter expert
- Production coordinator
- Production specialist
- Client
- Instructor/trainer

Let's look at the major components of each role.

PROJECT MANAGER

- Initiate/conduct/review the needs analysis.
- Develop a project plan based on the needs analysis.
- Brief each team member as to individual and group responsibilities.
- Coordinate resources.
- Review draft materials.

- Provide immediate feedback.
- Mediate any problems among team members or between the team and the client.
- Secure acceptance from client (internal or external) decision maker *at each step of the design process*.

INSTRUCTIONAL DESIGNER

- Conduct/review the needs analysis.
- Develop objectives.
- Develop a detailed performance strategy/design.
- Monitor development and production of materials to ensure that the strategy is followed.
- Test the materials to ensure that the strategy works.
- Evaluate the implemented program to determine the extent to which objectives have been met.

WRITER/DEVELOPER

- Develop the required materials (e.g., job aids, workbooks, scripts) according to the approved strategy design.
- Maintain the materials as required.

SUBJECT-MATTER EXPERT

- Provide the instructional designer and writer with content information.
- Ensure quality (as opposed to quantity) content information.
- Be available for answering questions throughout the PSD process.
- Review draft material for accuracy of content.

PRODUCTION COORDINATOR

- Get production requirements from project manager or instructional designer.
- Develop detailed production specifications.
- Monitor all aspects of production.
- Review interim and final materials (galleys, rough media, finished components) to ensure that production specifications have been met.
- Monitor distribution of finished material.

PRODUCTION SPECIALIST (ARTIST, DIRECTOR, PHOTOGRAPHER, CAMERAPERSON ETC.)

- Produce the specific media component(s) for which you are responsible.

INSTRUCTOR/TRAINER

- Provide input to the instructional designer on the audience and learning environment.
- Review design document and draft materials for relevance of strategy.
- Use the materials, exactly as specified, during the development/field test to confirm the strategy and content.
- Implement the materials as required.
- Maintain the materials as required.

CLIENT

- Request a solution to a performance problem.
- Provide a budget for project development.
- Provide a subject-matter expert(s).
- Ensure that the instructional designer/writer has priority access to the subject-matter expert(s).
- Coordinate review of draft materials.
- Provide consolidated feedback on draft materials.
- Approve materials at mutually agreed-on points in the development cycle.

HOW TO DEVELOP A PROJECT PLAN

You will require a project plan for any performance systems design project (see Figures 7–1 and 7–2). The complexity of the project plan should correlate directly to the complexity of the project itself. For a relatively simple project, your plan may be a single sheet of paper. A more complex project may include detailed charts, budgets, and resource allocations.

Keep your plan as simple as possible given your project requirements. At a minimum, any project plan should include

- Major requirements: Basic sales training
- Time line: Four phases: three days each—one per quarter
- Individuals responsible: Tom Duncan, Corporate Sales Staff
- Budget (optional): $18,000 plus staff salary costs

Figure 7–1 Project plan

Project Title _____ Project Manager _____ Start Date _____ Completion Date _____

Requirements	Time Line in Weeks/Months 1 2 3 4 5 6 7 8 9 10 11 12 13 14 15 16 17 18 19 20	Budget	Individual(s) Responsible

Analysis
  Performance Analysis
  Needs Analysis
  Task Analysis
  Report
  Review
Design
  Information Gathering
  Learning Objectives
  Instructional Strategies
  Content Outline
  Review
Materials Development
  Student Workbooks
  Job Aids
  Leader's Guides
  Scripts
  Reference Material
  Developmental Test
Production
  Art/Graphics
  Media Production
  Typesetting
Reproduction
  Print
  A/V Media
  Binders
  Tabs
  Collating Assembly
  Packaging
Implementation
  Train-the-Trainer
  Classes Scheduled
	Major requirements and subcategories will vary by type and complexity of project. Figure 7–2 provides an example of a project plan for a video training project for which the analysis has already been completed.		

Figure 7-2 Detailed project plan—video component excalibur training plan

PROJECT TITLE **EXCALIBUR END USER**   PROJECT MANAGER **K. McCLUNG**   START DATE **3/9/88**   COMPLETION DATE **6/1/88**

REQUIREMENTS	TIME LINE IN WEEKS	INDIVIDUAL(S) RESPONSIBLE
Design		
Start-Up Meeting	3/9/88	KM, GK, CZ
Data Gathering		GK, KM
Develop Treatment	In Conjunction With Development Excalibur Manual	GK
Internal Review		KM
Client Review	4/4/88	CZ
Develop Script Draft One		GK
Internal Review		KM
Client Review	4/16/88	CZ
Develop Script Draft Two		GK
Internal Review		KM
Client Review	4/21/88	CZ
Develop Final Script		GK
		KM, GK, CZ
Preproduction		
Casting		KM
Crew Arrangements		KM, JW, GK
Scout Location		PH
Design Set		PH
Develop Graphics		
Production		
Location Shoot	4/25 - 4/29/88	JW,KM,GK,CZ
Post-Production		
Pre-Edit		JW, KM
Off-Line Edit	5/6/88	JW, KM
Client Review		CZ
On-Line Edit	5/16/88	JW, KM
Deliver Masters		KM
Reproduction		
Graphic Concept	4/4/88	GK, KM
Client Review		CZ
Create & Produce Labels		IDG
Create & Produce Pkg.		IDG
Reproduce Tapes		IDG
Collate & Insert		IDG
Deliver Final Materials	6/1/88	IDG

Note: Several requirements may be concurrent.

Your plan should be developed in two stages. The first stage only includes the analysis phase. After analysis is complete, you should be able to identify the project requirements.

Very few plans survive an entire project without being changed. To make your plan workable, it should be easy to modify as factors change.

The project plan model shown in Figures 7–1 and 7–2 can easily be adapted to the needs of most sales training projects. Each of the requirement categories on this master plan can be further broken down into subplans: media production, for example, may require further delineation, into plans for audio and video production.

PDS STEP 1: ANALYZE THE PROBLEM/ REQUIREMENT

The analysis step is a critical part of the PSD process. It provides the basis from which any course, job aid, and so on, is developed. All too often, when analysis is neglected, the result is an excellent instructional product which, unfortunately, doesn't meet the real needs of the target audience. A good example is an excellent video sales training program developed by a major office equipment manufacturer. This program was intended for independent office supply dealers. Unfortunately, less than 5 percent of the dealers had VCRs at the time. Thus, the potential effect of an excellent instructional program was almost completely negated by delivery constraints. All of us have sufficient requirements that we can't afford to waste our time and money creating "instructional bric-a-brac."

On the other hand, you must also be careful to avoid "analysis paralysis." One should not become so involved in the analysis stage that the rest of the PSD process is neglected.

There are really three stages to analysis that may be conducted prior to development of materials. These are

- Performance analysis
- Needs analysis
- Task analysis

Performance Analysis: Finding the Causes—and Possible Solutions—to Problems

Performance analysis is an initial screening that isolates the cause of a performance problem and recommends a solution based on this cause. This stage is also referred to as front-end analysis. Sivasailam Thiagaijan (in a National

Society for Performance and Instruction presentation, October 1987) has identified 5 possible causes of performance problems:

- Lack of skill/knowledge—lack of product knowledge
- Lack of motivation—not motivated by commissions
- Lack of health—electronic lead tracking system has been misdirecting a significant percentage of leads.
- Lack of environmental support—slow response to inquiring for demonstration materials, letters to client
- Lack of organizational support—inability to deliver product to client/customer as promised

He has also identified 15 possible solutions which address performance problems.

- Job aids
- Training
- Education
- Interest
- Incentives
- Physical health
- Emotional health
- Tooling
- Ergonomics
- Facilities design
- Work flow
- Personnel
- Feedback systems
- Team building
- Culture change

If the solution identified by the performance analysis is one that is within the scope of your sales training organization (training, job aid, etc.), you're ready for the next analysis stage: needs analysis.

Needs Analysis: Five Key Areas to Consider

Once the performance analysis has identified a potential sales training requirement, further investigation is required. There are five major areas to be considered in your needs analysis:

- Organizational impact (what will the cost be if there is no solution)
- Audience
- Learning environment
- Recommended solution
- Constraints

The following checklist provides you with points to be considered within each area.

Sales Trainer. Needs Analysis Checklist

CLARIFICATION OF THE PROBLEM

_____Is the problem clearly described?
_____What is the potential impact of the problem on organization?
_____Are there associated problems or concerns (motivation, policy, etc.)?

AUDIENCE DESCRIPTION

_____What is the age range?
_____What is the distribution by sex?
_____What is the average education?
_____What is the reading level?
_____What is the motivational level?
_____Are there any physical limitations among the audience?
_____Does the audience possess skills/knowledge in related areas?
_____What is the total audience size?
_____What is the time available to take course, and is the course offered during the normal business day?
_____What is geographical dispersion of the audience?

LEARNING ENVIRONMENT DESCRIPTION

_____Where will the learning take place?

_____What type of facility is it?

_____What is the facility's availability?

_____What A/V equipment is available?

_____Is there a cost associated with using the facility?

_____Is there an above-average noise level within the facility?

_____Is there adequate ventilation within the facility? Does the ventilation shut down? If so, at what time?

_____What type lighting is available?

_____What is the physical size of the available facilities?

_____What is the proximity of food/lodging to the training facility?

RECOMMENDED SOLUTION

_____What are the overall performance objectives for the solution?

_____Is there a content outline?

_____What is the instructional approach (self-study, workshop)?

_____What media will be required?

_____What is the estimated length of the solution?

_____What is the estimated cost of the solution? (include development, implementation, and maintenance)

_____What is the estimated development time line?

_____Are there resource requirements (other than cost)?

CONSTRAINTS

Are there any specific constraints which could affect development or implementation of the solution? Be sure to consider each of the following:

- Time
- Budget
- People
- Product availability/stability
- Subject-matter expert availability

Task Analysis

If you discover during your needs analysis that the job(s) or procedure(s) being investigated is not completely documented, or is very complex, you may need to initiate the third, and most detailed, stage of analysis: task analysis.

A task analysis is conducted through a combination of observation, interview, and documentation review. A thorough task analysis requires a lot of time and resources.

The following checklist can be used when planning for the job task analysis as well as when developing your Job Task Analysis Report.

Task Analysis Checklist

_____Statement of scope

_____Names of people interviewed

    _____Position

    _____Length of experience with reported tasks

    _____Educational background

_____Description of observation procedure with sample worksheet (e.g., What were the criteria upon which you based your observation, and how did you consistently record your findings?)

_____Description of interview procedure with sample questionnaire

_____Description of work flow

_____Task listing, including

    _____Action verb (e.g., *telephones* client; *completes* client profile)

    _____Tools/aid required (e.g., telephone, computer, profile sheets)

    _____Level of proficiency required (e.g., obtains at least lead for every 20 calls)

PSD STEP 2: DESIGN A STRATEGY OR APPROACH TO SOLVE THE PROBLEM

Course design is begun in the needs analysis stage and completed after the initial recommendations of the needs analysis have been accepted. There are four major design components, all of which interact together:

- Objectives
- Learning approach

- Instructional strategy
- Design document

Establish Specific Objectives

Specific performance objectives should be established for all sales training. The number and specificity of these objectives is dependent upon the size and complexity of the project and the time available for development.

All objectives should be related to the attainment of a specific performance. Although there are many different ways to phrase a performance objective, each objective should contain the following four components:

- *Audience*—to whom is the objective directed?
- *Behavior*—what is to be accomplished?
- *Condition*—how will this behavior be accomplished?
- *Degree*—how is accomplishment to be measured?

This A-B-C-D format is easy to remember. Here's an example of an objective done in this form.

The sales trainees (audience) will demonstrate their knowledge of the office copier product line's competition (behavior) by responding with the appropriate office copier benefits to match any competitive product shown (condition). Sales trainees will respond correctly to at least 80 percent of competitive products (degree), for example, if the reduce/enlarge feature of the competition is mentioned, the sales trainee will present our zoom feature.

Five Major Learning Approaches Used in Sales Training

There are five major learning approaches generally used in sales training:

- Job aids
- Seminar (lecture)
- Instructor-led workshop
- Self-study
- Supervised self-study

For detailed explanations of these approaches, refer to the specific chapters devoted to these topics. As a general reference, the following checklist provides guidance in determining which learning approach is best for a particular application.

Learning Approach Checklist

APPLICATION	APPROACH
_____ Are human interaction skills needed (e.g., sales skills, customer service skills)?	Seminar Instructor-led workshop Supervised self-study
_____ Is there a need for live, immediate feedback?	Seminar Instructor-led workshop
_____ Is co-worker interaction needed?	Seminar Instructor-led workshop Supervised self-study
_____ Is the student population small?	Seminar Instructor-led workshop
_____ Does the request involve corporate changes in which there is little time to educate the field?	Seminar Job aids
_____ Is the content highly conceptual and abstract?	Seminar Instructor-led workshop
_____ Does the course involve hands-on practice with equipment (when equipment is readily available)?	Instructor-led workshop Supervised self-study Self-study Job aids
_____ Is the subject matter relatively stable?	Self-study Supervised self-study
_____ Is there large student population?	Self-study
_____ Is local control or responsibility for instruction desirable?	Self-study
_____ Is the course content procedure related (with no employee interaction need)?	Self-study
_____ Is the travel time and travel budget limited?	Self-study
_____ Does the course involve factual information such as product knowledge and forms completion?	Self-study

APPLICATION	APPROACH
_____ Is a high level of detail required?	Self-study
_____ Is workshop/seminar prework required?	Self-study
_____ Are small-group activities needed to enhance the learning (e.g., product knowledge applied to customer situations)?	Supervised self-study Instructor-led workshop
_____ Is scheduling flexibility required?	Self-study
_____ Do students need to learn at their own pace?	Self-study
_____ Are there relatively few qualified instructors to teach the content?	Self-study
_____ Are the students at varying experience and instructional levels?	Self-study
_____ Are predictable results and consistent quality a necessity?	Self-study
_____ Does the course involve a new product or performance with a long life cycle?	Instructor-led workshop
_____ Is a simulation required which has potential risks to equipment if mistakes are made?	Self-study

Note: Several requirements may be concurrent.

Instructional Strategy: How to Accomplish Your Objectives

After you've completed your objectives and determined your overall approach, you're ready to prepare your instructional strategy. An instructional strategy tells *how* you're going to accomplish your objectives. There are really two levels of strategy to be developed: a macro strategy and a micro strategy.

Macro Strategy: Looking at the Whole Picture. The macro strategy is usually one of the outputs of the needs analysis. The macro strategy looks at the entire project and determines that media and components are required, the format for each component, and how all components will interact to achieve the overall course objective.

A major part of macro strategy is determining what media to use. The following checklist should give you some guidance based upon your specific requirements. Other chapters of this book provide more details on the use of various media.

Media Selection Checklist

MEDIA	REQUIREMENT
Print	• Referability • Easy updates • A lot of detail • Portability • Low budget
Video/film	• Motivation • Visually oriented material • Role modeling • Motion
Audio	• Motivation • No visual requirement • Role modeling • Low budget
Slide/tape	• Visually oriented material • No motion • Long viewing requirements for specific visuals • Easy update • Low budget
Multiimage slide/tape	• Motivation • Large audience • High impact • Easy update
Computer-based training (CBT)	• Multiple levels of instruction • Simulation • Self-prompting • Tracking of responses • Large audience • Stable product/process

MEDIA	REQUIREMENT
Overhead transparencies	• Guide discussion
Flipcharts	• Summarize responses
	• Small-group activities
	• Quick
	• Easy
	• Low budget
Actual product	• Product demonstration

Micro Strategy: Looking at the Specific Objectives. The micro strategy should be developed concurrently with the content outline for each of the sales training objectives. It is a description of how each topic listed in the outline will be presented in the training. This strategy involves

• Intended media treatments (overhead video, slides, etc.)

• Brief description of exercises, summaries, and quizzes/test items that correspond to the objectives (case study, role play, paired practice, demonstration, discussion, lecture, etc.)

Design Document. The design document combines your objectives, strategies, and content outline in a form which you can present to your decision makers. A suggested format is:

• Introduction, including learning approach and macro strategy.

• Each performance objective (in order of presentation, if possible) with the micro strategy and content outline for that objective.

Get sign-off from your decision makers at this point before beginning actual materials development. This will ensure that the tone as well as the content is acceptable. It is much quicker, less expensive, and far less heart-breaking to change your approach at this stage than after the first draft has been completed.

Use the following checklist as a reference to help you evaluate a sales training course design.

Course Strategy Checklist

OBJECTIVES

_____Do objectives appear at the beginning of each section/module?

_____Are the objectives stated in performance terms?

_____Do objectives indicate the way to measure or identify the expected performance?

_____Do objectives define the major outcomes of the instruction?

_____Are the objectives practical within the parameters of the course?

_____Is each objective directly needed to achieve the desired learner outcomes?

CONTENT OUTLINE/LEARNING TASKS

_____Is the content outline consistent with the information in the front-end analysis?

_____Is the proposed course content organized in a logical fashion?

_____Are the topics presented in a sequence that will optimize, organize, and promote retention?

_____Is the level of sophistication of the topics consistent with the background and experience of the target population?

_____Do the learning tasks relate directly to the stated objectives?

TESTS

_____Do the test specifications and/or questions match the objectives and the proposed learning tasks?

_____Are clear and concise directions given for test completion?

_____Is test administration information provided for the instructor or course supervisor?

MEDIA

_____Do the proposed media presentations fulfill the learning objectives?

_____Does the strategy include descriptions of media interaction (when more than one medium is used)?

_____Are the media used at appropriate times within the proposed course?

_____Is each medium used in its area of strength?

PSD STEP 3: DEVELOP SUPPORT MATERIALS _____

After you've received approval of your course design, the next step in the PSD process is materials development. Although materials development generally takes more time than any other step, your detailed design document makes this step very straightforward.

Four Points to Consider When Developing Course Materials

There are four primary considerations in developing course materials:

- Subject-matter accuracy
- Instructional quality
- Writing style and clarity
- Format consistency

All these factors need to be matched against the time available for development. The old adage, "good, quick, cheap . . . pick any two," certainly applies here.

Since Chapter 10 is devoted to developing and writing sales training programs, it is not necessary to expand upon these considerations at this point. The following checklists will, however, assist you in evaluating the development of

- Instructor's guides
- Student workbooks
- Self-study materials
- Scripts

Instructor's Guide Checklist

OBJECTIVES

_____Does the Instructor's Guide provide step-by-step instructions for conducting the seminar/workshop?

_____Is an agenda provided with recommended time frames?

_____Is a course introduction included?

____Does the course introduction contain descriptions of the

- Purpose
- Audience
- Precourse requirements
- Test administration and scoring
- Equipment needed
- Preparation needed for course delivery?

____Is each module/section separated by tabs or packaged individually?

____Is the tabbed material referenced in the student notes?

____Are Student Workbook materials provided in the Instructor's Guide to allow instructors to see both instructor and student information at the same time?

____Do the directions for instructors contain enough detail for the instructor's experience level?

____Is the print for specific instructions large enough to allow it to be easily read at a glance while conducting the class?

Student Workbook Checklist

____Is an agenda provided with references to the various sections/modules?

____Is there a course introduction?

____Does the course introduction contain descriptions of the:

- Purpose
- Precourse requirements
- Equipment needed
- Organization and use of all student materials
- Objectives
- Testing information?

____Is there a table of contents?

____Does the material follow the original strategy?

____Do the activities relate to the objectives?

____Do the learning activities involve the students?

____Is there sufficient practice?

____Is there sufficient reinforcement?

____Does the text follow an easy-to-read format?

_____Are graphics/illustrations provided, when needed?

_____Is the writing clear and concise?

_____Is the required content covered?

_____Is the content presented in the appropriate sequence?

_____Is each module/section separated by tabs or packaged individually?

_____Are the materials within project specifications for

- Number of print pages
- Number of photographs
- Number of illustrations
- Print colors used?

Self-study Materials Checklist

_____Are the instructions to the students clearly defined?

_____Is there a table of contents?

_____Is there a course introduction?

_____Does the course introduction contain descriptions of course objectives, media interaction, and utilization of the self-study materials?

_____Does the text follow an easy-to-read format?

_____Is the writing clear and concise?

_____Are overviews included in the beginning of each section?

_____Do exercises and learning activities have clear and concise directions?

_____Is each objective covered by content?

_____Do the exercises and activities relate to the objectives?

_____Do solutions to exercises/activities repeat the questions and include the correct answers, with rationale, if needed?

_____Is there enough reinforcement and practice?

_____Are the directions for use of another medium (e.g., audio, video) explicit and consistent?

_____Are illustrations included to clarify and reinforce abstract concepts?

Scripts Checklist

_____Does the script indicate the content of both the audio and visual (if applicable) components?

_____Does the script include a description of the number and type of characters, and the estimated length of the presentation?

_____Are all sound effects indicated?

_____Is the location and type of music noted?

_____Is each audio segment labeled with a title and source?

_____Are the beginnings and endings of scenes indicated?

_____Does each page of the script end with a complete sentence?

_____Are directions to students indicated in the narratives (if appropriate)?

_____Is the appropriate content included in each segment/scene?

_____Is the content presented in a logical sequence?

_____Does the script support the related objective?

_____Is the scriptwriting clear, concise, and interesting?

_____Is the length of the script and number of characters consistent with course parameters?

PSD STEP 4: TEST THE MATERIALS

All instructional materials should be tested before they are finalized. This stage in the PSD process is referred to as the developmental test, field test, or formative evaluation.

The purpose of a developmental test is to "fine tune" the materials by determining

- How the sequence of content and activities work together.
- If there are any textual ambiguities.
- Student reactions to specific activities and the course as a whole.
- If students can pass course tests.
- If instructors' guides or self-study materials provide sufficient directions for conducting or completing the course.

The developmental test is *not* intended to be

- A test of the students.
- A time for clients to make their first detailed look at materials.
- A time for redefinition of the audience.

The developmental test is your final opportunity to improve the materials prior to implementation. For the test to be effective, it must be well planned and carefully executed. The following checklist will provide you with guidelines for planning and conducting a developmental test.

Developmental Test Guidelines Checklist

_____Use a representative sample of the actual student population.

_____Keep class size small enough for thorough debriefing of each student (maximum of 10–12).

_____Use one of the instructors designated to eventually teach the course.

_____Try not to use a course developer as instructor. This does not provide a fair test of the materials.

_____Replicate actual class environment to the greatest extent possible.

_____Use draft print materials and media scripts, rather than finished audio, video, or slide/tape wherever possible. This reduces cost and time in making changes.

_____Conduct the test as though it were a regular course with the exception of having students and instructors make notes about problems. A discussion of materials is not a valid test.

_____Ask students and instructors for the following feedback and encourage them to make notes on the materials.

    _____Did you have problems understanding the material?

    _____Does the material flow logically from one topic to another?

    _____Are terms adequately defined?

    _____Is wording appropriate, clear, and concise?

    _____Are exercises useful and realistic?

    _____Are directions clear and complete?

    _____Is there sufficient practice to enable you to answer test questions correctly?

    _____Does the media appropriately illustrate the narrative?

    _____Does the Instructor's Guide provide adequate instructions for conducting the course?

    _____Is the Instructor's Guide easy to follow while you are in front of the class?

_____If possible, debrief students and instructors after each section or module rather than waiting for completion of the course.

_____Debrief the students individually if possible.

_____Allow at least 1.5 times normal presentation time to allow for debriefing.

_____Allow the course developer to be an observer and debriefer.

PSD STEP 5: PRODUCE/REPRODUCE THE FINAL MATERIALS

Even the best instructionally designed materials can lose their effectiveness through problems in the production/reproduction process. Since the materials will usually leave your direct control at this point, it is very important that you plan and coordinate with the department(s) or vendor(s) who will produce/reproduce the materials.

Production is the stage in which a single master copy of all materials is made. For print this is camera-ready manuscript (referred to as "boards") with finished artwork, photographs, and cover and label designs. For audio, video, slides, or computer-based training, this is the creation of a completely edited and approved master of the audiotape, videotape, or computer program.

Reproduction is the stage in which multiple copies of the materials are made. These copies are made in the format in which they will be used. (For example, your video master may be in one-inch format for highest quality, while your reproduced copies may be in VHS format for convenience.) It is also in this stage that all print materials, including labels, are reproduced. Finally, all materials have to be collated and packaged in whatever final form is required (binders, slipcases, shrinkwrap, etc.).

There are many places to lose control in production/reproduction. For example, custom tabs and binders often take five to six weeks to produce; therefore they must be started while materials are being developed. Ask questions of someone with experience wherever possible. *Above all, start your planning early.*

You should do any required contracting and scheduling for production/reproduction as soon as you've completed your needs analysis. It is not unusual for production/reproduction to take one to two months. Having to rush this stage can double or triple your costs and often result in inferior finished materials regardless of cost.

PSD STEP 6: IMPLEMENT THE PROGRAM

Now that you've developed your materials, it's time for them to be implemented. This stage includes distribution of the materials, plus train-the-trainer and delivery for instructor-led training.

The train-the-trainer session ensures that instructors are fully qualified and prepared for a specific course. Requirements for a train-the-trainer can be broken into two phases: preparation and delivery.

Preparation Activities

When preparing for the program, the planner must be sure that

- Sufficient quantities of material/equipment have been shipped and are available at the class site.

- Instructor has reviewed materials and has rehearsed for train-the-trainer session.
- Instructor has list of attendees.
- Instructor has tested all equipment and checked materials for completeness.

Delivery Requirements

When delivering a train-the-trainer session the instructor should

- Introduce the course and explain design rationale, objectives and development constraints.
- Conduct sections of the course, explaining rationale and purpose of activities.
- Point out group dynamics needed to make course work.
- Have attendees practice, if possible.

PSD STEP 7: EVALUATE THE RESULTS

As sales trainers, we have a threefold responsibility:

- To diagnose sales performance problems.
- To develop and implement sales training and other interventions specifically designed to solve these performance problems.
- To determine the extent to which our solutions have solved the problems we originally diagnosed.

This third responsibility is evaluation. Since its purpose is evaluation of an entire program's effectiveness, it is also referred to as summative evaluation. This evaluation should provide diagnostic feedback to the participants and to management. This third responsibility is just as important as the other two, both to the participants and to those who pay the bills.

Four Major Levels of Sales Training Evaluations

There are four major levels of sales training evaluations:

- Student reaction
- Instructional effectiveness

- Job performance
- Bottom-line impact

Student reaction is the easiest to measure but doesn't necessarily correlate to the other three. Job performance and bottom-line impact are the most important and also are the most difficult to measure. Increased job performance and a corresponding impact on a company's productivity and bottom-line performance are the most significant indicators of any sales training program's success.

Succeeding chapters will provide you with specific techniques for evaluating sales training. These techniques vary widely depending upon which level of evaluation you choose to measure.

How to Achieve a Successful Evaluation

There are two keys to any successful evaluation. The first of these is the objectives around which the training program has been developed. These objectives must provide for specific, measurable behavior to be able to determine adequately whether that objective has been met.

The second key to successful evaluation is matching the evaluative instrument to the specified objectives. Be sure that your test or other evaluative instrument is actually meeting the conditions of the program's objectives. This seems self-evident, but it is very easy to develop an evaluative instrument in which the items bear little relevance to the program's objectives, especially when the objectives relate to attitudes and motivation, rather than specific skills and knowledge.

PSD STEP 8: MAINTAIN AND PERIODICALLY UPDATE

You are in a rapidly changing world. If your company is going to stay competitive, it must continually improve and expand its products and services. As a result, your sales training programs will usually require periodic updates to maintain their effectiveness.

This is a stage of the PSD process that is best initiated at the beginning of a project. During your needs analysis you should try to determine

- The life cycle of the course.
- How often, and to what extent, the contents are expected to change.
- How often these changes should be integrated into the materials.
- Who will be responsible for making the changes.

If you have this information, you can plan for, and budget, maintenance require-ments through the entire life cycle of your sales training materials. This keeps us from getting trapped with the forward-looking blinders of too many things to do and not enough resources with which to do them. Avoid having an excellent course sit on a shelf because of organizational or product changes.

It is also important that those facets of a program that are most likely to change be designed in such a way to facilitate change. For example, if you were developing a sales training curriculum for a company whose products were subject to rapid change, you might want to design any product demonstrations as sepa-rate, easily replacable, modules.

SUMMARY

This chapter has provided you with a brief introduction to how Performance Systems Design can help you provide your clients with the best sales training possible. The eight steps of PSD are:

1. Analyzing problems or requirements.
2. Designing an approach/strategy to solve the problem.
3. Developing materials to support the strategy.
4. Testing to fine-tune the materials.
5. Producing/reproducing the materials.
6. Implementing the program.
7. Evaluating the results.
8. Maintaining the program throughout its life cycle.

The checklists provided throughout this chapter will help you at each step of the PSD process; later chapters explain the steps in detail.

REFERENCES

Anderson, Robert H. *Selecting and Developing Media for Instruction*. New York: Van Nostrand Reinhold/ASTD, 1976.

Bloom, Benjamin S., ed. *Taxonomy of Educational Objectives*. New York: David McKay, 1956.

Briggs, Leslie J. *Instructional Design*. Englewood Cliffs, NJ: Educational Technology Publications, 1977.

Crosby, Philip B. *Quality Is Free*. New York: New American Library, 1979.

Davis, Robert T., Lawrence T. Alexander, and Steven L. Yelon. *Learning Systems Design*. New York: McGraw-Hill, 1974.

Dick, Walter, and Lou Carey. *The Systematic Design of Instruction*. Glenview, IL.: Scott, Foresman, 1978.

Gagne, Robert M., and Leslie J. Briggs. *Principles of Instructional Design*. New York: Holt, Rinehart and Winston, 1974.

Kearsley, Greg. *Training for Tomorrow*. Reading, MA.: Addison-Wesley, 1985.

LeFrancois, Guy R. *Psychology for Teaching*, 2nd ed. Belmont, CA.: Wadsworth, 1975.

Mager, Robert F. *Goal Analysis*, 2nd ed. Belmont, CA.: Pitman Learning, 1984.

Mager, Robert F. *Measuring Instructional Results*, 2nd ed. Belmont, CA.: Pitman Learning, 1984.

Mager, Robert F. *Preparing Instructional Objectives*, 2nd ed. Belmont, CA.: Pitman Learning, 1984.

Mager, Robert F., and Peter Pipe. *Analyzing Performance Problems*, 2nd ed. Belmont, CA.: Pitman Learning, 1984.

National Society for Performance and Instruction. *Introduction to Performance Technology*, Vol. I. Washington, DC: NSPI, 1986.

Tuckman, Bruce W. *Evaluating Instructional Programs*. Boston: Allyn & Bacon, 1979.

Zemke, Ron, Linda Standke, and Philip Jones, eds. *Designing and Delivering Cost Effective Training and Measuring the Results*. Minneapolis, MN.: Lakewood Publications, 1981.

chapter eight ———————————————

DETERMINING SALES TRAINING NEEDS

———————————— *ROBERT C. IMMEL*

Robert C. Immel is director of business and industrial training and development, Indiana Vocational Technical College, Richmond, Indiana. He also has his own training and consulting practice, specializing in communications and management development. For 20 years, he was a marketing training manager for two divisions of the Sperry Corporation and was formerly associated with Albion College and Michigan State University. He is contributing author to four books, and in 1983, McGraw-Hill published his training program on time management. A member of the National Society of Sales Training Executives since 1969, he has received seven awards for his editorial contributions, including two coveted gold awards for papers entitled "Managerial Needs Analysis" and "We Understand How Important It Is To Listen." He served as the society's president in 1980.

The needs determination process is basic to the success of sales training. The entire training effort depends on accurate needs analysis: the development of a training program and its materials, delivery, and measurement will only be as good as the training objectives as established. If the program objectives are not what's really needed, the training will fall into the same category. If no clear needs or objectives are set forth, it will not be feasible to assess whether the training has served a real purpose.

In this chapter, you'll learn how to conduct a needs analysis using any one or a multiple of methods to achieve your objective. Frequently the information required has already been recorded, and it is a matter of retrieval and interpretation of data to be of real value. In other instances, usable data may not be readily available, and it is up to you, the training professional, to collect and interpret data using the methods most applicable to the situation.

NINE BENEFITS OF CONDUCTING A NEEDS ANALYSIS

The reasons for conducting an analysis to determine the sales training needs of an organization are as follows:

1. By conducting a needs analysis, the training staff will be able to identify realistic needs and focus their training programs on them.
2. Knowing the training needs of the sales organization will help in utilizing the resources of your organization more effectively.
3. Knowing and communicating the training needs improves the coordination of other groups such as sales management, service, merchandising, and advertising.

4. You will have measures for assessing the effectiveness of the training.

5. Quantitative data about training needs will help justify requests to management for training resources.

6. Communications within your training and support groups will be aided. Well-defined training needs enable all members of the staff to contribute in a unified manner to the development and presentation of the program.

7. Involving the sales organization and others in the needs analysis and sharing the results of the analysis with them will cause them to be more supportive of the training effort.

8. A thorough needs analysis will enable the trainer to prioritize training efforts better.

9. By identifying the most critical needs, you are maximizing the probability of success of the training department and the organization as a whole.

Unfortunately, many smaller training departments do not have the resources, personnel, and time to conduct frequent or extensive needs assessments. However, experience dictates that in prioritizing tasks, needs analysis must place high to assure that the training efforts are applied to the areas of greatest need which will produce the greatest results.

Determining sales training needs need not be complex. You can employ many methods to ascertain your organization's training needs. The depth and/ or breadth of the needs analysis will depend on the magnitude of the problems to be resolved or the objectives to be achieved. To resolve a simple problem such as providing complete and accurate data on an order form should not require an in-depth survey. It may be as simple as assessing what causes inaccurate or incomplete orders.

Training needs analysis can originate from a number of sources such as objectives not achieved by a significant number of personnel, feedback picked up during field visits or while attending sales meetings, or the launching of new products or programs.

IS IT A TRAINING PROBLEM?

One of the greatest values of a needs assessment is to ascertain whether, in fact, there is a training problem or need. What may be perceived as a training problem by sales management may have no relationship to training but, rather, be the effect of other factors. Those factors could be compensation, product reliability, product availability, recognition, and so on.

EXAMPLE

Following an extensive introduction of a new product, the sales force failed to achieve objective. A refresher training program was offered six months later with similar results. The training department was assigned the task of finding out what training was required to resolve the problem. Field visits to various locations with a number of personnel clearly pointed out that the problem was not training. The new product had a lower cost and since the sales force was compensated by commission, they found greater personal reward in selling the higher-priced items. Consequently, little effort was placed on the new lower-priced product line.

EXAMPLE

After an in-depth, hands-on-training program to introduce a new industrial product, the sales force proceeded to write orders in excess of manufacturing's capacity to produce and deliver. In six weeks, orders for the first year's production were written. As a result, many customers did not receive delivery for eight or ten months, a consequence that resulted in disgruntled customers and harassed sales personnel. The following year, recounting the bad experiences of the previous year, sales personnel did not sell the product as aggressively, and objectives were not achieved.

Since pricing and product availability are not the domain of the sales training department, no amount of additional product, sales, or territory management training would resolve these situations.

If your needs analysis is in response to a performance discrepancy, your initial task will be to determine if the discrepancy is due to a skill deficiency. If it is, there is a training problem. When you know that the person has the skills to perform but is not performing, it is probable that something other than instruction is needed. As Mager and Pipe state:

> In general, the remedy is that of performance management. Rather than modify the person's skill or knowledge (since it's likely that the ability already exists), you will have to modify the conditions associated with the performance or the consequences or result of that performance. Rather than change what the person can do, change something about the world in which he or she does it so that it will be more attractive, or less repulsive, or less difficult.[1]

---

[1] Robert F. Mager and Peter Pipe, *Analyzing Performance Problems or You Really Oughta Wanna*, 2nd ed. (Belmont, CA: David S. Lake, 1984), pp. 59–60.

Take, for example, the sales representative who has put forth extra effort to succeed. This person has made confirming phone calls before leaving home in the morning, frequently drives extra miles to help an account with a major project, and is known to work three or four evenings a week during special program periods. The sales performance has been excellent; however, the individual has exceeded the telephone, travel, and meal allowance budgeted for a given period. Rather than being complimented for excellent performance, the individual is reprimanded for not operating within budget. Is this a skill deficiency or training problem? Not hardly. Rather than modify the person's skill or knowledge, you will have to modify the consequences or result of that performance.

SUGGESTIONS FOR COLLECTING AND REPORTING DATA

Three important considerations should be kept in mind while planning and carrying out the collection of needs data: (1) getting the optimal amount of data, (2) assuring that the data collected are accurate and reliable, and (3) collecting data in a manner so that those who are to support or receive training do not become hostile before training begins.[2]

What constitutes the optimal amount of data? When your continued analysis provides repetitive conclusions. Certainly with points (1) and (2), if you represent a large sales organization, you would want to collect data from several sales districts. In a smaller organization, you may include all sales personnel. In collecting data there should be no implication that what the sales personnel are doing is wrong. Your task is to collect information to help them perform duties better, easier, and more effectively with greater results.

Involve the learners in assessing their own developmental needs. It will provide them with insights about their strengths and weaknesses that can motivate them to engage in learning activities to meet their needs.[3]

It is important to remember that your role is that of a consultant. When you let the salespeople give you their input, by the time you implement the program, they will feel it is theirs, not yours. Doing this will gain credibility for your implementation.[4]

---

[2] Charles E. Watson, *Management Development Through Training* (Reading, MA: Addison-Wesley, 1979), p. 53.

[3] Alan B. Knox, *Helping Adults Learn* (San Francisco: Jossey-Bass, 1986), p. 59.

[4] Jarad F. Harrison, *The Management of Sales Training* (Orlando, FL: National Society of Sales Training Executives, 1983), p. 37.

Achieving and maintaining a high level of respect and trust by those you serve is paramount in getting honest, useful information. The role of the sales trainer is not to evaluate individual sales personnel performance as in a performance appraisal. That is the job of line management. Rather, your job is to assess their performance for those deficiencies which can be overcome through training.

At times, you may face a very challenging task in reporting your findings to line management along with your proposal for training to satisfy their needs. Often, you will be asked who said what or who are the deficient performers. To maintain your credibility in the field, it is better to respond that the issue is not the who but the what and treat the issue objectively, not personally.

Since you need top management support for training, it is important that those in your department who conduct the needs study gain management's trust and credibility. Focusing on the behaviors and not the individuals will keep your assessments objective and enable you to achieve this credibility.

Although each needs assessment method has its merits, none should be relied upon exclusively to provide the kinds and amounts of information that are required for an accurate and thorough assessment of needs. Competent training specialists draw upon a combination of methods selected on the basis of the information each method is designed to gather.[5]

How Postmeeting Questionnaires Can Help

Often, information is already available within the training or personnel departments which identifies training and development needs. Most training departments employ a postmeeting evaluation to obtain immediate feedback from participants prior to leaving the training site. Frequently, questionnaires contain such questions as "What recommendations do you have for improving this program in the future?" (See Figure 8–1 for the kinds of questions that are typically asked.) The answers are an indication of needs that were not satisfied or that still exist. The catch-all at the end of the postmeeting questionnaire is: "What other types of training do you desire?" Unfortunately, sometimes the trainer is tired, has to pack his or her training aids, and catch a plane to get back to prepare for the next program. A quick review may indicate that 80 percent of the participants said the program was good or excellent. The evaluations go in a file only to be discarded a year later. However, a couple hours spent analyzing the data may provide insight into other training needs or means of strengthening current programs.

---

[5] Watson, *Management Development Through Training*, p. 57.

Figure 8–1 Field trip questionnaire

1. What account(s) are you calling on today?

   _____

   _____

2. What are your call objectives for these account(s)?

   _____

   _____

3. Do you have a regular call schedule that you follow?

   _____

   _____

4. Do you have some accounts you call on more frequently than others?

   Yes _____ No _____ If yes, why? _____

   _____

5. In the most recent sales/product education program, what parts were:

   a. Most helpful _____

      Why _____

   b. Least helpful _____

      Why _____

6. What would have made the program of greater value to you?

   _____

   _____

7. What do you feel are your greatest strengths as a salesperson?

   _____

   _____

Figure 8–1 (Cont.)

8. What do you feel are your greatest weaknesses as a salesperson?

   _____

   _____

9. What would be of greatest help to you in overcoming your weaknesses?

   _____

   _____

10. Do you use the product education manuals? Yes _____ No _____

    If yes, how do you use the manuals? _____

    _____

11. What would make these manuals of greater value to you?

    _____

    _____

12. How often does your manager have district meetings? _____
13. What is the typical agenda at a district meeting?

    _____

    _____

14. How much time is spent on the following?
    Product knowledge?   _____
    Sales techniques?   _____
    Territory management?   _____
    Merchandising programs?   _____
    Credit/lease/pricing?   _____
    Other _____? _____
15. Approximately how much time does your manager spend traveling with you?

    _____

Figure 8–1 (Cont.)

16. What is normally discussed during manager visits?

_____

_____

17. Do you find these field visits to be helpful?

_____

18. How often do you have a performance review? _____

19. How much time is spent in the performance review? _____

20. Are the performance reviews helpful to you? _____
21. What would make the performance reviews more helpful?

_____

_____

22. In addition to new product introduction, what other subjects would you like to have covered at the next regional sales meetings?

_____

23. What other ways can the sales training department be of help to you?

_____

_____

_____

SEVEN METHODS FOR ASSESSING SALES TRAINING NEEDS

Conduct Performance Appraisals

Performance appraisals are an excellent source for identifying training needs, provided they are an assessment of how well individuals accomplish specific tasks and goals as opposed to performance appraisals that merely evaluate people on various traits.

If performance appraisals are assessments of how well people have accomplished predetermined objectives and they involve an analysis of what was done that was effective or ineffective, they can yield extremely useful information about training and development needs.[6]

EXAMPLE

Early in my career, I worked for a vice president of sales who asked that I review the performance appraisals of the 200-plus sales force. It was his contention that by being removed from the everyday pressures of bottom-line accountability, I might see something that others had missed. He was right. A tabulation of the data under "weaknesses" showed that 28 percent of the sales personnel were deficient in the areas of planning and organization (territory management). This startling figure revealed another fact. If 28 percent of the sales force showed weaknesses in these critical areas, it also implied that the district managers were not effective in coaching and counseling. Before training would be presented to the sales personnel, an intensive development program would first be offered to line management.

Design and Use Questionnaires

The most common means of assessing training needs is the use of the questionnaire. Questionnaires can be brief or extensive, depending on the objectives of the project. Questionnaires can be completed by the individual for which training is being planned or they can be completed by the immediate manager who has observed the performance of the individual over a period of time.

It is very important to survey all groups that would affect training. The first obvious advantage of the questionnaire is that you can reach large numbers of people in a short time. This ensures everyone a chance to be heard. A question-

---

[6] Ibid., p. 67.

Figure 8–2 Levels of learning—product

Sales Representative: ____________ Territory: _____

Sold in Territory: Yes	No	Product/Programs	A	K	U	S
X		Widgets		X		
X		Gidgets				X
X		Gizmos				X
	X	Do-Dads	X			
X		Co-op Advertising			X	

PRODUCT FAMILIARITY

A. *Awareness*. The individual is aware that the company has such a product. He or she does not know the specifications, capacities, and so on, of that product.

K. *Knowledge*. The individual has knowledge of the product; can quote models, capacities, features, and prices in general terms; does not have an understanding of why the product may have a particular part or the purpose of that part.

U. *Understanding*. The individual is beyond the A and K level and understands how the product functions and why the part is there. As an example, he or she understands and is able to interpret the value of a cam-action drive on a manure spreader or reverse knife grinding on a forage harvester. When a person "understands," he or she is able to explain the "how's" and "why's" of a concept as well as its implications. Cause-and-effect relationships, as well as advantages and disadvantages, are part of understanding.

S. *Skill*. The ability to perform—the ability to operate or demonstrate the product. It's not expected that the salesperson will have the degree of skill of an experienced operator; however, he or she should be able to show someone how to use the product if necessary.

Figure 8–3 Territory management needs

ANALYSIS OF TRAINING NEEDS

Individual's Name: T. O. Slow   Title: District Manager

Your Name: I. M. Tough   Title: Branch Manager

CIRCLE THE NUMBER IN FRONT OF ITEM IN WHICH YOU FEEL TRAINING
IS NEEDED

A. PLANNING

   1. Developing work objectives, modifying objectives to accommodate change.
   2. Working out performance objectives with each employee.
   3. Planning his or her own time.
   4. Getting his or her people to participate in goal setting.

  28. Handling selection interviews effectively.
  29. Helping others to relax at the beginning of an interview.
  30. Conducting orientation training with new employees.
  31. Ability to give on-the-job instruction.
  32. Giving advice on further education and training.

  65. Writing technical reports.
  66. Preparing performance appraisals.
  67. Being accurate in using company forms.
  68. Gaining acceptance and respect of employees.
  69. Motivating employees—encouraging, inspiring, or enforcing performance.
  70. Personally setting an example for the rest of the unit.

---

naire can be designed so that tabulating or compiling replies can be done easily. Responses to questionnaires can be anonymous, assuring more "honest" answers. A shortcoming is that clarification of responses or in-depth answers is impossible to get.[7]

Questionnaires and checklists can be used in the identification of training needs in product (see Figure 8–2) or territory management (see Figure 8–3). Data collected from the Analysis of Training Needs is useful in many ways: determining group needs, determining individual needs, and in one instance, determining candidates ready for promotion.

---

[7]Harrison, *The Management of Sales Training*, p. 24.

Determining group needs. Once the data from your group have been tabulated, you will see patterns from where several members have like needs. In such instances (depending on the size of your sales force), you would no doubt offer training to the entire sales force where general need has been identified.

Determining individual needs. There may be some individuals that have needs identified that are unique to them alone. In these instances, the needs may be answered through on-the-job coaching and counseling or by selecting an external program for the individual to attend since it's highly impractical that you would develop a program for two or three individuals.

Determining developmental needs. Most organizations have identified a cadre of high-potential promotable candidates or "fast trackers." Unfortunately, frequently the basis of selection has been subjective. "He graduated from State U the same as the boss" or "His dad was a fraternity brother of the boss" or "He's over 6'1", and everyone knows that tall men make better bosses" and, in one instance, "He has good judgment—he drinks Scotch." No doubt all the candidates have good qualities or they would not have been noticed initially. However, with the use of the Analysis of Training Needs, one is able to be both objective and specific in identifying needs. It then enables the trainer and line management to provide those developmental experiences that enable the individual to grow on the job.

Interpret Survey Feedback

Survey feedback is associated with large across-the-board changes in organization climate. A comprehensive view of the literature on survey feedback suggests that its primary impact appears to be on attitudes and perceptions of the situation. Studies suggest that survey feedback might best be viewed as a bridge between diagnosis of organizational problems and the implementation of active programs to solve those problems.[8]

Survey feedback is normally associated with the organization development practitioner. However, the professional sales trainer is also involved in organization development.

When one thinks of the survey feedback instrument, the terms that come to mind are "attitude" or "opinion" surveys. These surveys look at a much broader picture than do sales personnel skills alone. What is frequently found in the analysis of the data from such surveys is that factors external to the direct sales force are having an indirect impact on performance.

Your role as a training professional, and you may be the only such person in your organization, is to interpret the data and recognize how certain factors

---

[8] Edgar F. Huse, *Organization Development and Change*, 2nd ed. (St. Paul, MN: West, 1980), p. 224.

are having an impact on sales personnel performance. In essence, you become an internal consultant to the marketing organization or even beyond.

The development of survey questionnaires is a special field requiring skills few sales trainers have had the opportunity to learn or the time to employ. There are professional organizations that specialize in the development and implementation of these surveys. You can locate these organizations through the directories of services provided by organizations such as American Society for Training and Development or American Society for Personnel Administration. Another source may be a major university located in your area.

The typical survey covers opinions about the field's perception of customer relations, sales and marketing practices, sales support, communications, job satisfaction, sales incentives, managerial practices, and career development. Findings uncover a range of issues that impact sales performance and effectiveness.

For example, ineffective sales tools, high prices, poor supervision, and so on may be identified as hindering performance. In many instances, the sales training department has courses or can develop courses to overcome these obstacles to performance.

Opinion surveys conducted by an independent research company rather than in-house offer three major advantages:

1. *Anonymity.* Respondents are more honest in their appraisals when they are assured that no one at the home office will see their ratings. The fear of being identified is real. The independent firm usually guarantees confidentiality.

2. *Comparative Data.* Analyzing statistical data is not as easy as it seems. You need comparative data—what is a "typical" response to a particular question—in order to determine if your findings are above or below average. Independent research firms usually have this type of information. For example, Sales Staff Surveys, Inc., Danbury, CT, has surveyed over 25,000 sales representatives on over 150 questions directly related to sales effectiveness. The "norms" developed from their research provide valid, statistical comparative data.

3. *Objectivity.* The interpretation of data is often influenced by what one expects or wants the results to be. An independent research firm is more objective than are those close to a situation. In analyzing survey findings, data should be segregated by titles, levels of experience, and any other criteria that may influence response.

Opinion surveys can raise morale and encourage the field to support programs that result from survey findings. This occurs because of the participation in the decision-making process.

Seek Trainee Input

How active or passive are the participants in your programs? Active learning takes many forms in which participants help clarify preferences, make choices, ask questions, seek answers, select activities, practice procedures, and give feedback on their progress and satisfaction. How important is such active learning? How satisfactory is the extent of active participation in your program? What indications do you look for as you seek to maintain an optimal level of participant involvement?

An early way to be responsive to adult preferences for active participation is to include them in the needs assessment and objective setting process. You can do this by asking them their expectations and reasons for participation.

Include attention to both what they want to accomplish and why. Only some of their expectations may match the purposes and contribution of your program, but it is useful for both you and the participant to recognize the expectations that match and those that do not. They will appreciate the relevance of the program objectives that you value to the extent to which those objectives relate to their purposes. Because trying to change can be threatening, helping to set goals can give participants a sense of security, ownership, and responsibility for their learning.[9]

Develop Structured Interviews

You should include interviews in identifying training needs. Leave your office and ask fellow employees (e.g., in other departments within your division) what they perceive to be the needs of a sales training program.[10]

Interviews are usually most productive if they are structured to follow a predetermined plan to obtain specific information. They should not be overly structured, however, so as to be unable to pursue areas of importance that are unanticipated.[11]

The major advantage of the structured interview is that you can get in-depth information. Answers can be followed up with questions. Unclear responses can be clarified, and incomplete data can readily be updated. The disadvantage of this method of assessing training needs is that it takes a great amount of time and only small numbers can be surveyed. As with the questionnaire, extensive planning should go into developing your structured interview. The planning starts by identifying the kinds of information you need to know. Once you have

[9] Knox, *Helping Adults Learn*, p. 35.

[10] Harrison, *The Management of Sales Training*, p. 12.

[11] Watson, *Management Development Through Training*, p. 68.

done this, then start writing specific questions needed to get the answers.[12] Your questions should be open-ended, which leaves the interviewee free to respond and express opinions and needs. Closed questions tend to elicit short yes or no answers, which restricts the amount of data received.

Five Steps in Conducting a Structured Interview

1. *Set the stage*. Explain what you are doing, why you are doing it, and what will happen with the information. It's best to read directly from the list of questions. Explain this before the formal interview commences. Make it plain that you are not there to evaluate the individual or report to anyone your opinion of his or her performance or attitude. If trust is not developed, the responses will include only what the sales representative thinks you want to hear.

2. *Be relaxed*. Don't let your gestures evaluate responses. An approving or disapproving nod can cause the individual to edit future responses.

3. *Take notes*. Taking notes is important. If not handled correctly, however, note taking can inhibit the interview. If you write down only negative responses, all responses will soon be guarded. You might have an occasion to use a tape recorder in some of your interviews. If time permits, play it back so that the interviewee can hear it.

4. *Be positive*. Follow-up responses with positive statements such as "That's good," "I understand," "I can appreciate that." Remember, you're on their side! You are a recorder, not an evaluator.

5. *Get specific responses*. Probe general responses for specific answers with such statements as "Could you expand on that?" "What prompts you to feel that way?" "That's interesting, tell me more," and "How did that come about?"[13]

When to Use Structured Interviews. Structured interviews can be used at any time to assess training needs. The structured interview can also be used to get more in-depth feedback on your current training programs.

Due to the time restrictions in utilizing the structured interview with an entire sales organization, you may find it useful with a limited number in other ways. One helpful use is to conduct a limited number of structured interviews prior to developing a questionnaire that will be distributed to the entire sales

---

[12] Harrison, *The Management of Sales Training*, p. 32.

[13] Ibid., pp. 32–33.

Figure 8–4 Advisory committee guidelines

INDIANA VOCATIONAL TECHNICAL COLLEGE
REGION 09—RUSHVILLE

BUSINESS AND INDUSTRY TRAINING—ADVISORY COMMITTEE

The Advisory Committee shall operate in the specific area for which it has been appointed and shall limit its activities to advising on matters directly related to the development/procurement and offering of programs for business and industry.

The purpose and duties of the Business and Industry Advisory Committee are
  • Advise on current job needs of business and industry and the types of programs visualized as meeting these needs.
  • Advise as to skill development required to prepare people adequately for promotion within their own organizations.
  • Serve as a communication link between the educational and the business-industrial communities.
  • Suggest new workshops, conferences, or programs that will meet the ever-changing needs of the publics served.
  • Assist in studies and/or surveys to ascertain specific development needs.
  • Assist in maintaining a positive public image for the college with the community.
  • Aid the college in identifying emerging trends and changes in occupations and technologies.
  • Assist in the marketing of programs offered by the Business and Industry Training Department.
  • Review goals, objectives, and terminal behaviors of programs proposed/conducted by IVTC.
  • Assist in the procurement of competent resource personnel.
  • Assist in obtaining financial resources for pilot programs.
  • Assist in the evaluation of Business and Industry programs.

MEMBERSHIP

  • Members shall be selected and appointed to represent a cross section of the community, business, and industry served by the college.
  • The active advisory committee may suggest names of others whom they feel would make a positive contribution in achieving the goals and objectives.
  • Appointment of an advisory committee member shall be for three years; thus, one-third of the total membership shall be appointed each year. Exceptions to this include an appointment to fill an unexpired term and appointments in the first year, 1988.

Figure 8–4 (Cont.)

- An individual may be reappointed to the committee to serve a second successive term. An individual may not serve more than two successive terms.
- An appropriate representative of the college will be present at each meeting.

MEETINGS

- Regular meetings of the Advisory Committee will be held quarterly. Additional meetings can be called as required.
- Written notices of committee meetings shall be mailed to all members at least ten (10) working days before each meeting.
- Agenda for the meeting shall be prepared by the director, Business and Industry Training, and mailed to all members at least ten (10) working days before each meeting.
- Failure to attend 50 percent of the Advisory Committee meetings is cause for deletion of said individual from the committee.

---

organization. This will enable you to target specific questions related to training needs as perceived by sales personnel.

Another useful application is following the compilation of data from field survey questionnaires. The structured interview enables you to confirm your interpretation of data, to gather additional information, to clarify and verify. It also provides one more opportunity for the sales force to buy into, to accept ownership in, the results of your survey.

Establish a Team Approach to Needs Analysis

In some organizations, it is possible to establish training needs through a team approach. This calls for selecting representatives from various departments (or sales districts) to sit on a "training committee." They are asked to consider and suggest training needs as they relate to their particular department and the organization as a whole. The trainer's job is to pull together the various opinions and suggestions to help formulate a training program acceptable to all and incorporate the common characteristics.[14] Figure 8–4 illustrates guidelines for a typical advisory committee.

If we use the old acronymn, "No one of us is as smart as all of us," it stands to reason that the team approach is a valuable method to employ. I'll share three experiences; I initiated two and the third was assigned, but I used it to advantage for identifying training needs.

---

[14] Ibid., p. 16.

Three Examples of the Team Approach in Action

EXAMPLE 1

Faced with performance discrepancies in the field, I called on my staff of field training managers for help. Twice a year, meetings were held with the training managers from across the continent. The purpose of the meetings was to introduce new training programs and also spend a day and a half with new branch training managers on the subject of instructional techniques. The branch training manager's position for the most part was a passthrough one from sales representative to district manager. Most were only in the position for two years, some for three. I asked for volunteers to meet with me one evening to help identify training needs. Every person showed up.

First, we listed all the job responsibilities of a sales representative on one flipchart. Since most had been salespeople in the recent past, they were familiar with these duties. Next, on a second flipchart, we listed all types of training presently offered. We then crossed out those job responsibilities for which training was offered. What remained were the gaps in the present training agenda. We had done an excellent job in product education, selling skills, merchandising, and communications. What was lacking was the administrative responsibilities—how to manage the work. The problem was that the work was managing many of the salespeople. It seemed so simple and clear, yet I would not have been able to identify it without the help of the team.

EXAMPLE 2

Shortly after my transfer to another division, I recognized that there was no structured approach for training new sales representatives at the branch level. When they attended the home office training program for new sales personnel held annually, I found wide discrepancies in their initial training. With the approval of the sales director, I selected three relatively new salespersons. Their tenure was one year, one and a half years, and two years. Each was from a different branch. The four of us met at a hotel for two days, away from the home and district office. With flipcharts, felt-tip pens, and masking tape, we started out. What must the industrial salesperson know, and what are the job responsibilities? It wasn't long until every wall of the room was covered with sheets of paper. We then prioritized the tasks by importance and chronology. At the end of two days, a training guide for the new sales representatives had been developed. In the publication of the guide, credits were given to the three sales representatives who actually put it all together. As the trainer, I was only a facilitator.

EXAMPLE 3

The division for which I worked provided an annual opportunity for sales personnel to express their concerns. It was called the "Sales Feedback Panel." Each district would appoint a representative who would poll all other personnel in the district for their questions or concerns. Prior to my joining the staff, a representative of the personnel department received the input. It was the task of the facilitator to obtain answers to all questions raised while keeping the source (person or district) obscure. In many instances, it necessitated going to the vice president of personnel, the vice president of marketing, or even the vice president of engineering for some answers. Once the answers were received, a report was written listing the concern along with the reply. This was sent to the district representative who would then review it at the next district meeting with all district sales personnel.

A high degree of trust between the panel and the facilitator is required for this to be a success. Problems, some, but minor. Sometimes it started to turn into a gripe session but normally other panel members brought it back to objectivity.

The value from the training perspective is that we often get clues to training needs that we would otherwise never know. Examples could be delays by customer service in responding to customer inquiries. Does customer service have a training need? A breakdown in communications? A failure to communicate? Does management have a training need? Overall, it turned out to be very beneficial to the training department and, in the end, the sales department and organization as a whole.

Be Observant in Field Trips

If you are primarily concerned with the design of training programs for sales personnel, then it is essential to get a firsthand look at where the training will be used and how it will be used. Talk to as many salespeople as you can to get their perceptions as to what they feel is needed.

Field visits will give you positive feedback. This feedback should start with the customers. Find out what they expect of your company's representatives. Determine what the customers see as strong and weak points of the company reps who presently call on them. Ask the customers (dealer or distributor) what kinds of information or services they would like to have which they are not presently receiving.[15] Obtain permission to work with some of the salespeople (good and bad) as an *observer* of the sales calls. Don't go as a coach—that's the

---

[15] Ibid., p. 15.

line manager's job—but more as a scout. Watch the salespeople at work and observe how they conduct their business.

Be observant and ask questions about precall planning, sales techniques, sales aids, training, and activities during the sales call.[16]

SALES MANAGER NEEDS AND THEIR IMPACT ON YOUR ANALYSIS

Before solving an assigned problem, it is frequently necessary to resolve certain other problems, policies, or procedures. Such is the situation in assessing the training needs of the first-line field supervisor or district sales manager.[17]

References have already been made to the need for management training as expressed in 28 percent of sales personnel being weak in planning and organizing. Or the lack of or breakdown of communications. A number of indicators may dictate that management training is called for.

A careful review of performance appraisals of the sales personnel will soon identify those managers who are nit-pickers, those who are more concerned with activity rather than results. In reviewing appraisals for two or more years, you will identify those managers have been successful in helping people to overcome their deficiencies and those who have not.

After you have been with an organization for two, three, or more years, you will see patterns form. As an example, with one sales branch with two districts, I observed seven people promoted from one district, while with the other, none was promoted. It was evident that one manager was a developmental manager while the other was reductive. The same was reflected in the sales objectives achieved by the two sales districts.

The needs analysis survey identified in Figure 8–3 has proved most useful in identifying managerial development needs. The results achieved have shown universal needs for the management group as a whole, which dictates training be offered to all. The survey also indicates individual needs that, if unique to only one or two individuals, may prompt sending them to an external training seminar.

Do not discount reviewing the performance appraisals of the district managers. You will no doubt learn of training needs not discussed freely by line management.

Field visits with salespeople can reveal a lot about how their managers work with them, coach, and communicate. The performance of the sales representative is a reflection of the manager's style of operation. Travel with more than one representative from the district to get a true perspective. Select a cross

---

[16] Ibid., p. 42.

[17] Ibid., p. 58.

section of performers, stars, reliable and questionable. Prepare a structured interview (as referred to earlier in this chapter) and a checklist so that you are consistent in your questioning and observations while traveling with the salesperson. This will also make the tabulation of your observations much easier. The questions need not be complex. Frequently they are openers, and the sales representative will expand on the subject. Simple questions, such as

How frequently does the district manager travel with you?

How much time does he or she spend (hours or days)?

Describe a typical visit from your district manager.

Does the district manager review new products and product updates?

Do you receive coaching on your selling skills?

Does the district manager review new policies or procedures?

How often do you receive a performance review?

What actions result from these reviews to help you perform better?

You will no doubt adapt the questions to meet your organization and manner of selling. Try to keep the interview positive. Look for the strengths, as well as the deficiencies, as you may be able to incorporate some of the strengths in future training programs. Or better yet, use the district manager as an adjunct trainer.

DISTRICT MEETINGS

Attending district meetings can be very revealing. It will not take you long to determine those who have planned, prepared, and conducted excellent meetings.

Such indicators are that sales reps were notified in advance, so it does not disrupt their schedule. They also know the purpose of the meeting. Presentations are clear, discussion is open. There is demonstrated an air of trust and respect between manager and the sales personnel. Be alert of the managers who

Cut presentations short—"Here's the new literature, read it."

Resent the meeting—"Headquarters says we have to review this."

Disagree with others—"I don't care what you say."

Believe in putdowns—"Joe here really blew the Ajax account."

Listen to what others are saying during the meeting and at breaks. Observe the behavior of the attendees. Note the facial expressions and body language. They also communicate a message. The results of these observations may indicate any of a number of managerial development needs or it could be none. The most common are

Coaching and counseling

Performance appraisal

Planning and conducting effective sales meetings

Managing the development of people

Responsible listening

SUMMARY

These thoughts expressed on the analysis of sales training needs are the culmination of 30 years of experience in adult continuing education. Much has been learned from the publications referred to in this chapter as well as others currently out of print. However, my greatest resource for ideas has been other practicing sales training managers. The editorial papers of the National Society of Sales Training Executives continue to provide inspiration and new ideas in all aspects of sales training.

Formalizing needs assessment procedures contributes to agreement among leaders, yourself and other people associated with the training program, about the needs on which to focus. As the program is planned and conducted, participants meet some needs and discover others, and perspectives shift regarding the pertinence and importance of needs. Therefore, needs assessment should be an ongoing process, consisting in evaluation for the purpose of planning future training programs.[18]

REFERENCES

Harrison, Jared F. *The Management of Sales Training*. Orlando, FL: National Society of Sales Training Executives, 1983.

Harrison, Jared F. *The Sales Manager as a Trainer*. Orlando, FL: National Society of Sales Training Executives, 1983.

Huse, Edgar F. *Organization Development and Change*, 2nd ed. St. Paul, MN: West, 1980.

Knox, Alan B. *Helping Adults Learn*. San Francisco: Jossey-Bass, 1986.

Mager, Robert F., and Peter Pipe. *Analyzing Performance Problems or You Really Oughta Wanna*, 2nd ed. Belmont, CA: David S. Lake, 1984.

Valen, Carol, and Bill Valen. *Developing Sales Training, A Competency-Based Approach*. Cave Creek, AZ: William B. Vallen & Assoc., 1987.

Watson, Charles E. *Management Development Through Training*. Reading, MA: Addison-Wesley, 1979.

---

[18] Knox, *Helping Adults Learn*, p. 57.

chapter nine

THE BIG PICTURE: HOW TO CREATE A THEME FOR YOUR SALES TRAINING PROGRAM

DIANE HESSAN AND THOMAS C. KEISER

Diane Hessan is senior vice president of marketing for The Forum Corporation in Boston, Massachusetts. In that capacity, she is responsible for marketing all of Forum's capabilities in training, research, and consulting; market planning and research; corporate communications; promotion; and new product marketing. Hessan was formerly vice president of Forum's Sales Productivity business where she was involved extensively in Forum's Sales Competency Research project, a major study on sales success. Subsequently, she helped to pioneer "Face-to-Face Selling Skills," a sales improvement seminar based on the research. She has worked with hundreds of Forum clients in designing, implementing, and teaching sales, management, and customer-focus training programs. She is a frequent speaker at professional associations and conferences in the areas of relationship management and customer orientation. Hessan graduated summa cum laude from Tufts University with a double major in Economics and English and received an M.B.A. with distinction from the Harvard Business School.

Thomas C. Keiser is senior vice president of product development and research, The Forum Corporation, Boston, Massachusetts, where he is responsible for developing new products and capabilities for the firm. During his 15 years at Forum, he has been a principal architect of several packaged programs and many customized programs. Formerly, he was senior vice president of the management division of Forum which conducted all executive, management, and supervisory education. Earlier, Keiser was in charge of Forum's consulting practice. He has also been in sales and sales management and marketing at Forum. He has worked extensively on the design, development, and implementation of specialized programs for Forum clients, including Citicorp, Chemical Bank, CBS, the U.S. Department of Commerce, E. I. du Pont de Nemours, General Electric, Hercules, Mellon Bank, Prudential, Sun Oil, Trans World Airlines, and United Technologies. Keiser has published articles in *Harvard Business Review*, *The Journal of Services Marketing*, and *The ASTD Training and Development Handbook*. He speaks frequently on sales negotiations and customer focus. He graduated cum laude from the University of Pennsylvania with a double major in Honors Sociology and History. He received a master's degree in Sociology at the same time as his bachelor's degree.

Marketing people know the importance of positioning their products and services in market segments; salespeople are trained to position themselves and their companies as an integral part of the accounts they work with. In the same vein, it is useful to position sales training programs in a way that creates widespread organizational understanding of their purpose and contribution. As you will discover in this chapter, "positioning" involves concentration on the overall theme and mission for each program at both the design and implementation stages. Above and beyond the obvious roles of delivering training programs, most of these positions today require the ability to

- Consult with line divisions on sales productivity issues.
- Perform needs analyses for different sales problems and opportunities.
- Recommend training and nontraining solutions.
- Market programs internally.
- Design programs for fundamental and special needs.
- Provide follow-up and reinforcement solutions.
- Evaluate the impact of training.

Added to the foregoing list is the increased demand for sales training that is linked to changing corporate strategic direction, Thus, the tasks of "redesigning" and "reconsulting" are constant and complex.

In our work with hundreds of organizations, we have searched for key levers and ideas used by the most successful sales trainers in North American corporations. One important finding is that highly effective sales trainers understand and communicate the principles and concepts behind what they are doing: by creating the big picture.

This chapter describes what the big picture is, why it is important, and how to craft an effective message. We also share sample applications and some potential pitfalls to recognize and avoid.

WHAT IS THE BIG PICTURE?

The big picture is a statement, or set of statements, that summarizes the mission, concepts, philosophy, or message of your sales training program. It is used as a means of driving the marketing and design of a training program as well as tying it to the strategy of your business. It can appear in a variety of formats, from program titles and slogans to letters welcoming participants, to the course and executive overviews.

Think for a moment about how you read a magazine. The name and table of contents are similar to the name and program outline for a sales training program. The remainder of the magazine might be analogous to your program materials. The real key, however, to whether you'll purchase and read the publication is what's on the front cover underneath the title. How have the editors summarized the issue? What do they see as important? What do they think will capture the reader's interest? Although we are not suggesting that you literally need to summarize your message on the front of your program notebook, we are saying that carefully drafted linkages such as effective headlines are a key to successful sales training programs.

It can be very difficult to come up with a key message for a program. In fact, along with the design of the program itself, coming up with the key message is the most creative part of training program research, development, and design. We believe strongly that it is important to come up with a basic message after conducting needs analysis and research but before beginning to design the program. The reason is that the message needs to shape the fundamental design of the program. By keeping the key points in the front of your mind as you design, you reduce the risk of drifting in different directions.

TEN REASONS CREATING AN OVERALL MESSAGE CAN HELP IMPROVE YOUR SALES TRAINING DELIVERY

This section summarizes the primary reasons why creating an overall message can add to better development and delivery of sales training.

1. *To Retain Learning.* Sales trainers are continually looking for ways to reinforce program content. Having a key message helps participants to remember what is important in the same way that marketers use positioning so that consumers will remember key product attributes.

2. *To Build More than Skills.* Frequently, sales trainers are in situations where skill building alone is not sufficient. For instance, most salespeople know how to ask questions, but need to learn when and why to ask questions and how important the skill is. Thus, motivational and attitudinal dimensions come into play. A message summarizing a program's questioning philosophy can be as valuable as skill practice.

3. *To Link Training to Corporate Strategy.* To the extent that the big picture can be derived from your organization's strategy, you can elevate the importance of a training program. Beyond a program being seen as part of a standard curriculum, it can be seen as essential to, and consistent with, future company success.

4. *To Create Discipline for the Course Designer.* One pitfall in designing training programs is the danger of creating a fragmented design in which each segment works but has no unifying theme. An overall set of concepts, established beforehand, solves this problem because the designer has to ask "how does this segment relate to the overall message of the program?"

5. *To Create Continuity.* Although many sales trainers have a variety of responsibilities, most programs are created and implemented by a team of people. Agreement on the big picture helps to create clarity for each player at every step of the project.

6. *To Make Training More Issue Focused.* By their very nature, having a big picture is a means of saying "This is more than just training. This effort is a critical aspect of our company's future productivity."

7. *To Develop Organizational Value Statements.* Particularly under conditions of change, trainers need to address the culture and values of an organization. Salespeople will be more effective under changing circumstances if they know what is expected of them, what behaviors are rewarded, and so forth. A big picture simplifies these issues.

8. *To Help Market Training Internally.* There is great marketing value in an overall program message. Creating a program philosophy forces the trainer to think through what is important to the participant and what is needed to make a program unique. This usually creates and expands the interest level of potential participants.

9. *To Gain Management Support.* Ideally, sales managers would go through training whenever their people do, for maximum chance of reinforcement. When this is not practical, managers who know and understand the big picture can use it to reinforce behaviors even without knowing the specific skills covered.

10. *To Assure Consistency and Clarity of Direction.* In today's complex business environment, salespeople do not always feel they get a clear message about how they should behave. "Do I push for a fast close or watch out for the relationship?" "Do I get the order no matter how unprofitable?" and "Should

I do *anything* a customer wants?" are typical questions trainees will ask. Having a clear set of principles will help increase the value of training and people's confidence that they are doing what's right.

HOW TO CRAFT THE BEST MESSAGE FOR YOUR SALES PROGRAM IN FOUR STEPS

There is no single way to go about crafting an effective message for a sales training program. The following guidelines are gleaned from the school of hard knocks. We've tried to draw some general conclusions from both successful and less successful projects.

The basic methodology for creating a key message is to get the right people together first to diverge on different possible approaches and then to converge on the optimal alternative. Remember, this should take place before program design. For the best results, there needs to be a string of diverge/converge processes, so that each idea session builds upon the results of the former session. This process of iteration is extremely important to coming up with a high-quality message. The more time you have to go through different iterations, the higher quality the message.

It's also true that the people you select to do the brainstorming will greatly impact the quality of the message you ultimately craft. Consider inviting all those who have a stake in the outcome of the program—that is, the people who will be affected by its results. Consider also inviting outsiders, who bring a perspective different from the "prevailing wisdom." Consider inviting people from the target population as well as experts who really know what it takes to be successful in the topic area you're talking about. Try to use the same people throughout the process, so you don't have to get new people up to speed each time. Finally, try to select people who are conceptual by nature since the task you are undertaking requires a strong ability to understand and make sense of abstractions.

Step 1: Prepare for Your Brainstorming Session

To be ready for a session in which ideas will be generated, you need to do a complete *needs analysis*, including comprehensive interviews with the "stake-holders" as well as the target population for the training program. For the interviews with the target population, you will need to understand the language and norms of the group so you can use words that have meaning to these people.

Sample Questions to Ask Key Stakeholders

• What differentiates high from low sales performers? What skills? knowledge? attitude?

- What is the impact of a single ineffective sales call on the long-term relationship you have with a key customer?
- When it comes to developing in depth relationships with customers, which of our competitors does it best? How do they do it?
- How do you evaluate the effectiveness of a sales strategy?
- What would be the consequence of delaying this training initiative for a year? What would be the impact on next year's growth? What would be the impact on retention of high performers?

It is also important to conduct a *literature search* to discover what is already known about the topic. We have found that the bigger this literature search, the better. Look deep, for some of the best information is in relatively unknown publications. Often, an entire article yields only one key idea, but when put together with other key ideas, they can communicate an extremely impactful message. For example, in developing a sales training program we uncovered a publication which stated that when customers are listening to a salesperson, they discount by about 50 percent whatever the salesperson says. The reason is the natural skepticism most customers have about salespeople. This, plus statements from other publications, combined with our own research on sales effectiveness, led us to a key principle for our program called "persuade through involvement." The principle sums up the importance of using "pull" rather than "push" approaches in the selling process.

Original research is also an important part of the preparation process. The interpretation of research results plays a vital role in the development of a big picture. It is a mistake to leave the research in a form which only a trained statistician can understand. Most participants want to know that you've taken the time to do research, but they also want you to invest the time to cull from the data the essential message.

Two aspects of the preparation phase make or break the success of the project. One is touching the key stakeholders. If these people are not involved before you try to distill the basic message of the program, they will feel excluded and may derail the process later on. The second factor that matters is doing very broad research. The wider you cast your net, the more fish you will find and the better ideas you will ultimately incorporate into your program.

Step 2: Increase the Number of Options

Some particularly useful books on brainstorming and "lateral thinking" techniques are listed in the references for this chapter. Most such books state the importance of a formal process. We have been involved in a lot of brainstorming sessions where the leader took a casual approach. These meetings were far less productive than when the leader carefully followed the rules of brainstorming.

Simply put these rules are

- The facilitator does not participate in the substance of the discussion. His or her sole job is to keep the discussion going, to make transitions, and to record the output of the discussion.
- Team members should refrain from any form of criticism of each other's ideas.
- Any idea is accepted. Even if an idea is impractical, it may serve to spark a more practical suggestion.
- Team members are encouraged to build on each other's ideas.
- Set a time limit.
- Go beyond the first pause. The best ideas often emerge after the group gets stuck the first time.

Make sure at the beginning of the meeting to establish guidelines which demonstrate the importance of diverging, converging, and reiterating that process until the message has been perfected. If you carefully set expectations at the beginning, there will be less danger that someone will come in expecting to solve everything in one meeting or that all his or her ideas will be incorporated in the final solution. Consider using a facilitator from outside the project team. Since brainstorming requires the facilitator to record, not advocate ideas, using someone neutral means you won't have to "take yourself out of the game."

During the diverge phase, push for alternative frameworks so that you have different options available to you. Don't go down just one path because you may need a fallback after the testing.

One final suggestion is to set a time frame, but leave enough time so you can fully explore the issues. Usually the agenda of these meetings cannot be successfully accomplished in only a half day. They require much more thought and "soak time." A comfortable and relaxed setting, preferably away from the office, also helps because of decreased interruptions.

Step 3: Converge—Look for Common Themes

Converging takes place when you distill all the different ideas that come up during a brainstorming session. Basically, you're looking for the themes and clusters that the group agrees are central. There are several techniques:

- Content coding.
- Identifying key words which recur on many flipcharts.
- Cluster analysis.
- Identifying "what goes with what" and labeling the clusters.
- Rank ordering—having each person rank-order or rate the importance of

improving different factors, to identify which issues are most critical, and then coming up with a group average of the rankings or ratings.

It's important during the converge phase to maintain openness and to show a willingness to build alternative frameworks. If people begin to get into power battles or debates about one idea being better than another, the discussion can go downhill very quickly.

Step 4: Test Your Chosen Theme

After you have a candidate list for the key principles or topics of the program, it is imperative to go out and test it. You can do this through a focus group, where you put the idea forward in front of several participants and test their reactions, or you can do it one on one with participants and stakeholders. What is imperative is to listen to the feedback you are getting and be able to report back to your project team in a way they understand. The key is being flexible enough "to hear the feedback."

Review and Revise as Needed

As we mentioned before, the four steps just listed usually need to be reiterated on the basis of the tests you have done. By examining the question several times, you substantially improve its quality. Make sure you leave enough time to go through different reviews of the key concepts before you design. If you begin designing before you finalize the message, substantial rework (in the form of cost and schedule overruns and frustration) will be inevitable.

Case Study: How to Develop a Generic Sales Program

When we developed a generic sales program, we began by selecting a diverse project team consisting of some outstanding salespeople, several course designers, a few marketing people, and a writer. An outside video company was added later. All except the course designer and writer had substantial sales and sales management experience.

The program was built on the foundation of original research that we had conducted as well as an extensive literature search. The literature search was boiled down by one person to a 200-page stack of key article extracts and other information about the sales task. Each member of the project team began by reading through the stack individually and identifying what they believed were key themes. We then sorted through all our individual work and identified key recurring themes. By the end of our three-day meeting, we were confident of our understanding but felt our method of conveying the message was too verbose and confusing. We continued working on it with the core project team. A person from marketing and two people from R&D worked and reworked the definitions and the words through countless iterations. Since the program was one which would be delivered to many different companies, to literally tens of thousands of salespeople, we had to make sure that the basic message was absolutely right.

The degree of conflict among the team waxed and waned as we went from one version to another. In retrospect, one of our mistakes was that we began designing the program before we finalized the concepts we wanted to teach. While we were able to test our total design concept more thoroughly, we also had a lot more rework each time we made small changes to the fundamental concepts.

Four Measurements for Evaluating the Quality of Your Message

It's amazing how attached you can become to your first draft. It's helpful to have criteria against which to test that draft before you cast it in concrete. There are four important criteria to consider: *rigor*, *relevancy*, *simplicity*, and *taking a stand*.

Rigor. First is *rigor*. The message that you craft needs to have the appropriate level of rigor for the level of research which has been conducted and for the importance of the project that you are undertaking. If your needs analysis and research phase has been highly empirical, it's important to use a rigorous approach to coming up with the major themes for your program. Factor analysis is a technique which is commonly used to find out "what goes with what." Items that have high statistical correlation are put together and separated from those items that have a lesser affinity. To conduct highly empirical research and then slap together items that appear to look good together is not appropriate. The second important aspect of rigor is the importance of the program itself. As the consequences of being wrong get worse, the level of rigor should go up.

For example, when Forum conducted its Customer Focus research, it was important to take a rigorous approach in identifying the key organizational factors which lead to higher levels of customer satisfaction. Since the research would drive the design of several programs which would be delivered to thousands of people, it was essential to understand statistically the underlying structure of the data.

Relevancy. A second factor to consider is *relevancy*. If the program principles are seen as having little bearing on the participants' day-to-day work or are seen as simplistic, the trainer will have to overcome needless resistance on the part of participants and may alienate them simply through the choice of words.

The choice of words is particularly important when dealing with psychological factors which may be seen as threatening or "touchy-feely." For example, we had a valuable debate in our project team when we were developing a sales program for a multiindustry audience which had different levels of sales experience. We knew that building strong rapport with customers was a very important skill, but we needed to come up with a label for that skill which wasn't seen as trivial by the hardened, experienced salespeople who would attend. After a lengthy process, we came up with the word "connect" for the skill. Some of our design team felt that the word was still too "soft" for the audience. We market tested several alternatives but ultimately found that "connect" was the best alternative. Relevancy is of course extremely subjective, and one needs to approach the choice of words and labels as a marketing question since words mean so many different things to different people.

Simplicity. The third criterion is *simplicity*. If most people have trouble remembering more than a couple of seven-digit phone numbers, it is probably arrogant of us to assume they will retain a long list of learning points of our three-day training program. Although we may turn up lots of interesting and rich material during the needs analysis and research phase, this material must be boiled down into something people can retain. In fact, the whole point of coming up with the big picture is to identify a couple of key themes which you expect people to remember.

Taking a Stand. The fourth criterion is *taking a stand*. If your message conveys the obvious, it will not be heard in the first place. The message should take a bit of a controversial position for it to generate discussion and meaning for participants. For example, a sales management program which suggests that sales managers need to plan, organize, and control their sales forces will not capture the imagination or attention of most experienced sales managers. One which challenges sales managers that their role is not to maintain the status quo but to force change in the marketplace and within their own organization is a much more controversial statement that will trigger more discussion and be more memorable for participants.

HOW TO INCORPORATE THE BIG PICTURE THEME IN YOUR SALES TRAINING PROGRAM _____

Include the Big Picture in Program Design

The most successful sales training programs we have seen incorporate the big picture right into program design. Prior to introducing specific skills to be taught, a module explains the philosophy and beliefs behind the program.

Figure 9–1 shows the "driving principles" from a Forum Corporation sales

Figure 9–1 Three driving principles for salespeople

Focus on the Customer

High-performing salespeople view the customer as the center of the buying process. This attitude is more than just being concerned about a customer's needs.

The most successful salespeople are aware of how they themselves like to buy, which enables them to focus on the customer's point of view. This means that, as a salesperson, you should concentrate on the customer's buying cycle rather than on your agenda. Make sure that everything you say and do is vital to the sales interaction and is valuable to the customer. A good salesperson knows the answer to the question, "What's in it for the customer?" and knows how to make every minute of the sales call count.

Building the customer's trust is an important part of the buying process, and the customer cannot trust you if you are not aware of his or her state of mind, situation, and needs.

Earn the Right to Advance

High-performing salespeople earn the right to proceed at each step of the buying process. They understand that a customer's mild interest in a solution is not sufficient for the salesperson to "pitch" the product or to close an order. The customer's concerns have to be addressed at each step of the buying process before the salesperson can advance. For example, the salesperson should *not*:

- Ask questions until the customer understands why sharing information is important.
- Present a solution if the customer does not perceive his or her need as urgent.
- Close if the customer has insufficient information to make a decision.

High-performing salespeople approach selling as a series of business transactions in which the salesperson is a problem solver; that is, the salesperson is aware of the customer's needs and problems and is not just a "peddler" of a product or service.

Persuade Through Involvement

Persuasion is the process of involving the customer, not of giving the most dazzling sales pitch. People who make the best presentation do not necessarily make the best salespeople.

People believe more what they say themselves than what other people say. And salespeople who give customers the opportunity to become involved in the buying process are more persuasive than salespeople who do not.

Figure 9–1 (Cont.)

> High-performing salespeople also know that getting information from the customer and giving the customer a choice are at least as important as giving information. When the customer shares information with the salesperson, he or she is getting involved in the buying process. For this, listening skills are important: the best salesperson is someone who talks less and listens more.
>
> Opposition is a sign of involvement. If the customer opposes or disagrees with the salesperson, it should be considered a neutral or positive sign—not a negative one. Any involvement by the customer is better than no involvement at all—the toughest customer does not give any information about what he or she is thinking. When the customer raises objections, he or she is taking an active part in the buying process.
>
> Involvement can help both customer and salesperson: the more involved the salesperson is in helping the customer solve a problem or need, the more persuasive that salesperson will be.

Source: *Face-to-Face Selling Skills,* The Forum Corporation © 1984. Reprinted with permission of The Forum Corporation.

training program. These principles—Focus on the Customer, Earn the Right to Advance, and Persuade Through Involvement—are discussed, and are part of a small-group exercise in the first module of the program. Because they represent the philosophy of the program, they also are woven throughout each unit. Thus, when participants learn how to understand customer needs, they do it in a way that is explicitly consistent with the principles. Used in this way, the message drives the salesperson's behavior. Just as individuals learn the golden rule as a means of behaving in their daily life, so does this work with sales behavior.

The principles in Forum's program are based on five years of sales competency research; however, driving principles can come from basic research into critical behaviors or from other sources. Many sales trainers will create such principles by pulling them right from a corporate strategy statement or a similar document.

Participant Communications: Make It *Sound* Important

Here is an excerpt of a standard prework letter welcoming participants to a sales training program.

> Welcome to Acme's Basic Sales Training. The 2-day program in which you will be participating is designed to equip you with a range of skills

fundamental to selling. These include listening, questioning, selling features and benefits, understanding needs, closing, and handling difficult situations. Attached is a detailed outline of the program.

Contrast the preceding excerpt with the following:

Welcome to Customer-Centered Selling, the cornerstone of our sales training curriculum. The 2-day program is designed to equip you with the skills and principles critical to differentiating Acme—how we treat our customers. Although we will cover some key skills, we will also focus on how to look at selling from the customer's point of view, a perspective that should enrich your experience.

These two letters describe the same program, yet one *sounds* more important because of an "extra layer." The program objective and emphasis are more indicative of what is crucial to Acme and to what leads to success. This is especially important in prework and in other materials used to describe training.

Choose a Program Title Carefully

New product developers and marketers in industrial companies have been known to spend up to six figures to create the perfect name for a product. Yet, in training, the opportunity to link the program's philosophy to the title is frequently missed. Contrast "Customer-Centered Selling" with "Basic Selling Skills" or "Creating Customer Partnerships" with "Key Account Selling." Although subtle to some people, a small change can make a program seem much more relevant and important.

Create Slogans and Symbols

An easy and highly effective approach is to include corporate slogans in training programs. At one annual sales meeting, a sales force at a major equipment manufacturer created a motto for the year, "Obsessed to Serve the Customer." This principle was incorporated into every aspect of their sales training, from its revised title, "Customer Obsession Skills," to its internal brochures, to its follow-up job aids. The concept of using the language and philosophy that already exists helps tie program themes into what people have already bought into.

In another situation at a consumer products company, sales managers expressed a primary concern that salespeople were wasting good sales calls with unproductive accounts. An intensive program was developed by the sales training department to teach time and territory management to the sales force. Two years later, the participants find most memorable and useful the main message of the program: "If it's not worth doing, it's not worth doing well."

Creating a *visual* identity for a program can be a quick but effective way to convey your philosophy as well as to remind program participants of what they have learned. Such symbols are particularly applicable for posters, wallet cards, binder covers, and even letterhead for program communications. Most successful for us has been a logo used for a sales negotiations program that shows hands shaking, as if agreeing on a deal. We have given special clients permission to reproduce this logo in their program colors.

Keep Your Explanations Clear and Brief

It's no secret that training gets more support when line managment understands it. Unfortunately, line managers rarely have the time to hear details of a program. If, for instance, you only have five minutes to explain a sales training program, solely reviewing the schedule will not be memorable or exciting to the listener. Nor will a long laundry list of benefits. The big picture in this case is often a summary of what people will get out of the program. Being able to "net it out" with line managers is important, as exhibited by the following statement made by a highly successful sales trainer:

> You sales managers only need to know three things about this program. It says that
> a. Sales from new accounts cost 5 times as much to get as sales from existing accounts.
> b. Losing existing accounts has a huge negative impact on this company's profitability.
> c. The way you manage existing relationships has a lot to do with what you know about your customer's business.
> People learn how to do the last thing in this program better than anyone else.

The result, in this situation, was that all sales managers insisted on learning how they could get their people through the program as soon as possible and how they could support and reinforce it. They never asked for more detail because the big picture was enough to communicate the message.

Four Potential Pitfalls to Avoid

People tend to have four common problems in coming up with a big picture. One, and probably the most common, is closing down the discussion too early. When you have a high need for closure and tough deadlines staring you in the face, it's hard to listen open-mindedly to people suggesting alternatives. As we have alluded to earlier, however, you must allow enough time to do the necessary conceptualization before you do program design. Once the concept has been put together, it is relatively easy to come up with a course design and to develop materials. The cost of rework if you are constantly changing the basic concept is extremely high.

The second major problem people experience is analysis paralysis. This usually comes to pass when some members of your team want to see more data or do not feel comfortable with the current version of the basic concepts of the program. This can cause the project to drift and schedules to be missed. Conceptualization needs to be treated as an urgent item because so much development work follows it. It is difficult to create this sense of urgency when there are still several months left in the project, but it must be done. It also helps to get agreement on how to test the ideas with the target market or target population to arrive at a solution, and to have a clear guideline on who will make the decision if consensus is not possible.

The third typical problem is politics. It is impossible to take a stand if you are hesitant to take issue with those in your organization who feel differently about things. The best approach to minimize the politics is to seek input early and often, keep people informed about project progress, demonstrate what's in it for them, include the stakeholders, and make sure your sponsorship is clear and constant in its support.

The final common problem is thinking you can do it on your own. There are many heroic project managers who believe that they alone can distill all the information into a workable program. Even if it's possible to do this in terms of the substance of the program, it is impossible to do it in terms of the politics. Get a team involved early and keep the same team involved throughout the process.

SUMMARY

Just as organizations cannot operate at peak performance without a mission, a training program can seem fragmented without a carefully conceived set of principles and beliefs.

You will know you've succeeded when you have a top management team interested in, and committed to, your program—because they will view it as "more than just a training session." On a more practical level, you will have created a program that changes attitudes in addition to skills, that requires very few design changes down the road, and finally, that is linked directly to successful execution of your organization's strategy.

REFERENCES

DeBono, Edward. *Lateral Thinking*. New York: Harper & Row, 1970.

Gordon, W. J. J. *Synectics*. New York: Harper & Row, 1961.

Prince, George M. *The Practice of Creativity*. New York: Harper & Row, 1970.

von Oech, Roger. *A Whack on the Side of the Head*. New York: Warner Books, 1983.

chapter ten

DEVELOPING AND WRITING SUCCESSFUL SALES TRAINING PROGRAMS

WARREN KURZROCK

Warren Kurzrock is president of Porter Henry & Company, Inc. He joined the Porter Henry staff in 1966 and became president in 1977. He has comprehensive experience in every facet of sales force development, including needs analysis, sales training, sales management, incentive compensation, and long-range planning. Starting as a sales representative for Republic Steel Company, Kurzrock joined Anken Chemical and Film Corporation progressing to branch manager, regional manager, and then to division manager. In his last assignment, he built the most profitable sales force in the company. Kurzrock is a member of American Society for Training and Development and the Sales Executives Club of New York and is a frequent speaker at the National Society of Sales Training Executives. He holds an A.B. degree from Duke University and a Master's degree in marketing from New York University.

This chapter leads you through a series of guidelines to develop a training program from the point at which you have identified a need for it. It will start with formulating an overview of the content and continue with selecting appropriate research and training methods, identifying learning objectives, developing a content and a treatment outline, planning the reinforcement of training, and finally, giving tips and techniques for writing participant workbooks and leader's guides.

FORMULATING A CONTENT OVERVIEW

Based on your knowledge of the subject matter, the organization, and targeted training participants, you should develop a broad overview of the anticipated content. Your content overview should read like a table of contents, indicating chapter topics. The overview will enable you to focus your research to develop the specific content to be included in the program.

For a basic selling skills program, your content overview may look like this:

BASIC SELLING SKILLS PROGRAM CONTENT OVERVIEW

1. Precall Planning
2. Probing
3. Listening
4. Selling Benefits
5. Handling Resistance
6. Closing
7. Postcall Analysis

DEVELOPING CONTENT: FOUR KEY METHODS _____

Several methods can be used to develop content for a training program. Using a minimum of two to three methods can help to validate resources and obtain diverse data. Your immediate goal is to develop specific ideas and content for the training program. Secondarily, you need to build *support* for the training you will deliver. Talking with members of the sales force and sales force management will establish an early rapport that will benefit you later on. The training will be easier to implement if many people have an opportunity to contribute ideas and suggestions. Their input is a subtle but important endorsement of the training itself.

The following are several methods you can use to gather input on the content, delivery methods, product knowledge, and selling skills which ultimately comprise your program.

Focus Groups: Six Steps to a Productive Session

One of the most popular and efficient ways to develop content is to conduct a focus group. A focus group is an open discussion in which you, the trainer, probe the trainees—salespeople or sales managers—to uncover their learning needs. To ensure an effective session, you'll want to take certain steps.

- Set an agenda ahead of time. This is absolutely crucial to the success of your focus group. Review your experience, familiarity with the products, and selling skills to select key areas for discussion. For example, what *exactly* do the sales reps or managers do? What are typical customer needs, problems, or concerns?
- Organize specific questions and information needs before the session. Knowing what you want will greatly increase the likelihood of getting it. Sample questions might be
  - Who do you call on?
  - What are their concerns?
  - Who are the decision makers?
- Use visual aids such as a flipchart to indicate your objectives and agenda.
- Invite/select four to six participants. Include the most successful people so you can identify the best ideas.
- Schedule a half-day session. Normally, a concentrated two- to three-hour period will generate lively and informative discussion. Longer sessions become redundant as people exhaust their ideas and lose interest.
- Audiotape the session or appoint a note taker to ensure that you capture the essential information. You may want to have the group write down

their ideas as a change of pace and a means of capturing the content with a minimum of effort.

Individual Interviews

While a focus group is an excellent way to gather content from a cross section of people, pinpointing specific product information for a sales training program is probably best accomplished in individual interviews. These interviews are useful in gathering general concepts, techniques, and experience and are particularly efficient when directed at specific product or application knowledge. To zero-in on your targeted information, arrange to meet with product managers or technical experts individually to gather information for your program. These sessions are often exhausting because of the intensity of the questioning. You may be gathering unfamiliar technical information, so keep them focused and short, not more than an hour at a time.

Keep in mind that product knowledge and selling skills often overlap. Use the individual interview to identify which products the program will emphasize. As you review product or application information, begin to tailor your program to those special products. Focusing on specific products or applications optimizes trainee learning and reinforces management's objectives and sales goals.

Interview questions you might consider are

- What products should be emphasized in this program?
- How would you describe your selling cycle?
- Can you provide a list of product features and benefits?
- What are the most common objections you get from customers?
- How do you handle those objections?
- What supporting statistical or factual information do you have to back up your selling points?
- What visual aids do you use?
- How would you describe your customers?
- Who are your competitors?
- What are your product strengths and weaknesses?
- What are competitive product's strengths and weaknesses?
- How do you sell against the competition?

This is not a complete list, but the answers to these questions will provide you with a base of information from which you can develop clarifying questions and request more specific and detailed information.

Field Visits

It is critical when designing a sales training program to schedule sample field visits. These visits will familiarize you with the real-life situations, from which you will gather information for case histories and identify some of the problems and pitfalls firsthand. Whether you "came out of the field" or not, it's a good idea to broaden or update your experience by traveling with a few people. Depending on needs and objectives, this could involve traveling with one or two people for a day each, or significantly more. Obviously, the amount of time you spend on field visits depends on what you're trying to accomplish while gathering information and how much exposure you need from the field. Work from a questionnaire to solicit the specific information you need and save time. While you are interviewing the salespeople, you will also be able to observe their skills and behaviors on field calls.

If you are developing a time and territory management program, the information you ask for during a field visit should include

- How many accounts do you handle on a consistent basis?
- What constitutes a small, medium, and large account?
- How many days per quarter can you realistically devote to making sales calls?
- What obstacles do you encounter in terms of managing your time?
- How do you prioritize your accounts and sales calls?
- How do you currently plan your schedule?

Here are some examples of how to research content during field visits.

HOW TO	EXAMPLE
Determine needs and objectives for field visit.	Review objectives with field sales manager, reps.
Gather information for case histories.	Conduct focus group(s) with sales manager and/or reps and training manager.
Identify problems/pitfalls.	Interview and observe.
Develop questionnaire to obtain specific information.	Review original objectives; ask questions to uncover strengths/ areas to improve or to confirm information.
Observe skills and behavior.	Make mental and written notes during "ride-with" or "work-with."

Mail Questionnaires

The advantage of a mail questionnaire is that it can build wide support for your training program because you send it to many people quickly and at low cost. It's a good idea to keep your questionnaire very short and focused to really get the essential information. For example, if you are seeking answers or responses to typical sales objections you might identify these and ask the salespeople how they would respond. Generally, a questionnaire of this type should be limited to no more than 20 questions, with a variety of closed-end and open-end responses. The size of your sales force will dictate how many and to whom the questionnaires are sent. Aim for a representative sample. To ensure a greater degree of validity, send out at least 50, recognizing that you will certainly not get a 100 percent return. A portion of a sample mail questionnaire to assess selling skills is illustrated in Figure 10–1.

Assuming that you have used at least two or maybe all of the four methods, described—focus groups, individual interviews, field work, and mail questionnaire—you are now confronted with the challenge of assimilating and organizing this information so that you can select a training method. There are many methods to choose from: several are described in the sections that follow.

Figure 10–1 Portion of sample mail questionnaire to assess selling skills

Selling Skills

Frequency Values

Always Usually Often Sometimes Rarely Never

☐ ☐ ☐ ☐ ☐ ☐ 1. Have a specific objective for each sales call.

☐ ☐ ☐ ☐ ☐ ☐ 2. Prepare before the call—to review customer needs and problems, and to develop plans or recommendations.

☐ ☐ ☐ ☐ ☐ ☐ 3. Spend adequate time on amenities (being social, sharing personal and trade chit-chat) before getting down to business.

☐ ☐ ☐ ☐ ☐ ☐ 4. Bridge into business conversation or presentation, professionally, without excessive delay.

☐ ☐ ☐ ☐ ☐ ☐ 5. Ask relevant questions to learn about customer needs or to get update on changes, applications, and problems.

Figure 10–1 (Cont.)

Always	Usually	Often	Sometimes	Rarely	Never	
☐	☐	☐	☐	☐	☐	6. Listen effectively and retain the information provided.
☐	☐	☐	☐	☐	☐	7. Present benefits and concepts in an organized manner.
☐	☐	☐	☐	☐	☐	8. Select specific benefits and tailor the presentation to meet customer needs.
☐	☐	☐	☐	☐	☐	9. Reinforce important benefits with sales aids, data, and successful experiences of other customers.
☐	☐	☐	☐	☐	☐	10. Respond tactfully to questions, comments, or objections.
☐	☐	☐	☐	☐	☐	11. Provide complete answers to customer concerns and objections.
☐	☐	☐	☐	☐	☐	12. Encourage feedback from customer at critical points in the discussion.
☐	☐	☐	☐	☐	☐	13. Ask for the order, or a commitment, at the appropriate time.
☐	☐	☐	☐	☐	☐	14. Accept "turndowns" or delays gracefully.
☐	☐	☐	☐	☐	☐	15. Am genuinely concerned about customer's time, and keep sales presentations brief and to the point.
☐	☐	☐	☐	☐	☐	16. Have a good understanding of customer's business, products, needs, costs, and objectives.
☐	☐	☐	☐	☐	☐	17. Understand and respect the decision-making authority and chain-of-command in customer's organization.
☐	☐	☐	☐	☐	☐	18. Avoid "over selling" or making misleading claims and statements.
☐	☐	☐	☐	☐	☐	19. Handle group selling presentations effectively.
☐	☐	☐	☐	☐	☐	20. Refrain from "knocking" competition.
☐	☐	☐	☐	☐	☐	**Totals**

SELECTING A TRAINING METHOD _____

Prior to developing your sales training program, you will need to choose the methods you will use to deliver the training. Commonly used methods include

- Lectures
- Readings
- Preworkshop assignments
- Programmed instruction
- Structured discussion
- Brainstorming
- Case histories
- In-baskets
- Role plays
- Team workshops

We'll review each of these methods and give you suggestions for making each work for you. But, first, let's take a moment to explore some of their common characteristics. No matter which methods you use, they will fall either into the self-study category or the workshop category. (Several of these methods are described in more detail in other chapters of this book.)

Self-study

The major advantage of the self-study method is that it is cost-effective—it doesn't require a meeting leader or trainer to be with the participants throughout the instruction. It also allows the learner to study at his or her own pace. Self-study, done effectively, can also be much more efficient and effective than a workshop. It is more structured and the information in the material can be reinforced as often as required.

Self-study is more useful in imparting knowledge than in developing skills. The obvious reason is that people need practice and feedback from a "coach" to fully develop skills.

Self-study can be effective in introducing skills: why they are needed, when they are to be used, the actions they consist of, and the order in which those actions are to be performed. In sales training, self-study manuals sent as preworkshop materials can be used to address such skills as:

Establishing rapport—Psychological aspects of selling

Probing—Types of questions and their uses

Listening—Definition of active listening, benefits thereof and how-to's

Handling Objections—Types of objections and steps in handling each

Selling Benefits—Features, advantages, benefits defined; how to relate benefits to customers

Two benefits you can realize in your workshops from using self-study materials to introduce skills are:

1. You can replace a number of "introductory lectures" with shorter, more participative "reviews" of preread materials.
2. You can structure your training based on a universal, higher entry level of trainee knowledge.

Self-study may be used to present the knowledge required in such areas as:

Product knowledge

Orientation to corporation/division/sales structure

Systems

Competition

Customer profiles

Sales support and internal resources

Much of this knowledge is required for new-hire salespeople. Incumbent salespeople can learn about changes, new products, and new systems via self-study materials. A shortcoming of self-study is that you must literally test the participant to know how much information he or she has acquired. You generally cannot test through observation or discussion, as you can in a workshop—unless there is to be a workshop which requires application of the information, or workbook exercises are sent to the trainer for grading.

Self-study options are numerous and range from very simple reading assignments to sophisticated programs. The self-study methods we'll discuss here are

- Readings
- Preworkshop assignments
- Programmed instruction

Reading assignments are given to learners to introduce them to lots of information. Readings are most effective when followed by accountability exercises. You may want to announce a discussion or focus group on the readings as a way to motivate learners and evaluate their comprehension of the material. An outline accompanying the assignment will guide the learner to the salient points. Readings should give the learner information that will help solve real-world problems. Later we'll see why workshops are an ideal format in which to exchange feedback on the reading assignment.

Reading assignments can be appropriate to any training program as an introduction to the topic of the training, to state objectives and help set realistic expectations for the program, and to motivate and excite participants. A reading assignment could be as short as a one-page introduction and agenda or as long as a short book.

In preparation for a sales management program, you may send out copies of a book, such as *The One Minute Manager* as a premeeting reading. If you do use a book, be sure it is short, easy reading or you'll create a negative response from participants before they even enter the classroom.

Preworkshop assignments may include a reading assignment or having learners prepare specific information about their territories (number of accounts/potential accounts, types of accounts, etc.). Generally, the assignment prepares the learner to apply skills or knowledge learned in the workshop to real-life selling situations.

For example, in preparation for a sales presentation skills program, you may ask participants to select an upcoming real-life presentation to develop in the training session. Additionally, you may ask them to answer the following questions:

- What is the objective of your presentation?
- Who is your audience?/How large is your audience?
- What information does your audience need to know to do what you want them to do?

In general, preworkshop readings and assignments should take no more than 1 to 2 hours. Participants will be giving up valuable selling or managing time during the seminar. Don't ask them to give up much more than that.

Programmed instruction is step-by-step learning directed by a book, audiovisual tape, or computer program. The learner reads or listens to a small amount of material, called a "frame," and is immediately tested on it. The correct answer is always made available for comparison and review. Depending on how the learner answered the question, and depending on what type of instruction this is, the learner will be instructed to (1) move ahead or return to previous frames—

a "linear" program—or (2) access related frames—a "branching" program. You will decide on the content of the program, but the learner must complete the exercises alone.

Programmed instruction is an effective method for teaching product knowledge. Product knowledge is very technical and often involves memorizing product operations, features and benefits, restrictions and capabilities. Self-study materials allow learners to work at their own pace and to go back and review information if necessary.

While a certain amount of practice can be simulated with video or audio in a self-study program, it's essential for sales training programs involving selling skills to provide some type of workshop or practice. Therefore, if you are considering using self-study, note that it is most effectively used to *supplement* the workshop in the form of a preworkshop assignment or after-hours reading, or to build certain self-study segments into your overall program.

Workshop

Workshops, not self-study, will probably be the preferred format for most types of sales training. However, it is cost-efficient and -effective to integrate some self-study into the workshop format in the form of preworkshop reading assignments, self-study in the evening, or even follow-up material after the workshop is completed. The major advantage of the workshop is that it allows you to incorporate the skill practice that is vital to sales training. It permits the use of role practice as well as a variety of other learning methods to build participation and reinforce the knowledge and the selling skills.

The Advantages of Conducting a Workshop. There are many advantages of conducting a workshop. In addition to providing practice and enhancing the interest with participative exercises, it also enables an exchange of ideas between the participants. This is particularly useful if you are training or retraining people with field experience. Another advantage is that it can help shape positive attitudes. Enthusiasm is catching, and people working in a highly motivated environment will tend to get caught up in the momentum and experience a positive attitude change.

Size Limitations to Consider. Certain constraints need to be addressed in planning and conducting a workshop. First, you are limited in the number of people that can be trained in a group. Depending on the level of training, you may be dealing with as few as 1 or 2 people or as many as 15 or 18. The smaller the group, the most effective the learning. It takes a skilled trainer to handle a larger group and to make sure that everyone has an opportunity to participate. Most experts feel that you have to make a trade-off at some point in terms of utilizing group dynamics and building participation. Fewer than 10 participants in your group usually means that you will lose some of the group dynamics but will gain in individual learning. On the other hand, when you get more than 10 or 12 you will increase interaction, participation, and competition which will enhance training. There is no magic formula for the size of the group, but under no circumstances should it be so large as to limit participation. An additional requirement for workshop training is that you will be required to do intensive preparation so that all of the objectives can be reached and proper reinforcement takes place.

Some of the training methods used in workshops are

- Lectures
- Structured panel and open forum discussions
- Brainstorming
- Case histories
- In-baskets
- Role plays
- Team workshops

Lectures are discourses given to an audience and as such present a somewhat passive learning experience for the trainee. On way to liven up a lecture is to allow questions from the audience. A lecture can be appropriate for delivering a great deal of information to a large group at a relatively low cost. If you give a lecture, use dynamic, positive language. Illustrate your points with interesting examples. Speak in a pleasant, upbeat tone and talk loud enough to be heard by everyone. Organize your lecture on a single theme and offer a variety of self-contained and independent data to support it. Stick to real-world situations and avoid abstract or overly conceptual ideas. Use analysis, examples, statistics, and quotations to make your point. Wherever possible, add displays to vary and broaden your delivery. Another technique is to leave obvious areas open to your presentation and allow learners to comment on and question the material.

A *structured discussion* is a conversation among learners about very specific training needs and objectives. You should define these objectives ahead of time and draw up a list of questions to ask the group. If you want to create a highly structured discussion, pass around an outline or post one where everyone can easily see it. The key is to direct the learners to focus on one topic. Structured discussions are effective when learners are looking for guidance and not in need of a forum to express their ambivalence about the learning objectives.

A *panel discussion* is a series of short lectures by several professionals, often followed by a question and answer session. An *open-forum discussion* means that any member of the group may address any other member, in which case a moderator may guide the group.

Brainstorming is a discussion among learners about real-life solutions to real-life problems. This is a powerful means for the learner to practice holding off on making a decision or judgment until all or most of the possibilities have been aired. It also teaches learners to tolerate opposing ideas. You'll need at least five or six participants for the session. Preset a topic that involves solving a real-life problem and instruct the learners to come up with as many good ideas as they can. Discourage analytical or judgmental thinking while new ideas are being generated. Encourage learners to build on ideas that have already been shared. You'll want to keep track of everyone's suggestions via a background, flipchart, or overhead transparency. When you feel you've gotten the best ideas from the group, encourage discussion and analysis of each. Together with the group, evaluate the problem-solving potential of each idea. Establish criteria for successful solutions. Use these criteria to plan a course of action for the learners to take back with them.

Case histories are written models of the learner's real-life situations. They must be authentic to the last detail, believable, and they must describe a conflict that the learner must solve. If you present the case history in a group discussion, be sure to set specific, measurable objectives for the group. Ask for ideas, actions, decisions. Another option is to ask learners to write their own case histories based on current business. These can be shared in a workshop or group discussion. The key is to ask for action-oriented solutions to the problems outlined in the case history.

In-baskets are a collection of written materials that simulate the actual contents of the learner's in-basket on a typical workday. You'll want to instruct learners to process all paperwork exactly as they would in the office. Specify a time limit and then have them return the materials to you. You must review the materials and give feedback to the learners in a discussion. Reinforce the strong areas, point out where updating is in order, and allow the learners to evaluate their own performances. In-baskets are good pre- or postworkshop assignments and provide specific, measurable results.

Role plays or simulations are reenactments of situations learners have encountered, or expect to encounter. Role playing these scenes allows learners to evaluate how they acted/reacted and to try out other behaviors. Learners might initially feel embarrassed or uncomfortable with role playing, so ease into it gradually. You can initiate a simple role play with a learner as a means of answering a comment or question. For example, a learner tells you "One of my reps likes to ask for suggestions and then blatantly ignores everything I say." At this point, you should ask the learner a question that leads the learner to "act" the rep's part. For instance, "What was the last question he asked you?" The learner is role playing the rep's part, so ask for a volunteer or pick someone to play the learner's part. You can have everyone doing simultaneous role plays, but keep the number of participants to two or three. When assigning a preplanned role play, give everyone an objective to reach for. You may set time limits, but don't pressure learners to resolve every problem. The important thing is that the key exchanges were simulated. Feedback or a debriefing session is very important after a role-play session. Always give participants specific, measurable goals and work with the learners to evaluate the content of each role play.

Team workshops are units of three to four learners who break off into groups to confront an issue/idea and arrive at a tangible "solution," for example, a decision, recommendation, or report. These units work simultaneously and are called on to share their conclusions with other teams. It is crucial for you to plan these assignments ahead of time.

Here are some other terms with which you should be familiar. You probably know many of them already, but notice how they've been adapted here.

Clinics are sessions in which learners get together to solve a particular problem. A discussion format is common so that everyone has a chance to contribute. This is a good means to teach learners how to make decisions, solve problems, and become team players.

Conferences are generally gatherings of members of the same profession for the purpose of sharing information on a common theme. Activities may include lectures, discussions, clinics, and workshops.

Seminars are usually highly structured meetings using a combination of formats or a serial lecture with several speakers who each give a talk on the same theme or topic.

Symposia are gatherings, formal or informal, where specialists speak on a common topic. They often resemble panel discussions.

As we mentioned, the two preferred formats are self-study and workshop.

The information that follows will deal with the workshop format. This is quite often the most difficult format to structure. It may also incorporate some self-study elements.

LEARNING OBJECTIVES FOR STRUCTURING YOUR WORKSHOP

While you may have certain objectives in mind, based on your sales training needs analysis, it is essential that they be stated in behavioral terms, that is, include an observable action and be specific, measurable, and results-oriented. This is especially important when dealing with skill development. Behavioral objectives take time, effort, and a great deal of practice to develop. To create learning objectives, you should identify an overall objective for your program as well as learning objectives for each specific segment. Later, as we discuss outlines and organization, we will deal with segments and modules. However, for now, recognize that designing and developing a program means breaking it down into segments, most of which will be anywhere from a half-hour to 2 hours long.

We said that an effective sales training objective should be behavioral, that is, describe the intended results of the training, not the process itself. It should be stated in precise terms. It should describe what the learner will be able to do as a result of the training. Equally important, when the conditions under which the performance is expected to occur can be described, you will be off to a good start. Finally, if possible, you should also specify the quality or level of performance considered acceptable. The more specific and measurable you can make your objectives, the more effective the ultimate sales training will be. If you have a nebulous or general objective your training will be equally general and probably lack the necessary focus to get the job done. On the other hand, if you truly start out with a realistic and measurable objective, your training will be the better for it.

For example, an overall learning objective for a one- or two-day sales training program or workshop might be: "Each salesperson will be able to make a complete sales presentation on product X using our six-step selling process."

Another example that relates to a module of the same program might be: "Each salesperson will be able to handle the customer price objection by following our formula for handling objections and utilizing one of four responses to offset the objection itself."

It is important to recognize that this is not an exercise for form only and that the objectives become your stepping stones in the program. You should constantly refer to them while designing the program, and you should include them in the program itself so that the learners know what is expected of them. And equally important, the objectives should serve as guidelines for incorporating the training elements necessary to deliver the program. Once your major objectives are in order, you are ready to begin writing.

BEFORE YOU WRITE YOUR PROGRAM: TWO CRITICAL OUTLINES TO DEVELOP

Before you can begin serious writing, two critical outlines should evolve from your objectives and content research. The first is a straightforward content outline which is simply a list of necessary topics in order of learning sequence. The second is called a "treatment outline." It is somewhat more detailed and essentially describes the overall activity sequence, timing, and other elements in the program. It's important to do both, beginning with the regular outline and eventually evolving it into a treatment outline. This ensures that nothing important gets left out and that your learning objectives are achieved as the result of the training.

What to Include in Your Program: The Topic Outline

The topic or content outline is relatively easy to do. The main concern is ensuring that you've covered all the necessary topics and started the process of putting them in the right sequence. For example, you may have to decide whether to teach product knowledge or selling skills first. Next, you would develop a series of steps and subtopics which cover the necessary information and skills. There is no established format for doing this outline; use a system with which you are comfortable. Just remember to be consistent. If you start out with capital letters, don't suddenly switch to Roman numerals. A sample topic outline is illustrated in Figure 10–2.

If you follow this procedure, you will have a sense of your program's overall dimension. You should ruthlessly discard topics you feel are not necessary for

Figure 10–2 Sample topic outline

TOPIC OUTLINE
SECTION 1
PROGRAM INTRODUCTION

I. Introduction: Why an LPA Training Manual?
   A. Significance of LPA
      1. A "new" concept (new method for administering "old" drugs)
      2. A wide-open market with great potential
      3. Competitive efforts
   B. A more sophisticated sales approach
      1. Requires sales rep/LPA specialist to educate hospital personnel about entire LPA *system*
      2. Requires sales rep/LPA specialist to build credibility by exhibiting knowledge of pain and pain treatment
      3. Involves group decision at the hospital
II. Purpose of This Program
   A. As a training manual this program will help sales reps/LPA specialists:
      1. Build broad knowledge base about pain treatment and LPA
      2. Develop a strategic approach to selling LPA as a total system
      3. Increase sales
      4. Establish a competitive edge
   B. As a reference manual this program will help sales reps/PCA specialists:
      1. Add to existing knowledge base as further information about LPA-based pain management is developed

---

achieving your objectives. Keep in mind that while the primary importance of this outline is to develop a sequence, it's also an important step in gaining the big picture, and in providing a mechanism for feedback from other people within your organization. In some instances, this may be your immediate boss; in others, it may be people within the sales organization. In very large organizations, it may be a "client" who is actually another division for whom you are designing the program. By keeping the other people informed, whether they be subject-matter experts or clients, you will obtain more technical expertise, language, product features and benefits, and customer objections. It will also give you better insight into sequence and organization and will create buy-in or ownership.

It's a good idea when starting this topic outline to restate your objectives for each element. It's also helpful to add some "sell" in the training benefits to the trainee and state them strongly and clearly for motivation and clarity. Your other concern should be flow—it should be logical, starting with an overview of each topic, then getting down to subtopics.

Treatment Outline

The treatment outline is perhaps the most crucial element of your entire program and therefore cannot be skipped. It will guarantee the quality and learning effectiveness of your entire program, and essentially become the design document for the program itself. The treatment outline includes a detailed account of content as well as the method of presentation/learning to be used.

Determination of methods to be used depends on a number of factors. First, of course, is choosing a method that will achieve the learning objective. A directed discussion, for example, is not the preferred method for practicing probing skills. Other criteria relate to limitations of money, time, and space. Of greater importance, however, are criteria related to the nature of human beings and how they learn.

Methodology must take into consideration such issues as:

Is it early morning and trainees are at the top of their form or is it late afternoon and everyone is drooping?

What will trainees have done during the topic preceding this?

You can see how time of day can help you choose whether to do a role play, give a lecture, show a film, or have small group discussions.

You can also see that, if trainees will have just finished listening to a lecture, you will want to present the next topic in a highly participative manner.

Taking these criteria into consideration will help you make sound decisions regarding program design.

One simple format for you to follow involves the use of two columns. An example is shown in Figure 10–3. In this example, we have indicated the module number, the module title, the time required, and most important, the learning objective for that particular module. The two columns are labeled "content" and "method." This format enables you to rework your topics to fit them into a tight framework of a module that functions within a specified period of time. The first column lists the topics covered and is actually a reshaping of your earlier topic outline. Next to each topic indicate the method to be used. If you are providing the learner with knowledge, as opposed to skills, you can do so through lecture, video, films, reading, or other "nonparticipative" methods.

At the same time, you will also have to start thinking of skill elements and how to build and reinforce each one. You can develop skills through conference or discussion, team workshops, or role practice sessions. Skill development must be buttressed with a great deal of reinforcement and practice. Salespeople cannot learn the skills simply by listening or watching. They must do them. The closer you can come to a real-life situation, the more effective your results will be.

The treatment outline must be carefully designed to include all knowledge elements and adequate reinforcement of the skills themselves. Depending on the types of skills you are teaching and the other elements of your overall training, it may be desirable to build the skill in a sequence going from information about the skill, to discussion, to preliminary practice or application, to an ultimate reinforcement with role plays. These are considerations that you should, of course, build into the treatment outline.

Five Steps for Reinforcing Learning. In designing a sales training program, one of the most important elements is making sure that the learning material, whether knowledge or skill, is adequately reinforced. While there are no easy formulas for doing this, the guidelines that follow should be helpful to the neophyte who is designing a selling skills program.

In creating the treatment outline, the designer should follow the steps that are often found in linear programming:

- New information should be introduced in small doses or simple steps. By providing the information in incremental steps you allow the learner to digest one step at a time before going on.

- The information or skill needs to be reinforced immediately, either through feedback or in an exercise. It is critical that the learner have an opportunity to try out the knowledge or skill while it is fresh in his or her mind and get feedback or positive reinforcement.

- While it's important to reinforce information immediately, each element should also be buttressed perhaps three or four times during the program— as an isolated step, an opportunity to summarize, or in a try-out in a larger situation. For example, if you are teaching salespeople to handle objections

Figure 10–3 Sample treatment outline

Objective: Participants will be able to describe why physicians agree to prescribe specific products, differentiate between product features and benefits, describe the benefits of using visuals to support a presentation, list types of visuals, and use visuals effectively.	Module: 4 Title: Features and Benefits/ Using Visuals Time: 60 minutes class time

Content	Method

A. Introduction
  1. Objectives: Participants will be able to
    a. Describe why physicians agree to prescribe specific products
    b. Differentiate between product features and benefits
    c. Describe the benefits of using visuals to support a presentation
    d. List available sales visuals
    e. Use visuals effectively
	Meeting leader lecture and discussion

2. Agenda
    a. Why physicians prescribe specific products
    b. Features and benefits
    c. Benefits of visuals
    d. Types of visuals
    e. Using visuals effectively
	Overhead transparency 4–1

B. Audiotape	Audiotape 1 in which sales representative presents only features, not benefits. Discussion. Audiotape 2 in which same representative sells more effectively by selling features and benefits.

and are using a three-step process, it is desirable to provide them with each step and then reinforce that step with some exercise or practice. Later on, through role play, have them use all three steps together so they can synchronize the situation and use it realistically.

- It's desirable to use a variety of ways to apply or reinforce this information so that the group is constantly challenged. Learning can be improved if you make it interesting and challenging.

- Finally, follow a defined sequence. Don't provide new information until the prior material or step has been learned and practiced.

There are many ways to incorporate interaction and reinforcement. The trainer must be skillful in selecting those ways that best reinforce the appropriate knowledge or skill and, at the same time, use variety to enhance the learning. The following is a list of activities that reinforce learning. They are provided in two lists to indicate which are best for learning knowledge or information and which are best for developing skills. Keep in mind that there's frequently a cross-over in that one method can be used for both knowledge and skill.

Activities That Reinforce Learning

KNOWLEDGE

- Questions from the trainer
- Note taking
- Group discussion or conference
- Small-team discussions
- Individual application exercises
- Small-team application or written exercises
- Quizzes or tests

SKILL DEVELOPMENT

- Team workshops to solve problems and apply skills
- Role-practice sessions, either individual or in small group
- Games and simulations
- Interactive video applications, with teams of individuals responding to video-taped scenarios

In designing or writing programs it's important that the trainer have a good understanding of all these methods and devices. Particularly in the conference method of sales training, team workshops and role practice are "musts." Role play, because it is so important, has a chapter devoted to it elsewhere in this book.

SPECIFIC FORMATS FOR DEVELOPING AND WRITING PROGRAMS

As indicated earlier, you may be developing many different programs that require a variety of formats to support them. In the section headed "Selecting a Training Method," we examined several formats and classified them as either workshop or self-study. Because the workshop and self-study formats are two of the most popular and effective training devices, we'll use them to illustrate how to develop and write good programs.

Writing a Leader's Guide

You could be writing a Leader's Guide for your own use in a workshop or one that could be used by others. One effective way of translating your treatment outline into an active format, as described earlier, is to use the layout illustrated in Figure 10–4. This format consists of two columns: "Directions–Techniques–Aids" on the left and the "Instructor's Outline" on the right. The left-hand column is used to explain the things the instructor should be doing, such as showing an overhead transparency, asking the audience to do something, giving out an assignment, and so on. The right-hand column can either be a word-for-word script or an abbreviated outline of a script. If you are writing a leader's guide for someone else to use, or for a number of people to use, write it word-for-word and let them tailor it to their own needs and style. At worst, they will have sufficient examples and more words than they will actually need. On the other hand, if you are writing it for yourself, you probably would not spell out every concept, but would use an abbreviated style. You can further distinguish the two columns visually by using upper- and lowercase letters in the left-hand column and all caps in the right-hand column.

Writing a Participant's Workbook or Manual

There are a few differences between a participant's workbook used during a workshop and a manual designed for reading or self-study. The formats and style are basically similar. The major difference is that the self-study program is a complete piece designed to be read without interruption, whereas the participant's workbook would include a variety of interim exercises, case histories, and reading materials to recap what was covered in the workshop. A sample page of a participant's workbook is illustrated in Figure 10–5.

Figure 10–4 Sample leader's guide layout

DIRECTIONS-TECHNIQUES-AIDS	INSTRUCTOR'S OUTLINE

Discuss Sales Negotiating Opportunities

OVERHEAD TRANSPARENCY 1-4

SALES NEGOTIATING OPPORTUNITIES	Let's review some of the many negotiating opportunities and get examples.

> SALES NEGOTIATING
> OPPORTUNITIES
> • Obstacles
> • Requests for add-ons
> • Organization changes
> • Formal contract negotiations
> • Competitive activity

Ask for other examples as you define the five opportunities.

Let's review some of the many negotiating opportunities and get examples.

• An **obstacle,** for instance, might be an important objection such as "price."

• **Requests for add-ons** include free goods, extended warranties, and so on.

• **Organization changes** occur with both seller and buyer: if we raise prices, this is an organization change. On the account side, a change in buyers could trigger certain issues being renegotiated.

• **Formal contract negotiations** are another opportunity—they are often done annually and are usually lengthy and structured.

© Copyright 1986, Porter Henry & Co., Inc.

Figure 10–5 Sample workbook page

PLANNING THE KEY SALES CALL

The five major steps in planning the key sales call are described below:

1. The **lead-in** is just that—the step by which you begin the meeting. It consists of introductions, your key sales call objectives, a major benefit—what the account will gain by agreeing to your proposal—and the agenda. Introductions should be handled in the manner most appropriate to the situation. The choices are introduce each person individually, have each do a self-introduction, or orchestrate the process yourself as a "master of ceremonies." Your key sales call objectives are what you want the account group to do—take the next step, commit to a new concept, etcetera.

   The lead-in will be satisfactorily completed if it
   • Accomplishes introductions.
   • States sales call objectives.
   • Provides a major benefit.
   • Overviews agenda.

2. The **update** brings all the account participants "up to speed" so that everyone is at the same level of understanding. The update should briefly explain the facts, calls, relationships, account needs, and other historical events preceding the key sales call. The update can also be an effective way to correct misconceptions concerning the purpose of the meeting.

   The update will be satisfactorily accomplished if it
   • Describes what has happened to date.
   • Establishes the rationale for the meeting.

© Copyright 1986, Porter Henry & Co., Inc.

Developing and writing programs are skills that require practice, and more practice. However, anyone with sufficient desire and the willingness to follow proven guidelines can be successful. Here are the guidelines:

1. Get accurate input (reflecting your prior needs analysis) and buy-in from both field and headquarters salespeople and sales management.
2. Establish realistic, measurable, specific learning objectives.
3. Develop a tight content outline.
4. Write a complete treatment outline.
5. Write your program with good communication principles in mind.

chapter eleven ─────────────────

THE LEARNING ENVIRONMENT AND ITS IMPACT ON YOUR SALES TRAINING

ISABEL L. KERSEN

Isabel L. Kersen is director of sales training and communication, Cosmair, Inc. A native New Yorker, Kersen began her professional career as an English teacher in the New York City school system. In 1972, she was hired by the Xerox Corporation as an independent consultant-writer in training and development. When she left there, three years later, she was a professional sales trainer. Since then, Kersen has handled training assignments for the United States Navy, Hertz Truck Leasing, Hertz RentaCar, Warner Cosmetics, and Cosmair (the U.S. subsidiary of L'Oréal International). Although she specializes in sales training, Kersen also conducts workshops in supervision, presentation skills, delegation, time management, and other management areas. She has extensive experience in assertiveness training and instructor training and appears as a guest speaker at a variety of professional functions. Kersen lives in New Jersey and commutes to Cosmair's corporate headquarters on New York's Fifth Avenue.

This chapter explores the importance of creating a training environment that contributes to the effectiveness of training regardless of where it is done.

You will find here information about a number of environmental factors that can greatly affect the success of your efforts. Those factors include facilities, locations, room sizes and shapes, seating arrangements, catering choices, support materials, and both the physical and psychological climates.

Guidelines and suggestions are provided to assist you in making choices and taking actions to maximize your potential for conducting truly professional training activities.

The learning environment is defined as a combination of elements which surround the implementation of the training program and which make up the situation in which the learner must function. As with any other environment, there are two categories of elements which come into play: the physical and psychological.

The physical elements of the learning environment are all the factors you can perceive through the five senses. This encompasses everything from the room temperature to how hard the seat of the chair is to whether a participant can hear you.

The psychological elements are not as easy to see but are at least as important as the physical—perhaps even more important. This category includes such things as your expectations and the expectations of the participants, the perceived relationship of the program to the on-the-job needs of the group, the group dynamics, and your skills in dealing with people and groups.

HOW ENVIRONMENT AFFECTS THE LEARNING PROCESS

Since all these elements will have a direct, and often powerful, effect on the degree to which people can and will learn, it is important that you, as a trainer, are aware of them and that you do as much as you can to make the environment one which facilitates learning rather than hinders it.

The relationship of the learning environment to the training program itself can be defined as a hygiene factor. What this means is that, if you have a totally ineffective training program, the best learning environment in the world will not make it any better. On the other hand, if you have a superb training program, a number of negative elements in the learning environment will doom the program to failure. Thus, we pay attention to the learning environment not to make our training programs better, but to protect the effectiveness of our programs.

The environment affects the learning process in two ways. First, each element that the learners perceive sends them messages—messages about value of this experience and the value placed on them by the trainer. Dirty rooms, poor food, inferior quality of course materials all tell trainees a lot about how this program, and its participants, are perceived by you and the organization you represent. Audio/visual equipment that does not work or a handout you forgot to bring to class sends a loud message about your concern for the success of the program—and its participants.

Second, each element contributes to the degree to which trainees are psychologically and physically comfortable or uncomfortable. This is an important consideration because discomfort of any kind interferes with one's ability to learn. If the room is too hot, too cold, too airless; if learners are crowded too close together, if the way you handled the last answer volunteered by a participant has embarrassed the participant and made the group uncomfortable, learning will be seriously impeded.

Such discomforts distract the attention of learners, interfere with their ability to concentrate, and, in some cases, can cause a total withdrawal from the learning situation. If, for example, learners are kept too cold too long, they may become unwilling to participate at all; some have even been known to walk out.

On the other hand, a high level of physical and psychological comfort makes it easier for participants to concentrate and to endure whatever level of discomfort they may be experiencing as a result of the act of learning. (It is important, when considering the learning environment, to keep in mind that learning new ways of doing things always carries some measure of psychological stress and discomfort.)

As a trainer you are responsible for the effectiveness of training activities. Since you create a training environment every time you conduct a program, you might as well make sure that the environment in which you place your learners works *for* you, not against you. To accomplish this, you must pay attention to all the following elements.

SEVEN MAJOR ELEMENTS OF THE LEARNING ENVIRONMENT

Seven major elements determine the nature and quality of the training environment. They are

1. Geographic location—the city in which the training will be conducted
2. Facility—the type of building which houses the training
3. Meeting room(s)—the room(s) in which the training will occur
4. Seating arrangement—the positioning of the participants during the course of the training
5. Food and drink—all meals and coffee breaks during the training
6. Support materials—handouts, manuals, audio/visual equipment and any other tools used
7. People—the participants, the trainer, and anyone else who becomes part of the training situation

The following section on each element discusses the ways it can affect the learning process, some of the options you will have to choose among, the advantages and disadvantages of each, and some guidance for making your choices.

For the purpose of these discussions, it will be assumed that you are in a rather typical sales training situation with salespeople scattered around the country and few trainers on staff. This scenario was chosen not only because it is the most common, but also because it surfaces most of the challenges. If your situation is different, you should find it easy to adapt the information to meet your needs.

Geographic Location

Tips on Choosing a Location. Unless you have a corporate training center or other designated area for conducting training (regional offices, perhaps), you will have to choose a location every time you plan a program. There are a number of considerations involved.

The first choice you must make is whether the training will occur in one place or you will take it "on the road." To bring trainees from all over the country into your home city or to any other single location, some of the following factors should exist:

1. There is a local plant or operation the trainees must see.
2. There are a number of local people who will be part of the program or who the trainees must meet.

3. You have made arrangements with a local facility that keeps room and board and other charges low.

4. Large numbers of trainees are close enough to your location to keep travel costs low.

5. Your training involves the use of equipment which cannot be moved around the country.

If such factors do not exist, it is simply much more expensive to fly people from all over the country to one location than it is to fly a trainer to various locations while keeping travel costs for the sales force low. If you find two or three trainees living within driving distance of a likely site, you will probably also have found your least expensive location so far as travel is concerned.

In addition to considering costs, you also want to make selections based on what will work best for your participants. The issues related to geographic location which are of interest to and affect participants are climate, ease of access, and the activities that are available when class is out.

Climate Considerations. Taking people into a climate which is, in any way, worse than the one in which they live makes them unhappy. Don't take people from Washington, DC, to Chicago in February. Don't take people from Atlanta to Florida in hurricane season. In addition to having to deal with the increased potential for travel delays and hazards, people who are moved to a "worse" climate arrive with the feeling that this training experience is a form of punishment. They also tend to consider the trainer insensitive for having so arranged the program.

In most cases it is impossible to justify bringing your Chicago people to Florida in February for a sales training program. You should, however, be aware of the seasons and the climates as you look at site selection and planning for your training. Why not plan to train people from the southern territories in February in Florida and save Chicago training for the warmer weather? In the event you cannot avoid bad-weather programming, do everything possible to choose cities most people can fly into directly and use facilities nearest the airports.

Travel Accommodations. Salespeople, many of whom feel as if they spend their lives traveling, are resentful of what they perceive as poor travel planning. Selection of a city with poor access results in long delays, changes of planes, wasted time, and disgruntled trainees. When choosing a city, check to see how all your trainees will get there. If they all have to change planes in Atlanta, why not have the program there, and save everyone the extra waiting and flying time? (Otherwise, you can be sure that by breakfast the first morning, everyone will know that everyone had a layover in Atlanta.)

In some situations, the training department budget contains money to cover participants' travel and room and board. If this is your situation, you must decide how to best spend those dollars. Analysis of air fares and hotel costs will play a large part in the selection of your training sites. It is important, however, not to let concern about dollars overwhelm the need to deliver effective training.

Activities—Look for Entertainment and Exercise Opportunities. The need to be concerned about what people can do when they are not in class varies from situation to situation. In the event of a one- or two-day program, the trainees are pleased to have each other to talk to or have dinner with. Any location will work. Arrival night is no problem, and neither is the first night the group is together. When a program goes longer than this, people begin to feel captive in one building, and they need places that they can get to easily and inexpensively.

Facilities isolated in suburban areas can be difficult to get out of without advance arrangements. If, like many trainers, you prefer not to spend your evening with the trainees, you really need to consider what your chosen location offers them to do on their own.

In today's culture, you risk great resentment if you take people to a place that offers no opportunity for physical activity. Pools and workout areas are available in a large percentage of meeting sites. If you cannot find, or do not prefer, a site which offers this, you should be sure you can at least offer people a place to walk or run—even in the winter. Facilities in unsafe downtown areas or desolate airport locations or right on suburban highways, unless they offer extensive on-site workout facilities and space to walk, should not be considered for programs longer than a day or a day-and-a-half.

Training Impact. By selecting geographic locations with these criteria in mind, you will help ensure that your participants face the morning of the first training day with no resentment or annoyance related to where you have taken them. Instead, they will be appreciative of the carefully selected location.

There is, incidentally, nothing wrong with spending a few minutes on the first morning explaining how you made your choice.

Facility: Five Basic Types from Which to Choose

The term facility refers to the actual building in which your training is to be conducted. Your selection of the facility you will use may be closely related to the geographic location you have chosen since some locations will not offer all the various facility choices. You have five basic types of facilities from which to choose.

1. Corporate training centers
2. In-house spaces

3. Conference centers

4. Hotels

5. Resorts

Each type of facility offers enough of the same basic services to allow you to successfully conduct training in any one of them. There are, however, sufficient differences among them that you need to be aware of which is most ideally suited to your situation. (Also see summary in Table 11–1.)

Corporate Training Centers. A number of large corporations have built and maintain their own training centers. If you are part of such an organization, you are probably on the staff of this center and many of your training choices have been made for you. If you are not familiar with this type of facility, you may be interested in knowing something about them.

Corporate training facilities are generally built by large organizations which

Table 11–1 Likely benefits and liabilities of various types of facilities

Facility	Likely Benefits	Likely Liabilities
Corporate Training Center	Keeps dollars inside corporation	Costly investment to build
	Greatest control of environment	Heavy ongoing costs
	Favorable impression on trainees	Usually in hard-to-reach location
In-House Office Spaces	Lowest costs	Space not designed for training
	Access on short notice	Difficult to control interruptions
	Proximity of internal resources	Difficulty in providing for food and refreshment needs
Conference Centers	All-inclusive price	Heavy cancellation penalties
	Strong "image" environment may add to impact of training	"Image" environment may detract from training
	Varied space selections	Isolated locations
Hotels	Easily accessible	Meeting rooms may be oddly built
	Trained, experienced staff	Noise level in adjacent rooms
	Able to meet varied needs	May not be a "meeting" hotel
Resorts	Luxurious, attractive settings	Usually isolated locations
	Low "off-season" pricing	May not be a "meeting" facility
	Recreational facilities may add to the value of the training	Recreation facilities may distract from training

have a need to train large numbers of people. It is not likely such a training center deals only with sales training. The starting point for a training center is sometimes the need for training a large number of people who are based in one geographic (usually the headquarters) area. Or there may be a need for technical training requiring the presence of extensive equipment that creates a demand for a permanent home for training.

In such custom-designed training facilities, all the environmental issues discussed in this chapter should have been properly provided for.

A training center also keeps hundreds of thousands of dollars that would have been spent outside for public facilities inside the corporation. It is the careful assessment of these savings versus the costs of maintaining the facility that forms the basis for a decision to build a corporate training center.

Doing training at one's own center has positive psychological implications. Being in a corporate location, the participant feels very strong ties to the organization. The training, more in this kind of setting than in any other, bears the corporate stamp, and the organization is perceived as much more paternalistic/maternalistic than when training is conducted outside.

Some training centers are truly training "palaces" or "resorts." These large, attractive, and effective facilities make very positive impressions on trainees and often enhance their opinion of the organization. There is assumed to be a resultant increase in loyalty to the company with an accompanying reduction in turnover.

One of the risks in building a corporate training center is an inability to keep it full. A half-empty or underutilized facility is a financial drain. Since these huge locations do not go away when training needs do or shrink in size as work forces are reduced, a number of companies have ended up with white elephants on their hands. The decision to build one's own training facility is a long-term commitment that should be made very carefully in today's volatile business environment.

Sometimes these corporate locations have space available for rental to outside companies. Availability varies with the needs of the owning corporation, but, if you can book them, they offer excellent rates. Normally, they charge a per person rate that is all inclusive: room, board, meeting room, and all audio/visual equipment. These rates are usually a fraction of the going hotel rates in the area. The quality of such centers is usually quite high, but never book space at one you have not visited or stayed at overnight.

Most corporate training centers are located in suburban or rural areas. Many are self-contained communities with pools, bars, billiard rooms, bowling alleys, gyms, running tracks, shops, and other amenities.

Several major airlines have training centers. So do large hotel and fast-food chains. Companies doing extensive technical training also tend to have training facilities. The Xerox Corporation's International Center for Training and Management Development in Leesburg, Virginia, is a magnificent facility which is sometimes available. Contact the headquarters of likely corporations for the location of their center and a name and telephone number to call for information.

A drawback to using the training center of another corporation is the tendency of the trainees to see the owning corporation in a favorable light and to compare it to their own. If there is already a high degree of dissatisfaction among your trainees, this is not a good alternative for you to consider.

In-House Spaces. Even without a training center, your company may have spaces that you can use for sales training. Among these spaces could be conference rooms, auditoriums, or meeting halls. There may be training rooms used for other purposes that may also suit and be available to you.

If you can conduct training in-house, you can save money while providing your trainees with the advantages of being in a corporate location. There are, however, a number of drawbacks you must be aware of. Access to your training room is usually easy for anyone in the office, and it is not unusual for the vice president of sales or other upper-level managers to drop in to see what is going on or to say a few words to the "troops." You will have to deal with these interruptions and unscheduled observations.

At break times, salespeople tend to wander off to find telephones they can use or headquarters personnel they need to talk to. In the event your meeting room has a telephone, don't be too surprised if a sales rep actually gives the number to a customer or if some sales manager gets through with an urgent need to talk to one of your participants.

Food and drink are often difficult to handle in this kind of setting. Unless there is a corporate cafeteria to provide coffee in large quantities, you will need to break more often so people can get coffee. There will also be a need to either go out of the building for lunch or lunch will have to be ordered in. The first alternative is very time consuming, and the second leads to the "Prisoner of Zenda syndrome," which results in numerous brief trips out of the room. Also, as you do more and more training, you will notice that people who do not leave the training room for at least one extended period during the day are quite through learning by three or four o'clock.

Corporate space not designed for your purposes is rarely ideal for sales training. If you are considering using a conference room or any other space, be sure it allows you to deal properly with all the issues discussed later in the sections on meeting room, seating arrangement, and food and drink.

Before you decide you can save a lot of money by training in this kind of space, be sure to check all costs. If your headquarters is located in a major city, hotel and food costs may more than offset any savings on meeting rooms and other rentals.

Conference Centers. Conference centers are very similar to corporate training centers in that they tend to be all-inclusive in their pricing and they establish a self-contained community with a particular flavor.

Pricing is on a per person basis and includes room, board, and most, if not all, meeting room costs. Some conference centers charge for audio/visual rentals. You need to check on exactly what is covered by the basic rate. Since these are profit-making operations, prices will be higher than in corporate centers.

A great variety of settings is offered by conference centers. Some are as luxurious as any resort; some are as stark and austere as a monastery. Most conference centers are in suburban or rural areas and require rather lengthy and expensive trips from the airport. Some offer transportation, but many don't.

Selection of a conference center must be tied very closely to the message you want to send your participants about this training experience. Look carefully at the image the center is playing on in its literature. That message will be played out throughout your program. Is it appropriate for the people and the situation you will be dealing with?

Also consider the total package offered by the center. Does it fit your program? Do you want all that fishing or golf around if you are not going to give people a chance to enjoy it? Do you want people to not have television sets in their rooms if your program will leave them with time to kill in the evenings?

As with corporate training centers, check out a conference center very carefully. Look at what their rates include and how they carry out their theme or image. Watch very carefully for their cancellation penalties, which can be quite severe. If your programs are highly susceptible to change, such penalties may take a conference center out of consideration for you.

Hotels. Hotels are the easiest facilities to find—there is at least one in just about every community in the United States. They are also the easiest to get into trouble with. Many hotels are not truly prepared or equipped to meet all your training program needs. Be assured that this will probably not stop them from accepting your meeting arrangements.

One of the pitfalls in a hotel that is not really a meeting hotel is a continental breakfast for eight that consists of eight hard danishes, each covered with a one-inch thick layer of melted sugar, and eight cups of coffee in styrofoam cups. In such hotels, audio/visual equipment comes from an outside vendor. If you don't know how to work a piece of equipment, there is no one at the hotel to help you, and, if the bulb in your projector blows out, the delay can be 2 hours or more. The help is unaccustomed to the needs of meeting groups. Your room is not cleaned at break times or lunch time and, by the end of the day, is in danger of being condemned.

If you know of independent hotels or hotels in a small chain that meet your needs, by all means stick with them. If not, you are urged to consider the larger hotel chains. These chains have become specialists in dealing with meetings of all sizes. In addition to being able to handle meetings for hundreds of people, many of their hotels have personnel who pay particular attention to the needs of the smaller meeting.

This does not mean that nothing ever goes wrong in these hotels. You may still have to call to find out what happened to your 10:30 coffee break, but at least these people will know what a coffee break is, that it should have been there on time, and how to put yours together and deliver it in a matter of minutes.

Most of these hotels have their own audio/visual departments and well-trained personnel to help you. Because they host so many meetings, the staff is easy to work with. They understand your needs quickly and can even be good sources of guidance. At the least, they can tell you what they have seen work—or not work—for other meetings.

You should call the sales office of the hotel you wish to use. If there are to be no sleeping rooms—all your participants are local—call the hotel, but ask for the catering department. In many major cities, the chains offer you a choice of downtown, suburban, or airport locations.

Handling your bill is easiest if you arrange for master billing. You can put meeting room costs, audio/visual rentals, coffee breaks, and group meals on your master bill. In the event you are also responsible for the room charges for the participants, you may have them put on the same bill. Penalties for cancellation vary among hotels in a chain. Read your contracts carefully.

The major hotel chains work at presenting and maintaining an image suitable for corporate activities. Your participants will feel professional and will generally be well treated. Their rooms will be comfortable and clean, and you are likely to hear few complaints. There will probably be some provision made for exercising, working out, walking, or running. The food will be at least acceptable.

In most cases, your meetings at such hotels will go smoothly, but check every detail as carefully as possible. Even the finest hotels have meeting rooms with columns in the middle that will interfere with your seating plan, or rooms that are L-shaped and put some of your participants around a corner from the rest of the group. It is up to you to leave nothing to chance!

Resorts. Resorts are excellent for meetings that allow time for recreational activities. If your training program is filled with learning experiences that run from morning to night, it seems cruel to surround people with the many temptations resorts offer. Most corporations will tolerate the use of resorts for training only if top management is involved and if there will be some recreational time.

Some resorts offer very low prices for off-season meetings, but getting from the airport to the resort may be so costly that savings are offset. In the event the resort invites meetings only during their off-season period, you may run into the same problems as you would in any other nonmeeting hotel.

There are resorts that include what they call conference centers. This is usually a sign that you can expect professional meeting services and a professional, if relaxed, atmosphere. A definite resort atmosphere will prevail in such places.

Training Impact. Your choice of facility sends many messages to your participants. As soon as they receive notice of where the meeting will be held, they develop a set of expectations of what the experience will be like. It makes sense for you to choose a place that sends a fairly accurate image of your program and that supports that image through the kind of activities and quality of services it offers.

The Meeting Room—Comfort Is Critical

Although the selection of a geographic location and a facility are important, it is the meeting room itself in which the training takes place and in which you and the participants will spend the training day. There is virtually no aspect of this area that will not have an effect on your program.

The following discussion has validity for all training situations whether they are conducted on corporate premises or in an outside facility. In the event that you are creating a training environment, you have an opportunity to make each of these aspects as nearly perfect as possible.

Lighting. Lighting in the training room must be adequate, and there should be no dead spots where participants will sit. Poor lighting makes concentration difficult and, after a time, causes headaches. The most even light is found in rooms with large fluorescent fixtures. For best results, try to get meeting rooms with windows and fluorescent lights. The addition of daylight to the room makes for less eye strain, and people don't seem to get as sleepy in rooms where there is at least one window.

Ballrooms are often broken up by hotels to create smaller meeting rooms. These areas make notoriously poor training rooms for a number of reasons which will be introduced in later sections. As far as the lighting is concerned, the spots usually set in the very high ceilings are generally inadequate to work by, and the accompanying chandeliers seldom produce an even light. This means many people are literally left in the dark.

In addition to being interested in the quality of light in your meeting room, you must also be concerned with your ability to control the level of light. Most trainers use some form of audio/visual support during the course of a program. Whether you use video, film, overheads, or slides, you must be able to darken the area in which your screen is placed. In the same instance, you probably do not wish to plunge the room into total darkness, particularly if note taking is desired.

When you make your arrangements, ask about the light sources and the controls you will have in the room. If your meeting space is actually a divided section of a larger room, you may find that your light controls will be in the other section—which may even be someone else's meeting space. Resolve this situation before you book your space. Once you arrive, there is nothing you can do.

If you have a window, there will be blinds or drapes. You can turn out your lights and use the blinds or drapes to control the level of light in the meeting room.

Check your lights and your light controls immediately upon arriving at the meeting space. (Your arrival should be at least 30 minutes earlier than that of your participants.)

Temperature and Ventilation. Many hotels save money by not turning on the air conditioning until someone is in the meeting room. If the air flow is not on when you arrive, turn it on immediately. If there is a problem, you can try to have it resolved before the group arrives.

"Climate control" seems to be the biggest challenge for large facilities. Do everything possible to bring your meeting room temperature into a comfortable range. In most cases it is the engineering department that needs to be involved in your problem. Don't stop pressing for help until a member of that department has come to your room to see the problem firsthand. A temperature of 70 degrees Fahrenheit is ideal. Most people feel neither hot nor cold, so temperature is not an issue and does not distract from the learning process.

As the person in charge of ensuring a positive learning environment, you are required to set the rules about smoking during the program. There are two reasons you must become involved in this situation: first, because the rights of smokers and nonsmokers must be balanced to prevent smoking from becoming an issue that detracts from your program; second, because the presence of high levels of smoke in the air makes many people uncomfortable and *literally* puts people to sleep! Some trainers prohibit smoking totally. If you do this, you will find your smokers either keep popping out for a quick smoke or spend a good deal of time watching the clock and waiting for the next break. Either way, you tend to "lose" these people for a good part of the program.

In an effort to counteract the effects of this prohibition, you can plan for more frequent breaks. Assuring smokers that they will have a break every 90 minutes does cut down on the trips to the hallway. It does not, however, do much for the clock watching and may not fit the flow of your program.

Another approach is to allow smoking in the room, but limit it to just one cigarette at a time. Having permission to smoke, the smoker is less compulsive about the situation. There is never an issue about one's rights or about who should or should not be smoking at any particular time. The smokers work it out quite well among themselves, and both smokers and nonsmokers seem to feel this is a fair and reasonable alternative.

Floor Plan. The size and shape of your meeting room must allow comfortable seating of your participants, the presence and effective utilization of your equipment, and the implementation of the training activities you have planned.

It is not possible to simply state that if you have X number of participants, you need Y square feet. Factors you must take into consideration include

1. How will these people be seated? Some seating arrangements require more space than others.

2. How much equipment will be in the room? What kind? How will it be used? A setup for video-taping sales call role plays requires a great deal more space than the use of an overhead projector.

3. What activities are planned? Do they affect the space you need? Breaking a class up into small discussion groups requires additional space.

Discuss your needs with the salesperson from the facility. If you want comfort, add two or three to the count of people who will be using the room. You might also want to exaggerate a little the space you need for audio/visual equipment. These are defensive measures which are recommended because hotels have been known to cram meetings into the only space they have available as long as they can fit in the requisite number of chairs.

Remember, however, that it is also possible to find yourself in a training room that is too large for the group. The result is that the class tends to be overwhelmed by the room and participation drops off severely.

Some major hotel chains provide excellent tools for meeting planners. One that is particularly useful is the *Meeting Planners Guide* from Marriott. In addition to general information about each Marriott property, you are given a floor plan of every one of the meeting spaces, the square footage each contains, and the number of people the space can accommodate in a variety of seating arrangements. Studying this information can help you determine how much space you need for each program you plan. Call the Marriott sales office in your city for a copy of this book. Check with other chains to see what they can provide.

When you are setting up a room, make sure that there is enough space for people to get into and out of the chairs easily. As a rough estimate, you need about 3 feet between the wall and the back of the chair when it is pushed in.

When chairs are set at a table, they should not touch each other. If they do people seated there will feel cramped and uncomfortable and may even lack the necessary space to work efficiently.

Rooms with very high ceilings are difficult to work in. There is usually difficulty being heard, and you will really have work to project your voice even to a small group. High ceilings also seem to intimidate people. You will find that a group you expect to be actively participative becomes quiet and nonresponsive. People even tend to whisper in high-ceilinged training rooms just as they do in high-ceilinged banks and national monuments. Any ceiling over 8 feet high is a potential problem. If you have a ceiling over 10 feet, you must make adjustments to your program to compensate for lowered group participation.

Consider the shape as well as the size of your room. A long, narrow room may work well for a small group. As the number of participants grows, however, people must be pushed farther and farther back away from the trainer. The same number of square feet in a different shape works better for large groups.

The shape of the room can also limit your choice of seating arrangement. Since the nature of the program is the best determinant of how people should be seated, be sure the room layout is not infringing on your quality training.

Beware of columns! You may find that your seating plan is impossible or that your video monitor cannot be seen because there is a column right smack in the middle of your meeting room. Facility salespeople sometimes forget to mention these, so always ask if there are any in your room.

Avoid Distracting Decor. In most corporate training centers, conference centers, and major hotels, meeting room decor is acceptable and basically unobtrusive. If you are using smaller hotels with which you are unfamiliar, you might want to inquire about how your meeting room is decorated. What you want to avoid is decor so outstanding in some way that it is likely to be distracting.

If you are creating your own training space, you might want to consider the effects surrounding colors have on people. (Warm colors such as red and yellow make rooms feel warmer and keep energy levels up while cool colors such as blue and green make rooms feel cooler and people feel calmer.)

Many trainers use the walls of their training space to hang sheets of chart paper. This is a common device for allowing certain ideas, content, lists, objectives, and so on to remain in view during the course of the program. In most cases, masking tape does an excellent job without damaging the walls. It does not, however, work on oiled or waxed wood paneling. Some hotel meeting rooms have large oil paintings covering the wall making it unsuitable for hanging anything. If you want to use the walls of the training room, be sure to ask what they are covered with. If you are creating your own space, you can design your walls for maximum use.

Training Impact. The meeting room is the space you and your participants will be confined to for anywhere from one hour to a week or more. Although the room itself is not a part of your program, it will, to a great degree, determine the success or failure of your endeavors. It is well worth any time and effort required to make your meeting room the ideal training space.

Seating Arrangement: Five Possibilities Depending on Your Objectives

The manner in which you place people in a group is closely related to a number of your objectives.

1. The kinds of activities you want people to perform
2. How you want people in the group to relate to each other
3. How you plan to present yourself and work with the group

There are five basic seating arrangements commonly used for training purposes: conference style, U-shaped, classroom style, theater style, and small tables. We will examine each in relation to its advantages and limitation and the achievement of your objectives.

Conference Style. Conference-style seating consists of one large table with all members of the group seated around it. Usually only active participants are seated at the table. Observers are placed in additional seating behind the participants.

This seating arrangement can accommodate all kinds of training activities. People can read and write and sit comfortably at the table. Unless the room is too crowded, it is a simple matter to break up into smaller groups by simply moving chairs into clusters on either side of the table toward the back and front of the room. Pair work is done by turning neighboring chairs toward each other.

Conference-style seating allows the participants to look at each other and, therefore, to listen and speak directly to each other. People seated in this arrangement usually tend to feel equal to and in partnership with each other. Trainees are more likely to be supportive of and helpful to each other.

The trainer in a conference style room has a number of options. During the course of the program, you can stand at the front of the group, you can sit at the table, you can move around the room, you can stand at a podium at the front of the room. Each of these presents a slightly different image of the trainer to the group.

Standing in front of the group keeps you as the leader, the person in charge. This position allows you to move easily about from chart pad to wall to screen to projector as needed. Attention of the group will always follow you when you are standing in front of the room.

Sitting at the table allows you to function, for as long as you sit, as a member of the group. For some discussions, particularly those in which you hope people will lower their defenses, this may be your most effective place. As soon as you stand up again, you resume your leadership role.

Moving around the room allows you to establish greater contact with the people seated to the rear of the table. This action is perceived by the group as less formal and more energetic than standing in the front of the room throughout the program.

The use of the podium as a place to stand while speaking to the group, particularly with this basically open seating arrangement, is perceived as a very formal approach and puts the most distance between the trainer and the group.

U-Shaped. A U-shaped arrangement consists of tables set out in a U-shape: two long arms coming off a short end. Chairs are usually placed along the outside of the U. All types of training activities are possible with this room arrangement.

Similar in appearance to conference-style seating, this arrangement makes a big difference in the appearance of a room and the utilization of space. People are moved back toward the walls, and there is a great deal of open space left in the center. Thus, seating to accommodate the same number of people fills up a room much more than conference seating does. The room feels smaller and people have less space to move around. Participants are able to see, hear, and relate directly to each other. They are, however, farther apart and somewhat more distanced than in conference seating.

The trainer has less room to move around. As you move around the outside of a conference table, you never turn your back on anyone. If you wish to move closer to the people at the back of the U, as you go into the open space you leave other participants behind you. Should your training require that you be able to work or speak individually with each trainee, the U allows you simply to walk directly up to every person in the room. It is very difficult, perhaps even impossible, in this setup to step out of the leadership position unobtrusively and casually sit down to encourage more open discussion.

U-shaped seating can be used as an alternative to conference seating if there are too many people for the conference table or if the room is so large that there is a need to fill it up.

Classroom Style. Classroom-style seating consists of a number of long, desk-like tables arranged in rows with chairs placed behind them. The look and feel of this arrangement is very much that of the traditional schoolroom, and, for that reason, most trainers try to avoid it. In general, you will only see this seating for large groups.

Since the participants cannot see each other, group discussion does not work too well. All other activities, including small group work, are quite easy to implement. This kind of seating turns your group into an audience. Participation will, therefore, be difficult to achieve. Trainees will see themselves as playing a more passive role. Since all they will see of most of the other trainees will be the backs of their heads, there will be little group identity and little support for each other.

Classroom seating turns the person in the front of the room into an "expert." In some cases, when the trainer has all the information or knowledge and must quickly and easily provide it to the group, this is the best way to do so.

If, on the other hand, this seating arrangement is purely the result of the number of people who must be accommodated, the trainer may wish to work with the group in a less formal and authoritarian style. One way to achieve this is to maintain the classroom seating only as long as it is absolutely necessary and to break the class into smaller discussion groups at every opportunity.

Theater Style. Very similar to classroom seating in its effect, theater style truly turns the group into an audience. This arrangement consists solely of chairs placed in rows and is generally used only to accommodate very large groups.

Theater seating is also appropriate when high mobility within the room is desirable and tables are not necessary. Suppose, for example, you want to break a class into subgroups and have them spend several hours doing role plays or brainstorming. Some observers may be required to move their chairs from group to group. People may need to separate and regroup. In situations like this, tables may be totally unnecessary.

Without tables, however, participants cannot be expected to read, write, or take notes effectively. If you keep people in a theater-style arrangement, looking and listening will be all they can do, and your participants will be quite passive.

Again, since they will have only the backs of each others' heads to deal with, this seating plan does not encourage people to try to relate to each other. Anyone stepping to the front of the room will be expected to be a source of information, knowledge, and/or wisdom. As the trainer in this situation, you must be able to deliver valuable content. When you use less formal seating arrangements, the interest of the group is held, in part, by their opportunity to participate. When you use seating that makes you the "star," the entire burden for holding the attention of the audience, for keeping everyone awake and involved, for avoiding boredom and loss of interest rests squarely and solely on you.

Small Tables. An arrangement of small tables and chairs placed several feet apart in a meeting room is a rather unusual training environment. This setting, which is ideal for small-group work of all kinds, makes most other activities difficult. If people sit facing their tables, many of them will have their backs to the trainer in the front of the room. Most of them will have their backs to a number of other participants. This arrangement is, however, ideal if the trainer intends only to start the group going and will then leave them to work at their individual tables.

Groups seated this way are difficult to control, and you cannot expect them to pay careful attention to long presentations from the front of the room. The seating arrangement does much to encourage side conversations and peoplewatching. People will build relationships with those seated at their table, but will probably not have any particular sense of those seated elsewhere.

If you use a training process that consists totally of small-group participant activities and you are required only to facilitate, small-table seating will work for you.

The Front of the Room. No matter what seating arrangement you use, some area of the meeting room will be designated the "front"—the part where the trainer stands, where the screen is located, where the chart pad and easel are. This is the end of the room which people sitting classroom or theater style face, the end of the room where you place the open end of the U and where no one is seated at the conference table.

The front of the room should be on the side opposite the door so people don't have to walk in front of everyone else to go to the bathroom. The front should be opposite—or at least far away from—the area where the coffee table stands to minimize interruptions.

Don't let an overflow of screens, chart pads, handouts, video equipment, plugs, and cords turn the front of your room into an obstacle course. It will make you look disorganized and give you lots of opportunities to hurt yourself. You can always ask for tables to be set against the side walls in the front of the room to hold your materials until you need them.

Training Impact. How you use space, place people, and position yourself in front of a group are all reflected in the seating arrangements you choose for your programs. Once again, you must concern yourself with what will work best for the program you have designed and what messages you choose to send the people you are working with.

Food and Drink—Quality and Quantity Indicate You Care

While you are conducting training, you are also responsible for the needs of your participants for food and drink. Even in situations where you do not directly see to its provision, you must be sure that access to food and drink is made easily available.

Cold water and clean glasses should be on your training tables in the morning and replaced during lunch or earlier if needed.

If you are providing morning coffee, be sure to include tea and decaffeinated coffee since many people prefer them. A continental breakfast ideally contains foods as opposed to cakes. People are too health conscious today to make a breakfast of anything that is covered with icing or filled with custard. Juice or fresh fruit, muffins, croissants, rolls, or bagels will make participants feel that you gave them a reasonable breakfast.

Coffee and tea should be refreshed at the time you take your first break. This usually occurs between 10 and 10:30 A.M.

Lunch offers several options. If you are in a corporate location, you will probably use the company cafeteria. You can, if you wish, simply direct people there and tell them what time to return to class. If you are in a hotel, you can arrange for a group lunch to be served in a private room, or you can provide people with some direction as to where lunch may be found and send them out on their own, or you can make reservations for the group in the hotel restaurant or coffee shop. If you are paying for lunch, the group will have to stay together. If people are buying their own, you can let them choose where they want to go.

Many trainers avoid private lunches where they have to order food for everyone. Selecting a meal that everyone will like is getting harder as people develop very strong preferences and convictions. Trying to keep vegetarians, calorie counters, low-fat dieters, salt-watchers, and others happy is not easy. The best alternative when you're paying: reserve space for the group in the hotel coffee shop and let everyone order from the menu.

Even if you are not arranging for or paying for lunch, urge your participants to eat lightly. Since heavy meals will invariably lead to heavy eyelids and an accompanying high level of discomfort, suggest that people consider soup and either a salad or sandwich for the noon meal. For those who "live to eat," dinner will provide them with a more appropriate opportunity.

Discourage—strenuously—the drinking of alcohol at lunchtime. If people are persistent about having a drink with their meals, you may need to discuss the matter with them privately. Training is a business matter, and drinking during the program is no more desirable than it would be during a sales call.

Afternoon breaks are usually scheduled for about three o'clock. For years these consisted solely of coffee, tea, and soft drinks. Many hotels today offer more interesting breaks: cookies, ice cream, make-your-own-sundae, nuts and dried fruits, cookie and candy combinations are some of the choices you may have. Although they cost more, these breaks can help a group get through some particularly long and tough training days. You'll know when you really need that kind of assistance.

Dinner is handled in a manner similar to lunch. Your involvement with the dinner plans of the trainees is directly related either to your need to pay for the meal, your desire to dine with them, or both.

Training Impact. If anything sends a message to people about how you value them, it is the quantity and quality of the food and drink you offer them. While there is no need to overwhelm trainees with huge meals, providing them with adequate amounts of good food at appropriate times of the day lets them know that you—and the company you represent—are aware of and care about them as people.

Support Materials: What to Include As Training Aids

Everything you use to support your training—handouts, preschool materials, charts, movies, slides, transparencies, videos, workbooks—all can either enhance or damage the quality of your program. Pay attention to detail to make sure that each item has a look that is consistent with all other materials and with the nature, quality, and style of the program. In this way you can ensure that the integrity of your program and of your training image is protected.

Your copies must be clean, your forms attractively formatted, your equipment working, your slides and transparencies of high visual clarity and interest.

Make sure people have everything they need to function. You can safely assume that everyone will bring a pen and/or pencil to class. But you cannot assume that pens won't run dry or that pencils won't break or run out of lead. Bring spare pens or pencils with you.

A good supply of paper should be strategically placed on the table. You can't really expect people to take notes of any value on the tiny pads provided by hotels. Loose-leaf paper is preferred over plain bond for three reasons: it is cheaper, the lines make it easier to write clearly while hurriedly taking notes, and the three holes seem to indicate that you expect people to save these notes. As a matter of fact, all your handouts should be three-hole-punched for this same reason.

If your company has a sales training manual that is a ring binder, there should be a divider at the back marked *TRAINING*. This section should be reserved for the inclusion of all materials coming from the training function. Among these might be classroom handouts, notes taken during class, and other materials or news items sent to the field. Salespeople can also be encouraged to insert other magazine or newspaper articles they find useful.

If a training course does not include a three-ring workbook, it is important to provide participants with inexpensive—but attractive—folders in which to place class materials. The folders not only give people a convenient place to store the materials neatly, they also send a message: "This stuff is important! Hold on to it and treat it with respect!"

A tent card placed in front of each participant's seat with his or her name clearly printed on it makes the table more attractive and makes people feel more expected and more welcome.

You want the people in your programs to put forth effort and to give you their best. It is only fair every detail of the program show them that you have done the same.

People: How to Create a Relaxed Atmosphere in Which Learning Can Occur

The relationships among all the people in the training situation create most of the psychological environment which surrounds your program. Conflict and hostility between any two people can make everyone else so uncomfortable that learning comes to a stop.

The instructional staff has responsibility for creating a relaxed and relatively nonthreatening atmosphere in which learning can occur. Trainees must feel safe with you. How can you expect people to risk giving a wrong answer or to expose themselves to the perils of a videotaped role play if you have not gone out of your way to reassure them?

It is useful, at the beginning of the program, to describe the psychological environment you want and how you expect people to treat each other. Explain to the group that each person is here to learn—and to help the others learn.

(Unless you build it into the program, there is no need for participants to be competitive with anyone but themselves.) Remind them that each will be expected to grow in his or her own way and in relation to where each was when he or she arrived, not in relation to how well anyone else is doing.

You may also want to call to their attention the fact that learning information and trying new skills can be a risky, and sometimes scary, business and, since all of them are in the same boat, they really should be very supportive of each other. Now is the time to assure them that your role will be purely supportive. You can tell them that *here* is the place to try new things, to experiment, to make mistakes. After all, they can't lose any orders in the classroom, and everyone else is only going to help.

It is imperative that all your words and actions reflect what you are promising. This means no jokes at the expense of others, no sarcasm in dealing with wrong or slow answers, no signs of impatience or annoyance, and no criticism that is not tactful and constructive as well as honest and direct. In your actions, as in the rest of the environment you have created, nothing but your best!

It is paramount that, no matter what happens, you maintain your composure. Conflict between you and participants must be avoided. Head off hostilities as soon as you suspect there may be a problem. (This is best done privately—at break time, perhaps.) If you wait too long, you may have a major explosion to clean up after. There are a number of excellent techniques for controlling and/or resolving conflict among others. Learn them, and use them as soon as you see trouble coming.

By expecting good things from your trainees, your program, and yourself, you can open your training on a positive note. Keep that mood going by being patient and supportive and never saying or doing anything to humiliate or embarrass a trainee.

ON-THE-JOB TRAINING: CHOOSING THE BEST
LEARNING ENVIRONMENT POSSIBLE _____

Most on-the-job sales training consists of the trainer spending a day making calls with the salesperson. Although there is no way of predicting exactly where you will be in the course of the day, you still want to try to do your training in the most suitable environment possible. The only training activity performed in this kind of situation is one-on-one coaching and counseling.

Geography

There is probably limited choice where geography is concerned. The best place to observe and evaluate the job performance of salespeople is in front of their customers. This means you go into the territory of the trainee.

Setting

The training which goes on in these situations takes place in a number of settings—a number limited only by the places the sales rep takes you. Each place is a suitable environment for some aspect of your training activity.

Car. A lot of time is usually spent in an automobile. This is a good place to discuss the objectives of the upcoming call, to do demonstrations of sales skills, and to role play the behaviors you are encouraging.

Although the car is the most private location you will be in all day, it is not the best place for extensive serious discussion, extended coaching, or counseling. Car seats do become uncomfortable after a while, and the fact that you are not seated facing each other makes it difficult to go beyond a rather superficial level of rapport.

Restaurants. Restaurants make convenient places for discussions of selling skills, behaviors, and problem areas. Select a place that lets you sit, relax, and be served. Avoid noisy and/or crowded restaurants where private conversation is difficult. Overly "impressive" restaurants can be intimidating and can interfere with the need for openness in the training relationship.

Airports. If you must do your training in an airport, take the trainee to a restaurant or coffee shop. Waiting areas in terminals offer no privacy. Should you do this, keep the foregoing suggestions in mind. Many trainers belong to airline clubs and can take advantage of their facilities. The clubs are quite comfortable, but they can get overly crowded at peak travel periods.

Customer's office. When you and the salesperson are in a customer's office, the only training function you can properly fulfill is that of observation. Any attempt on your part to interrupt or interfere with a salesperson for purposes of correction, demonstration, or coaching will serve to undermine the salesperson's credibility with the customer. Also, there is nothing salespeople resent more than outsiders interfering with their customers. Such resentment certainly makes it more difficult for learning to occur.

Offices. If there is a local sales office, or if the salesperson has an office where you can have privacy, you will be in the best place for conducting in-depth training. The business setting adds to the professionalism of the situation, the physical comfort level is high, and you will not be disturbed—an excellent environment for an assessment of what you have seen and for a discussion of future developmental activities.

People. You must establish your relationships with salespeople in whatever way is best for you. The important thing to remember is that certain feelings facilitate learning. Ideally, salespeople should feel safe with you, secure in the knowledge that you will say and do nothing that will embarrass or humiliate them. They should also respect you as a professional so that they can accept your ideas, guidance, and suggestions. Without these elements in the psychological environment, one-on-one sales training cannot be effective.

Support materials. Just as in a classroom setting, the high quality of all materials you use for training purposes adds to the image and increases the acceptability of your field training. Have forms professionally typeset or done on a computer to avoid the "just-typed" look.

Everything you bring with you into this situation matters—even personal items. Carry an attractive briefcase or leather portfolio. Dress to the highest standards set for your salespeople. If your marketplace is such that your salesmen can wear sport jackets, you should still wear a suit when you travel with them. You should be your own best "support."

THE IDEAL CLASSROOM ENVIRONMENT: A
TEN-POINT CHECKLIST

The ideal classroom environment will have the following characteristics:

1. A good flow of air with the temperature evenly maintained at 70 degrees Fahrenheit
2. A combination of both fluorescent light and daylight that leaves no dark spots where people sit
3. Sufficient space to accommodate the group comfortably and for the audio/visual equipment and coffee service to be easily available, but out of the way
4. An 8-foot ceiling
5. No columns to interfere with seating or room usage
6. Attractive, clean, and unobtrusive room decor
7. Comfortable, roomy seating for participants
8. Seating that allows all necessary activities
9. Food and drink that is readily available, of good quality, in sufficient quantity, and appealing to participants
10. All materials of excellent quality
11. No telephone
12. A feeling among participants and trainers that this is a pleasant, productive experience for all

Unfortunately, perfection is often impossible to achieve. Despite your best efforts, you may end up with a too high ceiling or an air-conditioning system that keeps you alternating between sweating and shivering. The fact that you strive to avoid such negative factors will certainly minimize both the number of times you must deal with them and the number of them which can occur.

SUMMARY

Creating a training environment that takes into consideration every aspect that can either add to the prestige and impact of your program or detract from it takes a good deal of time, thought, and effort. Failure to do so can make the time, thought, and effort you will have put into program design a whopping waste.

Your objective is an environment that—*in no way*—makes the learner uncomfortable or interferes with the learning process. Put yourself in the place of each trainee sitting in your training room. What do you see, feel, hear that causes discomfort or distraction? You can be sure that, if something bothers you, it will bother others.

Hold to high standards when you work with the salespeople of public meeting facilities. They often give their best spaces and services to the customer who demands them. Make sure it's you!

If you can't get the best, you may be able to get "makeup" services at favorable prices. You can always try to negotiate a midday ice cream treat to make up for the column in your room. Be sure your participants know that the facility—and you—know and care about the potential problem. It really does help people deal with such things.

The payoffs for all this care and attention are successful training resulting in true learning and growth, a highly professional image, and people who want to come back and work with you again. Can you ask for more?

chapter twelve ———————————————

SALES TRAINING
ADMINISTERING

——————————————————————— *GERRY MARX*

Gerry Marx is president of Gamma Presentations, an oral communications and sales skills development consulting and training practice located in Colorado Springs, Colorado. Retired after 36 years with the Wm. Wrigley, Jr., Company of Chicago, Marx organized and directed Wrigley's Sales Training Center during his last 7 years before retiring. A member of American Society for Training and Development (ASTD) for 11 years, he was the recipient of the James R. Ball Memorial Award at the 1984 ASTD National Conference, an award presented by the Sales and Marketing Professional Practice Area of ASTD. Author of over three dozen business articles published regionally and nationally, he has been in demand as a lecturer and workshop facilitator. The bulk of his present practice is working with high-tech firms helping their technical people improve oral communication skills. His biography has been in *Who's Who in the Midwest* since 1982. Marx has been active in a number of ASTD national and regional and local committees and was the 1988 president of the Pikes Peak Chapter of ASTD.

The sales training department, or any other training department, is not merely a function of operations in a company. Its success relies largely on its ability to be more of a living organism, very people based, empathetic, and a model of corporate efficiency and effectiveness.

In this chapter we'll cover ways and means to effect such a training department. Logistics for daily operations, such as recordkeeping, site location, budgeting, computerizing the department, and detailed organization are only part of the functions. It is also necessary to structure the department as a role model of skill usage along with the strong bonds to those who have the ultimate responsibility for an employee's training, the first-line manager.

POSITIONING SALES TRAINING MANAGERS AS "COUNSELORS"

In her keynote address at a recent ASTD national conference, Jean Houston referred to trainers as the "conscience of the company." The sales training department is indeed the "conscience of the company." It has the ability to influence most departments in the company. In his article, *"HRD Takes Its Place in the Executive Suite,"*[1] management consultant Allan Cox makes a strong case for training managers/directors to position themselves and their departments in a "counselor role." In this counselor role they would be situated right along with other high-influence department heads in the executive suite. As a result of performing needs assessments at various levels, understanding and defining the needs of the company as well as the performance needs of the sales staff, the sales training department is automatically thrust into the role of the conscience of the company.

---

[1]*Training and Development Journal*, September 1984.

In addition to the strong influence that the training department can have on corporate activities, consider the influence on the sales staff itself. Most salespeople, by their widespread geographic locations, rarely witness the overall administrative functions of their companies. Consequently, the sales training department becomes a showcase of their firm's quality of operations, standards of excellence, and level of expected performance. How well we reflect a high company performance image will have a strong effect on operations in the field.

Trainers as Role Modelers

The old expression *"Do what I say, don't do what I do"* has no place in sales training. If any profession or any training function ever offered an opportunity to model skills, it is the sales training function.

In the development of sales training department activities, job descriptions must be prepared for trainers (and performance expectations for outside consultants brought in to train your people). These performance descriptions should model the very skills being learned. In a sense, we are selling trainees on buying into various behaviors that can improve performance.

For instance, considerable emphasis in customers' needs-based selling focuses on information gathering. This is a sales skill that continually needs improving as it involves the art of asking questions. Fortunately information gathering is one important skill that can be easily demonstrated, both in the training room and by the training department. In the training room, trainee attention becomes much keener when sessions start out with such questions as, "Have you ever considered. . . .?" "Why do you think . . .?" "Have you ever experienced . . .?" "Can you tell some of the problems that . . .?" And appropriate follow-up questions would be "Would you comment further on that?" "Why do you think they felt . . .?" These are the same questions that salespeople should be using to gather information from customers.

Going one step further, in the training room, summarizing conclusions implied by the trainee's answers serves as an excellent demonstration of summarizing for the close of a sale, another skill often lacking in salespeople. "Okay, we agree that . . ." is an approach to summarizing followed by, "Do you consider that to be a fair appraisal of our position?"

The training department also can demonstrate information gathering by determining various needs that the trainee will have when he or she comes in for the program. Later in this chapter we'll see how a prearrival form will identify certain needs, again serving as an information gathering tool.

Other demonstrated skills and arts cover presentations, proper use of visual aids, grammar and elocution, time and organization management, right through to attitudes, particularly empathy and enthusiasm.

What has this to do with administering a sales training department? The department's director or manager should have performance standards published for all the department's staff. Included in these standards should be the role modeling of all the skills taught, by the department's office staff as well as the trainers. After all, as Robert Louis Stevenson said, "Everyone lives by selling something."

SKILLS THAT CAN BE MODELED

Planning is a priority skill to be modeled, as too many salespeople fail in this area. The finite planning that goes into a productive trip to the sales training center/location can positively influence salespeople. In the training room, the effective use of questions, participation techniques, elocution, the use of props, illustrations, A/V techniques, presentation development, and platform abilities all present excellent role-modeling opportunities. In addition, empathy, attitude, and enthusiasm can set good personal standards.

Earlier, mention was made that we actually sell sales skill development ideas to trainees. In reality we are persuading trainees to reevaluate their skill behaviors and consider options available. The trainer's role in the persuasion process is akin to the role that salespeople have with their customers. By demonstrating effective oral communication skills, salespeople can witness their reaction to the persuasion process and even empathize with their customers' positions. It all boils down to *"What's in It for* Me?"

The very administration of the sales training department is another necessary role-modeling activity. Salespeople sometimes have difficulty managing and administering their office and territory responsibilities. Some training department managers actually have one or two members of their office staff conduct miniprograms on office organization, filing, effective "tickler file" systems, and even neat and orderly letter writing.

A perfect example of office staff involvement in the actual training function is in systems for filing. A good office clerk can put together a one-hour program that takes a salesperson step by step through sorting mail, systems for filing, developing the urgency follow-up file, systems to develop "tickler files," and even elements of good typing, that is, how to frame a letter.

Program developers and trainers can demonstrate proper use of visual aid equipment as well as developing effective visuals such as overhead transparencies, flipcharts, and so on. These are the same tools that salespeople use in the field.

Quite often salespeople are lacking in mathematical abilities. A program developed by someone in the financial department can help improve these skills. Other resources for such learning can be found at local community colleges, high schools, or various continuing education departments at schools.

Guest speakers from other company departments should be used to help salespeople be more aware of how they should administer their field work for cohesive functioning with such departments. Particularly important are the order and traffic departments.

When using guest speakers the sales training manager should provide them with guidelines and even coach them on good platform abilities, again modeling the work expected of trainees in their territories. Let trainees know right away that you and your trainers will be role modeling certain skills and activities. This not only makes the trainee more aware of the skills but further develops the professionalism of the trainer.

Four Key Support Levels for a Successful Training Department

A great portion of the success of a sales training department depends on the support from the various levels in the organization. For example, the sales training department can utilize needs assessment to gain strong support from four sectors that weigh heavily in the future and the success of the department, namely, senior management, the immediate superior, the trainee, and the customer. If properly tapped, these sectors can add tremendous support to the success of the department.

These sectors should be consulted as to their needs that can be satisfied through training. Not only can the training manager learn of what they are looking for in results and behavior adjustments, they can also allow these sectors to "buy into" the training program by helping to author it. Then, they can evaluate it by measuring future performance against their stated needs.

Senior Management. The needs of senior management must be incorporated into the training programs, be it sales management or selling skills. The company's mission, the corporate objectives, and the vision of sales support perceived by senior executives are acute concerns for the sales trainer. It is essential that the training manager be able to speak and understand the language of senior executives. In addition, much of the budgetary consideration for training rests in these senior offices. The very survival of the department can be tied to the ability to maintain senior management's support.

The Immediate Superior. This person is especially important as he or she has the "clout" to see to the application and results from the training. We'll cover some effective methods of how to bring the immediate supervisor into the picture under prearrival activities. A manager is generally responsible for training his or her immediate staff. The strategies used to bring managers into the training function can enhance the support from them.

The Trainee. Naturally there must be a strong desire on the part of the person being trained to want to learn and want to succeed. During the needs analysis with the trainee, the desire can be heightened if that person can help identify or "author" the material he or she would like to have available. Although they will not actually write the program, their input will help create "ownership" of the program and improve the chances of their applying it back on the job.

The Customer. Here is a greatly overlooked opportunity to enhance your company's image and effectiveness. A sales training department can have an immense influence on a customer's regard for the company by showing an interest in developing salespeople better to meet the customer's expectations of good sales behavior. In addition, the sales training manager becomes more aware of the marketplace and its ever-shifting needs, not to forget the competition lurking all about.

Successful administration of a sales training department depends on the proper positioning for those who can influence its success and who depend on its effectiveness. As the department weaves its way through the areas just covered, it earns the right to be called "the conscience of the company."

WHERE TO LOCATE THE SALES TRAINING DEPARTMENT/TRAINING ACTIVITIES _____

Formal sales training takes place either "on site" or "off site." The remainder is conducted through coaching on the job. We'll concern ourselves with the on-site (internal) and off-site (external) facilities and how to manage these activities.

On-site Training Facilities

This refers to sales training conducted on the premises of the department. This is usually a facility developed by a company for a centralized operation.

Location of Facility. Experience has shown that, although it is preferable to locate the training department close to the headquarters of the sales and marketing departments, it should not be too close. It is a convenience to be close enough to sales department decision makers so they can address trainees. It is also a convenience to have such decision makers accessible to training managers for corporate direction and input. On the other hand, being too accessible to a variety of corporate officials can bring unnecessary interruptions and "sitting in." The latter can cause considerable anxiety on the part of trainees and lessen their involvement in the program. If possible, keep your distance from home office interruptions. Stay in the neighborhood, but out of the house.

Site Considerations. If you are developing a center or moving into a new one, look for an office complex that has a motel/hotel associated with it as well as restaurants. This eliminates transporting the trainees back and forth.

Another consideration is national transportation. You should be close enough to an airport to which the hotel you select has limousine service. Having a permanent site at such a location allows you to make good arrangements with hotels, transportation, and restaurants.

Other considerations in choosing a site are health care, exercise clubs, and entertainment. In choosing a site with the hotel's facilities in mind, be sure the hotel's rooms are appropriate for studying.

Facilities Layout. Unless a lot of thought is given to the room needed, a training manager can encounter a number of complications and even run out of space.

The training facility should be developed in two separate but unified sections. The first is the administrative area. This includes space for reception, secretaries, and other office staff as trainers and the training manager. In addition design rooms, supply and inventory storage, office machinery, and refreshment areas are all to be considered. Most of this administrative area should be away from trainees' traffic pattern. Aside from interrupting the staff, trainees are inclined to read what's lying about, which may cause them to be conveyors of false information.

Training facilities are another matter. In general, it is best to "think maximum." Depending on how many will be in a given class (the maximum number you will allow) and the table/chair arrangements, provide a training room that will meet the needs of the future as well as the present. Then there are breakout rooms. Generally two are reasonable, with one serving as a conference room as well.

Try to have the audio/visual equipment in a separate, soundproof room. Audio/visual equipment also involves tapes, trays, films, cassettes, and so on that take considerable space. So take heed.

Breaks are necessary during training sessions, so refreshment and washroom facilities must be considered. Finally, sufficient storage space must be included.

In a training room environment, trainees are sitting most of the time. Comfortable chairs, preferably swivel so they can converse in participation sessions, and roomy desk space will greatly increase the trainees' comfort. Good color coordination (soft colors) avoids distraction; a professional office design consultant can help in these areas.

Off-site Training

Off-site locations are used by training departments that have no centralized training facilities. Off-site locations are used to take training sessions out to the field for geographic reasons, for example, when the training is applicable in only a certain region such as a product introduction. Also, the training center may be booked when another important program must take place.

Fortunately, hotel and conference center meeting planners understand the needs of and how to handle training sessions arrangements. Still, a checklist of needs for off-site training should always be handy so that your room size, equipment, transportation, requirements, and so on, can be filled. Working with a particular chain or a couple of hotel chains can be valuable. For instance, they can enter your needs on their computer files and satisfy those needs with consistency and minimum problems. Far more training sessions are held in hotels, motels, and conference centers than in training centers. Advanced reservations should be made to ensure space. Lining up a good catering manager at hotels/motels is also advised.

Reward Programs. While networking to learn of good off-site space, survey the opportunities to make training programs at a resort-type location. These are choice for high-achiever programs, reward programs, or morale boosters as well as training strategy sessions. Again, be sure that your logistics are in line with the needs and double-check your packing list to be sure that you take what you need. It's embarrassing to arrive in Hawaii and find that something has been left on the mainland. If shipping materials, do so two weeks ahead of time; if the materials get lost, then there will be time to reproduce them. This also applies to programs being done in foreign countries. In an unfortunate personal experience, materials shipped ahead to Manila got lost in Philippines Customs. Our team had to work until three in the morning reproducing everything that was then carried aboard. Earlier shipping could have precluded the emergency.

Taking Training Programs into the Field. Field trip training by sales trainers also affords good opportunities to see what's really happening out there. Rapid changes on the marketplace, problems confronting salespeople and their managers, the competition, and the opportunities to meet and discuss issues with customers are the bases of sales training program development. Abraham Lincoln once said, "A general's hindquarters belong in the saddle, not at headquarters."

When going on field visits, the manager in the area may want you to bring a half-day or an all-day training reinforcement program. After all, many salespeople may only get to the training center every few years. It's a good idea to have a "laundry list" of materials, training room needs and accommodation needs to furnish to the sales manager in the field. He or she can then make arrangements for such an off-site training session, with the certainty that everything that is necessary for a flawless get-together has been covered. Field managers have many more responsibilities as a manager than managers in other corporate departments. The more specific you can be to help them satisfy your training facilities needs, the better the results.

RECORDS, BUDGETS, AND PURCHASING
RECORDKEEPING

Sales training departments differ in certain respects from other training and HRD departments. One such difference in many companies is the association with the personnel department. Whereas most training departments come under the wing of a personnel department, sales training departments often report directly to the sales or marketing department.

However, sales training activities are closely allied to important information needed by the personnel department. Sales training managers should be aware of these communication links and maintain a close relationship with personnel.

Recordkeeping

Every instance of sales training (orientation, skills development, key account selling, sales management, in-field programs, self-help programs, adult education extension programs, etc.) should be recorded and posted in the employee's file. Systems vary on how this can be done. Most departments record this information in a computer. If a company maintains a master human resources file, the training must be entered to keep such a file alive and active. (See Figure 12–1.)

Figure 12–1 is a simple form that can be entered on a computer and linked to the personnel department. It can also be typed, as shown, adjusting it to the terminology used in the particular company. Most important, it feeds information so that accurate records are maintained for all training that has been completed by the salesperson.

Aside from knowing the training history for performance appraisal assessments, accurate training records are valuable for newly appointed or transferred supervisors. The accountability for a salesperson's training rests with the immediate supervisor. The supervisor relies heavily on the sales training department

Figure 12–1 Sales training record

RECORD OF SALESPERSON TRAINING				
NAME	CODE #	DISTRICT	TRAINING RECEIVED	DATE

to see that certain phases of training are performed. When a new supervisor takes over a sales area, he or she should have each salesperson's training history.

Another purpose of training history records should focus on the "training intervals." This refers to the amount of time between formal training periods. For instance, the training record may show that a salesperson has not attended a program on "Time and Territory Management" or "Territory Market Analysis" in eight years. The training world upgrades its abilities and access to the latest information almost daily; eight years between programs would show that the salesperson could be lacking in contemporary knowledge and behavior, aspects that a customer would notice also. Such records also help the training director to think in terms of "training systems" rather than training programs. (See Figure 12–2.)

For example, a key function of performance appraising by the immediate supervisor is to identify if any performance deficiencies can be corrected through training. Performance appraisal copies are stored in the personnel files. Maintaining a direct-line access to this information by way of the personnel department can alert the sales training people to needed training programs. Conversely, if the records show that the salesperson has had the training and still is not performing, then that information can help the supervisor assess the course of action to take.

Figure 12–2 Training recommendation form. (This form can be provided to the Personnel Department by the Sales Training Department.)

FROM: PERSONNEL DEPARTMENT   DATE _____

TO: SALES TRAINING DEPARTMENT

From information recently received on the PERFORMANCE APPRAISAL, please note the recommendation for training.

SALESPERSON: _____ SUPERVISOR _____
TRAINING RECOMMENDATIONS _____
_____
_____
_____
_____
DATE OF PERFORMANCE APPRAISAL _____

Equal Employment Opportunity Data. This is very important to the personnel department for current files.

Previous Training. Many newly hired salespeople may have had various sales training experiences with former employers. For instance, many companies offer "shelf" programs, computerized learning, and self-study. If a salesperson has already had any of these opportunities, it might be unnecessary for him or her to go through it again.

Another advantage of recording previous training experiences is that person may be a possible facilitator if you use a particular program. This is especially important with first-line field managers. As we will note, first-line managers can be excellent program leaders in certain training subjects. Candidates for manager/trainer assistance in the training room can easily be spotted in computerized records. Training departments have a history of frequent turnover of personnel. Successor managers will find all records on previous training history very helpful. (See Figure 12–3.)

Recording Training Costs. A sales training department should develop a system to keep track of hours and costs (per hour and per employee-trainee). In many companies, each sales unit (division, region, etc.) is responsible for a training budget. Costing out the programs into measurable amounts can help supervisors balance budgets as well as determine future ones. In addition, the sales training

Figure 12–3 Training history form

RECORD OF TRAINING HISTORY		
SALESPERSON		CODE #
ASSIGNMENT (DISTRICT/REGION)		
TRAINING HISTORY		
DATE	TRAINING PROGRAM(S)	TRAINER

department can compare these costs with those of other companies to make sure they are in line and getting the best return on their investment. This is akin to cost-per-call recording as done in the sales department.

Recording costs can be done in a variety of ways, generally depending on the needs and demands of the company's financial department. Tangible costs can include materials, hardware for the training, depreciation on equipment, office staff and training staff time and expense, as well as traveling and housing and meals. Intangibles include the time out of the field by the trainee.

Numerous training departments keep their records by "hours in training." Some companies require that X number of hours be spent by each salesperson or supervisor in formal training sessions.

Computerized Training Records

Not only has sales training gained recognition and support from their respective companies, thus demanding more and more records, but the vast amount of specialized training available behooves us to enter all such information on a computer. From invitations to follow-ups, as well as details of each sales employee's training history, entry on computer files is especially beneficial to the sales training department's responsibility for training moderate to large numbers of sales personnel. (See Figure 12–4 for a sample computerized course report.)

Fortunately, a variety of software is available for maintaining computerized records and for administering a sales training department. In all likelihood such software will seek you out before you have to seek it; the daily mail will bring a variety of options as your training center gets on the mailing lists.

Computers are also useful in reducing time and error in the registration of trainees—a form can be run off for registration in very much the same way hotels have guests preregistered all except for the person's signature. (See Figure 12–5.)

Computerized forms can be used to fill in the appropriate information for the welcome packet that the registrant will receive when checking into the hotel.

The registration form provides early information about the trainee and also shows a personal concern that helps to remove some of the anxiety that trainees often have prior to coming to a training session. The information should be adjusted for the needs of the particular training center; some of it may already be established such as the hotel assignment. Getting flight information can ensure that all attendees will arrive early enough for opening receptions, sessions, and so on. It can also assure that they did not book return flights before the program ends.

Information received from these prearrival forms can be sent to the hotels for their preregistration information, thus speeding up the hotel registration process.

Figure 12–4 Sample computerized course report and departments represented

```
STORAGE SYSTEMS TRAINING          ROSTER REPORT  Date:   9/21/89
Course Number    SQ111-01                        Page:      1

Course Title:    TEAMWORK SKILLS
Instructor:      R PAYNE & C CORMIER
Location:        MARBLE ROOMS
Begin Date:      09/20/89
End Date:        09/21/89
Begin Time:      8:00
End Time:        5:00

                                   Badge    Cost
Employee                           Number   Center   Dept.   Address      Mail Stop
Name                               ------    ------   ------   ------       ------
------


Total number on roster:   16


                          *** END OF REPORT ***
```

254

Figure 12-5 Registration form

```
┌─────────────────────────────────────────────────────────────────────┐
│                      FORM FOR PREREGISTRATION                         │
│                                                                       │
│  NAME _____  DATE _____   │
│  ADDRESS _____  │
│  TERRITORY ASSIGNMENT _____  │
│  IMMEDIATE SUPERVISOR _____  │
│  TITLE OF TRAINING PROGRAM _____   │
│  HOTEL ASSIGNMENT _____   │
│  IN EMERGENCY CONTACT _____   │
│  ANY SPECIAL DIETARY NEEDS _____   │
│  SMOKER _____ NONSMOKER _____                                   │
│  TRAVEL SCHEDULE:  ARRIVING DATE _____ TIME_____             │
│                    FLIGHT _____ AIRLINE _____                │
│                    DEPARTING DATE _____ TIME _____           │
│                    FLIGHT _____ AIRLINE _____                │
│  COMMENTS (PERSONAL INFO SUCH AS BIRTHDAY, HEALTH NEEDS): _____        │
│  _____    │
│  _____    │
│  _____    │
└─────────────────────────────────────────────────────────────────────┘
```

Other Computer Uses for the Sales Training Manager

A vast number of "packaged," "shelf," or "proprietary" training programs are available on the market today. With access to more and more specialized programs dealing with esoteric sales skills development (as well as sales psychology), a training manager should consider recording vendor information that might be of future value in the department's computer.

Often we learn of professionally developed programs through our network of peers in associations such as ASTD. Perhaps these programs do not apply to our current program menu, but could be great instructional material in the future. Cataloging them on the computer will mean that the information is available when the demand arises.

This is particularly important for training managers developing new departments or expanding an otherwise part-time training operation. Generally new departments start with orientation programs, basic interpersonal selling skills, and some sales management programs. Eventually key account selling, deeper intensity into basic interpersonal skills, and senior sales management administrative training and development could be in demand. Examples are budgeting/ forecasting and professional business presentation.

Perhaps the fastest and most reliable way to cover these high-skill topics is to contact a business college or a local university, whose professors are not only current in these topics but are willing to export their knowledge to the corporate world. Budgeting can also be covered by the corporation's planning and finance departments, as well as the marketing people.

Professional proprietary (or shelf) training programs are available for making business presentations. Quite often a company's major customers have their own guidelines on how to present, with the emphasis on actionable numbers and methods for getting the numbers across, such as graphs and charts. If arrangements can be made, engage the customer's own staff for information on making presentations to the customer. Having a purchasing agent, or merchandising executive, from a customer lay out his or her guidelines can be some of the best education the trainees can get.

A growing operation in many companies is the advanced curriculum of programs being offered. Based on past demand/acceptance, sales training centers send out a "catalog" of programs and dates for the following six months or even a year. This is better controlled when programs/dates/past records/attendees are kept on a computer. In a sense, the training manager can delegate this operation phase to the computer.

Program evaluation, either by trainees and/or by trainers, can also be computerized for reaction to certain programs. Finally, back-on-the-job results can also be recorded and associated with the training programs that had the behavior changes to produce those results.

Computerized development of training manuals has gained in popularity over the past decade. Word processors have enabled programmers to make the ever-constant changes needed in training manuals and make them with ease. Up-to-date training manuals and information are essential for sales training department credibility.

Computers are also helpful in administering the records of the sales training department staff. These include purchases for the department, safety inspections, remodeling, leases, and so on.

Library records, inventory of equipment and manuals, learning aids and follow-up kickers, when put into the computer, can keep the department's administration effective and current. (See Figure 12–6.)

If your department has certification to issue continuing education units, have those files on your computer. Some companies require training information to be recorded by "hours in training." This is related to budgetary control as well as personnel balance, that is, controlling the amount of time in essential programs.

Finally, if you are a computer-based operation, you may be able to tie into the personnel department's computer files. If not, personnel is sure to want all

Figure 12–6 Library record form

```
┌─────────────────────────────────────────────────────────────┐
│                                        DATE DUE _____   │
│                                                               │
│              TRAINING DEPARTMENT LIBRARY                      │
│                                                               │
│   BORROWER (Name and Address) _____ │
│   _____ │
│   ITEM:  Book _____ Cassette _____ Film _____ Other _____ │
│   TITLE _____ SHELF NO. _____ │
│   DATE OUT _____ DATE RETURNED _____   │
│                                                               │
└─────────────────────────────────────────────────────────────┘
```

vital training information needed for their human resource records. Arrangements should be made with personnel departments to assure information compatibility.

Personnel Department Resources for Training

Perhaps your company has a department for training office staff, factory workers, or other nonsales occupations. Even if there is no separate department, chances are someone in personnel does training. Many programs they have on line or access to can also be used by the sales training department. These include time management, work organization, administrative sciences, business writing, leadership, holding department meetings, management skills, financial/budgetary programs for the nonfinancial person, and so on. This training may also be useful to various people on the sales staff and their use would avoid reinventing the wheel.

Budgets

Who is responsible for paying for the training? Does the sales training department take all responsibility or is it prorated to various divisions in the sales organization?

A variety of budgetary methods are used for accounting for the training dollar. Some companies put the dollar responsibility for training on the particular field unit (region, district, etc.). Since the first-line manager has the clout to implement training and is in a position to measure results, the fiscal responsibility might then be built into his or her budget responsibilities. This would also include the "time spent" budgeting. As time is money, time can be used as a unit of measurement when examining results.

The computer is very helpful in assisting regional or district managers in "keeping books" of the training history and costs for their respective responsibilities. Expenses related to travel, lodging, feeding, and materials can be charged by the trainee to the training center, and the program can be transferred to a region or district manager's budget. If that is the operation procedure in your company, then the sales training department's budget can be developed strictly for departmental administration.

Software is available from a variety of sources to help department managers develop effective budget systems. A big difference in budgeting for a training department versus some other department is the purchasing of outside programs and bringing in outside trainers/consultants. Field trips by the training managers and trainers should be generously allocated as field/client exposure is imperative.

If training is taken into the field, the financial responsibilities for the training may again be in the budget of the manager who requested it. Travel accommodations for the sales training department staff would be in the training department's budget, but attendees would be on the standard field budgets established by the field managers. Be sure to have agreements in place as to where expenses for the hotel conference room, projectors, and other needed training equipment are to be charged.

Staffing, Administrating, and Purchasing

Depending on your training commitment, budgets, and space, the size of your staff will fluctuate. A receptionist/secretary and possibly a trainer may be the minimum. In small operations the sales manager may be the entire training function. Such small training units rely heavily on outside training resources.

The point to remember is that, no matter how many or how few in a sales training department, they should be attuned to the training function. They must be outgoing, positive, and empathetic. Many salespeople approach sales training with great anxieties. Staff personnel should realize this. In addition, sales training staff personnel should role model skills. Sales training departments are often considered "glass houses." Our behavior has to reflect our stated goals.

If there is one sales managerial skill that is often neglected or done poorly, it is appraising performance. Certainly the knowledge of and the training in various managerial skills is in the domain of the training department and should be demonstrated as an example to field personnel.

Most sales training managers arrive at their position by way of field sales. Back in the field, these sales training managers were on the "selling end" of business conversations. Now that they are sales training managers, they also become "buyers." Salespeople representing a wide variety of products from office equipment to training programs to audio/visual aids will be knocking on the door. These salespeople offer a variety of benefits in addition to what they are selling.

You can observe their selling behavior and judge their selling habits and styles. You can become more empathetic toward the buyers that your salespeople contact through the emotions and attitudes you develop as a "purchasing agent." You can visualize the decision-making process a buyer uses and gain greater respect for the "What's in It for Me" mindset.

As you administer the department operation, look upon all these activities as constant educational opportunities. See what can be learned, developed, and translated into effective training and operating practices.

External Trainers

When going outside for trainers, witness their performance personally first. When that is not possible, request a videotape of at least one hour of continuous platform activity. Request a detailed breakdown of the subject matter and how it will be covered (participation techniques, workbooks, etc.). Fill in the potential trainers on all necessary information about the company, product, why this session has been scheduled, the needs to satisfy results desired, the entry level of trainees, the time, facilities, and all other logistics a staff trainer would know.

Also, have a contract with the external trainers. When a training manager goes outside for trainers, the chances of losing control of the learning session are greater.

There are numerous reasons for going to external practitioners. For example, business mathematics may not be in the "respect realm" of the staff trainers. As well-qualified person in that field may be just down the road a few miles in a local college. A needed course in oral presentation skills might best be delivered by a well-qualified professional speaker.

Another source might be professional organizations. They represent every conceivable professional and are all over the country with local chapters. Another source readily availabe is the *ASTD Buyer's Guide and Consultant Directory*. Furthermore, many supplier companies will call on the sales training center to inform the administrator of their products and services. Record their information in your files or computer to have it accessible if the need arises. The ASTD's *Trainet* is an excellent source.

In-Company Guest Speakers

The ability to have a variety in platform speakers adds more zest to a program, especially one lasting for several days. It not only alters the energy but brings fresh thinking into the learning experience. We'll cover various guest speakers in the next section. When asking someone inside the company to do some of the training or to address the group on company outlook, be sure you specify the time limits that he or she has and provide the names of the trainees so they can be as personal as possible.

Many companies have found that the use of in-company personnel is more effective after hours, at evening sessions, or at round table discussions.

ACTIVITIES TO CONSIDER PRIOR TO TRAINEE ARRIVAL

As *role modeling* is a requirement for a successful sales training department, it is imperative that all activities be well thought out and demonstrate real state-of-the-art departmental management. This modeling not only influences the trainee's perceptions of a smooth, results-oriented training unit, it can also serve as a model to other departments in the organization.

Let us run through a hypothetical training program that has been approved for a group of sales trainees. This could be for line sales people or for managers. The "run-through" presented here can be extracted and developed into a checklist for the office staff and can be put on your computer. It serves as a good training instrument in case of department and office staff turnover.

Who Will Attend?

Selection of trainees in sales departments is mostly done by the managers or supervisors. On a national or a regional basis, supervisors do not want to clean out their districts/regions/divisions (whatever terminology they use geographically) by having too many salespeople gone at once. Second, the manager or supervisor is in a position to ensure the eventual application of the skills. He or she is also aware of the particular needs of the salespeople. Finally, the first-line manager has the ultimate responsibility for training his or her staff (it should be in the manager's job description).

Keeping the immediate superior right in the middle of the training decisions will pay off in many benefits, from good support of the sales training function to ensuring the application and confirming the results. *Timing* is also essential at this point. The manager may have promotional plans, new product introduction plans, his or her own sales meeting and reviews, and conventions or conferences that the salesperson is scheduled to attend. Allow plenty of time prior to scheduling sales training so the manager can ensure account coverage while the salespeople are away.

Where Will They Stay?

Hotel reservations, including eating and transportation considerations, must be secured before invitations are mailed. If doubling up room occupants, be sure they are compatible, smokers/nonsmoker; personal interests; varying parts of the country/state, and so forth.

Selecting Guest Speakers

When aligning your sales training program, consider a variety of guest speakers. Increasing a trainee's perspective can add a lot of support to the skills, techniques, and developmental material being covered. Some of those who can be invited are the following:

Company Officers and Department Heads. These creditable people can explain the company's thrust and emphasize the support needed from the sales staff. In many companies salespeople are seldom, if ever, introduced to or get to meet the top decision makers. Company senior managers can offer a variety of topics such as the reasons for certain procedures, promotional plans, growth opportunities within the organization; explain acquisitions; and squash or support rumors. At the same time, well-conducted meetings between trainees and senior management can provide feedback, so management also obtains firsthand information of what is going on in the field. Use such opportunities advantageously.

You may want to consider inviting in-company personnel for after-hours get-togethers. Otherwise, they may want to sit in on part of the skill development sessions; as mentioned earlier, this has a detrimental effect.

Field Managers. If it is geographically convenient, using a field manager or two has benefits for the manager, the trainee, and the training department. The manager can benefit by learning/practicing platform skills as well as refreshing his or her material. Trainees can benefit by getting the personal side of a manager's responsibilities as well as experiences. The training department can benefit from the increased support such managers can give to the overall training function.

A word of caution: Some managers may be awkward on the platform and as a result may mortgage their credibility. Guidelines set forth by the training manager can serve as a performance standard for managers who share in the training.

Perhaps the greatest benefit of all is the adult training experience gained by the manager. The responsibility for a salesperson's training and development rests with the immediate superior. Coaching on the job is a major role in this training. Exposure to the professional approach to sales training will help the manager's knowledge of how to interrelate with salespeople when training them.

Buyers, Customer Staff Personnel. Here is an often overlooked excellent opportunity. Buyers and other customer personnel, such as their training department personnel, make excellent luncheon and dinner speakers. Training staff personnel from customer's operations may even be invited into the training rooms. Perhaps reciprocal arrangements can be made where you or your training staff is invited to a customer's training session. The solidifying of good relations with customers through training cooperation greatly enhances the sales training department's credibility in affecting the company's overall sales effort.

Inviting the Trainees

Training department staff should approach invitations to trainees in the same way they would plan a trip for themselves. Thorough information, clear communications, and "everything you always wanted to know about coming to a training program" should be covered in the materials packet sent to the invitees.

An appropriate and warm letter of invitation is the heart of this packet. In addition to the letter of invitation you may also want to consider the following:

An Objectives Form. The purpose of this particular program, the format, and so on are covered in the invitation letter. Many sales training departments have found it expedient for trainees to think out and set particular objectives. These objectives would include bringing up certain issues that have bothered the salesperson; selling situations that have had negative or no results; skill deficiencies that the person realizes exist; and skill development areas suggested by the superior. We'll note later how the immediate supervisor can have a strong influence on establishing the trainee's objectives.

One simple approach to help the trainee develop objectives is to have the trainee ask himself or herself: "What do I dislike most about (a particular skill, activity, or practice)?" "What would I like to gain or develop most while at this program?" "Complete this sentence—'I hope that the program leader will. . . .'" (See Figure 12–7.)

Recommended Pre-Arrival Reading. Reading material related to the subject can always be made available to the trainee. This eliminates going over something in the training room that can easily be read and thus speeds up the training process. This, too, can be a ROLE MODELING experience. A pet peeve of buyers concerns salespeople who give them material to be read and then sit there and read it word for word to them. The buyer just wants it explained or questions answered. The trainer may use this as an example of improved sales performance. If the program being presented is continuing from a previous one, the trainee can refer back to the original manual. Another preparation would be suggested library reading.

Some training departments have a standard reference list of materials on how adults learn in a corporate training environment which is sent with every invitation, no matter what the program. It sets a mental conditioning for the learner, who may rarely be in an adult learning environment and helps adapt to the training course.

Logistics for Complete Information. Again, this is akin to going on a trip to a far-off location. Just as we would want to have all the information we need ahead of time, so does the person being invited. Some considerations are:

Figure 12–7 Sample prearrival objectives questionnaire

(Title of Training Program)

To assure that you are getting the maximum from this workshop and that your time is respected, please answer the following so your needs will be recognized and fulfilled.

1. The things that bother me most about (reference to the training program's subject) are _____

   _____

   _____

   _____

2. The things I want to learn most during this program are _____

   _____

   _____

   _____

3. I hope that the program's facilitator will _____

   _____

   _____

   _____

Transportation arrangements (including surface service from the airport to the hotel)

Program dates and days

Billing, expenses information

Hotel arrangements, roommates, smokers versus nonsmokers

All schedules, start-up time, reception get-together

List of other attendees and their origins

Dress code (so they properly pack)

Weather to expect

Exercise facilities (swimming pool, gym, jogging, etc.)

Business items to bring (account/territory listing, hand calculator, etc.)

Maps of local area (airport to hotel/training center, city maps for those coming by car, etc.)

Names of training office staff that they will be seeing.

A form to return to the training center letting the office staff know flight arrangements and any special request (such as medical, dietary, avoiding smokers, etc.)

Assorted do's and don'ts (avoid carrying excess cash; if you meet certain senior officers refer to them as "Mr." and not first name; be security conscious especially in the hotel; etc.)

Contacting the Immediate Supervisor—A Three-Phase Process

The trainee's immediate superior is the safest bet to see that results are gained on the job after the training experience. Chances are the immediate superior recommended or approved the trainee's registration in the first place. A three-phase communication program with the immediate supervisor can enhance the trainee's involvement and ensure better results.

A few weeks prior to the trainee's trip to the training program, you may want to send a letter to his or her superior outlining what will be covered and some expectations for results. Then the supervisor can sit down with the trainee, and together they can outline several objectives. The objectives could include a particular skill that is weak and was brought to the salesperson's attention during a performance appraisal. As mentioned earlier, performance appraisals are not performed consistently by sales managers and are often poorly done. Offering an opportunity for salesperson and manager to discuss these issues presents an opportunity for appraisal as well.

The manager may also want the salesperson to network with another attendee, one who has had a certain experience or one who has an account responsibility that affects the trainee's activity. The manager may suggest taking a problem that has bothered all the region/district and seeing if the trainee can get helpful answers. Having the trainee "author" his or her own objectives also helps to assure that the objectives will be addressed by the trainee. By having the immediate superior assist the trainee, follow-up and results should be improved.

We will cover the other two communication phases in the next two sections.

Arrival Packet

A welcome packet with the hotel registration adds warmth and information. This packet can include a brief letter of welcome, information on schedules, maps of the local area, dining facilities, time and location of the welcome reception, safety measures (hotel security), and medical facilities.

Medical Emergencies

When managing a training department you are dealing with a vast array of public, and medical emergencies are bound to arise. If a person has dietary considerations, handicaps, or other health problems, access to a medical clinic or hospital emergency service should be automatic. A first-aid kit should also be in the office.

Anniversaries, Celebrations, and Other Festive Occasions

Perhaps a trainee is away from home on his or her birthday, or has an employment anniversary while in training, or is celebrating some other special event. Calling attention to a personal celebration can add warmth and comradery to the group.

W.I.I.F.M.

Just as the salesperson's customers are interested in knowing "What's in it for me," so is the invitee. Have all invitation materials drawn up in a stimulating manner to create interest and desire. The very completeness and personal warmth of the sales training invitation is a model of how salespeople should be communicating with their customers.

OPERATIONS DURING THE TRAINING PROGRAM

While trainees are in the training center or at an external location, the sales training manager/director should take precautions to maintain control. He or she should also make certain that the environment is conducive to learning.

Environment Considerations

When salespeople enter a training environment, they are "fish out of water." Salespeople are used to action all day. They are very mobile, active, and often high strung. In the training room they face long hours of sitting down; the very opposite of their normal routine. In a very short time they can become restless and sleepy and take mental excursions.

Phone Calls

The bane of sales trainers is the use of telephones by trainees whenever there is a break. One call leads to another, the 10-minute break becomes 25 or 30 minutes, and control begins to crumble. One professional training manager has said that if a manager-trainee has to call the office even once during the training session, then the company made a mistake promoting him or her to that position.

Set rules about telephone calls. When holding the program in a hotel, it is more difficult to enforce the rules; trainees can go to their hotel room. In a training center such rules are more enforceable.

Incoming calls can be screened by the switchboard operator or secretary. Most incoming calls or outgoing calls are to the salesperson's office or immediate superior. Ground rules should be spelled out with the superior before the program. Arrangements should be made by trainee and superior for coverage of accounts and territory.

Refreshments

In addition to coffee and soft drink service, ice water and a dish of mint candies are good to have on the training desks.

Smoking

Set the smoking regulations according to company and training center policies.

Training Department Privacy

If programs are being held in the training center near the center's administrative offices, encourage trainees to keep out of this area. They can be distracting to the office staff. If they become curious about memos being written and read over the shoulder of the office personnel, they can misinterpret what they are reading and start rumors.

Private Consultation

It is advisable for trainers to hold private conversations with the trainee at the trainee's request. Many times trainees feel reticent about discussing a particular skill deficit in front of their peers. Private consultations are very helpful to understanding better how to control some concerns. Caution must be taken that conversations do not drift into personal conflicts; just stick with the behavioral aspect of the person's skills.

Emergencies

Aside from medical emergencies (covered earlier), others can pop up, especially in the off-site environment. The electricity may go out during a thunderstorm. Even with emergency lighting, which is not bright enough for reading, other activities can be initiated. Break groups into triads to discuss particular issues; war stories are often disdainful program draggers under normal situations, but can be a filler in such emergencies. Snowstorms can delay flights in or prevent trainees from getting home. In any event, be prepared for noncontrollable emergencies and have options available.

Communicating with the Immediate Superior

It was mentioned earlier that a three-phase system to keep the trainee's supervisor in the picture can abet the training results. The first issue covered was establishing objectives, a consensus system between the salesperson and the supervisor. While the trainee is in the program, a second letter can inform the supervisor what can be expected to result from the program and that an action plan will be developed by the trainee with a copy for the supervisor. The purpose is to rally the support of the supervisor so that effective and measurable implementation of learning will take place back on the job.

FOLLOW-UP/EVALUATION—POSTPROGRAM ACTIVITIES

As soon as possible, a staff conference with the trainers should be held. Perhaps the program has to be modified using numerous ideas that came from the field. Equipment may have had problems or new equipment needed. Revisions may be necessary in the training materials. At the same time rating systems in use should be recorded.

Certificates of Accomplishment

Many training centers issue a certificate to the trainees recognizing their attendance at the program. Salespeople enjoy posting certificates of accomplishment on their office walls.

More and more training centers are now requiring that the salespeople really "earn" these certificates. Rather than arbitrarily issuing them, they are held back until the trainee's action plan has been accomplished. Once the action plan activity has been reviewed by the immediate superior and progress has been recorded, the certificate is sent from the training center. Sometimes the certificate is enclosed in portfolio or some other useful business gift. This puts far more meaning into the recognition process.

Responding to Trainee's Needs

Many requests are made by trainees during sessions: copies of an article, information on local training opportunities, information they asked the trainers to pass on to other departments, and so on. Be sure to follow through as soon as possible. This has a strong effect on department credibility.

Training Center Library

A lending library system can easily be developed, even put on the department's computer. Several copies of a popular book can be stocked along with less popular and more esoteric subjects.

An audio cassette library is especially popular with salespeople as most have players in their automobiles. Get your company to issue a series of "product knowledge" cassette tape for new products, for upcoming promotions, or for modifications and problem solving. General monthly cassette newsletters can also be developed and sent out by the training department.

Audio cassette tapes made with a strong benefit reminder to salespeople have a training reinforcement effect.

Gaining in popularity in sales training department libraries are video cassettes and disks. Training films can also be useful, but videos are more popular (especially in the ½-inch size) as the popularity of video players increases in homes. Sometimes the library media items can be loaned to the training department of a customer, thus making further thrusts toward good customer relations.

Communication with Management

If action plans are set, go to the third phase of communicating with the immediate supervisor. Advise how to handle the follow-up process when the trainee returns to the territory. Were the objectives accomplished? What is the action plan and how can the supervisor help in its accomplishment? What controls are established?

Also, get back to senior management. If they had a hand in the development of the program, let them know what was accomplished. Perhaps you or your trainers have learned something that has to be passed on. The purpose is not to overlook this important sector which makes decisions on the training department's future.

Recording

Be sure all program information is recorded and passed on to the personnel department as necessary.

Reinforcement

Some sales training centers send a follow-up letter to trainees a few weeks after the program. It informs them of reinforcement cassettes, business magazine articles, and so on that are available to help change behavior styles and requests that the trainees set up their own measurement for results.

Also, and very important, a follow-up letter explains to the trainees that training is ongoing; it is not a one-shot program at the training center. It is really the responsibility of the employee to attend to his or her development. Training is a continuous system, not a break in the normal routine.

SUMMARY _____

The sales training department should be a state-of-the-art operation. Many sales training department managers have complained that they are victims of corporate budget chopping and have difficulty getting quantitative assessments from their corporate masters. The material discussed in this chapter is aimed at developing the department into a functional sales builder, not only through thorough and professional development of the sales staff but as a customer- and marketplace-oriented entity. Being a "model" of company standards and a spokesperson for the values and morals of the company can only enhance the strength and need for the department.

The department has the obligation to serve as the conscience of the company, exploring, discovering, and presenting the training and development needed to fulfill the corporate goals.

The rewards in the business world are for results, not efforts. The professionally administered sales training department can eventually find its reward when it becomes a counselor to the senior executive offices. And that's a reward well worth the efforts.

_____ **REFERENCES**

Berry, D. R. *A Guide for the Company Instructor.* Washington, DC: International Correspondence Schools, 1969.

Davis, Kenneth, and Frederick Webster. *Sales Force Management.* New York: Ronald Press, 1968.

Hanan, Mack. *Key Account Selling.* New York: AMACOM, 1985.

Harrison, J. F. *The Sales Trainer as a Manager.* Orlando, FL: National Society of Sales Training Executives, 1983.

Harrison, J. F. *The Management of Sales Training.* Orlando, FL: National Society of Sales Training Executives, 1983.

Macali, Paul. *The Field Managers Problem Solver.* New York: Hawthorn Press, 1978.

Mager, Robert. *Preparing Instructional Objectives.* Belmont, CA: Pitman Press, 1975.

Mager, Robert. *Analyzing Performance Problems or Ya Really Oughta Wanna.* Belmont, CA: Pitman Press, 1970.

Mager, Robert. *Developing an Attitude Toward Learning.* Belmont, CA: Pitman Press, 1968.

McLagan, P. A. *Getting Results Through Learning.* St. Paul, MN: M & A Press, 1982.

Mills, H. R. *Teaching & Training*. New York: Halstead Press, 1970.

Scannel, Edward, and Les Donaldson. *Human Resource Development*. Reading, MA: Addison-Wesley, 1978.

Stroh, T. F. *Training & Developing the Professional Salesman*. New York: AMACOM, 1973.

section three

IMPLEMENTING SALES TRAINING

Chapter 13   Coaching and On-the-Job Sales Training

Chapter 14   Self-study and Self-development, Flexibility, and Learner Initiative for Your Sales Training System

Chapter 15   Lecture Method and Classrooms: The Most Frequent Method in Sales Training

Chapter 16   Using Role Playing for Sales Training

Chapter 17   The Case Method: Teaching Sales Managers to Make Better Decisions

Chapter 18   Gaming and Simulation in Sales Training

Chapter 19   Effective Audiovisual Media in Sales Training

Chapter 20   How Video and Film Can Improve Your Sales Training Presentations

Chapter 21   Computer-Based and Interactive Video Training Technology

Chapter 22   How to Improve Your Sales Training Meetings

chapter thirteen

COACHING AND ON-THE-JOB SALES TRAINING

NICHOLAS H. WARD AND KATHRYN S. WOLFSON

Nicholas H. Ward is executive vice president, MOHR Development, in Stamford, Connecticut, where he is responsible for sales and sales management programs and the Financial Services Division. He has many years of experience with MOHR and a division of the Xerox Corporation. He is a frequent speaker at meetings and conventions and is the author of many articles on sales and sales management. He has a Bachelor of Arts degree from Williams College and a Master of Arts degree from Brown University.

Kathryn S. Wolfson is a senior consultant and director of product development at MOHR Development, Inc., Stamford, Connecticut. Her work at MOHR has involved development of programs such as Financial Services Selling (client-centered selling) and Financial Services Sales Management (sales leader for banking). She has eight years of experience as corporate and operations training, management, and technical training at Home Insurance Company and as director of training at Saks Fifth Avenue. She has a Bachelor of Arts degree from Smith College.

The purpose of this chapter is to provide guidelines and concrete examples of how sales managers can bring structure and efficiency to on-the-job learning. If your organization has a training department, this chapter will suggest ways to provide training and support for sales managers to allow them to function as on-the-job extensions of the training department.[1]

HOW TOP-PERFORMING SALESPEOPLE LEARN

Selling is one of the most highly compensated professions in the world—the high compensation is a direct reflection of the complexity and difficulty of achieving consistent sales success. If selling were simply a matter of following prescribed steps and levels of activity, then the task of building new salespeople would be altogether a different matter and sales would not be such a highly rewarded field.

In fact, however, because selling involves such a complex constellation of knowledge, skills, judgment, and personal motivation, it is not something that can be adequately taught through books or classroom learning alone. The evidence is clear that top-performing salespeople learn most of what they do from three key sources beyond the corporate classroom:

1. On-the-job trial and error
2. Observation of peers
3. Coaching and feedback of sales managers

---

[1] Much of the material in this chapter—specific procedures, forms, and so on—has been taken directly from a MOHR Development program, *Sales Leadership Systems*, copyright 1981.

Regardless of the quality of the formal learning experience that is provided to salespeople, these three learning sources will continue to be the largest contributors to job knowledge. The problem, of course, is that too often this learning is inefficient and random.

At worst, salespeople "try on behaviors" that they observe from peers and managers that are simply ineffective, and they develop patterns of behavior that will be the seeds of their failure. Too often, salespeople have managers who are untrained and unequipped to provide the basics of knowledge and skill development. The result is a salesperson who tries to hide his or her real activity from the manager, fearing that the manager won't understand and may interfere.

Leading sales organizations recognize the tremendous potential that can be realized by harnessing the power of on-the-job learning. Structure, preparation, and well-trained sales managers are the keys to converting random and inefficient learning into a powerful developmental process. At the hub of this process is the front-line sales manager.

It is the sales manager's role to function as an on-the-job training department, translating product knowledge and sales skills into field reality. Ultimately the most successful sales managers instill in their people the ability to function as their own coach.

Organizations without adequate front-line sales management or with sales managers who function as administrators or supersalespeople are at risk. It is their challenge to provide, in some other manner, the support systems outlined in this chapter.

The most consistent and reliable approach for upgrading the quality of on-the-job learning hinges on providing sales managers with the tools they need to:

1. Monitor and assess performance, knowledge, and skill.
2. Provide consistent on-the-job coaching, focusing on individual accounts and sales skills.
3. Provide strategic performance management coaching, focusing on territory management, overall activity, and progression toward goals.

PERFORMANCE ASSESSMENT AND MONITORING

Basic to all on-the-job learning is the ability to determine existing levels of skill and knowledge and then build development plans for growth. The complexity of the sales process demands an ability to assess accurately every aspect of sales performance. A crucial ingredient of the sales manager's role is to help the salesperson develop the ability to assess his or her own performance. Research indicates that the most successful sales performers are rigorous assessors of their own performance. Developing this ability is an essential sales management function.

Establishing a Skills Inventory

Before a salesperson and a sales manager can begin to speak about skill development, it is helpful to establish a basic vocabulary of the essential skills, why they are important, and the salesperson's current level of development.

The best way to arrive at a common viewpoint is to use a basic skills inventory form listing the underlying skills and knowledge necessary to do the job and then list key selling activities that you or your company have identified as important to the job. Examples of key skills and knowledge could be "Understand Target Market's Business" or "Listening Skills." Examples of key selling activities could be "Prospecting" or "Presenting Proposals to Groups." If possible, the skills, knowledge, and activities should be consistent with the content of any internal manuals or formal sales training programs used by your company.

A simple form can be devised to be filled out by both the salesperson and the manager. The forms should then be discussed in a meeting or by phone to establish a base point for future coaching and on-the-job training. A sample basic skills inventory form is shown in Figure 13–1.

It is also possible to augment the sales manager/salesperson profiles with a client rating form which most clients are happy to complete.

Whether or not you choose to complete a formal skill inventory process, it is important that you use some vehicle to establish a common frame of reference and language from which to begin on-the-job training.

Seven Questions to Ask When Reviewing Call Reports

Coaching and on-the-job training require accurate information about sales activity and time and territory management. Call reports and expense reports are the basic vehicles for providing this information.

The purpose of this chapter is not to give guidance on building or creating reporting systems, but to suggest some fundamental areas to help managers use these reports as sales management information. Approached diagnostically, they can reveal important data about sales behavior that will help managers pinpoint areas needing attention and provide direction for coaching and on-the-job training.

Timely call reports can provide a manager with critical information about basic activity patterns and should answer these questions:

Figure 13–1 Basic skills inventory form

Name _____	**Knowledge or Skill**	**Rating**
Rating Scale		
5 – A great strength	_____	_____
4 – A strength	_____	_____
3 – Adequate	_____	_____
2 – Not a strength	_____	_____
1 – Opportunity for real growth	_____	_____

Sales Activity		**Rating**
1. _____		_____
2. _____		_____
3. _____		_____
4. _____		_____
5. _____		_____
6. _____		_____

List the three knowledge and skill areas most in need of strengthening:

1. _____   2. _____   3. _____

List the three sales activities most in need of strengthening:

1. _____   2. _____   3. _____

- Are enough calls being made?
- Are the calls consistent with preplanned activities?
- Are they in the right markets and geographic areas?
- Do the calls reflect the right balance of new prospects, sales in progress, and add-on business?
- Is office or internal follow-up time planned appropriately?
- Do calls reflect appropriate product mix?
- Is activity taking place at appropriate levels in the client organizations?

Asking these seven questions while reviewing each salesperson's call reports will help uncover training needs. For example, too many calls in one stage of the selling cycle such as servicing, or initial calls, would indicate a need for training in how to prepare for, conduct, and get beyond that particular step in the cycle.

By interviewing several seasoned salespeople and managers or by assembling a task force of representatives of the sales organization, sales trainers should be able to assess existing reports and suggest modifications if reports are inadequate. Samples of completed reports may then be used to help managers learn to analyze activity using the seven questions. Basic sample reports are shown in Figure 13–2.

HOW SALES DATA ANALYSIS HELPS TO PINPOINT TRAINING NEEDS

Analysis of basic activity reports can reveal underlying behavior patterns; long-term consistent results depend upon the salesperson doing the "right things" consistently. Ultimately, however, it is bottom-line results that count in sales, and unless these activities are done *well*, results will suffer.

Monitoring and diagnosing the quality of sales activity can come from two sources:

1. Analysis of results (sales reports)
2. On-the-job observations

Much of this chapter is devoted to on-the-job coaching and observing sales effectiveness. The purpose of this section is simply to highlight the importance of using sales data in conjunction with on-the-job coaching. Straightforward analysis of sales data will help to pinpoint training needs by revealing where the salesperson is experiencing success or failure:

- Is volume sufficient?
- Does sales volume reflect best possible pricing?
- Is success appropriate to market and product mix goals?
- Is there an appropriate mix of repeat products to existing customers?
- Is there an appropriate mix of add on products to existing customers?
- Is there an appropriate mix of new customers?
- Does the volume per customer indicate too much reliance on a few customers or too little penetration with many?

Asking these seven questions while reviewing sales data is an essential complement to the analysis of activity reports and on-the-job observation. It should be remembered, however, that while all training needs will show up in sales results, by the time problems are reflected in results, there have already been costs to the organization. Analyzing data can help to confirm training needs, but it is not a substitute for one-on-one management.

Using samples of existing data in brief case study workshops and adopting these seven questions to highlight the critical factors for their own business should help trainers build sales managers' analytical skills.

ALTERING COMFORTABLE PATTERNS OF SALES ACTIVITY

Often, sales behavior patterns that are inappropriate or inconsistent with goal achievement may be attributed to a "lack of motivation" or an "attitude problem." Research suggests, however, that this is more often than not a convenient managerial excuse. Forceful exhortations or powerful motivations to do something—like sell more of a particular product—will not result in more of the appropriate behavior, unless:

1. The salesperson knows why the behavior is important.
2. The salesperson knows how to do that sales behavior.
3. The salesperson feels confident that the behavior can be executed competently with predictable success.

Salespeople *will not do*, in front of customers, those things that they don't feel *comfortable* doing. Left to their own devices, they will establish patterns of activity with products, customers, and markets where they feel most comfortable.

It is the sales manager's job as coach and trainer to use reporting systems

Figure 13–2 Sample reports

WEEKLY CALL REPORT

NAME _____
WEEK ENDING _____

Company Name/Contact Name/Title	PRODUCT	LEAD SOURCE	PHONE CONTACTS		FACE-TO-FACE CONTACTS							Results/Follow-up Comments
			COLD CALLS	FOLLOW-UP CALLS	COLD CALLS	INITIAL MEETINGS	IN-DEPTH INTERVIEW	PRESENT RECOMMENDATION AND CLOSING	REVIEW AND EXPAND BUSINESS	SERVICE CALLS	FOLLOW-UP CALLS	
M O N												
T U E												

280

| W |
| E |
| D |
| T |
| H |
| U |
| F |
| R |
| I |

TOTAL FACE-TO-FACE CALLS BY CATEGORY

TOTAL, F-F CALLS
TOTAL DAYS AVAIL
F-F CALLS/AVAIL DAY
TOTAL CC-F-F/PHONE
TOTAL INITIAL MEET

281

Figure 13-2 (Cont.)
FOUR-WEEK PLANNING FORM

	MONDAY	TUESDAY	WEDNESDAY	THURSDAY	FRIDAY
W K 1					
W K 2					
W K 3					
W K 4					

MAILING LIST UPDATE

Contact Name	Client Name	Title	Address	Telephone	Delete	Add	Change

to diagnose inappropriate patterns and provide sufficient training and coaching to help develop levels of comfort and competence so that the salesperson will end up doing more of the right things. When the diagnostic approach is used in conjunction with the coaching and performance management models suggested later in this chapter, the salesperson will begin to develop the fundamental disciplines of self-diagnosis and self-coaching. Sales trainers can help managers recognize that every on-the-job experience has the potential to result in on-the-job learning.

THE TRAINER'S ROLE IN BUILDING ONE-ON-ONE COACHING EFFECTIVENESS

The diagnostic part of sales management is, in many ways, also the easiest. Diagnosis helps the manager define some of the areas where a salesperson needs improvement. Translating the diagnosis into action, and pinpointing sales skill deficiencies, require face-to-face sales management. It is here that most coaching and on-the-job learning breaks down.

The remaining material in this chapter outlines how to help sales managers improve their success and comfort in face-to-face situations. It is broken down into two sections:

- *Day-to-day coaching*—meetings with salespeople focusing on individual sales calls and accounts
- *Performance management coaching*—meetings with salespeople focusing on strategic territory management and goal achievement.

Each section provides background, rationale, and concrete steps for handling each face-to-face sales management situation. Sales trainers may use this material in conjunction with any of the commercially available sales management and training programs, or as a basis for creating their own internal programs. If trainers are constructing their own training programs, it is recommended that for each face-to-face situation identified in the text, trainers could:

- Create a sample case study that would illustrate the learning points.
- Allow sales managers in a workshop format to analyze how the steps would apply to the case study.
- Then have managers select a salesperson who actually works for them and prepare to have a discussion with that salesperson following the suggested steps.

- Conduct role plays, allowing managers to practice using the steps with "real-life" situations.
- Have managers go back to their jobs, conduct the planned discussion with the salesperson, and report back to the trainer on the results of the discussion.

It is not recommended that trainers follow this process with too many face-to-face situations in one training session. Rather, the trainer should select two or three situations from day-to-day coaching or performance management coaching that would have the most relevance and impact for their sales organization. The objective should be to conduct a number of sessions over a period of time to begin to build sales management comfort and to institutionalize their role in on-the-job training.

Additionally, the trainer should make every effort to formalize the manager's coaching and on-the-job training responsibility. The coaching and performance management forms suggested in the chapter should be modified to reflect the critical issues of the sales organization and should be consistent with existing terminology and existing operational requirements.

The trainer should work closely with senior sales executives to enlist their support and to make certain what is taught in workshops is consistent with sales management job descriptions and accountabilities. In the event that the sales manager's job and span of control does not permit the kind of one-on-one diagnosis and coaching described in this chapter, the trainer could propose alternatives. For example, a buddy system could be worked out where senior salespeople function as surrogate coaches.

Regardless of how the link is made to the sales organization, the trainer who is interested in building on-the-job coaching effectiveness will recognize the importance of building sales management skill and comfort in the face-to-face situations described in the rest of this chapter. The truly effective trainer will succeed in integrating the coaching process and on-the-job learning into the very fabric of the way their organization does business.

DAY-TO-DAY COACHING OF THE CALL PROCESS

No aspect of coaching is more crucial than the sales manager's ability to help the salesperson with specific sales situations. Certainly, this day-to-day coaching has always been considered an important part of the manager's job. However, there is evidence to suggest that far too little coaching actually takes place in most field sales organizations. Lack of time and geographical distance are often cited as excuses for not conducting pre- and postcall reviews or going on joint calls with salespeople. In many cases, though, the real reason is that the sales manager simply feels uncomfortable with how to handle the coaching process.

Whether or not the sales manager is able to observe the actual call, coaching of the call process *is* critical to individual sales success. The effective sales manager can help the salesperson plan and control the sales process as well as develop appropriate call strategies. In addition, with regular coaching on relevant sales skills, market and product knowledge can be strengthened. Ultimately, the objective of this kind of coaching is to develop the employee's ability to self-critique when the sales manager is not available.

Coaching the call process involves several key activities:

- *Precall coaching*—helping the salesperson think through the call and develop strategies to meet call objectives.
- *Joint-call coaching*—assisting the salesperson in implementing the precall strategy as well as observing, assessing, and gathering critical information to provide meaningful postcall feedback.
- *Postcall coaching*—reviewing the call by using the salesperson's perceptions to evaluate the interaction and identify sales skill strengths and limitations.

Precall Coaching

The primary purpose of a precall review is to increase the chances of the salesperson's success by building confidence and competence.

Coaching discussions before the call help the salesperson to think through call objectives and strategies to meet those objectives. Should particular skills or knowledge (e.g., product, market) need to be strengthened for the call, the manager can demonstrate and practice critical elements of the call with the salesperson. Additionally, if the manager and salesperson will be going on the call together, this precall discussion allows them to share information about the account and define their roles (see joint-call coaching for more information).

Although the precall discussion is an opportunity for the sales manager to help the salesperson achieve the best strategy for any individual call, it should also be viewed as helping with the salesperson's overall development. The salesperson is being coached to think through the sales process, to develop and refine critical selling skills, to make optimal use of background information, and to develop account development strategies.

It is very tempting to convey either a combination cheerleader/drill sergeant or supersalesperson approach when conducting these precall discussions. Most salespeople however, do not respond well to these roles, but would prefer a range of coaching tips from which they can develop their own successful strategies.

Therefore, the sales manager should make certain that the discussion is constructive and positively motivating. The emphasis is on reinforcing the salesperson's skill and maintaining the salesperson's self-esteem, while helping to ensure that the salesperson has developed the most effective call strategy.

Depending on logistics, precall discussions can take place in person or by phone. The frequency and extent of these discussions will depend upon the salesperson's tenure and level of development and the importance of a specific call to the individual account. It is suggested, however, that any joint sales call be preceded by an effective precall discussion.

The call coaching guide in Figure 13–3 is a helpful tool sales managers can use to both plan for and review the call with their salespeople. Sections I and II should be completed by the salesperson and reviewed with the sales manager before the call. Section III should be completed by the manager immediately following a call to provide feedback.

Six Action Steps for a Successful Precall Review. Though each discussion will vary, the sales manager can help to ensure success by following several action steps when conducting a precall review:

1. *Review account history, needs, and potential.* How much background is reviewed will depend on how familiar the sales manager is with the account. In any case, since the salesperson has more in-depth and up-to-date information, it is important that he or she briefly position the account for the manager. Having to verbalize this information will also help the salesperson think through key issues.

2. *Review salesperson's call objective and come to agreement.* The sales manager should use the information obtained in step 1 to assess the appropriateness of the salesperson's call objective. The manager should also make sure the objective is specific and measurable. For example, "getting the business" is a vague objective, and it will be difficult to know to what extent it was achieved (as opposed to an objective of "closing the account on 500 units of item 302").

3. *Review salesperson's call strategy.* In reviewing the call strategy (i.e., the steps the salesperson intends to use to achieve the call objective), the salesperson should be asked for a step-by-step outline rather than a verbatim projection of expectations. The sales manager should step in to help the salesperson anticipate any objections that may be raised.

4. *If necessary, role play key steps of call, reinforce effective skills, and demonstrate skills to be strengthened.* The sales manager should choose elements of the call to role play based on his or her anticipation of key or pivotal points in the call, such as a specific objection or customer concern. After the role play, feedback to the salesperson should be specific. "You did a

Figure 13–3 Call coaching guide

Salesperson: _____   Date: _____
Sales Manager: _____   Customer/Company: _____
                                Contact: _____

General Information

1. Customer [ ]   2. Telephone [ ]   3. This Call: 4. Manager's
   prospect [ ]   in-person [ ]   role in call:

                                         Qualifying   Presentation
                                         Call   [ ]   of recommend-
                                         Initial call [ ]   ations   [ ]
                                         In-depth   Reviewing and ex-
                                         Interview   [ ]   panding the re-
                                                                    lationship [ ]

                                                                   _____
                                                                   _____
                                                                   _____
                                                                   _____
                                                                   _____

Precall Planning

1. Account History: _____
   _____

   Needs: _____   Potential: _____
   _____   _____

2. Call Objective: _____   3. Call Strategy: _____
   _____   _____
   _____   _____
   _____   _____

Figure 13–3 (Cont.)

Postcall Review

	Exceptional	Satisfactory	Needs improvement
PREPARATION			
Understands customer's business	—	—	—
Decision making process	—	—	—
Call objectives	—	—	—
Overall account strategy	—	—	—
SALES SKILLS			
Opening			
Built rapport	—	—	—
Set appropriate tone	—	—	—
Focusing on Customer Needs			
Gained interest	—	—	—
Developed needs we can satisfy	—	—	—
Questioning			
Used open and closed questions	—	—	—
Used questions to develop needs	—	—	—
Encouraged customer to express needs	—	—	—
Selling Benefits			
Linked features and benefits to needs	—	—	—

Figure 13–3 (Cont.)

	Exceptional	Satisfactory	Needs improvement
Handling Objections			
Understood customer concern		___	___
Introduced appropriate information	___	___	___
Placed concern in proper perspective	___	___	___
Closing			
Gained evidence of commitment	___	___	___
Gained clear understanding of next step	___	___	___

FOLLOW-UP PLAN

	Exceptional	Satisfactory	Needs improvement
Appropriate timing and technique	___	___	___
PRODUCT KNOWLEDGE			
Features and benefits			___
Application to customer's business	___	___	___
Competiton	___	___	___
Market	___	___	___
Use of documentation and support tools	___	___	___

SPECIAL STRENGTHS

_____

_____

AREAS/PLAN FOR IMPROVEMENT

_____

_____

289

good job" won't help the salesperson as much as saying, "You overcame the objection on shipping costs well by explaining the benefits of air freighting. That really will help the customer feel more comfortable about spending the extra dollars."

One good way to demonstrate skills to be strengthened is to have the manager first handle the situation as a salesperson, then let the salesperson try it until he or she feels more confident.

5. *Clarify your role during the sales call*. If the manager will be going on the call with the salesperson, this step is critical to be clear about how the manager will participate in the call. (More about this will be discussed in joint call coaching.)

6. *Express confidence in salesperson's ability to implement call strategy*. Finishing the review by enhancing the salesperson's self-esteem will help ensure success. Research has shown that high self-esteem going into a call is key to making the sale.

Joint-Call Coaching

Historically, joint sales calls are one of the most abused aspects of a sales manager's job. But, potentially, they can be one of the most useful. The overall objective of a joint call program should be to improve the salesperson's productivity by:

- Providing specific skill-related feedback.
- Helping the salesperson to develop a critical understanding of the call process and enabling the person to self-critique the call.
- Helping the salesperson learn to develop efficient sales strategies for each account.

The joint call is a crucial event in the relationship between sales manager and salesperson. It is a time when the performance of both the sales manager and the salesperson are most clearly observable; it takes place under the performance pressures of a real customer interaction and is frequently surrounded by anxiety. Because the joint call is such a critical aspect of the sales manager's role, it is essential that anxieties be minimized through the establishment and maintenance of a trusting relationship. When sales managers know how to relate constructively to the salesperson on joint calls, their presence, especially in critical selling situations, is welcomed.

For the joint call to be most effective, a call strategy must be developed prior to the customer interaction. Whenever possible, the sales manager should make sure the salesperson has both one major business objective and one knowledge/skill development objective for each call. The roles of both the sales manager and the salesperson must be clearly understood and appropriate to these call objectives. Over time, the salesperson must learn to trust that the sales manager will maintain his or her predetermined role, and anticipate situations in which exceptions may occur.

Three Types of Joint Calls and When to Use Them. There are three types of joint calls:

1. Modeled call
2. Shared call
3. Observed call

Each type of call has a different purpose and its own strategy. The *modeled call* is conducted entirely by the sales manager and provides the salesperson with an opportunity to observe an effective role model. This is often used to demonstrate how to sell a new product or address a particular market segment. Another reason for doing a modeled call may be to demonstrate a specific skill that the manager and salesperson have identified as an area for the salesperson to strengthen.

To have the most learning impact for the salesperson, the modeled call should be followed up as soon as possible by a similar call where responsibilities are shared between the manager and the salesperson, or the salesperson conducts the entire interview on his or her own.

The *shared call* should be used as follow-up to a modeled call or in high-risk situations where the customer is important and the manager does not feel that the salesperson will be comfortable handling a tough situation. In this call, the sales manager has two objectives: (1) to model or assist in implementing a portion of the predetermined strategy and (2) to observe, assess, and gather critical information during the employee's portion of the call that can be used to provide meaningful postcall feedback. It is critical that the different roles and a system of manager/salesperson cues be worked out in advance.

The *observed call* is the one fully conducted by the salesperson. These calls should be used in fairly low-risk situations and/or when the manager and salesperson decide that the salesperson can handle the entire call (e.g., after a modeled or shared call). Even though the manager does not take an extremely active position in this call, strategies and roles should be worked out in advance; however, the manager should always be prepared to step in when necessary. And, of course, the manager has the best opportunity to observe performance for postcall feedback.

Guidelines for Maximizing the Effectiveness of a Joint Call. Although the sales manager must be flexible in approaching each of these types of calls, there are specific guideposts for behavior which can help to maximize the effectiveness of any joint call. These are:

1. Build rapport and establish role.
2. Reinforce and keep focus on salesperson.
3. Help salesperson stay on track.

A look at each of these guidelines in more detail follows.

Build Rapport and Establish Role. The joint call is an opportunity to reinforce your company's interest in the customer and the account. The sales manager's very presence indicates that this is not an ordinary meeting and may create some feelings of ambivalence in the customer. When conducting a joint call, it is important that the sales manager be sensitive to the following potential areas of concern:

- While it is advisable to let the customer know in advance that it will be a joint call, in many cases the additional person is a surprise that may make the customer ill at ease.
- The customer may fear that the sales manager is there to apply pressure.
- The customer may feel that the salesperson is being "tested" and so may try to protect the salesperson and avoid expressing real needs.
- The customer may feel threatened by the sales manager and feel the need to establish competence by proving his or her expertise.
- The customer may try to enhance his or her own esteem by establishing a relationship with the sales manager at the expense of the relationship with the salesperson.

It is, therefore, essential that the sales manager begin the call by setting the proper tone through building rapport and establishing a clear role. The manager can do this by following these steps:

1. *Indicate how pleased you are to meet the customer and reinforce some aspect of the customer's business.* Being specific about an aspect of business deserving of praise (growth, merger, etc.) is much more effective than general compliments.
2. *If appropriate, comment on an area of personal interest to the customer.* Personal comments may relate to the individual's job, personal life, or mutual

interests. Personal style will be an important element in choosing the appropriateness of this step. To avoid anything too personal, the manager should check with the salesperson about what might be appropriate. In any case, this should be a brief, yet sincere and natural expression of interest.

3. *Clarify your role.* In the case of an observed interview, the sales manager should emphasize his or her role as a resource who may be of some assistance if needed. In a shared interview, the salesperson should still be positioned as the major contact. Regardless, the manager should not detract from an expression of confidence in the salesperson's ability to satisfy customer needs. After clarifying the manager's role, the focus of the presentation should, if appropriate, be returned to the salesperson.

Reinforce and Keep Focus on Salesperson. The quality of the relationship between the salesperson and the customer should be the fundamental link. Effective sales managers are careful never to intrude upon the salesperson's role as the primary resource for customer contact. By keeping the focus of the call on the salesperson, the sales manager reinforces the salesperson's role.

Even when the call strategy identifies the sales manager as a primary participant in the interview, every opportunity should be taken to build in references to the salesperson as the primary contact for follow-up. Skillful use of reinforcement and strict adherence to preestablished roles should enable sales managers to share their experience and expertise with the customer without shifting the focus away from the salesperson.

Possible Ways of Keeping Focus on the Salesperson
- Use body language such as eye contact and turning to salesperson.
- Refer questions to the salesperson.
- Build the salesperson into follow-up action plans.
- Ask the salesperson for comments or reactions.
- Reinforce the salesperson's role and behavior selectively.

The sales manager on a joint call is in an ideal position to reinforce the salesperson's role selectively. During the call, reinforcement can serve such strategic purposes as:

- Strengthening the customer's perception of the salesperson as a valuable resource.
- Increasing the salesperson's confidence in his or her ability to handle the call.

- Helping to "shape" the customer's sense of need and build confidence in your company's ability to meet those needs.

Reinforcement techniques are many and varied. They can range from a nonverbal nod of the head to a verbal underscoring of a special benefit of your product or service already mentioned by the salesperson. Activities that win approval or are rewarded or supported tend to be repeated. Reinforcement is a behavior that can achieve powerful results when used skillfully by the sales manager in the joint call.

In individual discussions with the salespeople, "excessive reinforcement" is not typically a problem with which sales managers need to be concerned. During a joint call, however, it is wise to be sensitive to the impact of your reinforcement upon the customer. Rapid-fire reinforcement of each of the key issues raised by the salesperson may have the impact of putting too much pressure on the customer and could result in defensive behavior by the customer. Selective and skillful reinforcement, on the other hand, will enhance the salesperson's role and strengthen the customer's commitment to doing business with your company.

Help Salesperson Stay on Track. If the sales call is clearly off track (for instance, the "social" conversation is running on or perhaps the salesperson has made an error on company policy or procedure), the sales manager should assume responsibility and give necessary direction to correct the error or refocus the call on objectives. The sales manager should do only what is necessary to help the salesperson stay on track. This does not mean that the sales manager should use it as an opportunity to abandon the preestablished role and take over. Rather, it means that the manager should use his or her best judgment in determining when to help the salesperson stay on track.

Before exerting control and stepping into the call, sales managers should always ask themselves the following questions:

- Is the call sufficiently off track that waiting for the salesperson to redirect the call would substantially reduce the probability of achieving the call objective?
- Is the situation clear enough for the salesperson to recognize and learn from the situation during the postcall review?
- Can the call be redirected and the focus returned to the salesperson in the customer's eyes?

If, in the sales manager's judgment, these questions can be satisfactorily answered in the affirmative, the sales manager should intervene and help the salesperson stay on track. When intervening, the sales manager should be sure not to erode the salesperson's self-esteem.

Postcall Coaching

Whether or not the manager was present on any given sales call, the skillful sales manager uses the salesperson's perception of the call to sharpen insight and strengthen specific skills. The objectives of an effective postcall discussion include reinforcing the salesperson's strengths, improving future performance, and developing ongoing account strategies.

Managers who avoid the "report card" approach will create a more open environment for postcall reviews. The sales manager should offer ideas about the call only after the salesperson is first asked for input. In addition, even though there may be several aspects of the salesperson's sales approach that may require correction or strengthening, the sales manager should focus on no more than one or two.

The postcall review form (Figure 13–3) can be used to provide feedback after a call. Immediately after the call, the manager should complete this using as many specific examples as possible. This should be reviewed with the salesperson. Even when the call is not observed, the manager and salesperson can complete the follow-up plan, special strengths, and areas/plan for improvement.

Six Steps for Maintaining a Constructive Approach to Problem Solving. Sales managers can maintain a constructive problem-solving approach to coaching the salesperson by following these steps:

1. *Review extent to which call objectives were met.* The manager should first explore the extent of the call's success from the salesperson's perspective. Measuring the call's success against the predetermined objectives builds a more analytic approach to selling than will simply relying on overall feelings.

2. *Discuss key factors that led to call outcome.* In this step, the manager looks for the salesperson to discriminate those aspects of the call that were critical to the outcome. The sales manager should discourage blow-by-blow review of the call by directing the salesperson's attention to the key factors. In many cases, due to prior conditioning, the salesperson may want to discuss what was wrong with the call immediately. However, the sales manager's guidance should help the salesperson avoid the tendency to focus on the negatives at this point.

3. *Reinforce skills used effectively during the call.* When reinforcing effective skill use, the manager should give specific examples whenever possible. Letting the salesperson also know *why* the skill was effective will help to ensure continued use.

4. *Ask for salesperson's ideas on how the call could have been strengthened and offer your own.* This is the opportunity to focus on less effective skills or strategies and the consequences for the call. By first asking for the

salesperson's review of limitations as well as suggestions for improvement, solutions are likely to be more valued. Managers should keep the focus on no more than one or two areas for improvement.

5. *If appropriate, demonstrate the skills discussed and have salesperson practice demonstrated skills.* This demonstration and role play of sales skills should be kept brief and focused on a limited skill objective.

6. *Summarize discussion and account follow-up activities.* As the one-salescall close becomes less frequent, helping the salesperson to develop a future strategy for the account protects the value of the sales development investment.

PERFORMANCE MANAGEMENT COACHING

The image of the salesperson as impulsive, disorganized, and desirous of a "freewheeling" work environment is today's myth. Recent research points to the fact that the best salespeople expect to be professionally managed and to participate in their own management. As a result, sales organizations with well-executed performance planning produce high levels of job satisfaction. Companies with articulated job standards and solid performance appraisal programs are seen as most attractive.

In the past, it may have been sufficient simply to assign salespeople to a territory, hand them a quota, and make sure they made enough calls to bring in the orders. Today's salesperson knows that a well-developed strategy is required to realize the full potential of a territory.

Attention to market share, profitability, and other measures of sales success are forcing sales managers to look beyond volume as the sole determinant of success. There is an increasing recognition that effective sales management requires an integrated approach to performance planning. Salespeople report increased confidence and competence when their sales managers skillfully communicate performance standards, establish goals, conduct progress reviews, and engage in performance appraisals objectively and developmentally. Sales managers must be able to engage skillfully in such interactions as:

- **Setting goals**—gaining commitment to meaningful, challenging goals, and establishing the strategies to achieve overall performance targets such as sales or product quotas.

- **Holding progress reviews**—conducting periodic, straightforward discussions that assess progress toward goals.

- **Handling performance problems**—identifying and improving sales performance deficiencies as they occur.

- **Conducting performance reviews**—reinforcing performance and behavior that met or exceeded expectations while clearly identifying areas needing improvement.

Setting Meaningful and Challenging Goals

Most sales organizations have some form of target against which salespeople can measure their performance. Research shows that the more people feel a sense of direct participation in establishing goals, the more commitment they will feel about achieving them. Additionally, salespeople involved in the goal-setting process are likely to arrive at more challenging goals than are those set by their sales manager alone.

In some sales organizations it may not be practical to involve salespeople in determining specific quotas or overall performance targets. Yet, even where overall goals are distributed from the top down, effective sales managers know that success will depend upon their ability to establish meaningful subgoals and action plans with their salespeople that will lead to the accomplishment of overall performance goals.

It is recommended that the sales manager conduct goal setting as a two-part process. During an initial meeting, the manager and salesperson establish the salesperson's overall performance goals as challenging but achievable targets.

Then, in a subsequent meeting, the sales manager helps the salesperson think strategically about breaking the performance target into subgoals and action plans that can be monitored throughout the planning period. This allows both people time in between to develop the most appropriate approach to reaching goals. These goal-setting meetings can even happen in conjunction with other coaching sessions (e.g., pre-, joint, or postcall). If this is impossible, combining the meetings into one is feasible as long as the salesperson knows the objective of the session in advance.

Effective sales managers know that the way in which goals are communicated and agreed to is in many ways as important as the goal itself. By listening to the salesperson and reinforcing the salesperson's successes, the sales manager can ensure that goals are perceived as neither too easy nor impossible to attain. In this way, the sales manager is able to maximize employee motivation while helping the salesperson to establish practical strategies that will lead to goal attainment.

The following action steps will provide sales managers with a logical approach to the goal-setting meetings.

How to Conduct the Initial Goal-Setting Meeting

1. *Outline group sales goals and strategies for the period.* The sales manager should briefly review the period sales goals and strategies of the group (the district, region, etc., to which the salesperson belongs). This gives the

salesperson the "big picture" which enhances self-esteem as well as provides a perspective for what the individual salesperson is being asked to do.

2. *Together decide on salesperson's goals that will help to achieve group sales goals.* By asking for suggestions for goals before adding his or her own, the sales manager increases the likelihood that the salesperson will commit to the achievement of the goals.

3. *Ask salesperson to prepare subgoals and action plans.* The sales manager will want to ask the salesperson to prepare and write down subgoals and action plans that will enable overall goals to be reached. For example, if the goal was an 18 percent increase in total sales, a subgoal could be to increase cross selling to establish a better product mix with accounts. Specific action plans, or steps to achieve this subgoal, could include targeting key accounts and putting together a mailing to existing customers mentioning new products.

4. *Set a follow-up date.* This date should allow the salesperson adequate time to prepare the subgoals and action plans.

The second meeting enables action plans to be fleshed out.

Action Planning: Strategic Steps for Monitoring the Goals

1. *Together agree on and write down salesperson's subgoals and action plans that will lead to achievement of performance goal.* By writing down the salesperson's ideas for the action plan, the sales manager demonstrates that the salesperson's input is valued.

2. *Ask salesperson for suggestions on self-development goals for the period and offer yours.* Establishing self-development goals in addition to performance-based subgoals enriches and develops the salesperson. These goals can include completing external or internal courses related to work objectives or becoming involved in activities outside the job's scope.

3. *Decide on dates to review progress.* Short-term follow-up date(s) should be linked to time frames specified when the action plan is developed.

4. *Express confidence in the salesperson's ability to achieve all goals and encourage him or her to call on you for help as needed.* The sales manager should stress that it is the salesperson's responsibility to go to the sales manager for help.

The performance management form (Figure 13–4) enables the sales manager and salesperson to record and keep copies of the salesperson's subgoals and action plans (as well as progress toward those subgoals). The form also provides an opportunity to record self-development goals.

Figure 13–4 Performance management form

Name: _____
Date: _____

Group Goals: _____

GOAL SETTING	PROGRESS REVIEW (Date): _____	PROGRESS REVIEW (Date): _____	PERFORMANCE REVIEW (Date): _____
SALESPERSON'S OVERALL GOAL(S)	PROGRESS TOWARD GOAL(S)	PROGRESS TOWARD GOAL(S)	ATTAINMENT OF GOAL(S)

SUBGOAL(S) AND ACTION PLANS

GOAL SETTING	PROGRESS REVIEW	PROGRESS REVIEW	PERFORMANCE REVIEW
SUBGOAL (1)	PROGRESS TOWARD SUB-GOAL (1)	PROGRESS TOWARD SUB-GOAL (1)	ATTAINMENT OF SUB-GOAL (1)
Action Plan	Action Plan	Action Plan	Comments
SUBGOAL (2)	PROGRESS TOWARD SUB-GOAL (2)	PROGRESS TOWARD SUB-GOAL (2)	ATTAINMENT OF SUBGOAL (2)
Action Plan	Action Plan	Action Plan	Comments

Figure 13–4 (Cont.)

SUBGOAL (3)	PROGRESS TOWARD SUB-GOAL (3)	PROGRESS TOWARD SUB-GOAL (3)	ATTAINMENT OF SUB-GOAL (3)
Action Plan	Action Plan	Action Plan	Comments

SELF-DEVELOPMENT

SELF DEVELOPMENT GOAL	PROGRESS TOWARD GOAL	PROGRESS TOWARD GOAL	ATTAINMENT OF SELF DEVELOPMENT GOAL
Action Plan	Action Plan	Action Plan	Comments

Holding Progress Reviews

Too often neglected as a part of the performance planning process, progress reviews are in some respects the most important element. Skill in conducting periodic straightforward discussions that assess progress toward goals has been recognized by effective sales managers as a powerful performance management tool.

During a progress review meeting, performance deficiencies can be diagnosed and addressed before they become serious problems. The progress review should enable the sales manager to assess the effectiveness of individual action plans. If market conditions or underlying assumptions have changed, goals may have to be revised or action plans may have to be changed. Progress reviews should also be seen as an opportunity to help sharpen the salesperson's strategic thinking and maintain the salesperson's perspective of each element of the performance plan.

It is the progress review process that gives life to sales performance planning and keeps it a dynamic process. Especially in today's selling environment, where the trend is toward long-term system selling with fewer closings and longer sales cycles, skillfully conducted progress reviews may make the difference between success and failure.

How to Conduct a Progress Review. To conduct these reviews most effectively, these action steps can be followed:

1. *Ask salesperson to review performance against sales goals, reinforcing all on-target performance.* Before this discussion, the sales manager should ask the salesperson to be prepared to discuss his or her progress against goals. During the progress review, you can both refer to the individual performance summary for a record of existing goals. The sales manager should be sure to keep the focus on specific behaviors and accomplishments by asking for and giving examples whenever possible.

   Each goal should be reviewed one at a time, recognizing any that the salesperson has met or exceeded. By reinforcing accomplishments, the manager maintains the salesperson's confidence and commitment. Identifying successful behavior encourages continuation of that behavior.

2. *For below-target performance, discuss both of your ideas to ensure all goals will be met and agree on steps to be taken.* After all goals have been reviewed, the manager should return to any that are below target. In discussing steps to be taken to make sure these goals are ultimately achieved, the focus should remain on the future rather than the past. Asking the salesperson for his or her ideas first will increase commitment. Any agreed-upon additional action plans to reach goals can be recorded on the performance management form.

3. *Revise goals where there has been a major change in underlying assumptions.* If original assumptions have changed, there may be a need to revise (e.g., increase, decrease) those goals affected by the changes. Changes that may impact goals include shifts in your company's or major clients' objectives, the economy, or the market. It is critical, however, that goals not be altered unless the changes are major enough to warrant. For example, if your company is not able to produce enough units of a particular item due to unanticipated operational limitations, the salesperson's goal may have to be lowered during this period.

4. *Summarize the discussion and set a date for the next progress review.* Conducting these reviews regularly is critical. How often managers have progress discussions will vary depending on factors such as the salesperson's experience level. What *is* important is that there are no surprises for either the manager or the salesperson at performance review time.

Handling Performance Problems

Even when performance is managed on a regular basis, performance problems may arise. Rather than waiting for a regularly scheduled progress review, sales managers should discuss the situation before the problem becomes a barrier to goal attainment and in such a way that motivates the salesperson to improve.

Many sales managers find handling performance problems to be an unpleasant part of their management responsibility. Some ignore the problem until it becomes chronic and others overreact and make the problem worse. This may be a result of most salespeople's initial defensive reaction to performance problem feedback.

When a salesperson is criticized, the salesperson may respond as if a personal negative judgment were being made. In an effort to protect self-esteem, the salesperson becomes defensive. In turn, this prevents the salesperson from using the information provided to improve performance.

Despite the potentially touchy nature of the situation, managers who focus on the problem rather than on the salesperson will be most successful. By refusing to become defensive, managers will be able to relate the problem to the salesperson in a way that increases the likelihood of improved performance. By acknowledging the merits of the salesperson's work, involving the salesperson in solving the problem, and then gaining the salesperson's commitment to action, the manager can reduce defensiveness and effectively motivate the salesperson to improve performance.

Tips on Handling Performance Problems

1. *Focus on the problem and its consequences.* The focus should be kept on the specific performance problem area and how the problem impacts the company, customers, and the salesperson himself or herself. By not attacking

the salesperson personally, defensiveness will be reduced and the salesperson will clearly understand.

2. *Ask for the salesperson's ideas on how to correct the problem, and add your own.* The salesperson is more likely to value those solutions he or she thinks of than ones offered by the sales manager. Let the salesperson suggest ideas first.

3. *Come to agreement on the steps each of you will take.* The manager and salesperson should both write down the steps to reinforce commitment. Goals and short-term follow-up dates for each step should be defined.

4. *Express confidence that the salesperson will correct the problem.* When expressing confidence the sales manager should relate it to some specific positive salesperson behavior.

5. *Set a specific follow-up date.* When setting this follow-up date, the sales manager will want to express the expectation that the problem will be resolved by then.

6. *Praise the salesperson when the desired behavior is first observed.* Positive reinforcement is called for at the very *first* sign of improvement to motivate the salesperson.

Conducting Performance Reviews

Sales managers have long been responsible for appraising performance. Changes in the nature of the sales function are demanding that sales managers more closely examine performance. The effective sales manager has to know how to integrate performance review and performance planning, to "go beyond the numbers" and assess performance across a broad range of activities.

Often the performance review is either uncomfortable for both the salesperson and the sales manager or unproductive in that the review process doesn't seem to motivate or help the salesperson. By making the performance review into a natural step in the performance management cycle, both the sales manager and the salesperson should be able to approach the session with positive, realistic expectations. In effect, the performance review should be seen as a formal tying together of day-to-day feedback and progress review sessions. It should clearly identify and reinforce factors leading to successful performance, while specifically addressing performance areas that can be improved. The successful performance review forms a solid basis for the next discussion on setting goals.

In preparation for the review, the sales manager will want to confirm the time and place for the discussion and remind the salesperson to complete the final results section of the performance management form.

Six Steps for Appraising Performance. The performance review should be an expanded version of the progress reviews held during the period.

1. *Review achievement of overall goal and come to agreement on the degree to which each subgoal was achieved.* The manager should ask the salesperson to review to what degree the overall goal was achieved as well as agree on the extent to which each subgoal was achieved. Both parties can refer to their performance management forms.

   The sales manager has the responsibility to encourage an expression of feelings and a discussion about the ratings. This should be done through "open-ended" types of questions (tell me how you feel about this, what are your thoughts about this, etc.).

2. *Identify all factors that contributed to goal attainment as well as those that got in the way.* The sales manager should be referring to a sampling of behavior over time and not just the most recent events or some unusual one-time event. As many behavioral examples as possible should be cited to support a given rating. This applies to both negative and positive performance.

3. *Ask for salesperson's ideas on how the causes for missed goals can be more effectively dealt with and offer your ideas.* The input the salesperson has in the action plan the more it will be valued. In many cases the salesperson will have the best suggestions for his or her own development.

4. *In the same way, review performance for other areas of responsibility.* This step focuses on any activities not assessed on the performance management form but that are still important in the performance of job responsibilities.

5. *Postively reinforce all attained goals and indicate your confidence in the salesperson's ability to attain future goals.* If relevant, this will also be an opportunity to have the salesperson sign the review form and/or set goals for the next performance period.

SALES MEETINGS: GROUP COACHING OPPORTUNITIES

Individual day-to-day and performance management coaching are essential to each salesperson's success. However, the effective sales manager also realizes the benefit of periodically bringing salespeople together for group coaching sessions. Although salespeople function independently, these sessions are key to building confidence and understanding situations shared in common. In addition to giving sales managers the chance to communicate important information, sales meetings

- Provide a forum for reinforcing group priorities and goals, as well as individual successes.

- Give salespeople an opportunity to share and learn from each other's successes and problems.
- Give salespeople an opportunity to discuss mutual problems and to determine more effective ways of handling accounts and using selling skills.
- Motivate and generate commitment to goals by helping salespeople feel good about themselves, their group, and the company.

Upcoming challenges like a new product introduction can be discussed, with goals and action plans developed and skill practice conducted.

TRAINING SALES MANAGERS TO BE COACHES AND ON-THE-JOB TRAINING REINFORCEMENT

Research shows that today's sales force is more highly educated and achievement oriented than ever before. Salespeople expect a close and candid relationship with a manager. They expect to be professionally managed and to participate in their own management. Self-esteem is a strong need along with tangible rewards. As a result, this condition creates a special challenge in training sales managers to be effective coaches.

The evidence is clear that regardless of how much formal training is provided in selling, the vast majority of sales learning is on-the-job learning. Whether an organization is able to harness the power of that on-the-job learning process—or simply leave it to chance—is largely dependent upon the organization's willingness and ability to make sales managers into effective coaches and on-the-job trainers.

chapter fourteen ─────────────────

SELF-STUDY AND SELF-DEVELOPMENT: FLEXIBILITY, AND LEARNER INITIATIVE FOR YOUR SALES TRAINING SYSTEM

───────────────────────── *GRETA KOTLER*

Greta Kotler is Manager, professional product development, at the American Society for Training and Development. Prior to coming to ASTD in 1985, Kotler was director of training for the CRG Training Institute, Washington, D.C. She designed and presented training programs in sales and marketing skills. She was also director of special projects at Beacon College for five years, which included extensive work with self-study and self-development.

This chapter addresses self-study and self-development in sales training and has the following purposes:

1. To examine the features, benefits, and use of self-study and self-development in sales training.
2. To develop "how to" processes for working with each approach.
3. To examine the trainer's and learner's role in each approach.

As defined, self-study and self-development have unique characteristics. This chapter examines each approach in several areas: description, summary of benefits, application in sales training, designing programs, and the sales trainer's role.

"Self-study" is a generic term which is often differentiated from instructor-led learning. In self-study, participants engage in learning activities on their own. An instructor may act as a resource person, facilitator, or administrator, but does not provide direct instruction as in a classroom setting. In self-study, the learning activities are designed to provide both instruction and feedback. They are also often portable, which allows the learner to complete them at a distance from the training site.

Several learning terms are similar in definition to a self-study methodology, for example,

1. *Self-instruction*—"an all-inclusive term used to describe any teaching situation where students take responsibility for their own learning."[1]
2. *Autoinstruction*—self-study designed for specific behavioral change.

---

[1] Marjorie L. Budd, "Self-Instruction," in *Training and Development Handbook*, 3rd ed., Robert L. Craig, ed. (New York: McGraw-Hill, 1987), p. 489.

3. *Distributed learning*—that learning in which place and schedule are determined by the learner.

4. *Programmed instruction*—popular during the 1960s, it includes highly structured individualized instruction and was sometimes completed by students using teaching machines.

5. *Distance education*—programs of study which are completed by students who are removed from both the learning site and teachers.

6. *Self-paced learning*—programs which allow the learners to determine the speed and schedule for completion.

7. *Home study*—specific learning programs which trainees complete at home (or elsewhere) for formal evaluation by a workplace or college staff. Correspondence courses with colleges are one form of home study.[2]

In this discussion, self-study and self-instruction will be used interchangeably and will focus on learning designs that have the specific goal of changed behavior on the job. In addition, self-study and self-instruction under discussion for sales training have the following characteristics.

1. They are professionally designed by sales training staff or purchased as off-the-shelf materials.

2. They allow sales professionals to complete them on his or her own schedule, but within a specified time.

3. The materials are designed to stand alone and to include feedback to the sales professional as he or she progresses.

4. The final evaluation of changed behavior on the job will be left to the trainer or sales manager.

"Self-development" has a broader definition. Its goals may be related to present or future work or can go beyond work altogether. The learners have greater control over the purpose, goals, learning methods, selection of resources, and evaluation standards and procedures. Self-development is particularly appropriate to career planning, defined as "an individual's deliberate process for becoming aware of self, identifying career goals, and taking steps to attain career goals; accomplished alone or with the organization's help."[3] In this discussion,

---

[2] For additional information on self-study approaches, see ibid., pp. 488–499; Greg Kearsley, *Training for Tomorrow* (Reading, MA: Addison-Wesley, 1985); Derek Rowntree, *Teaching Through Self Instruction* (London: Kogan Page, 1986), pp. 9–17; and Wendell I. Smith and J. William Moore, *Programmed Learning* (Princeton, NJ: D. Van Nostrand, 1962).

[3] Ellen Segalla, "Best Ideas for Career Development Systems," *Infoline* (Alexandria, VA: American Society for Training and Development, 1987), p. 2.

self-development is limited to acquiring new knowledge, skills, or attitudes for present or future work.

SELF-STUDY

How Self-Study Is Used in Sales Training

Self-study can be used for teaching knowledge, skills, and attitudes. The most widely used area, however, has been for knowledge and skills which are straightforward and are not expected to change extensively over time. In sales training, this includes product knowledge, company processes and procedures, and basic selling steps.

EXAMPLE

Company A wants its entire sales force to learn about a new product that has just been introduced. By creating a self-study module on the product, the company can get the information out without having to bring the sales force together as a group. Self-study is less effective when knowledge, skills, or attitudes require extensive amounts of imitative behaviors.

This is changing, however, as the technology of self-study expands rapidly to include more interactive video and computer-based training. These technologies are increasing the ability of self-study to include complex simulations, analysis of successful and unsuccessful applications, self-analysis, and behavior modeling.[4]

As all or part of a program. Self-study can be used as an entire program or for part of a program. As an entire program, it is designed to provide orientation, instructional activities, feedback, reinforcement, and evaluation.

EXAMPLE

The sales professional is given an entire package on "building a prospect list." He or she is expected to read all materials, listen to tapes, complete all exercises and tests, check answers against sample answers, and evaluate his or her learning based upon these sample responses.

---

[4] Kearsley, *Training for Tomorrow.*

EXAMPLE _____

Where self-study is used as a part of a program, certain modules are included which sales professionals are required to complete before the course. For example, all participants might be asked to do background reading and exercises on "building a prospect list" and to come to the training sessions with these completed. These exercises are corrected, discussed, or used as part of the formal group training program. These are often problem situations or cases of current practice and are formally built into the program.

Self-study modules can also be assigned in the middle of a program. A group may come together for part one of the program and then self-study assigned to be completed prior to a second group session. The approach allows the trainee to analyze cases or experience and receive feedback at the subsequent session. Or self-study can be assigned at the end of a group program. In this case, it can be used for remedial training or for reinforcement.

Number of trainees. For self-study to be cost effective, the number of trainees must be large. Once designed, self-study can provide consistent content and delivery for many sales professionals in different locations.

EXAMPLE _____

An insurance company has hundreds of sales professionals located throughout the United States. Each needs training on the company's procedures for closing the sale. With a self-study format, one program can be developed and then made available to the sales professionals in their diverse locations.

Learner motivation. Self-study programs require a high degree of learner motivation. This may be either internal or external. Learners who are internally motivated to complete programs on their own time without the benefit of the social interaction and learning from instructors and peers do very well with self-study methods. Or external motivators, such as deadlines or required tests, can be imposed on learners. Sales professionals often do particularly well with self-study because they are used to working alone and motivating themselves.

Learner ability. Self-study materials must be consistent with the learner's ability and background to complete the materials and to use any necessary software or hardware.

Location of learning. Self-study offers flexibility in where the learning takes place. Print and audio materials can be completed on multiple sites or transported to a convenient place by the learner. This feature is an advantage to sales training as the learning programs can be completed while on travel or away from the office.

Machine-based materials require hardware, but when that is available through an organizational learning center or at home, the same benefits of flexibility in study locations apply.

Administrative functions. Self-study requires administrative systems for the following:

1. Tracking materials.
2. Maintaining equipment.
3. Accommodating learners' schedules.
4. Maintaining evaluation records.
5. Involving sales trainers and sales managers.

These administrative systems, because they involve large numbers of learners on individual schedules and often geographically dispersed, require careful attention. Computerized administrative systems will assist as they become more available in organizations.

Cost. Assessing the cost of self-study requires analysis on an organizational basis. It can be stated generally, however, that up-front design, production, or purchase costs are greater than for instructor-led training. However, if the usage period and numbers of users are substantial, the costs over the life of the programs should be lower than those of instructor-led programs.

EXAMPLE

Company D is implementing extensive customer service training for thousands of employees located in fourteen different cities throughout the United States. Costs (including travel, lodging, facilities, instructors, and materials) necessary to bring these people together in several groups were estimated. These costs were compared to the cost of developing a uniform self-study print and audio package which could be produced and sent to the sales professionals needing the program. In preparing the analysis, the sales manager concluded that although the initial costs of producing the packages were high, the costs would be lower in the long run than the group trainings. In addition, she felt that she could control the quality of the information better by producing the program once and having it available for new hires as well.

Summary of Self-study Benefits

The major benefits of self study are:

1. Provides consistent content in format and delivery to large numbers of trainees.

2. Requires fewer instructors and facilities than classroom teaching.

3. Allows for flexible scheduling.

4. Allows for flexibility in location of learning.

5. Reduces travel time and expense for trainees who are not required to attend group sessions.

6. Is cost effective over time.

Application in Sales Training

In 1978, the Conference Board examined sales training objectives, content, and methods of 156 companies. In exploring methods, they considered on-the-job training, classroom, coaching, observation, home study, and special outside courses. Home study is a self-study approach; for purposes of this discussion, the findings are revealing.

> On-the-job training employed by virtually all the firms represented ranks first in relative importance by a clear majority. In addition to such training, most companies ensure that trainees receive formal instruction of one kind or another. Classroom training ranks second in importance, followed at some distance by coaching and by observation, which are also widely used training approaches. Home-study training, whose importance rarely ranks high in comparison to other approaches, is nonetheless used in about two out of three companies. Similarly, special outside courses are a supplementary approach for approximately half the companies surveyed.[5]

In 1987, a follow-up status report found lecturing to be the presentation method used most frequently. Role playing, particularly with video equipment, was found to be the fastest-growing approach.[6] This is consistent with the great need in sales training to practice skills and analyze performance.

In 1988, Executive Knowledge Works published a report on "Sales Training in America," including sales training practices from 235 companies in seven industries. They found that 71 percent of the companies surveyed used self-paced programs, either alone or in combination with other methods. Methods with more widespread use included role plays (96 percent), instructor lecture and video (93 percent), and case studies (86 percent).[7]

---

[5] David S. Hopkins, *Training the Sales Force: A Progress Report* (New York: The Conference Board, 1978), p. 6.

[6] Earl D. Honeycutt, Jr., Clyde E. Harris, Jr., and Stephen B. Castleberry, "Sales Training: A Status Report," *Training and Development Journal* (May 1987), pp. 42–45.

[7] Executive Knowledge Works, *Sales Training in America* (Palatine, IL: Anthony J. Fresine and Associates, 1988), p. V.15.

EXAMPLE

A telecommunications company develops a catalog with many sales courses available to employees. Some of these courses are available at the training centers only and require that employees arrange their schedules to attend. These courses include many opportunities to develop skills through role playing with critique from professional trainers. In addition, many courses in sales skills are now available through self-study. Some of these courses use print materials, others are supplemented with audio tapes. These do not provide opportunities for role playing. However, the catalog now lists a few computer and interactive video programs that do require sales practice.

Self-study, then, although not the most widespread method, continues to result in both flexibility and efficient learning. Computer and interactive video technologies, too, point to increased possibilities for simulation, practice, and response in the areas of selling techniques and communication. Greg Kearsley presents an example of a rookie salesperson facing 20 different kinds of customers or sales situations in a short period of time. With on-the-job training, this same experience might require a month or two.[8]

Designing a Self-study Training Program in Six Steps

Many of the principles that apply to designing any training program apply to self-instructional design. However, there are concerns unique to this approach because of the goal to provide instruction outside of the classroom. Each design step will be discussed with guidelines for self-study:

1. Analyzing needs
2. Developing learning objectives
3. Planning and developing learning
4. Validating design and materials
5. Implementing the instruction
6. Evaluating results

Step 1: Analyzing Needs. Needs analysis for self-study is not different from other needs analysis. Self-study is appropriate when the following performance gaps are found.

---

[8] Kearsley, *Training for Tomorrow*, p. 23.

1. Sales force performance would improve with uniform information about product service or policy.
2. Sales force performance requires uniform information about each step within the sales process.
3. Sales force needs skills to operate specific machinery.
4. Sales force needs better skills in communication, listening, negotiating, closing the sale.
5. Sales force needs to be motivated.

A needs analysis for self-study should also examine learner ability and motivation for self-study.

Step 2: Developing Learning Objectives. Learning objectives for self-study can address knowledge, skills, and attitudes. Regardless of area, they should include specific learner behaviors to be demonstrated. As the learners undertake self-study, they must understand

1. What are the conditions of learning?
2. What behaviors will they be expected to demonstrate at the end of the self-study?
3. What performance standards must be met?
4. How will evaluation be done?

Examples of learning objectives include

1. After reviewing written materials on new products (A, B, C, and D), trainees will describe in writing the benefits, features, cost, and application of these products for two different categories of users.
2. After completing self-study modules, trainee will pass test on sales procedures with 90 percent accuracy.
3. After listening to tape series on communication, trainee will correctly analyze three communication case studies.
4. After completing self-study modules on closing, trainee's sales results will increase by 25 percent.

Step 3: Planning and Developing Learning. Self-study materials should be organized in short, self-sufficient modules or segments. Learners should be able to complete each module in a time period of under two hours.

EXAMPLE

A forty-hour program in self-study on "introduction to sales skills" to be used by a manufacturing company includes eight different sections from "building a prospect list" to "follow-up after closing the sale." Some of these sections are further broken down into parts A and B so that the background materials, exercises, and evaluation can be completed by learners in no more than two hours. This is helpful to the participants for both maximum learning as well as planning their time. They know that if they schedule two hours for self-study, they will be able to complete a module.

Three sequencing methods for sales training materials. The sequencing of the modules and the materials within the modules can influence the overall success of the training materials. Sequencing can also influence both the learner's motivation to complete the materials and learning outcomes. The sequencing of materials has to make sense to the learner who is proceeding on his or her own and may be making selections as to which modules to complete and in which order.

Possible sequencing methods for sales training materials are

By chronology

By difficulty

By situational problems

1. *Chronology.* Material is organized in the order that it occurs on the job. For example, modules on the sales steps might include developing contacts, prospecting, getting an interview, preparing a presentation, closing, and follow-up beyond closing. Chronological sequencing is also appropriate for presenting the procedural steps in completing company reporting requirements. Chronological sequencing is appropriate for new sales trainers or for new material.

2. *Difficulty.* Materials can be presented in order of difficulty, requiring mastery of one module before moving to the next. If preparing modules on product knowledge, the first module might include description of features and benefits, followed by modules on different applications. The learner would need to be able to describe specific features and benefits before applying a product to a variety of customer settings. Again, the learner must know that the sequencing is "by difficulty" and that modules should not be skipped.

3. *Problem.* Content is presented by cases or problems. Each requires analysis, allowing the trainees several possible responses.

EXAMPLE _____

A new computer company has developed a course on "overcoming obstacles," which includes a number of situations to be analyzed by the trainee.

- a company is not ready to make the commitment to buy computer equipment;
- a company has already invested in another system and is not able to change systems now;
- managers cannot meet the pay schedule as outlined by the sales representative.

After each situation, the sales professional is asked to describe how he or she might overcome these obstacles. Answers are recorded and then compared with prepared answers for those objections.

This kind of problem-centered sequencing is appropriate for more experienced salespeople who have mastered basic steps, have gained some experience and are ready to apply analytical skills to situations.

Skills in problem-centered analysis are becoming increasingly important in sales training as the focus for sales professionals moves toward meeting customer needs. Customers have problems which sales professionals have the opportunity to solve. Executive Knowledge Works reports that in seven areas of skills, knowledge, and attitudes needed for future sales professionals, general skills (including problem solving) were ranked first by 34 percent of respondents. This was higher than personal skills (17 percent), sales skills (10 percent), product knowledge (10 percent), environment knowledge (10 percent), company knowledge (5 percent), and attitudes (14 percent).[9]

Appropriate learning activities. In planning self-study, all principles of adult learning apply. Learning activities should actively engage the salesperson. He or she should understand their purpose and how they will help back on the job. Possible methods are

1. Reading or listening to content followed by worksheets requiring recall or response to closed or open-ended questions. Model answers can follow to allow the learner to compare his/her responses.
2. Case studies asking the learner to analyze a situation and prescribe the next action. Branching allows the learner to move to the next appropriate case if the response demonstrates mastery of the topic.

_____

[9] Executive Knowledge Works, p. IX.4.

3. Practicing a skill in simulated or real-life situation with self-assessment first and another's assessment following. This is difficult to design into self-study, but advanced salespersons can be asked to have sales sessions videotaped for his or her and other analysis later.

4. Watching/listening to materials for motivation and attitude development. These materials should include trainee response and feedback—written, audio, or video or computer.

Reinforcing activities should be included; in addition, cumulative reviews are recommended following every two or three modules. To keep the interest of the learner, review activities should be varied.

Each training activity should be assessed based upon such questions as

1. Can this activity be completed by the learner without assistance? If assistance is needed, is it available?

2. Are the purpose and steps of the activity clearly stated?

3. Will completion of this activity lead to the intended outcomes?

Types of media that can be used. Several media are appropriate for self-study—print, audio, video, and computer—or a combination of these. The following chart lists examples of appropriate media for self-study in sales training followed by the uses and advantages and disadvantages of each.[10,11]

PRINT Uses

1. To present factual information on product, policies, procedures.
2. To present step-by-step information on sales steps.
3. To present cases of successful and unsuccessful sales for anlaysis.

ADVANTAGES	DRAWBACKS
Relatively inexpensive to produce	Difficult to update
Learner can take with him or her	Can be lost or damaged
Learner can work at own pace	Learner is passive

---

[10] Information on print, audio, and video was adapted for sales training from Ronald H. Anderson, *Selecting and Developing Media for Instruction*, 2nd ed. (New York: Van Nostrand Reinhold, 1983). For additional information on appropriate media for self-instruction, see Ronald H. Anderson, Robert A. Reister, and Robert M. Gagne, *Selecting Media for Instruction* (Englewood Cliffs, NJ: Educational Technology Publications, 1983), pp. 76–80.

[11] For additional discussion on videodisk and computer-based technologies, see Kearsley, *Training for Tomorrow.*

ADVANTAGES	DRAWBACKS
Learner can be directed to other sources	Often little visual stimulation
Learner can skip around; does not have to follow sequentially	Often writing or design not motivating
Learner familiar with print materials	

AUDIO Uses

1. To present factual information, for example, policies procedures, and sales steps with live voice or different voices.
2. To present cases of successful/unsuccessful sales work.
3. To motivate salesperson.

ADVANTAGES	DRAWBACKS
Material is standard	Material is difficult to update
Equipment is widespread and easy to use	When used for long periods of time, can be monotonous to listener
Can be used during travel or downtime	Material is sequentially presented; it is difficult to skip material
Can be timed to stop and allow the participant to react orally	For best use, often requires other instructional materials to be used with it

SLIDE TAPE Uses: As listed with the added benefit of being able to demonstrate materials presented visually.

VIDEO Uses

1. To demonstrate successful and unsuccessful sales tactics in any of the steps.
2. To demonstrate interpersonal behaviors and skills.
3. To motivate learner.
4. To present large amounts of material in short time.
5. To allow the learner opportunity to practice skills for individual and outside critique.

ADVANTAGES	DRAWBACKS
Familiar medium	High developmental costs
Holds learners' attention	Requires hardware not often available while traveling
Can capture interpersonal behaviors critical to sales training	Sequential; learner cannot easily skp around
Can motivate with experts and authorities	Relatively high developmental costs
Good off-the-shelf programs	
Easily played back	
Many organizations have the equipment: cassettes, monitors	

VIDEODISK Uses

Appropriate for same areas of instruction as video, but has several advantages for self-instruction.

ADVANTAGES	DRAWBACKS
Provides random access, allowing learners to find any audio or video sequence	Requires use of videodisk player which is not widely available
Can mix variety of audio and video screens	Cannot be taken easily on travel
Provides several interactive features—such as access to other microcomputers during the instruction	
Provides additional branching possibilities to meet individualized instructional needs	
Provides a variety of learning materials within a given program	

COMPUTER-BASED Uses

1. To present simulations and tutorials.
2. To present complex information in a variety of ways.
3. To provide for testing with learner feedback.
4. To provide "embedded" training on necessary equipment.
5. To administer and track training progress.

ADVANTAGES	DRAWBACKS
Provides feedback to learner	Appropriate equipment necessary
Provides branching to meet learner needs	Learner often needs training in use of equipment
Allows rapid updating of course materials	Difficult for salesperson to take on the road. For some, it is an unfamiliar medium.

Step 4: Validating Design and Materials. Materials should be tested during development by learners with characteristics similar to those who will use them. This is particularly important with self-study because the design and materials provide the actual instruction. Neither other participants nor an instructor will be available to interpret them.

Since sales training is often decentralized and therefore removed from the company headquarters, it is particularly important to have sales professionals review the materials before they are finalized. To do this, distinct groups of potential users should be identified, and representatives from each group should be asked to work through the materials.

The number in the validation test should be large enough for reliability of results. If it is not possible to have representatives from each potential user group in the validation test, it is important to ask a few potential sales professionals to test the materials.

EXAMPLE

John Freemont is director of sales training for a large retail company. His department has just completed a draft self-study program on customer service. The program consists of several modules including "understanding customer needs," "handling the angry customer," "procedures for returning items," and "company standards for customer service." John has a limited training budget and cannot run a full validation test on the materials. He decides, however, to ask six sales managers to each designate one sales professional to review the materials, completing all exercises and tests.

After the six complete the package, he sends them an open-ended evaluation form which they complete. He is trying to find out:

- is there relevant information for each learning objective?
- will completion of this self-study program lead to the learning objectives on customer service?
- are the sales professionals' backgrounds appropriate for this design?

Even with the small number of reviewers, he learns that some of the exercises were not clearly understood; and some of the examples were too limited to a particular situation. In addition, he found that completion of the self-study module did not lead to achievement of the stated learning objective on handling the angry customer. He did find that the materials were appropriate for the backgrounds of the sales professionals. They all had the current level of knowledge and skill to complete the program. He was glad that he had made this qualitative effort at getting feedback on the design and materials. It enabled him to improve the design in several areas.

1. Is there relevant information for each learning objective?
2. Will completion of this self-study program lead to achieving the learning objectives?
3. Are the trainees' backgrounds appropriate for this design and these specific materials? Consider prior experience, current level of knowledge, current level of skill, ability to use the equipment and even reading level.

Step 5: Implementing the Training. Before beginning a self-study program, the sales professional should be oriented to the specific objectives, methods, administrative procedures, evaluation, and medium of instruction. Because the sales professional often works alone and has to be self-directed, completing self-study packages may be easier for him or her than for other learners. Places where the materials can be completed are

1. *A resource center at the work site*. Individualized resource centers are available in many organizations where the learner can spend a few hours on a particular program.
2. *At home*. If interruptions can be held to a minimum and if necessary hardware is available, home can be a good place for completing self-instructional modules.
3. *In the car*. For sales professionals, the car can be a relaxing place for listening to tapes. Where this is applied, activities to reinforce the information on tape needs to be completed at another site.
4. *At a library or other quiet place*, particularly for print materials.

Step 6: Evaluating the Results. Evaluating self-study can follow the four evaluation steps outlined by Kirkpatrick: reaction, learning, behavior, and results.[12]

1. *Reaction*. If trainees liked self-study and felt comfortable learning alone, it is important for the sales trainer to know this because of its potential cost effectiveness as a training strategy. Therefore, it is important to ask the trainee such questions as, Did he or she like learning through self-study? Was he or she motivated to complete the self-study in the time frame outlined? Did he or she find it an appropriate way to learn? Would the trainee engage in self-study in the future?
2. *Learning*. Kirkpatrick defines learning as understanding of principles, facts, and techniques separate from their application. Methods for evaluating learning include assessing performance on learning activities, be they written, taped, or computer activities; paper and pencil tests; and oral, written, or taped summaries by the trainee.
3. *Behavior*. Changed behavior on the job is the goal of self-study as of all other training. Behaviors need to be looked at over time—two weeks, three months, six months—to be sure that behavior changes are consistent.
4. *Results*. Sales results because of changed behaviors are the most difficult to measure and attribute directly to training. However, it is the quantifiable

---

[12] This discussion is drawn from Donald L. Kirkpatrick, *Evaluating Training Programs* (Alexandria, VA: American Society for Training and Development, 1975), pp. 1–17.

results in sales increases, number and percentage of sales closed, and increased dollar value of sales that are most important to measure and monitor.

EXAMPLE

Susan Stevens completed a forty-hour self-study program on "principles of selling." She found it a very appropriate way to learn because it was on her time (reaction). Blocking out twenty two hour segments was difficult, but she managed to do it. She scored very high on all the review tests throughout the program (learning). She and her supervisor noticed immediate changes in her attitude and confidence, particularly in the area of "making sales calls" (behavior).

Increasing her actual sales did not follow immediately and required additional discussion, critique and coaching from her supervisor. She clearly understood the principles, but getting results took additional time.

Susan's situation is typical. Formal training, whether or not through self-study, often needs to be reinforced through practice, critique and coaching.

Purchasing Self-study Materials

Self-study materials can be purchased or tailored by outside consultants or companies to meet specific training requirements. In either case, the preceding discussion provides guidelines for evaluating the materials. Several additional questions can be useful in determining the appropriateness of off-the-shelf materials to an organization's specific needs.

1. Do these materials meet our training objectives?
2. Is the message of these materials consistent with our standards, procedures, and culture?
3. How will they be used? As a complete program? As a part of another program? For beginning salespeople? For advanced salespeople?
4. How many separate packages do we require?
5. How will the materials be distributed?
6. How does the cost-benefit of the developed materials compare with those developed in-house?

Two sources of information on consultants and vendors are *ASTD Buyer's Guide and Consultant Directory* (annual publication available from ASTD, 1630 Duke Street, P.O. Box 1443, Alexandria, VA), and *The Marketplace Directory* (annual publication of *Training Magazine*, Lakewood Publications, 50 South Ninth Street, Minneapolis, MN 55402).

Sales Trainers' Roles in Self-study Approaches

The trainer has several roles in self-study approaches. In 1983, 15 roles were defined for trainers in *Models for Excellence*.[13] Eight roles describe the trainer's function in self-study.

1. Needs analyst—the role of defining gaps between ideal and actual performance.

2. Task analyst—identifying activities tasks, subtasks and human resource and support requirements necessary to accomplish specific results in a job or organization.

3. Program designer—the role of preparing objectives, defining content, and selecting and sequencing activities for a specific program.

4. Instructional writer—the role of preparing written learning and instructional materials.

5. Media specialist—the role of producing software and using audio, visual, computer, and other hardware-based technologies for training and development.

6. Manager—the role of planning, organizing, staffing, controlling training and development operations or training and development projects, and linking training and development operations with other organization units.

7. Program administrator—the role of ensuring that the facilities, equipment, materials, or event are present and that materials' coordination program logistics run smoothly.

8. Evaluator—the role of identifying the extent of a program, service, or product's impact.

SELF-DEVELOPMENT

Self-development discussed here is defined as that learning which enhances the skills, knowledge, or attitudes for present or future work. In self-development, learners also have substantial control over the purpose, goals, learning methods, resources, and evaluation procedures and criteria. Self-development will become an important strategy as the requirements for a flexible, educated work force increase and is particularly appropriate for highly motivated individuals, a characteristic of many in the sales force.

---

[13] Patricia A. McLagan and Richard C. McCullough, study directors, *Models for Excellence: The Conclusions and Recommendations of the ASTD Training and Development Competency Study* (Alexandria, VA: American Society for Training and Development, 1983), p. 29. The roles are presented here in a different order.

Self-development can be loosely structured and completely in control of the individual or can be more structured and used as an employer strategy for increasing worker productivity and competitiveness. It can be initiated and monitored by the sales trainer or by the individual. It can be organized with specific measurable short-term individual and organizational goals or for longer-range individual career development and organizational staffing.

Although the focus here is on self-development in which the trainer or sales manager has a role and interest, the information provided here could also be used by an individual without involvement from members of the organization.

How Self-development Is Used in Sales Training

Because self-development is an individual strategy and process, its uses are harder to quantify than are other types of training practices. Currently, self-development in sales training may be included as part of the sales manager's informal coaching, as well as part of a formal mentoring system. Executive Knowledge Works reports that 61 percent of the companies surveyed use informal coaching and 24 percent have formal mentoring systems.[14] In addition, formalized self-development strategies are also becoming more possible through resources offered by company learning centers.

Even with these increases, self-development as a formalized, systematic approach in sales training is not widely used at the present time, although less formalized approaches to self-development are widespread. Formalizing self-development is an important strategy for future development of sales professionals however, because self-development is designed to meet individual needs within an organizational structure.

Self-development approaches are also particularly relevant to sales professionals because they require the same attitudes as those for successful selling. Executive Knowledge Works identified several key factors of successful salespersons. Forty-eight percent of those factors are categorized in the area of "attitudes" and include such characteristics as "self-starter, persistent, innovative, hard worker, motivated, committed, self-confident."[15] These are the same characteristics necessary for success in formalized self-development, which is why this approach is particularly relevant for sales professionals.

Learner motivation. In self-development, internal learner motivation is critical. The learner usually initiates or demonstrates interest in participation. In addition, the learner must be able to work with little or no direction once the initial outline has been developed and approved.

Variation in learning styles is more easily accommodated in self-development approaches because the learner can select the method. For example, if the sales

---

[14] Executive Knowledge Works, p. VI. 2.

[15] Executive Knowledge Works, p. IV. 26.

professional learns best in groups, he or she can select a class, seminar, or workshop. A professional who learns best through reading and reflection can select individual study. If the individual spends time driving or at hotels, audio tapes may give the person a feeling of using time wisely and therefore be motivating.

Learner Ability. Success in self-development requires many of the same skills and attitudes as does success in selling. Drawing from Executive Knowledge Works' skills and attitudes for success in selling, self-development requires the following:[16]

> *Skills:* Planning, problem solving, time management, detail orientation, prospecting, entrepreneurial, and consulting approach.
>
> *Attitudes:* Self-starter, persistent, innovative, hard worker, enthusiastic, versatile, motivated, positive attitude, commitment to quality, integrity, and self-confidence.

Again, many of the skills and attitudes required for successful self-development are identical to those required for successful selling. For this reason, self-development approaches are particularly relevant for sales professionals.

Location of learning. The specific learning design will determine where the learning should occur. However, self-development can be used in a wide variety of places, for example, libraries; learning centers; classes; seminars; at home; or in workplace meetings with mentors, consultants, or colleagues.

Administrative functions. Self-development requires administration by the learner. However, if the sales manager or trainer has included these programs as part of a company program, organizational administration of tracking, resource coordination, and resource referrals will be necessary.

Cost. The cost of self-development depends upon the quality of the learning and its resulting impact to both the individual and the organization. From the sales training manager's point of view, the costs are usually low, since it requires no instructors, no materials development, and no additional facilities. The costs of the resources may be borne by the organization or by the individual. The benefits to the participating individuals are usually great in both learning results and increased self-confidence.

Summary of Self-development Benefits

The major benefits of self-development approaches are

1. Provides opportunity to meet specific learner needs.
2. Requires no instructors and limited facilities.

---

[16] Executive Knowledge Works, p. IV. 26.

3. Allows for flexibility in purpose, goals, methods, resources, scheduling, and evaluation.

4. Provides opportunity for development of motivated staff.

5. Provides structure to meet organization's needs for staff development.

Designing a Long-Term Self-development Program

Two ways to use self-development processes are presented here. A longer-term self- or career development plan should include a self-assessment, goals statement, and several individual learning projects. Shorter-term learning projects can also be developed without the self-assessment and goals statement.

The model presented here is optimal and should be adapted to the learner's needs, interest, time, and resources available. It is also recommended that learners test out the process with a short-term learning project before attempting a full-scale self-development plan.[17]

Self-assessment. The first step is to develop an accurate assessment of the current situation. Learners need to assess their present level of knowledge, skills, and attitudes.

1. What knowledge have they acquired previously and what knowledge needs to be developed?

2. In what skill areas are they already competent?

3. What new skills need to be developed and what present skills levels need improvement?

4. What behaviors do they now demonstrate and what behaviors do they need to demonstrate and at what competency level?

In addition, it is important to assess and predict individual attitudes and reactions in a variety of professional situations.

In preparing a self-assessment, it is helpful for the learner to have assistance from the trainer or manager in forming and adequately answering these questions. An assessment should be written to allow review and tracking over time.

Goals Statement. This statement carefully outlines where the learner would like to be at the end of this development activity. It can cover either level or function. For example, a salesperson may want to demonstrate the competencies of a sales manager. These specific competencies and evaluation standards would

---

[17] Based on an educational model used at Beacon College, Boston. This model was adapted from Malcolm Knowles, *Self-Directed Learning: A Guide for Learners and Teachers* (Chicago: Follett, 1975).

need to be identified. Or a sales trainer may want to develop professional competency in computer-based applications to training. Again, competency and evaluation standards would need to be defined.

Individual Learning Projects. Individual learning projects are the third part of the plan. They can also be developed independently. In either case, the following should be included:

1. Overview
2. Learning goals and objectives
3. Methods
4. Learning resources
5. Time frame
6. Cost
7. Evaluation/evidence of learning

Overview. The learner should state why this particular project is important to his or her development. This statement is important not only to define an area, but it also disciplines the learner to ask, "Why?" It may be "nice to know," but is it really necessary?

Learning goals and objectives. In this section an overall goal which directs the project is developed. In addition, learning objectives or steps to reach the overall goal are outlined for the specific project. The major goal can either be written behaviorally or quantitatively or more generally using terms "to learn" or "to understand." The learning objectives should be more specific and should describe behaviors or outcomes.

Self-development learning projects are appropriate for various kinds of learning and knowledge, for example,

1. To describe the steps in the sales process.
2. To describe benefits and features of a specific product.
3. To identify strategies to overcome specific objections.
4. To develop personal selling style. (This would provide a personalized approach, especially appropriate to an experienced salesperson.)
5. To develop positive attitude toward telephone selling.

Methods of learning. Methods of learning answer the question of "how" this learning will take place. Self-development methods include but are not limited to the following:

1. Tutorials
2. Guided and independent reading
3. Professional or employer-based training programs
4. Self-instructional approaches outlined in last section (workbooks, computer-based training, interactive video simulations)
5. Workshops/seminars/conferences
6. Internships
7. Apprenticeships
8. Action projects
9. Independent research
10. College or correspondence coursework

The learning methods are diverse and must be examined closely as to their availability and feasibility for meeting specific learning objectives.

Resources. After selecting the method of learning, the specific resources should be identified. Therefore, if the method is guided reading, each book selected should be listed under resources. This helps the learner to be specific and provides real directions for getting started. Specific sources of learning resources are discussed later.

Time frame. To make this approach successful, it is necessary to develop a time frame and schedule. If time frame is not included, there is a danger that the plan will be too ambitious and require more time than is possible for the learner.

Cost. The costs of learning resources need to be estimated as closely as possible.

1. How much will each learning resource cost?
2. When will payments have to be made?

Evaluation. Evaluation should answer the concerns addressed in self-study:

1. *Reaction*: Did I like this method of learning? Was it successful for me? Would I do it again?
2. *Learning*: What did I learn? What evidence of that learning can I demonstrate?
3. *Behavior*: Can I demonstrate and quantify changed behavior over time—two weeks, three months, six months?
4. *Results*: What results in my work life—present and future—can I measure?

In addition to self-evaluation, another person—the sales manager or trainer—
might be asked to provide an evaluation. Or the learner might prefer a person
not associated with the workplace. This is appropriate if the learner will be
threatened by the participation of the manager. The purposes and criteria of
the evaluation as well as its format should be understood from the beginning.

Sample Learning Project for Self-development in Sales Training

OVERVIEW

I have difficulty selling by telephone. I am used to face-to-face selling where I can interact
personally and read the body language of the customer. Since my company is requiring
additional telephone sales work, I want to improve my skills and my results in telephone
selling.

GOALS AND OBJECTIVES

OVERALL PROJECT GOAL

To increase my telephone selling results by 20 percent in six months.

OBJECTIVES

1. To assess my current telephone selling skills and identify specific areas for improvement.
2. To be able to describe the sales applications to telephone selling, including call strategy,
   customer motivation, gaining prospects and contacts, effective listening, opening, timing,
   handling objections, and closing the sale.

METHODS

1. Self-study: Study and complete all exercises in "Selling on the Phone: A Self-Teaching
   Guide."
2. Coaching: Ask Dan Jones, who is very successful in telephone sales, to work with
   me in analyzing my practice.
3. Practice: Practice telephone sales over six-month period, analyze results, identify additional strategies, and practice these skills.

RESOURCES

1. James D. Porterfield, *Selling on the Phone: A Self-teaching Guide* (New York: John
   Wiley, 1985).
2. Dan Jones
3. My own practice
4. Telephone sales analysis sheet (to be developed by me after completion of self-teaching
   guide)

TIME FRAME

This project will be completed over a six-month period: January 1–June 30, 1989.

1. January 1–February 28: Work with self-teaching guide, reading, studying in depth, and completing all exercises (2 hours per day, Tuesday–Saturday).
2. March 1–June 30: Biweekly meetings with coach, Wednesdays, 7:00–9:00 P.M.

    Complete and analyze 120 calls a week.

COST

Book: Approximately $13
Coach: Dan Jones has agreed to provide me with biweekly meetings for $50.00 a session × 8 sessions: $400
Miscellaneous materials: $25
Total: $438

EVALUATION/EVIDENCE OF LEARNING

My goal will have been met if my selling by telephone increases my sales results by 20 percent over a six-month period.

Evidence of learning will also include

1. Completion of exercises in self-study guide.
2. Oral descriptions of selling process to coach.
3. Analysis of own sales calls with plan for improvement.

Notes:
    Methods: A variety of methods could have been selected to reach the same learning goals. For example, a seminar or course on "telephone selling" could have been used. In addition, it would have been possible to complete this project less expensively without the coach; however, the learner should assess the importance of the project and his or her willingness to invest the money to get additional assistance in reaching a goal. Because successful completion of this project would result in increased sales and income for the learner, the extra time and money could be justified.
    Time frame: The time frame could also have been shortened if the coaching and practice period were eliminated or streamlined.

Types of Resources

Many resources are available for self-development. This section will provide information for sales trainers in finding their own resources and/or directing others to resources.

Books/Periodicals. Extensive written materials are accessible for self-development. Because there is so much written material available, it is necessary for the learner to take care in selecting and using these materials. Decisions have to be made to limit material and to determine the purpose for using each selection. How thoroughly must each selection be understood? Is the purpose familiarity with overall concepts, or is it in-depth study? How will one demonstrate understanding or mastery of subject matter?

For information on how to find written materials through libraries and database searches, see

Olivetti, L. James. "Information Resources." In *Training and Development Handbook*, 3rd ed., Robert L. Craig, ed., pp. 787–789 and 794–798. New York: McGraw-Hill Book Company, 1987.

Bard Ray, Chip R. Bell, Leslie Stephen, and Linda Webster. *The Trainer's Professional Development Handbook*, pp. 215–226. San Francisco: Jossey-Bass, 1987. This book also describes over 100 of the best books in the training and development field. Although not specifically addressing sales training, this book provides valuable information for all trainers.

College and University Classes. For self-development, the participant may decide upon formal sales or marketing courses through a local college or university. In using this learning resource for self-development, the learner should do more than enroll in the course; he or she should attain information on

Course objectives (What will I know or be able to do as a result of taking this course?)

Method of teaching (Will this method allow me to meet my learning objectives? Does it agree with my learning style?)

Background of instructor

Number of students

Required assignments

Class schedule

Method of evaluation

In addition to local colleges and universities, correspondence courses and external degree programs are available. These programs allow the learner a self-study option, namely, to complete courses without attending traditional classes. For information, see

Ready, B. C., and R. D. Sacchetti. *Peterson's Independent Study Catalog, NUCEA's Guide to Study Through Correspondence, 1986–1988*. This resource lists over 12,000 high school, college, graduate, and noncredit courses offered by over 70 colleges and universities.

Sullivan, Eugene. *Guide to External Degree Programs in the United States*. New York: American Council on Educatin and Macmillan Publishing Company, 1983.

Training Programs/Seminars/Courseware. This includes films and videos, CBT software, packaged programs, assessment instruments, and others. Training programs and seminars are effective learning resources for self-development. There are several sources of information to find the best seminar. Again, it is essential for the learner to assess the seminar's goals and methods to be sure that this seminar will meet individualized learning goals.

The following are some sources of training programs and seminars:

TRAINET, a computerized service for national members of ASTD. Trainet lists over 100,000 training seminars and workshops by topic. For additional information on this service, contact the ASTD Information Center, 1630 Duke Street, P.O. Box 1443, Alexandria, VA 22313.

Fetteroll, E., L. Nadler, and Z. Nadler. *The Trainer's Resource*, 1987. Amherst, MA: Human Resource Development Press, 1987. This book presents information on "packaged" training programs in 20 categories, including sales management and sales training. Training programs include self-study programs or segments to be used as a part of the overall training program.

Wasserman, Steven, Jacqueline Wasserman O'Brien, and Edmond L. Applebaum, eds., *Learning Independently*, 3rd ed. Detroit: Gale Research, 1987. Approximately 4,000 self-instruction resources are included in this catalog (correspondence courses, programmed learning products, audio and video cassettes, and others). Sales management, sales techniques, sales training, and salesmanship are included as topics.

Associations. Associations and their subgroups provide an excellent way to meet, exchange information, and learn from others in the field. In addition, professional opportunities to write and make presentations provide ways to organize materials and gain professional feedback from peers. Three appropriate associations for sales trainers are

American Society for Training and Development—Sales and Marketing Professional Practice Area—a subgroup of 2,500 professionals with the purpose of fostering the "professional development of those ASTD members with a special interest in the training of sales and marketing personnel" (1630 Duke Street, P.O. Box 1443, Alexandria, VA 22313).

National Society of Sales Training Executives—An organization of approximately 125 corporate directors and managers of sales and marketing training and human resource development. The society publishes a quarterly journal, *Sales and Marketing Training*, and holds meetings for sales trainers (NSSTE Headquarters, 1040 Woodcock Road, Orlando, FL 32803).

Sales and Marketing Executives International provides several services, including a Graduate School of Sales Management and Marketing, conventions and leadership conferences, local chapter activities, and professional certification programs and a variety of products and awards (Statler Office Tower, Cleveland, Ohio 44115).

There are a wide variety of other associations, some aimed at a certain profession, some industry specific. To locate appropriate associations, see:

Gruber, Katherine, ed., and others. *Encyclopedia of Associations*, 21st ed. Detroit: Gale Research, 1987. This is a guide to over 23,000 associations, including information on purpose, membership services, products, and addresses. This encyclopedia is also available on computer tapes and diskettes through Gale Research and online through Dialog Information Services, Inc.

Professional Associates. Coaches, mentors, and other professionals are excellent resources for self-development. The learner should select and negotiate an agreement with them, outlining their relationship: goals, process, schedule, and any cost. This provides a more formal agreement outlining the commitment of both resource person and learner.

Locating mentors and coaches requires knowledge of potential resources. These relationships usually work best when they are close in proximity or in the same organization. It is important that the coach or mentor understand the culture and environment in which the learner is working.

For locating professionals with which to discuss issues less frequently, ASTD's Member Information Exchange (MIX) is available to national members. MIX is a computerized networking system which identifies individuals who can be contacted to informally discuss items of professional interest. Included in the MIX index are the following areas of sales skills:

1. The selling process
2. Key account management
3. Territory management
4. Product knowledge training
5. Sales management; managing field managers

Resource Centers. Organizational resource and information centers can provide access to training materials, databases, and hardware for use in self-study and self-instructional programs. Resource centers usually have staff available to assist in the selection and location of learning resources.

SUMMARY

This chapter has addressed self-study and self-development approaches for sales training.

Self-study is designed for trainee learning outside the classroom for the purpose of changed behavior on the job. Major benefits of self-study are consistent content in format and delivery can be provided to large numbers of trainees; few instructors and limited facilities are needed; there is flexibility in scheduling and place of learning; and time and expense for trainee travel are eliminated. Designing self-study emphasizes methods and activities which include instruction, feedback, review, and evaluation. Self-study materials can be developed in-house or purchased and can be used for all or part of a training program.

Key roles for sales trainers using self-study include needs analyst, task analyst, program designer, instructional writer, media specialist, manager, and program administrator.

Self-development is a structure for developing long- or short-term learning addressing present or future work. Self-development allows participants control over purpose, goals, methods, resources, and evaluation of their learning. Major benefits are creative opportunities to meet learner needs for development, no requirement for instructors or facilities, administration is limited and performed predominantly by the learner, and the structure can meet the needs of motivated employees as well as the human resource planning needs of organizations. Two designs are presented: a plan for long-term self- or career development and a plan for shorter-term learning projects. Resource listings for self-development were also included.

Self-study and self-development are not the predominant methods presently used for sales training. However, they provide opportunities for flexibility and learner initiative in an overall sales training system. Both approaches are consistent with adult learning theory and salesforce motivation. Over the coming decade, they should grow in use as technology is expanded and competition requires a flexible, highly developed and motivated work force.

REFERENCES

Anderson, Ronald H. *Selecting and Developing Media for Instruction*. New York: Van Nostrand Reinhold, 1983.

Bard, Ray, Chip R. Bell, Leslie Stephen, and Linda Webster. *The Trainer's Professional Development Handbook*. San Francisco: Jossey-Bass, 1987.

Budd, Marjorie L. "Self-Instruction." In *Training and Development Handbook*, 3rd ed., Robert L. Craig, ed. pp. 488–499. New York: McGraw-Hill, 1987.

Cox, John H. "A New Look at Learner-Controlled Instruction." *Training and Development Journal*, Vol. 36(3) (March 1982), pp. 90–94.

Dejoy, Judith Klippel, and David M. Dejoy. "Self-directed Learning: The Time Is Now." *Training and Development Journal*, Vol. 41(9) (September 1987), pp. 64–66.

Ellington, Henry. *Producing Teaching Materials: A Handbook for Teachers and Trainers*. New York: Nichols, 1985.

Executive Knowledge Works. *Sales Training in America*. Palatine, IL: Anthony J. Fresina and Associates, 1988.

Frenzel, Louis E. "Hands-on Training at Home." *Training and Development Journal*, Vol. 34(9) (September 1980), pp. 60–66.

Hahne, C. E. "Sales Training." In *Training and Development Handbook*, 3rd ed., Robert L. Craig, ed., pp. 659–680. New York: McGraw-Hill, 1987.

Hartley, James. *Designing Instructional Text*, 2nd ed. New York: Nichols, 1985.

Honeycutt, Earl D., Jr., Clyde E. Harris, Jr., and Stephen B. Castleberry. "Sales Training: A Status Report." *Training and Development Journal* (May 1987), pp. 42–45.

Hopkins, D. S. *Training the Sales Force: A Progress Report*. New York: The Conference Board, 1978.

Kearsley, Greg. *Training for Tomorrow: Distributed Learning Through Computer and Communications Technology*. Reading, MA: Addison-Wesley, 1985.

Knowles, Malcolm. *Self-directed Learning: A Guide for Learners and Teachers*. Chicago: Follett, 1975.

Knowles, Malcolm, and Associates. *Andragogy in Action: Applying Modern Principles of Adult Learning*. San Francisco: Jossey-Bass, 1984.

Lambert, Michael L. and Sally R. Welch. *Home Study: Course Development Handbook*. Washington, DC: National Home Study Council, 1988.

Niemi, John A., and D. D. Gooler, eds. *Technologies for Learning Outside the Classroom*. San Francisco: Jossey-Bass, 1987.

Olivetti, James L. "Information Resources." In *Training and Development Handbook*, 3rd ed., Robert L. Craig, ed. New York: McGraw-Hill, 1987.

Pedler, Mike, J., Burgoyne, and T. Boydell. *A Manager's Guide to Self-Development*. New York: McGraw-Hill, 1978.

Reiser, Robert A., and Robert M. Gagne. *Selecting Media for Instruction*. Englewood Cliffs, NJ: Educational Technology Publications, 1983.

Romiszowski, A. J. *Developing Auto-Instructional Materials*. New York: Nichols, 1986.

Rowntree, Derek. *Teaching Through Self-instruction*. London: Kogan Page, 1986.

Sales and Marketing Professional Practice Area. *Sales and Marketing Training Resources in Print*. Alexandria, VA: American Society for Training and Development, 1987.

Schuttenberg, Ernest M., and Saundra J. Tracy. "The Role of the Adult Educator in Fostering Self-directed Learning." *Lifelong Learning,* Vol. 10(5) (February–March 1987), pp. 4–7.

Smith, Martin E. "Self-paced or Leader-led Instruction?" *Training and Development Journal* (February 1980), pp. 14–18.

chapter fifteen

LECTURE METHOD AND CLASSROOMS: THE MOST FREQUENT METHOD IN SALES TRAINING

ROBERT W. PIKE

Robert W. Pike is president of Resources for Organizations, Inc., Eden Prairie, Minnesota. He has developed and implemented training programs for business, industry, and government since 1969. He leads sessions over 150 days per year on topics of leadership, attitudes, motivation, communication, decision making, problem solving, personal and organizational effectiveness, conflict management, team building, and managerial productivity. More than 15,000 trainers have attended his Creative Training Techniques Workshop. As a consultant, Pike has worked with organizations such as AT&T, American Express, Pfizer, Upjohn, McGraw-Edison, Caesars Boardwalk Regency, Minnesota Title Insurance, University of Georgia, Vanderbilt University, and IBM. A member of American Society for Training and Development since 1972 he has served as director of special interest groups and was on the national board. Pike has presented at national and regional conferences for ASTD, *Training* magazine, and the National Safety Congress. He is a featured speaker in the convention and certificate programs of the American Society of Association Executives. He has been in *Who's Who in the Midwest* since 1980 and is in the current edition of *Who's Who in Finance and Industry*. He has contributed to *Training*, *The Personnel Administrator*, and other magazines.

Lecture has long been a staple in sales training. But lecture no longer means just standing up and talking. It means taking advantage of the one-to-many delivery process in every way possible.

In this chapter you'll learn how to use the three alternatives to the traditional lecture: the symposium, the debate, and the question and answer session. You'll explore the two different purposes for questions in a lecture—information and opinion—and when, where, and why to use them. You'll discover five different ways to assess an audience so that the information you present is on target. You'll find some simple, yet powerful, ways to use the most common visual supports to lecture—slides, flipcharts, and transparencies—along with 22 tips for communicating effectively. Next, you'll examine ways to develop a lecture that can be delivered with impact by developing strong openings, strong closes, and strong support in the body of the presentation. Finally, you'll explore some room arrangements that can spice up your lecture.

HOW THE LECTURE METHOD IS USED IN SALES TRAINING

Lecture is probably the first form of training ever developed. Picture the first human who discovered how to climb a tree to escape a wild animal. Probably one person was told first, but then a group was formed to hear about it again: story-telling—and learning from the lesson learned—the first form of lecture.

And it's just as common in sales training. A salesperson discovers how to make a particular sale. The natural step is for the sales manager to have that individual share the technique at the next gathering of salespeople—the sales training lecture is born. It's simple, straightforward—and in its original form it was delivered by people who had done it talking to people who wanted to know how to do it.

Lecture can generally be defined as one person systematically presenting information. The lecture's major advantage is that it allows a maximum amount of information to be presented in a minimum amount of time. Since there is little participation, it allows materials to be presented in a systematic fashion—even when those materials and ideas may be quite diverse.

The major disadvantage of lecture is that it represents one person's point of view. It is strongly influenced by the personality of the presenter. It provides for only one-way communication and also limits group participation.

Three Types of Lecture Variations Useful to Sales Trainers

Today, more than ever before, variety is important. We live in the age of entertainment. With 68 channels of cable we flip through them with our wireless remote controls and then say "there's nothing on." Even when lecture is a primary communication method we can still apply the method with some variety. Here are three:

Symposium

DESCRIPTION

Three or more people present different viewpoints in short presentations. After a question and answer session is conducted by a moderator.

ADVANTAGES

Offers multiple viewpoints and provides opportunity for clarification through questions and answers.

LIMITATIONS

Depends on having speakers of equal ability and a good moderator and allowing questioners to be candid with their questions.

For example, XYZ Company is holding its annual sales conference. Its three top producers use three different prospecting methods. Fred uses center of influence prospecting, Gail uses nest prospecting, and Chris uses referral prospecting. Each makes a 10-minute presentation on what the method is, how they use it, and why they like their method best. The moderator leads the question and answer session afterward.

DEBATE

DESCRIPTION

An equal number of speakers, directed by a moderator, presenting different sides of an issue.

ADVANTAGE

Holds a group's interest by presenting two sides of an issue.

LIMITATION

Can become too emotional. Requires a good moderator to maintain control.

For example, you're holding a training course for sale people. The subject is personal computer product knowledge. The debate proposition: developing in-depth knowledge of an industry is the best way to sell and serve a client base. Jack takes the pro side, arguing that it takes time to understand an industry and once you understand it, you can transfer the knowledge gained from serving client a to client b and so on.

Jan argues against it, citing economic instability. She says it is unwise to put all your eggs in one basket (one industry), and so on. You can't protect the basket, and so on. Jan states that having a broader industry base protects against a downturn in any one industry.

Question and Answer

DESCRIPTION

The presenter generating response through inquiry, generally from more than one person.

ADVANTAGES

Can be combined with other methods. Clarifies information to meet specific needs.

LIMITATIONS

Can become too formal. May not be of interest to the group. Easy for several participants to dominate a question and answer session.

For example, Tim presents an overview of five different closing methods. He then asks the group to volunteer which they've used and with what results. He follows this by having people form groups of three and gives each group two minutes to generate any clarifying questions they would like to ask. At the end of the two minutes, he takes one question from each group in turn and answers them until all questions or the time available have been exhausted.

HANDLING QUESTIONS AND ANSWERS

How to Prepare Yourself for Asking Informative Questions

Here are some general considerations before asking questions:

1. *Plan* your questions. Think through what you're going to ask, why, and where in your presentation. Will you have covered the information? Will the group have sufficient experience to answer the question?

2. Know your *purpose*. Are you looking for information or opinions? Information can be right or wrong. Opinions are just that—opinions—and need to be valued as such. If you're looking for information are you sure that you're going to get what you're looking for?

    For example, asking the selling price of a product is an information question, that is, there's a right or wrong answer. But asking which of several prospecting methods one uses, or asking how many prefer one method over another, are opinion questions. There is no right or wrong answer.

3. *Show* how the question relates. Make sure your audience has a frame of reference or context for the question. Can they see how your question at this particular time makes sense? For example, asking for more information when sales drop in a traditionally strong market, that is, were there any unusual circumstances during the past quarter that related to the drop in sales?

4. Move from *general* to *specific*. Ask broader questions first; then focus on clarifying questions that are more specific about individual points.

    For example, you might ask participants: "What are the six methods for closing sales?" (That's general). Then, after getting an answer to the question, follow up with: "Describe the sixth method you mentioned, the impending event close." Or, to use another example, you might ask:

    "What is the biggest challenge in selling?" is a general question. On finding that closing is most often given by the group, the follow-up question "Why is closing your biggest challenge?" is more specific.

5. Keep the questions *short*. Some questions are so long the audience has forgotten what was asked by the time the question is finished.

6. Whenever possible avoid "yes," "no," or answer implied questions. Questions like "You don't want any, do you?" "You'd like an evening session wouldn't you?" or "Have you used group involvement?" may get you an answer, but not very much stimulating discussion.

   A better approach to each would be to ask: "Would you like one or two?" "What advantages could you see in an evening session?" "What experience have you had in using group involvement?"

7. Don't interrupt the person answering. Once you've asked the question allow the person to answer. Don't use the first part of their answer as a springboard to your next monologue.

Three Formats to Use When Asking or Answering Questions

When answering questions from participants, listen for *both* content and intent—not only what people say, but what they mean by what they say. Three formats can be used to either ask or answer questions: the group as a whole, individuals, and subgroups.

Addressing the Group As a Whole. When asking questions of the group as a whole, it's important to consider first whether or not the group is likely to know the answer to the question. Sounds simple, doesn't it? Yet asking a question that somebody answers incorrectly can be a devastating experience for both you and the individual. And it may create an environment that causes others not to want to even attempt an answer in the future. So ask yourself first, "How likely is it that the group will know the answer?" And remember, when in doubt, don't ask. Or better yet, make your question an open book question. For example, you might say, "The product features and benefits are found in Chapter 5 of the reference manual. I have ten questions about model 12. I'll ask a question and then you'll have 30 seconds to find the answer in the manual. Here's question 1 . . . go!"

This approach has several advantages:

1. It involves your entire group.
2. It ensures the correct answer.
3. It familiarizes the entire group with the reference manual.

Addressing Individuals. When asking questions of individuals, some of the same cautions apply as when asking questions of the group: You don't want to embarrass people. You want to avoid telling people they're wrong.

Here are some helpful guidelines:

1. Consider allowing people to guess. This indicates that you don't necessarily expect a right answer—you want participation. Remember that you can ask for opinions as well as information—and opinions aren't right or wrong.

   For example, in a recent survey, car buyers were asked to name the three most important features they were looking for in their next car. John, what's your *guess* as to what one of the features might be?

2. As you ask the question, make eye contact with the entire group. This will not only keep everybody included, but will also allow you to pick up nonverbal cues as to people who might be appropriate to call on. People who avoid making eye contact with you are probably not good candidates if you're looking for a right answer. Individuals, on the other hand, who nod, smile, and so on are fairly good candidates.

3. If the individual gives an incorrect response, check your audience for nonverbals. If you see someone who appears to disagree (frowning, shaking the head no, and so on are cues for this), you can call on him or her by saying, "There seem to be other opinions, what's yours?" This can enable you to correct information without directly telling the person they're wrong.

Addressing Subgroups. Perhaps one of the most effective methods to use is asking subgroups to respond to a group of questions that you've developed. You can set a time limit. You can create enough questions to keep people completely involved in the time available.

When the time limit is up, you can ask the first group to respond to question one and check with other groups for concurrence. Group responds to question and so on, rotating through all the questions and all the groups.

This method has some distinct advantages:

1. It creates more involvement.
2. It gives people a chance to respond without pressure.
3. It can allow people to either use or not use other resources.

It also requires caution. Your experts in the group can take charge and answer everything. Unless you're careful, this can minimize participation. Consider generating enough questions that it requires each subgroup to break into smaller groups to respond to all the questions.

How to Delay Answering With the "Capture the Question" Technique

Generally the best time to answer a question is when the question is asked. However, there are times when answering a question will confuse some of your participants. You can avoid this with a technique called "capture the question."

Let's say that Ann says something like: "What arguments do you think are most persuasive in a preapproach letter to buyers of our competitor's model?" Ann's question is a good one, but answering it now would leave some of the group confused.

Respond to Ann's question by saying, "That's a good question, but so that everyone can have the framework for understanding my response, there's some ground I need to cover. With your permission I'd like to capture your question on this flipchart. (Here you take a piece of flipchart paper and post it on a wall. You write down the question, Ann's name, and the time by which you will have answered it, on this flipchart.) If I haven't answered it by 2 P.M. this afternoon I'd like you to ask it again. On that basis will you allow me to defer the question?"

This enables you to stay on track, but posting the questions so they're visible at all times is your signal to the group that you will answer the question. You also create a sense of personal responsibility for the questioner to remind you if you haven't. And you can check off questions as you've answered them. Always be sure that you get the questioner's agreement that the question has been answered satisfactorily. Perhaps the best way to make sure questions are asked and answered is to use the subgroup approach. After covering some content area, break up the group into subgroups of five to seven participants. Give them some time (generally 3 to 5 minutes is plenty) to review the material and brainstorm two or three questions they'd like answered.

If you have covered, in the past three hours, appointments and presentations, you would say: "This morning we've focused on appointments and presentations. Take three minutes in your groups to look at what we've covered. Then, as a group, generate two or three questions you'd like to ask on anything that's been covered."

Once the time is up you can announce that there is 15 minutes available (or whatever your time limit may be) and that you'll answer as many questions as possible in the time available. You then rotate from table to table until your time limit has expired.

There are a number of distinct advantages to this method:

1. It generates a higher quality of question.
2. Many questions will be answered within the groups.
3. Greater involvement of more participants is achieved.

THE "KILIS" METHOD FOR ASSESSING THE AUDIENCE

The two key questions to ask in preparing for any lecture are: What do people need to know? and What do people need to be able to find? If we know what people need to know, we can make that the focus of the lecture, discussion, skill practices, and anything else that we may have planned. We can spend less time, then, on what people need to be able to find by providing appropriate reference materials, job aids, and so on.

The question is: How do you assess your audience? For years we've been teaching the *KILIS formula* in creative training techniques: Knowledge, Interest, Language, Influence, and Situational Elements.

Knowledge. The old journalistic law applies here: never underestimate the intelligence of your listeners, never overestimate their need for information. For example, you may take it for granted that most businesspeople are familiar with the concept of equipment depreciation, but you can't necessarily expect them to know what depreciation formula applies in their organization.

Interest. The success of your presentation is 50 percent assured if you have the audience's interest working with you from the beginning. Even if interest is initially lacking, it can be created through skillful presentation techniques. Remember your audience is turned to radio station WII-FM, "What's *in it for me*?" How will this help me do my job faster, better, easier? What are the benefits that I will gain? What are the losses I will avoid? Answer these questions and you'll know just what to prepare to make your presentation powerful.

Language. This can be tricky even if you're an expert on the topic. How expert is your audience? What terms are they likely to know? Would a new salesperson understand the terms "nest-prospecting" or "impending event close?" Would an experienced salesperson appreciate a detailed definition of what an objection is?

Remember this general rule of thumb: "When in doubt, check it out." Ask your audience to define terms. And always err on the side of conservatism. Better to explain too much than too little. Define jargon, at least briefly, the first time it's used. Nothing loses an audience faster than terms it doesn't understand. The problem is more complex if several disciplines are involved.

For example, chemical terms used to describe air pollution would be familiar to somebody selling in that industry. However, an audience that includes not only pollution control salespeople, but also real estate and automobile salespeople might find itself lost.

Influence. Ask yourself these questions:

Who are the decision makers in this group?

Who are the "natural" leaders?

What are their problems and personal interests?

What can you tell them to make their jobs easier?

How can you challenge them?

Does your message threaten their prestige and influence?

Situational elements. These include size of the audience, time of day, length of course, special nature of the occasion, and so on.

1. *Size of the audience.* An audience of 500 people will make it difficult to use a flipchart as a visual aid, require that handouts be given to participants in advance, require the use of a microphone by the speaker as well as audience microphones if a question and answer period is planned. A group of 20 eliminates the special considerations just mentioned.

2. *Time of the day.* Generally, adults need more physical movement and energizers in the afternoon and evening as opposed to mornings. Also helpful: two shorter breaks in a presentation, stretch breaks, some stand-up activities (for example, "Jot down your answers to questions 1 and 2 on the overhead transparency. As soon as you have your answer, please stand." When everyone's standing, you reveal questions 3 and 4, saying "Great, now please jot down your answers to questions 3 and 4. When you're finished, please be seated.")

3. *Length of course.* A 45-minute presentation that stands alone may not require much variety. But if you're running a five-day course, you'll probably want to consider changing the room arrangement—where the front is located, exchange round tables for rectangular tables, use small-group discussion activities, and so on as changes of pace to keep the participants energized.

4. *Special nature of the occasion.* Training delivered as part of the company's annual incentive trip to Hawaii has to be different from training done in the home office. People are going to expect more free time, the opportunity to share their experiences (after all, they were good enough to win the trip), and so on. Training is only one reason for the meeting (and probably not the major one), and the program should reflect that.

DEVELOPING YOUR PRESENTATION USING THE FOUR-STEP AIDA FORMULA

The AIDA formula is one that is based on the opera *Aida*. It forms the four major steps of a presentation or segments of a training program: *A*ttention, *I*nterest, *D*esire, and *A*ction.

Step 1: Secure the Audience's Attention

In the beginning of any presentation, you need to get the attention of the group. So in preparing, you need to think about what you can do to get attention.

One of the most obvious suggestions is to ask a question. People have been conditioned to respond to questions, so asking a question is a good way to break preoccupation. When you begin any initial presentation, you will find that the members of your audience or participants, members of the training group, will be preoccupied with their own conversations, and asking a question is a great way to break through the preoccupation—particularly if you ask it of the audience and begin eliciting responses.

For example, if you were running a program on decision making, you might ask a group of sales managers this question: "What kinds of problems do sales managers have because they don't make decisions effectively?" Or, if it were problem solving, you could say, "What kinds of problems do sales managers have because they don't solve problems effectively?" Or "What kinds of problems have you seen come into the lives of people because they don't handle money well?" You can then use a chart pad to begin recording participant responses. This is just an open audience brainstorming process where you're going to elicit their responses.

As the audience begins hearing, as your participants begin responding what you'll find is that the side conversations will stop and they'll stop much more quickly than if you simply begin lecturing. This particular technique has several benefits:

1. You break through preoccupation. You get favorable attention.
2. You gain immediate audience involvement.
3. The question also begins to build a case for why your topic or why this particular part of the training program is important. Because as people list these problems, they are not necessarily saying they have this problem, but they're saying that it is a problem, and as the list grows longer, there will certainly be some of these problems that each person in your group can identify with. So you're not only gaining their attention; you're also demonstrating why the particular topic that you're presenting to them is important.

Step 2: Maintain Interest

The interest step is where you begin to answer the question: "What's in it for me?" It's helpful in our presentation to realize that nearly every person to whom we will ever make a presentation is constantly tuned to radio station WII-FM: "What's *in it for me?*" So as we look at the information we want to present, and as we remember that there are things we want them to know, feel, and do as a result of our presentation, then we have to answer in the interest step the question: "What's in it for Me?" How will the participant benefit from knowing what you want them to know, feeling what you want them to feel, doing what we want them to do?

EXAMPLE _____

You are teaching a group negotiating techniques. And to appeal to their sense of "what's in it for me," you might show that what they are learning is applicable in all phases of their life. Not only can they use these skills while negotiating deals for the company, but they also can use them when buying a car, a house, and other large priced items. Acquisition of transferable skills is always a powerful motivation when learning new skills.

Step 3: Develop Desire in Action

This is where we begin to develop the content that says in the attention and interest steps we showed you some end results. In the desire step we share with you the vehicle(s) that will help you get the end results. So in the desire step we begin to share the content, the how-to's, if you will, of how I can get benefits or results that we talked about in the attention and interest steps, how I can solve the problem, how I can close more sales, how I can give more effective decisions, how I can close more sales, how I can give more effective performance appraisals, how I can train more creatively. This is covered in the desire step.

Step 4: Move Audience to Take Action

This step is the wrap-up of the presentation which asks: "What action are we going to take?" "What did you learn and how are you going to put it into practice?" This is where we honestly ask people to evaluate, "How have you done it in the past?" (if in fact they have done whatever it is that we're training them to do or making a presentation on). You get their feedback that says this new way is more effective and they are willing to give it a try.

EXAMPLE _____

We talked about the need to become very sensitive to other people's body language. When you leave the session today, I want you to do three things:

1. I want you to begin observing what the other person is doing as well as saying at the beginning of each other's sales call.
2. I want you to quickly determine if what the person's saying and what the person's doing match. If not, use more intensive questioning to flush out further information.
3. Take the time after each call to do a summary of the call which includes key observations on the other person's body language. Such documentation will be invaluable should you turn your territory over to a new person.

TECHNIQUES FOR BUILDING SUPPORT IN YOUR PRESENTATION _____

Types of Support to Use in Reinforcing Your Sales Training. Be careful to build support for points that the audience may not fully understand or fully agree with. There are at least two times when it's important to support your ideas: first, when the audience is skeptical of its truth or its value (even though they may understand the idea), and, second, when the concept that you're presenting is difficult to understand. Here are key types of support that you might use:

1. *Figures*—numerical representations of facts.
2. *Statistics*—data that express factual relationships based on counting.
3. *Facts*—statements about present or past realities that are verifiable, either by third-party support or by direct observation.
4. *Definition*—an inquiry into understanding the nature of something, usually by going from general to specifics, for example, "a ranch-style house (term) is a type of building (general class) that contains only one level (particular qualities)."
5. *Anecdote*—a story or an experience that is used to demonstrate a point but not necessarily prove it.
6. *Example*—a representative object or incident usually brief in nature that proves or clarifies a general statement.
7. *Illustration*—a more detailed example that generally offers more specific clarifications, point by point.
8. *Authority*—citing a reliable, recognized source other than yourself to support your point.
9. *Analogy*—describing a set of similar conditions that by similarity and familiarity shed additional light on the subject being discussed.

Remember that people will remember illustrations, examples, anecdotes, and so on longer than even the key points that they are intended to illustrate. If you speak in broad, general terms, you may be indicating to your audience, first, that you're not sure of your facts; second, that you did not thoroughly research your material; and third, that you really didn't care enough about your presentation or your audience to prepare thoroughly.

Almost everyone is guilty at one time or another of bypassing points. However, if we do not support what we say with evidence, then we may miss the goal of persuading or influencing the audience completely. Have you used things like:

"Research shows . . .".

What research? Most of the time your audience will not know.

It is convenient to use terms like "they say" or "everyone knows," but who is "they" or who is "everyone?" Be sure that there will be at least one or more listeners asking those questions, and this can reduce the credibility of your presentation.

Be accurate, as well as factual. Avoid vague generalities.

"A recent survey . . .". How recent? Whose survey?

"Statistics prove . . .". Whose statistics and do they really prove?

Train yourself to speak and think clearly and provide specific facts.

When citing examples, use specific names of people, places, things, when possible and permissible. Effective support is effective only when:

1. It's relevant.
2. It's clear.
3. It's accurate.
4. It's strong; that is, it can stand against counterarguments.
5. It's easy to explain.

Transitions. Transitions are used to move from one key thought in a presentation to another. Here are some kinds of transitions:

- Minisummary: This can take two forms. In the first, you summarize what's been covered so far. For example, so far we've seen that the first step to handling an objection is to acknowledge it as real in the prospect's mind. And the second step is to connect it into a question that can be answered. So when a prospect says, "I want to think about it," we might say, "What you're wondering is whether you've given the proposition adequate consideration, is that right?"

- Refocus: This is excellent if a participant sidetracks the group. For example, if your discussion is on prospecting methods a participant, Fred, goes off on a tangent talking about how he organized his prospecting system. You refocus the group by saying, "Just before Fred's comment, what was our general topic?" You're now back on track without wasting valuable time working your way back from where Fred led you—and you haven't put Fred on the spot.

- Question and answer: Allow a short period of time for participants to ask their general questions on a topic. At the end, move to the next topic.
- Pauses: Just stop, look at your group and say nothing for 5 to 10 seconds.
- Physical movement: We're going to do a short brainstorming exercise. Divide into three groups. Be sure to pair yourself with some people you haven't spent a lot of time with today. I'll give you 10 minutes to come up with as many ideas for expanding your sales territory this year as possible. Because we've been sitting all day, I'm going to have you stand around the flip charts for this task.
- A new visual aid: This cartoon really summarizes what I've been saying about good customer service.
- Change of media from overhead to blackboard, blackboard to flipchart, flipchart to slides, and so on.

Audience Contact. There are a number of ways of maintaining contact with your audience:

- Listening carefully: "What didn't work during your cold calling?" (New salespeople.) Allow the people plenty of time to talk about what didn't work. Also be sure to create a safe atmosphere so they can talk freely without being viewed as a failure.
- Asking open-ended questions of individuals in the audience: "If you had to identify factors that helped you increase your sales, what factors would you select?"
- Asking questions that generate a general audience response, such as applause or show of hands: "How many people sitting here today average more than two calls a day?" (Show of hands.) "Three?" (Show of hands.)
- Transmitting your personal enthusiasm for the subject: "I love talking about body language because I always have eager listeners and it is one of the fastest growing areas of interest in sales."
- Eye contact: Be sure to establish eye contact with every member of the audience if it's not too large. People feel included.

OTHER TECHNIQUES FOR SUPPORTING THE LECTURE

Remember the Chinese proverb, "What I hear I forget, what I see I remember, what I do I understand." It is important that we support lecture with visual aids and other types of support. What can we do to help the audience *see* what we're communicating? And even though we're lecturing, what can we do to involve them?

Technology proliferates. We have more media choices than ever before to support the lecture format. Yet the most common are slides, overhead transparencies, and flipcharts. Here are some tips and guidelines for preparing and using these common choices.

Preparing the Room and Equipment

1. The screen must be visible to every viewer with the bottom of the screen 42 inches off the floor.
2. Place the screen in the corner and angle it toward the center of the room. This improves visibility, especially when using transparencies.
3. Whether using slides or transparencies, position the projector so that it does not block the audience's view of the screen.
4. A matte surface improves visibility and seating breadth allowable.
5. Remember the 2 × 6 rule: Distance from screen to first participant equals 2 times screen width; distance from screen to last participant equals 6 times screen width.
6. Tape power cords to floor.
7. Don't block any projector fans.

General Slide Transparency Guidelines

1. Choose your words carefully. Never use more than 6 words per line or 6 lines per visual.
2. Only one idea per visual.
3. Think visually; 80 percent of all slides made are text slides. They are not visual. For example if there are four principles for handling objections—acknowledge, restate, answer, and reclose—consider putting those four principles on a scroll to help visualize the concept of "principles." Or change the word "principle" to the word "key" and have a key next to each point.
4. Use, but don't overuse, color. Color enhances and highlights, but for most visuals (other than photographs) two or three colors are better than five or six.
5. Don't show all your points at once. Whether slides or transparencies, design them so you reveal only one point at a time. This keeps the audience focused on where you are right now.
6. Keep your presentation moving. Reveal a new point or go to a new visual every 30 to 45 seconds.

7. Avoid vertical lettering on your graphics. They're too difficult to read.

8. Use a maximum of two type faces on any visual. More than that makes them too cluttered.

9. When listing nonsequential items, use checkmarks, bullets, or boxes rather than numbers.

10. Make sure your colors contrast, that is, light colors on dark backgrounds or dark colors on light backgrounds.

11. Use upper- and lowercase letters on your visuals—they are easier to read.

Guidelines for the Overhead Projector

1. Make sure that you stand to the side of the overhead—to avoid blocking your projected image.

2. Never point to the screen. If you want to point something out, point to it on the visual itself using a pencil, pointer, and so on.

3. Always face your audience. Let them look at the visual. You make eye contact with them.

4. Turn the projector off when you are changing visuals—and when talking for an extended period of time between points. This puts the focus back on you and eliminates distractions.

Guidelines for the Flip Chart

1. The flip chart should be used with smaller groups, generally 25 or less, for the sake of readability.

2. Don't write on the bottom one-third of the flipchart. It's this part that can be unreadable to some of your audience.

3. Use water-based markers; this eliminates bleed-through onto other sheets of your flipchart.

4. Print with large letters, generally 3 inches high or higher. And use a marker with a broad enough tip that the letters are clearly legible.

Guidelines for Using Handouts

Most of us have been in sessions where it was the instructor, us, notepads, and pencils—and that's it. We began to write as soon as the first visual appeared—and never stopped writing. We didn't hear much—it took too much concentration to copy the visual before it was replaced by the next one. And we often heard frustrated groans and gasps as the visual disappeared before we could finish copying it.

But is this what training is all about—a roomful of participants copying visuals? No, no, no. Yet we give credence to what we see. We must have a copy of what we see. So we become copyists. Appropriate handouts can eliminate this dilemma.

At the very least a handout should include copies of the material displayed by the visuals. Consider reproducing the visual, but with a few key words missing. This encourages the participant to complete the handout, but doesn't detract from the participant's involvement with the presentation.

Handouts can also include additional support materials that go beyond the visuals. For example, if the visuals are about getting client testimonial letters, several sample letters might be included.

The handout is also a good place to include an extensive biographical sketch of the presenter(s). This can convey much more information than an introduction can (or should) and can build the presenter's credibility with participants, as well as satisfy natural curiosity.

Guidelines for Using Resource Manuals

Resource manuals can be a big plus when offering programs that go beyond a few hours. Here are just a few of the materials that can be included in a resource manual:

1. Samples of forms—both blank and completed
2. Samples of tables
3. Charts and graphs
4. Glossary of terms
5. Bibliography
6. Article reprints
7. Copies of transparencies
8. Copies of slides
9. Notepaper
10. Instructor biographical sketch
11. Course outline
12. Housekeeping items (location of restrooms, phones, restaurants, break times, meal times, social functions, if any, and so on)
13. Evaluation forms

TIPS FOR COMMUNICATING SALES TRAINING
IDEAS EFFECTIVELY _____

1. *Present single ideas.* Present your ideas to your audience one at a time time. If you don't most of them will get lost. Encourage participants to react to an idea once you've presented it. For example, you might say: "How can you apply open-ended questions to your qualifying your prospect?"

2. *Get people to "buy off" on one idea before presenting another.* Ask people to respond or react to your ideas so that you're sure in your own mind that they understand and accept the idea before going on to another point. For example, "write down three applications of this technique and how and when you can use it in the next month."

3. *Be specific.* Communicate as accurately as possible. The more specific and the clearer you can be, the better. Use examples, analogies, illustrations. Avoid overgeneralizing. For example, "What's the most useful idea you've gained from the program so far?"

4. *Respond to emotions.* Encourage people to share not only their thoughts, but also their feelings. When a person expresses emotion, try to draw that feeling out. Express appreciation for the other person's feeling. Provide praise. Provide understanding. Accept the negative emotions that people may have. Look for the signs of irritation, confusion, or frustration that may indicate that your participants are not listening, understanding, or accepting the content the way they could be. For example, the participants may have tried open-ended questions before with poor results. As an instructor you may have to explore what happened and why.

5. *Share yourself.* Be open to giving *all* that you have and *all* that you know. When you are open to your attendees, the attendees will be open to you. You'll have better, deeper, richer communication from other people when you open yourself. For example, "I've always had a fear of asking what I consider a great question and have the person just glare at me. As a matter of fact that's happened before."

6. *Know what you want to say.* Make sure that your thoughts and ideas, the points that you want to cover, are clear in your mind. Make sure that you have a total grasp of the content that you want to communicate. If it's not clear to you, it is almost impossible to make it clear to somebody else.

7. *Use a logical sequence.* Put your thoughts and ideas in a sequence that makes sense to your listeners; it may be chronological, it may be topical, it may be from more important to less important, but put your ideas together in a way that will make sense to your attendees and enable them to apply what you have to offer. For example, open with an overview of the day or

lecture and what will be covered: 1. cold calls, 2. objections management, and 3. closings.

8. *Communicate when people are in the mood to listen.* If people are worried, frustrated, angry, upset, or irritated, it's no time to try to communicate. Those feelings must be dealt with first, and eliminated, ventilated, and so on. For example, if a group is hostile, ask why. If they've been ordered to attend, ask what's not getting done and how your session can be productive to improve their skills to do their job more efficiently.

9. *Use a language common to your listeners.* Avoid jargon, unfamiliar terms, difficult words. These may prove to your audience that you're smart, but make them far less likely to be able to use what you have to offer. For example, new sales recruits rarely know the meaning of simple sales terminology like "territory," "closings," and so on.

10. *Involve your audience.* To reach the minds of your audience, you need to draw your audience out. Ask your participants for input. Listen to them as they speak. For example, "Who is willing to share with the group their worst sales call ever?"

11. *Give feedback.* When someone says something that can be interpreted several ways, give your interpretation to make sure that it matches the speaker's. Don't merely parrot back what was said. Communicate what you think was meant, and see if the speaker agrees. For example, "What I hear you saying is that you realized you had a basic personality conflict from day one because he liked the other sales person from your company better?"

12. *Create interest.* Interest can be created by asking questions, by demonstrating the importance of the topic to the attendees. Answer the question, "What's in it for me?" for the audience. Be brief, be specific, but make sure you find common ground by thinking about the feelings, the viewpoint, the attitudes your attendees will have about the subject. For example, "When I attend a workshop like this, I hope to take away one new idea. If I do, the time is worth it because I know I'll be a better sales person."

13. *Think first, talk second.* A car won't run if there's no gas in the tank, and a speaker can't communicate effectively if there aren't clear thoughts in the brain.

14. *Know your aim.* Ask yourself: "What is my purpose in communicating this information?" Be specific. Know what you want attendees to know or feel or do when you finish.

15. *Take into account the total environment whenever you communicate.* Be aware of lighting, room arrangement, physical setting, time of day, the circumstances under which people came to the meeting (i.e., voluntarily requested, ordered), and so on.

16. *Get the opinions of others.* Ask other people for their feedback and their

interpretation of your ideas. Make sure that you get a clear perspective of how other people actually see the ideas you're trying to communicate. Avoid assuming that you're being clear; get feedback that tells you you're clear. For example, "What would you think about sending out a team to negotiate the larger sales contracts?"

17. *Be aware of intent as well as content.* Your tone of voice, body posture, facial expressions, how you dress, your openness to the input of others, all communicate your respect or lack of respect for your content and for your audience.

18. *Remember that individuals are tuned to radio station WII-FM.* Individuals are asking, "What's in it for me?" And so be sure to point out the personal benefits involved. Focus on the person, not on you, not on the organization, but on the individual.

19. *Follow up.* Feedback tells you that your message has been received. Check to make sure you're actually putting your content across. Ask questions. Encourage feedback and input during the presentation, and then follow up afterward to make sure you did an excellent job. For example, "What advantages do you see in rotating sales people through other territories over a three year period?"

20. *Communicate for long term as well as short term.* People respond negatively to change. Be sure that people not only see the short-term effects, but also the long-range benefits of new ideas, new actions, and new directions. Provide information in a way that instills confidence rather than fear. For example, "Once you have learned to use your portable computers out in the field, you will wonder how you ever did without them. I realize, however, learning how to use them is a real challenge."

21. *Make sure actions and attitudes support your presentation.* We've all heard the cliche, "What you are speaks so loudly, that I can't hear what you say. Make sure that your attitudes are compatible with your words."

22. *Be a good listener.* Very few people are looking for a good talker, but they are looking for someone who will listen, and they give a lot of power to a good listener.

HOW TO PRESENT AN EFFECTIVE OPENING ⎯⎯⎯⎯⎯⎯⎯⎯⎯⎯

1. *Open with energy, enthusiasm, and animation.* No one is interested in listening to another dull, dry, boring, listless presentation. Project your interest and enthusiasm. Bring vitality and intensity to the presentation. Challenge your participants. For example, many trainers are using "muppet" movies to grab people's attention. Sales people are used to glitz.

2. *Don't apologize.* Do everything you can by way of preparation to be sure that you don't need to. Remember: If you feel the need to apologize, probably 85 percent of your audience will not know what it is you're apologizing for, unless you point it out, and the other 15 percent won't be affected by it. For example, if the overhead projector is not working and you've created flip charts instead, say nothing. Few, if any, will know you had planned to use fancy transparencies. Let the problem begin and end with you.

3. *Make eye contact.* There is power in eye contact, and if you want to build that power, focus on a single eyeball of an individual. Look at a person eyeball to eyeball and move about the room establishing the same intense eye contact. Don't stare at the floor, the ceiling, the walls, or the back of the room. Make eye contact with the people that you want to communicate with. For example, some trainers have greeted people at the entry way, shaken hands, and established opening eye contact—sales people place high value on good eye contact.

4. *Be "others" oriented.* We've all heard the phrase, "self-conscious"; you should be "others-conscious." If you focus on putting your message across, delivering your content, and persuading and influencing your listeners as to the importance of what you're communicating to them, you won't have to worry about whether or not you're putting on a "good performance." People will appreciate you and respect you if you make a sincere, honest effort to benefit them. Thomas Carlyle said over 100 years ago, "Care not for the reward of your speaking, but simply, and with undivided mind, for the truth of your speaking." For example, sales people may be used to glitz, but if there is no substance then they will soon forget a speaker.

5. *Give the audience an overview.* Remember the cliche, "Tell them what you're going to tell them, tell them, tell them what you told them." Right now, tell them what you are going to tell them. Set out the groundwork for the material that you'll cover. Define key terms. Build a common ground with your audience. Review everything that led up to the present situation. For example, excellent sales people are good time managers. They expect their speakers to be well organized, direct and good communicators of information.

6. *Focus attention.* What are the key questions that need to be answered? What makes them urgent? What's the first question? For example, "our sales have dropped 30 percent this past quarter and management feels there are some controllable factors, but they need your help."

7. *Be open.* For the most part, the audience should know exactly who you are, what your attitudes are about the subject, and the fact that you've got the confidence to handle it. They should be able to feel your sincere interest in communicating and sharing your ideas with them. For example, "Though I have not worked in our western territory, I have worked with those sales people for the last 20 years."

8. *Appearance*. Consider the following:

   Clothing: Are you dressed professionally? Are you properly groomed?

   Gestures: Do you use your hands and your head comfortably? Are your gestures compatible with what you are saying?

   Facial expressions: Is your face animated? Does it communicate an interest in your audience and your subject?

   Posture: Do you stand alert and erect, without being stiff?

   Body movement: As you move and as you change body position, do these movements serve a communication purpose? Do they focus attention on the subject at hand?

   Eye contact: Do you establish eye contact with individuals in the audience? Remember, focus power in the eye contact by focusing on only one eye.

9. *Voice*. Note the following:

   Tone: Do you communicate enthusiasm, seriousness, joy, laughter, mirth, excitement, in your tone of voice?

   Enunciation: Do you pronounce or enunciate each word clearly, or do you slur or skip certain sounds?

   Pace, speed: Do you time your pauses? Do you fill those vocal pauses with a's and ah's? Do you speak fluently or haltingly? Are you too fast or too slow?

   Word choice: Do you use the appropriate words to convey your thoughts?

10. *Audience contact*. Do you reach out and touch the audience? Do you establish rapport that will help them be interested in identifying with your thoughts, with your ideas? For example, most sales people like to have a speaker physically close, so sales trainers have a lot of contact activities, such as informal exercises where the instructor circulates. With all of the points we've just mentioned in mind, try watching a television presentation with the sound turned down. Look to see how effectively the speaker communicates without the power of the voice. Then, try him or her without seeing the presentation—listen, but don't look.

CLOSING THE LECTURE

It doesn't do a lot of good to have spent your time influencing, persuading, informing, and so on if the presentation just fizzles out at the end. Just as there are a logical opening and support that is built in the body, there needs to be a close. What type of close depends on the purpose of the lecture. If the purpose is to inform, then some sort of testing or review might be in order. A general question and answer period followed by a brief summary might also be appropriate. For example, a lot of sales groups do action plans for the next 30 days after sales call training.

If the purpose is to persuade, then a call to action may suit the occasion. This might involve a show of hands, filling out and handing in a goal card, having people share specific actions they plan to take as a result of the meeting, and so on. For example, a sales group may commit to increasing sales 10 percent in the next quarter and decide how the group will reward itself.

Make sure that your opening and your close support one another. It may pay you to develop your opening first, your close second, and then make sure that your support material in the middle leads your participants from one to the other.

ROOM ARRANGEMENT FOR THE LECTURE
METHOD

When people think of training, particularly sales training, the thought that enters the minds of many is lecture. Yet more than lecture we may think of the dynamic motivational sales trainer who is going to inspire us to reach new heights. Yet that model, that picture, may very well be a picture of the past, not the future. Lecture is simply one of a large variety of methods that may be used—and used effectively.

At the same time the classroom arrangement that we use may significantly affect the training in ways that we never understood in the past. For example theater style, schoolroom style, and the U-shape are the three most common arrangements used in most training programs. What do they communicate about the kind of communication that is going to take place? For the most part the training is going to revolve around the instructor. There may be participant to instructor dialogue, but the room setup minimizes the idea that participants will interact among themselves.

Five Room Arrangements to Help in Delivering Your Training Message

Here are five room arrangements that can be used in the delivery of training. One of the first considerations to the delivery of training ought to be the physical environment—specifically taking into account the kind of interaction that is desired.

Theater or Auditorium Style

1. Arranges audience with chairs only—normally in rows of 10 with an aisle between rows.

2. Minimizes interaction when it is important to convey a lot of information in a short period of time. Can accommodate a large number of people.

3. Reinforces only the presenter's point of view. Only one channel of communication. Limits group participation. Strongly dependent on the speaker's personality.

Classroom Style

1. Seats participants on one side of a table. Usually 3 to a table. Normally 6 to 9 to a row. Generally an aisle between rows.
2. Allows participants to take notes more readily. Communicates a somewhat more formal instructional setting than theater style. Can be used with moderately large groups.
3. Minimizes interaction among participants since you only readily make eye contact with the person on either side.

U-Shape

1. Arranges seating to accommodate the group around the outside of a U-shaped arrangement of tables.
2. Allows eye contact and interaction with more participants. Is good for whole group dialogue.
3. Does not readily allow for subgroups. Can become too large for total participation. Difficult to accommodate more than 20 to 25 participants.

Herringbone or Chevron Style

1. Arranges 4–6 participants on one side of 6-foot by 30-inch tables. A number of tables are arranged in a "herringbone," "chevron," or "V" with the front of each table angled toward the center and front of the room.
2. Allows focus to shift from the participants at the table to the presenter at the front of the room. Participants can interact in small groups for discussion, problem solving, and so on.
3. Not as effective for "information-only" sessions since it encourages participation.

Round

1. Arranges groups of 5 to 7 around one-half to two-thirds of a round table. The part of the table toward the front of the room is left open.

2. Encourages participation. Allows more eye contact and close physical proximity than the other arrangements.

3. Rounds occupy the most space of all the methods used—three times the space of theater style. Requires an instructor who wants participation and who knows how to refocus the whole group at the appropriate time.

Room arrangements can maximize or minimize participation. One factor to consider is that either rounds or herringbone have the advantage of group dynamics to a greater degree than other methods. In small groups there is an unspoken accountability to other members. Small groups can foster participation and allow peer teaching to take place. Don't underestimate the experience that a group may have available to share.

Environmental Factors

While the physical arrangement of the room is important, we must not overlook other elements that can also significantly impact a training program.

Color. The trend in decorating today is toward light tones, warm tones, earth tones. Think of the last meeting room you were in. Were the walls dark? Did the colors make you feel warm and welcome? That's part of what the psychology of color is all about. If you have a choice, consider color as a part of your classroom strategy, in the posters you may use or the wall hangings you display.

Sound/Music. The Bulgarian psychologist Lozanov has spent years perfecting "superlearning" techniques that involve the use of 60-cycle music (Baroque being the most common, e.g., Vivaldi's "Four Seasons") to synchronize the physical, mental, and emotional cycles that individuals have. Music can help set the mood at the beginning of a session. Music can be used in the background during small-group discussions. And it can be an excellent support to beginning and ending breaks. Music should be used to complement if used during your sessions, not to compete.

Lighting. Another overlooked factor is lighting. Make sure that you check any room that you are going to use for training to make sure that the lighting is both sufficient and controllable. Is there enough light for participants to take notes and interact? There should be. Can lights be dimmed/turned off independently to allow good contrast for an overhead projector or video projector and at the same time light for participants to take notes? There ought to be. Nothing is more tiring than trying to see washed-out visuals—except possibly trying to take notes when there's insufficient light.

Seating. The cliche is "The mind can absorb what the seat can endure." And it's true. Here are some quick guidelines for participant seating:

1. Adequate back support
2. Comfortable padding for the seat
3. Arm rests for multiple day seminars

In a word, seating should be "ergonomic"—that is, designed to keep the participant comfortable and alert. If you think most hotel chairs won't qualify, you're right. But it doesn't hurt to ask and you want to keep it in mind for your own facility.

SUMMARY

The lecture has come a long way from that first person explaining how to climb a tree to escape the wild beast. If we want to present with impact, we must pay attention to dozens of details the audience may not even notice, from the way the room is arranged, decorated, and lighted, to the support material the participants will be given, to the way we'll support our presentation visually, and, oh, yes, right down to thinking through what we're going to say. The simple lecture isn't so simple. Yet properly prepared, it can be a powerful tool in any trainer's instructional resources.

chapter sixteen —————————————————

USING ROLE PLAYING FOR
SALES TRAINING*

SCOTT B. PARRY

Scott B. Parry is chairman of Training House, Inc. He has conducted workshops for educators and training directors in Africa, Europe, South America, Asia, and Australia for UNESCO, the Ford Foundation, and private corporations. He pioneered the development of the ASTD-NYU program for professional trainers. His background includes associations with Harcourt Brace and World, Hill & Knowlton, and Sterling Institute before establishing Training House. His clients include AT&T, IBM, Ford, GTE, Philips, Mobil, Kodak, Coca-Cola, Air France, Aetna, and Pfizer. He holds an A.B., Princeton; M.S., Boston University; and Ph.D., New York University. He is author of four books and dozens of training courses and has run more than 200 "Train the Trainer" workshops.

---

*Permission to reprint various materials throughout this chapter is granted by Training House, Inc.

This chapter will prepare you with the concepts and skills needed to use role play effectively in the training of salespeople and their support staff (service reps, sales managers, customer relations specialists, tel-sell, etc.). This chapter

- Describes two major purposes of role play and shows how to achieve each.
- Relates role play to the development of knowledge, attitudes, skills.
- Explains the three-stage learning model and the place of role play.
- Tells how to write or edit role play assignments effectively.
- Cites some practical guidelines for selecting players.
- Describes at least four variations to the two-person format.
- Introduces role play to a class so that it is fully effective.
- Shows how to follow correct procedures for conducting a role play.

ROLE PLAY—A MISUNDERSTOOD
OPPORTUNITY FOR TRAINING PROGRAMS _____

The lecture and discussion have just ended. The trainees have just seen a videotape that demonstrated the do's and don'ts of effective selling. The instructor announces: "Now, we're going to practice and apply what you've been learning by doing some role plays. Do I have any volunteers?"

What is the typical reaction of the trainees? Do their faces reflect joy? Do they welcome this opportunity for "hands-on" learning? Do they look forward to taking their turn in the role plays?

In most organizations, they do not. Role play is a four-letter word (two of 'em, in fact) for many people, conjuring up a picture of innocent lambs being brought to slaughter at the front of the room. Or perhaps the scene is reminiscent of the Roman coliseum, where the Christians were fed to the lions to the mixed emotions of a spectator mob.

Why is role play so feared by many trainees? Why is it misunderstood, misused, and mistrusted?

Unfortunately, role play has not always been used effectively by instructors. Postenactment critiques were allowed to become destructive. Trainees have been put into enactments prematurely, before the desired behaviors were understood. Trainees have been given role assignments that were not sufficiently structured to be real, leading to enactments where participants "ham it up" or fabricate information or go off on tangents or "play to the gallery" or try their hand at method acting. All these are symptoms that a role play has been poorly written, poorly positioned, or both.

The picture need not be dismal. Many instructors report that their trainees enjoy role playing, and their end-of-course evaluation sheets often mention this method as one of the most beneficial aspects of the course.

Hence this chapter. We believe that the potential for sharpening interpersonal skills is greater with role playing than in other instructional methods. Whether this potential is realized or not rests with the trainer.

TWO MAIN SALES TRAINING PURPOSES OF ROLE PLAYS

The two major purposes to which role play can be put in a training session are application and discovery. The first purpose, application, is present whenever the trainees have just been taught how to perform and are then placed in role plays that give them an opportunity to practice, refine, and apply what they have just learned. The second purpose, discovery, is present whenever we use role play to discover more about the player's values, mannerisms, and patterns of behavior.

Let's look at some examples of each type. In the training of salespeople, role playing is used primarily for *application* purposes rather than discovery. The sales manager or trainer will first instruct the new salesperson in such subjects as product knowledge, how to handle objections, how to convert features into benefits, how to listen for cues that reveal the prospective buyer's needs, and how to close. Frequently these will be demonstrated by film, videotape, or live presentation. Thereafter, role playing is used as a means of giving trainees a chance to practice and apply their new knowledge and skills.

For example, many sales training programs teach the four behavioral styles of Carl Jung: sensor, intuiter, thinker, and feeler. The point is to sell the "way the customer wants to buy," using the style of the customer. After discussing the characteristics of each style and how to recognize a sensor, an intuiter, and so on, the class might practice giving a sales presentation four different ways: to a sensor, an intuiter, a thinker, and a feeler.

In the *discovery* type of role play, the trainees are not given prior instruction in how to behave. This would defeat our purpose, which is to give them and us the opportunity to discover the assumptions, habits, and biases that are present when they interact with others. For example, an instructor who wanted to give trainees a chance to identify their "natural" or predominant style on Carl Jung's behavioral types might start class with a role play and then follow it with a discussion of the characteristics of each of the four types. The person in the client (prospect, customer) role might then give feedback to the salesperson, indicating whether he or she was selling like a sensor, intuiter, thinker, or feeler. To be sure, some discovery is always present in any use of role play. Salespeople discover what it's like to deal with rejection (when the customer says no), what effort is required to remain in control and not respond emotionally to an irrational customer, what it's like to be in the customer's shoes.

HOW IT ALL STARTED

It is interesting to note that the first use of role playing in the United States was as a form of psychotherapy. An Austrian psychiatrist, J. L. Moreno, came to the United States in 1925 and made use of a technique known as psychodrama. The patient was expected to act out aggressions, frustrations, inhibitions, phobias, and so on as a means of achieving catharsis (release) and of giving both therapist and patient an understanding of the nature of the problem.

Although Moreno's primary activity was as a therapist, he did make use of psychodrama in a training program that he conducted for R. H. Macy's Department Store in 1933. This is the first known use of role playing in the classroom in business and industry.

Today's widespread acceptance of role playing as a training tool is heavily attributable to the work of Norman R. F. Maier at the University of Michigan. Dr. Maier took role playing out of its clinical origins and shifted the emphasis from discovery to application of skills in such areas as problem solving, human relations, sales, supervision, and communications.

The Behavioral Triangle and Its Impact on Sales Training

Norman Maier's use of role play combined for the first time the three components of human behavior with which trainers are concerned: knowledge, attitude, and skills. They are shown in our "behavior triangle" as a hierarchy. Skills are at the apex, easy to see and teach. Knowledge supports these skills, and attitudes, in turn, underlie knowledge. To develop the desired behavior in class, trainers must address all three. Let's look at each in turn, as they relate to sales training:

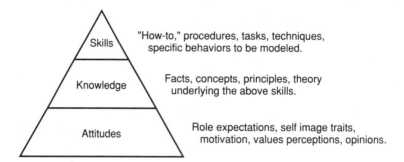

- *Knowledge.* The acquisition of knowledge consists of learning facts, concepts, principles, rules, and procedures and the relationships among these. Salespersons must learn product information, pricing policy, procedures for processing an order, delivery schedules, policy on returns, credit, and so on.
- *Skills.* This arena is broad, including manual skills (e.g., how to operate or demonstrate merchandise), analytical skills (e.g., how to analyze the buyer's needs, financial position, buying style), and communication skills (e.g., how to listen, use persuasive language, talk benefits).
- *Attitudes.* Here we are concerned with values, perceptions, beliefs, and factors that are closely related to personality. Salespeople are taught to believe in the integrity of the customer, the importance of service, long-term customer satisfaction versus short-term gains, consultative selling versus hard-sell tactics, and so on.

In examining these three components of human behavior in general and the development of salespeople in particular, we can now view role playing as a technique that is uniquely qualified to enable trainer and trainee alike to recognize and refine the knowledge, skills, and attitudes needed to sell effectively. Its dual qualities of application and discovery make role play especially effective both in diagnosing and in shaping behavior.

DEMONSTRATION: KEY TO THE THREE-STAGE LEARNING MODEL

The first stage of learning anything new is *acquisition*: the trainee must acquire new concepts, terms, procedures, and so on. This is the information "input"—the knowledge, skills, and attitudes that must be imparted to the learner.

The second stage of learning is *demonstration*. As soon as trainees have acquired new information, they want to see how it works. They need examples, cases, situations, illustrations that bring the new input to life. A good demonstration will give the trainees an opportunity for vicarious application—they should be able to identify with the demonstration.

The third stage of learning is *application*, where trainees have a hands-on opportunity to practice. It is here that knowledge is applied, skills are refined, and attitudes are reinforced. This stage occurs in class via simulations and on the job via the everyday opportunities that challenge the trainee.

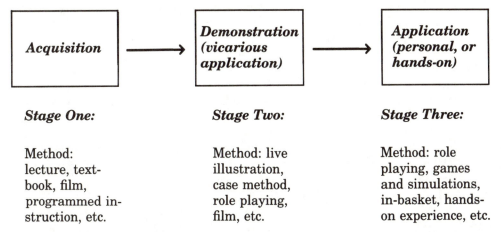

| **Acquisition** | → | **Demonstration (vicarious application)** | → | **Application (personal, or hands-on)** |

| *Stage One:* | | *Stage Two:* | | *Stage Three:* |

| Method: lecture, text-book, film, programmed instruction, etc. | | Method: live illustration, case method, role playing, film, etc. | | Method: role playing, games and simulations, in-basket, hands-on experience, etc. |

As can be seen in the learning model, role play is useful in both the demonstration and the application stage. A case could also be made for its use in the acquisition stage as a means of diagnosing the trainee's needs by identifying the "entering behavior" that the trainee brings to class.

A common error of trainers is that they go directly from acquisition to application with little or no demonstration. We must give trainees the opportunity to observe and critique the desired behavior. It is during this critique that trainees and instructor receive feedback that indicates whether or not they are ready to move into the role plays in the application stage. If we throw them in the water before they are ready to swim, the uncertainty of the situation and the threat of failure consumes much of their energy and distracts them from applying their new knowledge and skills—they have only been told what to do, and must experience it vicariously and safely (demonstration) before they practice it in role plays (application).

HOW TO HANDLE THE DEMONSTRATION IN SALES TRAINING

The common way of handling demonstration in sales training is for the instructor to fill the salesperson role and demonstrate the do's and don'ts of an effective presentation. This can be live or on videotape. Alternatives to this include the use of a respected, experienced, successful salesperson doing the demo, or the use of commercially available films or videotapes.

There are some problems here. For one thing, the trainees should be analyzing each behavior—whether it was appropriate, what effect it had on the buyer, how it might have been worded differently, and so on. Usually the demo is given "straight through," thus depriving the trainees of in-depth analysis and discussion. The moral, of course, is to get the demo on videotape or film, so that it can be stopped at each critical step and analyzed. Trainees can be asked, "What would you have done? What do you think the salesperson should say next? Should they try to close the sale?"

An alternative form of demonstration makes use of a technique known as script analysis. Trainees are given a script showing correct and incorrect behavior. Their job is to analyze the script, indicating what they do and don't like about the way the salesperson handled the interaction, and why. After they have discussed and processed the script to their satisfaction and the instructor's, they are then ready for role playing.

An example of script analysis is shown in Figure 16–1. As can be seen, trainees make their evaluative comments in the "Notes" column beside the script. Each line is numbered to make their references easier. A sales presentation can be scripted to show only correct actions (behavior modeling) or to illustrate a mixture of appropriate and inappropriate behaviors. The advantage of the latter is that trainees must show their ability to discriminate and not merely to label what is being done.

This means that the trainees need feedback on the appropriateness of their critique. An example of a feedback sheet is shown in Figure 16–2. Perhaps the most useful form of feedback is to give trainees a script of the "correct" or improved sales presentation that they just analyzed. An example is shown in Figure 16–3.

Figure 16–1 Sample script analysis

Dianne's Presentation to Hal Kramer: Example A

In the script that follows, you will see Dianne Rogovin making a presentation to Hal Kramer, who heads a 320-person department. Dianne could be a sales representative, or outside consultant, or an internal specialist in administrative support systems. Her goal is to convince Mr. Kramer to install the Lenke System 175, a word processing system that will improve the speed and ease of getting typed words on paper. Several days ago she asked for 20 minutes of Hal Kramer's time to make her presentation. As we look in, she is entering his office.

Your job is to indicate in the column to the right the things you do and don't like about the way she is handling the presentation.

Figure 16–1 (Cont.)

NOTES

1. Kramer: Yes, Dianne, come on in. (***To secretary: "Mary, would you hold all calls?"***) Please have a seat. What can I do for you?

2. Dianne: I'd like to talk about what I can do for you . . . and for everyone in the department who is responsible for getting out letters, memos, and reports, whether at the input or output stage.

3. Kramer: I take it you're talking about word processing equipment.

4. Dianne: Yes, and from what I've gathered in talking with secretaries and typists, a new system is overdue. They feel that there is little opportunity for advancement in being a secretary and working for one or two people. A word processing center would give us much more career mobility. Why, one girl I talked to who has been here going on nine years was telling me about a friend of hers whose company put in a Commugrafix 1800B System that was able to . . .

5. Kramer: (***Interrupting***) Dianne, I'm interested in exploring word processing and examining all the trade-offs. But the implications of our talk go beyond this department, for three reasons. First, any decision is likely to be made upstairs on a companywide basis, with equipment installed where it will give support service to all who need it. Second, I've been reading about some exciting advances in linking word processing equipment with the computer and our leased phone lines, which could tie other locations into a communication network. So maybe Joe Ames up in MISD should be in on our talks. And third, have you talked with Mrs. Bennis? Her group has been providing secretarial support service and doing a great job. Her insights relating to the needs of different departments and managers that call on her group would be useful to have.

6. Dianne: Well, I felt it would be premature to talk to the others before getting your feelings. Are you acquainted with the Lenke System 175 that has instant CRT display, storage on floppy disks, visual scan, and text editing capability? (***She shows him a four-color brochure.***)

Figure 16–1 (Cont.)

7. Kramer: Selecting the right systems has got to come much later. First, we have to study the present flow of work to see how our needs are and are not being met. Then, we have to talk with managers and look at their projects to see what capability they need—whether it's statistical typing or book copy with justified right-hand margins or many rewrites of the same document, as with a legal contract. Then we probably ought to see what some other companies have done and what history they have had.

8. Dianne: I've already talked with some of the managers. They would like to be able to shorten the turnaround time on present typing. With only one secretary for every three or four managers, the process is slowing down. They can't find someone to take dictation, and when they do, it takes a half day or more to get the finished letter back. Another feature of a word processing center is that managers could phone the center at any time, day or night, and dictate.

9. Kramer: You mean we'll have to run shifts to have the center covered at all times?

10. Dianne: No . . . a recording unit is activated and the dictation is taped. However, it's often a good idea to have two shifts. Employees can then pick the hours they want to work, say, from 7 A.M. to 3 P.M. or from 11 A.M. to 7 P.M.

11. Kramer: Hmmm. That gives us 12-hour coverage—7 to 7—with double-strength staffing during the peak hours, 11 A.M. to 3 P.M. Look, Dianne, why don't I have a talk with some of the others who should be in on the planning of our secretarial support system. I'll feel them out on their needs and the direction they think we should go. Then you and I can get together. Okay?

12. Dianne: How long do you think that will take?

13. Kramer: Depends on their schedules and availability. But as soon as I can talk with them, I will.

14. Dianne: There's something I can do to help get ready for our next talk. I've been wanting to do a line count analysis of what the actual typing demand is throughout the department. I realize you're talking about a company-

Figure 16–1 (Cont.)

	NOTES
wide installation, but by starting with your own department, we'll have a model that other departments can follow in calculating the potential savings and the type and amount of equipment needed. I can have that ready in two weeks . . . by the 17th. Maybe we could talk then . . .	

Figure 16–2 Sample feedback sheet

Dianne's Presentation to Hal Kramer: A Critique

Here are some of the key mistakes we spotted where Dianne (Figure 16–1) failed to make a successful presentation.

- Dianne failed to establish the need. In paragraphs 2 and 4, she should have stated her objectives from Kramer's point of view. She talks about advancement and career mobility of secretaries, which can improve when word processing systems (WPS) are installed, but they are hardly the main reason for installing WPS.

- Dianne is dealing with a well-organized, impatient buyer who is looking for a detailed presentation. Her rambling disorganized approach is not taking the buyer's personality into account.

- In paragraph 6 Dianne is premature in getting into the features of one particular system. Also, she never translates these features into benefits for Kramer.

- In paragraph 8, Dianne now seems to be addressing the buyer's needs and his interest in details (e.g., "one secretary for every three to four managers"). However, she has missed a chance to use his "one-two-three" outline style of speech (see paragraph 5). She might have said: "Results: (1) not available when needed, (2) delayed turnaround time, (3) dictation is limited to office hours."

- Dianne was not equipped to handle Kramer's objections (paragraphs 5 and 7). She might have acknowledged the value of his suggestions instead of countering them. She should be working with him . . . not opposing him. As a result, he takes matters into his own hands (paragraph 11). She does handle this well, getting a timetable established (two weeks, paragraph 14), despite his unwillingness to set a date for their next talk.

- Lacking a well-planned presentation, Dianne lost control to Kramer early in their meeting.

Figure 16–3 Sample improved sales presentation

Dianne's Presentation to Hal Kramer: Example B

In Figure 16–1, you examined a script of Dianne's poorly organized attempt to persuade Hal Kramer. In the script that follows, we see Dianne doing a much better job of making her presentation. Study the script and make notes in the margin to the left indicating what you do and don't like about the way Dianne handled her second presentation to Mr. Kramer.

1. Kramer: Yes, Dianne, come in. (**To secretary: "Mary, would you hold all calls."**) Have a seat. What can I do for you?

2. Dianne: I've been taking a close look at our secretarial support system. It should be **extending** the effectiveness of our managers and administrative people. But I have to report that at present the system is **limiting** our effectiveness as a department and preventing managers from performing efficiently.

3. Kramer: How did you reach this conclusion? Do you have evidence that our secretaries aren't doing their job?

4. Dianne: The problem lies not with the secretaries but with the system and the way they are assigned. Let me explain. (**At this point, she handed Kramer a typed list of problems.**)

5. Kramer: Are our managers complaining about their secretarial support? Can't they get help from Mrs. Bennis and her group when they need it?

6. Dianne: I don't think anyone is to blame. And the problems we just looked at are being faced by any other large organization that turns out a lot of typed communications.

7. Kramer: What are they doing about it?

8. Dianne: Let me answer that in a minute, but first I'd like to make some additional remarks since it's important for us to understand the roots of the problem before we look at possible solutions. I think there are three major causes of the problem. Let's look at each. (**She takes a blank sheet of paper and writes "1. Quality of Education."**)

   First, the quality of education in the public schools has gone down greatly in the decades since you went to school. Today's graduates haven't been taught to spell, punctuate, or apply basic grammer rules very effectively.

9. Kramer: I don't know if that applies here. My secretary is a whiz. I think my secretary is a product of the present-day schools.

Figure 16–3 (Cont.)

10. Dianne: Your secretary is one of the best in the department. And that's part of the problem. The best people have been assigned to the top brass for reasons of status more than need. Yet there are probably a number of people in the department whose daily output of correspondence is greater than that of the upper echelon.

11. Kramer: Are you advocating taking my secretary away to become available to someone else who writes more than I do?

12. Dianne: Certainly not . . . that would be the quickest way to alienate you I could think of. (**Both laugh**) But I am suggesting two kinds of functions a secretary performs: correspondence and administrative chores. You should have an administrative assistant, with the corresponding secretaries—those whose grammer skills are good—assigned to a word processing center. Did you read the article I sent you several days ago?

13. Kramer: Yes . . . the reprint from **Nation's Business.**

14. Dianne: Good. Then you're familiar with the distinction between administrative secretaries and corresponding secretaries.

15. Kramer: And which would my secretary become?

16. Dianne: I'm not going to fall into the trap of answering that. (**Laughs**) But it relates to the second cause of the problems we were looking at. (**Dianne takes the paper and writes: "No. 2. No Career Appeal."**) There was a day when the secretary's role was glamorous, as depicted in movies and novels of the 1940s and 1950s. But today it is seen as a dead-end job, low paying, low visibility, and people with talent are simply not attracted to it. Well, a career in word processing gives one a lot better pay, visibility, and mobility.

17. Kramer: Those are pretty words. But aren't you simply upgrading their titles and giving more pay? I don't see how that will address the problems we identified?

18. Dianne: You're right. That wouldn't change a thing, any more than calling a garbage handler a sanitation inspector doesn't get the garbage picked up any better. The critical difference is that we have long since outgrown the hardware limitations of a simple typewriter. An effective secretary is limited in output by having to proofread, make corrections, and retype documents.
    The text editing typewriter that is at the heart of a word processing system makes correcting and retyping virtually automatic. The typist need only indicate where changes are to be made and the machine retypes the document at a speed three times faster than a person

Figure 16–3 (Cont.)

can type. That's the savings at the **output** end. (*Dianne pulls out a folder showing a schematic flow diagram of the parts of a word processing system, and points to the output equpment.*)

Now let's look at the savings at the *input* end. (*Dianne points to input equipment.*) The person originating a letter, memo, or report can simply pick up the phone, dial the center at any hour of the day or night, and dictate. The next available person will prepare the document. This means that there is no delay for our people waiting to give dictation to a secretary who is busy on another project. Moreover, it's possible to have two shifts working the word processing center . . . from 7 A.M. to 3 P.M. and from 11 A.M. to 7 P.M. This means that managers who come in early or stay late can get secretarial support.

19. Kramer: From 7 to 7. That's 12-hour coverage, with double-strength staffing during the peak hours, 11 A.M. to 3 P.M. I like that.

20. Dianne: That's why I said that the biggest thing word processing brings us is the ability to multiply efficiency by a factor of two to three times. (*She takes the paper and writes "No. 3. WPS Multiplies Efficiency."*) It's amazing when you stop to think about how the paper output of a growing company can double every 5 to 10 years, yet the efficiency of one person at one typewriter has remained about the same ever since the typewriter was invented.

21. Kramer: Up until the advent of word processing equipment.

22. Dianne: Exactly. Now we're ready to answer the question you raised earlier, namely, if the problems we're facing are common to other large companies, what are they doing about it?

23. Kramer: And the answer, you're about to tell me, is that they are buying word processing equipment. Right?

24. Dianne: In due time, right. However, a buying decision is the last step in the sequence. Here are the steps that have to be taken first. And I'd like your permission to make this my project. (*Dianne hands him the sheet listing the "Action" steps. She allows time for him to read them.*)

25. Kramer: The list looks well thought out. How long will it take you?

26. Dianne: Working an average of two days per week, about four to five weeks. And I have a question in that connection: Will that give you enough time to get budget approval in next year's budget if you reach a purchasing decision?

Figure 16–3 (Cont.)

---

27. Kramer: Oh, yes. We don't have to have that in 'til November. And we'd probably want a lease anyway, if that's an option.

28. Dianne: It is. Two questions, and then I'll get out of your hair. First, you'll see on this list that I don't get back to you until the eighth step. Do you want me to check with you any earlier?

29. Kramer: Let's see . . . (**He reads list again.**) No, on step 1 you seem to know what kinds of issues to get at without my help. But why don't we talk again after step 5?

30. Dianne: Good. I'll feel better if I can give you a verbal progress report. And my second question is whether you have anyone whom you particularly want me to talk to—or to avoid talking to—during step 2.

31. Kramer: No. I don't think of anyone offhand. And I wouldn't want to bias your sample anyway.

32. Dianne: (**Laughs**) Okay, then. I'm off and running. Thanks for your time.

33. Kramer: Let me know if you run into any snags that I can help with.

GUIDELINES FOR SELECTING ROLE PLAYS

Since our primary concern with role plays is their use in providing learners with an opportunity to apply the skills that they have previously acquired, it follows that the process of selecting a given role play for workshop use will be based largely on its effectiveness in eliciting those skills that learners have already been taught. In this regard, role plays should be field-tested to assure that they are effective in producing the desired behavior.

It is not uncommon for the writer of a role play to expect directions to the players to focus on a particular issue, only to discover later in actual use that the players concentrated on another issue—a minor detail that the writer included only to give reality to the assignment—and missed the main point altogether. If the writer uses the role play in subsequent sessions with other trainees and discovers the same situation to be true, it is clear evidence that revision is necessary.

The "Critical Incident" Method

The writing of effective role plays is not an easy task, and instructors should first examine what has been published and use existing material wherever possible. This is not always feasible, however . . . especially if you are teaching procedures that are unique to your organization. Role plays designed to develop skills that relate to your organization's policies and procedures must be written by the instructor, since published material is not likely to be available.

In doing this, the "critical incident" technique is often helpful. Let's illustrate this with an example. Suppose that you are responsible for training the sales managers in your organization. Periodically you bring them together for training and for briefings on new policy, changes in the product line, and so on. At these training meetings you would like to use role play to help them develop skill in supervising their salespeople. You could, of course, sit down and write several role plays dealing with different aspects of the sales manager's relationship with the company's salespeople, but you frankly have not been in touch with field operations for several years and are quick to admit that you don't really know the kinds of day-to-day problems that your sales managers are encountering. How do you find this out? By asking them, using the "critical incident" technique as your means of gathering data. A sample is shown in Figure 16–4.

Figure 16–4 Sample request for critical incidents (to serve as input to the writing of role plays)

MEMO TO: Regional Sales Managers

FROM: Director of Sales Training

SUBJECT: Preparation for Sales Meeting on November 27

To help us prepare for our meeting next month, I would like you to recall the events of the last months and select one incident that occurred between you and one of your salespeople that you feel is unusual and that taught you a lesson (either a "how-to-do-it" success story, or a "how-not-to-do-it" failure story). The incident you select may be used to develop a case history or role play for our forthcoming meeting; all names will be changed.

You will probably need several paragraphs to describe the incident or series of incidents. Please indicate what made the situation "critical." That is, describe what it was that made this situation different from the usual, and what you learned from it.

Please have this back to me by November 15.

From the responses you receive, you will be able to identify the kinds of problems or situations that are on the minds of your sales managers, and their relative effectiveness in dealing with these situations. Armed with these data, you should be able to prepare realistic role plays for your meeting. Another advantage to the critical incident technique is that there will probably be greater interest and ego involvement by your trainees at the November 27 meeting because they have previously given some thought to the problems that you are dealing with in the role plays you have created.

Ask Your Salespeople or Managers to Write Their Own Role Plays

In Figure 16–4, the director of sales training took on the responsibility of creating the role plays from the "critical incidents" that the sales managers turned in. An alternate is to ask your salespeople (and/or their managers) to write the role plays themselves, submitting them to you for editing and duplication. These can then be made available to the sales managers and anyone else who does sales training.

An example of this is shown in Figure 16–5. This food manufacturer scheduled a half-day of training in how to write role plays for its district sales managers (DSMs) and had each create a role play as part of the workshop. These were then edited, typeset, reproduced, packaged, and distributed to the DSMs. The cover sheet of instructions is shown in Figure 16–5 and the first role play in the package is reproduced in Figure 16–6.

How to Sort the Information

In writing a role play, it's important to sort the information needed by players into three categories: information (facts and feelings) that both parties have, information known only to the salesperson, and information known only to the customer. A common mistake is to tell only one player an essential piece of information that both would know in real life, but the writer forgot to include it in both roles. The two role plays are like a picture puzzle. Each player has some of the pieces. Neither can construct an acceptable picture alone; they must interact and work together to create a picture that both find acceptable. Incidentally, this is why the observers of a role play are usually given both role assignments to read prior to the enactment. Theirs is the job of analyzing how effective the salesperson was in eliciting the missing pieces from the customer and blending the two parts of the picture puzzle. (And here the analogy breaks down, for in role playing there is usually no single "correct picture" but rather a number of acceptable or appropriate outcomes.)

Figure 16–5 Role play instructions

Suggestions for Using These Role Plays

INDIVIDUAL TRAINING

If you wish to prepare one of your salespeople for a forthcoming sales call of a type dealt with in one of the role plays, use it to help your person "tune up." Similarly, if you suspect a weak area and wish to assess strengths and weaknesses, select the role play that will best diagnose the individual's performance. In using role plays individually, you will probably want to play the role of the buyer (store manager, chain supervisor, etc.) yourself.

GROUP TRAINING

When used at sales meetings or training sessions, role playing can be an excellent stimulus to get your salespeople discussing what works for them. In groups it is best not to play the buyer role yourself, but to pick one of the members. This enables you to be more objective in your role as discussion leader. Here's a simple procedure to follow:

1. Select the two players and tear one of the role sheets in half giving each player his or her part. They are not to collaborate or read one another's role sheets.

2. Give the observers both roles (untorn sheets) and allow time for everybody to read them.

3. Meet privately with each player to clarify any questions they might have.

4. Let the role play begin. Don't interrupt the players. If they step out of character and ask for help, give it (briefly) but return them to their role and the situation.

5. As soon as the interaction has concluded, ask the salesperson, "How do you feel about the way you handled it? Looking back would you do anything differently?"

6. Ask the buyer (manager, dealer, etc.) how he or she felt about the meeting and whether anything might have been handled better.

7. Then open the discussion to the group. Where appropriate, summarize on the blackboard or flipchart the things they like or dislike, bringing out the do's and don'ts of this type of selling situation.

8. Give the players the other person's role, so that everyone has a complete set. Conclude the session.

Figure 16–6 Role play

Role: Store Manager—Fred Rollins
Place: High-Volume Chain Store

You are the store manager of one of the highest volume stores in your chain. Every day, one sales representative or another approaches you for something, and you have promised yourself that today you won't be disturbed; you must make out next week's employee work schedule which already should have been posted. The job is made more difficult by the fact that one of your competitors across the street is closing and you expect a significant increase in volume. You are having difficulty in scheduling hours during peak volume periods.

Your chain follows rigid shelf-management policies published by headquarters. These policies are enforced by your supervisor who expects compliance with all company policies but does have the authority to make adjustments. The thought of doing any resetting at the present isn't too appealing, though, because of the way business is going to pick up, and you are understaffed.

The girl in the courtesy booth just called upstairs and told you that Vic Bell would like a few minutes. You reluctantly give him permission to come up.

Role: Sales Representative—Vic Bell
Place: High-Volume Chain Store

The other day you called on the supervisor of one of your key chain accounts and presented your company's "shopping mall" concept for merchandising your line (i.e., locate your products and your competitors at the opposite ends of the section, with the smaller brands in between). He liked the plan, but would not accept it for all his stores at the present time. Permission was received to institute the plan in three different stores, but only if the respective store managers consent.

You are waiting to present the program to Mr. Rollins, manager of one of the chain's largest stores. It has been several weeks since your last call on this store and you notice that business is considerably better than usual. Relations with this particular manager are pretty good, but he is known for not being overly cooperative. This is the first store to be approached—a key location. Needless to say, you are eager to sell this store.

Presently, this account follows the plan-o-gram of your competitor. In your preparation for the presentation, you have noticed similarities with our "shopping mall" theory in their dairy and product departments.

The girl in the courtesy booth tells us that you can go upstairs now.

THE THREE OBJECTIVES IN
   WRITING ROLE PLAYS _____

Set the Stage for Players to "Act Naturally"

In writing role plays, your job is to set the stage rather than to prepare the script. Role playing is not a theatrical exercise for trainees to show their dramatic talent. Rather, our objective is for the players to be themselves and to "act naturally." This means, for one thing, that the situations you create must be realistic. That is, the situation you describe in the assignment that goes to each player must lie well within his or her framework of experience as "something that has happened or might easily happen."

Wherever possible, you should avoid writing instructions on how to act during a role play. For example, suppose we are teaching customer courtesy to bank tellers and are creating role-play assignments for the teller and the customer. It would not be appropriate to tell the customer to "be angry" or to tell the teller that "Mrs. Smith is a very finicky customer who has made you angry on previous occasions." Instead, your job is to set the stage in such a way that the opportunity for anger is strongly present. You might, for example, tell the customer that "you have been waiting in line for ten minutes, during which other persons who arrived at the bank long after you did have already been taken care of at the other teller windows."

Promote Interaction

A role play is successful to the degree to which it promotes interaction between the two players. One of the shortcomings of role-play assignments prepared by a novice writer is that they "let the players off easily" . . . that is, the players can reach a mutually satisfactory solution with a minimum of interaction. Often this is the result of giving one of the players most of the pieces of the puzzle, so that very little interaction is needed to get the remaining pieces from the other player.

For example, consider the role play reproduced in Figure 16–7. The first version requires very little interaction, since (1) there are very few facts given to either player (i.e., the puzzle is too simple, requiring too few pieces to be fit together to solve it); and (2) there is really only one acceptable solution to this role play: the supervisor must give Sally enough time off to go to the dentist. On the bottom half of the page, we have rewritten this role play in a way that will produce much more interaction, thus giving the players a better opportunity to practice and apply the skills they have learned.

Figure 16–7 Interaction design

FIRST VERSION:

ROLE: Sally, Retail Sales Associate   **PLACE:** Supervisor's Desk

You have just called your dentist for an appointment. He has told you that the best time for you to come would be tomorrow at 3:00 P.M. You have agreed to come at that time.
You now realize that you must ask Mr. Watkins, your supervisor, for time off tomorrow to go to the dentist. The dentist's office is about 45 minutes away from work, so you'll probably need the entire afternoon—at least from 2:00 P.M. on. You find your supervisor at his desk and ask him.

ROLE: Mr. Watkins, Supervisor   **PLACE:** Your Desk

You are seated at your desk doing some administrative work when Sally, one of your sales associates, approaches you.

REWRITTEN VERSION:

ROLE: Sally, Retail Sales Associate   **PLACE:** Supervisor's Desk

You've been having a toothache for the past three days. Today you've hardly been able to take care of customers, the pain has been so great. So you've just called the dentist for an appointment. The earliest he can see you is tomorrow at 3:00 P.M. In fact, you got this appointment only because another of his patients had canceled. (He's a busy dentist—booked up for at least three weeks in advance.) You now realize that you must ask Mr. Watkins, your supervisor, for time off tomorrow to go to the dentist. The dentist's office is about 45 minutes away from work, so you'll probably need the entire afternoon—at least from 2:00 P.M. on. You find Mr. Watkins at his desk and ask him.

ROLE: Mr. Watkins, Supervisor   **PLACE:** Your desk

Your boss just called. The special inventory count he wanted from you by next Monday has been moved up, and now he wants it by tomorrow at 6:00, when he leaves for the airport, report in hand! You are going to need everyone in the department. In fact, you might need some overtime. You are seated at your desk, trying to figure how to get the work done by tomorrow at 6:00, when Sally, one of your best sales associates, approaches you. Sally is the group leader . . . whatever she does, the other associates usually go along with. You're hoping she will agree to stay 'till 6:00 tomorrow, if necessary.

Provide Enough "Ammunition" to Practice Sales Skills

In some types of training, you want the role play to provide the learners with an opportunity to practice their skills in dealing with a variety of problems. For example, in the teaching of selling skills, we might want the salesperson to have a chance to deal with a number of common objections. In writing the customer's or prospect's role, we might arm them with three or four objections. Thus, the customer's role might contain something like this:

> On the one hand, you like what you've heard about XYZ, the new salesperson's company, and its line. However, you've always gotten good service from ABC, whom you have dealt with for some time now. It wouldn't be fair to them. And, though XYZ is offering you a better price, you're familiar with that old trick . . . once they make the sale, the price would be right up there with ABC's price. As for XYZ's service contract, you've heard that it is better than ABC's. However, ABC has given you prompt service whenever you needed it—without contract. And their charges for service calls have been less, you suspect, than XYZ's charge for the service contract.

As can be seen in the example, the salesperson is likely to encounter three objections during the sales presentation: loyalty to present vendor, price, and no need for service contract. This will make the interaction much more meaty and realistic, for the customer has been given enough "ammunition" to give the salesperson lots of skills practice.

PREPARING THE TRAINEES: TWO TRAINING TOOLS FOR PRACTICING WHAT IS LEARNED

Before going into a role play, you want to be sure that your trainees are ready. That is, have they had enough acquisition and demonstration? Are they ready to practice what they have just learned? Do they know what constitutes appropriate and inappropriate behavior in the situations they are about to encounter? There are two training tools that can help you to answer these questions, at the same time preparing your trainees to get the most out of their enactments. Let's look at each—the rating sheet and the planning sheet.

Rating Sheets

These serve as a checklist of criteria to be met during the role play. An example is shown in Figure 16–8. Perhaps the best time to distribute the rating sheet is before the Demonstration, so that your trainees can use it to evaluate the demo presentation. Then, when you introduce the role plays, they are already familiar with the criteria.

Figure 16—8 Rating sheet

What kind of job did the presenter do in:	Ex-cellent	Very Good	Good	Fair
OPENING				
1. Defining the need and the objective clearly in opening statements				
2. Presenting objective in terms of how it would benefit the other person				
PRESENTING				
3. Establishing a friendly climate				
4. Presenting one idea at a time, using evidence or an example to bring it to life				
5. Stating the facts and features (especially dollars) as benefits				
6. Giving costs of not doing it (if appropriate)*				
7. Empathizing by relating to the other person's interests				
8. Involving the other person(s) by asking questions				
9. Listening attentively				
10. Giving feedback to show understanding				
HANDLING OBJECTIONS				
11. Re-stating benefits				
12. Probing with questions to get real reason for objection (if appropriate)*				
13. Using questions to clarify buyer's concerns				
14. Referring to past success (sale) or experience (if appropriate)*				

385

Figure 16–8 (Cont.)

What kind of job did the presenter do in:	Ex-cellent	Very Good	Good	Fair
ACTIONS 15. Giving an outline for implementation, explaining who will do what, and when 16. Outlining plans for follow-up, with who, what, when				
CLOSING 17. Asking appropriate questions to get commitment or desired action 18. Restating advantages and disadvantages 19. Soliciting other person's opinion for solution (if appropriate)* 20. Seeking a course of action that was beneficial to both parties				
Total number of checks				
Multiply by	4	3	2	1
To get				

Directions for scoring:
Allow four points for every checkmark in the "Excellent" column, three points for every "Very Good," two points for every "Good," and one point for "Fair" checks. Thus, a perfect score would be 80 points.

* If these steps are not appropriate, award the presenter an "Excellent."

Some instructors prefer to develop the rating sheet themselves, distributing it as a handout. Others prefer to have the group generate its own list of criteria which are recorded on the flipchart for subsequent reference during the postenactment critique.

Four Benefits of Using Rating Sheets

1. They are a useful way of summarizing the knowledge and skills imparted during the acquisition and demonstration stages of the course.
2. They give the instructor a tool to help focus the postenactment discussion on relevant issues.
3. They make the ratings more specific and objective, thereby minimizing the typical "safe" comments (e.g., "You looked nervous," "I thought you had a nice, easygoing manner," etc.).
4. The numerical scores at the bottom of each observer's sheet give the instructor feedback on how ready the group is for role playing. (A wide range of scores indicates that further instruction on the desired performance is needed, since all observers saw the same behavior but rated it very differently.)

Planning Sheets

These are forms that the trainees complete in preparation for an enactment. A sample four-page planning sheet for sales presentations is shown in Figure 16–9. It is based on the script of Dianne's presentation to Hal Kramer (Figure 16–3) and shows how she might have prepared for the presentation.

Figure 16–9 Presenting and selling—A planning sheet

1. ESTABLISH NEED   What do you want to accomplish? What needs, problems, opportunities are you addressing, and why?

I want to get company to install word processing centers wherever they are needed (as indicated by needs analysis which I want permission to do). These centers will help us overcome the present problems of

- shortage of competent secretarial help who can spell, punctuate, etc.
- managers can't find secretary to take dictation when needed.
- turnaround time (between dictation and signing of complete letter) too long.

Figure 16–9 (Cont.)

- people traveling can't dictate field reports—will be able to phone in.
- present pay and career/growth opportunities are weak for secretaries.

2. MEASURING SUCCESS   How will you know if you've succeeded? What action do you expect as a result of your presentation?

Want to get Mr. Kramer's permission to let me do a line count analysis of our typing needs, along with interviews with managers who dictate a lot. He can pave the way for these interviews. By getting him involved, he will be more receptive to initialing the internal request to purchase the new equipment.

3. THE BUYER   What do you know about the other person(s)? List their interests, peculiarities, etc. How can you use this in your presentation?

Mr. Kramer is well-organized, logical, methodical, impatient, dynamic. He likes facts and details. I'll prepare a sheet outlining the problems with the present system (like the list in No. 1 above, only longer and expressed in terms that make sense from Mr. Kramer's point of view). I'll also prepare a sheet listing the steps to be taken (like the list in No. 8—Actions).

4. OPENING   What will you say to establish your objective? How can you state this in terms of how it will benefit your buyer?

I'll start by pointing out that secretarial support systems should be designed to *extend* the effectiveness of those they serve; at present our system is *limiting* our effectiveness and preventing managers from performing efficiently.

5. FACTS/FEATURES   What evidence will you use to support the need, etc? Costs in time and money.

Word processing system makes our best secretaries available to everyone on an "as needed" basis.

Center can receive dictation 24 hours a day via phone, including night lines.

A two-shift (7 A.M.–3 P.M. and 11 A.M.–7 P.M.) arrangement gives greater coverage.

Storage capacity of editing typewriter permits many rewrites; final draft is prepared automatically once changes are "programmed" in.

Form letters can be originally typed and personally addressed.

6. BENEFITS   What benefits do you see as a result? Spell out probable consequences, immediate and long-range. What are benefits for the other person(s)?

Quality secretarial service—the typed word—can be had by everyone, regardless of rank or status.

Figure 16–9 (Cont.)

People in the field, at home, or in hotel can call in and dictate reports, memos, proposals.

Managers who arrive early or stay late can have secretarial service.

Corrections can be made quickly without concern about retyping time.

Our monthly letter to the 22-member Advisory Committee can be personal.

7. HANDLING OBJECTIONS   What objections or questions do you anticipate? List them and indicate for each how you will respond. What benefits can you offer as an alternative?

''We don't need it yet''	I'll show what it's costing to *not* have it. Break this down to cost per day of *not* having it, compared with cost per day of having it.
''Too large a capital expenditure''	We can lease on monthly basis with option to buy.
''Should be done on a company- . . . . wide basis, and not at this level.''	We've got to start somewhere, our department is one of the largest initiators of letters, memos, reports. We need it now. Let us take leadership for the company.

8. ACTIONS   What steps are you proposing for implementation and for followup? Outline in brief your plan, listing timetable and who should do what.

1. Prepare a list of what we want to know from present managers regarding strengths/weaknesses of present system and needs for tomorrow.

2. Identify the top users of secretarial services and interview them.

3. Talk with Mrs. Bennis to see what is/isn't working and to win her support.

4. Carry out a word count analysis for a sample period (e.g., one week).

5. Summarize results of steps 2–4 and determine needs.

6. Talk with 2–3 vendors and get proposals and names of users.

7. Check out several users . . . other firms with installations similar to our needs.

8. Make final presentation (and report) to Mr. Kramer.

9. Purchase and install equipment; train operators and users.

9. CLOSING   How will you close your presentation? Indicate what questions you will ask to increase commitment and arrive at a decision or desired action.

Figure 16–9 (Cont.)

- Ask Mr. Kramer if he wants me to check with him after taking any of the steps above (e.g., Steps 1 and 5).
- Tell him it will take me 4–5 weeks to complete Steps 1–8 (working at the rate of two days per week on the project). Ask if that allows us enough time to get budget approval in next year's budget if a decision to buy it is reached.
- Ask him for his advice on whom I should talk to in Step 2 of the Action steps.

10. SELLING AIDS   What materials (printed, handout, A/V) do you want the other person(s) to have in advance, during, or as follow-up to your presentation?

Advance	During	Follow-up
Send via interoffice mail a copy of the article from *Nation's Business* on word processing as tomorrow's office today. With cover note suggesting he look it over before our meeting.	Sheet outlining problems with the present system (Like my list in No. 1 on planning sheet). Sheet listing my Action steps (No. 8 above). Folder showing schematic flow diagram.	Memo summarizing what we agree to at meeting.

11. TIME AND PLACE   Select the best time and place, and determine the amount of time needed.

When (day, date, time) <u>this Thurs. at 10:00</u>   How long? <u>20 minutes</u>

Where <u>Mr. Kramer's office</u>   Appointment made <u>Yes</u>

12. SELF-PREPARATION   Have you prepared yourself psychologically and physically?

Psychologically
- ✓ Believe in self and idea
- ✓ Know my own motivations
- ? Analyze past successes & failures (none with Mr. Kramer)
- ✓ Eliminate false assumptions
- ✓ Have single-minded goal
- ✓ Empathize

Physically
- ✓ Appearance suitable to audience
- ✓ Organized all materials
- ✓ Have all equipment

- ✓ Have prepared presentation
- ✓ Have practiced presentation

Figure 16–9 (Cont.)

CHECKLIST FOR PRESENTING

OPENING: State the need and your objective in terms of how it will benefit the other person.

PRESENTING: Establish a friendly climate. Present one idea at a time, and present the facts and features as benefits. Use empathy. Maintain rapport and involve the other person with questions, attentive listening and feedback to show understanding.

HANDLING OBJECTIONS: Restate benefits and probe to get the real reason for an objection. Clarify your buyer's concerns. Refer to past experiences or successes, when appropriate.

ACTIONS: Present a plan for implementation and follow-up, spelling out who will do what, and when. Involve the other person(s) in some of the action planning.

CLOSING: Ask appropriate questions to get commitment or desired action. Restate the advantages and disadvantages. Seek the other person's opinion if you need it. Compromise if necessary to get action that is beneficial to both parties.

Like rating sheets, planning sheets might best be introduced during the demonstration to acquaint trainees with the form and to illustrate its correct use. If "hands-on" experience is desired, have each trainee complete a planning sheet immediately after the demo sales presentation, showing how the salesperson must have prepared for it.

Planning sheets are also a good way of helping new salespersons to prepare for their first real sales after training, and trainees should be encouraged to take a few blank ones with them as a bridge to help them transfer their new learning from class to the job.

CONDUCTING THE ROLE PLAY

In most role plays, one person is in the "critical" role (the role we are training everyone to fill), while the other person is in the "foil" role (the person with whom our trainees must interact in their jobs). Thus, in sales training, the critical roles are those of the salesperson, bank teller, retail associate, and so on. The foil roles are those of the customer, prospect, and so on.

Tips on Selecting Role Players

When you are using role play for the first time, or if you are using participants for the demonstration, you should pay special attention to selecting the participants. They must meet your two objectives: to model the proper application of knowledge and skills and to show others that role play is safe, beneficial, and even fun.

Select players who are generally well liked by the others, who are well prepared to do a better than average job, who are not going to be embarrassed by the experience, and who are willing to go first. In short, you are looking for confidence and competence.

Allow Time for Players to Become Comfortable

Give each player only the copy of his or her role. The remaining trainees are observers and should be given both roles and the rating sheet, if you are using one. After they've had time to read their roles, check with each player. See if they have any questions. Find out if they are comfortable in the role, or if there are any personal reasons why they would rather not enact this particular role play. Ask a few questions to make sure they understand the situation (e.g., "Why do you think X wants to see you? What do you think will happen? How do you plan to deal with it?").

Then set the stage for the role play to begin. You might want a table in the front of the room, indicating that it is a sales counter, or X's desk, or the teller's window. Make sure the necessary materials are on hand and in the appropriate person's hands—sales literature, catalog, merchandise, price list, and so on. Have the trainees take their places and begin the role play.

Don't Interrupt Too Soon

The role plays used in training classes typically last for 5 to 10 minutes. Do not interrupt unless the enactment should be aborted (e.g., it's becoming a farce, the assignment is unclear, a player is losing emotional control). Many role plays will get off to a poor start, but the person in the critical role is able to turn it around and end successfully. Give them the chance to rescue themselves.

Help Players End If Necessary

Sometimes the players reach an impasse but don't know how to end the interaction. After going around the merry-go-round several times, you might interrupt with, "Well, I think that we've got a pretty good sample of many techniques. Even though we didn't make the sale (resolve the problem, etc.), we still have a lot of meat for discussion, so let's get started."

LEADING THE POSTENACTMENT DISCUSSION

Some instructors have a tendency in leading the postenactment discussion of putting themselves in the position of the one with all the answers. When anyone asks a question, they always have an answer. When another participant answers, they tend to say, "that's correct" or "not quite . . . let me explain." This kind of behavior will make trainees more hesitant to express their opinions or enter into the discussion.

Ask Open-Ended Questions

A more effective technique is to get the group to come to conclusions, to decide what was appropriate, to resolve differences among themselves. During the acquisition and demonstration stages of our learning model, the instructor's role is that of expert—the one with the answers. However, once the imparting is over and the practicing begins, the instructor's role during application is that of a catalyst—the one with the questions that lead participants to evaluate their own behavior.

Your most useful tool for leading discussion is the open-ended, nondirective type of question. Here are some examples: "How did you all feel about the way X handled . . . ?" "What was the customer's reaction to . . . ?" "Did any of you see other alternatives to . . . ?" "Why do you think the customer reacted that way?" And so on.

Encourage Self-evaluation from the Participants

Before opening the discussion, give your participants a chance to complete the rating sheets or to study the list of criteria that you placed on the flipchart. You want each trainee to evaluate the behavior of the person in the critical role without being influenced by the perceptions and opinions of others. As soon as the rating sheets have been filled out, begin the discussion.

The first person to be called on should be the trainee who filled the critical role: "How do you think you did? What had you hoped to accomplish? Where did you feel most comfortable? Where were you least comfortable?" And so on. By having the chance for self-appraisal before the others comment, the trainee is likely to be more objective and less defensive. Also, insight is usually a more powerful (and certainly more palatable) form of learning than is criticism.

Hold a Group Discussion

A soon as the self-evaluation is over, open the discussion to the group. In the interest of time, you probably should not discuss every one of the criteria on the rating sheet—only those where the behavior was especially good or poor.

Where behavior was not appropriate, ask the person in the critical role (or someone else, if you sense potential embarrassment) to replay a portion of the enactment, interacting with the foil or with you in the foil role.

End the discussion by collecting the rating sheets. Some instructors give them directly to the person who filled the critical role; others keep them as a form of evaluation and feedback. You can accomplish both objectives by taking a minute to spread the sheets across a table so that you can scan them and give the group a quick summary on what was handled well and on what areas need further work. Of course, you also want to see how similar or dissimilar the ratings are from different trainees who all saw the same role play. A wide range indicates that they do not agree on, and therefore do not understand, the criteria listed on the rating sheet. Some clarification as to what constitutes the desired performance is appropriate before doing the next role play. After your brief analysis of the rating sheets, you should turn them over to the trainee who earned the ratings.

The discussion is as valuable a learning experience as the role play itself, if not more so. In general, you should expect to spend about twice as much time on the follow-up discussion as was spent on the role play. If you have a choice between spending time on another role play and completing the discussion of a role play, it is wisest to use the time for discussion. By observing and critiquing the role play, and refining the criteria for good performance, both observers and players get a clearer idea of how to perform their job. They "internalize" the standards. This is especially true when they arrive at standards themselves through discussion.

GROUP VARIATIONS: SIX TECHNIQUES _____

Up to now, our discussion has focused on role playing in which the majority of the class serves as observers. This is appropriate if the role play is being used to demonstrate (stage 2 of our three-stage learning model), since the instructor wants to retain control of the group and lead the discussion with its analysis and critique of the outcome. However, when we reach stage 3 (application), our concern is in giving **all** our learners as much opportunity as class time will allow for them to apply what they have learned. This cannot be done by putting two people on stage. Rather it must be done by giving as much "hands-on" experience to as many participants as possible.

The six techniques described in the paragraphs that follow will help you to meet the objectives that are always present when your learners are ready for stage 3—Application.

Here are those objectives, in brief:

- To give learners enough practice in applying new concepts and skills that will carry over into the real world ("transfer of training" from class to job).

- To develop self-confidence and the feeling that "this isn't so difficult after all; I can do it."
- To give the instructor feedback from as many participants as possible, so that additional (corrective) instruction can be given where needed.
- To enable participants to learn more techniques by observing one another (learning from one's peers).
- To simulate as nearly as possible the conditions back on the job (and minimize the limitations and conventions of the workshop setting).

Technique 1: Baseball

In playing baseball, many batters take their turn against one pitcher. Applied to role playing, you can have one person play the foil role (preferably an outsider to the group whom you've briefed on how to play the role) and let many participants take their turn at bat against the foil. You can interrupt every 3 to 5 minutes, sending in another participant to carry on where the last person left off. You might want to discuss and analyze each interaction before sending the next person in, or you might want to preserve continuity and wait until the role play has come to its end before discussing **all** participants' performances.

Technique 2: E Pluribus Unum

This Latin phrase that graces the quarter in your pocket means "From many, one." In role playing it is sometimes appropriate to assign many to the same role and ask them to act as one. That is, anyone can speak and pick up the ball at any time. It is useful when using this technique to add another rule: anyone who isn't pleased with the way things are going can call "time" and stop the action. This forces the player in the foil role to leave the group so that the many players in the critical role can analyze how the interaction is going and what they should do different. Then, when they call "time in" the foil player returns. They then tell the foil whether they are starting over, picking up where they left off, or perhaps backing up to an earlier point.

Technique 3: The Eternal Triangle

In some role plays, the person in the critical role is competing with other parties for business (e.g., a salesperson bidding against competitors, a bank officer of one bank determining the rates and conditions of a loan to a customer who is also "shopping" two other banks, officials of one union trying to organize employees who are being courted by another union, etc.). In these "three-way" (triangular) relationships, you might find it useful to create a role play that has two parties filling the critical role and competing against each other to win over the foil role. An example of this technique is found in the Price–Bender role play shown in Figure 16–10.

Figure 16–10 The Price–Bender role play

INSTRUCTOR GUIDELINES FOR: Price–Bender Role Play

TIME: 60 Minutes

OBJECTIVE: At the end of the cassette and workbook lesson, participants are told to read the enclosed role play and prepare to enact it at the workshop. Only half the group will enact the role they've studied; the other half will prepare to enact the role of George Bender, the customer. Several rounds of negotiation will enable everyone to practice their skills, applying the different techniques taught in the preworkshop materials.

PROCEDURE: FIVE MINUTES

1. Begin by breaking the participants into four-person groups (and three-person groups as needed if your total number is not evenly divisible by four).

2. Explain that "Although most role plays involve two parties, this one will require three: a buyer (George Bender) and two competing salespersons (each named "D. W." Price), who will take turns presenting their "best deal" in sequence." Let each group decide who the two sales representatives will be. Have one of them go to the left side of the room, and the other to the right. Tell them to take their price schedules with them, and encourage them to confer with the others who will be playing the same role, though with different George Benders.

3. Announce that the participants who are seated will play the role of George Bender. In four-person groups, two people will play the role, and either can speak at any time. In three-person groups, of course, there is only one George Bender. Distribute the George Bender role play to those who are seated.

4. Explain that "Bender wants the best furniture he can get at the lowest price. And each sales representative wants the order. We'll find out after all groups have finished the role play which party came out ahead . . . the left side of the room, the right side, or the Mr. Benders."

FIVE MINUTES

5. Explain that "Everyone will have five minutes to compare notes and get ready for the first call. Then the "D. W." Prices on the left side of the room will present, with those on the right going out of the room. You'll have five minutes for your sales call. Then you must leave so that the others will have their five-minute turn. You will continue to take five-minute turns until a contract is reached between one of the representatives and George

Figure 16–10 (Cont.)

Bender. The sales representatives who are outside the room may confer. Okay, take five-minutes to get ready."

6. Address this next comment to the seated George Benders. Tell them that "It might be a good idea for you to look through your workbook and review the table that summarizes the different negotiation strategies. Then confer with one another and see what ideas you can pick up to help you achieve your objective."

FORTY MINUTES

7. After five minutes, announce that it's time to begin the role play. Tell the representatives that they can take the price schedule with them to use as a sales aid. Tell the George Benders not to refer to their own copy of the price schedule during the negotiations. Explain that you will announce the time at the end of every five-minute negotiating session. Ask the left side of the room to go to Mr. Bender's office, introduce themselves, and begin the role play.

8. During each round of negotiations, announce "Four minutes are up" and then "Five minutes. End of this round. Time to switch sales representatives." You will also want to make sure that the waiting representatives are not listening in on their competitor's negotiations. And you'll want to see how the negotiations are progressing in each group so that you can give them the Questions sheet the moment they reach agreement on a price and package.

9. As each group finishes, invite the "losing" sales representative to come back into the group. Give each member of the group a copy of the Questions sheet, and tell the participants to spend their time discussing the issues raised.

10. If any groups are still negotiating after 30 minutes, tell them to wind up the role play in another five minutes or so. Then give them five minutes to evaluate their success (or lack of it) in light of the issues on the Questions sheet.

TEN MINUTES

11. Now get the attention of the entire group. Find out which sales representative got the order in each group, and why. See if there were any groups in which the negotiations failed and there was no sale. Call on some of the George Benders to explain their strategy.

12. Conclude the exercise by pointing out that a high win-win is not always possible. In this case George Bender did not want to spend more than $2,500,

Figure 16–10 (Cont.)

> though the furniture he required would cost about $1,000 more than that. Either he would have to settle for less furniture (or cheaper furniture), or the sales representative would have to give up some of the 40 percent commission. To the degree that both took place, there was compromise and a partial win-win. But to the degree that the "give-and-take" was not bilateral, to that end the session resulted in a win-lose outcome which can have adverse effects in the future.
>
> **SAMPLE ROLE PLAY**
>
> **Role of "D. W." Price, Manufacturer's Rep.**
> **Place: Mr. Bender's office**
>
> You are employed by a major supplier of office equipment, furniture, and printed forms. Your firm represents a number of manufacturers who ship directly to your customers. You are on salary plus commission, with a low salary and high (40 percent) commission.
>
> Yesterday your boss gave you a "hot lead"—a Mr. George Bender is one of the three partners in a newly established law firm that will open their offices next month. You phoned him yesterday and arranged to meet today to determine his needs. However, you did establish by phone that there are three offices (the three lawyers, two secretaries) and a conference room to be furnished. Mr. Bender also mentioned that he has invited one other firm to quote . . . one that is new in town and has not built up the reputation your firm enjoys.
>
> Your firm's policy is to allow a 12 percent discount on all orders over $3000, and an additional 3 percent for payment in full advance (i.e., upon signing of purchase order). These discounts are taken on the gross selling price. There are also individual discounts on some items (e.g., chairs, as shown on your price schedule). Beyond these discounts, any further reduction of price comes out of the 40 percent commission you receive.
>
> Although you and the other reps do take a reduced commission from time to time in order to make a major sale or land an account that will lead to future business, you try to keep this to a minimum, since it is your bread and butter.
>
> Now it is time to prepare for your meeting with Mr. Bender. Study your price schedule and come up with a presentation that will give Mr. Bender a feel for the "ball-park cost" of furnishing their new offices. You don't want a specific proposal, of course. That would be premature, since you don't know the details of his requirements yet. But you do want to be prepared and to show him that you've done your homework.

Figure 16–10 (Cont.)

PRICE SCHEDULE—OFFICE FURNITURE LINE

	RETAIL PRICE
DESKS:	

Executive Desk . . . five drawers, two with locks, Formica top in simulated wood, top measures 24″ by 66″. Delivery time: 3 weeks from Nashville $160.00

Standard Desk . . . same design as above, but without locks and with top that measures 24″ by 60″. Delivery time: 2 weeks from Chicago 135.00

Pedestal Desk . . . two drawer (one file, one 5″), with Formica top that measures 30″ by 48″. Typing height; can attach to secretary's desk or stand alone. Delivery time: 3 weeks from Nashville. Specify drawers on left or right side 105.00

CHAIRS:

Executive Swivel with Molded Arms . . . black plastic arms sculptured on tubular chrome frame, four ball casters, adjustable two ways, foam padding, one piece vinyl covering in four colors. From Mobile, 2 weeks 185.00

Side Chair with Arms . . . same as above but with floor protective glides rather than casters. Deduct 10% on four or more. From Mobile, 2 weeks 155.00

Value Contour . . . one piece molded construction with foam padding and vinyl upholstery on arms, back, and seat. Four casters, adjustable two ways, two colors. From Detroit, 3–4 weeks 125.00

Value Contour Side Chair . . . same as above, but with protective floor guides rather than casters. Deduct 15% on six or more. Detroit, 3–4 weeks 100.00

Typing Chair . . . two piece (seat and back) in vinyl-covered foam, with three-way adjustment. Four colors. From Mobile, 2 weeks 85.00

CONFERENCE TABLES:

Senior Table . . . tapered (surf-shaped) with Cordova woodgrain laminated top 48″ by 96″, seats 8–10. Chrome pedestal at each end is inset so as not to interfere with legroom. Two weeks from Chicago 320.00

Junior Table . . . same as above, with top 36″ by 72″, seats 6–8 . . . . 240.00

Mini Table . . . same as above, but 42″ round, seats 4 . . . . 120.00

Figure 16–10 (Cont.)

BOOKCASES:

Four-Shelf Model . . . one fixed, two adjustable, of laminated particle board with look of walnut. Measures 60"h by 36"w by 12"d. From Chicago, 2 wks 95.00

Three-Shelf Model . . . one fixed, one adjustable. Same as above, but with 45"h 75.00

Two-Shelf Model . . . center shelf fixed. Same as above, but with 30"h 55.00

SAMPLE ROLE PLAY

Role of George Bender
Place: Your office

You are a recent graduate of law school. Two of your fellow graduates and you are in the process of opening a law practice and have acquired an office on the east side of town. You have scheduled an open house for five weeks hence.

Yesterday you called two local firms that handle office furnishings, and have arranged for their sales representatives to call on you today. You happen to know that office furniture has a high mark-up, and sales representatives typically receive 35 to 50 percent commission, so you plan to negotiate for the best price possible. Your dad was in the furniture business before he retired, so your two partners have left the purchase of furniture to you. But you did agree that the new furniture should not exceed $2,500. This is a tight budget, but you plan to play one supplier against the other to get the price down to under $2,500.

You and your two partners will each have a private office and will need a desk, a desk chair, and two or three "side chairs." Each of you should also have a bookcase (although you are prepared to drop this later in your negotiation if it will help you to stay within your budget).

You also have two secretaries and need for each a desk, a typing desk (at right angles to the main desk), and a typing chair. Finally, you have a conference room and will need a table and chairs for eight persons.

You plan to see each representative individually and get their best price. Then you'll call each back to tell them the better features of their competitor's bids (so as to get their price down). It may take two or three visits from each, but you have an obligation to your two partners. You plan to apply your best skill in negotiating so as to get the offices furnished for under $2,500.

Figure 16–10 (Cont.)

EVALUATION SHEET

1. What behavior (strategy, deal, etc.) did the winning D. W. Price offer that led you to select him or her?

2. Was there anything that either salesperson did that caused them to ''lose ground'' or sour the negotiation?

3. At what point (which round) in the negotiation did Bender decide who the winner was? How many rounds took place thereafter? With what additional gain?

4. What strategies were employed? What concessions on each party's part (e.g., Bender's dropping of bookcases, agreeing to use some of the office chairs in the conference room, Price's willingness to reduce his commission)?

5. What final package and price did Bender select? That is, what furniture and price is in the final contract?

6. Did any other Bender do better?

Technique 4: Round Robin

To give everyone an equal chance to perform, you might create three equivalent role plays and select three players to serve as critical role, foil role, and observer. After going through the first role play, you then rotate the players (whoever played the critical role now serves in the foil role, the foil player now becomes observer, and the observer now fills the critical role). By going "round robin" over the three role plays, each person will have served in all three capacities.

Technique 5: Multiple Roles

When you are teaching skills and concepts that can best be practiced in a group setting (such as group dynamics or conference leadership), you may want to create a role play that has a number of different roles to be filled—usually four or five. You can thus create "hidden agendas" and differences in perception among the players, and greatly increase the opportunities for interaction. An example familiar to many trainers is the "survival" exercise in which the trainees are put in the position of being marooned (on the desert, at sea, in the Arctic, etc.) and are faced with the task of agreeing on which items on their list of supplies are important to their survival. The exercise teaches group decision making and related skills.

Technique 6: Syndicates

By breaking the class into subgroups, or syndicates, of three to five persons, you can have everyone enacting the same role play simultaneously, thus giving all participants "hands-on" experience. This technique can be used in conjunction with any of the other techniques just discussed. It frees the instructor to circulate during the enactments, listening in on each briefly and collecting examples of good and poor behavior. The instructor thus has more material to draw from in leading the postenactment discussion and helping the full group to process and interpret their behavior.

SUMMARY

Not too many years ago the words "role playing" bore something of a mystical aura. The idea of having people act out how they would behave in different situations was thought to be too personal, too clinical for the classroom. Such methods had their place, perhaps, in the office of a psychiatrist or marriage counselor. But certainly not in training, where there is danger of a person being embarrassed or held up to ridicule in front of fellow workers!

Today there are very few companies that have not made use of role playing to train their employees at all levels. We have discovered that role playing is a most effective way to learn and that virtually every trainee can benefit significantly from it, both by watching and by taking his or her turn as player. Textbooks on educational psychology devote much space to the concept of "transfer of training." Role playing is, without a doubt, one of the most effective teaching techniques for helping trainees to *transfer* their training by giving them a chance to exercise new skills and develop behavior patterns (habits) that they will continue to practice back on the job.

REFERENCES

Material from the following publications has been used in this chapter.

Baker, Richard. *Skills for Negotiation and Conflict Management.* Princeton, NJ: Training House, 1978.

Gallitano, Kathryn. *Making Presentations and Selling Ideas.* Princeton, NJ: Training House, 1980.

Jones, J. W., and J. W. Pfeiffer, "Role Playing." In *The 1979 Annual Handbook for Group Practitioners,* J. E. Jones and J. W. Pfeiffer, eds., pp. 182–193. San Diego, CA: University Associates, 1979.

Parry, Scott B. *Using Role Playing for Group Instruction.* Princeton, NJ: Training House, 1980.

Shaw, M. E., R. J. Corsini, R. R. Blake, and J. S. Mouton. *Role Playing: A Practical Manual for Group Facilitators.* San Diego, CA: University Associates, 1980.

chapter seventeen —————————————

THE CASE METHOD: TEACHING SALES MANAGERS TO MAKE BETTER DECISIONS

————————————— *THOMAS J. NEWMAN*

Thomas J. Newman is director of training, Johnson Wax, Racine, Wisconsin. He plans and conducts training programs for managers and supervisors both in the United States and in many of the 45 overseas subsidiaries of Johnson Wax. He has been handling training assignments at Johnson Wax for nearly 20 years. Earlier he established a training division in an international temporary help firm and spent eight years as the training director of a large insurance company. Newman has a B.S. degree from the University of Wisconsin and an M.Ed. from Marquette University. In 1976, he was named the international trainer of the year by the American Society for Training and Development.

The goal of this chapter is to discuss sales training applications of the case method. The chapter will describe what it is, when and how to use it, how to develop cases, and ways to measure its effectiveness and identify sources for case method materials and information.

The case method of teaching businesspeople how to handle typical problems was developed by faculty members of the Harvard Graduate School of Business Administration after World War I. The faculty recognized that the case method was a break from traditional educational practice. It is speculated that for this reason the system did not have immediate acceptance. At the time, many professors at Harvard and other universities looked at their profession as one of imparting knowledge to students: a one-way form of teaching. Thus, it was criticized by professors who felt strongly about their traditional mission.

Because professors (and trainers) are known for their love of talking, the case method—which turns most of the talking over to participants—ran into heavy resistance when it was introduced. Arguments ensued between the "case" and "noncase" instructors. (There is on record at least one public fist fight between two professors arguing the merits and demerits of the case method.)[1]

We now see the case method widely used as a productive method of teaching sales and business managers how to make better decisions. Two parts of the case method have had a particular impact on sales and management training. First, the case method arouses the interest of the learner by making him or her an active rather than passive participant. Second there is a heavy emphasis on the decision-making process instead of philosophical or theoretical concepts alone.

C. I. Gragg said succinctly that the "mere act of listening to wise statements

---

[1] Dwight R. Ladd, *Notes on Use of the Case Method* (Boston: Intercollegiate Case Clearing House, Soldier Field, n.d.), p. 7.

and sound advice does little for anyone—if the learning process is to be effective, something dynamic must take place in the learner."[2]

THE CASE METHOD: A SYSTEM OF INSTRUCTION BASED ON "GUIDED DISCOVERY"

What is a case and what is the case method?

A *case* is or should be based on a real business situation which includes not only the facts but generally some additional information on the company and possibly the industry with sufficient data to reflect the decision to be made and the setting in which the decision maker finds himself.[3]

The *case method* is a system of instruction built on the premise that people are most likely to retain and use what they learn if they reach understanding through "guided discovery." It sets up a problem situation for a group of participants to solve. Along the way, trainees refine analysis skills, develop the will to take risks in the face of uncertain outcomes, and come to know their own strengths, weaknesses, and talents.

Objectives for using the case method in business training sessions vary. Some of the most often stated objectives are

- To improve cold calling skills.
- To improve customer problem-solving to enhance skills.
- To develop negotiation skills.
- To increase sales through effective territory management.

Advantages

The advantages of using the case method to teach sales management are many. It

- Provides experimental learning in real-life situations with limited time and resources.

---

[2] Charles I. Gragg, *Because Wisdom Can't Be Told* (Boston: Intercollegiate Case Clearing House, Soldiers Field, n.d.), p. 4.

[3] J. Hatcher, M. R. Vaghefi, and W. J. Arthur, *The Case Method—Its Philosophy and Educational Concept* (Boston: Intercollegiate Case Clearing House, Soldiers Field, n.d.), p. 2.

- Develops a "skill" through doing, not just fantasizing about an unreal situation.
- Illustrates that the training experience applies to the real world, and thus is more credible.
- Creates a sounder understanding of human behavior through insights and people's feelings and beliefs.
- Develops a procedure for analyzing problems that can be transferred back to the "real world."

Drawbacks

However, as with any teaching technique, the case method has its critics. For example, it

- Encourages too many conversational side trips of little value.
- Encourages superficial thought rather than developing analytical decision-making skills.
- Does not provide participants with an opportunity to continue working through their decisions; thus a tendency exists to make unrealistic recommendations.
- Causes participants to experience frustration because there are no "right" answers, resulting in a reduced level of enthusiasm for cases.

In expressing his feelings about cases one training consultant says, "Cases are antidotes for the tendency to avoid real issues by talking about theory rather than about its application."[4]

DESIGNING YOUR OWN CASES WITH IN-HOUSE RESOURCES

Finally, and perhaps most important, are your own in-house resources for real cases in your business setting. At Johnson Wax, to create real-life settings in our district sales manager week-long seminars, we require that each manager submit three actual case situations on paper to the instructor *before* the seminar is conducted. These cases do not have to be 30 pages long; 2 or 3 is the average. Each must spell out a performance issue with which the district sales manager

---

[4] Dugan Laird, *Approaches to Training and Development* (Reading, MA: Addison-Wesley, 1978), pp. 144–145.

is currently working or has handled recently. These cases are edited, retyped, and used during the "Handling Sales Performance Problems" portion of the seminar, a day-long event.

Armour-Dial's Method: Use Your Managers' Problems When Planning a Program

Armour-Dial's Ray Higgins, director of sales training and development, comments in a National Society of Sales Training Executives editorial (1977) that he learned early in his career the value of developing a workshop based on the problems of the managers themselves. In a course titled "Getting Work Done Through People," a five-day course, Higgins required, as the price of admission, that each participant submit one written case problem he or she was currently (or recently) experiencing in getting the work done through people.

These cases were edited, reproduced, and placed in each participant's notebook before the workshop. Then the workshop program was built around the problems submitted. Higgins has followed this procedure—of starting with manager's problems, then planning the program—ever since, and with *all kinds* of manager groups. He says editing actual case problems gives the richest kind of education to the trainer (he now has 2,500 participant cases) and turns around the whole process of program development.

The five-day course at Armour-Dial is a success! Higgins recommends that you "Try it, you'll like it" (and so will the trainees)![5]

DDI Cases: Interaction Management

Development Dimensions, International's well-known behaviorally based supervisory course titled "Interaction Management" relies on short prepared cases to get across the use of the principal steps of supervision. But soon into the course, participants write a "Describe Your Own" situation (DYOs), based on a real situation they're currently facing on the job. These DYOs then form the basis of the practice when participants try out the critical steps of a skill. For example, in the module "Improving Work Habits," the six critical steps are

1. Describe in detail the poor work habit you have observed.
2. Indicate why it concerns you.
3. Ask for reasons and listen openly to the explanation.
4. Indicate that the situation must be changed and ask for ideas for solving the problem.

---

[5] Ray Higgins, *Using the Participant-Prepared Case Method to Teach Management*, Unpublished editorial for the Naional Society of Sales Training Executives, Phoenix, 1977, pp. 1–14.

5. Discuss each idea and offer your help.

6. Agree on specific action to be taken and set a specific follow-up date.

This realistic skill practice on actual work situations provides outstanding results. DDI's research into the usage and impact of these modules is dramatic in its success.[6]

The "Jamaica Johnson" Case

Johnson Wax recently introduced Block-Petrella's workshop on "Staff Consulting Skills." This workshop is aimed at staff people who need to get things done through others over whom they have no authority or control. It is a four-, five-, or six-day workshop, depending on the needs of the participants. Over 60 percent of the time during the workshop is spent in working through a case situation. Block-Petrella provides excellent cases from their experiences as starters for the workshop.

It quickly became apparent that the non-Johnson–related cases were adequate, but left something to be desired. Johnson Wax now has two outstanding cases based on sales and technical issues that form the base for the workshop. These cases are very well received, practiced, and played out during the workshop. The following is a fictional case based on a real situation.

EXAMPLE

Introduction

Jamaica Johnson (not a real company) is a medium-sized company, with sales annually of about $20 million. Recent conplaints, turnover, and an opinion survey show that morale is at an all-time low in the Sales Division.

You are a member of an internal sales development staff group (internal consultants) responsible for analyzing problems and recommending changes to improve sales performance. Your boss has asked you to investigate the Jamaica Johnson Sales Division problem and to help them to upgrade morale and performance.

The People Involved

Lou, Regional Director. Quiet, but when he sets his mind to a task, moves quickly to resolve issues. Has used staff consultants in the past to tackle tough line assignments alone.

---

[6] William C. Byham and James C. Robinson, "Interaction Management: Supervisory Training That Changes Job Performance," *The Personnel Administrator* (February 1976).

Marty, Jamaica Johnson President. Likes to keep information close to the chest, tends to be reflective, but not communicative. Knows Lou is anxious to resolve morale and other issues in the Sales Division. Feels something will need to be done in Sales, but is reluctant to "rock the boat."

Kim, Jamaica Johnson Sales Director. Abrasive, abrupt, easily set off temperamentally. Becomes defensive when staff consultants want to interfere. Knows sales people are not happy, but feels they have a job to do, and he is not responsible for their "happiness."

Jerry, Jamaica Johnson Sales Representative. Been with the company two years; is rated highly. Likes Sales and is anxious to do a good job. Likes Lou, thinks Marty is "okay." Thinks Kim is good at sales but is a poor supervisor, hard to work for, and is not to be trusted.

Your Task

You have a meeting with Marty to contract for your visit to Jamaica Johnson. At this meeting you will need to find out

- Who is the client?
- Will Lou support your visit to Jamaica Johnson?
- What really are your objectives on this assignment?

The Jamaica Johnson case has served as an excellent base for new sales development appointees and other staff people who have to work through others over whom they have limited authority and control. In fact, now when a trainee is about to fall into an avoidable situation, friends are apt to warn the trainee, "This looks like a potential Jamaica Johnson case."

Three other real cases have been fictionalized and are used in management development workshops. The case method is serving a vital role in the broadening of management skills at Johnson Wax.

USING SPECIALIZED PROFESSIONAL WRITERS TO DEVELOP CASES

Cummins Engine in Columbus, Indiana, has been using the case method in its internal management development program, designed to broaden its managers and executives. Cummins uses outside university specialists to research and design their internal cases. The cases were developed by having the people who were involved in the situations write a narrative of their perception of the situation. Then the university specialists used this background material to develop the actual cases.

The experience at Cummins was so powerful that all the management training programs include at least one case study. Incidentally, these internal cases were taught by the outside university specialists who were well trained in the case teaching techniques, not an easy skill to learn.

At Blue Bell, Inc., only Harvard cases were used in its marketing linkage training seminar. Both Harvard and Wharton business professors conduct these case discussions. Three cases were assigned and read in advance. Then after small-group discussions, the total class of 50 regroup for a longer discussion, led by the lead professor. Afterward, the leader explains what *really happened* in the actual case. This discussion is always rated highly by participants.[7]

One consideration to keep in mind when employing outside case writing specialists is the *cost*. If you truly are searching for a high-quality, challenging, provocative final product, be prepared to pay for this work. Get a clear estimate in advance rather than let the meter run for six or nine months, only to be surprised by a bill that breaks your budget for the next two years.

MAINTAINING QUALITY IN CASE SELECTION _____

As you explore the case method for possible use in your company, don't do all the leg and brain work yourself. Enlist a small committee of sales managers typical of those to whom you will be teaching the case. Ask them to read each potential case and be prepared to evaluate each as a possible teaching tool in your program. In this way, you will not only get invaluable assistance from users, but the method will aid in securing their "buy in" on the cases which "they selected." Finally, if a particular case appeals to one of the committee members, ask him or her to help in facilitating the case during the seminar. This strategy should give you a reasonable amount of quality control over your case selection.

WHEN TO USE THE CASE METHOD _____

The case method can be appropriate for a number of training purposes. Besides the three uses mentioned here, you may have unique needs in your organization which will fit the case method style of training.

First, cases can be used to *improve executive and managerial decision-making skills*. Quality decision making is a key skill in business. To make high-quality decisions, a manager must be able to sort out the facts, figure out what needs to be done, and arrive at a reasonable plan of action. Applying analytical

---

[7] Thomas J. Newman, *Case Studies: To Use or Not to Use*, Unpublished editorial for the National Society of Sales Training Executives, Orlando, Fl, 1980.

thinking effectively is the mark of an astute manager. When the case is based on a true business situation, the executive will be faced with making a decision in a realistic setting as if he or she was sitting in the desk of the main character of the case.

EXAMPLE

Sales executives are often faced with the tremendous challenge of creating a valid data base on which to make decisions. Decision making cases can (1) emphasize the key skills needed to develop and work from an information source (data base), (2) allow the participants to work with raw data, and (3) show participants how circumstances and personalities become important variables in the decision making process.

Second, the case method is an excellent strategy for *improving communication and interaction skills*. One of the most tested of the many "survival" exercises, "Desert Survival"[8] is a masterful training experience in improving both participant communication behavior and interpersonal skills, particularly sensitivity and empathy. It is a group decision-making exercise for increasing individual and team effectiveness. If the participants in a case method experience are broken in small groups to discuss the case preliminarily, these discussions inevitably aid in communication skills enhancement. Interacting with others who have varying points of view on the case provides participants with a perfect opportunity to work on their interaction skills in a safe, nonthreatening environment.

EXAMPLE

Salespeople also find the case method can help them tackle communication and interaction issues frequently faced with customers. Analyzing the various personality types through the cases helps salespeople develop strategies for coping with difficult people.

Finally, cases can assist in *solving complex issues for management teams*. The case method has been used by management groups to discuss and come up with solutions to difficult management problems. Usually these problems are of the type that require heavy employee and management input. For example,

---

[8] J. Clayton Lafferty, Patrick M. Eady, and John M. Elmers, *The Desert Survival Problem* (Plymouth, MI: Experimental Learning Methods, 1974).

- Should the company discontinue selling products in a particular territory due to low demand and high delivery costs.
- Should the company hire additional sales people to handle a new product line.
- Should the company reorganize sales territories based on new sales data.
- To explore the problems frequently encountered by sales people when a new buyer is assigned to an account.
- To develop strategies for establishing and maintaining a good working relationship.

Issues similar to these have been approached and solved through the skilled use of the case method and experienced case facilitators. Because cases, when well researched or written, are from real events, they can be rich resources in sales training or management workshops.

HOW TO WRITE A CASE IN FIVE STEPS _____

"If you don't know where you are going, you'll only end up where you should by a happy accident."

Step 1: State Your Objectives

Start with the *objectives* of the case method sesssion. This is a critical step, as it provides the foundation for the case. Think through the skills you want the participants to develop in the session. Until you define what you want the case to do for you, you will only waste time writing a case.

Step 2: Research the Case Possibilities

This step can be handled through interviews with appropriate managers in the area, reading files for historical data that will add interest to the case, and gathering generic case ideas from other cases written for the subject.

EXAMPLE _____

Your long-term, top salesperson can be a rich source of information on coping with new buyers. To succeed, this person has had to establish a number of new relationships. Failure to do so would have meant a loss of business, so if the account has grown, you have an excellent indication that the salesperson has the skills to develop new relationships.

Step 3: Choose the Most Appropriate Case for Your Needs

To *select the appropriate case* for your objectives, keep in mind that the best cases are related to the participants' industry. The most valuable cases present characters and problems with which the participants are familiar. High levels of interest and learning are achieved through accurate portrayal of the case situation. Listen to people describe their bosses, peers, and employees. We are surrounded with good case characters.

EXAMPLE _____

Development of the characteristics of new buyers is key. Good salespeople often educate new buyers and are able to size them up quickly. Most salespeople will tell you that new buyers have "universal" characteristics that can be easily incorporated in a case text.

Step 4: Strive to Capture the Learners' Attention

To *capture learners' attention*, describe the situation, the attitudes, and the characters, but don't bury the reader in too much detail. Assemble a disguised version of the situation by changing names, titles, places, and times. Work hard to establish a believable scenario in the case. Don't try to create imaginary facts or incidents in the case. If you do, and the participants want to know more about the case, you may lose credibility and interest if you cannot come up with accurate answers to their questions.

Case length and depth is purely a judgment call on your part. We have found that cases with over 15 pages can be discouraging to participants to read unless the case is a perfect fit for them. Unless it is a financial training seminar, too many charts and graphs can turn off readers, too. Occasionally, a 2- or 3-page case can contain sufficient elements to cover all the principles you are trying to teach.

EXAMPLE _____

A salesperson would be apt to share the following buyer as one of the most difficult. Don Williams has recently become a buyer for PT Manufacturing. He has a drafting background, worked as a supervisor on the manufacturing floor, and is attending college classes in the evening. He's come up through the ranks. He is replacing Hank Strunk who had been in the job for years. Don's perception is that Hank dealt with people he liked rather than those who would deliver quality parts. There were many days on the floor when Don hated his company's buyers because he was coping with poor quality parts. He swore when he became a buyer, he'd pursue quality. A sales group might be asked how to establish a good relationship with Don, knowing he's lumped all previous vendors in his "poor quality" category.

Step 5: Customize a Commercial Case

To *customize a commercial case*, such as the type available from the Harvard Intercollegiate Case Clearing House, proceed carefully. If participants perceive that the case situation was "too close to our business," they might get too immersed in intricate technical aspects of the setting. However, if you decide to make changes to a commercial case, don't stop with a name change or two, hoping that this will suffice.

Start with the basic lesson the case is designed to bring out. If there is no connection to your learning objectives, toss it out and start over.

When you find a relatively close fit for your goal, skim the case to be certain there are no major elements too overwhelming to make it adaptable to your changes. For example, manufacturing cases are not acceptable in banking or insurance circles, nor is the reverse situation palatable. If yours is a heavily unionized company, a nonunion case would not be a comfortable setting for the participants. And, finally, if the company you're working with is a foreign company, try to find a case that took place in the correct culture. Difficulties arose in our Brazilian subsidiary when we used uncustomized situations. All cases were then customized to the Brazilian environment.

This case in point really illustrates the need when customizing a case to compare the basic elements of the case to your group's learning objectives and then tailor the setting with this matchup in mind. Always take the final draft to three or four typical participants (or their bosses), have them read it, and get their reaction to realism of your version. Customizing cases can save you time, but needs to be handled carefully.

HOW TO USE THE CASE METHOD IN AN IN-HOUSE TRAINING SEMINAR

A master case method instructor teaching an in-house seminar at Johnson Wax was asked how he had learned the technique of teaching cases so expertly. Perhaps with a bit of tongue-in-cheek, he said, "While I was at Harvard, I made up my mind that I wanted to teach using the case method. So I taught the case method at Harvard for nine years, learning along the way, and now that I've been at Stanford for three years, still teaching cases, I'm getting more and more comfortable with the technique."

This professor's comments should not deter you from trying out and learning to instruct using the case method. He was referring mainly to a university setting, where a typical semester runs for 36 meetings of $1\frac{1}{2}$ hours each. Thus the development of instructor/student rapport and understanding builds over an extended period of time. And the instructor has an advantage in watching the students mature in the quality of handling increasingly complex cases during the semester.

However, in a sales management or business seminar, the time span is necessarily reduced, and the instructor must condense the learning time through limiting the amount of case details, setting rather rigid time limits, and clearly describing the end result desired (a recommendation, decision or action plan), and perhaps even providing a list of questions for the participants to answer on their way to the final product.[9]

When instructing using the case method, the teacher has a lower profile than in the standard seminar setting. This may mean some adjusting on the part of the instructor. Often a new case instructor finds that he or she must work as hard as the participants at using thinking skills necessary for effective business leadership.

What to Do When the Discussion Strays

Occasionally, the discussion may head down a path which the instructor feels is not in line with his or her objectives for the case. When this occurs, the best tactic is to listen to what the participants are saying. It could well be that a strategy or thought is developing that the instructor had not visualized in the case.

Watching a skilled instructor during a case discussion, you may get the impression that he or she is making very little contribution to the seminar. A new instructor may be surprised to learn that preparing for a case class is more rigorous than preparing for a lecture class. As a lecturer, the instructor determines the material to be delivered and the sequence of the delivery.

As a case instructor, he or she has given up the autocratic style permissible in a lecture class. And yet the instructor realizes that it is critical to maintain control over the class discussion.

How to Prepare to Teach a Case

Thus, preparing to teach a case requires more than a brief skim through the case with an eye toward a few questions to begin the discussion. Instead, the instructor needs to become thoroughly acquainted with the case to handle any oblique comments, questions, or solutions which the participants may introduce.

---

[9] Laird, *Approaches to Training and Development*, p. 145.

The *first step* in instructor preparation is to master the facts. It is critical to make an outline and marginal notes and to recheck any calculations which participants will be looking at critically. Knowledge of all the issues is important, both real and imagined.

The *next step* is to consider the specific learning objectives toward which the case is pointed. Some issues may be more heavily emphasized than others. *Next*, probing, thought-starting questions need to be assembled for the case discussion. How these questions are worded requires considerable thought.

To introduce a case discussion to a seminar, assume that at least some of the participants have not had case discussion experience. Thus a brief description of the Case Method process is to explain how it presents a problem to be solved by a group of managers. Explain the participants' role, the instructor's role, and the ground rules of the discussion. Make clear to the group that the main purpose of the discussion is not to have all agree, but rather to help them to think through their version of the case. In effect, you expect them to refine their thinking.

Five Strategies for Controlling a Case Discussion

To direct and control the participants' discussion, the instructor must do many things:

1. *Establish a quality standard.* This can be done by insisting on the use of case facts, by requesting supporting analysis, and by asking questions with obvious answers.

2. *Keep the discussions orderly.* By nailing down each point before allowing the discussion to move on, the instructor can assure that a point is agreed to, changed, or argued by the group.

3. *Ask relevant and pivotal questions.* This forces the participants to think deeper on an issue or can move the direction of the discussion to more productive subjects.

4. *Manage participant input.* Through calling on quieter people, or bringing in an "expert" opinion from someone's relevant past experience or utilizing the particularly adept participant in complex problems, the instructor can include everyone in the "action" of the case.

5. *Summarize frequently.* This tactic can add significantly to the learning process. A summary is especially helpful at the end of a discussion or when the opportunity arises to explore other facets of the case.

By utilizing these five strategies, a case method instructor will productively lead the group toward the learning objectives of the case.

The Participant's Role

One important item frequently missed in case method discussions is the role of the participant. The participants must be prepared to do the talking, thinking, and the decision making. Thus they will need to

1. Read the case thoroughly.
2. Understand major ideas and make notes on these ideas.
3. Stand up for their logic and beliefs.
4. Participate in the discussions.
5. Recognize the rights of others in the group.[10]

Use Humor to Get Your Point Across

Humor can be used effectively to generate reference points during a case discussion, for example, by pointing out specific personalities in a joke or dramatic situations in a story that can be referred to later on in the discussion.

In a performance appraisal seminar which uses cases in Johnson Wax, I have frequently used the story of the husband whose wife was complaining that he didn't tell her often enough that he loved her. The husband said, "I told you when I married you 20 years ago that I loved you, and if anything changes, I will let you know!" I refer back to the anecdote frequently while emphasizing the need for managers to tell their employees how they feel about their performance—a critical skill in effective performance appraisals.

Also, humor can be used to diffuse potential problem situations and to evaluate discreetly the quality of individual participant contributions.

Pacing the Discussion to Maintain Interest

The pacing of a case discussion often is not varied enough by the instructor to maintain a high level of interest. Wheelwright and Sasser[11] suggest a "short drill" as a solution to this problem. After taking 10 to 15 minutes to identify the major issues, the instructor might move the class toward shorter comments. This method reduces comments to 30 to 60 seconds each, allowing more individuals to talk, gets them committed to the topic, and can further develop the concept

---

[10] Merwin M. Hargrove, *The Case Approach* (Boston: Intercollegiate Case Clearing House, Soldiers Field, n.d.), pp. 6–7.

[11] W. Earl Sasser and Steven C. Wheelwright, *Ideas on Planning Teaching Tactics for Case Courses*, Unpublished paper, Stanford University Graduate School of Business, Stanford, CA, 1980, p. 9.

or decision of the case. The pace can be slowed as new topics and issues are introduced. This strategy gives a larger number of participants a chance to introduce their opinions and can help to maintain attention and high interest in the case.

Option for Concluding a Case Method Discussion

At the conclusion of a case method discussion, the instructor has several options. One, he or she can risk asking one or more of the high-quality contributors to summarize the conclusions the groups have reached. Two, the instructor can "stop the action" for a minute, to think through the groups' conclusion for the summary statement. Or, finally, the instructor can just stop completely, after all the comments have ended, and say "Thank you!" This third method holds the most risk, because inevitably the participant will want to know "What really happened in the actual case?"

Method 2 is recommended, "stopping the action" for a brief timeout, making a few summary notes, then giving the group your summary statement, with the update of what really happened in the case (assuming this information is available; if not, explain before the case discussion begins that the conclusion is not available). However, one writer suggests that "what actually happened, or what any one person thought ought to be done is of no great significance; what is significant is that *you* know what *you* would do in a specific situation."[12]

Finally, list the objectives achieved during the session as your conclusion to the case.

FOUR METHODS OF MEASURING THE IMPACT OF THE CASE METHOD

One case method consultant[13] suggests that "the Case Method often fails because the trainer is waiting for the trainees to solve the case, and the trainer doesn't know where the sticky issues are and how to guide people through the hard ones."

In a university setting, the impact of the Case Method can be evaluated by the professor (who grades the students' performance) or by the students (who rate the professor's teaching ability). However, in a business or sales training seminar, the impact of the case method must necessarily be measured in ways other than college grades.

---

[12] John S. Hammond, *Learning by the Case Method* (Boston: Intercollegiate Case Clearing House, Soldiers Field, n.d.), p. 3.

[13] Helen Kelly, *Get Results with the Case Method* (Alexandria, VA: Info-line, ASTD, March 1987), p. 10.

Figure 17–1 Sample case method seminar evaluation form

Date _____

1. My overall satisfac-   1   2   3   4   5   6   7   8   9
   tion with the day is: Very Low   Very High

2. As far as my job is   1   2   3   4   5   6   7   8   9
   concerned, the   Irrelevant   Highly
   skills I learned to-   Relevant
   day were:

3. The pace today   Much Too   Too   Just Too   Much
   was: Slow   Slow Right Fast   Too Fast

4. The extent to which   1   2   3   4   5   6   7   8   9
   my interest level was   Very Low   Very High
   maintained during
   the day was:

5. Was any part of the
   day confusing? If
   yes, please give de-
   tails.

6. My satisfaction   1   2   3   4   5   6   7   8   9
   with the trainer   Very Low   Very High
   was:

7. The trainer could
   improve the learn-
   ing experience by:

8. The extent to which   1   2   3   4   5   6   7   8   9
   my expectations   Not At All   Completely
   were met was:

Permission granted by Johnson Wax.

Several methods of seminar evaluation are possible. First, participants' *opinions* (did they like the method?) can be easily solicited immediately after the seminar concludes. (See Figure 17–1.) Second, the participants' *gain in knowledge* (did they learn from the seminar?) can be evaluated through final questionnaires which measure their gain in knowledge as a result of the learning experience. Third, the transfer of knowledge to the *use on their jobs* (will they use it?) can be tested through observation on the job (salespeople call it "work with" days), boss interviews, self-questionnaires, or subordinate interviews—risky, but extremely helpful. And finally, the bottom-line measure of "Did it favorably affect the company's sales?" can be measured through increased sales by the participant, with an accompanying favorable effect on the company year-end results.

Each of these four methods of evaluation has strengths and limitations. As you progress from like, to learn, to use, to bottom line, you'll see that it takes more time, money, and effort to collect each succeeding measure, but also you'll gain more valuable data as you progress from 1 to 4. So the final judgment is up to you to decide how far you want to go to discover the results.

SOURCES OF CASE MATERIAL FOR YOUR TRAINING PROGRAM

If you are intrigued enough with the case method to consider using it in a sales management workshop or seminar, there are a number of fine sources in which you can find excellent cases.

First, begin keeping a file of case studies that cross your desk accidentally in the course of a week. Within a year, you'll have to open a second file drawer just to hold the volume of cases that will accrue. This personal library will serve as a perfect reference as you research possible cases for your courses.

Second, get in touch with Harvard's Intercollegiate Case Clearing House, Soldiers Field, Boston, MA 02163. It is a veritable gold mine of widely varied business cases developed through 50 years of research and use in Harvard Business School classes. Also, the Clearing House provides valuable articles and short papers about the case method. The prices are reasonable, the quality and depth outstanding.

SUMMARY

To succeed in business, we know that a series of pat answers are not going to bring ultimate resolution to complicated problems. Each business situation is a new one, requiring imaginative understanding leading to sound judgment and finally, action. As we ponder the business seminar trainee who has not had the advantage of studying through the case method, an anonymous writer contributed this appropriate limerick:

> "A student of business with tact
> Absorbed many answers he lacked.
> But acquiring a job,
> He said with a sob,
> How does one fit answer to fact?"[14]

In this chapter, we have introduced you to the case method for use in sales or management training. You now know where to find excellent case materials, when to use the case method, and have even had an exposure to how to write cases.

Additionally, the information on conducting a case discussion should prove helpful as you begin to introduce this powerful training technique into your repertoire of skills. Unless you measure the results of your training efforts, you may be wasting valuable time, effort, and resources of both you and your organization.

The case method leads participants into innovative thinking and critical judgments. They face real situations, not philosophical ones. They become active, not silent bystanders. Properly taught, the case method can enhance participant learning tenfold, and ultimately lead to stronger, more dynamic business skills through the active learning process.[15]

---

[14] Gragg, *Because Wisdom Can't Be Told*, p. 6.

[15] Ibid., p. 9.

chapter eighteen ————————————

GAMING AND SIMULATION
IN SALES TRAINING

J. CLAYTON LAFFERTY,
—*WITH TRACY L. RANGE*

As founder and CEO of Human Synergistics, J. Clayton Lafferty, Ph.D., is concerned with helping people work together more effectively. His pioneering simulation, the Desert Survival Situation, demonstrates that when people are supportive of one another and follow a rational sequence in dealing with a problem, they are able to perform beyond the sums of their individual resources. In other words, the whole is greater than the sum of its parts. This is the meaning of "synergy" and the philosophy of Human Synergistics, a consulting firm dedicated to tapping the wealth of resources available to individuals and groups.

Tracy L. Range is a technical writer for Human Synergistics, Inc., a consulting firm which specializes in the publication of management development training materials for corporations and other organizations. Ms. Range is responsible for all writing connected with the development of the company's products, as well as the revision of its established products. In addition, she works with the company's director of development and research in designing new training materials to add to Human Synergistics' product line.

The sales game is largely an interpersonal one defined by two polar opposites: the purchaser and the seller. As sales training professionals, you are faced with the often difficult task of developing fresh, effective approaches to training in this volatile business. This chapter will address how gaming and simulation as used in sales training programs can add a major dimension to the development of a superior sales force.

This chapter begins with some general background information; it then discusses using prepared simulation materials, how to create your own exercises, the future of this valuable learning tool, and, last, where to go for more information.

HOW GAMING AND SIMULATION ARE USED IN EVERYDAY LEARNING EXPERIENCES

Role Play as Children and as Adults

Life itself can easily be considered a game: important, but not always serious. As children, we rehearse for our roles in life by pretending on the playground. We invent scenarios of good guys fighting bad guys, of a happy household where "mommy" cooks dinner and "daddy" comes home from work. By playing these roles, we explore difficult concepts and discover the range of our human behavior. Indeed, play is an essential learning experience in our development. Because these roles exist only in our imagination, we can explore all life's possibilities while assuming none of its risks.

As adults, we too can learn by incorporating the tools contained within these early learning experiences. By using role play and simulation, we can create a manageable version of our world where it is possible to practice our behaviors and correct our mistakes, generally without harmful consequences. The use of simulations prepares us to function more effectively in the real world.

Simulations in Occupational Training

Simulations serve an important training function in many occupations, because people learn best through direct experience. Before they are qualified for space travel, astronauts are tested on sophisticated electromechanical equipment which closely imitates situations they will encounter beyond the earth's gravity pull. Similarly, through the use of specially designed aircraft simulators, commercial airline pilots qualify to fly certain types of aircraft without ever leaving the ground. In much the same way, salespeople can become proficient in fundamental sales situations before ever calling on a prospect.

SIMULATION: THREE KEY TYPES TO CONSIDER IN THE TRAINING ENVIRONMENT

For our purposes, a simulation is a problem-solving learning activity that attempts to imitate, or "simulate," a real-life situation. A form of drama, it may involve the acting out of a story, the playing of roles, or participation in a game. But unlike a short drama or skit, the simulation has learning objectives, involves many people playing roles, and often has unpredictable outcomes.

EXAMPLE

A sales seminar simulation may have as its primary objective to teach the sales people win/win negotiating techniques by placing them in a traditionally adversarial negotiating situation: keeping an account when there has been a price increase.

The activity is centered on a problem of concern to those involved and is both controlled and open ended. The scenario or description of roles is carefully designed by the trainer or facilitator, but both the unraveling of the problem and the outcome can be somewhat unpredictable for the trainer or the participants. The learners interact as though the problem situation were real and, in their involvement, they become aware of real complexities. What they learn through their participation encourages them to reflect upon past experience and more accurately anticipate future situations. Their discussion of the experience, guided by the trainer or facilitator, raises questions of how to meet anticipated problems. Thus, the simulation gives participants the opportunity to experience indirectly what they may experience on the job in the future.

EXAMPLE

In the price increase situation the sales person may be surprised by a purchasing agent who announces that the increase is enough to warrant his or her company putting the contract out for bid. The salesperson then faces the challenge of having to try and deal with the situation effectively.

There are essentially three basic types of diagnostic simulations for trainers to consider: the direct, true-to-life simulation such as overcoming objections to a price increase; the strategic, insightful simulated or game situation such as how to reorganize a sales territory after key indicators show serious problems; and the role-play simulation such as when you take the role of a salesperson on a call dealing with a difficult buyer. Each will be described with some advantages and limitations of each.

True-to-Life Simulation: The In-Basket Exercise

An example of a direct, true-to-life simulation is the widely used "In-Basket Exercise." An individual is asked to set priorities quickly, make decisions, and analyze the letters and organizational materials found in a carefully arranged in basket. Typical problems built into this exercise include unexpected phone calls and incomplete information.

Typically, this type of simulation is used to measure an individual's ability to delegate, schedule, plan, and carry out tasks. It can also be used to assess more specific skills such as typing, composing letters, and answering the telephone. As such, this exercise is an accurate tool for examining an individual's analytical skills and judgment by placing him or her in a reality-based situation. This type of simulation is often used in assessment centers, where the assessors provide the participant with feedback on his or her performance.

The in-basket type of simulation can be enhanced by building in an "emergency" to test the strength of the participant's managerial talents. As one example, a leaky roof in an unoccupied warehouse was added to gauge participants' reaction to a "surprise" development. The important question is the degree to which the participant acts on this insignificant item based on the simple fact that someone asks about it. Consequently, we were able to measure the qualities of outer-directedness, extreme concern with pleasing people, and taking the conventional route in each participant. In viewing the results of this exercise, we were able to discover something about a participant's thought patterns in reference to a reality-based situation. Building a level of sophistication into the in-basket exercise (such as the example provided) will greatly increase its value to trainers and managers.

Strategic/Insightful Simulation

The strategic or insightful simulation or game attempts to expose participants to a situation which requires overall decisive thinking. In this type of simulation, participants seek solutions to an oral or written account of a conflict situation and process their experience through group discussion. A typical example of this kind of simulation is Human Synergistics' Desert Survival Situation, which is described in detail shortly.

Strategic, insightful simulations are used to provide a shared experience that will foster interaction and provide a basis for discussion and analysis. They also enable participants to develop analytical skills as well as gain insight into approaches to problem solving from other points of view.

Strategic, insightful simulations are both fun and challenging for participants. They are centered on active learning, which encourages participation and sustains interest throughout the simulation. Participants are able to use the knowledge they have and are stimulated to obtain new knowledge to advance through the steps of the exercise. Because participants must commit themselves to specific actions as designated by their decisions, they become more proficient at understanding and evaluating their own motivations.

To achieve the training objectives, this type of simulation must be planned and administered *very carefully*. No designer can anticipate every problem which may arise in the actual simulation. Participants' intense enthusiasm and preoccupation with "winning" the simulation can create problems for the trainer. Consequently, in this type of simulation, it is very easy for participants to overlook the training objectives set forth by the trainer or facilitator.

Role-Play Simulation

The third type of simulated activity—role play—can help trainers to accomplish a wide variety of training objectives, ranging from providing information to changing attitudes. Role-play simulations enable participants to learn through imitation, observation, feedback, analysis, and conceptualization. These simulations are also useful for testing and practicing new behaviors participants can use in their jobs. Perhaps most important, role play allows participants to experience and understand a variety of problem situations from others' points of view.

WHEN TO USE A SIMULATION OR GAME _____

A simulation is an appropriate learning activity when the trainer desires a high degree of participation from group members. It is particularly effective as a warm-up activity at conferences and workshops. Because it encourages participants to focus on the problems at hand, the simulation is a good means of energizing them and creating interest and enthusiasm for a project. Since a simulation involves human interaction, it is an excellent way for participants to relax and get to know each other. In forms such as the interactive video, simulations are also useful for solo practice and skill reinforcement.

The strengths of simulation learning easily override the weaknesses. A simulation is learner centered and often learner directed to the extent that the outcome and the unraveling of the exercise are up to the learners. As a controlled activity, a simulation can only replicate or imitate what the trainers and participants choose to act out. Because of this, a simulation is a problem-solving situation that enables participants to come as close as possible to learning within the "real" situation.

Simulation Advantages and Limitations

SIMULATION ADVANTAGES

- Active learning encourages involvement and sustains interest throughout the simulation. Simulation games are fun and challenging.
- Decisions produce consequences as immediate feedback. Participants can immediately correct errors and receive positive feedback for good performances.
- Players must commit themselves to specific actions as designated by their decisions. This helps players understand and evaluate their motivations.
- The game medium helps to simplify complicated issues or concepts.
- Simulation games broaden understanding of problems and solutions.
- Simulation participants are more apt to accept new ideas and to change their attitudes as they become more personally involved in the game.
- Simulations increase managers' confidence and ability to handle problems.
- They help individuals and organizations to identify their weaknesses and plan for the future.
- They make theory and strategy tangible through performance.
- They help participants understand the relationships between a system's component parts.
- Simulations are shared experiences that are both the basis for discussion and for continued learning.
- Participants are able to use knowledge they have and are challenged to obtain new knowledge.

LIMITATIONS

- Participants sometimes become overconfident about their abilities to deal with real situations.
- Because designers of simulations must choose from a broad range of variables and options in real-life situations, some simulations may foster a simplistic view of reality.

- To achieve the training objectives, the simulation games must be *very carefully* planned and administered.
- Participants' intense interest in the games can create problems for the trainer. They may become bored by less entertaining learning activities.
- Participants may overlook training objectives because they are too interested in "winning."
- No designer can anticipate every problem that may arise in the actual use of simulations.
- Simulations can be time consuming.
- Commercially prepared simulations can be expensive.
- Trainers frequently must adjust or modify some commercial simulations to suit the particular needs of the group.
- It takes additional time to train staff and participants to administer and play simulations.
- Simulations tend to be highly structured and, therefore, too restrictive—especially computer-assisted versions.

HOW TO USE PREPARED SIMULATION MATERIALS

The Desert Survival Simulation: Working Together to Solve a Common Problem

The original Desert Survival Situation has been used with great success for years. Since its publication in 1971, it has received a great deal of attention from training departments, and its widespread popularity continues. It is presented here as an example of simulations that are currently available to trainers.

When developing this simulation, our first objective was to expose a group of people to a situation with which they are relatively unfamiliar. For this reason, we selected the remote desert as the ideal "location," reasoning that very few people have ever found themselves stranded there. Using the desert as the setting virtually eliminated the possibility of an "expert" being present in any group. Therefore, since the problem is one about which participants know little, everyone starts on equal ground.

How the Desert Survival Scenario Is Presented. Rather than relying upon the arbitrary judgment of a trainer, we sought the advice of world-renowned survival experts to help formulate the scenario and to provide us with a rank-ordered list of items, some of which would significantly contribute to a group's ability to "survive." The input of an expert is important because it helps to reduce greatly any disputes regarding the ranking of items which participants may direct toward the facilitator. First, participants read a short scenario which describes a desert plane crash and explains the parameters of the situation. Finding that they are "survivors" of this crash, they must then rank-order items which they have found scattered about the crash site. These items range from a mirror and a water supply to salt tablets and food. Participants are told that an expert has ranked each of these items according to how they might be used to increase the odds of survival. However, before they are able to rank-order the items, participants must formulate a survival "strategy" (e.g., go for help or stay at the crash site).

After each individual has formulated his or her survival strategy and rank-ordered the items accordingly, the facilitator separates participants into groups. Next, each group must perform the same task, ultimately arriving at a consensus plan of action and subsequent ranking of items.

The main purpose of the Desert Survival Situation is to observe how people work together to solve a common problem. It is designed so that information is sometimes unclear or unavailable. Therefore, participants must deal with some ambiguity and are forced to use their assumptions and a few known facts to solve a problem—a situation not unlike ones they face daily on the job.

The Importance of Being Flexible. As groups begin their discussion and try to reach a consensus, the subtle design of the learning process begins to emerge. Of course, each person has his or her own solution and has committed himself or herself to a course of action. One of Desert Survival's instructional objectives is to help participants recognize that it is possible to be flexible, even though they are seemingly devoted to their own courses of action. This process illustrates for participants that backing down from an idea can indeed involve severe personal strain. For example, an extremely dominant individual will demonstrate this style, even in a situation where he or she is relatively ignorant of the consequences. We have observed that some individuals, although lacking in information, have successfully imposed their views, values, and ideas on others using the sheer force of persuasion. In contrast, other better informed individuals with less assertive personalities were successfully talked out of their ideas.

Reaching a Consensus Can Increase a Team's Performance. Research on the Desert Survival Situation shows that many things become apparent to groups as the experience unfolds. Charting the progress of over 205 carefully observed teams, we discovered that up to 85 percent of the teams have better scores than any individual on those teams. This fact demonstrates for teams that openness of discussion, rational disagreement with one another, and a tendency to avoid voting result in optimum performance. In fact, reaching a consensus on the best and most rational ideas can increase a team's performance by hundreds of percentage points.

To summarize our discussion on the Desert Survival Situation, there is value in using a versatile, professionally validated simulation to accomplish your training objectives. By examining the difference between their score, their team score, and the expert ranking, individuals can learn a great deal about their skills in problem solving, interpersonal communication, rational analysis, objective setting, and ability to differentiate between objectives and problems. Once learned, these skills can be easily and directly applied to professional decision-making situations.

Guidelines for Purchasing Simulation Exercises

Thousands of commercially prepared simulation exercises and numerous simulation-related products exist to help you plan your training program. To discover what is available, contact organizations such as ASTD or the Association for Business Simulation and Experiential Learning (see references at end of chapter). These groups and others like them can provide you with the names of consulting firms and other marketeers of simulations.

First Priority: Establish Simulation Objectives. When deciding on which professionally prepared simulation materials to use, it is important to examine your objectives. What skill or skills are you seeking to teach? Do you require a simulation that teaches skills that explore feelings, behaviors, or attitudes? It is also crucial to examine the age range, interests, and abilities of your group. Does the simulation appropriately suit its members?

Experience a Simulation for Yourself First. When choosing among professionally prepared simulations, try, first, to experience the exercise for yourself. Self-participation will provide you with much insight into the power that can exist within a well-designed simulation. Trying a simulation on for size also enables you to check the instructions: Are they clear and easily understandable? Other factors to investigate include the length of the exercise (preparation, play, and discussion times) and its versatility (Is the exercise easy to adapt to group needs, or more difficult, as with some computer simulations?). Obviously, the cost of the simulation will also be a determining factor. Is your choice the most cost-effective way of accomplishing your training goals?

Choose High-Quality Professional Simulations. When choosing among professionally prepared simulations, it is vital that the purchased materials have been extensively and thoroughly validated. Reliable suppliers have a great deal of experience in the construction of simulations, and they continually test them. Testing simulations for validity clearly requires the input and skills of a trained professional.

Simulation Selection Checklist. Whether you are purchasing a professionally prepared simulation or are designing your own, use the following guidelines to help you:

- What are the training objectives? Does the simulation teach skills or explore feelings, behaviors, and attitudes?
- What does the exercise represent? What social system, situation, or problem? How does it relate to your objectives?
- What is expected of your players? Does the simulation suit the abilities and interests of your group? What age group is the exercise geared toward?
- How many players will you need to play roles, observe, and assist?
- Are the instructions clear? Will you need a supplementary instructions sheet for players?
- What equipment, room size, and seating arrangements are required?
- How long is the exercise? (Consider preparation, play, and discussion times.)
- How easy is the simulation to modify?
- Is this the most cost-effective way of accomplishing your training goals?

Using Simulations and Games in Your Training Program

Once you have chosen simulation materials for your group, you must learn the most effective way to facilitate them. Whether you are using a prepared simulation or one you created yourself, it is important to prepare for the actual simulation by practicing how to run the game well. If you don't have time actually to play, mentally rehearse both administering and playing. Scrutinize the materials and read the instructions from both the facilitator's and the player's points of view. It is important to understand all the player's moves, constraints, and possible strategies. A leader's guide should accompany prepared simulations. These guides contain step-by-step information for the maximum ease and convenience of the facilitator. In addition to leader's guides, companies that sell prepared simulations often provide useful companion products. As an example, to help provide the Desert Survival Situation facilitator with maximum flexibility and creativity, Human Synergistics recently developed a three-part, 18-minute video as a companion product to the exercise. Through the use of actual desert film footage, this tape creates a true-to-life experience for participants. This "you are there" effect heightens the level of group interaction, which greatly increases the value of the Desert Survival Situation.

Tips on Introducing the Simulation. Introduce the game by stating its purpose. Novice simulation participants will need a clear definition of "simulation," a general explanation of the exercise's content and value, and a short, accurate description of how it works. Don't overexplain the simulation! Participants learn more by making mistakes and discovering aspects of the simulation on their own.

Answer questions only about simulation rules. It is important to remember that the simulation is a learning experience which forces participants to find their own solutions. As a facilitator, you are responsible for observing and helping participants. Take cues from them that will tell you how to help. Look for facial expression which indicates confusion or negative responses to feedback.

How to Conclude the Simulation. At the conclusion of the simulation, discuss the actual events by analyzing the simulation from different points of view. Expect participants to argue and vent their emotions before they can begin a more analytical discussion.

Structure the postsimulation discussion by first asking participants to critique the game. Allow all to articulate their thoughts and feelings regarding the simulation and to challenge others' impressions. Next, introduce a discussion of the underlying reality of the model by referring to the participants' experience during the game. This will draw out the participants' genuine feelings about the real situation and give the facilitator a chance to suggest new ways of perceiving it.

Last, direct participants to explain in specific terms their involvement in the simulation and to apply this to the real situation. Ask how the simulation experience compares and contrasts with reality. During this discussion, it is also important to solicit feedback from participants on the value of the simulation. Did the group feel that it was clear, interesting, and relevant?

Measuring Simulation Results

What results do we have a right to expect from simulated or gamed experiences? These results generally rely on adequacy of design, professionalism of conduct, and the degree to which the simulation approximates reality. As a general rule, after conducting a simulation, you can expect a higher degree of enthusiasm, a greater degree of involvement among group members, and an elevated degree of thinking by participants as they put the learned principles into their own language. If your measurement techniques are sophisticated enough, you should also be able to show that participants spread these learned principles to the general workplace population, to be absorbed and then used daily.

EXAMPLE

A salesperson encountering price increase objections will be capable of countering them because he has had practice and is confident due to facing such a challenge in a simulation situation.

The negotiation simulation exercise has several purposes:

1) To teach sales people a win/win approach.

2) To give sales people practice taking the other side.

3) To allow sales people to see themselves on video tape as the simulation is used.

Three to four hours of preparation is allowed. Usually a team of two to four people is assigned to each side. The groups create and define the roles. Negotiation sessions usually last one to two hours and are video taped in full.

Evaluation comes through the use of several forms:

1. Observable body language.

2. Overall reaction to the session.

3. Negotiation offer and debrief sheet.

The next example will illustrate the results of using a simulation directly with a sales force. This case involved a large manufacturer of specialty items for business and industry.

The company's marketing department had decided to reorganize its very large sales force to create cross-selling between what were essentially isolated divisions. Because this decision was made with no input from the sales representatives who would be affected, it was met with much resistance. Without the key support of those directly responsible for sales, it was reasonable to expect that the new program would fail, resulting in profit losses for the company.

Recognizing its error, the marketing department sought to correct it using a specially designed simulation. The purpose of the simulation was to demonstrate the benefits of reorganization to the sales department and attempt to gain their acceptance of the reorganization plan. The principles used in designing this simulation involve the equation ED (effective decisions) = Q (quality) $\times A$ (acceptance). This means that effective decisions are the products of quality thinking multiplied by the acceptance of the decisions by the people who have to implement them. In this company's case, while the quality of the marketing department's decision may have been high, the people who must implement the decision did not accept it. Therefore, making the reorganization plan acceptable to those who must make the plan work was the ultimate goal of the simulation used by the company.

The company's entire sales force of over 1,200 people participated in the simulation. Through its use, they were shown a specific benefit of the reorganization plan: through their own initiative, the company's sales representatives could now contact other reps selling different products in the same geographic area who were previously unknown to them, and share sales opportunities and experiences.

The use of this simulation resulted in the sales force's unanimous acceptance of the marketing department's reorganization plan. This acceptance occurred because the simulation enabled the sales force to discover the benefits of the plan *for themselves*. Indeed, a major strength of simulations is that they allow participants to explore possibilities and eventually reach decisions based on the facts presented and their own interests, thoughts, and feelings.

This example illustrates the extensive possibilities for using simulations in sales training. Because salespeople work largely independent of both management and one another, they often lack opportunities to improve their interpersonal skills, learn to work synergistically with others, and come to consensus decisions. Additionally, the isolated nature of direct sales provides no insight into how well the sales manager functions with the people who are the lifeblood of a sales organization: the sales force. As the preceding example effectively illustrates, using simulations in sales training is a simple, nonthreatening method for getting everybody moving in the same direction.

Costs of Prepared Simulations

Before concluding this section, the issue of cost must be considered. The adage "You get what you pay for" is very relevant to the training industry. Before purchasing professionally prepared simulations, thoroughly examine the product as well as the consulting firm's reputation in the industry. Obviously, the most cost-effective simulations are those that have been thoroughly researched and proven to solicit validated, well-documented results.

For information on professional suppliers of simulations, refer to the organizations listed in the References for this chapter.

CREATING YOUR OWN SIMULATION MATERIALS

The effectiveness of any simulation is predicted by the degree to which it forces participants to deal with a basic issue. Therefore, the adequacy of design and the insight of the person creating the design are key variables. The construction of an effective simulation is not an easy task and requires a great deal of work. It is very possible to design a well-structured simulation, but one with totally unrealistic content and learning objectives which have not been well thought out.

To create simulations that guard against this, it is extremely important that you make fertile and extensive use of conceptual design experts in the area of simulation when attempting to custom design training materials.

Design a Simulation to Suit Participants

In line with a simulation's credibility is its level of sophistication. Many simulations have a tendency to insult the intelligence of participants. Your simulation will more than likely result in a true learning experience for participants if you remember to structure it to their level of experience.

The Value of Exploring Different Prepared Simulations

Before designing your own simulation, take every opportunity to participate in existing ones. This participation stimulates creativity and is the best reference for designing simulation games. Next, define your training objectives precisely, but broadly enough to include the participants' personal objectives.

The next step is to outline your scenario. You may choose as a context for your problem a social system such as a company, organization, government agency, or cultural group. After establishing your scenario, choose the amount and types of roles participants will play. Too many roles will tend to complicate the game and confuse participants. Write role descriptions outlining role players' goals and available resources.

Then, establish criteria for "winning" and "losing." It is also important to formulate rules and expected patterns of interaction between players. Decide, also, on what kinds of materials you will need (manuals, contracts, or play money, for instance).

Last, and perhaps most important, design procedures for communicating feedback to participants. Proper debriefing of a simulation is crucial to its success. Always pretest and validate any new simulation before using it to fulfill a training objective. Solicit colleagues' opinions, particularly in the early phases of design when you are just beginning to conceptualize and build simulations and games. Be aware that you can save time and trouble by redesigning an existing simulation. After identifying your central objective, find a game that treats the same or similar problem.

Adapting Existing Sales Training Methods to Simulated Experiences

Existing sales training methods such as the dual sales call can easily be reduced to simulated experiences which greatly increase the value of the learning experience. During a dual sales call, much of the learning which occurs is passive. Developing an interactive video to simulate a sales call allows this training area to come alive and creates a feedback learning situation. A plus in developing such a video is, obviously, direct sales experience. In fact, when developing your own sales training materials, two heads may be better than one. Combining your sales experience with that of a qualified designer of simulations is, perhaps, the key to developing the most effective sales simulations.

SUMMARY

The success of any simulation depends, to a very large extent, upon the quality of the concept and the development of the model. Because the process of training salespeople is so complex, gaming and simulation can add a major dimension to the training and development of a sales force. Indeed, the methods outlined in this chapter can make a sales trainer's role much more effective.

Today, the media and other tools at our disposal are allowing us to make major advances in the development of simulations compared to what was available just a few years ago. However, the conceptualization of possible simulations and the nature of the strategic learning process behind them has yet to match the capability of this new equipment. The need, then, is not for better equipment. Rather, the training industry needs more sophisticated concepts and clearer instructional objectives before it can use these technological advancements most effectively. The future of gaming and simulations clearly rests in the forward strides of the training industry.

REFERENCES

Books

Greenblat, C. S., and R. D. Duke. *Principles and Practices of Gaming Simulation.* Beverly Hills, CA.: Sage, 1982.

Horn, R. E., and A. Cleaves. *The Guide to Simulation Games for Education and Training.* Beverly Hills, CA.: Sage, 1980.

Jones, G. T. *Simulation and Business Decisions.* Middlesex, England: Penguin, 1972.

Oridone, G. *Training by Objectives: An Economic Approach to Management Training.* New York: Macmillan, 1970.

Journals

Simgames: The Canadian Journal of Simulation and Gaming
Champlain Regional College
Lennoxville, Quebec, J1M 2A1
Canada

Simulations and Games: An International Journal of Theory
Design and Research
Sage Publications
275 South Beverly Drive
Beverly Hills, CA. 90212

Organizations

American Society for Training and Development
1630 Duke Street
P.O. Box 1443
Alexandria, VA 22313
(703) 683–8150

Association for Business Simulation and Experiential Learning
J. Bernard Keys
Department of Management
Landrum Box 8152
Georgia Southern College
Statesboro, GA 30460–8152
(912) 681–5216

Human Synergistics
39819 Plymouth Road
Plymouth, MI 48170
(313) 459–1030

International Simulation and Gaming Association
Cathy Greenblat
34 Bayard Lane
Princeton, NJ 08540

North American Simulation and Gaming Association
Bahram Farzanegan
Department of Political Science
UNC Asheville
Asheville, NC 28804–3299
(704) 258–6422

chapter nineteen

EFFECTIVE AUDIO/VISUAL MEDIA IN SALES TRAINING

RICK WILLS

Rick Wills is manager of sales and sales supervisory training, Growmark, Inc. In his present position, he leads the development and implementation of training programs for the 1,500 sales and marketing, 250 sales supervisory, and 2,250 sales support personnel in the Growmark System. He is also a member of the Growmark marketing committee, working with all product divisions on the implementation of their marketing plans. Wills was teacher, coach, and salesperson before accepting the position of sales trainer with FS Services in 1977. In 1981, he accepted his present position. He has been active locally and regionally in the American Society for Training and Development. Nationally, he is active in the National Society of Sales Training Executives. He has won several NSSTE honor awards for his editorials on sales and supervisory training and has authored several articles for marketing and training periodicals. Wills is chairperson, ongoing faculty member, and frequent director of NSSTE's Sales Trainer Clinic. In his work with the Sales Trainer Clinic, he has helped hundreds of new trainers to develop their skills and abilities in becoming more professional.

Since Confucius' suggestion that "one picture was worth a thousand words," people have been aware of the need to show, as well as tell, information to better communicate. As we move ever farther into the information age, the effective transfer of information becomes more critical. Today's trainers and communicators realize that the use of audio/visual media can aid in developing knowledge, skills, and attitudes of people in the workplace.

In this chapter on audio/visual use in sales training we will be looking at

- Why use audio/visuals.
- Deciding which audio/visuals to use.
- Preparing and using audio/visuals.
- Selecting and working with audio/visual suppliers.
- Advantages, disadvantages, and user tips on commonly used audio/visuals.

Some specific aids such as film and video and computer-based and interactive video training will be omitted because they are covered in greater depth in other chapters.

AUDIO/VISUAL MEDIA: SUPPORT MATERIALS THAT APPEAL TO SALES TRAINEES

As we discuss audio/visual media in this chapter, we consider them to be *any support materials used to help an audience increase its understanding of information by appealing to the audience's sense of sight and/or hearing*. Audio/visual media vary in technology and sophistication from rough hand-drawn diagrams to modern state-of-the-art computer interactive systems. The intent of all A/Vs, however, is the same—to increase understanding by the salespeople on the receiving end of the information.

Audio/visual media may be used to support information presented by an instructor or they may be materials which present information that is supported by the instructor. They also can be materials which communicate the intended message without the assistance of an instructor.

HOW AUDIO/VISUAL MEDIA HELP SALES TRAINEES TO LEARN

Sales trainers use audio/visual support materials to increase understanding and retention of information, such as product application or market trends, by their trainees. Most people remember things better when a variety of senses are used in receiving the material. The role which various senses play in learning was determined in a survey conducted by Socony-Mobil Oil Company.[1] That study concluded that people learn through

Sight	83%
Hearing	11%
Smell	3%
Touch	2%
Taste	1%

Their tests went on to conclude that people generally remember

Of what they read	10%
Of what they hear	20%
Of what they see others do	30%
Of what they hear and see others do	50%

These data alone should be ample evidence of how we can be more effective sales trainers through the use of audio/visual aids. However, as trainers there are additional reasons for taking the time and trouble to incorporate good A/Vs into your presentations. These reasons include

---

[1] B. Y. Auger, *How to Run Better Business Meetings* (St. Paul, MN: Business Services Press, 1966), p. 44.

- Helping to increase and maintain audience interest in the subject.
- Helping to simplify complex issues.
- Reinforcing of the presenter's message.

One significant additional reason for using effective audio/visuals can be seen in a 1981 report prepared by the Wharton Applied Research Center at the Wharton School, University of Pennsylvania.[2] This independent study, sponsored by the Audio Visual Division of 3M, assessed the effects of overhead projection in business meetings. The results of the study showed that audio/visual aids have a significant influence on the outcome of meetings.

The study involved 123 M.B.A. candidates who participated in 36 group meetings to decide whether or not to introduce a fictional new product. The results of the meetings showed that when no overhead projections were used there was a 50/50 split in the group on whether to introduce the "new product." When overhead projection was used to promote a "Go" position, the result was 2 to 1 in favor of introducing the "new product." When the overhead projection was used to promote the "No-Go" position the results were 2 to 1 opposed to the new introduction.

Why the varying results? According to the participants, the presenters using the visual aids appeared to be "significantly better prepared, more professional, more persuasive, more highly credible and more interesting."

These tests indicate the use of audio/visual aids can be critical in presenting effective training. There are more things to consider than just "using" an audio/visual, however. Unless the appropriate A/V aid is used correctly, the message presented to the group may be different from the message intended. The importance of planning for audio/visual use, including how to select the proper medium and develop it to get maximum effectiveness, will be looked at later in this chapter.

TYPES OF AUDIO/VISUAL MEDIA TO USE IN SALES TRAINING

Entire books have been written on audio/visual aids, so we will not cover the advantages and disadvantages of every possible audio/visual here. We will, however, explore the more commonly used A/Vs and look at what can be accomplished with each.

---

[2] *How to Present More Effectively and Win More Favorable Responses from More People in Less Time* (St. Paul, MN: Audio Visual Division/3M, 1981).

Flipcharts or Chart Pads

The chart pad or flipchart is one of the most commonly used visual aids. It consists of an easel which has attached a large (typically around 28 inches × 34 inches) pad of paper. The paper is written on with felt-tip markers or crayons. Flipcharts can be used to present to a group such things as the features and benefits of a new product. They can also be used to record information from a group. The easel can also be used to present prepared charts, such as sales trends, to a group.

Advantages. Flipcharts are relatively inexpensive to purchase and use. Since they are both common training tools and relatively inexpensive, they are normally available for rent or loan in most facilities which may be used for off-location training. Flipcharts are also relatively portable, so they may be taken with you from your location. Flipcharts are used with the lights on and are especially effective with groups of fewer than 50 people.

Flipcharts are quite versatile. They can be prepared before a session to visualize information you wish to present, or they can be used to record the information developed by the group during the session. Flipcharts are especially effective when used to emphasize key information by writing or drawing it for the audience.

Because of their portability, flipcharts are very useful for workshops where salespeople work in small groups then come together to make reports to the joint group. For example, a group of salespeople could be brought together to discuss sales strategies for specific target markets. Each group can be given a flipchart, sent off to work, and have their information recorded and readily available for the final report.

The sales trainer must plan how he or she is going to use the chart pad, even when the group will develop the topic to be discussed. This assures that the key points the trainer wishes to discuss will be emphasized visually on the chart.

One key advantage of the flipchart is that the information recorded on chart pad pages is permanent. The pages can be taped to the wall for future reference or even if not taped up, the pages can be "flipped back to" to review points made earlier. Pertinent information can even be added to flipchart pages long after they have been developed to keep the points relevant and in focus.

Disadvantages. While the chart pad is a useful tool, many a sales trainer's "right hand," there are potential disadvantages.

While more permanent than chalkboards, once pages have been developed and posted, they tend to have a short life span. Any recorded information that is posted must consciously be removed from the wall and saved or rewritten to a more portable, manageable size. Also, because of their limited size, chart pads are not normally useful with large audiences.

Instructors who use flipcharts need good printing and spelling abilities. It is important that prepared information or inputs from the group be in a legible, readable form. Additionally, it is difficult for most instructors to talk and write on charts at the same time.

Prepared Charts versus Charts Developed with the Group

When evaluating whether to prepare charts prior to a session or develop the charts with inputs of the group, several factors must be considered by the instructor:

1. Is the experience level of the group on this topic area so low that they will be unable to provide the information that needs to come out?
2. Do I need to improve my writing and spelling abilities for use in front of the group?
3. Is there a specific order in which this material must be developed?
4. Are complex drawings needed that would be time consuming to do in front of the group?
5. Will this information be presented to several groups within a fairly short time?
6. Will writing this information in front of the group be overly time consuming or distracting?
7. Am I uncomfortable developing a chart in front of the group?

The more of these questions the sales trainer answers "yes," the more advantageous it is to prepare the flipchart prior to the training session.

User Tips

- With pre-prepared charts, pencil-in notes on side of sheet to act as your leader guide.
- Pretear masking tape in 4-inch strips and place on the side of the easel for taping pages to wall. To simplify taping pages to the wall, place the pretorn tape on the edges of the paper prior to tearing it off the pad.
- Use broad-tipped, *water-based* markers, not alcohol based, to reduce chance of bleedthrough. This is especially important if you add information to pages

already posted on walls. Hotel and motel managers are not thrilled with "bled-through" notes left on their wallpaper. Accidental marks on clothing are also easier to remove.

- Check, before the session, to be sure markers are not dried out or points mashed. If they are, discard them.

- Use blue, black, or purple broad tipped markers to record most information. Other colors, such as red, yellow, green, or orange, while not as visible, are effective for highlighting and emphasis.

- When preparing chart pads for a presentation, it is important to prevent the information from following pages showing through the page being used. This can be prevented by either leaving a blank page between each prepared sheet or by starting with the back page and using the chart pad in reverse order. When used in reverse order begin by flipping all of the pages over the back of the chart to hide them from the view of the audience. Each individual page can then be flipped from the back to the front to share its information.

- On individually prepared pages, sections of the page can be hidden from the group by folding the page up from the bottom to cover the information. Information can also be covered with paper strips held in place by masking tape or paper clips. When you are ready to discuss these points, merely remove the specific paper strip or unfold that section.

- If you have trouble writing on a straight line, purchase prelined chart pads.

- Be sure, before the session, that there is sufficient paper on each pad for that session.

- Simple pictures or cartoons can help to emphasize your message and illustrate points quickly that take tremendous amounts of time to describe verbally. These can be predrawn in marker, or in light pencil to be copied in marker in front of the audience.

- Post on the wall pages with key points or items to which you want to return at a later time. This keeps them in sight so you don't have to rely totally upon your memory.

- Write large enough so the last person in the last row of your room can read it comfortably.

- Use two chart pads with a group, one with pre-prepared material and the other to record information from the group.

- Use a piece of paper folded lengthways and taped to the side of the chart stand. It can act as a leader's guide. This can identify the main areas to be covered and keep the leader free of the lectern.

Overhead Projectors, Transparencies, and Computers

The overhead projector is a versatile audio/visual training aid which is normally available for rent or loan at most off-site training locations. Overhead projectors shine light through a glass stage to a mirror above, which projects the image to a screen. These projectors are placed in front of the audience with the image projected onto a screen behind the instructor. Images can be projected from premade transparencies or now from computers. New technology allows a presenter to bring in a computer, hook it up to a special device and project screen images on an overhead projector for presenting information. Information which will be shared several times, such as new product introduction details or territory management techniques, is ideal for development into a series of overhead transparencies. The overhead can be used much as a flipchart by recording inputs or responses from the class on blank acetate sheets with grease pencil or transparency markers, or items can be typed into a computer and projected on-screen.

EXAMPLE _____

In the past sales trainers have been frustrated in their efforts to use a variety of colored graphics. Now with the new technology, sophisticated sales and marketing materials can be used in training without having to use expensive multicolored transparencies.

Overhead projectors allow the sales trainer to face the audience and maintain eye contact while utilizing the equipment. In addition, overhead projectors can be used in rooms without turning down the room lights. (Note: Sales Trainers must be sure there is no direct light shining on the screen which will "wash out" the projected image.)

Advantages. Overhead projectors allow high presenter flexibility by using both high-quality prepared transparencies with prepared information or blank transparencies to make lists or diagrams as the discussion proceeds.

Prepared transparencies can help keep training sessions flowing as the instructor planned. An entire program on goal setting can easily be developed on transparencies and presented with just an overhead projector for A/V equipment. Such prepared transparencies can be easy and inexpensive to develop. Good-quality transparencies can be made utilizing the copying equipment found in most offices. Consequently, transparencies can be easily produced and duplicated. Prepared transparencies can be reused many times if reasonable care is taken.

Transparencies tend to project with good visibility and clarity. These overhead images work well for audiences of up to 100 people, even larger if screen size, projector distance, and room lighting are suitable for increasing the image size.

Transparencies with overlays can be developed to allow for cumulative presentations. A transparency of the human body that starts with the skeletal system and then adds overlays of the internal organs, blood system, muscular system, and skin can be useful in discussing general physiology with pharmaceutical salespeople.

These overlays add professionalism to any presentation. Also, by setting the overhead projector beside the lectern, the presenter can work easily from prepared notes or lesson plans.

An instructor can effectively keep the sales trainees' focus on individual concepts of a transparency by covering the transparency with a sheet of heavy paper or thin cardboard and moving the paper downward line by line.

EXAMPLE

The new computer overhead systems are revolutionizing sales and marketing presentations. Long and complex programs can be prepared and presented with ease. All the sales trainer needs to do is move the computer program from screen to screen and all the information is available.

Disadvantages. Overhead projector lenses may distort the image projected. To avoid this, the screen may be set with the top of the screen approximately 6 inches closer to the projector than the bottom.

While most transparencies can be developed inexpensively, the commercial development of transparencies can be quite expensive.

The projector arm holding the lens can cause interference with the field of vision for people directly in front of the projector.

EXAMPLE

Sales trainers need to make arrangements to have both a computer with a colored screen and the computer overhead available for the training session. These items are not easily rented in a lot of smaller locations. Special advanced arrangements need to be made.

Criteria for Developing Prepared Transparencies. Prepared transparencies can add a touch of professionalism and credibility to any presentation. The effectiveness of the overhead and prepared transparencies was shown in the 1981 Wharton report referred to earlier.[3]

Certain criteria may affect your decision to develop the transparencies yourself or have them commercially produced, for example,

- *The group to whom you will be presenting.* Generally the higher the level of the group, the higher the quality of the visual. (This is more for the image of the presenter than improving of the information transferring capability of the aid.)
- *The number of times the transparencies will be used.* The higher the frequency of use, the less costly commercially prepared transparencies become on a per use basis.
- *Your budget.* Self-made transparencies are less costly than commercially made ones.
- *The amount of time before the meeting.* Commercially developed transparencies can take weeks to produce while self-developed may take only minutes.
- *The number of technical drawings to be used.* The more technical information needing to be shown, the greater the benefit of a commercial transparency to utilize color and other emphasis techniques to highlight key areas.

User Tips. *Never* use regular typing for transparencies; it will seldom project large enough to be read comfortably.

- Whenever possible, use frames around transparencies to reduce the glare of light shining from sides of the transparencies and make them easier to handle.
- If it is not possible to frame transparencies, place paper between them to help keep them separated.
- Make notes on the frames of the transparency; this enables you to present your information without having to fumble with extra cards or papers.
- Always focus the projector before the session begins.
- To focus an overhead projector without a transparency, place a coin with a serated edge, such as a dime or quarter, or a key on the stage and turn the focus knob until the serations of the coin or key are clearly in focus.
- If the projector is to be moved in during the session, prefocus the projector where it will be used. Then mark the floor with pieces of tape to indicate

---

[3] Ibid.

where the wheels of the cart holding the overhead should be placed for the overhead to be in focus during use.

- If the overhead will not be moved during a session, tape the cord to the floor with wide masking tape or duct tape to reduce the chances of someone stumbling over the cord.

- Turn the projector off between each transparency. If many transparencies are to be used sequentially, tape a flap of cardboard to the lens head of the projector. This can be flipped down to cover the lens and black out the light to avoid continually turning the projector off and on.

- To highlight points on a transparency, use a pointer on the transparency itself rather than turning around and facing the screen.

- Use large print and not too many words for each transparency. Generally, six lines with six words per line is most effective.

- A quick rule of thumb to check the projected size of a transparency is to place the transparency on a piece of white paper on the floor. If you can read the transparency *easily and comfortably* while standing above it, it should be appropriate for projection.

- Utilize different color transparency stock, printing, and highlight to add variety and interest to your presentation.

- Use lowercase letters; they are more legible than capitals. Capitals help provide contrast.

- Cover material on the transparency to focus group attention on specific information. This can be done more effectively by placing the paper between the transparency and the projector stage. This will keep the paper from being blown off the stage by the projector fan. It will also allow the instructor to easily read the upcoming lines of the transparency and any notes on the frame.

- Utilize colored washable pens to highlight key areas. When using or developing permanent transparencies on which you will repeatedly write or highlight, add a blank transparency over the prepared transparency so accidental writing by a permanent marker will not damage the prepared transparency.

- Keep a spare bulb available at all times, even with machines which have a built-in spare bulb. Murphy's law is still in effect.

- Number all the frames in the same corner for any presentation. This will help keep your visuals in sequence and your presentation flowing well.

- Keep your transparencies to be used face up in a pile on the side of the projector closest to you. After you have used a transparency, place it *face down* in a pile on the *opposite side* of the projector. This will reduce the chances of getting transparencies mixed up, and when finished, your transparencies will be in correct order for your next presentation.

- Utilize cartoons to emphasize your points. Cartoons taken from newspapers and periodicals can be enlarged on most copiers to create an image suitable for use as a transparency. For extra effectiveness, develop your own caption relevant to the point you are making with your verbal presentation, possibly substituting names familiar to the group.

- When developing word transparencies, use only key words or phrases, not entire sentences. "Bullets" are helpful in identifying each separate idea.

- Use blank transparencies or an acetate roll in developing responses from the group. Blank transparencies are easier to refer back to than the actetate roll, but the acetate roll is much easier to use if a considerable amount of writing is to be done.

EXAMPLE

Criteria for Preparing Computer Programs

A combination of word processing, graphics, and desktop publishing programs are being used.

Popular word processing programs include:

Microsoft *Word*

Wordstar

Word Perfect

Popular graphics packages include:

Harvard Graphics

Popular desktop publishing packages include:

Page Maker

Publish It

Ventura

Programs can be developed and transferred onto a single disk for presentation. The disk simply needs to be loaded into the computer and a page by page sequence can be presented. Such a program allows both a sophisticated visual program and a set of hard copy sheets to be used as handouts to be produced.

In addition the disks can be duplicated inexpensively and sent to their sites for use quickly and easily.

Tips

- Carefully select your software programs to ensure they can produce what you need.
- Use someone who has an in-depth knowledge of those programs to help produce your program.
- Produce a program with variety.
- Order equipment well in advance.
- Take 2 back-up program disks in case something happens to your original program disks.
- Think in terms of supplying copies of the disks to all sales training participants so that they can use them for further reference on their computers.
- Understand that color copying machines are in the early stages of technology so hard copy will be in black and white.
- Allow enough time to develop your program.

35mm Slides

The 35mm slide projector is reasonable in cost, simple to use and relatively commonplace. Slides provide excellent quality images and graphics with relatively small expense for development or reproduction. They can be effectively used with small to large groups. Used with cassette tapes, they can add an effective audio/visual dimension to a presentation.

EXAMPLE

Many companies produce slide presentations for ongoing training of sales personnel. Slides provide a polished but relatively inexpensive way of presenting the year's sales, goals, projected sales figures, special sales incentive programs, and product training.

Advantages. 35mm slides can be inexpensive to produce and duplicate. They can be produced quickly; in most cities slides can be developed commercially in 24 hours or less. They can also be used many times. When presentations are put together for long-term use, such as a company orientation program, slides provide additional flexibility in that new ones can be produced and inserted to update information, such as specific new products, without changing the entire presentation.

Commercially developed "generic" slides are readily available at reasonable cost to help develop professional looking presentations quickly. 35mm slides also allow a simple, quick, and easy opportunity to use photographs of people and places familiar to your group.

The 35mm slide projector is portable and readily available at most off-site locations. It can be run remotely by the leader or an assistant, or many can be programmed with an audiotape to advance the slides. Most slide projector manufacturers provide information and training on the use and maintenance of their equipment. Impressive, professional presentations utilizing multiple projectors can be very effective in grabbing audience interest.

Disadvantages. The main disadvantage of 35mm slides is that they require a darkened room. Darkening the room can reduce the leader's interaction with the group and ability to read the group's reactions. In general, slides tend to reduce the amount of interaction taking place within a group.

Longer lead time is necessary for slide production than for flipcharts or overhead transparencies. Additionally, multiple projector presentations generally require professional support and can become quite expensive.

Using Locally Produced versus Commercially Developed Slides. When looking at developing slides for a presentation, several factors should be evaluated in making your decision whether to produce your own slides or to have them professionally produced:

- The group to whom you will be presenting
- The number of times the slides will be used
- Your budget
- The amount of time before the meeting
- The amount of technical drawing to be used
- The availability of a quality commercial firm to produce your materials

In general, if your answers to any or all of these questions tend to be "higher rather than lower," commercially developed materials may be more appropriate.

User Tips

- Design slides to be read from the farthest seat of the last row.
- Edit slides before your presentation to assure they are in proper sequence, rightside-up and forward. Then recheck your slides immediately before the session.

- Start with a black slide to keep from blinding the audience with the bright projector light.
- Be sure to have a script for the leader and an assistant with a description of what slide appears at what point in the dialogue. This will help to ensure that the right slide is projected at the right time.
- Put the lectern close enough to the screen so both the slide and the narrator are in the participants' natural field of vision.
- Use a shaded light on the lectern so the presenter can read the narrative without causing a distraction to the group.
- Since most projection screens are wider than they are high, use only the horizontal format for slides.
- Limit word slides to five to six words per line with no more than five lines per slide to produce the best image.
- Use both capitals and lowercase letters on word slides.
- Use only a limited narration with each slide; slides that are on the screen for over 30 seconds will generally lose their interest to an audience.
- Change slides any time there is a thought change. This will visually help your audience focus on the transition.
- Provide variety in the amount of time slides are on the screen to help reduce monotony and predictability, even for short presentations.
- Keep participant interest by providing variety and using slides with other audio/visual media.
- Use 35mm slides to accurately show product use. This can be extremely useful for presenting technical information on your own product or the competitor's.

Chalkboards and White Boards. Chalkboards, tools most of us know from school days, are still used as training aids. Most chalkboards are attached to the wall; however, they are also made on portable rolling wooden or metal frames. White boards are smooth white-surfaced boards that can be written on with erasable alcohol-based markers. These boards, like the flipchart, are best used with groups of fewer than 50 participants and should be no more than 30 feet from the audience. Because of the inability to store information, chalkboards appear to have been replaced by chart pads for many trainers. Some white boards can even photocopy the contents of the board before it is erased.

Advantages. Most of the advantages of the flipchart are true also of these boards. The key ones are

- Chalkboards are relatively inexpensive and very versatile.
- They are a good tool for recording inputs of the group.

- They can be used in regular room lighting.
- They are effective for the spontaneous diagramming of a concept or mechanism.

Additionally, both the chalkboard and white board provide large writing surfaces on which considerable information can be placed. Changes to information can be made simply and quickly by erasing the old information and inserting the new.

Disadvantages. One of the main disadvantages of these boards is that once information is erased, it is permanently gone. Therefore, instructors must allow participants sufficient time to copy written material before erasing it. If the instructors want to save any of the information, they must recopy it themselves.

Chalkboards can also be messy from chalk dust. While using colored chalk can help in a presentation, it is not easily erased without washing the board. This may add to the mess during a presentation.

The chalkboard and white board are rigid and somewhat inflexible in the way they can be used, and most are not very portable. If there is a considerable amount of information to share, these boards are not easily prepared in advance. Also, unless instructors are very adept, it is easy to talk to the board rather than to the group.

One unique potential disadvantage is that chalkboards may induce negative connotations with some trainees because of their relationship to school. This factor is difficult to measure but is worth considering as instructors evaluate the use of chalkboard.

User Tips

- Print, don't write, and do so very carefully.
- Print large enough so the member of the group farthest from the board can read it easily. A rule of thumb is to have print height 1 inch high for each 10 feet the farthest person is from the chalkboard:

Feet Away	Print Size Should Be
10 feet	1 inch
20	2
30	3
40	4

- On large boards, segment the board into smaller areas. This will help keep the ideas within the participants' field of vision.

- On a blackboard, use white chalk since it tends to show up best. On a green chalk board, use yellow chalk since it seems to be most visible.
- Utilize colored chalk or markers to highlight ideas and functions, and for variety.
- As the instructor you must be aware of where you stand as you use the board so as not to block the field of vision of the participants.
- To best use the board, if right handed, stand on the right side of the board at a 45° angle, as you face the audience.
- When referring to information on the board, use a pointer to reduce the amount of the information blocked by your arm and body.
- When finished writing, put the chalk or marker down so it does not become a distraction to participants.
- Be aware of possible glare on the board from room lights.

Flannel/Magnetic/Display Boards

These devices tend to be display stands for charts or models. These devices are convenient for developing step-by-step presentations, in that items can be placed on them one at a time as the instructor talks with the group and items can be shifted or changed easily.

Flannel boards are made of flannel over a wood, metal or plastic frame. The stick-on pieces are backed with flannel, flock, masking tape, or sandpaper to hold them in place on the slightly tilted board.

Magnetic boards have a frame containing a thin metallic surface upon which stick-ons can be placed. Stick-ons for magnetic boards are magnetic or backed with magnetic strips. Many magnetic boards can also be written on with special markers and used as a white board.

Display boards or cork boards are made of a soft substance into which tacks can easily be inserted and removed.

Advantages. These types of aids are reusable by simply changing the information placed upon them. Changes in the information or order can be easily made by merely moving or changing the pieces. Stick-ons can be prepared in advance or left blank to be written on as ideas develop from the group.

These tools provide a simple and graphic method of model building. A step-by-step building of the selling process, with discussion of each step as the model is built, can be very effective. As such they are good attention getters.

Disadvantages. These boards may not be readily portable. They are seldom found as normal training materials provided by off-site locations.

User Tips

- The flannel board has drifted from popularity over the past years. This may make it an advantageous medium to use for the emphasis it may add through its uniqueness.
- Building models or diagrams step by step is especially effective on these tools.
- Use of color and shape can help to spice up these presentations.
- Styrofoam models can be constructed to use on these tools to add a third dimension to the presentation.

Models and Demonstrations

The materials the trainee will be actually using can be effective audio/visual media. Showing or modeling how to do a task using the tools and materials trainees will be using provides very effective training and may at times be simpler for the instructor. Using the actual product to practice a salesperson's product demonstration will help increase both their knowledge and skills.

When the materials the trainees will use in their actual job are too large or too small to be used in training, models of those objects can be useful. These models are scale reproductions and can help to bring realism to classroom training.

Advantages. Modeling is a powerful, effective form of teaching. Models are as close to real life as possible and can be very useful for investigating complex subjects and materials.

Models and demonstrations provide actual hands-on application of ideas being taught. This hands-on training will increase retention of the knowledge or skill. Models work extremely well with small groups.

Training models can be made so details not pertinent to ideas at hand can be omitted while still allowing for realistic demonstrations.

Disadvantages. Use of models can result in oversimplification that may cause additional questions when trainees use the actual object. Also, if the model or demonstration does not go exactly as planned, it could have an impact more negative than other media.

User Tips

- When using a model, color parts differently to make them more identifiable by trainees.

- Don't pass a model around the group. It distracts from what you as leader are saying. Have one for each group member, or allow the group to look at it after your presentation.
- Don't present an object or model until you are ready actually to use it; this will keep focus on you as leader.

Written Materials

Manuals, books, and handouts are a common form of visual aid used by many trainers. These provide a lasting resource which can be beneficial during the training session and even more valuable when used back home on the job.

These materials have excellent flexibility in that handouts can be developed to cover exactly the information you wish to include. Books can be used in their entirety or by individual section, chapter, page, paragraph, sentence, or phrase. Manuals, especially product manuals and pricing manuals, are valuable learning aids in that they allow trainees to use tools they will be using on the job.

Advantages. Written materials are inexpensive and extremely flexible. The material they contain is highly relevant to the training sessions and useful for later reference. Most written materials are readily available or reproducible on short notice.

One of the greatest advantages of written materials is that they can be used to build a uniform level of knowledge within a group of trainees prior to attending training. This can allow valuable classroom time to focus on difficult or highly technical information, or to work on skill building.

Disadvantages. One of the biggest drawbacks of written materials is that they require high trainee self-motivation. This can be especially true if there is a significant amount of written material. A "mountain of paper" can be demotivating to individuals preparing for a training session in addition to performing their regular job.

For trainers, developing written materials requires good writing skills, an ability that not all trainers possess. Additionally, even though the material will be available as a reference tool, many trainees will not utilize the information after the training session.

User Tips

- Illustrations, pictures, and cartoons help improve readership of your materials.
- Use large type to make reading easier.

- Reading should be for pre- or postsession assignments or evening homework. Don't invest valuable classroom time for reading.
- A prechapter overview and postchapter summary can help to increase readership and retention of material.
- Application exercises within the written material help to increase retention.
- Printing on both sides of pages can help to reduce size of large volumes.
- If sections of a workbook or manual can be separated, send or give them to trainees one at a time with instructions for their use.
- Provide reading lists for after session references that include sections or chapters, not just entire books.

Teleconferencing

Teleconferencing is a method of providing audio/visual inputs from people at sites other than where the training session/meeting is located. Teleconferences can be an excellent way for product managers or corporate executives to talk with the entire sales force at one time. The teleconference can be an audio-only presentation utilizing a telephone hookup amplified for the entire audience to hear. It may also be a telephone hook-up that has support of 35mm slides or a fax machine to provide visualization of the information being presented by the person on the phone hook-up. A third method of teleconferencing is satellite television hook-up combined with telephone hook-up to the sending location. This provides a live visual image to the group and allows for input from the group via the telephone.

EXAMPLE

Many sales teams that are spread out all over the U.S. and other countries have found this a cost effective way to conduct meetings and training. Training on new products, new forms, a new marketing approach have been accomplished very successfully. The salespeople don't lose valuable time traveling to headquarters. Instead they have contact with others, ask questions, receive valuable information, are trained on a quick turnaround time and can return to their sales territory, often the same day.

Advantages. Teleconferencing can be a very cost-effective way to reach many people fairly quickly. It reduces travel time, travel expenses, and time-out-of-field costs of participants. A teleconference provides for two-way communication between a presenter and group that would not be available through video cassette or 35mm slide presentation. This two-way communication allows participants an opportunity to clarify or expand on presented information.

Disadvantages. Television satellite teleconferencing sessions can be expensive. They require highly technical people at all stages of development and implementation. These hook-ups also require a great deal of long-range planning. Time is critical. In video teleconferences especially, presentations must be well rehearsed and timing well established.

While teleconferencing is a two-way medium, most teleconferences involve a knowledge-sharing process—little "real" two-way communication actually occurs.

User Tips

- Plan a time that will be acceptable to all potential participants.
- Select strategic sites easily accessible to a majority of the potential participants in an area.
- Some hotels and motels, local TV stations, and public broadcasting systems stations have systems which can be utilized, for a fee. These facilities have people with the technical expertise necessary to administer such a program professionally.
- To increase chances of a meaningful question and answer period, "plant" a few key questions in the group to start the interaction.

Audio Tapes

Audio cassette tapes are an excellent method of providing information, such as product knowledge, to trainees. These tools, though providing only one-way communication, allow the transfer of information at a very reasonable cost. In addition, if the trainees spend significant travel time on their jobs, tapes can be a productive way to utilize that "windshield" time.

Advantages. Audio tapes can be extremely inexpensive to produce and duplicate. They, therefore, can be used with the entire sales force. Audio tapes are very useful in distributing technical and product information to the sales force. They can also be used to develop such things as verbal and language skills. Audio tapes are portable, durable tools that can help to produce a positive, professional image about training. Many professional programs are available for purchase or loan through public or private libraries.

Disadvantages. While audio tapes can present a very positive message, there is no visualization to utilize sight as a learning sense. Tapes which are locally produced or contain a great deal of highly technical information can become boring, which can reduce retention. While bordeom can be a problem, likewise an audio cassette that gets people too involved can be dangerous if used in an automobile. Additionally, it is difficult to move to specific portions of cassette tapes with speed or accuracy.

User Tips

- Keep tapes short and fast moving; 10 to 12 minutes is best.
- Utilize local talent only if they have good speaking skills. An exception to this is the use of an individual who commands respect of the sales force such as the vice president of sales or marketing.
- Be creative; audio tapes provide the opportunity to create many special effects through vivid language and background sounds.
- Prepare the script to do exactly what you want it to do and have the talent follow it exactly.
- Prepare a "Recommended Listening List." Many public libraries carry business audio tapes which can be checked out.
- Work with a commercial tape rental library to develop a rental library designed to meet the needs of you, your company, and your sales force.
- With a group, use a top-quality player and sound system so the message is easy to hear and understand.
- Encourage salespeople to use blank tapes to practice sales calls and self-coach.
- Give work assignments with tapes to make tape more relevant to the listener.

The characteristics of each of the foregoing types of audio/visual media are summarized in the accompanying chart.

Computer-Generated Graphics

These tools are essentially a developmental process for creating visuals which can be utilized on overhead transparencies, 35mm slides, or video tapes. When developing computer-generated graphics, trainers should use the same basic rules as in developing slides or transparencies.

Instructional Carrels

Carrels allow trainees to utilize an audio/visual aid for individual application. They can be utilized with almost any of the aids mentioned earlier in this chapter as well as with video or computer-assisted instruction.

Audio/visual guidelines

Audio/ Visual	Group Size	Advantages	Disadvantages	Tips
Flip chart	Up to 50	• Inexpensive • Portable • Lights stay on • Versatile • Can be prepared or spontaneous • Can be posted on wall • Useful for work groups • Emphasize key info • Pages permanent • Info can be added	• Short page life span • Need good printing, spelling skills • Difficult to talk, write at same time • Not useful with large audiences	• Water color markers • Blue, black to record, red to highlight • Colored markers to highlight • Check markers before session • Pencil in notes on page • Start from back for prepared charts • Use prelined paper • Post to wall for reference • Pretear tape for posting • Check quantity of paper before session • Use pictures to illustrate • Write large enough • Use two pads, one prepared, one for inputs
Overhead projector	Up to 100	• Transparencies easy, cheap to develop • Reusable • Flexible • Help produce good flow • Cumulative presentations • Good visibility	• Image may distort • Commercial ones expensive • Lens arm interferes with field of vision	• Never type transparency • Use frames on transparencies • Make notes on frames • Prefocus projector • Use key or coin to focus if transparency not available • Tape down cord • Turn off or flap lens to darken between transparencies • Large print, 6 × 6 words • Use color transparency stock for attention • Use upper- and lower-case letters • Cover information on transparency to focus attention • Keep spare bulb handy • Highlight with color markers • Number frames in same place • Place used transparencies face down on opposite side from those to be shown.

461

Audio/visual guidelines (Cont.)

Audio/ Visual	Group Size	Advantages	Disadvantages	Tips
33mm slides	Up to 250	• Cheap, quick to produce and copy • Reusable • Easy to update presentations • Easy to use own products, people in presentations • Generic commercial slides available • Portable • Available off site • Can be run by leader, assistant or be programmed • Manufacturers provide training	• Required darkened room • Reduce amount of group interaction • Longer lead time than transparencies • Multiprojector presentations expensive	• Use cartoons to make points • Use bullets • Use blank acetates for recording group inputs • Edit and check before presentation • Start with black slide • Have lectern, with shaded light, in field of view • Have script for leader and assistant • Use horizontal format • 5 × 5 word slides • Not on screen over 30 seconds per slide • Change slide with thought change • Vary amount of time slides are on screen • Use with other media • Use capital and lower-case letters • Use to show products in action
Chalkboards/ White boards	Up to 50	• Relatively cheap • Versatile • Good for recording group inputs • Use with lights on • Good for diagramming • Easy to change • Large writing space	• Once erased, information is lost • Messy • Inflexible • Not portable • Not easily pre-prepared • May induce negative schoolroom feelings	• Print • Segment large board into smaller areas • Colored chalk to highlight • Stand at 45° angle from board • Use pointer in referring to information • Beware of glare • Put marker down when done
Magnetic/display boards	Up to 50	• Simple • Easily changed • Can be prepared in advance • Reusable	• Not portable • Not found off site	• Build models step by step • Use colors, shapes • Styrofoam models add third dimension

Method	Audience	Advantages	Disadvantages	Guidelines
Models/demon-strations	Up to 50	• Close to real life • Simplify complex subjects • Hands on application • Useful with small groups • Can omit unnecessary details	• May oversimplify • If doesn't work, very negative impact	• Color parts differently • One model for each group member • Keep out of view until ready to use
Written materials	Individuals, small group	• Inexpensive • Flexible • Relevant, real life • Easily reproducible • Build uniform knowledge level	• Require high self-motivation by user • Requires good writing skills for developer • Often not used	• Use illustration and pictures • Use large type • Use application exercises • Have prechapter overview and postchapter summary • Duplex pages • Provide reading lists
Teleconference	Large groups	• Cost-effective • Reduce travel time • Two-way communication	• Expensive • Highly technical • Planning critical • Timing critical • Little real two-way communication	• Plan time acceptable to all participants • Select accessible sites • Some hotels/motels can be utilized • Plant starter questions for two-way interaction
Audio tapes	Individuals, small group	• Inexpensive • Many uses • Portable • Durable • Professional ones available for purchase or loan	• No visuals • Technical tapes boring • Difficult to move to specific part of tape	• Keep them fast moving and short • Utilize people with good speaking skills • Be creative • Use prepared script • Develop "Recommended Listening List" • Work with commercial tape rental businesses • Add assignments • With group, use good-quality player and sound system

SIX CRITERIA FOR DECIDING WHICH AUDIO/VISUAL AID TO USE IN YOUR TRAINING PROGRAM

When evaluating which audio/visual aids to use for your meeting, several criteria should be used:

1. *The objective of the training session.* This is a key factor. Too often audio/visual aids are utilized because they are the latest thing or the equipment is new to the organization. The key question should be "How will this help the trainees meet the objectives set?"

2. *The number of trainees in a session.* For individual training, study carrels can be very effective. For a group of 250, the same tool would be impossible and implausible. By the same token if you wish to visualize something to a group of 250, 35mm slides will be your best medium, or possibly a video projected upon a large screen. For groups of 50 or less, chartpad, chalkboard, and overhead projector can be effective visual aids.

3. *The experience level of the group.* When the group is low in experience in the task being focused on, it is important that the instructor be prepared to present most of the information to the group. In these cases audio/visual aids must be prepared ahead of time. When the group has a high level of experience with the subject matter, aids that are more spontaneous can be of benefit in letting the group develop ideas to increase knowledge and skill levels.

4. *The environment of the meeting room.* The physical layout and environment of the meeting room can help to determine the type of visual to use. If it is inconvenient to change the level of lighting, stay away from aids that require reduced light, such as films or 35mm slides. More logical choices would be video through several small monitors, overhead transparencies, or chartpads.

   In a room where there is insufficient light, stay away from the chartpads or chalkboards and focus on the use of overhead, slides, or videos.

   In small rooms, where there is no wall-mounted screen, the addition of a standing screen may reduce the size of the room even more. If the wall is a light color and slightly textured, you may be able to project directly onto the wall. If you cannot use the walls or a screen effectively, you may stay away from projected media and focus on video and/or chartpads.

5. *Availability of resources.* Most commercial hotels and motels which do training business have access to common audio/visual aids. These typically include chartpads, overhead projectors, and video players. Some facilities carry even more sophisticated equipment including sound systems. Often facilities which don't have the more sophisticated materials on premises

can refer you to local firms which can provide them. If you are conducting sessions off site and have questions about the availability of equipment, contact the facility regarding the availability of the aids you require.

6. *Instructor experience with specific audio/visual tools.* If an instructor does not have adequate experience with a certain audio/visual aid and does not have time to become familiar with it prior to a session, they would be well advised to change to a more familiar tool. If it is not possible to switch media, the instructor must find someone to assist. A poorly used aid is much worse than no aid at all!

GUIDELINES FOR PREPARING AND USING AUDIO/VISUAL AIDS

Audio/visual aids can be a critical portion of your total instructional message. Therefore, it is important to keep several things in mind as you prepare to use audio/visual aids.

1. Make sure the audio/visual you are considering is appropriate to the objective you are trying to achieve. Any and all support materials should reinforce the point of your message.

2. Keep your audio/visuals simple, concise, and to the point. This helps you focus the aid on the key point you want to stress. Word visuals should be key words or short phrases.

3. Use pictures, cartoons, graphs, and charts in place of words whenever possible on visuals. Ideas can be transmitted more simply and more effectively through picture images.

4. Visual aids should be colorful. Color helps to attract attention, represent an idea, or just keep things interesting. Different colors have different effects on people.
   • Red tends to have a warm, stimulating effect.
   • Blue produces a cool, sedative, depressing effect.
   • Purple gives the effect of dignity and reserve.
   • Green conveys the effect of pleasant messages.
   • Yellow indicates a warm, sunny message.
   • Orange produces a stimulating effect.

5. Plan when in your presentation you will use each A/V. Each should be built into the lesson plan to assure that it is used when most beneficial and for the purpose intended.

6. Use a variety of audio/visual aids. The variety provided by changing the types of audio/visuals can help reduce boredom and predictability of presentations.

7. Audio/visuals must be correct. The message transmitted by the audio/visual can be stronger than the words used by the instructor. People do remember more of what they see than what they hear!

8. Evaluate every A/V you plan to use. Does the aid truly help focus on the point in question, or is it merely a restatement of it? If it can be best presented in text, do so and save the impact of the other aids you will use.

9. Make sure all members of the audience will be able to receive the message the audio/visual is designed to send. Can the visual be seen and read by everyone in the room? Can everyone hear the message from tapes or video you are planning to use? If participants cannot adequately see and hear, the aids may be more frustrating than useful.

10. Familiarize yourself with the equipment well before time to utilize it.

11. Be sure the audio/visuals you plan to use are functioning well *immediately* before your session starts.

12. Remove or turn off the A/V equipment when you have finished. This will remove possible distractions to your audience.

13. Talk to your trainees, not to the audio/visuals.

WORKING WITH AUDIO/VISUAL SUPPLIERS TO CHOOSE THE BEST EQUIPMENT FOR YOUR NEEDS

When evaluating audio/visual equipment for purchase, it is important to find a dealer who can assist in making the buying decision. The dealer should also provide the service to keep your equipment operating effectively. A dealer should provide total support, including advice, consultation, training on the equipment, supplies, and accessories.

To evaluate equipment, you must first identify how you are going to use it. A good dealer will assist you by asking such questions as

- Who will be using the equipment?
- How will the equipment be used?
- Will usage of the equipment change in the foreseeable future?

Your dealer should also help evaluate whether to buy or rent equipment. Factors that should be evaluated include

- The frequency of use of the equipment over a given period.
- The portability of the equipment, if sessions will be held outside of a central location.
- The total cost of the equipment versus the accumulated rental costs over three to four years.
- The availability of rental equipment of the quality necessary for your presentations.
- Potential technological advances during the next few years in that equipment and their importance to your use of that equipment.

A useful tool when evaluating potential equipment is *The Equipment Directory of Audio-Visual, Computer, and Video Products*[4] published by the International Communications Industries Association. This book lists up-to-date information about current audio/visual products from over 450 manufacturers and 1,500 dealers. Each product has illustrations and specifications, including suggested list price, model number, weight, capacity, features, applications, accessories, and other technical details.

SUMMARY

Audio/visual aids can be beneficial tools for the effective sales trainer, which should result in better reception and retention for your trainees and a more professional image for the presentation and presentor.

REFERENCES

Allen, Sylvia. *The Manager's Guide to Audiovisuals*. New York: McGraw-Hill, 1979.

Anderson, Ronald H. *Selecting and Developing Media for Instruction*. Florence, KY: Van Nostrand Reinhold, 1983.

The Audiovideo Marketplace, 1985. New York: R. R. Bowker, 1985.

Auger, B. Y. *How to Run Better Business Meetings*. St. Paul, MN: Business Services Press, 1966.

The Best of Audiovisual Notes from Kodak. Rochester, NY: Eastman Kodak, 1983.

Broadwell, Martin. *The Supervisor as an Instructor, A Guide for Classroom Training*. Reading, MA: Addison-Wesley, 1968.

Brown, James W., Richard B. Lewis, and P. Harcleroad. *AV Instruction Technology, Media, & Methods*. New York: McGraw-Hill, 1983.

---

[4] *The Equipment Directory of Audio-Visual, Computer and Video Products*, 33rd ed. (Fairfax, VA: The International Communications Industries Association, 1987–88).

Bullard, John R., and Calvin E. Mether. *Audiovisual Fundamentals: Basic Equipment and Simple Materials Production*. Dubuque, IA: William C. Brown, 1974.

The Communicators Catalog. Rochester, NY: Eastman Kodak, 1984.

Craig, Robert L., ed. *Training and Development Handbook, ASTD*, 3rd ed. New York: McGraw-Hill, 1987.

Gropper, George L. *Criteria for Selection & Use of Visuals in Instruction*. Engelwood Cliffs, NJ: Educational Technology Publications, 1971.

Effective Training, A Guide for the Company Instructor. Scranton, PA: International Textbook Company, 1969.

The Equipment Directory of Audio-Visual, Computer and Video Products, 33rd ed. Fairfax, VA: The International Communications Industries Association, 1987–88.

Goudket, Michael. *An Audiovisual Primer*. New York: Teachers College Press, 1974.

Halas, John, and Roy Martin-Harris. *Audiovisual Techniques for Industry*. New York: United Nations Sales Section, Publishing Division, 1978.

Harrison, Jared F., ed. *The Sales Manager as a Trainer*. Reading, MA: Addison-Wesley, 1977.

How to Present More Effectively and Win More Favorable Responses from More People in Less Time. St. Paul, MN: Audio Visual Division/3M, 1984.

How to Run Better Business Meetings, A Reference Guide for Managers, The 3M Meeting Management Team. New York: McGraw-Hill, 1987.

Instructional Aids—How to Make and Use Them. Dubuque, IA: William C. Brown, 1974.

Kemp, Jerrold E. *Planning and Producing Audiovisual Materials*. New York: Harper & Row, 1985.

Kinder, James S. *Using Instructional Media*. Florence, KY: Van Nostrand Reinhold, 1973.

Laird, Dugan. *A-V Buyers Guide*; *A User's Look at the Audio-Visual World*, 3rd ed. Fairfax, VA: The International Communications Industries Association, 1980.

Munson, Lawrence. *How to Conduct Successful Seminars*. New York: McGraw-Hill, 1984.

Staton, Thomas F. *How to Instruct Successfully*. New York: McGraw-Hill, 1960.

A special thanks to the many members of the National Society of Sales Training Executives for their contributions to this effort; to Cindy Ploense, Larry Hoyt, and the people of North Star Audiovisual in Bloomington, IL, for their assistance.

chapter twenty

HOW VIDEO AND FILM CAN IMPROVE YOUR SALES TRAINING PRESENTATIONS

JAMES A. BAKER

James A. Baker is vice president of Baker Consulting Services and president of Baker Communications, Inc. (BCI), a leading training and career development firm. Established in 1979, BCI's ongoing clients list includes over 75 *Fortune* 500 companies and numerous medium and small organizations. Baker is a co-founder of the National Center for Dispute Settlement of the American Arbitration Association, an instructor in business communications at the Jesse H. Jones Graduate School of Business at Rice University, a member of the National Speakers Bureau, on the board of directors of Cenikor Foundation, an Executive Committee Member of Houston's Drug-Free Business Initiative, and a charter member of the Houston Chapter of Sales and Marketing Executives Society of Houston. He is a Ph.D. candidate in Psychology at Saybrook University, San Francisco, California, and holds a Master's degree in Business Administration from the State University of New York and a Bachelor's degree from the University of Rochester.

This chapter deals with subjects that relate to the visual aspects of selling. While both film and video are covered, technological advances, ease of use, and decreasing costs make video the method of choice for the presentation of visual materials. Much useful training material is still available on film, but increasingly, these films are being transferred to video tape.

This chapter covers:

- The purposes of training materials.
- The advantages and disadvantages of video and film.
- Developing a training program with video tape.
- Buying video-taped training materials.
- Making video-taped training material.
- Marketing your own training program.

Scientific research has proven that 90 percent of a sales representative's success is a result of both the *visual* and *vocal* image imparted. Persuasive selling depends largely on the face-to-face interchange and "chemistry" of the sales message. Studies have shown that only 10 percent of buying decisions are based on the words that the sales representative uses, and most decisions are made in the first three minutes of the sales call. The decision is based on three elements: nonverbal communication (60 percent); tone, pace, and pitch of voice (30 percent); and words (10 percent).[1]

One informal study conducted at the Harvard Negotiations Institute in 1986 compared response to training sessions in which lecture comprised more

---

[1] Albert Mehrabian, *Silent Messages* (Belmont, CA: Wadsworth, 1971), p. 43.

than 80 percent of class time with response to similar sessions where videotaped exercises and role plays comprised 80 percent of class time. Measured learning nearly doubled in the "video-enhanced" group, which helps explain why training often fails to achieve long-term results when teaching is dominated by lecture.

Video modeling and practicing on video tape can improve sales representatives' ability to use to their advantage all the psychological and situational reasons that clients buy, as well as their ability to improve the "chemistry" between themselves and prospective buyers. Words in a lecture cannot show salespeople how they appear to buyers. Nor can words show how to recognize body language and other signals from buyers who are evaluating the sales message and analyzing options.

When we train sales representatives to accommodate their style to match the customer's behavioral cues and to control their *own* body language and tone of voice, it increases their ability to recognize nonverbal positive and negative signs from the buyer. Everything we know about the visual nature of the art of selling supports the need for greater use of visual materials in training and far less use of the "lecture" method.

INSTRUCTIONAL GUIDELINES FOR USING TRAINING MATERIALS

Different types of instructional materials and methods are suited for each part of the training process:

1. Describing the selling concept—notebooks, written materials, and lectures
2. Skill demonstration—video modeling
3. Skill transfer—videotaped sales call and role playing

Written materials and lectures are most useful in describing specific aspects of a product or service. There is a vast difference between detailing the benefits of a pharmaceutical to a doctor and promoting a long-term consulting service in a chemical plant. The purely conceptual need will vary. In addition, the delivery of the sales presentation and the behavioral change sought in the buyer will differ somewhat, although much is common to all good sales calls, regardless of industry.

Effective sales training programs must develop the sales representative's knowledge of the sales process and evaluate whether or not he or she retains that knowledge. As a first step, the training program uses words to build concepts. This step should be accomplished quickly and effectively, and studies have proven that learning retention increases when video is added to lectures used in classrooms.[2]

Cognitive training does not mean that we have brought the salespeople to an application level. Though we have taught them what to do, we have not taught them how to do it. We have taught concept, but not practical application.

Visual and Verbal Persuasion

A sales presentation must be visually as well as verbally persuasive. This is where video and film become the most effective tools, with their ability to demonstrate skills. A short video vignette modeling a skill increases the understanding of the participants substantially and enables them to move on to role plays or applications with a stronger understanding of the skill.

The Motivational Impact of Video and Film

In addition to alienating the monotony of a lecture, video and film can be very motivational; in fact, they can create the perception of the need for the training. It is easy to show pictures of the actual sales environment, whether a chemical laboratory, retail setting or an oil field. This immediacy also makes training more relevant.

"Right Way" Versus "Wrong Way" Examples

As part of the video training process, it is critical to show incorrect ways in which things may be handled. Surprisingly, many sales representatives, as well as their managers, do not actually *recognize* what has been done wrong, despite many training experiences. In several tests, we asked experienced sales personnel to evaluate a deliberately flawed sales call as a program benchmark. Virtually unanimously, they agreed it had been a "great call" in each experiment and could therefore see no need for the training. "Right way" versus "wrong way" examples can illustrate the specific flaws in a sales presentation and show how salespeople may be causing some of their own problems with difficult customers.

THE ADVANTAGES AND DISADVANTAGES OF VIDEO AND FILM

Many trainers like to use film or video because they are effective for modeling behavioral skills, and the combination of sight and sound can accelerate learning.

---

[2] Walter Arno Wittich and Charles Frances Schuller, *Audiovisual Materials*, 4th ed. (New York: Harper & Row, 1967), p. 31.

In fact, one study indicated that learning visual identification, concepts, principles, procedures, skilled perceptual motor acts, as well as developing desirable attitudes and opinions, increased significantly with motion pictures.[3] With film or video, you can demonstrate and model individual skills separately or together. In either case, films or tapes can be replayed as often as needed. They can carry part of the instructional load as well as being a vehicle for skill recognition.

Film and video provide the excitement of continuous action, which slides, transparencies, or filmstrips cannot. For example, imagine how much the children's show *"Sesame Street"* would lose as a slide show.

Another general advantage of film/video is that it can help to ensure high-quality instruction. Many sales trainers have no formal background in either sales or training and must rely heavily on professionally prepared visual productions.

Visual productions can also demonstrate models of effective body language and tone of voice, which is important since up to 90 percent of a sales representative's success depends upon his or her visual image. Video or film can provide a large number of delivery skills that project confidence and believability, illustrate techniques for being aggressive in a strange territory, and demonstrate how to produce results. Specifically, we can demonstrate the right ways to:

- Assume a natural yet authoritative stance.
- Use natural hand gestures, including both what to do and what not to do with the hands.
- Control vocal delivery with pauses or inflection for reinforcement of a point or for emphasis.
- Handle discussions with clients.
- Listen for customer cues.
- Maintain eye contact.
- Restate questions.
- Answer questions succinctly without rambling.

The possibilities of video or film are as limitless as the needs of the sales staff.

Film: The Preferred Choice for Large Groups and "Top-Quality" Productions

Film is the natural choice when the audience or training group is large. Obviously, 50 to 100 people cannot conveniently view one TV monitor, and having

---

[3] Jack Tanzman, and Kenneth J. Dunn, *Using Instructional Media Effectively* (West Nyack, NY: Parker, 1971), p. 157.

several monitors in one meeting location may be impossible. Even when several monitors are available, the inevitability of chair movement while each participant locates a spot with a clear view is disruptive to the flow of the presentation. The need for a single screen may dictate the use of a film presentation.

Another strength of film over video, though this may not remain true in the future, is quality. Film has a wider "contrast range." That is, it can capture more detail in dark areas. It has higher resolution, more sharpness, and more color saturation. If, for example, you need depth of field for an outdoor shot in the shade, film gives better results. The technical sophistication of the viewing audience might be a consideration when you make a choice between film or video.

A final consideration is standardization of size. Film and film projectors come in relatively few sizes.

The disadvantages of film, however, deserve careful consideration. Film is generally more expensive to produce than video, by as much as four times. It takes separate crews to handle the film and the sound. If cost is no concern, filming will produce better quality in the master which can then be converted to video. Special effects can be added at that point; putting words on video is cheaper than on film.

Film, too, is less convenient to stop for stills since it was not designed to "freeze" as a frame. The ability to stop is critical in either role modeling or in replaying an individual sales representative's role playing. If you want to discuss a part of a film, you have to wait for the mechanical winding down of a machine, which is not as immediate as stopping a video player. If a film breaks during a presentation, it is not as convenient to have an extra reel of film as it is to have an extra video cartridge. It is also more expensive to duplicate film than video. Further, it is almost comical to reverse a film. Replaying a certain segment, over which there may have been confusion or questions, is also somewhat difficult.

After lectures (which we believe still occupy too much of the time in typical training programs), video is the most popular training method. In a study compiled by *Training* magazine, video surpassed films in actual use by more than 80

Table 20–1 Training method use in organizations of more than 50 employees

Lectures	81.8
Video tapes	80.5
One-on-one instruction	62.2
Role plays	45.1
Case studies	41.5
Films	47.4
Self-assessment instruments	35.8
Noncomputerized self-study	27.6

percent of the trainers in organizations with more than 50 employees. The amount of use and percentages are shown in Table 20–1.[4]

While each method has value, we argue for greater use of active methods over passive and, in particular, for greater use of video in role playing, case studies, and self-assessment.

Video: Seven Reasons for Its Popularity with Training Managers

There are many reasons to choose video over film: convenience, cost, sophistication, creativity, stability, flexibility, and the playback and recording features.

Convenience. Most offices now have a television set and a VCR; even in prevideo days when film was the only choice for movement and sound, most offices did not have film projectors, screens, or rooms to accommodate them. Video takes a limited amount of space compared to film. And video is more transportable. You can carry a camcorder, video player, and a few tapes on and off planes and in and out of cars much more easily than film reels and projector equipment.

Cost. Video is usually less expensive. Duplication of a video tape, often a necessity in a large training program, is also more economical than film, since blank VHS tapes sell for under $5.00. In addition, mistakes are much easier and less expensive to correct on video than on film.

Sophistication. After years of commercial television, people are accustomed to the jazzy, artful, dynamic quality of video. Video is a part of everyday living, while going to see a film may only be a once-a-month or once-a-year event. People are comfortable dealing with video, and they have come to expect the special effects that video makes possible.

Creativity. Much is being done with video character generation (producing words on the screen), and the process is becoming cheaper. Words can precede or follow action; words can appear adjacent to or overlaid on action. Graphics can accompany narrative. Many of the benefits of the lecture method still so popular with trainers can be combined with the added benefits of subliminal reinforcement. The combination of words, visual images, and graphics can dramatically increase retention.

---

[4] Jack Gordon, "Where Training Goes," *Training*, Vol. 23 (10) (October 1986), p. 60. Reprinted with permission from the October 1986 issue of *Training*, The Magazine of Human Resources Development. Copyright 1986, Lakewood Publications Inc., Minneapolis, MN (612) 333–0471. All rights reserved.

Stability. On the whole, video withstands the abuse of travel, weather, and use. It stays in its own case. It is not fed through a machine loaded with mechanical gears. Video can be stopped and restarted with ease, where film cannot. Over a period of time, film may break and video may stretch, but stretching is much easier to deal with in the midst of a training session. If this happens, usually the tracking adjustment on the VCR will correct the situation until the tape can be replaced. If a film breaks, it must be removed from the projector, spliced, and then rewound.

If the heads on the VCR damage a tape during a presentation, the problem still isn't as great as having a film break. And it is much easier to have a spare tape than a spare film.

Flexibility. A key advantage of video over film is the ability to reverse and fast forward to a specific spot with ease. Most VCRs can search or scan quickly with the image on the screen. In a training situation, the relatively short time with which this can be accomplished is a great advantage, especially when the subject is behavioral skills. Trainers can repeat and review demonstrations of sales skills as often as they need. Trainers can use the scan feature of video to resequence course material to fit particular needs, even something as drastic as trimming a three-day course to one day. That's tough with film.

Recording and Playback. As training moves into practice, video has its greatest benefit—on-site recording. Most of us do not have a whole sound stage handy, which is necessary for film. Video is more mobile, convenient, and immediate.

Video facilitates practice of skills. You can tape a sales representative's practice session and play it back immediately as part of several different training methods: self-study, assessment, group discussion, feedback, and suggestions for improvement. Fast-forwarding with the image on screen can emphasize the simplest unconscious gesture or mannerism; an image is worth a thousand words in making sales representatives aware of needed improvements. Studies show people learn more by observing body language and tone of voice. In this case, the sales rep learns from his or her own image. The immediate reinforcement provided by video is central in changing behavior patterns.

Along these lines, video can be beneficial in focusing on improvement of behavioral skills in general. By using a simple evaluation form that parallels the skills being taught, you have the perfect vehicle for discussion. A sample is shown in Figure 20–1. The form becomes a guide for self-improvement.

Video taping is also very effective for evaluating the organization and completeness of a sales presentation. Depending on the skills being taught, an evaluation form might rate a sales representative on his or her *delivery* alone (without regard to content). A form for these purposes is shown in Figure 20–2.

Again, when a form like this is completed by both the sales representative

Figure 20–1 Sample video evaluation form

5 = excellent	1 = poor	Comments
Eye contact (strong/minimal)	5 4 3 2 1	_____
Language (persuasive/weak)	5 4 3 2 1	_____
Posture (excellent/poor)	5 4 3 2 1	_____
Appearance (professional/careless)	5 4 3 2 1	_____
Voice (energetic/boring)	5 4 3 2 1	_____
Self-confident (poised/uneasy)	5 4 3 2 1	_____
Gestures (expressive/minimal)	5 4 3 2 1	_____

Figure 20–2 Sample sales representative delivery evaluation form

5 = excellent	1 = poor	Comments
Opening (focused/scattered)	5 4 3 2 1	_____
Body (coherent/illogical)	5 4 3 2 1	_____
Closing (forceful/inconclusive)	5 4 3 2 1	_____
Strategy (persuasive/argumentative)	5 4 3 2 1	_____
Enhancement (memorable/boring)	5 4 3 2 1	_____
Visuals (reinforce/distract)	5 4 3 2 1	_____
Time (appropriate/lengthy)	5 4 3 2 1	_____
Audience (strong/minimal)	5 4 3 2 1	_____

and the trainer, you have a very useful guide for self-study, discussion, further practice, and assessment. You can also measure and document improvement over time by using checklists periodically.

Potential Drawbacks. In spite of all its advantages, video does have some drawbacks. For one, it tends to have a passive quality—it can produce an impersonal one-way communication that lulls both trainer and trainee to sleep. Viewers are also at the mercy of the taped information, which is often targeted to the slowest possible student. Although the tape doesn't have to be professionally produced, good production can markedly increase communication.

Like film, video is harder to update (for example, if the company changes its name) than are slides, handouts, and transparencies.

Video should be an aid, not a substitute. Perhaps we should view video as a very tasty dish on the cafeteria line of instructional tools. It is really up to us to decide whether it is the main course or dessert.

DEVELOPING A SALES TRAINING PROGRAM WITH VIDEO TAPE

Most organizations still rely on in-house training staff to develop and deliver training. In 1983, 42.5 percent of organizations surveyed preferred in-house development and delivery of sales training, and in 1986, 30.4 percent still preferred this method. The preference for sales training developed by outside vendors but delivered by in-house staff was much lower: 16.0 percent in 1983 and 14.8 percent in 1986. Vendor- or consultant-developed sales training seminars or workshops consisted of only 9.8 percent of companies' total training effort in 1983, growing by only .4 percent in 1986.[5] The implications appear to be clear: in-house trainers carry the primary responsibility for the development of sales training programs and must be prepared to select the best materials, be they slides, transparencies, handouts, video, or film. They may have to produce their own video tapes or films while considering *value received* for dollars spent. And they must construct sales training programs that are effective in improving that crucial 90 percent of sales representatives' visual and vocal message.

Before you can make final decisions about whether to buy or make your own instructional materials, you must consider the size of the total training staff, as well as the size of the company. *Training* magazine's study in its October 1986 issue (Table 20–2) offered interesting insight concerning the "Development and Delivery of Training as a Function of Organization Size."[6]

---

[5] Ron Zemke, "Development and Delivery: Classroom Training Still Most Common Option," *Training*, 23 (10) (October 1986), p. 58.

[6] Ibid., p. 59.

Table 20–2 Development and delivery of training as a function of organization size

Number of Employees

50–99
*********************25.9
++++++++++++++27.0
100–499
********************************34.1
+++++++++17.4
500–999
********************************34.1
+++++++13.3
1,000–2,499
***39.8
++++++11.3
2,500–9,999
**39.4
++++10.9
10,000+
************************************37.2
+++8.3

* = Mean percent of all training designed and delivered by in-house staff.
\+ = Mean percent of all training obtained via public workshops, seminars, and conferences.

Developing your own training programs depends on the amount of in-house expertise available as well as the number of employees requiring training and the number of times and locations the programs will be used.

Six Questions to Ask When Considering Media and Film for Your Program

To help evaluate the best approach, ask yourself the following questions:

- What is the purpose of this training program and how will it meet the organization's strategic goals? Do you need to convey information or do you need to improve skills? This is critical to your choice of training media.
- How many times will we use this sales training program? Your company may be able to justify spending more on a sales program that will be used dozens of times, and have a direct return on their investment.
- How will this sales training program be presented? This will affect the simplicity or complexity of the program. You may have to build more prepack-

aged material into a sales system that will be conducted by office sales managers or others who lack formal training skills.

- Where will the sales training occur? What audio/visual and duplicating equipment already exist at each location? This will affect the mobility of the program and the machinery that must be carried to each location.
- Who is the audience? This will affect content. Are you developing a program for both sales representatives and sales managers, both sales engineers and technical service representatives who interface with customers?
- What is already available? Do you have the core of the sales training program and simply need to update it? Or are you starting from scratch?

The following sections discuss in more detail the problems of creating a totally new sales training system and deciding whether to buy off-the-shelf programs or to make your own.

PURCHASING VIDEOTAPED SALES TRAINING MATERIALS

There are still some very poorly made videos being sold. One recent example is a set of training films for telephone sales. Each tape in the set was edited from a two-day live presentation and was condensed into a total of two hours. The editing was sloppy and the segue was missing. The recommendations to "stop and discuss the tape . . . now" were amateurish and redundant. The producers of the tapes apparently confused their audience with an audience who lacked any TV exposure.

Mechanical stops for discussion are necessary, especially when the trainer is not knowledgeable or is merely a facilitator, but the machine should not insult us by telling us to do it. Some videotapes overcomplicate, including virtually everything on the tape when some of it could be left for the training manual. Other videotapes oversimplify to the point that they are hardly any use at all.

In general, however, off-the-shelf video training programs have changed considerably since the middle 1970s. Early videotapes were often awkwardly staged, terribly simplistic, and set in unbelievably stark office settings with actors who sounded like actors. Too many obvious mechanical stops were built in. Role playing and modeling seemed far removed from the practical world. The examples used often had limited application—technical people were unimpressed with how to sell washing machines in a department store. Other tapes may have had great selling ideas, but little understanding of the business milieu. Some well-packaged programs used good concepts and skill applications but did not include the rationale for when or why to use them. Some vendors of training tapes tried to put everything on the tape, instead of putting some of the program into an instructor's guide. Most of these problems have disappeared today.

The entire business of producing instructional video materials has undergone tremendous growth, and competitiveness has bred new levels of sophistication. As with any infant industry, trial and error and new entrants have improved the early products. Today, Harvard case studies, behavioral data, and statistical quality information have all been added to the skill-based programs to keep up with the complexity of competitive world markets. Currently, off-the-shelf training materials are more informative, more practical, more tailored to individual needs, and more cognizant of how people learn; further, the number of vendors and video tapes has multiplied many times over. Magazines for human resource specialists contain hundreds of advertisements for sales training programs, and universities stock extensive corporate training films. In larger cities, telephone Yellow Pages list local vendors and video production companies.

How to "Preview" Video Tapes

Previewing video tapes can be a quick method to locate possible training materials at either nominal or no cost. Some need merely to be rented for a few days. However, don't expect the panoramic extravagance of *Gone with the Wind* on a local news bulletin budget. Just because it's video does not mean that it's sophisticated or complete.

Searching through the mass of video tapes available can be very disappointing and potentially fruitless. Previewing video tapes by ordering from vendors' brochures may take several weeks, which is no problem only if you have plenty of time. Instead, try visiting a university with a large library of tapes (an economical and practical solution) or use personal contacts within your industry. Or check with local professional associations to see what might be available.

However you begin your search, eventually you will have one or more tapes to evaluate. Figure 20–3 is a checklist to use to compare video tapes for the particular characteristics you want.

Even though evaluating video tapes can be time consuming, it must be considered a serious effort. Budgets can be defended for effective training programs.

HOW TO MAKE YOUR OWN SALES TRAINING VIDEO IN SIX STEPS

Even with the proliferation of corporate sales training video tapes, you may not be able to find exactly what you want. One option is to buy or borrow tapes that can be used in part and simply work them into the program. This may be acceptable if you are in control of the training and have the time to develop the rest of the program.

Figure 20–3 Checklist for video tape selection

Desired Characteristic	Tape A ( )	Tape B ( )	Tape C ( )
PURPOSE			
New information			
Concepts			
Skills			
Drills, practice			
Supplemental information			
ADVANCE			
Sophistication (rate 1–5)			
Need (rate 1–5)			
Preparation (rate 1–5)			
FORMAT			
Interactive			
Motivational			
Controllable			
Sound instructional design			
Clear information			
Provides reinforcement			
Assessable			
PRODUCTION			
Color and graphics (1–5)			
Music and narrative (1–5)			
Appropriate setting (1–5)			
LIFE EXPECTANCY			
Dated video information			
Dated audio information			
COMPLETENESS			
Instructional booklets			
Practice vehicles			
TONE			
Too condescending			
Too technical			
TIME REQUIREMENTS			
Mastery time			
Organizational deadline			
COST FACTORS			
Absolute cost			
Useful life			

Figure 20–3 (Cont.)

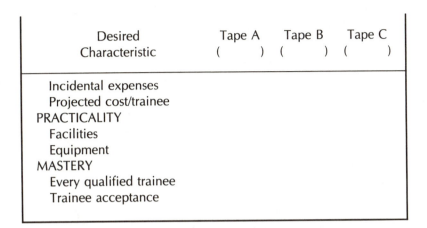

Desired Characteristic	Tape A ( )	Tape B ( )	Tape C ( )
Incidental expenses			
Projected cost/trainee			
PRACTICALITY			
Facilities			
Equipment			
MASTERY			
Every qualified trainee			
Trainee acceptance			

Another option, if budget allows, is to hire a vendor to custom-make a video tape that exactly suits your needs. This can either be the most or the least expensive option. At this writing, a complete customized program can cost from $100 to $1,000 per course participant, depending on the number of people to be trained.

The final option is to make your own sales training video tapes. This may be the most economical option if you have the equipment, can hire professional talent, and will use the tape enough times for it to pay for itself. One training manager received bids ranging from $15,000 to $30,000 for an entire training tape. She ended up making the tape herself for $1,300 after assessing her own resources.

Her company already had all the recording equipment she would need. She had in-house talent to handle cameras, microphones, lights, sets, and direction. Through a friend, she located actors at a local community theater who worked without compensation, except for a personal copy of the finished tape. She wrote her own script, took a course on video production on her own time, and spent many hours developing the specific plan.

In the beginning, she had no idea how much time was involved in planning, in shooting and reshooting, or in building and tearing down sets. Further, since a script is never shot in sequence but rather broken down into segments that will all be shot on a certain set, she discovered the importance of labeling each tape with the exact page of the script so that editing would not be a nightmare.

On the whole, the video tape suited her organization's needs exactly. The actual cost, not counting salaries of employees involved, included only the blank tapes and final editing by a video production company. She and her boss felt that the time and effort were well spent.

Another trainer in a natural gas industry also made her own video tape using company officers as actors. Two of the officers took early retirement shortly after production of the tape, which automatically dated it. Although she felt that her tape was more specific to her needs than anything off the shelf, she now strongly recommends hiring professional actors.

While we could not hope to present an entire course on video production within these pages, we can share the techniques and procedures that have worked for us.

Building a sales training video tape has several important sequential steps: developing the idea, determining the treatment, gaining authorization, developing the script and story board, production, and postproduction.

Step 1: Developing the Idea

Clarify Specific, Immediate Need for the Video. Start with long discussions including sales management, sales personnel, and company officers, and determine the specific need the video will address. Training becomes more effective and appealing if it is based both on sales representatives' statements of what they need (how to handle price increases) and what management says it needs (why price increases are needed). This is much more effective than basing a course solely on what training personnel thinks the sales team needs (to improve sales). When the sales team has a sense of "ownership" in the project, it will support the training program and help to achieve management goals.

Consider Long-Range Factors. Equally important at this stage is the future direction the company plans to take. What are its needs and wants? Is it selling a long-term relationship in a tight-supply market or selling a commodity product at a higher price in a competitive market? Assessing these factors is crucial; you must take into account where your company is going, what the industry and competitors are doing, and what long-term company goals are. The final training program will only be as effective as the amount of effort and planning that occurs at this first stage, and everyone who will be affected by it should have input.

At this point, too, establish a budget, draw up a time frame for completion, and gain commitment and political buy-in from everyone who must be involved in the ensuing steps.

Step 2: Determining the Treatment

The first step in drawing up the preliminary plan for your video is to answer these questions:

- Who is your target audience? What are their ages, educational backgrounds, and socioeconomic levels?
- Is your production service a one-call close, or are you required to make several sales calls to gain the order?
- What is the message? What information and impression do you want your target audience to have? How much of the video should be narrative and how much should be examples? What kind of examples will be most useful? Are you attempting to persuade, educate, or inform your audience?
- What script material is available? Are there articles, brochures, and press releases that would be helpful? How should material be presented—"bottom up" or "top down"? Should an overview be presented first?
- What type of production do you envision? Will you use a studio or customer-simulated location shots, and what additional problems and expenses are associated with each location? Will you video tape your "sales" environment at restaurant locations, customer offices, or in sales in situations where people are driving in their cars.
- What type of talent do you need? You can use a sales manager, or members of your marketing organization, actors as spokespersons, actors in roles, and voice-over narration.
- What types of props and costumes are required? Is typical business dress appropriate?
- Do you require special effects, such as animations, computer graphics, illustration art, and so on?

While you consider these above questions, keep in mind the following guidelines:

- Create the need for the training film.
- Use a familiar setting, whether it is a lab, office, conference room, or an outdoor location.
- Hold the content to no more than five skills in any two-day training session.
- Don't oversimplify or overcomplicate the skills; keep the video tape simple without being condescending.
- Use a minimum of language; let the visual material carry the load.
- At some point, break down each skill into its parts without being overly repetitious.
- Build in stops for discussion, but make sure they are subtle and natural.
- Limit stops to no more than four or five stops per 10 minutes of video tape.

- Build in some form of test or assessment, but make it nonthreatening.
- Plan how to enhance video with words, but keep in mind that words on film are still rather expensive.

Start by preparing a four- to six-page overview of the idea. Include as many varying opinions as possible from all those who contributed to the project. Does the planned execution correspond with their needs? Do they agree with the examples? Do they feel the content is complete? Here again, make an effort to create a sense of ownership in the project in everyone who is involved.

Step 3: Gaining Authorization

This is a very critical stage. The treatment you develop must be kept short, so that each party involved will read it. And all participants must be available promptly for the project to stay within the predetermined time frame. To control program development, it is best to have each person indicate approval by initialing and dating each page. Suggestions for additions, deletions, and changes in the treatment can then proceed smoothly. Be sure to keep everyone informed of any changes made.

Step 4: Developing the Script and Storyboard

Your writer (whether in-house or outside) must be business-oriented, creative, humorous, human, and never condescending. The writer must be able to:

- Identify with the culture and sophistication of the audience and understand adult learning styles.
- Understand the training process enough to write a good overview, set clear objectives, develop good models of the skills being taught, and structure the sequence of learning.
- Understand the business and corporate style enough to use models or examples that relate well.
- Understand video as a medium well enough to write believable dialogue rather than straight narrative, using video to titillate, teach, and test.

The wording of the narrative and the examples of selling skills must show the people-side of the business. Any stiffness in wording should be changed. If the writer is not equally strong in all the foregoing areas, an unbalanced learning curve will occur, and the video will be less effective.

It's generally a good idea to have the producer work with you as you develop the script and storyboard, if possible. (See the next section for some suggestions in finding a producer.) Everything you write into the script costs money, and the producer can help you budget the production as you go along. Many companies, in fact, pick the producer before they choose a script writer.

The cost of producing your video will depend on many factors: locations of shoots, crew and equipment required, talent used, and incidental expenses. Avoid surprises by writing your production to fit within your budget. Asking a producer "How much will it cost to produce a 30-minute video?" makes as much sense as asking a builder "How much will it cost to build a three-bedroom house?"

As the script is being written or immediately after it is complete, you must add visuals to the words or create the "storyboard." How will skills be illustrated? What action is taking place? What exactly should the graphics look like? What should the set look like?

The presence of the producer is an absolute must for storyboarding, because their knowledge of visual methods of presenting information is invaluable. In 90 percent of cases, a two-column script with the words on the left and a written description of the visual image on the right is sufficient. In some cases where the action is very complex or multiple cuts are involved, you may have to illustrate the action with pictures and diagrams. This is quite time consuming and expensive, however, so avoid it if you can.

When you have the final script in hand, check back with the various sales representatives, sales managers, and company officers to get final approval. The best way to reach final script approval is to conduct a sort of "round table" reading with the writer, sales personnel, and management all present. Again, suggest that each person initial and date each page of the script. If the script needs a significant number of changes, you may need another "round table" discussion to approve the rewording.

Step 5: Producing the Video

The right producer will be one with a sound business sense, creativity, and the ability to collaborate with you, your staff, and the writer. Contact local film and video associations or locate producers through referrals from others in your field. View some of their productions. Check their references. Check their equipment. There are three categories of equipment to choose from: "consumer" format (simple and inexpensive VHS equipment), "industrial" format (¾-inch), and "broadcast" format (1-inch or high-speed, ½-inch BetaCam or M2). The difference among the image quality of these three types of equipment is similar to the difference among Super-8 home movies, the 16mm movies you saw in high school, and the 35mm movies you see at a movie theater. The preferred format for quality and ease of editing is the ½-inch BetaCam. Most TV stations now use this format for their local news features. Check the rate cards. The busier the producer, the better the quality of the finished product, as a rule.

Establish regular meetings with your producer at important checkpoints: script and storyboard approval, budgeting, logistical planning, and postproduction editing. You want your producer to develop his or her own sense of "ownership" in the project. Before any filming proceeds, make sure you negotiate a contract to preclude the possibility of unauthorized price changes.

Selection of actors comes next. Always request final approval of the actors that the producer will use. The whole video will become ineffectual if the actors are not professional or are not suitably chosen. We recommend you use members of the Screen Actors' Guild. On one occasion, a producer wanted to substitute an actress when the one we had approved could not fulfill her commitment. The character we needed was a strong, aggressive businessperson, and the new actress looked like a frail, victimized woman who had never worked outside the home. We knew that this substitution would create a casting mistake that could ruin the video tape. To avoid this kind of problem, make sure all members of the cast will be available on schedule.

Preproduction planning involves logistics, getting the shoot accomplished most expeditiously in terms of locations and talent. If you use expensive talent, you may want to shoot all the scenes with that talent first, even if you have to travel to several locations. Or you may find it less expensive to shoot all scenes at each location and keep talent hanging around, even if they're not doing anything. Working out the shooting schedule may be quite a juggling act, because in nearly every case, videos are shot out of sequence. The cost of production will depend on the cost of crew, equipment, and talent (priced at hourly or daily rates), plus incidental travel and other expenses.

Shooting for graphics, special video effects, and studio art usually takes place during the same time frame as the action sequences, so that when you get to the editing step, you have all the raw material you need to work with.

Carrying out the actual shoot may be done in a number of different ways. In many cases, video segments will be only 30 seconds or so long, and actors can learn their lines while the lights and sets are being readied. A few runthroughs without taping is generally enough, after which the scene can be taped as many times as necessary to get everything right.

For longer segments, memorization can also be used, although it may be equally effective (and less expensive) to use cue cards, a teleprompter, or audio prompting. Many professional spokespersons use a pocket recorder with an earplug that cannot be seen on camera. They record their dialogue into the recorder and play it back in their ear, parroting their own lines. This is very effective, because it is easy to change on the set (unlike teleprompting), it is suitable for long narrations, and the actor does not have to maintain eye contact with the camera. It also avoids the vocal inflections that typically go with reading.

Graphics are done in the studio, and may comprise character generation, computer graphics, digital video effects (DVE), and straight photographic art. Character generation is a simple method of "typing" words on the screen, and the cost is generally included in editing. Computer graphics may run the gamut from fairly simple to the incredibly exotic, depending on your budget. DVE include such devices as squeezing an image into a box and tumbling it around, flying through an image, and other phenomena familiar to those who watch the TV news. Photographic art refers your company logo or other artwork. Slides and film can also be transferred to video. Except for character generation, graphics are all priced separately, so be sure to include them when developing your budget. If you can't include every special effect you want, take heart—this kind of thing can be overdone.

Step 6: Postproduction Editing

The first step is "off-line" editing, a simple procedure using two machines and a controller that is much less expensive than an "on-line" editing suite. Off-line editing is where all the creative decisions are made regarding which segment of which tape to place where, with what kind of cut, dissolve, fade, wipe, and so forth to the next segment.

First, the raw footage is window-dubbed with a time code. This means that a duplicate tape is made which shows a time code and frame number in a small box in the corner of each screen. This tape is used for the rough off-line edit: segments are extracted and combined to produce a finished program with all the numbers that describe the footage. This is the tape used to make revisions and obtain the sign-offs you need for final production.

Once your rough edit tape is approved, you move on to the on-line suite. The on-line edit is a fairly mechanical process in which a computer takes the original tape, plus the numbers from the rough edit, plus information on what kinds of cuts and dissolves you want, and produces the final tape. Usually there are two options (with different prices) for on-line edit: two machine and three machine. With a two machine edit, you have one source machine and one recording machine, so only simple cuts between segments are possible. With a three-machine edit, you have two source machines, which make it possible to do fades, wipes, and other special effects involving dual image transitions.

Once your final tape is produced, the only remaining steps are transferring the tape to the distribution format (usually VHS) and making enough copies to distribute.

Generally speaking, if your planning is sound and involves the right people, if your treatment and script are well thought out and approved by the same people, and if your producer is creative and willing to work both for you and with you, you should have a final video tape that will meet your needs successfully and effectively.

MARKETING YOUR SALES TRAINING VIDEO
PROGRAM _____

We all know what it's like to conduct a training program where everything goes as planned and each participant feels uplifted or renewed or more informed because of our efforts. Regretfully, many participants in training programs come because their supervisor tells them to, or they come out of curiosity, or they need a break from the routine, or they need a course to include in their performance appraisal review. This is just reality.

David Dayton, a management consultant, offers some advice on drawing the right crowd. He suggests marketing the program carefully. Make sure the participants need the knowledge you're offering and know that they need it. Further, he suggests marketing what it is you have the answer for, that is, "selling the ends, not the means." Finally, he says, "cite specific changes" that will result from the training, both in the short and the long term.[7] You will achieve greatest success if video is used to its full potential to improve visual and vocal image.

SUMMARY

When we look at how far video has come and how universally it is used, we can be certain that video technology will enhance training programs far into the future. Today with the development of the laser disc and the computer-driven interactive teaching technologies utilizing it, the horizon seems still broader.

Yet, on a practical scale, video has more to achieve in industrial training programs. Far too many businesses still rely too heavily on lecture and too little on role playing, visual feedback, and reinforcement. Ninety percent of selling is a body language art. And video is far and away the best method of teaching it.

BIBLIOGRAPHY

Allesandra, Anthony J. *Non-Manipulative Selling*. Reston, VA: Reston, 1981.

Argyle, M., F. Alkema, and R. Gilmour. *"The Communication of Friendly and Hostile Attitudes by Verbal and Nonverbal Signals."* Unpublished manuscript, Institute of Experimental Psychology, Oxford University, 1971.

Berscheid, E., and E. H. Walster. *Interpersonal Attraction*. Reading, MA: Addison-Wesley, 1969.

Bettger, Frank. *How I Multiplied My Income and Increased My Happiness in Selling*. Englewood Cliffs, NJ: Prentice Hall, 1982.

Bettger, Frank. *How I Raised Myself from Failure to Success in Selling*. Englewood Cliffs, NJ: Prentice Hall, 1975.

Elsea, Janet G. *The Four Minute Sell*. New York: Simon & Schuster, 1984.

Evered, James F. *A Motivational Approach to Selling*. New York: American Management Association, 1982.

---

[7] "Training Today," *Training*, 23(10) (October 1986), p. 10.

Gayle, Willie. *Power Selling*. Englewood Cliffs, NJ: Prentice Hall, 1959.

Girard, Joe, and Stanley H. Brown. *How to Sell Anything to Anybody*. New York: Warner Books, 1979.

Herman, Fred. *Selling Is Simple—Not Easy, But Simple*. New York: Vantage Press, 1970.

Hill, Napoleon, and W. Clement Stone. *Success Through a Positive Mental Attitude*. New York: Pocket Books, 1977.

Hopkins, Tom. *How to Master the Art of Selling*. New York: Warner Books, 1982.

Kimble, G. A. *Hilgard and Marquis' Conditioning and Learning*. New York: Appleton-Century-Crofts, 1961.

Kinder, Jack, and Garry Kinder. *The Selling Heart*. Indianapolis: R & R Newkirk, 1974.

Linkletter, Art. *How to Be a Supersalesman*. Englewood Cliffs, NJ: Prentice Hall, 1974.

Mehrabian, A. "Verbal and Nonverbal Interaction of Strangers in a Waiting Situation." *Journal of Experimental Research in Personality*, 5 (1971), pp. 127–138.

Mehrabian, A. "Significance of Posture and Position in the Communication of Attitude and Status Relationships." *Psychological Bulletin*, 17 (1969), pp. 359–372.

Mehrabian, A. "Orientation Behaviors and Nonverbal Attitude Communication." *Journal of Communication*, 17 (1967), pp. 324–332.

Molloy, John. *Dress for Success*, New York: Warner, 1976.

Riesman, D. *The Lonely Crowd*. New Haven, CT: Yale University Press, 1950.

Roth, Charles B., and Roy Alexander. *Secrets of Closing Sales*. Englewood Cliffs, NJ: Prentice Hall, 1982.

Schuller, Robert H. *Tough Times Never Last, but Tough People Do*. Nashville: Thomas Nelson, 1983.

Schwartz, David J. *The Magic of Thinking Big*. St. Louis: Cornerstone, 1962.

Shook, Robert. *Winning Images*. New York: Macmillan, 1977.

Sommer, R. *Personal Space*. Englewood Cliffs, NJ: Prentice Hall, 1969.

Wallace, Joanne. *Dress with Style*. Old Tappan, NJ: Fleming H. Revell, 1982.

Walters, Dottie. *The Selling Power of a Woman*. Englewood Cliffs, NJ: Prentice Hall, 1962.

Wheeler, Elmer. *Sizzlemanship*. Englewood Cliffs, NJ: Prentice Hall, 1983.

Willingham, Ron. *The Best Seller: The New Psychology of Selling*. Englewood Cliffs, NJ: Prentice Hall, 1984.

Ziglar, Zig. *See You at the Top*. Gretna, LA: Pelican, 1975.

Zimmer, Allen E. *The Strategy of Successful Retail Salesmanship*. New York: McGraw-Hill, 1966.

chapter twenty-one ─────────────

COMPUTER-BASED AND INTERACTIVE VIDEO TRAINING TECHNOLOGY

─────────── *PATRICK A. CATALDO, JR.*
AND DANIEL S. COOPER

Patrick A. Cataldo, Jr., is the vice president of educational services for Digital Equipment Corporation. He is responsible for the design, development, delivery, and measurement of training products, systems, and services for both internal employees and Digital customers. Prior to assuming corporate responsibilities for educational services in 1985, he was the corporate sales training manager. Pat has held previous line management positions as the U.S. area training manager and the general international area training manager, where he began his career with Digital in 1974. Cataldo has completed the Executive Development Program at the University of Virginia and holds an M.B.A. from Boston College and a B.A. from St. Francis College in Loretto, Pennsylvania. Most recently he received an honorary doctor of science degree from Grand Valley State University, Michigan. He has been recognized for his achievements in training by the American Society for Training and Development and is a past recipient of the ASTD International Trainer of the Year award. He is a member of the board of directors of the National Society of Sales Training Executives.

Daniel S. Cooper is the training plan systems manager within the educational services department of Digital Equipment Corporation. He has extensive experience in the development, implementation, and evaluation of computer-based and interactive video training programs, including several years in sales training at Digital. He received his Ph.D. in Education from Stanford University, specializing in computer-based training and his A.B. from Harvard University.

Two of the most exciting new sales training technologies are computer-based training (CBT) and interactive video training (IVT). Both are uniquely powerful ways to learn interactive skills, such as qualifying a customer. However, both technologies are in their infancy, and their use can be risky if not carefully managed.

This chapter identifies the special strengths and weaknesses of CBT and IVT for sales training. It describes their use for several different sales training objectives and audiences. Highlights include a discussion of practical requirements and a preview of technology advances that are altering the uses of CBT and IVT. Also included are suggestions for introducing these technologies smoothly into an existing sales training program.

COMPUTER-BASED AND INTERACTIVE VIDEO TRAINING: FAST, EFFECTIVE, AND FLEXIBLE

Suppose you are the manager of a sales force that is facing intense competitive pressures. You are racing to get a new product to market before a competitor's product is unveiled. The time is very short for introducing your new product to your sales force, and this new product will require your sales force to use special procedures for qualifying customers. Sales representatives also will have to deliver a very precise positioning message. If the sales force does not follow the new procedures, customers may have unrealistic expectations, leading to disappointment and the loss of valued customers. In summary, a poor selling job—or a selling job done in old ways—will cause more harm than good.

Under these circumstances, you would probably seek a very effective training method, which could deliver training rapidly, in widely dispersed locations with

---

The views expressed here are not necessarily those of Digital Equipment Corporation.

flexible schedules. An ideal training method might include the capacity for interactive simulation exercises. Simulations would verify that the sales representatives had learned correct sales procedures and messages for the new product. Simulation exercises would also reassure you that sales representatives had practiced their ideas and seen the consequences in private, rather than in front of your customers.

Finally, you might be very interested in a method of delivery that provided good record keeping and information tools to manage the training. For example, the computer-generated records and tools might let you know when most of the sales representatives in each region had received training and alert you to any unusual problems that developed in particular locations.

Computer-based training and interactive video training offer the potential for just this type of training.

HOW CBT AND IVT DIFFER FROM OTHER FORMS OF MEDIA-BASED INSTRUCTION

Computer-Based Training

CBT, which has been evolving gradually since the late 1950s, refers to training that is delivered through computer terminals or personal computers. It uses the computer screen (or paper printout) to present most of the instructional material. It also relies on the keyboard or some other computer input device (such as a touch-sensitive screen) to get responses from students. Occasionally, computer-based training may be supplemented by printed materials like workbooks or by audio/visual materials like video and audio cassettes. However, the distinctive characteristic of computer-based training is the use of a computer terminal or personal computer as the primary medium of instruction.

Interactive Video Training

IVT first appeared in the late 1970s. The invention of computer-controlled videodisk players was the key ingredient in its evolution. In practice, its distinctive characteristic is the use of video in close combination with a computer terminal or personal computer.

To the student, it typically appears to be a combination of television and computer-based training, together on the screen at once. For example, it usually involves motion pictures with sound, as in television, or sometimes still pictures with sound. However, the pictures and sound stop from time to time to present problems or choices to the student. Based on the student's responses, the system presents different follow-up materials.

In other words, interactive video differs from traditional, "linear" video by adding the capability to "branch" to different follow-up sequences, depending on the student's actions. This quality of "interactive" stop-and-go dialogue with the student, which encourages an active participation by the student and modifies the training for different students' needs, is shared with computer-based training. It is the quality that most distinguishes these two training methods from all other forms of media-based instruction.

FIVE EXAMPLES OF CBT AND IVT PROGRAMS USED IN SALES TRAINING

A few examples may clarify the types of computer-based training and interactive video training that are presently used in sales training programs. (These are composites of several programs, not specific real programs.)

Computer-Based Training with Minimal Branching

This form of computer-based training is functionally similar to printed workbooks, except that it has "unskippable" exercises. A minimal-branching computer-based training program might be used to communicate simple information, such as straightforward facts and procedures for a new product. Typically, standard computer terminals or personal computers would be used, and the program would appear as a series of "pages" (individual screens with text and possibly graphics).

At appropriate points, quizzes would appear, requiring a multiple-choice or fill-in-the-blanks response from the student. Depending on a student's responses, the program might offer congratulations or suggestions for further review.

The primary advantages of minimal-branching computer-based training over a workbook are that (1) exercises are almost always completed (because they are difficult to skip), providing important feedback to learners and performance reports to managers; and (2) the screen pages can be kept up to date more easily than printed pages. One common use of minimal-branching computer-based training is for inexpensive training on using computer software, like an order-entry system.

Sometimes, this type of computer-based training consists primarily of a single, long example that requires the student to enter software commands at appropriate points. In this case, it is sometimes called a "tutorial." Tutorials offer knowledgeable users a way to learn rapidly by seeing the new software in action. Their value for novices depends on whether the material is easy to learn by simply following an example.

Computer-Based Training with Extensive Branching

This variety of computer-based training differs from the preceding one by offering many branches, or paths, through a course. Typically, the different paths are aimed at learners with different backgrounds or learning needs. For instance, a program to prepare many sales representatives for selling a new type of product might have different branches for representatives with little experience in their company, much experience in their company, much experience in a competitive company, and so forth. To direct various learners among the branches, the program might use tests, self-choice menus, or a combination of both.

Computer-Based Training Simulation Games

Programs employing computer-based training simulation games usually have a competitive element. For example, they may put the learner in the position of a new sales representative trying to achieve a difficult sales quota. Then they may pose a series of challenging decisions, keeping track of all the learner's responses. At the end of a simulation problem, they usually provide a summary score with commentary on the individual's choices and suggestions for future learning. In some forms of computer-based training simulation games, lea ners participate in teams, making collective decisions and vying for the best team scores.

Interactive Video Training with Minimal Branching

This form of training is comparable to the "workbook" form of computer-based training, except that pages of computer text and graphics are replaced with video pictures and sound, much like a video taped course. In some cases, still pictures with sound are used instead of full-motion video, to reduce costs. As in computer-based training with minimal branching, presentations of new material usually are mixed with tests or exercises, to verify comprehension.

Common applications include the introduction of new products or selling practices. In this situation, photographic images with accompanying sound may be essential for generating excitement and demonstrating appropriate sales techniques. On the other hand, the higher cost of interactive video training with extensive branching (which requires more interactive video training material to be developed for all the branches) may not be justified. If the course is to be seen by relatively few learners during its lifetime, minimal branching may be more cost effective.

Interactive Video Simulation Training

Another type of interactive video training that has proven particularly well suited for sales training is the interactive video simulation game. These games are used to develop interpersonal skills by simulating difficult sales or sales management situations in which the student must make rapid decisions. Later, the student is given an opportunity to see the consequences of those decisions.

HOW CBT AND IVT COMPARE WITH OTHER SALES TRAINING METHODS

One of the key problems facing managers of sales training programs is to make the best possible use of each available training method. Given the relatively recent appearance of CB and IV training, it can be difficult to position them appropriately in relation to other innovative methods, such as case studies, video cassettes, and role playing.

However, the essential consideration in positioning these methods is not complex: it depends on how and where an organization conducts its training. If the sales force is all located in one location, and it is not difficult to get them together for training, then the relatively high cost of developing an interactive video or computer-based training program would not be recovered quickly. In this situation, it is important to use interactive video training and computer-based training judiciously, focusing on types of training that are difficult to deliver with any other method.

Two Advantages of IVT and CBT

Unique Opportunity for Immediate Feedback. Interactive video and computer-based training have at least two special niches in a sales training curriculum. First, they can provide a type of effective practice with feedback that is difficult to match in anything but the most professional role-playing situations—or by on-the-job trial and error. Because it can put people into a dramatic interpersonal situation, interactive video offers simulation of a quality that is unmatched in other forms of self-paced instruction and rarely matched in live instruction.

Computer-based training also offers excellent simulation opportunities for skills that are normally performed either with pencil and paper or on a computer terminal. It (and interactive video training) can provide stimulating and interesting practice in decision-making skills. For example, it can provide practice in choosing an appropriate sales strategy. It can also help improve skills in timing the closing of a sale, or even in interpreting a customer's moods based on verbal cues.

High Motivation in Self-paced Instruction. The second special niche of computer-based and interactive video training is to provide exceptionally high motivation in self-paced instruction. Compared to self-paced instruction with books or standard ("linear") video, interactive video and computer-based training can more effectively stimulate the student by posing challenging problems and providing informative feedback. A student's attention is not likely to wander in the middle of a lively interactive dialogue. If the interactions are well designed, they can increase the chances that a student will complete a course, even in the face of distractions.

Cost-effectiveness of CBT and IVT

Compared to live instruction (lecture or case study), these interactive methods may be either more or less expensive, depending on the circumstances. The costs of developing computer-based training and particularly interactive video training are often considerably higher than are those of a live lecture course or text-based self-paced course (although they may be only slightly higher than a well-developed case study or linear video course).

If the salespeople to be trained are distributed over a wide geographical area, and bringing them together for live instruction would be expensive, then computer-based training and interactive video training may offer considerable savings. Interactive video training and computer-based training are like most other forms of self-paced instruction in this regard. However, they are likely to offer much higher completion rates (percentage of students completing a self-paced course) than many other forms of self-paced instruction.

On the other hand, if the salespeople are located centrally, and the costs of travel for training are minimal, then the delivery costs of interactive video training and computer-based training may be higher than those of live training. In this case, the primary advantages of computer-based and interactive video training may be their effectiveness for simulation practice. (See Figure 21–1.)

PRACTICAL APPLICATIONS _____

Three Major Benefits of CBT and IVT for Sales Training

The benefits of computer-based training and interactive video training fall into three general categories. (See Figure 21–2.) First, they offer improved effectiveness for certain kinds of training. The effectiveness is heightened because it is easy to customize the objectives and topics for the needs of a particular student. In addition, these methods offer powerful visual presentation modes

Figure 21–1 Cost-effectiveness of computer-based and interactive video training

		Importance of high-quality simulation practice	
		LOW	HIGH
Costs of travel for live training	HIGH	CBT promising IVT possible	CBT promising IVT promising
	LOW	CBT possible IVT unlikely	CBT promising IVT possible

Key: "promising" = likely to be cost effective
"possible" = may be cost effective
"unlikely" = unlikely to justify costs

(in the case of interactive video training, these include audio as well as visual presentation modes). Furthermore, all these methods offer sophisticated practice with feedback, with consistent quality (compared to such training methods as role playing by amateur role players, whose skills may vary widely).

By customizing topics and objectives, interactive video training and computer-based training can save time and hold attention for the student with a

Figure 21–2 Major benefits for sales training of CBT and IVT

Features	Benefits
1. Customizing presentations to individual salespeople's needs	Increased attention and decreased learning time
2. Powerful visual presentations	Increased attention and low-risk learning
3. Sophisticated practice with feedback	Fluency, confidence

relatively strong background. For the student with a weaker background, interactive video and computer-based training can provide needed introductory information or prerequisite skills, along with the primary training.

Powerful visual (or audio/visual) methods offer two types of benefits. First, they can take advantage of motion pictures' well-established potential for holding adults' attention over long periods of time. Second, they can provide realistic simulation of interpersonal situations that would otherwise be inconvenient or risky to use for training. For example, real situations involving an unhappy customer would be risky training grounds for practicing the critical skills of answering customers' objections. However, practice in front of a screen, watching a realistic motion picture portrayal, can provide a high level of skill development—without the risk.

Using CBT and IVT in a Complete Sales Training Program

For New Sales Representatives with Little Experience Selling. For new sales representatives, interactive methods of training offer several advantages. One is that computer-based and interactive video training provide a rich variety of experience in a short time. They can expose the student to interpersonal situations where selling skills must be practiced—skills such as qualifying, effective listening, presentation of a proposal, handling objections, and so forth. Each of these skills depends on carefully observing the actions and mood of the customer and making decisions about one's own behavior based on the observations. The skills of making those observations and decisions accurately and rapidly, like most skills, require practice, and the ideal practice might include exposure to a wide variety of situations. In this way, the student can learn to detect the key signals that are common to all of the varied situations.

Computer-based training and particularly interactive video training (because of its ability to show live human characters) can provide varied and compressed practice on interpersonal skills, with highly individualized feedback, which can support the rapid development of observation and decision-making skills.

In addition, both computer-based and interactive video training can be highly effective supplements to on-the-job training. Because these methods are suitable for on-demand training without elaborate prior scheduling, inexperienced sales representatives can undergo on-the-job training in local sales offices while taking a variety of courses through these electronic media. This strategy allows them to alternate between acquiring knowledge, on the electronic system, and applying the knowledge on the job.

For Experienced Sales Representatives. Experienced sales representatives need training on new products and sales policies. For them, the feasibility of on-demand training is particularly important, since they may have particularly full schedules with few openings for training. Often their training must be brief, and brief training sessions typically provide little time for realistic, effective practice. However, practice may be a key ingredient in new product training. It may be necessary for mastering a new way of positioning the product, for example, and it might be difficult to provide in any form of self-paced training other than computer-based or interactive video training.

These advantages would apply both for the training of an existing sales force and for the training of a new sales representative with previous sales experience in another company.

Costs and Other Resource Requirements

The costs and other resource requirements of computer-based and interactive video training fall into three categories: equipment acquisition and support, courseware acquisition or development, and program operation.

Equipment Acquisition and Support. The first category of costs is acquiring equipment on which to play the computer-based or interactive video courses. In the case of computer-based training, the necessary computer terminals or personal computers may already be on hand. However, in the case of interactive video training, few interactive video systems are in widespread use, and frequently it will be necessary to purchase them. Typical equipment consists of a personal computer with a special video monitor plus a videodisk player; in some cases, an existing personal computer can be adapted.

The cost of the special monitor and the videodisk player is approximately the same as that of the personal computer itself. Consequently, for many sales organizations, it will not be practical to put interactive video equipment on the desk of every sales representative. However, the fact that the interactive video training system includes a personal computer or terminal means that the equipment is readily usable for other purposes when it is not being used for interactive video training. In most cases, having one or two interactive video training systems in a local sales office will prove adequate for many kinds of training.

Courseware Acquisition or Development. The second major category of costs is the acquisition or development of computer-based and interactive video courses. At present, most programs are custom developed. Relatively few off-the-shelf programs for sales training exist, although a few large sales training organizations have developed or announced the intention to develop off-the-shelf interactive video programs. It is reasonable to assume that many more off-the-shelf programs will be available in the future.

Typically, off-the-shelf programs are priced equivalent to live training for two to five students. On the other hand, the costs of custom programs, while they vary widely, have been known to approximate the costs of live training for 500 to 2,500 students. (Computer-based training typically costs less to develop than interactive video training, and costs equivalent to 200 to 1,000 times the costs of live training are common at the present time.)

Given the high costs of custom development, it is wise to plan the use of computer-based training and especially interactive video training carefully. Custom development is usually cost justified when the number of students is large. In those cases, the development costs are spread over many students and may be recovered in savings on delivery and operation costs. In other cases, custom-developed courseware is expensive in the beginning but cost justified over the long run, because it provides substantial reductions in student learning time.

The development costs of computer-based and interactive video training can be held down in many ways. The first and most important is careful planning and instructional design. Frequently it is possible to provide some portions of a training program through live or printed materials, which usually have lower development costs. Restricting the use of expensive materials, like custom interactive video courseware, to just those applications for which they are best suited tends to minimize overall development costs.

Also, discretion is wise in the use of high-cost presentation techniques. Full-motion video, with professional lighting, sets, and acting, can be quite expensive—as much as several thousand dollars per minute. Furthermore, these costs are relatively stable. Unlike many costs associated with computers, video costs are not dropping rapidly, because of the relative maturity of the video production industry. On the other hand, the costs of software development in computer-based training and graphics in interactive video training are dropping, as better techniques are developed for creating and editing computer graphics. These trends can be expected to continue in the future.

Furthermore, traditional animation techniques, which often cost as much or more than full-motion video, are being gradually replaced by sophisticated computer graphic development tools. Consequently, simple forms of on-screen animation may be expected to drop in cost. In addition, the costs of audio (whether recorded on disk or generated by a computer speech synthesizer) are relatively modest compared to those for video and animation.

In summary, by carefully planning the mixture of high-cost and low-cost techniques, it may be possible to significantly reduce the total cost of custom-developed courseware in the future.

"Authoring" Approaches. Another consideration in courseware development is the choice of "authoring" methods. Authoring is the process of developing both (1) text and images for a CBT or IVT course and (2) the accompanying computer programs. Authoring involves designing, writing, entering, and testing all the information and computer instructions of the course.

There are several approaches to authoring, each with its own strengths and weaknesses. Figure 21–3 describes the main approaches.

Figure 21–3 Comparison of approaches to authoring

Computer Software	Development Done by	Ease of Development	Flexi-bility	Ease of Support[1]
Standard programming language	Team of writer(s) and developer(s)	Low	High	High
Special-purpose CAI or IVT lang.	Team of writer(s) and developer(s)	Moderate	Moderate	Low
Special "fill-in-the-blanks" software	Single individual can do all	High	Low	Low

[1] Ease of finding skillful programmers or authors who can update a course in the future (relative to other approaches).

As the figure shows, no one approach is best for all situations. For example, when flexibility to handle novel requirements is critical, the approach that uses a standard programming language may be best. On the other hand, the approach that uses special, "fill-in-the-blanks" software may be more appropriate when ease of development by a single individual is paramount, and the need for flexibility is secondary.

Unfortunately, there is little compatibility at this time between courses developed under different approaches—or even under similar approaches on different types of equipment. This means that custom-developed courses frequently run only on one type of equipment. When it is important for courses to be "portable" among various types of equipment, experienced development consultants may be able to make special adjustments in the courses.

Program Operation. A final cost category, for computer-based and interactive video training, is ongoing program operation. These costs include equipment maintenance, courseware updates and distribution, and the labor of support personnel. For computer-based training, maintenance and updates are typically no more expensive than for other kinds of software. For interactive video training, these expenses may be somewhat higher, due to the specialized nature of the equipment used.

For both, costs of support personnel sometimes include administrators, tutors, or other assistants. Although such support people are not always required, the availability of a support person by phone may make it possible to overcome some natural anxieties among new users of the system.

Introducing CBT or IVT into a Sales Training Program

Stages in the Introduction of a New System. There are several stages in an organization's response to the introduction of a new information system, including a computer-based or interactive video training system. Although there are several different theories about these stages, there are certain stages that seem to be virtually universal. (See Figure 21–4.)

For example, it is commonly observed that an initial period of interest in the new system is soon replaced by a period of anxiety and resistance among staff members who fear the possible impact of the new system on their jobs. For staff members in a sales training organization, anxieties may focus on the replacement of existing activities, which may be valued by some members of the staff, or the inconvenience of learning to deal with the new system.

Although such concerns may be widespread initially, they can be overcome with careful planning. For instance, concerns about the replacement of established programs may be addressed by workshops or printed materials that identify the benefits of the new system and seek the participation of the staff in decisions about it. Staff members who are involved in significant decisions, such as choosing the first courses to go on the new system, tend to view the system with more enthusiasm than do uninvolved staff members.

Obviously, staff members who will experience a negative impact of the system on valued portions of their jobs need an opportunity to discuss possible changes in their jobs. In some situations, these discussions can be facilitated through "coaching" by a central implementation staff. However, these concerns are not likely to be more traumatic than the introduction of any other significant changes in the workplace.

Sometimes it is easy to forget that training systems also require training. Trainers need to learn how to use the new system, as well as how to introduce

Figure 21–4 Stages in the introduction of a new CBT or IVT system

---

Stage 1: Initial interest and curiosity about the new system
Stage 2: Anxiety and resistance to change
Stage 3: Learning to use new system and incorporate it into ongoing programs
Stage 4: Overcoming obstacles and accommodating special needs
Stage 5: Acceptance as "normal" operation

---

it to their students. Furthermore, they may need extra access to support people, as questions or problems arise during the early phases of the system introduction. For this reason, it is often wise to schedule introductions of the new system in one location at a time. This way, a central support team can focus on the special requirements of individual locations and provide a high level of support during the introduction period.

ADVANCES IN CBT AND IVT TECHNOLOGY

Developments that are underway today are likely to revolutionize CBT and IVT—indeed all forms of industrial training—in the near future. Today's CBT and IVT systems may become obsolete quickly, unless they are planned with these developments in mind.

Networking: Encouraging the Evolution of "Information Warehouses"

The advancing technology of networking will have a profound effect on instructional uses of computers. Low-cost, powerful networks will encourage the evolution of "information warehouses," from which widely dispersed workstations will draw course materials. Course information and software will be "downloaded" to workstations as needed. This arrangement will remove one of the classical economic constraints of education: the need to bring students to a central location, in order to have the most up-to-date information and learning experiences.

Subscription Agreements. Costs of high-speed, reliable connections between computers should drop rapidly, as the technology evolves. With low-cost connections, frequent updates to electronic training materials will be much less expensive than with traditional printed or audio/visual materials. In this environment, both creators and users of instructional materials are apt to seek "subscription" agreements, giving users access to the most current version of courses. Subscribers will want customized training, to keep them abreast of new developments. Customized training will be inexpensive since (1) subscribers' preferences can be stored in a computer, allowing automatic generation of appropriate materials, and (2) distribution of many customized versions of materials will be easy. Today's mass production of materials, driven by economies of scale in duplication, should give way to tailored materials, produced and delivered as needed.

Networking may mean more self-instruction, but it will not necessarily mean less human contact. In fact, convenient, powerful networks may make possible extensive tutoring and interaction with students, at times and places that would have been impractical in the past.

Electronic Mail. Electronic mail (both printed and voice) may be an important medium for instructional communication, thanks to networks. An important advantage of electronic mail is that it is delivered even when the recipient is away, or is busy, unlike live communication. Yet unlike postal mail, responses can usually be expected in a matter of minutes or hours, in a well-engineered system. The popularity of current, primitive mail and "electronic bulletin board" systems suggests that this combination of advantages satisfies many communication requirements. One important new application will be "collaborative learning," in which computers are used to share information among geographically dispersed learners. For example, top sales representatives could share their success stories instantly, via electronic "conferences."

Person-to-Person "Live" Communication. However, networks will also facilitate "live" communication, when the time is right. Telephone calling will be merged with data communication, in the coming networks, and person-to-person exchange of pictures or typed messages will be easy. We are even likely to see "training windows" appear on screens in some situations, allowing direct assistance during a real application of some newly learned skills.

Also, remote tutors can provide a key "person-to-person" form of motivation. They can provide some of the social and emotional structure that has helped students in live courses complete their training more frequently than students in comparable self-instruction courses. For example, a remote tutor might help a student create a personal plan for completing a course, anticipating the most difficult sections of a course, as well as the sections that could be skipped. Later, the tutor could answer questions that arose while the student was taking the course.

Workstation Systems: New Techniques at Modest Prices

Powerful workstations with dazzling graphics, comparable with today's most advanced computer-aided design (CAD) engineering workstations, will soon become widely available at modest prices. Vivid images will be nearly as easy to create and manipulate on these workstations as text is on a word processor today. High-fidelity sound will also be available. Unfortunately, unsophisticated presentations (as in many present CBT programs) are apt to lose their appeal quickly, as black-and-white television programs did, after the advent of color television.

Developers of training programs for these systems will be able to make "training windows" appear when needed, such as during a confusing task. With this technique, training will appear at times when learners are highly motivated to pursue it—when they need it to solve a problem.

These workstations will have considerably more computing power than comparable systems today. One use of the power will be to improve the apparent intelligence shown by the systems during simulations or tutorial dialogues. Intelligent responses depend on sophisticated models of students' mental processes, and these models require powerful computers. Model-equipped workstations will be able to detect various types of learning "breakdowns" and automatically adjust for them. Adjustments might include changing the pace of instruction for a student who is confused or providing memory tips for a student who is "rusty" from lack of practice.

New audio/visual media and techniques will be incorporated into these workstations as they become available. The new techniques should allow more interesting and concise presentations of concepts, along with more effective practice simulations. They may combine the attention-holding power of video games with large computer databases concerning various sales techniques, for example.

Software: Powerful and Customized

Powerful new software will become available to take advantage of the workstations, networks, and other developments. The software should benefit not only the student but also developers and tutors.

For example, planning software will allow tutors or students to create sophisticated plans for individual learning—plans that help students use their time effectively and also keep their motivation high. The software will suggest customized combinations of course sections, practice exercises, and even instructional methods (e.g., loosely structured exploration of a simulation versus highly structured materials), depending on a student's past experience and preferences. Other kinds of new software will assist a student while actually taking a course. It will verify progress and readiness for the next course section by generating appropriate checkpoint exercises and analyzing the results. Based on the analysis, it will recommend remedial work, outside assistance, or a change in the planned sequence of topics. It will even help connect the student to a live tutor (or an automated tutoring system) for special assistance.

To assist the developer of instructional materials, the software will analyze group data about students' progress. It will identify points in a course where many students were having problems, and it may even suggest possible corrections.

The evolution of electronic publishing software is also likely to change sales training significantly. Paper materials are frequently important adjuncts to interactive video and computer-based training, because of their portability, freedom from special equipment requirements, and convenience for reference. Unfortunately, the costs of books and other printed materials have risen dramatically in recent years. One reason is the large amount of labor involved in handling centrally printed materials. Central printing has created economies of scale, in the past, but much of the economy is later lost in storage, distribution, and the eventual disposal of obsolete materials.

New electronic publishing technologies will allow on-demand printing of materials with virtually magazine quality. As a result, it will be possible to publish high-quality materials with no inventories and with minimal handling. These changes will lead to cost reductions, constantly current versions of printed materials, and wider availability of customized materials. Furthermore, local "just-in-time" printing will make printed materials available at times and places that would not have been practical in the past.

Human Interaction Techniques: Models for More Sophisticated Systems

Advances in our understanding of educational and communications processes will probably make the largest impact on future interaction techniques. Better theoretical models of how students learn from different types of presentations and practice exercises will enable the creation of more sophisticated instructional systems. The models will allow systems to predict accurately the consequences of different choices a student may make about topic sentences or methods of instruction. For example, the model-based system might be able to predict that a detailed presentation of some topic would be too boring or confusing for a student and recommend a longer but more entertaining path through the material for that particular student. By accurately forecasting outcomes, the system would help the student get the most out of the available learning opportunities.

The "human interface" of instructional workstations—the display formats, keyboards, and other means of moving information into and out of the workstation—will also become markedly easier to use. Software developed by different companies will adopt a standard layout in screen windows. Input devices like keyboards and "mice" (small, roll-around boxes that move pointers on a screen) are also apt to become standard, making it easier for users to move from one type of workstation to another.

THE IMPACT OF COMPUTER-BASED AND INTERACTIVE VIDEO TRAINING ON SALES TRAINERS

Computer-based and interactive video training are very important techniques for sales training, because they are uniquely powerful ways to teach interactive skills such as qualifying a customer or answering objections. In fact, a sales organization might gain a significant competitive advantage by using these techniques to develop an exceptionally skillful sales force.

However, both these techniques are in their infancy at the present time. Many CBT and IVT courses of today—as well as the systems on which they run—are likely to become obsolete in the future. This poses a dilemma for the sales trainer. Ignoring these techniques runs the risk that a competitor could master them and gain a competitive edge. On the other hand, a premature investment in courses or systems that might be obsolete soon (through incompatibility with future technology) would be a costly mistake. Resolving the dilemma will generally require expert consulting advice. The ideal consultants would be highly competent in modern instructional design techniques as well as current developments in CBT and IVT. Consultants who are strong in both areas could help a sales training organization find courses and systems that will be cost-effective and capable of surviving the technology revolution.

More importantly, expert consultants could make crucial contributions to the development of custom courseware. They should help sales trainers design powerful courses, while holding down costs and minimizing obsolescence. Their help would make it possible to benefit from early investments in CBT and IVT with low risk of obsolescence.

SUMMARY

Two of the most distinctive requirements of sales training are (1) the need for a continuous stream of new courses, as new products and marketing strategies are created and (2) the importance of practice, particularly interactive simulations (or role playing), to develop fluency and confidence in selling skills, before the skills are applied in a real sales situation.

Computer-based and interactive video training offer unique advantages in meeting both requirements. However, they are relatively new and complex forms of training, with the risks of high costs and technological obsolescence, if they are employed in an ill-planned manner. The most important guideline for using them is to ensure that their use is consistent with an appropriate instructional design, centered on the particular requirements of specific learners.

REFERENCES

Currier, Richard L. "Interactive Videodisc Learning Systems." *High Technology* (November 1983), pp. 51–59. A limited but well illustrated and readable introduction to the topic.

Dalton, David W. "How Effective Is Interactive Video in Improving Performance and Attitude?" *Educational Technology* (January 1986), pp. 27–29. An example of a research study on the effectiveness of CBT and IVT.

Gagne, Robert M., and L. J. Briggs. *Principles of Instructional Design.* New York:

Holt, Rinehart and Winston, 1974. A classic text on principles of instructional design; useful for working with course development consultants.

Gropper, George L., and Paul A. Ross. "Instructional Design." In *Training and Development Handbook*, 3rd ed., ed. Robert L. Craig. New York: McGraw-Hill, 1987. A thorough discussion of the principles of instructional design, intended for professional trainers and their clients.

Hart, Frank A. "Computer-Based Training." In *Training and Development Handbook*, 3rd ed., ed. Robert L. Craig. New York: McGraw-Hill, 1987. An authoritative summary of CBT practices, benefits, and principles of effective use.

Heines, Jesse M. *Screen Design Strategies for Computer-Assisted Instruction*. Bedford, MA: Digital Press, 1984. Profusely illustrated introduction to design principles for CBT (as well as IVT); useful for understanding what makes well-designed courseware effective.

Huntley, Joan Sustik, and Stephen M. Alessi. "Videodisc Authoring Tools: Evaluating Products and a Process." *Optical Information Systems*, 7(4) (July–August 1987), pp. 259–281. A report on evaluations of current authoring tools.

Kearsley, Greg. "Instructional Videodisc." *Journal of the American Society for Information Science*, 34(6) (November 1983), pp. 417–423. A brief overview of different IVT approaches, with useful examples.

Lee, Chris. "*Training* Magazine's Industry Report 1987," *Training* (October 1987), pp. 33–35. One of an annual series of reports, including survey results on CBT and IVT usage in U.S. organizations.

chapter twenty-two

HOW TO IMPROVE YOUR SALES TRAINING MEETINGS

HOMER SMITH

Homer Smith is president of Marketing Education Associates and past president of the National Society of Sales Training Executives. He is editor of the biweekly publication, *Master Salesmanship*, published by Clement Communications, Inc., and is a columnist on sales meetings for *Sales & Marketing Management* magazine. He is the author of numerous books and articles on selling and meetings, including the recent *Selling Through Negotiation*. His firm is located at 4004 Rosemary St., Chevy Chase, MD 20015.

The ability to hold an effective sales meeting has become an important measurement of sales managers at any level. Ever since selling was recognized as a science that calls for planning, training, and mental conditioning, the sales meeting has been accepted as an important management tool for accomplishing those objectives.

FIVE REASONS FOR CONDUCTING A SALES TRAINING MEETING

Since planning and conducting sales meetings demands time and effort that could be devoted to other phases of management, why do the more progressive businesses consider it worthwhile?

While there is some overlapping of purpose between them, here are five reasons generally given for holding a sales meeting:

1. *To instruct and train.* A major benefit of a sales meeting is the opportunity it provides for group instruction and development of the sales force. It makes possible the use of valuable training techniques that are not possible on an individual basis, particularly those that allow the sharing of views and experiences of other salespeople.

2. *To improve communications.* Modern business calls for good communications between management and the selling team. Regular sales meetings provide a medium for announcements and information of interest and importance to the salespeople, but they add two important ingredients not present in other methods: the opportunities for clarifying questions and for discussion.

3. *To introduce new products or policies.* A meeting attaches more importance to the new policy or product, permits exchange of ideas, and minimizes misunderstandings.

4. *To solve problems*. The sales meeting makes it possible to bring together those involved in a problem and those who can contribute solutions.

5. *To motivate the sales staff*. Motivation is a vital ingredient in any sales plan. The meeting program may be designed to encourage self-motivation. When conducted properly, the sales meeting itself can be a motivating factor by contributing to the salesperson's feeling of importance and a member of the team. A poor meeting, unfortunately, has the opposite effect.

HOW TO MAKE YOUR SALES TRAINING MEETING A "GROUP SALE" USING THE AIDA FORMULA

All five of the reasons given for holding a sales meeting could be considered a part of a single basic purpose: to instruct and develop salespeople. Since that is also the theme of this book, let's look at ways in which the sales meeting can be more effectively used for that purpose.

Sales meetings vary in their mechanical details, but they all have a common purpose—to sell the salespeople on an idea, policy, or procedure and get them to take enthusiastic action. Planning a sales meeting, therefore, can be approached in much the same way a professional salesperson plans an important sales presentation.

Think of your favorite formula for selling. Wouldn't it work equally well in planning a good sales meeting? Let's take the most universal and simple formula as an example: AIDA. Here's how it might function in planning a meeting.

1. *Get ATTENTION*. How you start the meeting sets the mental attitude toward the rest of the meeting.

2. *Arouse INTEREST*. Say or do something to show how the salespeople will personally benefit by what you are going to do in the next few minutes.

3. *Create DESIRE*. Here is your plan, your presentation or demonstration. Like any sale, the key is convincing the customers (the salespeople) of the benefits to them personally. These might include ways to increase sales, to sell more easily, to beat competition, to get fewer customer problems, or to improve the salesperson's image with customers.

4. *Get ACTION*. Too many meeting leaders, like too many salespeople, stop short of this important step: closing the "sale." They consider they have accomplished the meeting objective when they have explained a product or program and have pointed out its benefits. But a better objective for the meeting is getting a commitment from the salespeople to put the new information and ideas into action. Examples of commitments after introducing a new product are make X number of demonstrations and report results,

quotas for sales, ask for names of salespeople's customers who might be interested in the new product, or role play a demonstration of the new product.

SETTING THE MEETING OBJECTIVE

If you agree that a sales meeting is a "group sale," your first step in planning any meeting is to decide what it is you want to sell. If you had unlimited time and a belief that the sales meeting could solve all your problems with your salespeople, you might try to handle all of them at one meeting. Since this obviously can't be done, you must select those objectives that can be accomplished with reasonable success with a sales meeting and within the time allowed. Use other times and other media for the rest.

Don't Try to Cover Too Much

Specific objectives for sales meetings can soar in numbers, so it is easy for a manager to end up with an all-purpose meeting that tries to cover too much. When this happens, salespeople come away loaded with information but with confused or incorrect interpretation and no plan for putting the information to use.

Single out just those objectives that are most important, that can be successfully handled by your "sales" formula, and within the time frame allowed.

Too many sales meetings present the opposite problem. Sales meetings are held as company policy on a certain day and time every week. The problem becomes one of filling the meeting hour with something and everyone suffers as a result.

Sales meetings are costly in time and salaries for managers and salespeople. If another method of educating or communicating would be more effective than a meeting, use it!

Basic Objectives to Cover in Your Sales Training Meeting

While the specific objectives for a sales meeting are almost endless, the basic objectives are to

1. Inform.
2. Instruct.
3. Enthuse.
4. Sell.

It would be difficult to think of any sales meeting that was not intended to accomplish these four goals. But here are some more specific objectives that are often considered:

- To inform and instruct on a new product or service.
- To stimulate greater sales effort on an old product or line.
- To present a plan for getting more sales.
- To uncover problems the sales force might be having.
- To change an attitude on a product, policy, or procedure.

FOUR BASIC TRAINING METHODS FOR SALES MEETINGS

After establishing the objectives for your sales meeting, you select the methods for accomplishing these objectives through group training. Group training techniques fall into one or more of these four basic methods for group training:

1. *Telling.* Information is given to salespeople by lecture, a panel of experts, charts, dramatizations, audio/visuals, and so on. This method is used when the objective of the meeting is to *inform* the salespeople about a product or plan.
2. *Showing.* Most training, particularly when there is a skill involved, requires showing as well as telling. The demonstration, skits, films, and video tapes are examples.
3. *Tryout.* In good training procedures, telling and showing must be followed by a third step, tryout by the trainee to establish what has been learned. During the sales meeting, role play is commonly used for this purpose. But the best tryout phase of the training is held in the field under supervision.
4. *Conferring.* Here the salespeople work out their own answers to selling problems under the leadership of a qualified director. The conference and buzz groups are examples of this training technique.

The Conference Method

Socrates and his fellow philosophers knew the value of group discussion and used it for training. But the use of the discussion or conference method for sales meetings, seminars, and clinics is fairly recent in modern business. In this method of training, the conferees, with a qualified conference leader, work out for themselves the answers to selling problems. Aside from its value in solving selling and training problems, it has a special psychological value to recommend it in that actual participation sharpens the salesperson's interest and stimulates thinking.

Because the training that takes place comes from the experience of the group, the pure conference is reserved primarily for experienced salespeople. When coupled with the other forms of training, however, the discussion or conference can be used for clarification and plans for taking action. Frequently, the first portion of the sales meeting consists of a presentation, a demonstration, tape recordings, visual aids, role playing, and so on. The last portion can then go into controlled discussion of the points brought out earlier. The key to a successful conference is a competent leader who has learned to draw solutions from the participants, to challenge their views, and to stimulate discussion. There are books on conference leadership that are helpful. A special section in the book, *Organizing for Better Meetings*,[1] is devoted to ways to open, control, and guide a discussion, key words and questions that help develop discussion, and how to generate charts during the conference.

Buzz Groups Come Up with Answers

When the group is too large for proper individual participation, or when there is not enough time to cover the subject matter by discussion before the entire group, the leader can break the group into smaller "buzz groups." One popular form of the buzz session is called the "Phillips 66" method. It takes its name from Dr. J. Donald Phillips who started it at Michigan State University. It consists of breaking the audience into committees of six and giving each buzz group six minutes to work on a problem.

With groups seated at round tables, such as following a meal, each table can be designated a buzz group or team. In a classroom setting, alternate rows of persons can turn around to form table buzz groups.

While the leader can appoint the team captain, it seems to work better if each table elects its own. The leader checks to see that all have done so, then asks each captain to appoint a recorder. The leader than assigns a topic to each table, or all groups can work on the same topic, like identifying the main problems in selling a particular product. The leader announces that they have five or six minutes to come up with their suggestions, and at the end of the time, calls on each group recorder in turn to report on what the group has suggested. These suggestions are tabulated on a board, pad, or overhead slide by the leader or an assistant, eliminating duplications.

Depending on the time and purpose, the buzz groups can go back for another session on another problem. In the example of coming up with a list of problems, the next buzz session could be to come up with solutions for the problems. Each

---

[1] Marketing Education Associates, 4004 Rosemary Street, Chevy Chase, MD 20815.

group would be assigned one or more of the problems on the list. After time was called, the solutions could be recorded as before.

Because the buzz session gets a lot of thinking screened and summarized in a hurry, it can be used for a variety of purposes, such as getting a list of questions to ask a panel, a list of nonprice benefits to offer a client in negotiations, or favorite methods for saving time on sales calls.

Techniques for using the various group training methods are discussed in other chapters of this book so we will not cover them here.

SEVEN STEPS FOR PLANNING A SUCCESSFUL SALES MEETING

No matter how large or complex the meeting, there are just seven basic steps that must be covered if the meeting is to succeed. The stages may be very fleeting in planning the simple meetings, and poorly executed in others, but they will be there.

Step 1. Determine the Objectives

What should those attending do after the meeting that they aren't doing now? We have already discussed specific meeting objectives.

Step 2. Consider Alternatives

Is a meeting the best way to achieve the objectives? Would a letter, a phone call or some other means of communication be just as effective?

Step 3. Plan the Meeting

This is the most important step after determining the objectives. Considerations within the framework of planning include

- Who will be responsible for planning, developing and executing the meeting?
- Who should attend the meeting? Who can benefit? Who can contribute?
- When and where should it be held?
- What will it cost? How will the budget be met?
- What will be the agenda? Program content? How long?
- Who will conduct the sessions?
- What methods of presentation will be used?

Step 4. Prepare for the Meeting

Preparing for the meeting involves refinement of decisions made in the planning stage.

- Communicating with participants and performers.
- Handling publicity and reporting.
- Matching site facilities to the program needs. (Are they adequate?)
- Preparing a timetable.
- Developing program content and coordinating performers.
- Developing and collecting meeting materials.
- Making staff assignments.
- Arranging for physical needs: meals, lodging, entertainment, transportation.

Step 5. Conduct the Meeting

This is it! The meeting is started, the agenda goes into gear and is carried to completion.

Step 6. Follow-up

What did the meeting provide that should be put into action? Is the action in operation? What needs were uncovered in the meeting that require further implementation? Is this being done? If another meeting is indicated, begin the cycle again.

Step 7. Evaluate the Success of the Meeting

How close did the meeting come to fulfilling the objectives? What changes should be made if the meeting is to be repeated? What was learned that can be applied to future meetings?

NATIONAL VERSUS REGIONAL MEETINGS

Companies with national distribution are faced with the choice of holding a single, national sales meeting or holding a series of regional meetings throughout the territories. The national meeting involves bringing the sales force or distributors into a central spot for one grand affair. Regional meetings have a smaller version for key distribution centers. The tradition of the company and its marketing program affect the choice of a national or regional meeting, but here are some additional factors that are considered in making the choice.

The National Meeting

1. Costs are greater because of the tendency to use more expensive talent, entertainment, and facilities and the costs of bringing in all the sales force and distributors rather than sending a few executives around the country. Costs *could* be kept comparable to regional meetings, but this is not usual.
2. It takes less time for executives involved but more time for sales force.
3. Because of the larger budget, the facilities, program, talent, and entertainment can be more spectacular.
4. A larger pool of experience for sharing ideas is created. Sales force and company personnel who do not ordinarily meet during the year can exchange experiences and make friendships.

Regional Meetings

1. Travel expenses for sales force are lower, executive travel costs are higher. Regional meetings tend to be shorter than national, thus entail lower on-site costs.
2. Shorter travel and meeting time means that salespeople take less time away from their work.
3. Total budget may be lower than one national meeting, but not necessarily. Individual meetings tend to be less pretentious and shorter in length than the national meeting, but total costs will have to be compared.
4. Climate for individual training will be better. Smaller groups permit greater individual participation and exchange of ideas. Common experiences within the same market area allow for more specific training.
5. There is a greater opportunity for improvement. Any errors uncovered in the first meeting can be corrected before the next ones are held.

While there is a trend away from the national spectacular in favor of the more intensive regional meetings, the national image can be built into the regional format to get the best of both systems.

Publicity that ties all regional meetings into one program helps to build the national concept. Having top executives at each regional meeting permits an even better personal association with the sales force. A national contest or similar types of competition between the regions helps to strengthen the one-team concept.

With proper planning and an adequate budget that permits quality programming and production, it is possible to have the advantages of the regional meeting concept and still have the prestige of the national meeting.

TIPS FOR PLANNING LARGE SALES MEETINGS

The companywide sales meeting, whether one national meeting or a series of regional meetings, is basically just a collection of small meetings or sessions. But the greater size and longer duration of such meetings presents some additional problems not found in the typical local, short sales meeting.

The companywide sales meeting obviously requires careful and detailed planning, since it involves greater expense and more people and puts management into the national spotlight. Here are some guidelines that will help in the development of a successful national sales meeting program.

1. Involve top management. Get its active support and participation.
2. Involve sales supervisors. Keep them informed. Use special talents on the program and committees.
3. Budget carefully. Be ready to answer top management questions on costs and time problems like lost production.
4. Plan the meeting agenda carefully. Use a select committee to review and approve it.
5. Test the meeting agenda. Watch for conflicts with other company programs, projects, production schedules, space availability, holidays, vacations, union plans, and so on.
6. Project the anticipated results. Justify approval of the budget by top management by showing how results will be worth it.
7. Choose a general chairperson. He or she will preside at the meeting and be accountable to top management for it. The chairperson will delegate responsibility for production.
8. Choose a project manager. He or she will be responsible for the overall direction of the meeting, planning, organizing, and coordinating.
9. Choose a program director. He or she will be responsible to the project manager for the program content, schedules, speakers, and publications involved in the program.
10. Choose a facilities manager. He or she will be responsible to the project manager for preparing and maintaining the site for the meeting. The facili-

ties manager will see that all equipment, supplies, and materials are on hand when needed and are properly disposed of after the meeting.

11. Choose an entertainment director. He or she will be responsible to the project manager for all food, refreshments, and entertainment required at the meeting.

12. Choose a promotion director. He or she will be responsible to the project manager for promotion and publicity both internal and external. This will include relations with the press and the public.

Depending upon the size of the meeting, the availability of personnel within the company, and the authorization for their use, the duties of these individuals may be combined or greatly expanded by the addition of more assistants. One person might be responsible only for food service and another for entertainment, for example.

Here are major meeting activities that may require individual committee chairpersons.

Program

Budget and Finance

Hotel and Physical Arrangements

Props and Decoration

Promotion and Publicity

Transportation

Registration

Entertainment

Banquet and Meals

Awards and Prizes

SOURCES FOR MEETING SPEAKERS

The meeting program planner's job is to find speakers who have the greatest odds of success, have a message that fits the meeting objectives, and come at a price that fits the budget.

The best way to select speakers for your meeting is to use those that you yourself have used or heard and feel they would do a good job at your meeting. The next best source is a trusted meeting planner you know who has used speakers successfully that you can also use. Exchange speaker information with other meeting planners on a regular basis. Maintain a file on speakers you have heard and those who have been recommended. Add comments to help you in future planning such as address and phone, background, specialty, topics, fees, where performed, and recommendations.

While these leads for speakers offer the least chance for error, they are not foolproof. They may also offer you a limited scope of possible speakers for a particular meeting. Here are some additional resources for speakers that professional meeting planners use:

1. *Executives and employees of your company.*

2. *Leaders in your industry outside your firm.*

3. *Customers and clients.*

4. *Specialists in your meeting's objectives* (experts on financial planning, personnel, retirement, sales and marketing, and so on). Caution: Be sure that in addition to being an expert they can also make a good presentation.

5. *Trade associations and professional societies* (your own and affiliated industries). They may have speakers available or the staff can make recommendations.

6. *Sales and management training societies.* For speakers on sales and management training and development, contact National Society of Sales Training Executives, 203 E. Third Street, Sanford, FL 32771, and American Society for Training and Development, 1630 Duke St., Alexandria, VA 22313.

7. *Trade press and other media.* Editors and reporters know about your industry, can discuss trends, and give examples of how others in your industry resolved problems.

8. *Government agencies.* Civil servants can give expert analysis of government regulations, trends, and so on.

9. *Politicians.* Big names can make an interesting program feature.

10. *The mail.* Get on mailing lists as a meeting planner by subscribing to meetings magazines (see item 15) and joining organizations like the Meeting Planners International. Speakers will begin sending you literature and proposals.

11. *Bird dogs.* Ask your company executives and others who attend meetings to tell you about outstanding speakers they hear at other functions.

12. *Local business clubs* (Chamber of Commerce, Board of Trade, Kiwanis, Rotary, Lions, etc.). These may have rosters of members who will speak to groups. Officers or staff can also give advice on speakers they have used.

13. *Schools, colleges, and universities* (faculty available for general speaking engagements, conducting and developing seminar sessions, or leading discussions on special topics).

14. *National Speakers Association* (largest organization of professional speakers). Its directory of member speakers available to meeting planners contains

photos, data, contact information. Cross section lists speakers by subjects like sales or motivation, by special interest like humor or seminar leaders, and geographically like those living in Chicago, Toronto, or Sydney, Australia. (Contact NSA: 4747 N. 7th St., Phoenix, AZ 85014. 602–265–1001.)

15. *Meetings magazines directories.* The popular meetings magazines publish annual directories of meeting sites and other data including speakers. Ask for the latest directory from these magazines: *Successful Meetings*, 633 Third Ave., New York, NY 10017. *Meetings and Conventions*, 500 Plaza Dr., Secaucus, NJ 07094. *Meeting News.* 1515 Broadway, New York, NY 10036.

16. *Professional speaker bureaus.* These are agents for big-time speakers as well as local celebrities. Check the classified telephone directories and advertising in meetings publications like those listed in item 15.

Determine the Speaker's Talents Before You Commit Yourself

Before you settle on hiring a speaker you have located through one of the leads suggested, check with one of his or her former clients personally as a safety precaution. Ask the client about recent engagements. Get more than just an answer to "Did he or she come through okay?" Ask enough questions so you can be sure the speaker's topic will fit your objectives and your audience will feel you made a good choice. Remember the two major requirements: the content and quality of the material covered and the speaker's effectiveness before an audience. A good check question is: "Would you hire the speaker again?"

The best way to check out a potential speaker is to witness an actual presentation. Ask the speaker for his or her schedule and ask for permission to attend a meeting where he or she will speak. If you can't attend yourself, ask another person you trust to attend and report back. The next best thing is to listen to a tape made at an actual meeting.

PREPARING AN EFFECTIVE MEETING ROOM ENVIRONMENT

Meeting planners rightfully give a lot of thought to the program content and the quality of speakers for the important meetings. They understand the importance of these factors to the ultimate success of any meeting. Unfortunately, the selection of the meeting site and the preparation of the surrounding environment does not get equivalent planning and evaluation. Yet the quality of the meeting room environment and its relationship to the objectives of the meeting can have a considerable influence on the ultimate success of any meeting.

Selecting the Meeting Site

A number of factors enter into the determination of the meeting site. They include, but are certainly not limited to, these twelve:

1. The objectives of the meeting
2. Who will attend
3. Size of the group
4. The meeting program
5. Physical requirements of the program
6. Length of the total meeting and individual sessions
7. Housing requirements and quality desired
8. Meals and refreshment requirements and quality desired
9. Recreation and entertainment desired
10. Prestige image desired
11. How participants will travel to the meeting
12. Budget limitations

The objectives for the meeting are the first consideration in site selection just as they are for the program content. The site environment has to contribute to the attainment of these objectives. The meeting program and its physical requirements along with the size and expectations of the group that will attend give specific checkpoints for the site selection.

How important is prestige and entertainment? Is the length and intensity of the program unusual? The balance of these considerations will affect your choice of a resort location, a college campus, a conference center, or a hotel or motel with the desired facilities.

The resort environment encourages relaxation and fun. The businesslike environment of the university or conference center suggests that you're expecting to get a lot of work accomplished. The choice of a hotel or motel can go either way depending on the location and program scheduled.

The transportation of the participants should be a consideration in site selection, too. Cost of transportation may be a budget problem, reducing registrations. Getting to and from the meeting site should be convenient for the majority of the participants. If the distance is conducive to driving, a motel or hotel with adequate parking is a good choice. If flying is necessary, an airport meeting site should be considered.

Budget limitations, while at the bottom of our list, can unfortunately rule out the selection of sites that would otherwise meet the planner's dream. Then it becomes a matter of balancing the essentials against the nice-to-have's.

Inspecting the Meeting Room

The meeting room environment is very important because the participants will spend the most time there. The conditions should maximize the learning and retention results. Here are some factors to consider when inspecting the meeting rooms of any facility.

1. *Is the room the right size*? The room size should be adequate for the largest number of people involved plus space for equipment, displays, podium or stage, audio/visuals, and whatever else you require. The room should not be too large, either, or you will make a small group look out of place, lose the value of intimacy, and make listening more difficult. The choice of seating arrangements will affect the size of the room required. We will discuss the various seating arrangements later.

2. *Are there any obstructions*? Check for pillars, partitions, low chandeliers.

3. *How is visibility*? Other things can affect visibility besides obstructions. Could you see well if you were in the last row?

4. *Are there visual distractions*? Do windows look out onto the swimming pool or busy street? Is there a public passage through the area? Can entrance be at the rear so as to minimize distractions of coming and going?

5. *Are there sound distractions*? Are rooms adequately soundproof to avoid noise from adjoining rooms and hallways? Listen for noises from the hall, street, elevators, air conditioning, music.

6. *How is the sound amplification*? How much will the facility furnish? Is the amount and quality adequate for your purpose? What will you have to bring or rent?

7. *How is the air quality*? The internal climate is more important than the one outside. Does the air conditioning do a good job? Will the heat be adequate if you need it?

8. *Is the furniture adequate*? Are there sufficient chairs? Are they in good condition and comfortable? Are there tables for displays, projection equipment, registration, coffee breaks? Will the tables be covered with cloths?

9. *Is the room suitable for audio/visual aids*? Can it be darkened if needed? How? Can lights be controlled easily? Are there adequate electrical outlets with enough power for your equipment?

10. *Where are the elevators, stairs, and restrooms*? You can't change the locations if they aren't convenient, but you can choose the meeting room with these in mind.

11. *How will the room be serviced*? Get agreement with the site management on the room setup and service.

It will be unusual if you find a meeting site and particular room that will meet all these tests perfectly. Your chances of choosing the best room available for your meeting will be improved, however, if you take the time to run these tests on your options. More important, you are more apt to avoid the serious problems in environment that can jeopardize an otherwise perfectly planned program.

Choosing the Meeting Room Layout

Successful meeting planners pay particular attention to the layout of seating and accessories used by the attendees. They know that the comfort and effective communication level of the participants are important to the ultimate success of any meeting.

The choice of the meeting layout plan depends primarily upon the purpose of the meeting, the presentation methods used, the size of the group, and the shape and size of the meeting room.

Nine basic styles for meeting room seating. The two most popular meeting layout plans for larger groups are auditorium style and classroom style. (See Figures 22–1 and 22–2.) Auditorium seating is the most common method for large groups. Classroom style is the preferred seating for meetings that last longer than two hours or when the participants are required to take extensive notes or work with papers and materials. Narrow tables, usually 18 inches wide and 6 or 8 feet long are normally used.

Figure 22–1. Auditorium-Style Seating   **Figure 22–2 Classroom-Style Seating**

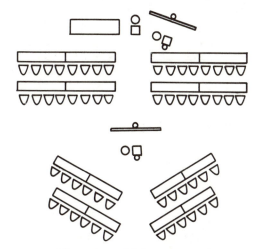

Chevron or V Arrangement

Figure 22–3 Tables for Buzz Groups

Figure 22–4 Boardroom Seating

Figure 22–5 Conference Meetings

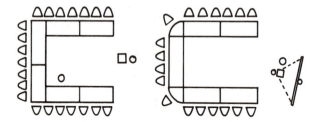

Figure 22–6 V-Shape Seating

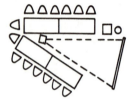

Figure 22–7 Hollow Square or Hollow Circle Seating

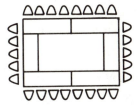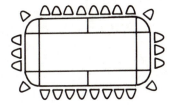

Figure 22–8 T-Shape Seating

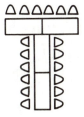

Figure 22–9 E-Shape Seating

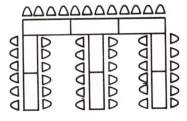

The other seating plans are used for special types of meetings. The illustrations which follow are shown with permission from the booklet *Guidelines for Effective Meeting Room Environment* published by Marketing Education Associates, 4004 Rosemary St., Chevy Chase, MD 20815. It comes with a plastic template of standard meeting room furniture for making layout sketches in common ¼- and ⅛-inch scales. (See Figures 22–3 through 22–9.)

Provide layout sketches along with a list of the desired equipment, meeting date, room, and time for the site setup crews to make sure they end up with the room layout and equipment at the time you desire.

Don't Stop Now!

Because of space limitations, we have had to be selective of the meeting problems and techniques discussed. If meetings become an important part of your operation, you will appreciate help in other areas like site selection and negotiation, budgeting, meal planning, transportation, recreation, registration, audio/visuals, exhibits, special types of meetings, and more suggestions for meeting programs. For these, get a comprehensive meetings handbook, especially one that provides a wide variety of checklists that can save you time, problems and embarrassment. Two comprehensive meeting planning handbooks are recommended: *Organizing for Better Meetings* concentrates on meetings for group training, especially sales meetings, and has checklists for all meeting activities. It is available from Marketing Education Associates, 4004 Rosemary St., Chevy Chase, MD 20815. *The Meeting Planner's Guide to Logistics and Arrangements* covers the many details of planning larger meetings and conventions and is available from Institute for Meeting and Conference Management, P.O. Box 14097, Washington, D.C. 20044. Your library may have these and other books on meeting techniques.

Consider the sales meeting as an important part of your total training arsenal. Your reward will be increased sales, of course. But there's nothing that boosts the morale of a sales manager or trainer so much as hearing a salesperson say, "That was a great meeting! I learned a lot from it."

section four

MEASURING AND EVALUATING SALES TRAINING

Chapter 23 Practical Guidelines for Measuring What Trainees Learn

Chapter 24 Measuring Training Results: Behavioral Change and Performance Improvement

chapter twenty-three

PRACTICAL GUIDELINES FOR MEASURING WHAT TRAINEES LEARN

THOMAS R. CURRIE

Thomas R. Currie is director of human resource development for the Reynolds Metals Company of Richmond, Virginia. Following more than a decade of experience with
IBM Corporation and the 3M Company in sales, sales training, and sales management, Currie joined the Reynolds
organization in 1963. Named to his present position in 1974,
he previously served as the company's director of sales
training and corporate director of training and development. He was a founding director of ASTD's Richmond
Chapter. He has been a member of the National Society
of Sales Training Executives since 1968. He served on that
association's board of directors for eight years and as its
president in 1977. On three occasions, Currie has been
the recipient of coveted NSSTE awards for editorial excellence.

The purpose of this chapter is to share with you, the trainer of sales and marketing personnel, some *practical* thoughts on how to measure the extent to which participants in your training programs have learned what you intended for them to learn.

As used here, learning is defined as the acquisition and retention of knowledge and skills. The measurement of behavioral change and performance improvement as *results* of learning are covered in subsequent chapters.

FIVE REASONS FOR MEASURING LEARNING IN SALES TRAINING

The measurement of learning surely ranks as one of the most perplexing and frustrating challenges confronting the sales trainer. It can be costly, time consuming, and more often than not, less than totally successful. So, why bother? There are a number of valid reasons; some of the more compelling of which are discussed here.

To Determine Whether Learning Objectives Are Being Met

Ideally, you begin each of your sales training sessions with a clear concept of what we expect to accomplish. You have a "game plan" complete with well-defined objectives that, if met, will bring our learners from where they are to where they need to be. As with any set of plans however, Murphy's law ("If anything can go wrong, it probably will") sometimes intrudes. How do we shield ourselves from Mr. Murphy? The best way is to devise a system of controls and checkpoints that will alert us if any aspect of our program deviates from the course we have set for it.

EXAMPLE _____

If key skills are being taught to new salespeople, points in the training can be planned where those participating are forced to define and demonstrate the ability to apply what they have learned. If the key skill is developing the ability to use open-ended questions when qualifying a new prospect, then the salesperson will have to develop a set of open-ended questions to ask when dealing with a new prospect.

If the new salesperson is unable to develop such questions, then the training session design would accommodate some remedial work with this person. Computer based training often has this automatically built into the program. The learner isn't allowed to go on to the next skill area until the present skill on the screen is mastered.

With respect to learning objectives, this means taking measurements *before* and *after* each major segment of the program to make certain that the learners are acquiring and retaining the knowledge and skills explicit in the objectives. If they aren't, corrective action can be taken to modify the design, the delivery system, or even the learning objectives themselves if they are proving to be unrealistic.

EXAMPLE _____

Learning objectives for a territory management program:

- To teach the participants what a sales territory is as defined by our company.
- To introduce all the IBM computer based marketing data available on each territory through actual use of their portable computers in the session.
- To teach participants how to gather usable sales data on their territory's sales force and report it to headquarters.
- To introduce the home office staff, their functions, and how to most efficiently use the services they have to offer by having them trace the lines they would follow in using the system to solve a proposed problem.

To Determine Whether Learning Is Transferable to the Job

Whereas learning for the sake of learning can be a virtue, it is not very cost-effective to burden a training program for sales and marketing personnel with sessions that have no practical, on-the-job application. Therefore, we need to ensure that what we are teaching is transferable to the job in a way that can produce tangible business results, for example, more sales calls, more closes, increased sales.

One way to test application validity is by measuring learning in a simulated job environment. For example, before-and-after practice sales presentations based on carefully constructed scenarios that *faithfully* mirror real-world market conditions can be very useful in measuring the acquisition of most selling skills.

To Strengthen Future Programs

As much as we might enjoy reading them, participant evaluations (sometimes called "smile sheets") usually do little more than give us general reactions to our training programs. While learner comments can be helpful in improving design and delivery systems, they rarely offer tangible evidence of *how much* was learned. Knowing, in quantitative terms, what our learners have acquired from a given program is one of the keys to strengthening future sessions.

To Evaluate Instructor Effectiveness

"If the learner hasn't learned, the teacher hasn't taught." An old saw to be sure, and one subject to considerable debate. However, none of us would questions the fact that the competency of the instructor is a critical element of the learning process. If we determine, through measurement, that our learners aren't learning, we must consider the instructor as a *possible* "villain" in the plot. If we fail to measure, we are left with little more than subjective criteria upon which to evaluate instructor effectiveness.

To Survive in Today's Business Environment

In today's tough-minded business environment, there is a growing intolerance with any function that will not or cannot show a contribution to the achievement of the organization's objectives. Unless we, as sales trainers, evaluate our efforts and prepare ourselves to demonstrate a beneficial impact on the bottom-line, we deservedly become prime targets when the inevitable cost-containment axe falls. While it may not be possible to prove that learning per se makes a *direct* contribution, the well-advised sales trainer will be able to show that knowledge and skill enhancement actually occur as a result of the training effort. Only to the *degree* that such enhancement can be shown are we able to establish a distinct linkage between the training and any subsequent performance improvement.

LEARNING OBJECTIVES: THE BASICS FOR MEASUREMENT

It has been said that, "If you don't know where you want to go, any road will take you there." Parenthetically, if you don't know what you want to measure, any measuring device will do. Too often, we trainers don't know what we want to measure because we haven't taken the time or made the effort to develop sound learning objectives for our training programs. Attempting to measure learning without specific, predetermined objectives is like trying to decide which team won a football game played without goal lines—it simply cannot be done!

Early in the program planning stage, and *before* design work begins, we should ask ourselves the questions, "What do our learners need to know that they don't already know?" and "What do they need to be able to do that they can't already do?" Obviously, in some cases neither question can be answered until pretesting has been done to determine existing knowledge and skill levels. In other instances, such as training the field sales force to introduce a new product, pretesting may not be practical. In any event, the answers to the questions should lead logically to precise, well articulated learning objectives.

Example of Developing Preset Objectives

You are Sally Jones of the XYZ Company. You intend to modify one of your existing products and that modification will mean additional user benefits to your customers and prospects. You have determined that your field sales personnel are unfamiliar with the design changes and that a training program is needed to bring them "up to speed." As a result of the training, you want your sales representatives to have a full *understanding* of the changes and to be able to *demonstrate* what the changes mean in terms of added benefits to the user. In other words, you want them to have acquired specific knowledge and skills. Your learning objectives might be stated as follows:

1. To be able to show a complete understanding of the Model B-100 design changes by scoring 100 percent on a 20-question written test.
2. To be able to demonstrate the ability to convert the design changes into added user benefits in a role-playing exercise.

These objectives, then, will become the basis for determining whether your sales people actually learned what you wanted them to learn. The important point here is that unless you have preset your learning objectives, *no* measurement method is going to answer the question, "Are my sales and marketing personnel acquiring *needed* knowledge and skills from my training programs?"

Methods of Measuring Learning

As stated earlier, attempts to measure learning often produce results that are less than totally satisfactory in that they are not always completely objective or quantitative. We cannot allow ourselves to become discouraged, however, by the realization that some of our efforts to measure will render results that are more subjective and therefore less credible than others. The goal should be to evaluate *all* training, using a measurement method that promises the *best possible* outcome.

The selection of an appropriate method is largely a matter of matching a method to a particular kind of training. For example, written tests might prove best for gauging acquisition of knowledge, while role play could be the better approach to measuring skills development.

Following is a general discussion of the more popular methods. It is not meant to be an exhaustive treatment of each method nor is it intended to be all inclusive.

Written Tests: Two Main Types

Of all methods discussed here, written tests are probably the most objective and lend themselves more readily to quantitative analysis. This will be particularly true if you follow the steps suggested by D. L. Kirkpatrick in the *Training and Development Handbook* listed in the bibliography at the end of this chapter.

Dr. Kirkpatrick recommends:

1. The learning of *each conferee* should be measured so that quantitative results can be determined.
2. A before-and-after approach should be used so that any learning can be related to the program.
3. Where practical, a control group (not receiving the training) should be compared with the experimental group which receives the training.
4. Where practical, the evaluation results should be analyzed statistically so that learning can be proved in terms of correlation or level of confidence.

Generally speaking, there are two kinds of written tests: standardized and "tailored." Standardized tests are relatively easy to obtain for a variety of subjects and usually have the further advantage of having been constructed by experts who understand how to rid tests of bias and subjectivity. A possible disadvantage is that you must design your instruction around the test content. Otherwise, you may end up with incompatibility between course content and what the test is intended to measure.

Tailored tests are those that you design yourself. Presumably, these will have the advantage of built-in compatibility since they will be based on the exact material you intend to cover. Because most of us are not test design experts, the tailored variety may be somewhat flawed from an objectivity standpoint and may not facilitate statistical analysis as readily as standardized tests.

At Reynolds Metals we have used both types with equally satisfactory results. For instance, with our field sales managers, we have found Don Kirkpatrick's "Management Inventory on Managing Change" to be useful as an aid to analyzing training needs (pretest) and as an objective way to measure learning (posttest). An extract from that instrument is shown in Figure 23–1.

Managers are asked to either agree (A) or disagree (DA) with each item

Figure 23–1 Management inventory on managing change

	Agree	Don't Agree
Answer all questions even if you are not sure of the correct answer.		
1. Managers should constantly be looking for changes that will improve department efficiency.	A	DA
2. People doing a particular job are one of the best sources of ideas to improve that job.	A	DA
3. In order to get a large number of suggestions from people, you must give money or prizes for ideas that are accepted and implemented.	A	DA
4. People will automatically accept changes decided on by experts.	A	DA
5. People who don't understand the reasons for a change will always resist it.	A	DA
6. People are always anxious to move from an old office to a new one.	A	DA
7. "You can't argue with success."	A	DA
8. Decisions to change should be based on opinions as well as facts.	A	DA
9. Explaining the reasons for a change will always turn resistance into acceptance.	A	DA
10. When a change has been decided on, it is a good idea to get subordinates involved in helping you implement the change.	A	DA

in the inventory, and their answers are graded, using an accompanying scoring key. By incorporating Kirkpatrick's rationale for each inventory item in our instructional design, we are reasonably assured of covering the same material over which the learners are tested.

Reynolds industrial products sales representatives need to acquire a great deal of technical product knowledge. Along with other methods, we determine their pre-post knowledge through a series of in-house developed written tests. An example appears in Figure 23–2.

Compatibility between the tests and the material presented is achieved by having the product expert who will function as the instructor for a particular session prepare the questions for that session. Occasionally, some minor editing by the professional training staff is necessary for clarification and consistency, but always with the knowledge and consent of the instructor.

Figure 23–2 Sample knowledge test

Written Examination Number 5
ALUMINUM PROPERTIES AND ALLOYS

1. Four advantages of aluminum's low density (light weight) are

   (1) _____ (2) _____
   (3) _____ (4) _____

2. Another characteristic of aluminum alloys is _____ corrosion resistance.

3. The naturally occurring film of aluminum oxide on the surface of aluminum may be replaced with a thicker oxide film by an electrochemical treatment known as _____ .

4. The use of aluminum in cooking utensils and in heat exchangers are illustrations of the use of aluminum because of its _____ .

5. T F In the 2000 series (sometimes called the 2XXX Series) the main alloying constituent is copper.

6. T F The 3000 series (sometimes called the 3XXX series) has manganese as the principal alloying ingredient.

7. T F The 4000 series (sometimes called the 4XXX series) has silicon as the principal alloying ingredient.

8. T F The 5000 series (sometimes called the 5XXX series) has magnesium as its major alloying ingredient.

Role Play

Role play, while perhaps more subjective as a measurement method than written tests, offers several distinct advantages over the latter. Unlike written tests, it is dynamic in nature and more closely simulates real-world conditions under which the learner's knowledge and skills will ultimately be challenged. Whereas written tests are normally administered in a static environment and allow relatively protracted periods of time for the learner to think through and formulate a response, role play demands much the same kind of spontaneity that the sales person must rely on when confronted with a customer's questions or objections. Also, role play can require physical as well as mental demonstration of knowledge and skills. Finally, most sales and marketing personnel seem to *enjoy* the interaction and immediate feedback inherent in role play. (Also see Chapter 16, "Using Role Playing for Sales Training.")

As with any method of measurement, role play must focus on and be relevant to what you want to measure. If, for instance, you wish to evaluate *closing* ability, the role-play situation or scenario should be designed to determine the extent to which the learner can convincingly summarize key buyer benefits and skillfully apply whatever closing techniques you are teaching.

In terms of its ability to measure learning, role play will be effective *only* to the extent that competent evaluators are used. We sometimes make the mistake of relying on the learner's peers to make these judgments. A mistake, because peers usually have no training in this area and because they are often less than totally candid since they realize that they will eventually be subjected to the same scrutiny. This is not to say that your learners should be denied feedback from their contemporaries. To the contrary, peer comments and observations can be quite useful in developing and maintaining the interaction and esprit so important to training programs. But peer evaluations usually fall short of telling you *how much* your learners have learned. This should be the domain of you and your professional colleagues if your objective is *accurate* measurement of learning. Of course, even the professional must know *exactly* what is to be measured and will need tools to facilitate measurement.

Following is a hypothetical illustration of how role play might be used to measure learning.

> Seller Scenario: You have just given an impressive demonstration of the Model B-100 to a prospect who appears to be interested but complains that your price is too high. Convince the prospect that B-100 is fairly priced.

> Buyer Scenario: You have just witnessed an impressive demonstration of the Model B-100. You believe it would fill a need in your organization, but you feel that it is overpriced by about $500.00 compared to a competitive product.

In this illustration, the trainer wants to determine the extent to which the learner (seller) understands the objection-handling techniques being taught and the degree to which the learner can apply the concept of price versus value.

As the scenario is role played, the trainer observes and uses the evaluation form depicted in Figure 23–3 to make judgments about how much the learner has acquired in terms of specific knowledge and skills.

Assuming the learner in this illustration role played the same scenario before and after the training, and assuming that the trainer is competent as an evaluator, there should result a reasonably valid measurement of learning.

Figure 23–3 Role-play evaluation form

```
                          Role-Play Evaluation

Subject: _____

Participant: _____ Date: _____

Technique Used: _____
Level of Understanding Demonstrated:

           Lo |  |  |  |  |  |  |  |  | Hi |
            0   1  2  3  4  5  6  7  8  9  10


Price Versus Value Concept:
   Did participant:  Amortize price?              Yes ☐    No ☐
                     Sell quality?                Yes ☐    No ☐
                     Sell service?                Yes ☐    No ☐
                     Sell total system?           Yes ☐    No ☐
Level of Effectiveness:

           Lo |  |  |  |  |  |  |  | Hi |
             1   2  3  4  5  6  7  8  9  10


Comments: _____
_____
_____
```

INDIVIDUAL AND GROUP PROBLEM SOLVING

Those of us who train sales and marketing managers usually offer a session in problem solving. Typically, we cover brainstorming techniques, situation analysis, problem definition, alternative courses of action, and finally, decision making. These are much easier subjects to teach than they are to measure in terms of learning acquired. But there are instruments available that can help us to determine how well individuals and groups of learners can apply the principles taught. One that we have used rather successfully is Human Synergistics' "The Desert Survival Problem II." (Also see the gaming and simulation in Chapter 18.)

Following the training session on problem-solving principles, managers individually study the situation posed by "Desert Survival" and then define major problems to be overcome. Next, they select a basic strategy from 6 alternative courses of action and then make 16 decisions regarding items needed to survive in a hostile environment.

After the individual tasks are completed, the managers are formed into teams of five or six. Individual solutions to the problems are discussed and consensus is reached on team answers.

Both individual and team solutions are scored against an expert's answers, providing us with a quantitative measurement of the degree to which the principles taught were applied.

An obvious shortcoming of this approach is the lack of a baseline measurement taken before the training occurred. It isn't practical to use the same instruments as a pre-post test. Different, though similar, instruments could be used but the result would be somewhat suspect since we would, in a sense, be comparing "apples to oranges."

ACTION PLANNING: HOW TO MEASURE SPECIFIC AREAS OF LEARNING

Action planning provides a way of measuring learning in specific subject areas. It lends itself quite well, for example, to determining the extent to which learners can apply principles taught in sessions on setting performance objectives.

Suppose that you are training sales managers to write performance objectives, to set standards of performance against which achievement of those objectives will be measured, and to develop action plans for accomplishing the objectives. At the end of the session you can test their understanding and ability to apply the principles by having them write an action plan similar to the one shown in Figure 23–4.

This form can be used as a pretest as well as a posttest. And, aside from

Figure 23–4 Action plan form

Action Plan for (Learner's Name) _____

Performance Objective: _____

_____

_____

Standard of Performance (describe the conditions that will exist when the objective has been met).

_____

_____

_____

Action Steps

   What   Who   When

_____

_____

_____

_____

_____

Resources Needed _____

_____

_____

Possible Barriers _____

_____

_____

_____

its value as a measuring device, action planning can be used as a way to get learners to formulate personal performance improvement objectives based on any or all parts of your training program. An objective to improve communication skill is an example.

LEARNER EVALUATIONS: ENCOURAGING PEOPLE TO THINK ABOUT WHAT THEY'VE BEEN TAUGHT

As mentioned earlier, learner evaluations must be viewed as suspect in terms of measurement validity. They can, however, provide more than subjective reaction if formatted in such a way as to require the learner to *think* about what has been taught. Figure 23–5 offers an approach that combines a subjective value assessment with a requirement to reflect on the training and make a judgment about how it can be *applied* in specific ways.

Figure 23–5 Learner evaluation form

Evaluation Form

Value Rating Scale

Excellent = 5
Very Good = 4
Good = 3
Fair = 2
Poor = 1

Session: *Improving Communication Through Better Listening*
1. In terms of its usefulness to me, I give this session a value rating of ___ .
2. I believe this session would have been more helpful to me if _____
_____ .
3. I will apply the learning acquired in this session in the following way(s).

_____

_____

_____

_____

The third item on the evaluation form should give us a reasonably reliable measurement of *what* was learned even if it falls short of being quantifiable. Obviously, this method cannot be used to pretest.

SUMMARY

The measurement of learning is unquestionably difficult and often frustrating. Yet there are sound reasons why trainers of sales and marketing personnel must make every effort to do so. Among these is the ever-increasing abhorrence toward any function which cannot prove that its activities contribute to the achievement of the organization's goals and objectives. Unless we find ways to evaluate the results of our training efforts, we will be ill-prepared to show that we are cost-effective, contributing members of the corporate citizenry.

This chapter offers several practical approaches to measuring learning. Some will produce more objective, quantifiable results than others. You will have to decide which method best suits your purpose. Other than to recommend strongly *before* and *after* testing and the use of control groups when feasible, no attempt has been made to deal with the so-called "rigorous experimental design" concepts. Such is beyond the scope and purpose of this work.

One should remember that measuring learning is only one phase of the overall process of training evaluation. But it is an extremely important part that will become critical when we later try to link classroom learning to on-the-job behavioral change and performance improvement.

BIBLIOGRAPHY

Bass, B. M., and J. A. Vaughn. *Training in Industry: The Management of Learning.* Belmont, CA: Wadsworth, 1966.

Davies, I. K. *Competency-Based Learning.* New York: McGraw-Hill, 1973.

Hamblin, S. C. *Evaluation and Control of Training.* London: McGraw-Hill (UK) Ltd., 1974.

Harrison, J. L., ed. *The Management of Sales Training, A Publication of the National Society of Sales Training Executives.* Reading, MA: Addison-Wesley, 1977.

Kearsley, G. *Costs, Benefits, and Productivity in Training Systems.* Reading, MA: Addison-Wesley, 1982.

Kirkpatrick, Donald L. In *Training and Development Handbook*, 3rd ed., Robert L. Craig, ed. New York: McGraw-Hill, 1987.

Kirkpatrick, Donald L. *Evaluating Training Programs, a Compilation.* Alexandria, VA: American Society for Training and Development, 1975.

Laird, Dugan. *Approaches to Training and Development.* Reading, MA: Addison-Wesley, 1978.

Mager, Robert F. *Preparing Instructional Objectives*, 2nd ed. Belmont, CA: Fearon, 1975.

Merwin, S. *Effective Evaluation Strategies and Techniques*. San Diego, CA: University Associates, 1986.

Odiorne, George S. *Training by Objectives*. New York: Macmillan, 1970.

Rae, Leslie. *How to Measure Training Effectiveness*. New York: Nichols, 1986.

Warren, M. W. *Training for Results*. Reading, MA: Addison-Wesley, 1969.

chapter twenty-four ——————————

MEASURING TRAINING RESULTS: BEHAVIORAL CHANGE AND PERFORMANCE IMPROVEMENT

C. E. (GENE) HAHNE, ROBERT E. LEFTON, ——————— V. R. BUZZOTTA

C. E. (Gene) Hahne is manager of training—human resources, Shell Oil Company, Houston, Texas. His group is responsible for training the 20,000 employees in manufacturing, pipeline, technical operations, and sales and operates four retail learning centers to train jobbers, dealers, and distributors. Hahne has received Shell Oil's Excelsior Award, ASTD's Distinguished Contribution to Employer Award, and ASTD's 1984 Torch Award. He served on both the old ASTD board and the board of directors under the new governance structure; as treasurer of the society; chair, budget and finance committee; and on many national committees and task forces. He was director of the ASTD Sales and Marketing Division and received the 1981 James R. Ball Award. His contributions to the community include work with Texas A&M University, University of Texas, Houston Baptist University, and the University of Houston. He helped design and implement a management development program for the state of Texas and has conducted workshops throughout the world. He has been named to *Who's Who in the South and Southwest*, *Who's Who in Finance and Industry*, *Personalities in the South*, *Personalities in America*, and *Men of Achievement in the World*.

Robert E. Lefton, a co-founder and president of Psychological Associates, Inc., is a leading U.S. consultant in sales, management, and organization development. Since earning his Ph.D. at Washington University, he has been a consultant to many *Fortune* 500 companies and served on the faculty of Washington University, The Presidents Association of the American Management Associations, Motorola Executive Institute, and CEO International, Inc. A developer of dimensional training, he has conducted seminars throughout the United States and Europe. He is co-author of *Improving Productivity Through People Skills*, *Effective Selling Through Psychology*, and *Effective Motivation Through Performance Appraisal*. He has contributed to journals such as *Sales Meetings*, *Training and Development Journal*, *National Productivity Review*, and *Marketing*. He has been quoted or referred to in *St. Louis Post Dispatch*, *The Wall Street Journal The Daily Telegraph of London*, *Forbes*, *Fortune*, *Time*, and *Psychology Today*. A major interest of his is the link between leadership skills and productivity. He frequently addresses groups such as ASTD, the Security Industries Association, Bankers Administrative Institute, Young Presidents Club, United Kingdom's Education and Training Conference, and the American Chamber of Commerce in London.

V. R. Buzzotta is co-founder, chairman of the board, and chief operating officer of Psychological Associates, Inc. Since the firm's inception in 1958, he has pioneered new concepts in sales, management, and appraisal training and has worked as a consultant with many major corporations in the United States and abroad. Dr. Buzzotta received his Ph.D. in Psychology from Washington University and is co-author of numerous articles on training and development, as well as three books on applied behavioral science: *Improving Productivity Through People Skills*, *Effective Motivation Through Performance Appraisal* and *Effective Selling Through Psychology*. Psychological Associates, Inc., develops and markets training programs in management, sales, and performance appraisal under the trademark *The Dimensional Training System*. These "people-skills" programs are used by a broad range of clients in the financial, consumer, and industrial fields in the United States and abroad, through affiliates in the United Kingdom, Canada, France, and South Africa. The company's Assessment Division helps clients with the selection and promotion of personnel.

Let's begin with a common occurrence:

A company puts a group of salespeople, say, 15 of them, through a sales training course designed to improve their sales behavior. At the end of the course, the salespeople are supposed to be more proficient at opening sales calls, examining customer needs, presenting their products persuasively, handling objections, and closing. To do all these things, they're supposed to have learned a number of "behavioral" skills: how to create rapport with customers, how to raise receptivity, how to ask resultful questions, how to adapt to different kinds of customers, how to resolve differences, and so on. The purpose of the training program, in short, is to change, or at least modify, their sales behavior, and thereby improve their sales results.

The question is: How can management ascertain if the program did what it was supposed to? Behavioral change, after all, is an intangible. Can something so immaterial be measured? Has management any way to determine if it's spent its money wisely?

Yes, management has a way. Skill at asking questions, say, or at motivating, cannot, of course, be measured by paper-and-pencil testing (as skill at reading the price book, for example, can). But behavioral change can be measured. It can be measured if the behavior is linked to measurable events—things that happen because of the behavior. This chapter explains how it's done.

HOW BEHAVIORAL-CHANGE TRAINING WORKS

Behavioral-change training for salespeople is training designed to change what salespeople say and do when they interact with prospects or customers. *Change* can mean a number of things: elimination of bad habits or hurtful behaviors, a decrease or increase in the frequency of certain behaviors, the instilling of new behaviors or the modification of certain behaviors. In every case, however, *behavior* refers to *interactional* behavior—to patterns of speech and action that affect the give-and-take between salesperson and customer. The fundamental purpose of behavioral-change sales training is to make salespeople more persuasive, more proficient at influencing a broader range of buyers and potential buyers.

This last point gets to the heart of the matter: the purpose of behavioral-change training is to enable salespeople to produce more sales. Companies don't spend money on such training for its own sake; they spend it because they expect the training to bring in more business than they'd otherwise get. If the efficacy of such training is to be measured, then, it must be measured in terms of practical results.

THREE KINDS OF OBJECTIVES FOR A SUCCESSFUL SALES-BEHAVIOR TRAINING PROGRAM

A well-designed sales-behavior training program—a program, that is, that conduces to measurement—must be designed with three kinds of objectives in mind, each seeking different results. We'll call these business objectives, application objectives, and learning objectives.

Business Objectives

These are the measurable business outcomes to be achieved as the end result of the training. An increase in dollar volume, an increase in units sold, deeper penetrating of a given market, higher closing ratios, lower cost of sales—these are examples of business objectives. They're the measurable—and ultimate—effects sought by the company as a result of sales training.

Application Objectives

These are activities that must be carried out on the job if the business objectives are to be attained and that can be carried out effectively only if the sales training is itself effective. Better prospecting, more efficient sales calls, fewer nonproductive calls, improved ability to contend with competition, more closes on first calls—these are examples of application objectives. They're things that must happen *before* the business objectives can be reached, and they're more likely to happen *after* sales training.

Learning Objectives

These are the selling skills (the sales *behaviors*) that must be mastered before the application objectives can be met. These are the skills that the sales training program is designed to teach. The ability to create rapport with prospects, ask questions that uncover needs, link benefits to needs, make persuasive presentations, handle objections—these are examples of learning objectives. It's skills such as these that must be learned—and that behavioral training is designed to teach—*before* the application objectives can be attained.

We see, then, that the three kinds of objectives are inextricably linked. As in the old song ("The hip bone's connected to the thigh bone, the thigh bone's connected to the knee bone"), the business objectives and the application objectives are connected to the learning objectives. It's the business objective that companies are ultimately interested in; it's to attain them that companies spend money on training. But the business objective cannot be achieved directly; they can be achieved only by attaining the application objectives; and these, in turn, can only be achieved by attaining the learning objectives. That, in a nutshell, is why companies do sales training.

THREE QUESTIONS FOR DETERMINING IF YOUR SALES-BEHAVIOR OBJECTIVES HAVE BEEN MET

To determine whether these three kinds of objectives have been met, sales trainers must answer three questions after any sales training program:

1. Did the participants learn the selling skills they were expected to learn? (Were the learning objectives met?)

2. The answer to the first question hinges on the answer to the second: Are the participants doing things differently and better in the field? (Are they meeting their application objectives by applying their new skills?) This is the acid test. A participant may seem to have "learned" many skills in a training program, but not until he or she uses them successfully in the field can anyone be sure of it.

3. As a result of things being done better and differently in the field (that is, as a result of the application objectives being met), are the company's business objectives being met? This is the "payoff" question. Unless the answer is "yes," the training may be deemed a failure even though the learning objectives have been met. Obviously, though, if the three kinds of objectives were originally set with one another in mind, the achievement of the first two should ensure the achievement of the third—the business objectives.

We should note here that circumstances beyond anyone's control—for example, a severe decline in the market—could make the business objectives unreachable. On the other hand, if both the learning and application objectives are achieved and a windfall occurs—for example, the competition goes out of business—it would be difficult to determine whether the business objectives were accomplished because of the training or because of the windfall. These uncontrollable factors should always be considered in the evaluation of results.

TWO BASIC MEASURING PRINCIPLES

It's Not How Much You've Learned, It's What You've Learned That Counts

We're ready to state a basic principle of measurement: the acquisition of skills does not guarantee changes in behavior, or, in our terms, the attainment of learning objectives doesn't ensure the attainment of application objectives. Even if a training manager measures the attainment of learning objectives, and finds (by carefully observing role plays, say) that the participants have indeed become more proficient at, say, probing, he has learned little of any practical value. To a trainer who says, "The people in this group are now better at probing than they were before being trained," a perfectly fair rejoinder would be, "So what?" What *difference* does it make? The answer, plainly, is that it makes no difference at all unless the newly learned probing skills are used in the field and contribute to the attainment of certain business objectives. What's ultimately being measured is not the amount learned, but the practical value of what was learned. Management, which after all pays the bills, isn't interested in buying improved behavior; it's interested in buying improved results.

Sales Trainers Need the Cooperation of Upper Management to Succeed

A second, related principle is: sales trainers alone cannot set the three kinds of objectives. They need the help of people at higher levels. Just who those people are will of course vary from company to company, but very rarely are they in the training department. In most companies, sales manager or even managers at higher levels must tell the sales training manager what their business objectives are. In many cases, the sales manager also sets forth the application objectives. The training manager then sets learning objectives that will ensure the attainment of the application objectives and, by extension, of the business objectives.

None of this, however, is likely to happen as a matter of course; very few companies are so well coordinated that all these pieces fall into place effortlessly. Generally, it's the sales training manager (whose formal title may be director of sales training or something similar) who must initiate the setting of objectives, usually in response to some precipitating event.

PRECIPITATING EVENTS: WHY SALES TRAINING PROGRAMS DON'T "JUST HAPPEN" _____

In well-run companies, sales training programs don't "just happen." They happen in response to a felt need. Something happens that makes management feel "we'd better do something to help our salespeople do a better job." Usually, this "something" is (1) an organizational change, (2) an internal problem or (3) external pressure.

Organizational Change

The sudden hiring of large numbers of new salespeople, the acquisition of such salespeople through acquisition or merger, the addition of new products or services—occurrences such as these frequently cause management to feel that "we must augment the skills of our salespeople."

Internal Problems

Observed deficiencies in the sales force, sudden losses of business, rising cost of sales, unsatisfactory closing ratios—any such problem may be a precipitating event.

External Pressures

Successful incursions by competitors, the entry of new competitors or competing products into the marketplace, abrupt changes in the market, a decline in the economy—any such pressure from the outside may also be a precipitating event.

What all such incidents have in common is that they may cause management to feel "something's wrong (or about to go wrong), and we'd better try to fix it." We say "feel," because, at this early stage, nobody may yet have articulated a need for training (or any other remedy). Management may have nothing more than a vague, unformed feeling that things aren't quite right. Sooner or later, someone in management may vocalize the need in concrete language: "We need this-or-that kind of sales training." But, most often, it's the training manager who must sense the need, probe it, and give it shape. That's what professional trainers are paid to do.

HOW SALES TRAINERS DETERMINE THE NEED FOR SALES-BEHAVIOR TRAINING _____

The trainer can find out by either informal or formal means what must be done to counteract the bad effects of the precipitating event.

Informal Means

Proficient trainers are "tuned in" to what goes on in their companies. They make a point of listening and observing, of keeping up with things, of staying in touch with a wide array of co-workers, especially those in a position to know "what's happening." As a result, they usually learn of precipitating events as they occur, or soon after. If they notice, or even guess at, a need for sales training, they can pursue the matter with the appropriate people in management.

Formal Means

A number of more systematic methods are available to training managers who want to know if a need for sales training exists. These are need surveys, attitude or climate surveys, and research into company records.

Need surveys are the most direct technique, but they too often provide only limited information; they tell the trainer much more about deficiencies in skills than deficiencies in application. They're usually more useful in setting learning objectives than application objectives.

Attitude and climate surveys may be less direct, but the information they disclose may be more useful. While they may not explicitly reveal a need for training, they may imply it. They may also give the training manager a good idea of shortcomings in application.

Finally, companies keep records that, if properly interpreted, may reveal a need for sales training. Records of turnover, trends in sales volume, trends in sales costs, performance appraisal results—any of these may indicate a need for more or different training.

Together these formal and informal methods may suggest to the training manager that the company has, or soon will have, problems that might be solved by more or better sales training. His next step is to check out this assumption with those people in management who are capable of verifying it, and who can authorize that something be done about it. In doing this, the training manager must get the answers to these questions:

1. There's evidence to suggest we have such-and-such a problem. Do we?
2. What must we (the company) make happen so as to overcome the problem? In other words, what objectives ought we set for ourselves to solve it?
3. Which of these objectives can be furthered by more or better sales training? (These will ultimately become the business objectives.)
4. What must our salespeople do in the field to ensure, or at least advance, the attainment of these objectives? (These activities will become the application objectives.)

Usually there's no need to ask management about the training objectives. The training manager can work these out on his own.

The answers to the foregoing questions can be worked out, of course, only through give-and-take between the training manager and other representatives of management. The former must not go into the meeting as a supplicant, seeking permission to "do some sales training." He or she must go in as a specialist with valuable experience and expertise to contribute.

HOW TO WIN MANAGEMENT'S COMMITMENT BY LINKING VARIOUS CORPORATE INTERESTS

One problem confronting the sales training manager is that not everyone in the company is interested in all three kinds of objectives, or even sees the connection between them. While it's risky to generalize about this matter, it's probably safe to say that the following interests manifest themselves in most companies:

- Senior management is typically interested in broad business and application objectives.
- Middle management is likely to concentrate on narrower business and application objectives (designed, of course, to meet the broader objectives of senior management).
- Salespeople are likely to focus most of their attention on application and learning objectives. They'll be interested in business objectives too, of course, *if* they know what they are.
- Sales trainers are, as a rule, mainly interested in learning objectives and to a lesser extent, in application objectives.

None of this is surprising. At every level, people are most interested in those objectives that most directly impinge upon them. The job of the sales training manager, then, is to link all three kinds of goals. It's his or her responsiblity to make the connections known to everyone involved, from top management to sales trainer. Senior managers should never have to ask such questions as, "Why in the world are we spending money to teach our sales people how to probe?" Sales trainers should never have to wonder, "Once we've taught our salespeople how to probe, what good will the company get out of it?"

Neither senior- nor middle-management can be expected to be automatically committed to any sales training program. In most cases, such commitment must be won. Sales training managers can win management commitment if

- They involve management in the objective-planning process.
- They demonstrate the connection between the program's learning objectives and its application objectives and its business objectives.

The message to management must be: "We're not doing sales training for its own sake; we're doing it to further the attainment of this company's business objectives." Then, after the training, if the commitment is to be sustained, the sales training manager must show that the learning and application objectives were indeed achieved, and that, as a result, the company has moved closer to its business objectives.

DESIGNING MEASURES OF BEHAVIORAL AND PERFORMANCE CHANGE

In designing measures of behavioral and performance change, the training manager must answer two sets of questions: when and how?

When to Measure

1. *Postmeasures.* Should the evaluation focus on behavior after the training? If so, when is "after"? Immediately after? A month after? A year after? Several of these?
2. *Pre- and postmeasures.* Should evaluations be made both before and after the training, so that comparisons can be drawn?
3. *Comparison groups.* Should evaluations be made of salespeople who have not been trained as well as those who have, so that comparisons can be drawn? Should a matched control group be evaluated alongside the trained group?

How to Measure

1. *Qualitative versus quantitative.* Should "anecdotal" evaluations or testimonials be used? Or only numerical evaluations? Should other-person evaluations (say, by a sales manager who is making calls in the field with a newly trained salesperson) be included? How about a combination of these techniques?
2. *Self versus other.* Should subjective evaluation (by the trainee) be allowed? Or only objective (presumably) evaluations by other people (sales managers or customers)?

3. *Audits of company records, questionnaires, surveys.* Which of these will provide the most useful and valid data? Does anyone on the staff know how to design questionnaires or surveys?

4. *Interviews.* Are the time and resources available to do these? Who will be interviewed? Who will do the interviewing?

5. *Observation.* How objective will these reports be? How detailed? How structured?

Each of these techniques of measurement has its advantages and drawbacks. Which is best depends upon the expertise available, the people, the budget, and the time. We'll examine each technique in the light of these factors.

WHEN TO MEASURE

In most companies, *when* to measure is largely affected by the availability of the resources needed to do it. Generally, where money, people, and time are very limited, measurement is restricted to posttraining results; where they're less limited, measurement may include pretraining results as well; and where they're ample, measurement may also include comparison groups. No matter how many resources are available, however, sales trainers should be aware of the following considerations:

Two Types of Postprogram Measurement

There are two kinds of postprogram measurement: that done immediately after the program, utilizing program critiques filled out by participants, and that done a month or more (perhaps a year) after, utilizing self- or other-questionnaires, or actual calls on prospects. We'll call the first "instant" measurement and the second "delayed" measurement.

"Instant" Measurement. The major advantage of instant measurement is that it provides immediate feedback. The sponsoring company doesn't have to wonder if participants "liked" the program or considered it "beneficial"; it can find out at once, simply by asking. Program critique forms used in instant measurement can disclose two things: (1) how well, in the opinion of the participants, the learning objectives were met and (2) how useful the participants expect the learning to prove. In other words, application objectives cannot be measured at this early time, but application *intentions* can.

Instant measurement offers a number of *dis*advantages as well, and trainers should be alert to these. For one thing, program critiques are often "love letters" rather than hard-headed evaluations; sales training programs often end on a high note, with salespeople singing the praises of the program, and ignoring its flaws. Even worse, the ratings are highly subjective, since most participants have no standards by which to measure the program. About all any participant can honestly say is: "I *think* the program was well done, I *think* it'll prove useful, and I *intend* to put it to use." That's something, of course, but it's hardly enough to justify the money the company spent on the training.

Delayed measures, on the other hand, put time and distance between the measurement and the program. The participant may be asked to fill out a questionnaire, evaluating the program in the light of later experience. Or someone else, say, the sales manager, may be asked what he's observed the program has done for the salesperson. Or, best of all, a sale planned during the program may be tracked to determine if the plan worked and if the sale was made.

"Delayed" Measurement. Delayed measurement offers a number of advantages missing in instant measurement. Questionnaires are much less likely to become love letters; by the time of the measurement, some low notes have probably mingled with the high notes. More than that, the opinions of people other than the participants can be solicited; by the time the measures are taken, the sales manager in particular should have had an opportunity to observe the participants in action. Finally, delayed measures yield more information; learning objectives, application objectives, and (if enough time has elapsed) business objectives can all be measured.

Nevertheless, there are a couple of disadvantages to delayed measures. For one thing, a self-questionnaire is still likely to yield highly subjective information; participants still have no independent criteria for evaluating the program. Moreover, delayed measurements are harder to compile, since the participants have scattered. On balance, however, delayed measures are much to be preferred to instant. Figure 24–1 is a sample delayed measurement survey.

Pre- and Postprogram Measures

In this kind of evaluation, postprogram results are measured and compared with preprogram results. This is obviously a much more objective way to ascertain whether anything has really changed as a result of the training.

The advantages of pre-post comparisons are numerous. To begin with, these measures provide a valid standard for comparison: whatever happened *before* the training becomes the baseline. Such a line makes it fairly easy to determine if performance has improved since the training, worsened, or stayed the same. Pre-post comparisons make it possible to measure all three kinds of objectives: learning, application, and business.

Figure 24–1 Sample delayed measurement survey

This survey is used as a *postonly* instrument. Dimensional Sales Training®—I (DST-I) is a skill-building program that teaches new or untrained salespeople the basics of successful selling.

DIMENSIONAL SALES TRAINING®—I (DST-I)
FOLLOW-UP APPLICATION SURVEY

Some time ago you participated in a DST-I seminar. Now that some time has passed, we would like to find out if you have had a chance to apply your skills and what happened when you did. The survey focuses on application of the DST-I skills to the "on-the-job" tasks which you carry out as a salesperson. By taking a few minutes to respond to this, you will help refine DST-I and help your company evaluate the DST-I program.

Your answers will be combined with those of other participants. Your responses will remain completely confidential, so please respond with high candor.

The survey will take ten (10) minutes or less to complete. We need your questionnaire by the date specified in your cover letter if your ideas are to count. Thank you for responding.

DST-I POSTSEMINAR SURVEY

1. Have you applied your DST-I selling skills since the program?

   ☐ YES   ☐ NO
   If NO, please go on to question 8.

2. Below is a list of tasks frequently assigned to salespeople. Please check all those (*one or more*) to which you've applied your DST-I skills.

   ☐ Persuade customer to renew present contract   ☐ Qualify prospect
   ☐ Persuade customer to place new order   ☐ Uncover additional customer needs
   ☐ Improve profitability of product or service mix   ☐ Demonstrate product/explain service
   ☐ Get customer to provide referral   ☐ Train distributor salespeople
   ☐ Introduce new product   ☐ Gather sale-related information

Figure 24–1 (Cont.)

☐ Handle complaint   ☐ Notify customer of price increase
☐ Handle postsale service   ☐ Notify customer of change (in product, service)
☐ Get approval for new promotional program   ☐ Get proposal approved
                                             ☐ Activate inactive account

☐ Other (please describe): _____

3. Approximately how many customer contacts do your check marks reflect? _____

4. How much benefit to you and your company do you expect from these calls in the 12 months following the seminar?

No Benefit	Minimal Benefit	Moderate Benefit	Significant Benefit	Very Significant Benefit
☐	☐	☐	☐	☐

5. What do these calls represent in gross revenue to your company? (Try to compute this figure accurately and realistically.) $ _____

6. How important has each of the following DST-I skills been to you on these calls? (Please check one box on each line across.)

	No Importance	Some Importance	Moderate Importance	Great Importance	Very Great Importance
a. DST-I FOUR-STEP SALES PROCESS	☐	☐	☐	☐	☐
b. BENEFIT STATEMENTS	☐	☐	☐	☐	☐
c. PROBING	☐	☐	☐	☐	☐
d. MANAGING OBJECTIONS	☐	☐	☐	☐	☐
e. SPECIFIC SKILLS FOR MANAGING Q1, Q2, OR Q3 BARRIERS	☐	☐	☐	☐	☐

7. How much of your success on these calls would you assign to your DST-I skills?
☐ 0% ☐ 10% ☐ 20% ☐ 30% ☐ 40% ☐ 50% ☐ 60% ☐ 70% ☐ 80% ☐ 90% ☐ 100%

Figure 24–1 (Cont.)

8. If you have not yet applied your DST-I selling skills, please tell us what's kept you from doing so.

_____

9. How much have you used your **Postseminar Application Manual** since the seminar?

    No Use   Moderate Use   Extensive Use
     ☐   ☐   ☐

10. Has your manager used the **Sales Manager's Coaching Manual** with you since the seminar?
   ☐ YES   ☐ NO

11. a. Would you recommend the DST-I course to other salespeople? ☐ YES ☐ NO
    b. Why or why not? _____

_____

12. In view of your experience, how would you respond to this statement: "DST-I training for our sales staff is a sound investment for our company."?

Strongly Disagree   Disagree   Uncertain   Agree   Strongly Agree
     ☐   ☐   ☐   ☐   ☐

13. In view of your DST-I experience, would you be interested in taking a more advanced Dimensional Sales or Sales Management course?

   a. Advanced Sales Training-II (DST-II)

     Strong Interest   Some Interest   No Interest   Doesn't Apply To Me
        ☐   ☐   ☐   ☐

   b. Advanced Sales Management-II (DSM-II)

     Strong Interest   Some Interest   No Interest   Doesn't Apply To Me
        ☐   ☐   ☐   ☐

14. How many months ago did you take DST-I? _____

Thank you very much for your candid responses.

There are, however, a couple of disadvantages. One is that the collection of data is significantly more difficult when the data must be collected at different times. The other, and by far the more troubling, is that there's not "experimental control," no way of knowing for sure that a change in the postprogram results is attributable to the program itself. Other factors having nothing to do with the training (for example, the closing of a competitor's business or an abrupt change in foreign demand) may account for the change. True, in most instances, sales trainers can reasonably assume that the training made the difference, but they cannot be sure. See Figure 24–2 for a pre- and postprogram measurement survey.

Comparison Groups

Comparison groups provide the experimental control that's lacking in pre-post measures. Other variables may still intervene, but, with the setting up of control groups, they're less likely to "contaminate" the measures.

Let's use a simple example to illustrate the point. Suppose group A (the experimental group) goes through a week of sales training, and group B (the matched control group, with characteristics similar to group A's) does not. In the month before the training, each group produces sales volume of $1 million. In the month after the training, group A produces $2 million and group B produces $1 million. Unless some other factor can be discovered that affected one group but not the other, it's a safe assumption that group A's superior performance can be attributed to the training. For instance, if a sudden recession hit group B's territory but not group A's, that would be an external factor that might override the effects of training. But, barring the existence of such a factor, the training would seem to be the variable that made the difference.

The advantages of using pre- and postmatched groups are plain. This is a rigorous technique, and an objective one; nobody writes or sends "love letters." Furthermore, it sets a clear standard for comparison: what both groups did *before* the training. It's excellent for measuring all three kinds of objectives, and it yields the most definite conclusions.

Nevertheless, there are some serious disadvantages. The matching of groups may not be easy, and the collection of data is sure to be more difficult than in other kinds of measurement. Moreover, the technique is comparatively costly in both time and money. Worst of all, it will work only if one group—the control group—is *deprived* of training. This can be costly. In our earlier example, group B, the control group, produced $500,000 less volume in one month than did group A. If the training enabled group A to produce $2 million, similar training given to group B might have enabled it to sell $2 million too. If so, the technique of measurement required the sacrifice of $500,000 in volume.

To summarize, in determining "when" to measure the results of sales training, trainers should keep in mind that preprogram and postprogram measures

Figure 24–2 Sample pre- and postprogram measurement survey

The following is used as a *pre-* and *post*instrument. It measures the effectiveness of the manager before and after the manager has been trained as evaluated by that manager's subordinate.

YOUR *LAST*
PERFORMANCE APPRAISAL CYCLE

HOW *EFFECTIVE*
WAS IT?

OBJECTIVE

This questionnaire will help you size up your experience with your organization's Performance Appraisal Cycle. It will help you evaluate both the benefits you derive from it and any problems which impair what you and your organization get out of it. We've designed this survey so that you can give us . . . in strictest confidence . . . your *candid* opinions.

INSTRUCTIONS

On the next few pages are eight statements about the Performance Appraisal Cycle. Please read each and then rate your reaction, using the 5-point scale following each statement. Base your rating *on your own experience with your own boss* in your last full cycle. Remember: a cycle covers a full year of work, from goal setting to progress reviews to the year-end appraisal interview. (If there's no full Performance Appraisal Cycle in your organization, or if you've changed bosses, please answer the questions to the best of your ability.)

1. *Clarity of Goals.* Think about your job goals—or your job objectives—what you were expected to accomplish on the job. The goals should have been realistic, not too easy, not too hard. They should have been specific—quantity, time, quality, and so on should have been spelled out. They should have been comprehensive and clearly understandable, so you knew what you were to achieve in all areas. They should also have been in writing. In the light of these criteria, please rate your goals.

Figure 24–2 (Cont.)

Overall, my goals were (**circle** *only one number*):

1   2   3   4   5

(Nonexistent or   (Okay. Some were clear   (Excellent. High
vague and incomplete.)   some were not.)   clarity.)

2. **How Your Goals Were Established.** Ideally, you should have had some say in setting your goals. They should have been worked out jointly with your boss, who should have listened to your ideas respectfully and considered them on their merits. If, because of company policy, the goals were handed to you without prior discussion, you should at least have had a voice in setting subgoals, acceptable performance levels, and developing plans for achieving the goals. In the light of these guidelines, please rate how your goals were established.

My goals were established (**circle** *only one number*):

1   2   3   4   5

(Without my input   (With my partial   (With my full involvement
or very little say.)   involvement.)   according to the
                                                                    foregoing criteria.)

3. **Commitment to Goals.** Ideally, as a result of your goal-setting meeting, you should have been fully committed to your goals—willing to work hard to achieve them. They should have (a) made sense to you—seemed fair, reasonable, and useful; (b) appealed to you—seemed interesting, challenging, and suited to your talents; (c) motivated you—you should have understood and valued the benefits that would accrue to you if you attained them. With these criteria in mind, please rate your commitment to your goals.

My commitment to my goals was (**circle** *only one number*):

1   2   3   4   5

(Nonexistent or   (So-so. I was fairly   (Strong and compelling. I
minimal. I did what   committed to some of   was eager to do what was
I had to.)   my goals, but not all.)   needed to achieve them.)

Figure 24–2 (Cont.)

4. *Progress Reviews.* Think about what's happened after your goals were set at the start of the performance year. Ideally, periodic progress reviews should have been held, with you and the boss checking to see if you were moving toward your goals on schedule or, if not, why. These reviews should have given you constructive feedback on performance. In these reviews, the two of you should have candidly discussed questions like: Was there good reason to change any of the goals? If you weren't on schedule, what could be done about it? Was your progress being hindered by factors beyond your control? Was the boss a help or hindrance? Data should have been systematically collected and recorded. With these ideas in mind, please rate your progress reviews.

My progress review meetings were (*circle* only one number):

1	2	3	4	5
(Nonexistent or very inadequate.)		(Okay, but not thoroughly or frequently enough.)		(Excellent. They met all the foregoing criteria.)

5. *Preparation for the Performance Appraisal Meeting.* Think back to the preparation for your last (year-end) performance appraisal interview. Ideally, *both* you and the boss should have planned it well in advance, to allow time to do it right. Performance data should have been systematically collected and documented throughout the year and available to you for planning your appraisal. Before the appraisal meeting, you should have been told how to prepare, and what would be expected from you in the meeting. With these points in mind, please rate the overall preparation for your last performance appraisal.

The preappraisal planning the boss and/or I did was (*circle* only one number):

1	2	3	4	5
(Nonexistent or very inadequate.)		(We planned some but not very thoroughly.)		(Excellent. It met all the foregoing criteria.)

6. *The Performance Appraisal Meeting.* Recall your last (year-end) performance appraisal interview. It should have been an open, candid exchange of information and opinions between you and your boss. The two of you should have worked toward the same end: not only a valid evaluation of your performance, but a clear understanding of its causes and how it could be made even *better*. You should have felt free to express any relevant views, which should have been attentively heard and seriously considered. The exchange should have been businesslike and realistic, with both of you facing all the facts, positive and negative. In the light of these criteria, please rate how your last appraisal meeting was conducted.

Figure 24–2 (Cont.)

My last appraisal meeting was (*circle* only one number):

1	2	3	4	5
(Nonexistent or very inadequate.)		(Okay. We seemed to do an adequate job.)		(Excellent. It met all the foregoing criteria.)

7. **Results of the Appraisal Meeting.** Think about the end results of your last performance appraisal meeting. Ideally, it should have produced complete understanding by you of the quality of your performance, and strong commitment to setting new goals. You should have left with a firm resolve to do what's necessary to strengthen your performance. You should have felt the appraisal was honest, fair and useful. With these points in mind, please rate the results of your last appraisal.

My last appraisal meeting produced (*circle* only one number):

1	2	3	4	5
(None or very few of above results.)		(Some of the above results.)		(All of the foregoing results.)

8. **Payoff from Performance Appraisal Cycle.** Summarize how your last Performance Appraisal Cycle (goal setting through progress reviews through appraisal interview) benefitted you, the boss, and the organization. Ideally, the cycle should pay off in higher productivity. Also (a) you should know where you stand and why, and that your efforts are suitably recognized and rewarded; (b) your boss should manage more effectively than before, thanks to your improved efforts; (c) the organization should have reliable information for making decisions (about compensation, promotion, transfers, etc.). In other words, the appraisal should pay off for everyone involved. With this in mind, please rate the overall benefits from your last appraisal.

The benefits from my last appraisal have been (*circle* only one number):

1	2	3	4	5
(Minimal to everyone: me, my boss, the organization.)		(So-so. There have been some benefits, but not as many as there should have been.)		(Maximal. The last cycle has paid off in all the ways mentioned.)

© 1981 Psychological Associated, Inc.

568

utilizing matched control groups will usually yield the most thorough, objective, and valid data. The technique, however, requires extensive resources and probably should not be undertaken unless the necessary money, time, and people are available.

HOW TO MEASURE

When it comes to the "how" of measurement, we find, once again, that the resources available usually have a decisive effect upon the kind of measuring that's done. Companies with tightly limited resources (time, money, and people) mostly use qualitative and other "soft" measures. Companies with more extensive resources are free to make more use of quantitative and other "hard" measures. In the long run, quantitative measures are more desirable, but qualitative measures should not be spurned. Both kinds offer certain advantages.

Qualitative

These are usually *anecdotes* (reports of something that happened during the training or, presumably, as a result of training, and that shed light on the value of the training) or *testimonials* (statements affirming the worth of the training, not necessarily accompanied by anecdotal reports). Qualitative measures should not be thought of as mere "puffery." They can and often do disclose information of real value in judging the effectiveness of training. They can be used to measure learning, application, and business objectives.

Qualitative measures have several obvious advantages. One is that they're relatively easy to obtain; to get them, all you need do is ask for them, either in interviews or in writing. Another is that they bring the results of training to life; exuberance and enthusiasm or disappointment and frustration cannot be captured in numbers, but they can be captured in anecdotes. And, quite obviously, either set of emotions may affect the end results of the training. Finally, anecdotes and testimonials (especially anecdotes) may help explain things that are otherwise hard to understand. A trainer may, for instance, feel that she did "everything right" in a seminar, and yet be aware that, for some "mysterious" reason, the seminar didn't "click." A few candid qualitative reports may explain why.

Still and all, qualitative measures have some serious drawbacks. For one thing, they're difficult to tabulate and compare; the trainer cannot always be sure he or she is dealing with apples and apples. Even more troubling is the vagueness and subjectivity of the reports. Statements like "I think the program was great" (and such statements are common) aren't very useful. In fact, any thoughtful trainer is bound to ask: "How do you *know*? What makes you—a salesperson—such an authority on sales training? And what do you mean by *great*? What are your standards of judgment? Do you really mean *great*, or merely *good?*" And so on. Qualitative reports sometimes raise more questions than they answer.

None of this means that qualitative measures should be disregarded. But they should be used with caution.

Quantitative

Quantitative reports, which are essentially numerical, impress most trainers and other businesspeople as "scientific" and "empirical," both of which are words with strong positive connotations. Indeed, quantitative measures do offer several advantages lacking in qualitative measures.

One advantage, of course, is that quantitative measures are relatively easy to tabulate and compare. Whatever's being measured (e.g., units, dollars, calls, closes), the trainer can be certain she's dealing with apples and apples. Another advantage is the objectivity of the measures: If group A produced $2 million volume in one month, that's that, and nobody can reasonably ask, "Says who?"

But there are disadvantages. One is that quantitative measures don't "just happen"; somebody must design the system of measurement. This takes time and know-how. Even more serious, however, is the lack of critical incidents, that is, reports of events that clarify otherwise puzzling matters. Numbers reveal a lot, but they're not likely to reveal the sudden insight, the sharp perception, that "makes everything fit."

To summarize, any company with the necessary resources will want to use qualitative measures. But these measures, while valuable, are not perfect. They may leave certain big questions unanswered.

OBTAINING QUALITATIVE OR QUANTITATIVE MEASUREMENTS

Self- and Other Judgments

The sales trainees themselves, their managers, and their customers are all potential sources of anecdotes and testimonials.

The trainees are, of course, the easiest source to tap. Their reports have one unique advantage: they come from the "inside." If anyone *should* know whether the training has paid off, it's the person who was trained. But sometimes this very specificity makes these "insider" reports undependable, especially if the reports are requested immediately after the training. Some trainees may be exhilarated; others may be downhearted. In either case, their reports may be so colored as to be virtually valueless. Most sales trainers who have relied on self-reports by trainees will admit, as one of them put it to the authors, that "these things always make me queasy—I never know how much of what I'm reading is fact and how much is fiction."

Other-person judgments, usually rendered by either the sales manager or the customer, are generally more objective. In effect, they say, "Here's what I've observed since this trainee (or group of trainees) went through the training program. I have no ax to grind, so I have no qualms about reporting the facts as I see them." Reports by customers can be especially instructive, but they're frequently difficult to obtain (and many companies are reluctant to ask for them).

It would be wrong, however, to leave the impression that other-person judgments are always fair and objective. They're sometimes based on very inadequate data; a sales manager, say, may have had very little opportunity to observe a given salesperson since the training program, but may nevertheless render a dogmatic judgment. Even worse, other-person reports are sometimes skewed by bias; a customer, for instance, may be so eager to be a "good guy" that he reports things that simply aren't so ("This salesperson has been doing a terrific job," when the truth is that he's been doing an adequate job).

The best approach, then, is to combine self- and other reports. This is more difficult and time consuming than an either-or approach, but it does something the either-or approach cannot do: it offers *different* perspectives on the value of the training. If the "true story" can be obtained at all, it can best be obtained by hearing the versions of the salespeople, their managers, and at least a sample of their customers.

Program and Company Records

Training program records are usually of two types: (1) postprogram questionnaires and (2) reports on sales calls planned during the training and carried out afterward.

Questionnaires have one of the same drawbacks as self-reports (the objectivity of the responses is open to question), but they're usually more focused, and they lend themselves to quantification. A salesperson who's asked to rank, on a scale of 1 to 10, say, the knowledgeability of the instructor cannot ramble and equivocate; he must come up with a number. This doesn't mean his response will necessarily be objective or informed, but it is easy to compile and to compare with the numbers on other questionnaires. Such questionnaires are commonly used at the conclusion of seminars to see if participants think the learning objectives were met. They can also be used later to determine if the application objectives are being met.

Reports on sales calls are much more objective and instructive. Typically, here's what happens: during the sales training seminar, each salesperson picks a hard-to-sell account, designs a strategy for making a sale to the account (using the skills learned in the seminar), and practices and polishes the strategy. Very soon after the seminar, the salesperson calls on the hard-to-sell account and puts the strategy to work. The results, of course, are easy to track. Immediately after the call, or series of calls, each salesperson can file a report answering such questions as: Did you make the sale? Did you sell the amount you projected? If the strategy works, and if a hard-to-sell account becomes a *sold* account, it's evident that the sales training has paid off.

Company records may also provide useful information, although the link between the information and the sales training is not always easy to establish. The records may show, for example, that sales have increased 5 percent in the six months since the entire sales force concluded a sales training program, but, without controls, the connection between the increase and the training will remain tenuous. Records of other things (turnover, closing ratios, etc.) may be equally difficult to correlate with training. Still and all, such records are usually easy to collect and reasonably objective.

Of the various kinds of records reviewed, reports on sales calls planned during training and carried out afterward are usually the most useful.

Audits, Surveys, Interviews, Observation

Audits can be used to measure sales volume, closing ratios, calls per salesperson, or profitability. Their major advantage is their relevance: they can be designed to measure very specific kinds of data. And they can be used to determine if various business objectives are being met. They are not, however, very helpful when it comes to measuring learning or application objectives.

Surveys and questionnaires are usually easy to administer and yield quantitative data. They can be used to measure all three kinds of objectives: learning, application, and business. But it's sometimes difficult to construct reliable quantitative scales, and without such scales, the value of the surveys diminishes considerably.

Interviews, if well done, can provide a wealth of useful qualitative data. In the experience of the authors, most salespersons enjoy being interviewed; under skillful questionings, they're likely to provide a number of critical incidents and insights. Unfortunately, interviews are time consuming and, especially when the sales force is scattered over a wide area, costly, they sometimes yield highly subjective and unquantifiable results, and they're often conducted by sales trainers who are less than effective questioners. Still and all, at their best, interviews are a useful measuring tool.

Finally, observation can also yield useful data. The report of, say, a sales manager who's closely observed a salesperson making calls, can provide quite a lot of relevant information. But such reports are often difficult to obtain or to quantify. And, unless a company's sales managers are concentrated in one area, such reports are also expensive.

Figure 24–3 summarizes our discussion of when and how to measure.

Figure 24–3 Training measurement dimensions

Resources:	Few	Moderate	Many
1. What:	Learning objectives	Learning and application objectives	Learning, application, and business objectives
2. When:	Postprogram	Pre and post	Pre and post Comparison groups
3. How:	Self-ratings Anecdotes Questionnaires (qualitative)	Self- and other measures Questionnaires (quantitative) Statistical tabulations	Self- and other ratings Audits Questionnaires Interviews Observations Comparisons
	Tentative conclusions	Firmer conclusions	Definitive conclusions

HOW NUMBERS HELP IN COMPILING RESULTS

In compiling the results of measurement, keep in mind that qualitative information can often be rendered quantitatively. Suppose, for instance, that a postseminar questionnaire asks: "What was your overall reaction to the seminar?" If 12 participants respond with answers like "Great seminar," "Very productive," "Terrific experience," "Time well spent," and so on, these responses can all be classified as "favorable." And if three participants respond "A waste of time" or "A disappointment," these responses can be classified as "unfavorable." The training manager can then report the responses as 75 percent favorable and 25 percent unfavorable. For reporting purposes, qualitative data have been rendered in quantitative form.

Of course, the use of scales rather than questions lends much greater precision to the results. Suppose that instead of asking, "What was your overall reaction to the seminar?," the trainer in our previous example had included the following item on the questionnaire:

**Circle the number that best describes your
overall reaction to the seminar:**

1	2	3	4
Useless	Some value	Definite value	Great value

If five participants circle "1," five "2," seven "3," and three "4," the training manager can then report that 25 percent of the participants considered the program useless, 25 percent of some value, 35 percent of definite value, and 15 percent of great value, and he can do this *without* making "judgment calls" (e.g., should a particular response be classified as favorable or unfavorable?). In our example, the judgment of the trainer need not intervene.

Why bother to render qualitative data quantitatively? For two reasons:

1. Numbers are easier to communicate. A report in which information is rendered qualitatively is always longer, and frequently more ponderous, than is one in which it's rendered quantitatively.

2. Numbers permit the application of statistics. Statistics make comparison much easier, and comparisons, as we've seen repeatedly, make it possible to draw much more valid conclusions. Moreover, statistics permit much more intensive and sophisticated analysis of the data. With statistics, information can be analyzed for frequencies, percentages, means, and ranges.

DESIGNING PRE- AND POSTMEASURES _____

We've already explained the desirability of pre-post measures. Before undertaking such measures, however, the training manager should

- Make sure the predata are available. It will prove futile, for example, to attempt to measure pre- and postclosing ratios if no records of closing ratios were kept before the training seminar.

- Make sure the postmeasures are designed to measure exactly the same thing as the premeasures.

- Make sure line management understands the benefits to accrue from the measures. This is important because the cooperation of line management may be needed to get the predata. If access to the records is controlled by line management, and if line management thinks the measurement project is unimportant, it may assign a low priority to the training manager's need for obtaining the records.

In comparing pre- and postresults, the training manager may have to call upon expert statisticians. A difference that may look significant to the inexpert eye may be statistically insignificant, or vice versa. There are only two ways to decide on the importance of any difference: the training manager can make a subjective judgment ("My hunch is that this is a really important change"), or he can apply statistical checks (T-tests, Chi squares, F-ratios, etc.). If he lacks the skill to do the latter, he'll have to call for help. The alternative, which is to rely on subjective judgment, is too risky.

THE NEED FOR STATISTICAL CHECKS

Statistical checks, which can also be called "tests of significance," are *essential* for doing controlled comparisons. Differences in performance between trained (experimental) groups and untrained (control) groups can be evaluated only if rigorous statistical checks are applied.

A statistician (or someone with good statistical skills) can also determine if "matched" groups are really matched. Suppose a training manager wants to compare two sales districts; in one, the average education is 13.5 years, the average experience 10 years, the average age 28, and the average group sales volume before the training $200,000 a month; in the other district, the average education is 14.5 years, the average experience 15 years, the average age 35, and the average group sales volume before the training $215,000 a month. The question for a statistician would be: Are these groups matched? How significant are the differences?

How sophisticated should a statistical analysis be? That depends upon the availability of resources. Where money, time, and people are all quite limited (and where, for that reason, measurement is probably limited to postmeasures), companies often settle for frequencies, percentages, means, and ranges. Where somewhat more resources are available, many companies (which are probably doing pre- and postanalysis) add T-tests and correlations. And where resources are extensive (and experimental controls are probably in place), some companies add such techniques as variance analysis and regression analysis. A qualified statistician can advise on all these techniques and their costs.

PRESENTING THE FINDINGS

Once you have answered the original questions (Did the sales training change behavior? Did it produce results? Did it achieve its learning, application, and business objectives?), you have one task left: to communicate the answers to the appropriate members of management. Before you can do that, you must answer four additional questions:

- To whom will I make my report?
- What will I attempt to achieve?
- What will I report?
- How will I do it?

Whom. The first thing to determine is: Who will be in the audience? What are their concerns? What do they want to know?

Purposes. The report (written or oral) will of course concentrate on the results of the training. But which results? Those having to do with business objectives? Or application objectives? Or learning objectives? Or some combination? The answer, obviously, will depend upon the audience and its interests.

What. Several questions must be answered here. Will the report be highly detailed or a summary? Will it draw a full range of firm conclusions or only tentative initial conclusions? Will it stop at that point or will it recommend further action?

How. Will the report be written or oral or some combination of the two? Will it use visual aids (slides, charts, graphs)? If so, what kinds? If the presentation will be oral, should reading matter be sent out ahead of time? How will audience participation (very important) be obtained? How much detail will be "enough"? How much will be "too much"? What props (microphones, flipcharts, etc.) will be needed?

It may be a good idea to tie up the main points graphically. Figure 24–4 depicts the sequence of events we've recommended.

Figure 24–4 Training measurement sequence

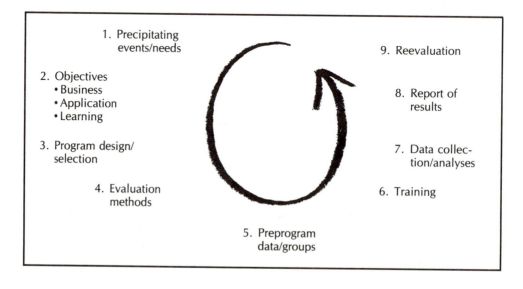

SUMMARY

When companies send their salespeople to sales training programs, one of the questions they ask is: "Can management ascertain if the program did what it was supposed to?" In other words, can the results of the training be measured and, if so, how? The answer to the first question is "yes." To answer the second, we must look at three kinds of inextricably linked objectives.

Business objectives are the measurable business outcomes to be achieved as the end result of the training. *Application objectives* are activities that must be carried out if the business objectives are to be met. And *learning objectives* are the selling skills or sales behaviors that must be mastered before the application objectives can be achieved.

Two basic principles training managers should keep in mind are: (1) the attainment of learning objectives doesn't ensure the attainment of application objectives, and (2) sales trainers alone cannot set the three kinds of objectives—they need the help of people at higher levels.

In well-run companies, sales training programs occur in response to a need triggered by one or more of the following factors: (1) organizational change, (2) internal problems, and (3) external pressures. The training manager must sense the need by formal means (needs, attitude, or climate surveys) or informal means (asking questions, staying tuned in). He or she must also confront two issues: (1) people at different levels of the company focus on different objectives, and (2) management cannot be expected to commit automatically or easily to any sales training program.

Training managers must make the point to management that sales training's purpose is to help achieve the company's business objectives. Then, after the training, they must demonstrate that both the learning and application objectives have been achieved, which should lead to the attainment of business objectives. In determining *when to measure* the results of sales training, trainers should keep in mind that preprogram and postprogram measures utilizing matched control groups will usually yield the most thorough, objective, and valid data. Including all three elements, of course, depends on the availability of resources.

Availability of resources also determines *how to measure* results—qualitatively or quantitatively. Any company with the necessary resources should seriously consider using quantitative measures.

Techniques for obtaining either qualitative or quantitative measurements include self-judgments, on the part of the trainees themselves; other-person judgments, from the sales manager or customer; program records, which include postprogram questionnaires and reports on sales planned during the training and carried out afterward; company records; audits; surveys; interviews; and observation.

In compiling the results of measurement, training managers should remember that qualitative information can often be rendered quantitatively for two good reasons: (1) numbers are easier to communicate, and (2) numbers permit the application of statistics, which, in turn, makes comparison easier.

Comparison between pre- and postseminar results requires statistical checks and, if possible, a statistician to determine what data are statistically significant or insignificant and if "matched" groups are indeed matched.

Before a sales training manager communicates the results of the training to management, he or she must decide (1) to whom the report will be made, (2) what is to be achieved with the report, (3) the content and complexity of the report, and (4) how the report will be presented.

REFERENCES

American Telephone & Telegraph Co. *The Trainer's Library: Measurement and Evaluation.* Reading, MA: Addison-Wesley, 1987.

American Vocational Association. *The Effects of Types of Training Evaluation on Support of Training Among Corporate Managers.* Alexandria, VA: AVA, 1986.

Birnbrauer, Herman. "Evaluation Techniques That Work." *Training and Development Journal*, 41(7) (September 1987), pp. 52–55.

Brinkerhoff, Robert O. *Achieving Results from Training.* San Francisco, CA: Jossey-Bass, 1987.

Brinkerhoff, Robert O. "An Integrated Evaluation Model for HRD." *Training and Development Journal*, 42(2) (February 1988), pp. 66–68.

Campbell, D. T., and J. C. Stanley. "Experimental and Quasi-Experimental Designs for Research on Teaching." In *Handbook of Research on Teaching*, pp. 171–246. N. L. Gage, ed., Chicago: Rand McNally, 1963.

Chabotar, Kent J., and Lawrence J. Lad. *Evaluation Guidelines for Training Programs.* Lansing, MI: Dept of Civil Service, State of Michigan, 1974.

Cook, T. D., and D. T. Campbell. *Quasi-Experimentation: Design and Analysis Issues for Field Settings.* Chicago: Rand McNally, 1979.

Deming, Basil S. *Evaluating Training.* Alexandria, VA: ASTD, 1982.

Fetterol, Eugene. "Did the Training Work?" Part II. *Training News*, 9(3), 1987, pp. 11–12.

Harris, Joan L., and Chris Bell. *Evaluating and Assessing for Learning.* London: Kogan Page, 1985.

Herman, Joan L., Lynn Lyons Morris, and Carol Taylor Fitz-Gibbons. *Evaluator's Handbook.* Newbury Park, CA: Sage, 1987.

Kirkpatrick, Donald L. *More Evaluating Training Programs.* Alexandria, VA: ASTD, 1987.

Kirkpatrick, Donald L. "Evaluation." In *Training and Development Handbook*, 3rd ed., Robert L. Craig, ed. New York: McGraw-Hill, 1987.

Kirkpatrick, Donald L. *Evaluating Training Programs.* Alexandria, VA: ASTD, 1975.

Lord, F. M. "Statistical Adjustments When Comparing Pre-existing Groups." *Psychological Bulletin*, 72 (1969), pp. 336–337.

May, Leslie S., Carol A. Moore, and Stephen J. Zammit, eds. *Evaluating Business and Industry Training.* Norwell, MA: Kluwer Academic Publishers.

Mayo, Douglas G., and Philip M. Dubois. *The Complete Book of Training: Theory, Principles and Techniques*. San Diego: University Associates, 1987.

Newstrom, John G. "Confronting Anomalies in Evaluation." *Training and Development Journal*, 41(7) (September 1987), pp. 56–60.

Parker, Barbara L. *Summative Evaluation in Training and Development: A Review and Critique of the Literature, 1980 Through 1983*. St. Paul, MN: Unviersity of Minnesota Press.

Patton, Michael Quinn. *Creative Evaluation*. Beverly Hills, CA: Sage, 1987.

section five

SELECTING AND USING SALES TRAINING RESOURCES

Chapter 25   Sources of Sales Training Information

Chapter 26   Sales and Marketing Education: Programs Offered by Colleges and Universities

Chapter 27   Consulting with External Sales Trainers: A Four-Phase Process

Chapter 28   When and How to Use Packaged Sales Training Programs

Chapter 29   Public Seminars as a Sales Training Resource

Chapter 30   Professional Networking for Sales Training

chapter twenty-five ——————————

SOURCES OF SALES TRAINING INFORMATION

EDWARD R. DEL GAIZO
————————————AND DERWIN A. FOX

Edward R. Del Gaizo, Ph.D., is director of research in the Product Development Department for Learning International. He has responsibility for the overall research and evaluation of Learning International products and services. Prior to joining Learning International in 1983, Dr. Del Gaizo was employed at Equitable Life, where he was personnel research director before becoming director of training and development. Prior to this, he was project manager for a consulting firm specializing in organizational research and effectiveness. In addition, he has been a featured speaker at several professional conferences throughout North America and has authored numerous published articles in the performance areas.

Derwin A. Fox is senior vice president of marketing and development for Learning International. He is responsible for the planning, development, and marketing of Learning International's programs and services in the areas of management, supervisory, service, and sales training. He has conducted seminars for many professional organizations such as ASTD and the National Society for Sales Training Executives and has published numerous articles in national publications. Prior to joining Learning International, Fox was manager of instructional technology at the Xerox Research and Development Center. Prior to this, he was manager of training for General Motors.

In corporations, sales training refers to one or two disciplines: teaching product knowledge and training in selling skills. Since product knowledge is industry specific, as well as company specific, this chapter will cover some techniques used for product training then concentrate on sources of information for developing selling skills.

Information for sales trainers comes in many forms: print, video, workshops, classrooms, and electronic media. It is not our intent to recommend any one vendor, author, publisher, or provider over a competitor. To the extent possible, the sources we list include organizations, directories, catalogs, distributors, or other services that provide a broad range of information about sales training materials, which come in a variety of media. The lists of specific products are not meant to be exhaustive, but rather the personal choices of a sample of sales training professionals.

Even if we could objectively identify the five or ten best sources of information available in each medium, the list would be outdated before you read it. Yet, gathering information can be very expensive, and before you buy anything, you will want to assess, from all that is available, what information source will be most helpful to you. Most trainers look for recommendations from colleagues who have had similar needs and fulfilled them satisfactorily. This chapter points you toward sources of recommendations you can trust.

Because prices change often, we have not included any in our descriptions of products and services.

PRODUCT TRAINING

When you are responsible for teaching salespeople about your company's product—the product they will sell—no off-the-shelf training materials can do that job completely for you. Your program has to be designed specifically for your product. Your choices include not only the instructional design of the product training but also the medium that best suits your needs.

Should you present a workshop, provide self-study materials, use video, or employ computer-based instruction (CBI)? Whether you want helpful information on how to design your own product training or need help in finding an expert to do it for you, professional organizations are a good place to start.

Instructional Design

The following is an organization through which you can contact experts in instructional technology.

National Society for Performance and Instruction (NSPI)
1126 Sixteenth Street, NW
Suite 214
Washington, DC 20036
(202) 861-0777

NSPI is an international organization of some 2,500 members, with local chapters in most major cities. Members include instructional technologists as well as professionals in a variety of fields whose objective is to improve human performance. NSPI publishes the *Performance and Instruction Journal* ten times annually and the annual *Official Membership Directory* that identifies members' expertise. The society also sponsors an annual conference. (For other sources of instructional design information, see Chapter 7, "Designing Instructional Systems.")

Computer-Based Instruction

The following association and publications will provide you with a basis for obtaining CBI sources.

Association for the Development of Computer-Based
Instructional Systems (ADCIS)
409 Miller Hall
Western Washington University
Bellingham, WA 98225
(206) 676-2860

ADCIS members include courseware designers, researchers, evaluators, vendors, communications experts, courseware distributors, programmers, and audio/visual specialists, as well as users of CBI. The association sponsors an annual conference, offers workshops, and publishes the quarterly *Journal of Computer-Based Instruction*, which reviews books and courseware and prints research studies. ADCIS has a growing network of local chapters and a number of special interest groups including one in computer-based training and one in interactive video/audio.

The Guide to Computer-Based Training
Weingarten Publications, Inc.
38 Chauncy Street
Boston, MA 02111
(617) 542-0146

Published annually, the guide discusses authoring systems and includes a directory of courseware vendors, listing the specialties of each.

Data Training
Weingarten Publications, Inc.
38 Chauncy Street
Boston, MA 02111
(617) 542-0146

Published monthly, this magazine primarily provides information on training in data processing, but it also publishes articles about computer-based training and publishes semiannually updates of *The Guide to Computer-Based Training*.

American Micro Media Microcomputer Training Catalog
19 North Broadway
Red Hood, NY 12571
(914) 758-5567

This catalog of training software represents about 60 publishers and includes both authoring and sales training programs.

For more information, see Chapter 21, "Computer-Based and Interactive Video Training."

Video Tape and Video Disk Instruction

The following organizations will be helpful in obtaining information on video instruction.

International Television Association (ITVA)
6311 N. O'Connor Road, Suite 110-LB51
Irving, TX 75039
(214) 869-1112

ITVA's more than 7,000 members are practitioners, suppliers, and users of nonbroadcast video. ITVA publishes a bimonthly magazine called *Corporate Television* and a membership directory. ITVA has chapters in most major cities and a number of national groups, some professional and others who are industry special interest groups. The association sponsors an annual international conference at which it presents awards in categories such as training and sales/marketing. The association also holds national institutes that are one- or two-day seminars which focus on a particular area of expertise.

Interactive Video Industry Association
c/o Future Systems, Inc.
P.O. Box 26
Falls Church, VA 22046
(703) 241-1799

The 90 companies that are members of this trade association include providers of programs, training, and hardware. The association offers an information package that includes a list of its members.

Society for Applied Learning Technology
50 Culpeper Street
Warrenton, VA 22186
(703) 347-0055

The society concentrates on sponsoring conferences on technology in education. It holds two major videodisk conferences a year, the Conference and Exhibition on Interactive Instruction Delivery and the Conference on Interactive Videodisc in Education and Training, as well as a number of smaller related meetings.

Nebraska Videodisc Symposium
P.O. Box 83111
1800 North 33 Street
Lincoln, NE 68501
(402) 472-3611

Held annually in early fall, this symposium, sponsored by the Nebraska Videodisc Design/Production Group, brings together designers, producers, entrepreneurs, and equipment manufacturers from all over the world.

Institute for Graphic Communication
375 Commonwealth Avenue
Boston, MA 02115
(617) 267-9425

For more information, see Chapter 20, "Film and Video."

SELLING SKILLS: PROFESSIONAL,
COMMERCIAL, AND PRACTITIONER SOURCES _____

If you are about to put together a sales training program, you will want information on how to sell, and information on how to train salespeople. Sources to help you meet both these needs are included in the following sections.

Professional Organizations

These organizations provide members with opportunities to focus specifically on selling techniques and sales training. They provide networking opportunities at local chapter meetings of special interest groups; sponsor annual conferences and workshops for members and nonmembers; publish periodicals, books, and directories; and offer a variety of other services indicated within the descriptions of each.

For information seekers, another advantage of membership in a professional association is that your name will be put on mailing lists of suppliers who will send you promotional materials of their products and services. What is junk mail to some can develop into an extensive file of up-to-date supplier information, which is often the first place trainers turn when they need help.

American Society for Training and Development (ASTD)
1630 Duke Street, Box 1443
Alexandria, VA 22313
(703) 683-8100

ASTD provides opportunities for networking with members in the sales and marketing professional practice area, local sales and marketing special interest groups and networks, and industry groups of your choice. The society also offers products and services that provide information for sales trainers, such as the *Training & Development Journal*, a monthly magazine. (See the following section entitled "Professional Journals.")

ASTD National Conference is an educational program that includes topics and skills important for sales trainers. The concurrent Expo showcases hundreds of suppliers.

Who's Who in Training and Development is the society's membership directory that cross-references names by professional practice, so that sales trainers can easily pick out members with similar interests.

ASTD Buyer's Guide & Consultant Directory is also published by ASTD. (See the section entitled "Directories of Vendors and Consultants.")

Info-Line is a monthly information series that deals with a single training topic in each issue. ASTD provides a complete list of back issues, which can be purchased separately. A recent one, *Train Your Sales People for Success*, includes an extensive bibliography of books and articles.

ASTD Information Center provides members with data on organizations that supply information, publications, programs, or training tools in a specific area, as well as bibliographies of articles and books on the subject. The Center also conducts searches via TRAINET, ASTD's database; however, this service requires an additional fee.

For the sales trainer who is looking for a trustworthy recommendation, one of the Information Center's most valuable services can be the Master Information Exchange. The exchange is a computer-based professional networking system that puts members in touch with colleagues who have agreed to share information, resources, and advice.

For nonmembers, the Information Center will provide a list of recent *Training & Development Journal* articles and ASTD-offered books on the subject of inquiry. It also provides an "Inventory of Performance Aids," an annotated list of buyer's guides, program directories, seminar directories, practical application sourcebooks, handbooks, media reviews, journal article indexes and abstracts, and on-line databases.

National Society of Sales Training Executives (NSSTE)
1040 Woodcock Road
Orlando, FL 32803
(305) 894-8312

NSSTE is the only association whose sole objective is to improve sales training effectiveness. It maintains strict membership requirements: persons considered for membership must have been trainers for at least a year, spending a majority of their working time in training and development activities; have attended at least one NSSTE meeting; and currently manage a sales training function. Further, members are required to attend at least one of the two society meetings held each year and submit one editorial annually. These editorials are reports on solutions to training problems that are distributed to all other members.

Nonmembers are welcome to attend all NSSTE functions and purchase all NSSTE publications except the editorials. However, nonmembers can read the editorials at the Indiana State University where a library of them is maintained.

NSSTE products and services include *Sales and Marketing Training*, a quarterly journal (see the section called "Professional Journals"); semiannual society meetings, with featured speakers and concurrent educational sessions, and local meetings; a series of "How to" booklets and other books on sales training and sales management; train-the-trainer clinics and sales management seminars; and a membership directory.

Sales and Marketing Executives-International (SME-I)
Statler Office Tower
Cleveland, OH 44115
(216) 771-6650

SME-I has some 10,000 members, most of them associated with local SME-I chapters. The organization provides a Marketing Information Center, which is a research service for members; a Sales and Marketing Support Center, which evaluates new training materials, motivational materials, and home-study courses and offers discount prices on selected materials to members; a staff counsel, which provides market contacts, career contacts, and market information; a Speaker Information Center, which contains up-to-date information on hundreds of speakers; *Marketing Times*, SME-I's bimonthly professional journal, which carries news and commentaries on the marketplace; and *SME-I Digest*, a bimonthly newsletter that contains sales and marketing news briefs.

SME-I also sponsors montly meetings with featured speakers; sales seminars and workshops; specialized clinics and roundtables; the Sales Management Institute, a four-day course held in key cities across the country; the Graduate School of Sales Management and Marketing, at Syracuse University, Syracuse, NY, a two-week session held once a year for two years that is the equivalent of six M.B.A. credits; and International Marketing Congress, SME-I's annual convention with speakers and workshops.

Sales and Marketing Executives of Greater New York
114 East 32 Street, Suite 1301
New York, NY 10016–5543
(212) 638-9755
(914) 604-6996

This club organizes conferences, lectures, and seminars; holds training sessions; and sponsors research. It publishes a monthly news magazine, *The Sales Executive*. Members must be managers.

Professional Journals

The following periodicals are dedicated to publishing articles on selling or sales training. To indicate the type of information each provides, we have included sample article titles in some of the descriptions.

Sales and Marketing Training
8101 Hinson Farm Road
Suite 201
Alexandria, VA 22306
(703) 360-9395

This quarterly magazine is sponsored by The National Society of Sales Training Executives. *Sales and Marketing Training* contains articles on sales training methods and resources, selling techniques and society activities, as well as reviews of current books, off-the-shelf training packages, videos, and electronic resources. Recent articles include "The Timidity Factor in Selling" and "How to Work with a Training Consultant."

Personal Selling Power
P.O. Box 5467
Fredericksburg, VA 22403
(800) 752-SELL
(703) 752-7000

Published eight times a year in newspaper tabloid format, the magazine carries advice from leading sales experts; announcements of new audio cassette and video cassette programs and computer software; reports on new training courses, training packages, and sales surveys; and reviews on sales and marketing books. Recent titles include "On Closing the Sale" and "Stop Telling—Start Selling." *Personal Selling Power* provides an index, by subject, of all articles printed since it began publication. The publisher also offers a number of books on selling.

Training & Development Journal
1630 Duke Street, Box 1443
Alexandria, VA 22313
(703) 683-8100

It is a monthly magazine, published by the American Society for Training and Development, that contains articles on all aspects of training and development. Many issues include features related to sales training in particular. Recent titles include "Sales Training: Changing Roles, Changing Needs" and "What Makes a Successful Sales Person."

Training, The Magazine of Human Resources Development
Lakewood Publications
Lakewood Building
50 South Ninth Street
Minneapolis, MN 55402
(612) 333-0471

It is a monthly magazine that has articles on all aspects of training and development. Many issues feature articles related to sales training. Recent titles include "Who Should Do the Training" and "Poof! You're in Sales."

Training News
Weingarten Publications, Inc.
38 Chauncy Street
Boston, MA 02111
(617) 542-0146

Training News is a monthly newspaper that contains advice and news about generic training programs, with some articles focused specifically on sales training. Recent titles include "Interpersonal Skills in Sales Training."

Sales and Marketing Management
Bill Communications
633 Third Avenue
New York, NY 10017
(212) 986-4800

The magazine, published monthly, is available at newsstands in New York City and by subscription elsewhere. Recent articles include "How to Murder Your Sales Trainer: Shortcomings in Training Programs" and "Sales Meetings and Sales Training."

Sales and Marketing Digest
10076 Boca Entrada Boulevard
Boca Raton, FL 33431
(305) 483-2600

Published monthly, *Sales and Marketing Digest* carries selected reprinted articles from other published sources, regularly including articles on sales techniques.

HRD Review
105 Berkely Place
P.O. Box 6
Glen Rock, NJ 07452
(201) 445-2288

HRD Review is a monthly newsletter that presents evaluative reviews of books, audio cassettes and video cassettes, three-ring binder-based workshops, games, and simulations in all areas of human resource development. The editors try to include at least one review of sales training or sales management material in each issue. The reviews are conducted by HRD professionals, whose backgrounds are included with each review.

Bibliographies, Indexes, and Abstracts

You can identify what the preceding publications, as well as other periodicals, have printed recently on your subject in a number of bibliographies, indexes, and abstracts. Some of these sources reference only periodicals; others include both articles and books.

Training and Development Alert and
HRD-in-Brief Series: Sales Training
Advanced Personnel Systems
P.O. Box 1438
Roseville, CA 95661
(916) 781-2900

Published bimonthly, *Training and Development Alert* is a service that annually abstracts approximately 750 journal articles and books on training and development. This information is distributed on single sheets of paper. The service also can provide copies of most of the abstracted articles.

HRD-in-Brief Series: Sales Training is a collection of every printed abstract on sales training since the service began publishing in 1979. The collection is revised quarterly so that any purchaser gets an up-to-date edition.

Sales and Marketing Training Resources in Print
American Society for Training and Development
1630 Duke Street, Box 1443
Alexandria, VA 22313
(703) 683-8100

This bibliography lists more than 75 books, as well as tapes, journals, films and videos, and abstracts of articles. The source of the abstracts is ABI/Inform. (See the Databases section on pages 601–604.)

Business Periodicals Index
H. W. Wilson Company
950 University Avenue
Bronx, NY 10452
(212) 588-8400

Available in most libraries, this index is published monthly, cumulated quarterly, and bound annually. It indexes approximately 300 periodicals, many of which occasionally contain articles on selling and sales training. The articles are listed alphabetically by subject. This index can lead you to articles such as

"Sales: Ten Do's and Ten Don'ts" in *Mortgage Banking* and "Laughing All the Way to the Bank: Humor as an Aid in Selling" in *Industrial Distributor*.

Current Index to Journals in Education (CIJE)
The Oryx Press
2214 North Central at Encanto
Phoenix, AZ 85004–1483
(602) 254-6156

This index is published monthly and cumulated semiannually. Available in most libraries, it includes abstract of articles listed by subject and author. Pertinent subject descriptors include sales occupations and retailing. CIJE can be accessed on-line through the ERIC data base. (See the Databases section.)

Resources in Education *(RIE)*
The Oryx Press
2214 North Central at Encanto
Phoenix. AZ 85004–1483
(602) 254-6156

Available in most libraries, RIE contains indexes and abstracts of books related to education in the same way CIJE does with journal articles. It can also be accessed on-line through ERIC.

Books

Each year scores of new books on how to sell are published. You don't have to search them out; rather you need to narrow the choices. You can begin to do that by asking respected colleagues which books they find helpful and by keeping up with the book review sections of some of the periodicals listed in the Professional Journals section of this chapter.

The following list represents the favorites of a sample of sales trainers. It is not meant to be all inclusive or even an objective "best" list. An indication of how personal such choices are, is that we found very little overlap in books named by the trainers surveyed.

Alessandra, A. *Non-manipulative Selling*. Reston, VA: Reston, 1981.

Bettger, F. *How I Raised Myself from Failure to Success in Selling*. Englewood Cliffs, NJ: Prentice Hall, 1958. (Paperback: Cornerstone, 1975.)

Buzzotta, V. R., et al. *Effective Selling Through Psychology: Dimensional Sales & Sales Management Strategies New Edition*. Cambridge, MA: Ballinger, 1985.

Hanan, M., J. Cribben, and H. Heiser. *Consultative Selling*, 3rd ed. New York: American Management Association, 1985.

Johnson, S., and L. Wilson. *The One-Minute Salesperson*. New York: William Morrow, 1984. (Paperback: New York: Avon, 1984.)

Liebling, H. E. *Handbook for Personal Productivity*. Penfield, NY: Skill Builders, 1983.

Mandino, O. *The Greatest Secret in the World*. New York: Bantam, 1978.

Qubien, N. *Professional Selling Techniques*. New York: Berkley, 1985.

Roth, C. B., and R. Alexander. *Secrets of Closing Sales*, 5th ed. Englewood Cliffs, NJ: Prentice-Hall, 1982.

Valen, C., and W. Valen. *Training Salespeople for Mastery Performance*. Cave Creek, AZ: Valen, 1986.

Willingham, R. *The Best Seller! The New Psychology of Selling & Persuading People*. Englewood Cliffs, NJ: Prentice Hall, 1984.

For additional books, refer to the section Textbooks and Other Support Materials under the head of "Selling Skills: The Academic Approach."

Publishers that Specialize in Books on Selling

The following publishers exclusively publish in the sales and marketing field or are business publishers with a significant sales/marketing line.

Dartnell
4660 Ravenswood Avenue
Chicago, IL 60640
(800) 621-5463
(312) 561-4000

Dartnell publishes the *Dartnell Sales Management Catalog* and provides it free of charge. It also publishes subscription items and videotapes.

Prentice Hall
Business & Professional Books Division
Prentice Hall Building
Englewood Cliffs, NJ 07632
(201) 592-2000

Prentice Hall publishes an extensive line of practical handbooks for field sales representatives, sales managers and telemarketing managers. It also publishes sales audio cassette programs.

The Salesman's Guide, Inc.
1140 Broadway, 12th Floor
New York, NY 10001
(800) 223-1797
(212) 684-2985

The Salesman's Guide publishes directories of retailers that include the buyer's name and items that person purchases.

John Wiley and Sons
605 Third Avenue
New York, NY 10158
(212) 850-6000

This company prints a catalog of training books called *The Art of Training*, as well as numerous business publications.

American Management Associations (AMA)
135 West 50 Street
New York, NY 10020
(212) 586-8100

The association publishes video tapes and presents seminars. Members receive discounts and additional services.

The Research Institute of America
90 Fifth Avenue
New York, NY 10011
(212) 337-4100

A free booklet, *Sales Management . . . Plus: How to Uncover and Use Hidden Motivators to Increase Sales*, is available from the institute.

Directories of Vendors and Consultants

You may be considering the purchase of a sales training program. If so, you must decide whether to have the vendor conduct the training in your organization or have the vendor train your in-house trainers to do the facilitating themselves. Another option is to contract a vendor to tailor the program specifically to your organization. Your fourth choice is to create your own program. Even if you create your own program, you may wish to hire a consultant to help with the design and development. To find out who offers the kind of service you need, check in one of the following directories. This is a list of custom developers as well as vendors of packaged programs, videos, CBI programs, and training materials. Your next step will probably be to interview a number of vendors/consultants and get references from people you trust before you make a final decision.

ASTD Buyer's Guide & Consultant Directory
American Society for Training and Development
1630 Duke Street, Box 1443
Alexandria, VA 22313
(703) 683-8100

This is a cross-referenced directory of training services, equipment, and supplies that is published annually. It lists more than 100 vendors and consultants in the categories of sales/management and sales marketing. Descriptions of their services are provided by the vendors.

Training Marketplace Directory
Lakewood Publications
50 South Ninth Street
Minneapolis, MN 55402
(612) 333-0471

Published annually, this directory is a guide to a selection of consultants, programs, supplies, and meeting sites for training. This directory obtains its information by sending questionnaires to about 6,000 suppliers and compiling the results. Under the categories of sales management and sales training, the directory lists some 600 suppliers of classroom training, packaged and custom programs, cassettes, books, CBI, films, and videos.

The Trainer's Resource
HRD Press
22 Amherst Road
Amherst, MA 01002
(800) 882-2801
(413) 253-3488

It is a guide to commercially produced training programs that can be delivered at a company on-site. The directory lists approximately 50 programs under the category of sales training. Prices, appropriate audiences, recent users, and a description of the product supplied by the vendor are included in the directory.

Training and Development Organizations Directory
Gale Research Co.
Book Tower, Department 7748
Detroit, MI 48277–0748
(800) 223-GALE
(313) 961-2242

This directory lists firms, institutes, specialized university programs, and other agencies that offer training. Descriptions of them are provided by the sponsor. The information is cross-referenced by subject, including sales management and sales training, and by location, including state and city.

Instructional Systems Association
P.O. Box 1196
Sunset Beach, CA 90742
(714) 846-6012

Instructional Systems Association is a trade association of companies that produce packaged training programs. The association provides a free directory of its member firms, the subjects they cover, the type of product they produce, and the kinds of clients they serve. Sales training is the association's second largest category, next only to management training.

For additional sources of information on vendors, see the following sections entitled Seminars and Workshops, Conferences, and Databases

Seminars and Workshops

There are a number of reasons why you might want to attend a public seminar or workshop or identify such a workshop for someone else in your organization. Perhaps you would like to test a vendor's program before bringing it in-house, or maybe you need a particular type of training for just one employee or for too few to make an in-house program practical. Several services and directories exist to help you find the right outside program to meet your needs.

To learn more about how to get the most from this sales training resource, refer to Chapter 29 on "Public Seminars."

Conferences

All the professional organizations previously listed sponsor annual conferences. Both the conference sessions and the accompanying exhibits by vendors offer a wide variety to interest sales trainers. Membership in the associations is not generally required for conference attendance, although nonmembers pay a higher fee. In addition to the ASTD annual conference and NSSTE's biannual meeting, another major training conference is

TRAINING Annual Conference and Expo
Lakewood Conferences
50 South Ninth Street
Minneapolis, MN 55402
(612) 333-0471

This conference, sponsored by *Training* magazine, is a five-day program of concurrent workshops that usually includes sales training programs, special events, and an exposition of vendors.

Video Tapes and Films

If you are thinking of creating your own sales training video tape, refer to the section called Videotape and Videodisk Instruction under the earlier head of "Product Training" for information on how to find production experts. More likely, you will look for a video tape or film that meets your needs from among the hundreds of commercially produced products. The following directories are available to help you choose.

> *Media Profiles: Career Development Edition* and
> *The Sales Trainer's Program Handbook*
> Olympic Media Information
> 550 First Street
> Hoboken, NJ 07030
> (201) 963-1600

Media Profiles is published six times a year. In one issue each year, reviews of 200 films and videos are published, a third of which focus on sales training and customer service. Training professionals prepare the reviews, which include audience identification, content summary, and evaluation.

The Sales Trainer's Program Handbook is a collection of almost 300 reviews, reprinted from *Media Profiles*, devoted solely to sales training.

> *ASTD Training Video Directory*
> ASTD Publishing Service
> P.O. Box 4856
> Hampden Station
> Baltimore, MD 21211
> (301) 338-6949

It is a two-volume directory of more than 14,000 video tapes that are geared to training and human resource development. Each volume contains a subject heading outline, a subject section, a title section with descriptions, and a directory of producers and distributors.

Video Tape and Film Distributors

Film and video distributors represent many producers. Some have their own catalogs; others send brochures and catalogs from appropriate producers. Most distributors have screening rooms, where you can view as many videos or films as you wish at no charge. The companies also make recommendations to you over the phone and ship preview copies of films or tapes to you at a fee established by the producer. The fee is deducted from the purchase price if you decide to buy.

Distributors are listed in the *ASTD Training Video Directory*. (See the Video Tapes and Films section.)

Computer-Based Sales Training

For information on computer-based sales training, refer to the section entitled Computer-Based Instruction under the head of "Product Training."

Other Sales Training Materials

Self-instructional Materials. A number of self-instructional programs for salespeople and sales managers are included in this directory.

> *Learning Independently*
> Gale Research Co.
> Book Tower, Department 7748
> Detroit, MI 48277–0748
> (800) 223-GALE
> (313) 961-2242

Assessment Instruments. You may want to use a validated questionnaire to determine your trainees' aptitude and current skill level. Instruments related to sales training are included in the following directory, which provides information on their intended audience, theory base, validity, and reliability.

> *Directory of Human Resource Development Instrumentation*
> University Associates
> 8517 Production Avenue
> San Diego, CA 92121
> (619) 568-5900

Databases

There are a number of advantages to doing your research via computer. Enter a few appropriate descriptor words, and the computer bursts forth with a volume of references. Depending upon the database, the references may be abstracted or the database may even allow you to access the original work. The process is fast and thorough and saves you from wading through numerous indexes and bibliographies. Furthermore, databases are much easier to update than printed materials, so you can count on getting the most recent information.

If you have a personal computer and a modem, you can subscribe to your chosen databases and conduct your own searches. Two database vendors from which you can access most of the important training-related databases are

> BRS Information Technologies
> 1200 Route 7
> Latham, NY 12110
> (518) 783-1161

> Dialog
> 3460 Hillview Avenue
> Palo Alto, CA 94304
> (800) 334-2564
> (415) 858-3785

If you are not equipped to do your own searches, you can have them conducted for you in many libraries or by the database producer.

Databases that provide useful information for sales trainers include

> TRAINET/The Seminar Database
> American Society for Training and Development
> 1630 Duke Street, Box 1443
> Alexandria, VA 22313
> (703) 683-8100

Available to ASTD national members only, TRAINET is a comprehensive database of training seminars, workshops, and conferences, including sales and marketing entries. Members can access TRAINET from their own computers, or the ASTD Information Center will do general searches.

> EdVENT
> Timeplace, Inc.
> 460 Totten Pond Road
> Waltham, MA 02154
> (800) 544-4023
> (617) 890-4636

EdVENT provides the same data as TRAINET but is available publicly. It also has report-making capabilities. EdVENT maintains information on 120,000 scheduled seminars, conferences, and workshops, including offerings by universities, businesses, associations, governments, and private consultants. It is updated weekly.

ABI/Inform
Data Courier, Inc.
620 South Fifth Street
Louisville, KY 40202
(800) 626-2823
(502) 582-4111

This database contains references with abstracts from 650 business and management journals, including all articles from these journals since 1971. Data Courier provides document retrieval.

WILSONLINE—Business Periodicals Index
H. W. Wilson Company
950 University Avenue
Bronx, NY 10452
(212) 588-8400

This is the on-line counterpart of the *Business Periodicals Index*. (See the section on "Bibliographies, Indexes, and Abstracts.") It is not available from Dialog or BRS.

ERIC
U.S. Department of Education
Office of Educational Research and Improvement
555 New Jersey Avenue, NW
Washington, DC 20208
(202) 357-6289

ERIC contains references with abstracts to books, journals, and research reports in all areas of education, including adult skills training. Salesmanship is one of the ERIC descriptors. Documents can be ordered through the ERIC Document Retrieval Service. Print counterparts are *Research in Education* and *Current Index to Journals in Education*. (See section called "Bibliographies, Indexes, and Abstracts.")

DATRIX II
University Microfilms, Inc.
300 North Zeeb Road
Ann Arbor, MI 48106
(313) 761-4700

This database identifies doctoral dissertations written on a particular topic. Such a search could reveal resources such as a recent dissertation entitled "An

Investigation of Presentation Style in the Rapport Building Stage of the Personal Selling Process," submitted by W. C. Chase of the University of Houston. The print equivalent is the annual *Dissertation Abstracts International*, available in many libraries. Complete dissertations can be ordered.

SELLING SKILLS: THE ACADEMIC APPROACH

The line between business and academia continues to blur as academics become consultants for companies, executives lecture at colleges and universities, and schools offer business seminars and certificate and degree programs for businesspeople. For a complete discussion of the links between higher education and sales training, see Chapter 27, "Colleges and Universities." The following section contains sources that can lead you to colleges in your area with related programs; associations of academics who are conducting research, teaching, or working in your field; current research that is being published; and the printed source materials being used to teach sales and marketing in schools.

Colleges and Universities

You can locate schools that offer two-year programs in sales, marketing, and sales training by utilizing the following sources.

American Association of Community and Junior Colleges
National Center for Higher Education
One Dupont Circle, NW, Suite 410
Washington, DC 20036
(202) 293-7050

This association can guide you to schools that offer programs in sales/marketing in specific industries.

The College Blue Book
Macmillan Publishing Company
866 Third Avenue
New York, NY 10022
(212) 702-2000

Available in many libraries, this directory lists degrees offered by subject and college. Related subject headings are sales, sales and marketing, sales and retailing, and training and development.

Other Sources

Professional organizations, such as the ones listed here, also can be of assistance to you.

Pi Sigma Epsilon (PSE)
145 North Avenue, Suite G
Hartland, WI 53029
(414) 367-5600

This is a national, professional fraternity in selling, sales management, and marketing. Undergraduate chapters are supported by educators, alumni, and professionals. Sales and Marketing Executives-International supports PSE through affiliated associations that sponsor PSE chapters. (See the section entitled "Professional Organizations" for more information on SME-I.)

The Journal of Personal Selling and Sales Management
155 East Capitol Road
Hartland, WI 53029
(414) 367-5600

Published three times a year by Pi Sigma Epsilon, this journal encourages professional and academic research and exchange of ideas in selling and sales management.

Textbooks and Other Support Materials

The range of available textbooks compares to that of books in the trade market, and the choices of educators is as varied as that of sales trainers. The three listed have multiple editions but are only a sample of the good-quality textbooks available.

Churchill, G., N. Ford, and O. Walker, Jr. *Sales Force Management*: *Planning, Implementation & Control*, 2nd ed. Homewood, IL: Richard D. Irwin, 1985.

Manning, G. L., and B. L. Reece. *Selling Today*: *A Personal Approach—An Extension of the Marketing Concept*, 3rd ed. (Text edition and instructor's manual.) Dubuque, IA: William C. Brown, 1987.

Russell, F., F. Beach, and R. H. Buskirk. *Selling*: *Principles and Practices*, 11th ed. New York: McGraw-Hill, 1982.

In addition to the previously mentioned professional organizations, Distributive Education Clubs of America is an organization that caters specifically to students.

Distributive Education Clubs of America (DECA)
1908 Association Drive
Reston, VA 22091
(703) 860-5000

DECA provides teaching materials and cooperates with colleges and high schools in the development and presentation of retail sales and management programs. It also works directly with some major retail companies.

SUMMARY

The information sources this chapter includes are not, by any means, all inclusive. However, they are a good place to begin your research. Sifting through the possibilities will not be an easy task, but much of the legwork has been done for you here.

chapter twenty-six ————————————

SALES AND MARKETING EDUCATION: PROGRAMS OFFERED BY COLLEGES AND UNIVERSITIES

———————————— *STEWART W. HUSTED*

Stewart W. Husted is a professor of business and director of the National Society of Sales Training Executives–sponsored Institute for Sales and Marketing Education at Indiana State University. He earned his Ph.D. degree from Michigan State University in 1977 and in addition holds degrees from Virginia Polytechnic Institute and State University (B.S.) and the University of Georgia (M.Ed.). Dr. Husted has held a variety of training and development positions at the Army Institute for Personnel Management, 5030th USAR School, Command and General Staff College and the U.S. Military Academy. He has co-authored four popular business and marketing texts and numerous articles on training and development and marketing education, including an article in the *Training and Development Journal* which provided an academic model for universities to use in establishing competency-based curriculums for training and development.

The role of colleges and universities in sales and marketing training is changing rapidly. Prior to 1982, there were few universities offering degrees in sales and even fewer offering degree programs in marketing education with an emphasis in training and development. Marketing education, as discussed in this chapter, includes sales training. Today there are a wide variety of formal degree programs offered and a growing number of continuing education programs offering courses in sales and marketing. Many of these courses are offered by leading academic institutions.

This chapter describes sales and marketing education on campus today and evaluates the many programs and services offered by colleges and universities to corporations. Programs reviewed will include both those offering degree programs and continuing education courses in sales, and those programs offering degrees which prepare graduates to become sales and marketing trainers and developers.

HOW SALES AND MARKETING EDUCATION ON CAMPUS BEGAN

The first formal sales training conducted in the United States was in 1905 at Filene's department store in Boston. The course was initiated by Lucinda W. Prince, a certified high school teacher and a member of the Women's Educational and Industrial Union of Boston.[1]

The success of the course led to the creation of the Union School of Salesmanship. The school was sponsored by six Boston stores that helped provide a class of 30 salesgirls. In 1908, Prince established the Prince School for Store Service

---

[1] Susan S. Schrumpf, *The Origin and Development of Distributive Education* (New York: McGraw-Hill, 1972).

Education. By 1910 three graduates from Wellesley College were enrolled in Mrs. Prince's school. Five other college graduates also attended the school between 1911 and 1912. All were quickly placed in stores or schools to teach salesmanship. The Prince School also drew the favorable attention of Simmons College, and in 1913 Prince's teacher-training program was absorbed into the college as a graduate school. In 1915 in addition to her duties as director of the school, Prince became education director of the National Retail Dry Goods Association. She remained as director of the school until her death in 1937.

Prince's work is important because of her efforts to initiate sales and marketing training programs, not only in stores but also in high schools and colleges across the nation. The success of her programs led to the development in 1918 of the Research Bureau for Retail Training at the Carnegie Institute of Technology and in 1919 the School of Retailing at New York University. To many, the most important part of her legacy was probably the institution in 1936 of the "Distributive Education" program created by the George-Deen Act. This federally funded program, now called "Marketing Education," has educated and trained millions of young people in high schools and postsecondary institutions for careers in sales and marketing.

TYPES OF PROGRAMS OFFERED ON CAMPUS TODAY

Today there are approximately 100 colleges and universities offering a major in Marketing Education. While many of these programs exist solely for the purpose of preparing high school marketing education teachers, there is a movement to diversify the program and to offer a training and development option or concentration within the major. A 1987 survey completed by Indiana State revealed that 11 universities offered T & D options at the bachelors, masters, doctoral, or certificate levels. Several other universities noted they were developing such programs. It should also be noted that over 250 institutions offering HRD/T & D degree programs are listed by ASTD in its Directory of Academic Programs in *Training and Development/Human Resource Development*. Except for the schools in Table 26–1, none offers a degree in Marketing Education.

Along with the growth in training and development programs, there is also a corresponding growth in the variety of training and development courses offered. The following list is a sampling of courses offered by the 11 schools in Table 26–1 with a Marketing Education major.[2]

---

[2] *Publisher's Note*: According to the author of this chapter, "There are no specific sales training courses offered. Many of these courses include sales training."

Table 26–1 University marketing education: T & D Programs

University	Bachelors	Degree Masters	Doctorate	Certificate
Indiana State University[1]	X			
Illinois State University[1]		X		
Auburn University	X	X		
University of Minnesota	X	X	X	X
University of Houston	X	X		
Winthrop College	X	X		
Old Dominion University	X	X		
University of Nebraska	X	X		
Utah State[1]	X			
Ohio State University		X	X	
Virginia Polytech Inst	X			

[1] Offers a Business degree.

> Training in Industry and Business
> Introduction to Marketing Education and Training
> Organizational Development in Industry and Business
> Training and Development: Cost-Benefit Analysis
> Critical Review of Research in Training and Development
> Strategic Planning in Training and Development
> Consulting in Industry and Business
> Internship: Training and Development
> Computers in Education and Industry
> Instructional Systems Design for Marketing Education
> Methods of Instructing Marketing Education
> Training Program Management
> Foundations of Adult Education
> International Training and Development
> Career and Life Planning

Resource Centers: Two Valuable Sources

On college and university campuses there are two valuable resource centers available to sales and marketing trainers.

Institute of Sales and Marketing Education. The Institute is co-sponsored by the National Society of Sales Training Executives and Indiana State University. The Institute houses the annual editorial contributions (from 1958) of NSSTE members. Each of the 3,300 editorials is related to a sales, marketing, or management training program conducted by the member's corporation. In total, there are 44 different training program categories. The Institute has editorials cataloged from 1980 and sells copies of editorials to NSSTE members. Because the NSSTE membership views the editorials as a member benefit, the Institute is not currently permitted to sell editorials to nonmembers; however, the NSSTE library, located in the ISU School of Business, is available to students, faculty, and any training professional. NSSTE does permit the sale of editorials to other colleges and university programs. In addition to offering the NSSTE library, Institute personnel are available for consulting on sales and marketing management topics, seminars, curriculum development, and other training-related services.

MarkED Resource Center. MarkED is a consortium from approximately 30 states which was founded in 1969. It is housed on the campus of Ohio State University and has a full-time staff of professionals who offer a variety of education and training services. The consortium, founded as a way of quickly developing competency-based marketing curriculums for high school and postsecondary students, has developed into the leading research-based curriculum developer for Marketing Education. MarkED assisted in the development of a new core curriculum and curriculum framework for use in high schools, adult education programs, and postsecondary institutions across the nation.

In addition to providing materials to educational institutions, the Board of Trustees voted in 1983 to expand MarkED's mission. Today MarkED also provides many of its services to corporations and other organizations who need competency-based training materials and tests. Furthermore, MarkED offers a series of professional sales training seminars. These include

PFS   Contact: A hands-on workshop to help sales trainers learn how to teach "positive" first impressions in a marketing environment.

PFS   Sales: Provides specific techniques for teaching selling, using a counselor-style selling technique.

PFS   Supervision: Ideas to help trainers teach new or experienced supervisors a balanced style, blending people orientation with task orientation.

For more information, contact

MarkED, The Ohio State University,
1375 King Avenue, P.O. Box 12226,
Columbus, OH 43212–0226

CONTINUING EDUCATION ON CAMPUS

In the mid-1970s colleges and universities began looking at continuing education for the corporate world as a vast untapped market area that would help revitalize sagging higher education enrollments. Again in the 1980s, Peters and Waterman in their best seller, *In Search of Excellence*, encouraged business schools to refocus their efforts to career development programs from entry-level degree programs.[3] The result is that many business schools are discovering that the getting "close to the customer" in the Peters and Waterman paradigm includes not only delivering the recruits to the door, but the complete servicing of the employee's and company's educational and developmental needs.[4] In addition, Peters and Waterman note that many of the companies are heavy users of continuing education.

A growing number of universities have made commitments to career development programs. These schools include Harvard, University of Michigan, University of Virginia, University of Pennsylvania (Wharton), Stanford, Columbia, Penn State, Duke, and Northwestern. In reality, these programs should be viewed as a cost-effective service being provided to the business community. These services include executive-development programs, management seminars, workshops, short courses, and customized programs and reading materials.

Seminars and Short Courses

Literally hundreds of colleges and universities are offering a variety of seminars and short courses which can serve the needs of marketing and sales trainers. These types of courses create the bulk of continuing education courses offered by colleges and universities.[5] One example is the highly successful sales management programs offered by the University of Wisconsin-Madison's Management Institute. Like many schools they offer several sales management seminars such as "Sales Management: Skills Update," "Teaching Salespeople to Sell," and "How to Use Direct Marketing to Cut the High Cost of Selling."

Custom Programs. While generic sales seminars of one to three days in length are popular, these seminars are being supplemented by seminars tailor-made for individual corporations. For example, when General Motors reorganized in 1981, it turned to the graduate business schools at Northwestern, Wharton,

---

[3] Thomas J. Peters and Robert H. Waterman, Jr., *In Search of Excellence* (New York: Harper & Row, 1982).

[4] Charles E. Brunner, "Changing the Focus of Business Schools," *The Collegiate Forum* (Fall 1983), p. 15.

[5] Richard Broderick, "Using Colleges and Universities to Meet Your Training Department Needs," *Training/HRD* (March 1982), pp. 20–28.

Indiana University, and UCLA to design curricula and reading materials specifically customized for their needs. A 1986 *Bricker's Directory* survey of 46 leading business schools found that more than half were offering custom programs. For example, at Northwestern, over one-half of the 3,000 corporate students attending seminars were in custom programs sponsored by 24 corporations.[6]

Cost Considerations. While the purpose of these seminars varies, most are designed to keep corporate personnel abreast of the latest developments. However, apart from that the similarities are few. For example, training managers should always be aware of what they are getting for their money. In 1987, one-week sales management seminars ranged in price from $650 to $2,850. A close examination of fees will reveal whether the cost of meals and housing is included. If housing is included, are participants paying for single rooms? Are rooms in a hotel/motel, university center, or in college dorms? These factors greatly affect price. In addition, discounts are often offered for two or more participants.

For instance in 1987 the one-week School of Sales Management at Arizona State (sponsored by the American Marketing Association) cost $900 (nonmember price). This price included luncheons and breaks but no housing. The faculty was largely composed of business practitioners, authors, and a few faculty members from other universities. On the opposite extreme was Columbia University's Sales Management Seminar which was taught primarily by Columbia faculty. The seminar was taught at Columbia's Arden House, the former country estate of Edward Henry Harriman. Included in the cost were all meals/breaks and a full recreation program and physical fitness program. The cost was $2,850. See Table 26–2 for more information on leading university sales and marketing seminars.

How to Find the Right Seminar for Your Employees. When trying to find the right seminar for your employees, there are several sources. These include *Seminar Directory: Industries Guide to Short Courses, Seminars, Conferences and Workshops*. This directory is published quarterly by Stemm's Information Systems and Indexes, P.O. Box 42576, Los Angeles, CA 90050. Other sources include ASTD's electronic database, TRAINET, ads placed in *Sales and Marketing Management* magazine, and the American Marketing Association's biweekly *Marketing News*. Other directories include *Bricker's Directory of University-Sponsored Executive Development Programs*[7] and the *International Directory of Executive Education*.[8]

---

[6] Kenneth Dreyfack, William J. Hampton, and John A. Byrne, "When Companies Tell B-Schools What to Teach," *Business Week*, February 10, 1986, pp. 60–61.

[7] George W. Bricker, *Bricker's Directory of University-Sponsored Executive Development Programs* (South Chatham, MA: Bricker Publications, 1986).

[8] Nancy G. McNulty, *International Directory of Executive Education, Pergamon Press, Inc.* (Elmsford, NY: Maxwell House, 1980).

Table 26–2 One-week short course in sales and marketing at selected universities

University	Housing	Meals	Methods of Instruction	Targeted Audience	Cost
University of Michigan	Executive Conference Center	All	Lecture/simulation	Experienced sales and marketing managers	$2,100
Wharton School	N/A	Partial	Lecture/case	Experienced sales managers	1,985
Columbia University	Executive Conference Center	All	Lecture/case	Regional and national sales managers	2,850
University of Wisconsin	N/A	Partial	Lecture/activity	Sales personnel District/regional sales managers	850
Arizona State University	N/A	Partial	Lecture	District/regional sales managers	900

While most companies do not have a policy on the selection of program participants, the overriding consideration is to select only those employees who can and will benefit from the program. Furthermore, selection should be based on demonstrated past job performance which represents a prediction concerning the quality of future performance. Under no circumstance should "problem" employees or those near retirement be selected.[9]

Executive Education Programs

The first executive education program was the Sloan Fellowship Program at M.I.T. in 1931. Today there are more than 50 such programs, typically ranging in length from 2 to 14 weeks. These programs should not be confused with Executive-MBA programs which are offered by universities as degree-granting programs for experienced, midlevel managers.

While all sales and marketing managers could benefit from general executive education programs, there is presently only one program specifically designed for sales and marketing managers. Founded in 1953, the Graduate School of

---

[9] Gerald H. Whitlock, "The Role of Universities, Colleges and Other Educational Institutions in Training and Development," in *Training and Development Handbook*, 2nd ed., Robert L. Craig, ed., p. 45-10. (New York: McGraw-Hill, 1986).

Sales Management sponsored by Syracuse University in cooperation with Sales and Marketing Executives-International offers a four-week (over two summers) program to meet the professional needs of senior field sales managers as well as recently appointed product or marketing managers. To date over 7,000 executives have attended this program. Those completing the course earn six graduate credits from Syracuse and earn 16 points (8 per year) toward the 35-point eligibility requirements of the Professional Certification Program of the Accreditation Institute of SMEI. The 1987 cost of this program was $5,000 for both years.

For more information on the sales and marketing education programs discussed, write to the following addresses:

Dean, Graduate School of Sales Management & Marketing
School of Management
Syracuse University
Syracuse, NY 13244–2130 (315) 423-3292

Executive Education Center
School of Business Administration
The University of Michigan
Ann Arbor, MI 48109–1234 (313) 763–1000

School of Sales Management
American Marketing Association
250 S. Wacker Drive
Chicago, IL 60606–5819 (312) 648–0536

The Office of Executive Education
The Wharton School
University of Pennsylvania
Vanee Hall
Philadelphia, PA 19104–6359 (215) 898–1776

Columbia Executive Programs
Graduate School of Business
324 Uris Hall
Columbia University
New York, NY 10027 (212) 280–3395

Institute for Sales and Marketing Education
School of Business
Indiana State University
Terre Haute, IN 47809 (812) 237–3232

Campus Services for In-House Training

When planning sales meetings or conferences, training managers should always keep in mind the excellent conference services and classroom facilities at many colleges and universities. Not only are many schools located in strategic spots that might offer central locations for your meetings, but many schools are located at scenic spots as well. For example, one corporation which has frequently used colleges is Hartz Mountain. One regional sales meeting was held at the University of San Diego, a beautiful campus that spreads across a palm-covered mesa overlooking Mission Bay, the home of Sea World. Other schools used by Hartz Mountain include Northwestern, University of Colorado, Texas Christian, University of Pittsburgh, University of Massachusetts, Drew University, and Monmouth and Robert Morris colleges.

Two Choices When Selecting Campus Facilities. There are two basic routes that can be selected when deciding to utilize campus facilities. The first is using dorms and regular classroom facilities vacated by students during summer months. Because colleges do not like laying off housekeeping and food service personnel every graduation, many schools, especially smaller liberal arts colleges, are more than willing to rent you dorm and classroom space. To stay in business, all schools need to keep resources productive, especially during these times of declining enrollments.

One major advantage of using the college campus is the relatively low cost. Normally, the cost of having a meeting on campus, using dorm and classroom facilities, is about one-third of the motel/hotel route. Many colleges can provide you double rooms (dorm) and three meals a day for less than $100 per week per person. For example, one midwestern state college provides air-conditioned suites (five persons per suite) and three meals per day for $84 per person (weekly rate). On a daily basis, most colleges charge around $30 to $40 per person (American plan). Other benefits include the use of libraries and recreation facilities like tennis and racketball courts and swimming pools. At an extra cost are classrooms, duplicating services, audiovisual equipment, and even the occasional university professor as a guest speaker. Keep in mind that most colleges and universities have a few well-known campus personalities.

For those who cannot get along without a telephone and TV in every sleeping room, private bath, and wake-up calls, the second route—university conference centers—offers a good alternative. University conference centers are usually first-class facilities designed specifically to handle the needs of varying types of organizations. While much of their business is university related, many corporations also use their services. Some of the more popular conference centers are those at Indiana University, Wake Forest, Columbia University, University of Georgia, University of Alabama, University of Michigan, University of Wisconsin, and the University of Arkansas. Some conference centers, like Columbia and Wake Forest, use estates donated to them by W. Averell Harriman, R. J. Reynolds, or other wealthy benefactors or alumni.

Each of these facilities is first class and offers everything found in a good hotel or outside conference center and much more. For example, the Paul W. Bryant Conference Center at the University of Alabama is a complex of four buildings located on 12 acres. The buildings include the Sheraton Capstone Inn, the Conference Center, Alumni Hall, and the Paul W. Bryant Museum. For conferences from 20 to 1,000 participants, the Bryant Conference Center offers maximum meeting space facilities. Spacious conference rooms include a multipurpose 1,000-seat auditorium, a 400-seat assembly room divisible into two 200-seat rooms, and 12 additional break-out rooms for groups of 2 to 100. These rooms are supported by the latest media technology to enrich the teaching-learning process: downlink teleconferencing, audio recording and high-speed duplicating, closed-circuit television, advanced video system for on-site sending, multiimage projection, and much more. In addition, a full range of registration and administrative services are provided.

Table 26–3 is a sample checklist that can be used when planning to use campus facilities.[10]

Table 26–3 Sample planning checklist

1. Greater need to plan ahead (winter to early spring) than with motels/hotels.

2. Contact housing office or summer conference coordinator.

3. Concentrate on dormitory, dining, and meeting facilities.

4. Inquire about newer or more desirable dorms (air conditioned). There are better dorms; students enter lotteries for them. No A/C needed in Boulder, Colorado, or San Diego; don't take A/C for granted in areas of need. Ask about A/C.

5. Housekeeping and linen arrangements are necessary.

6. Special female accommodations necessary.

7. Lounge area available with TV in dormitory? (Social center of group.)

8. Visit campus personally or have responsible field manager evaluate.

9. Build the scenic triangle (dorm to dining to class) walk to draw attendees to and through more of campus.

10. Room key deposits and forfeitures for lost keys?

11. Parking stickers required for vehicles?

12. Purchase and mail campus postcards in advance as "teasers."

13. Summer is campus renovation time. Inquire about jackhammers or bulldozers under class windows.

14. Buy college bookstore binders for agenda, maps, and meeting passouts (inexpensive but memorable).

15. Secure telephone numbers for emergencies during day and night.

16. Plan extracurriculum: pool, beach, tennis, golf, gymnasium, fields.

17. Discuss media needs—certain classrooms have more advanced facilities.

---

[10] Donald B. Reynolds, "We're Working Our Way Through College." Idea Mint presented at National Society of Sales Training Executives in San Diego, CA, December 4–7, 1983.

Table 26–3 (Cont.)

18. Any celebrity on faculty for possible "drop in"?

19. Secure certificate of your company's liability insurance for college needs.

20. Meal planning. Full room and board, three meals, about $30–35 daily.

21. Special barbecue or cookout arrangement for one night?

22. Example: breakfast and lunch daily, one evening barbecue, balance of dinners in representative, up-scale location near campus.

23. Bring camera for class picture (mail to all). Pictures make good slides for National Sales Meeting and house organ use.

24. Survey bookstore for end-of-season summer clearance during your meeting. (Making room for fall merchandise, big savings.)

    a. Participants can purchase college items for family members.

    b. Purchase a souvenir for each attendee.

25. Clothing extremely casual, the way summer students dress.

26. Ask for ash trays and water; professors in regular classes are not provided these meeting "standards."

27. Plan for coffee in lounge or drink out of doors in scenic spot; vending machines ever present.

28. Plan ahead on billing procedures. Will they advance credit?

29. Ascertain beer and liquor regulations. We found no institution that prohibited moderate use.

30. Don't "test" the campus police or parking restrictions.

31. Give plenty of advance notice to participants; most take summer vacations, but colleges are ready for you in June.

CONTINUING EDUCATION AND SERVICES OFF CAMPUS

Correspondence Courses

Presently there are 71 colleges and universities that offer over 12,000 correspondence courses through the National University Continuing Education Association. Thirty-three of these colleges offer either undergraduate credit or noncredit courses in the general marketing and marketing management areas. Furthermore, 25 of these schools offer management courses in principles and theory, personnel management, and organizational development and behavior. Other courses which may be of interest to trainers include Business Communication and Business Report Writing. These two courses are offered at 25 colleges and universities. At this time, Louisiana State University is the only school offering a specific course in training and development.

Seven Questions to Ask When Choosing a Correspondence Course. Correspondence courses are not for everyone. Students seeking graduate degree credit will find few courses from which to choose; however, students who desire to sharpen or brush up on their skills will find a variety of noncredit courses. Students seeking course credit toward the completion of an undergraduate degree need to ask several questions.[11]

1. Is this course acceptable for credit applied toward graduation?
2. How many total credits by correspondence will be accepted toward graduation?
3. If you are taking a course by correspondence from one institution, can it be transferred to another?
4. Will credits earned by correspondence be accepted in the major?
5. If courses are being used for certification purposes, are they acceptable to the certifying agency?
6. Is a possible pass-fail option acceptable to your college or certifying agency?
7. If the course is being used to meet company education requirements, is it acceptable?

More information can be obtained from Peterson's *The Independent Study Catalog*. It can be ordered from Peterson's Guides, Dept. 6608, Princeton, NJ 08543–2123 (telephone 800–225–0261).

Faculty Consulting Services

Colleges and universities offer a potential gold mine of outside assistance to corporations. Faculty members serving as consultants are frequently used by large and small companies to deliver professional training, advice, and/or service.

Methods for Choosing a Faculty Consultant. When seeking a faculty consultant, there are several methods used to discover campus talent. One common method is simply to ask for recommendations from other training professionals. Using this method you can often get a letter of recommendation from a previous user of the faculty member's services. The letter should provide the degree of satisfaction with the faculty consultant and what he or she accomplished. A second method is to make note of an especially good faculty presentation at a conference or meeting which you attend. After the presentation you can introduce

---

[11] *The Independent Study Catalog, 1986–1988.* (Princeton, NJ: Peterson's Guides, 1986).

yourself, and see if the professor is available for consulting. A third method is to contact the university directly. Depending on your needs, you could call the department chairperson of marketing, management, or the appropriate department within the School of Education. Usually these individuals are happy to make recommendations of qualified faculty.

Another university contact can be found in the Business School. Most accredited business schools have a center for research and economic development. Normally universities encourage their faculty to do consulting through these centers. The center acts as a broker bringing clients and faculty together. They can also offer other university services such as audio/visual, secretarial support, and aid from graduate assistants. These services are not available from faculty who "moonlight" on their own as there would be a conflict of interest. In addition, faculty who work through a center are often given release time from their teaching schedules to consult in the name of the university. In this case all funds generated go to the university. However, in cases where the faculty member doesn't receive release time, the fee is usually split. For example, one university charges clients $750 per day for in-house training sessions or $100 per hour for faculty time. The university keeps $250 and gives the faculty member $500.

Cost for Faculty Consultants. If you choose to seek a faculty member who operates his or her own consulting practice, be prepared to pay. While an individual from a lesser known college or university may be willing to consult for a $300 fee per day plus expenses, the average is probably closer to $800 plus expenses. As with any product, it pays to know what you are buying. Most faculty have doctorates and are local or regional experts in their field. If you want a name with a national reputation, the fee will be much greater. For instance, the fee for giving a keynote talk to management by one nationally known marketing professor from a Big 10 school is $6,000 per day plus expenses. On the average, however, fees will be pretty much on par with nonacademic consultants.

On-Site Programs by University Faculty

Besides offering seminars on campus, a large number of schools conduct their programs on-site. For example, Indiana State has delivered a wide variety of custom on-site seminars. One faculty member delivered a technical report writing seminar to a group of data processing managers at CBS's Columbia House Record Club. Another team of faculty conducted outplacement seminars for over 1,200 workers at a closed IBM distribution center. Another team of faculty members delivers supervisory training seminars for Snacktime Foods. This type of on-site training is taking place all over the nation. It is being conducted not only by major colleges and universities, but in many areas by community college and vocational-technical college faculty. In the latter case much of the cost of training the work force can often be picked up by state grants and/or private industry councils (PICs).

Sales and Marketing Training Research

While any university with qualified faculty can conduct research related to corporate training and development or marketing and management, three universities have established centers for specifically conducting research in training and development. They are the University of Minnesota (Training and Development Research Center), Peabody College at Vanderbilt University (The Corporate Learning Institute), and Indiana State University (Institute for Sales and Marketing Education).

The Training and Development Research Center at the University of Minnesota engages in research and support for the improvement of employee training and organizational development in business and industry. It is the mission of the Training and Development Research Center to promote, conduct, and disseminate inquiries about the practical problems associated with human performance in organizations.[12]

The Training and Development Research Center program of work includes:

1. Identification and clarification of problems that call for employee training and organizational development.
2. Investigation of methods for making human capital investment decisions.
3. Validation and improvement of training theories and methodologies.
4. Validation and improvement of organization development theories and methodologies.

The program of work is accomplished through cooperative efforts with industry and business in the form of contracted research internships, seminars, workshops, and consultations.

Peabody College is a part of a national resource for the development and sharing of new knowledge about how learning can be fostered more cost effectively in the workplace. A further focus is concerned with how and why people learn within business organizations.

The Institute for Sales and Marketing Education at Indiana State University is also unrivaled in that it focuses on research related to the training needs of sales and marketing personnel. As previously discussed, the ISU Institute con-

---

[12] "What Is the Training and Development Research Center?" *Training and Development* (St. Paul: University of Minnesota, 1987).

tains an extensive library of corporate training programs. This allows ISU researchers the unique opportunity to explore the various types of training taking place in the corporate world. For instance, a corporation might request that ISU researchers provide all available data on what types of role playing exercises are used by sales trainers in the chemical industry. Or perhaps a company is beginning a team selling effort, and the sales training manager is directed to develop a team selling training course. By contacting ISU faculty, the Institute will develop a report on all available information (but not the actual editorials) on team selling training programs contained in the library.

Needless to say, information like this is very valuable to a sales training manager. Furthermore, university research centers such as those discussed provide a very cost-effective alternative to obtaining data which is often not available through company resources.

INDUSTRY INVOLVEMENT WITH MARKETING ED/HRD ACADEMIC PROGRAMS

Thus far in this chapter, we have discussed what services you can find at colleges and universities. Now, the roles will be reversed, and various methods of becoming involved with colleges and universities will be discussed. In all likelihood, you will find your personal involvement or that of your corporation to be just as valuable for you as the university.

Executive-in-Residence Programs

E-I-R programs spread nationwide during the 1970s to bridge the gap between campuses and business. Today colleges and universities are eager to invite active business managers to campus. The programs are found not only in leading business schools but also at liberal arts colleges. Many liberal arts colleges find executives for on-campus programs through organizations such as the Woodrow Wilson National Foundation (Box 642, Princeton, NJ 08542, 609–924–4666). This foundation annually sends around 85 figures from business, journalism, government, law, medicine, and the arts for stays of up to a week as visiting fellows at smaller colleges.[13]

In addition, trade and professional associations are also involved in the selection of executives. For instance, the American Council of Life Insurance (1850 K Street, NW, Washington, DC, 20006, 202–862–4000) sends eight to ten senior life insurance executives to liberal arts campuses for three weeks.[14]

---

[13] Marilyn Wellemeyer, "Executives on Campus," *Fortune*, April 18, 1983, pp. 137–148.

[14] Ibid., p. 137.

Another example is the National Society of Sales Training Executives which annually sends two or more member executives to Indiana State University. NSSTE will soon expand this successful program to other colleges and universities.

While some E-I-R programs provide housing and meals for executives and their spouses, this is not always the case. Executives selected by NSSTE are expected to pick up their own housing and travel expenses, while the university pays for meals. It should also be noted that some foundations provide stipends to executives who are selected, but most executives decline. While most do not need the stipend, the typical executive attitude is that selection for an E-I-R program is a great honor and a time for "personal payback." Most executives also admit that the experience is a great "ego trip." Executive experiences indicate that students hang on every word and are most eager to learn real-world approaches to business management. Personal contact with students is the reward for such involvement.

Executives contemplating an invitation to campus usually ask what a typical day is like. The answer would certainly vary depending on the school and the individual, but most E-I-Rs agree they have few days at work that are more hectic than theirs on campus. If there is a typical day, it would probably begin with early breakfast followed by three to four class sessions spread out over the day. In between, E-I-Rs would most likely counsel students in their private offices or perhaps give a faculty seminar. Evenings are also packed with activities. These can range from dinners at the homes of faculty and administration, dining at exclusive clubs, addressing a crowd at a town-and-gown convocation, or perhaps even meeting with a business or social fraternity to talk about the job market. The possibilities are unlimited, but be sure to count on long days and plenty of conversation.

Of course not all E-I-R programs involve intense one- to three-week sessions. Many colleges actively seek newly retired executives to spend a semester or even a full academic year on campus. These programs offer many of the same benefits at a more relaxed pace.

Exchange Programs (Industry Personnel/Faculty)

While this type of program is relatively new to the field of sales and marketing, the exchange of faculty and industry personnel is common in other areas such as the sciences, technology, and even government. While it is not necessary for an actual one-on-one exchange to take place, the basic idea is for either businesspeople to come to campus as instructors or researchers or for faculty to work in a business setting for an extended period of time (most commonly six months to a year). In either case, the purpose is to exchange ideas and information which will sharpen the skills of both parties.

Exchanges frequently take place as a result of faculty sabbaticals/leaves or industry research contracts with universities. Faculty sabbaticals are earned every seven years or longer. They are awarded to faculty who submit proposals which outline how the sabbatical will improve their teaching or current research. Because faculty members are paid (full pay for a quarter/semester or one-half pay for a full year), many companies can utilize university faculty at bargain prices. Often companies give the full pay earned by the faculty member as a direct payment to the university. This way the school and faculty member benefit from such an arrangement.

Two companies which have used faculty personnel with success are Eastman Kodak and Wickes Lumber Companies. Eastman Kodak normally brings education professors in to their Rochester Marketing Education Center to teach and to work on special projects in curriculum design and media technology. Wickes recently utilized a marketing education professor from Virginia Tech to teach selling courses and to assist in the design of a management development program. Other companies have also utilized faculty exchange participants as instructors in management and marketing communications courses.

Adjunct Faculty Participation

Nearly every college and university has a need for adjunct faculty members. This is especially true in business schools where 15–20 percent of job openings are annually going unfilled.

The best way to find out about openings is simply to call the department chairperson at a nearby college or university and let them know you are available. You could also write them a letter with an enclosed resume. While they may not be able to use you immediately, your name will be kept on file for future opportunities.

Like executives in residence, training professionals involved in college teaching as adjuncts also find many personal rewards. However, it should be noted that financial rewards are once again not normally there. Payment for one three-hour semester course ranges from approximately $750 to $2,500. These payments vary according to experience and degrees held. Obviously, it is much cheaper to staff a department with part-time faculty members versus employing costly full-time faculty. The increasing costs of higher education have forced many community colleges to use part-time adjuncts as the bulk of their faculty. This way they not only keep salaries down, but also avoid paying most benefits.

Those training professionals nearing retirement may find college teaching to be a nice supplement to retirement. While many colleges and universities will not employ faculty for tenure-track positions who have not earned their doctorates, many will hire retired businesspeople for nontenured-track positions. These individuals are hired full time for periods of one to seven years with the understanding that they are not required to earn the doctorate, nor is the university obligated to provide continuing employment.

Why colleges and universities do not normally employ persons without doctorates is an extremely controversial issue. The issue is usually raised by a training professional who has spent 20 to 30 years in the corporate world and often teaches the same courses as Dr. Doe at a nearby university. The odds are that Dr. Doe has far less business experience than the seasoned training professional, and thus, it is hard for the businessperson to understand why the college won't employ him or her on a full-time basis.

The answer is simple but not obvious. First, four-year colleges and universities do not employ faculty to just teach. As a matter of fact at most large state and private universities, tenured faculty spend a small portion of their time actually in the classroom. The university expects the professor to conduct research in his or her field which not only will expand their knowledge and contribute to the profession, but in most cases will also financially benefit the university. Whether we like it or not, research contracts supervised by faculty are necessary to fund universitywide activities that benefit everyone on campus.

Individuals possessing the doctorate (Ph.D., Ed.D., or D.B.A.) have a different type of experience from the business person. In addition to extensive courses in their field, they also possess the ability to conduct the research that the college or university demands. Holders of such degrees must possess proficiency in one or more research "tools" such as a foreign language(s), statistics, or computer science. Thus, it is this ability to conduct advanced research that makes the "Ph.D." a valuable and sought-after faculty member. This individual is then expected to pass his or her trade in the classroom and/or lab to graduate students who are paying top dollars for their advanced degrees.

Cooperative Education/Internships

As previously mentioned in this chapter, cooperative education got its start in the early 1900s. Cooperative education at the collegiate level was begun in 1906 at the University of Cincinnati. Prior to 1960, only 65 colleges (primarily engineering schools) utilized the cooperative program. Today the program has expanded into nearly all academic areas and into over 1,100 colleges and universities. Two universities, Cincinnati and Boston's Northeastern, have built the co-op plan into all their undergraduate programs. In addition, 100 colleges offer graduate cooperative education programs.

The goal of cooperative education is to help orient students to the work world by providing a solid foundation for career planning and vocational guidance. There are three modes of co-op which can provide these opportunities for students. The most traditional is the alternating mode. Students exercising this option spend a minimum of two semesters as a full-time employee, separating this work with a semester/quarter of full-time studies on campus. In the parallel mode, students maintain their full-time status on campus, but hold part-time jobs in the local community. The third mode, the single field experience, requires one semester off-campus. This could possibly include internships, summer jobs, or special projects.

It should be noted that there are a few basic differences in regular co-op and internships. First, students performing internships are expected to already have the skills necessary to work along side a regular employee, and thus, the internship is often conducted in the student's last semester/quarter. It could be compared to student teaching. Like student teaching, most interns are not paid. Co-op students by comparison are paid ($800 to $1,500 per month is the common range) and are expected to learn on the job.

Co-op is becoming a primary recruiting tool of many companies such as General Motors and others, who see the value in employing students who they can observe earlier at lower pay rates. While co-op in sales and marketing is not widespread, there are many companies employing students on co-op in the areas of retail sales/buying, marketing research, and advertising. In the training area, all the universities previously listed are seeking appropriate internships for their students. Thus far, interns in those programs have been successfully used to telemarket training programs, perform task analysis, develop job descriptions and training manuals, develop management development courses, establish training libraries, and administer continuing education seminars. Regardless of your marketing or training area, the odds are that co-ops/interns could be a valuable asset to your organization.

Advantages to Hiring Co-Op Students. A U.S. Office of Education study revealed many other advantages to hiring co-op students.[15] These include:

1. Co-op graduates receive higher performance appraisal ratings than other college graduates.
2. Co-op graduates received higher raises in salary more frequently than other college graduates.
3. Recruitment costs averaged much less for co-op students than for recent college graduates.
4. The percentage of minority group members listed was twice as high among co-op students as among recent college graduates.

Sources for Additional Information. More information can be obtained from the following:

---

[15] Richard A. Hayes and Jill H. Travis, *Employer Experience with Cooperative Education: Analysis of Costs and Benefits* (Detroit: Detroit Institute of Technology, Cooperative Education Research Center, 1975).

National Commission for Cooperative Education
360 Huntington Avenue
Boston, MA 02115

Cooperative Education Research Center
Northeastern University
401 Stearns Center
Boston, MA 02115

Cooperative Education Association
221 North LaSalle Street, Suite 2026
Chicago, IL 60601

National Society for Internships and Experimental Education
1735 Eye Street, NW
Suite 601
Washington, DC 20006

Adopt-a-School Program

During the 1980s cooperation between business and education is a key strategy in achieving excellence in our school systems and colleges and universities. One example of this strategy is the popular "Partnership in Education" program which has been implemented nationwide at all levels of educational institutions. At the university level, there are several examples of excellent programs. In 1978, the University of Minnesota's School of Management and its Institute of Technology formed a Partners Program that linked them with 22 "partner" companies that donate a minimum of $20,000 annually. In return, they receive priority consideration in course enrollment and recruitment of university graduates. The Partners Program now generates over $3 million annually for the School of Management.[16]

Another example is the Marketing Studies Center at the University of Georgia. The center was formed with a $1 million contribution from the Coca-Cola Company. The president of Coca-Cola, Donald R. Keough, stated, "As marketers of the world's most popular consumer product, the Coca-Cola Company has a vested interest in the ability of the state of Georgia to continue to produce the highest caliber of marketing executives in the country." The center has expanded its program to include a Master of Marketing Research which offers an emphasis in brand management and sales management.

Endowed chairs also offer another method of assisting a university. With this method a company or organization normally donates $100,000 or more to a specific department for use in supplementing (from the interest earned) an

---

[16] Broderick, "Using Colleges and Universities," p. 21.

outstanding faculty member's salary. A person can hold a chair indefinitely, but they are often rotated on a three- to five-year basis and are used to attract people from outside the school. The best known chair in sales is the Memphis State Sales and Marketing Executives Chair in Sales. Because Memphis State University is one of the few universities in the nation with a sales major, the chair holder is expected to provide national academic leadership to this area of study. In 1989 UARCO, Inc., endowed a chair at Northern Illinois University. The chairholder will be responsible for developing a selling concentration at NIU and coordinating off-campus sales training seminars.

Other Linkages Between Business and Sales Training Personnel

There are several other ways that business and industry sales training personnel can become involved in academic programs. These include assisting in the placement of graduates from marketing and sales training programs; serving on advisory committees of marketing or sales training programs; acting as speakers for student business organizations like DECA, Delta Sigma Pi, Beta Gamma Sigma, and others; providing scholarships to deserving students who are preparing for careers in sales and marketing or training and development; or assisting employees to attend college by offering tuition aid programs.

EVALUATION OF COLLEGE AND UNIVERSITY PROGRAMS _____

It would be extremely difficult to come up with a method for evaluating the wide variety of programs available at colleges and universities. Research has indicated that most colleges and universities make little if any attempt to evaluate their programs systematically. What evaluation that does take place is usually done informally by examining participants' remarks (oral or written) or test results when appropriate.

The nature of the problem is influenced by the difference in education and training. Trainers can expect immediate feedback from those attending training sessions in any of several ways. The most obvious is better job performance, which translates into increased profits. Internal training course participants can be tracked and organizational influences noted. Unfortunately, college and university programs are attended by a few participants from any one company. Thus, because of small numbers and unknown organizational influences, meaningful evaluation is difficult at best. However, when the nature of a program lends itself to evaluation, every effort should be made to collect data systematically and to determine effectiveness. This is probably easiest in the case of custom-designed programs conducted for your company by a college or university. Regardless, it is not enough simply to leave a costly seminar just "feeling better." The end result should be a change in individual or organizational behavior which was created by a well-run college or university program. If the program is well run, it will probably offer cost savings to your company and will benefit both the corporation and the college or university.

SUMMARY

While there are few colleges and universities offering degree programs in sales, sales management, or sales training, there are many college and university programs and services relating to sales education that can benefit corporations. These programs and services include:

- resource centers
- seminars and short courses
- executive education programs
- in-house training facilities
- correspondence courses
- consulting
- on-site training, and
- sales and marketing research.

Each of these programs and services has advantages and disadvantages and thus must be carefully evaluated. On whole, however, colleges and universities are an excellent resource and can be a tremendous benefit to the sales training manager who is seeking ways to supplement his or her corporate sales training program.

chapter twenty-seven

CONSULTING WITH EXTERNAL SALES TRAINERS: A FOUR-PHASE PROCESS

CHIP R. BELL

Chip R. Bell is an independent consultant headquartered in Charlotte, North Carolina. His practice focuses on service management consulting and leadership training. A nationally known trainer and consultant, he was formerly a partner with LEAD Associates, Inc., and vice president and director of Management Development for NCNB Corporation. Dr. Bell is the author of seven books, including *Influencing: Marketing the Ideas that Matter* (University Associates, 1982), *The Trainer's Professional Development Handbook* (Jossey-Bass, 1987), *Instructing for Results* (University Associates, 1985), and *Clients and Consultants*, 2nd ed. (Gulf, 1985). His articles have appeared in numerous professional journals. He consults with a wide range of clients, including several *Fortune* 500 companies.

CONSULTING WITH EXTERNAL SALES TRAINERS ⎯⎯⎯⎯⎯⎯⎯

The complexity of the selling encounter coupled with the requirement for first class learning experiences have led many human resource development practitioners to look to external sales trainers to fill and/or amplify their sales training needs. But the myriad consultant jokes telegraph the need for caveat emptor—let the buyer beware.

This chapter is a fast-paced, pragmatic look at tips and tools important in effectively finding, selecting, managing, and evaluating an external sales training consultant. Within the chapter you will encounter numerous perspectives and cautions aimed at helping you maximize the client-consultant relationship, saving you time, money, and sleepless nights!

SALES CONSULTANT—SPECIALIST IN THE SELLING PROCESS ⎯⎯⎯⎯⎯⎯⎯

Sales consultant! Some would argue that the two words do not belong in the same title—a conflict in terms like the oft-quoted "military intelligence" or "postal service." *Selling* is about influencing another to accept, buy, or support a product, service, or idea that he or she might not have "bought" sans influence. *Consulting* is the practice of perceptual engineering—enabling the consultee to *see* a problem in a certain way with free choice in picking a solution.

The selling role implies persuader, convincer, energizer of acceptance. Sales professionals are charged with qualifying prospects, matching product to need, and moving inventory, not giving advice without closing the deal. Consultants, on the other hand, are charged with offering "wisdom" and heady but useful advice, not overcoming objections or touting benefits. So, we are back to the opening dilemma. How do we get past this wolf-in-sheepskin quagmire?

We can wade through the quagmire by first defining terms—particularly as they will be used in this chapter. To start, a true sales consultant is an advisor who specializes in the sales process—one armed with a depth of knowledge in the components needed to create, maintain, and manage a sales culture. Sales consultants serve as resources in assessing gaps in the sales process, recommending solutions targeted at closing the gaps, and keeping them closed. Their purview spans the sales process with emphasis on

- How salespeople are selected and prepared for their roles.
- The tools and training used to build confidence and competence in selling.
- The sales management process—goal setting, call preparation, coaching, follow-up, feedback, and the like.
- The organizational systems used to reinforce and reward appropriate sales behavior and sales results, including incentive tools, promotion criteria, and sales promotion efforts.
- The strategic planning effort and its impact on effective selling throughout the organization.
- The relationship among advertising, public relations, and selling.
- The interface between market research, marketing planning, and the selling effort.
- Assessment of barriers or obstacles which inhibit, retard, or frustrate the selling effort.
- The relationship between sales and how the customer or client is served after the sale.

The list is much longer! The sales consultant may not be a specialist in all solutions, but he or she has sufficient training and experience to be competent in gap identification and how gaps are best filled.

There are other roles often labeled "sales consultant." A frequent visitor to the corporate reception area is the sales trainer—a professional who designs and/or delivers sales training programs. Despite their calling, sales trainers are learning specialists, not consultants. There *are* sales consultants who can also design and teach sales training. However, in doing so, they are temporarily changing roles.

Finally, there are sales reps. Essentially, these are vendors or salespeople in the true sense of the word. They represent one or more companies which produce products or perform specific services related to some aspect of the selling process. One may be a sales representative for a packaged sales training program. Another may be selling a compensation program for salespeople or tools for an effective sales calling effort. For purposes of this chapter, we will make a distinction between sales consultants and sales reps.

Clarifying labels is more than a "meeting of the minds" semantics activity. By reducing ambiguity about roles, perspectives, and boundaries, we gain clarity about expectations and potential outcomes. Abraham Maslow's popular line, "If the only tool you have is a hammer, then every problem risks becoming a nail," is a caveat emptor word of caution.

This chapter will focus primarily on the external sales trainer, the second type of "consultant" described. Many of the comments, cautions, and caveats to follow could also apply to the true sales consultant. We will not discuss the sales training rep or vendor here since the area is covered in Chapter 28.

External sales trainers typically specialize in the design and/or delivery of a sales training program uniquely crafted for the client organization. Their entry into the organization and work while there typically follows the consulting pattern, although they are not true consultants in the manner earlier described.

Keep in mind that the call for their expertise is grounded in the premise that some type of learning activity is deemed to be in order. Either independently or with the assistance of a sales consultant, a solution (or at least a partial solution) has been chosen. Expectations for this solution should be bordered by measures of improved competence, not necessarily improved performance. There have been many participants who emerged with an "A+" in sales training but failed to sell more widgets. The missing link was not about competence, but rather some other factor in the participant's work world which deterred improved performance.

HOW AN EXTERNAL SALES TRAINER CAN HELP YOU _____

There are advantages and disadvantages to using an external sales trainer. Reviewing those that fit a particular organizational situation can be extremely beneficial when deciding whether to use an external sales trainer or not.

The most common reason external sales trainers are hired is the conclusion that an external resource can offer skills, perspectives, and/or credibility not available internally. There may be no one internally with the needed competence to design and/or deliver sales training. Or competent internal trainers may be busy with other important activities so it may be easier to bring in an outsider on a project basis than free up a qualified person internally.

Sometimes the issue is one of internal standards. Decision makers may conclude that, while internal trainers may have the capacity and availability to design and/or deliver a sales training program, it would not be up to the standards of quality required. External sales trainers often bring a broad perspective borne out of working with a wide range of clients. As one training manager stated, "My trainers teach communication, leadership, effective writing, and an assortment of other topics. They could probably do sales training as well. But I wanted someone who lives, eats, and sleeps sales training—a true specialist."

What are the cons? External is external—consultants don't know the organizational culture, norms, and values. To be effective, an external resource must work extra hard to gather an understanding of the informal ropes important to program credibility.

In addition, external resources can be less committed to the outcome since they will exit the organization, leaving the internal person to "clean up the mess" and manage the followthrough, reinforcement, and maintenance. And the more tailored the program, the more expensive it can be to bring the external trainer up to speed on aspects an insider may have devoted years attaining.

One mythical disadvantage is cost. Inexperienced clients who hear the tab for an external resource may react in shock. Clearly, there *are* consultants who overprice for their efforts and for the client's return on investment. However, the external resource called in to design a sales training program is likely to be an expert in design *and* selling content. Consequently, they can produce in days what might take an internal person months. Likewise, the external trainer called in to teach typically requires far less "prep" time to facilitate an effective selling course.

On the other hand, external training resources can be great assets, bringing refreshing perspectives, innovative ideas, and new concepts. Uncluttered by historical baggage, they can offer important insights and objectivity. Because of this, they can speak with credibility that insiders might envy but find difficult to duplicate.

External training resources can be great liabilities. They can create dependence instead of fostering competence. They can subtly usurp the stature and popularity that rightfully belongs to the internal person. They can promote guruism or hero worship instead of making heroes of their client. Like all professions, the world of the external sales trainer has its share of snake oil, charlatans, and ripoff artists.

THE CONSULTING PROCESS AND EXTERNAL SALES TRAINERS

Getting advice and assistance from an external training resource occurs through a consulting process. Consultation is fundamentally the act of helping. It is not simply the mechanical tossing of expertise and advice toward a needy client; it is an experience in shared resources. The core of the act of helping is the interaction of two or more people relating to improve some person, situation, or thing. The quality of those dynamics determines if the encounter results in more help than harm, more joy than sorrow, and more growth than stagnation.

The ingredients in an ideal client-consultant relationship are like those in an ideal marriage. Each is not easy to come by. Even if attained, they remain ethereal; all parties involved must work constantly to keep them at hand. As sensitivity to the dynamics of *helping* increases, the chances for effectiveness likewise climb.

There are numerous ways of perceiving the consulting process. Writers differ on their words for process steps and on the models or conceptual frameworks they utilize. Most agree there is entry, some type of diagnosis or assessment, a response or action, and, at a point, termination of the consultation. The quality of the client-consultant relationship (in our case the relationship between the user-client and an external training resource) lies in how each phase of the consultation process is managed.[1]

Since we have organized this chapter around a four-phase consulting process, it will be useful to say a few words about each one. *Entry* begins with the initial contact between client and resource. Through this phase, problems are jointly explored, perceived needs and symptoms are discussed, relationships are clarified, goals and roles are defined, resource parameters are identified, methodology is clarified, and ultimately a contract (written or verbal) is negotiated.

We will spend a major part of this chapter on the entry phase. As Peter Block has so eloquently stated, "Without proper entry, the remainder of the consultation will likely miss the mark."[2]

Diagnosis or *assessment* is the process of examining the cause of the problem and/or scope of the need. It generally begins with an understanding of the situation as perceived by the client and includes an appreciation of the client's goals and resources, as well as the client system's culture, values, norms, and beliefs.

Questionnaires, interviews, observation, and analysis of previous performance data are the most common methods used in obtaining a diagnosis of the situation and need.

Response is the action phase of the consulting process. It includes the planning and engagement of structured activities employed to correct the problem or meet the need. The *closure* phase includes evaluation of the results and planning for continuous process maintenance to ensure permanent integration and lasting change. Now we will examine the selection and management of an external sales trainer through the lens of the consulting process.

PHASE 1: ENTRY—THE INITIAL CONTACT

You have sorted through the decision-making process and come to the conclusion that an external sales trainer is needed. The next hurdle to traverse is selecting the right trainer. "What" is the question wedged between "why" and "where." Too often, trainers work through the rationale (the why) for hiring an

---

[1] Chip R. Bell and Leonard Nadler, *Clients and Consultants* (Houston: Gulf, 1985), p. 2.

[2] Peter Block, *Flawless Consulting* (San Diego, CA: University Associates, 1981).

external resource and immediately leap to the search (the where) with little thought for the selection criteria.

Four Criteria for Selecting an External Sales Trainer

Like choosing art, the "eye of the beholder" is at work when picking the right external sales trainer. However, the following criteria may be helpful considerations. Acceptance is not a prerequisite to success. Together, these criteria are merely a function of role clarity—being clear on what you expect the trainer to do. Excellent designers don't always have superior "platform skills," and "silver tongued" sales pitches may belie inadequacies as a creator of learning programs. Be clear on what you need the external resource to do and then build critical success factors around that role.

1. *Allegiance to professional growth*: A criterion many hiring professionals have found telling is the degree the "applicant" demonstrates an allegiance to growth. Answers to interview questions about recent books read, courses completed, and conferences attended can provide useful insight into the role which personal growth plays in the life of the potential external sales trainer.

2. *Role-modeled behaviors*: How well does the sales trainer model the skills, behaviors and attitudes imbued in the sales training course? The "applicant" who recommends a solution without identifying a need may be failing to model the approach you ultimately hope your future training participants acquire.

3. *Compatible philosophies*: You will avoid later grief if the person you select shares your general philosophy about how people learn and how best to facilitate growth. This does not mean you must see eye to eye on everything; part of a successful consulting relationship lies in the capacity to learn from each other. However, if you are an ardent subscriber of behavior modification and you hire an external resource who is wedded to discovery learning, you may waste valuable time (and money) working through philosophical conflicts.

4. *Relevant experience*: The issue of how much actual experience one must have to be credible is an area which has been argued for many years. Some sales trainers with little or no selling experience are superb. Highly experienced salespeople can be ineffective in conveying their experience in a learning environment. You might want to take risks with a great internal trainer who has limited sales experience. Contracting with a no-experience external sales trainer is high risk. Sure, there are world-class sales trainers and terrific sales training designers who have never personally closed a sale or handled an objection. But, if you are buying a temporary talent,

why take the risk? Credibility with the subject—particularly the subject of selling—is a crucial commodity. The "years logged" or "sales quotas met" is less of an issue than the external resource's capacity to stand in the shoes of the learner and create or facilitate from a position of depth of understanding. Such depth is more likely acquired through actual experience in a selling role.

Generic Versus Tailor-Made Training Program. There is a side of the experience issue which wires the "what" back to the "why." You should have arrived at this point in the process with a clear understanding of the need for sales training and the reasons an external sales trainer would be uniquely qualified to assist in meeting that need. A part of that needs analysis should have included a determination of whether your need was for a generic sales training program or one crafted for your specific requirements.

Even the most generic of sales training programs requires some adjustment. Generally, if the diagnosis suggests generic, your examination of external resource experience should include a look at presentation and facilitation skills. Find a way to observe candidates in action. If this proves impossible, ask for a video tape—one with more than a few excerpted clips of their "best stuff." Call references, including participants. References who hired external sales trainers may be tempted to color their recommendations to justify their decisions. Participants have little need to be less than candid.

If the analysis calls for a tailor-made program, examine each candidate's design prowess. The external sales trainer will need to interview people in the organization to gather the data which precisely focuses on the need. Will the person make the appropriate impression on those interviewed?

Often the experience of needs analysis will be a critical component in collecting the right information in an efficient (i.e., cost-effective) manner. A clue to the sales trainer's skill may be the questions he or she asks. Ask candidates for examples of other sales programs they have designed. Again, check references, zeroing in on whether the program worked—was there a link between the program and skills acquired by participants? Sample questions are

QUESTIONS TO ASK A SALES CONSULTANT

1. How would you describe your last sales training assignment?
2. What did you find most challenging? Most personally rewarding? Most difficult?
3. If you had the assignment to do over again, what are some things you would do differently?
4. Think back over the most difficult consulting assignment you have had. Outline some of the highlights and describe your role in the process.

5. If you were a client selecting a sales consultant, what are the qualities you would consider most important? Which of these qualities do you believe you possess? Which ones do you work hardest to improve?

6. Describe a professional growth experience you have had in the last two years of which you are particularly proud.

7. What is a personal growth experience you feel positive about?

8. What do you enjoy most about sales training? What do you like least?

9. If you are selected for this assignment, what are the steps you would take in carrying it out?

10. As a sales training consultant, what do you expect of a client? What are client actions which most often disappoint you?

11. If you were the client instead of the consultant, what are some ways you would evaluate the effectiveness of a sales training consultant?

12. What are the factors which contribute most to sales training success? What most often causes it to fail? What would be your role in preventing failure?

13. If you get this assignment, how would you begin the consulting assignment?

14. What are the parts of the consulting assignment which you would want to subcontract to someone else?

15. How do you typically price a consulting assignment like this one?

Locating External Resources. We are now ready to think about "where" to find the supertrainer you need. There are many sources for finding external sales trainers. The best is to call training professionals in other organizations whose opinions you value. Seek first those whose need has been similar and/or those whom you would like to emulate. External consultants in the field can also be important sources of information about external trainers particularly if they have no vested interest in the outcome and can respond objectively.

If you are unable to identify people you know, call some you don't know. Pick out several organizations you admire and contact your counterpart. Identify favorite books on sales training and contact the author(s). Review back issues of *Training and Development Journal, Training Magazine, Sales Management*, and so on and contact the authors. The American Society for Training and Development (ASTD) and *Training Magazine* both publish buyer's guides that identify types of resources. Also, the Sales and Marketing Professional Practice Area of ASTD can be a valuable resource in locating the right external training consultant.

Once the need has been identified, you may know precisely which consultant or consulting firm would be the perfect match. If you are less certain, you should consider more than one option. Even consultants or consulting firms which appear very much alike may be quite different. While "love at first sight" can work in sales training as well as courtship, a certain amount of "playing the field" is generally advisable.

Small Versus Large Consulting Firms. One consideration in choosing the best external resource is whether to go with the independent consultant (sole proprietor) or consulting firm (small, medium, or large). Your judgment will be shaped by many factors—need, timing, costs, and so on. Such a judgment can be made by considering the advantages and limitations of each.

The independent consultant or small consulting firm may be less expensive. Absent the overhead of a large firm, you may get more bang for your buck (all things being equal). They can be more responsive to tight time frames and able to assemble subcontractors for the assignment with less red tape. You typically will be working with the person who sold you on the contract. The small firm may be more concerned about ensuring that you get exactly what you need since they are more vulnerable to failure than a large firm which may be able to absorb the loss more easily.

However, the independent consultant or small firm typically has little backup. If the consultant gets sick, there is often no one to "pick up the slack." Independent consultants or small firms often divide their energies between marketing and consulting, thus not devoting their entire energies to "doing consulting." The small firm, dependent on subcontractors, may be less able to control results than a larger firm which does most of the work in-house.

The large firm often has the ability to do more field testing or research and development on a product before it is sold. (The small firm is more likely to be using you as a part of their research and development.) The larger consulting firm may be able to offer a range of options which a small firm could never provide. Also, the larger firm can often bring a more diverse experience to bear on your need—they are more likely to have dealt with a situation very close to yours.

The small firm (or independent consultant) may take an assignment which is outside their niche because of their month-to-month billing vulnerability. On the other hand, the large firm may recommend an in-house associate less suited to your needs and requirements because they have to promote their associates.

Stated differently, the small firm is tempted to say, "We can do that," when what they really mean is, "If I burn the midnight oil reading and preparing, I may be able to bluff my way through it." The large firm is tempted to say, "I'm not the one for that, but I have a colleague in our firm who can . . ." when what they really mean is, "You really should call another consulting firm better suited for that than we are, but I'm not about to let potential consulting work get away from us and go down the street." Pick your poison.

I trust you noticed the frequent use of words such as "may," "possibly," and "potentially" in the last few paragraphs. There are very fine external sales trainers and sales training consulting firms (small, medium, and large) which are exceptions to all the stereotypes outlined. My goal is to raise a few cautions and increase your awareness of a few "piles not to step in" (as they say on the farm).

Requests for Proposals: Experience Tells More than a Written Document. Some readers may look to this chapter as a resource for useful guidance on RFPs (Requests for Proposals). I intentionally give the issue limited exposure. There are many fine books on preparing RFPs. However, I will state my bias. RFPs are designed to solicit consulting proposals, all complying with a uniform format set of requirements to aid in selection objectivity. For instance, it keeps the voluminous proposal from automatically winning over the thin one (or vice versa) and the elaborately written document from being a shoe-in over the more blandly crafted one. RFPs need to communicate as much about the "why" as the "what." A successful outcome often has less to do with cost than value. Too often, the seekers of the lowest bid get far less than they pay for in terms of quality growth and durable change. The choice of an external resource should be based more on the deliverer's experience with outcomes than with inflated language of a written document.

Tips on Contracting

The second stem in the entry process is contracting. One of the all-time best references for this phase is Peter Block's *Flawless Consulting*—a must-read for all training professionals.[3] Among his nuggets of wisdom having to do with entry are

- All wants are legitimate. Ask the external sales trainer for all that you need, even if you have some reservations about it's being doable. The dialogue which ensues may result in a rich, realistic solution. Remember you can say no to any aspect which fails to meet your needs. While you don't always get what you want, at least if wants are on the table, you have a much higher chance of negotiating an agreement all are confident will be successful.

- The consulting relationship should be a 50/50 relationship. There is no need for you to feel dependent on the consultant; there is no need for the consultant to feel dependent on you. Watch out if either feeling occurs.

- You cannot contract with someone who is not in the room. While this is intended more for consultants than clients, it applies equally well. Agreements should be made with the person who will actually be doing the work. Watch for the quasi-bait-and-switch in which the silver-tongued partner sells the work and then passes the baton to a junior member of the firm after the ink has dried on the contract.

- Put it in writing! This is not a plea for formality and legalese. Even if your "contract" is a letter outlining your understanding of the arrangement,

---

[3] Ibid.

writing out expectations will aid in reducing ambiguity, and preventing later disappointment from unrealized hopes. Make certain your contract includes starting dates, deadlines, check points, deliverables, criteria for success, role clarification, and boundaries of involvement.

Guidelines for Assessing Cost

The sticky part of most consulting contracts is price—the tab which shows up as a negative variance on your budget. There are several ways to deal with the price problem.

The "Cost-Plus" Method. The simplest way is to agree on a per diem rate (plus routine travel expenses) and then try to scope the project to ascertain an approximate number of days.

All additional expenses are billed and receipted as actual—without overrides. This is typically the format used by independent consultants and small firms. The disadvantage of this method is that, unless there is a package price, the project can run longer than anticipated.

The Total Package Approach. The total package approach is typically arrived at through the same "cost-plus" method, with a few additional twists. Subcontracted work (printing, artwork, video production, etc.) is often added in at an "actual cost plus some percentage" basis. The view is that the effort involved in locating the subcontractor—coordination effort and lost billing time—is "worth" something. The cost is passed on to the buyer. The independent consultant using the cost-plus method sometimes "eats" this cost as the price of doing business. The advantage of the total package approach is that the costs to the buyer are finite and predictable. The disadvantage is that the buyer may not know exactly how much "padding" he or she is paying for. The other disadvantage is the risk that an external consultant (e.g., sales training resource) who has underbid a contract will, in the latter phases, begin to cut corners to recoup losses.

Nine Questions to Ask About Pricing. The pricing game is one which can be an awkward scene. Keep in mind that as the buyer, you are entitled to know what you will be paying for. Aside from the issues impacting your budget you might want to also learn

1. What are the deliverables? What tangible outcomes can the consultant "guarantee"?
2. Who owns the outcome? If the consultant develops a sales training program, will you have the copyright? If not, will you have unlimited reproduction rights? Can you alter a program owned by the consultant without permission from the consultant? Does it matter to you that a product developed for

you but owned by the consultant is marketed to your major competitor? Does it matter to you that the consultant works with your biggest competitor while consulting with you? Remember, all wants are legitimate.

3. How will the end product be evaluated? Will the consultant be involved in helping you create the evaluation criteria and process? Is "satisfaction guaranteed" important to you? If so, how will "satisfaction" be determined?

4. What escape clauses are important to you? What if you get halfway through the effort and discover you selected the wrong resource? Can you get out of the arrangement gracefully? What assurance does the consultant have? If you hired the consultant to teach several sales training sessions, and one is canceled a few days before it was scheduled, will the consultant invoice you anyway? How much notice of cancellation is appropriate? If you cancel too late but the consultant is able to book the time with another client, is it appropriate for you to be invoiced anyway?

5. What is covered in the costs? How do you feel about paying for phone calls, laundry while on site, first-class air fare, administrative overhead incurred by the consultant, rental cars which would exceed the cost of a taxi, taxi expense from the consultant's home to the airport, and so on? What invoice expenses do you want to see receipts for? (The most common practice is all items in excess of $25.00 must be accompanied by a receipt.) How much descriptive detail is required on the invoice? What is the mileage allowance for use of a personal automobile?

6. What happens if the consultant cannot deliver what was agreed? What provisions are made for backup trainers? What if (heaven forbid) the consulting company goes bankrupt during the consultation with you?

7. What provision has been made for the consultant to work with in-house staff? Is a team teaching arrangement appropriate? How will sensitive and/ or confidential information be handled?

8. Who is the client? If you are the client, what happens if the consultant begins to treat your boss as the client? Who can approve the assignment? Is there provision for all who will be involved in the approval process to be a part of the selection effort? If the scope of the project needs to be expanded after it is begun, who will have to sign off on the extra funds required? Are decision makers from the target area involved in the entry effort? You can bolster commitment with early participation by those who will be ultimately involved. Is the person selling the project to you the one who will deliver it? What happens if you experience "bait and switch" during the consultation?

9. You will be hiring a sales training resource. As such, this person will probably be a "supersalesperson." Does it matter to you that the consultant uses your audience to "sell more work" or sell his or her books, tapes, and so

on? What will be your response if, at the end of the consultation, the consultant asks to use you as a reference or asks you to write a "To whom it may concern" letter on his or her behalf?

All the questions appropriate to using an external sales training resource are not included in this chapter. The best way to handle the entry issues is to handle them. The best way to prepare for entry is by networking with colleagues who have been down the same road. Pick their brain on what worked and what didn't. The references at the end of this chapter will be useful in helping you maximize your experience by vicariously walking the same trails as others before you. The most experienced contract negotiator has had that first experience. Find out the lessons they have learned. Figure 27–1 is a letter of agreement from the client which could serve as a contract.

Figure 27–1 Sample letter of agreement

January 15, 19xx

Dear Mr. Boyd:

I enjoyed our time together yesterday and look forward to working with you on the Consumer Sales Academy. This is to outline my understanding of how we will be working together.

You will design a four-day module on effective selling targeted for consumer bank managers. Attached are the module objectives based on the needs analysis we conducted in the Fall. You will teach the pilot program plus conduct a three-day "train-the-trainer" session aimed at preparing our in-house trainers to teach the module you will design. All trainers will observe the pilot program. We will also video tape the pilot for use in preparing trainers.

The deliverables we agreed upon included

1. A detailed study guide to aid trainers in learning the program. I envision a document close to 100 pages single spaced.
2. A trainer's guide from which trainers will teach the program complete with overhead transparencies for use in the lecture portions. Lectures should not exceed about 20 minutes.
3. A camera-ready participant workbook (about 50 pages).

You will spend no more than five days interviewing our consumer bank managers, customers, and supervisory management unless we agree that more is required

Figure 27–1 (Cont.)

for you to design a tailored sales training program. I agreed to send you a product manual, performance review materials, call reports, strategic plan of the consumer division, and a brief bio on the people you will be interviewing.

The cost for this four-day module will be $18,000. While you will retain the copyright, we will have unlimited rights to reproduce and alter the program. You agreed not to market the program to any of our major competitors (see attached list) for one calendar year from the date of the pilot. In addition to the program costs, you will invoice for the five days of interview, four days for the pilot program, and three days for the train-the-trainer work at $1,250 per day plus routine travel expenses. All air travel will be coach class and will be apportioned if another consulting assignment is included on the same air receipt. Automobile travel should be invoiced at 25 cents per mile. Your invoice will have receipts for any travel expense which exceeds $25. Consulting days canceled by us less than 10 working days prior to their planned occurrence should be billed to us as if consulting had occurred.

We are excited about our work together. We anticipate beginning this project on Monday, February 8, with the pilot program scheduled for the first week in April. You should plan for the entire project to be completed by the end of April.

Boyd, please give me a call if this fails to reflect our understanding. Otherwise, sign at the bottom and return the original to me.

Hope you enjoyed the book I sent you! Let's plan to have dinner on the evening of February 8.

Sincerely,

Nancy Marie Rainey, VP
Director of HRD
NMR/sk
enclosures

_____
                                                    Harold Boyd

_____
                                                    Date

PHASE 2: DIAGNOSIS—EXAMINING THE PROBLEM

We began this chapter with the presumption that some diagnosis or assessment (i.e., needs analysis) has occurred which precipitated the request for sales training. Even the most thorough diagnosis will still require some assessment by the external sales training resource. While the principal focus may not be ascertaining need, familiarity with the culture is acquired via need diagnosis. It is very important that you provide the sales training resource with access to the "movers and shakers" of your organization. Their association can be beneficial in creating or implementing a sales training effort tailored to your needs.

Interview Key People

Most sales training consultants will want to interview key people who impact selling effectiveness. Arrange for the sales training resource to acquire information on

1. The products and services sold by the typical participant in the sales training program.
2. The aids, supports, and tools used by sales practitioners in their selling efforts.
3. Information on how sales practitioners are selected and trained for their role.
4. The manner in which sales practitioners are managed, coached, rewarded, and supervised.
5. Information about the competition and how they impact selling effectiveness.
6. Market research data regarding the demographics and psychographics on the target market.
7. The typical objections encountered by sales practitioners and information important in overcoming resistance.
8. Data related to calls per closings, the life cycle of a typical sales call, call reports, performance reviews of sales practitioners, and marketing plans for units involved in sales training.

It is very important that your hired resource enter the design and/or delivery effort armed with all the information needed for success. While the "criteria for success" vary with company and project, at a minimum they are those data which the external resource needs to gain internal insight or "stand in the shoes of the participant." Don't be concerned with data overload. Let that be the problem of the consultant. If the consultant reaches a point of saturation, he or she will (should) cry "uncle."

Most external resources experience some resistance from the "keepers of knowledge" within the organization. While those effective in their roles as data collectors will be adroit in overcoming resistance, even the most expert can benefit from your empowerment. Let those interviewed know why the external sales training resource will be contacting them. Encourage the interviewee to be open and candid in their response to questions. All benefit from a fully informed external resource. Sometimes a "to all employeees" memo from an executive in "mahogany row" can help pave the way to introducing an external resource, their rationale for being in the organization, and strong encouragement for candor in their data collection and supportiveness in their efforts.

PHASE 3: RESPONSE—THE ACTION PART OF THE CONSULTING PROCESS

Peter Block refers to "Response" as the main event, but he is quick to acknowledge that its implementation is specific to the situation and the expertise of the consultant. There are, however, several key factors to consider in the response phase of consulting.

1. Contract negotiation is a never-ending process. Characteristic of most consulting relationships is the "moving target" dimension. Needs, wants, and expectations change as the consultation unfolds. Therefore, it is incumbent on both client and external sales trainer to be flexible in their relationship and open in their expression of requirements. Remember, the relationship should be 50/50.

2. Data collection is also a never-ending effort. As talents change, the effective external resource is perpetually adjusting his or her response to be in sync with the target audience and their needs.

3. The professional external sales training resource remembers that sales training is a means to an end, not the end. What is of preeminent importance is that successful selling occurs—resulting in a positive impact on the bottom line of the organization. Naive sales trainers often get caught up in the activity and lose sight of the purpose or mission of sales training. Such myopia can result in the tail wagging the dog—all attention given to sales training rather than sales results.

4. There is an old adage that goes, "It is easier to turn a mule after the mule is moving." This pithy aphorism has its meaning in consulting in the following way—start (and accept) where the client is and get them

moving in the direction of improvement. Turning the client (i.e., changing the client) will come easier if you have gained his or her credibility by acknowledging (not necessarily accepting) their reality. The powerful consultant is the one who begins where the client is and helps them move as fast and as far as they are capable of moving toward effectiveness and productivity. Such a consultant (external sales training resource) is vitally interested in improvement—helping people be strong rather than just feeling better about being weak.[4]

A part of your job during the response phase is to monitor the training process to make certain the effort is proceeding according to the plan or contract. This includes ensuring that the effort stays within the boundaries established at the outset or "renegotiated" during the consultation.

While few consultants are easily controlled or coerced, most are capable of being directed and managed. Provide early warning on major changes. Keep the external sales training resource up to speed on what's happening within the organization. Send copies of strategic and "to all employee" memos. Make certain they are on the mailing list for the company newsletter or house organ. Brief them on key meetings which impact their work; invite them to attend discussions related to their work. Continually reassess the focus of the consultation to ensure it is on target with the need. Examine ways to internalize the results and offload the consultant's expertise to others within the organization. Managing the response or intervention can be simple and painless if you have negotiated a set of agreements up front and then keep your agreements. Here are several tips which may be useful in the response phase. While some are written from the perspective of the external sales trainer, all are important to a successful client-consultant relationship.

1. Most often resistance is due to relationship issues rather than the task or content being discussed.

2. Change is least likely to occur when there is defensiveness.

3. Do not fight battles which cannot be won, nor battles which should not exist. Effective consultants are less warrior and more mirror, focusing on facilitating growth not on drawing defense lines or winning arguments.

4. Insistence upon "being right" is self-defeating; aim for what works.

5. The client's enthusiasm often gets dissipated by the consultant's wisdom.

6. Some sales problems can be corrected through sales training alone, but very few.

---

[4] Chip R. Bell, "Building a Reputation for Training Effectiveness," *Training and Development Journal*, 38(5) (May 1984), p. 50.

7. Work in the most promising arena. If the probability of success is less than 50 percent, avoid it if you can; delay if you must.

8. Authority is the last resort of the inept.

9. You are only eligible to change another's view if they believe you understand their view.

10. As the late Herb Shepard said, "Innovation requires a good idea, initiative, and a few friends. Light many fires. Keep an optimistic bias and capture the moment."

PHASE 4: CLOSURE—EVALUATION AND FOLLOW-THROUGH

In consulting, there is too much wooing and wowing and not enough weaning and walking. So much energy goes into getting the project started; so little devoted to getting it ended. The primary tasks in the closure phase of the project are to evaluate the effort, plan for continuity and follow-through, celebrate successes, and disengage from the external sales trainer.

Evaluation is the process of judging what worked or did not work when compared to the expectations and objectives outlined in the entry phase. The "look back to learn forward" is critical if follow-through is to be focused and pragmatic. The evaluation process varies with each project. Some evaluations are only at the reaction level. ("Did participants enjoy the training program and believe it was potentially useful to them.") Other evaluations seek to tie sales training programs to sales results. Keep in mind that the external sales trainer was hired to design and/or teach a learning program. Holding them accountable for bottom-line impact may not be fair since the training was only one of many variables affecting performance and results.

Tony Putman, in an article entitled "Pragmatic Evaluation," outlines several questions which may be useful in approaching the evaluation of the sales training effort.[5]

1. What decisions are going to be made or questions answered by conducting an evaluation?

2. Who will be making the decisions (or answering the questions)?

3. What type of information counts (has credibility) with the person identified in question 2?

---

[5] Anthony O. Putman, "Pragmatic Evaluation," *Training and Development Journal,* 34(10) (October 1980), pp. 36–48.

4. What and how will you evaluate to get this type of information?

5. What are the resources needed to get the information?

6. What are the constraints which need to be overcome to get the information?

7. How will the information be presented to aid decision making?

The largest issue in closure is the disengagement—"weaning and walking." If the project has been successful, celebrate and then focus on parting ways. Let the external sales trainer know how they contributed to the success of the project. The critique should also include the "If we were to do it again, this is how it could be improved" focus. Select some means to telegraph the ending. These might include a final report, memos to those involved thanking them for their contribution, or announcement of an in-house person who will be assuming primary teaching responsibility for the sales training program.

We began this chapter by examining the consulting process and its use in engaging the assistance of an external sales trainer. Success will come if sales training is based on an accurately assessed need; if the consulting relationship is based on candor, authenticity, and clear agreements; and if there are actions taken to ensure support and follow-through for the sales training effort.

_____ **REFERENCES**

Bell, Chip R. *Influencing: Marketing the Ideas That Matter*. San Diego, CA: University Associates, 1982.

Bell, Chip R., and Leonard Nadler. *Clients and Consultants*. Houston: Gulf, 1985.

Block, Peter. *Flawless Consulting*. San Diego, CA: University Associates, 1981.

Kilmann, Ralph H. *Beyond the Quick Fix*. San Francisco: Jossey-Bass, 1984.

Ginsburg, Sigmund G. "Selecting and Managing Management Consultants." *Training and Development Journal* (January 1983), pp. 76–80.

Gorovitz, Elizabeth S. "Consultants: The Buyer's View," *Training and Development Journal* (January 1983), pp. 67–68.

Steele, Fritz. *Consulting for Organizational Change*. Amherst: University of Massachusetts Press, 1975.

Zemke, Ron. "How Training's Buyers and Sellers See Each Other," *Training Magazine* (August 1983), pp. 20–26.

chapter twenty-eight

WHEN AND HOW TO USE PACKAGED SALES TRAINING PROGRAMS

RICHARD C. McCULLOUGH

Richard C. McCullough is executive director of the Instructional Systems Association, a trade association of more than 80 firms which produces generic and custom-designed training packages. He has been in the training field about 35 years, for a long while with the federal government in positions such as assistant director for training, U.S. Office of Personnel Management; deputy director of the bureau of training, U.S. Civil Service Commission; and director of training, Internal Revenue Service. After leaving the government, he was vice president for professional development with the American Society for Training and Development, where he was staff project director for the ASTD Competency Study. He received his Master of Arts degree from The George Washington University with a specialty in Employee Training.

The objectives of the chapter are to

- Discuss how sales trainers decide whether it is better to develop a training program internally or to purchase it from outside sources.
- Explore the resources that can help the sales trainer find the packaged sales training programs which might meet the company's need.
- Identify the critical points sales trainers need to consider when deciding among various vendors and their products.
- Suggest ways to establish mutually productive working relationships between you, the sales trainer, and the vendor of your choice.

It is *not* the purpose of this chapter to "sell" packaged training programs. It is to present another option sales trainers should consider in meeting the challenges of their jobs.

THE TREND TOWARD PACKAGED SALES TRAINING PROGRAMS

An ever-increasing amount of sales training is being conducted through the use of packaged training programs. It is estimated that there are more than 1,000 major suppliers of such programs in the United States,[1] many of whom have programs specifically designed to help those in the sales training business.

By packaged training programs we mean a session, course, or entire program developed by a vendor for the purpose of providing training for employees within

---

[1] Vincent W. Hope, Mary B. Hope, and Thomas W. Hope, "Packages and Seminars," in *Training and Development Handbook*, 3rd ed., ed. Robert L. Craig (New York: McGraw-Hill, 1987), p. 828.

a business or company. They can be as simple as a one-hour film with leader's guide or as elaborate as a lengthy, multimedia program requiring extensive training and certification of instructors. Many are generic or "off-the-shelf" programs that can be used within a company with little or no tailoring. Others are custom designed for a particular company to meet their specific needs. In recent years there seems to be a trend toward increased customization to the point where annual expenditures have reached about $1.04 billion each for generic and custom-designed programs.[2]

The "training industry" which produces both generic and custom-designed programs has had a remarkable history of growth. Before 1970 there were fewer than 100 firms in the packaged training business. As the interest in training and development has grown over the past two decades, so has the use of packaged programs. It is estimated that by 1990 a total of $4.6 billion will be spent on training *not* developed in-house but rather purchased from outside sources.[3] Much of this training will be in the form of packaged training programs.

In addition to being varied in their complexity, length, and type, packaged training programs also vary in the subject areas they cover, from executive development to welding. Within the area of our interest—sales training—a listing of some of the titles of packaged programs available on the market today should illustrate the breadth of offerings:

"Consultation Selling Skills"

"Courtesy, Sales and Service"

"Face-to-Face Selling Skills"

"Interpersonal Selling Strategies"

"Managing Exceptional Sales Performance"

"Negotiating to Yes"

"Professional Telephone Selling Skills"

"Situational Selling"

"Successful Cold Call Selling"

"The Versatile Salesperson"

These represent only a small fraction of the generic programs available in the sales training area. When you add to the generic programs the custom-designed programs tailored to meet specific company needs, there is no limit to the types of sales training packages available for use.

---

[2] Beverly Geber, "Training Budgets Still Healthy," *Training*, 24(10) (October 1987), p. 40.

[3] Robert L. Craig, "The Future of HRD: It's Going to Get Tougher," *Public Personnel Management*, 15(4) (Winter 1986), p. 470.

GUIDELINES FOR MAKING YOUR BUILD OR BUY DECISION _____

The title of this section is absolutely the wrong place to begin, as any good trainer is aware. Without belaboring the obvious, good training practice tells us we should start a few steps back.

Most "sales training problems" uncovered by line or staff managers and brought to your attention are not susceptible to a training solution. Therefore, either building or buying a training program to solve them would be a mistake.

While now nearly 20 years old, Mager's little paperback of analyzing performance problems[4] is still an extremely useful tool in helping us to find the real causes for problems on the job. Step-by-step he asks you to

- Describe in detail exactly what is, or is not, happening that you feel is causing a problem.
- Estimate its importance.
- Determine whether it is caused by employees not knowing how to do what is expected of them.

And then, and only if the answers gotten so far warrant it.

- Decide what sort of training effort is needed.
- Establish measures for determining the effectiveness of the training.

This is all very basic, but sometimes we forget.

So, if a situation exists where, through careful analysis, you have determined that training is required, then there are a number of critical considerations that need addressing before you decide whether it is better to build or buy.

Cost: A Relative Matter

Packaged training programs range in price from a few hundred dollars for a generic session of a couple of hours on video tape with a few student handouts to many thousands of dollars for a customized series of courses using a variety of media and requiring extensive instructor preparation.

Cost is a relative matter influenced by each company's particular circumstances. If sales trainers can find a packaged program that truly meets their organization's identified needs, usually it will be less costly to buy it from an outside vendor than to develop it themselves. And some believe that "even

---

[4] Robert F. Mager and Peter Pipe, *Analyzing Performance Problems or 'You Really Oughta Wanna'* (Belmont, CA: Fearon, 1970).

'tailoring' these 'packaged' programs to meet special internal needs can be more efficient than spending money to develop the whole program from scratch."[5]

While it is true that the buyer pays for the vendor's generic program developmental costs, this is only a portion of the total, since the entire cost is spread among all of the purchasers. In addition, producing printed material and media in volume is far less costly than if it had been done for a single course or program.

Another advantage in purchasing a packaged program is that you know the total cost in advance. Programs developed internally have a lot of hidden costs (administrative overhead, employee benefits). Disregarding these costs, there is a mistaken idea sometimes that in-house programs are "free."

On the other hand, any money spent on a sales training packaged program that does not meet the objectives for which it was purchased is wasted.

Budget: Know Your Company's Limits

Aside from actual cost, another consideration in deciding on whether to make or buy is the manner in which your training money is budgeted. This will differ from company to company.

With "downsizing" there may be cases of fewer on-board staff to do the work and more money for contracting outside for services and products. It is often thought that during tight money situations one of the first restrictions is on outside training expenditures. The fact is that sales of generic training programs increased considerably during the 1980 and 1982 recessions.[6]

The bottom line is to know the budget policy in your particular organization and to act accordingly. But, as mentioned before, training managers and others should not delude themselves into believing that developing a sales training program using internal staff and facilities does not impact on the budget. The company pays, one way or another.

Time: Know Your Constraints

Obviously, if a generic sales training packaged program is found that meets a company's need and there is no problem finding money in the budget to pay for it, a great amount of time can be saved by buying rather than making. Course development, particularly if it involves media production, is extremely time consuming. There are a few cautions.

In some organizations, particularly agencies of the federal government,

---

[5] Craig, "The Future of HRD," p. 470.

[6] Hope, "Packages and Seminars," p. 831.

the procurement process can take many months. Announcements must be placed in the *Federal Register*. Requests for proposals are made. These must be carefully evaluated before a decision is reached. Even then, nonsuccessful bidders can instigate an appeal if they feel their proposal had not been given fair treatment. While not as complicated and time consuming, some large corporations have adopted similar procedures.

If any changing or tailoring of the generic training package is required, this will lessen the advantage gained from purchasing outside. This is particularly true if media changes are involved. If time is a critical consideration, careful thought should be given before requests for any changes are sought.

And, as was said concerning costs, if the package program does not fully meet the organization's objectives for which it was purchased, then getting it quickly is of no advantage.

Expertise and Experience: Judge For Yourself

This is another factor that needs careful consideration in light of what you know about your own organization. Some company's sales training emphasizes product or service knowledge while others put their emphasis on selling skills. In a recent study[7] the average time spent in sales training courses was divided among product information (35 percent), sales techniques (30 percent), market information (15 percent), company information (10 percent), and other (10 percent).

If the sales training program you need requires a great deal of product and company knowledge, it might be better for the program to be developed by those within the company who have that expertise and experience. However, it is possible to provide that knowledge to the vendor for incorporation into the packaged course.

If the sales training program is to emphasize selling skills rather than specific product or company knowledge, then the vendor who has specialized in producing that type of sales training program probably has more experience and expertise than even your most knowledgeable employee. They also have the experience and expertise for putting their knowledge into a training mode.

There is no clear-cut answer to where the type of expertise and experience needed for your particular situation exists. It will depend on the objectives of the course. If they are broad and general, the generic package may well fit the bill. If they are very product specific, then a customized package program, or one internally developed, may be more appropriate. In either case, it is a matter that needs careful thought.

---

[7] Earl D. Honeycutt, Jr., Clyde E. Harris, Jr., and Stephen B. Castleberry, "Sales Training: A Status Report," *Training and Development Journal*, 41(5) (May 1987), p. 43.

Quality of Program Materials

This is an area where the outside vendor usually has an advantage over the in-house training staff, particularly if media is involved. By producing sales training materials for sale to many companies, the vendor can afford to invest a great deal in films, video, student materials, instructor guides, and packaging. The typical internal training shop does not have this luxury.

If a package training program has been on the market for awhile, it also has the advantage of being thoroughly field tested. A commercially successful sales training package will already have the "bugs" worked out of it before you bring it in to your company, thus saving you the potential embarrassment of "having to go back to the drawing board." Remember that the packaged training industry is highly competitive. Sales training programs produced by vendors have to be of a high quality to stay on the market for any length of time.

There is a downside to this. Once vendors have invested considerable money in producing a particular training package, and it has been well received by their clients, it is unlikely that you can get them to make major changes in their content or approach. This is especially true when high-priced instructional media are involved. For example, it should not be expected that a vendor would want to make any changes in an interactive video-based sales training course to meet the needs of a single client. And if willing to do so, it would be at great expense to the client.

Because the vendor has the advantages of the latest in packaging techniques, you need to look beyond all the "bells and whistles" to the objectives of the program, the content being presented, and the instructional design being used. Be as critical of these polished vendor-produced programs as you would the ones produced in-house without the fancy trimmings.

And, while not dealing specifically with quality of materials, another advantage of purchasing materials from a vendor is that it can be done on an "as-needed" basis. You, the user, do not have to provide the storage space for large amounts of training materials. You do not have to worry about materials becoming obsolete. Often, when material is developed internally, there is a desire to produce in quantity to reduce costs. This leads to problems of storage and obsolesence.

Audience

Like other considerations, the nature of the individuals for whom the training is directed should be looked at when a make or buy decision is needed. Certainly, if your potential number of students is very small, the cost of an expensive packaged program would be hard to justify. It would be better to find a good public seminar or program with objectives as similar as possible to yours in which to enroll your few participants. Their training could be augmented with company-specific information prior to or after their attendance. Just as there are many fine packaged programs, there are many good public offerings on sales topics by academic institutions, professional associations, independent consultants, and by some of the same vendors who produce packaged programs. You should take the same care in choosing one of these programs as you would a packaged one.

If your potential number of students is large and easily assembled, a packaged approach makes more sense. Generally speaking, the larger the number of users of a program, the smaller the cost per unit. Many vendors give sizable discounts for volume purchases and for materials which will be repurchased over time. However, before entering into any long-term contracts, check to be sure there are no major internal changes anticipated that might obsolete the training materials you are considering.

If your audience is large, but widely separated geographically as individuals or in small groups, a packaged program with self-study or "stand-alone" qualities would seem most appropriate, whether purchased from a vendor or developed in-house. There seems to be a trend in packaged programs to make them less instructor dependent.[8] This provides more flexibility in administration, but less opportunity for exchange and exploration of ideas among students.

Company Culture

Companies have different histories, mores, and customs. In some, there is a belief that unless something was conceived, developed, and produced within the company, by company employees, then it cannot be worth very much. Some very large organizations often have the feeling that they are such leaders in the field that no one else could possibly teach them anything.

In other companies there is the reverse, particularly in regards to training and development. In these firms, line sales managers have a hard time believing that anything worthwhile can come from the company's internal sales training group. They are usually looking for a "name," or a "guru," or the latest "star" in the selling field to come up with the answers to the company's problems. They are often willing to spend thousands of dollars for a one-hour talk on the latest selling technique by the current superstar on the lecture circuit.

Astute sales training managers learn to deal with the culture prevalent in their organization. In the first instance, any vendor being considered must be able to demonstrate that their product is fully compatible with the way things are done within the company. The sales approach presented must be one that fits and builds on what is already viewed within the company as successful and "our way of doing things."

---

[8] H. Stanley Connell III, "Sales Programs," in *The Handbook of Human Resource Development,* ed. Leonard Nadler (New York: John Wiley, 1984), p. 14.12.

In the latter case, the sales trainer needs to try to get management to apply the same thoughtful analysis to any outside proposal in the training area as they would to other management decisions (i.e., what is the performance problem facing us, how important is it, are we sure it is a deficiency that can be helped by training, how will we measure to determine if any change occurred as a result of the event or activity being considered). It is always a good idea for the trainer to have alternative ways to meet a perceived need and encourage the line managers to consider them carefully before making a decision.

This may sound a bit Machiavellian, but in a climate where failures are not tolerated and yet there is a desire to try a risky training venture, it may be to the advantage of the sales training manager to encourage the purchase from outside just in case it does not produce the results expected. The vendor can go on to another client; the internal sales training staff has to live with their mistakes.

These are a few of the factors to be considered when making a decision whether it is better to build a sales training program or course internally or to go on the outside market and purchase one. It is not a black and white decision. There is no consistently right way to go. Companies differ greatly, circumstances change rapidly, demands come and go. Your job as sales trainer is to weigh the options—make it yourselves, buy a generic package, contract for a customized package—and then thoughtfully make the decision that appears the best for your company at that particular point in time.

IDENTIFYING PACKAGED PROGRAMS FOR SALES TRAINING

Let us assume that you have decided that a sales training effort is required to meet specific needs within your company, and further, you have determined that the best approach to be taken is to purchase such a program from an outside source. The next job you have is to identify the vendors and programs available for your consideration. There are various resources you can call upon.

Word of Mouth

Probably the best way to determine what really good packaged programs are available on the market today is to "ask the man who owns one." Through contacts in your local or national professional associations you should query your fellow sales trainers as to what programs they have used successfully and what vendors have they had the most success with.

The American Society for Training and Development has a Professional Practice Area of 2,500 members devoted entirely to sales and marketing training and the National Society of Sales Training Executives is made up of 150 individuals who head the sales training activities of their respective companies. Being a part of these networking groups enables you to get firsthand, practical, evaluative information about packaged training programs and the vendors who produce them.

Publications

Another way of identifying potential sales training programs to meet your company's needs is through specialized publications. For example,

The Trainer's Resource: A Comprehensive Guide to Packaged Training Programs
Human Resource Development Press
22 Amherst Road
Amherst, MA 01022
(413) 253-3488

This annual publication has been in existence since 1983. The 1988 edition lists nearly 500 different packaged programs available for purchase or lease from some 120 vendors. For the sales area, the programs are divided into two major categories: sales management with 9 programs listed and sales training with 52 programs listed. For each packaged program, the one-page entry includes

Audience. Describes the participants for whom the pack was developed (e.g., sales managers, new sales employees, experienced sales representatives in the medical or pharmaceutical field).

Objectives. What the program is intended to accomplish and the students learn.

Topics. Subjects covered by the material.

Delivery system. Length of program, format, number of participants, media used, and materials provided.

Instructional strategies. Instructional methods employed and train-the-trainer requirements.

Recent users. When the program was first introduced and the names of companies who have used it.

Cost. Includes the cost of purchase, rental, preview, and instructor training, if appropriate.

Contact. The name, address, and telephone number of the vendor and the specific name of the contact person to be called for more information.

The Marketplace Directory
Training magazine
Lakewood Publishers
50 South Ninth Street
Minneapolis, MN 55402
(612) 333-0471

Made available free to subscribers of *Training* magazine (and for $15 for nonsubscribers), this annual directory lists over 1,500 firms that produce products and programs for the training and development field. One section of the directory deals specifically with prepackaged, off-the-shelf, and customized programs.

In the 1987–88 edition, 141 firms indicated they could provide packaged programs and seminars in the sales management area, and 326 indicated they could do the same for the sales training area. Each company's entry is keyed to indicate off-the-shelf programs, consulting and custom-designed programs, and the media used.

ASTD Buyers' Guide & Consultant Directory
American Society for Training and Development
1630 Duke Street
Alexandria, VA 22313
(703) 683-8100

This directory is made available free each year to the national members of ASTD (and for $35 for nonmembers). The 1987 edition lists 83 firms who provide programs and services in "sales management" and 59 in "sales marketing." Individual entries for these firms provide a short description of the firm and the type of services and products they specialize in.

Directory of Member Firms
Instructional Systems Association
10963 Deborah Drive
Potomac, MD 20854
(301) 983-0783

ISA is a trade association of more than 80 firms which develop and market a wide variety of generic and custom-designed training programs. This directory describes the member firms, their size, how long they have been in business, their products, their services, and their clients by major industry groups. The index of the 1987 edition lists 34 member firms that specialize in training at sales and sales management skills.

By careful use of these four publications, the sales trainer can identify potential vendors of sales training packaged programs and begin the process of narrowing in on the few who may have the product or services to meet their needs.

Conferences and Trade Shows

Another rich source of information on sales training packaged programs is through attendance at professional conferences and their accompanying exhibitions. Many of the firms listed in the publications just referred to exhibit at trade shows associated with conferences like those sponsored by ASTD and Lakewood Conferences (*Training* magazine).

In each of these conferences, there are "tracks" developed specifically to sales training. At "Training '87" sponsored by Lakewood Conferences, the "Sales Management and Sales Training/Customer Service" track included 11 three-hour seminars. At the 1987 ASTD National Conference, its sales and marketing track included 18 concurrent sessions on sales training matters. Some of these sessions are conducted by senior members of companies in the packaged training business. It is an opportunity for you to see and hear them in person.

Also, when you are attending sessions such as these, the majority of your fellow participants are likely to be in the sales training business like yourself. This is a wonderful way to meet other sales trainers and expand your network. It is also a good opportunity to find out what sales training packaged programs are being used by your peers and with what success.

Most of the major packaged training vendors exhibit at one or both of these national conferences. Here is the opportunity for you to see sales training packages firsthand and to compare the programs of several companies. If you see something of interest, do not hesitate to ask for a more in-depth discussion of the product than can be had on the busy exhibit floor. Appointments can be made for discussions with vendor representatives at times the exhibition is not in operation or arrangements can be made for later meetings at your place of business.

By careful planning, attendance at professional conferences and trade shows can be extremely valuable in assembling information on the availability, quality, and cost of sales training packaged programs.

Advertising and Direct Mail

Vendors spend countless dollars each year to sell us on their packaged programs. Through advertisements in the trade journals (e.g., *Training* magazine, *Training and Development Journal, Training News*) and through mail sent to your home or office, they try to get the message across that their's is the package to meet your training needs. Do not overlook these as another source of information.

We often skip over ads in magazines and quickly discard the "junk mail" that seems to inundate us daily, but, if you have a sales training need identified and are clear on the objectives to be met, you can quickly and critically look at these materials as leads for you to follow up on. Some sales trainers keep a file of those ads or brochures that look the most promising for future use, even though they may not be in the current market for a packaged program.

By using several sources for identifying sales training packaged programs you decrease the chances you will be taken in by "the first pretty face to come along" and increase the possibility of finding the perfect fit between your company's sales training needs and a packaged program's proven solutions.

CHOOSING THE RIGHT SALES TRAINING PACKAGED PROGRAM

Having made the decision that your company's sales training needs can be best met by purchasing a packaged program from an outside vendor rather than developing within the company, and having identified several sources for such programs, now you need to evaluate the ones available to determine which to choose. Many of the considerations that came into play during your make or buy decision process will again be important.

Objectives

Foremost, you must keep in mind that you have identified needs for which training has been determined is the solution. Any program you eventually purchase must be based on learning objectives that will meet those needs. In other words, the stated objectives of the packaged sales training program you are considering must be compatible with your company's needs. Regardless of how well organized, designed, packaged, media supported, and performance proven a packaged program might be, unless it is aimed directly at solving your particular problem then it should not be chosen.

Always keep in mind *your* objectives. Ask vendors to provide specifics as to how their program will meet your objectives. Ask to see evidence that they have been successful in meeting similar objectives in the past. This evidence can be in the form of evaluation results or testimonials from former clients.

Cost Considerations

In discussing the cost of a program with vendors, be sure to determine the complete cost your company will incur. The cost includes such items as student workbooks, instructor guides, handouts, supplementary texts, films, video tapes, other support media, instructor training, pilot programs conducted by the vendor, and shipping charges.

If not totally generic, the development costs of a customized program are likely to be the largest single cost items in a major program effort. Building a program from scratch, or customizing an existing program, to meet the specific needs of a company may be the best way to go, but it is also expensive.

You probably already have the standard audio/visual equipment needed to conduct most programs (e.g., 35mm slide projectors, 16mm film projectors, overhead projectors, flipcharts), but, if the program being considered requires equipment not already in-house (e.g., disk players, computers, individual audio cassettes, rear screen projection), then the cost of purchasing or leasing the needed equipment must be factored in.

Regarding printed materials, you need to know what your reproduction rights are. Some vendors, particularly those providing custom-designed programs, give the customer complete reprinting authority. Other vendors give limited authority for a specified number of copies to be reproduced for internal training purposes only. And yet other vendors retain tight control over all materials and will permit no reproduction of anything, even for file copy purposes.

Vendors have a perfect right to protect their materials from unauthorized reproduction, and many of them are becoming more assertive of this right. The Training Media Distributors Association commissioned a study of the attitudes and practices of trainers concerning copyrighted material.[9] It showed that 34.6 percent of the organizations in the United States with 50 or more employees that use prerecorded videocassettes are copyright violators and that 75 percent believe that all printed materials provided with a packaged training program should be reprintable without additional cost. TMDA has an active educational and enforcement program for ensuring the proper usage of their members' products in light of the current copyright laws in our country.

You need to be clear as to what your rights are to reprint or reproduce items included with the packaged sales training programs you are considering. It will have a considerable impact on the eventual cost of your purchase.

Some vendors require that any one who will present their training packages be certified as an instructor. This is an essential part of some programs. If a train-the-trainer component is needed, be sure to factor in its prime cost (materials, vendor-provided instruction, facilities, etc.) and any secondary costs (travel and per diem for trainees).

Frequently, the nature of the packaged program requires that the vendor supply the instructors, or possibly your company is not able to supply them. This can be a significant expense; however, when weighed against taking a sales trainer off other duties or an in-house account executive out of production, it may be the most cost beneficial way to go. It sometimes comes down to a

---

[9] *Attitudes and Practices Concerning Pre-Recorded Videocassettes and Current Copyright Law in Corporate America* (Minneapolis, MN: Lakewood Research, 1987), p. 43.

trade-off between the need for expert skill in presenting the course material versus an in-depth knowledge of the company and product.

Take time to investigate the total costs of the programs you are considering, immediate and long range; be careful of hidden costs; weigh your options like discounts for volume purchases, in-house versus vendor-supplied instructors, and media alternatives; and then decide which program offers the most at the best price.

Time Considerations

Often, time can be the most critical factor. Management sometimes uncovers a crisis that needs correcting yesterday, or a project is nearly ready for implementation without allowing for the necessary sales force training. In such cases, a deadline date may be the deciding factor when choosing between two vendors' packaged programs. But even in more normal times, how long it will take to get a program into operation is an important consideration.

You should have a firm date in mind as to when it is essential for your sales force to have completed its first training on the newly purchased package. This allows you and prospective vendors to back away from that date and determine a time schedule to meet your needs.

Obviously generic programs are more readily available and can be delivered in a shorter period of time than can those that are custom-designed for your company. But even with a generic program a timetable needs to be worked out with the vendor allowing for necessary production, delivery, instructor preparation, prereading by students, purchase or rental of audio/visual equipment, and so on.

While custom-designed programs do require more time, it may not be as much more as you might imagine. Vendors in the custom business have a good grasp, based on experience, of how long it will take to develop a particular type of package and can provide accurate estimates of delivery time. Unlike most internal training shops, outside vendors have more flexibility in hiring additional employees to meet a deadline. They also can contract out portions of the work, particularly the production stages.

The important thing is to know your time constraints and to be clear about them with your prospective vendors.

Quality of Product: A Three-Part Consideration

This is really a three-part consideration: content, instructional methodology, and packaging. Determining ahead of time the quality of custom-designed programs is more difficult than with generic packages. With the latter you can review the objectives of the program and evaluate whether the content covers them. You, and other subject matter experts from your company, can carefully examine the materials before deciding on one vendor's program or another's to see if the content is compatible with the needs of your company.

With custom-designed material, you have to review materials the vendors have produced for other customers and decide whether the content is adequate for the stated objectives.

Next you need to evaluate the instructional technology used in the packaged program, whether custom-designed or generic. We certainly know enough today to realize that a combination of methods and techniques are essential to effective training. There needs to be a proper balance of lecture, discussion, interaction, practice, and feedback to meet the requirements of the subject matter, the differences in learning styles of various students, and the need to produce specifically desired behavioral changes. When comparing various packages, see which ones offer the best variety that will lead to the most likely positive results.

And, while "you can't tell a book by its cover," the quality of the packaging of a training program can be an indication of its overall quality, and often the degree of initial acceptance by the participants. Materials should be attractive, easy to read and understand, and user friendly.

In considering quality, you should never be satisfied with hearing about the product or with reading brochures describing them. Insist on reviewing actual materials, either at your site or in the vendor's place of business. More and more, vendors are charging a review fee. This is money well spent.

Expertise and Experience Considerations

In a general sense, the reputation that a vendor has earned over a period of time is the most important factor to consider. A good track record of accomplishment is the best indicator of future positive results. This is not to say that there are not a number of fine, new packaged training companies who have come on the scene in the last two or three years, but, in a field where customer satisfaction is such a critical issue, those vendors who have been around for ten or more years must be doing something right.

If you believe that your particular sales training situation needs specific product or industry knowledge, you should be sure to find out if the vendors you are considering have had that type of experience. There are some vendors who have limited their market by serving only the banking, or pharmaceutical, or automobile industries. If you can find a vendor who has had experience in your industry, that certainly is a mark in their favor.

If you are considering a customized training package, you may wish to meet those training analysts and developers who would be working on your program. In this way you can judge for yourself the level of their expertise and experience. This is also a way to find out if the vendor will be using in-house staff or will be subcontracting some of the work.

FORMER CLIENT SATISFACTION ————————————————————

Every vendor will provide you with information on how well its programs have met the sales training needs of other firms. These may be in the form of statistical analyses indicating postprogram increased sales, higher level of renewals, fewer customer complaints, and less customer turnover. All these will be directly attributable to the packaged training the firm had purchased from the vendor.

You are also likely to receive testimonial letters from previous clients expressing their satisfaction with program materials, administration, and results; and with the productive relationship between them and the vendor. All this information should be reviewed carefully.

If not offered, you should ask for the names and telephone numbers of contacts within the firms who have used the vendor's programs recently. These individuals should be contacted to verify the facts you have been told and to substantiate or refute your impressions of the vendor and their product. After gaining as much as you can from the contact provided by the vendor, ask that contact for the names and numbers of in-house instructors who had taught the program and of several participants in recent programs. Tell the contact that you would like to talk with these people if it is all right with the company.

Also, ask around at conferences, professional society meetings, and industry get-togethers to gather additional data on the vendors you are considering and their particular sales training packaged programs.

If it can be done, it is very useful to go beyond the references provided by the vendors.

"Gut Reaction"

While hard to describe, another consideration is your visceral reaction to those representing the vendor and to the materials presented to you. While very subjective, how you feel about the fit between your company's needs and the vendor's product can be very meaningful.

You are in the best position to know your organization and how its members react to certain people and to certain ways of presenting material. Even if successful somewhere else, you may sense that your company would have difficulty with certain concepts and approaches. You should listen to those feelings and check them out with other knowledgeable persons in your organization.

Training Package Checklist

When choosing one sales training packaged program over another, many considerations need to be taken into account—too many for most people to keep in mind without help. Some people find a checklist is helpful.[10] A sample checklist is shown in Figure 28–1.

In Column A of that figure are listed the seven considerations in choosing the right training package for your situation. Under each consideration there are a few words to remind the user of what is included. There is also room at the end of the column to add considerations overlooked in this chapter.

The importance of each consideration will differ company to company. In some, cost may be the overriding factor, in others it may be time, and in still others it might be the level of experience and expertise possessed by the vendor. Column B in Figure 28–1 gives you an opportunity to weight the importance of each consideration to your company's situation. The rating can be from a "1" meaning of no importance to a "5" meaning of critical importance.

Figure 28–1 Training package checklist

RANKING SHEET:	Vendor _____ Program _____		
(A) Considerations	(B) Importance to Your Company (1 = low, 5 = high)	(C) Vendor/Product Evaluation (1 = low, 5 = high)	(D) Rating (col. B × col. C)
Objectives			
• Clearly stated			
• Compatible			
• Measurable			
Cost			
• What covered			
• What not			
• Extras			
Time			
• How long			
• How sure			

---

[10] Ross Tartell, "What to Look for When You Buy Training," *Training and Development Journal,* 41(1) (January 1987), p. 30.

Figure 28–1 (Cont.)

Quality
  • Content
  • Instruction
    technique
  • Packaging

Expertise/ex-
perience

  • In subject
  • In industry
  • In methodology

Client satis-
faction

  • Level 1
  • Level 2

"Gut reaction"

Other

  •
  •
  •
  •

                                                         TOTAL SCORE =

Column C is where you would rate the particular vendor being evaluated against each consideration, again with "1" for poor or low and "5" for outstanding or high. By multiplying columns B and C you can get a weighted evaluation on each criteria which can then be totaled to get an overall score. In this way you can keep track of your opinions of several vendors on a number of different considerations.

Care should be taken not to let the evaluation form make the decision for you. It is only a tool. The total score of one vendor may not be as high as that of another, but one consideration is so important that it overshadows all the others. The final decision on which packaged sales training program to purchase will be based on your estimate of which program will most meet the perceived training needs of your company at the best cost within the time frame.

Is it an easy decision to make? Not usually. Can it be done? Yes, and it is every day. What we want to do is to improve the decision process.

WORKING WITH THE VENDOR OF YOUR CHOICE

Purchasing a packaged sales training program from an outside vendor in no way relieves you of accountability for its success, anymore than if you had developed it in-house.

You still have the responsibility for working with the vendor to ensure that the company's needs and the program's objectives are in line with one another; for assuring that the actual program is presented in the most efficient, effective, and professional manner possible; and for evaluating the results of the program to determine if objectives were achieved and appropriate behavioral changes were made.

Whether generic or custom-designed, you and the vendor of that program have entered into a temporary partnership that should be covered both by the formal and informal contractual procedures.[11]

The formal contract should clearly cover what is expected from the vendor and from you, representing the company. Try not to leave anything important out of the contract on the assumption that surely everyone knows that was what was intended. Specify exactly what products and services are being contracted for. This is where ownership, copyright, reproduction authority, instructor licensing, and follow-up activities need to be detailed.

A definite completion date for delivery of the training should be clearly stated, and, if appropriate, a timetable giving dates when interim actions will be completed leading up to delivery. This is particularly important when purchasing a custom-designed package that takes a considerable time to complete. Interim benchmarks provide you with a means for monitoring progress on your program.

Costs covering all aspects of the program need to be clearly stated along with a schedule of payments. Some vendors require a certain amount of up-front money, others schedule payments through the life of the contract, some want payment in full when the program is delivered, and others will allow you to stretch the payments out (particularly if there is the possibility for a continuation of the contract).

Provision should be provided for amending the contract. Conditions do change. This is protection for you as well as for the vendor. It is better to anticipate this and allow for it in the written contract rather than to be surprised later on and having either party feel "trapped."

---

[11] Richard C. McCullough, "To Make or Buy?" *Training and Development Journal*, 41(1), (January 1987), p. 27.

When it comes to contractual matters, the vendor's representative you have been working with may call in someone from their accounting/legal office to work out the details. You probably will want to do the same, but do not lose control. You are in the best position to know what was discussed, what was agreed upon, and what it was about the vendor's product, program, or service that made them your choice. Let the accountants and the lawyers make it correct, legal, and binding, but retain your responsibility for ensuring that the contract covers the items that you considered critical in making the decision to buy from this vendor.

In addition to the formal contract, it is important for you and the vendor to develop an informal contract also. In it you will want to agree on what each of you expects to get from the "partnership," what each one of you will do to make sure the experience is a success. You will want to spell out what levels of trust and communication are expected and agree upon ways they can be maintained. "Red flags" should be identified as ways each of you will know of problems so that you can alert the other as quickly as possible. A means for surfacing and resolving any potential conflict should be mutually agreed to.

While not part of the formal, official, binding contract, these informal understandings should be documented so that they are clear, understood, and accepted by both parties. It is far better to take the time before you enter into a working relationship to clarify these issues than it is to try to do it after a crisis has been reached. You have gone to a lot of trouble identifying and selecting the best sales training package program for your company. Now take the time to solidify your relationship with the vendor who has become your partner in meeting your company's sales training needs.

SUMMARY

In this chapter, we have made a number of points:

1. Packaged training programs—generic or custom-designed—are another option to consider in meeting your company's sales training needs.

2. Before anything else, make sure that training is the answer to your company's problem.

3. There are valid reasons why at times it is best to build your sales training programs internally with your own staff, while at other times it is better to go to an outside vendor.

4. Use a variety of resources to identify the many sales training packaged programs available for your examination.

5. Do not rush to a decision. Evaluate each of the programs you have identified in light of your particular company's training needs.

6. Establish good contractual arrangements with the vendor of the program you have selected.

7. Be as thorough and careful in selecting and implementing a sales training program purchased from an outside vendor as you would if you had developed it internally.

8. Once purchased, the sales training packaged program is no longer the vendor's. It is yours and your company's in the sense that you will be held accountable for its success or failure in meeting the objectives set for it. As a professional sales trainer, you cannot dodge this responsibility.[12]

Sales training packaged programs are not *the* answer to all your sales training problems; however, they are a viable alternative resource you should keep in mind. The vendors of these packaged programs should be seen as your auxiliary staff on whom you can call when the appropriate time arises. But you are the person who must decide on what is "the appropriate time" to call in that auxiliary staff.

---

[12] John A. Cantwell, Donald J. Hosterman, and Howard R. Shelton, "Using External Programs and Training Packages," in *Training and Development Handbook*, 2nd ed., ed. Robert L. Craig (New York: McGraw-Hill, 1976), p. 47–2.

chapter twenty-nine ──────────

PUBLIC SEMINARS AS A SALES TRAINING RESOURCE

────────────────────── *DON M. SCHRELLO*

Don M. Schrello, chief executive of Schrello Direct Marketing, Long Beach, California, is a Ph.D. with 30 years' experience and a leader in the training community. He is past president and chairman of Schrello Associates, Inc., creators and marketers of seminars and training programs, which was acquired by The Forum Corporation in 1981. He was a founder of the Instructional Systems Association and a member of its executive board for many years. Dr. Schrello has authored more than 50 books, seminars, and other publications, including *The Product Evaluation and Planning Seminar, Improving Your Competitive Position, The Seminar Market, Marketing Training Within Your Own Organization*, and *How To Market Training Programs*. He is a regular speaker at major national and regional meetings, including ASTD, National Society for Performance and Instruction, Instructional Systems Association, University Associates, and *Training* magazine. He is a frequent consultant to some of the nation's top companies. Before starting his own business, he held senior management positions in long-range planning, marketing, and new products with one of the country's largest high-technology companies.

A vital part of every sales trainer's inventory of training resources is the open-to-the-public seminar or short course. These programs are widely available and inexpensive and can be identified to match virtually every training need. This chapter provides what you need to know to identify, select, and get the most from public seminars for sales training.

PUBLIC SEMINARS AND THEIR ROLE IN SALES TRAINING

Clearly, the essence of a public seminar is that it is a *meeting* of those from *many affiliations* with a *seminar leader* for the purpose of *learning or development*.

While public seminars existed thousands of years ago (as any of the early religious books quickly proves), the modern public seminar industry began more than 60 years ago with the American Management Association. Since that time the public seminar business has grown dramatically, and today represents a $12–15 billion industry for seminar fees alone. The related costs for travel, lodging, and other expenses double and perhaps even triple this estimate.

Annual attendance at U.S. public seminars exceeds 40 million participants, with some sources estimating an even higher number. And attendance is growing at a rate of more than 10 percent annually, well above the rate of growth of the U.S. gross national product. Because the needs met by public seminars are so deeply rooted in business and economic trends, this growth rate is expected to continue for the foreseeable future.

Types of Providers of Public Seminars

To make sense out of the vast public seminar marketplace, it is helpful to have a framework such as is shown in Figure 29–1, adapted from *The Seminar Market*.[1]

---

[1] Don M. Schrello, *The Seminar Market* (Long Beach, CA: Schrello Enterprises, 1980).

Type of Providing Organization		Skills Training	Knowledge Acquisition	
			Current and Evolving	Established and Accepted
Specialized training organizations and individuals		HI	HI	LO
		Public Seminars		
Business, trade, and professional associations		MED	HI	LO
Schools, colleges, and universities	Non-Degree Programs	LO	MED	MED
	Degree Programs	LO	MED	HI

Figure 29–1. Primary outside providers of business educational products and services

Two broad types of courses are offered: knowledge acquisition and skills training. Knowledge acquisition, in turn, is divided into current and evolving knowledge (dealing with new products, services, customer needs/wants, laws and regulations, etc.), and established and accepted knowledge (dealing with established principles, practices, art, science, etc.). Skills training encompasses topics such as management skills, selling skills, presentation skills, time management, planning, as well as personal skills such as speed reading, foreign languages, and so on. These major divisions are shown across the top of Figure 29–1.

Down the left of Figure 29–1 are shown the three major types of outside providers of training: (1) specialized training organizations and individuals; (2) business, trade, and professional associations; and (3) schools, colleges, and universities. The notations of low, medium, or high indicated in the squares of Figure 29–1 refer to the extent to which each of these sources is responding to the needs for the different types of training.

The shaded portion of Figure 29–1 shows the share of the total market generally covered by public seminars. It focuses on skills training, and on the current and evolving portion of knowledge acquisition. The established and accepted body of knowledge, traditionally conveyed through the degree programs of schools, colleges, and universities, is usually not provided by public seminars.

The skills training of interest here deals with those skills of interest to the sales trainer. These seminars typically have long useful lifetimes, with only routine updating of the course content. Typical titles include "Time and Territory Management," "Professional Selling Skills," "Sales Management for the New District Manager," "The Art of Negotiating," and so on. Skills training is offered mainly by specialized training organizations and individuals and, to a lesser extent, by business, trade, and professional associations. Schools, colleges, and universities have some offerings in this area, but generally concentrate their efforts elsewhere.

The second major category of public seminars covered in this chapter concerns current and evolving knowledge, again primarily knowledge of interest to the sales trainer. These seminars often deal with specific types of products and/or markets and may involve product sales information, product service information, product use information, important developments in technology or the law, and so on. This category of seminars includes titles such as "Managing in a Downsizing Economy," "Increasing Foreign Sales," "How to Sell Health Care Services," "Selling to the Government After Gramm-Rudman," and similar topics.

In general, public seminars dealing with current or evolving knowledge are best presented by the authorities closest to—and participating in the advance of—the state of the art. For this reason, these programs often involve well-known authorities as speakers. Participants come to listen to these experts and ask them questions or perhaps to hear a panel of experts offer opposing views.

Four Major Types of Programs

It is also useful to examine where public seminars fit in the spectrum of business-related education and training. Figure 29–2 shows the relative proportion of four major types of training programs as a function of the size of the

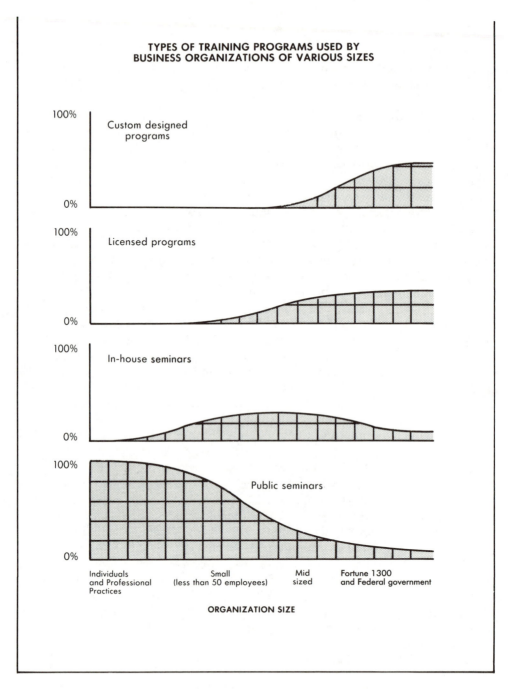

Figure 29–2. Types of training programs used by business organizations of various sizes

customer's organization.[2] These four types of programs are (1) public seminars, (2) in-house presentations, (3) licensed programs, and (4) custom-designed programs.

Public seminars are generally held in different geographic locations at convenient hotels or conference sites, are open to anyone desiring to attend, and are conducted by a representative of the providing organization.

In-house presentations (sometimes called on-site programs) are those in which the providing organization travels to the customer's site (or where the customer makes all the facility arrangements) and in which all the participants customarily are employees of a single company or a group of cooperating companies (sometimes called consortium training).

Licensed programs are those which are developed by an outside provider and then presented in-house at the client company, using the client's own employees as instructors.

Finally, *custom-designed programs* are the same as licensed programs, but are developed "from scratch" by either an outside provider or by the company's own program development staff.

Figure 29–2 shows that public seminars are used proportionately more often by the smaller-sized company, even though representatives of the very large companies may participate. Such participation by those from large organizations allows them to acquire their own skills and knowledge, or to preview the course for possible adaptation on an in-house, licensed, or customized basis.

The public seminar portion of Figure 29–2 is somewhat misleading because it suggests that there are more attendees in public seminars from smaller organizations than from larger ones. Actually, the reverse is true! The reason that the *proportion* of participation in public seminars declines as company size increases is that large organizations divide their training among all four types of presentation, whereas public seminars are almost the only means by which the smallest organizations obtain their training.

Seminar Versus Session

Throughout this chapter the word "seminar" refers to a given topic, presented in a course of a given duration, and offered by a given sponsor. The word "session" is used for an individual presentation of a seminar, at a specific location, on a particular date at a particular time.

The distinction between these two words is important because as a general rule each seminar is presented many times—that is to say, in many sessions.

---

[2] Don M. Schrello, "Sales and Marketing of Seminars, Training Programs and Instructional Systems," Keynote address to the Instructional Systems Association 1987 Sales and Marketing Conference, New York.

As we review the statistics and other information about public seminars, failure to observe this distinction can easily lead to confusion.

The Advantages of Public Seminars over More Traditional Teaching Methods

Public seminars have inherent advantages over more traditional forms of education and training. Among the advantages are the following (adapted from *The Seminar Market*[3]):

- Many seminars are conducted on a frequent, recurring basis, allowing people to attend as new education and training needs arise.
- Most seminars convey concentrated information in a relatively short time, ranging from less than a day to at most a few weeks.
- Except for the investment of time, fees, and other costs for the seminar itself, little or no additional commitment is required of the individual participant or their organization.
- Public seminars are often available locally and therefore require minimum travel. In those instances where travel is necessary, frequent schedules and varied locations help attendees combine seminar attendance with other travel plans.
- Seminars are generally open to anyone; no prior education, degree, or other prerequisites are required.
- Unlike books, audio or video tapes, computer-based training (CBT), and other self-instructional media, public seminars offer an opportunity to ask questions and to interact with other participants.

Changing economic conditions, technology, and the increased competitiveness of worldwide business combine to create a continuing demand for expanded sales knowledge and selling abilities. Public seminars provide a rapid and effective way to gain this knowledge and to build these skills.

In addition to market pressures, there continues to exist the requirements for new employee training, and for advanced level training of employees with new assignments and/or recent promotions. The availability of public seminars on topics such as "General Sales Management Skills," "Finance for Non-Financial Managers," "Sales Compensation," "Leadership," and others helps prepare and reeducate the salesperson for new roles and responsibilities.

Yet another use of public seminars is to meet training needs in the development of the sales organizations. The fact that an organization can participate

---

[3] Schrello, *The Seminar Market*.

at public seminars individually *or* as a group eliminates the necessity of drawing many salespeople out of the field at the same time. Public seminars that focus on such sales training organizational needs include titles such as "Building Teamwork and Commitment," "Strategic Planning," "Fundamentals of Customer Service," and "Product Evaluation and Planning."

Public Seminar Criticisms

The positives—no matter how many—do not automatically cancel the negatives. Unfortunately, a few public seminars have invited criticism toward the entire seminar industry by virtue of poor performance on the part of the sponsor. Many of these substandard programs involved the "soft" interpersonal skills and include "fad" courses that are in vogue for a brief but noisy term.

Notwithstanding these transitory problems, the most commonly recurring criticisms of public seminars are

- They are often expensive, especially including the costs of related travel, lodging, and so on.
- Sessions are sometimes canceled at the last minute by the sponsor, adversely impacting those attendees who combined attendance at the seminar with other business commitments.
- Seminar quality varies; even well-known sponsors have been accused of promising more than their programs can deliver.
- The initial enthusiasm of participants immediately following some programs may not be matched by permanent on-the-job application of the information or skills learned.

But, on balance, the advantages seem to outweigh the problems significantly, because in spite of these criticisms, public seminars are thriving. More important, public seminars have gained new esteem and an increasingly important role as a vehicle for providing needed skills and knowledge to business organizations around the world. The successful resolution of whatever problems do exist, combined with a strengthening of the advantages inherent in public seminars, should result in an even stronger and healthier group of seminar sponsors supplying the needed programs for the foreseeable future.

WHEN TO USE PUBLIC SEMINARS IN SALES TRAINING

From the perspective of the training director, line manager, or other individual responsible for sales training, the four main trade-offs determining the type of training employed to meet a particular need are

1. Public versus in-house program
2. Individual versus group instruction
3. Media-based versus a live trainer
4. Off-the-shelf packages versus custom-developed programs

Of course, these four variables can occur in various combinations. For example, a seminar could employ self-paced, media-based segments interspersed with group sessions led by a live instructor. Or it could include a mixture of off-the-shelf as well as custom-developed modules.

The fourth of these variables—off-the-shelf versus custom-developed—is important only for in-house programs. Regularly offered public seminars are usually considered to be off-the-shelf.

The experienced trainer recognizes that there is no one "best" way to deliver training. Under different circumstances, each of the combinations listed can be superior to the rest . . . while still other conditions could make any one inferior to others.

So how does one decide? The following sections offer some ideas.

Public Seminars: The Cornerstone of Training for Most Organizations

The large number and rich variety of public seminars, their ready accessibility, their comparatively low cost, and the relatively modest commitment needed to get started using them all combine to make such courses the cornerstone of training for all but the very largest organizations.

Moreover, the organizations and institutions that supply public seminars can spread the often considerable development costs over many users, thereby furnishing very-high-quality training at reasonable prices. As an example, one computer-based, interactive sales skills training program is estimated to have required more than a quarter-million-dollars' development investment. But the finished product is available to participants in a public seminar for a few hundred dollars.

Public seminars are frequently the approach of choice when one or more of the following circumstances exist:

- There are comparatively few people from the organization to train at any one time.
- Those to be trained are geographically dispersed.
- A suitable public seminar is already offered.
- It's desirable for trainees to meet participants from other organizations and exchange ideas with them.

- An existing seminar is to be previewed or "tried on" for eventual customizing or licensing in the organization.
- A long-term commitment to the training cannot be made.

In-House Programs

A larger variety of training program types is available for in-house programs than for public seminars.

In-house presentation of off-the-shelf programs by *outside trainers* is the preferred choice when:

- The training is best presented by an "outside expert."
- It saves money and/or time over sending the participants to public sessions of the same course.
- There are team-building/or proprietary application objectives requiring "at-home" participation.
- There are at least 8 to 10 persons to be trained at one time.
- It's desired to "pilot" test a new or customized program which will later be presented by in-house personnel.

Presentation by an *in-house trainer* of an off-the-shelf program can often be arranged via licensing and/or a train-the-trainer certification program. Suppliers of such programs can also frequently provide low-cost "tailoring" of these programs to meet the in-house needs more closely, while preserving the tested and proven elements of the course. Such adaptations may involve case studies, workshop applications, incorporation of the organization's own policies and procedures, or the preparation of special audio/video tapes, workbooks, and other materials.

Off-the-Shelf versus Custom Developed

Obviously, a short step is all that's needed to move from tailoring an off-the-shelf program to a wholly customized course, with the development done either by an outside resource or internally by the user organization. As mentioned earlier, this major undertaking is expensive and time consuming and could produce uncertain results if the developers are not thoroughly experienced. Especially where advanced technology (such as computer-based or interactive training) or media (such as video disks) is involved, using an off-the-shelf course, as-is or with some adaptations or tailoring, can be preferable to a custom course developed solely for the purpose.

In any case, in-house presentations of a group instruction program (whether off-the-shelf, tailored, or custom-developed) by an *in-house* resource can be best when

- There are large numbers of people to be trained.
- Greater flexibility in scheduling/timing is desired than can be accommodated by outside resources.
- An adapted, tailored, or otherwise special course is needed to meet the organization's needs.

ESTIMATING THE COSTS AND PAYBACK OF A PUBLIC SEMINAR

Seven Cost Factors to Consider

The total cost of attending a public seminar is broken down into a number of elements both direct and indirect:

The Cost of Preseminar Arrangements. This includes the management, staff, or clerical time devoted to locating, evaluating, and enrolling a participant in the selected course, and making the necessary travel arrangements (if any). While these costs are sometimes substantial, they are usually ignored in estimating the total cost of a seminar.

Participant Advance Preparation Time. A good public seminar—like any training—should be proceeded by suitable preparation in which the training objectives and results to be achieved are clearly defined. Often this involves meetings between the trainee and his or her supervision, co-workers, and others. Sometimes extensive written preparation is required such as completing questionnaires. In other cases the seminar sponsor will request that advance reading be done to speed the conduct of the course. Such time, if expended, needs to be figured in the total cost of training.

Travel and Away-From-Home Living Expenses. These are almost always identified as a major source of public seminar cost, often averaging two or three times the seminar fee. However, the growth of local seminars has considerably reduced these costs, especially for participants located in major metropolitan areas.

Seminar Registration/Cancellation Fees. These fees too are usually identified clearly in any training budget. Ideas on how to reduce these costs are presented elsewhere in this chapter.

Additional Materials Purchased. Often overlooked in the cost of seminar attendance is the purchase of additional references, books, job aids, and other materials either at the seminar or as a result of references suggested by the seminar leader. Such materials can become a permanent part of your training library, but may need to be included in the estimated cost of seminar attendance.

The Wages and Salaries of Those Being Trained. This is a controversial topic that is especially important for highly paid managerial and sales personnel, especially if their salaries while being trained will come from a different budget center.

Lost Productivity While at Training. Taking sales personnel "out of the territory" for training may result in lost sales at least in the short term, even though the long-term result is increased sales. Highly productive sales organizations always weigh time out of the territory in terms of overall effect of sales.

Each organization seems to take a different view of how the indirect (or "overhead") costs of the training function are handled. Smaller organizations usually lump them into a general or administrative account, and don't allocate these costs to specific training events. But larger organizations frequently do apportion these costs, and hence they must be included in estimating the total cost of attending a public seminar.

Tracking the "Payback" or Results of Training

Estimating the results of training—the payback—is much more elusive. In the ideal case, sales training would directly produce measurable increases in sales, orders, customer satisfaction, reduced callbacks and order cancellations, or other tangible measures of increased sales productivity. Unfortunately, this is usually not convenient or even possible, although sales training is the one area where reasonably objective standards for measuring productivity routinely exist.

In this respect, the ability to "demonstrate" the cost effectiveness of sales training is not different for public seminars than for any other type of training. The costs are all too evident, but the benefits (sadly) are much harder to measure clearly.

How Public Sales Training Seminars Operate

This section deals with the major elements of a public seminar, and some recent changes that have occurred. Examined are the currently popular sales training seminars and their sponsors, course duration, the locations most often scheduled, and public seminar prices (including cancellation provisions and guarantees).

This added insight into how public seminars operate will enable a more enlightened evaluation of competing seminars and a better selection of the one most likely to meet your sales training needs.

What's Available: A Look at Popular Sales Training Seminars and Sponsors

Using the same data sources and methods outlined later in this chapter, a survey was made of the most popular public sales training seminars currently offered. Table 29–1 summarizes a representative sample of the hundreds of separate programs identified in this survey.

Table 29–1. Representative public seminars of interest to sales trainers

Sponsor	Seminar Title	Days	Price
Dun & Bradstreet	How to Find/Sell New Customers	0.5	$ 65
Zig Ziglar	How to Win Friends	0.5	65
CareerTrack	Image and Self-projection for Pro-fessional Women	1	48
CareerTrack	Power Negotiation Skills	1	48
CareerTrack	How to Get Results with People	1	48
CareerTrack	Getting Things Done	1	48
Dun & Bradstreet	How to Find/Sell New Customers	1	86
UC Davis	The Art of Selling	1	90
UC Berkeley	A Career in Sales: Is It for You?	1	95
Dun & Bradstreet	Effective Sales Techniques	1	97
Dun & Bradstreet	Basic Selling Skills for Women	1	97
Pryor, Fred	How to Keep Customers	1	98
Performance Group	Getting Qualified Appointments	1	99
Learning Dynamics	Mastering the Art of Closing Sales	1	99
Sales Skills Institute	Essential Skills of Salesmanship	1	99
Performance Group	Becoming a Master Salesman	1	99
UCLA	Sales Management	1	135
Associated Management Institute	Building Teamwork and Commit-ment	1	165
Associated Management Institute	How to Close More Sales	1	165
Dartnell Institute of Management	Managing the Sales Force	1	195
Dartnell Institute of Management	Effective Sales Management	1	195
American Management Associa-tion	Selling to Health Care Industry (5208)	1	395
Washington Researchers	Practical Competitor Analysis Tech.	1	495
Karrass Seminars	Effective Sales Negotiating	1	550
Nicholas & Associates	Invitation to Selling	2	200
Blessing/White	Managed Sales Development	2	215
Paracommunications	Competitive Selling Strategies	2	295
Florida S&M Institute	Sales Strategies	2	325
Paracommunications	How to Sell Health Care Services	2	325
California CPA Foundation	Selling Skills for the Professional CPA	2	325

Table 29–1. (Cont.)

Sponsor	Seminar Title	Days	Price
Mountain Bell	Sales Skills Workshop	2	360
Selin Corporation	Increased Sales Through Cross-selling	2	395
S.E.T. Incorporated	The Sales Process	2	395
Effectiveness Institute	Effective People Skills	2	395
Bureau National Affairs	Self-motivation in Selling	2	420
Forum Corporation	Sales Negotiation	2	440
Tracom Corporation	Improving Sales Presentations with Social Style	2	475
Century Planning Association	How to Prepare Winning Proposals	2	550
American Management Association	Fundamentals of Customer Service (5270)	2	580
NTP	Master Sales Skills	2	595
Systema Corporation	The Sales Relationship	2	595
Frederick Knapp Association	Sales-Image	2	595
American Management Association	Selling for the New Sales Representative (5510)	2	595
American Management Association	Using Independent Manufacturers Representatives (5275)	2	635
American Medical Care Association	Sales Management-Highly Competent. Managed Care	2	650
American Management Association	Power Gets Results (2262)	2	695
Boston University	Peak Performance Salesmanship	2	695
Boston University	Self-presentation for Professional Women	2	695
Wilson Learning	The Versatile Sales Person	2	700
American Management Association	Sales Planning (5246)	2	715
American Management Association	Sales Compensation (5522)	2	725
Federal Publications	Franchising	2	725
Burke Marketing	Sales Forecasting Techniques	2	800
Wilson Learning	Connecting with People	2.5	500
Wilson Learning	Social Style Sales Strategies	2.5	600
Systema Corporation	Systematic Selling Techniques	2.5	625
American Management Association	Value-added Services: Competing (5226)	2.5	695
American Management Association	Successful Sales Call	2.5	750
American Compensation Association	Compensation of Salespersons	3	520
Weber Associates	Fundamentals of Professional Selling	3	525
Mountain Bell	Excellence in Sales Performance	3	540

Table 29–1 (Cont.)

Sponsor	Seminar Title	Days	Price
Weber Associates	Strategies for Successful Selling	3	575
Zig Ziglar	Born to Win	3	595
Forum Corporation	Face-to-Face Selling Skills	3	595
National Society of Sales Training Executives	Sales Management Seminar	3	650
B. J. Chakiris Corp.	Consultative Selling	3	675
Forum Corporation	Exceptional Sales Performance	3	680
Max Sacks, International	Professional Selling Clinic	3	695
American Management Association	Hi-Performance Teams (2523)	3	695
American Management Association	National Accounts Management (5206)	3	755
American Management Association	Sales Management—New District Manager (5227)	3	760
American Management Association	Managing Distributor Network (5590)	3	795
Learning International	Professional Selling Skills	3	1,000
Systema Corporation	Consultative Selling Skills	3.5	695
MOHR Development	Retail Negotiating Skills	4	395
Pace Organization	The Pace Seminar	4	550
Scientific Methods	Sales Grid Seminar	4	675
Century Planning Association	How to Prepare Winning Proposals	4	750
American Management Association	Principles of Professional Selling (5520)	4	850
Wilson Learning	Counselor Selling	4	850
American Management Association	Field Management of Salespeople (5598)	4	860
American Management Association	Executive Course for General Sales Manager (5509)	4	895
American Management Association	Fundamentals of Marketing (5512)	4.5	895
Sales Systems Development	Developing an Effective Field Sales Organization	5	600
Psychological Association	Dimensional Sales Training	5	670
American Marketing Association	The School of Sales Management	5	800
Psychological Association	Dimensional Sales Management	5	950
Wharton School	Perspectives in Sales Force Management	5	1,945
Dale Carnegie	The Dale Carnegie Sales Course	6	795
Columbia Executive Programs	Sales Management	6	2,850
Dale Carnegie	The Dale Carnegie Course	7	795

Table 29–1 does not show all available sales training seminars or all active sponsors. The seminar marketplace is so vast and unstructured that it is virtually impossible to be sure that important seminars and/or sponsors were not omitted, particularly ones that are primarily local, or that concentrate on a particular industry or service.

Nonetheless, Table 29–1 gives a reasonable cross section of active seminar sponsors, the titles offered, the duration of these seminars and their current single-seat, undiscounted prices. The table is arranged in order of increasing duration and increasing price.

Clearly the American Management Association (AMA) sponsors most of the titles in Table 29–1. And, with more than $100 million annual revenue, AMA may actually be the largest provider of public sales training seminars. But Table 29–1 is not a reliable guide to the largest sponsors, because it does not show how often each seminar is offered or the average number of participants in each session.

For example, many of the largest training organizations, who offer primarily in-house programs, also offer public sessions that they do not actively advertise. Organizations such as Learning International, Wilson Learning, and The Forum Corporation—among the largest providers of sales training—appear in Table 29–1 with only a few titles. In fact, all three organizations are major suppliers of public seminars which are sold primarily by their sales forces and by word-of-mouth. As one illustration, the number of attendees at public sessions of Learning International's popular three-day Professional Selling Skills seminar, which is not actively advertised, places them in the top 15 percent of all public seminar sponsors.

Course Duration: The Trend Toward Shorter Seminars

The table 29–1 entries are listed in order of increasing course duration, from the shortest to the longest. Not surprisingly, most public seminars today are one week or less, with the most popular duration being two days. The Dale Carnegie courses, which seem in the table to be longer duration, are actually offered one evening per week and therefore take less time off-the-job than their total duration suggests.

This trend toward shorter seminars is comparatively recent. In the late 1970s and early 1980s, many one week and even multiple-week seminars were popular. Today, the emphasis is on shorter seminars, and many sponsors have repackaged their longer seminars into several shorter ones which can be attended independently and in any order.

Even more prominently, the one-day or less seminar—which fell into a short period of disfavor in the early 1980s—has returned as a very popular format. Because seminar sponsors typically find much higher response rates to their advertising for seminars of one-day duration or less, they are able to conduct more sessions and to hold them in more widely scattered locations, thereby bringing the benefits of the seminar to an even wider audience.

Public Seminar Prices

As suggested in *Marketing In-House Training Programs*,[4] the prices shown in Table 29–1 can be displayed in a more informative way as a graph of the single-seat, undiscounted price as a function of seminar duration. Figure 29–3 presents this graph for 130 of the public sales training seminars identified in the survey. The three lines represent the minimum, average, and maximum prices found for each length course. Each symbol represents a single public seminar.

Prices in Figure 29–3 are for a single attendee and are undiscounted. However, discounts are widely available for multiple enrollments, for membership in the sponsor's organization, or (occasionally) for early registration and/or payment.

Group enrollment discounts are handled in a number of ways. Some levy an added "organization registration charge" on the first participant from an organization, while others offer a discount for the second and subsequent enrollments in a session. Still others offer discounted "tickets," "subscriptions," or similar advance purchase arrangements for those with many personnel to train. This is, of course, mutually beneficial, furnishing the attending organization reduced costs while furnishing the sponsor advance bookings.

Guarantees, Advance Payments, and Cancellation Refunds

Two additional factors related to public seminar prices are the offering of a money-back guarantee of satisfaction and a complete refund of any fees paid in the event that the participant needs to cancel at the last minute.

A growing number of seminar sponsors now offer a money-back guarantee to anyone dissatisfied with the seminar. Most sponsors find such guarantees rarely if ever used, since most participants feel that they get value from seminar attendance. When a money-back request is received from a participant, it usually signals difficulty with the seminar leader: often a new instructor not yet able to deliver the seminar to the specifications promised in the advertising.

For those concerned about whether a seminar is truly appropriate for them,

---

[4] Don M. Schrello, *Marketing In-House Training Programs* (Long Beach, CA: Schrello Direct Marketing, 1984).

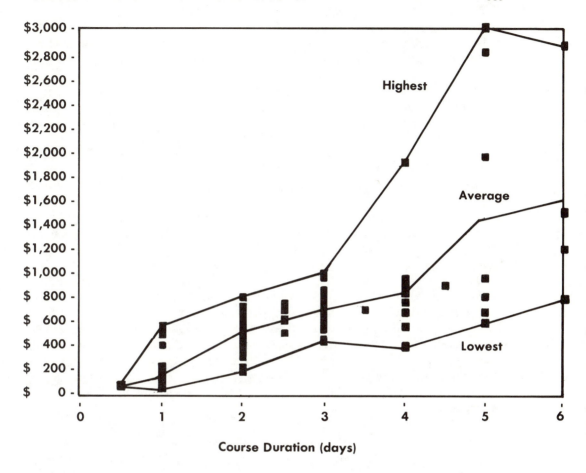

Figure 29–3. The range of prices charged for public sales training seminars

the money-back guarantee can provide reassurance. But remember that the fees to be refunded represent only a small part of the total cost of seminar attendance. So picking a seminar wisely is always a better strategy than counting on a money-back guarantee.

Incidentally, changes in consumer protection regulations now dictate that most seminar sponsors must refund your fees if you are unable to attend and do give reasonable notice. However, larger customer organizations may find it more convenient to simply "roll over" the enrollment fee to another course attendance rather than duplicate the paperwork connected with getting approval for the course in the first place.

Most seminar sponsors request—and receive—advance payment. Only a small fraction of participants' organizations (frequently the largest companies and the federal government) fail to pay before the seminar begins.

SOURCES OF INFORMATION ON PUBLIC SEMINARS

The large number of different seminars available—and the many different locations at which they are presented—present a formidable challenge for the sales trainer attempting to find the best possible program for his or her employees. However, with a little preparation you will not only find the right program, but also construct a system for updating your files so that your future training needs will take advantage of the widest possible base of information.

This section discusses the major sources of information on public seminars. These sources will enable you to identify many possible seminars by subject, title, sponsor, duration, location, cost, and other important factors. Once having identified the candidate seminars that could meet your needs, your next challenge is to select the best one.

There are three types of information sources to consult to identify public seminars:

1. National (and international) sources of information, which include (a) published directories, (b) on-line computer databases, and (c) locator/enrollment/registration services.
2. Networking with others, especially your organization's managers, other sales trainees, professional and trade associations of which your firm may be a member, colleges and universities, and others in your local area.
3. Your own public seminar files, which should be started at once and maintained in a systematic manner indefinitely.

National Sources of Public Seminar Information

Table 29–2 summarizes 15 major national sources of information on public seminars. At the present time, no single source is exclusively dedicated to sales training seminars; however most sources report 10 to 20 percent of the seminars they list are sales training. And, of course, if you have nonsales training needs, any and all seminars included in these sources could be of interest.

Table 29–2 shows the providing organization and the name of the publication or service they provide, the person to contact for information and their telephone number, the estimated number of seminar sponsors they represent, the approximate cost to you for using the resource, and remarks describing the resource, its currency, and so on. There are three different kinds of resources represented in Table 29–2.

Published Directories. Nine of the resources in Table 29–2 are printed listings which you can purchase or examine at a well-stocked library. Perhaps the most up-to-date and comprehensive of these are Seminar Information Service (SIS) and *Seminars Directory*. SIS lists seminars from the largest number of sponsors, in a loose-leaf, workbook format. *Seminars Directory*, the oldest publication of its type, publishes in a magazine format several times a year. Both sources list only seminars for which sponsors have furnished adequate information.

A common characteristic of the printed sources is that they do not charge the sponsor for listing their seminars. Costs are paid by the user, either through annual subscription fees or for each copy of the publication. As a result, the number of seminars listed is only limited by how vigorously the publisher solicits the training community.

If cost is a problem, there are two, essentially free, printed listings that offer a reasonable cross section of suppliers: the ASTD *Buyer's Guide* and the *Training* magazine *Marketplace Directory*. Both directories are published annually, and both are furnished without added charge as part of the membership and/or subscription fee. Unfortunately, neither list specific public seminar programs, nor does either adequately differentiate between sponsors providing only in-house programs, and those offering public seminars.

One other directory listed in Table 29–2 may also be useful for selecting public sales training seminars: Human Resource Development Press's *The Trainer's Resource*, which includes public seminars as well as programs which are only available in-house.

On-Line Computer Databases. There are at least three resources that allow real-time access to databases of currently offered seminars. Most of these services require an initial subscription charge (for which a confidential password is issued) and then charge per minute of time actually connected to the database. The user also pays the applicable telephone charges, although many services offer local or toll-free numbers.

These on-line services are shown in Table 29–2. At present only two of the three are truly independent. ASTD's Trainet is really an alternate name for The TimePlace/Edvent Service. Notice also that Executive Telecom System's Human Resource Information Network makes available the SIS and Seminar Clearing House databases.

The number of seminar titles, locations, and dates available through these on-line databases is truly staggering, and careful planning of your search and

Table 29–2 Resources for locating publis sales training seminars

Type	Organization & Service	Contact	Approx. No. of Sponsors	User's Cost	Remarks
P	American Society for Training and Development *Annual Buyer's Guide & Consultant Directory*[1]	Marketing Representative (703) 683–8129	525	Included with membership; $35 separately	Lists training providers classified by type of training offered; no course information provided
O	American Society for Training and Development Trainet On-Line service	Information Center Manager (703) 683–8190	(See Time-Place below)	60 cents per min. connect charge	PC accessible on-line database; free password for ASTD members; otherwise same as Time-Place EdVENT database below
P	Bill Publications *Sales & Marketing Management Magazine*[2]	Publisher (212) 986–4800	—	$38 per year	Monthly magazine
O	Executive Telecom System (a division of BNA) The Human Resource Information Network (AKA The Training & Development Network)	Director of Marketing (800) 421–8884	2–3,000	$890 per year plus connect charge	PC accessible on-line database; accesses Seminar Information Service and Seminar Clearing House databases, as well as others
S	The First Seminar Service	President (617) 452–0766	500	None (paid by seminar sponsor)	Locates seminars to meet specified criteria; furnishes brochures on prospective courses; handles all enrollment details
P	Gale Research Company *Training & Development Organizations Directory*[3]	Marketing Director (313) 961–2242	2,000	$270	Last edition 1983; "New Organization" supplement available; next edition expected May 1988; available in many public libraries
P	Hope Reports *Training Business Directory (Vols. 1 and 2)*[4]	Chairman (716) 458–4250	3,800	$60	Last edition 1984; next edition expected 1988; may be available in some public libraries
P	Human Resource Development Press *The Trainer's Resource*[5]	Publisher (413) 253–3488	200	$59.95	650-page directory published annually; contains packaged programs as well as public seminars

				Cost	No.	Description
P	National Society of Sales Training Executives	Administrative Manager (305) 894–8312	—	$225 initiation fee; $150 per year		Professional society; publishes the quarterly journal *Sales & Marketing Training*
S	National Training Network	President (619) 549–0214	10	None (paid by seminar sponsor)		Serves as a broker for selected sales training and other training programs; works primarily with in-house programs
S	Seminar Clearinghouse, International (formerly MANTREAD Seminary Registry)	President (612) 293–1044	1,500	$170 to $1,100 per year based on number of employees		Finds programs to meet specified criteria; also furnishes evaluations; on-line database available through Executive Telecom System (see above)
P	Seminar Information Service (SIS)	President (714) 261–9104	400	$275 per year		Workbook and monthly newsletter; on-line database service available through Executive Telecom System (see above)
P	*Seminars Directory*, The Directory of Professional Seminars, & Continuing Education Programs	Publisher (608) 231–3070	300	$60 per year		50–60 page directory published twice per year; some phone help available to locate specific programs
O	TimePlace, Incorporated EdVENT	Vice President, Marketing (617) 890–4636; (800) 544–4023	5,896	$200–10,000 per year		PC accessible on-line database; can be searched by many different criteria; course descriptions can be downloaded or read on screen
P	*Training: The Magazine of Human Resources Development*	Publisher	2,700	Included with subscription		Lists training providers classified by type of training offered; no course information

P, published directories; O, on-line computer databases; S, locator/enrollment services.
[1] *ASTD Buyers Guide & Consultant Directory* (Alexandria, VA: ASTD, annually).
[2] *Sales & Marketing Management* magazine (New York: Bill Publications, monthly).
[3] Paul Wasserman and Janice McLean, eds., *Training and Development Organizations Directory*, 3rd ed. (Detroit: Gale Research Company, 1983).
[4] Thomas W. Hope, *Hope Reports Training Business Directory*, Vols. 1 and 2 (Rochester, NY: Hope Reports, 1984).
[5] Leonard Nadler, Eugene Fetteroll, and Zeace Nadler, eds., *The Trainer's Resource* (Amherst, MA: Human Resource Development Press, 1987).

experience is required to avoid aimless wandering and excessive connect charges. Before you call, be familiar with the system commands (especially those for obtaining help and for ending the session) and also have available the phone number to secure "live" help if you find yourself at a dead-end. This is especially necessary for those unfamiliar with or just getting started with computers.

Locator/Enrollment Services. Three resources listed in Table 29–2 offer the service of locating public seminars that meet the criteria you specify and help you evaluate the various candidates they find. Two of these services will even enroll your attendees in sessions of your choice!

Of the three, Seminar Clearinghouse International has been in existence since the early 1960s and previously operated under the name *Mantread Seminar Registry*. It is the only one of the three to charge the user for the information it provides.

The other two services, First Seminar Service and National Training Network (not connected with the J. C. Penney Company's satellite teleconferencing service of the same name), charge the user nothing, but primarily supply information on seminars whose sponsors have agreed to pay commission on the registrations they secure. As a result, these organizations function as brokers for the seminar sponsors.

As long as you are aware of this arrangement, and are agreeable to any bias it may introduce, these services may be quite useful in helping you quickly identify the public seminars you need.

Networking with Others

For most public seminars, the best information is obtained directly from a recent participant. Such "real-world" information cuts through advertising hype and may reveal benefits the seminar sponsor never mentioned in their advertising. The best way to obtain names of recent participants is directly from the sponsor (more on that in the next section). A close, second best way is to establish a network of training colleagues and associates, either in your industry or geographic area, you can call upon to recommend public seminars they've had success with. It doesn't take long for word to spread about a good seminar and/or a good seminar sponsor.

Even the smallest organizations need some kind of networking as soon as definable training needs begin to surface.

Other common sources for networking are trade associations and professional societies. While associations are often reluctant to endorse a particular seminar sponsor, many offer public seminars themselves, and their selection of a seminar or sponsor may suggest ones you want to consider as well.

And don't overlook local colleges, universities, or even community colleges.

Many of these institutions are doing an excellent job of finding and developing seminars and short courses. Usually under the department of continuing education or an extension program, and often part of the business school, colleges and universities are responding to the changing demographics of continuing education. Because of their nonprofit status, institutions of higher learning can spend less money promoting and more money on the seminar, often enabling them to feature famous seminar leaders at rates well below what the same speakers would otherwise charge.

Finally, take advantage of your local Chamber of Commerce, Executive or Sales Manager Roundtable, Young President's Association, or other management group in your community that might have members interested in sales training. Again, this part of your network will be invaluable when you have specific training needs and want to know who really has the best program on a given topic.

Look at Your Own Files for Training Sources

As suggested before, your mail box or in-basket is one of your best sources of training information—if only you don't throw away all those brochures and catalogs!

Anyone with sales training responsibilities is well advised to set aside a file cabinet drawer (or even a cardboard box) and start dating and saving *all* public seminar announcements and other catalogs related to training. At first this may seem an insurmountable task. But in a short time you'll find that a pattern establishes itself; you get lots of material at regular intervals from a fairly small number of seminar sponsors, and considerably less material at very *in*frequent intervals from many others.

Organize the information by making a file folder for each sponsor whose material you receive, and keep the files in alphabetical order. When you receive something from a new organization, create a new file folder. This will usually be a fine start on your own resource library.

TIPS ON SELECTING THE RIGHT SEMINAR _____

With so many seminars available, selecting the right seminar can be difficult if you go at it in a haphazard fashion. On the other hand, carefully evaluating the available information in a number of clearly defined areas can often pinpoint exactly the *right* seminar, and—more important—can reveal what would turn out to be the *wrong* seminar if you attended without thorough investigation.

This section identifies those key evaluation elements, and suggests how to get the additional information you need to answer the important questions in each of these areas.

Thirteen Key Factors to Evaluate

There are 13 major areas in which information should be collected and evaluated for making the decision of whether or not to attend a particular seminar or to select which seminar should be attended from among a number of candidates. The 13 factors are

1. Content/program outline (includes the course title, subtitle, detailed outline, prerequisite courses, if any, and all related content information)
2. Sponsor (time in business, type of organization, number of seminars offered, reputation)
3. Presenter(s) (credentials, experience, method of compensation, familiarity with this course, number of times they've presented this course, past participants' evaluation of them)
4. Schedule (including duration and course dates)
5. Location (city, facility, availability of transportation, etc.)
6. Course format (method of presentation, class size, auxiliary media employed, such as computers, refreshments, meals or receptions included, seating arrangements, etc.)
7. Audience/participants likely to attend (organization sizes, positions, experience levels, prior attendees' organizations, "who should attend")
8. Testimonials of past participants (including the individual's name, position title, organization/division)
9. Materials provided (course notes, books, worksheets, audio or video tapes, job aids, etc.)
10. Price/discount(s)/payment/refund terms
11. Guarantee (if offered)
12. Continuing education units (CEUs) (including approval by certifying boards for professional credit where applicable)
13. Cancellation history (for this course and sponsor)

If your organization has its own evaluation factors, make your own list starting with these 13 factors, adding your own points and changing the terminology throughout to fit your situation.

How to Use These Evaluation Factors

To evaluate a number of different seminars, list the evaluation factors down the left side of a sheet of paper and the seminars to be evaluated across the top of the sheet. Then rate each seminar on each evaluation factor using a numerical scale of, say, 0 to 10, where 10 is best. Finally, add up the ratings for each seminar. The seminar with the highest total score is likely to be the best choice across all the factors.

Even if you choose not to be so formal in your evaluation, the 13 evaluation factors—plus any you added—will give you a much more objective way to examine the available information on each seminar.

Finding the Information You Need

The broadest sources of information described in the previous section—published directories, on-line computer databases, and locator/enrollment services—may not provide detailed information on all the evaluation factors you've identified. For example, directories and on-line computer services usually include only the barest essentials and don't contain detailed program outlines, cancellation histories, testimonials, and other information you will need.

In these cases, the information you need to make the final selection decision will be obtained primarily from the detailed brochure, catalog, or other materials furnished by the sponsor. But often even this material will not be adequate, and you will need to call the sponsor for still more specific, current information.

Besides, the additional information furnished by the sponsor provides yet another way to evaluate them. Are their materials professional in appearance and well-designed? Are they clear, easy to read and understand? Did the sponsor handle your request promptly and professionally?

Although the treatment of most of the key evaluation factors is self-evident, a few additional considerations are in order for some of them.

Content/Program Outline. Of course the content is the key reason for attending the seminar, and the closer the content conforms to your needs, the more likely the seminar will be of value. Experience suggests that industry specialization goes a long way toward increasing course relevance. So if one seminar is "generic," while another is tailored for your particular industry or sales situation, the latter is likely to be more valuable.

Many sponsors have found the two-level or three-level course outlines, which describe in detail both *what* will be covered and *when* in the program, are more effective than less detailed outlines. But be wary of detailed outlines that aren't connected with the time line. It's easy to include lots of topics in an outline which will not receive the corresponding proportion of time during the seminar.

And guard against another common problem: enrolling in a seminar because one or two items in a long outline are "exactly what we're looking for." We know, intellectually, that the course will probably not deal with our favorite topic for more than a short time, but we are attracted because it hit our "hot button."

Seminar Leader(s). Although many seminar sponsors show the name, photo, and brief biography of the seminar leader, often they are total unknowns to us.

And when we do see a familiar name, will that person be at every session? Such considerations are important because differences between seminar leaders often account for large variations in course quality, style of presentation, and even content.

Another note: When two or more leaders are involved in a seminar, participants invariably identify with one of the leaders over the others. This can result in apparently poor ratings for a truly excellent seminar leader who just isn't an "entertainer." If you get such adverse comments on the seminar leader scheduled for your session, probe further to determine if it's because they often work with someone who's a real dynamo and "steals the show." Try to determine how they are rated when they work alone; often it's a better indication of their true ability.

And become aware of how the seminar leader is compensated. Are they being paid hundreds of dollars a day to conduct the seminar? Are they working "free" in the expectation of getting business for themselves or their firm from participants in the seminar? Or are they merely building their résumé? If it's one of the first two, chances are good they'll do a professional job, since this is the way they earn their living. If it's the latter, you might be getting only their discretionary time, which may not produce the kind of seminar you expect.

Of course, if you are attending a seminar conducted by a well-known seminar leader, it's likely that enrollment fees, along with "back-of-the-room sales" of their books, tapes, and other materials, is what is paying their way. These seminars can be quite valuable because they are also "the product" which is being sold. And if they've been offered for a while, the product is probably valuable.

Schedule and Location. Many organizations select the location of the seminar so that other business can be transacted in conjunction with it. This practice spreads training costs and may also increase the number of available training opportunities.

It follows that resort locations will generally only be selected when one of the objectives is to provide a relaxed environment for the participant, or where that's the only place a desired seminar is offered. Sometimes such public seminars are used as a "reward," in which case a fancy location may be preferred. But a better solution is often to attend the seminar at a businesslike, city center or airport location and allow the participant recreational time at the beginning or the end of the seminar.

Another consideration for popular, limited class-size seminars is the possibility of a waiting list. Maximum class size is set either by 1) facilities limitations, 2) instructional requirements, or 3) staff limitations. Since meeting rooms are often reserved far in advance, facilities limitations are the most frequent reason for a waiting list, especially for new seminars which have been well-received.

At the other end of the spectrum are lightly enrolled seminars offered by sponsors who have a minimum class size, below which they won't run the session.

Such minimum class sizes are set by either 1) economic considerations, or 2) by the number of persons required for class activities such as role play, case studies, team assignments, and/or discussions.

Participant Testimonials. The most reliable way to check out a seminar is by speaking with those who have recently participated in it. This is easy since most sponsors will gladly furnish names and telephone numbers of individuals in positions similar to yours who have recently attended the seminar you're considering. Again, the more sophisticated seminar sponsors are even careful in this regard, so that they do not "overuse" any given participant as a reference, but will furnish their name only a few times, and only on a specific request.

Calling or Writing the Seminar Sponsor

In the course of collecting information from the different seminar sponsors you're going to find that direct contact with them furnishes another important perspective to help evaluate many other key elements. This perspective starts the moment the sponsor answers their phone, as you sense the professionalism with which your call is treated, how your call is routed, how many tries it takes to get to the right person, and how your request for information is handled.

What to Look For in the Confirmation Package

One piece of visible evidence on the seminar sponsor is the confirmation package sent to those who enroll. The better confirmation packages should contain at least the following:

1. A personal confirming letter signed by the registrar
2. Premeeting preparation materials (including specific instruction on how to use them)
3. The invoice or credit card charge slip for the seminar fee
4. Hotel reservation forms (for those programs involving an overnight stay) along with literature on the hotel or meeting facility
5. A fact sheet detailing transportation, restaurants, points of interest, dress codes, and other helpful information for the attendee
6. The latest copy of the promotional brochure on the program (which should be checked against the one you have already to be sure that there haven't been any course changes made). And this material should be mailed in a timely fashion, essentially immediately upon receipt of the registration.

GETTING THE MOST OUT OF A PUBLIC SEMINAR

O.K.! You've found the seminar that's just right, you've enrolled your attendee in the session that's most convenient, you've got a great deal on the hotel room and airline ticket, and you've found a way to combine the seminar with other business so the net cost in time and dollars is a rock-bottom minimum. What's more, the program is exactly what you were looking for, as personally confirmed by several recent participants with experience profiles similar to your attendee. Now all that's left is to make sure that you get what you paid for: a superb training experience.

Well, just signing up for the seminar isn't enough. There are some activities which you can do *before*, *during*, and *after* attendance at the seminar that can greatly enhance the value of the program to your participant and ensure that you get your money's worth from the event.

What to Do Before Leaving for the Seminar

To be sure your participant gets the most from a public seminar, here's a short list of activities that should be completed by all participants before they leave for the course:

1. Write out the reasons for attending: what you really want to get out of the program both on-the-job and personally.

2. Reread the seminar program, and mark the key topics you want to be certain are covered. (Be prepared to check these off as the program progresses or even to ask the seminar leader specifically to cover these topics because they're important to why you selected this course.)

3. Prepare a list of the questions you want to get answered at the seminar. A typewritten list is always in order. Consider making a copy so you can give the questions to the seminar leader before the course starts (or at the first break), telling him or her you're attending the seminar to get answers to the questions.

4. Have a discussion between the participant and their immediate superior, and/or co-workers, to identify the kinds of questions they want answered, and the kinds of information they would like brought home from the seminar. Add these to the list of questions started in entry 3.

5. Make sure to complete all the preseminar work, advance reading, and other preparation that's been assigned by the seminar sponsor. Some seminars require questionnaires to be completed and mailed to the sponsor for computer scoring before the program. Be sure to get these in on time. If there's reading to be done, be sure you allow enough time to complete it *before*

the seminar begins. And if there are books, software, supplies, or other items to be purchased in advance, make sure that's done as well.

6. Make your travel plans so that you do *not* have to leave the seminar early. Many seminar programs present important information at the very end, and leaving early not only causes you to miss this information, but may also be disruptive to the class as a whole.

7. Try to stay at the hotel where the seminar is being held. This cuts down on unnecessary commuting, enables you to get more work done during the breaks between sessions, and makes it easier for you to spend additional time with the seminar leader or other participants.

8. Beware of last-minute cancellations of the session by the seminar sponsor. Unfortunately, cancellations are a fact of seminar life, with experienced sponsors running only a few cancellations while others may cancel as many as one-fourth of their sessions. Such cancellations by the sponsor can be the result of a last-minute decline in enrollments, transfer of a large group of attendees to another sessions, accident or illness of a key seminar leader, or some other event outside the sponsor's control. But a frequent cause of seminar cancellation—especially for new courses—is the failure by the sponsor to promote the program adequately, thereby not getting a sufficient number of participants to hold the session.

    You can help alert yourself to these "surprises" by calling the seminar sponsor three to four weeks before the program and asking for the number of participants currently enrolled, as well as the names of the companies they represent. Be suspicious if most of the participants came from just a few organizations, because withdrawal of one organization might cause the session to be canceled. Ask the sponsor if *they* have any concerns about canceling the course, because you have other travel connected with it and don't want to have to cancel at the last minute. Most sponsors will be very open about their current state of enrollments.

9. Double-check all meeting and travel arrangements. It's hard to believe, but a surprising number of seminar attendees make travel arrangements to the wrong city, arrive on the wrong day, check into the wrong hotel, take the wrong ground transportation, forget their advance materials, and otherwise foul-up what should be a successful course attendance. Don't let it happen to you!

10. Finally, what if you must cancel your enrollment and will not be able to attend another session in the foreseeable future or send a substitute. If you are entitled to a full refund of any fees paid, make sure you request that refund when you cancel by phone, and confirm your conversation immediately by writing. If you do not receive your refund in a reasonable time (say, 30 days) ask for the refund again. Because of paperwork problems some seminar sponsors routinely wait for the second or third refund request

before paying it. Such delay often saves the unnecessary paperwork of checks crossing in the mail, travel authorizations, cancellations, reauthorizations and other turmoil for an individual who is going to sign up for another course a few weeks or a month later. But if it's clear you want a refund, ask for it—and follow-up until you get it!

What to Do at the Seminar

Public seminar participants can increase their benefits by the following actions:

1. Arrive at least one-half hour before the program is scheduled to begin. If the seminar advertises registration starting at 8:30 A.M. and the course beginning at 9:00 A.M., try to arrive at 8:15 A.M. This will enable you to get the best seat and also may provide an opportunity to chat informally with the seminar leader. Establishing such a person-to-person rapport will often get more personal attention by the leader during the seminar than will someone who arrives late and never talks to the instructor.

2. Sit near the front of the room where you can see the visual aids and demonstrations directly, and where you can talk informally with the seminar leader during breaks and quiet moments. While sitting at the back of the room may give a sense of anonymity and "easy escape," it will not yield the most from the seminar.

3. Be prepared to ask questions (from the list prepared in advance, or others that come up during the seminar) at every appropriate opportunity. Of course, it's important to strike a balance between "dominating the session" by asking too many questions (and irritating the seminar leader and perhaps other participants) and asking enough questions to be sure you get what you paid for. Contrary to common belief, good public seminar leaders actually like questions because they help get the audience participating actively and they set a direction for the course. Since seminars typically include far more material than can be presented in a single session, the questions that are asked often determine what will be covered.

4. If you have a short written list of questions (say, a half page or so), make an extra copy and give it to the seminar leader. Most leaders are conscientious and will go out of their way to be sure your questions are answered, even if it means extra time on their part at the breaks.

5. Try to get some personal time with the instructor, particularly if you have specific questions that you think he or she could answer. If lunches are not included in the program, offer to take the seminar leader to lunch at your expense. Tell them you'd like to have them answer a few questions

that wouldn't be of interest to the rest of the group. The same thing applies to a drink after the seminar, and again if you offer to pay you'll get more attention. Of course, if your questions are being adequately answered during the seminar (it is hoped that that's the norm) then you don't need to cut into the instructor's recuperation time outside of class hours.

6. Look carefully at "back-of-the-room" materials for sale at the seminar. Such materials are often purchased on impulse and may or may not have lasting value. Of course, if they supplement your reason for going to the seminar, or are what you've been looking for anyway, they can often be purchased at lower cost at the seminar, and by hand-carrying them you will have use of them without delay.

7. Try to ignore the normal "rough edges" which can be a part of any seminar. The more professional seminar sponsors will have worked out most of these "bugs" before you attend the course, but some problems arising from last-minute meeting room changes, failure of materials to arrive, schedule conflicts within the hotel, last-minute substitution of seminar leaders, and so on are totally unavoidable. Many times an inexperienced seminar attendee will rate the seminar down for congestion or noise in the hotel, expensive ground transportation, or excessive crowds in the city, none of which are really under the control of the course sponsor, who may have booked the session years in advance.

8. The last thing you need to be prepared for is what to do if, in spite of everyone's best efforts, the seminar is definitely *not* what you expected. If the seminar sponsor offers a money-back guarantee after the course is completed, you may want to consider "sticking it out," particularly if there is material later in the course that will meet at least some of your needs. If the sponsor only offers to refund fees up to the first break, you need to make a fairly quick decision.

For conferences and major sponsors, you may want to tell the instructor at the first break that the course is not what you expected. There have been cases where the instructor knew of another session running at the same time and place that *was* appropriate and helped get the transfer accomplished on the spot.

What to Do After the Seminar

It is rare that attendance at a seminar produces "magical" change in anyone. It's also true that the highest level of enthusiasm and excitement is immediately following the seminar. For this reason, many seminar sponsors call the postcourse evaluations "smile" sheets because the majority of participants are very excited and positive immediately upon conclusion of the seminar. You can turn this to your organization's advantage by taking the following actions upon completing the seminar:

1. Ask each seminar attendee to report back either verbally or in writing immediately upon the course's completion. This might include filling out a form, writing a letter, or having an interview with someone in the training organization. Such postseminar interviews can serve the important function of building the training department's files, so future participants can benefit from the immediate reactions to past participants.

2. Have the participant identify others in your organization who would benefit most from attending the same seminar. This would ideally be part of the previous item, but it needs to be given separate thought, especially if there are portions of the seminar that would be relevant for people outside the participant's own department.

3. If others are to attend, this is the time to consider whether the total population to be trained is large enough to warrant contacting the seminar sponsor to arrange either (a) discounted fees for additional public seminar attendees or (b) an in-house presentation of the program at your own facilities.

    In-house presentations are usually less expensive than public seminars for as few as 8 to 10 participants, because the travel costs and travel time are eliminated, and because sponsors frequently offer in-house seminars at a greatly discounted price compared to their public seminars.

4. Consider asking the attendee to make a short presentation for others in the organization on what was learned at the course. This may be arranged by the training department, or by the attendee. The purpose is to share with others the principal insights and "bottom-line" conclusions reached, *not* to try to reproduce the seminar experience (which is usually not possible). Be careful about the unauthorized reproduction of course notes or other copyrighted materials.

5. Finally, what do you do if, after the seminar is over, you realize that the program *was* a disappointment? First, take time to write down a clear description of why, separating, if possible, issues from real failures in the seminar. Then call the seminar sponsor and talk to one of their officers (not a registrar or customer service person). Tell them your concerns and disappointments frankly and unemotionally. Most seminar sponsors will go out of their way to make things right. And, remember, a refund of fees is not always the best way to handle these situations. Often an opportunity to attend one (or more) additional sessions at no charge is preferable because it gives a second chance to evaluate the course favorably—and to get additional training at low cost.

SUMMARY

Public seminars play a major role in sales training and represent a significant resource for sale trainers. This chapter has described the kinds of programs available, types of providers, when to use them, where to locate information about them, how to select those seminars most suitable for particular needs, and how to utilize them effectively.

ADDRESSES FOR TABLE 29–2 INFORMATION SOURCES

ASTD
1630 Duke St.
P.O. Box 1443
Alexandria, VA 22313

Sales and Marketing Management Magazine
633 Third Avenue
New York, NY 10017–6706

Executive Telecom System
9585 Valparaiso Ct.
Indianapolis, IN 46268

The First Seminar Service
88 Middle St., 2nd. fl.
Lowell, MA 01852–1813

Gale Research
1249 Washington
Detroit, MI 48226–1822

Hope Reports, Inc.
1600 Lyell Ave.
Rochester, NY 14606–2324

Human Resources Development Press
22 Amherst Rd.
Amherst, MA 01002

National Society of Sales Training Executives
1040 Woodchuck Rd.
Orlando, FL 32803–3588

National Training Network
10335 Thanksgiving Lane
San Diego, CA 92126

Seminar Clearinghouse International
630 Bremer Tower
St. Paul, MN 55101

Seminar Information Service
17752 Skylark Circle
Suite 210
Irvine, CA 92714–6419

Seminar Directory
1402 E. Skyline Dr.
Madison, WI 53705–1133

Timeplace, Inc.
460 Totten Pond Rd.
Waltham, MA 02154–1906

Training Magazine
50 S. 9th
Minneapolis, MN 55402–3165

chapter thirty ——————————————————————

PROFESSIONAL NETWORKING FOR SALES TRAINING

—————————————— *JUDITH C. QUINN*

Judith C. Quinn is corporate manager for training and development, Informix Software, Inc., Menlo Park, California. Formerly, she was manager of training and development for ITT Information Systems, San Jose, California, and a sales trainer specialist with the General Electric Medical Systems Division, Milwaukee, Wisconsin. Her sales career began in the industrial lighting sales ranks in Honolulu, and she now has ten years of experience in the microcomputer industry. She is a dedicated networker, who has created and used her contact base as she has started up and managed training departments for both manufacturers and marketers. She developed and delivered the first computer-aided instruction sales training involving the introduction and rollout of medical CAT scanners. She is a grandmother, scuba diver, pilot, and award-winning archer. She has a degree in adult education from the University of California, Berkeley. Professional networking continues to play a vital part in her professional life.

Professional networking arises out of the needs and visions that cannot be fulfilled in isolation. In fact, isolation is the very antithesis of networking. This chapter examines the nature of networks, and why they are becoming plentiful in the sales training field as well as in many other professional endeavors. The establishment, maintenance, and growth of networks, by networkers, is one of the most encouraging developments in the sales training field today. Networks need to be elevated as a consistent model for making and keeping contacts.

Professional networking is an exciting worldwide movement among professional sales trainers and in most other professional fields as well. Networking develops and uses contact with others who share similar fields of interest and expertise. Networkers deliberately make contacts to seek inside information, advice, and ideas; to elicit sales leads and referrals; to garner moral support when the going gets tough; and often just to talk in confidence with someone about job-related concerns and achievements.

Professional networking is a sophistication of one-on-one communication. Networking is personal. Networking is not mass communication. It is not a shallow or feckless exercise. Above all, networking empowers the individual. People in networks help, nurture, and enable one another to learn and to grow in their chosen professional fields.

Anytime you reach beyond yourself, and to another person, to obtain information, solve a problem, accomplish a task, understand yourself and your environment better, you are *networking*. Learning to network, and becoming a successful practitioner, will allow you to acquire a skill critical to success in your chosen career. In the field of sales training, networking is not an option. It is essential.

HOW TO NETWORK SUCCESSFULLY IN SALES TRAINING

In general, successful networkers appear to share a significant characteristic: a well-honed sense of intuition, that is, the ability to feel what is going on around themselves. They know where to get information, whom to trust, and how to share. Professional networkers know precisely where to go to find information, quite often operating on pure hunches. You can do it too.

Just as you build your personal relationships with care, you need to network with care. Your reward is enrichment of both your personal and professional life. Truly valuable networking is built on a foundation of sincerity, integrity, competence, and considerable hard work. Professional networking is not a matter of making a few quick phone calls, exchanging business cards, or "doing lunch." To network successfully, you must meet four primary responsibilities:

1. *Do Quality Work.* One of the most lasting impressions you make on another person is the quality of the work you produce. If you excel at your work, word will spread, and others whose work equals yours in quality—in other words, networkers—will judge you worth having as a contact and seek you out. In the sales training field, countless executives are placed solely by word of mouth, not by executive search firms. The best candidates for the senior positions come as referrals from others who have seen their work, the quality of their contributions to their professions. And that's what moves them up the ladder.

2. *Maintain the Highest Standards of Integrity and Sincerity.* It is indeed sad to see people in the professional world convinced that the only way they can get ahead is to compromise their integrity by lying, cheating, and taking outright advantage of others. Sooner or later, everyone is judged by the principles he or she lives by. The principles that endure, and serve to further enduring careers are honesty, sincerity, integrity. Colloquially put, what you put out comes back. My vote always goes for the professional who will pass up a profit, or anything else that would infringe on his or her principles.

3. *Choose People You Really Want To Be With.* In a network, choose the people you respect and like. The best networking contacts are also friends, with whom you can forge relationships based on values you share, and on the time you enjoy spending together. Some of the best networkers are often those who train in fields entirely different from your own. But because you share a mutuality of interests, and because you understand each other as friends, the benefits of your friendship carry over into the business arena.

4. *Become a Giving Person.* Always be willing to give something of yourself. Professional networkers are those who continually give of themselves without expecting a return on the investment. Some of the most valued relationships in networking spring up when you go out of your way to do a favor for another. When you are asked to participate in a network, check your own motivations. Make sure your generosity quotient is up to snuff. When everybody in a network asks themselves that question, the network is successful.

WHAT ARE NETWORKS? _____

To understand networking, you have first to understand the structure of networks. Contrary to some conventional wisdom, bureaucracies and hierarchies are not the only forms of social, political, or business organization.

Slowly, but surely, experts in business and the professions are beginning to see the wisdom of the network, bearing out the authenticity of the work of organizational theorists like Tony Judge,[1] who has written, "the network is not coordinated by anybody. The participating bodies coordinate themselves, so that one may speak of auto-coordination." In other words, everyone in the network participates equally to build a structure based on common concern and common need.

In their work, *The Networking Book: People Connecting with People*, Jessica Lipnack and Jeffrey Stamps[2] provide one of the best definitions of the structure and processes of a network. Paraphrased, their premise is this: *a network arises out of the needs and vision that cannot be fulfilled in isolation.* The tension between remaining in isolation and reaching beyond oneself is the linchpin of the phenomenon of professional networking. When one reaches out, one breaks the barrier that might exist, and careerwise, nothing is ever the same again. Look at the structure of a network. How it is built illuminates how it works. In a well-built network, the members use the network to their mutual and individual advantage.

Networks rely on the actuality of a connected, integrated *inter*connected design. The sense of the design is that everyone needs everyone else, and therefore, a network foundation is laid with cooperation.

A network structure obviates the need for a pyramid. A circle might be interpreted as a pyramid loosened a bit—quite a bit. A hierarchical model is instantly obsolete in a network because everyone participates as an equal, without regard to rank or position.

A network structure is flexible. There is no rigid or centralized control in a network. Nobody has all the answers, which relieves everyone the curse of being an "expert." It also eliminates the necessity of having someone in a power position. The power in a network derives from the interconnected whole looking out for each other, creating and building upon leads to advance each other's career.

Network structures are built of many perspectives, not unlike the view through a kaleidoscope. Rich in experience, as its members are rich in experience, no single viewpoint is obtainable from a network. This kaleidoscopic vision allows

---

[1] Tony Judge, in Michael Maccoby, *The Gamesman* (New York: Bantam Books, 1982).

[2] Jessica Lipnack and Jeffrey Stamps, *The Networking Book: People Connecting with People* (London: Routledge & Kegan Paul, 1986).

members the luxury of hooking into the right idea at the right time, rather than feeling obligated to listen to one source who has "all the answers." Diversity is the soul of the network, and it certainly derives its resilience from this feature.

Network structure relies on leadership diversity. If there is no one "source," leadership is a shared prerogative, dispersed among the members of the network who choose to assume it. This shared leadership, dispersed among many members at different times, and fitting different needs, enables growth and support and advancement. It doesn't control it. The difference between enable and control is key to understanding network leadership.

Combine the structure with the process that grows out of it—a process that relies on relationship, loosened boundaries, no specific individual exerting control while all members are involved in enabling action, a common concern for the individual spread among individuals who share common values, and a quest for each other's welfare—and you have a resilient structure that liberates success and reward for all who participate in it.

HOW TO START NETWORKING

Networking is not a pastime. It's a business, and thus involves your valuable time and energy. In *Networker's Guide to Success*,[3] author Jan Triplett points out, "Networking is costly in terms of energy, time, and money. No one can afford to throw those things away." The truly valuable and enduring network is built on sincerity, integrity, competence, and hard work. That's what successful sales training is built on as well.

When you network with care, as you build your relationships with care, your reward is enrichment of both your personal and professional life. Before you begin to network, and after you've decided to network, ask yourself three questions. Who? What? How? Who do you want in your network? What do you believe they can contribute? How can you keep them involved and interested in you and in the network? The answers have ramifications far beyond making a few phone calls.

Finding the Networkers

They're everywhere you are. They're young, old, middle-aged; lavishly experienced, not very experienced. Some are detail oriented. Others are broad generalists. They're men, women, creatively oriented, technically oriented. They're sales trainers and professionals in fields quite different from your own. They are knowledgeable professionals eager to share their knowledge with other professionals—you. Find them in all kinds of places.

---

[3] Triplett, *Networker's Guide to Success.*

Since they're your peers, they can be found at seminars, ASTD meetings, and a host of professional gatherings. When your company introduces a product at a rollout, there is usually a potential networker present. Don't be hesitant to walk up to an author with whose work you are familiar and talk about networking possibilities. Very often, nobody approaches these experts, fearing they're overbooked. Don't believe it. Many are flattered and more than needful of a network.

Another underused networking contact base is the world you live in—your social circle, your church, synagogue, educational and cultural roots, the community activities in which you are involved. Networkers are thick in these kinds of settings, often just waiting for an open door, or a friendly inquiry—from you. Within your own company, watch the shifting management levels. When a new person is promoted, become familiar with the person's background. See if this individual, or another on the way up the ladder is someone you want to make contact with, someone who belongs in your network.

Finding the people isn't difficult. What is sometimes difficult is making sure you are choosing the right person for your network. Ask yourself these questions.

Do I like this person?

Can I trust this person?

Is it possible this person will become "clingy," or overdependent on me?

Is this person honest and straightforward?

Will this person contribute to the network, or just use it?

These questions call for highly subjective conclusions, but unless you can use your intuition to answer them in the affirmative, you don't have the right person for the network. Networks are not enormously fragile, but why engage someone who is a potential weak link to erode the connections that bond you. Networkers may be everywhere, but sometimes, the selection process is a case of "many are called, but few are chosen." Don't be afraid to make the judgment. You'll have to live with it. It's part of the risk of networking that its greatest attraction—the cooperative action of human beings—is also one of its pitfalls. Try not to be deterred by the risk of connection. You'll find the ratio of suitable to unsuitable potential networkers is about 90 to 10.

The Five Rules of Professional Networking

1. *Listen*. Actively practice the hardest earned lesson of your sales experience. Most people are saying much more than is coming out of their mouths. Learn to listen for the subtext.

2. *Ask questions*. Remember to ask a question before you spew out advice. Good questions are many times more useful than terrific, and often premature, answers.

3. *Don't make assumptions*. This is one of the oldest and most routinely ignored rules of business. Before you proceed with an action, be sure you have *all* the information. It is never safe to assume you have all the information unless you really have it.

4. *Concentrate on quality not quantity*. How good the information is, is always more important than the number of people supplying it. It's far more valuable to have 4 or 5 active contacts in your net than 50 you seldom hear from.

5. *Be useful*. Get the information out succinctly. Put it all in a nutshell. You don't need to record volumes of detail. You simply have to know how to tell another networker where to find information, where to obtain the lead.

If you abide by these five rules and practice the elements of success mentioned earlier in this chapter—do quality work, maintain the highest standards of integrity and sincerity, choose the people you really want to be with, and become a giving person—you will be a successful networker.

HOW TO USE A NETWORK

If you use your intuition, the ability you possess to glean reality from what goes on around you, you will know who to trust, where to find information, and how to share it.

Here are a few simple tips to use as tools as you begin:

1. Devise a method of storing information. The method should allow you to retrieve what you need when you need it, simply and rapidly.

2. Keep it simple. Simplicity is the key to good networking. If you feel you are being overloaded, stop immediately, and learn to say "No."

3. Use the telephone for local networking, computer-assisted networks for long distance, and the mail for lengthy documentation.

4. Use postcards if you don't have your own personal computer. It is amazing how this technique teaches you how little you really need to say to get your thoughts and information across.

Networking can be a blessing and a burden at the same time. Professional networking is given an excellent description in Cooper Edens' book, *The Caretak-*

ers of Wonder.[4] He describes it as the "Great Rainbow Balancing Act." Think of it as trying to balance a rainbow on your hand—with its blending colors, its indistinct lines of delineated hue, its habit of being suddenly visible in certain lights but not in others. The rainbow is difficult to see, unless you are paying attention, holding your hand steady, so you can see how lovely it really is. That's the network for you.

THE BENEFITS OF NETWORKING IN SALES TRAINING

Networking is replacing the hierarchical management forms associated with frustration, impersonality, inertia, and failure in business. This failure of top-down power structures to solve today's human resource problems has forced people simply to talk to each other.

A major benefit of networking in the sales training field is putting different people together in ways that they make significant contributions to a wider range of decisions and do things that would ordinarily be slowed by a structured line organization. The new business networks are not elitist, as the "Good Old Boy" networks were. Every member is a peer. Networks are three dimensional, involving people from all levels with information being the great equalizer. Networking expert John Naisbitt says, "There are three fundamental reasons why networks have emerged as a critical social form now: (1) the death of traditional structures, (2) the din of information overload, and (3) the past failure of hierarchies."[5] Anyone who questions the value, or the future of networks should keep Naisbitt's words in mind.

A Sales Training Business Scenario

It is 4:00 P.M. on a Friday. Your manager has just relayed a bit of information down the line. He wants your recommendation for a top-level executive "Negotiating Skills" course on his desk by Monday morning. Before panic sets in and you are reduced to a quaking mass, you ask for a couple of details: How much does he want to spend per executive? How many is he planning to "invite" to attend the course? Is there any time constraint? (There may be other questions.) Finally, fortified with a little more information, you go to your computer and access your training informational network. ASTD's TRAINET puts an excellent database at your fingertips for complete seminar information, including detailed course descriptions which are now available with CourseWARE Database, as

---

[4] Cooper Edens, *The Caretakers of Wonder*, 2nd ed., (San Diego, CA: Green Tiger Press, 1987).

[5] John Naisbitt, *Megatrends* (New York: Warner Books, 1982).

well as on-line ordering ability. But before you hit the "o" (ordering key) and input the requested billing information, put your own "Professional Network" to work.

Your desk Rolodex address file has the telephone numbers of your personal networkers. A call to a training colleague is placed. You share your problem with this networker, asking her or him questions: Has she or he used this vendor before? What were the reactions of trainees and management to the content and the instructors? Did the company provide adequate visuals/documentation? What type of follow-up was used? Was there adequate experiential linkage? Was the course cost-justified? Would he or she use the vendor again? And a bottom-line question: Does she or he believe this course would appropriate for my company?

Based on the response from the personal network, you would then be in a position to make a decision. So it's back to the computer to write a short memo to the boss detailing the course recommendation you are making, prices, and available dates/locations. You attach a copy of the printout of the ASTD TRAINET detail, and it's ready for the boss's desk.

On Monday morning when the boss has made a decision based on your recommendations, he praises the detail information and your meeting the schedule deadline. It's now time to get back into ASTD TRAINET, make the course selection by hitting the "o" key to place the order, and request phone confirmation. Now you can notify the trainees (executives throughout the United States) and within an hour have all the necessary information on the computer "E MAIL" for enrollment.

The successful completion of this business scenario would not have been possible without the involvement of both a computer network and a professional network. Computer networks are one of the most powerful tools for change in professional training and development. They allow access to high-powered communication technology that can be utilized in research, data transmission, teleconferencing, and most important, experience sharing.

Professional networkers are natural outgrowths of such sharing-oriented organizations as NSSTE (National Society of Sales Training Executives). The recognition of different expertise among professionals, getting to know the personalities of the individuals, and an honest desire to share a knowledge base all contributes to a successful network. Someone once said, "It's not what you do, it's who you know and the smile on your face. It's contacts, contacts, contacts. . . ."

SUMMARY

In making the transformation from the industrial to the information age in a world of big powers, major media and monopolies, in a world approaching a population of 5 billion individuals, there is a critical need to network. Willis

Harman of the Stanford Research Institute has said, "We can choose either to understand and move with the tides of history, whatever they may be, or try to resist them. Upon that choice may rest in great measure the state of business in 1990 and beyond."[6]

Networking is replacing the hierarchical dinosaur forms of top-down bureaucracy that now tread the business landscape in decidedly decreasing numbers. Inertia, frustration, impersonality, and failure are their offspring, and they are dying off too. Only networking is really growing as a way of managing and support. Only networking is really putting people together in ways that make tremendous contributions to a wider range of decisions. Accomplishments are made that would normally be slowed or eliminated by the normal bureaucratically structured form. Because networking empowers the individual, people in networks tend to nurture one another. This kind of nurturing keeps everybody going. And when everybody is kept going, so is a healthy business climate.

Is networking a risk? Yes, but then all sharing is a risk. It is also rewarding. Early in my sales career, after a "bad day," a sales manager and personal mentor took a crumpled piece of paper out of his wallet and gave it to me. No author was credited. It read as follows:

> To laugh is to risk appearing the fool. To weep is to risk appearing sentimental. To reach out for another is to risk involvement. To expose feeling is to risk exposing true self. To place your ideas, your dreams before the crowd is to risk their loss. To love is to risk not being loved in return. To live is to risk dying. To hope is to risk despair. To try is to risk failure. But risk must be taken, because the greatest hazard in life is to risk nothing. The person who risks nothing does nothing, has nothing, and is nothing. He may avoid suffering and sorrow, but he simply cannot learn, feel, change, grow, love, live. Chained by his certitudes, he is a slave, he has forfeited freedom. Only a person who risks is truly free.

Networking is a new form of freedom, a new path to happiness and success. Networking is basic to organizing, managing and training people as we approach the twenty-first century.

<hr>

BIBLIOGRAPHY

Byrne, John A. "Office Pals: It Pays to Stay in Touch," *Business Week*, April 6, 1987.

Ehrenreich, Barbara. "In Praise of 'Best Friends,'" *Ms*, January 1987.

Ferguson, Marilyn. *The Aquarian Conspiracy, Personal and Social Transformation in the 1980s*. Los Angeles: J. T. Archer; New York: St. Martin's Press, 1980.

<hr>

[6] Willis Harman, Stanford Graduate School, Summer Session, 1986.

Helgesen, Sally. "The Truth About Networking: How to Be Helpful to the People with Whom You Work, and Ask for Help in Return," *Glamour*, September 1985.

Konrad, Walecia. "Influence Brokers for Women in the Corporation," *Working Woman*, October 1986.

Lloyd, Kate Rand. "The Old Girl Network Goes National," *Working Woman*, November 1985.

Kleinman, Carol. *Women's Networks*. New York: Ballantine Books, 1980.

Much, Marilyn. "The Disco Connection (Networking 500 Club Inc.'s Weekly Party)," *Industry Week*, September 29, 1986.

Nilson, William P. *Achieving Strategic Goals Through Executive Development*. Reading, MA: Addison-Wesley, 1987.

Sanoff, Alvin P., and Miriam Horn. "The Old School Ties That Bind (Harvard Graduates Help Each Other)," *U.S. News & World Report*, August 25, 1986.

Segal, Troy. "Ambitious and Single? Now There's a Club for You," *Business Week*, October 20, 1986.

Smith, Carol Cox. "Networking: Getting Your Money's Worth," *Savvy*, May 1, 1987.

Tichy, Noel, and Mary Anne Devanna. *The Transformational Leader*. New York: John Wiley, 1986.

Toffler, Alvin. *The Third Wave*. New York: Harcourt Brace Jovanovich, 1981.

Toffler, Alvin. *Future Shock*. New York: William Morrow, 1970.

Wickham, Dewayne. "Network to Network: A Variety of Professional Organizations Are Pooling Their Resources to Gain Economic Power and Foster Career Development," *Black Enterprise*, February 1987.

Wilson, Larry. *Changing the Game: The New Way to Sell*. New York: Simon & Schuster, 1987.

Wunderman, Lester. In Michael Maccoby, *The Gamesman*. New York: Bantam Books, 1982.

INDEX

ABI/Inform database, 603
Accommodations, 260
Adjunct faculty participation, 624–25
Administration of training department, 243–70
  budgets, 257–58
  computerized training records, 253–54
  computer programs for manager, 255–58
  external trainers, 259
  in-company guest speakers, 259–60
  invitations to trainees, 262–64
  location of department/training activities,
    247–49
    off-site training, 248–49
    on-site training, 247–48
  personnel department resources for training,
    257
  purchasing, 258–59
  recordkeeping, 250–53
  skills that can be modeled, 245–46
  staffing, 258–59
  support levels, 246–47
    customer, 247
    immediate supervisor, 246
    senior management, 246
    trainee, 246
Adopt-a-school program, 627–28
Adult learning skills, applied to sales training
  program, 123–24
Advance payments, public seminars, 688–90
Advisory committee guidelines, 170–71
AIDA formula, 346–48
  developing desire in action, 348
  maintaining interest, 347–48
  moving audience to take action, 348
  sales training meeting, 514–15
  securing audience's attention, 346–47
*American Micro Media Microcomputer Training
  Catalog*, 587
American Society for Personnel Administration,
  167
American Society for Training and Development (ASTD), 16, 20, 70, 94, 102, 167,
  255, 332, 589–90
  competency study, 70–72
  Member Information Exchange (MIX), 333
  Sales and Marketing Professional Practice
    Area, 16
Andragogy
  instructional guidelines, 120–23
    compared to teaching children, 122–23
    in sales training, 120–22
Application objectives, behavioral-change train-
  ing, 552

Arrival packet, 264
Association for the Development of Computer-
  Based Instructional Systems (ADCIS),
  586
ASTD Buyers' Guide & Consultant Directory,
  660
ASTD Training Video Directory, 600
Attendee accommodations, 260
Audience
  assessing the audience, 344–46
  audience contact, 351
  KILIS method and, 344–46
  lecture method and, 344–46
  packaged programs and, 656–57
Audio tapes, 459–60, 463
  advantages, 459
  disadvantages, 459–60
  user tips, 460
Audio/visual media, 439–68
  guidelines, 461–63
  as learning aids, 441–42
  preparation/use guidelines, 465–66
  suppliers, working with, 466–67
  types of, 442–63
    audio tapes, 459–60, 463
    chalkboards/white boards, 453–55, 462
    computer graphics, 460
    flannel/magnetic/display boards, 455–56,
      462
    flipcharts/chart pads, 443–46, 461
    instructional carrels, 460
    models/demonstrations, 456–57, 463
    overhead projectors, 446–48, 461
    teleconferencing, 458–59, 463
    35mm slides, 451–53, 462
    written materials, 457–58, 463
  which aid to use, 464–65
Auditorium-style seating, meeting room, 527
Audits, behavioral-change training and, 572

Baker, James A., 469–91
Basic competencies, 72
Behavioral-change training, 551–52
  designing measures of change, 558–70
  determining need for, 555–57
    by informal means, 556–57
    linking corporate interests, 557–58
  how to measure change, 569–70
    audits, 572
    interviews, 572
    numbers and, 573
    observation, 572

Behavioral-change training (*cont.*)
    program/company records, 571–72
    qualitative measures, 569–70
    quantitative measures, 570
    self-/other judgments, 570–71
    surveys/questionnaires, 572
    training measurement dimensions, 573
  measuring principles, 554
  objectives for, 552–53
    application objectives, 552
    business objectives, 552
    determining whether objectives have been
      met, 553–54
    learning objectives, 552–53
  precipitating events, 555
  pre- and postmeasure design, 574–75
  pre- and postprogram measurement survey,
    565–68
  presenting findings, 573–75
  statistical checks, need for, 573
  when to measure change, 558, 559–69
    comparison groups, 564–69
    postprogram measurement, 559–60
    pre- and postprogram measures, 560–64
Behavioral triangle, 367–68
Bell, Chip R., 630–49
Bennis, Warren, 47
Bibliographies/indexes/abstracts, on selling
    skills, 594–95
Big picture, *See* Theme of training program
Boardroom seating, meeting room, 528
Books, on selling skills, 595–96
Brainstorming, workshops, 205
BRS Information Technologies, 602
Budgets, 58–59, 257–58
Burns, James MacGregor, 67
Business objectives, behavioral-change train-
    ing, 552
Business Periodicals Index, 594–95
Buyer's Guide and Consultant Directory
    (ASTD), 259
Buzz group tables, meeting room, 528
Buzzotta, V. R., 548–79

Call coaching guide, 287–89
Campus service
  for in-house training, 615–18
    planning checklist, 617–18
    selecting campus facilities, 616–17
Cancellation refunds, public seminars, 688–90
Case histories, workshops, 205
Case method, 403–21
  advantages, 405–6
    case method seminar evaluation form, 419
    case selection, maintaining quality in, 410

designing your own cases, 406–9
  Armour-Dial's method, 407
  Development Dimensions International
    (DDI) cases, 407–8
  Johnson Wax case, 408–9
disadvantages, 406
how to write a case, 412–14
  capture learners' attention, 413–14
  choose appropriate case for your needs, 413
  customize commercial case, 414
  research possibilities, 412
  state objectives, 412
measuring impact, 418–20
pacing the discussion, 417–18
  concluding, 418
source materials, 420
use in in-house training seminar, 414–18
  controlling the discussion, 416
  discussions that stray, 415
  participant's role, 417
  preparation, 415–16
  use of humor, 417
using specialized professional writers, 409–
    10
when to use, 410–11
Cataldo, Patrick A. Jr., 492–511
Catalogs, for audio/video cassette sales/sales
    management programs, 25–26
Certificates of accomplishment, 267
Chalkboards/white boards, 453–55
  advantages, 453–54
  disadvantages, 454
  user tips, 454–55
Classroom-style seating, meeting room, 527
Climate considerations, geographic location,
    221
Climate control, large facilities, 229
Clinics, 206
Closing, lecture, 359–60
Coaching, 283–305
  day-to-day coaching of call process, 284–96
    joint-call coaching, 290–94
    postcall coaching, 295–96
    precall coaching, 285–90
  group coaching opportunities, 304–5
  performance management coaching, 296–304
  trainer's role, 283–84
  training sales managers for, 305
  See also Performance management coaching
College/university classes, 331
Communication
  computer-based training, definition of, 495
  with management, 268
  of sales training ideas, 355–57
Comparison groups, behavioral-change train-
    ing, 564–69

Computer-based training, 494–510
  advantages of, 498–99
  benefits of, 499–501
  compared to other forms of media-based instruction, 495–96
  cost-effectiveness of, 499
  courseware acquisition/development, 502–4
  equipment acquisition and support, 502
  example programs, 496–98
  with extensive branching, 497
  human interaction techniques, 509
  impact on sales trainers, 509–10
  introduction of new system, 505–6
  with minimal branching, 496
  practical applications, 499–506
  program operation, 504–5
  selling skills, 601
  simulation games, 497
  software, 508–9
  technological advances, 506–7
    electronic mail, 507
    "person-to-person" live communication, 507
    subscription agreements, 506
  used in a complete sales training program, 501–2
  workstation systems, 507–8
  See also Interactive video training; Video/film
Computer-generated graphics, 460
Computerized training records, 253–54
  sample of, 254
Computer programs
  criteria for preparation of, 450–51
  for managers, 255–58
Conferences, 206
  conference centers, 225–26
  selling skills, 599–600
Confirmation package, public seminars, 699
Continuing education, 612–22
  campus service for in-house training, 615–18
  correspondence courses, 618–19
  executive education programs, 614–15
  faculty consulting services, 619–20
  on-site programs by university faculty, 620–21
  seminars/short courses, 612–14
Contract instructors, 78–79
Cooperative education/internships, 625–27
  advantages to hiring co-op students, 626
    sources of additional information, 626–27
Cooper, Daniel S., 492–511
Correspondence courses, 618–19
  questions to ask about, 619
Course strategy checklist, 142–43
Critical competency skills
  instructor/program designer skills, 98
  manager/needs analyst skills, 98–101
  pretraining assessment of, 100–101
  training personnel, 97–101

Current Index to Journals in Education (CIJE), 595
Curriculum plan, 56–57
  elements of, 57
Currie, Thomas R., 533–47
Custom programs, seminars, 612–13

Databases, on selling skills, 601–4
Data Training, 587
DATRIX II database, 603–4
Debate, as lecture method, 340
Decor, meeting room, 231
Delayed measurement
  of behavioral change, 560
    delayed measurement survey, 561–63
Del Gaizo, Edward R., 583–606
Demonstration, 368–77
  definition of, 368–69
  handling demonstration in sales training, 369–77
  sample feedback sheet, 373
  sample improved sales presentation, 374–77
  sample script analysis, 370–73
Demonstration contests, 39–40
  case studies, 39
  competitive presentations, 39
  role play, 39–40
Desert Survival simulation, 429–31
Design team
  Performance Systems Design
    client, 130
    instructional designer, 129
    instructor/trainer, 130
    production coordinator, 129–30
    project manager, 128–29
    subject-matter expert, 129
    writer/developer, 129
Developer-trainers, 78
Dialog database, 602
Directory of Human Resource Development Instrumentation, 601
Directory of Member Firms, 660
Display boards, *See* Flannel/magnetic/display boards

Education in sales/marketing, 607–29
  continuing education, 612–22
  current program offerings, 609–11
  history of, 608–9
  resource centers, 610–11
    Institute of Sales and Marketing Education, 611
    MarkED Resource Center, 611
  T & D programs, 609–10
  See also Continuing education

EdVENT database, 602
Emergencies, during sales training programs, 266–67
Enhanceable competencies, 73
Equal Employment Opportunity data, record-keeping and, 252
Equipment
audio/visual equipment, selection of, 466–67
preparation of, 352
ERIC database, 603
E-shape seating, meeting room, 529
Essential competencies, 72–73
Evaluating sales training, 532–46
action planning, 543–45
action plan form, 544
individual/group problem solving, 543
learner evaluations, 545–46
learner evaluation form, 545
learning objectives, 537–42
developing preset objectives, 537
methods of measuring learning, 538
reasons for, 534–36
role play, 541–42
role-play evaluation form, 542
written tests, 538–50
standardized tests, 538
tailored tests, 529
See also Follow-up, postprogram activities
Evaluation
of college/university programs, 628–29
Performance Systems Design, 150–51
achieving successful evaluations, 151
levels of, 150–51
Evered, James F., 3–26
Exchange programs (industry personnel/faculty), 623–24
Executive education programs, 614–15
Executive-in-residence (E-I-R) programs, 622–23
Executives International, 332
Experienced practitioners, 78
External sales trainers, 259
advantages/disadvantages to use of, 633–34
assessing cost, 641–43
"cost-plus" method, 641
questions to ask about pricing, 641–43
total package approach, 641
consulting process, 634–49
closure, 648–49
diagnosis, 644–45
initial contact, 635–44
response, 645–48
contracting tips, 640–41
criteria for selection of, 636–37
generic versus tailor-made training programs, 637–38
interviewing key people, 645–46

letter of agreement, 643–44
locating external resources, 638
requests for proposals, 640
sales consultants, 631–33
small versus large consulting firms, 639

Facilities, 222–28
benefits and liabilities of various types of, 223
conference centers, 225–26
corporate training facilities, 223–25
in-house space, 225
resorts, 227
training impact, 228
types of, 222–23
See also Learning environment; Meeting room
Faculty consulting services, 619–20
cost of, 620
selection of, 619–20
Field trips
budgeting of, 11–12
questionnaire, 160–62
Field visits, 197
Flannel/magnetic/display boards, 455–56
advantages, 455
disadvantages, 455
user tips, 456
Flipcharts/chart pads, 443–46
advantages, 443
disadvantages, 443–44
general guidelines for, 353
prepared charts, 444–45
user tips, 444–45
Floor plan, meeting room, 229–31
Focus groups, 195–96
Focusing, by leaders, 48
Follow-up
managers, 40
postprogram activities, 267–69
certificates of accomplishment, 267
communication with management, 268
recording, 268
reinforcement, 268–69
responding to trainee's needs, 267
training center library, 268
sales manager, 40
Food and drink, 235–36
training impact, 236
Four-week planning form, sample of, 282
Fox, Derwin A., 583–606

Gaming/simulation, 422–38
advantages, 428
conclusion of, 433
creating your own materials, 435–37

introduction of, 432–33
journals, 437
limitations, 428–29
measuring results, 433–35
organizations, 438
prepared materials, 429–32
role-play simulations, 427
simulations in occupational training, 425
strategic/insightful simulation, 426–27
true-to-life simulation, 426
use in everyday learning experiences, 424–25
when to use, 427–29
Geographic location, 220–22
activities, 222
climate considerations, 221
tips on choosing, 220–21
training impact, 222
travel accommodations, 221–22
Goal setting, as management process, 52
Group coaching opportunities, 304–5
Group techniques
role plays, 394–402
baseball, 395
E Pluribus Unum, 395
The Eternal Triangle, 395
multiple roles, 401
round robin, 401
syndicates, 402
Guarantees, public seminars, 688–90
Guest speakers, 261–62
buyers and other customer personnel, 261–62
company officers/department heads, 261
field managers, 261
Guide to Computer-Based Training, 587

Hahne, C. E. (Gene), 548–79
Handouts, guidelines for use of, 353–54
Hessan, Diane, 177–92
Hiring model, 73–74
Hollow square/hollow circle seating, meeting room, 529
HRD-in-Brief Series: Sales Training, 594
HRD Review, 593
Husted, Stewart W., 607–29

Immediate supervisor, contacting of, 264
Immel, Robert C., 154–76
Implementation
Performance Systems Design, 149–50
delivery requirements, 150
preparation activities, 149–50
strategy formation, 50
In-baskets, workshops, 205

In-company guest speakers, 259–60
Information sources, public seminars, 705–6
Innovation, of leaders, 48–49
The Inquiring Mind (Houle), 111–12
Institute for Graphic Communication, 588
Institute of Sales and Marketing Education, 611
Instructional carrels, 460
Instructional guidelines, 107–24
adult learning characteristics, 109–20
andragogy, 120–23
experience brought to classroom, 112–13
learning styles, 117–20
participation in adult learning activities, 113–16
perceived barrier to learning, 115
pragmatism/learning orientation, 111–12
reasons for learning, 114
self-directed learning, 116–17
Instructional materials
Performance Systems Design
development test guidelines checklist, 148
testing of, 147–48
Instructional Systems Association, 599
Instructors, types of, 78–79
Instructor's guide checklist, 144–45
Interactive Video Industry Association, 588
Interactive video training
definition of, 495–96
with minimal branching, 497–98
Internal consultants, 79
International Center for Training and Management Development (Xerox Corp.), 224–25
International Television Association (ITVA), 587–88
Intern programs, 77–78
See also Cooperative education/internships
Interviews, 196
behavioral-change training and, 572
with key people, for external sales trainers, 645–46
See also Structured interviews
Invitations to trainees, 262–64
arrival packet, 264
logistics for complete information, 262–64
prearrival objectives questionnaire, 262
recommended prearrival reading, 262

Job competency approach
to assessing prospective personnel, 70–76
competency models, 71–74
key training and development roles, 70–71
Job positions/staffing alternatives, 76–81
instructors, 78–79
internal consultants, 79
intern programs, 77–78

Job positions/staffing alternatives (*cont.*)
  permanent versus temporary personnel, 77
  training coordinators, 79
Joint-call coaching, 290–94
Joint calls
  maximizing effectiveness of, 292–94
  types of, 291
Journal of Personal Selling and Sales Management, 605

Keiser, Thomas C., 177–92
Kersen, Isabel L., 217–41
KILIS method, of assessing the audience, 344–46
Kinney, Charles H., 27–41
Knowles, Malcolm, 120–21
Kotler, Greta, 306–35
Kurzrock, Warren, 193–216

Lafferty, J. Clayton, 422–38
Law, William E., 42–67
Leaders, characteristics of, 48–49
Leaders (Bennis), 47–48
Leader's Guide
  sample layout, 214
  writing of, 213
Leadership (Burns), 67
Learning
  barriers to, 115
  reasons for, 114
Learning environment, 217–41, 265
  effect on learning process, 218–20
  elements of, 220–38
  facility, 222–28
  food and drink, 235–36
  geographic location, 220–22
  ideal classroom environment, checklist for, 240
  meeting room, 228–31
  on-the-job training, 238–40
  people, 237–38
  seating arrangement, 231–35
  support materials, 236–37
Learning How to Learn—Applied Theory for Adults (Smith), 118
Learning Independently, 601
Learning objectives, behavioral-change training, 552–53
Lecture method
  assessing the audience, 344–46
  closing, 359–60
  developing your presentation, 346–48
  opening, 357–59
  room arrangement, 360–63

  types of, 339–41
    debate, 340
    question and answer, 340–41
    symposium, 339
  use in sales training, 338–41
  See also AIDA formula; KILIS formula
Lectures, workshops, 204
Lefton, Robert E., 548–79
Letter of agreement, external sales trainers, 643–44
Lighting, meeting room, 228–29
Long-term self-development programs, 326–29
  goals statement, 326–27
  individual learning projects, 327–29
    sample project, 329–30
  resources, 330–33
    associations, 332–33
    books/periodicals, 330–31
    college and university classes, 331
    professional associates, 333
    resource centers, 333
    training programs/seminars/course-ware, 331–32
  self-assessment, 326

McClung, Kenneth A. Jr., 125–53
McCullough, Richard C., 70, 94, 650–71
McGurer, Dennis S., 68–89
McLagan, Patricia A., 70, 94
Macro strategy, Performance Systems Design, 140–41
Magnetic boards, *See* Flannel/magnetic/display boards
Mailing list update form, 282
Management processes, 51–62
  gaining commitment of others, 53–54
  goal setting, 52
  mission statement, development of, 51–52
  stating objectives, 52–53
  strategic tools, 54–59
    business planning, 54–56
    curriculum planning, 56–57
    training budget, 58–59
    training schedule, 57–58
Managers
  computer programs for, 255–58
  as counselors, 243–45
  defining needs, 31–32
  establishing agreements between trainers and, 30–31
  follow-up, 40
  job responsibilities, 28–29
  levels of, 59–62
  performance evaluation, 59–62
  reasons for, 59
  reviewing needs of, 174–75

as role modelers, 244–45
strategic skills, 44–49
    doing right things versus wrong things, 44
    questions to improve leadership skills, 45–47
    recognizing training group characteristics, 47
    solving critical performance problems, 44–45
training leadership skills, 47–49
training program, preparing for, 32–36
training role versus manager role, 29–30
training sessions, 36–40
    demonstration contests, 39–40
    new product introduction, 38
    probing, 38
MarkED Resource Center, 611
Marketing plan, 65–66
    for training, 66
    for training managers, 66
The Marketplace Directory, 660
Marx, Gerry, 242–70
Measuring sales training, See Evaluating sales training
Media Profiles: Career Development Edition (video), 600
Media selection checklist, 141–42
Medical emergencies, 264–65
Meeting Planners Guide (Marriot), 230
Meeting room, 228–31, 360–63, 231–35
    auditorium-style seating, 360–61, 527
    boardroom seating, 528
    buzz group tables, 528
    classroom-style seating, 233, 361, 527
    conference meetings, 232, 528
    decor, 231
    environmental factors, 362–63
        color, 362
        lighting, 228–29, 362
        seating, 362–63
        sound/music, 362
    E-shape seating, 529
    floor plan, 229–31
    front of, 234–35
    herringbone/chevron style, 361
    hollow square/hollow circle seating, 529
    inspection of, 526–27
    layout of, 527–30
    preparation of, 352, 524–30
    round-style seating, 361–62
    selecting site, 525
    small tables, 234
    temperature/ventilation, 229
    theater-style seating, 233–34, 360–61
    training impact, 231, 235
    T-shape seating, 529

U-shape seating, 232–33, 361
V-shape seating, 528
Meetings, 513–30
    AIDA formula and, 514–15
    group coaching at, 304–5
    large sales meetings, planning tips, 521–22
    meeting room, 524–30
    national versus regional meetings, 519–21
    planning for success, 518–19
    reasons for conducting, 513–14
    speakers, sources of, 522–24
    training methods, 516–18
        buzz groups, 517–18
        conference method, 516–17
Micro strategy, Performance Systems Design, 142
Mission statement, development of, 51–52
Modeled calls, 291
Models/demonstrations, 456–57
    advantages, 456
    disadvantages, 456
    user tips, 456–57
Models for Excellence (McLagan/McCullough), 70, 94
The Modern Practice of Adult Education: Andragogy versus Pedagogy (Knowles), 121

National sales meetings, 520
    versus regional sales meetings, 519–21
National Society for Performance and Instruction (NSPI), 586
National Society of Sales Training Executives (NSSTE), 16, 20, 176, 332, 590
Nebraska Videodisc Symposium, 588
Needs analysis, 154–76
    benefits of conducting, 155–56
    collect/reporting data, 158–62
    district meetings, 175–76
    field trip observations, 173–74
    performance appraisals, 163–67
    postmeeting questionnaires, 159–62
        field trip questionnaire, 160–62
    purpose of, 156–58
    seeking trainee input, 168
    structured interviews, development of, 168–69
    team approach to, 171–73
Networking
    benefits of, 714–15
    by leaders, 48
    guidelines, 713–14
    how to start, 711–13
        finding networkers, 711–12
    network, defined, 710–11

Networking (*cont.*)
  public seminars and, 694–95
  rules of, 712–13
  use in sales training, 708–9
Newman, Thomas J., 403–21
New sales trainers
  preparation of, 102–4
    preparation options, 102–3
    training plan, 103–4

Objections, outline for handling, 34–35
Observation, behavioral-change training and, 572
Observed calls, 291
On-line computer databases, public seminars and, 691–94
On-site training, 247–48
  facilities
    facilities layout, 248
    location of facility, 247
    site considerations, 247–48
On-the-job training
  learning environment, 238–40
    geography, 238
    setting, 239–40
Opening, of lecture, guidelines for, 357–59
Organization structure, 62–66
  achieving balanced staff makeup, 64–65
  centralized versus decentralized structure, 62–64
  customer convenience/fragmentation, 62
  delivery mix strategy, 62–63
  marketing plan, 65–66
  reporting, 63–64
  staff, 64–65
Outlines, 207–13
  sample outline for handling objections, 34–35
  topic outline, 208–9
  treatment outline, 209–13
Overhead projectors, 446–48
  advantages, 446–47
  disadvantages, 447
  general guidelines for, 353

Packaged programs
  build/buy decision guidelines, 653–58
    audience, 656–57
    budget, 654
    company culture, 657–58
    cost, 653–54
    expertise/experience, 655
    quality of program materials, 656
    time, 654–55

former client satisfaction, 666–69
  "gut reaction", 666
  training package checklist, 666–69
identifying vendors/programs, 658–62
  advertising/direct mail, 661–62
  conferences/trade shows, 661
  publications, 659–61
  word of mouth, 658–59
selection of program, 662–65
  cost considerations, 662–64
  expertise/experience considerations, 665
  objectives, 662
  quality of product, 664–65
  time considerations, 664
trend toward use of, 651–52
when and how to use, 650–71
working with vendor of choice, 669–70
Panel discussions, workshops, 204
Parry, Scott B., 364–402
Performance appraisals, 163–67
  design/use of questionnaires, 163–66
  interpreting survey feedback, 166–67
Performance evaluation, 59–62
  call reports, questions to ask on review of, 276–78
  department, 86–88
  managers, 59–62
  monitoring and, 275–78
  skills inventory, 276–77
  staff, 86–88
    appraisal systems, 86–87
Performance management coaching, 296–304
  action planning, 298
  conducting initial goal-setting meeting, 297–98
  conducting performance reviews, 303–4
  handling performance problems, 302–3
  performance management form, 299–300
  progress reviews, 301–2
  setting meaningful goals, 297
Performance Systems Design, 124–53
  design team, 128–30
  effect on company, 127–28
  evaluation, 150–51
  implementation, 149–50
  instructional materials, 147–48
  periodic updates, 151–52
  problem/requirement analysis, 133–37
  producing/reproducing final materials, 149
  project plan, 130–32
  strategy, 137–43
  support materials, 144–47
Permanent instructors, 78
Persona Selling Power, 592
Pike, Robert W., 336–63

Planning sheets
  role play, 387–91
    sample of, 387–91
Positioning, by leaders, 48
Positioning sales training, checklist for, 46
Postcall coaching, 295–96
Postprogram measurement of behavioral
    change, 559–60
  comparison groups, 564–69
  delayed measurement, 560
    delayed measurement survey, 561–63
  instant measurement, 559–60
Prearrival objectives questionnaire, 262
Precall coaching, 285–90
Prepared materials
  gaming/simulation, 429–32
    costs of, 435
    Desert Survival simulation, 429–31
    purchase of, 431–32
Presentation
  building support in, 349–51
    audience contact, 351
    transitions, 350–51
    types of support to use, 349–50
  development of, 346–48
Pretraining assessment, of sales trainer compe-
    tency, 100–101
Probing, 38
Problem/requirement analysis
  Performance Systems Design
    needs analysis, 135–36
    performance analysis, 133–34
    task analysis, 137
Product training, resources, 585–88
Professional journals, on selling skills, 591–93
Professional networking, *See* Networking
Professional organizations, to improve selling
    skills, 589–91, 605
Program development/administration, 105–270
  administering sales training, 242–70
  creating a theme for sales training program,
    177–92
  developing/writing successful training pro-
    grams, 193–216
  instructional guidelines, 107–24
  learning environment, 217–41
  performance systems design, 125–53
  training needs, determination of, 154–76
Programmed instruction, self-study, 202–3
Project plan
  Performance Systems Design, 130–32
    development of, 130–33
    illustration of, 131–32
Prospective personnel
  assessment of, 70–76
    alternative approaches, 75–76
    job competency approach, 70–72

  multiple interviews, 75
  references, 76
  resumes, 75
  samples of work, 76
  telephone screening, 75–76
  transcripts, 75
Public seminars, 672–706
  actions at seminar, 702–3
  advance payments, 688–90
  advantages of, 678–79
  calling/writing sponsor, 699
  cancellation refunds, 688–90
  confirmation package, what to look for in,
    699
  criticisms of, 679
  duration of course, 687–88
  estimating costs/payback of, 682–83
    cost factors, 682–83
    prices, 688
    tracking payback, 383
  guarantees, 688–90
  information sources, 705–6
  national sources, 690–95
    checking your own files for, 695
    locator/enrollment services, 694
    networking, 694–95
    on-line computer databases, 691–94
    published directories, 691
  operation of, 683–88
  postseminar actions, 703–4
  pre-seminar preparation, 700–702
  providers of, 673–75
  representative seminars/sponsors, 684–86
  role in sales training, 67379
  selection tips, 695–700
    content/program outline considerations,
      697
    factors to evaluate, 696–97
    information sources, 697–99
    participant testimonial considerations, 699
    schedule/location considerations, 698–99
    seminar leader(s) considerations, 697–98
  seminar versus session, 677–78
  types of programs, 675–77
  when to use, 679–82
    in-house programs, 681
    off-the-shelf versus custom-developed, 681–
      82
Publisher's addresses, 24–25
Purchasing, 258–59

Qualitative measures of change, behavioral-
    change training, 569–70
Quantitative measures of change, behavioral-
    change training, 570–71

Questionnaires, 198–99
  behavioral-change training and, 572
  performance appraisals, 163–66
Questions and answers, 340–41
  formats for asking/answering questions, 342–
    43
    addressing the group as a whole, 342
    addressing individuals, 342–43
    addressing subgroups, 343
    "capture the question" technique, 343–44
    general considerations, 341–42
Quinn, Judith C., 707–17

Range, Tracy L., 422–38
Rating sheets
  role play, 384–87
    benefits of use, 387
    sample of, 385–86
Reading assignments, self-study, 202
Reading list
  for sales managers, 23
  for sales personnel, 22
Recordkeeping, 250–53
  Equal Employment Opportunity data, 252
  previous training experiences, 252
  training costs, 252–53
Refreshments, during sales training programs,
  266
Regional sales meetings, 520
  versus national sales meetings, 519–21
Registration form, sample of, 255
Reinforcement, as postprogram activity, 268–
  69
Resource manuals, guidelines for use of, 354
Resources, 583–606
  case method, 420
  cooperative education/internships, 626–27
  external sales trainers, 638
  long-term self-development programs, 330–
    33
  product training, 585–88
    computer-based instruction, 586–88
    instruction design, 586
  public seminars, 705–6
  selling skills, 589–606
    assessment instruments, 601
    bibliographies/indexes/abstracts, 594–95
    books, 595–96
    colleges/universities, 604
    computer-based sales training, 601
    conferences, 599–600
    databases, 601–4
    directories of vendors/consultants, 597–99
    professional journals, 591–93
    professional organizations, 589–91, 605
    publishers specializing in books on selling,
      596–97

  self-instructional materials, 601
  seminars/workshops, 599
  textbooks/support materials, 605–6
  videotape/film distributors, 600–601
  videotapes/films, 600
  speakers, 522–24
Resources in Education (RIE), 595
Role play, 39–40, 364–402, 541–42
  behavioral triangle, 367–68
  conducting of, 391–92
    allowing players to be comfortable in their
      roles, 392
    avoiding interruptions, 392
    helping players end interaction, 392
    selecting players, 392
  demonstration, 368–77
  group techniques, 394–402
  history of, 367–68
  instructions, 380
  interaction design, 383
  planning sheets, 387–91
  postenactment discussion, 393–94
  Price-Bender role play, 396–401
  purposes of, 366–67
  rating sheets, 384–87
  role-play evaluation form, 542
  role-play simulations, 427
  sample role play, 381
  selection guidelines, 377–81
    asking salespeople/managers to write role
      play, 379
    "critical incident" technique, 377–79
    sorting information, 379
  use in sales training, 365–66
  writing objectives, 382–84
Room arrangement, *See* Meeting room

Sales activity
  altering comfortable patterns of, 279–83
  setting objective of, 515–16
Sales data analysis, training needs and, 278–
  79
Sales managers, *See* Managers
Sales and Marketing Digest, 593
Sales and Marketing Executives of Greater New
  York, 591
Sales and Marketing Executives-International
  (SME-I), 591
Sales and Marketing Management magazine,
  593
Sales and Marketing Training, 591–92
Sales/marketing training research, 621–22
*Sales and Marketing Training Resources in
  Print,* 594
Sales meetings, *See* Meetings; Meeting room
Sales Staff Surveys, Inc., 167

Sales trainers, *See* Training personnel
Sales training
  evolution of, 8–9
  external resources, use of, 18–19
  field trips, budgeting of, 11–12
  follow-up, importance of, 14
  future of, 19–21
  implementation, 271–530
  internal resources, use of, 17–18
  manager's role in, 27–41
  objectives of, 14–15
  revolving programs, 13–14
  role of, changes in, 9–10
  staffing, 68–89
  trainee evaluation, 15–16
Sales training management, 42–67
Sales training programs
  attendee accommodations, 260
  communicating with immediate supervisor, 267
  developing content, 195–99
    field visits, 197
    focus groups, 195–96
    individual interviews, 196
    mail questionnaires, 198–99
    outlines, 207–13
  emergencies, 266–67
  Leader's Guide, writing of, 213, 214
  participant's workbook/manual, 213–16
  phone calls during, 265–66
  private consultation, 266
  refreshments, 266
  smoking, 266
  training department privacy, 266
  training method selection, 200–207
    self-study, 200–203
    workshops, 203–7
  who will attend, 260
  See also Program development/administration
Sales training record, 250
Schrello, Don M., 672–706
Scripts checklist, 146–47
Seating arrangements, *See* Meeting room
Self-development, 323–33
  benefits of, 325–26
  long-term self-development programs, 326–29
  use in sales training, 324–25
  See also Long-term self-development programs
Self-instructional materials, on selling skills, 601
Self-study, 200–203
  availability of materials, 17
  benefits of, 311–12
  characteristics of, 308

definition of, 307
  materials, purchase of, 322
  methods, 201–3
    preworkshop assignments, 202
    programmed instruction, 202–3
    reading assignments, 202
  program design, 313–22
  sales trainers' roles in, 323
  similar learning terms, 307–8
  use in sales training, 309–13
Self-study materials checklist, 146
Seminars, 206, 612–14
  cost considerations, 613
  custom programs, 612–13
  finding the right seminar for your employees, 613–14
  on selling skills, 599
Services off campus, *See* Continuing education
Shared calls, 291
Skills inventory, 276
  sample form, 277
Slide transparencies, general guidelines for, 352–53
Smith, Bob, 118
Smith, Homer, 512–30
Smoking, during sales training programs, 266
Society for Applied Learning Technology, 588
Software, computer-based/interactive video training, 508–9
Special/festive events, celebration of, 265
Spikes, W. Franklin, III, 107–24
Staff, 64–65, 258–59
  achieving balanced staff makeup, 64–65
  encouraging new career path opportunities, 86
  evaluating staff size, 64
  performance evaluation, 86–88
  training/development, 85–86
    in-house sources, 85–86
    outside sources, 85
Staffing alternatives, *See* Job positions/staffing alternatives
Strategic/insightful simulation, 426–27
Strategic training management model, 61
Strategy
  Performance Systems Design, 137–43
    course strategy checklist, 142–43
    design document, 142
    establishing objectives, 138
    learning approach checklist, 139–40
    macro strategy, 140–41
    major learning approaches, 138–39
    media selection checklist, 141–42
    micro strategy, 142
Strategy formation, 49–51
  identifying where training can positively affect performance, 50–51

Strategy formation (*cont.*)
  implementation, 50
  looking for changes that may impact on sales
    effort, 50
Structured discussions, workshops, 204
Structured interviews
  development of, 168–69
  how to conduct, 169
  when to use, 169
Student workbook checklist, 145–46
Subscription publications, 23
Suppliers, audio/visual media, 466–67
Support materials
  Performance Systems Design, 144–47
    development of, 144–47
    instructor's guide checklist, 144–45
    scripts checklist, 146–47
    self-study materials checklist, 146
    student workbook checklist, 145–46
  selling skills, 605–6
Surveys, behavioral-change training and, 572
Symposia, 206
  as lecture method, 339

T & D programs, 609–10
Teleconferencing, 458–59
  advantages, 458
  disadvantages, 459
  user tips, 459
Telephone screening, of prospective personnel,
    75–76
Temperature/ventilation, meeting room, 229
Textbooks/support materials, on selling skills,
    605–6
Theme of training program, 177–92
  crafting of message, 182–87
    case study, 185–86
    converge, 184–85
    increase number of options, 183–84
    prepare brainstorming session, 182–83
    review/revision, 185
    test chosen theme, 185
  incorporating theme in program, 187–92
    brevity of explanations, 191
    participant communications, 189–90
    potential pitfalls, 191–92
    program title selection, 190
    slogans/symbols creation, 190–91
    theme in program design, 187–89
  quality of message, 186–87
    relevancy, 186–87
    rigor, 186
    simplicity, 187
    taking a stand, 187
  reasons for creation of, 180–82

35mm slides, 451–53
  advantages, 451–52
  disadvantages, 452
  locally produced versus commercially devel-
    oped slides, 452
  user tips, 452–53
Topic outline, 208–9
Top-performing salespeople, learning skills,
    274–75
Trainable competencies, 73
*Trainer's Resource: A Comprehensive Guide to
    Packaged Training Programs*, 659
Trainer's Resource, 598
TRAINET/The Seminar Database, 103, 602
Training & Development Journal, 592
TRAINING Annual Conference and Expo, 599–
    600
Training center library, 268
Training coordinators, 79–80
  quality control checklist, 79–80
  vendor selection checklist, 80–81
Training department
  administration of, 83–85
  future of, 88–89
Training and Development Alert, 594
*Training and Development Organizations Di-
    rectory*, 598–99
Training history form, 252
Training management system, establishment
    of, 49
Training Marketplace Directory, 598
Training News, 593
Training personnel
  assessment of prospective personnel, 70–76
  corporate planning and, 10–11
  critical competency skills, 97–101
  impact on corporation, 5–6
  maintaining state-of-the-art programs, 16
  management of, 81–85
    keeping staff to a minimum, 83
    proactive versus reactive departments, 82
    separation of departmental functions, 83
  position on organizational chart, 4–5
  positive influence on organization, 91–92
  preparation of, 90–104
  responsibilities, 6–7
    recognition of, 21
    to company, 6–8
    to students, 6
  role of, 3–26, 94–97
  sales of, 12–13
  selection of
    career versus "passthrough" trainers, 93–
      94
    qualities to look for, 93
    tips, 92–94

special tips for, 41
trends in, 69
Training recommendation form, 251
Training, The Magazine of Human Resources Development, 592
"Train the Trainer" course, 102
Transitions, in presentations, 350–51
Transparencies
  criteria for developing prepared transparencies, 447–48
  user tips, 48–49
Travel accommodations, 221–22
Treatment outline, 209–13
  activities that reinforce learning, 212–13
  steps for reinforcing learning, 210–12
True-to-life simulation, 426
T-shape seating, meeting room, 529

U-shape seating, meeting room, 361

V-shape seating, meeting room, 528
Vendors
  directories of vendors/consultants, selling skills, 597–99
  packaged programs
    identification of, 658–62
    working with vendor of choice, 669–70
  selection checklist, 80–81
Video/film, 469–90
  advantages/disadvantages of, 472–78
    film, 473–75
    video, 475–78
  instructional guidelines, 471–72
    visual and verbal persuasion, 472
  making your own video, 481–89
    determining the treatment, 484–86
    developing the idea, 484
    developing the script/storyboard, 486–87
    gaining authorization, 486
    marketing, 489–90
    postproduction editing, 489
    producing the video, 487–89
  motivational impact of, 472
  "right way" versus "wrong way" examples, 472
  sales representative delivery evaluation form, 477

sales training program, development of, 478–80
video evaluation form, 477
video sales training programs
  preview of, 481
  purchase of, 480–81
videotape selection, checklist for, 482–83
Videotape/film distributors, 600–601
Videotapes/films, on selling skills, 600
Visioning, by leaders, 48

Ward, Nicholas, 273–305
Weekly call report, sample of, 280–81
Wenschlag, Roger E., 90–104
White boards, *See* Chalkboards/white boards
Wills, Rick, 439–68
Wilson, Larry, 43
WILSONLINE—Business Periodicals Index, 603
Wolfson, Kathryn S., 273–305
Workbook/manual
  sales training programs, 213–16
    sample page, 215
    writing of, 213–16
  sample page from, 215
  student workbook checklist, 145–46
Workshops, 203–7
  advantages of conducting, 203
  learning objectives for structuring of, 206–7
  role plays/simulations, 205–6
  on selling skills, 599
  size limitations, 203–4
  team workshops, 206
  training methods, 204–6
    brainstorming, 205
    case histories, 205
    in-baskets, 205
    lectures, 204
    panel discussions, 204
    structured discussions, 204
Written materials, 457–58
  advantages of use, 457
  disadvantages of use, 457
  user tips, 457–58
Written tests, 538–50
  standardized tests, 538
  tailored tests, 529